Greenland
(Kalatdlit-Nunat)
Alaska
Yukon
N.W.T.
Nunavut
[Labrador]
B.C.
Alta.
Sask.
Man.
Ont.
Que.
Nfld.
St.Pierre
and Miquelon
P.E.I.
N.B.
N.S.
Wash.
Mont.
N.Dak.
Minn.
Vt.
Maine
N.H.
Mass.
Oreg.
Idaho
Wyo.
S.Dak.
Wis.
Mich.
N.Y.
R.I.
Conn.
Iowa
Pa.
N.J.
Nev.
Utah
Colo.
Nebr.
Ill.
Ind.
Ohio
W.Va.
Del.
Md.
Va.
Kans.
Mo.
Ky.
Calif.
N.C.
Tenn.
Okla.
Ark.
S.C.
Ariz.
N.Mex.
Ala.
Ga.
Miss.
Tex.
La.
Fla.
Mexico

Flora of North America

Contributors to Volume 14

Charles M. Allen
Wendy L. Applequist
Daniel F. Austin†
John E. Averett†
Lynn Bohs
Hayden Brislin
Richard K. Brummitt†
Robert A. Bye
J. Richard Carter
Mihai Costea
Lawrence J. Davenport
Ellen A. Dean
W. Hardy Eshbaugh
Mark Fishbein
Kathryn L. Fox
Philip D. Jenkins†
Jordan C. Jones
C. Lee Kimmel
Sandra Knapp
Alexander Krings
David E. Lemke
Rachel A. Levin
Zheng Li
Katherine G. Mathews
Angela McDonnell
Jill S. Miller
Thomas Mione
Alexandre K. Monro
Guy L. Nesom
Linh Tô Ngô
Eliane Meyer Norman
James S. Pringle
Casie L. Reed
Kenneth R. Robertson
Patricia Sabadin
Tiina Särkinen
David M. Spooner†
John L. Strother
Janet R. Sullivan
Michael A. Vincent
Maggie Whitson

Editors for Volume 14

Tammy M. Charron
Managing Editor

Kanchi Gandhi
Nomenclatural and Etymological Editor

David E. Giblin
Taxon Editor for Gentianaceae and Loganiaceae

Dale E. Johnson
Taxon Editor for Gentianaceae

Robert W. Kiger
Bibliographic Editor
Lead Editor and Taxon Editor for Gelsemiaceae and Hydroleaceae

Alexander Krings
Taxon Editor for Apocynaceae

Geoffrey A. Levin
General Editor
Taxon Editor for Apocynaceae

Jay A. Raveill
Taxon Editor for Sphenocleaceae

Mary Ann Schmidt
Senior Technical Editor

John L. Strother
Reviewing Editor and Taxon Editor for Convolvulaceae

Janet R. Sullivan
Taxon Editor for Solanaceae

James L. Zarucchi†
Editorial Director (to 2019)

Volume 14 Composition

Tanya Harvey
Layout Artist and Editorial Assistant

Spigelia marilandica

Flora of North America

North of Mexico

Edited by FLORA OF NORTH AMERICA EDITORIAL COMMITTEE

VOLUME 14

Magnoliophyta: Gentianaceae to Hydroleaceae

NEW YORK OXFORD • OXFORD UNIVERSITY PRESS • 2023

Oxford University Press is a department of the University of Oxford.
It furthers the University's objective of excellence in research,
scholarship, and education by publishing worldwide.

Oxford New York
Auckland Cape Town Dar es Salaam Hong Kong Karachi Kuala Lumpur
Madrid Melbourne Mexico City Nairobi New Delhi Shanghai Taipei Toronto

With offices in
Argentina Austria Brazil Chile Czech Republic France Greece Guatemala Hungary Italy
Japan Poland Portugal Singapore South Korea Switzerland Thailand Turkey Ukraine Vietnam

Published by Oxford University Press, Inc.
198 Madison Avenue, New York, New York 10016
www.oup.com

Library of Congress Cataloging-in-Publication Data
(Revised for Volume 14)
Flora of North America North of Mexico
edited by Flora of North America Editorial Committee.
Includes bibliographical references and indexes.
Contents: v. 1. Introduction—v. 2. Pteridophytes and gymnosperms—
v. 3. Magnoliophyta: Magnoliidae and Hamamelidae—
v. 22. Magnoliophyta: Alismatidae, Arecidae, Commelinidae (in part), and Zingiberidae—
v. 26. Magnoliophyta: Liliidae: Liliales and Orchidales—
v. 23. Magnoliophyta: Commelinidae (in part): Cyperaceae—
v. 25. Magnoliophyta: Commelinidae (in part): Poaceae, part 2—
v. 4. Magnoliophyta: Caryophyllidae (in part): part 1—
v. 5. Magnoliophyta: Caryophyllidae (in part): part 2—
v. 19, 20, 21. Magnoliophyta: Asteridae (in part): Asteraceae, parts 1–3—
v. 24. Magnoliophyta: Commelinidae (in part): Poaceae, part 1—
v. 27. Bryophyta, part 1—
v. 8. Magnoliophyta: Paeoniaceae to Ericaceae—
v. 7. Magnoliophyta: Salicaceae to Brassicaceae—
v. 28. Bryophyta, part 2—
v. 9. Magnoliophyta: Picramniaceae to Rosaceae—
v. 6. Magnoliophyta: Cucurbitaceae to Droseraceae—
v. 12. Magnoliophyta: Vitaceae to Garryaceae—
v. 17 Magnoliophyta: Tetrachondraceae to Orobanchaceae—
v. 10 Magnoliophyta: Proteaceae to Elaeagnaceae—
v. 11 Magnoliophyta: Fabaceae, Parts 1 and 2—
v. 14 Magnoliophyta: Gentianaceae to Hydroleaceae

ISBN: 9780197691465 (v. 14)
1. Botany—North America.
2. Botany—United States.
3. Botany—Canada.
I. Flora of North America Editorial Committee.
QK110.F55 2002 581.97 92-30459

1 2 3 4 5 6 7 8 9
Printed and bound in the U.S.A. by Sheridan, Chelsea MI

Contents

This volume is dedicated to Sandra Knapp (b. 1956),
eminent student of Solanaceae systematics
and enthusiastic promoter of botany.

Founding Member Institutions

Flora of North America Association

Agriculture and Agri-Food Canada
Ottawa, Ontario

Arnold Arboretum
Jamaica Plain, Massachusetts

Canadian Museum of Nature
Ottawa, Ontario

Carnegie Museum of Natural History
Pittsburgh, Pennsylvania

Field Museum of Natural History
Chicago, Illinois

Fish and Wildlife Service
United States Department of the Interior
Washington, D.C.

Harvard University Herbaria
Cambridge, Massachusetts

Hunt Institute for Botanical Documentation
Carnegie Mellon University
Pittsburgh, Pennsylvania

Jacksonville State University
Jacksonville, Alabama

Jardin Botanique de Montréal
Montréal, Québec

Kansas State University
Manhattan, Kansas

Missouri Botanical Garden
St. Louis, Missouri

New Mexico State University
Las Cruces, New Mexico

The New York Botanical Garden
Bronx, New York

New York State Museum
Albany, New York

Northern Kentucky University
Highland Heights, Kentucky

Université de Montréal
Montréal, Québec

University of Alaska
Fairbanks, Alaska

University of Alberta
Edmonton, Alberta

The University of British Columbia
Vancouver, British Columbia

University of California
Berkeley, California

University of California
Davis, California

University of Idaho
Moscow, Idaho

University of Illinois
Urbana-Champaign, Illinois

University of Iowa
Iowa City, Iowa

The University of Kansas
Lawrence, Kansas

University of Michigan
Ann Arbor, Michigan

University of Oklahoma
Norman, Oklahoma

University of Ottawa
Ottawa, Ontario

University of Louisiana
Lafayette, Louisiana

The University of Texas
Austin, Texas

University of Western Ontario
London, Ontario

University of Wyoming
Laramie, Wyoming

Utah State University
Logan, Utah

For their support of the preparation of this volume,
we gratefully acknowledge and thank:

Fondation Franklinia

The Philecology Foundation

The Andrew W. Mellon Foundation

The David and Lucile Packard Foundation

an anonymous foundation

Christopher Davidson and Sharon Christoph

Jennifer H. Richards

Chanticleer Foundation

The William and Flora Hewlett Foundation

The Stanley Smith Horticultural Trust

Hall Family Charitable Fund

WEM Foundation

William T. Kemper Foundation

Members of the Flora of North America Board
and other colleagues

Illustration Sponsorship

For sponsorship of illustrations included in this volume, we express sincere appreciation to:

Ellen Dean, Ventura, California
Lycianthes asarifolia, Solanaceae

Julian P. Donahue, Tucson, Arizona
Amsonia palmeri, Apocynaceae
Asclepias elata, Apocynaceae
Funastrum cynanchoides, Apocynaceae
Haplophyton crooksii, Apocynaceae
Metastelma arizonicum, Apocynaceae
Pherotrichis schaffneri, Apocynaceae

Hardy Eshbaugh, Oxford, Ohio
Capsicum annuum var. *glabriusculum*, Solanaceae

Patricia A. Harris, Columbia, Missouri
Volume 14 frontispiece, *Spigelia marilandica*, Loganiaceae

Julie Kierstead, Ashland, Oregon
Datura wrightii, Convolvulaceae

Chris Kreussling (Flatbush Gardener), Brooklyn, New York
Asclepias incarnata subsp. *incarnata*, Apocynaceae

Scott Peterson, Grass Valley, California
Asclepias cryptoceras subsp. *cryptoceras*, Apocynaceae

Nancy R. Morin, Point Arena, California
Araujia sericifera, Apocynaceae—in memory of Christopher Davidson
Calystegia spithamaea subsp. *spithamaea*, Convolvulaceae—in memory of Richard K. Brummitt
Solanum jamesii, Solanaceae—in memory of David Spooner
Turbina corymbosa, Convolvulaceae—in memory of Helen Jeude

Janet Sullivan and Tom Laue, Lee, New Hampshire
Physalis crassifolia, Solanaceae
Physalis pubescens, Solanaceae
Physalis virginiana, Solanaceae
Physalis walteri, Solanaceae

Flora of North America Editorial Committee

(as of October 2022)

Project Staff — past and present involved with the preparation of Volume 14

Barbara Alongi, *Illustrator*

Mike Blomberg, *Imaging*

Ariel S. Buback, *Assisting Technical Editor*

Tammy M. Charron, *Managing Editor (2017–)*

Trisha K. Distler, *GIS Analyst*

Tanya Harvey, *Layout Artist and Editorial Assistant*

Suzanne E. Hirth, *Editorial Assistant*

Cassandra L. Howard, *Senior Technical Editor*

Ruth T. King, *Editorial Assistant*

John Myers, *Illustration Compositor*

Kristin Pierce, *Editorial Assistant*

Andrew C. Pryor, *Assisting Technical Editor*

Heidi H. Schmidt, *Managing Editor (2007–2017)*

Mary Ann Schmidt, *Senior Technical Editor*

Yevonn Wilson-Ramsey, *Illustrator*

Contributors to Volume 14

Charles M. Allen
University of Louisiana at Monroe
Monroe, Louisiana

Wendy L. Applequist
Missouri Botanical Garden
St. Louis, Missouri

Daniel F. Austin†
Arizona-Sonora Desert Museum
Tucson, Arizona

John E. Averett†
Georgia Southern University
Statesboro, Georgia

Lynn Bohs
University of Utah
Salt Lake City, Utah

Hayden Brislin
North Carolina State University
Raleigh, North Carolina

Richard K. Brummitt†
Royal Botanic Gardens
Kew, Richmond, Surrey, England

Robert A. Bye
Jardín Botánico
Universidad Nacional Autónoma de México
Mexico City, Mexico

J. Richard Carter
Valdosta State University
Valdosta, Georgia

Mihai Costea
Wilfrid Laurier University
Waterloo, Ontario

Lawrence J. Davenport
Samford University
Birmingham, Alabama

Ellen A. Dean
University of California Davis Center for Plant Diversity
Davis, California

W. Hardy Eshbaugh
Miami University
Oxford, Ohio

Mark Fishbein
Oklahoma State University
Stillwater, Oklahoma

Kathryn L. Fox
University of New Hampshire
Durham, New Hampshire

Philip D. Jenkins†
University of Arizona
Tucson, Arizona

Jordan C. Jones
Valdosta State University
Valdosta, Georgia

C. Lee Kimmel
North Carolina State University
Raleigh, North Carolina

Sandra Knapp
The Natural History Museum
London, England

Alexander Krings
North Carolina State University
Raleigh, North Carolina

David E. Lemke
Texas State University
San Marcos, Texas

Rachel A. Levin
Amherst College
Amherst, Massachusetts

Zheng Li
The University of Arizona
Tucson, Arizona

Katherine G. Mathews
Western Carolina University
Cullowhee, North Carolina

Angela McDonnell
Oklahoma State University
Stillwater, Oklahoma

Jill S. Miller
Amherst College
Amherst, Massachusetts

Thomas Mione
Central Connecticut State University
New Britain, Connecticut

Alexandre K. Monro
Royal Botanic Gardens, Kew and The Natural History Museum
London, England

Guy L. Nesom
Academy of Natural Sciences of Drexel University
Philadelphia, Pennsylvania

Linh Tô Ngô
University of Missouri
Columbia, Missouri

Eliane Meyer Norman
Stetson University
DeLand, Florida

James S. Pringle
Royal Botanical Gardens
Hamilton, Ontario

Casie L. Reed
North Carolina State University
Raleigh, North Carolina

Kenneth R. Robertson
Illinois Natural History Survey
Champaign, Illinois

Patricia Sabadin
North Carolina State University
Raleigh, North Carolina

Tiina Särkinen
Royal Botanic Garden Edinburgh
Edinburgh, Scotland

David M. Spooner†
University of Wisconsin
Madison, Wisconsin

John L. Strother
University of California
Berkeley, California

Janet R. Sullivan
University of New Hampshire
Durham, New Hampshire

Michael A. Vincent
Miami University
Oxford, Ohio

Maggie Whitson
Northern Kentucky University
Highland Heights, Kentucky

Regional Reviewers

ALASKA / YUKON

Bruce Bennett
Yukon Department of Environment
Whitehorse, Yukon

Justin Fulkerson
Alaska Center for Conservation Science
University of Alaska Anchorage
Anchorage, Alaska

Steffi M. Ickert-Bond, Regional Coordinator
University of Alaska, Museum of the North
Fairbanks, Alaska

Robert Lipkin
Alaska Natural Heritage Program
University of Alaska Anchorage
Anchorage, Alaska

David F. Murray
University of Alaska, Museum of the North
Fairbanks, Alaska

Carolyn Parker
University of Alaska, Museum of the North
Fairbanks, Alaska

Mary Stensvold
Sitka, Alaska

PACIFIC NORTHWEST

Edward R. Alverson
Lane County Parks Division
Eugene, Oregon

Curtis R. Björk
University of British Columbia
Vancouver, British Columbia

Mark Darrach
University of Washington
Seattle, Washington

Walter Fertig
Washington State University
Pullman, Washington

David E. Giblin
University of Washington
Seattle, Washington

Richard R. Halse
Oregon State University
Corvallis, Oregon

Linda Jennings
University of British Columbia
Vancouver, British Columbia

Aaron Liston, Regional Coordinator
Oregon State University
Corvallis, Oregon

Frank Lomer
New Westminster, British Columbia

Kendrick L. Marr
Royal British Columbia Museum
Victoria, British Columbia

Jim Pojar
British Columbia Forest Service
Smithers, British Columbia

Peter F. Zika
University of Washington
Seattle, Washington

SOUTHWESTERN UNITED STATES

Tina J. Ayers
Northern Arizona University
Flagstaff, Arizona

Walter Fertig
Washington State University
Pullman, Washington

G. F. Hrusa
Lone Mountain Institute
Brookings, Oregon

Max Licher
Northern Arizona University
Flagstaff, Arizona

Elizabeth Makings
Arizona State University
Tempe, Arizona

James D. Morefield
Nevada Natural Heritage Program
Carson City, Nevada

Nancy R. Morin, Regional Coordinator
Point Arena, California

Donald J. Pinkava†
Arizona State University
Tempe, Arizona

Jon P. Rebman
San Diego Natural History Museum
San Diego, California

Glenn Rink
Northern Arizona University
Flagstaff, Arizona

James P. Smith Jr.
Humboldt State University
Arcata, California

Gary D. Wallace†
California Botanic Garden
Claremont, California

WESTERN CANADA

Bruce A. Ford,
Regional Coordinator
University of Manitoba
Winnipeg, Manitoba

Lynn Gillespie
Canadian Museum of Nature
Ottawa, Ontario

A. Joyce Gould
Alberta Environment and Parks
Edmonton, Alberta

Vernon L. Harms
University of Saskatchewan
Saskatoon, Saskatchewan

Elizabeth Punter
University of Manitoba
Winnipeg, Manitoba

Sara Vinge-Mazer
Saskatchewan Conservation Data Centre
Saskatchewan Ministry of Environment
Saskatchewan, Canada

ROCKY MOUNTAINS

Jennifer Ackerfield
Denver Botanic Gardens
Denver, Colorado

Walter Fertig
Washington State University
Pullman, Washington

Ronald L. Hartman†
University of Wyoming
Laramie, Wyoming

Bonnie Heidel
University of Wyoming
Laramie, Wyoming

Robert Johnson
Brigham Young University
Provo, Utah

Ben Legler
University of Wyoming
Laramie, Wyoming

Peter C. Lesica
University of Montana
Missoula, Montana

Donald H. Mansfield
The College of Idaho
Caldwell, Idaho

B. E. Nelson,
Regional Coordinator
University of Wyoming
Laramie, Wyoming

Leila M. Shultz
Utah State University
Logan, Utah

NORTH CENTRAL UNITED STATES

Anita F. Cholewa
University of Minnesota
St. Paul, Minnesota

Neil A. Harriman†
University of Wisconsin Oshkosh
Oshkosh, Wisconsin

Bruce W. Hoagland
University of Oklahoma
Norman, Oklahoma

Craig C. Freeman,
Regional Coordinator
The University of Kansas
Lawrence, Kansas

Robert B. Kaul†
University of Nebraska
Lincoln, Nebraska

Gary E. Larson†
South Dakota State University
Brookings, South Dakota

Deborah Q. Lewis
Iowa State University
Ames, Iowa

Stephen G. Saupe
College of Saint Benedict/ St. John's University
Collegeville, Minnesota

Lawrence R. Stritch
Martinsburg, West Virginia

George Yatskievych
The Universtiy of Texas at Austin
Austin, Texas

SOUTH CENTRAL UNITED STATES

Jackie M. Poole,
Regional Coordinator
Fort Davis, Texas

Robert C. Sivinski
University of New Mexico
Albuquerque, New Mexico

EASTERN CANADA

Sean Blaney
Atlantic Canada Conservation Data Centre
Sackville, New Brunswick

Luc Brouillet,
Regional Coordinator
Institut de recherche en biologie végétale
Université de Montréal
Montréal, Québec

Jacques Cayouette
Agriculture and Agri-Food Canada
Ontario, Ottawa

Frédéric Coursol
Jardin botanique de Montréal
Montréal, Québec

William J. Crins
Ontario Ministry of Natural Resources
Peterborough, Ontario

Marian Munro
Nova Scotia Museum of Natural History
Halifax, Nova Scotia

Michael J. Oldham
Natural Heritage Information Centre
Peterborough, Ontario

NORTHEASTERN UNITED STATES

Ray Angelo
New England Botanical Club
Cambridge, Massachusetts

David E. Boufford,
Regional Coordinator
Harvard University Herbaria
Cambridge, Massachusetts

Tom S. Cooperrider†
Kent State University
Kent, Ohio

Allison Cusick
Carnegie Museum of Natural History
Pittsburgh, Pennsylvania

Arthur Haines
Canton, Maine

Michael A. Homoya
Indiana Department of Natural Resources
Indianapolis, Indiana

Robert F. C. Naczi
The New York Botanical Garden
Bronx, New York

Anton A. Reznicek
University of Michigan
Ann Arbor, Michigan

Kay Yatskievych
Austin, Texas

SOUTHEASTERN UNITED STATES

Mac H. Alford
University of Southern Mississippi
Hattiesburg, Mississippi

J. Richard Carter Jr.
Valdosta State University
Valdosta, Georgia

L. Dwayne Estes
Austin Peay State University
Clarksville, Tennessee

W. John Hayden
University of Richmond
Richmond, Virginia

Wesley Knapp
NatureServe
Arlington, Virginia

John B. Nelson
University of South Carolina
Columbia, South Carolina

Chris Reid
Louisiana State University
Baton Rouge, Louisiana

Bruce A. Sorrie
University of North Carolina
Chapel Hill, North Carolina

Dan Spaulding
Anniston Museum of Natural History
Anniston, Alabama

R. Dale Thomas†
Seymour, Tennessee

Lowell E. Urbatsch
Louisiana State University
Baton Rouge, Louisiana

Alan S. Weakley,
Regional Coordinator
University of North Carolina
Chapel Hill, North Carolina

Theo Witsell
Arkansas Natural Heritage Commission
Little Rock, Arkansas

B. Eugene Wofford
University of Tennessee
Knoxville, Tennessee

FLORIDA

Loran C. Anderson†
Florida State University
Tallahassee, Florida

Alan R. Franck,
Regional Coordinator
Florida Museum of Natural History
Gainesville, Florida

Bruce F. Hansen
University of South Florida
Tampa, Florida

Richard P. Wunderlin
University of South Florida
Tampa, Florida

Preface for Volume 14

Since the publication of *Flora of North America* Volume 11, Parts 1 and 2 (the twenty-third volume in the *Flora* series) in 2023, the membership of the Flora of North America Association [FNAA] Board of Directors has not changed. As a result of a reorganization finalized in 2003, the FNAA Board of Directors succeeded the former Editorial Committee; for the sake of continuity of citation, authorship of *Flora* volumes is to be cited as "Flora of North America Editorial Committee, eds."

Most of the editorial process for this volume was done at the Hunt Institute for Botanical Documentation, Carnegie Mellon University, Pittsburgh, Pennsylvania. Art panel composition was done by John Myers, Gaston, Oregon. Occurrence map generation was carried out by Geoffrey A. Levin, Canadian Museum of Nature, Ottawa, Ontario. Pre-press processing as well as final editing and production took place at the Missouri Botanical Garden. Typesetting, layout, and panel correction was done by Tanya Harvey, Fall Creek, Oregon.

Illustrations published in this volume were executed by three very talented artists: Barbara Alongi, John Myers, and Yevonn Wilson-Ramsey. Barbara Alongi prepared illustrations for most of Apocynaceae and all of Gelsemiaceae, Gentianaceae, Hydroleaceae, and Loganiaceae; she also created the frontispiece depicting *Spigelia marilandica* (Loganiaceae). Yevonn Wilson-Ramsey prepared illustrations for all of Convolvulaceae, Solanaceae, and Sphenocleaceae and *Vinca minor* in Apocynaceae. John Myers illustrated *Pentalinon luteum* in Apocynaceae and also composed and labeled all of the line drawings that appear in this volume.

Starting with Volume 8, published in 2009, the circumscription and ordering of some families within the *Flora* have been modified so that they mostly reflect that of the Angiosperm Phylogeny Group [APG] rather than the previously followed Cronquist organizational structure. The groups of families found in this and future volumes in the series are mostly ordered following E. M. Haston et al. (2007); since APG views of relationships and circumscriptions have evolved, and will change further through time, some discrepancies in organization will occur. Volume 30 of the *Flora of North America* will contain a comprehensive index to the published volumes.

Support from many institutions and by numerous individuals has enabled the *Flora* to be produced. Members of the Flora of North America Association remain deeply thankful to the many people who continue to help create, encourage, and sustain the *Flora.*

Introduction

Scope of the Work

Flora of North America North of Mexico is a synoptic account of the plants of North America north of Mexico: the continental United States of America (including the Florida Keys and Aleutian Islands), Canada, Greenland (Kalâtdlit-Nunât), and St. Pierre and Miquelon. The *Flora* is intended to serve both as a means of identifying plants within the region and as a systematic conspectus of the North American flora.

The *Flora* will be published in 30 volumes. Volume 1 contains background information that is useful for understanding patterns in the flora. Volume 2 contains treatments of ferns and gymnosperms. Families in volumes 3–26, the angiosperms, were first arranged according to the classification system of A. Cronquist (1981) with some modifications, and starting with Volume 8, the circumscriptions and ordering of families generally follow those of the Angiosperm Phylogeny Group [APG] (see E. M. Haston et al. 2007). Bryophytes are being covered in volumes 27–29. Volume 30 will contain the cumulative bibliography and index.

The first two volumes were published in 1993, Volume 3 in 1997, and Volumes 22, 23, and 26, the first three of five volumes covering the monocotyledons, appeared in 2000, 2002, and 2002, respectively. Volume 4, the first part of the Caryophyllales, was published in late 2003. Volume 25, the second part of the Poaceae, was published in mid 2003, and Volume 24, the first part, was published in January 2007. Volume 5, completing the Caryophyllales plus Polygonales and Plumbaginales, was published in early 2005. Volumes 19–21, treating Asteraceae, were published in early 2006. Volume 27, the first of two volumes treating mosses in North America, was published in late 2007. Volume 8, Paeoniaceae to Ericaceae, was published in September 2009, and Volume 7, Salicaceae to Brassicaceae, appeared in 2010. In 2014, Volume 28 was published, completing the treatment of mosses for the flora area, and at the end of 2014, Volume 9, Picramniaceae to Rosaceae was published. Volume 6, which covered Cucurbitaceae to Droseraceae, was published in 2015. Volume 12, Vitaceae to Garryaceae, was published in late 2016. Volume 17, Tetrachondraceae to Orobanchaceae, was published in 2019. Volume 10, Proteaceae to Elaeagnaceae, was published in 2021. Volume 11, Parts 1 and 2, Fabaceae, was published in 2023. The correct bibliographic citation for the *Flora* is: Flora of North America Editorial Committee, eds. 1993+. Flora of North America North of Mexico. 24+ vols. New York and Oxford.

Volume 14 treats 624 species in 105 genera contained in 8 families. For additional statistics please refer to Table 1 on p. xx.

Contents · General

The *Flora* includes accepted names, selected synonyms, literature citations, identification keys, descriptions, phenological information, summaries of habitats and geographic ranges, and other biological observations. Each volume contains a bibliography and an index to the taxa included in that volume. The treatments, written and reviewed by experts from throughout the systematic botanical community, are based on original observations of herbarium specimens and, whenever possible, on living plants. These observations are supplemented by critical reviews of the literature.

Table 1. *Statistics for Volume 14 of Flora of North America.*

Family	Total Genera	Endemic Genera	Introduced Genera	Total Species	Endemic Species	Introduced Species	Conservation Taxa
Apocynaceae	36	2	15	175	73	23	16
Convolvulaceae	18	1	5	167	52	32	23
Gelsemiaceae	1	0	0	2	1	0	0
Gentianaceae	18	2	2	112	74	4	18
Hydroleaceae	1	0	0	5	4	0	0
Loganiaceae	3	0	1	11	6	1	3
Solanaceae	27	3	11	151	30	50	5
Sphenocleaceae	1	0	1	1	0	1	0
Totals	**105**	**8**	**35**	**624**	**240**	**111**	**65**

Basic Concepts

Our goal is to make the *Flora* as clear, concise, and informative as practicable so that it can be an important resource for both botanists and nonbotanists. To this end, we are attempting to be consistent in style and content from the first volume to the last. Readers may assume that a term has the same meaning each time it appears and that, within groups, descriptions may be compared directly with one another. Any departures from consistent usage will be explicitly noted in the treatments (see References).

Treatments are intended to reflect current knowledge of taxa throughout their ranges worldwide, and classifications are therefore based on all available evidence. Where notable differences of opinion about the classification of a group occur, appropriate references are mentioned in the discussion of the group.

Documentation and arguments supporting significantly revised classifications are published separately in botanical journals before publication of the pertinent volume of the *Flora.* Similarly, all new names and new combinations are published elsewhere prior to their use in the *Flora.* No nomenclatural innovations will be published intentionally in the *Flora.*

Taxa treated in full include extant and recently extinct or extirpated native species, named hybrids that are well established (or frequent), introduced plants that are naturalized, and cultivated plants that are found frequently outside cultivation. Taxa mentioned only in discussions include waifs known only from isolated old records and some non-native, economically important or extensively cultivated plants, particularly when they are relatives of native species. Excluded names and taxa are listed at the ends of appropriate sections, for example, species at the end of genus, genera at the end of family.

Treatments are intended to be succinct and diagnostic but adequately descriptive. Characters and character states used in the keys are repeated in the descriptions. Descriptions of related taxa at the same rank are directly comparable.

With few exceptions, taxa are presented in taxonomic sequence. If an author is unable to produce a classification, the taxa are arranged alphabetically and the reasons are given in the discussion.

Treatments of hybrids follow that of one of the putative parents. Hybrid complexes are treated at the ends of their genera, after the descriptions of species.

We have attempted to keep terminology as simple as accuracy permits. Common English equivalents usually have been used in place of Latin or Latinized terms or other specialized terminology, whenever the correct meaning could be conveyed in approximately the same space, for example, "pitted" rather than "foveolate," but "striate" rather than "with fine longitudinal lines." See *Categorical Glossary for the Flora of North America Project* (R. W. Kiger and D. M. Porter 2001; also available online at http://huntbot.andrew.cmu.edu) for standard definitions of generally used terms. Very specialized terms are defined, and sometimes illustrated, in the relevant family or generic treatments.

References

Authoritative general reference works used for style are *The Chicago Manual of Style,* ed. 14 (University of Chicago Press 1993); *Webster's New Geographical Dictionary* (Merriam-Webster 1988); and *The Random House Dictionary of the English Language*, ed. 2, unabridged (S. B. Flexner and L. C. Hauck 1987). *B-P-H/S. Botanico-Periodicum-Huntianum/Supplementum* (G. D. R. Bridson and E. R. Smith 1991), *BPH-2: Periodicals with Botanical Content* (Bridson 2004), and *BPH Online* [http://fmhibd.library.cmu.edu/HIBD-DB/bpho/findrecords.php] (Bridson and D. W. Brown) have been used for abbreviations of serial titles, and *Taxonomic Literature*, ed. 2 (F. A. Stafleu and R. S. Cowan 1976–1988) and its supplements by Stafleu et al. (1992–2009) have been used for abbreviations of book titles.

Graphic Elements

All genera and 25 percent of the species in this volume are illustrated. The illustrations may show diagnostic traits or complex structures. Most illustrations have been drawn from herbarium specimens selected by the authors. Data on specimens that were used and parts that were illustrated have been recorded. This information, together with the archivally preserved original drawings, is deposited in the Missouri Botanical Garden Library and is available for scholarly study.

Specific Information in Treatments

Keys

Dichotomous keys are included for all ranks below family if two or more taxa are treated. More than one key may be given to facilitate identification of sterile material or for flowering versus fruiting material.

Nomenclatural Information

Basionyms of accepted names, with author and bibliographic citations, are listed first in synonymy, followed by any other synonyms in common recent use, listed in alphabetical order, without bibliographic citations.

The last names of authors of taxonomic names have been spelled out. The conventions of *Authors of Plant Names* (R. K. Brummitt and C. E. Powell 1992) have been used as a guide for including first initials to discriminate individuals who share surnames.

If only one infraspecific taxon within a species occurs in the flora area, nomenclatural information (literature citation, basionym with literature citation, relevant other synonyms) is given for the species, as is information on the number of infraspecific taxa in the species and their distribution worldwide, if known. A description and detailed distributional information are given only for the infraspecific taxon.

Descriptions

Character states common to all taxa are noted in the description of the taxon at the next higher rank. For example, if sexual condition is dioecious for all species treated within a genus, that character state is given in the generic description. Characters used in keys are repeated in the descriptions. Characteristics are given as they occur in plants from the flora area. Characteristics that occur only in plants from outside the flora area may be given within square brackets, or instead may be noted in the discussion following the description. In families with one genus and one or more species, the family description is given as usual, the genus description is condensed, and the species are described as usual. Any special terms that may be used when describing members of a genus are presented and explained in the genus description or discussion.

In reading descriptions, the reader may assume, unless otherwise noted, that: the plants are green, photosynthetic, and reproductively mature; woody plants are perennial; stems are erect; roots are fibrous; leaves are simple and petiolate; flowers are bisexual, radially symmetric, and pediceled; perianth parts are hypogynous, distinct, and free; and ovaries are superior. Because measurements and elevations are almost always approximate, modifiers such as "about," "circa," or "±" are usually omitted.

Unless otherwise noted, dimensions are length × width. If only one dimension is given, it is length or height. All measurements are given in metric units. Measurements usually are based on dried specimens but these should not differ significantly from the measurements actually found in fresh or living material.

Chromosome numbers generally are given only if published and vouchered counts are available from North American material or from an adjacent region. No new counts are published intentionally in the *Flora.* Chromosome counts from nonsporophyte tissue have been converted to the $2n$ form. The base number (x =) is given for each genus. This represents the lowest known haploid count for the genus unless evidence is available that the base number differs.

Flowering time and often fruiting time are given by season, sometimes qualified by early, mid, or late, or by months. Elevations 100 m and under are rounded to the nearest 10 m, between 100 m and 200 m to the nearest 50 m, and over 200 m to the nearest 100 m. Mean sea level is shown as 0 m, with the understanding that this is approximate. Elevation often is omitted from herbarium specimen labels, particularly for collections made where the topography is not remarkable, and therefore precise elevation is sometimes not known for a given taxon.

The term "introduced" is defined broadly to refer to plants that were released deliberately or accidentally into the flora and that now are naturalized, that is, exist as wild plants in areas in which they were not recorded as native in the past. The distribution of introduced taxa are often poorly documented and changing, so the distribution statements for those taxa may not be fully accurate.

If a taxon is globally rare or if its continued existence is threatened in some way, the words "of conservation concern" appear before the statements of elevation and geographic range.

Criteria for taxa of conservation concern are based on NatureServe's (formerly The Nature Conservancy)—see http://www.natureserve.org—designations of global rank (G-rank) G1 and G2:

G1 Critically imperiled globally because of extreme rarity (5 or fewer occurrences or fewer than 1000 individuals or acres) or because of some factor(s) making it especially vulnerable to extinction.

G2 Imperiled globally because of rarity (5–20 occurrences or fewer than 3000 individuals or acres) or because of some factor(s) making it very vulnerable to extinction throughout its range.

The occurrence of species and infraspecific taxa within political subunits of the *Flora* area is depicted by dots placed on the outline map to indicate occurrence in a state or province. The Nunavut boundary on the maps has been provided by the GeoAccess Division, Canada Centre for Remote Sensing, Earth Science. Authors are expected to have seen at least one specimen documenting each geographic unit record (except in rare cases when undoubted literature reports may be used) and have been urged to examine as many specimens as possible from throughout the range of each taxon. Additional information about taxon distribution may be presented in the discussion.

Distributions are stated in the following order: Greenland; St. Pierre and Miquelon; Canada (provinces and territories in alphabetic order); United States (states in alphabetic order); Mexico (11 northern states may be listed specifically, in alphabetic order); West Indies; Bermuda; Central America (Belize, Costa Rica, El Salvador, Guatemala, Honduras, Nicaragua, Panama); South America; Europe, or Eurasia; Asia (including Indonesia); Africa; Atlantic Islands; Indian Ocean Islands; Pacific Islands; Australia; Antarctica.

Discussion

The discussion section may include information on taxonomic problems, distributional and ecological details, interesting biological phenomena, and economic uses.

Selected References

Major references used in preparation of a treatment or containing critical information about a taxon are cited following the discussion. These, and other works that are referred to in discussion or elsewhere, are included in Literature Cited at the end of the volume.

CAUTION

The Flora of North America Editorial Committee **does not encourage, recommend, promote, or endorse** any of the folk remedies, culinary practices, or various utilizations of any plant described within this volume. Information about medicinal practices and/or ingestion of plants, or of any part or preparation thereof, has been included only for historical background and as a matter of interest. Under no circumstances should the information contained in these volumes be used in connection with medical treatment. Readers are strongly cautioned to remember that many plants in the flora are toxic or can cause unpleasant or adverse reactions if used or encountered carelessly.

Key to boxed codes following accepted names:

- [C] of conservation concern
- [E] endemic to the flora area
- [F] illustrated
- [I] introduced to the flora area
- [W] weedy applies to taxa listed as weeds on the State and Federal Composite List of All U.S. Noxious Weeds (https://plants.usda.gov/java/noxComposite) and the Composite List of Weeds from the Weed Science Society of America (https://wssa.net/wssa/weed/composite-list-of-weeds), with input from treatment authors and reviewers.

Flora of North America

GENTIANACEAE Jussieu • Gentian Family

James S. Pringle

Herbs [**shrubs, trees**], annual, biennial, or perennial, autotrophic, with green stems and leaves, or mycotrophic; when strongly mycotrophic, stems and leaves weakly chlorophyllous (only in *Bartonia*) or yellowish, whitish, purplish, or buff, or lacking chlorophyll (only in *Voyria*). **Leaves** cauline, often also basal, opposite, whorled, or rarely alternate, sessile or petiolate, simple; stipules absent [rarely present as ocreae]; blade margins entire. **Inflorescences** cymes (sometimes racemoid, spicoid, or capitate), thyrses, or verticillasters, or solitary flowers; flowers pedicellate or sessile. **Flowers** bisexual or occasionally some unisexual [all unisexual on some or all plants], homostylous [heterostylous], protandrous and outbreeding or less often homogamous and autogamous, radially [somewhat bilaterally] symmetric, 4–12(–14)-merous [rarely 3-, 6-, or 16-merous] except for carpels; perianth hypogynous, calyx and usually corolla persistent; calyx green or occasionally ± hyaline (absent in *Obolaria*), sepals connate or some [or all] nearly distinct, lobes imbricate in bud, often ± unequal, colleters often present adaxially near base; corolla petaloid, petals connate, lobes contorted in bud or rarely imbricate (*Obolaria*, *Voyria*), spurs present only in *Halenia*, 1 per petal; stamens epipetalous, isomerous and alternate with petals, all fertile [rarely some sterile], equal [unequal]; filaments free or connected by a corona; anthers 2-locular, dehiscing longitudinally [with terminal pores], remaining straight, recurving, or coiling helically or circinately, distinct or (only in some spp. of *Gentiana*) coherent; pistil 1, 2-carpellate; ovary 1[or 2]-locular; placentae 2, parietal [axile]; style present or absent, erect or initially deflexed to one side [declinate], uncleft, shallowly 2-cleft, or deeply cleft (*Sabatia*); stigmas 1 or 2, coiling only in *Sabatia*, decurrent on ovary (only in *Lomatogonium*, sometimes slightly so in *Bartonia*). **Fruits** capsular, dehiscence septicidal or rarely rupturing irregularly (*Obolaria*) [indehiscent capsules, berries]. **Seeds** few–very many, usually sessile; endosperm abundant and embryo small in autotrophic species, endosperm scant and embryo undifferentiated in completely mycotrophic species (*Voyria*).

Genera ca. 100, species ca. 1800 (18 genera, 112 species in the flora): nearly worldwide.

In the tribal classification by L. Struwe et al. (2002), genera 1–7 in this flora are in tribe Chironieae Dumortier, subtribe Chironiinae G. Don. Species of Chironieae generally lack nectaries, although *Sabatia* reportedly has indistinct nectaries at the base of the ovary. Genera 8–17 are in tribe Gentianeae Dumortier. *Gentiana*, in which the nectaries are on the gynophore, is in subtribe Gentianinae G. Don; the remaining genera of the Gentianeae in the flora area, all of which have epipetalous nectaries, are in subtribe Swertiinae Grisebach. *Voyria* constitutes the monogeneric tribe Voyrieae Gilg, in which the nectaries (when present) are on the ovary or the gynophore.

Pedicel lengths given here refer to the true pedicels, between the most distal pair of bracts or bractlets and the calyx. In some genera, notably *Centaurium*, *Sabatia*, and *Zeltnera*, a flower terminating the ultimate branch of an inflorescence, directly subtended by bractlets, although sessile by this definition, may appear pedicellate. Corolla lengths as given are from the receptacle to the apices of the lobes (or plicae in *Gentiana andrewsii*).

The Gentianaceae include many species esteemed in ornamental horticulture. In addition to those noted under the respective genera, the more important species in North American horticulture include *Exacum affine* Balfour f. ex Regel, Persian-violet, native to the island of Socotra, Yemen, which is widely grown as a florists' pot plant.

SELECTED REFERENCES Mansion, G. 2004. A new classification of the polyphyletic genus *Centaurium* Hill (Chironiinae, Gentianaceae): Description of the New World endemic *Zeltnera*, and reinstatement of *Gyrandra* Griseb. and *Schenkia* Griseb. Taxon 53: 719–740. Struwe, L. and V. A. Albert. 2002. Gentianaceae: Systematics and Natural History. Cambridge. Struwe, L. and J. S. Pringle. 2019. Gentianaceae. In: K. Kubitzki et al., eds. 1990+. The Families and Genera of Vascular Plants. 15+ vols. Berlin etc. Vol. 15, pp. 453–503. Wood, C. E. Jr. and R. E. Weaver. 1982. The genera of Gentianaceae in the southeastern United States. J. Arnold Arbor. 63: 441–487.

1. Corollas with 4 spurs . . . . 8. *Halenia*, p. 34
1. Corollas without spurs.
    2. Leaves and stems yellowish, whitish, purplish, or buffy, without chlorophyll or weakly chlorophyllous; leaves scalelike, blades to 5 mm.
        3. Stems and leaves yellowish green or purplish; corollas narrowly campanulate, lobes 4, longer than tube; widely distributed in c, e North America (including Florida) . . . . 12. *Bartonia*, p. 49
        3. Stems and leaves white to pale buff; corollas salverform, lobes 5, shorter than tube; Florida only . . . . 18. *Voyria*, p. 92
    2. Leaves and usually stems green; some leaf blades 5+ mm.
        4. Flowers subtended by 2 separate, leaflike bracts but without calyx. . . . . 13. *Obolaria*, p. 52
        4. Flowers with calyx of 4–12(–14) sepals, some or all connate at least near base.
            5. Corollas with projecting summits of plicae between lobes, or (in *G. sceptrum*) with the summit of the plicae forming a truncate gap between the lobes . . . . 17. *Gentiana*, p. 72
            5. Corollas without plicae or truncate gaps between lobes.
                6. Stigmas decurrent along sutures of ovary; corollas rotate to widely campanulate . . . . 9. *Lomatogonium*, p. 36
                6. Stigmas not decurrent on ovary; corollas salverform, funnelform, campanulate, subrotate, or rotate.
                    7. Nectaries in pits prominent adaxially on corolla lobes, pit openings (and sometimes adjacent areas on corolla lobes) with fringed or projecting margins.
                        8. Nectaries 1 per corolla lobe, or if 2 with both opening into single fringe-rimmed area on corolla lobe . . . . 11. *Frasera* (in part), p. 39
                        8. Nectaries 2 per corolla lobe, with completely separate openings each surrounded by fringed rim.
                            9. Cauline leaves opposite or alternate; corollas blue or violet-blue (rarely greenish white), 4- or 5-lobed . . . . 10. *Swertia*, p. 38
                            9. Cauline leaves whorled; corollas yellowish green with purple spots and occasionally purple suffusion, 4-lobed . . . . 11. *Frasera* (in part), p. 39
                    7. Nectaries, if present, not in pits with fringe-rimmed openings (corolla may be fringed at throat, but fringes do not surround nectary-pit openings).

[10. Shifted to left margin.—Eds.]

10. Corollas rotate; styles cleft 1+ mm, style branches and stigmas often helically coiled; anthers coiling circinately or remaining nearly straight . . . . . 5. *Sabatia*, p. 21
10. Corollas tubular, salverform, funnelform, or campanulate; styles not cleft or cleft to 1 mm, neither style branches nor stigmas coiling; anthers remaining straight or coiling helically.
    11. Corollas widely campanulate, lobes 2+ times as long as tube . . . . . 6. *Eustoma*, p. 32
    11. Corollas funnelform or salverform, lobes shorter than 1.5 times tube.
        12. Corollas with fringes of trichomes or fringed scales on adaxial surface near base of lobes; margins of lobes not fringed.
            13. Pedicels longer than subtending internodes; nectaries 2 times as many as corolla lobes. . . . . 14. *Comastoma*, p. 54
            13. Pedicels mostly shorter than subtending internodes; nectaries same number as corolla lobes. . . . . 15. *Gentianella* (in part), p. 55
        12. Corollas without fringes or scales on adaxial surface near base of lobes; margins of lobes fringed or not.
            14. Corollas tubular, funnelform, or campanulate, or if ± salverform then with fringes or conspicuous teeth on margins of lobes, lobes not abruptly spreading horizontally at summit of tube, margins entire, dentate-serrate, or fringed.
                15. Corolla lobe margins entire, corollas 0.4–3 cm, lobes 4 or 5, shorter than tube . . . . . 15. *Gentianella* (in part), p. 55
                15. Corolla lobe margins dentate-serrate or fringed, corollas (1.2–)2–8 cm, lobes 4, ± as long as tube, or if shorter then with margins as above. . . . . 16. *Gentianopsis*, p. 62
            14. Corollas salverform, with lobes abruptly spreading ± horizontally at summit of slender tube (corollas often closing in specimen preparation), margins entire or minutely erose near apex only.
                16. Corollas yellow; anthers remaining straight, not coiling . . . . . 7. *Cicendia*, p. 33
                16. Corollas pink to rose-violet or occasionally white; anthers coiling helically.
                    17. Inflorescences largely spicate, only proximally, if at all, dichasial . . . . . 2. *Schenkia*, p. 8
                    17. Inflorescences dichasially or partly monochasially cymose (distally sometimes racemoid or subcapitate).
                        18. Stigmas 2, elliptic to ovate or orbiculate; capsules cylindric . . . . . 1. *Centaurium*, p. 5
                        18. Stigmas 2, fan-shaped, or 1, 2-lobed (sometimes appearing subcapitate in *Zeltnera trichantha*); capsules ovoid to ellipsoid.
                            19. Stigmas 2, fan-shaped, or 1, with 2 ± fan-shaped lobes . . . . . 3. *Zeltnera*, p. 9
                            19. Stigma 1, shallowly 2-lobed with hemispherical lobes . . . . . 4. *Gyrandra*, p. 20

1. CENTAURIUM Hill, Brit. Herb., 62, plate 9 [upper left]. 1756 • Centaury [Greek *kentauros*, centaur, alluding to plant's supposed medicinal use by Chiron in Greek mythology] [I]

James S. Pringle

Herbs annual or biennial [perennial], chlorophyllous, glabrous [stems papillose-puberulent]. **Leaves** cauline, opposite, often also basal. **Inflorescences** dichasial or partly monochasial cymes. **Flowers** (4–)5-merous; calyx deeply lobed nearly to base; corolla pink to rose-violet or occasionally white [salmon, yellow], with whitish eye, salverform, glabrous, lobes abruptly spreading, elliptic-oblong, shorter than tube, margins entire or erose-tipped, plicae between lobes absent; stamens inserted in distal 1/2 of corolla tube, all initially deflexed to one side [not deflexed]; anthers free, coiling helically at dehiscence; ovary sessile; style deciduous, distinct, in most species initially deflexed away from stamens, shallowly cleft; stigmas 2, ovate, elliptic, or orbiculate; nectaries absent. **Capsules** cylindric. $x = 10$; polyploidy frequent, aneuploidy occasional.

Species ca. 15 (3 in the flora): introduced; Eurasia, n Africa, Pacific Islands (New Zealand), Australia; introduced also in Mexico, West Indies, Central America, and South America; temperate to dry-mesic tropical regions.

Erythraea Borkhausen is a name formerly widely used but has been determined to be illegitimate.

G. Mansion (2004) and Mansion and L. Struwe (2004) concluded from molecular phylogenetic studies that the native North American species that had been included in *Centaurium*, with the exception of *C. capense* C. R. Broome in Baja California Sur, are more closely related to *Eustoma*, *Sabatia*, and the European genus *Exaculum* Caruel than to *Centaurium* in the narrow sense, and that *Centaurium* in the narrow sense is more closely related to the African genera *Chironia* Linnaeus and *Orphium* E. Meyer. They segregated the native North American species as *Zeltnera*.

The introduced *Centaurium* species often colonize disturbed sites, where populations may be of short duration. All of the species in the flora area are likely to be found in additional provinces and states and may disappear from others.

Plant size varies greatly within the species, being affected by environmental factors and by the interaction of photoperiod with the time of seed germination. The smallest plants often do not exhibit the characteristics of branching pattern, basal-leaf persistence, and flower size by which the respective species are usually distinguished.

Both outbreeding and autogamy probably occur in the species of *Centaurium* in the flora area. The style is at first deflexed in one direction and all stamens in the opposite. Later, both the style and the stamens become erect, at which stage autogamy may occur.

SELECTED REFERENCES Mansion, G. and L. Struwe. 2004. Generic delimitation and phylogenetic relationships within the subtribe Chironiinae (Chironieae: Gentianaceae), with special reference to *Centaurium*: Evidence from nrDNA and cpDNA sequences. Molec. Phylogen. Evol. 42: 951–977. Mansion, G., L. Zeltner, and F. Bretagnolle. 2005. Phylogenetic patterns and polyploid evolution within the Mediterranean genus *Centaurium* (Gentianaceae-Chironieae). Taxon 54: 931–950. Melderis, A. 1972. Taxonomic studies on the European species of the genus *Centaurium* Hill. J. Linn. Soc., Bot. 65: 224–250. Zeltner, L. 1970. Recherches biosystématique sur les genres *Blackstonia* Huds. et *Centaurium* Hill. Bull. Soc. Neuchâtel. Sci. Nat. 93: 1–164.

1. Flowers with distinct pedicels 1–5(–11) mm, all plants with some pedicels 2+ mm .... 3. *Centaurium pulchellum*
1. Flowers sessile or subsessile, pedicels to 2 mm.
   2. Basal rosettes generally well developed and present at anthesis; corolla lobes (3–)4–8 mm .... 1. *Centaurium erythraea*
   2. Basal rosettes generally poorly developed or absent at anthesis; corolla lobes (1–)2–4.5 mm .... 2. *Centaurium tenuiflorum*

1. Centaurium erythraea Rafn, Danm. Holst. Fl. 2: 75. 1800 • European centaury, petite-centaurée commune F I

Herbs annual or biennial, 3–60 cm. **Stems** 1–several, branching mostly above middle. **Leaves:** rosette of basal leaves usually present at flowering; blade elliptic to spatulate-obovate, 15–70 × 5–20 mm, apex rounded to subacute; cauline blades elliptic (proximal) to lanceolate or linear (distal), 8–50 × 1–8 mm, apex obtuse to acute. **Inflorescences** dense, ± corymboid, dichasial cymes; flowers sessile or occasionally on pedicels to 2 mm. **Flowers:** calyx (2–)5–7 mm; corolla (6–)10–17 mm, lobes (3–)4–8 mm; anthers (1–)2–2.5 mm; stigmas ovate-elliptic. **Seeds** brown. $2n$ = 40 (Europe), 42 (Spain).

Flowering summer–early fall. Fields, roadsides, other open, disturbed habitats; 0–1000 m; introduced; Man., N.S., Ont., Que.; Calif., Idaho, Ind., La., Md., Mass., Mich., Mont., N.Y., N.C., Ohio, Oreg., Pa., Vt., Va., Wash.; Eurasia; introduced also in Central America, South America, Pacific Islands (New Zealand), Australia.

The name *Centaurium minus* Moench is a synonym of *C. littorale* (Turner) Gilmour but has often been misapplied to *C. erythraea*. *Centaurium minus* Garsault and *C. umbellatum* Gilibert are invalidly published names that have been applied to *C. erythraea*.

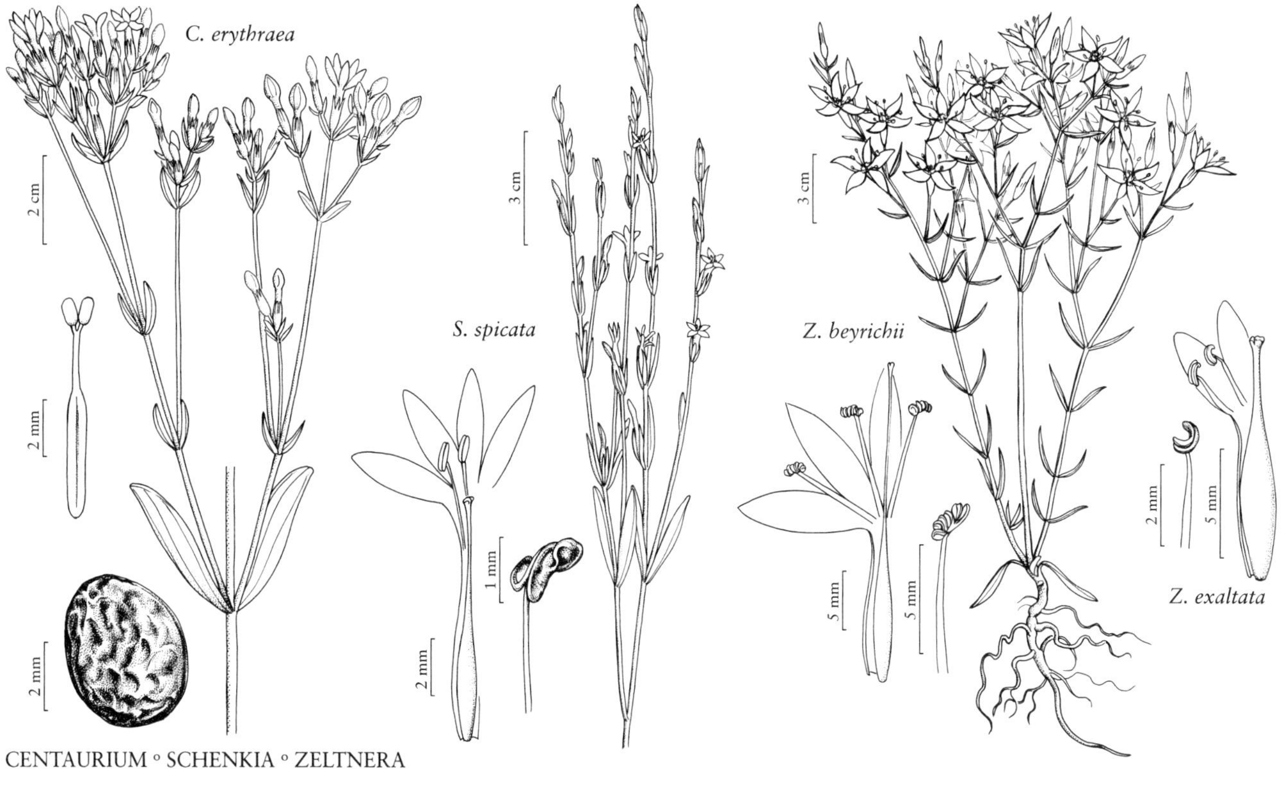

CENTAURIUM ∘ SCHENKIA ∘ ZELTNERA

In the flora area, *Centaurium erythraea* appears to be represented only by what is called subsp. *erythraea* in more inclusive circumscriptions of the species, for example, A. Melderis (1972). The one chromosome count for a North American population is consistent with this interpretation.

2. **Centaurium tenuiflorum** (Hoffmannsegg & Link) Fritsch, Mitt. Naturwiss. Vereins Univ. Wien, n. s. 5: 97. 1907 (as Centaurion) • Slender or June centaury [I]

Erythraea tenuiflora Hoffmannsegg & Link, Fl. Portug. 1: 354, plate 67. 1820

Herbs annual, 2–75(–90) cm. **Stems** usually 1, simple or branching above or near middle or occasionally with few, slender branches from base. **Leaves:** rosette of basal leaves present or absent at flowering but not forming well-developed rosette; blade obovate to oblong, 15–25 × 3–15 mm, apex rounded to obtuse; cauline blades obovate to elliptic-oblong, ovate, or lanceolate (proximal) to narrowly lanceolate (distal), 10–30(–42) × 2–9(–18) mm, apex obtuse (proximal) to acute. **Inflorescences** dense, ± corymboid, dichasial cymes; flowers sessile or occasionally on pedicels to 2 mm. **Flowers:** calyx 5–9(–11) mm; corolla 8–14(–17) mm, lobes (1–)2–4.5 mm; anthers 0.7–1.7 mm; stigmas elliptic. **Seeds** yellowish to reddish brown. **2*n*** = 40.

Flowering summer–fall. Along streams, edges of marshes, seasonally damp meadows, bluffs, and roadsides; 0–1800 m; introduced; Ala., Calif., La., Miss., Okla., Oreg., Tex.; Eurasia; n Africa; introduced also in West Indies, South America, Pacific Islands (New Zealand), Australia.

Pending the availability of a more satisfactory classification, the name *Centaurium tenuiflorum* is used here in a broad sense. According to G. Mansion et al. (2005), *C. tenuiflorum* in the broad sense includes a diploid entity, *C. tenuiflorum* subsp. *acutiflorum* (Schott) L. Zeltner [sometimes treated as *C. acutiflorum* (H. W. Schott) Druce]; a probable autotetraploid, *C. tenuiflorum* subsp. *tenuiflorum*; and an unnamed entity believed to be an allotetraploid derivative of diploid *C. tenuiflorum* × *C. erythraea*. The first two are native in Europe and not known from North America. The entity that has become naturalized outside its native range, including the flora area, is the allotetraploid.

Centaurium tenuiflorum has often been assumed to be native in California and has incorrectly been called *C. floribundum* (Bentham) B. L. Robinson or *C. muehlenbergii* (Grisebach) W. F. Wight ex Piper (J. S. Pringle 2010b). Those names are correctly associated with *Zeltnera muehlenbergii*.

Exceptional plants of *Centaurium tenuiflorum* are more diffusely branched and have longer primary branches than is usual. In the most extreme cases, the plants may be branched from near the base, with large numbers of flowers borne in noncorymboid, witch's-broomlike inflorescences. These plants may represent the results of injuries, virus infections, or unusual environmental conditions.

3. **Centaurium pulchellum** (Swartz) Hayek ex Handel-Mazzetti, Stadlmann, Janchen & Faltis, Oesterr. Bot. Z. 56: 70. 1906 • Lesser or branching centaury, centaurée élégante [I]

Gentiana pulchella Swartz, Kongl. Vetensk. Acad. Nya Handl. 4: 85, plate 3, figs. 8, 9. 1783

Herbs annual, 2–25(–30) cm. **Stems** 1–5 (sometimes appearing more numerous because of near-basal branching), branching throughout or in small plants often only above middle. **Leaves:** basal leaves withered or occasionally persistent at flowering; blade obovate to elliptic-oblong, 5–25 × 2–6 mm, apex usually acute to acuminate, occasionally obtuse; cauline blades elliptic-oblong (proximal) to lanceolate (distal), 5–15(–25) × 1–5(–12) mm, apex usually acute to acuminate, occasionally obtuse. **Inflorescences** dense to ± open, dichasial or occasionally distally monochasial cymes, usually not distinctly corymboid; pedicels 1–5(–11) mm. **Flowers:** calyx (3–)5–9(–11) mm; corolla (5–)10–15(–17) mm, lobes (1–)2–5 mm; anthers 0.7–1.1 mm; stigmas widely ovate to elliptic or orbiculate. **Seeds** dark brown or reddish brown. $2n = 36$ (Europe, w Asia).

Flowering spring (southward)–fall. Moist fields, other moist to wet, open places; 0–1500 m; introduced; N.B., N.S., Ont., Que.; Colo., Conn., Ill., Ind., La., Maine, Md., Mass., Mich., Minn., Miss., Nebr., N.J., N.Y., Ohio, Pa., S.Dak., Tex., Vt., Va., Wash., Wis., Wyo.; Eurasia; introduced also in West Indies, South America, Australia.

There are historical records of *Centaurium pulchellum* from California, Delaware, and North Carolina.

In relatively large plants of *Centaurium pulchellum* in open habitats, the first branching is usually no more than three (rarely four) nodes above the base, with the branches being strongly divaricate. Each branch usually comprises a single internode, terminating in a central flower and two lateral branches, so that most of the plant above ground is a dichasial cyme. In *C. tenuiflorum,* the first branching is usually (four or) five or more nodes above the base, near or above mid-stem, with the branches diverging at narrower angles, forming a corymboid inflorescence more compact than that of *C. pulchellum.* In the flora area, however, *C. pulchellum* is usually represented by small plants that branch only above the middle. Such plants can be recognized as *C. pulchellum* by the consistently present although short pedicels and more slender stems, and sometimes by their occurrence with larger plants having the more usual branching pattern of that species.

Infraspecific taxa are sometimes recognized in *Centaurium pulchellum* in the broad sense in Asia. In the flora area, *C. pulchellum* is represented only by subsp. *pulchellum.*

2. SCHENKIA Grisebach, Bonplandia (Hannover) 1: 226. 1853 • Centaury [For Joseph August Schenk, 1815–1891, (Austrian-) German botanist and palaeontologist] [I]

James S. Pringle

Herbs annual or biennial, chlorophyllous, glabrous. **Leaves** basal and cauline, opposite. **Inflorescences** spikelike [racemoid], monochasial cymes, sometimes dichasial at base. **Flowers** 5-merous; calyx lobed nearly to base; corolla pink to rose-violet, salverform, glabrous, lobes abruptly spreading, elliptic-oblong, shorter than [± as long as] tube, margins entire or erose-tipped, plicae between lobes absent; stamens inserted in or near corolla sinuses, diverging radially; anthers distinct, coiling helically at dehiscence; ovary sessile; style deciduous, erect, distinct, not cleft; stigmas 2, [1, 2-lobed]; nectaries absent. **Capsules** ellipsoid. $x = 11$.

Species 5 (1 in the flora): introduced, Massachusetts; Eurasia, n Africa, Pacific Islands, Australia; temperate to dry-mesic tropical regions.

G. Mansion (2004) and Mansion and L. Struwe (2004) inferred from molecular studies that *Schenkia* is more closely related to *Zeltnera* and the European *Exaculum* Caruel than to *Centaurium* in the narrow sense. The stigma morphology of *Schenkia* is similar to that of *Zeltnera*.

Schenkia differs from *Centaurium* and *Zeltnera* most distinctly in its spicate inflorescences, although the inflorescence of *S. spicata* is often dichasial at the one or two most proximal divisions. Conversely, the inflorescences of some *Zeltnera* species, such as *Z. muehlenbergii*, are distally monochasial with the pedicels short or none. Although *Schenkia* generally differs from *Zeltnera* in having the stamens inserted in rather than below the sinuses of the corolla, *S. spicata* varies in this respect, sometimes having the stamens inserted slightly below the sinuses (G. Mansion 2004).

1. **Schenkia spicata** (Linnaeus) G. Mansion, Taxon 53: 726. 2004 • Spiked centaury [F] [I]

Gentiana spicata Linnaeus, Sp. Pl. 1: 230. 1753; *Centaurium spicatum* (Linnaeus) Fritsch

Herbs 4–55 cm. **Stems** 1–several, branching throughout. **Leaves:** basal often withered or absent by flowering; blade widely ovate to elliptic, 6–30 × 2–17 mm, apex obtuse to acute; cauline blades elliptic (proximal) to lanceolate (distal), (6–)10–30(–45) × 3–8(–12) mm, apex acute. **Inflorescences** with central flower at proximal 1 or 2 divisions, otherwise all or most flowers sessile or subsessile. **Flowers:** calyx (4–)7–11 mm; corolla 10–15 mm, lobes 3.5–5.5 mm; anthers 1–1.5 mm; stigmas 2, widely fan-shaped. **Seeds** dark reddish brown to black. $2n = 22$ (Mediterranean region and Russia).

Flowering summer–early fall. Sandy borders of salt marshes; 0 m; introduced; Mass.; Eurasia.

In the flora area, *Schenkia spicata* is established only on the island of Nantucket, records from which span more than 150 years. It was formerly known from Maryland and Virginia.

Taxa sometimes treated as varieties or subspecies of *Schenkia spicata* (as *Centaurium*) were treated as distinct species of *Schenkia* by G. Mansion (2004), whose treatment is followed here, and by some earlier authors. The plants in the flora area are *S. spicata* in the narrow sense.

3. ZELTNERA G. Mansion, Taxon 53: 727, figs. 4, 5I–N. 2004 • Centaury [For Louis Zeltner, b. 1938, and his wife, Nicole Zeltner, b. 1934, Swiss botanists]

James S. Pringle

Herbs annual, biennial, or perennial, glabrous throughout except in *Z. glandulifera*. **Leaves** cauline, opposite, often also basal. **Inflorescences** dichasial or partly to largely monochasial cymes. **Flowers** 4- or 5-merous; calyx deeply lobed, lobes narrowly linear, usually ± ridged, distinctly keeled only in *Z. davyi*; corolla pink to rose-violet (white), with ± distinct white or occasionally pale green or pale yellow eye, salverform, tube generally constricted above ovary (scarcely so in *Z. nudicaulis*), lobes abruptly spreading, shorter to slightly longer than tube, margins entire or erose-tipped, neither adaxial scales or trichomes nor plicae between lobes present; stamens inserted in distal 1/2 of corolla tube, in some species all initially curved to one side; anthers distinct, coiling helically at dehiscence; ovary sessile; style deciduous, erect or initially deflexed away from stamens, distinct, not cleft or cleft to 1 mm or less; stigmas 2 or, if 1, 2-lobed, stigmas or lobes generally ± fan-shaped and divergent, lobes connivent and sometimes appearing as a subcapitate stigma only in *Z. trichantha*; nectaries absent. **Capsules** narrowly ellipsoid to ovoid. $x = 17, 21$; disploidy frequent.

Species ca. 25 (14 in the flora): North America, Mexico, West Indies, Central America, South America; temperate to tropical regions.

Plant size varies greatly in *Zeltnera* species, especially in *Z. venusta* and others that sometimes grow in intermittently moist habitats such as the edges of vernal pools. Small, one- or two-flowered plants, some only 3–10 cm, do not exhibit the characters of branching pattern, basal-leaf numbers or persistence, leaf and flower size, pedicel length, and orientation of styles and stamens by which the species are usually distinguished. It is scarcely feasible to identify such plants strictly from their own morphology; their provenance and the identity of larger plants in the vicinity must be considered. Branching pattern appears also to be affected by the density of the surrounding vegetation.

Some *Zeltnera* species are primarily outbreeding. They have large, showy corollas with conspicuous, sharply defined eyes. The style is at first deflexed in one direction and the stamens in the opposite. Later, the style becomes erect, and the stamens become radially disposed and erect to incurved. In these species, the stigma or stigmas are borne on a slender style often as long as or longer than the ovary at anthesis and much exceeding the stamens, although the latter are also distinctly exserted. Style length may vary among the flowers on a single plant, although true heterostyly is not known to occur in the genus. Smaller, presumably autogamous flowers may be present along with larger flowers. Other species are primarily, although probably not obligately, autogamous. They have smaller, often less brightly colored corollas with poorly defined eyes. The styles are shorter and generally erect, and the filaments may be erect or incurved throughout the life of the flower. *Zeltnera arizonica*, *Z. beyrichii*, *Z. calycosa*, *Z. maryanniana*, *Z. trichantha*, and *Z. venusta* exhibit the floral syndrome associated with outbreeding; the other taxa in the flora area, to various degrees, exhibit the morphology associated with autogamy (C. R. Broome 1973).

G. Mansion and L. Zeltner (2004) divided *Zeltnera* informally into a Californian group including species 1 through 6, a Texan group including species 7 through 14, and a Mexican group represented in the flora area only by *Z. nudicaulis*. Within the respective species-groups, especially the Californian group, some of the species are poorly defined morphologically. Some species, especially *Z. venusta*, vary greatly in the size and shape of the corolla lobes as well as in habit and vegetative characters and may thus resemble other species. Hybridization, introgression, and allopolyploidy are believed to account for some of the variation that has made *Zeltnera* problematic taxonomically, although some variation suspected of being due to hybridization may merely represent variability within species. Some otherwise well-differentiated species tend to be more similar to each other in morphology where their ranges overlap; in such cases, similarity has been interpreted as being due more to convergent responses to selective pressures than to interspecific hybridization (C. R. Broome 1981). In some localities, populations of intermediates appear to be more or less stabilized. In the classification presented here, plants treated as or suspected of being hybrids and introgressants may be more numerous and more widely distributed than one would consider pragmatic. Nevertheless, this treatment seems to present a truer picture of the evolutionary situation than would the recognition of intermediates between sympatric species as additional species or subspecies. In California, intergradation has been reported to involve, in various combinations, *Z. davyi*, *Z. exaltata*, *Z. muehlenbergii*, *Z. namophila*, *Z. trichantha*, and *Z. venusta*. In Texas, *Z. calycosa* appears to hybridize with *Z. beyrichii* and *Z. texensis*.

SELECTED REFERENCES Mansion, G. 2004. A new classification of the polyphyletic genus *Centaurium* Hill (Chironiinae, Gentianaceae): Description of the New World endemic *Zeltnera*, and reinstatement of *Gyrandra* Griseb. and *Schenkia* Griseb. Taxon 53: 719–740. Mansion, G. and L. Zeltner. 2004. Phylogenetic relationships within the New World endemic *Zeltnera* (Gentianaceae-Chironiinae) inferred from molecular and karyological data. Amer. J. Bot. 91: 2069–2086.

1. Stems and midveins of leaves and sepals scabridulous to papillate-puberulent . . . . . . 12. *Zeltnera glandulifera*
1. Stems, leaves, and calyces glabrous.
   2. Pedicels mostly 10–75 mm, all or most of them longer than closed corollas.
      3. Corolla 6.5–10(–12.5) mm, tube scarcely constricted above ovary; distal cauline leaf blades usually narrowly linear to filiform, less than 2 mm wide, occasionally lanceolate to ovate or elliptic, to 4 mm wide . . . . 14. *Zeltnera nudicaulis*
      3. Corolla 9–20 mm, tube distinctly constricted above ovary; distal cauline leaf blades linear to lanceolate or ovate, mostly (1–)2–6(–9) mm wide.
         4. Flowers more than 3 per stem except on unusually small plants; corollas usually 4-merous, lobes ca. 1/2 as long as tube or less . . . . 1. *Zeltnera exaltata*
         4. Flowers usually 1–3 per stem; corollas 5-merous, lobes more than 1/2 as long as tube . . . . 7. *Zeltnera multicaulis*
   2. Pedicels 0–30(–60) mm, all or most of them shorter than closed corollas.
      5. Mid-stem and distal cauline leaf blades filiform to narrowly linear, narrowly lanceolate, or narrowly oblanceolate, most less than 2 mm wide and ± as wide as stem diam.
         6. Primary stems (except those of small plants) and larger branches with well-developed central axis, with dichasial branching if any only near summit and in lateral cymules . . . . 2. *Zeltnera namophila*
         6. Branching largely or entirely dichasial, with a central flower at each fork, or distally monochasial.
            7. Basal leaves absent; corolla lobes 3–7 mm . . . . 10. *Zeltnera texensis*
            7. Basal leaves or withered remains usually numerous and conspicuous at flowering; corolla lobes 7–16 mm.
               8. Primary stems 3–30; basal leaves generally green at flowering; corolla lobes ovate-elliptic . . . . 11. *Zeltnera maryanniana* (in part)
               8. Primary stems 1, although often branching near base; basal leaves often withered or absent at flowering; corolla lobes narrowly oblong-lanceolate to linear . . . . 13. *Zeltnera beyrichii*
      5. Mid-stem and distal cauline leaf blades linear to lanceolate or wider, most usually 2+ mm wide and/or distinctly wider than stem diam.
         9. Pedicels mostly less than 6 mm, distal flowers often sessile.
            10. Apices of corolla lobes acute to acuminate; stigmas 2, closely appressed and sometimes appearing as 1 . . . . 4. *Zeltnera trichantha*
            10. Apices of corolla lobes obtuse to subacute; stigmas 2, divergent . . . 5. *Zeltnera muehlenbergii*
         9. Pedicels (2–)4–30(–60) mm.
            11. Calyx lobes with distinct keels proximally 0.3–0.6 mm wide; corolla lobes 3–7 mm . . . . 6. *Zeltnera davyi*
            11. Calyx lobes ± sharply ridged but not distinctly keeled; corolla lobes (5–)6–20 mm.
               12. Basal leaves usually numerous and green at flowering; stigmas 1, lobes tardily diverging . . . . 11. *Zeltnera maryanniana* (in part)
               12. Basal leaves few, present or absent at flowering; stigmas 2, lobes ± divergent.
                  13. Leaves usually cauline only; corolla lobes widening ± abruptly above base (except in forms with very narrow lobes), proportions variable . . . . 3. *Zeltnera venusta*
                  13. Basal leaves usually present at flowering; corolla lobes widening gradually above base, less than 1.5 times as wide distally as at base.
                     14. Plants (4–)9–30 cm; primary stems usually 1; corollas 12–23 mm . . . . 8. *Zeltnera calycosa*
                     14. Plants (10–)20–50(–60) cm; stems often several from base; corollas (closed), (15–)18–25 mm . . . . 9. *Zeltnera arizonica*

1. Zeltnera exaltata (Grisebach) G. Mansion, Taxon 53: 731. 2004 • Great Basin or tall or desert centaury [F]

Cicendia exaltata Grisebach in W. J. Hooker, Fl. Bor.-Amer. 2: 69, plate 157, fig. A. 1837; *Centaurium exaltatum* (Grisebach) W. Wight; *C. nuttallii* (S. Watson) A. Heller

Herbs annual, (3–)10–60 cm. **Stems** 1–10, simple below inflorescence (small plants) or branching variously, sometimes ± throughout, but branches usually few. **Leaves:** basal present or occasionally ± withered by flowering, similar to proximal cauline leaves or larger; cauline blades oblong-elliptic to lanceolate (proximal or occasionally all) to linear (distal), 10–30(–50) × 1–10(–17) mm, apex acute or proximal leaves obtuse. **Inflorescences** proportionately narrow cymes, proximally dichasial, distally monochasial (on larger plants) or completely monochasial; pedicels (2–, on ultimate branches)10–70 mm. **Flowers** 4-merous; calyx (4–)6–11 mm; corolla 10–20 mm, lobes lanceolate to oblong or narrowly elliptic-obovate, 2.5–6 × 0.8–2.5 mm, that is, ca. 1/2 as long as tube or less, apex truncate to rounded or obtuse; stigmas 2, fan-shaped. **Seeds** dark reddish brown to nearly black. **2*n*** = 40, 74.

Flowering spring–early fall. Stream banks, marshes, lakeshores, margins of hot springs and vernal pools, other wet, alkaline places often surrounded by desert; 200–3100 m; B.C.; Calif., Colo., Idaho, Nev., N.Mex., Oreg., S.Dak., Utah, Wash., Wyo.; Mexico (Baja California, Baja California Sur, Chihuahua).

There is a historical record of *Zeltnera exaltata* from Montana. An old record from Nebraska is considered to be based on an adventive occurrence of short duration.

G. Mansion and L. Zeltner (2004) reported that plants compatible with descriptions of *Zeltnera exaltata* and similar in molecular characters included some populations with 2*n* = 40 and others with 2*n* = 74. They considered the latter likely to be of allopolyploid origin, derived from the hybridization of a diploid component of *Z. exaltata* in the narrow sense with a species having 2*n* = 34. Plants in the South Coastal Ranges from Baja California north to Monterey County, California, from which region Mansion and Zeltner reported 2*n* = 40, have smaller flowers, with the closed corollas 10–14 mm, subglobose seeds 0.25–0.33 mm in diameter, and pollen ca. 22 µm in diameter (C. R. Broome 1973). Plants (except those less than 10 cm) from localities farther inland and northward, from regions in which Mansion and Zeltner found 2*n* = 74, have corollas 14–20 mm, ellipsoid seeds 0.5–0.75 mm long, and pollen grains ca. 30 µm in diameter. By typification, the name *Z. exaltata* in the narrow sense is applicable to the entity occurring in the more northern and inland localities, with 2*n* = 74; if the species should be divided, it is the populations of the South Coastal Ranges, with 2*n* = 40, that should be treated as new. None of the names listed in the synonymy of *Z. exaltata* by Broome and Mansion (2004) is typified by specimens from the South Coast Ranges.

Zeltnera exaltata varies greatly in the number of stems arising from the base, the presence or absence of basal leaves at flowering time, and, especially in its easternmost populations, leaf width. The narrow angle of branching and the long pedicels generally give plants of this species a distinctive appearance. Its corollas are usually four-merous, whereas five-merous corollas prevail in the other *Zeltnera* species in the flora area (although four-merous corollas are not uncommon in *Z. nudicaulis*), and the corolla lobes, being about half as long as the tube, are proportionately shorter than those of the other *Zeltnera* species in the flora area. The four-merous corollas are useful in identifying very small plants of this species. The small corolla lobes are useful in distinguishing *Z. exaltata* from *Z. multicaulis*, which likewise has long pedicels but corolla lobes nearly as long as the tube.

In Nevada and adjacent regions of California, *Zeltnera exaltata* appears to intergrade with *Z. namophila*. As these species differ in chromosome number, additional chromosome counts and other techniques appropriate for the study of hybridization will be necessary for a satisfactory interpretation of apparent intermediates.

2. Zeltnera namophila (Reveal, C. R. Broome & Beatley) G. Mansion, Taxon 53: 732. 2004 • Spring-loving centaury [C] [E]

Centaurium namophilum Reveal, C. R. Broome & Beatley, Bull. Torrey Bot. Club 100: 353, fig. 1. 1974

Herbs annual, (5–)15–45 cm. **Stems** often 1, occasionally 2–5 (–10), branching ± throughout. **Leaves** cauline; blade narrowly lanceolate (proximal) to linear or filiform (distal or all), 10–50 × 1–5 mm, apex subacute to acute. **Inflorescences** thyrses of largely dichasial cymules; central pedicels in divisions of proximal cymules 5–20 mm, other pedicels 0–9 mm. **Flowers** generally 5-merous; calyx 6–9 mm; corolla 13–18 mm, lobes lanceolate, 5–8 mm, apex acute to acuminate except for blunt extreme tip; stigma 1, shallowly 2-lobed, lobes fan-shaped. **Seeds** black. **2*n*** = 34.

Flowering spring–fall. Wet meadows near springs and streams; of conservation concern; 600–1400 m; Calif., Nev.

Zeltnera namophila has been reported from Utah, but documentation has not been found in studies for this flora.

Although plants of *Zeltnera namophila* sometimes have been identified as *Z. exaltata*, these species are usually distinct in morphology. The larger plants of *Z. exaltata* are often several-stemmed at the base, with strongly ascending branches. Branching is largely dichasial proximally, monochasial distally. Most of the pedicels (distal to all bracts) are longer than the flowers. *Zeltnera namophila*, regardless of plant size, is often although not invariably single-stemmed at the base. The primary stem or stems of well-developed plants of *Z. namophila* usually maintain a distinct central axis for much of their height, the inflorescence thus being a paniclelike thyrse. The proximal primary branches spread relatively widely, often at more than 45°. The pedicels are generally shorter than the flowers, although solitary flowers borne above small bracts on slender branches may appear to be long-pedicellate. All leaves of *Z. namophila* are linear or nearly so, whereas in *Z. exaltata* the basal leaves, which are often persistent, and the cauline leaves to mid stem are usually oblong-elliptic to lanceolate. The flowers of *Z. namophila* are generally five-merous, whereas those of *Z. exaltata* are four-merous, and the fresh corollas of *Z. namophila* are deeper pink. The corolla tubes of *Z. namophila* are 6–9 mm, and the lobes are 5–7 mm, distinctly more than half as long as the tube; the corolla tubes of *Z. exaltata* are 7–15 mm, and the lobes are 3–5 mm, about half as long as the tube or shorter.

3. **Zeltnera venusta** (A. Gray) G. Mansion, Taxon 53: 733. 2004 • California or charming or beautiful centaury

Erythraea venusta A. Gray in W. H. Brewer et al., Bot. California 1: 479. 1876; *Centaurium venustum* (A. Gray) B. L. Robinson

Herbs annual, (3–)8–30(–50) cm. **Stems** 1 or occasionally several, usually branching only near or above middle, occasionally with slender lower branches. **Leaves** usually all cauline; blade ovate to oblong or lanceolate, (3–)5–25 × (1–)2–6 mm, apex obtuse or acute. **Inflorescences** open, proximally dichasial, distally monochasial or completely monochasial cymes; pedicels (2–)4–25(–45) mm. **Flowers** 5-merous; calyx (3–)6–14 mm; corolla (7–)16–30 mm, lobes usually elliptic or elliptic-obovate, occasionally lanceolate to narrowly oblong or linear, (2–)5–20 mm, apex usually rounded to obtuse, less often acute, usually erose; stigmas 2, fan-shaped. **Seeds** nearly black. **2*n*** = 34.

Flowering spring–summer. Dry scrub, grasslands, open woods, forest openings; 0–1800 m; Calif.; Mexico (Baja California).

Although the maximum flower size in *Zeltnera venusta* is the largest for any *Zeltnera* species in the flora area, both plant and flower size are highly variable in this species, and corolla size is not reliable for the identification of this species. Small, one-flowered plants and plants intermediate in stature and flower size may be present near larger plants in microhabitats where less time has elapsed between seed germination and flowering. The whitish corolla eye of *Z. venusta* is especially prominent, sometimes with rose-purple spots in the throat aligned with the midveins of the corolla lobes.

Plants of *Zeltnera venusta* with relatively large flowers and proportionately wide corolla lobes are readily identifiable. Corolla lobes, however, range from linear or narrowly oblong to widely elliptic, with apices ranging from rounded (usually associated with proportionately wide corolla lobes) to acute. The largest flowers and the proportionately widest corolla lobes tend to occur in populations near the Pacific coast, from Los Angeles County, California, to Baja California, Mexico. Variation in corolla-lobe shape is especially pronounced in Fresno County, California, where plants with wide, elliptic lobes rounded at the apex and plants with narrowly lanceolate, acute-tipped lobes are sympatric and occur with many intermediates. Plants and flowers tend to be smaller in populations northward and toward the interior, especially at higher elevations, but plants with relatively small, narrowly lobed corollas occur as far south as Riverside, San Bernardino, and San Diego counties in California. Plants approaching all extremes in length-width proportions and other aspects of shape sometimes occur within a local population, as indicated by herbarium specimens at DUKE and UC that represent sampling from single populations. In studies for this work, no distinct entities were discerned that could be segregated from *Z. venusta* or recognized as infraspecific taxa.

Plants of *Zeltnera venusta* with relatively wide corolla lobes may resemble the similarly large-flowered species *Z. arizonica* and *Z. calycosa*. The corolla lobes of such plants of *Z. venusta* widen abruptly above the base, whereas those of the other large-flowered species widen more gradually.

In parts of California, plants of *Zeltnera venusta* with narrow corolla lobes resemble *Z. trichantha* to various degrees. Relatively densely branched plants of *Z. venusta* are highly similar to *Z. trichantha* in aspect and branching pattern. Such plants are more common north of Yosemite National Park, especially in Shasta County, but occasionally occur south to western Riverside and San Bernardino counties.

Zeltnera trichantha and *Z. venusta* are largely allopatric, but their ranges overlap northward. Although C. R. Broome (1973) was unsuccessful in attempts to cross *Z. trichantha* with other California species, morphology suggests that natural hybridization occurs, especially in Lake, Sonoma, and Tehama counties, California, and that introgression may be extensive. Some suspected hybrids resemble *Z. trichantha* but combine the elliptic leaves and subsessile flowers usual in that species with wider, obtuse corolla lobes and/or shorter styles and with stigmas approaching those of *Z. venusta* in morphology. Other plants differ from typical *Z. trichantha* in their more open inflorescences and longer-pedicelled proximal flowers, or in having very small stigmas, like those usually occurring in *Z. trichantha*, on distinctly cleft styles, like those of *Z. venusta*.

The type specimens of the names *Centaurium venustum* subsp. *abramsii* Munz and *Zeltnera abramsii* (Munz) G. Mansion, and most of the other specimens that have been so identified, are interpreted here as probable hybrids between *Z. trichantha* and *Z. venusta*. These names are typified by plants from the vicinity of Redding, Shasta County, California, that combine traits of these two species. Subsequent reports of this taxon have largely been limited to scattered localities in this part of northern California. The corolla lobes of the type are narrow and lanceolate, as in *Z. trichantha*, rather than oblong, but are obtuse at the apex, as in *Z. venusta*. The stigmas are small, about 0.3 mm wide, as in *Z. trichantha*, but the style divides into two branches about 1 mm below the stigmas, as in *Z. venusta*. The largest seeds are about 0.4 mm long and dark brown, but seed size varies and some appear malformed. Other specimens from the vicinity of Redding are more characteristic of *Z. venusta* in the narrow sense, and both *Z. trichantha* and *Z. venusta* occur in Shasta County. Reports of *Z. abramsii* from elsewhere in California (G. Mansion 2004) may have been based on variants of other species, or plants representing other hybrid combinations.

In molecular studies by G. Mansion and L. Zeltner (2004), plants identified as *Zeltnera abramsii* appeared in a basal position in the clade comprising the Californian group, including *Z. trichantha* and *Z. venusta*. The apparently basal position of *Z. abramsii* is not irreconcilable with a hypothesis of its hybrid origin, however, as molecular phylogenetic studies using nrITS or chloroplast DNA are not well suited for the detection or confirmation of hybridization.

In southeastern California, *Zeltnera venusta* and *Z. arizonica* are sometimes similar in appearance, but their respective ranges are separated by a part of the Sonoran Desert. Generally, *Z. venusta* is single stemmed at the base and *Z. arizonica* is multistemmed, although occasional specimens of *Z. venusta* are multistemmed, perhaps as a response to injury, and the smallest plants of *Z. arizonica* are sometimes single stemmed. The corolla lobes of *Z. venusta* are usually conspicuously and abruptly narrowed at the base, thus differing from those of *Z. arizonica*, but the plants of *Z. venusta* with the narrowest corolla lobes show little difference from *Z. arizonica* in this respect.

Plants from Los Angeles and San Diego counties, California, with pedicels to 50 mm and sometimes with several stems from the base and/or with more or less persistent basal leaves, may be hybrids with the southwestern form of *Zeltnera exaltata*.

4. Zeltnera trichantha (Grisebach) G. Mansion, Taxon 53: 733. 2004 • Alkali centaury [E]

Erythraea trichantha Grisebach, Gen. Sp. Gent., 146. 1838; *Centaurium trichanthum* (Grisebach) B. L. Robinson

Herbs annual, 5–35(–45) cm. **Stems** 1(–5), branching near or above 1/3 of height, occasionaly lower. **Leaves** cauline; blade ovate to lanceolate, distal sometimes linear, 10–40 × 1–11 mm, apex acute. **Inflorescences** ± dense, proximally dichasial, distally sometimes monochasial, usually corymboid cymes; proximal flowers sessile or pedicels to 3(–6) mm, distal flowers usually sessile. **Flowers** 5-merous; calyx 8–14 mm; corolla 12–22 mm, lobes lanceolate, (3–)5–10 mm, apex acute to acuminate; stigmas 2, cuneate, closely appressed and sometimes appearing as 1. **Seeds** dark brown. $2n = 34$.

Flowering spring–summer. Alkaline and saline flats, moist sites in chaparral and open woods, often in serpentine soils; 0–800 m; Calif.

The inflorescence of *Zeltnera trichantha* is generally corymboid, but the level at which plants first branch varies. Plants branching at or near the base, so that the whole plant above ground is obconic, plants branching only in the upper fifth or less, and a complete range of intermediates may occur within a single population.

Unequivocal *Zeltnera trichantha* occurs in the Inner North Coast Range and San Francisco Bay region of California, from San Mateo and Stanislaus counties north to Tehama County, often on alkaline flats and in serpentine soils. This species can generally be recognized by a distinctive combination of relatively dense, obconic inflorescences; all flowers sessile or on true pedicels usually less than 4 mm, rarely to 6 mm; narrow, acute to acuminate corolla lobes, notably contrasting with the obtuse or abruptly acute corolla-lobe apices prevalent among the other *Zeltnera* species in the flora area; styles that are longer and more slender than those of most *Zeltnera* species, often extending 6–11 mm beyond the throat of the corolla; and cuneate stigmas 0.2–0.3 mm wide at the summit, which remain more or less appressed to each other throughout much of the life of the flower. Basal leaves are absent at flowering time. Usually, but less consistently, the cauline leaves are elliptic, widest near the middle, tapering toward the base, and acute to acuminate at the apex. Elsewhere in California, plants of *Z. venusta* with narrow corolla lobes often resemble *Z. trichantha* to various degrees. These species are contrasted in the discussion of *Z. venusta*. Other species, notably *Z. muehlenbergii*, occasionally approach *Z. trichantha* in habit, but, like *Z. venusta*, differ in having two distinctly separate, fan-shaped stigmas on a shallowly cleft style.

5. **Zeltnera muehlenbergii** (Grisebach) G. Mansion, Taxon 53: 731. 2004 (as muhlenbergii) • Muhlenberg's centaury [E]

Erythraea muehlenbergii Grisebach, Gen. Sp. Gent., 146. 1838; *Centaurium curvistamineum* (Wittrock) Druce; *C. floribundum* (Bentham) B. L. Robinson; *C. muehlenbergii* (Grisebach) W. Wight

Herbs annual, 3–30(–40) cm. **Stems** usually 1, occasionally 2–4, simple or variously branching. **Leaves:** basal present or absent at flowering but not forming well-developed rosette; blade obovate to oblong, 15–25 × 3–15 mm, apex rounded to obtuse; cauline blades narrowly elliptic-ovate to elliptic-lanceolate, 10–25 × 2–9 mm, apex obtuse or distal leaves of large plants acute. **Inflorescences** ± open, distally or completely monochasial cymes; proximal flowers in center of cyme divisions sessile or pedicels to 12 mm, distal flowers generally sessile or pedicels to 4 mm. **Flowers** 5-merous; calyx 8–13 mm; corolla 12–19 mm, lobes oblong-lanceolate to narrowly elliptic, 2–7 mm, apex obtuse to subacute; stigmas 2, fan-shaped. **Seeds** brown.

Flowering early summer–early fall. Wet meadows, marsh edges, wet openings in woods, seeps, often in serpentine soils; 0–1600 m; B.C.; Calif., Idaho, Oreg., Wash.

The name *Centaurium muehlenbergii* has often been misapplied to *C. tenuiflorum* in the flora area (J. S. Pringle 2010b). This error accounts for reports of *C. muehlenbergii* in Louisiana, Mississippi, and Texas. The name *C. floribundum*, although typified by specimens referable to *Zeltnera muehlenbergii*, has also been misapplied to *C. tenuiflorum*, hence comments by some authors that *C. floribundum* should perhaps be included in *C. tenuiflorum*. This complex nomenclatural history has caused confusion with regard to the conservation status of plants called *Centaurium* or *Z. muehlenbergii*. From studies for this flora work, it appears that the true *Z. muehlenbergii* is appropriately of conservation concern.

The inflorescences of *Zeltnera muehlenbergii* are relatively open, in contrast to the dense, corymboid dichasia of *Centaurium tenuiflorum*. The proximal branching of the larger plants of *Z. muehlenbergii* is usually dichasial, with the central flower sessile or on a pedicel to 5 mm long or occasionally to 12 mm. The distal branching is mostly monochasial, with a branch developing only on one side of each flower; the inflorescences of small plants are often monochasial throughout. The flowers at the distal divisions are sessile or on pedicels to 4 mm. Both *C. tenuiflorum* and *C. pulchellum*, which are sometimes similar in aspect, differ from *Z. muehlenbergii* in the style and stigma characters noted in the descriptions of the respective genera (J. S. Pringle 2010b).

Some plants of *Zeltnera davyi* and exceptionally small plants of *Z. exaltata* look much like *Z. muehlenbergii* but consistently have proximal pedicels more than 4 mm long and often more than 12 mm. Even on small plants the pedicels of *Z. exaltata* are often longer than the flowers.

The name Monterey centaury is sometimes associated with this species, but that vernacular name came into use when the name *Centaurium muehlenbergii* was being misapplied to *Zeltnera davyi*. True *Z. muehlenbergii* has rarely been found as far south as the northern shore of Monterey Bay (J. S. Pringle 2010b).

SELECTED REFERENCE Pringle, J. S. 2010b. The identity and nomenclature of the Pacific North American species *Zeltnera muhlenbergii* (Gentianaceae), and its distinction from *Centaurium tenuiflorum* and other species with which it has been confused. Madroño 57: 184–202.

6. **Zeltnera davyi** (Jepson) G. Mansion, Taxon 53: 730. 2004 • Davy's or Monterey centaury [E]

Centaurium exaltatum (Grisebach) W. Wight var. *davyi* Jepson, Man. Fl. Pl. Calif., 762. 1925; *C. davyi* (Jepson) Abrams

Herbs annual, (2–)5–30(–50) cm. **Stems** 1–10, simple (small plants) or few-branched ± throughout. **Leaves:** basal absent or occasionally persisting at flowering, similar to cauline; cauline blades elliptic-oblong to ovate, 8–26 × 3–8(–13) mm, apex obtuse to acute. **Inflorescences** completely monochasial or occasionally proximally dichasial, ± racemoid cymes; pedicels (2–)4–25(–55) mm. **Flowers** 5-merous; calyx 8–10 mm; corolla 12–17 mm, lobes ovate-oblong, 3–7 mm, keeled (uniquely in this species in the flora area), apex obtuse; stigmas 2, widely fan-shaped. **Seeds** dark brown.

Flowering spring–summer. Moist coastal bluffs, interdunal depressions, open woods, sometimes in ultramafic soils; 0–1000 m; Calif.

The name *Centaurium muehlenbergii* has sometimes been misapplied to *Zeltnera davyi*, with true *Z. muehlenbergii* then being called *C. floribundum* (J. S. Pringle 2010b).

The distinctly keeled calyx lobes cause the calyces of *Zeltnera davyi* to appear greater in diameter than those of related species and ovoid to ellipsoid rather than nearly cylindric. The combination of this calyx morphology and the proportionately wide, relatively deeply pigmented corolla lobes (usually evident in herbarium specimens) gives the flowers of *Z. davyi* a distinctive aspect.

Zeltnera davyi and *Z. muehlenbergii* are sometimes similar in habit. Medium-sized plants of *Z. davyi* are usually several-stemmed from the base, whereas that pattern is much less common in *Z. muehlenbergii*. *Zeltnera davyi* usually differs from *Z. muehlenbergii* in the presence of elliptic to ovate leaves over 5 mm wide (except on the smallest plants) well into the inflorescence; consistently present pedicels 4–30 mm long; calyx lobes with keels proximally 0.3–0.6 mm wide; and ovate-elliptic corolla lobes 3–7 × 2–3 mm. In *Z. muehlenbergii*, elliptic to narrowly ovate leaves, when present, are usually limited to the proximal one-third or less of the plant, with the distal leaves being narrower, and the corolla lobes are elliptic-oblong, 2–7 × 1–2 mm (J. S. Pringle 2010b).

7. **Zeltnera multicaulis** (B. L. Robinson) G. Mansion, Taxon 53: 734. 2004 • Tufted or many-stemmed centaury

Centaurium multicaule B. L. Robinson, Proc. Amer. Acad. Arts 45: 396. 1910

Herbs annual, 4–38 cm. **Stems** (1–)3–12, branching throughout. **Leaves:** basal and near-basal present or more often ± withered at flowering; blade obovate to oblanceolate, 6–20 × 4–10 mm, apex rounded to subacute; cauline blades narrowly lanceolate, 6–33 × 1–9 mm, apex acute. **Inflorescences** racemoid, monochasial cymes, usually 1–3-flowered; pedicels 5–63 mm. **Flowers** 5-merous; calyx 5–10 mm; corolla 9–20 mm, lobes lanceolate, 3.5–7.5 × 1–2.7 mm, apex acute; stigma 1, shallowly 2-lobed, lobes club- to fan-shaped. **Seeds** reddish brown. $2n = 42$ (Mexico).

Flowering late winter–summer. Moist meadows; 1300–1500 m; Tex.; Mexico.

Zeltnera multicaulis is widespread in Mexico, although the populations are widely scattered, occurring there at both higher and lower elevations. In the flora area, it is known only from Culberson and Presidio counties, Texas. From the similarly long-pedicelled *Z. exaltata*, which does not occur in Texas, *Z. multicaulis* differs in that its corolla lobes are distinctly more than half as long as the tube.

8. **Zeltnera calycosa** (Buckley) G. Mansion, Taxon 53: 734. 2004 • Buckley's centaury, rosita

Erythraea calycosa Buckley, Proc. Acad. Nat. Sci. Philadelphia 14: 7. 1862; *Centaurium breviflorum* (Shinners) B. L. Turner; *C. calycosum* (Buckley) Fernald; *C. calycosum* var. *breviflorum* Shinners; *Zeltnera breviflora* (Shinners) G. Mansion

Herbs annual or biennial, (4–)9–30 cm. **Stems** usually solitary, occasionally 2–5, usually branching ± densely near or above middle, stems of smallest plants usually sparsely or not branched below inflorescence. **Leaves:** basal often present at flowering; blade ovate to lanceolate, oblanceolate, or occasionally linear, 9–40 × 3–12 mm, apex rounded to subacute; cauline blades elliptic to lanceolate, or distal or all cauline leaves linear, 7–40 × 1–9(–13) mm, apex acute. **Inflorescences** diffuse, proximally dichasial, distally monochasial cymes; pedicels (2–)4–30(–40) mm. **Flowers** 5-merous; calyx 7–12 mm; corolla (12–)14–23 mm, lobes lanceolate to lanceolate-ovate or elliptic, 5–12 × 1–5 mm, apex acute; anthers 1.2–3.5 mm; stigmas 2, fan-shaped. **Seeds** light brown. $2n = 40$.

Flowering spring–summer. Stream banks, prairies, roadsides, beaches, and edges of salt marshes; 0–2000 m; Tex., Utah; Mexico (Coahuila, Nuevo León, Tamaulipas).

Zeltnera calycosa is known from Missouri only from a historical introduction.

Plants formerly identified as *Centaurium calycosum* from localities west of Texas have been segregated as *Zeltnera arizonica*. A report from Louisiana was based on a misidentified specimen of *C. tenuiflorum* (J. S. Pringle 2010b).

Zeltnera calycosa varies greatly vegetatively, especially in the spacing, size, and proportions of its leaves. This variation is correlated in part with plant size but appears also to be due to introgression of genetic material from *Z. texensis* and perhaps from other species.

Small-flowered plants of *Zeltnera calycosa* were distinguished as *Centaurium calycosum* var. *breviflorum* by L. H. Shinners. Although the floral dimensions given for *C. calycosum* var. *breviflorum* and the autonymic variety overlap appreciably, this variety was accepted by D. S.

Correll and M. C. Johnston (1970), who gave its range as central Texas and also salt marshes along the Gulf Coast. B. L. Turner (1993d) raised this taxon to species rank but restricted its circumscription to plants in Texas south of 28°30'N and east of 100°W and a few sites in northern Nuevo León and Tamaulipas, Mexico, in which range he said that it completely replaced *C. calycosum* in the strict sense. He excluded central Texas from its given range, although he acknowledged that small-flowered plants occurred sporadically throughout the range of *C. calycosum*. C. R. Broome (1973), in contrast, reduced the name *C. calycosum* var. *breviflorum* to synonymy under *C. calycosum* var. *calycosum*, and included plants from the Gulf Coastal region in that variety. She thus indicated that some plants from that area were relatively large-flowered, as in her classification small-flowered plants of this species were treated as *C. calycosum* var. *nanum* (A. Gray) B. L. Robinson. In studies for this flora, specimens from sites well within the range given by Turner for *C. breviflorum*, notably those seen at PAUH, included plants with floral dimensions fully conforming to those given for *C. calycosum* in the strict sense and similar to those of specimens from elsewhere in range of *Z. calycosa*, indicating that *breviflorum* forms do not completely replace larger-flowered forms in that area. D. J. Pinkava, in notes on a manuscript for this flora, likewise expressed doubt that species rank was appropriate for the plants that Turner had treated as *C. breviflorum*. Research to date has not sufficed to indicate whether in all cases plants with smaller flowers represent phenotypic plasticity, or whether an ecotype adapted and restricted to the periphery of salt marshes has become genetically differentiated.

The type specimen of the name *Centaurium calycosum* var. *nanum* appears to be derived from *Zeltnera calycosa* × *Z. beyrichii*. Other plants to which that name has been applied are *Z. calycosa* × *Z. texensis* and small-flowered plants of *Z. calycosa* occurring throughout the range of the species (B. L. Turner 1993d; studies for this flora).

Chromosome counts for *Zeltnera calycosa* in the broad sense, that is, *Z. arizonica* and both varieties of *Z. calycosa* as circumscribed in this flora, indicate that significant variation exists in this complex, but counts remain too few for interpretation in relation to classification. In addition to those cited here for the respective taxa, C. R. Broome (1978) reported $2n = 84$ for a plant from Nuevo León, Mexico, identified as *Centaurium calycosum* var. *calycosum* by her and as *C. arizonicum*, that is, *Z. arizonica* of this flora, by B. L. Turner (1993d). Some of the plants identified as *C. calycosum* var. *calycosum* that were the source of seeds from which plants were raised for chromosome counts by Broome were obtained where *Z. texensis* was also present, and some of the voucher specimens appear intermediate in morphology.

9. Zeltnera arizonica (A. Gray) G. Mansion, Taxon 53: 733. 2004 • Arizona or marsh centaury

Erythraea calycosa Buckley var. *arizonica* A. Gray in A. Gray et al., Syn. Fl. N. Amer. 2: 113. 1878; *Centaurium arizonicum* (A. Gray) A. Heller; *C. calycosum* (Buckley) Fernald var. *arizonicum* (A. Gray) Tidestrom

Herbs annual or biennial, (10–)20–50(–60) cm. **Stems** 1–10, usually branching ± sparsely near or above middle. **Leaves:** basal usually present at flowering, sometimes numerous; blade oblanceolate to lanceolate, (7–)15–70 × 4–10 mm; cauline blades lanceolate to oblanceolate, (13–)25–70 × (2–)5–8(–13) mm, apex obtuse to acute. **Inflorescences** predominantly dichasial or distally monochasial cymes; pedicels 4–40(–60) mm. **Flowers** 5-merous; calyx 7–12 mm; corolla (15–)18–25 mm, lobes (linear to) lanceolate to lanceolate-ovate or elliptic, 7–12 × 1–5 mm, apex acute; anthers 2.5–3.5 mm; stigmas 2, fan-shaped. **Seeds** dark reddish brown. $2n = 24, 40$.

Flowering spring–fall. Stream banks, marshes, other moist, open habitats; 50–2800 m; Ariz., Calif., Colo., Nev., N.Mex., Tex., Utah; Mexico (Chihuahua, Coahuila, Nuevo León, San Luis Potosí, Sonora).

Zeltnera arizonica and *Z. calycosa* appear to intergrade in western Texas and Coahuila, Mexico, as noted by C. R. Broome (1973), and have often been treated as varieties of a single species. *Zeltnera arizonica* was subsumed in undivided *Centaurium calycosum* by N. H. Holmgren (1984b), who attributed its allegedly distinguishing features largely to environmental effects, whereas B. L. Turner (1993d) considered the resemblance between these taxa to be superficial and *Z. arizonica* (as *Centaurium*) appropriately recognized at species rank. From studies for this flora, acceptance of this species seems warranted.

In *Zeltnera arizonica*, the relatively sparse branches generally spread at 10–20°, whereas in *Z. calycosa* the usually denser branches spread at 20–60°.

Zeltnera arizonica is highly variable in the proportionate width of its corolla lobes. Some plants in the western part of its range resemble *Z. exaltata* vegetatively but differ in having corolla lobes much longer in proportion to the tube.

10. Zeltnera texensis (Grisebach) G. Mansion ex J. S. Pringle, Rhodora 113: 514. 2012 • Texas or Lady Bird's centaury

Erythraea texensis Grisebach, Gen. Sp. Gent., 139. 1838; *Centaurium texense* (Grisebach) Fernald

Herbs annual, 3–30 cm. **Stems** 1(–5), dividing below middle into ± equal branches, these further branching throughout or, in the smaller plants, branching only distally. **Leaves** usually all cauline; blade narrowly elliptic to linear (proximal) to filiform (distal), 7–25 × 0.5–4.5(–8) mm, apex acute. **Inflorescences** open, much-branched, dichasial, with a central flower in each fork, or distally sometimes monochasial cymes; pedicels 4–20 mm. **Flowers** 5-merous; calyx 6–12 mm; corolla 12–20 mm, lobes narrowly oblong to elliptic-oblong, 3–7 × 0.8–1.5 mm, apex acute; stigmas 2, fan-shaped. **Seeds** light brown. $2n$ = 40, 42.

Flowering summer. Prairies, barrens, open woods, other open, usually rocky sites, in calcareous soils; 100–500 m; Ark., Kans., Mo., Okla., Tex.; Mexico (Nuevo León).

Zeltnera texensis has been reported from Nuevo León, Mexico (J. N. Mink et al. 2011b), remote from its range in the United States, but specimens have not been examined in studies for this flora. Reports of *Z. texensis* (as *Centaurium*) from Louisiana have been based on misidentified *C. pulchellum* and *C. tenuiflorum* (J. S. Pringle 2010b). Reports from New Mexico have been based on *Z. arizonica* and *Z. maryanniana* (studies for this flora).

Zeltnera texensis differs from *Z. maryanniana* most conspicuously in being proximally single-stemmed without persistent basal leaves.

The name Lady Bird's centaury is for Claudia Alta Taylor "Lady Bird" Johnson, former First Lady of the United States, who promoted the conservation of native plant species and their use in beautifying highway verges. She especially liked this species and had seeds of it sown around the airstrip at the Johnson ranch in Texas.

11. Zeltnera maryanniana (B. L. Turner) G. Mansion, Taxon 53: 734. 2004 (as maryanna) • Gypsum centaury [E]

Centaurium maryannianum B. L. Turner, Phytologia 75: 269, fig. 4. 1993 (as maryannum)

Herbs annual, 3–20 cm. **Stems** 3–30, branching throughout. **Leaves:** basal present at flowering; blade narrowly oblanceolate to obovate, 20–50 × 2–8 mm, apex obtuse to acute; cauline blades oblanceolate (proximal) to narrowly oblong-lanceolate or linear (distal), 15–25 × 1–4 mm, apex obtuse (proximal) to acuminate (distal). **Inflorescences** dichasial cymes; pedicels 5–20 mm. **Flowers** 5-merous; calyx 7–10 mm; corolla 14–22 mm, lobes ovate-elliptic, 6–11 × 1–4.5 mm, apex acute; stigma 1, 2-lobed, lobes fan-shaped, tardily diverging. **Seeds** black. $2n$ = 42.

Flowering spring–summer. Roadsides, sand hills, rocky ridges, other open sites in gypsum soils; 900–1700 m; N.Mex., Tex.

Zeltnera maryanniana is known only from gypseous habitats in southeastern New Mexico and trans-Pecos Texas, but might be expected in northern Chihuahua, Mexico.

Prior to its recognition as a species in 1993, specimens of *Zeltnera maryanniana* were identified as *Centaurium beyrichii*, *C. calycosum* in the broad sense, or *C. texense*, or as hybrids between species that do not occur in the range of *Z. maryanniana*. In aspect, *Z. maryanniana* is similar to *Z. beyrichii*, from which its range is separated by about 650 km. Its flowering stems arise from a dense tuft of basal leaves and rosettes that still are green at flowering time, whereas such basal leaves are usually more or less withered when plants of *Z. beyrichii* are in flower. Nearly all of the leaves of *Z. maryanniana* are distinctly wider than the stem diameter, whereas the mid-stem and distal leaves of *Z. beyrichii* are often only about as wide as the stem. *Zeltnera maryanniana* differs from *Z. texensis* in its tufted stems, and from both *Z. beyrichii* and *Z. texensis* in its single stigma and black seeds. The corolla lobes of *Z. maryanniana* are elliptic and proportionately wider than those of *Z. beyrichii* and *Z. texensis*. *Zeltnera maryanniana* further differs from all other *Zeltnera* species in having filaments expanded at the base (G. Mansion 2004).

Plants of *Zeltnera maryanniana* sometimes flower again after the fruits have matured on the first stems, but no specimens seen in studies for this flora show evidence of a truly perennial habit. Plants of some Mexican species of *Zeltnera* likewise sometimes flower a second time before dying (C. R. Broome 1973).

12. Zeltnera glandulifera (Correll) G. Mansion, Taxon 53: 734. 2004 • Rough-stemmed centaury [E]

Centaurium beyrichii (Torrey & A. Gray) B. L. Robinson var. *glanduliferum* Correll, Wrightia 4: 76. 1968 (as Centaureum); *C. glanduliferum* (Correll) B. L. Turner

Herbs annual, 5–20 cm. **Stems** 1–30+ from crown, branching throughout, densely papillose-puberulent. **Leaves:** basal present or ± withered at flowering, similar to cauline; blade linear, 10–40 × 1–3 mm, apex obtuse to acute, abaxially papillose-puberulent along midvein. **Inflorescences** dense, much-branched, dichasial cymes; pedicels 2–10 mm. **Flowers** 5-merous; calyx 5–9 mm, papillose-puberulent along midveins; corolla 13–17 mm, lobes narrowly oblong-lanceolate, 5–7 mm, apex acute; stigmas 2, fan-shaped. **Seeds** nearly black. **2*n*** = 40.

Flowering summer–fall. Roadsides, other open, often rocky sites, in calcareous soils; 800–1100 m; Tex.

Zeltnera glandulifera is endemic to Brewster, Pecos, and Terrell counties in trans-Pecos Texas. Its range is about 200 km west of that of *Z. beyrichii.*

Zeltnera glandulifera is similar to *Z. maryanniana*, from which its range is wholly southeast. The species differs most distinctly in that its stems, leaf and calyx midveins, and especially its pedicels and inflorescence branches are copiously papillose-puberulent, mostly in lines.

13. Zeltnera beyrichii (Torrey & A. Gray) G. Mansion, Taxon 53: 733. 2004 • Rock or mountain centaury, mountain-pink, quinineweed [E] [F]

Erythraea beyrichii Torrey & A. Gray in R. B. Marcy, Explor. Red River Louisiana, 291, plate 13. 1853, based on *E. trichantha* Grisebach var. *angustifolia* Grisebach in A. P. de Candolle and A. L. P. P. de Candolle, Prodr. 9: 60. 1845; *Centaurium beyrichii* (Torrey & A. Gray) B. L. Robinson

Herbs annual or perennial, 10–45 cm. **Stems** 1–20 from crown, if 1, dividing near base into ± equal branches, these much-branched distally. **Leaves:** basal usually withered or absent by flowering, similar to cauline; cauline blades narrowly linear-oblanceolate or linear (proximal) to filiform (distal), 10–30 × 1–2(–3) mm, apex acute. **Inflorescences** dense, much-branched, dichasial cymes; pedicels 3–15 mm. **Flowers** 5-merous; calyx 7–11 mm; corolla 15–30 mm, lobes narrowly oblong-lanceolate to linear, 7–16 × 1–2 mm, apex acute; stigmas 2, clavate to narrowly fan-shaped, ± tardily diverging. **Seeds** dark brown. **2*n*** = 40, 82.

Flowering spring–summer. Wet, gravelly or rocky places, usually in calcareous soils; 100–700 m; Okla., Tex.

Reports of *Zeltnera beyrichii* (as *Centaurium*) from Arkansas have been based only on the provenance of the type specimen, given as Arkansas Territory but probably in present-day Oklahoma. New Mexico plants have been separated as *Z. maryanniana.* A specimen at ASU from Coahuila, Mexico, appears to represent this species but requires further study.

Plants of *Zeltnera beyrichii* often have 20 or more stems from the base. These stems are much-branched. The larger plants usually bear over 200 flowers and sometimes over 1000. *Zeltnera beyrichii* is often described as an annual, but C. R. Broome (1973) characterized it as a true perennial, which perennates by the production of vegetative shoots from the roots.

Although the style of *Zeltnera beyrichii* is initially deflexed like those of other large-flowered *Zeltnera* species and the stigmas are well separated from the anthers when the latter dehisce, the stamens are radially divergent at anthesis rather than all being curved to one side (C. R. Broome 1973).

C. R. Broome (1973, 1978) considered *Zeltnera beyrichii* a probable allopolyploid derivative of *Z. calycosa* and *Z. texensis.*

14. Zeltnera nudicaulis (Engelmann) G. Mansion, Taxon 53: 735. 2004 • Sonoran or Santa Catalina Mountain centaury

Erythraea nudicaulis Engelmann, Proc. Amer. Acad. Arts 17: 222. 1882; *Centaurium nudicaule* (Engelmann) B. L. Robinson

Herbs annual, 4–27 cm. **Stems** 1–several, usually simple or branching only above middle. **Leaves:** basal often present at flowering; blade elliptic to obovate, 3–26 × 2–10 mm, apex obtuse to acute; cauline blades elliptic-oblong to lanceolate or rarely ovate, (proximal, sometimes extending distally into inflorescence) to narrowly linear or filiform (distal), 3–31 × 0.7–6 mm (distal leaves usually 1–2 mm wide, occasionally to 4 mm), apex subacute to acuminate. **Inflorescences** diffuse, proximally dichasial, distally monochasial or completely monochasial cymes; pedicels (8–)15–75 mm. **Flowers** 5-merous; calyx 4–8 mm; corolla scarcely constricted above ovary, 6.5–10(–12.5) mm, lobes lanceolate, 2.5–4.5 × 0.5–3 mm, apex obtuse; stigmas 2, fan-shaped. **Seeds** reddish brown. **2*n*** = 42 (Mexico).

Flowering winter–spring. Stream banks in open woods, wet meadows; 900–1500 m; Ariz., N.Mex., Tex.; Mexico (Sonora).

Zeltnera nudicaulis is the only species of this genus in the flora area in which the corolla is not distinctly constricted above the ovary, and it has leaves thicker than those of the other *Zeltnera* species with which it is sympatric. It differs distinctly from *Z. exaltata* in its widely separated cauline leaves, which contrast with its much wider basal and near-basal leaves.

4. GYRANDRA Grisebach in A. P. de Candolle and A. L. P. P. de Candolle, Prodr. 9: 44. 1845 • Centaury [Greek *gyros*, circle, and *andros*, of male, alluding to helically twisting showy anthers at dehiscence]

James S. Pringle

Herbs perennial [annual], chlorophyllous, stems, leaves, and calyces scabridulous to papillate-puberulent in lines. **Leaves** basal and cauline, opposite. **Inflorescences** partly dichasial, partly monochasial [completely monochasial] cymes. **Flowers** 5-merous; calyx lobed nearly to base; corolla [pale violet to] deep rose-pink without distinct white eye, salverform [subrotate], glabrous, lobes abruptly spreading ± as long as [longer than] tube, margins entire, plicae between lobes absent, spurs absent; stamens inserted in distal 1/2 of corolla tube, all initially curved to one side; anthers distinct, coiling helically at dehiscence; ovary sessile; style deciduous, initially deflexed away from stamens, distinct, not cleft, not coiling; stigma 1, shallowly 2-lobed [subcapitate]; nectaries absent. **Capsules** ellipsoid. $x = 18$.

Species 6 (1 in the flora): Texas, Mexico, Central America; warm-temperate to dry-mesic tropical regions.

G. Mansion (2004) and Mansion and L. Struwe (2004) inferred from molecular studies that *Gyrandra* is more closely related to *Eustoma* and *Sabatia* than to *Zeltnera*, although it resembles *Zeltnera* in its helically coiling anthers and in its calyx with a short tube and strongly ascending lobes.

Gyrandra differs from most species of *Zeltnera* in having corolla lobes about as long as or longer than the tube. Some Mexican species of *Gyrandra* have large, nearly rotate corollas with the lobes two to three times as long as the tube. In *G. blumbergiana* and the Mexican *G. brachycalyx* (Standley & L. O. Williams) G. Mansion, however, the lobes are about as long as the tube; conversely, in *Z. multicaulis*, *Z. trichantha*, and *Z. venusta* the corolla lobes are occasionally as long as or longer than the tube.

Gyrandra further differs from *Zeltnera* in its abaxially viscid corolla tubes, best observed in fresh material [C. R. Broome 1973, as *Centaurium* sect. *Gyrandra* (Grisebach) A. Gray]. The capsule walls of all species of *Gyrandra* are relatively thick and woody at maturity, whereas those of *Centaurium* and *Zeltnera* are variable in this respect, often being very thin. The angles of the stems of *Gyrandra* are minutely serrulate or papillose, whereas the stems of all *Zeltnera* species except *Z. glandulifera* are smooth. The leaves of *Gyrandra* are generally linear to narrowly lanceolate, more or less densely crowded proximally but not forming a distinct rosette, gradually more widely spaced distally, all persistent at flowering time.

1. **Gyrandra blumbergiana** (B. L. Turner) J. S. Pringle, Rhodora 115: 99. 2013 • Blumberg's centaury C E F

Centaurium blumbergianum B. L. Turner, Sida 21: 87, figs. 1, 2. 2004

Herbs 10–40 cm. **Stems:** rosettelike vegetative and elongated flowering stems present concurrently, to 100+ from crown, flowering stems branching throughout. **Leaves:** basal and near-basal present at flowering; blade narrowly linear, 10–30 × 0.4–0.6 mm, apex acute. **Inflorescences** proximally dichasial, distally monochasial cymes; pedicels 15–30 mm. **Flowers:** calyx 8–12 mm; corolla 15–18 mm, tube constricted above ovary, lobes elliptic, 8–9 × 3.5–4.5 mm, strongly constricted at base, where 1/3 as wide as at mid length, apex abruptly short-acuminate; anthers 2–3 mm; stigma lobes hemispherical. **Seeds** brown.

Flowering summer. Along sulfur streams in open sites; of conservation concern; 800–900 m; Tex.

Gyrandra blumbergiana is known only from a single canyon in Presidio County but should be sought in northeastern Chihuahua, Mexico.

The case for placing *Gyrandra blumbergiana* in *Gyrandra* was presented by J. S. Pringle (2013). The corolla lobes of *G. blumbergiana*, being strongly constricted at the base, differ in shape from those of most *Zeltnera* species in the flora area, although they are approached by some of the variants of *Z. venusta*. This species differs from all other species of *Gyrandra* and from most species of *Zeltnera* in its perennial, tufted habit, with short, more or less rosettelike vegetative stems accompanying the flowering stems. Since this species was placed in *Gyrandra* only on the basis of its morphology, molecular studies of its relationships would be desirable.

5. SABATIA Adanson, Fam. Pl. 2: 503. 1763 • Marsh- or sea-pink, rose-gentian [For Liberato Sabbati, ca. 1714–ca. 1779, Italian botanist and physician]

James S. Pringle

Lapithea Grisebach

Herbs annual, biennial, or perennial, perennials sometimes stoloniferous, individual crowns then sometimes biennial; chlorophyllous, glabrous. **Leaves** cauline, opposite, often also basal. **Inflorescences** cymes, thyrses, heads, or paired or solitary flowers, cymes dichasial in species with opposite branching, monochasial in species with alternate branching. **Flowers** (4–)5–12(–14) -merous; calyx tube urceolate, campanulate, or obconic, lobes longer or shorter than tube; corolla pink, purplish pink, cream, or white, often with an adaxial yellow, yellowish green, or white eye encompassing the corolla tube and extending into the proximal parts of the lobes, rotate, glabrous, lobes much longer than tube, entire, plicae between lobes absent; stamens inserted in or immediately below (species 1–14, 17–20) or slightly but distinctly below (species 15 and 16) the corolla sinuses; anthers straight, recurved or recoiling circinately or curving or twisting helically, distinct; ovary sessile; style deciduous, initially deflexed to one side or less often erect, distinct, 2-cleft; stigmas 2, styles and stigmas often coiling helically; nectaries 5 at base of ovary, not clearly differentiated. **Capsules** cylindric to ovoid or globose. $x = 7$; aneuploidy common.

Species 21 (20 in the flora): North America, n Mexico, West Indies; temperate to subtropical areas.

The orthographic variant *Sabbatia* has been widely used for those names cited here that were published during the nineteenth century and still later by G. C. Druce and J. K. Small. The spelling *Sabatia* was an intentional Latinization by M. Adanson and must be retained.

Sabatia capitata and *S. gentianoides* have sometimes been segregated as the genus *Lapithea* or as *Sabatia* sect. *Pseudochironia* Grisebach, distinguished by sessile rather than pedicellate flowers and merely curved rather than coiling anthers. These species further differ from the other species of *Sabatia* in having the filaments inserted slightly but distinctly below, rather than in, the sinuses of the corolla. Molecular phylogenetic studies by K. G. Mathews et al. (2015) confirm the close relationship of these species to each other but support their retention in *Sabatia*.

Most species of *Sabatia* are protandrous and outbreeding. In most of the outbreeding species the styles are deeply cleft and the stigmas are linear, appearing as continuations of the style branches. Initially, these combinations of style branches and stigmas are helically coiled around each other and are deflexed to one side, while the stamen filaments are erect and the anthers are straight. Later, the style becomes erect and its branches uncoil and diverge, exposing the stigmatic surfaces, and the stamens diverge radially. The anthers coil or at least recurve circinately upon or following dehiscence. In *S. capitata* and *S. gentianoides* the style branch-stigma combinations are widely oblanceolate and are scarcely coiled at any stage, and the anthers remain straight or are slightly coiled helically after dehiscence. *Sabatia arenicola* and *S. calycina* have relatively small, homogamous flowers and are generally autogamous. In these species the style is not deflexed at anthesis. It and the stigmas, which are already receptive when the pollen is shed, are scarcely or not coiled until after pollination occurs. The anthers remain straight or nearly so after dehiscence (J. D. Perry 1971).

Great variation in plant and flower size occurs within *Sabatia* species, especially in the annuals and biennials. This variation is due to environmental factors, including shade, soil moisture, and the interaction of photoperiod with receding water levels or other conditions permitting seed germination (J. D. Perry 1971).

Stems of *Sabatia* species may be terete or four-angled. If the stems are angled, narrow wings may extend from the apex of the angles. Calyx tubes described here as campanulate or hemispheric are rounded at the base, whereas those described as obconic (or turbinate in some literature, for example, R. L. Wilbur 1955) taper at the base. Some species are intermediate in this respect. Pedicel length in this treatment refers to the true pedicels beyond the most distal pair of bractlets, which may be minute, and the calyx, rather than to the ultimate divisions of the inflorescence.

SELECTED REFERENCES Blake, S. F. 1915. Notes on the genus *Sabatia*. Rhodora 17: 50–57. Mathews, K. G., M. S. Ruigrok, and G. Mansion. 2015. Phylogeny and biogeography of the eastern North American rose gentians (*Sabatia*, Gentianaceae). Syst. Bot. 40: 81–85. Perry, J. D. 1971. Biosystematic studies in the North American genus *Sabatia* (Gentianaceae). Rhodora 73: 309–369. Wilbur, R. L. 1955. A revision of the North American genus *Sabatia* (Gentianaceae). Rhodora 57: 1–33, 43–71, 78–104.

1. Flowers 7–12(–14)-merous.
    2. Flowers sessile above bractlets, usually in heads.
        3. Basal leaf blades oblong-spatulate, sharply differentiated from linear cauline leaves; cauline leaf blades 1–3 mm wide . . . . 15. *Sabatia gentianoides*
        3. Basal and cauline leaf blades similar, blades oblong to elliptic, 7–20(–25) mm wide . . . . 16. *Sabatia capitata*
    2. Flowers pedicellate.
        4. Basal leaves usually present at flowering, blade oblanceolate to spatulate, sharply differentiated from linear to lanceolate cauline leaves . . . . 20. *Sabatia decandra*
        4. Basal leaves absent at flowering or similar to cauline leaves, blade linear to oblong-lanceolate.
            5. Primary branching mostly opposite; calyx tube not or obscurely ridged, lobes linear-filiform . . . . 17. *Sabatia kennedyana*
            5. Primary branching alternate (rarely all opposite in *S. foliosa*); calyx tube distinctly ridged, lobes linear to narrowly spatulate or foliaceous.
                6. Stolons poorly developed or absent; internodes mostly 1.25+ times as long as subtending leaves; salt brackish (rarely fresh) marshes along the Atlantic Coast . . . . 18. *Sabatia dodecandra*
                6. Stolons well developed; internodes generally less than 1.25 times as long as subtending leaves; inland habitats, not brackish . . . . 19. *Sabatia foliosa*

1. Flowers (4–)5(–7)-merous.
   7. Calyx lobes with lateral veins (submarginal in lobes) more prominent than midvein, tubes with prominent, commissural ridges.
      8. Leaf blades elliptic to ovate or obovate; corolla lobes 4–10(–13) mm; beaches, interdunal depressions, salt flats along the Gulf Coast . . . . 3. *Sabatia arenicola*
      8. Leaf blades linear to ovate; corolla lobes 8–25 mm; widely distributed, generally not in coastal saline habitats.
         9. Basal leaves usually present at flowering; calyces usually shorter than or as long as corollas; corolla lobes elliptic-rhombic. . . . . 2. *Sabatia formosa*
         9. Basal leaves absent at flowering; calyces usually as long as or longer than corollas; corolla lobes obovate or spatulate-obovate.
            10. Leaf blades lanceolate-elliptic to ovate; corolla lobes obovate, less than 1.8 times as long as wide, apex abruptly ± acute; widely distributed . . . 1. *Sabatia campestris*
            10. Leaf blades linear to narrowly lanceolate; corolla lobes narrowly spatulate-obovate, 1.8+ times as long as wide, apex rounded; c Arkansas. . . . . . . 4. *Sabatia arkansana*
   7. Calyx lobes with midvein as or more prominent than lateral veins or, if linear-filiform, subulate, or setaceous, without discernible lateral veins, tubes not ridged below sinuses, or at most with low ridges from base to sinuses.
      11. Flowers subsessile above bractlets or true pedicels to 10(–15) mm.
         12. Stems terete; calyx lobes 0.1–3 mm; corolla lobes 4–7(–9) mm . . . . .5. *Sabatia macrophylla*
         12. Stems 4-angled at least distally; calyx lobes (2–)3–8(–14) mm; corolla lobes 4.5–21 mm.
            13. Stems 4-angled throughout, the angles with wings 0.1–0.5 mm. . . . . .6. *Sabatia quadrangula*
            13. Stems at least proximally terete or nearly so (distally ± 4-angled, wings absent).
               14. Plants annual, biennial, or occasionally short-lived perennial; stems with all branching opposite or distal branching all or partly alternate; corollas pink or rarely white . . . . 7. *Sabatia brachiata*
               14. Plants perennial; stems with all branching opposite; corollas white . . . . 8. *Sabatia difformis*
      11. Pedicels above bractlets 10–100(–150) mm.
         15. Stems 4-angled with wings distally 0.2–0.3 mm wide. . . . . 11. *Sabatia angularis*
         15. Stems terete or nearly so, wings absent.
            16. Calyx lobes oblanceolate to spatulate or foliaceous . . . . 12. *Sabatia calycina*
            16. Calyx lobes setaceous, filiform, or subulate to linear.
               17. Calyces generally less than 1/2 as long as corollas, lobes 3–8 mm; corollas white . . . . .9. *Sabatia brevifolia*
               17. Calyces generally 1/2+ as long as corollas, lobes (4–)6–25(–30) mm; corollas pink or rarely white.
                  18. Mid-stem leaves generally less than 2.5 mm wide, distal leaf blades filiform; corolla lobes (13–)17–30 mm . . . . .10. *Sabatia grandiflora*
                  18. Mid-stem leaf blades generally 2.5+ mm wide, distal leaf blades linear; corolla lobes 5–24 mm.
                     19. Plants perennial, stems clustered; calyces usually 0.8+ times as long as corollas. . . . . 13. *Sabatia campanulata*
                     19. Plants annual or biennial, single-stemmed; calyces usually less than 0.8 times as long as corollas . . . . 14. *Sabatia stellaris*

1. Sabatia campestris Nuttall, Trans. Amer. Philos. Soc., n. s. 5: 197. 1836 (as Sabbatia) • Western marsh-pink, prairie or Texas rose-gentian, meadow-pink, Texas-star [E]

Herbs annual. **Stems** generally single, 4-angled, sometimes with wings to 0.2 mm wide, 0.6–5 dm, branching all or mostly alternate. **Leaves** all cauline at flowering time; blade lanceolate-elliptic to ovate, that is, widest proximal to or near middle, 0.8–4 cm × 5–20 mm. **Inflorescences** open cymes; pedicels (10–)20–100 mm. **Flowers** 5-merous; calyx tube shallowly campanulate, 3–8 mm, commissural veins more prominent than midveins, strongly ridged, distally keeled, lobes linear, (6–)10–22(–32) mm; corolla pink or occasionally white, eye often absent or not sharply defined, when present greenish yellow, projections of eye into corolla lobes oblong to narrowly triangular with a reddish border alternating with shorter white or paler yellow zones, tube 4–9 mm, lobes obovate, that is, widest distal to middle, 10–25 × (5–)9–15 mm, apex abruptly ± acute; anthers coiling circinately. $2n = 26$.

Flowering mid spring–early fall. Dry or wet open woods, prairies, fields, roadsides, sandy soils; 0–600 m; Ark., Ill., Kans., La., Miss., Mo., Okla., Tex.

Sabatia campestris occurred as an introduction in Connecticut and Maine in the past. Reports from North Carolina have been derived from the misreading of the label of a specimen actually from Missouri. Reports from Iowa were probably derived from a combination of an ambiguous description and misinterpreted geographic data.

Sabatia campestris often resembles *S. angularis* in aspect, and its stems may be distinctly although narrowly winged. The species can readily be distinguished from *S. angularis* by the prominent keeled ridges below the sinuses of its calyx. Also, in the corollas of *S. campestris,* the yellow projections of the eye are oblong to narrowly triangular, alternating with shorter, pale yellow to white zones; those of *S. angularis* the yellow or yellowish green projections of the eye are widely triangular, contiguous, not alternating with pale yellow zones.

2. Sabatia formosa Buckley, Proc. Acad. Nat. Sci. Philadelphia 14: 7. 1862 (as Sabbatia) • Stately rose-gentian, Buckley's sabatia [E]

Herbs annual. **Stems** 4-angled, sometimes with wings to 0.2 mm wide, 0.6–3 dm, branching all or mostly alternate. **Leaves** basal and cauline present at flowering time; blade lanceolate to ovate, that is, widest proximal to middle, 0.8–2.5 cm × 3–13 mm. **Inflorescences** open cymes; pedicels 20–70 mm. **Flowers** 5-merous; calyx tube campanulate, 2–8 mm, commissural veins more prominent than midveins, strongly ridged, lobes linear, (4–)8–22 mm; corolla purplish pink, eye greenish yellow, projections of eye into corolla lobes with a red border alternating with shorter white or paler yellow zones, tube 3–8 mm, lobes elliptic-rhombic, that is, widest near middle, 9–20 × 4–19 mm, apex ± acute; anthers coiling circinately.

Flowering early–mid spring. Prairies, fields, beaches, roadsides; 0–600 m; La., Okla., Tex.

Sabatia formosa closely resembles *S. campestris*, within which it often has been included. By its earlier flowering, however, *S. formosa* is to some degree reproductively isolated from *S. campestris* where the two species are sympatric. N. B. Bell and L. J. Lester (1980) provided molecular as well as morphological evidence supporting the recognition of *S. formosa* as a species, and further morphological support has been found in studies for this flora. In *S. formosa,* the largest leaves are generally at and near the base of the stem, with the basal rosette usually persisting at flowering time. In *S. campestris,* the proximal leaves are generally smaller than those at mid stem, and the basal rosette is absent. The corollas of *S. formosa* are more deeply pigmented than those of *S. campestris*. The corolla lobes of *S. formosa* tend to be elliptic-rhombic, widest near the middle, whereas those of *S. campestris* are obovate, widest distally.

3. Sabatia arenicola Greenman, Proc. Amer. Acad. Arts 34: 569. 1899 (as Sabbatia) • Sand or coast rose-gentian

Sabatia carnosa Small

Herbs annual. **Stems** generally single, 4-angled with wings to 0.3 mm wide, 0.3–2(–3.2) dm, branching alternate or proximally occasionally opposite. **Leaves** all cauline at flowering time; blade elliptic to ovate or proximal occasionally obovate, 0.6–2(–2.7) cm × 2–9(–13) mm. **Inflorescences** open cymes; pedicels 2–40(–70) mm. **Flowers** 5-merous;

calyx tube campanulate, 3.5–8.5 mm, commissural veins more prominent than midveins, ridged, lobes oblong-lanceolate to narrowly ovate-triangular, 3–20 mm; corolla white to pink, eye white to pale yellow, projections of eye into corolla lobes oblong, without a contrasting border, tube 2–5 mm, lobes spatulate-obovate, 4–10(–13) × 2–8(–11) mm, apex obtuse to subacute; anthers remaining straight or slightly coiling circinately. **2*n*** = 28.

Flowering summer. Beaches, interdunal depressions, salt flats; 0 m; La., Tex.; Mexico (Tamaulipas).

Plants of *Sabatia arenicola* are densely leafy throughout. The vegetative parts are more succulent than those of the other species of *Sabatia* and darken upon drying (a useful character in identifying herbarium specimens). The mid-stem leaves of *S. arenicola* are mostly elliptic, widest near the middle, whereas those of *S. campestris* and *S. formosa*, to which it is most closely related, are mostly lanceolate to ovate, widest near the base.

Although *Sabatia arenicola* appears to be largely autogamous, it is now quite well isolated temporally from *S. formosa*. Molecular evidence indicates introgression of genetic material from *S. formosa* into part of the range of this species (N. B. Bell and L. J. Lester 1978).

4. **Sabatia arkansana** J. S. Pringle & Witsell, Sida 21: 1250, figs. 1, 2, 3[right], 4, 5. 2005 • Pelton's or Arkansas rose-gentian [C] [E]

Herbs annual. **Stems** 4-angled with wings to 0.2 mm wide, branching alternate or proximal rarely opposite. **Leaves** all cauline at flowering time; blade linear to narrowly lanceolate, 0.7–3 cm × 1–4.5(–6) mm. **Inflorescences** open cymes; pedicels (2–)10–40 mm. **Flowers** 5-merous; calyx tube campanulate, 2.8–5.5 mm, commissural veins more prominent than midveins, ridged, lobes linear, 9–13 mm; corolla purplish pink or rarely white, eye yellow or yellowish green, projections of eye into corolla tube oblong to narrowly triangular, alternating with shorter white or paler yellow zones, tube 3–7 mm, lobes narrowly spatulate-obovate, 8–18 × 3–6 mm, apex rounded; anthers coiling circinately.

Flowering summer. Seasonally wet sites in glades; of conservation concern; 100–200 m; Ark.

Sabatia arkansana is known only from shale and nepheline syenite glades in Saline County.

5. **Sabatia macrophylla** Hooker, Compan. Bot. Mag. 1: 171. 1836 (as Sabbatia) • Large-leaved sabatia [E]

Herbs perennial, not stoloniferous. **Stems** several, clustered, terete, 5–14 dm, branching opposite throughout. **Leaves** all cauline at flowering time; blade lanceolate to ovate-oblong or ovate, 2.5–6(–8.5) cm × 5–30 (–45) mm. **Inflorescences** corymboid dichasia of compact cymules; pedicels 1–5 mm. **Flowers** 5-merous; calyx tube campanulate, 1–2 mm, mid- and commissural veins about equally prominent, not ridged or with low, narrow ridges, lobes triangular to linear-subulate, 0.1–3 mm; corolla white or cream throughout, tube 2–4 mm, lobes oblong-oblanceolate, 4–7(–9) × 2–3(–4) mm, apex rounded to obtuse; anthers recurving.

Varieties 2 (2 in the flora): se United States.

The inflorescences of *Sabatia macrophylla* are nearly flat-topped and usually contain more flowers than those of other *Sabatia* species. The relatively small, closely spaced flowers give this species a distinctive aspect. It further differs from *S. difformis* in its glaucous stems and leaves.

Sabatia macrophylla is restricted to central and southern Georgia, northern Florida, and southern Alabama, Louisiana, and Mississippi, mostly but not exclusively (in Georgia) near the Gulf Coast. The range of var. *macrophylla* extends farther west than that of var. *recurvans*.

1. Calyx lobes erect or spreading, shorter than or ± as long as tube . . . . 5a. *Sabatia macrophylla* var. *macrophylla*
1. Calyx lobes recurved, longer than tube . . . . 5b. *Sabatia macrophylla* var. *recurvans*

5a. **Sabatia macrophylla** Hooker var. **macrophylla** [E]

Calyx lobes erect or spreading, narrowly triangular to linear-subulate, 0.1–1.5(–2) mm, shorter than or ± as long as tube. **2*n*** = 38.

Flowering summer. Wet, open pine woods, savannas, grasslands, bogs; 0–200 m; Ala., Fla., Ga., La., Miss.

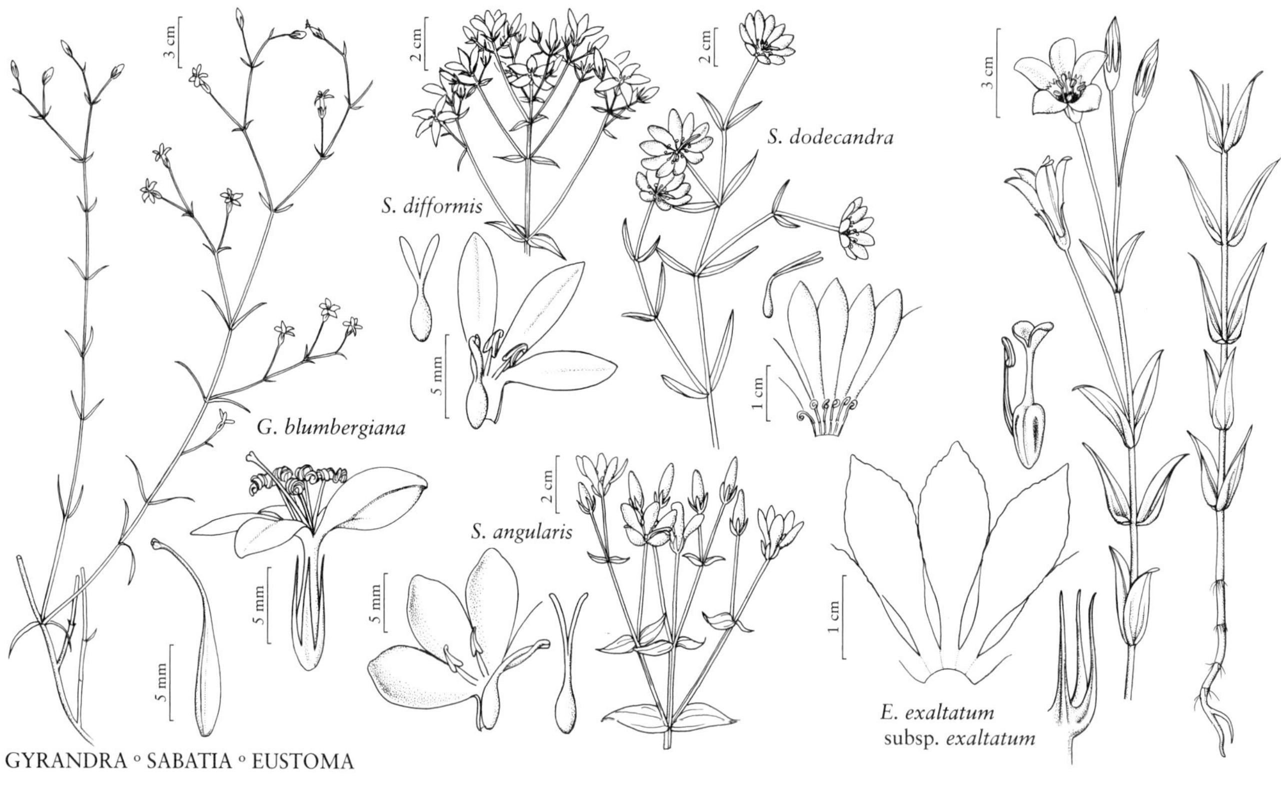

GYRANDRA ◦ SABATIA ◦ EUSTOMA

5b. Sabatia macrophylla Hooker var. **recurvans** (Small) Wilbur, Rhodora 57: 17. 1955 [E]

Sabatia recurvans Small, Man. S.E. Fl., 1049. 1933 (as Sabbatia)

Calyx lobes recurved, linear-subulate, (1–)1.5–3 mm, longer than tube. **2*n*** = 38.

Flowering summer. Wet, open pine woods, savannas, bogs; 0–100 m; Fla., Ga.

6. Sabatia quadrangula Wilbur, Rhodora 57: 22. 1955 • Branching or four-angled sabatia [E]

Herbs biennial. **Stems** usually single, occasionally 2–several, 4-angled with wings 0.1–0.5 mm wide, 1.5–5(–7.5) dm, branching all opposite or secondary and/or tertiary occasionally alternate. **Leaves** basal and cauline or only cauline present at flowering time; basal blades spatulate-obovate; cauline blades linear-oblong to ovate-lanceolate, 0.8–2.5(–6) cm × 3–8(–18) mm. **Inflorescences** cymes of compact cymules; pedicels 1–2(–4) mm. **Flowers** 5-merous; calyx tube widely obconic to campanulate, 1.5–3(–3.5) mm, mid- and commissural veins about equally prominent, low-ridged, lobes linear-subulate or rarely wider, 2–8(–11) mm; corolla white throughout (sometimes drying yellow or salmon) or occasionally with a yellow eye, projections of eye into corolla lobes, when present, triangular, without a contrasting border, tube 2.5–7 mm, lobes oblanceolate to narrowly spatulate-obovate, 4.5–15 × 1.5–6 mm, apex rounded to obtuse; anthers coiling circinately. **2*n*** = 32, 34.

Flowering late spring–summer. Fields, open pine woods, granite outcrops; 0–300 m; Ala., Fla., Ga., N.C., S.C., Va.

The name *Sabatia paniculata* (Michaux) Pursh was misapplied to this species for many years but is typified by a specimen of *S. difformis* (R. L. Wilbur 1955).

7. Sabatia brachiata Elliott, Sketch Bot. S. Carolina 1: 284. 1817 (as Sabbatia) • Elegant or narrow-leaved rose-gentian [E]

Herbs annual, biennial, or occasionally short-lived perennial. **Stems** usually single, rarely 2 or 3, proximally terete, distally 4-angled but not winged, 1.5–7 dm, branching all opposite or distally all or partly alternate. **Leaves** cauline and usually also basal present at flowering time; basal blades spatulate; cauline blades oblong to lanceolate, 1–3(–5) cm × 3–10(–16) mm. **Inflorescences**

open cymes or thyrses; pedicels 1–8(–13) mm. **Flowers** 5-merous; calyx tube turbinate to campanulate, 1–4 mm, mid- and commissural veins about equally prominent, with low, narrow ridges, lobes linear-subulate, (4–)6–10(–15) mm; corolla pink or rarely white, eye greenish yellow, projections of eye into corolla lobes triangular, usually with a dark red border, tube 3–6 mm, lobes oblanceolate to narrowly spatulate-obovate, 5–20 × 2–8 mm, apex obtuse to subacute; anthers coiling circinately. $2n = 32$.

Flowering late spring–summer. Open, wet or occasionally dry sandy woods, savannas, fields, roadsides; 0–500 m; Ala., Ark., Ga., La., Miss., Mo., N.C., S.C., Tenn., Va.

8. Sabatia difformis (Linnaeus) Druce, Rep. Bot. Exch. Club Soc. Brit. Isles 3: 423. 1914 (as Sabbatia) • Lance-leaved or white sabatia [E] [F]

Swertia difformis Linnaeus, Sp. Pl. 1: 226. 1753; *Sabatia lanceolata* Torrey & A. Gray

Herbs perennial, not stoloniferous. **Stems** 1–several, clustered, proximally terete, distally sometimes ± 4-angled but not winged, 2.5–10.5 dm, branching opposite throughout. **Leaves** all cauline at flowering time; blade linear-lanceolate to narrowly or occasionally widely elliptic-ovate, 1–4(–6) cm × 3–14(–22) mm. **Inflorescences** corymboid dichasia of compact cymules; pedicels 1–8(–15) mm. **Flowers** 5(or 6)-merous; calyx tube shallowly campanulate, 1–2(–3) mm, midveins somewhat more prominent than commissural veins, low-ridged, commissural veins scarcely ridged, lobes narrowly lanceolate to filiform, (2–)4–9(–14) mm; corolla white throughout (sometimes drying cream to yellow), tube 2.5–6 mm, lobes oblanceolate, (5–)7–21 × 2.5–8 mm, apex rounded; anthers recurving. $2n = 36$.

Flowering late spring–summer. Wet, open pine woods, savannas, bogs, clearings, ditches; 0–100 m; Ala., Del., Fla., Ga., N.J., N.C., S.C., Va.

A historical record of *Sabatia difformis* from Maryland is documented, but no recent collections or reports are known from that state. Old reports of *S. lanceolata* from New York and Tennessee, for which no documentation was found in studies for this flora, are believed to be erroneous, probably based on misidentifications or misapplications of the name.

The name *Sabatia paniculata* (Michaux) Pursh is typified by a specimen of *S. difformis* but has often been misapplied to *S. quadrangula* (R. L. Wilbur 1955).

9. Sabatia brevifolia Rafinesque, Atlantic J. 1: 147. 1832 (as Sabbatia) • Elliott's or narrow-leaved or short-leaved sabatia [E]

Sabatia elliottii Steudel

Herbs annual. **Stems** single, terete, 1.5–7 dm, branching all or mostly alternate. **Leaves** all cauline or basal occasionally persistent at flowering time; blade linear to oblong-lanceolate, 0.5–3 cm × 1–5(–7) mm. **Inflorescences** open cymes or solitary flowers at ends of branches; pedicels (10–)20–40(–50) mm. **Flowers** 5-merous; calyx tube obconic, 1–3 mm, mid- and commissural veins about equally prominent, not or low-ridged, lobes filiform, 3–8 mm; corolla white, eye greenish yellow, projections of eye into corolla lobes without a contrasting border, tube 1–3 mm, lobes oblanceolate, 6–18 × 2–7 mm, apex obtuse to acute; anthers coiling circinately. $2n = 32$.

Flowering late summer–fall. Open pine woods, savannas, bogs; 0–70 m; Ala., Fla., Ga., S.C.

Reports of *Sabatia brevifolia* from Louisiana were based on a specimen of questionable provenance and are considered probably erroneous by students of that state's flora.

In some older literature, the name *Sabatia difformis* was misapplied to *S. brevifolia*.

10. Sabatia grandiflora (A. Gray) Small, Fl. S.E. U.S., 928. 1903 (as Sabbatia) • Large-flowered marsh-pink or sea-pink or rose-gentian

Sabatia gracilis Michaux var. *grandiflora* A. Gray in A. Gray et al., Syn. Fl. N. Amer. 2(1): 115. 1878 (as Sabbatia)

Herbs annual. **Stems** single, terete, 1.5–9(–11) dm, branching alternate. **Leaves** all cauline at flowering time; blade mostly linear, 1–5 cm × 0.5–2 mm or those near base to 5 mm wide, distal leaves filiform. **Inflorescences** open, few-flowered cymes or solitary flowers; pedicels (20–)40–120 mm. **Flowers** 5-merous; calyx tube campanulate, 6–25(–30) mm, midveins slightly more prominent than commissural veins, veins not ridged or midveins low-ridged, lobes subulate to linear; corolla pink or occasionally white, eye yellow, projections of eye into corolla lobes oblong, usually with red border, tube 3–8 mm, lobes narrowly to medium-widely obovate, (13–)17–30 × 5–15 mm, apex rounded to subacute; anthers coiling circinately. $2n = 36$.

Flowering year-round. Marshes, shores, and wet, open pine and cypress woods; 0–60 m; Ala., Fla.; West Indies (Cuba).

11. Sabatia angularis (Linnaeus) Pursh, Fl. Amer. Sept. 1: 137. 1813 (as Sabbatia) • Common marsh-pink or rose-gentian, square-stemmed rose-gentian or sabatia, bitterbloom [E] [F]

Chironia angularis Linnaeus, Sp. Pl. 1: 190. 1753

Herbs biennial. **Stems** single, 4-angled with wings 0.2–0.3 mm wide, (0.5–)3–7.5(–9) dm, branching proximally mostly opposite, distally mostly alternate. **Leaves** all cauline at flowering time or basal sometimes persistent; basal blades oblong-spatulate to ovate-orbiculate; cauline blades lanceolate to widely ovate, 1–4 cm × 5–30(–40) mm. **Inflorescences** open cymes; pedicels 10–60 mm. **Flowers** 5(or 6)-merous; calyx tube shallowly campanulate, 1–2 mm, mid- and commissural veins about equally prominent, low-ridged, lobes linear to narrowly oblong-lanceolate or occasionally ± foliaceous, 4–15(–18) mm; corolla pink or occasionally white (sometimes drying orange), eye greenish yellow, projections of eye into corolla lobes triangular, usually with dark red border, tube 4–7 mm, lobes ± narrowly spatulate-obovate, 6–22 × 2–9(–11) mm, apex rounded to subacute; anthers coiling circinately. 2*n* = 38.

Flowering late spring–summer. Open pine and mixed woods, prairies, fields, marshes, shores, granite outcrops, roadsides; 0–800 m; Ala., Ark., Del., Fla., Ga., Ill., Ind., Kans., Ky., La., Md., Mich., Miss., Mo., N.J., N.Mex., N.Y., N.C., Ohio, Okla., Pa., S.C., Tenn., Tex., Va., W.Va.

Sabatia angularis has been reported as weakly naturalized in New Mexico (K. W. Allred 1999). Historically, *S. angularis* has also been found introduced in Ontario, Connecticut, the District of Columbia, and Massachusetts, and can be expected elsewhere. An old report from Maine is not implausible, but no documentation has been located. A report from Wisconsin likewise is also plausible, but the provenance of the specimen is doubtful.

12. Sabatia calycina (Lamarck) A. Heller, Bull. Torrey Bot. Club 21: 24. 1894 (as Sabbatia) • Large-sepaled sabatia, coastal rose-gentian

Gentiana calycina Lamarck in J. Lamarck et al., Encycl. 2: 638. 1788

Herbs perennial, not stoloniferous. **Stems** usually single, occasionally 2–several, terete or distally ± 4-angled but not winged, 0.8–5 dm, branching all or mostly alternate. **Leaves** all cauline at flowering time; blade elliptic to widely spatulate, 1–6(–10) cm × 4–30 mm. **Inflorescences** open, few-flowered cymes; pedicels (10–)30–60 mm. **Flowers** 5–7-merous; calyx tube shallowly campanulate, 1.5–5 mm, midveins slightly more prominent than commissural, veins not ridged or midveins with low, narrow ridges, lobes oblanceolate to spatulate or ± foliaceous, 8–25 (–32) mm; corolla pale pink proximally, distally white, or white throughout except for eye, eye yellow, projections of eye into corolla lobes triangular, without a contrasting border, tube 3–6 mm, lobes oblanceolate to narrowly spatulate-obovate, 6–15 × 2–6 mm, apex rounded to obtuse; anthers becoming slightly recurved. 2*n* = 64.

Flowering summer. Marshes, swamps, wet woods, riverbanks, ditches; 0–60 m; Ala., Fla., Ga., La., Miss., N.C., S.C., Tex., Va.; West Indies (Cuba, Dominican Republic).

13. Sabatia campanulata (Linnaeus) Torrey, Fl. N. Middle United States 1: 217. 1824 (as Sabbatia) • Slender marsh-pink or rose-gentian, savannah or campanulate sabatia [E]

Chironia campanulata Linnaeus, Sp. Pl. 1: 190. 1753; *C. gracilis* Michaux; *Sabatia campanulata* var. *gracilis* (Michaux) Fernald

Herbs perennial, not stoloniferous. **Stems** 1–many, clustered, terete or distally 4-ridged but not angled or winged, 1.5–6 (–9) dm, branching all or mostly alternate. **Leaves** all cauline at flowering time; blade narrowly lanceolate or oblong (proximal) to linear (all or distal), 1–4 cm × 1–7(–12) mm. **Inflorescences** open, few-flowered cymes or solitary flowers at ends of branches; pedicels (20–)40–70(–90) mm. **Flowers** 5-merous; calyx tube turbinate to shallowly campanulate, 1–3 mm, mid- and commissural veins about equally prominent, not or low-ridged, lobes setaceous to narrowly linear; corolla pink or rarely white, eye yellow, projections of eye into corolla lobes oblong, usually with a red border, tube 2–6 mm, lobes oblanceolate, 6–24 × 3–9(–11) mm, apex obtuse; anthers coiling circinately. 2*n* = 34.

Flowering summer–early fall. Freshwater marshes, bogs, wet pine savannas, wet fields, ditches; 0–700 m; Ala., Ark., Del., D.C., Fla., Ga., Ky., La., Md., Mass., Miss., N.J., N.Y., N.C., S.C., Tenn., Tex., Va.

Sabatia campanulata formerly occurred in Indiana and Pennsylvania, but there are no recent records from those states.

The differences upon which varieties of *Sabatia campanulata* have been based appear largely to be phenotypic responses to seasonal phenomena and conditions of the habitat. They exhibit less correlation with geographic distribution than some authors have indicated (R. L. Wilbur 1955).

There is a record of a hybrid of *Sabatia campanulata* with *S. kennedyana*.

14. Sabatia stellaris Pursh, Fl. Amer. Sept. 1: 137. 1813 (as Sabbatia) • Annual marsh-pink or sea-pink, saltmarsh-pink

Herbs annual or biennial. **Stems** single, terete or distally 4-angled, 0.2–5(–8) dm, branching alternate. **Leaves** all cauline at flowering time; blade linear to elliptic or obovate, 0.5–6(–9) cm × (1–)2–10(–15) mm. **Inflorescences** open, few-flowered cymes or solitary flowers; pedicels (10–)40–100(–150) mm. **Flowers** (4- or)5-merous; calyx tube obconic to ± campanulate, 1.5–6 mm, midveins slightly more prominent than commissural veins, veins not ridged or midveins shallowly ridged proximally, lobes setaceous to linear, (4–)6–11(–22) mm; corolla pink or rarely white, eye yellow, projections of eye into corolla lobes 3-lobed, usually with a red border, tube 3–8 mm, lobes oblanceolate or narrowly to medium-widely spatulate-obovate or elliptic, 5–20 × 2–10 mm, apex rounded to obtuse; anthers coiling circinately. $2n$ = 36 + 0–4B.

Flowering summer–fall, year-round in Fla. Saltwater and brackish marshes, swales, ditches, and (in Fla.) sand barrens, restricted to coastal habitats north of Fla.; 0–30 m; Ala., Conn., Del., Fla., Ga., La., Md., Miss., N.J., N.Y., N.C., R.I., S.C., Va.; Mexico; West Indies (Bahamas).

The range of *Sabatia stellaris* extends to higher elevations in Mexico.

Sabatia stellaris is known only historically from Massachusetts; although there is at least one correctly identified specimen, some reports from that state were based on misidentified specimens of *S. angularis* and perhaps *S. campanulata* (studies for this flora). Specimens from Pennsylvania have been reidentified in studies for this flora as *S. campanulata*. The basis for an old, undocumented report of *S. stellaris* from Maine is unknown.

Specimens of *Sabatia campanulata* and *S. stellaris* are difficult to distinguish if lacking the basal parts, and misidentifications have led some botanists to consider these species as not distinguishable. The mid-stem leaves of *S. campanulata*, below the first branching, are usually narrowly oblong to oblong-lanceolate, widest near or proximal to the middle, and rounded at the base; those of *S. stellaris*, although variable in proportions, are usually widest distal to the middle and cuneate at the base. The calyces of *S. campanulata* are usually more than 0.8 times as long as and often longer than the corollas; those of *S. stellaris* are usually about 0.75 times as long as the corollas, but both species occasionally deviate from these proportions. The corolla lobes of *S. stellaris* are usually obovate and less than twice as long as wide. The two style branches (including the stigmas) of *S. campanulata* are slightly longer than the uncleft style below them; those of *S. stellaris* are much longer than the very short uncleft portion. In *S. campanulata*, the ultimate lateral (later-flowering) branches of the inflorescence usually bear a pair of bracts below a solitary true pedicel; in *S. stellaris*, the ultimate lateral branching usually, but not invariably, gives rise directly to true pedicels without bracts.

15. Sabatia gentianoides Elliott, Sketch Bot. S. Carolina 1: 286. 1817 (as Sabbatia) • Pinewoods or spider rose-gentian [E]

Lapithea gentianoides (Elliott) Grisebach

Herbs annual. **Stems** single, terete or slightly 4-ridged but not angled or winged, 1.5–5 (–6.5) dm, branching opposite or alternate. **Leaves** cauline and often also basal present at flowering time; basal blades widely oblong-spatulate; cauline blades abruptly differentiated, linear, 1–10 cm × 1–3 mm. **Inflorescences:** flowers solitary or in dense, few-flowered clusters, sessile. **Flowers** 7–12-merous; calyx tube widely campanulate, 3–8 mm, not ridged, lobes setaceous, 3–17 mm; corolla pink, eye greenish yellow, projections of eye into corolla tube oblong, without a border, tube 6–10 mm, lobes oblanceolate to narrowly spatulate-obovate, 12–30 × 4–11 mm, apex rounded to obtuse; anthers slightly twisting helically, not coiling circinately. $2n$ = 28.

Flowering late spring–fall. Open wet pine woods, pine savannas, wet meadows, roadsides; 0–200 m; Ala., Ark., Fla., Ga., La., Miss., N.C., S.C., Tex.

The name spider rose-gentian is derived from the appearance of the involucre subtending each solitary flower or cluster of a few flowers, which comprises two to four or more closely spaced pairs of narrowly linear leaves.

16. Sabatia capitata (Rafinesque) S. F. Blake, Rhodora 17: 54. 1915 • Cumberland or Appalachian rose-gentian, upland sabatia [C] [E]

Pleienta capitata Rafinesque, Fl. Tellur. 3: 30. 1837; *Lapithea capitata* (Rafinesque) Small

Herbs annual. **Stems** single, terete or slightly 4-ridged but not angled or winged, 1.5–4.5(–7) dm, branching opposite or alternate. **Leaves** basal and cauline present at flowering time; blade oblong to elliptic, 2–5(–7) cm × 7–20 (–25) mm. **Inflorescences** heads, sessile. **Flowers** 7–12-merous; calyx tube widely campanulate, 3–6 mm, not ridged, lobes linear, 4–10 mm; corolla pink or rarely white, eye pale yellow, projections of eye into corolla lobes semicircular, without a contrasting border, tube 5–7 mm, lobes narrowly spatulate-obovate, 12–25 × 5–13 mm, apex rounded; anthers remaining straight or nearly so, not coiling. **2*n*** = 76.

Flowering summer–early fall. Open dry or mesic oak-hickory woods, sandstone regions; of conservation concern; 200–900 m; Ala., Ga., Tenn.

Sabatia capitata is endemic to the southernmost portions of the Cumberland Plateau and the Ridge and Valley Province in northern and central Alabama, northwestern Georgia, and southeastern Tennessee. An old specimen was labeled by a later recipient as being from North Carolina, but its provenance is uncertain.

17. Sabatia kennedyana Fernald, Rhodora 18: 150, plate 121, figs. 1–3. 1916 • Plymouth gentian or rose-gentian, large sabatia [C] [E]

Sabatia dodecandra (Linnaeus) Britton, Sterns & Poggenburg var. *kennedyana* (Fernald) H. E. Ahles

Herbs monocarpic but generally requiring 3+ years to flower or occasionally short-lived perennials, stoloniferous. **Stems** 1–several, scattered or occasionally in small clusters, terete, 1.5–6.5 (–8) dm, branching mostly opposite, distally (or on small plants all) sometimes alternate. **Leaves** basal and cauline or only cauline present at flowering time; basal blades linear-oblong to narrowly oblanceolate, 1.5–10 cm × 2–10(–16) mm; cauline blades lanceolate to linear, 1.5–6 cm × 2–7 mm. **Inflorescences** open cymes; pedicels (5–)10–50(–70) mm. **Flowers** 7–12-merous; calyx tube shallowly campanulate, 2–4.5 mm, mid- and commissural veins about equally prominent, obscurely or not ridged, lobes linear-filiform, 5–18 mm; corolla pink to pinkish violet or rarely white or patterned pink and white, eye yellow, 3-lobed, with or without red border, tube 5–8 mm, lobes spatulate to narrowly spatulate-obovate, (9–)12–27 × 4–14 mm, apex rounded or occasionally nearly truncate or emarginate; anthers coiling circinately. **2*n*** = 40.

Flowering summer–fall. Nonsaline pond shores, wet woods, often in shallow water at least early in the season; of conservation concern; 0–10 m; N.S.; Mass., N.C., R.I., S.C., Va.

Sabatia kennedyana is endemic to a few disjunct localities all near the Atlantic coast, although the habitats are nonsaline. Occurrences in Virginia may be derived from introductions.

Prior to its recognition in 1916, this taxon was included in *Sabatia dodecandra* (usually as *S. chloroides*). According to R. L. Wilbur (1955), *S. kennedyana* is best distinguished from *S. dodecandra* and *S. foliosa* by the combination of stems almost completely devoid of ridges; leaves thin, smooth, and brittle when dried; primary branching generally opposite in well-developed specimens; terminal flowers generally much overtopped by the first internode of the lateral branches; calyx tube wide, thin, not ribbed; calyx lobes linear-subulate, less than 0.8 mm wide (0.8+ mm wide in *S. dodecandra* and *S. foliosa*), hyaline-margined, thin and flat in cross section; and corolla lobes spatulate-obovate, that is, widest near the apex. *Sabatia kennedyana* further differs from *S. dodecandra* in the much greater frequency and size of its stolon-borne rosettes.

Although habitat destruction, including drainage, eutrophication, and other disturbances of the coastal-plain ponds constitute much of the basis for conservation concern, the picking of *Sabatia kennedyana* for bouquets is also significant. Because of the similarity of the generic name to Sabbath, this species has traditionally been used for decorating churches in some localities. In picking, the crowns and stolons are readily even if unintentionally ripped out of their oozy substrate. The life history of this species as related to conservation concerns was discussed by L. C. Orrell Ellison (2006).

There is a record of a hybrid of *Sabatia kennedyana* with *S. campanulata*.

Sabatia kennedyana is in the Center for Plant Conservation's National Collection of Endangered Plants.

18. Sabatia dodecandra (Linnaeus) Britton, Sterns & Poggenburg, Prelim. Cat., 36. 1888 (as Sabbatia) • Perennial or large marsh-pink, perennial sea-pink, giant or marsh rose-gentian [E] [F]

Chironia dodecandra Linnaeus, Sp. Pl. 1: 190. 1753; *Sabatia chloroides* (Michaux) Pursh

Herbs perennial; stolons absent or weakly developed. **Stems** 1–several, clustered, terete or distally 4-ridged but not angled or winged, 0.8–6 dm, branching all or mostly alternate. **Leaves** basal absent at flowering time, internodes between cauline leaves mostly 1.25+ times as long as subtending leaves; blade elliptic- or oblong-lanceolate, 1.5–7 cm × 4–12(–16) mm. **Inflorescences** open, few-flowered monochasia or solitary flowers at ends of branches; pedicels 10–90(–110) mm. **Flowers** 7–12(–14)-merous; calyx tube obconic to campanulate, 1.5–4 mm, mid- and commissural veins about equally prominent, 4-ridged; lobes linear to oblong-lanceolate or occasionally narrowly spatulate or ± foliaceous, 4–20 mm; corolla purplish pink or rarely white, eye yellow, projections of eye into corolla lobes oblong, sometimes shallowly 3-lobed, usually with a red border, tube (3–)4–8 mm, lobes oblanceolate to narrowly spatulate-obovate, (10–)12–25 × 3–11 mm, apex rounded to subacute; anthers coiling circinately. $2n = 34 + 8B$.

Flowering summer–fall. Saltwater, brackish, or rarely freshwater marshes; 0–10 m; Conn., Del., Fla., Ga., Md., N.J., N.Y., N.C., S.C., Va.

There are historical records of *Sabatia dodecandra* from Connecticut and New York. Reports from west of the range given here have been based on a concept of the species that included *S. foliosa*.

19. Sabatia foliosa Fernald, Bot. GaZ. 33: 155. 1902 (as Sabbatia) • Leafy marsh-pink or rose-gentian [E]

Sabatia dodecandra (Linnaeus) Britton, Sterns & Poggenburg var. *foliosa* (Fernald) Wilbur; *S. harperi* Small; *S. obtusata* S. F. Blake

Herbs perennial, stoloniferous. **Stems** several–many, scattered or loosely clustered, terete or distally 4-ridged but not angled or winged, 0.8–7(–10) dm, branching generally all or mostly alternate, rarely all opposite. **Leaves** basal absent at flowering time, internodes between cauline leaves generally less than 1.25 times as long as subtending leaves; blade ovate-lanceolate to elliptic or linear, 1.5–6 cm × 4–14(–20) mm. **Inflorescences** open, few-flowered monochasial cymes or solitary flowers at ends of branches; pedicels 10–70(–100) mm. **Flowers** 7–12(–14)-merous; calyx tube shallowly campanulate, 1.5–4 mm, mid- and commissural veins about equally prominent, 4-ridged; lobes linear to narrowly spatulate or ± foliaceous, 10–20 mm; corolla purplish pink or rarely white, eye yellow, projections of eye into corolla lobes oblong, sometimes shallowly 3-lobed, usually with a red border, tube (3–)4–8 mm, lobes oblanceolate to narrowly spatulate-obovate, 12–30 × 3–10 mm, apex rounded to subacute; anthers coiling circinately. $2n = 38$.

Flowering summer. Swamps, wet pine woods, shores, riverbanks, ditches, inland, nonsaline habitats; 0–100 m; Ala., Fla., Ga., La., Miss., S.C., Tex.

Sabatia foliosa has often been treated as a variety of *S. dodecandra*. Its recognition at specific status follows R. L. Wilbur (1970b) and J. D. Perry (1971). As the internodes of *S. foliosa* are generally less than 1.25 times as long as, and often shorter than, the subtending leaves, whereas those of *S. dodecandra* are mostly 1.25–3.5 times as long, this species has a more leafy appearance than *S. dodecandra*. The leaves of *S. foliosa* are thinner in texture than those of *S. dodecandra*; the apices of the mid-stem leaves of *S. foliosa* are usually obtuse and those of the distal leaves merely subacute, whereas the apices of the mid-stem and distal leaves of *S. dodecandra* are usually acute; and the corolla lobes of *S. foliosa* are proportionately narrower than those of *S. dodecandra* (M. L. Fernald 1902; Wilbur 1955). Because of its stoloniferous habit, *S. foliosa* often forms dense colonies, which *S. dodecandra* does not. Natural hybrids between these taxa are unknown, and artificial hybrids are sterile (Wilbur 1970b; Perry).

20. Sabatia decandra (Walter) R. M. Harper, Bull. Torrey Bot. Club 27: 432. 1900 • Bartram's marsh-pink or rose-gentian [E]

Chironia decandra Walter, Fl. Carol., 93. 1788; *Sabatia bartramii* Wilbur; *S. dodecandra* (Linnaeus) Britton, Sterns & Poggenburg var. *coriacea* (Elliott) H. E. Ahles

Herbs perennial, not stoloniferous. **Stems** single, terete, 2.5–8(–10) dm, branching all or mostly alternate. **Leaves** basal and cauline usually present at flowering time; basal blades oblanceolate to spatulate, 4–10 cm × 9–25 mm; cauline blades abruptly differentiated, linear to lanceolate, 1.5–5(–6.5) cm × 1–8(–15) mm. **Inflorescences:** flowers solitary or paired at ends of branches; pedicels (30–)80–120 mm. **Flowers** 8–12(–14)-merous; calyx tube shallowly campanulate, (2–)3–4(–8) mm, mid- and commissural veins about equally prominent, not ridged; lobes narrowly linear to subulate, 4–20 mm; corolla pink or rarely white, eye yellow, projections of eye into corolla lobes oblong, usually with a red border, tube 5–9 mm, lobes narrowly spatulate-obovate, 16–35 × 5–12 mm, apex rounded to obtuse; anthers coiling circinately. 2*n* = 36.

Flowering summer–early fall. Wet pine savannas, cypress woods, pond margins, ditches, sometimes in shallow water; 0–100 m; Ala., Fla., Ga., Miss., S.C.

As well as differing as indicated in the descriptions, *Sabatia decandra* differs from sympatric related species in its thickened calyx lobes, which are semicircular in cross section rather than thin and flat, and in having cauline leaves usually no wider than the diameter of the stem.

Sabatia decandra has generally been called *S. bartramii* in recent years. No specimen associated with the original description of *Chironia decandra* by Walter is known to exist, and R. L. Wilbur (1955) concluded that Walter's description of *C. decandra* did not suffice to indicate unequivocally to which species he had applied the name. D. B. Ward (2007) neotypified the name *C. decandra* with a specimen of the species to which the name *S. decandra* is applied here.

6. EUSTOMA Salisbury, Parad. Lond. 1: plate 34. 1806 • Catchfly- or bluebell- or prairie-gentian [Greek *eu*-, fine, and *stoma*, gap, alluding to showy, open-mouthed corolla]

James S. Pringle

Herbs annual, biennial, or short-lived perennial, chlorophyllous, glabrous. **Leaves** basal and cauline, opposite. **Inflorescences** open, proximally dichasial or completely monochasial cymes; flowers pedicellate. **Flowers** 5-merous; calyx lobed nearly to base; corolla widely campanulate, glabrous, lobes 2+ times as long as tube, margins entire or inconspicuously erose, plicae between lobes absent; stamens inserted near middle of corolla tube; anthers distinct; ovary sessile; style proximally persistent, erect, distinct; stigma 2-lobed; nectariferous ring at base of ovary. **Capsules** compressed ovoid-ellipsoid. *x* = 18.

Species 1: United States, Mexico, West Indies, Central America; introduced in n South America and Pacific Islands (Guam); temperate to tropical areas.

This showy-flowered genus is popular in horticulture, where it is often called lisianthus, that being an incorrect spelling of the name of a genus from which these plants have long been segregated. True *Lisianthius* P. Browne is a Central American and West Indian genus that is very different in appearance and doubtfully in cultivation.

SELECTED REFERENCES Parke, M. 1986. The new look of lisianthus. Horticulture (Boston) 64(3): 32–34. Shinners, L. H. 1957. Synopsis of the genus *Eustoma* (Gentianaceae). SouthW. Naturalist 2: 38–43.

1. Eustoma exaltatum (Linnaeus) Salisbury ex G. Don, Gen. Hist. 4: 211. 1837 [F]

Gentiana exaltata Linnaeus, Sp. Pl. ed. 2, 1: 331. 1762

Herbs 1.5–10 dm. **Leaves** glaucous; basal blades spatulate-obovate to elliptic-oblong, 2–10 cm × 5–20 mm, apex obtuse; cauline blades elliptic-oblong to lanceolate or narrowly ovate, 1.5–9(–14) cm × 4–30(–50) mm, distal blades with apices acute to acuminate. **Inflorescences:** pedicels 3–10 cm. **Flowers:** calyx 10–25 mm, lobes subulate to linear, apex acuminate; corolla predominantly pale to deep violet-blue or rarely rose-violet, with successive zones of greenish yellow, darker and paler shades of the predominant color in the throat, or rarely white or pale yellow, 1.8–7 cm, lobes elliptic to obovate, apex rounded to subacute, often mucronate.

Subspecies 2 (2 in the flora): United States, Mexico, West Indies, Central America; introduced in n South America and Pacific Islands (Guam).

Whether plants of *Eustoma exaltatum* are annual, biennial, or perennial depends on the amount and distribution of rain during the growing season.

Although the two subspecies recognized here have often been treated as distinct species, observations and measurements in studies for this flora indicate that variation in plant size, leaf shape, flower size, and corolla-lobe shape is essentially continuous and exhibits less correlation with geographic distribution than has been indicated in some publications. Many specimens from New Mexico and Oklahoma, especially, are intermediate in morphology.

1. Corollas 1.8–4(–4.5) cm . . . . 1a. *Eustoma exaltatum* subsp. *exaltatum*
1. Corollas (3.5–)4–7 cm. . . . . 1b. *Eustoma exaltatum* subsp. *russellianum*

1a. Eustoma exaltatum (Linnaeus) Salisbury ex G. Don subsp. **exaltatum** • Alkali chalice, West Indian-bluebell [F]

Eustoma barkleyi Standley ex Shinners

Corollas 1.8–4(–4.5) cm, lobes elliptic to narrowly obovate. $2n$ = ca. 72.

Flowering year-round. Prairies, beaches, levees, roadsides, fresh to brackish or saline marshes, other open, moist to wet, often alkaline sites; 0–1900 m; Ala., Ariz., Ark., Calif., Fla., La., Miss., Nev., N.Mex., Okla., Tex.; Mexico; West Indies; Central America; introduced in South America (Venezuela) and Pacific Islands (Guam).

1b. Eustoma exaltatum (Linnaeus) Salisbury ex G. Don subsp. **russellianum** (Hooker) Kartesz in J. T. Kartesz and C. A. Meacham, Synth. N. Amer. Fl., nomencl. innov. 11. 1999 • Texas-bluebell, lira de San Pedro [E]

Lisianthius russellianus Hooker, Bot. Mag. 65: plate 3626. 1838 (as Lisianthus); *Eustoma grandiflorum* (Rafinesque) Shinners; *E. russellianum* (Hooker) G. Don

Corollas (3.5–)4–7 cm, lobes narrowly obovate to widely spatulate-obovate. $2n$ = 18 (cultivated in China), 72.

Flowering spring–fall. Prairies, meadows, other open, usually moist to wet sites; 10–2000 m; Colo., Kans., Mont., Nebr., N.Mex., Okla., S.Dak., Tex., Wyo.

The Montana record may represent an escape from cultivation, of short duration. A report of this taxon from Louisiana was based on a specimen collected in 1806, only three years after the Louisiana Purchase, and "Louisiana" probably did not refer to the present state of Louisiana.

Cultivars of subsp. *russellianum* have been selected for plant size and habit, corolla size and shape (usually 6–11 cm in diameter in cultivated plants), diversity in corolla color and markings, supernumerary petals, and other horticulturally significant traits. Tetraploidy has been induced through colchicine treatment. Plants with supernumerary petals often show other floral irregularities, such as tricarpellate pistils.

7. CICENDIA Adanson, Fam. Pl. 2: 503. 1763 • [Italian *kikenda*, ancient name used in Tuscany for Gentianaceae taxon, probably *Gentiana lutea*]

James S. Pringle

Herbs annual, chlorophyllous, glabrous. **Stems** erect or decumbent. **Leaves** cauline, opposite, sometimes also basal. **Inflorescences:** solitary flowers or 2–5-flowered, irregular or ± dichasial cymes. **Flowers** 4-merous; calyx tubular, lobes longer than tube, connate, 2 outer lobes wider than inner [not connate, all equal]; corolla yellow, salverform, glabrous, lobes shorter, abruptly spreading at summit of tube, margins entire, plicae between lobes absent; stamens inserted in corolla sinuses; anthers distinct, remaining straight; style deciduous, erect, distinct, cleft 1 mm or less [not cleft]; ovary short-stipitate; stigmas 2; nectaries absent. **Capsules** widely ellipsoid. x = 13.

Species 2 (1 in the flora): United States, South America, Europe; introduced in Australia.

The name *Cicendia* was long misapplied to the European genus *Exaculum* Caruel, and the genus described here was called *Microcala* Hoffmannsegg & Link.

1. **Cicendia quadrangularis** (Dombey ex Lamarck) Grisebach, Gen. Sp. Gent., 157. 1838 • Timwort, American or square cicendia, tiny yellow gentian [F]

Gentiana quadrangularis Dombey ex Lamarck in J. Lamarck et al., Encycl. 2: 645. 1788; *Microcala quadrangularis* (Dombey ex Lamarck) Grisebach

Herbs 1–9 cm. **Stem** 1, branching at various levels in larger plants. **Leaves** basal often withered at flowering time, cauline spreading; basal blades spatulate-obovate to oblanceolate or elliptic-oblong, 3–15 × 1–3 mm; cauline blades narrowly elliptic-oblong to oblanceolate, 3.5–7 × 0.5–2 mm, base connate-sheathing 0.5 mm, apex obtuse to subacute. **Inflorescences** 1–5-flowered; pedicels 3–60 mm. **Flowers:** calyx hemispheric, 2–7 mm, much exceeding corolla tube in diam., outer lobes proximally rhombic, distal portion triangular, apex acute, inner lobes proximally oblong, distal portion with strongly invaginated flanges, abruptly narrowed into ovate-triangular, cucullate, acuminate apex; corolla 3–10 mm, lobes ovate-triangular to ovate-elliptic, 1–4 mm, apex rounded or abruptly acute. $2n = 26$.

Flowering spring–early summer. Grassy places, edges of vernal pools, other open habitats; 0–2700 m; Calif., Oreg.; South America; introduced in Australia.

In the latter part of the life of the flower, the stamens of *Cicendia quadrangularis* arch over toward the stigmas, permitting self-pollination if outbreeding has not already occurred. The corollas close at night and in cloudy weather.

8. HALENIA Borkhausen, Arch. Bot. (Leipzig) 1(1): 25. 1796 • Spurred gentian [For Jonas Petersson Halenius, 1727–1810, Swedish physician and student of Linnaeus]

James S. Pringle

Herbs annual [perennial], chlorophyllous, glabrous. **Leaves** basal and cauline, opposite [whorled], sessile, basal and proximal cauline sometimes ± petiolate. **Inflorescences** cymes (often umbelloid) or thyrses. **Flowers** 4-merous; calyx lobed nearly to base; corolla yellow or pale green to violet [white], campanulate or ± ovoid and scarcely opening, glabrous, lobes longer than tube, margins entire or nearly so, with 4 spurs, 1 diverging from tube below each lobe [merely protuberance], plicae between lobes absent; stamens inserted in corolla sinuses; anthers distinct; ovary sessile; style persistent, erect, short or indistinct; stigmas 2; nectaries at ends of spurs, 1 in each spur. **Capsules** compressed-ovoid. $x = 11$.

Species ca. 40 (2 in the flora): North America, Mexico, Central America, South America, Asia; temperate to alpine areas.

Halenia brevicornis (Kunth) G. Don, a widespread neotropical species, occurs in northwestern Chihuahua, Mexico, and should be sought in pine woods and along arroyos in southeastern Arizona and southwestern New Mexico.

SELECTED REFERENCE Allen, C. K. 1933. A monograph of the American species of the genus *Halenia*. Ann. Missouri Bot. Gard. 20: 119–222.

1. Corollas pale green to violet; spurs proximally divergent at ca. 45°, 3–5 mm (occasionally absent on small plants) . . . . 1. *Halenia deflexa*
1. Corollas yellow; spurs proximally nearly horizontal, (3–)5–8(–16) mm. . . . . 2. *Halenia rothrockii*

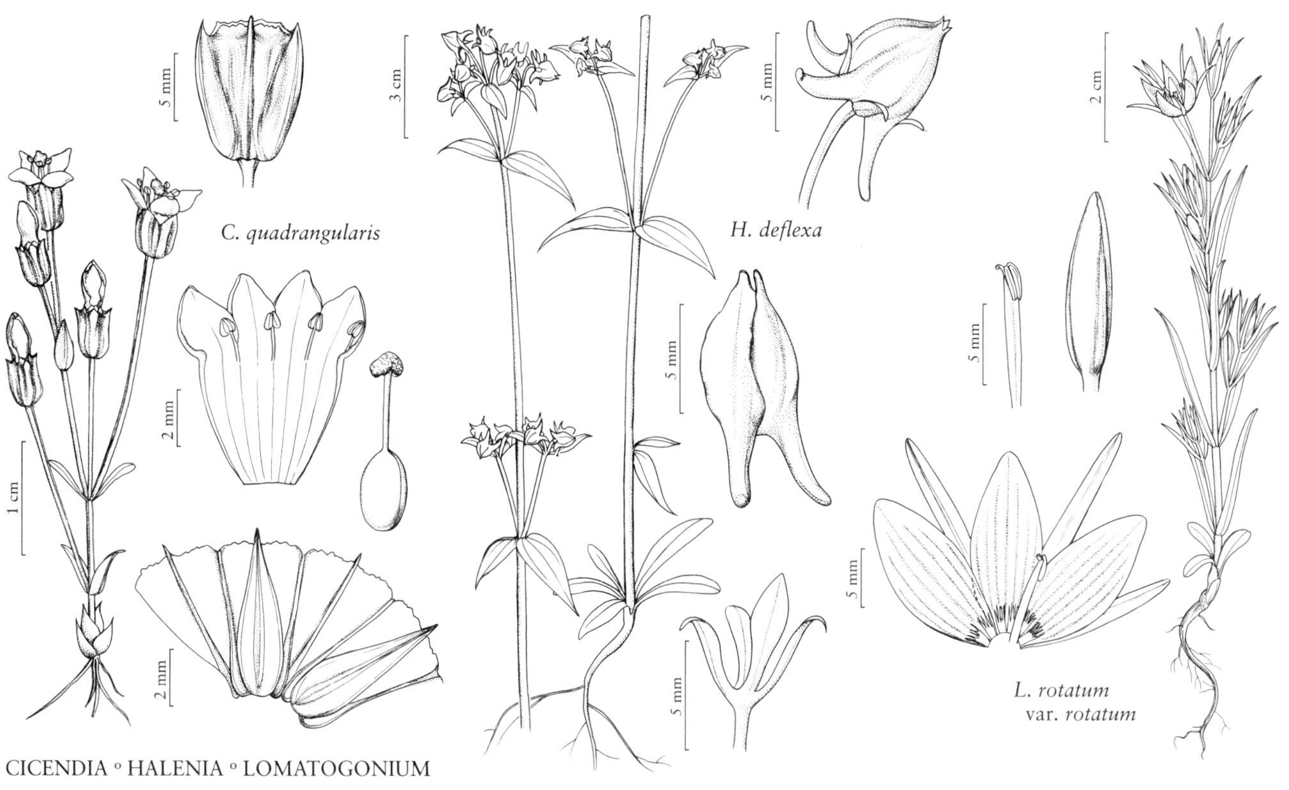

CICENDIA ◦ HALENIA ◦ LOMATOGONIUM

1. **Halenia deflexa** (J. E. Smith) Grisebach in W. J. Hooker, Fl. Bor.-Amer. 2: 67. 1837 • Northern spurred gentian, halénie défléchie [E] [F]

Swertia deflexa Smith in A. Rees, Cycl. 34(II): Swertia no. 8. 1816; *Halenia deflexa* subsp. *brentoniana* (Grisebach) J. M. Gillett; *H. deflexa* var. *brentoniana* (Grisebach) A. Gray

Herbs annual (rosettes sometimes appearing in fall the year before flowering), 0.2–9 dm. **Stems** usually 1, erect, simple or short-branched. **Leaves:** basal and proximal cauline leaves sometimes ± petiolate; basal blades spatulate to elliptic; cauline blades elliptic (proximal) to ovate or lanceolate (distal), 1–7 cm × 5–30(–40) mm. **Inflorescences** 1–75-flowered cymes or thyrses of umbelloid clusters, often with smaller inflorescences on branches. **Flowers:** calyx 4–8 mm, lobes lanceolate to narrowly ovate with narrowed base, apex acuminate; corolla pale green to violet, 8–15 mm, lobes ovate, apex acute to acuminate, spurs proximally divergent at ca. 45°, 3–5 mm, distally ± upcurved or occasionally reduced or absent on flowers on lowest branches, or in stressed environments on most or all flowers. 2*n* = 22.

Flowering summer–early fall. Damp woods, white-cedar swamps, rail edges, shaded stream banks and lakeshores, usually in sandy, calcareous soils, also in rocky or sandy sites near the sea; 0–2000 m; St. Pierre and Miquelon; Alta., B.C., N.B., Nfld. and Labr., N.W.T., N.S., Ont., Que., Sask.; Maine, Mass., Mich., Minn., Mont., N.H., N.Y., N.Dak., S.Dak., Vt., Wis., Wyo.

Mexican plants formerly included in *Halenia deflexa* have been excluded from this species by R. L. Wilbur (1984).

Plants 0.2–1.5 dm, with densely spaced leaves and relatively small corollas deep violet throughout or with the lobes greenish only toward the apex, occurring in damp to wet, exposed, rocky or sandy sites near the sea in St. Pierre and Miquelon, Newfoundland and Labrador, Nova Scotia, and Quebec, have been called var. *brentoniana* or subsp. *brentoniana* but probably represent a phenotypic response to cold, exposed habitats. Intergradation between such plants and larger plants in less extreme habitats is extensive.

2. **Halenia rothrockii** A. Gray, Proc. Amer. Acad. Arts 11: 84. 1876 • Southwestern or Mount Graham or Rothrock's spurred gentian

Herbs annual, 1–5 dm. **Stems** usually 1, erect, simple or short-branched. **Leaves:** basal blades spatulate to elliptic-lanceolate; cauline blades linear, 1.5–4 cm × 2–5 mm. **Inflorescences** 4–27-flowered cymes or thyrses, terminal cluster often umbelloid, plus smaller inflorescences sometimes present on branches. **Flowers:** calyx 5–14 mm, lobes lanceolate, apex acuminate; corolla yellow, 5–12 mm, lobes ovate, apex acute to acuminate, spurs proximally nearly horizontal, distally ± upcurved, (3–) 5–8(–16) mm.

Flowering late summer–fall. Open coniferous woods, mountain meadows; 1900–3000 m; Ariz., N.Mex.; Mexico (Chihuahua, Coahuila, Durango, Sonora).

The name *Halenia recurva* (J. E. Smith) C. K. Allen has often been applied to this species, but its basionym, *Swertia recurva* J. E. Smith, is typified by a specimen of the South American species *H. asclepiadea* (Kunth) G. Don (J. S. Pringle 2008).

9. LOMATOGONIUM A. Braun, Flora 13: 221. 1830 • Felwort [Greek *lomatis*, fringe or hem, and *gonium*, diminutive of *gone*, pistil, alluding to decurrence of stigma along ovary margin]

James S. Pringle

Herbs annual [perennial], chlorophyllous, glabrous. **Leaves** basal (often withering early) and cauline, opposite. **Inflorescences** diffuse cymes, often also axillary. **Flowers** 4, 5 (or 6)-merous; calyx lobed nearly to base; corolla white or blue [violet-blue], rotate [to widely campanulate], glabrous, lobes much longer than tube, margins entire or nearly so, plicae between lobes absent, spurs absent; stamens inserted near base of corolla tube; anthers distinct; ovary sessile; style absent; stigma 2-lobed, ± decurrent along sutures of ovary, not coiling; nectaries in pits on corolla tube near summit, 2 [1] per petal, rim of openings with [without] fringed, scalelike, or tubular prolongations. **Capsules** compressed-ellipsoid. $x = 8$ (see discussion under *L. rotatum*).

Species ca. 20 (1 in the flora): North America, Eurasia; montane to alpine and boreal to arctic areas.

Prior to the mid-twentieth century, this genus was generally known by the illegitimate name *Pleurogyne* Eschscholtz ex Grisebach.

Lomatogonium and *Comastoma* are closely related (R. Wettstein 1896; K. B. von Hagen and J. W. Kadereit 2002). Further transfers in both directions might be appropriate, but the taxa in the flora area would probably not be affected. *Lomatogonium* is also closely related to *Swertia* and has occasionally been included in that genus. Borderline cases with regard to the position of the stigmas occur in some Asiatic species.

SELECTED REFERENCE Liu, S. W. and Ho T. N. 1992. Systematic study on *Lomatogonium* A. Br. (Gentianaceae). Acta Phytotax. Sin. 30: 289–319.

1. Lomatogonium rotatum (Linnaeus) Fries, Summa Veg. Scand., 554. 1849 • Marsh felwort, lomatogone rotacé [F]

Swertia rotata Linnaeus, Sp. Pl. 1: 226. 1753; *Pleurogyne rotata* (Linnaeus) Grisebach

Herbs 3–50(–60) cm. **Leaves:** proximal blades spatulate to elliptic-oblanceolate, middle and distal blades narrowly ovate to lanceolate or linear, 4–40 × 1–10 mm, apex obtuse to acuminate. **Flowers:** calyx 5–15 mm, lobes linear, apex acute or acuminate; corolla white to blue with darker veins, widely campanulate to subrotate, 5–20 mm, lobes elliptic-ovate, apex acute to apiculate, margins of nectaries forming a deeply fringed tube, 0.6–1.2 mm; stigmas strongly decurrent on the sutures of the ovary.

Varieties 3 (2 in the flora): North America, Eurasia.

The report of $2n = 10$ for *Lomatogonium rotatum* from the Manitoba shore of Hudson Bay by Á. Löve and D. Löve (1982b) is in accord with their earlier reports for this species in Iceland (Löve and Löve 1956; D. Löve 1953). More recently, however, $2n = 16$ has been reported for *L. rotatum* in Iceland by Löve and Löve (1986) and in China by Yuan Y. M. and P. Küpfer (1993). No counts have been reported for var. *fontanum*. Reports of $2n = 10$ for this species in any part of its range should be considered questionable.

1. Branches, if any, and pedicels moderately ascending; basal leaves usually persistent, basal blades spatulate to elliptic-oblanceolate, cauline blades narrowly ovate to narrowly lanceolate, 1–3.8 cm × 1.5–6(–10) mm; corollas usually blue, rarely whitish . . . . . . 1a. *Lomatogonium rotatum* var. *rotatum*
1. Branches and pedicels appressed-ascending; basal leaves usually absent at flowering, cauline blades narrowly lanceolate to linear, 1.5–4 cm × 0.7–3(–4) mm; corollas white, often suffused with pale blue, or pale blue throughout . . . . 1b. *Lomatogonium rotatum* var. *fontanum*

1a. Lomatogonium rotatum (Linnaeus) Fries var. **rotatum** [F]

Herbs 3–25(–45) cm. **Stems** usually dark purplish to reddish brown; branches, if any, and pedicels moderately ascending. **Leaves:** basal usually persistent, blades spatulate to elliptic-oblanceolate, 1–2 cm × 2–5 mm, apex obtuse; cauline blades narrowly ovate to narrowly lanceolate, 1–3.8 cm × 1.5–6(–10) mm, apex obtuse to acute. **Flowers:** calyx lobes acute; corollas pale to medium violet-blue, rarely whitish.

Flowering summer. Wet seaside meadows, occasionally around salt springs, consistently near coast in the flora area, estuarine marshes, rock crevices at high-tide mark, moist sand or silt substrate in tidal estuaries; 0–20 m; Greenland; Man., N.B., Nfld. and Labr., N.W.T., Nunavut, Ont., Que., Yukon; Alaska, Maine; Eurasia.

1b. Lomatogonium rotatum (Linnaeus) Fries var. **fontanum** (A. Nelson) J. S. Pringle, Rhodora 115: 107. 2013 [E]

Pleurogyne fontana A. Nelson, Proc. Biol. Soc. Wash. 17: 177. 1904

Herbs 10–50(–60) cm. **Stems** (especially as seen in herbarium, greener when fresh) usually yellow or amber, occasionally reddish brown; branches and pedicels appressed-ascending. **Leaves:** basal usually absent at flowering; cauline blades narrowly lanceolate to linear, 1.5–4 cm × 0.7–3(–4) mm, apex acute to acuminate. **Flowers:** calyx lobes acuminate; corollas white, often suffused with pale blue abaxially and along darker blue veins, or occasionally pale blue throughout.

Flowering summer–fall. Alkaline or saline wet meadows, fens, stream banks, drainage ditches; 400–3000 m; Alta., B.C., Man., N.W.T., Sask., Yukon; Alaska, Colo., Idaho, Mont., N.Mex., Utah, Wyo.

The name *Lomatogonium rotatum* subsp. *tenuifolium* (Grisebach) A. E. Porsild has often been applied to this taxon but was based on Asiatic plants with which those of interior North America are not equated taxonomically here. Plants of this species with strongly ascending branches like those of Rocky Mountain plants do occur in eastern Asia, but such Asian specimens seen in studies for this flora had larger flowers with more deeply blue-pigmented corollas and proportionately wider corolla lobes. Pending molecular comparisons, it is assumed here that the Rocky Mountain plants have a separate origin from those in Asia.

10. SWERTIA Linnaeus, Sp. Pl. 1: 226. 1753; Gen. Pl., ed. 5, 107. 1754 • [For Emanuel Sweert, ca. 1552–1612, Dutch gardener and illustrator]

James S. Pringle

Herbs perennial [monocarpic], chlorophyllous, glabrous [stems and leaves puberulent]. **Leaves** basal and cauline, opposite or all or distal leaves subopposite or alternate. **Inflorescences** thyrses or verticillasters, occasionally racemoid [flowers solitary or few]. **Flowers** 4- or 5-merous; calyx lobed nearly to base, lobes lanceolate; corolla violet-blue or pale green to white, usually with violet-blue markings and/or suffusions [rarely brownish red], rotate or nearly so, lobes much longer than tube, entire, plicae between lobes absent; stamens inserted near base of corolla tube, usually connected by corona consisting of low ridge from which trichomes [scales] arise between filaments; anthers distinct; ovary sessile [short-stipitate]; style persistent, erect, short [absent]; stigmas 2; nectaries in [1] 2 foveae per corolla lobe, rim of openings raised [rarely scarcely so], fringed. **Capsules** compressed-cylindric to ovoid. x = [7, 8, 9, 10, 12, 13] 14.

Species ca. 120 as commonly circumscribed (1 in the flora): North America, Eurasia, Africa, Pacific Islands (New Zealand); temperate to high-altitude tropical regions.

There are differences of opinion about the appropriate circumscription of *Swertia*. Some, for example, V. V. Zuev (1990) and J. Shah (1984), included all of the taxa treated here as *Swertia* and *Frasera*, along with many Asian and African species, within a broad concept of *Swertia*. Others, provisionally followed in this flora, segregate all of the species endemic to North America, including Mexico, as *Frasera*, but retain many Asian species similar in morphology in *Swertia*. Still others, for example, H. Toyokuni (1963), emphasizing chromosome numbers, advocated the restriction of *Swertia* to *S. perennis*, broadly circumscribed, this being the only species in the complex known to have n = 14 or any multiple of 7. This topic is further discussed under 11. *Frasera*.

SELECTED REFERENCES St. John, H. 1941. Revision of the genus *Swertia* (Gentianaceae) and the reduction of *Frasera*. Amer. Midl. Naturalist 26: 1–29. Shah, J. 1984. Taxonomic Studies in the Genus *Swertia* (Gentianaceae). Ph.D. thesis. University of Aberdeen.

1. Swertia perennis Linnaeus, Sp. Pl. 1: 226. 1753 • Star or mountain bog swertia, star-gentian [F]

Swertia perennis var. *obtusa* (Ledebour) Grisebach

Herbs 0.8–6.5 dm. **Stems** usually 1 + 1–few rosettes. **Leaf blades** obtuse to subacute, not white-margined; basal and rosette blades spatulate-obovate to elliptic, 3–22 cm × 10–32 mm; cauline alternate, subopposite, or distal often opposite, blades elliptic-oblanceolate. **Inflorescences** narrow, open, often few-flowered. **Flowers** (4- or)5-merous; calyx 4–8 mm; corolla bluish white to violet-blue, sometimes with darker spots, or rarely greenish white, veins darker, 7–16 mm, lobes lanceolate-oblong, apex acute; corona low, ± fringed ridge; openings of foveae oval to round, rims fringed all around. $2n$ = 28.

Flowering summer–early fall. Wet meadows, thickets, bogs, stream banks; 0–3900 m; B.C.; Alaska, Ariz., Calif., Colo., Idaho, Mont., Nev., N.Mex., Oreg., Utah, Wash., Wyo., restricted to high elevations southward; Eurasia.

A specimen considered perhaps to be from the Yukon Territory is more likely from British Columbia, but *Swertia perennis* should be expected in the Yukon.

11. FRASERA Walter, Fl. Carol., 9, 87. 1788 • Green-gentian [For John Fraser Sr., 1750–1811, Scottish botanical and horticultural explorer]

James S. Pringle

Leucocraspedum Rydberg; *Swertia* Linnaeus sect. *Frasera* (Walter) Zuev; *Tesseranthium* Kellogg

Herbs perennial or long-lived monocarpic, chlorophyllous, glabrous or with stems and leaves puberulent; stems stout and hollow, proximally over 1 cm diam. in *F. caroliniensis* and *F. speciosa*, in other species firm and more slender. **Leaves** basal and cauline, opposite or whorled in 3s to 5s. **Inflorescences** thyrses, verticillasters, or occasionally racemoid cymes. **Flowers** 4-merous; calyx lobed nearly to base, lobes lanceolate; corolla violet-blue or pale green or pale yellow to white, usually with violet-blue markings and/or suffusions, nearly rotate or rarely campanulate (distinctly campanulate only in *F. tubulosa*, somewhat so in *F. fastigiata*), lobes much longer than tube, margins entire, without plicae between lobes; stamens inserted near base of corolla tube, usually connected by a corona consisting of a low ridge from which, in most species, trichomes or scales arise between the filaments; anthers distinct; ovary sessile; style erect, gradually or abruptly differentiated from ovary, persistent; stigmas 2; nectaries in 1 or 2 foveae on each corolla lobe, rim of fovea openings in some species also surrounding a differentiated area on the adaxial corolla surface distally adjacent to the opening, raised and fringed, the fringe components distinct to the base or nearly so except in *F. tubulosa*. **Capsules** compressed-cylindric to ovoid. $x = 13$.

Species 15 (15 in the flora): North America, Mexico.

The division of *Swertia* in the broad sense, which is sometimes circumscribed so as to include *Frasera*, is supported by molecular phylogenetic studies by P. Chassot et al. (2001), but as of this writing, the matter of the most appropriate segregation of genera is unsettled. The acceptance of generic status for *Frasera*, which has become more frequent in recent years, is provisional in this flora. Relatively few of the North American species were included in the study by Chassot et al., and its usefulness for generic delimitations is further impaired by the molecular techniques available at the time. Satisfactory morphological characterizations of *Frasera* and other genera that might be segregated from *Swertia* in the broad sense remain elusive. Some character states that have sometimes been said to distinguish *Frasera* from *Swertia* in the narrow sense, including the presence or absence of connate-sheathing leaf bases, four- versus five-merous flowers, and whether or not a distinct, slender style is present, are not consistently applicable in the flora area, and some extralimital species that clustered closely with *S. perennis* in the phylogenetic study cited above are morphologically similar to species treated as *Frasera* in the flora area. Further generic realignments in the complex are to be expected.

Characters associated with the nectaries are important in distinguishing among the species and varieties of *Frasera*, so herbarium specimens should be prepared so that the adaxial surfaces of some corollas are visible. The nectaries, which may be one or two on each corolla lobe, are located in pits, called foveae, which open adaxially. Two kinds of foveal morphology are present in North American *Frasera*. In species 1 through 5, the opening is immediately adaxial to the nectary, and a raised, fringed rim surrounds the fovea opening only. This foveal morphology is present in the type species of both generic names *Frasera* and *Swertia*. In species 6 through 15, including all of the species that have been placed in *Leucocraspedum*, the fovea is pocketlike, with the adaxial opening distal to the nectary or nectaries. The raised, fringed rim extends beyond the opening of the fovea, so as to surround both the opening and a distally adjacent portion of the adaxial corolla surface differentiated in color and texture. These structures associated with the nectaries are to be distinguished from the androecial coronas of some species, which consist of fimbriate ridges or entire, serrate, or laciniate scales between the bases of the filaments.

In the first group of species distinguished above, the ovary in species 1 and 2 tapers toward the stigmatic lobes with scarcely any differentiation of a style, as also occurs in *Swertia perennis*, whereas in species 3 through 5, including *Frasera caroliniensis*, the type of the generic name *Frasera*, a distinct, slender style is present. All species in the second group have distinct, slender styles. White leaf margins prevail in the second group of species but in the first group are present only in *F. tubulosa*.

The inflorescences of *Frasera* are thyrses or verticillasters, consisting of a central axis along which smaller thyrses (proximally) and/or dichasial or modified cymules are borne in pairs or whorls. The cymules of the larger plants are usually accompanied by flowers on pedicels that arise directly from the main axis. In *F. coloradensis* and *F. parryi*, the primary branches of the inflorescence may be nearly as long as the central axis. Small inflorescences, which occur frequently in *F. ackermaniae* and *F. gypsicola* and occasionally in *F. albicaulis*, may be racemoid. Inflorescences are described here as dense if the primary branches are closely spaced, less than 2 cm apart except sometimes at the proximal nodes, and relatively short, so that the inflorescences are usually less than 6 cm wide and/or more than five times as long as wide, with crowded flowers; as narrow but not dense in *F. gypsicola*, in which the inflorescence width is similar but the flowers are few and are not closely spaced; or as diffuse, if the branches are more widely separated and both the branches and the pedicels are longer and generally strongly divergent, so that the inflorescences are generally more than 6 cm wide and most of the flowers are well separated.

Plants of at least some of the monocarpic species of *Frasera* remain in a rosette stage for several to many years. Studies of *F. caroliniensis* and *F. speciosa* have shown that, within any one population, after several years in which all or most plants have remained in the rosette stage, all or most of the larger plants are likely to flower in the same year, while the smaller plants remain vegetative. The environmental factors that induce flowering in such species are not well understood. In some species, the plants are monocarpic with respect to the original crown, which dies after flowering, but prior to flowering may produce new crowns from adventitious buds on the roots. These may persist as independent plants and flower in later years, after having attained larger size (D. W. Inouye and O. R. Taylor 1980; P. F. Threadgill et al. 1981).

SELECTED REFERENCES Card, H. H. 1931. A revision of the genus *Frasera*. Ann. Missouri Bot. Gard. 18: 245–280, plate 14. St. John, H. 1941. Revision of the genus *Swertia* (Gentianaceae) and the reduction of *Frasera*. Amer. Midl. Naturalist 26: 1–29. Shah, J. 1984. Taxonomic Studies in the Genus *Swertia* (Gentianaceae). Ph.D. thesis. University of Aberdeen.

1. Openings of foveae directly adaxial to nectaries and surrounded by a raised, fringed rim, with no differentiated areas on the adaxial corolla surface; leaves with or without white margins.
    2. Stems proximally 1+ cm diam.; calyces 6–25 mm.
        3. Foveae 2 on each corolla lobe, with separate openings; w North America . . . . 3. *Frasera speciosa*
        3. Foveae solitary; e, c North America . . . . 4. *Frasera caroliniensis*
    2. Stems to 1 cm diam.; calyces 4–15 mm.
        4. Rim of fovea opening prolonged into a tube 1+ mm, which divides into 2 oblong projections; leaf blades white-margined, proximal blades oblanceolate, distal blades linear-oblong . . . . 5. *Frasera tubulosa*
        4. Rim of fovea opening fimbriate, scarcely or not prolonged into a tube; leaf blades not white-margined, all or at least basal and mid-cauline blades widely elliptic to ovate.
            5. Corollas light to medium blue or violet-blue with darker veins, often spotted; androecial corona of sparse hairs or absent; Idaho, Washington . . . . 1. *Frasera fastigiata*
            5. Corollas predominantly white to pale greenish yellow, occasionally with a violet-blue tinge, not spotted; androecial corona of many dense hairs; California, sw Oregon . . . . 2. *Frasera umpquaensis*

1. Openings of foveae distal to the nectaries, usually with a differentiated area on the adaxial corolla surface extending distally from the opening of each fovea (differentiated area absent in *F. gypsicola*), with a raised rim surrounding the combined foveal opening and differentiated area on the corolla lobe; leaf blades with white margins.
   6. Nectaries 2 per corolla lobe, foveae paired or, if single, 2-lobed at base.
      7. Nectaries in closely paired foveae with separate openings into a single differentiated area shaped ± like the spade on playing cards, rim of differentiated area ± sparsely short- to long-fringed all around, no part scalelike; Arizona, Colorado, New Mexico, Utah . . . . 6. *Frasera paniculata*
      7. Nectaries each at the base of a 2-lobed fovea with a single opening into an oblong to elliptic differentiated area on the corolla surface, rim proximally projecting as a scale, distally deeply fringed; California, w Nevada . . . . 7. *Frasera puberulenta*
   6. Nectaries 1 per corolla lobe, foveae not 2-lobed at base.
      8. Raised, completely fringed rim surrounding a round fovea opening only, no differentiated area on corolla surface . . . . 12. *Frasera gypsicola*
      8. Raised, proximally or completely fringed rim surrounding the fovea opening and a differentiated area on the corolla surface.
         9. Leaves proximally whorled on main stem, opposite on branches . . . . . . 8. *Frasera albomarginata*
         9. All cauline leaves opposite.
            10. Inflorescences diffuse, 5–30 cm wide.
                11. Plants 1.5–2.5 dm; stems and adaxial leaf surfaces puberulent; foveae opening into an orbiculate to elliptic-oblong differentiated area . . . . 9. *Frasera coloradensis*
                11. Plants 5–16 dm; stems and leaves glabrous; foveae opening into a U-shaped differentiated area on the corolla surface . . . . 10. *Frasera parryi*
            10. Inflorescences narrow, ± dense, 1.5–4.5(–6) cm wide.
                12. Only a low, fringed ridge between bases of filaments; mountains of s California . . . . 11. *Frasera neglecta*
                12. Corona scales 1–6 mm between bases of filaments; widely distributed in California and/or elsewhere.
                    13. Corollas unmarked; foveae opening into an elliptic-obovate to suborbiculate differentiated area, rim ± evenly fringed all around . . . . 15. *Frasera montana*
                    13. Corolla with dark blue or purple dots (except in *F. albicaulis* var. *idahoensis*); foveae opening into an oblong, elliptic-oblong, or lance-ovate differentiated area, fringe at distal end of this area absent or distinctly shorter than at base.
                        14. Lowest cauline internode generally longer than all basal leaves; corolla lobes oblong-obovate, widest distal to midlength, tapering ± abruptly to an acute or short-acuminate apex . . . . 13. *Frasera albicaulis*
                        14. Lowest cauline internode generally shorter than some basal leaves; corolla lobes narrowly proximally oblong, widest near midlength, tapering gradually to an acuminate apex . . . . 14. *Frasera ackermaniae*

1. **Frasera fastigiata** (Pursh) A. Heller, Bull. Torrey Bot. Club 24: 312. 1897 • Clustered frasera or green-gentian [E]

Swertia fastigiata Pursh, Fl. Amer. Sept. 1: 101. 1813

Herbs monocarpic, 3–14 dm, glabrous. **Stems** 1. **Leaf blades** not white-margined; basal spatulate-obovate, 15–60 (including proximal petiolar portions) × 3–17 cm; proximal cauline leaves whorled, distal often opposite, blades widely elliptic to ovate, apex obtuse to acute (proximal) to acuminate (distal). **Inflorescences** dense, sometimes interrupted proximally. **Flowers:** calyx 4.4–13 mm; corolla light to medium blue or violet-blue with darker veins, often spotted, 6.7–13.4 mm, lobes elliptic-ovate, apex short-acuminate; androecial corona a sparse fringe of hairs to 5 mm, occasionally absent; style short, ± stout and indistinctly differentiated from summit of ovary; nectaries and foveae 1 per corolla lobe, foveae round or nearly so, opening directly adaxial to nectary, without a differentiated area on the corolla surface, rim raised, with long, incurved fringes all around.

Flowering late spring–summer. Mountain meadows, open woods; 1600–2000 m; Idaho, Wash.

2. **Frasera umpquaensis** M. Peck & Applegate, Madroño 6: 12. 1941 • Umpqua frasera or green-gentian [C] [E]

Swertia umpquaensis (M. Peck & Applegate) H. St. John

Herbs monocarpic, 3–14 dm, glabrous. **Stems** 1. **Leaf blades** not white-margined; basal spatulate-obovate, 15–60 (including proximal petiolar portions) × 3–17 cm; cauline leaves generally whorled, blade widely elliptic to ovate, apex obtuse to acute (proximal) to acuminate (distal). **Inflorescences** dense, sometimes interrupted proximally. **Flowers:** calyx 7–15 mm; corolla white or pale greenish yellow, pale yellow near midvein, sometimes with violet-blue tinge, not spotted, 5.3–11 mm, lobes ovate-oblong, apex acuminate; androecial corona a dense fringe to 5 mm; style short, ± stout and indistinctly differentiated from summit of ovary; nectaries and foveae 1 per corolla lobe, foveae round or nearly so, opening directly adaxial to nectary, without a differentiated area on the corolla surface, rim raised, with long, incurved fringes all around. $2n = 78$.

Flowering late spring–summer. Mountain meadows, open woods; of conservation concern; 1200–2000 m; Calif., Oreg.

Former uncertainty as to the distinctness of *Frasera umpquaensis* from *F. fastigiata* has been cleared up by B. L. Wilson et al. (2010).

3. **Frasera speciosa** Douglas ex Grisebach in W. J. Hooker, Fl. Bor.-Amer. 2: 66, plate 153. 1837 • Giant or showy frasera, monument plant, deer's-ears, elkweed, showy green-gentian

Frasera macrophylla Greene; *Swertia radiata* (Kellogg) Kuntze, Revis. Gen. Pl. 2: 430. 1891; *Swertia radiata* (Kellogg) Kuntze var. *macrophylla* (Greene) H. St. John; *Tesseranthium radiatum* Kellogg; *T. speciosum* (Douglas ex Grisebach) Rydberg, not *S. speciosa* Wallich ex D. Don 1836

Herbs monocarpic, 5–20 dm, glabrous or stems and leaves puberulent. **Stems** 1. **Leaf blades** not white-margined; basal spatulate or oblanceolate to elliptic-obovate, 7–50 × 1–15 cm, apex rounded to acute; cauline leaves whorled, blade oblong-lanceolate. **Inflorescences** elongate, open proximally, ± dense distally. **Flowers:** calyx 10–25 mm; corolla pale yellowish green, purple-dotted, occasionally suffused with purple distally, 12–25 mm, lobes elliptic-oblong to obovate, apex [obtuse or] acute to short-acuminate; androecial corona scales 7–9 mm, deeply multicleft; style slender, distinct; nectaries and foveae 2 per corolla lobe, foveae narrowly elliptic, opening directly adaxial to nectary, each opening with a ± even fringe all around, the pair opening into a green but not rimmed area on the corolla surface. $2n = 78$.

Flowering summer. Open woods, montane to subalpine meadows; 1500–3500 m; Ariz., Calif., Colo., Idaho, Mont., Nev., N.Mex., Oreg., S.Dak., Tex., Utah, Wash., Wyo.; Mexico (Nuevo León, Tamaulipas).

Swertia radiata var. *maderensis* Henrickson, endemic to Mexico, is conspecific with *Frasera speciosa*, but the necessary combination in *Frasera* has not been published. If that is done, our material will become var. *speciosa*.

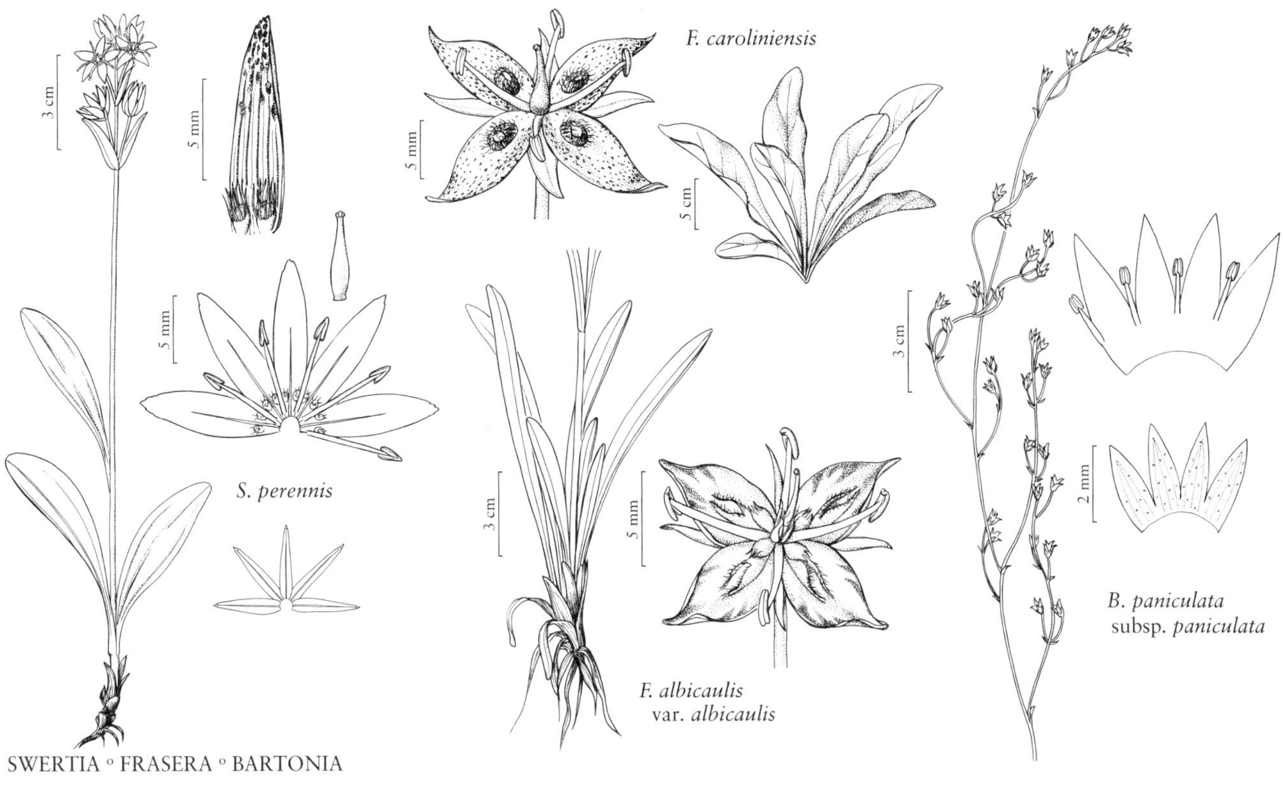

SWERTIA ◦ FRASERA ◦ BARTONIA

4. **Frasera caroliniensis** Walter, Fl. Carol., 88. 1788 • American columbo [E] [F]

Swertia caroliniensis (Walter) Kuntze

Herbs monocarpic, 10–25 dm, glabrous or stems and leaves puberulent. **Stems** 1. **Leaf blades** not white-margined; basal spatulate to elliptic or narrowly obovate, 20–45 × 3–10 cm, apex rounded to acute; cauline leaves whorled, blade oblong-lanceolate, 10–32 × 3–7 cm. **Inflorescences** elongate, open proximally, ± dense distally. **Flowers:** calyx 6–16(–25) mm; corolla pale yellowish green, purple-dotted and sometimes suffused with pale purple, 10–21 mm, lobes elliptic-oblong, apex short-acuminate; androecial corona of trichomes and deeply fringed scales to 3 mm; style slender, distinct; nectaries and foveae 1 per lobe, foveae round, opening directly adaxial to nectary, without a differentiated area on the corolla surface, rim ± evenly fringed all around. $2n = 78$.

Flowering late spring–early summer. Deciduous, ± open woods, often ravines, calcareous soils; 100–700 m; Ont.; Ala., Ark., Ga., Ill., Ind., Ky., Mich., Miss., Mo., N.Y., N.C., Ohio, Okla., Pa., S.C., Tenn.

An old report of *Frasera caroliniensis* from Louisiana is plausible, but no documentation is known.

5. **Frasera tubulosa** Coville, Proc. Biol. Soc. Wash. 7: 71. 1892 • Kern frasera [E]

Swertia tubulosa (Coville) Jepson

Herbs perennial, (0.6–)2–11 dm, glabrous. **Stems** generally 1, with 0–few rosettes. **Leaf blades** white-margined; basal oblanceolate, 2–9 × 0.3–1(–1.5) cm, apex abruptly pointed; cauline leaves whorled, distal blades linear-oblong. **Inflorescences** elongate, dense, branching near base of main stem, nearly continuous distally. **Flowers:** calyx 6–11 mm; corolla white to pale blue, veins darker, campanulate, 8–13 mm, lobes elliptic-oblong to obovate, apex short-acuminate; androecial corona scales narrowly oblong, 2–3 mm, margins fringed; style slender, distinct; nectaries and foveae 1 per corolla lobe, foveae round at base, opening directly adaxial to nectary, without a differentiated area on the corolla surface, rim prolonged into an appendage ca. 3 mm, proximally tubular, dividing distally into 2 oblong, fringed lobes.

Flowering summer. Open pine woods; 1800–2700 m; Calif.

Frasera tubulosa is endemic to Inyo and Tulare counties.

6. Frasera paniculata Torrey in War Department [U.S.], Pacif. Railr. Rep. 4(5): 126. 1857 • Utah or tufted frasera [E]

Frasera utahensis M. E. Jones; *Leucocraspedum utahense* (M. E. Jones) Rydberg; *Swertia utahensis* (M. E. Jones) H. St. John

Herbs monocarpic, (4–)7–15 dm, glabrous. **Stems** usually 1. **Leaf blades** white-margined; basal linear-lanceolate to narrowly oblanceolate, 5–20 × 0.7–2(–3) cm; cauline leaves opposite, distal blades lanceolate. **Inflorescences** diffuse. **Flowers:** calyx 3–6 mm; corolla greenish white to greenish yellow, dark green- or purple-dotted, 7–12 mm, lobes elliptic-ovate, apex abruptly acuminate; androecial corona ± fringed; style slender, distinct; nectaries 2 per lobe, in closely paired, ± separate foveae, foveae opening distal to nectaries, into a single widely ovate-triangular differentiated area on the corolla surface shaped ± like the spade on playing cards, rim ± sparsely short- to long-fringed.

Flowering late spring–summer. Desert scrub, dry, open pinyon-juniper woods, often on pink sand dunes; 1200–2200 m; Ariz., Colo., N.Mex., Utah.

The only specimen from Nevada identified as *Frasera paniculata* has been reidentified as *F. albomarginata* var. *albomarginata* in studies for this flora.

7. Frasera puberulenta Davidson, Bull. S. Calif. Acad. Sci. 11: 77, plate 1. 1912 • Inyo frasera [E]

Swertia albomarginata (S. Watson) Kuntze var. *purpusii* Jepson; *S. puberulenta* (Davidson) Jepson

Herbs monocarpic, 1–3(–5) dm; stems and adaxial leaf surfaces puberulent. **Stems** 1–several. **Leaf blades** narrowly white-margined; basal oblanceolate or narrowly obovate to elliptic-oblong, 2–12 × 0.6–2.2 cm; cauline leaves opposite or rarely some whorled, distal blades oblong to lanceolate. **Inflorescences** diffuse, branching near base of main stem. **Flowers:** calyx 5–12 mm; corolla greenish white, copiously blue-purple-dotted, 7–13 mm, lobes oblong-obovate, apex abruptly acuminate; androecial corona poorly developed, ± fringed; style slender, distinct; nectaries 2 per corolla lobe in a single, basally 2-lobed fovea, fovea opening distal to nectaries, into a single differentiated area on the corolla surface that is oblong to elliptic, ± widened at distal end, proximal side of rim prolonged into a scale 2 mm, fringed ± 1/2 its length, distal side of rim long-fringed.

Flowering summer. Dry, open coniferous woods; 1700–3400 m; Calif., Nev.

Frasera puberulenta is endemic to high elevations in Inyo and Mono counties in California and Mineral and Esmeralda counties in Nevada.

8. Frasera albomarginata S. Watson, Botany (Fortieth Parallel), 280. 1871 • White-margined or desert frasera [E]

Leucocraspedum albomarginatum (S. Watson) Rydberg; *Swertia albomarginata* (S. Watson) Kuntze

Herbs perennial, (1–)2–6 dm, glabrous. **Stems** 1–few. **Leaf blades** white-margined; basal oblanceolate, 2–10 × 0.5–1 cm; cauline leaves whorled at least proximally on main stem, usually opposite on branches, distal blades linear-lanceolate. **Inflorescences** diffuse, often branching at or near base of main stem. **Flowers:** calyx (2–)5–12 mm; corolla greenish white, often purple-dotted, 8–14 mm, lobes elliptic-obovate, apex abruptly acuminate; androecial corona a fringe, sometimes scarcely developed; style slender, distinct; nectaries and foveae 1 per corolla lobe, foveae opening distal to nectary, into a differentiated area on the corolla surface that is proximally oblong with a raised rim fringed ± 1/2 its height, distally wider with rim fringed to base.

Varieties 2 (2 in the flora): w United States.

1. Inflorescence branches and calyx lobes glabrous; distal end of differentiated area on corolla lobes cordate to deeply 2-lobed . . . . 8a. *Frasera albomarginata* var. *albomarginata*
1. Inflorescence branches and calyx lobes puberulent; distal end of differentiated area on corolla lobes rounded, truncate, or shallowly notched . . . . 8b. *Frasera albomarginata* var. *induta*

8a. Frasera albomarginata S. Watson var. **albomarginata** [E]

Inflorescence branches glabrous. **Flowers:** calyx lobes glabrous; distal end of differentiated area on corolla lobes cordate to deeply 2-lobed.

Flowering late spring–summer. Dry, open, rocky, mostly pinyon-juniper woods; 1300–2500 m; Ariz., Calif., Colo., Nev., N.Mex., Utah.

8b. Frasera albomarginata S. Watson var. **induta** (Tidestrom) Card, Ann. Missouri Bot. Gard. 18: 275. 1931 [E]

Frasera induta Tidestrom, Proc. Biol. Soc. Wash. 36: 183. 1921

Inflorescence branches densely puberulent. **Flowers:** calyx lobes puberulent abaxially except near midvein; distal end of differentiated area on corolla lobes rounded, truncate, or shallowly notched.

Flowering late spring–summer. Dry, open, rocky pinyon woods; 1500–2200 m; Calif., Nev.

Variety *induta* is restricted to the Clark Mountains in San Bernardino County, California, and the Spring (Charleston) Mountains in southern Nevada.

The name *Swertia albomarginata* var. *purpusii* has been applied in part to *Frasera albomarginata* var. *induta* but is typified by a specimen of *F. puberulenta.*

9. Frasera coloradensis (C. M. Rogers) D. M. Post, Bot. GaZ. 120: 3. 1958 • Colorado frasera [C] [E]

Swertia coloradensis C. M. Rogers, Madroño 10: 108. 1949

Herbs perennial, 1.5–2.5 dm; stems and adaxial leaf surfaces puberulent. **Stems** 1–several. **Leaf blades** narrowly white-margined; basal narrowly oblanceolate, 4–17 × 0.4–2 cm; cauline leaves opposite, distal blades linear-oblanceolate. **Inflorescences** diffuse, 5–30 cm wide, branching near base of main stems. **Flowers:** calyx 7–13 mm; corolla white to cream, usually rather sparsely purple-dotted, 7–10 mm, lobes oblong-obovate, apex abruptly acuminate; androecial corona poorly developed, usually comprising erose scales 0.2–1 mm and few trichomes; style slender, distinct; nectaries and foveae 1 per corolla lobe, foveae opening distal to nectary, into an orbiculate to elliptic-oblong differentiated area on the corolla lobe, rim ± fringed all around.

Flowering summer. Grasslands, open pine-juniper woods, often around sandstone or limestone outcrops; of conservation concern; 1200–1700 m; Colo., Okla.

Frasera coloradensis is endemic to Baca, Bent, Las Animas, and Prowers counties in southeastern Colorado and Cherokee County in adjacent Oklahoma.

To a greater degree than other *Frasera* species, *F. coloradensis* forms relatively large clusters of divergent rather than erect stems. Otherwise, it is somewhat similar to *F. puberulenta*, from which it is separated by about 1500 km. *Frasera coloradensis* has corymboid inflorescences, about as wide as long, with branches or pedicels arising in pairs at each node of the main axis, and many long leaves in the inflorescences; *F. puberulenta* has more elongate inflorescences, usually with branches and/or pedicels arising in whorls from some nodes of the main axis, and few if any large leaves in the inflorescences. In *F. puberulenta* the nectary is more deeply pocketed. *Frasera coloradensis* is also somewhat similar to *F. albomarginata* var. *induta.* It differs in its opposite rather than proximally whorled leaves, its larger leaves within the inflorescence, and the highly dissimilar shape of the differentiated areas on the corolla lobes.

Frasera coloradensis is in the Center for Plant Conservation's National Collection of Endangered Plants.

10. Frasera parryi Torrey in War Department [U.S.], Pacif. Railr. Rep. 4(5): 126. 1857 • Parry's frasera

Swertia parryi (Torrey) Kuntze

Herbs perennial, 5–16 dm, glabrous. **Stems** 1 or 2. **Leaf blades** white-margined; basal strap-shaped to elliptic-oblanceolate, 5–25 × 0.8–4 cm; cauline leaves opposite, blades ovate to lanceolate-oblong. **Inflorescences** diffuse, 5–30 cm wide, primary branches ± elongate and sometimes racemoid. **Flowers:** calyx 8–17 mm; corolla pale green, dark purple-dotted, 9–20 mm, lobes oblong-obovate, apex short-acuminate; androecial corona a low fringe; style slender, distinct; nectaries and foveae 1 per corolla lobe, foveae opening distal to nectary, into a U-shaped differentiated area on the corolla surface, rim ± evenly fringed all around.

Flowering spring–summer. Dry, open woods, chaparral; 100–1900 m; Calif.; Mexico (disjunct in Baja California Sur).

Inclusion of Arizona in the range of *Frasera parryi* appears to have been derived from speculation. The vernacular name Coahuila elkweed, sometimes applied to *F. parryi*, is inappropriate as this species is not known from Coahuila.

11. Frasera neglecta H. M. Hall, Bot. GaZ. 31: 388, plate 10. 1901 • Pine frasera or green-gentian [E]

Swertia neglecta (H. M. Hall) Jepson

Herbs perennial, 2–5.5 dm, glabrous. **Stems** 1–several, with several rosettes. **Leaf blades** narrowly white-margined, basal 2–20 × 0.3–0.9 cm, linear to narrowly oblanceolate; cauline leaves opposite, blade linear to narrowly oblanceolate. **Inflorescences** narrow, 1.5–4 cm wide, dense, interrupted. **Flowers:** calyx 5–8 mm; corolla greenish white, purple-streaked, 7–15 mm, lobes oblong-obovate, apex abruptly acuminate; androecial corona a low fringe; style slender, distinct; nectaries and foveae 1 per corolla lobe, foveae opening distal to nectary, into a round to nearly square differentiated area on the corolla lobe, rim ± evenly fringed all around.

Flowering late spring–summer. Dry, open woods; 1400–3500 m; Calif.

Frasera neglecta is endemic to the San Bernardino, San Emigdio, and San Gabriel mountains of southern California.

12. Frasera gypsicola (Barneby) D. M. Post, Bot. GaZ. 120: 3. 1958 • White River or Sunnyside frasera [C] [E]

Swertia gypsicola Barneby, Leafl. W. Bot. 3: 155. 1942

Herbs perennial, 1–3.5 dm, glabrous. **Stems** 1–few from each division of the caudex, often with several rosettes. **Leaf blades** white-margined, basal 3–8 × 0.1–0.3 cm, narrowly linear; cauline leaves opposite, blades similar to basal. **Inflorescences** narrow but not dense, few-flowered. **Flowers:** calyx 3–4 mm; corolla cream, dark purple-dotted, 5–9 mm, lobes lanceolate, apex acute to short-acuminate; androecial corona scales oblong, ca. 2 mm, margins subentire to ± lacerate; style slender, distinct; nectaries and foveae 1 per corolla lobe, foveae opening ± round, distal to nectaries but without a differentiated area on the corolla surface, rim deeply, evenly fringed all around.

Flowering summer. Valley bottoms, in white-barren soils; of conservation concern; 1500–1700 m; Nev., Utah.

Frasera gypsicola is endemic to two small calcareous mountain areas in northeastern Nye County and adjacent White Pine County, Nevada, and western Millard County, Utah.

Because of its multicipital caudex with the divisions more strongly divergent than in other *Frasera* species except *F. coloradensis*, *F. gypsicola* has a distinctive cespitose habit.

13. Frasera albicaulis Grisebach in W. J. Hooker, Fl. Bor.-Amer. 2: 67, plate 154. 1837 • White-stemmed frasera [E] [F]

Leucocraspedum albicaule (Grisebach) Rydberg; *Swertia albicaulis* (Grisebach) D. Douglas ex Kuntze

Herbs perennial, 1–6.5 dm. **Stems** 1–few, with several rosettes. **Leaf blades** narrowly white-margined; basal oblanceolate, 4–23 × 0.3–1.2(–2) cm, basal leaves generally longer than lowest internode; cauline leaves opposite, proximal blades oblanceolate to oblong, distal linear-oblong. **Inflorescences** narrow, 1.5–4(–5) cm wide, interrupted proximally, ± continuous distally. **Flowers:** calyx 3–7(–12) mm; corolla greenish white to pale or medium blue, usually dark blue- to purple-dotted and/or with a dark blue, purple, or green central stripe (except in var. *idahoensis*), 6–12 mm, lobes oblong-obovate-elliptic, widest near midlength, abruptly tapering to acute to short-acuminate apex; androecial corona scales present, variable among the varieties; style slender, distinct; nectaries and foveae 1 per corolla lobe, foveae distal to nectary, opening into an elliptic-oblong to lance-ovate differentiated area on the corolla surface, rim fringed all around but with fringes shorter distally, or not fringed toward distal end.

Varieties 5 (5 in the flora): w United States.

Frasera albicaulis varies in vegetative and abaxial corolla puberulence and in the size and dissection of the androecial corona. It has sometimes been divided into several species and additional varieties, but, from the material available for study at the time of this writing, it appears appropriate to recognize only five varieties, although *F. ackermaniae* might be considered for inclusion at varietal rank. Each of these varieties has a more or less distinct geographic range, but some intergrade.

1. Stems and abaxial or both leaf surfaces puberulent; rim of differentiated area on corolla surface not fringed at distal end.
   2. Corona scales oblong, margins lacerate or deeply toothed, ultimate divisions threadlike . . . . 13a. *Frasera albicaulis* var. *albicaulis*
   2. Corona scales ovate to obovate or oblanceolate, margins entire or shallowly 2–3-lobed, rarely absent . . . . 13b. *Frasera albicaulis* var. *modocensis*
1. Stems glabrous; leaf blades glabrous or puberulent only near base and abaxially along midvein; rim of differentiated area on corolla surface fringed all around, fringes shorter near distal end.
   3. Corona scales ovate to elliptic, margins entire or shallowly undulate-erose or toothed . . . . 13c. *Frasera albicaulis* var. *cusickii*
   3. Corona scales ovate to elliptic to oblong-lanceolate, margins lacerate or deeply lobed.
      4. Corollas greenish white or pale to medium blue, generally with dark blue to purple spots; corona scales 1–4 mm, oblong-lanceolate . . . . . 13d. *Frasera albicaulis* var. *nitida*
      4. Corollas pale blue, usually without darker spots; corona scales 2–6 mm, widely ovate to elliptic. . . 13e. *Frasera albicaulis* var. *idahoensis*

13a. Frasera albicaulis Grisebach in W. J. Hooker var. **albicaulis** [E] [F]

Frasera pahutensis Reveal

Stems densely puberulent. **Leaf blades** abaxially puberulent. **Flowers:** calyx lobes adaxially puberulent; corolla pale to medium blue or greenish white to cream, generally with dark blue to purple spots; corona scales oblong, deeply 2-lobed, often further cleft into threadlike segments, 1–3(–4) mm, margins lacerate or deeply toothed or lobed; rim of differentiated area on corolla fringed proximally, not fringed toward distal end. **2*n*** = 26 (as *F. pahutensis*).

Flowering late spring–early summer. Dry or moist open sites; 500–2600 m; Idaho, Mont., Nev., Oreg., Wash.

A report of *Frasera albicaulis* from British Columbia is unsubstantiated. California plants that have previously been included in var. *albicaulis* are treated here as var. *modocensis*.

The Nevada plants that were called *Frasera pahutensis* are disjunct by about 1200 km from other populations of *F. albicaulis* var. *albicaulis*. Although they occur nearer the ranges of var. *modocensis* and var. *nitida*, they resemble the more northern var. *albicaulis* in the narrow sense, differing, respectively, from the two varieties in California in their white or scarcely blue-tinged corollas, relatively small, deeply lacerate corona scales, and in their densely puberulent stems and leaves. Examination of numerous specimens by D. M. Post (unpubl.) and in studies for this flora has disclosed no morphological features by which the Nevada plants could be differentiated from var. *albicaulis*, although further study would be appropriate. The Nevada plants that have been called *F. pahutensis* may be diploid whereas some other taxa here included in *F. albicaulis* may be hexaploid, but with the ploidy level of var. *albicaulis* elsewhere being unknown and only one count having been published for any of the other components of *F. albicaulis*, it is not known which ploidy level prevails in which of the varieties. Additional studies of ploidy levels in this complex might indicate that further refinements in classification would be appropriate.

13b. Frasera albicaulis Grisebach in W. J. Hooker var. **modocensis** (H. St. John) N. H. Holmgren in A. Cronquist et al., Intermount. Fl. 4: 22. 1984 • Modoc frasera [E]

Swertia modocensis H. St. John, Amer. Midl. Naturalist 26: 18. 1941

Stems densely puberulent. **Leaf blades** abaxially puberulent. **Flowers:** calyx lobes abaxially puberulent; corolla greenish white or pale to medium blue, generally with dark blue to purple spots; corona scales ovate to obovate or oblanceolate, (0.8–)1.5–2.5(–3.3) mm, margins entire or shallowly 2–3-lobed, rarely absent; rim of differentiated area on corolla surface fringed proximally, not fringed toward distal end. **2*n*** = 52 (as *Swertia albicaulis* var. *albicaulis*).

Flowering summer. Dry, brushy sites; 900–1900 m; Calif., Nev., Oreg.

Variety *modocensis* has sometimes been included in var. *albicaulis*, but var. *albicaulis* in the narrow sense has been characterized by N. H. Holmgren (1984b) as being more robust than var. *modocensis*, as well as differing in the corona-scale character noted in the descriptions here.

13c. Frasera albicaulis Grisebach in W. J. Hooker var. **cusickii** (A. Gray) C. L. Hitchcock in C. L. Hitchcock et al., Vasc. Pl. Pacif. N.W. 4: 60. 1959 • Cusick's swertia [E]

Frasera cusickii A. Gray, Proc. Amer. Acad. Arts 22: 310. 1887; *F. caerulea* Mulford; *Leucocraspedum coeruleum* (Mulford) Rydberg; *Swertia albicaulis* (Grisebach) D. Douglas ex Kuntze var. *cusickii* (A. Gray) J. S. Pringle; *S. nitida* (Bentham) Jepson subsp. *cusickii* (A. Gray) Abrams

Stems glabrous. **Leaf blades** glabrous or puberulent only near base and abaxially along midvein. **Flowers:** calyx glabrous; corolla greenish white or pale to medium blue, generally with dark blue to purple spots; corona scales ovate to elliptic, 2.5–3(–4.5) mm, margins entire or shallowly undulate-erose or toothed; rim of differentiated area on corolla surface fringed all around, fringes shorter near distal end.

Flowering early summer. Dry, rocky slopes, open pine woods; 800–2200 m; Idaho, Oreg.

Variety *cusickii* is endemic to the mountains of western Idaho and the Blue Mountains of eastern Oregon.

13d. Frasera albicaulis Grisebach in W. J. Hooker var. **nitida** (Bentham) C. L. Hitchcock in C. L. Hitchcock et al., Vasc. Pl. Pacif. N.W. 4: 60. 1959 • Shining frasera [E]

Frasera nitida Bentham, Pl. Hartw., 322. 1849; *F. albicaulis* var. *columbiana* (H. St. John) C. L. Hitchcock; *F. nitida* var. *albida* Suksdorf; *Swertia albicaulis* (Grisebach) D. Douglas ex Kuntze var. *nitida* (Bentham) Jepson; *S. nitida* (Bentham) Jepson

Stems glabrous. **Leaf blades** generally glabrous, except sometimes for puberulent sheathing bases of proximal leaves in Oregon and Washington plants. **Flowers:** calyx glabrous; corolla greenish white or pale to medium blue, generally with dark blue to purple spots; corona scales oblong-lanceolate, 1–4 mm, margins deeply 2–several-lobed, rim of differentiated area on corolla surface fringed all around, fringes shorter near distal end.

Flowering late spring–early summer. Dry, open woods, rocky slopes, chaparral, prairies; 50–1900 m; Calif., Oreg., Wash.

Variety *modocensis* and var. *nitida* intergrade in northern California.

Plants from southeastern Washington and northeastern Oregon, treated as *Frasera albicaulis* var. *columbiana* by C. L. Hitchcock (1959), were said to be well isolated from var. *nitida* but scarcely separable. As variations overlap both in leaf-sheath puberulence and in the length of the corona scales, the characters by which these varieties were distinguished, these taxa are treated here as a single variety. If the plants in northeastern Oregon and southeastern Washington are distinguished taxonomically from var. *nitida*, the correct varietal epithet would be *albida*, based on its use in the combination *F. nitida* var. *albida*, which has priority at that rank over *columbiana*. As of this writing, the epithet *albida* has not been published at varietal rank under either *Frasera* or *Swertia albicaulis*.

13e. Frasera albicaulis Grisebach in W. J. Hooker var. **idahoensis** (H. St. John) C. L. Hitchcock in C. L. Hitchcock et al., Vasc. Pl. Pacif. N.W. 4: 60. 1959 • Idaho frasera [C] [E]

Swertia idahoensis H. St. John, Amer. Midl. Naturalist 26: 24. 1941; *S. albicaulis* (Grisebach) D. Douglas ex Kuntze var. *idahoensis* (H. St. John) J. S. Pringle

Stems glabrous. **Leaf blades** glabrous. **Flowers:** calyx glabrous; corolla pale blue, usually without darker spots; corona scales widely ovate to elliptic, 2–6 mm, margins deeply lacerate; rim of differentiated area on corolla surface fringed all around, fringes shorter near distal end.

Flowering early summer. Dry to moist, rocky slopes, open pine woods; of conservation concern; 900–2100 m; Idaho, Oreg.

Variety *idahoensis* is endemic to the Seven Devils Mountains in Adams County, Idaho, and the Wallowa Mountains in northeastern Oregon.

14. Frasera ackermaniae C. C. Newberry & Goodrich, W. N. Amer. Naturalist 70: 415, figs. 1, 2. 2010 (as ackermanae) • Ackerman's frasera [C] [E]

Herbs perennial, 0.5–2.5 dm. **Stems** 1–several, with several rosettes, puberulent. **Leaf blades** narrowly white-margined, glabrous or abaxial surfaces puberulent proximally; basal narrowly oblanceolate, 1.4–14 × 0.2–0.7 cm, basal leaves usually shorter than lowest internode; cauline leaves opposite, distal blades nearly linear. **Inflorescences** narrow, 1.5–4 cm wide, ± continuous throughout or interrupted proximally. **Flowers:** calyx 4–8 mm; corolla white or slightly suffused with blue, dark blue-dotted, abaxially with a green central stripe, 6–9 mm, lobes proximally oblong, distally triangular, apex acuminate; corona scales deeply cleft into distally threadlike segments, 2 mm; style slender, distinct; nectaries and foveae 1 per lobe, fovea narrowly oblong, opening into an oblong differentiated area on the corolla surface, rim fringed all around but fringes shorter distally.

Flowering summer. Semibarren clay hillsides and wash bottoms; of conservation concern; 1700–1800 m; Utah.

Frasera ackermaniae is known only from Uintah County.

Frasera ackermaniae is similar to *F. albicaulis*, especially var. *albicaulis*, and its treatment as another variety of *F. albicaulis* might be considered appropriate. As noted in the original description, in *F. ackermaniae* the lowest branching of the inflorescence is near the base of the plant, with the lowest internode usually exceeded by the basal leaves, whereas in *F. albicaulis* var. *albicaulis* the lowest branching is higher, with the lowest internode generally longer than the basal leaves. *Frasera ackermaniae* also differs from *F. albicaulis* in its narrower leaves and in its ovate corolla lobes, which narrow gradually from midlength toward the acuminate apex, rather than more abruptly as in *F. albicaulis*.

15. **Frasera montana** Mulford, Bot. GaZ. 19: 119. 1894 • White frasera [E]

Swertia montana (Mulford) H. St. John, Amer. Midl. Naturalist 26: 16. 1941; *Leucocraspedum montanum* (Mulford) Rydberg

Herbs perennial, 2.5–8 dm, glabrous or stems and leaf bases minutely puberulent. **Stems** 1–several, with several rosettes. **Leaf blades** white-margined; basal narrowly spatulate-oblanceolate to lanceolate, 7–30 × 0.5–1.5 cm; cauline leaves opposite, blades narrowly lanceolate to linear. **Inflorescences** narrow, 1.5–4 cm wide, dense. **Flowers:** calyx 3–6(–8) mm; corolla white to cream, unmarked, 5–9 mm, lobes elliptic-ovate, apex rounded, apiculate; androecial corona scales obovate-oblong, 1–2 mm, margins nearly entire to deeply lacerate or fringelike; style slender, distinct; nectaries and foveae 1 per lobe, foveae opening into an elliptic-obovate to suborbiculate differentiated area on the corolla surface, rim ± evenly fringed all around.

Flowering late spring–summer. Dry mountain meadows, sagebrush slopes, open pine woods; 1200–2000 m; Idaho.

Frasera montana is endemic to the mountains of western Idaho.

Frasera montana appears to be closely related to *F. albicaulis* and might be treated as another variety of that species, but its proportionately wide corolla lobes with rounded rather than acute to acuminate apices, reminiscent of the petals of apple blossoms, give the flowers of this attractive species a distinctive appearance. The differentiated areas on the corolla surface into which the foveae open are elliptic to nearly round and are proportionately shorter than those of any variety of *F. albicaulis*. *Frasera montana* might most readily be confused with *F. albicaulis* var. *idahoensis*, which likewise usually has unspotted corollas, but it can be distinguished not only by the shape of its corolla lobes but also by the differentiated areas on its corolla lobes with rims that are more or less evenly long-fringed all around; those of all varieties of *F. albicaulis* are distally more shallowly or not fringed. The androecial corona scales of both species are variable, but those of *F. montana* are generally cleft more or less longitudinally, if at all, whereas those of *F. albicaulis* var. *idahoensis* (but not all varieties of *F. albicaulis*) usually bear lateral as well as terminal lobes or fringes.

12. BARTONIA Muhlenberg ex Willdenow, Ges. Naturf. Freunde Berlin Neue Schriften 3: 444. 1801, name conserved • [For Benjamin Smith Barton, 1766–1815, American botanist] [E]

James S. Pringle

Herbs annual or perennial, weakly chlorophyllous, with stems and leaves yellowish green or purplish, glabrous. **Leaves** cauline, alternate, subopposite, or opposite, scalelike. **Inflorescences** dichasial or racemoid cymes, reduced thyrses, or solitary flowers. **Flowers** 4-merous; calyx lobed nearly to base, or some or all lobes proximally connate; corolla white or greenish white to pale yellow, yellowish green, to green, sometimes purple-tinged, narrowly campanulate, glabrous, lobes longer than tube, margins entire or erose, plicae between lobes absent, spurs absent; stamens inserted in corolla sinuses; anthers distinct; ovary sessile or subsessile; style short and stout or absent; stigma 2-lobed or decurrent along distal portion of sutures; nectaries absent. **Capsules** compressed-cylindric. x = 11, 13.

Species 3 (3 in the flora): c, e North America.

Plants cultivated as *Bartonia* are species of *Mentzelia* (Loasaceae), to which the homonym *Bartonia* Sims of 1804 was formerly applied.

Bartonia has a reduced root system, minute leaves, and stems that are sometimes weakly chlorophyllous, and is partially mycotrophic (D. D. Cameron and J. F. Bolin 2010). *Bartonia* species have been assumed to be annuals, from the appearance of the basal and underground parts, but this has been questioned.

SELECTED REFERENCES Gillett, J. M. 1959. A revision of *Bartonia* and *Obolaria* (Gentianaceae). Rhodora 61: 43–62. Mathews, K. G. et al. 2009. A phylogenetic analysis and taxonomic revision of *Bartonia* (Gentianaceae: Gentianeae), based on molecular and morphological evidence. Syst. Bot. 34: 162–172.

1. Leaves ± evenly spaced; corollas 4.8–11 mm, lobed nearly to base, lobes narrowly spatulate-obovate to elliptic; flowering winter–spring . . . . 3. *Bartonia verna*
1. Leaves crowded near base of stem, distally widely separated; corollas 1.9–6.2 mm, lobed 0.5–0.8 times their length, lobes oblong or oblong-lanceolate; flowering summer–fall.
    2. Corolla lobe margins distally erose-serrate, apices rounded to abruptly acute, mucronate . . . . 1. *Bartonia virginica*
    2. Corolla lobe margins entire, apices acute to acuminate, not mucronate . . . . 2. *Bartonia paniculata*

1. Bartonia virginica (Linnaeus) Britton, Sterns & Poggenburg, Prelim. Cat., 36. 1888 • Yellow or Virginia or late-flowering bartonia, yellow screwstem, bartonie de Virginie [E]

Sagina virginica Linnaeus, Sp. Pl. 1: 128. 1753; *Bartonia tenella* Willdenow

Herbs ± erect, yellowish green, often purplish proximally or occasionally more extensively, 3–45 cm. **Leaves** all opposite, proximal alternate and distal opposite, or rarely all alternate, closely spaced near base of stem, gradually more widely spaced distally; blade 0.9–4.7 mm. **Inflorescences** racemoid cymes or small thyrses with strongly ascending branches. **Flowers:** calyx lobed nearly to base, lobes lanceolate, 2–4.5 × 0.4–1.1(–1.4) mm, apex acuminate; corolla white to greenish white or yellowish green, distally or occasionally more extensively often purple-tinged especially in age, 2.3–4.4 mm, lobes oblong, 1.6–3.2 × 0.7–1.4 mm, that is, 0.6–0.8 times the length of the corolla, margins usually erose-serrate distally, apices rounded to abruptly acute, mucronate; anthers often recurving in age but not coiling, yellow or purple, 0.5–1.2 mm, apex mucronate to short-acuminate; style absent; stigmas decurrent along sutures of ovary. **Capsules** dehiscent medially. 2n = 52.

Flowering late summer–fall. Bogs, shores, mesic to wet, open woods, usually peaty soils; 0–500 m; St. Pierre and Miquelon; N.B., N.S., Ont., Que.; Ala., Conn., Del., D.C., Fla., Ga., Ill., Ind., Ky., La., Maine, Md., Mass., Mich., Minn., Miss., N.H., N.J., N.Y., N.C., Ohio, Pa., R.I., S.C., Tenn., Vt., Va., W.Va., Wis.

Inrolling of margins during drying may cause the corolla lobes of herbarium specimens of *Bartonia virginica* to appear to taper more gradually than they actually do, leading to misidentification as *B. paniculata*.

2. Bartonia paniculata (Michaux) Muhlenberg, Cat. Pl. Amer. Sept., 16. 1813 • Branched or twining bartonia, panicled screwstem, bartonie paniculée [E] [F]

Centaurella paniculata Michaux, Fl. Bor.-Amer. 1: 98, plate 12, fig. 1. 1803

Herbs decumbent to erect or ± twining, yellowish green to purplish, 3–52 cm. **Leaves** all alternate or distal or occasionally most leaves opposite or subopposite, closely spaced near base of stem, gradually more widely spaced distally; blade 0.5–3 mm. **Inflorescences** racemoid cymes, or thyrses with branching variable, often arcuate-ascending. **Flowers:** calyx lobed nearly to base or some or all lobes proximally connate, lobes lanceolate to ovate, 1–3.2 × 0.3–1.1 mm, that is, 0.5–0.8 times the length of corolla, apex acute to acuminate; corolla white or occasionally pale yellow to green, distally often purple-tinged, 2–6.2 mm, lobes oblong-lanceolate, 1.5–4 × 0.7–2 mm, margins entire, apices acute to acuminate, not mucronate; anthers not recurving or coiling, yellow or purple, 0.3–0.9 mm, apex rounded to obtuse, sometimes mucronate; style absent; stigmas decurrent along sutures of ovary. **Capsules** dehiscent from apex.

Subspecies 3 (3 in the flora): c, e North America.

Bartonia paniculata is less well differentiated morphologically from *B. virginica* than some references indicate. The stems of some plants of subsp. *paniculata* are sinuous or occasionally twine around grass culms, but many are as straight and erect as those of *B. virginica*. Conversely, the stems of *B. virginica* are often decumbent and occasionally somewhat sinuous. Both species vary in the extent of purple suffusion in the stems. The leaves of some plants of *B. paniculata* are all widely spaced and distinctly alternate, but other plants have mostly opposite or subopposite leaves and/or closely spaced proximal leaves, as *B. virginica* has been characterized. The inflorescence branches of *B. paniculata* are often relatively long and divergent or distally arcuate-ascending but are sometimes short and/or strongly ascending their whole length as in *B. virginica*. Most such specimens represent variability in vegetative morphology within the respective species rather than true intergradation or hybridization. Corolla, anther, and stylar characters reliably distinguish these two species and indicate that they generally retain their distinctness where sympatric. Occasional plants have been found, mostly on the Atlantic plain, that are intermediate in corolla and vegetative morphology and at least in some cases are sterile. K. G. Mathews et al. (2009) found a few plants combining traits of both *B. paniculata* and *B. virginica* but no plants exhibiting intermediate character states.

Bartonia paniculata subsp. *paniculata* and subsp. *iodandra* intergrade. Plants combining deeply four-parted, narrowly lobed calyces with various degrees of purple suffusion in the anthers are relatively frequent along the Atlantic coast from New Jersey to Nova Scotia and occasionally occur south to Maryland. Plants variously combining traits attributed to subsp. *paniculata* and subsp. *iodandra* have been called var. *intermedia*. Most such plants are included in subsp. *iodandra* here, following J. M. Gillett (1959). Plants intermediate between subsp. *paniculata* and subsp. *texana* have been reported from Caddo Parish, Louisiana (K. G. Mathews et al. 2009).

1. Stems purplish throughout, erect; some or all calyx lobes proximally connate, forming sheath or tube to 3 mm; anthers purple or occasionally yellow, 0.4–0.9 mm. . . . . .2b. *Bartonia paniculata* subsp. *iodandra*
1. Stems yellowish green, proximally sometimes purplish, erect, decumbent, or ± twining; all calyx lobes distinct nearly to base; anthers yellow, 0.3–0.5 mm.
    2. Stems decumbent to erect or ± twining; calyx lobes 1.5–2.9 mm; corollas 2.9–6.2 mm . . . . .2a. *Bartonia paniculata* subsp. *paniculata*
    2. Stems erect; calyx lobes 1–1.5 mm; corollas 2–3.1 mm . . . .2c. *Bartonia paniculata* subsp. *texana*

2a. Bartonia paniculata (Michaux) Muhlenberg subsp. **paniculata** E F

Bartonia lanceolata Small

Stems decumbent to erect or ± twining, yellowish green, proximally sometimes purplish, 10–52 cm. **Flowers:** calyx lobes distinct nearly to base, 1.5–2.9 × 0.5–1 mm; corolla 2.9–6.2 mm, lobed 0.6–0.8 times its length; anthers yellow, 0.3–0.5 mm, apex rounded. $2n = 52$.

Flowering summer–fall. Bogs, fens, white-cedar swamps, wet, open woods, peaty lake margins; 0–500 m; Ont.; Ala., Ark., Conn., Del., D.C., Fla., Ga., Ill., Ind., Ky., La., Maine, Md., Mass., Mich., Miss., Mo., N.H., N.J., N.Y., N.C., Okla., Pa., R.I., S.C., Tenn., Tex., Va., Wis.

A specimen from Newfoundland identified as this subspecies was reidentified as subsp. *iodandra* in studies for this flora.

Molecular studies (C. Ciotir et al. 2013) of plants identified morphologically as subsp. *paniculata* disclose greater genetic divergence than expected between plants from the Atlantic coastal plain in New Jersey and the disjunct populations in the Great Lakes region, the latter clustering more closely with subsp. *texana*. If populations in the interior were distinguished taxonomically, the epithet *paniculata*, whether at specific or infraspecific rank, would remain for the plants of the Atlantic coastal plain.

2b. Bartonia paniculata (Michaux) Muhlenberg subsp. **iodandra** (B. L. Robinson) J. M. Gillett, Rhodora 61: 54. 1959 E

Bartonia iodandra B. L. Robinson, Bot. GaZ. 26: 47. 1898; *B. paniculata* var. *intermedia* Fernald; *B. paniculata* var. *iodandra* (B. L. Robinson) M. L. Fernald; *B. paniculata* var. *sabulonensis* (Fernald) Fernald

Stems erect, purplish, 3–25 cm. **Flowers:** calyx lobing variable, some or all lobes usually distinctly connate proximally, forming sheath or tube, 1.5–3.2 × 0.5–1.1 mm, sheath or tube 0.5–3 mm; corolla 3–6.2 mm, lobed 0.5–0.7 times its length; anthers purple or occasionally yellow, 0.4–0.9 mm, apex rounded or mucronate.

Flowering summer–fall. Bogs, peaty lake margins; 0–400 m; St. Pierre and Miquelon; N.B., Nfld. and Labr. (Nfld.), N.S.; Conn., Maine, Mass., R.I., restricted to coastal sites southward.

In some respects, subsp. *iodandra* is intermediate between subsp. *paniculata* and *Bartonia virginica*, and for that reason it was treated by A. Haines (2011) as a species derived from past hybridization of those taxa. The range of subsp. *iodandra* extends well beyond that of either subsp. *paniculata* or *B. virginica*, completely displacing subsp. *paniculata* in the northern part of the range of the species as circumscribed here. It appears that although some plants that have been called subsp. *iodandra* may be derived from hybridization, most plants of subsp. *iodandra* are sufficiently similar to subsp. *paniculata* and sufficiently distinct from *B. virginica* to justify their retention in *B. paniculata*. Morphologically, the more abruptly tapering corolla lobes of subsp. *iodandra* might be assumed to be attributable to such introgression, but the more or less spathaceous calyces and the greater extent of purple pigment in its vegetative parts presumably would not.

2c. Bartonia paniculata (Michaux) Muhlenberg subsp. **texana** (Correll) K. G. Mathews, Dunne, E. York & Struwe, Syst. Bot. 34: 168. 2009 [C] [E]

Bartonia texana Correll, Wrightia 3: 191, fig. 61(6–10). 1966

Stems erect, yellowish green, 15–23 cm. **Flowers:** calyx lobed nearly to base, lobes all distinct nearly to base, 1–1.5 × 0.3–1 mm; corolla 2–3.4 mm, lobed 0.6–0.8 times its length; anthers yellow, 0.4–0.5 mm, apex obtuse or obscurely mucronate.

Flowering fall. Around seeps on wooded slopes, stream banks, often in clumps of moss; of conservation concern; 20–100 m; La., Tex.

Subspecies *texana* is endemic to eastern Texas and northern Louisiana, occurring mostly in the Piney Woods region.

3. Bartonia verna (Michaux) Rafinesque ex Barton, Fl. Virgin. 51. 1812 • White or spring bartonia [E]

Centaurella verna Michaux, Fl. Bor.-Amer. 1: 98, plate 12, fig. 2. 1803

Herbs erect, purplish or rarely yellowish, 2–23 cm. **Leaves** alternate or distal leaves subopposite, ± evenly spaced; blade 0.6–3.5 mm. **Inflorescences** ± open, racemoid cymes or solitary flowers. **Flowers:** calyx lobed nearly to base, lobes lanceolate, 0.9–2.8 × 0.5–1.5 mm, apex obtuse to subacute; corolla white, 4.8–11 × 1.5–4 mm, lobed nearly to base, lobes narrowly spatulate-obovate to elliptic, margins usually undulate-erose, apices rounded to subacute, not mucronate; anthers sometimes coiling, yellow, 0.5–1.1 mm, apex rounded; style short and stout; stigmas decurrent along its length. **Capsules** dehiscent medially. $2n = 44$.

Flowering winter (southernmost part of range)–spring (northward). Bogs, shores, moist savannas and meadows; 0–50 m; Ala., Fla., Ga., La., Miss., N.C., S.C., Tex.

13. OBOLARIA Linnaeus, Sp. Pl. 2: 632. 1753; Gen. Pl. ed. 5, 280. 1754 (as Obularia) • Pennywort [Greek *obolos*, coin, and *aria*, possession, alluding to leaf shape] [E]

James S. Pringle

Herbs perennial, chlorophyllous, glabrous. **Leaves** cauline, opposite. **Inflorescences:** flowers terminal and axillary, solitary or in cymules of 3. **Flowers** 4-merous, subtended by 2 separate, leaflike bracts; calyx absent; corolla white to pale violet, narrowly campanulate, glabrous, lobes imbricate in bud, ascending, slightly longer than tube, entire or erose, with 2 minute scales per petal on proximal part of tube, plicae between lobes absent, spurs absent; stamens inserted in corolla sinuses; anthers distinct, remaining straight; ovary sessile; style persistent, erect, short; stigmas 2, remaining straight; nectaries in ring at base of ovary. **Capsules** compressed-ovoid, rupturing irregularly. $x = 28$.

Species 1: c, e United States; temperate areas.

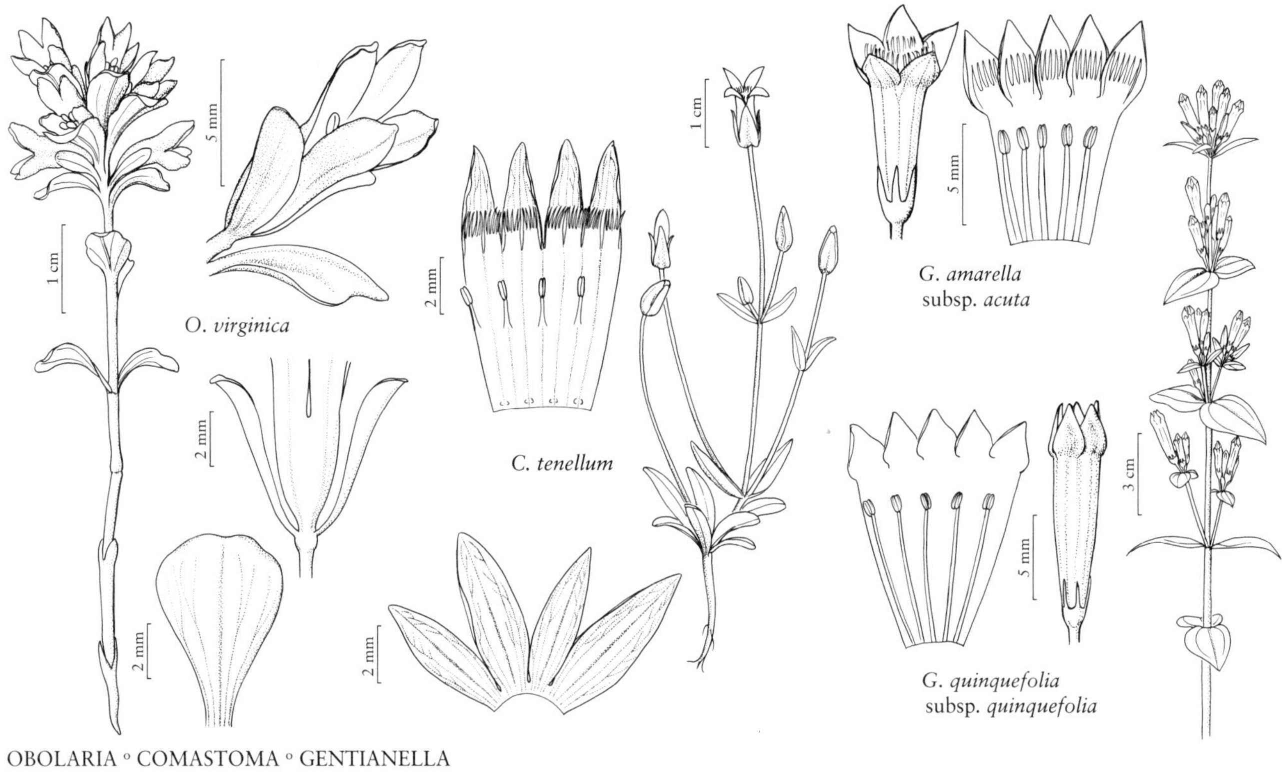

OBOLARIA ◦ COMASTOMA ◦ GENTIANELLA

Obolaria has a reduced root system and is probably strongly mycotrophic, although its leaves are green. The paired bracts subtending each flower have often been interpreted as two separate sepals, following T. Holm (1897). As noted by A. Gray (1848b) and as illustrated by Holm, these bracts are sometimes parallel with the carpels with their midveins aligned with the midveins of the carpels, and sometimes at right angles to the carpels, with their midveins aligned with the sutures of the ovary. The case presented by Gray for interpreting these structures as bracts is more convincing.

SELECTED REFERENCES Gillett, J. M. 1959. A revision of *Bartonia* and *Obolaria* (Gentianaceae). Rhodora 61: 43–62. Gray, A. 1848b. *Obolaria virginica*, Linn. Mem. Amer. Acad. Arts, ser. 2, 3: 21–31. Holm, T. 1897. *Obolaria virginica* L.: A morphological and anatomical study. Ann. Bot. (Oxford) 11: 369–383.

1. Obolaria virginica Linnaeus, Sp. Pl. 2: 632. 1753 [E] [F]

Stems 1–3, simple or few-branched, 4–17(–25) cm. **Leaf blades** in inflorescence fan-shaped to spatulate-obovate or orbiculate, 4–16 × 3–11 mm; all or most leaves below inflorescence minute, scalelike. **Flowers:** corolla 6–15 mm, lobes obovate-oblong, apex acuminate. **2*n*** = 56.

Flowering late winter (southward)–spring. Moist deciduous woods, soils rich in leaf mold; 0–400 m; Ala., Ark., Del., D.C., Fla., Ga., Ill., Ind., Ky., La., Md., Miss., Mo., N.J., N.C., Ohio, Pa., S.C., Tenn., Tex., Va., W.Va.

14. COMASTOMA (Wettstein) Toyokuni, Bot. Mag. (Tokyo) 74: 198. 1961 • [Greek *kome*, hair of head, and *stoma*, mouth or gap, alluding to fimbriate scales at corolla throat]

James S. Pringle

Gentiana Linnaeus sect. *Comastoma* Wettstein, Oesterr. Bot. Z. 46: 174. 1896

Herbs annual [biennial or perennial], chlorophyllous, glabrous. **Leaves** cauline, opposite, sometimes also basal. **Inflorescences:** flowers solitary at end of primary stem and branches (if any) [in small, sometimes racemelike cymes]. **Flowers** 4(–5)-merous; calyx deeply lobed, lobes often unequal; corolla violet-blue or greenish white [pale yellow], tubular-salverform [narrowly campanulate], adaxially with 2 fringed scales [1 continuous scale] near base of each lobe, lobes shorter than tube, margins entire or nearly so, plicae between lobes absent, spurs absent; stamens inserted near middle of corolla tube; anthers distinct, remaining straight; ovary sessile; styles persistent, erect, short or indistinct; stigmas 2, remaining straight; nectaries on corolla tube near base, 2 per petal. **Capsules** compressed-cylindric. $x = 5$, [6, 8, 9].

Species ca. 25 (1 in the flora): North America, Eurasia; temperate to arctic and alpine areas.

Although *Comastoma* has much in common morphologically with *Gentianella*, molecular phylogenetic studies indicate that the origin of *Comastoma* was separate from that of *Gentianella* and place it closer to *Lomatogonium* (Yuan Y. M. and P. Küpfer 1995; K. B. von Hagen and J. W. Kadereit 2002). Palynological and embryological studies also indicate a separate origin (S. Nilsson 1967; Liu J. Q. and Ho T. N. 1996).

Comastoma tenellum differs distinctly from *Gentianella* species in the flora by its pedicels, which are generally longer than the subtending internodes.

1. **Comastoma tenellum** (Rottbøll) Toyokuni, Bot. Mag. (Tokyo) 74: 198. 1961 • Samiland or Danes' or slender or one-flowered gentian, spæd ensian, gentiane délicate [F]

Gentiana tenella Rottbøll, Skr. Kiøbenhavnske Selsk. Laerd. Elsk. 10: 436, plate 2, fig. 6. 1770; *G. tenella* var. *monantha* (A. Nelson) J. Rousseau & Raymond; *Gentianella tenella* (Rottbøll) Börner; *G. tenella* subsp. *pribilofii* J. M. Gillett; *Lomatogonium tenellum* (Rottbøll) Á. Löve & D. Löve

Herbs 1–15(–26) cm. **Stems** decumbent to erect, solitary or clustered, simple or branched from base or throughout. **Leaves:** basal blades elliptic-oblong to spatulate, 3–20 × 1–5 mm; cauline blades elliptic to ovate or lanceolate, 4–9 × 1–3 mm. **Inflorescences:** solitary flowers at ends of main stem and branches (if any); peduncles 2–10 cm. **Flowers:** calyx 4–11 mm, base minutely saccate, 2 outer lobes ovate-triangular to lanceolate, inner lobes lanceolate, shorter; corolla pale violet-blue to white or greenish white, 6–17 mm, lobes spreading, ovate-oblong, 2–4.5 mm, apex obtuse, with 2 short scales deeply fringed. $2n = 10$, also reported from Eurasia.

Flowering summer–early fall. Sea and lakeshores, tundra, dry to wet, rocky montane to alpine meadows; 0–3900 m; Greenland; N.W.T., Nunavut, Ont., Que., Yukon; Alaska, Ariz., Calif., Colo., Idaho, Mont., Nev., N.Mex., Oreg., Utah, Wash., Wyo., restricted to high elevations southward; Eurasia.

A report of *Comastoma tenellum* from British Columbia was based on a collection reidentified as *Gentianella amarella* subsp. *acuta* in studies for this flora. Other reports from that province remain unsubstantiated.

Plants from the Pribilof Islands in the Bering Strait were distinguished as *Gentianella tenella* subsp. *pribilofii* on the basis of more branching, shorter internodes, and wider outer calyx lobes rounded to truncate at the base rather than cuneate or more narrowly rounded. However, in studies for this flora, similar specimens have been seen from interior Alaska, the Yukon Territory, Banks Island (Northwest Territories), and elsewhere. Some plants from the Rocky Mountains have more slender calyces than those of plants from the arctic coasts, with the sepals less strongly unequal, but many specimens from interior localities show no such tendencies or are intermediate in morphology. Other alleged differences between interior and coastal North American populations, or between North American and Eurasian populations, have not been substantiated in this study.

15. GENTIANELLA Moench, Methodus, 482. 1794, name conserved • Gentian

[Genus name *Gentiana* and Latin *-ella*, diminutive, alluding to resemblance]

James S. Pringle

Aloitis Rafinesque; *Amarella* Gilibert, name rejected; *Arctogentia* Á. Löve

Herbs annual or biennial [perennial], chlorophyllous, glabrous. **Leaves** cauline, opposite [whorled], sometimes also basal. **Inflorescences** cymes or solitary flowers; pedicels mostly shorter than surrounding internodes. **Flowers** 4- or 5-merous; calyx with tube cylindric to narrowly campanulate or sometimes very short, sometimes with 1 or 2 sepals distinct to base or nearly so, or rarely cleft on 1 side and spathiform (*G. wislizeni*); corolla blue, blue-violet, rose-violet, pink, pale yellow, or white, [red, bright yellow, green, bicolored], tubular, funnelform, campanulate, or nearly salverform [nearly rotate], adaxially with or without a fringe of separate trichomes or a deeply fringed scale near the base of each lobe, lobes shorter than tube or ± as long (*G. tortuosa*) [longer than tube], margins entire or nearly so, without plicae between lobes, spurs absent; stamens inserted near or below [above] middle of corolla tube; anthers distinct; ovary sessile, subsessile, or short-stipitate; style persistent, erect, short or indistinct; stigmas 2; nectaries 1 [rarely 2] per petal, on corolla tube near base. **Capsules** compressed-cylindric to compressed-ovoid (*G. aurea* and *G. tortuosa*). $x = 9, 12$.

Species ca. 250 (10 in the flora): nearly worldwide except in Africa; temperate to arctic and alpine regions.

Gentianella in the flora area includes some species having vascularized fringes of trichomes or deeply fringed scales near the throat of the corolla and other species lacking any such fringes, vascularized or not. Molecular studies indicate that *Gentianella* circumscribed to include both of these groups (but excluding Asiatic species with paired nectaries) is monophyletic and that each of the two groups as traditionally defined is very largely monophyletic. *Gentianella microcalyx* is an exception, evidently being more closely related to the species with fringed corollas although it lacks fringes itself (K. B. von Hagen and J. W. Kadereit 2001). These two groups have occasionally been distinguished at generic rank, species 1 through 7 being retained in *Gentianella* and species 8 through 10 being placed in *Aloitis*. When both groups are retained in *Gentianella,* they have sometimes been designated sect. *Gentianella* and sect. *Arctophila* (Grisebach) Holub, respectively.

The corolla fringes normally present in species 1 through 6 may be reduced or absent on small flowers on the proximal branches or on flowers produced late in the season, or on all flowers of the smallest plants.

SELECTED REFERENCES Gillett, J. M. 1957. A revision of the North American species of *Gentianella* Moench. Ann. Missouri Bot. Gard. 44: 195–269. Hagen, K. B. von and J. W. Kadereit. 2001. The phylogeny of *Gentianella* (Gentianaceae) and its colonization of the Southern Hemisphere as revealed by nuclear and chloroplast DNA sequence variation. Organisms Diversity Evol. 1: 61–79.

1. Corollas without adaxial fringes of trichomes or scales.
   2. Calyx tubes deeply cleft, spathaceous; lobes minute, linear, 0.5–1.5 mm. . .6. *Gentianella wislizeni* (in part)
   2. All sepals united proximally to ± the same level, tube not deeply cleft or spathaceous; lobes usually 2–20 mm (occasionally some smaller only in *G. microcalyx* and *G. quinquefolia* subsp. *quinquefolia*).
      3. Stems usually branched only above base if at all; calyx lobes subequal or ± irregular but not distinctly dimorphic.
         4. Corolla lobe apices obtuse to acute but not attenuate; calyces 2–3(–5) mm; corollas 6–13 mm; Arizona . . . . .7. *Gentianella microcalyx*
         4. Corolla lobe apices short-attenuate; calyces 2–15 mm; corollas 10–25 mm; c, e North America. . . . . 8. *Gentianella quinquefolia*

3. Stems (except of smallest plants) branched from near base; 2 outer calyx lobes distinctly larger than inner.
   5. Flowers (4- or)5-merous; outer calyx lobes oblong to narrowly obovate, larger than inner but not foliaceous; inflorescences proximally ± enveloped by subtending leaves; Greenland. . . . . 9. *Gentianella aurea*
   5. Flowers 4-merous; outer calyx lobes elliptic to ovate-oblong, often ± foliaceous; inflorescences not proximally enveloped by subtending leaves; widely distributed but not in Greenland . . . . 10. *Gentianella propinqua*

1. Corollas of all or all but smallest flowers with fringes of trichomes or deeply fringed scales adaxially near base of lobes.
   6. Calyx tubes deeply cleft, spathaceous; all calyx lobes becoming distinct at the summit of the tube, minute, linear, 0.5–1.5 mm. . . . . .6. *Gentianella wislizeni* (in part)
   6. Calyx tubes not spathaceous, composed of either all sepals united proximally and all lobes becoming distinct from each other at ± the same level, or with 1 or 2 lobes distinct nearly to the base but with the other sepals united to a higher level and becoming distinct at the summit of the tube; lobes diverse in shape, 2–20 mm.
      7. Calyces with 1 or 2 lobes distinct nearly to base of calyx, these ovate, foliaceous, proximally enveloping the other lobes; fringes on adaxial corolla surface basally connate, forming deeply cleft scales . . . . 4. *Gentianella heterosepala*
      7. Calyces with all lobes proximally connate, diverging distinctly above base of calyx, ± equal or unequal but none foliaceous; fringes on adaxial corolla surface consisting either of trichomes separate to base or of fringed scales.
         8. Plants 2–10(–16) cm; corolla lobes ± as long as tube . . . . .2. *Gentianella tortuosa*
         8. Plants 2–50(–80) cm; corolla lobes distinctly shorter than tube.
            9. Plants 2–25(–30) cm; calyx lobes, ovate-triangular to suborbiculate, shorter than tube; fringes on adaxial corolla surface basally connate, forming a deeply cleft scale at the base of each lobe; Alaskan islands . . . 1. *Gentianella auriculata*
            9. Plants (2–)4–50(–80) cm; calyx lobes linear or linear-oblong to lanceolate or narrowly elliptic, ± as long as or longer than tube; fringes on adaxial corolla surface consisting of trichomes separate to base; widely distributed in the flora area, including (*G. amarella*) Alaskan islands.
               10. Corolla lobes 3–5.5 mm; corollas blue to purplish pink, pale yellow, or white, 7–21 mm . . . . 3. *Gentianella amarella*
               10. Corolla lobes (5–)7–10 mm; corollas white or pale yellow, (15–)20–30 mm . . . . .5. *Gentianella wrightii*

1. Gentianella auriculata (Pallas) J. M. Gillett, Ann. Missouri Bot. Gard. 44: 261. 1957 • Eared or auricled gentian

Gentiana auriculata Pallas, Fl. Ross. 1(2): 102, plate 92, fig. 1. 1789

Herbs annual or winter annual, 2–25(–30) cm. **Stems** erect, usually simple or branched distally only, occasionally at or near base. **Leaves:** basal sometimes withered by flowering, blades spatulate-obovate to oblanceolate or elliptic, 4–30 × 1–13 mm; cauline blades elliptic to ovate-lanceolate, (5–)14–40 × 3–10(–15) mm. **Inflorescences** terminal and axillary, few-flowered, dichasial cymes or solitary flowers; pedicels 0–35 mm. **Flowers** 4(or 5)-merous; calyx 7–16 mm, lobes all ovate-triangular to suborbiculate, similar in length but the outer lobes wider, 2–8 mm; corolla pale to dark violet-blue [white], narrowly funnelform-salverform, 15–30 mm, lobes spreading, narrowly ovate, 3–8 mm, apex rounded or obtuse, adaxial corolla surface with a scale fringed ca. 0.6 times its length at base of each lobe; ovary subsessile. 2*n* = 48 (Siberia).

Flowering late summer–fall. Gravelly, open slopes; 0–300 m; Alaska; Asia (n Japan, Kamchatka, coastal Siberia).

In the flora area, *Gentianella auriculata* is known only from Attu in the Aleutian Islands and St. Lawrence Island in the Bering Strait.

The two larger calyx lobes of *Gentianella auriculata* are abruptly narrowed at the base, which therefore appears cordate or more or less auriculate, hence the specific epithet.

Gentianella auriculata is the only species of *Gentianella*, as the genus is circumscribed here, reported to have a chromosome number other than a multiple of 9.

Although its $2n = 48$ might suggest a closer relationship to *Comastoma* and *Lomatogonium*, molecular studies indicate that this species is appropriately included in *Gentianella* (K. B. von Hagen and J. W. Kadereit 2001).

2. Gentianella tortuosa (M. E. Jones) J. M. Gillett, Ann. Missouri Bot. Gard. 44: 248. 1957 • Curly-stemmed or Utah or Cathedral Bluff gentian [E]

Gentiana tortuosa M. E. Jones, Proc. Calif. Acad. Sci., ser. 2, 5: 707. 1895

Herbs annual, 2–10(–16) cm. **Stems** decumbent to erect, branched from base, often with shorter branches throughout. **Leaves:** basal leaves present at flowering time, blades oblong-elliptic to spatulate, 5–25 × 2–6 mm; cauline blades narrowly elliptic to lanceolate, 5–35 × 1–5 mm. **Inflorescences** terminal and axillary, diffuse, few-flowered cymes or solitary flowers; pedicels 3–15 mm. **Flowers** 5-merous; calyx 5–12 mm, lobes oblanceolate to linear, subequal, 4–10 mm; corolla white, sometimes blue-tinged, narrowly campanulate, 4.5–8.5 mm, lobes spreading, ovate, 2.5–4 mm, ± as long as tube, apex obtuse, adaxially with a fringe of trichomes, often ± divided into 2 clusters, at base of each lobe; ovary sessile.

Flowering late summer. Moist, open pine, spruce-fir, or aspen woods, rocky, calcareous soils, often on talus slopes; 2000–3400 m; Colo., Nev., Utah.

Gentianella tortuosa is largely restricted to the Utah Plateaus, with outlying populations in the Spring (Charleston) Mountains in southern Nevada and on Roan Plateau in western Colorado.

The plants of *Gentianella tortuosa*, having several stems or branches from the base, are almost spherical above ground, with the closely spaced larger leaves usually ascending well above the flowers, suggesting that it may sometimes be dispersed as a tumbleweed. The long, slender taproot, often twice as long as the height of the aerial portion of the plant, is also unique among the *Gentianella* species in the flora area. The epithet *tortuosa* refers to the distal branches and pedicels, which are often conspicuously curved in various directions. The calyx lobes are also often curved. Uniquely among the species of *Gentianella* in the flora area, this species produces six or fewer seeds per capsule, in contrast to the large numbers produced by most species in the genus (J. M. Gillett 1957).

3. Gentianella amarella (Linnaeus) Börner, Fl. Deut. Volk., 543. 1912 • Northern gentian, gentiane amarelle, smalbægret ensian [F]

Gentiana amarella Linnaeus, Sp. Pl. 1: 230. 1753

Subspecies ca. 5 (1 in the flora): North America, Eurasia.

3a. Gentianella amarella (Linnaeus) Börner subsp. **acuta** (Michaux) J. M. Gillett, Ann. Missouri Bot. Gard. 44: 253. 1957 • Marsh or rangers' gentian, rose-gentian [F]

Gentiana acuta Michaux, Fl. Bor.-Amer. 1: 177. 1803; *G. amarella* Linnaeus var. *plebeia* (Chamisso ex Bunge) Hultén; *G. amarella* var. *acuta* (Michaux) Herder; *G. strictiflora* (Rydberg) A. Nelson; *Gentianella acuta* (Michaux) Hiitonen; *G. strictiflora* (Rydberg) W. A. Weber

Herbs annual, (2–)4–50(–80) cm. **Stems** erect, simple or branched distally, sometimes also with long branches from near base. **Leaves:** basal present or withered by flowering, blades spatulate to elliptic-oblong, 6–50 × 3–20 mm; cauline blades ovate to oblong-lanceolate, 8–60 × 2–20 mm. **Inflorescences** terminal and axillary, dichasial cymes, lateral flowers in smaller cymes or axillary flowers solitary; pedicels (0–)8–25(–50) mm. **Flowers** (4- or)5-merous; calyx 4–13(–18) mm, lobes all connate proximally to ± the same level, linear to lanceolate, subequal or occasionally distinctly unequal but not foliaceous, 2–10(–15) mm; corolla deep to pale violet-blue to purplish pink (often bluer dried than fresh), pale yellow, or white, tubular to narrowly campanulate, 7–21 mm (proximal flowers often much smaller than distal), lobes ascending to spreading, ovate-triangular, 3–5.5 mm, apex obtuse to acute, with a fringe of trichomes near base of each lobe (fringes occasionally poorly developed, or absent in small flowers); ovary sessile. $2n = 18, 36$.

Flowering summer–early fall. Mesic to wet meadows, fens, prairies, clearings, roadsides, open woods, cliffs, beach ridges, tundra, generally calcareous soils; 0–4100 m (restricted to higher elevations southward); Greenland; St. Pierre and Miquelon; Alta., B.C., Man., N.B., Nfld. and Labr., N.W.T., N.S., Nunavut, Ont., Que., Sask., Yukon; Alaska, Ariz., Calif., Colo., Idaho, Maine, Minn., Mont., Nev., N.Mex., N.Dak., Oreg., S.Dak., Utah, Vt., Wash., Wyo.; Mexico (Durango, Nuevo León, Sinaloa); e Asia.

An outlying population of subsp. *acuta* in Vermont, historically much collected, is no longer extant. Records for Nunavut are from Akimiski Island in James Bay only.

Subspecies *acuta* and its homotypic synonyms were formerly associated only with the North American (including Mexican) representatives of *Gentianella amarella*, but in recent decades, the taxon has generally been circumscribed to include the plants of this complex in eastern Asia south to central China.

Some authors do not separate subsp. *acuta* from subsp. *amarella*, whereas others treat it as a distinct species. Morphologically, subsp. *acuta* is weakly differentiated from subsp. *amarella*. The flowers of the plants in much of Europe average larger when flowers in similar positions on plants of similar size are compared, but in all such comparisons there is extensive overlap in flower size. Other distinctions reported in earlier literature have been found not to be consistent upon examination of specimens from throughout the range of the species in the broad sense (M. L. Fernald 1917b; studies for this flora). Specimens from Scandinavia examined in studies for this flora appeared indistinguishable from North American plants. Presumed differences in ploidy level have probably influenced some authors' acceptance of the taxon *acuta* at specific or subspecific rank, but chromosome counts remain too few to indicate adequately the ranges of plants with the respective ploidy levels. Counts of $2n$ = 18 have been reported from Québec and California, and $2n$ = 36 has been reported from Manitoba, Iceland, Spain, and Sweden.

Historically, the North American representatives of *Gentianella amarella* have variously been divided into several species, as well as subspecies and varieties, based largely on plant stature and branching, but also considering four- versus five-merous flowers, corolla color, and other characters, as well as habitats. J. E. Weaver and F. E. Clements (1929) and J. M. Gillett (1957) concluded from their studies of this species that much of the morphological diversity in North America represented responses to interactions of seasonal phenomena with the time of seed germination and with light and shade, density of surrounding vegetation, and other conditions of the diverse macro- and microhabitats in which this taxon is found. Their conclusion was subsequently accepted by N. H. Holmgren (1984b), who included all of the North American taxa that had been segregated from this species or treated as subspecies or varieties of it in the synonymy of undivided *G. amarella*. In studies for this flora, it was found that plants of comparable height but differing in branching pattern, some with long branches from the base and others with only short, distal branches, were often present in the same population and included in the same collections. Plants with relatively closely spaced nodes and strongly ascending branches have been given the epithet *strictiflora* in various combinations, but complete intergradation appears to connect the extremes and little if any correlation is evident between plant habit and geographic distribution. The epithet *plebeia* has been applied to plants relatively low in stature, with few branches and wide leaves, but this distinction has generally been rejected in recent literature. Such plants often occur in the same areas as subsp. *acuta* in the narrow sense. They have sometimes been assumed to be more or less isolated reproductively through genetically determined earlier flowering, but most such specimens seen in studies for this flora had been found in flower no earlier than specimens not exhibiting such morphology. Experiments involving transplants or similar techniques would be of interest, but no such studies have been reported.

4. Gentianella heterosepala (Engelmann) Holub, Folia Geobot. Phytotax. 2: 117. 1967 • Engelmann's gentian

Gentiana heterosepala Engelmann, Trans. Acad. Sci. St. Louis 2: 215, plate 8. 1863; *Gentianella amarella* (Linnaeus) Börner subsp. *heterosepala* (Engelmann) J. M. Gillett; *G. amarella* var. *heterosepala* (Engelmann) Dorn

Subspecies 2 (1 in the flora): w United States, c Mexico.

Subspecies *durangensis* Villarreal is known from central Mexico.

4a. Gentianella heterosepala (Engelmann) Holub subsp. **heterosepala** [E]

Herbs annual, (2–)10–40(–60) cm. **Stems** erect, simple or branched above base. **Leaves:** basal sometimes withered by flowering, blades spatulate to obovate, 10–35 × 3–9 mm; cauline blades oblanceolate to ovate or lanceolate, 20–65 × 3–17 mm. **Inflorescences** terminal and axillary, dichasial cymes; pedicels 5–75 mm. **Flowers** (4- or)5-merous, (calyx usually 5-lobed); calyx 7–20 mm, with 1 or 2 lobes distinct nearly to base, these ovate, foliaceous, proximally enveloping the others, with 3 or 4 inner lobes, 1 usually distinct nearly to base, the others variably connate, inner lobes lanceolate to narrowly linear, 6–10 mm; corolla pale violet-blue to pinkish white, white, or pale yellow, tubular to narrowly campanulate, (9–)14–20(–24) mm, lobes spreading, ovate, 4–8 mm, apex acute, with a deeply fringed scale at base of each lobe; ovary sessile.

Flowering mid–late summer. Moist meadows, stream banks, open woods; 1900–3500 m; Ariz., Colo., N.Mex., Utah, Wyo.

In *Gentianella heterosepala*, one or more often two sepals are separate to the base or nearly so; these are distinctly longer than and at least twice as wide as the others, which they envelop proximally. According to N. H. Holmgren (1984b), plants considered intermediate between this taxon and *G. amarella* subsp. *acuta* had generally been so interpreted only because of their having unequal calyx lobes. Other aspects of their floral morphology were characteristic of *G. amarella* subsp. *acuta*. Following Holmgren, whose observations were supported in studies for the present work, such plants from the Intermountain Region are considered to represent variability within *G. amarella* subsp. *acuta* rather than true intergradation. Specimens previously identified as *G. heterosepala* from sites north to Saskatchewan, Idaho, and Montana are likewise included within *G. amarella* subsp. *acuta*, as the variation among the lobes of individual calyces of these plants does not reach the extremes that characterize *G. heterosepala*. Specific status for *G. heterosepala*, following G. L. Nesom (1991g), is thus supported, although some hybridization with *G. amarella* subsp. *acuta* may occur. These taxa also differ in seed diameter, which is about 1 mm in *G. heterosepala* and 0.75 mm or less in *G. amarella* subsp. *acuta* (J. M. Gillett 1957).

5. **Gentianella wrightii** (A. Gray) Holub, Folia Geobot. Phytotax. 2: 118. 1967 • Wright's gentian

Gentiana wrightii A. Gray in A. Gray et al., Syn. Fl. N. Amer., ed. 2, 2: 118. 1886; *Gentianella amarella* (Linnaeus) Börner subsp. *wrightii* (A. Gray) J. M. Gillett

Herbs annual, 25–50(–70) cm. **Stems** erect, simple or branched throughout. **Leaves:** basal often withered by flowering, blades ovate to elliptic or spatulate, 10–20 × 6–12 mm; cauline blades ovate to lanceolate, 20–50 × 8–17 mm. **Inflorescences** terminal and axillary, dichasial or ± umbelloid cymes; pedicels 3–5(–15) mm. **Flowers** (4- or)5-merous; calyx (6–)8–15 mm, dividing into lobes distinctly above base, narrowly elliptic to linear-oblong, subequal, (5–)6–10 mm; corolla white or pale yellow (often drying deeper yellow), rarely with short violet-blue streaks distally, narrowly funnelform, (15–)20–30 mm, lobes spreading, ovate, (5–)7–10 mm, apex obtuse to acute, with a fringe of trichomes near base of each lobe; ovary sessile.

Flowering late summer–fall. Wet meadows; 2100–2400 m; N.Mex.; Mexico (Chihuahua, Sinaloa, Sonora).

Gentianella wrightii may no longer occur in the flora area, from which it is known only from nineteenth-century collections by E. L. Greene in one locality in Grant County, New Mexico.

6. **Gentianella wislizeni** (Engelmann) J. M. Gillett, Ann. Missouri Bot. Gard. 44: 235. 1957 • Wislizenus's or Chiricahua Mountain gentian [C]

Gentiana wislizeni Engelmann, Trans. Acad. Sci. St. Louis 2: 215, plate 7. 1863

Herbs annual, 10–50 cm. **Stems** erect, branched above base. **Leaves:** basal often withered by flowering, blades ovate to elliptic or spatulate, 5–20 × 3–6 mm; cauline blades ovate to lanceolate, 15–40 × 4–15 mm. **Inflorescences** terminal and axillary, dichasial or partly monochasial cymes; pedicels 3–20 mm. **Flowers** (4- or)5-merous; calyx 3–6 mm, tube deeply cleft, spathaceous, lobes minute, linear, 0.5–1.5 mm; corolla pale violet to pinkish white (often drying yellowish), tubular to narrowly funnelform, 6–14 mm, lobes spreading, narrowly ovate-elliptic, 3–4 mm, apex mucronate, with fringe of trichomes at base of each lobe (occasionally absent, especially in small flowers); ovary subsessile.

Flowering fall. Rocky sites in open pine and pine-oak woods; of conservation concern; 2000–2600 m; Ariz.; Mexico (Chihuahua, Durango).

In the flora area, *Gentianella wislizeni* is known only from a few localities in southeastern Arizona, the northernmost being in southern Apache County.

The corolla trichomes of *Gentianella wislizeni* are relatively short and in the undissected flowers of herbarium specimens are often concealed.

7. **Gentianella microcalyx** (Lemmon) J. M. Gillett, Ann. Missouri Bot. Gard. 44: 246. 1957 • Eggleaf or Chihuahua gentian

Gentiana microcalyx Lemmon, Pacific Rural Press 23: 129. 1882

Herbs annual, 15–60 cm. **Stems** decumbent to erect, branched distally, often also at or near base. **Leaves:** basal often withered by flowering, larger leaves ± petiolate, blades elliptic to spatulate or oblanceolate, 10–25 × 2–8 mm; cauline blades cordate-ovate to lanceolate, 5–50 × 3–30 mm. **Inflorescences** terminal and often axillary umbelloid cymes; pedicels 3–25 mm. **Flowers** (4- or)5-merous; calyx 2–3(–5) mm, lobes triangular to oblong or elliptic, ± unequal, 1.5–3.5 mm; corolla yellowish white to pale rose-violet or pale violet-blue, occasionally with red-violet spots distally, narrowly funnelform to nearly salverform, 6–13 mm, lobes spreading, lanceolate-ovate, (2.5–)3–6.5 mm, apex obtuse to acute (in dried specimens sometimes appearing

± acuminate through inrolling), without adaxial scales or fringes; ovary short-stipitate.

Flowering late summer–fall. Moist, rocky sites in ± open pine and pine-oak woods; 600–2100 m; Ariz.; Mexico (Chihuahua, Coahuila, Durango, Sonora).

Gentianella microcalyx is primarily of Mexican distribution, where its range extends to higher elevations than in the flora area, where it is known only from southern Arizona.

8. Gentianella quinquefolia (Linnaeus) Small, Fl. S.E. U.S., 929. 1903 • Stiff gentian, ague-weed, gentiane de cinq feuilles [E] [F]

Gentiana quinquefolia Linnaeus, Sp. Pl. 1: 230. 1753; *G. quinqueflora* Hill; *Aloitis quinqueflora* (Hill) Rafinesque

Herbs annual or biennial, 2–80 cm. **Stems** erect, usually branched distally but without long branches near base. **Leaves:** basal usually withered by flowering, blades spatulate to oblanceolate, 5–35 × 2–12 mm; cauline blades ovate, 5–60(–80) × 2–35 (–45) mm. **Inflorescences** terminal and often axillary, dichasial or partly umbelloid cymes; pedicels 1–17 mm. **Flowers** 5-merous; calyx 2–15 mm, lobes subulate to lanceolate or linear-oblong, usually subequal, (1–)2–8 (–10) mm; corolla violet, violet-blue, blue, or occasionally pale yellow or white, narrowly funnelform, opening narrowly, 10–25 mm, lobes incurved, ovate-triangular, 3–8 mm, apex short-attenuate, without adaxial scales or fringes; ovary stipitate.

Subspecies 2 (2 in the flora): c, e North America.

The differences between the subspecies of *Gentianella quinquefolia* often have been overstated. Although the subspecies are largely separated geographically, plants intermediate in morphology are common throughout much of the range of the species. To some degree, the extremes probably represent phenotypic responses to local environmental conditions.

According to C. T. Mason and H. H. Iltis (1966) *Gentianella quinquefolia* is at least sometimes a biennial in Wisconsin, whereas elsewhere it has been reported to be an annual. Of the many flowers often present on a single plant, only a few may have open corollas at any one time.

1. Calyces 2–6(–8) mm, lobes subulate to linear-oblong, 1–4(–6) mm; corollas 10–23 mm . . . . 8a. *Gentianella quinquefolia* subsp. *quinquefolia*
1. Calyces 5–15 mm, lobes linear-oblong to lanceolate or occasionally subfoliaceous, 3–8 (–10) mm; corollas (15–)18–25 mm . . . . .8b. *Gentianella quinquefolia* subsp. *occidentalis*

8a. Gentianella quinquefolia (Linnaeus) Small subsp. **quinquefolia** [E] [F]

Herbs: larger plants usually with extensive primary and secondary branching. **Flowers:** calyx 2–6(–8) mm, lobes subulate to linear-oblong, 1–4(–6) mm; corolla 10–23 mm, lobes 3–7 mm. **2*n*** = 36.

Flowering late summer–fall. Open woods, grassy balds, stream banks, roadsides; 0–1800 m; Ont.; Conn., Ga., Md., Mass., N.J., N.Y., N.C., Pa., S.C., Tenn., Vt., Va., W.Va.

There are historical records of subsp. *quinquefolia* from Québec, Maine, and New Hampshire.

8b. Gentianella quinquefolia (Linnaeus) Small subsp. **occidentalis** (A. Gray) J. M. Gillett, Ann. Missouri Bot. Gard. 44: 245. 1957 [E]

Gentiana quinqueflora Lamarck var. *occidentalis* A. Gray, Manual, 359. 1848; *Aloitis quinqueflora* (Hill) Rafinesque subsp. *occidentalis* (A. Gray) Á. Löve & D. Löve; *Gentianella occidentalis* (A. Gray) Small; *G. quinquefolia* var. *occidentalis* (A. Gray) Small

Herbs: larger plants usually with short primary branching only. **Flowers:** calyx 5–15 mm, lobes linear-oblong to lanceolate or occasionally subfoliaceous, 3–8(–10) mm; corolla (15–)18–25 mm, lobes 4–8 mm. **2*n*** = 36.

Flowering fall. Moist open woods, prairies, wet bluffs, open woods, roadsides; 200–600 m; Ont.; Ark., Ill., Ind., Iowa, Kans., Ky., Mich., Minn., Miss., Mo., Ohio, Tenn., Wis.

9. Gentianella aurea (Linnaeus) Harry Smith ex Hylander, Uppsala Univ. Årssk. 1945(7): 259. 1945 • Golden gentian, bleg ensian

Gentiana aurea Linnaeus, Syst. Nat. ed. 10, 951. 1759; *Aloitis aurea* (Linnaeus) Á. Löve & D. Löve; *Arctogentia aurea* (Linnaeus) Á. Löve

Herbs annual, 2–30 cm. **Stems** decumbent to erect, usually with branches from near base. **Leaves:** larger leaves ± petiolate; basal blades elliptic to spatulate, 3–21 × 1–10 mm; cauline blades elliptic to ovate, 9–26 × 4–12(–15) mm. **Inflorescences** terminal and axillary, dichasial cymes, proximally ± enveloped by subtending leaves; pedicels

0–13 mm. **Flowers** (4- or)5-merous; calyx 3–8 mm, lobes narrowly obovate to linear-oblong, 2.5–6 mm, 2 outer lobes usually oblong to narrowly obovate, larger than the linear inner lobes; corolla pale yellow, occasionally suffused with pale violet-blue, tubular to narrowly funnelform, 6–11 mm, lobes erect or nearly so, ovate-oblong, 2–4 mm, apex mucronate, without adaxial scales or fringes; ovary subsessile. $2n = 18$.

Flowering late summer. Tundra meadows; 0–300 m; Greenland; Eurasia.

In the flora area, *Gentianella aurea* occurs only in Greenland. It is widely distributed in Eurasia, where its range extends to much higher elevations.

10. Gentianella propinqua (Richardson) J. M. Gillett, Ann. Missouri Bot. Gard. 44: 236. 1957 • Four-parted or small-flowered gentian, gentiane fausse-amarelle

Gentiana propinqua Richardson in J. Franklin, Narr. Journey Polar Sea, 734. 1823; *Aloitis propinqua* (Richardson) Á. Löve & D. Löve; *Arctogentia propinqua* (Richardson) Á. Löve & D. Löve

Herbs annual, 2–25(–40) cm. **Stems** decumbent to erect, all but smallest plants usually branched from near base. **Leaves:** basal blades elliptic to spatulate-obovate or oblanceolate, 5–35 × 2–8 mm; cauline blades elliptic-ovate to lanceolate, 5–35 × 2–10 mm. **Inflorescences** terminal and axillary, dichasial cymes, subcapitate in subsp. *aleutica*; pedicels 0–30 mm (lateral flowers sessile or on shorter pedicels than central). **Flowers** 4-merous; calyx 5–12 mm, lobes 3–8 mm, 2 outer lobes elliptic to ovate-oblong and ovate, larger than the inner and often ± foliaceous, inner narrowly lanceolate to linear-subulate; corolla blue-violet or rose-violet to white, tubular to narrowly funnelform, 7–22 mm, lobes spreading, lanceolate-ovate, 3–5 mm, apex obtuse, acuminate, or mucronate to bristle-tipped, without scales or fringes (flowers often smaller or rudimentary on basal branches); ovary short-stipitate.

Subspecies 2 (2 in the flora): North America, Asia (ne Siberia).

Plants of *Gentianella propinqua* usually have an erect central stem from which several more slender, proximally curved or decumbent branches arise at or near the base. A similar branching pattern and general aspect sometimes occur in the highly variable *G. amarella* subsp. *acuta*, so one should make sure that the corollas are devoid of adaxial trichomes before identifying plants as *G. propinqua*.

The subspecies of *Gentianella propinqua* are weakly differentiated. Some specimens from the Aleutian Islands approach subsp. *propinqua* in morphology. Conversely, although subsp. *propinqua* has often been described as having merely mucronate or abruptly bristle-tipped corolla lobes, in contrast to the acuminate corolla lobes of subsp. *aleutica*, many specimens from the mainland of Alaska, the Yukon Territory, and elsewhere have more or less long-acuminate corolla lobes.

1. Central flowers conspicuously larger than lateral; corollas blue-violet or occasionally white; cauline leaf blades narrowly elliptic to lanceolate . . . . 10a. *Gentianella propinqua* subsp. *propinqua*
1. Central flowers ± same size as lateral; corollas pale violet to rose-violet to white; cauline leaf blades elliptic-ovate to ovate. . . . . 10b. *Gentianella propinqua* subsp. *aleutica*

10a. Gentianella propinqua (Richardson) J. M. Gillett subsp. **propinqua**

Gentiana arctophila Grisebach; *G. propinqua* Richardson subsp. *arctophila* (Grisebach) Hultén; *Gentianella propinqua* subsp. *arctophila* (Grisebach) Tzvelev

Herbs 2–25(–40) cm. **Cauline leaf blades** narrowly elliptic to lanceolate. **Inflorescences** dichasial cymes; pedicels 0–30 mm. **Flowers:** central distinctly larger than lateral; corolla blue-violet or occasionally white, those of central flowers 12–22 mm, lobe apices long-acuminate or abruptly tapering to mucronate or bristle-tipped or occasionally obtuse. $2n = 36$.

Flowering summer–early fall. Beaches, stream banks, borders of bogs and muskeg, moist alpine and montane meadows, thickets, tundra, roadsides, clearings, other gravelly disturbed sites; 0–3000 m; Alta., B.C., Man., Nfld. and Labr., N.W.T., Nunavut, Ont., Que., Yukon; Alaska, Idaho, Mont., Wyo., restricted to higher elevations southward; Asia (ne Siberia).

Plants of subsp. *propinqua* with obtuse corolla lobes, to which the epithet *arctophila* has been applied, occur in the same populations throughout much of the range of the species as plants with bristle-tipped corolla lobes and plants intermediate in this respect. Branched and small, unbranched plants likewise occur within the same populations (J. M. Gillett 1957). From specimens now available, the character states alleged by E. Scamman (1940) to divide *Gentianella propinqua* in mainland Alaska into two groups, most readily distinguished by plant size and anthocyanin concentration in the leaves and stems, appear neither well marked nor consistently associated in syndromes.

10b. Gentianella propinqua (Richardson) J. M. Gillett subsp. **aleutica** (Chamisso & Schlechtendal) J. M. Gillett, Ann. Missouri Bot. Gard. 44: 241. 1957 • Aleutian gentian C E

Gentiana aleutica Chamisso & Schlechtendal, Linnaea 1: 175. 1826; *G. propinqua* Richardson var. *aleutica* (Chamisso & Schlechtendal) B. Boivin; *Gentianella propinqua* var. *aleutica* (Chamisso & Schlechtendal) S. L. Welsh

Herbs 2–10 cm. **Cauline leaf blades** elliptic-ovate to ovate. **Inflorescences** subcapitate cymes; pedicels 0–5 mm. **Flowers:** central and lateral similar in size; corolla pale violet to rose-violet to white, 7–13 mm, lobe apices acuminate but neither mucronate nor bristle-tipped.

Flowering late summer. Mountain meadows, rocky or gravelly soils; of conservation concern; 100–500 m; Alaska.

Subspecies *aleutica* is endemic to the Aleutian Islands and a few localities near the coast of mainland Alaska.

16. GENTIANOPSIS Ma, Acta Phytotax. Sin. 1: 7. 1951 • Fringed gentian [Genus name *Gentiana* and Greek *-opsis*, resembling]

James S. Pringle

Gentiana [unranked] *Crossopetalae* Froelich, Gentiana, 109. 1796

Herbs annual, biennial, or perennial, chlorophyllous, glabrous or with peduncles and calyces papillate-scabridulous. **Leaves** basal (sometimes withering before flowering) and cauline, opposite. **Inflorescences:** flowers solitary or occasionally paired (*G. barbellata*), terminating main stem and each of any branches; peduncles or pedicels generally shorter than stem except in *G. holopetala*. **Flowers** 4-merous; calyx in most species with 2 outer lobes lanceolate, longer, narrower, and more strongly acute or acuminate than ovate inner lobes (nearly equal in size and shape in *G. macrantha* and *G. simplex*), margins of lobes narrowly hyaline at least in proximal 1/2, apices acute to acuminate, and with tube in most species ± as long as inner or all lobes (all lobes longer than tube in *G. macrantha* and sometimes in *G. holopetala*), not cleft and spathaceous; corolla blue or in most species occasionally rose-violet or white, rarely pale yellow, tube widely tubular or tubular-campanulate, adaxially glabrous or with minute trichomes near insertion of stamens, lobes spreading, often tardily reflexed, ± as long as or shorter than tube, margins usually distinctly toothed and/or fringed, rarely entire or nearly so, plicae between lobes absent, spurs absent; stamens inserted in proximal 1/2 of corolla tube; anthers distinct, not coiling; ovary subsessile or stipitate; style persistent, erect, short, indistinct or rarely slender and distinct; stigmas 2; nectaries 1 below each corolla lobe, on corolla tube near base. **Capsules** compressed-ovoid. x = [11], 13.

Species ca. 15 (8 in the flora): North America, Mexico, Eurasia.

Gentianopsis has sometimes been treated as a subgenus or section of *Gentianella*, but molecular and other cladistic studies indicate that its distinctness and phylogenetic position are comparable to those of widely accepted related genera (Yuan Y. M. and P. Küpfer 1995; K. B. von Hagen and J. W. Kadereit 2001).

The taxonomic treatment of North American *Gentianopsis* has varied greatly in recent as well as in older literature. The combined history of misidentifications, different circumscriptions of taxa, and consequent difficulties in determining synonymy has resulted in erroneous or confusing statements as to the distribution of several of the taxa recognized here. This has also caused problems in determining their appropriate conservation status.

Some authors have assumed that North American *Gentianopsis* includes one group of taxa (the *G. crinita* complex) with $x = 13$ and another (the *G. detonsa* complex) with $x = 11$, but the occurrence of chromosomes in multiples of 11 in any North American *Gentianopsis* is unsubstantiated and should be considered doubtful. Reports of $2n = 44$ for taxa treated here as subspecies of *G. detonsa* and as *G. thermalis* have been based only on an early report of $2n = 44$ for *G. detonsa* [subsp. *detonsa*] in Iceland, subsequently retracted (Á. Löve and D. Löve 1986); all of the more recent counts for *G. detonsa*, including those for Icelandic plants, are $2n = 78$.

The subspecies of both *Gentianopsis detonsa* and *G. virgata* have often been treated as full species, but those within each of the respective more broadly defined species are so similar that specimens are sometimes difficult to identify except through consideration of their geographic origin. Even taxa placed in different species in the treatment accepted here are less well differentiated than some references indicate and are occasionally difficult to distinguish. Differences are sometimes overstated, especially with regard to the shapes and proportions of the calyx lobes. Although isolated occurrences of one geographically restricted subspecies within the range of another would not be inconceivable, in this genus it is much more likely that the specimens on which such reports have been based represent morphological convergence due to environmental factors.

Gentianopsis taxa differ in the sizes of the largest plants and the largest flowers, but there is extensive overlap among taxa in the sizes of the smaller plants and their flowers. Likewise, the diagnostic value of the numbers and spacing of leaf pairs on the main axis is limited by variation in plant size and growing conditions. The lengths of the gynophores continue to elongate during the life of the flowers and as the fruits develop.

Gentianopsis species are generally restricted to more or less calcareous habitats. This is especially evident in *G. virgata* subspp. *macounii* and *virgata*.

SELECTED REFERENCES Gillett, J. M. 1957. A revision of the North American species of *Gentianella* Moench. Ann. Missouri Bot. Gard. 44: 195–269. Iltis, H. H. 1965. The genus *Gentianopsis* (Gentianaceae): Transfers and phytogeographic comments. Sida 2: 129–154. Pringle, J. S. 2004. Notes on the distribution and nomenclature of North American *Gentianopsis* (Gentianaceae). Sida 21: 525–530.

1. Flowers sessile or occasionally on peduncles to 0.7 cm; plants perennial . . . . 7. *Gentianopsis barbellata*
1. Flowers on peduncles 0.7+ cm; plants annual or biennial, or seemingly biennial but with new stems arising from spreading roots in *G. simplex*.
    2. Margins of corolla lobes entire or merely erose-undulate; peduncles mostly much longer than stem . . . . 6. *Gentianopsis holopetala*
    2. Margins of corolla lobes dentate, serrate, erose, or fringed; peduncles often shorter than stem.
        3. Stems always simple even at base, 1-flowered; basal leaves withered at flowering; seeds striato-reticulate. . . . . 8. *Gentianopsis simplex*
        3. Stems of most plants branched, 2–many-flowered (larger plants of some of these species occasionally accompanied by small, 1-flowered plants; such plants usually have basal leaves persistent at flowering time); seeds papillate.
            4. Mid-stem and distal leaves lanceolate to ovate, apex acute; corolla lobes fringed on sides and around apex. . . . . 1. *Gentianopsis crinita*
            4. Mid-stem and distal leaves linear or narrowly lanceolate to oblong-elliptic, elliptic, or oblanceolate (rarely ovate in *G. thermalis*, if so, apex ± obtuse); corolla lobes not fringed or fringed only on sides, apex merely dentate.

[5. Shifted to left margin.—Eds.]

5. All or at least outer calyx lobes usually 1.5–2.5 times as long as or longer than calyx tube; branches, if present, from above base of primary stem rather than from base; s Arizona . . . . 5. *Gentianopsis macrantha*
5. All or inner (shorter) calyx lobes less than 1.5 times as long as tube; branching from base of primary stem and/or higher; widely distributed in flora area, including Arizona.
   6. All calyx lobes similar in size and shape, keels often ± undulate-crisped distally . . . . 4. *Gentianopsis thermalis*
   6. Outer calyx lobes narrower and more strongly acute or acuminate than inner, keels not undulate-crisped distally.
      7. Branches or peduncles rarely arising from base; basal leaves usually withered at flowering; apices of mid-stem and distal leaves acute; calyx keels usually papillate- or granular-scabridulous proximally (seen at 50×), occasionally nearly smooth . . . . 2. *Gentianopsis virgata*
      7. Branches and/or peduncles often arising from base as well as distally; basal leaves usually persistent and green at flowering; apices of all leaves obtuse, acute in subsp. *detonsa*; calyx keels smooth . . . . 3. *Gentianopsis detonsa*

1. **Gentianopsis crinita** (Froelich) Ma, Acta Phytotax. Sin. 1: 15. 1951 • Eastern fringed gentian, gentiane frangée [E] [F]

Gentiana crinita Froelich, Gentiana, 112. 1796; *Anthopogon crinitus* (Froelich) Rafinesque; *Gentianella crinita* (Froelich) Berchtold & J. Presl

Herbs annual or biennial, (0.3–)1–6(–10) dm. **Stems** except those of smallest plants with branches or peduncles arising from nodes distinctly above base, rarely from base. **Leaves:** basal often withered by flowering, blades spatulate to oblanceolate, 0.8–4 cm × 1–11 mm, apex rounded to acute; cauline blades (narrowly to) widely lanceolate to widely ovate, 1–8 cm × (4–)7–25 mm, apex acute. **Peduncles** 1–15(–20) cm. **Flowers** 1–many; calyx 14–40(–50) mm, keels slightly papillate-scabridulous proximally, all or at least inner lobes less than 1.5 times as long as tube, outer lobes lanceolate, apices acuminate to attenuate, inner lobes lance-ovate to ovate, apices acute to acuminate; corolla deep blue or rarely rose-violet or white, 25–60(–75) mm, lobes elliptic-obovate, 10–25 × 5–15 mm, margins with fringes to 6 mm laterally and around apex, apex rounded; ovary distinctly stipitate. **Seeds** papillate, not winged. $2n = 78$.

Flowering late summer–fall. Wet meadows, prairies, savannas, alvars, stream banks, roadsides, other moist, open sites, usually in ± calcareous soils; 0–1400 m; Man., Ont., Que.; Conn., Ga., Ill., Ind., Iowa, Maine, Md., Mass., Mich., Minn., N.H., N.J., N.Y., N.C., N.Dak., Ohio, Pa., R.I., Vt., Wis., restricted to higher elevations southward.

There are historical records of *Gentianopsis crinita* from Delaware, Tennessee, Virginia, and West Virginia. Reports from states and provinces west of the range given here were based on misidentifications or on circumscriptions of the species that included *G. virgata*.

Occasional plants or populations of *Gentianopsis crinita* approach *G. virgata* subsp. *virgata* in leaf shape and, perhaps less often, plants or populations of *G. virgata* approach *G. crinita* in this respect. Although in most populations, such plants represent variation within the respective species, as indicated by corolla morphology, hybridization evidently does occur, notably in disturbed sites in northern Ohio.

2. **Gentianopsis virgata** (Rafinesque) Holub, Folia Geobot. Phytotax. 2: 120. 1967 • Slender fringed gentian, gentianopsis élancé [E]

Anthopogon virgatus Rafinesque, Fl. Tellur. 3: 25. 1837 (as virgatum)

Herbs biennial or perhaps sometimes annuals, 0.2–5(–7) dm. **Stems** except those of smallest plants with branches or peduncles arising from nodes distinctly above base, infrequently from base (subsp. *macounii*). **Leaves:** basal usually withered by flowering, blades spatulate to oblanceolate, 0.8–6 cm × 1–18 mm, apex rounded to acute; cauline blades linear to linear-lanceolate, 1.5–9 cm × 2–9(–12) mm, apex acute. **Peduncles** 1–20 cm. **Flowers** 1–many; calyx 12–60 mm, keels usually minutely granular-scabridulous to strongly papillate-scabridulous proximally, sometimes smooth, all or at least inner lobes less than 1.5 times

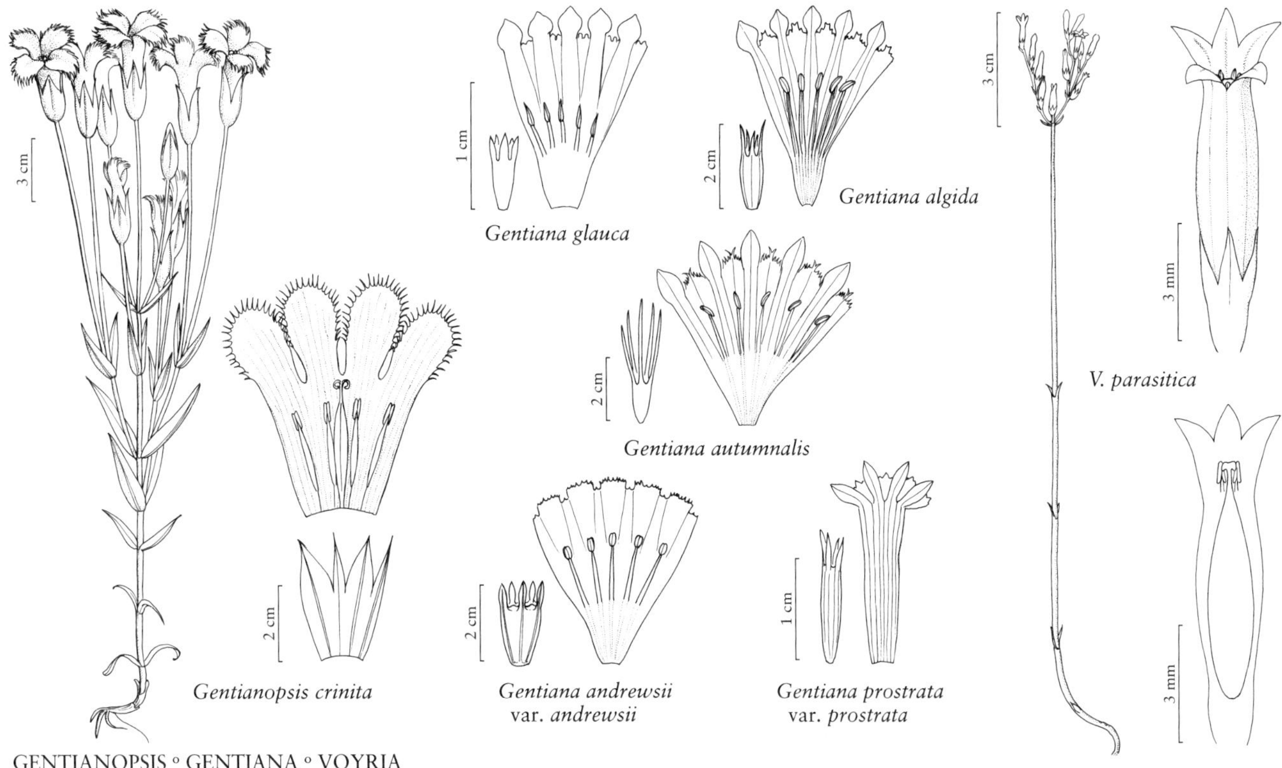

GENTIANOPSIS ◦ GENTIANA ◦ VOYRIA

as long as tube, outer lobes linear-lanceolate, apices acuminate to attenuate, inner lobes lance-ovate to ovate, apices acute to acuminate; corolla deep blue or rarely pale blue, rose-violet, white, or pale yellow, 18–60(–75) mm, lobes oblong-obovate to ovate-elliptic or orbiculate, 10–30 × 3–18 mm, margins proximally fringed, distally dentate or erose, or sometimes entire (subsp. *victorinii*), apex rounded; ovary subsessile to long-stipitate. **Seeds** papillate, not winged.

Subspecies 3 (3 in the flora): North America.

1. Calyx keels usually strongly papillate-scabridulous, occasionally minutely or obscurely so or smooth; corolla lobes with lateral fringes usually 2–6 mm . . . . 2a. *Gentianopsis virgata* subsp. *virgata*
1. Calyx keels minutely granular-scabridulous near base or nearly smooth; corolla lobes with sparse lateral fringes to 2 mm or merely erose, dentate to serrate, or entire or nearly so.
   2. Corolla lobes deeply dentate or serrate toward apex, teeth mostly 1+ mm; ovary at flowering with slender gynophore ± as long as body; estuary of St. Lawrence River, Quebec . . . . 2b. *Gentianopsis virgata* subsp. *victorinii*
   2. Corolla lobes erose or shallowly dentate toward apex, teeth mostly less than 1 mm; ovary at flowering subsessile or with a thick gynophore much shorter than body; widely distributed in c, w North America, disjunct on Gaspé Peninsula, Quebec . . . . 2c. *Gentianopsis virgata* subsp. *macounii*

2a. Gentianopsis virgata (Rafinesque) Holub subsp. **virgata** • Narrow-leaved fringed gentian, gentiane grande [E]

Gentiana procera Holm; *Gentianella crinita* (Froelich) Berchtold & J. Presl subsp. *procera* (Holm) J. M. Gillett; *Gentianopsis crinita* (Froelich) Ma subsp. *procera* (Holm) Á. Löve & D. Löve; *G. procera* (Holm) Ma

Herbs 1.5–5(–7) dm. **Peduncles** 2–20 cm. **Flowers** often 1–many; calyx 15–45(–60) mm, keels usually strongly papillate-scabridulous, rarely obscurely scabridulous or smooth; corolla deep blue or rarely pale blue, rose-violet, white, or pale yellow, 20–60(–75) mm, lobes orbiculate-obovate, 20–30 × 9–18 mm, margins proximally with fringes 2–6 mm, distally dentate; ovary subsessile to distinctly slender-stipitate. 2*n* = 78.

Flowering late summer–fall. Fens, interdunal depressions, alvars, wet prairies, moist ditches, calcareous soils; 200–400 m; Man., Ont., Sask.; Ill., Ind., Iowa, Mich., Minn., N.Y., N.Dak., Ohio, Wis.

Subspecies *virgata* is believed to be extirpated from Pennsylvania. Reports of this taxon, under various names, as occurring northwest to the Yukon Territory and Alaska have been based on misidentified specimens or on unusually inclusive circumscriptions of the species.

Most representatives of *Gentianopsis virgata* from Saskatchewan can be assigned to subsp. *macounii*, but occasional plants exhibit the relatively long lateral corolla-lobe fringes and prominently papillate calyx keels of subsp. *virgata*. In Saskatchewan populations, such plants usually occur with intermediates. Specimens of subsp. *virgata* from Minnesota generally approach subsp. *macounii*.

The common names smaller fringed gentian and lesser fringed gentian, often given for this taxon, are misleading, as the range of variation in plant and flower size is similar in subsp. *virgata* and *Gentianopsis crinita*. In both taxa, flower size is more or less correlated with plant height, but plants of subsp. *virgata* tend to have fewer but larger flowers than plants of *G. crinita* of similar stature.

Subspecies *virgata* appears to be the most variable in corolla color of the *Gentianopsis* taxa in the flora area. Although the great majority of plants have blue corollas, white or rose-violet corollas occur in some localities; plants with pale yellow corollas have been reported in Wisconsin; and a bicolored form has been found in Sleeping Bear National Lakeshore, Michigan, in which the corolla tube and the proximal part of the lobes are white except for a central blue stripe.

2b. Gentianopsis virgata (Rafinesque) Holub subsp. **victorinii** (Fernald) Lammers, Michigan Bot. 42: 163. 2004 • Victorin's gentian, gentiane de Victorin [C] [E]

Gentiana victorinii Fernald, Rhodora 25: 87, plate 139. 1923; *Gentianella crinita* (Froelich) Berchtold & J. Presl subsp. *victorinii* (Fernald) J. M. Gillett; *Gentianopsis procera* (Holm) Ma var. *victorinii* (Fernald) H. H. Iltis; *G. victorinii* (Fernald) H. H. Iltis

Herbs (0.2–)1–5 dm. **Peduncles** (1–)2–13 cm. **Flowers** 1–5; calyx 12–30 mm, keels minutely or obscurely granular-scabridulous near base; corolla deep blue or rarely white, 20–45 mm, lobes widely ovate-elliptic to orbiculate, 10–20 × 10–12 mm, margins proximally with a few fringes to 1.5 mm or merely dentate to entire, distally dentate to serrate with teeth mostly 1+ mm; gynophore distinct, slender, as long as body of ovary at flowering. **2*n*** = 78.

Flowering late summer–fall. Upper limits of tidal zone of St. Lawrence River; of conservation concern; 0 m; Que.

Subspecies *victorinii* is endemic to the estuary of the Saint Lawrence River, along about 150 km centered approximately on Quebec City. The subspecies is restricted to a zone where tidal action causes inundation but where the water is fresh. Its habitat is inundated for two to three hours per day during periods of relatively high tides and not inundated during periods of lower tides (J. Rousseau 1932; F. Courson 1998; Environment Canada 2011).

Subspecies *victorinii* differs from the other subspecies of the species in its widely elliptic to orbiculate corolla lobes, which are rounded at the apex, less than twice as long and often about as long as wide, with the teeth at and near the apex mostly more than 1 mm long. The corolla lobes of the other subspecies are more nearly oblong, more or less truncate, and usually proportionately narrower, with teeth less than 1 mm long. Subspecies *victorinii* also differs in its distinct, slender style 1–2 mm long and in its relatively slender gynophore about 5 mm or longer at flowering, that is, 0.5–1 times as long as the body of the ovary.

2c. Gentianopsis virgata (Rafinesque) Holub subsp. **macounii** (Holm) J. S. Pringle, Sida 21: 529. 2004 • Macoun's fringed gentian, gentiane de Macoun [E]

Gentiana macounii Holm, Ottawa Naturalist 15: 110. 1901; *G. gaspensis* Victorin; *G. tonsa* (Lunell) Victorin; *Gentianella crinita* (Froelich) Berchtold & J. Presl subsp. *macounii* (Holm) J. M. Gillett; *Gentianopsis crinita* (Froelich) Ma subsp. *macounii* (Holm) Á. Löve & D. Löve; *G. macounii* (Holm) H. H. Iltis; *G. procera* (Holm) Ma subsp. *macounii* (Holm) H. H. Iltis

Herbs 0.5–4 dm. **Peduncles** 5–20 cm. **Flowers** often 1 per primary stem, occasionally 2–5; calyx 13–25 mm, keels minutely granular-scabridulous near base or nearly smooth; corolla deep blue or rarely pale blue to white, 18–40 mm, lobes oblong-obovate, 10–20 × (3–)5–13 mm, margins with sparse lateral fringes to 2 mm, merely erose or shallowly dentate, with teeth less than 2 mm, distally and around the apex; ovary at flowering subsessile or with a short, thick gynophore much shorter than body. **2*n*** = 78.

Flowering summer–early fall. Wet meadows, riverbanks, margins of sloughs, fens, beach ridges, lakeshores, brackish meadows; 0–1900 m; Alta., B.C., Man., N.W.T., Ont., Que., Sask.; Minn., Mont., N.Dak., S.Dak.

A report of subsp. *macounii* from Nebraska is erroneous, having been based on a specimen actually from North Dakota (studies for this flora). At least one report of this taxon from the southern part of the Northwest Territories is correct, but most specimens from the Northwest Territories so identified have been reidentified as *Gentianopsis detonsa* subsp. *raupii* in studies for this flora.

A disjunct population in brackish meadows along the estuary of the Bonaventure River, Gaspé Peninsula, Quebec, has been called *Gentiana gaspensis* but appears to constitute only a relatively uniform population, the morphology of which falls within the range of variation exhibited by plants called *Gentianopsis virgata* subsp. *macounii* (H. A. Gleason 1952, as *Gentiana tonsa*; J. M. Gillett 1957; A. Dutilly et al. 1958).

Subspecies *macounii* and subsp. *virgata* intergrade in Minnesota, southern Manitoba, and occasionally as far west as southern Saskatchewan.

3. Gentianopsis detonsa (Rottbøll) Ma, Acta Phytotax. Sin. 1: 15. 1951 • Sheared or serrate gentian

Gentiana detonsa Rottbøll, Skr. Kiøbenhavnske Selsk. Laerd. Elsk. 10: 435, plate 1, fig. 3. 1770; *Gentianella detonsa* (Rottbøll) G. Don

Herbs annual or biennial, 0.2–6 dm. **Stems:** branching diverse, often from base as well as distally. **Leaves:** basal usually persistent and green at flowering, blades narrowly obovate to spatulate, oblanceolate, oblong, or linear (subsp. *raupii*) 0.5–3.5(–6) cm × 3–18 mm, apex rounded or obtuse, sometimes subacute or acute (subsp. *detonsa*), at least distal cauline blades oblanceolate, narrowly elliptic, lanceolate, or linear, (0.5–)1.5–6.5 cm × 1–7(–15) mm, apex rounded to obtuse, or usually acute (subsp. *detonsa*). **Peduncles** (0.3–)1–15 cm. **Flowers** often 1 per primary stem, occasionally 2–5; calyx 7–30 mm, keel smooth, all or at least inner lobes less than 1.5 times as long as tube, lobes ovate to narrowly lanceolate, varying with subspecies; corolla pale to deep blue or occasionally pale yellow or white, 12–50 mm, lobes oblong, oblong-lanceolate, oblong-triangular, oblong-orbiculate, or proximally oblong, distally obovate to suborbiculate, varying with subspecies, 5–20 × 4–12 (–15) mm, margins proximally fringed or merely erose to dentate, distally dentate; ovary ± short-stipitate. **Seeds** papillate, not winged.

Subspecies 4 (4 in the flora): North America, Eurasia.

References to *Gentianopsis detonsa* in the Rocky Mountains from Montana south to New Mexico have been based on circumscriptions of the species that included *G. thermalis*. Specimens formerly so identified from Illinois, Indiana, and Ontario south of the James Bay region are *G. virgata* subsp. *virgata*. Those from Minnesota are *G. virgata* subsp. *macounii*.

1. Margins of corolla lobes dentate or erose, not fringed; halophytes of northern seacoasts (including Mackenzie Delta).
    2. Basal leaf blades oblanceolate, apex obtuse to acute; cauline leaf blades (if present) linear, apex usually acute; corolla lobes 0.5+ times as long as tube; seed coat papillate only toward ends . . . . 3a. *Gentianopsis detonsa* subsp. *detonsa*
    2. Leaf blades obovate to spatulate or oblanceolate or distal cauline leaf blades (if present) elliptic to lanceolate, apex rounded or obtuse; corolla lobes to 0.5 times as long as tube; seed coat completely papillate . . . . 3d. *Gentianopsis detonsa* subsp. *nesophila*
1. Margins of most corolla lobes fringed proximally, dentate toward apex; plants of the interior, approaching Arctic coast only in Mackenzie Delta (subsp. *raupii*).
    3. Distinct rosette of basal leaves present at flowering, separated from cauline leaf blades; corolla lobes acute apically . . . . 3b. *Gentianopsis detonsa* subsp. *yukonensis*
    3. Basal leaves generally persistent but rosette not well developed, transitional to cauline leaves; corolla lobes rounded apically . . . . 3c. *Gentianopsis detonsa* subsp. *raupii*

3a. Gentianopsis detonsa (Rottbøll) Ma subsp. **detonsa** • Seaside or arctic serrate gentian, salt-ensian

Gentiana detonsa Rottbøll var. *groenlandica* Victorin; *G. richardsonii* A. E. Porsild

Herbs 0.2–2 dm. **Stems** simple or those of larger plants with branches or peduncles arising from base. **Leaf blades:** basal oblanceolate, apex obtuse to acute; cauline leaves linear, apex usually acute, occasionally absent. **Flowers:** outer calyx lobes distinctly longer than inner, apex acuminate, inner ovate to lanceolate, apex acute to short-acuminate; corolla 12–36 mm, lobes oblong-triangular, 0.5+ times as long as tube, margins dentate, not fringed, narrowed apically, apex ± acute; gynophore short, thick. **Seed coat** papillate only toward ends. $2n = 78$ (Iceland).

Flowering summer. Wet seaside meadows, brackish mudflats, north of tree line; 0–10 m; Greenland; N.W.T., Nunavut; Alaska; Eurasia (arctic regions).

The corolla lobes of subsp. *detonsa* and subsp. *yukonensis* are proportionately narrower than those of the other *Gentianopsis* taxa in the flora area and taper to a more nearly acute apex than those of any other taxa except *G. macrantha*.

Plants from arctic Canada and Greenland do not appear to be separable from those from Iceland, the provenance of the type, nor from plants from Norway. If representatives of *Gentianopsis detonsa* in any other part of continental Europe are considered taxonomically separable, subsp. *detonsa* in the flora area is appropriately retained in the autonymic subspecies.

3b. Gentianopsis detonsa (Rottbøll) Ma subsp. **yukonensis** (J. M. Gillett) J. M. Gillett, Canad. J. Bot. 57: 186. 1979 • Yukon fringed gentian [E]

Gentianella detonsa (Rottbøll) G. Don subsp. *yukonensis* J. M. Gillett, Ann. Missouri Bot. Gard. 44: 215. 1957

Herbs 0.5–6 dm. **Stems** except those of smallest plants with branches or peduncles arising from nodes distinctly above base, occasionally also from base. **Leaf blades:** basal leaves in a distinct rosette, blades oblong to oblanceolate, apex obtuse to subacute; cauline lanceolate to linear, apex acute. **Flowers:** outer calyx lobes distinctly longer than inner, lance-ovate, apex acuminate to attenuate, inner lanceolate, apex acute to short-acuminate; corolla 12–50 mm, lobes oblong to elliptic-lanceolate, 0.5+ times as long as tube, margins distinctly although sometimes sparsely fringed proximally, dentate toward apex, apex acute; gynophore short, thick. **Seed coat** papillate only toward ends.

Flowering summer. Tundra, wet meadows and subalpine heathland, open woods, sandy or gravelly soils; 100–500 m; Yukon; Alaska.

Subspecies *yukonensis* is endemic to the watershed of the Yukon River and its tributaries.

Subspecies *yukonensis* combines character states associated with both subsp. *detonsa* and subsp. *raupii*, but it cannot satisfactorily be included in either. Differences in branching pattern have been overstated. Branches or peduncles frequently arise from the base in subsp. *yukonensis* as well as in subsp. *raupii*, and plants not branched from the base, although often one-flowered, are frequent in subsp. *raupii* as well as in subsp. *yukonensis*. Occasional plants from the range of subsp. *raupii* resemble plants from the Yukon River basin, but subsp. *raupii* usually has a stouter aspect than subsp. *yukonensis*, and plants of subsp. *raupii* of comparable stature usually have more basal branches that closely approach the main axis in height and stem diameter.

Subspecies *yukonensis* usually has a well-developed rosette of oblanceolate leaves wider than the cauline leaves, the rosette being separated from the cauline leaves by a distinct internode, whereas the basal rosette of subsp. *raupii* is usually less well developed, and its leaves are gradually transitional to the proximal cauline leaves. The corolla lobes of subsp. *yukonensis* distally taper gradually to an acute apex, whereas those of subsp. *raupii* are shaped more like an old-fashioned keyhole, proximally oblong, rather abruptly widening to an ovate to suborbiculate distal portion, rounded at the apex.

Because of its widely separated, linear cauline leaves, subsp. *yukonensis* is more similar in aspect to the taller plants of subsp. *detonsa* than to subsp. *raupii*. Subspecies *yukonensis* differs from subsp. *detonsa* in its obtuse cauline leaf apices and proximally fringed rather than merely dentate corolla-lobe margins and usually has more basal leaves.

Reports of *Gentiana barbata* Froelich from Alaska (E. Hultén 1968) were based on specimens of *Gentianopsis detonsa* subsp. *yukonensis*. The Eurasian taxon formerly called *Gentiana barbata* has sometimes been included in *Gentianopsis detonsa* but is usually treated as a distinct species, *Gentianopsis barbata* (Froelich) Ma. Were these taxa considered conspecific, the epithet *detonsa* would have priority over *barbata*.

3c. Gentianopsis detonsa (Rottbøll) Ma subsp. **raupii** (A. E. Porsild) Á. Löve & D. Löve, Bot. Not. 128: 517. 1976 • Northern fringed or Raup's gentian [E]

Gentiana raupii A. E. Porsild, Sargentia 4: 60. 1943; *Gentianella detonsa* (Rottbøll) G. Don subsp. *raupii* (A. E. Porsild) J. M. Gillett; *Gentianopsis barbata* (Froelich) Ma subsp. *raupii* (A. E. Porsild) Elven

Herbs 0.5–6 dm. **Stems** except those of smallest plants usually with branches or peduncles arising from base, rarely only distally. **Leaf blades:** basal leaves usually present at flowering but few, transitional to cauline leaves; basal and cauline blades oblanceolate to linear, apex obtuse. **Flowers:** calyx lobes subequal in length, outer lobes narrowly ovate to lanceolate, apex acuminate to attenuate, inner lobes lanceolate, apex acute to acuminate; corolla 12–50 mm, lobes proximally oblong, distally obovate to suborbiculate, 0.5+ times as long as tube, margins of most lobes distinctly although sometimes sparsely fringed proximally, dentate toward apex, apex rounded; gynophore intermediate among subsp. of *G. detonsa* in length and thickness. **Seed coat** completely papillate.

Flowering summer. Alluvial meadows, riverbanks; 100–300 m; Alta., N.W.T., rarely approaches Arctic coast in Mackenzie Delta, otherwise not on Arctic coast.

Reports of subsp. *raupii* (or of *Gentianopsis detonsa* exclusive of subsp. *nesophila*) as a taxon rare in or extirpated from Ontario, known only from the shores of James and southern Hudson bays, have been based on specimens that are here included in subsp. *nesophila*

(J. S. Pringle 2004 and references cited therein; subsequent identifications in studies for this flora). Subspecies *raupii* is interpreted here as being endemic to the Slave and Mackenzie River basin in northern Alberta and the Northwest Territories; all plants of *G. detonsa* in the broad sense in the saline coastal meadows on the James and southern Hudson Bay shores are included in subsp. *nesophila*. Some of the latter plants resemble subsp. *raupii* in having fewer leaves in proportion to plant size than is usual in populations of subsp. *nesophila* farther east, but they have the proportionately short, unfringed corolla lobes characteristic of subsp. *nesophila*. Conversely, occasional plants of subsp. *raupii* from the Mackenzie Valley approach subsp. *nesophila* in numbers of leaves.

3d. Gentianopsis detonsa (Rottbøll) Ma subsp. **nesophila** (Holm) J. S. Pringle, Sida 21: 527. 2004 • Island gentian, gentiane des îles [E]

Gentiana nesophila Holm, Ottawa Naturalist 15: 111. 1901; *Gentianella detonsa* (Rottbøll) G. Don subsp. *nesophila* (Holm) J. M. Gillett; *Gentianopsis nesophila* (Holm) H. H. Iltis

Herbs 0.2–2.5 dm. **Stems** except those of smallest plants with branches or peduncles arising from base. **Leaf blades:** basal and at least proximal cauline obovate to spatulate or oblanceolate, distal cauline (if present) sometimes elliptic to lanceolate, apices of all rounded to obtuse. **Flowers:** calyx lobes all similar in shape and length, ovate to lance-ovate, all with apex acute or outer short-acuminate; corolla 12–50 mm, lobes oblong-orbiculate, to 0.5 times as long as tube, margins dentate or erose, not fringed, apex rounded; ovary subsessile or gynophore stout, to 5 mm at anthesis. **Seed coat** completely papillate.

Flowering summer–early fall. Coastal marshes, gravelly beaches, saline habitats; 0–20 m; Nfld. and Labr. (Nfld.), Nunavut, Ont., Que.

Subspecies *nesophila* occurs only along the shores of the Gulf of Saint Lawrence and Hudson and James bays. Records for Nunavut are from islands in James Bay only.

Subspecies *nesophila* generally has more pairs of leaves in proportion to plant height than the other subspecies of *Gentianopsis detonsa*, and spatulate to oblanceolate leaves prevail at more nodes above the basal rosettes. The basal leaves, which are generally numerous, tend to be both absolutely and proportionately wider than those of the other subspecies of *G. detonsa* and all subspecies of *G. virgata*. In contrast to those of all other *Gentianopsis* taxa in the flora area, the corolla lobes of subsp. *nesophila* are less than half as long as the tube. Occasional plants of subsp. *nesophila*, especially on the western shores of Hudson and James bays, approach subsp. *raupii* in some aspects of their morphology and vice versa. The papillae of the seed coats of subsp. *nesophila* are mostly smaller than those of the other subspecies of *G. detonsa*.

Where both taxa are present on the southwestern shore of James Bay, subsp. *nesophila* grows in supratidal meadow-marshes, whereas *Gentianopsis virgata* subsp. *macounii* is found at slightly higher elevations, on low raised beach ridges (J. L. Riley and S. M. McKay 1980).

4. Gentianopsis thermalis (Kuntze) H. H. Iltis, Sida 2: 134. 1965 • Rocky Mountain fringed or western fringed or feather gentian [E]

Gentiana thermalis Kuntze, Revis. Gen. Pl. 2: 427. 1891; *G. detonsa* Rottbøll var. *unicaulis* (A. Nelson) C. L. Hitchcock; *G. elegans* A. Nelson; *Gentianella detonsa* (Rottbøll) G. Don subsp. *elegans* (A. Nelson) J. M. Gillett; *G. detonsa* var. *elegans* (A. Nelson) Dorn; *Gentianopsis detonsa* (Rottbøll) Ma var. *elegans* (A. Nelson) N. H. Holmgren

Herbs biennial or perhaps annuals in some areas, 0.5–5(–9) dm. **Stems** except those of smallest plants usually with branches or peduncles arising from base. **Leaves:** basal usually persistent at flowering, blades spatulate to oblanceolate or narrowly obovate, (0.5–)1–7 cm × (2–)5–12(–25) mm, apex rounded or obtuse; cauline blades narrowly to widely elliptic-oblong to ovate or lanceolate, rarely linear, 1.5–5 cm × (1–)3–12(–23) mm, apex obtuse to acute. **Peduncles** 2–13(–18) cm. **Flowers** 1–many; calyx 15–35(–40) mm, keel often undulate-crisped distally but not papillate-scabridulous, all or at least inner lobes less than 1.5 times as long as tube, outer lobes narrowly ovate, apex acute to short-acuminate, inner lobes ovate, apex acute to short-acuminate; corolla deep blue to blue-violet or rarely white, (20–)30–70(–75) mm, lobes proximally oblong, widening into ovate to orbiculate distal portion, 10–25(–32) × 3–15 mm, margins proximally sparsely to copiously fringed, distally dentate; ovary distinctly stipitate. **Seeds** papillate, not winged.

Flowering summer–early fall. Around hot springs, wet mountain meadows, stream banks, roadsides; 1700–3600 m; Ariz., Colo., Idaho, Mont., Nev., N.Mex., Utah, Wyo.

Gentianopsis thermalis occurs near the California state line in Washoe County, Nevada, and should be sought in California. This species is known only from one historical record from Arizona.

The range of *Gentianopsis thermalis* approaches that of *G. virgata* subsp. *macounii* only in Montana. Although occasional plants of *G. thermalis* have linear leaves similar to those that prevail in *G. virgata* subsp. *macounii*, the leaves of *G. thermalis* are usually elliptic-oblong and often more than 5 mm wide. The basal leaves of *G. thermalis* are usually larger and more numerous than those of *G. virgata* subsp. *macounii* and are more likely to persist until flowering.

The four calyx lobes of *Gentianopsis thermalis* are nearly equally wide. The keels are narrowly but distinctly winged along the midvein all the way to the apex and are often more or less undulate-crisped distally, with the wing usually sharply defined in dark purple. Some specimens from the southern part of the range have keels that are scarcely papillate-scabridulous distally but are otherwise characteristic of the species. The calyx keels of the other western *Gentianopsis* taxa are smooth rather than papillate-scabridulous, although the less prominent, usually noncrisped keels of *G. detonsa*, *G. holopetala*, and *G. macrantha* may also be narrowly defined in dark purple. The calyx keels of all subspecies of *G. virgata* are papillate- or granular-scabridulous proximally rather than on the lobes. As compared with those of other *Gentianopsis* species in the flora area except for *G. detonsa* subsp. *raupii* and *G. macrantha*, the corolla lobes at least of the larger flowers of *G. thermalis* are more distinctly divided into an oblong proximal portion and an abruptly expanded distal portion, reminiscent of an old-fashioned keyhole. The style of *G. thermalis* is slender and one-third to nearly as long as the ovary at anthesis, thus being proportionately much longer than that of any other *Gentianopsis* taxon in the flora area except *G. virgata* subsp. *victorinii*. The gynophore is also proportionately long and slender at anthesis, as long as or longer than the style. The tendency among some *Gentianopsis* taxa for the margins of the stigmas to be petaloid, more or less recurved and crisped, and erose to shallowly lacerate is especially pronounced in *G. thermalis*.

Gentianopsis thermalis is sometimes treated as a subspecies or variety of *G. detonsa* and has been designated simply as *Gentiana*, *Gentianella*, or *Gentianopsis detonsa* in some field guides. Molecular studies support the acceptance of *G. thermalis* at species rank, indicating that it and *G. macrantha* may be more closely related to *G. crinita* and *G. virgata* than to *G. detonsa* (B. A. Whitlock, pers. comm.).

Gentianopsis thermalis is the official flower of Yellowstone National Park.

5. **Gentianopsis macrantha** (D. Don ex G. Don) H. H. Iltis, Sida 2: 135. 1965 • Large-flowered fringed or grand fringed gentian

Gentianella macrantha D. Don ex G. Don, Gen. Hist. 4: 179. 1837; *Gentiana grandis* (A. Gray) Holm; *G. superba* Greene; *Gentianella detonsa* (Rottbøll) G. Don subsp. *superba* (Greene) J. M. Gillett

Herbs annual, 1–9 dm. **Stems** simple or with branches or peduncles arising from nodes distinctly above base. **Leaves:** basal usually persistent at flowering, blades spatulate to oblanceolate, 0.8–8 cm × 2–20 mm, apex rounded or obtuse; cauline blades elliptic to narrowly lanceolate or linear, 2–9 cm × 3–23 mm, apex obtuse (proximal leaves) to acute. **Peduncles** 6–35 cm. **Flowers** 1–many; calyx 18–55 mm, keels minutely granular-scabridulous, all or at least outer lobes 1.5–2.5 times as long as tube or longer, lobes nearly equal in shape and size, ovate-lanceolate, apex of outer lobes acuminate to attenuate, apex of inner lobes acuminate; corolla proximally white with violet-blue to violet veins, distally violet-blue to violet, 40–80 mm, lobes proximally oblong, expanding into rhombic distal portion, that is, with apex ± acute, 18–53 × 11–25 mm, margins with fringes to 5 mm on the sides, dentate around apex; ovary distinctly stipitate. **Seeds** papillate, not winged.

Flowering fall. Open woods, roadsides; 1500–2800 m; Ariz.; Mexico (Chihuahua, Durango, Sonora, Zacatecas).

In the flora area, *Gentianopsis macrantha* is known only from extreme southern Arizona. Specimens from New Mexico have been reidentified as *G. thermalis*.

Gentianopsis macrantha is distinguished from large-flowered plants of *G. thermalis* by its combination of stems that are simple or branched only well above the base; all four calyx lobes distinctly longer than the tube, usually more than 1.5 times as long; a more gradually flaring corolla tube, which is obconic rather than more nearly cylindric; rhombic distal portions of the corolla lobes, which widen above the base and taper from below the middle to a nearly acute apex, being shaped somewhat like the blade of a traditional mason's trowel rather than being ovate to orbiculate; and the pale portion of the corolla that extends to the proximal part of the lobes, in which the veins are sharply outlined in deep violet-blue or violet, whereas in *G. thermalis* the corolla lobes and upper part of the tube are usually deep blue throughout. A few specimens of *G. thermalis*, mostly from Arizona and New Mexico, are similar to *G. macrantha* in flower size or have calyx lobes that are relatively long in proportion to the tube and might be considered intermediate between these species. In occasional specimens of *G. thermalis*, mostly from Utah and Arizona, the pale zone of the corolla extends into the proximal portion of the lobes as in *G. macrantha*. Such specimens have sometimes been identified as *G. macrantha* but can generally be distinguished by the branching pattern and/or the corolla-lobe shape.

J. M. Gillett (1957) has suggested that the name *Gentiana macrocalix* Lexarza may have been applied to *Gentianopsis macrantha*, in which case, at species rank, the epithet *macrocalix* would have priority over *macrantha*. No type specimen of the name *Gentiana macrocalix* is known, but from studies for this flora, this name is assumed to have been applied to a Mexican taxon not occurring in the flora area.

6. **Gentianopsis holopetala** (A. Gray) H. H. Iltis, Sida 2: 136. 1965 • Sierra or tufted gentian [E]

Gentiana serrata Gunnerus var. *holopetala* A. Gray in W. H. Brewer et al., Bot. California 1: 481. 1876; *G. holopetala* (A. Gray) Holm; *Gentianella detonsa* (Rottbøll) G. Don subsp. *holopetala* (A. Gray) J. M. Gillett

Herbs annual, 0.3–3(–4.5) dm. **Stems** simple or those of larger plants with branches or peduncles arising from base. **Leaves:** basal persistent, blades spatulate to oblanceolate, 2–10 cm × 3–15 mm, apex rounded to acute; cauline blades lanceolate-elliptic to linear, 1.5–6 cm × 3–10 mm, apex acute. **Peduncles** 2–22 cm, generally much exceeding stem. **Flowers** solitary; calyx 14–36 mm, keels smooth, all or at least inner lobes less than 1.5 times as long as tube, outer lobe lanceolate to narrowly ovate, apices acute to acuminate, inner lobes lanceolate to ovate, apices acute to acuminate; corolla deep blue, 20–55 mm, lobes ovate, 8–15 × 3–9 mm, margins entire or distally erose-undulate, apex obtuse to subacute; ovary long-stipitate. **Seeds** papillate, not winged. $2n = 78$.

Flowering summer–fall. Wet subalpine meadows; 1800–4000 m; Calif., Nev.

Gentianopsis holopetala is endemic to the central and southern High Sierra Nevada and San Bernardino Mountains in California and the Inyo and White mountains of California and adjacent Nevada.

Gentianopsis holopetala has been said to comprise two cryptic species, in that the populations in the Inyo and White mountains differ genetically from those in the Sierra Nevada to an extent not paralleled in their morphology (B. A. Whitlock, pers. comm.). Misidentifications of *G. barbellata* or *G. simplex* as *G. holopetala* appear to have been the basis for statements that this species perennates by means of shoots arising from the roots.

Gentianopsis holopetala and *G. simplex* have often been confused. The calyx of *G. simplex* is completely green or sometimes slightly suffused with purple, without the sharp purple outlining of the midribs that is common in both *G. holopetala* and *G. thermalis*. The peduncles of *G. holopetala* are usually more than 2.5 times as long as the subtending internodes, regardless of plant size, whereas those of *G. simplex*, the only other *Gentianopsis* species in California, are usually less than 2.5 times as long. The anthers of *G. holopetala* are 2.2–3.7 mm, and those of *G. simplex* are 1–1.5 mm (N. H. Holmgren 1984b). When seeds are available, the differences in their morphology also distinguish small plants of *G. holopetala* from *G. simplex*. The corolla lobes of *G. barbellata* are usually more than 2 times as long as wide, and those of *G. holopetala* are usually less than or about 2 times as long as wide, although both species are somewhat variable in this respect.

7. **Gentianopsis barbellata** (Engelmann) H. H. Iltis, Sida 2: 136. 1965 • Perennial fringed or fragrant fringed or windmill gentian [E]

Gentiana barbellata Engelmann, Trans. Acad. Sci. St. Louis 2: 216, plate 11, figs. 1–6. 1863; *Gentianella barbellata* (Engelmann) J. M. Gillett

Herbs perennial, 0.5–1.5 dm. **Stems** arising from buds on roots, forming clonal colonies; individual stems indeterminate, forming an erect caudex from which 1+ lateral flowering branches arise each year, generally unbranched. **Leaves:** basal persistent, blades narrowly spatulate to oblanceolate, 1.5–10 cm × 3–14 mm, apex rounded or obtuse; cauline blades (when present) oblanceolate (proximal) to linear (distal), 1–6 cm × 1–10 mm, apex obtuse to acute. **Peduncles** usually absent, occasionally to 0.7 cm. **Flowers** usually 1 per primary stem, occasionally paired; calyx 11–25 mm, keels smooth, outer lobes lanceolate, apices acuminate, inner lobes lanceolate to narrowly ovate, apices acute to short-acuminate; corolla bright to deep blue, 25–45(–50) mm, lobes ovate-oblong, 15–25 × 2–6 mm, margins proximally with fringes 1–2 mm, distally dentate; ovary distinctly stipitate. **Seeds** papillate, not winged.

Flowering late summer–fall. Rocky slopes, open coniferous and aspen woods, wet alpine meadows, calcareous soils; 1000–3800 m; Ariz., Colo., N.Mex., Utah, Wyo.

The perennating habit of *Gentianopsis barbellata* is unique within the genus in the flora area. Stems arise from root buds, forming clonal colonies. These stems are perennial and indeterminate. They generally persist as vegetative rosettes for a few years, then each year thereafter produce a flowering branch from an axillary bud. Especially robust plants may produce more than one flowering branch per year (G. Engelmann 1879; P. A. Groff 1989).

The flowers of *Gentianopsis barbellata* appear to have longer peduncles than they actually do when they are closely subtended by narrow bracts that may be confused with the calyx lobes. Although a few minute trichomes are sometimes found on the corolla tube near the level of stamen insertion in other species of *Gentianopsis* in the flora area, these are more numerous and more prominent in *G. barbellata*, forming distinct rows flanking the adnate portions of the stamens, as in the European *G. ciliata* (Linnaeus) Ma (G. Engelmann 1863; J. M. Gillett 1957).

8. **Gentianopsis simplex** (A. Gray) H. H. Iltis, Sida 2: 136. 1965 • Hikers' or one-flowered gentian [E]

Gentiana simplex A. Gray in War Department [U.S.], Pacif. Railr. Rep. 6(3): 87, plate 16. 1857; *Gentianella simplex* (A. Gray) J. M. Gillett

Herbs with individual stems biennial, but perennating with stems arising from buds on spreading roots, forming clonal colonies, 0.4–4 dm. **Stems** unbranched. **Leaves:** basal usually withered by flowering, blades spatulate, 0.5–1.5 cm × 1–6 mm, apex rounded to obtuse; cauline blades elliptic to ovate or lanceolate, 5–22 cm × 1–7 mm, apex obtuse to acute. **Peduncles** 1.3–5(–11) cm. **Flowers** 1 per plant; calyx 9–27 mm, keels smooth, all lobes lanceolate, apices acute to short-acuminate; corolla deep blue, (16–)28–35(–45) mm, lobes oblong, (5–)9–15(–22) × 3–6 mm, margins proximally entire to erose or with few fringes to 1.5 mm, distally shallowly erose to prominently dentate, apex rounded to obtuse; ovary distinctly stipitate. **Seeds** striato-reticulate, winged, especially at ends.

Flowering late summer–fall. Wet mountain meadows, bogs; 1200–3400 m; Calif., Idaho, Mont., Nev., Oreg.

Reports of *Gentianopsis simplex* from Yellowstone National Park, Wyoming, have been based on small plants of *G. thermalis*. When plants from localities outside the range given above are thought to be *G. simplex*, attempts should be made to obtain seeds, as the seed coats of this species are distinctively different from those of all other North American *Gentianopsis*.

Gentianopsis simplex forms clonal patches as new flowering stems develop from root buds. In contrast to those of *G. barbellata*, these stems are determinate and biennial, with terminal flowers (P. A. Groff 1989). Consequently, if the underground parts are not seen, the plants appear to be monocarpic.

Gentianopsis simplex has relatively strongly petaloid, crisped stigma margins.

17. GENTIANA Linnaeus, Sp. Pl. 1: 227. 1753; Gen. Pl. ed. 5, 107. 1754 • Gentian [For Gentius, d. 168 BC, king of Illyria, discoverer of medicinal use of *G. lutea*]

James S. Pringle

Calathiana Delarbre; *Chondrophylla* A. Nelson; *Ciminalis* Adanson; *Dasystephana* Adanson; *Gentianodes* Á. Löve & D. Löve; *Pneumonanthe* Gleditsch

Herbs annual, biennial, or perennial, chlorophyllous, glabrous or stems and calyces puberulent; stems solitary or clustered from a single crown, arising at intervals from a horizontal rhizome only in *G. glauca*. **Leaves** cauline, opposite [whorled], sometimes also basal. **Inflorescences** terminal and sometimes axillary cymes, often condensed into heads, or flowers solitary; terminal heads or dense cymes usually subtended by 1–3 pairs of involucral leaves; in species of sect. *Pneumonanthe* (specified in discussion below) that generally have flowers in dense heads or cymes, the flowers are individually subtended by paired small bracts (some bracts may be absent when flowers are in dense, many-flowered heads), bracts absent in *G. autumnalis* and *G. pennelliana*. **Flowers** usually 5-merous, sometimes 4-merous (*G. prostrata*) [*G. cruciata*], rarely to 7-merous; calyx tube cylindric to narrowly campanulate or less often cleft and spathaceous (only in *G. platypetala*, occasionally in *G. affinis*), calyx tube (except *G. fremontii* and *G. prostrata*) extending above bases of the lobes, which thus appear to diverge below the summit, calyx lobes ascending to nearly erect (except *G. clausa*, *G. flavida*, and *G. latidens*, in which they spread nearly horizontally); corolla blue, greenish blue, violet, rose-violet, pale yellow, greenish yellow, or white [deeper yellow, orange, red], campanulate, funnelform, tubular, or salverform, glabrous within, lobes shorter than tube, margins entire or minutely erose-serrate, alternating with projecting or rarely truncate plicae (defined below), spurs absent; stamens inserted in proximal $^1/_2$ of corolla tube; anthers distinct or connate; ovary stipitate; style persistent, erect, short or not clearly differentiated from ovary; stigmas 2; nectaries as many as corolla lobes, on gynophore. **Capsules** ovoid-ellipsoid to cylindric (short-obovoid in *G. fremontii*), dorsiventrally compressed, stipitate. *x* = 6, 7, 9, [10], 13.

Species ca. 400 (28 in the flora): North America, Mexico, Central America, South America, Eurasia, nw Africa, Atlantic Islands (Iceland), sw Pacific Islands, Australia.

Gentiana nivalis is in sect. *Calathianae* Froelich. *Gentiana glauca* is in sect. *Kudoa* (Masamune) Sataka & Toyokuni. *Gentiana algida* is in sect. *Frigida* Kusnezow. Species 4 through 25, plus *G. septemfida* Pallas, are in sect. *Pneumonanthe* Gaudin. Species 26 through 28 are in sect. *Chondrophyllae* Bunge. *Gentiana lutea* Linnaeus is in sect. *Gentiana*, and *G. cruciata* Linnaeus is in sect. *Cruciata* Gaudin (A. Favre et al. 2016).

In most of the perennial species, flowering stems arise directly from the caudex, with the basal and near-basal leaves reduced and scalelike, the cauline leaves gradually increasing in size distally. *Gentiana algida*, *G. newberryi*, and *G. setigera* non-flowering stems, with the leaves generally so closely spaced as to form rosettes, arise from the caudex, and flowering stems, the following year, although they may appear to arise from the caudex, arise from axillary buds of the previous year's rosettes. The proximal cauline leaves, rather than being reduced, are as large as or larger than the distal leaves.

Except in *Gentiana lutea*, the primary components of the corolla tube in *Gentiana* alternate with distally widening, and in most sections, more or less infolded components of similar texture, which are vascularized only by branches from the lateral veins of the adjacent petals on each side. In most species, these extend into free portions, which alternate with the corolla lobes. These components of the corolla, historically called appendages, are now generally called plicae. (In *G. nivalis,* the plicae may be infolded so that the free portions appear superficially to adaxially at the base of the lobes rather than between them.) In nearly all of the species in the flora area, the sinuses on each side of each plica are equally deep or nearly so. Only in *G. flavida*, *G. linearis*, and *G. villosa* are these sinuses distinctly unequal, the sinus nearer the apex of the obliquely triangular free portions of the plicae being conspicuously less deep than the other. The free portions of the plicae are generally more or less erect, except that in *G. decora* the smaller of the two divisions, to the left as viewed adaxially, is often inflexed. The form of the free portions of the plicae is very useful in distinguishing among some of the species in the flora area. Herbarium specimens should be prepared so that these floral parts are visible, preferably by including a corolla slit lengthwise and unrolled.

In most of the species of sect. *Pneumonanthe* in the flora area, the exceptions being *Gentiana villosa* in the East and *G. calycosa* and *G. parryi* in the West, the distal cell-wall corners of the marginal leaf cells extend outward into cilia (seen at 10×). In the exceptions, the cell-wall corners extend outward only far enough to appear as minute teeth.

In all species except those in sect. *Chondrophyllae*, the calyx tube extends above the bases of the lobes, so that the lobes appear to diverge from the abaxial surface slightly below the truncate summit, with gaps between the bases of the lobes. This intracalycular membrane is interpreted as an invagination of the adaxial surface. In sect. *Chondrophyllae*, the calyx lobes diverge at the summit of the tube, with only V- or U-shaped sinuses between them, although in *Gentiana fremontii*, in sect. *Chondrophyllae*, the longitudinal infolding of the hyaline portions of the tube below the sinuses may give the appearance of an intracalycular membrane.

In *Gentiana*, as in some other Gentianaceae, some species have corollas that open only in sunshine; in some other *Gentiana* species, the corollas remain essentially closed. In species with corollas that open, the portions of the corolla lobes exposed in bud and the tube below each lobe are often suffused abaxially with green and/or purple. In species 4 through 24, relatively dark continuous or fragmented stripes commonly extend downward adaxially between the central and lateral veins of each petal into the paler proximal portion of the tube, and their presence does not serve to distinguish among species. The gynophores elongate as the fruits mature, variably so even within a species. Elongation is often especially prominent in *Gentiana glauca*, *G. newberryi*, and *G. prostrata*, elevating all or much of the capsule above the persistent corolla.

In some species in sect. *Pneumonanthe*, the distal corners of the marginal cell walls of the leaves and calyx lobes are prolonged to the extent that they appear as short, firm cilia when seen at 10×; in other species they are shorter, merely forming minute teeth. The presence or absence of such cilia is useful in identifying small or otherwise exceptional specimens of some species, especially in the Coast Ranges and the Rocky Mountains.

In those species in sect. *Pneumonanthe* in which most or all of the flowers are clustered in a terminal head, this terminal inflorescence is directly subtended by leaves, usually in two or three densely spaced pairs, that are about as large as the distal cauline leaves and of the same color, texture, and, in most species, shape. These are called "involucral leaves" here. The term "bracts" is used here for much smaller structures, mostly less than 2 cm, that directly subtend the individual flowers. These bracts are usually paired, but in large inflorescences, their development may be somewhat irregular.

Species in sect. *Pneumonanthe* are extensively interfertile, with at least 18 interspecific hybrid combinations known to have occurred in nature (J. S. Pringle 1967; studies for this flora). These species, all of which have $2n = 13$, are relatively large-flowered and are pollinated primarily by bumblebees. The hybrids, which are fertile, represent occasional events involving otherwise well-differentiated species that are usually more or less isolated geographically and/or ecologically and in some cases are highly dissimilar in floral morphology. Hybrids are most frequent among the prairie species *Gentiana andrewsii*, *G. flavida*, and *G. puberulenta*.

Gentiana lutea, great yellow gentian, native to Europe, is adventive in Alberta. Plants of this species are 0.5–2 m tall, with solitary, stout, unbranched stems and widely ovate to elliptic leaves to 30 cm. The flowers are borne in dense, terminal and axillary clusters and have bright yellow, deeply cleft corollas 10–25 mm, with five (to nine) spreading, linear lobes. As noted above, *G. lutea* lacks the plicae between the corolla lobes that are otherwise characteristic of the genus. *Gentiana cruciata*, crosswort or star gentian, also native to Europe, has escaped from cultivation in Manitoba and Massachusetts. It has a basal rosette of strap-shaped leaves, decumbent stems with numerous ovate-elliptic leaves, and flowers in small axillary and terminal clusters. The mostly four-lobed corollas are bright blue, 20–25 mm, with spreading lobes alternating with small, nearly symmetrically triangular or few-toothed plicae. *Gentiana septemfida* in the narrow sense, crested gentian or summer gentian, native to the Caucasus Mountains of Eurasia, has escaped in Ontario, Illinois, and Wisconsin. It resembles *G. plurisetosa* but has more numerous, lance-ovate leaves and usually more flowers per stem, in a terminal cluster. The corollas are deep blue, 30–50 mm, with spreading lobes and prominent, finely divided free portions of the plicae.

In addition to those mentioned above, some other European and Asiatic species of *Gentiana* are grown as ornamentals in North American gardens, although none has become really common in cultivation. Several species in sect. *Ciminalis* (Adanson) Dumortier and sect. *Kudoa* are cultivated in alpine gardens in the cooler parts of the flora area, and other species, mostly in sect. *Cruciata* and sect. *Pneumonanthe*, are cultivated in perennial borders.

SELECTED REFERENCES Favre, A. et al. 2020. Phylogenetic relationships and sectional delineation within *Gentiana* (Gentianaceae). Taxon 69: 1221–1238. Halda, J. J. 1996. The Genus *Gentiana*. Dobré. Ho, T. N. and Liu S. W. 2001. A Worldwide Monograph of *Gentiana*. Beijing and New York. Ho, T. N., Liu S. W., and Lu X. F. 1996. A phylogenetic analysis of *Gentiana* (Gentianaceae). Acta Phytotax. Sin. 34: 505–530. Köhlein, F. 1991. Gentians (English translation by D. Winstanley). Portland. Pringle, J. S. 1967. Taxonomy and distribution of *Gentiana*, section *Pnumonanthae* [sic] in eastern North America. Brittonia 19: 1–32.

1. Plants 0.1–2.5(–3) dm; corollas 7–25 mm; subarctic to arctic and/or subalpine to alpine regions.
   2. Flowers solitary or in dense terminal heads of 2–5(–7); calyx tube extending adaxially above base of lobes.
      3. Annuals, with a single flowering stem; rhizomes absent; corollas salverform, with spreading lobes, deep blue or rarely white . . . . 1. *Gentiana nivalis*
      3. Perennials, with flowering stems arising at intervals from horizontal rhizomes; corollas tubular, with ascending lobes, greenish blue, greenish yellow, or rarely white. . . . . 2. *Gentiana glauca*
   2. Flowers solitary on an unbranched stem or at ends of branches, in the latter case the inflorescence forming a diffuse cyme; calyx tube not extending above base of lobes.
      4. Flowers in a diffuse cyme; corollas predominantly white, adaxially with purple dots; free portions of corolla plicae divided to base into 2 equal segments . . . . 28. *Gentiana douglasiana*
      4. Flowers solitary; corollas white or blue, without adaxial dots; free portions of corolla plicae triangular, merely erose or at most shallowly notched.
         5. Leaf blades obscurely or not white-margined; all cauline leaves ± spreading; corollas adaxially medium to deep blue or occasionally white; capsules compressed-cylindric . . . . 26. *Gentiana prostrata*
         5. Leaf blades conspicuously white-margined; distal cauline leaves strongly ascending; corollas adaxially white or pale (rarely deeper) blue; capsules short-obovoid . . . . 27. *Gentiana fremontii*
1. Plants (0.1–)0.4–12 dm; corollas (12–)18–65 mm (if plants less than 2.5 dm, corollas more than 23 mm); species widely distributed.
   6. Calyx tubes deeply cleft into 2 spathaceous segments, calyx lobes minute; corolla lobes 6–11 mm; free portions of corolla plicae low-triangular, less than 1 mm, merely notched, otherwise entire . . . . 11. *Gentiana platypetala*
   6. Calyx tubes not deeply cleft, with well-developed lobes, or occasionally cleft with some lobes much reduced in *G. affinis*, in which the free portions of the corolla plicae are larger and laciniate.
      7. Flowers usually solitary on unbranched stems, occasionally on stems with 1 or 2 branches each bearing a solitary flower; peduncles (0.3–)1–8 cm, with neither involucral leaves nor bracts at base of flowers; corollas 30–65 mm, open, with spreading lobes; corolla lobes 10–25 mm; species of Atlantic and Gulf coastal plains.
         8. Corollas usually blue, rarely rose-violet or white; Atlantic Coastal Plain, New Jersey to South Carolina. . . . . 4. *Gentiana autumnalis*
         8. Corollas white, abaxially suffused with purplish green on and below lobes; nw Florida . . . . 5. *Gentiana pennelliana*
      7. Flowers usually in terminal heads or compact cymes of 2–25, rarely solitary, sometimes also in distal axils of primary stem or on short lateral branches, sessile above paired bracts; corollas (12–)18–60 mm, opening or remaining closed; corolla lobes 0.5–17 mm; species widely distributed, in diverse habitats.
         9. Non-flowering rosettes of leaves present at flowering time; flowering stems arising laterally below summit of a rosette, and/or basal and near-basal leaves of flowering stems, densely spaced, persistent; at least some rosette or basal leaves 2–14 cm, larger than distal leaves.
            10. Free portions of corolla plicae divided nearly to base into 2 or 3 threadlike parts . . . . 7. *Gentiana setigera*
            10. Free portions of corolla plicae less deeply or finely divided, segments threadlike only toward apex if at all.

11. Corolla lobes ascending; plicae erose, not cleft into triangular segments; rosette and basal leaf blades narrowly spatulate to oblanceolate, 6+ times as long as wide . . . . 3. *Gentiana algida*
11. Corolla lobes spreading; plicae ± bifid, with 2 triangular, serrate to lacerate segments; rosette and basal leaf blades widely spatulate to obovate or oblanceolate, less than 6 times as long as wide . . . . . . 6. *Gentiana newberryi*

[9. Shifted to left margin.—Eds.]

9. Non-flowering rosettes absent at flowering time, flowering stems arising directly from caudex; basal and lowest cauline leaves reduced, ± scalelike.
   12. Corollas remaining tightly closed; corolla lobes reduced to a mucro or 0.5–2(–3) mm, distinctly shorter than plicae . . . . 21. *Gentiana andrewsii*
   12. Corollas remaining closed or opening; corolla lobes 0.7–15 mm, ± as long as or longer than plicae.
      13. Corolla plicae truncate and entire at the summit, not or scarcely extending above base of lobes . . . . 12. *Gentiana sceptrum*
      13. Corolla plicae triangular to ± oblong, bifid, toothed, or fimbriate at the summit, with free portions extending above base of lobes.
         14. Free portions of corolla plicae divided nearly to base into a fringe of 5–20 threadlike parts . . . . 8. *Gentiana plurisetosa*
         14. Free portions of corolla plicae not or less deeply or finely divided, segments threadlike only toward apex if at all.
            15. Free portions of corolla plicae obliquely triangular, that is, confluent with the adjacent lobe higher on the left side when the adaxial surface is viewed, wider than long, sometimes with a few low teeth but not cleft or lacerate.
               16. Distal leaf blades linear to lanceolate, generally less than 15 mm wide and 6+ times as long as wide . . . . 22. *Gentiana linearis*
               16. Distal leaf blades lanceolate to ovate, elliptic, obovate, or spatulate, 15+ mm wide and/or less than 6 times as long as wide.
                  17. Calyx lobes lanceolate to ovate-triangular, spreading; corollas white, sometimes tinged pale yellow . . . . 24. *Gentiana flavida*
                  17. Calyx lobes linear to oblong-lanceolate or oblanceolate, ascending; corollas usually violet-blue, violet, or greenish white ± suffused with violet, occasionally rose-violet or white.
                     18. Involucral leaves ascending and proximally conduplicate, ± enveloping base of inflorescence; proximal and mid-stem leaf blades linear to oblong or ovate. . . . . 23. *Gentiana rubricaulis*
                     18. Involucral leaves spreading, not enveloping base of inflorescence; proximal and often mid-stem leaf blades elliptic or obovate to spatulate . . . . 25. *Gentiana villosa*
            15. Free portions of corolla plicae shallowly to deeply bifid, with the summit distinctly erose, toothed, or lacerate, not oblique, that is, with the sinuses on each side of a plica not distinctly unequal.
               19. Leaf blades widely ovate to orbiculate, mostly less than 2 times as long as wide; marginal leaf and calyx-lobe cell walls forming minute teeth (seen at 10×) but not extending into cilia; corollas open; seeds not winged.
                  20. Involucral leaves similar to or smaller than distal cauline leaves, ± spreading from the base, not enveloping bases of inflorescences . . . . 9. *Gentiana calycosa*
                  20. Involucral leaves wider than distal cauline leaves, proximally conduplicate, ascending and enveloping bases of inflorescences . . . . 10. *Gentiana parryi*
               19. Leaf blades linear, oblong-lanceolate, or lanceolate to elliptic or ovate, generally 2+ times as long as wide (rarely proportionately shorter in *G. catesbaei*); marginal leaf and calyx-lobe cell walls extending into cilia (seen at 10×); corollas opening or remaining closed; seeds winged.

[21. Shifted to left margin.—Eds.]

21. Corollas tubular-funnelform to narrowly campanulate, opening with spreading or reflexed lobes.
    22. Corollas (12–)18–40(–45) mm, lobes 3–7(–10) mm, less than 2 times as long as free portions of plicae . . . . 13. *Gentiana affinis*
    22. Corollas (30–)35–60 mm, lobes 6–15 mm, 2+ times as long as free portions of plicae . . . . 14. *Gentiana puberulenta*
21. Corollas tubular, opening to various degrees or remaining closed.
    23. Corolla lobes 0.7–3 mm, ± as long as plicae, if lobes 2+ mm (*G. austromontana),* then lobes ca. 1/2 as wide as free portions of plicae; corollas remaining completely closed.
        24. Corolla lobes ca. as wide as free portions of plicae; calyx lobes widely elliptic to obovate or orbiculate; stems and calyx tubes glabrous . . . . 18. *Gentiana clausa*
        24. Corolla lobes ca. 1/2 as wide as free portions of plicae; calyx lobes lanceolate to narrowly ovate or elliptic; stems and calyx tubes puberulent . . . . 20. *Gentiana austromontana*
    23. Corolla lobes 2.5–10 mm, in most species exceeding plicae, if lobes less than 3 mm or ± as long as plicae (*G. latidens*), then lobes ca. as wide as free portions of plicae; corollas remaining loosely closed or slightly to widely opening.
        25. Calyx lobes spreading nearly horizontally when fresh, mostly ovate, obovate, orbiculate, or rhombic; corolla plicae ± as long as lobes; mountains of w North Carolina . . . . 19. *Gentiana latidens*
        25. Calyx lobes ± erect, linear-subulate, lanceolate, elliptic, or oblanceolate; corolla plicae shorter than lobes; more widely distributed in e, c United States.
            26. Calyx lobes subulate to linear or rarely oblanceolate; calyx tubes and stems puberulent; corolla plicae unequally bifid at summit, narrower segment usually deflexed . . . . 17. *Gentiana decora*
            26. Calyx lobes narrowly elliptic to oblanceolate; calyx tubes glabrous, stems glabrous or occasionally ± puberulent; corolla plicae subequally bifid at summit, both segments erect.
                27. Leaf blades linear to elliptic, widest near middle; calyx lobes shorter than or ± as long as tube; corolla lobes usually less than 2 mm longer than plicae; widely distributed in e, c United States . . . . 15. *Gentiana saponaria*
                27. Leaf blades usually ovate, widest near base; calyx lobes mostly longer than tube; corolla lobes usually 2–4 mm longer than plicae; Atlantic Coastal Plain . . . . 16. *Gentiana catesbaei*

1. Gentiana nivalis Linnaeus, Sp. Pl. 1: 229. 1753
• Snow gentian, gentiane des neiges, sne-ensian

Herbs annual, 0.1–1.5(–3) dm, glabrous, not rhizomatous. **Stems** 1–10, ascending to erect, simple or few-branched. **Leaves** basal and cauline; cauline ± evenly spaced; basal blades elliptic to obovate, 0.5–1.2 cm × 2–8 mm, apex obtuse; cauline blades ovate to lanceolate, 0.3–1.2 cm × 2–6 mm, apex obtuse to acute. **Inflorescences** solitary flowers at ends of main axis and branches and in axils, branches of largest plants thus forming cymes. **Flowers:** calyx (8–)11–16 mm, lobes linear-lanceolate, 2–6 mm, margins not ciliate; corolla deep blue or occasionally white, salverform, open, (11–)17–25 mm, lobes spreading, ovate-elliptic, 3–6 mm, plicae small, ± triangular, distally bifid; anthers distinct. **Seeds** not winged. **2*n*** = 14 (Iceland, Europe).

Flowering summer. Meadows, exposed, rocky slopes; 0–700 m; Greenland; Nfld. and Labr. (Labr.), Que.; Eurasia; Atlantic Islands (Iceland).

Gentiana nivalis is found in arctic and alpine regions of Europe and western Asia.

2. Gentiana glauca Pallas, Fl. Ross. 1(2): 104, plate 93, fig. 2. 1789 • Glaucous or blue-green or inky gentian [F]

Gentiana glauca var. *paulensis* Kellogg

Herbs perennial, 0.2–1.6 dm, glabrous. **Stems** erect, arising singly at intervals from slender, elongating, horizontal rhizomes, forming patches. **Leaves** basal and cauline; cauline ± abruptly more widely spaced; blade elliptic to spatulate-obovate, apex obtuse; basal and rosette leaf blades 0.8–2(–2.7) cm × 4–12 mm; cauline blades

0.5–1.7 cm × 3–8(–12) mm. **Inflorescences** reduced, ± dense cymes, 2–5(–7)-flowered, also often a pair at most of the distal node, rarely solitary flowers. **Flowers:** calyx 5–7 mm, lobes ascending, lanceolate-triangular, 2–4 mm, margins not ciliate; corolla greenish blue or greenish yellow, rarely white, tubular, opening narrowly, (8–)12–20 mm, lobes ascending, triangular, 1.8–4 mm, free portions of plicae rounded, minutely erose; anthers distinct. **Seeds** winged. $2n$ = 24.

Flowering summer. Arctic and alpine tundra slopes and meadows; 0–2500 m; Alta., B.C., N.W.T., Yukon; Alaska, Mont., Wash., restricted to high elevations southward; Asia (n Japan, Kamchatka, coastal Siberia).

The description by Kellogg of var. *paulensis* does not indicate clearly how this variety was believed to differ from *Gentiana glauca* elsewhere in its range. Specimens from the Pribilof Islands examined in studies for this flora, including an isotype and other plants from Saint Paul Island, do not appear to be taxonomically separable.

3. **Gentiana algida** Pallas, Fl. Ross. 1(2): 107, plate 95. 1789 • Arctic or whitish gentian [F]

Gentiana romanzovii Ledebour ex Bunge; *Gentianodes algida* (Pallas) Á. Löve & D. Löve

Herbs perennial, 0.4–2 dm, glabrous. **Stems** 1–few, arising laterally below rosettes or seemingly from center of rosettes because of large basal leaves, decumbent to erect. **Leaves** basal and cauline; blades of basal and rosette leaves narrowly spatulate, 2–10(–14) cm × 1–7 mm, transitional to cauline leaves with blades oblanceolate to lanceolate or linear, 2–5 cm × 2–5 mm, blades of all leaves 6+ times as long as wide, apices obtuse to acute. **Inflorescences** 1–3-flowered. **Flowers:** calyx 15–28 mm, lobes linear to lanceolate, (5–)7–12 mm, margins not ciliate; corolla abaxially suffused with blue-violet on and below lobes, adaxially white or yellowish white with dark purple and/or greenish spots and streaks, funnelform, open, 35–50 mm, lobes ascending to spreading, ovate-triangular, 2–5 mm, summit of plicae nearly truncate, erose; anthers distinct. **Seeds** winged. $2n$ = 24.

Flowering summer–early fall. Arctic tundra, alpine meadows; 0–4000 m; Yukon; Alaska, Colo., Mont., N.Mex., Utah, Wyo., restricted to high elevations south of the Arctic; Asia.

Gentiana algida of North America and Asia is sometimes considered conspecific with *G. frigida* Haenke of Europe. Early reports of differing chromosome numbers have been superseded by more recent counts indicating that $2n$ = 24 prevails in both species. If these taxa are united, the North American plants can be called *G. algida* var. *algida*. This designation is also applicable if varieties are recognized among the Japanese or other Asiatic representatives of this species.

4. **Gentiana autumnalis** Linnaeus, Cat. Edwards's Nat. Hist., 11. 1776 • Pine-barren gentian [E] [F]

Dasystephana porphyrio (J. F. Gmelin) Small; *Gentiana porphyrio* J. F. Gmelin; *G. stoneana* Fernald

Herbs perennial, 1.5–5.5 dm, glabrous. **Stems** 1(–3), terminal from caudex, decumbent to erect. **Leaves** cauline, gradually more distantly spaced distally; blade linear to narrowly oblanceolate, 2–10 cm × 0.5–5 mm, apex obtuse (proximal leaves) to acute. **Inflorescences** solitary flowers, occasionally also terminating 1 or 2 branches, not subtended by bracts. **Flowers:** calyx 17–40(–53) mm, lobes linear, 10–25(–36) mm, margins not ciliate; corolla deep blue with greenish yellow dots adaxially on lobes or occasionally rose-violet or white, funnelform, open, 30–65 mm, lobes spreading, widely ovate, 10–20 mm, free portions of plicae shallowly to deeply divided into 2 subequal, lacerate, attenuate segments; anthers distinct. **Seeds** winged. $2n$ = 26.

Flowering fall–early winter (southward). Moist meadows, pine barrens; 0–100 m; N.J., N.C., S.C., Va.

Gentiana autumnalis is believed to have been extirpated long ago from Delaware and Maryland.

Variation in the number of floral parts, from four to seven per whorl, is more frequent in *Gentiana autumnalis* than in the other species of *Gentiana* in the flora area.

There is a record of a hybrid of *Gentiana autumnalis* with the highly dissimilar *G. villosa* in North Carolina.

5. **Gentiana pennelliana** Fernald, Rhodora 42: 198. 1940 • Wiregrass gentian [C] [E]

Diploma tenuifolia Rafinesque, Fl. Tellur. 3: 27. 1837, not *Gentiana tenuifolia* Petrie 1913; *Dasystephana tenuifolia* (Rafinesque) Pennell; *G. autumnalis* Linnaeus subsp. *pennelliana* (Fernald) Halda

Herbs perennial, 0.7–3.5 dm, glabrous. **Stem** 1, terminal from caudex, decumbent. **Leaves** all cauline, gradually more distantly spaced distally; blade linear to narrowly oblanceolate, 1–3.5 cm × 0.5–5 mm, apex obtuse (proximal leaves) to acute. **Inflorescences** solitary flowers. **Flowers:**

calyx 18–45 mm, lobes linear, 10–30 mm, margins not ciliate; corolla white with greenish purple lines abaxially on and below lobes, funnelform, open, 35–65 mm, lobes spreading, ovate, 15–25 mm, free portions of plicae deeply divided into 2 subequal, lacerate, attenuate segments; anthers distinct. **Seeds** winged.

Flowering fall–early spring. Moist, open pine woods; of conservation concern; 0–70 m; Fla.

Gentiana pennelliana is endemic to Bay, Calhoun, Franklin, Gadsden, Gulf, Leon, Liberty, Wakulla, and Walton counties in northern Florida. It usually grows in plant communities in which wiregrass, *Aristida stricta*, is a prominent component, hence the common name.

Gentiana pennelliana differs further from *G. autumnalis* in its longer, more gradually flaring corolla tube; the division of the lateral veins of the corolla near the base, so that each petal has five primary veins rather than three as in *G. autumnalis* and the other *Gentiana* species in the flora area; and stamens 7–12 mm above their insertion on the corolla tube, as contrasted with 13–30 mm in *G. autumnalis*.

6. Gentiana newberryi A. Gray, Proc. Amer. Acad. Arts 11: 84. 1876 • Newberry's or alpine gentian [E]

Herbs perennial, 0.1–1.5(–3.5) dm (below flowers), glabrous. **Stems** 1–5, arising laterally below rosettes, from a stout tap root, tufted, decumbent. **Leaves** basal and cauline; blades of basal rosette and proximal cauline leaf blades widely spatulate to obovate or oblanceolate, 0.8–5 cm × 2–25 mm, apex obtuse or mucronate, at least these leaves with blades less than 6 times as long as wide, distal cauline leaves few, with blades oblanceolate to lanceolate or linear, 2–5 cm × 2–5 mm, apices acute. **Inflorescences** terminal, flowers usually solitary, occasionally 2 or 3. **Flowers:** calyx 14–30 mm, lobes linear to narrowly ovate, (4–)6–12 mm, margins not ciliate; corolla white or blue, campanulate, open, 23–55 mm, lobes spreading, elliptic-obovate, 7–17 mm, free portions of plicae divided into 2 triangular, serrate to lacerate segments; anthers distinct. **Seeds** winged.

Varieties 2 (2 in the flora): w United States.

The two varieties of *Gentiana newberryi* intergrade extensively. The most distinctive form of var. *newberryi*, with relatively tall stems and medium to deep blue corollas, occurs in the northern part of the range of the species, from the Klamath and White mountains of California north into Oregon. Plants most clearly referable to var. *tiogana* prevail in the southern part of the range of the species, from Butte County south to Inyo and Tulare counties, California. In the central part of the range of the species, plant size and corolla color are less consistently correlated, with occasional plants combining low stature with deep blue corollas or tall stems with predominantly white or pale blue corollas. In that part of the range, corolla color may be highly variable within a single population.

The leaves of *Gentiana newberryi* are thick-textured and distinctively concave, usually spoon-shaped, when fresh. Narrower leaves sometimes occur in var. *tiogana*, but many plants of that variety have widely spatulate leaves like those of var. *newberryi*.

1. Corollas medium to deep blue with greenish to dark purple lines abaxially on and below lobes, usually 35–55 mm. . . . . 6a. *Gentiana newberryi* var. *newberryi*
1. Corollas white to pale blue except for greenish to dark purple lines abaxially on and below lobes, usually 23–42 mm. . . . . 6b. *Gentiana newberryi* var. *tiogana*

6a. Gentiana newberryi A. Gray var. **newberryi** [E]

Stems, leaves, and flowers usually in higher part of size ranges for species; corollas usually medium to deep blue with dorsal greenish to dark purple lines adaxially on and below the lobes, occasionally paler, usually 35–55 mm.

Flowering late summer–fall. Wet mountain meadows; 1200–2200 m; Calif., Oreg.

There is a record of a hybrid of var. *newberryi* with *Gentiana calycosa*.

6b. Gentiana newberryi A. Gray var. **tiogana** (A. Heller) J. S. Pringle, Sida 14: 186. 1990 • Tioga gentian [E]

Gentiana tiogana A. Heller, Leafl. W. Bot. 2: 221. 1940; *G. newberryi* subsp. *tiogana* (A. Heller) Halda

Stems, leaves, and flowers usually in lower to middle part of size ranges for species; corollas white to pale blue (sometimes drying deeper blue) except for greenish to dark purple lines abaxially on and below lobes, usually 23–42 mm. 2*n* = 26.

Flowering late summer–fall. Wet mountain meadows; 1500–4000 m; Calif., Nev.

7. Gentiana setigera A. Gray, Proc. Amer. Acad. Arts 11: 84. 1876 • Mendocino or elegant gentian [C] [E]

Gentiana bisetaea J. T. Howell

Herbs perennial, 2–4.5 dm, glabrous. **Stems** 1–12, arising laterally below rosettes, decumbent. **Leaves** basal and cauline; cauline leaves gradually more widely spaced distally; basal and rosette blades spatulate-obovate, 2.5–8.5 cm × 5–15 mm, apex obtuse; cauline blades elliptic, 1–3 cm × 5–17 mm, apex obtuse to acute. **Inflorescences** solitary flowers or 2–4-flowered heads. **Flowers:** calyx 14–23 mm, lobes ovate-oblong, 5–8 mm, margins not ciliate; corolla deep blue, campanulate, open, 25–50 mm, lobes elliptic-obovate, 10–16 mm, free portions of plicae divided nearly to base into 2 or 3 long, threadlike segments; anthers distinct. **Seeds** winged.

Flowering late summer–fall. Bogs and wet mountain meadows; of conservation concern; 300–1100 m; Calif., Oreg.

Gentiana setigera is endemic to Gasquet Mountain, Del Norte County, and Red Mountain, Mendocino County, California, and a small area in Josephine County, Oregon. At the Oregon site, it has been called *G. bisetaea* or Waldo gentian. Reports from other sites have been based on *G. plurisetosa*, with which *G. setigera* has often been confused.

K. L. Chambers and J. Greenleaf (1989) and C. T. Mason (1991) distinguished *Gentiana plurisetosa* from *G. setigera*, clarified the nomenclature of *G. setigera*, and included *G. bisetaea* in *G. setigera*. Prior to those studies, all components of this complex had been of conservation concern. With *G. plurisetosa* comprising only a part of this complex, and with *G. setigera* now being more narrowly circumscribed, conservation concern remains appropriate for both of these species.

8. Gentiana plurisetosa C. T. Mason, Madroño 37: 289, fig. 1. 1991 • Klamath or elegant gentian [C] [E]

Herbs perennial, 0.5–5 dm, glabrous. **Stems** 2–10, terminal from caudex, decumbent to erect. **Leaves** cauline, ± evenly spaced; blade elliptic to orbiculate, 1–6 cm × 7–38 mm, apex obtuse to acute. **Inflorescences** solitary flowers or 2–5-flowered heads, sometimes with additional flowers at 1–3 nodes. **Flowers:** calyx 17–35 mm, lobes lance-elliptic, 10–14(–20) mm, margins not ciliate; corolla deep blue, campanulate, open, 35–50 mm, lobes spreading, oblong-obovate to orbiculate, 7–14 mm, free portions of plicae divided nearly to base into 5–20 threadlike, often crisped segments; anthers distinct. **Seeds** winged.

Flowering late summer–fall. Wet mountain meadows; of conservation concern; 1200–2000 m; Calif., Oreg.

Gentiana plurisetosa is endemic to the Coast and Klamath ranges of northern California and southwestern Oregon. The name *G. setigera* has often been misapplied to this species.

Gentiana plurisetosa further differs from *G. setigera* in that the connate leaf bases at mid stem form a sheath around the stem 5 mm or longer. In all other *Gentiana* species in the flora area, the sheathing leaf bases are less than 5 mm.

9. Gentiana calycosa Grisebach in W. J. Hooker, Fl. Bor.-Amer. 2: 58, plate 146. 1837 • Explorers' or mountain bog or Mount Rainier or Rainier pleated gentian [E]

Gentiana calycosa subsp. *asepala* Maguire; *G. calycosa* var. *asepala* (Maguire) C. L. Hitchcock; *G. calycosa* var. *obtusiloba* (Rydberg) C. L. Hitchcock; *G. calycosa* var. *xantha* A. Nelson

Herbs perennial, 0.5–4.5 dm, glabrous. **Stems** 2–15, terminal from caudex, ± decumbent. **Leaves** cauline, ± evenly spaced; blade ovate to elliptic or orbiculate, 1–5 cm × 6–30 mm, margins glabrous, apex obtuse to acute. **Inflorescences** solitary flowers or 2–5-flowered heads, sometimes with additional flowers at 1–3 nodes. **Flowers:** calyx (5–)10–20 mm, tube uncleft, lobes lanceolate to widely ovate or elliptic, some or all occasionally ± foliaceous, 3–8(–10) mm, or tube deeply cleft and spathaceous, lobes reduced, linear, 0–3 mm, or some or all vestigial or absent, margins not ciliate; corolla deep blue, usually with greenish yellow dots adaxially on lobes, rarely violet or pale yellow, campanulate, open, 25–50 mm, lobes spreading, oblong-ovate or ovate-triangular to orbiculate, 5–13 mm, free portions of plicae divided less than 1/2 their length into 2 or 3 triangular segments threadlike only toward apex, rarely undivided; anthers distinct. **Seeds** not winged. $2n = 26$.

Flowering summer–fall. Wet mountain meadows, rocky slopes; 1000–3900 m; Alta., B.C.; Calif., Idaho, Mont., Nev., Oreg., Utah, Wash., Wyo.

In addition to the color forms noted in the description, a bicolored form of *Gentiana calycosa* in southwestern Washington has blue corollas with the center of the distal third of the corolla lobes, including the short-acuminate tip, white.

Gentiana calycosa varies greatly in stature and in the size of its leaves and flowers. Most plants of *G. calycosa* from the Pacific coastal region, including western Washington and Oregon and most of California, have tubular calyces with well-developed, ovate to elliptic lobes varying in size but usually 3–8 mm. In Canada such plants also prevail east to Alberta; southward, in the United States, they also prevail in Wyoming. From eastern Washington and Oregon to Idaho and parts of Montana, and in Nevada and Utah, the calyces are usually cleft and spathaceous, with the lobes much reduced, less than 3.5 mm, and linear, or vestigial or absent. Extreme forms of such plants have been called subsp. or var. *asepala*. Because of the sporadic rather than continuous distribution of such forms, the intergradation, especially in the vicinity of Mount Rainier, Washington, and in parts of Montana, where populations may include plants approaching both extremes along with intermediates, and the lack of correlation with other morphological variation, such plants are not distinguished taxonomically in this flora. To some extent their occurrence appears to be correlated with warmer, drier regions, and may be influenced by the habitat.

In most of its range, Gentiana calycosa grows in wet alpine meadows and similar moist habitats, but in the western Cascades it almost always grows in drier, north-facing sites on rocky slopes and cliffs. (T. Harvey, pers. comm.). Plants of *G. calycosa* in well-drained rocky slopes were distinguished as *G. saxicola* English (1934), not Grisebach (1838). According to C. S. English (1934), these plants differ from those of wet alpine meadows in having erect rather than decumbent stems; internodes mostly about as long as the leaves rather than distal internodes much longer than the leaves; and spreading rather than erect calyx lobes, which are larger and proportionately wider. According to Harvey, in contrast, the stems of the plants from the drier, rocky sites in the western Cascades are more likely to have decumbent stems. No consistent association of these morphological variations or correlation with geographic distribution was substantiated in the studies for this flora, but further study would be desirable, as molecular phylogenetic studies (A. Favre, pers. comm.) suggest that taxonomic recognition of these ecotypes might be appropriate.

There are records of hybrids of *Gentiana calycosa* with *G. affinis* var. *affinis* and *G. newberryi* var. *newberryi*.

10. Gentiana parryi Engelmann, Trans. Acad. Sci. St. Louis 2: 218, plate 10. 1863 • Parry's gentian [E]

Gentiana bracteosa Greene; *Pneumonanthe parryi* (Engelmann) Greene

Herbs perennial, 1–3.5(–4.5) dm, usually glabrous, occasionally minutely puberulent in lines on stems only. **Stems** 1–7(–14), terminal from caudex, decumbent to erect. **Leaves** cauline, ± evenly spaced; blade ovate, 1.5–4 cm × 8–21 mm, margins not ciliate, apex obtuse; involucral leaves wider than cauline, ascending and conduplicate, partially enveloping base of inflorescence. **Inflorescences** 2–7-flowered heads or occasionally solitary flowers, occasionally with additional flowers in 1 or 2 distal axils. **Flowers:** calyx 10–20(–27) mm, lobes linear to lanceolate, (1–)4–8 mm; corolla deep blue or violet-blue, campanulate, open, 33–50 mm, lobes spreading, obovate, 4–9 mm, free portions of plicae divided less than 1/2 their length into 2–5 triangular segments threadlike only toward apex; anthers distinct. **Seeds** not winged.

Flowering summer–fall. Mountain meadows; 1800–3900 m; Ariz., Colo., N.Mex., Utah, Wyo.

Reports of *Gentiana parryi* outside the range indicated here have been based on specimens of *G. calycosa* or *G. affinis* (studies for this flora). Because of its restriction to high altitudes, populations of *G. parryi* are widely scattered, especially in the southern part of its range.

Gentiana parryi has sometimes been included in *G. calycosa* and less often in *G. affinis*, but its larger, conduplicate involucral leaves, which are more or less sharply differentiated from the distal cauline leaves and largely envelop the calyces, give *G. parryi* an aspect distinctly different from that of either *G. calycosa* or *G. affinis*. Biometric studies by J. R. Spence (unpubl.) have supported its recognition as a species. As noted by N. H. Holmgren (1984b), these species also differ in anther length, which is 3.5–5 mm in *G. parryi* and 1.6–3.2 mm in *G. calycosa* and *G. affinis*. In the Rocky Mountains of central Colorado, *G. parryi* and *G. affinis* appear especially well differentiated. Where the ranges of *G. parryi* and *G. calycosa* approach each other in the Intermountain Region, *G. parryi* usually grows in drier habitats than *G. calycosa*. The distinctive involucre of *G. parryi* is less well developed in some Arizona plants otherwise identifiable as this species, which should be given further study.

11. Gentiana platypetala Grisebach in W. J. Hooker, Fl. Bor.-Amer. 2: 58. 1837 • Broad-petaled gentian [E]

Gentiana covillei A. Nelson & J. F. Macbride; *G. gormanii* Howell

Herbs perennial, 0.5–3.5 dm, glabrous. **Stems** 1–5, terminal from caudex, erect or nearly so. **Leaves** cauline, ± evenly spaced; blade widely ovate to elliptic, 1.5–4 cm × 8–22 mm, apex obtuse. **Inflorescences** solitary flowers or occasionally a terminal pair. **Flowers:** calyx 8–12 mm, tube cleft to base or nearly so into 2 spathaceous segments, lobes elliptic to ovate-lanceolate, 0.5–5 mm, margins not ciliate; corolla bright blue, campanulate, open, 30–38 mm, lobes widely ovate-triangular, 6–11 mm, free portions of plicae spreading, low-triangular, less than 1 mm, notched at apex, otherwise entire; anthers distinct. **Seeds** not winged.

Flowering late summer. Alpine and coastal mountain meadows, heathlands, rocky and boggy slopes; 0–1400 (–2100) m; B.C.; Alaska.

Gentiana platypetala is restricted to sites near the Pacific largely confined to the insular ranges of British Columbia and southern Alaska from northern Vancouver Island, Queen Charlotte Islands, and Alice Arm, British Columbia, northwest to Kodiak Island, Alaska, but occasionally on mainland coastal ranges.

The distinctive spathaceous calyces of this species are strongly suffused with reddish purple.

12. Gentiana sceptrum Grisebach in W. J. Hooker, Fl. Bor.-Amer. 2: 57, plate 145. 1837 • King's-scepter or staff or Pacific gentian [E]

Gentiana menziesii Grisebach; *G. sceptrum* var. *cascadensis* M. Peck; *G. sceptrum* var. *humilis* Engelmann ex A. Gray

Herbs perennial, 1–9 dm, glabrous. **Stems** 1–4(–10), terminal from caudex, decumbent to erect. **Leaves** cauline, gradually more widely spaced distally; blade lanceolate to ovate or elliptic, 1–9 cm × 5–15 (–20) mm, apex obtuse to acute. **Inflorescences** solitary flowers or 2–5-flowered heads or umbellate cymes, often also on peduncles to 8 cm or in cymules on branches from distal 1–4(–7) nodes. **Flowers:** calyx 13–27 mm, tube rarely deeply cleft, lobes lanceolate to elliptic-ovate or rarely foliaceous, 6–15 mm, margins not ciliate; corolla blue or rarely rose-violet or white, narrowly campanulate, opening only slightly with lobes ascending to incurved, 25–50 mm, lobes oblong-ovate to orbiculate, 5–10 mm, summit of plicae forming ± truncate, entire sinus, not or scarcely extending beyond bases of lobes; anthers distinct. **Seeds** winged toward ends only. $2n = 26$.

Flowering summer–fall. Bogs, wet meadows; 0–1300 m; B.C.; Calif., Oreg., Wash.

Plants of *Gentiana sceptrum* with ascending, narrowly lanceolate leaves conspicuously exceeded by the distal internodes have been called *G. menziesii*. Plants with spreading, elliptic to ovate, more closely spaced leaves have been called *G. sceptrum* var. *cascadensis*, although that variant might more appropriately be considered nomenclaturally typical. Studies for this flora have indicated that the extremes are connected by many intermediates, and that little correlation exists between leaf shape and geographic distribution, or between leaf shape and the other traits by which the segregates have been characterized, such as stature, erectness of stems, or numbers and sizes of flowers. Even within a relatively limited area, such as Vancouver Island, British Columbia, or Humboldt County, California, plants can be found with leaves ranging from narrowly lanceolate to widely elliptic.

Small plants with strongly decumbent stems, found at a few localities at or near the coast in California and southern Oregon, have been called *Gentiana sceptrum* var. *humilis*. Spathaceous calyces occasionally occur in these plants. Such plants are not recognized taxonomically here because larger plants approaching typical *G. sceptrum* have been found at the same localities or nearby, but they should be given further study.

13. Gentiana affinis Grisebach in W. J. Hooker, Fl. Bor.-Amer. 2: 56. 1837 • Rocky Mountain or oblong-leaved or marsh gentian [E]

Pneumonanthe affinis (Grisebach) Greene

Herbs perennial, 0.5–8 dm. **Stems** 1–10(–20), terminal from caudex, decumbent to erect, glabrous or puberulent in lines below leaf bases or more extensively. **Leaves** cauline, variably spaced; blade oblong or elliptic to ovate, lanceolate, or nearly linear, 1–5 cm × (2–)3–20(–25) mm, generally 2+ times as long as wide, margins ciliate, apex obtuse to acute. **Inflorescences** racemoid thyrses of ± dense 1–6-flowered cymules, terminating main stem and usually on short branches at distal 1–6(–12) nodes. **Flowers:** calyx 5–18(–23) mm, tube occasionally deeply cleft, lobes linear to narrowly elliptic-lanceolate or occasionally some rudimentary, (0–)1–13 mm, margins ciliate; corolla blue, sometimes with pale dots adaxially on lobes, or rarely pale violet or white, tubular-funnelform, open, (12–)18–40(–45) mm, lobes ± spreading, oblong-ovate, 3–7(–10) mm, free portions of plicae divided less than 1/2 their length into 2 ± triangular, lacerate segments; anthers distinct. **Seeds** winged.

Varieties 2 (2 in the flora): w, c North America.

Gentiana affinis is highly variable, and some authors have divided it into several species, or have recognized more varieties than the two accepted in this flora. N. H. Holmgren (1984b) speculated that further studies might disclose patterns that would warrant the recognition of additional infraspecific taxa in *G. affinis*. From studies for the present and earlier works, however, only two varieties appear to be well differentiated, and even these intergrade in Idaho, Montana, Oregon, and northern California, especially in Del Norte County, California. Despite the very different vegetative aspects of the extremes, which sometimes occur more or less sympatrically, intergradation is so extensive that the recognition of these taxa as distinct species seems inappropriate. The treatment of *G. affinis* presented here, although based on an independent approach involving the study of numerous specimens, is essentially in agreement with that of Holmgren and is supported by biometric analyses by J. R. Spence (unpubl.), which confirmed the intergradation between var. *affinis* and var. *ovata*.

Only var. *affinis* approaches the range of *Gentiana puberulenta*. Large plants of *G. affinis* generally differ from small plants of *G. puberulenta* in the proportionate lengths of the corolla lobes and the free portions of the plicae. In *G. affinis*, the lobes are less than two times as long as the free portions of the plicae; in *G. puberulenta* they are generally two or more times as long.

1. All leaf blades generally linear to lanceolate, narrowly oblong, or narrowly ovate; calyx lobes narrowly linear (occasionally some or all rudimentary); corollas (12–)18–30(–35) mm, lobes usually 3–5 mm. . . . . . . 13a. *Gentiana affinis* var. *affinis*
1. Proximal and mid-stem leaf blades elliptic to ovate; calyx lobes linear to elliptic-lanceolate (occasionally some rudimentary); corollas (12–)30–40(–45) mm, lobes usually 5–7 mm. . . . . .13b. *Gentiana affinis* var. *ovata*

13a. Gentiana affinis Grisebach in W. J. Hooker var. **affinis** E

Gentiana affinis var. *bigelovii* (A. Gray) Kusnezow; *G. affinis* var. *forwoodii* (A. Gray) Kusnezow; *G. affinis* var. *parvidentata* Kusnezow; *G. bigelovii* A. Gray; *G. forwoodii* A. Gray; *G. interrupta* Greene; *G. rusbyi* Greene ex Kusnezow; *Pneumonanthe bigelovii* (A. Gray) Greene

Stems 1–5 dm. **Leaves** ± evenly spaced or occasionally more closely spaced proximally; blades all generally linear to lanceolate, narrowly oblong, or occasionally narrowly ovate, usually 3–15 mm wide, occasionally linear, 2–3 mm wide. **Flowers:** calyx tube uncleft or occasionally ± deeply cleft, lobes narrowly linear or occasionally some or all rudimentary, (0–)1–7(–10) mm; corolla (12–)18–30(–35) mm, lobes 3–5(–7) mm. $2n = 26$.

Flowering late summer–fall. Mesic to wet meadows, prairies, river bars, open woods; 200–3900 m; Alta., B.C., Man., N.W.T., Sask.; Ariz., Colo., Idaho, Minn., Mont., Nev., N.Mex., N.Dak., Oreg., S.Dak., Tex., Utah, Wash., Wyo.

An old collection of var. *affinis* from Huron County, Ontario, presumably represents a chance introduction of short duration.

Variety *affinis* as circumscribed here includes some Colorado, New Mexico, and Wyoming plants with the distal leaves closely spaced, blades linear or nearly so, and relatively long. Such plants have sometimes been segregated as var. *bigelovii* or as *Gentiana bigelovii*. Studies for this flora, however, have indicated that although the extremes are distinctive in appearance, they intergrade with plants having the distal leaves more widely spaced and shorter, and do not appear to predominate in any part of the range of *G. affinis*. Reports that the *bigelovii* vegetative morphology is associated with paler corolla color should be given further study.

Other taxonomic segregation, including the recognition of *Gentiana interrupta*, has been based on variation in the density of the inflorescence, the number of nodes below the terminal inflorescence, if any, at which axillary cymules or flowering branches are borne, and the length of such branches. Studies for this flora have shown that there is extensive intergradation and little correlation of this variation with geographic distribution. Some authors have treated plants with spathaceous, one- or two-cleft calyces with reduced or rudimentary lobes as *G. affinis* var. *parvidentata*, or have emphasized such calyces in their characterization of *G. forwoodii*. Plants with spathaceous calyces and various degrees of reduction in the calyx lobes occur sporadically throughout most of the range of the species, in populations otherwise referable to either of the two varieties recognized here, and may represent responses to growing conditions.

Occasional plants as far southeast as Wyoming approach var. *ovata* in leaf size and shape. Some relatively large-flowered plants in Arizona, New Mexico, and southern Utah, although here included in var. *affinis*, also approach var. *ovata*. Such more or less intermediate plants have sometimes been called *Gentiana affinis* var. *major* A. Nelson & J. F. Macbride. Some may represent the results of introgressive hybridization with *G. parryi*. Flower size and stem vestiture have been cited in distinguishing *G. rusbyi* in Arizona and New Mexico, but the flowers of the type collection are not unusually large for *G. affinis* var. *affinis*. Plants from that region, whether or not relatively large-flowered, have at most only minute, papillate puberulence like that found among plants elsewhere in the range of *G. affinis*.

Variety *affinis* appears to have hybridized occasionally with *Gentiana calycosa* and *G. puberulenta*, although its variability makes the morphological recognition of hybrids uncertain.

13b. Gentiana affinis Grisebach in W. J. Hooker var. **ovata** A. Gray in W. H. Brewer et al., Bot. California 1: 483. 1876 • Oregon gentian [E]

Gentiana oregana Engelmann ex A. Gray

Stems (0.5–)1.5–7 dm. **Leaves** closely spaced proximally, widely spaced distally; blade elliptic to ovate (proximal), or ovate to lanceolate (distal), 4–20(–25) mm wide. **Flowers:** calyx tube usually not cleft, lobes linear to elliptic-lanceolate or occasionally some rudimentary, (0–)7–13 mm; corolla (12–)30–40(–45) mm, lobes (3–)5–7(–10) mm.

Flowering late summer–fall. Meadows, brushy places, open woods; 0–2300 m; Calif., Idaho, Mont., Nev., Oreg., Wash., Wyo.

Variety *ovata* differs from var. *affinis* in its corollas that are wider at the throat, with larger lobes. Plants of this variety, especially those with proportionately wide leaves, thus resemble *Gentiana calycosa* in aspect. Although some hybridization may have occurred, these taxa can generally be distinguished by details of morphology, especially the presence or absence of cilia on the margins of the leaves and calyx lobes and, in fruiting specimens, whether or not the seeds are winged.

14. Gentiana puberulenta J. S. Pringle, Rhodora 68: 213, plate 1334, figs. 3, 4. 1966 • Downy or prairie gentian [E]

Herbs perennial, 1–6 dm, puberulent on stems and abaxially on midveins of leaves and primary veins of calyx tubes. **Stems** 1–5(–20), terminal from caudex, erect or nearly so. **Leaves** cauline, ± evenly spaced; blade narrowly oblong-lanceolate, 1.5–7 cm × 4–18 mm, apex obtuse to acute. **Inflorescences** 1–6-flowered dense cymes or heads, sometimes with additional flowers at 1–3 nodes or on short branches. **Flowers:** calyx 11–36 mm, lobes linear, 4–18(–25) mm, margins ciliate; corolla deep blue or rarely rose-violet, narrowly campanulate, open, (30–)35–60 mm, lobes spreading or ± recurved, ovate, 6–15 mm, free portions of plicae divided less than 1/2 their length into 2 ± triangular, lacerate segments; anthers distinct. **Seeds** winged. $2n = 26$.

Flowering late summer–fall. Mesic to ± dry savannas and prairies, calcareous soils; 100–1300 m; Man.; Ark., Ill., Ind., Iowa, Kans., Ky., Mich., Minn., Mo., Nebr., N.Dak., Ohio, Okla., S.Dak., Tenn., Wis.

Gentiana puberulenta is evidently extirpated from Ontario, Louisiana, Maryland, and New York, where outlying prairie communities have largely been eliminated by agricultural and urban expansion.

The name *Gentiana puberula* Michaux 1803, not Franchet 1890, and the homotypic synonym *Dasystephana puberula* (Michaux) Small have long and often been misapplied to this species but are typified by a specimen of *G. saponaria*.

Some small plants of *Gentiana puberulenta* appear similar to *G. affinis* var. *affinis*, but only a few specimens appear actually to be hybrids between these species. Where their ranges approach each other, the flowers of *G. affinis* are generally much smaller than those of *G. puberulenta*, and the corolla lobes of *G. affinis* are generally less than twice as long as the free portions of the plicae, whereas those of *G. puberulenta* are more than twice as long. The flower size of *G. affinis* var. *ovata* more closely approaches that of *G. puberulenta*, but in that variety, the range of which does not overlap that of *G. puberulenta*, the leaves are usually ovate to elliptic rather than narrowly oblong-lanceolate, and the distal internodes are often about as long as or longer than the leaves, in contrast to the proportionately shorter internodes of *G. puberulenta*. For further guidance in distinguishing between *G. puberulenta* and *G. affinis*, see discussion under 13. *G. affinis*.

Hybrids of *Gentiana puberulenta* with the strikingly dissimilar *G. andrewsii*, constituting *G.* ×*billingtonii* Farwell (as species), and with *G. flavida*, constituting *G.* ×*curtisii* J. S. Pringle, occur in the tall-grass prairies. Hybrids with *G. saponaria* formerly occurred in western Maryland.

15. Gentiana saponaria Linnaeus, Sp. Pl. 1: 228. 1753 • Soapwort gentian [E]

Dasystephana latifolia (Chapman) Small; *D. saponaria* (Linnaeus) Small; *Gentiana cherokeensis* (W. P. Lemmon) Fernald; *G. elliottii* Chapman var. *latifolia* Chapman

Herbs perennial, 0.7–6.5 dm, usually glabrous, occasionally puberulent on stems only. **Stems** 1–5, terminal from caudex, decumbent to erect. **Leaves** cauline, ± evenly spaced; blade linear to widely elliptic, 1.5–12 cm × 3–30 mm, apex obtuse to acute. **Inflorescences** ± dense 1–8-flowered cymes or heads, sometimes with additional cymules on short branches.

Flowers: calyx 9–32 mm, lobes spreading nearly horizontally when fresh, narrowly oblanceolate, 4–17 mm, shorter than or ± as long as tube, margins ciliate; corolla blue or rarely rose-violet, tubular, loosely closed to slightly or (in southernmost part of range) almost fully but narrowly open, 30–50 mm, lobes ovate-triangular, 3–7 mm, usually less than 2 mm longer than plicae, free portions of plicae divided 1/2 or more of their length into 2 subequal, erect, ± triangular, lacerate segments; anthers connate. Seeds winged. $2n$ = 26 (including plants identified as *G. saponaria* and *G. cherokeensis*).

Flowering late summer–fall. Mesic to moist open woods, savannas, swamps, fens, roadsides; 0–900 (–1200) m; Ala., Ark., Del., D.C., Fla., Ga., Ill., Ind., Ky., La., Md., Mich., Miss., N.J., N.Y., N.C., Ohio, Okla., Pa., S.C., Tenn., Tex., Va., W.Va.

Gentiana saponaria is believed to be extirpated from the District of Columbia. Reports from west of the range given here have been based mostly on specimens of *G. andrewsii* × *G. puberulenta*, occasionally on *G. flavida* × *G. puberulenta* or other hybrids. Some reports from the northeastern United States, including all records from Vermont and upstate New York, were based on specimens of *G. clausa* that antedate the recognition of that species in standard floras. Other reports have been based on misidentified *G. linearis*.

The name *Gentiana puberula* Michaux is typified by a specimen of *G. saponaria* but has generally been misapplied to *G. puberulenta*.

Plants from the northern parts of the range of *Gentiana saponaria* tend to have corollas more nearly closed than those from the southern parts of the range, but their corollas are not so firmly closed as those of *G. clausa*, the corolla lobes are larger, the summits of the plicae are usually more or less visible in herbarium specimens, and the shape of the calyx lobes is distinctively different. Plants of *G. saponaria* in the southernmost part of its range tend to have somewhat larger and more open corollas, approaching *G. catesbaei* in these respects, but they differ in their elliptic rather than ovate leaves and calyx lobes mostly shorter than or about as long as the tube rather than longer. Plants from bog and lake-shore habitats in Watauga County, North Carolina, at 1200 m, above the usual altitudinal range of *G. saponaria*, have attracted interest because of their linear to narrowly elliptic leaves mostly 3–9 mm wide. Their calyx and corolla morphology strongly supports their inclusion in *G. saponaria*, as does the occurrence of occasional plants with wider leaves in the same populations. Plants with similarly narrow leaves occur elsewhere in the range of *G. saponaria* and include those that have been identified as *G. cherokeensis*.

The epithet *saponaria* refers to a resemblance of the stems and leaves of this species to those of soapwort or bouncing-bet, *Saponaria officinalis* (Caryophyllaceae). Soaplike substances were not obtained from the gentian, so the invention of "soap gentian" as a common name is not appropriate.

Gentiana saponaria hybridizes with *G. andrewsii* relatively frequently in the Ohio Valley and occasionally elsewhere. Hybrids with *G. catesbaei*, *G. clausa*, *G. decora*, and *G. puberulenta* are also known.

16. **Gentiana catesbaei** Walter, Fl. Carol., 109. 1788
• Catesby's or coastal plain gentian [E]

Dasystephana parvifolia (Chapman) Small; *Gentiana catesbaei* var. *nummulariifolia* Fernald; *G. elliottii* Chapman

Herbs perennial, 1–7 dm, usually puberulent on stems only, occasionally glabrous. **Stems** 1–5, terminal from caudex, erect or nearly so. **Leaves** cauline, ± evenly spaced; blade usually ovate, occasionally elliptic, 1.5–7.5 cm × 4–30 mm, apex acute. **Inflorescences** ± dense 1–10-flowered cymes or heads, sometimes with additional flowers at 1–4(–8) nodes or on branches. **Flowers:** calyx 17–55 mm, lobes erect, lanceolate, 10–35 mm, mostly longer than tube, often ± foliaceous, margins ciliate; corolla blue or occasionally rose-violet, tubular, slightly to fully but narrowly open, 35–55 mm, lobes ± erect to spreading, deltate-ovate, 5–10 mm, usually 2–4 mm longer than plicae, free portions of plicae divided 1/2 or more of their length into 2 subequal, erect, ± triangular, lacerate segments; anthers connate. **Seeds** winged.

Flowering fall(–winter in Fla.). Moist ± open woods, clearings, roadsides; 0–100 m; Del., Fla., Ga., Md., N.C., S.C., Va.

Gentiana catesbaei is believed to be extirpated from New Jersey and Pennsylvania. Specimens from Alabama have been reidentified as *G. saponaria* as all such specimens seen in studies for this flora had the elliptic leaves and short calyx lobes typical of *G. saponaria* rather than the ovate leaves and much longer calyx lobes that characterize *G. catesbaei*.

From the more widely distributed *Gentiana saponaria*, *G. catesbaei* differs most conspicuously in its ovate rather than elliptic leaves, widest proximal to rather than near mid-length; calyx lobes widest near mid-length and usually 1.5–3 times as long as the tube; and generally with spreading rather than incurved corolla lobes.

Gentiana catesbaei is almost entirely restricted to the Atlantic coastal plain, where it displaces the closely related *G. saponaria* south of northeastern North Carolina. In the northern part of its range, where the ranges of these species overlap, they generally remain distinct, although a few plants apparently of hybrid origin have been found. A hybrid with the much less similar *G. villosa* is also known.

17. Gentiana decora Pollard, Proc. Biol. Soc. Wash. 13: 131. 1900 • Mountain or Appalachian gentian [E]

Dasystephana decora (Pollard) Small

Herbs perennial, 1.5–6 dm, puberulent on stems and calyx tubes. **Stems** 1–4, terminal from caudex, decumbent to erect. **Leaves** cauline, ± evenly spaced or somewhat more widely spaced distally; distal blades lance-elliptic to ovate-elliptic, 3–10 cm × 7–40 mm, apex acute to acuminate; proximal blades oblanceolate to obovate, apex obtuse to acute. **Inflorescences** ± dense 1–15-flowered cymes or heads, sometimes with additional flowers at 1–3 nodes or on short branches. **Flowers:** calyx 10–20 mm, lobes ± erect, subulate to linear or occasionally oblanceolate, 2–8(–12) mm, margins ciliate; corolla white to pale or occasionally medium blue or violet, tubular, loosely closed to fully but narrowly open, 25–45 mm, lobes ovate-triangular, 3–6 mm, longer than plicae, free portions of plicae divided 1/2 or more of their length into 2 unequal, ± triangular, lacerate segments, narrower segment usuallly deflexed; anthers connate. **Seeds** winged. **2*n*** = 26.

Flowering fall. Mesic woods, roadsides; 600–1600 m; Ga., Ky., N.C., S.C., Tenn., Va., W.Va.

Hybrids of *Gentiana decora* with *G. austromontana*, *G. clausa*, and *G. saponaria* are known.

18. Gentiana clausa Rafinesque, Med. Fl. 1: 210. 1828 • Meadow closed or bottle gentian, gentiane close [E]

Herbs perennial, 2–8 dm, glabrous. **Stems** 1–10, terminal from caudex, erect or decumbent. **Leaves** cauline, ± evenly spaced; blade ovate, 3–15 cm × 10–45 mm, apex acuminate. **Inflorescences** 1–20-flowered heads, sometimes with additional flowers at 1–3 nodes, rarely on short branches. **Flowers:** calyx 8–22 mm, lobes spreading nearly horizontally, widely obovate or elliptic to orbiculate, 2–6(–10) mm, margins ciliate; corolla blue or occasionally violet or white, tubular, completely closed, 23–40 mm, lobes incurved, ovate-triangular to semicircular, 0.7–2 mm, free portions of plicae ± as long and as wide as lobes, oblong, deeply and unequally bifid, summit erose; anthers connate. **Seeds** winged. **2*n*** = 26.

Flowering late summer–fall. Moist, open woods, stream banks, roadsides, acid soils; 0–800 m; Que.; Conn., D.C., Maine, Md., Mass., N.H., N.J., N.Y., N.C., Ohio, Pa., R.I., Tenn., Vt., Va., W.Va., restricted to higher elevations southward.

The corollas of *Gentiana andrewsii*, *G. austromontana*, and *G. clausa* all remain completely and tightly closed but are pollinated by bumblebees, which force open the corollas. The fresh corollas of *G. clausa* are rounded at the summit, with the plicae concealed by the true lobes. In contrast, the fresh corollas of *G. andrewsii* and *G. austromontana* are more acute, with the plicae forming much or all of the visible summit. The corolla lobes of *G. clausa* are about as long and as wide as the free portions of the plicae, whereas those of *G. andrewsii* and *G. austromontana* are distinctly narrower than the plicae. Also, in contrast to those of other species of *Gentiana* in the flora area except for *G. flavida* and *G. latidens*, the calyx lobes of *G. clausa* when fresh spread almost horizontally rather than being nearly erect.

Reports of *Gentiana clausa* from Indiana to Missouri and elsewhere west of the range given here have been based on specimens of *G. andrewsii* var. *dakotica*, second- or later-generation plants derived from *G. andrewsii* × *G. puberulenta*, or other hybrids and introgressants. In these plants, in contrast to *G. clausa*, the sepals are lanceolate and nearly erect, and the lobes of the intact corolla do not entirely conceal the plicae.

Gentiana clausa is largely isolated ecologically and geographically, but a few hybrids with *G. andrewsii*, *G. austromontana*, *G. decora*, and *G. saponaria* are known.

19. Gentiana latidens (House) J. S. Pringle & Weakley, Rhodora 111: 394. 2009 • Balsam Mountain gentian [C] [E]

Gentiana saponaria Linnaeus var. *latidens* House, Muhlenbergia 6: 75. 1910

Herbs perennial, 2–10 dm, glabrous, rarely puberulent on stems and calyx tubes. **Stems** 6–100+, terminal from caudex, erect or decumbent. **Leaves** cauline, ± evenly spaced; blade ovate, 3–15 cm × 10–55 mm, apex acuminate. **Inflorescences** 1–20-flowered heads, sometimes with additional flowers at 1–3 nodes. **Flowers:** calyx 15–35 (–45) mm, lobes spreading nearly horizontally when fresh, obovate, elliptic, ovate, orbiculate, or rhombic, 3–25(–35) mm, often strongly unequal, margins ciliate; corolla blue, ± loosely closed, 30–55 mm, lobes ± incurved to nearly erect, ovate-triangular, 2.5–5 mm, free portions of plicae ± as long and wide as lobes, oblong, deeply and unequally bifid, summit erose; anthers connate. **Seeds** winged.

Flowering late summer–fall. Moist to wet rocky slopes, roadsides, acid soils; of conservation concern; 1300–1700 m; N.C.

Gentiana latidens is known only from the Plott Balsam and Great Balsam mountains and Pisgah Ridge in Haywood, Jackson, Macon, and Transylvania counties, and perhaps also in Clay County.

20. Gentiana austromontana J. S. Pringle & Sharp, Rhodora 66: 402, fig. 1. 1964 • Blue Ridge or Roan Mountain or Appalachian gentian [E]

Herbs perennial, 0.7–5 dm, puberulent on stems and calyx tubes. **Stems** 1–50, terminal from caudex, erect or nearly so. **Leaves** cauline, ± evenly spaced; blade lanceolate to ovate or elliptic, 3–12 cm × 10–30 mm, apex acute to acuminate. **Inflorescences** 1–15-flowered heads, sometimes with additional flowers at 1–3 nodes or on short branches. **Flowers:** calyx 8–25 mm, lobes lanceolate to ovate-triangular, narrowly ovate, or occasionally elliptic, 1.5–12 mm, margins ciliate; corolla violet-blue, usually deeply colored, tubular, completely closed, 30–50 mm, lobes erect, triangular to nearly semicircular, 1.5–3 mm, free portions of plicae ± as long as lobes and ± 2 times as wide, deeply divided into 2 nearly equal, triangular to ± oblong segments, margins minutely erose; anthers connate. **Seeds** winged.

Flowering fall. Grassy balds, open woods, acid soils; 600–2100 m; N.C., Tenn., Va., W.Va.

Because of the relatively late recognition of *Gentiana austromontana* as a distinct species, specimens of this species have been identified both as *G. clausa* and as *G. decora*. *Gentiana austromontana* is distinguished from *G. andrewsii*, *G. clausa*, and *G. latidens* by the combination of puberulent stems and calyx tubes with free portions of the corolla plicae that are about as long as the lobes and about twice as wide, divided into two more or less triangular segments each similar to a true lobe in size and shape. Because of its puberulence, it has been confused with *G. decora*, which differs in its more open, generally paler corollas with longer lobes and plicae, and usually narrowly linear calyx lobes.

Both *Gentiana austromontana* and *G. decora* occur in the higher elevations of eastern Tennessee and western North Carolina, although *G. decora* tends to occur in shadier habitats. These species are usually distinctly dissimilar, in each case bearing a greater resemblance to other species than to each other, but populations of plants variously intermediate between the two occur relatively frequently, especially in Greene and Unicoi counties, Tennessee. Plants otherwise typical of *G. austromontana* but with narrowly open corollas have been found in Mount Jefferson State Natural Area, Ashe County, North Carolina, and may be derived from introgression of genetic material from *G. decora*. *Gentiana austromontana* also hybridizes occasionally with *G. clausa*.

21. Gentiana andrewsii Grisebach in W. J. Hooker, Fl. Bor.-Amer. 2: 55. 1837 • Bottle or fringed bottle or prairie closed gentian, gentiane d'Andrews ou close [E] [F]

Dasystephana andrewsii (Grisebach) Small; *Pneumonanthe andrewsii* (Grisebach) W. A. Weber

Herbs perennial, 1–12 dm, glabrous or rarely puberulent. **Stems** 1–20, terminal from caudex, decumbent to erect. **Leaves** cauline, ± evenly spaced; blade elliptic-oblong to lanceolate or narrowly ovate, 3–16 cm × 10–50 mm, apex acuminate. **Inflorescences** 1–25-flowered heads, often with additional flowers at 1–6(–9) nodes or on short branches. **Flowers:** calyx 9–29 mm, lobes lanceolate to ovate or occasionally oblanceolate, 2–15 mm, margins ciliate; corolla blue, white, or rarely rose-violet, tubular, completely closed, 28–45 mm, lobes reduced to a mucro or ± triangular, 0.5–2(–3) mm, free portions of plicae oblong, shallowly and nearly symmetrically bifid, summit truncate, erose; anthers connate. **Seeds** winged.

Varieties 2 (2 in the flora): North America.

Gentiana andrewsii is the only species of *Gentiana* in which the plicae of the corolla are distinctly longer than the lobes.

Gentiana andrewsii has often been reported outside its actual range. Although the epithets of some of the species that have been confused with or considered inseparable from *G. andrewsii* have priority, the familiar name *G. andrewsii* is often misapplied, sometimes because it is assumed that any "closed gentian" is *G. andrewsii*. As *G. clausa* was not distinguished from *G. andrewsii* in standard floras prior to 1950, reports from the northeastern United States based on specimens identified before 1950 should be considered doubtful if the specimens have not been reexamined. Old reports from the southern Appalachians are also questionable because *G. austromontana* was not recognized until 1964. Some reports from the southeastern and south-central United States and along the Atlantic seaboard have been based on specimens of *G. saponaria*. True *G. andrewsii* is distinguishable as the only *Gentiana* species in which the corolla plicae distinctly exceed the minute lobes. The fringed tip of the completely closed corolla, at first white, soon turning reddish brown, is an excellent field mark for distinguishing *G. andrewsii* from *G. clausa*. In *G. clausa*, the summit of the intact corolla appears completely blue (in the typical color form), and the plicae are concealed. *Gentiana andrewsii* grows in calcareous soils and *G. clausa* in noncalcareous soils.

Because of this ecological separation, there are only a few records of hybridization between *Gentiana andrewsii* and *G. clausa*. In the tall-grass prairies, *G. andrewsii* hybridizes with *G. flavida*, producing *G.* ×*pallidocyanea* J. S. Pringle, and with *G. puberulenta*, producing *G.* ×*billingtonii* Farwell (as species). Northward, it occasionally hybridizes with *G. rubricaulis*, producing *G.* ×*grandilacustris* J. S. Pringle, and in the southeastern part of its range it hybridizes with *G. saponaria*.

1. Corolla lobes reduced to a mucro or at most minutely triangular, less than 1 mm . . . . 21a. *Gentiana andrewsii* var. *andrewsii*
1. Corolla lobes triangular or ± rounded, 1–2(–3) mm . . . . 21b. *Gentiana andrewsii* var. *dakotica*

21a. Gentiana andrewsii Grisebach in W. J. Hooker var. **andrewsii** E F

Stems glabrous. **Leaf widths** varying throughout range for species. **Corolla lobes** reduced to a mucro or at most minutely triangular, less than 1 mm. $2n$ = 26 [C. L. Rork 1949].

Flowering late summer–fall. Moist, open woods, swamps, savannas, mesic to wet prairies, calcareous soils; 0–1700 m; Man., Ont., Que.; Conn., Ill., Ind., Iowa, Ky., Md., Mass., Mich., Minn., Mo., N.J., N.Y., N.Dak., Ohio, Pa., Vt., Va., W.Va., Wis.

Isolated populations of var. *andrewsii* in Colorado and Virginia are no longer extant. This species has probably also been extirpated from Vermont, where it had been restricted to calcareous sites in the western part of the state, and may also have been extirpated from Massachusetts. Reports from Rhode Island have been based on specimens of *Gentiana clausa*, and the provenance of an old specimen labeled as being from New Hampshire is doubtful. More recent reports from New Hampshire have been based on misidentified specimens of *G. clausa* (D. E. Boufford and K. Porter-Utley, pers. comm.; studies for this flora).

21b. Gentiana andrewsii Grisebach in W. J. Hooker var. **dakotica** A. Nelson, Bot. GaZ. 56: 68. 1913 • Dakota gentian E

Stems glabrous or rarely puberulent. **Leaf widths** usually in narrower part of range for species. **Corolla lobes** triangular or ± rounded, usually 1–2(–3) mm but still exceeded by plicae.

Flowering late summer–fall. Mesic to wet prairies, savannas, calcareous soils; 100–1200 m; Man., Sask.; Ill., Iowa, Minn., Mo., Nebr., N.Dak., S.Dak., Wis.

Variety *dakotica*, which largely replaces var. *andrewsii* westward, may be derived from introgressive hybridization with *Gentiana puberulenta*. It differs from *G. clausa* in that the corolla lobes do not conceal the plicae in the intact corolla, as well as in the shape and orientation of the calyx lobes and in the length and shape of the free portions of the corolla plicae. Similar plants in the Ohio Valley are derived from *G. andrewsii* × *G. saponaria*. Other reports from outside the range given here are probably derived from various hybrid combinations (studies for this flora).

22. Gentiana linearis Froelich, Gentiana, 37. 1796 • Narrow-leaved or bog gentian, gentiane à feuilles linéaires E

Herbs perennial, 1–9 dm, glabrous. **Stems** 1–30, terminal from caudex, erect. **Leaves** cauline, nearly evenly spaced or somewhat more widely spaced distally; blade linear to lanceolate, 4–9 cm × 3–14 mm, apex acute. **Inflorescences** ± dense 1–7-flowered cymes, sometimes with additional flowers at 1–4 nodes, sessile or on branches to 12 cm. **Flowers:** calyx 8–28 mm, lobes linear to oblong, 2–12(–15) mm, margins not ciliate; corolla blue or occasionally violet or white, tubular, loosely closed or slightly open, 25–50 mm, lobes ± incurved, semicircular, 2.5–5 mm, free portions of plicae obliquely triangular, margins entire or shallowly erose, with a minute, deflexed second segment; anthers connate. **Seeds** winged. $2n$ = 26.

Flowering late summer–fall. Bogs, wet meadows, shores, generally strongly acid soils; 0–2000 m; N.B., Nfld. and Labr. (Labr.), Ont., Que.; Maine, Md., Mass., Mich., N.H., N.Y., Pa., Tenn., Vt., Va., W.Va., Wis.

Gentiana linearis is extirpated from New Jersey. Reports from Manitoba and Minnesota have been based on circumscriptions of *G. linearis* that included *G. rubricaulis*, mostly prior to the recognition of *G. rubricaulis* as a distinct species in standard floras. A report from North Carolina is incorrect, having been based on a misunderstanding as to where a photograph was taken (W. F. Hutson, pers. comm.). Narrow-leaved specimens of *G. saponaria* are occasionally misidentified as *G. linearis* but can be distinguished by their ciliate calyx lobes and by the shape of the calyx lobes and the free portions of the corolla plicae.

23. Gentiana rubricaulis Schweinitz in W. H. Keating, Narrat. Exp. St. Peter's River 2: 384. 1824 • Great Lakes or red-stemmed or purple-stemmed gentian [E]

Dasystephana grayi (Kusnezov) Britton; *Gentiana grayi* Kusnezov; *G. linearis* Froelich var. *latifolia* A. Gray; *G. linearis* subsp. *rubricaulis* (Schweinitz) J. M. Gillett; *G. linearis* var. *rubricaulis* (Schweinitz) MacMillan

Herbs perennial, 1–8 dm, glabrous. **Stems** 1–5, terminal from caudex, erect. **Leaves** cauline, gradually more widely spaced distally; blade linear to oblong-lanceolate (proximal) or lanceolate to ovate (distal), 3–9 cm × 8–30 mm, apex acute. **Inflorescences** dense 1–15-flowered cymes, basally ± enveloped by ascending, conduplicate involucral leaves, rarely with additional flowers at one node. **Flowers:** calyx 10–26 mm, lobes oblong, 2–14 mm, margins not ciliate; corolla grayish violet to violet-blue or occasionally rose-violet or white, tubular, loosely closed or slightly open, 30–45 mm, lobes ascending, ovate-triangular, 4–8 mm, free portions of plicae obliquely triangular, erose, with minute, deflexed second segment; anthers connate. **Seeds** winged. $2n = 26$.

Flowering late summer–fall. Fens, swamps, wet meadows, stream banks, interdunal depressions, calcareous soils; 0–700 m; Man., N.B., Ont.; Maine, Mich., Minn., Wis.

The name *Gentiana linearis* var. *lanceolata* A. Gray was applied originally to plants referable to *G. linearis*, although the name *G. rubricaulis* was cited in synonymy. The name *G. linearis* var. *latifolia* was applied originally only to *G. rubricaulis*, but both of these names were applied subsequently to both that species and relatively wide-leaved specimens of *G. linearis*. This confusion has been responsible in some cases for the rejection of specific status for *G. rubricaulis*, and for erroneous reports of *G. rubricaulis* in New York and Vermont. Reports from Nebraska were based on an old misidentification of *G. puberulenta*. Reports from Saskatchewan were also based on misidentified specimens. Reports of *G. rubricaulis* in Maine and New Brunswick are correct, although these populations are disjunct by about 775 km from the easternmost populations in Ontario.

In marked contrast to all other species of *Gentiana* in eastern and central North America, including *G. linearis*, the involucral leaves of this species are strongly ascending and somewhat conduplicate as well as being wider, and envelop the proximal portion of the flower cluster.

In the vicinity of Lake Superior, where the ranges of *Gentiana rubricaulis* and *G. linearis* overlap, these species maintain their distinctness, with *G. rubricaulis* occurring in calcareous soils and *G. linearis* in granitic and similar strongly acid soils (J. S. Pringle 1968). A few hybrids of *G. rubricaulis* with *G. andrewsii*, which is likewise a calciphile, are known. These hybrids have been designated *G.* ×*grandilacustris* J. S. Pringle.

24. Gentiana flavida A. Gray, Amer. J. Sci. Arts, ser. 2, 1: 80. 1846 • White prairie or white or cream or yellowish or pale gentian [E]

Dasystephana flavida (A. Gray) Britton

Herbs perennial, 3–10 dm, glabrous. **Stems** 1–10, terminal from caudex, decumbent to erect. **Leaves** cauline, ± evenly spaced; blade lanceolate to ovate, 5–15 cm × 15–50 mm, apex acuminate. **Inflorescences** dense 1–20-flowered cymes, often also with additional clusters at 1 or 2 nodes. **Flowers:** calyx 10–30 mm, lobes spreading, with bracketlike keels, lanceolate to ovate-triangular, 3–15 mm, margins not ciliate; corolla white, sometimes with yellowish or greenish tinge (drying yellowish), with veins outlined in green, tubular, loosely closed or slightly open, 30–55 mm, lobes incurved to nearly erect, widely ovate-triangular, 4–6 mm, free portions of plicae obliquely triangular, erose to shallowly lacerate, with minute, deflexed second segment; anthers connate or some sooner or later distinct. **Seeds** winged. $2n = 26$.

Flowering late summer–fall. Mesic prairies and savannas, calcareous soils; 100–800 m; Ont.; Ark., Ill., Ind., Iowa, Kans., Ky., Mich., Minn., Mo., Nebr., Ohio, Okla., Pa., W.Va., Wis.

The name *Gentiana alba* Muhlenberg has often been applied to this species. Uncertainty had long persisted, first as to whether the name *G. alba* was validly published by G. H. E. Muhlenberg in 1813, then, after that publication had been deemed invalid, whether it was validated by T. Nuttall in 1818. A group of nomenclatural authorities considered this issue on behalf of this flora and concluded that neither of those publications of the name *G. alba* had been valid, and that *G. flavida* A. Gray was the earliest validly published name for this species (K. N. Gandhi, pers. comm.).

Outlying eastern populations of *Gentiana flavida* in North Carolina, Pennsylvania, and West Virginia are no longer extant, and the continued existence of other peripheral populations in isolated prairie remnants is precarious. Reports from Manitoba have been based on misidentified *G. rubricaulis*. A report from Maryland was based on the misreading of a label of a specimen actually from Indiana (studies for this flora).

In contrast to those of the other species of *Gentiana* in the flora area, with the exceptions of *G. clausa* and

G. latidens, the calyx lobes of *G. flavida* spread widely, with keels like shelf brackets decurrent on the tube.

Morphological variation in *Gentiana flavida* should be given further study. According to J. T. Curtis (1959), plants of this species from the northern part of its range, as seen in the field, appear distinctly different in inflorescence form from plants native farther south.

In the tall-grass prairies, *Gentiana flavida* hybridizes with *G. andrewsii*, producing *G.* ×*pallidocyanea* J. S. Pringle, and *G. puberulenta*, producing *G.* ×*curtisii* J. S. Pringle. Reports of *G. flavida* with the corollas distally lilac have been based on plants derived from such hybridization, probably through backcrossing.

25. Gentiana villosa Linnaeus, Sp. Pl. 1: 228. 1753 • Striped or pale or straw-colored gentian E

Dasystephana villosa (Linnaeus) Small; *Gentiana deloachii* (W. P. Lemmon) Shinners; *G. ochroleuca* Froelich

Herbs perennial, 0.7–6 dm, glabrous. **Stems** 1–5, terminal from caudex, erect. **Leaves** cauline, ± evenly spaced; blade obovate or spatulate to elliptic, 2.5–10 cm × 10–40 mm, proximal blade apices retuse or truncate to obtuse, distal ± acute. **Inflorescences** ± dense 1–10-flowered cymes, often with additional flowers at 1 or 2(–4) nodes or on branches. **Flowers:** calyx 11–50 mm, lobes linear to oblanceolate, 5–35 mm, margins not ciliate; corolla largely white or greenish white with veins outlined in green, sometimes suffused with violet, or grayish violet ± throughout, tubular, narrowly open, 30–55 mm, lobes ascending, ovate-triangular, 4–10 mm, free portions of plicae obliquely triangular, erose, occasionally shallowly bifid; anthers connate or distinct. **Seeds** not winged. **2*n*** = 26.

Flowering fall(–early winter southward). Mesic woods; 0–800 m; Ala., Fla., Ga., Ind., Ky., La., Md., Miss., N.C., Ohio, Pa., S.C., Tenn., Va., W.Va.

Gentiana villosa is believed to be extirpated from Delaware, the District of Columbia, and New Jersey.

Although the name *Gentiana ochroleuca* is a heterotypic synonym of *G. villosa*, it was sometimes applied to *G. flavida* during the nineteenth century. Such a misapplication is responsible for reports of *G. ochroleuca* from Illinois. Reports of *G. villosa* from Arkansas are plausible but remain unsubstantiated.

The species name is a misnomer as plants of *Gentiana villosa* species are glabrous. The use of the translation "hairy gentian" as a common name is inappropriate and potentially confusing.

There is one record each of hybrids of *Gentiana villosa* with *G. autumnalis* and *G. catesbaei*.

26. Gentiana prostrata Haenke in N. J. Jacquin, Collectanea 2: 66, plate 17, fig. 2. 1789 F

Chondrophylla prostrata (Haenke) J. P. Anderson; *Ciminalis prostrata* (Haenke) Á. Löve & D. Löve

Varieties 2+ (1 in the flora): w North America, Eurasia, questionably in s South America.

Gentiana prostrata is variously divided into subspecies and/or varieties. At least two, perhaps more, varieties seem appropriately recognized at that rank (one in the flora).

26a. Gentiana prostrata Haenke in N. J. Jacquin var. **prostrata** • Pygmy or tundra gentian F

Chondrophylla nutans (Bunge) W. A. Weber; *Gentiana prostrata* subsp. *americana* (Engelmann) A. E. Murray; *G. prostrata* var. *americana* Engelmann; *G. prostrata* subsp. *nutans* (Bunge) Halda

Herbs biennial or annual, 0.1–1.5(–2.5) dm, glabrous. **Stems** generally 1, decumbent, from a tap root. **Leaves** basal and cauline, all ± similar; cauline leaves spreading, gradually more widely spaced distally; blade spatulate-obovate to widely oblanceolate or distal blades ovate, 0.2–1.2 cm × 1–4 mm, obscurely or not white-margined, apex mucronate. **Inflorescences** solitary flowers at ends of branches. **Flowers** usually 4-, occasionally 5-merous; calyx 3–15 mm, lobes triangular, 1–2.5 mm, margins not ciliate; corolla medium to deep blue or occasionally white, nearly salverform, open, 7–25 mm, lobes spreading, ovate, 2–6 mm, free portions of plicae triangular, entire or shallowly erose-serrate; anthers distinct. **Seeds** not winged. **2*n*** = 36.

Flowering summer. Moist to wet alpine and arctic tundra, meadows and heathland, wet ledges, lake and stream banks, along animal trails; 0–3900 m; Alta., B.C., N.W.T., Yukon; Alaska, Calif., Colo., Idaho, Mont., Oreg., Utah, Wyo., restricted to high elevations southward; Eurasia (Alps, Caucasus e to Afghanistan).

Because of its restriction to high altitudes in much of its range, populations of *Gentiana prostrata* are widely scattered. Reports of *G. prostrata* from Nevada, New Mexico, and elsewhere in North America outside the range given here have been based on circumscriptions of the species that included *G. fremontii*.

Molecular phylogenetic studies have indicated that the North American plants here included in *Gentiana prostrata* should perhaps be treated as a species distinct from European *G. prostrata* (A. Favre, pers. comm.). The North American plants have been said to differ from the European plants in having four-merous rather than five-merous flowers, but four-merous and five-merous flowers occur both in Europe and in North America.

Plants exhibiting the morphological extremes alleged to distinguish the taxa *nutans* and *prostrata* in the narrow sense frequently occur within the same population, along with intermediates, especially in the northern parts of the range of *Gentiana prostrata* in North America, but occasionally as far south as Colorado and Wyoming. The character states alleged to distinguish these taxa are not consistently associated in syndromes (studies for this flora, especially specimens at A). What appear to be nodding flowers are often late-stage buds, with the flowers on the same plants becoming erect at anthesis. The relationship between floral orientation in this species and the time of day or cloudiness has not been studied but may also be relevant. The length of the gynophore at maturity, emphasized by W. A. Weber and R. C. Wittmann (1985) in distinguishing *G. nutans* Bunge from *G. prostrata*, varies greatly within *G. prostrata*. No correlation between gynophore length and other traits alleged to distinguish *G. nutans* were observed in studies for this flora.

The corollas of *Gentiana prostrata* and *G. fremontii* close not only at night and in cloudy weather but also when the flower is touched or the stem is jarred.

27. Gentiana fremontii Torrey in J. C. Frémont, Rep. Exped. Rocky Mts., 94. 1843 • Moss or Frémont's or lone gentian [E]

Herbs biennial or sometimes annual, 0.1–1.3 dm, glabrous. **Stems** 1–10(–25), decumbent to erect. **Leaves** basal and cauline, cauline leaves gradually smaller, more widely spaced and more strongly ascending distally; blade conspicuously white-margined, apex acute; basal blades widely spatulate to ovate or orbiculate, 0.2–1.3 cm × 1.5–8 mm; cauline blades oblanceolate to linear, distal blades 4–7 × 0.6–2 mm. **Inflorescences** solitary flowers. **Flowers:** calyx 4–12 mm, lobes narrowly oblong-triangular, 1.5–3.5 mm, margins not ciliate; corolla white to pale blue or rarely deeper blue, often with dark blue lines abaxially, nearly salverform, open, 7–15 mm, lobes lance-ovate, 2–4 mm, free portions of plicae low-triangular with margins entire or shallowly erose-serrate or notched at apex; anthers distinct. **Seeds** not winged.

Flowering (late spring–)summer. Subalpine wet meadows; 600–3700 m; Alta., Sask.; Ariz., Calif., Colo., Mont., Nev., N.Mex., Utah, Wyo., restricted to high elevations south of Saskatchewan.

In contrast to the deep green stems and leaves of *Gentiana prostrata*, the vegetative parts of *G. fremontii* are much paler. G. Engelmann (1879) described plants of *G. fremontii* as having a pale, sickly appearance, and J. A. Ewan annotated specimens as having been yellowish when seen in the field, reminiscent of a fungus or broomrape (*Aphyllon* or *Orobanche*). This suggests that mycorrhizal symbiosis is especially significant in this species, but its trophic ecology has not been studied.

Gentiana fremontii differs further from *G. prostrata* in having obovoid capsules less than twice as long as wide, generally not fully exserted from the marcescent corolla, narrowly winged distally along the sutures, with valves that eventually separate nearly to the base, whereas the capsules of *G. prostrata* are compressed-cylindric, more than twice as long as wide, often fully exserted at maturity, not winged, with the valves separating only above the middle. Also, although both species vary in this respect, *G. fremontii* more often has the flower parts in fives.

The names *Gentiana aquatica* Linnaeus and *Chondrophylla aquatica* (Linnaeus) W. A. Weber have often been applied to this species. *Gentiana fremontii*, although similar to the Siberian and Chinese *G. aquatica*, appears to differ consistently in the wider, more conspicuous white margins of its leaves, longer and proportionately narrower mid-cauline leaves, usually white rather than blue corollas, and corolla plicae that generally have jagged rather than entire summits. The illegitimate name *G. humilis* Steven 1812, not Salisbury 1796, has also been applied to *G. fremontii*, but the North American plants are not now considered conspecific with the type from Azerbaijan.

28. Gentiana douglasiana Bongard, Mém. Acad. Imp. Sci. St.-Pétersbourg, Sér. 6, Sci. Math. 2: 156, plate 6. 1832 • Douglas's or swamp gentian [E]

Herbs annual, 0.5–2.7 dm, glabrous. **Stem** 1, often branched from near base and distally, main axis erect. **Leaves** basal and cauline, cauline ± evenly spaced; basal blades oblong-obovate to ovate, 0.4–2.3 cm × 2–9 mm, apex obtuse to acute; cauline blades ovate to elliptic, 0.3–1 cm × 2–7 mm on main axis, smaller on branches, apex generally ± acute. **Inflorescences** solitary flowers or open, 2–7-flowered cymes terminating main axis and branches. **Flowers:** calyx 4–7 mm, lobes linear-oblong, 1.5–3 mm, margins not ciliate; corolla adaxially white with purple spots near base of lobes distal to yellowish green throat, abaxially suffused with green and distally with deep blue, nearly salverform, open, 9–14 mm, lobes ovate-triangular, 3–5 mm, free portions of plicae symmetrically divided to base into 2 lanceolate, acuminate segments; anthers distinct. **Seeds** winged. $2n = 26$.

Flowering late spring–early fall. Bogs, boggy woodlands, wet meadows and tundra; 0–1500 m; B.C.; Alaska, Wash.

18. VOYRIA Aublet, Hist. Pl. Guiane 1: 208, plate 83. 1775 • [Garipons, spoken in French Guiana, *voyria*, local name alluding to edible roots of achlorophyllous heterotrophic plants, probably *Voyria rosea* Aublet]

James S. Pringle

Leiphaimos Schlechtendal & Chamisso

Herbs perennial, achlorophyllous, glabrous [variously puberulent]. **Leaves** cauline, opposite [proximal alternate], scalelike. **Inflorescences** small dichasial cymes [flowers solitary]. **Flowers** [4]5-merous; calyx small, tube campanulate [to cylinidric]; corolla white or pink [orange, yellow, pale violet-blue], salverform [funnelform], glabrous [tube pubescent on adaxial or both surfaces], lobes shorter than tube, imbricate in bud, entire, plicae between lobes absent, spurs absent; stamens inserted in distal 1/2 of corolla tube; anthers subsessile, distinct [connate], remaining straight; ovary stipitate [sessile]; style persistent [deciduous], erect, short [long], remaining straight; stigma 2-lobed [unlobed]; nectaries absent [2, on proximal part of ovary]. **Capsules** ellipsoid.

Species ca. 20 (1 in the flora): Florida, Mexico, West Indies, Central America, South America, Africa; warm-temperate to tropical areas.

Plants of *Voyria* are mycotrophic. The root anatomy and mycorrhizae of several species have been investigated by S. Imhof (1997, 1999, and references cited therein). The one species in the flora area is in subg. *Leiphaimos* (Schlechtendal & Chamisso) V. A. Albert & Struwe.

SELECTED REFERENCES Albert, V. A. and L. Struwe. 1997. Phylogeny and classification of *Voyria* (saprophytic Gentianaceae). Brittonia 49: 466–479. Maas, P. J. M. and P. Ruyters. 1986. *Voyria* and *Voyriella* (saprophytic Gentianaceae). In: Organization for Flora Neotropica. 1968+. Flora Neotropica. 121+ nos. New York. No. 41.

1. **Voyria parasitica** (Schlechtendal & Chamisso) Ruyters & Maas, Acta Bot. Neerl. 30: 143. 1981 • Ghost-plant F

Leiphaimos parasitica Schlechtendal & Chamisso, Linnaea 6: 387. 1831

Herbs white to pale buff, 1–1.5 dm. **Leaf blades** 3–5 mm, apex obtuse to acute. **Inflorescences** 1–30-flowered. **Flowers:** calyx campanulate, lobed about to middle, 2–4 mm, lobes lanceolate-triangular, apex acute; corolla 5–9 mm, lobes ovate, 1 mm, apex obtuse; stamens inserted in distal 1/4 of corolla tube.

Flowering year-round. Hammocks; 0–10 m; Fla.; Mexico; West Indies; Central America.

The illegitimate name *Voyria mexicana* Grisebach has often been applied to this species.

In the flora area, *Voyria parasitica* occurs only in Miami-Dade and Monroe counties. This species occurs to 1300 m, in more diverse habitats, outside the flora area.

LOGANIACEAE R. Brown ex Martius • Strychnos Family

Katherine G. Mathews

Herbs, subshrubs, shrubs, or trees [lianas], taprooted or rhizomatous. **Stems** sometimes with pseudodichotomous branching; wood often with internal phloem and vestured pits; colleters often present in axils of leaves and calyx; hairs unicellular or uniseriate. **Leaves** [in rosettes] usually opposite, sometimes whorled or pseudowhorled, simple, entire, penniveined or with ascending, arcuate lateral veins; stipules or interpetiolar lines absent or present, stipules interpetiolar, leafy, membranaceous, or reduced to ridge, or paired at leaf base. **Inflorescences** terminal or axillary, cymose, dichasial, monochasial, or thyrsoid [solitary flowers]; bracteoles (0–)2(or 3). **Flowers** bisexual [unisexual]; perianth [4 or]5-merous; sepals valvate or imbricate [contort], equal or unequal [calycophyllous, replaced by involucre]; corolla yellow, white, green, pink, red, or purple, tubular, funnelform, campanulate, or urceolate [rotate], glabrous or hairy outside and within; stamens [1–]5, alternating with corolla lobes; filaments glabrous [hairy], distinct [connivent or coherent]; anthers 2- or 4-celled, glabrous [hairy], dehiscence introrse [latrorse], pollen colporate to colpate; ovary superior or semi-inferior, 2-locular, syncarpous or apically apocarpous, glabrous or hairy, ovules [1–]5–50 per locule, anatropous or hemi-anatropous, placentae axile, frequently peltate; styles [0 or]1, persistent or deciduous, filiform, glabrous or hairy. **Fruits** capsules or berries [drupes], capsule dehiscence septicidal and usually loculicidal, separating partly or completely into 2 or 4 valves, also circumscissile in *Spigelia*, berries yellow or orange to green [red, brown, blue-black]. **Seeds** polyhedral, round, or lens-shaped, not winged [winged], endosperm fleshy or cartilaginous [starchy], embryo straight.

Genera 15, species ca. 400 (3 genera, 11 species in the flora): c, se United States, Mexico, West Indies, Central America, South America, Asia, Africa, Indian Ocean Islands, Australia; pantropical, temperate areas.

A monophyletic Loganiaceae has been delineated to contain 15 genera, including *Antonia* Pohl, *Bonyunia* M. R. Schomburgk ex Progel, *Gardneria* Wallich, *Geniostoma* J. R. Forster & G. Forster, *Labordia* Gaudichaud-Beaupré, *Logania* R. Brown, *Mitrasacme* Labillardière, *Mitreola*, *Neuburgia* Blume, *Norrisia* Gardner, *Phyllangium* Dunlop, *Schizacme* Dunlop, *Spigelia*, *Strychnos*, and *Usteria* Willdenow (C. L. Frasier 2008), as opposed to the broader, polyphyletic concept of A. J. M. Leeuwenberg and P. W. Leenhouts (1980). The North American genera *Buddleja* (including *Emorya*) and *Polypremum*, formerly included in Loganiaceae in the broad sense, are now commonly grouped together in Buddlejaceae (Lamiales); the most recent treatments place *Buddleja* in Scrophulariaceae (D. C. Albach et al. 2005) and *Polypremum* into Tetrachondraceae (S. J. Wagstaff 2004). *Gelsemium* has been excluded from Loganiaceae in the narrow sense and placed in Gelsemiaceae (L. Struwe et al. 1994).

In the past, segregate families have been proposed to segregate various groups from Loganiaceae in the narrow sense, including Spigeliaceae Berchtold & J. Presl and Strychnaceae de Candolle ex Perleb; because knowledge of intergeneric relationships is still provisional, they are not recognized here.

Several genera are known to contain bioactive compounds, including secoiridoids and indole alkaloids, and have been used as drugs, poisons, and medicines, particularly *Spigelia* and *Strychnos*.

SELECTED REFERENCES Frasier, C. L. 2008. Evolution and Systematics of the Angiosperm Order Gentianales with an In-depth Focus on Loganiaceae and Its Species-rich and Toxic Genus *Strychnos*. Ph.D. dissertation. Rutgers, The State University of New Jersey. Leeuwenberg, A. J. M. and P. W. Leenhouts. 1980. Loganiaceae. Taxonomy. In: H. G. A. Engler et al., eds. 1924+. Die natürlichen Pflanzenfamilien..., ed. 2. 26+ vols. Leipzig and Berlin. Vol. 28b(1), pp. 8–96. Mennega, A. M. W. 1980. Loganiaceae. Anatomy of the secondary xylem. In: H. G. A. Engler et al., eds. 1924+. Die natürlichen Pflanzenfamilien..., ed. 2. 26+ vols. Leipzig and Berlin. Vol. 28b(1), pp. 112–161. Rogers, G. K. 1986. The genera of Loganiaceae in the southeastern United States. J. Arnold Arbor. 67: 143–185.

1. Shrubs or trees . . . . 3. *Strychnos*, p. 100
1. Herbs or subshrubs.
   2. Corollas tubular or funnelform, 6–52 mm; capsules 2-lobed, dehiscing into 4 valves . . . . 1. *Spigelia*, p. 94
   2. Corollas urceolate, 1.5–3 mm; capsules 2-horned, dehiscing along medial line into 2 valves . . . . 2. *Mitreola*, p. 98

1. SPIGELIA Linnaeus, Sp. Pl. 1: 149. 1753; Gen. Pl. ed. 5, 74. 1754 • Pinkroot [For Adriaan van den Spiegel, 1578–1625, Paduan physician and author]

Herbs [subshrubs or shrubs]. **Stems** ascending to decumbent, sparsely to diffusely, sometimes divaricately, branched, glabrous or scabrous [otherwise hairy]. **Leaves** sessile or petiolate, 1 pair or 2 closely spaced pairs forming pseudowhorl proximal to inflorescence; blade ovate, lanceolate, elliptic, oblong, rhombic, linear, obovate, or oblanceolate, venation 1 or 2 pairs of secondary veins from near base, curved along margins, not reaching apex. **Inflorescences** terminal [axillary], monochasial, 2–45-flowered; bracts and bracteoles subulate, 1–3(–7) mm, each flower subtended by (0–)2(or 3) bracteoles. **Flowers:** sepals persistent, often accrescent in fruit, shortly connate at base, green, linear, linear-lanceolate, lanceolate, or lanceolate-subulate [deltate]; corolla white, yellow, pink, or scarlet, funnelform or tubular [salverform, campanulate, or urceolate], glabrous [villous]; ovary superior; stigmas conic, capitate, or bulbous, unlobed or scarcely 2-lobed. **Fruits** capsules, green or brown, 4-valved, 2-lobed, dehiscence septicidal and loculicidal, circumscissile near base. **Seeds** tan to dark brown, obliquely ovoid, usually obcompressed, tuberculate or reticulate. $x = 8$.

Species ca. 60 (7 in the flora): c, se United States, Mexico, West Indies, Central America, South America, se Asia, w Africa.

SELECTED REFERENCES Gould, K. R. 1997. Systematic Studies in *Spigelia*. Ph.D. dissertation. University of Texas. Gould, K. R. and R. K. Jansen. 1999. Taxonomy and phylogeny of a Gulf Coast disjunct group of *Spigelia* (Loganiaceae sensu lato). Lundellia 2: 1–13. Henrickson, J. 1996. Notes on *Spigelia* (Loganiaceae). Sida 17: 89–103. Hurley, H. 1968. A Taxonomic Revision of the Genus *Spigelia* (Loganiaceae). Ph.D. dissertation. George Washington University.

1. Corollas scarlet or pink (rarely white), 25–52 mm.
   2. Rhizomes stout; cauline leaf blades 4–12 cm; corollas scarlet (rarely white) outside, yellow to greenish yellow (rarely pink) inside . . . . 2. *Spigelia marilandica*
   2. Rhizomes slender; cauline leaf blades 2–4 cm; corollas light pink outside, light pink to white inside.
      3. Cauline leaf blades ovate, base rounded; corollas 25–30 mm, lobes barely opening at anthesis. . . . . 3. *Spigelia gentianoides*
      3. Cauline leaf blades usually lanceolate to elliptic, rarely narrowly ovate, ovate, or obovate, base cuneate to rounded; corollas 36–50 mm, lobes spreading to reflexed at anthesis. . . . . 4. *Spigelia alabamensis*
1. Corollas white or whitish, 6–17 mm.
   4. Plants annual; cymes 22–45-flowered; corollas 6–9 mm; s Florida . . . . 1. *Spigelia anthelmia*
   4. Plants perennial; cymes 2-flowered; corollas 7–17 mm; n Florida, Texas.
      5. Stems 3+; cauline leaves: proximals with blades (1.2–)1.5–3(–3.5) × 0.3–1 (–1.3) cm, distals usually opposite; open prairies and woods, thin, rocky soils; Texas . . . . 7. *Spigelia hedyotidea*
      5. Stems 1–3; cauline leaves: proximals with blades (1.4–)3–6.5 × (0.5–)0.9–1.7 (–2.2) cm; distals usually pseudowhorled; riparian woodlands and swamps, clay to sandy soils; e Texas or Florida.
         6. Corollas (7–)10–17 mm; Florida . . . . 5. *Spigelia loganioides*
         6. Corollas 8–11(–13) mm; Texas. . . . . 6. *Spigelia texana*

1. Spigelia anthelmia Linnaeus, Sp. Pl. 1: 149. 1753 • West Indian pinkroot [W]

Herbs annual, 15–40 cm; rhizome absent. **Stems** 1, sparsely to profusely branched. **Cauline leaves:** proximals 0 or 1 pair per stem, sessile or petiolate, blade narrowly lanceolate to linear, 1.5–6.3 × 0.4–1.5 cm, base narrowly cuneate; distals pseudowhorled, 2 unequal pairs. **Cymes** 22–45-flowered. **Flowers:** calyx lobes lanceolate-subulate, 2–3(–4) mm; corolla white, each lobe with 2 vertical purple stripes, funnelform, 6–9 mm, lobes spreading at anthesis. **Capsules** 4–5 × 4–7 mm, muricate distally. **Seeds** 2 mm. $2n$ = 32.

Flowering year-round. Disturbed and cultivated soils, roadsides; 0–200 m; Fla.; s Mexico; West Indies (Cayman Islands, Cuba, Dominican Republic, Haiti, Jamaica, Puerto Rico, Virgin Islands); Central America (Belize, Costa Rica, El Salvador, Honduras, Nicaragua, Panama); n South America (French Guiana, Guyana, Suriname, Venezuela); introduced in se Asia (se China, Indonesia), w Africa.

Spigelia anthelmia is widely distributed in the subtropics and tropics. In Florida, it is restricted to Miami-Dade and Monroe counties. From Costa Rica to South America, it overlaps in range and is commonly confused with its close relative *S. hamelioides* Kunth (K. R. Gould 1997; A. V. Popovkin et al. 2011).

2. Spigelia marilandica (Linnaeus) Linnaeus, Syst. Nat. ed. 12, 2: 734. 1767 • Indian pink, woodland pinkroot [E] [F]

Lonicera marilandica Linnaeus, Sp. Pl. 1: 175. 1753

Herbs perennial, 30–60 cm; rhizomes stout. **Stems** 2 or 3, rarely branched. **Cauline leaves:** proximals 3 or 4 pairs per stem, sessile, blade usually ovate to lanceolate, sometimes elliptic, 4–12 × 1–5 cm, base rounded to cuneate; distals opposite. **Cymes** 4–17-flowered. **Flowers:** calyx lobes lanceolate-subulate, (5–)9–13 mm; corolla scarlet (rarely white) outside, yellow to greenish yellow (rarely pink) inside, tubular, narrow at base, gradually widening and inflated just below throat, 45–52 mm, lobes reflexed at anthesis. **Capsules** 5–6 × 8–9 mm. **Seeds** 3–4 mm. $2n$ = 48.

Flowering (Mar–)May(–Jul). Rich, often circumneutral, soils, forest margins; 20–300 m; Ala., Ark., Fla., Ga., Ill., Ind., Ky., La., Miss., Mo., N.C., Okla., S.C., Tenn., Tex.

Despite the epithet, *Spigelia marilandica* is not known to occur in Maryland. The type specimen was allegedly collected in Virginia, where its occurrence is also not documented (T. Wieboldt, pers. comm.). It is closely related to both *S. alabamensis* and *S. gentianoides* (K. R. Gould and R. K. Jansen 1999; A. V. Popovkin et al. 2011) from which it is easily distinguished from the latter two species when in flower by its red and yellow (versus pink) corollas. Various alternate color forms of *S. marilandica* have been found throughout its range,

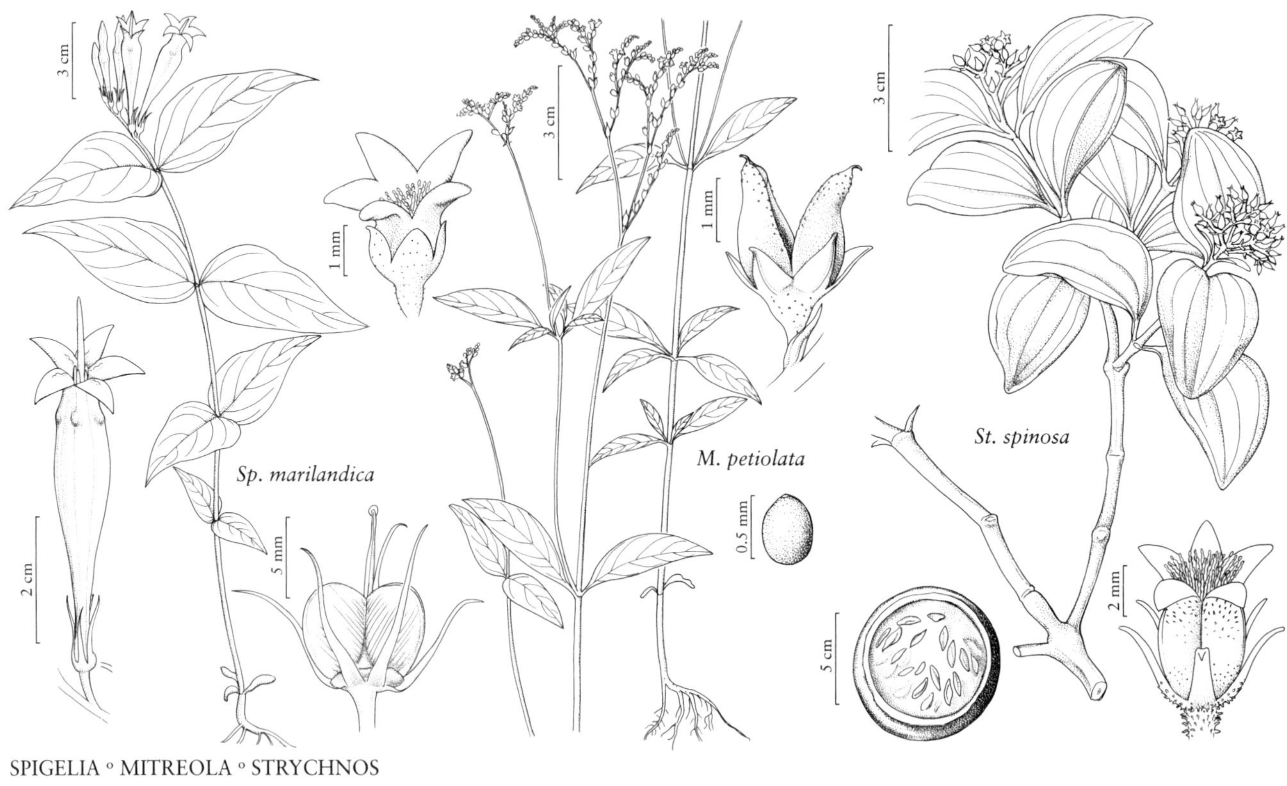

SPIGELIA ◦ MITREOLA ◦ STRYCHNOS

including plants with white corollas outside (green inside), and scarlet corollas outside and pink or pink with red stripes inside. *Spigelia marilandica* overlaps *S. gentianoides* in distribution and may be confused with it when in vegetative state or in fruit; *S. marilandica* is a generally taller (30–60 cm versus 25–40 cm), more robust plant with stout (versus slender) rhizomes, longer (4–12 cm versus 2–4 cm) leaves that are equal to or longer (versus sometimes shorter) than the internodes with acuminate (versus acute to acuminate) apices, and generally has longer inflorescences with up to 17 (versus up to eight) flowers per cyme.

3. **Spigelia gentianoides** Chapman ex A. de Candolle in A. P. de Candolle and A. L. P. P. de Candolle, Prodr. 9: 5. 1845 • Gentian pinkroot [C] [E]

Herbs perennial, 25–40 cm; rhizomes slender. **Stems** 1–3, rarely branched or branched near base. **Cauline leaves:** proximals 3–6 pairs per stem, sessile, blade ovate, 2–4 × 1–2 cm, base rounded; distals opposite. **Cymes** to 8-flowered. **Flowers:** calyx lobes lanceolate; corolla light pink outside, sometimes with 2 dark pink lines on each lobe, light pink to white inside, funnelform, 25–30 mm, lobes barely opening at anthesis. **Capsules** 5–6 × 8 mm. **Seeds** 3 mm.

Flowering (Apr–)May–Jul(–Oct). Sandy loam, upland longleaf pine-wiregrass and pine-oak-hickory woods; of conservation concern; 10–100 m; Ala., Fla.

Spigelia gentianoides is listed as an endangered species under both federal and Florida state law; five extant populations are known (V. Negron-Ortiz 2011). The primary threats to its existence are habitat loss and alteration of its native habitat mainly for pine silviculture and associated disruption of fire regimes (R. A. Kral 1983; Negron-Ortiz). *Spigelia gentianoides* is closely related and morphologically similar to *S. alabamensis*, formerly classified as a disjunct variety of *S. gentianoides*. It can be distinguished from *S. alabamensis* by its ovate (versus lanceolate to elliptic) leaf blades and by its shorter corollas with lobes that do not reflex at anthesis.

4. Spigelia alabamensis (K. Gould) K. G. Mathews & Weakley, J. Bot. Res. Inst. Texas 5: 444. 2011 • Alabama pinkroot [C] [E]

Spigelia gentianoides Chapman ex A. de Candolle var. *alabamensis* K. Gould, Sida 17: 418, fig. 1. 1996

Herbs perennial, 25–40 cm; rhizomes slender. **Stems** 1–3, rarely branched or branched near base. **Cauline leaves:** proximals 3–6 pairs per stem, sessile, blade usually lanceolate to elliptic, rarely narrowly ovate, ovate, or obovate, 2–4 × 1–2 cm, base cuneate to rounded; distals opposite. **Cymes** 2–4(–6)-flowered. **Flowers:** calyx lobes lanceolate, 8–11 mm; corolla light pink outside, sometimes with 2 dark pink lines on each lobe, light pink to white inside, broadly funnelform, 36–50 mm, lobes spreading to reflexed at anthesis. **Capsules** 5–6 × 8 mm, smooth. **Seeds** 3 mm.

Flowering May–Jun. Dolomite outcroppings; of conservation concern; 70–80 m; Ala.

Spigelia alabamensis was formerly classified as a disjunct variety of *S. gentianoides*, from which it can be distinguished by its mainly lanceolate to elliptic (versus ovate) leaves and by its longer calyx and corollas, with corolla lobes that reflex at anthesis. *Spigelia alabamensis* is endemic from 17 dolomite glades, known as the Ketona Glades, in Bibb County.

5. Spigelia loganioides (Torrey & A. Gray ex Endlicher & Fenzl) A. de Candolle in A. P. de Candolle and A. L. P. P. de Candolle, Prodr. 9: 4. 1845 • Florida pinkroot [C] [E]

Coelostylis loganioides Torrey & A. Gray ex Endlicher & Fenzl, Nov. Stirp. Dec. 5: 33. 1839

Herbs perennial, 7.5–27 cm; rhizomes slender. **Stems** 1–3, sparsely branched. **Cauline leaves:** proximals 2–7 pairs per stem, attenuate-petiolate, blade ovate, lanceolate, elliptic, or rhombic, (1.4–)3–6.5 × (0.5–)1–1.7(–2.2) cm, base narrowly cuneate; distals sometimes pseudowhorled. **Cymes** 2-flowered, subsessile. **Flowers:** calyx lobes linear-lanceolate, 3–5 mm; corolla white, suffused with pink or yellow and with 2 pale lavender lines on each lobe outside, white inside, funnelform, (7–)10–17 mm, lobes spreading at anthesis. **Capsules** 2–3 × 4–5 mm. **Seeds** 1–2 mm.

Flowering (Apr–)May(–Jul). Wet woodlands, hydric hammocks, floodplain swamps, limestone-based soils; of conservation concern; 0–20 m; Fla.

Morphologically, *Spigelia loganioides* is nearly identical to *S. texana* of eastern Texas, from which it can be distinguished by its slightly longer (10–17 versus 8–11 mm) corollas and by its inconsistent production of a pseudowhorl of four leaves below the inflorescence (the pseudowhorl is present in about 50% of the specimens examined). Specimens from one possibly extirpated population (Ocala, Marion County), including the type specimen, differ strongly from *S. texana* in having relatively small, mostly broadly ovate, obtuse to acute, strictly opposite leaves and in being relatively short in stature (see J. Henrickson 1996, for illustration). In a molecular phylogenetic analysis including multiple populations of *S. hedyotidea*, *S. loganioides*, and *S. texana* (K. R. Gould and R. K. Jansen 1999), *S. loganioides* was sister to the other two species combined and is therefore recognized as a distinct species from *S. texana*.

Spigelia loganioides is endemic in Levy, Marion, Pasco, Sumter, and Volusia counties.

6. Spigelia texana (Torrey & A. Gray) A. de Candolle in A. P. de Candolle and A. L. P. P. de Candolle, Prodr. 9: 5. 1845 • Texas pinkroot [E]

Coelostylis texana Torrey & A. Gray, Fl. N. Amer. 2: 44. 1841

Herbs perennial, (10–)20–45(–50) cm; rhizomes slender. **Stems** 1–3, seldom branched. **Cauline leaves:** proximals 2 or 3 pairs per stem, attenuate-petiolate; petiole 3 mm; blade ovate, lanceolate, oblong-elliptic, elliptic, or rhombic, 3–5.5 × 0.9–1.7 cm, base narrowly cuneate; distals usually pseudowhorled. **Cymes** 2-flowered. **Pedicels** 1–3 mm. **Flowers:** calyx lobes linear-lanceolate, 2–5 mm; corolla white, suffused with pink or yellow and with 2 pale lavender lines on each lobe outside, white inside, funnelform, 8–11(–13) mm, lobes spreading at anthesis. **Capsules** 2–3 × 4–5 mm. **Seeds** 1–2 mm.

Flowering Apr–Jul. Black-clay soils in thickets, woods along streams, riparian forests; 0–500 m; Tex.

Spigelia texana, of eastern Texas, is most closely related to *S. hedyotidea* of central-western Texas and northern to central Mexico (K. R. Gould and R. K. Jansen 1999). It can be distinguished from *S. hedyotidea* by its taller growth habit, its larger, more membranaceous leaves that dry flat (versus coriaceous and drying wrinkled), its tendency to be glabrous throughout, except at the nodes where stiff papillae occur (versus scabrous throughout), its tendency to produce a pseudowhorl of four leaves (versus paired) that subtend the inflorescence, and its slightly wider corollas.

7. **Spigelia hedyotidea** A. de Candolle in A. P. de Candolle and A. L. P. P. de Candolle, Prodr. 9: 7. 1845 • Prairie pinkroot

Coelostylis lindheimeri (A. Gray) Small; *Spigelia lindheimeri* A. Gray

Herbs perennial, 5–15(–19) cm; rhizomes slender. **Stems** 3+, usually branched. **Cauline leaves:** proximals 3 or 4 pairs per stem, sessile to attenuate-petiolate; blade oblanceolate, lanceolate, ovate, or elliptic, (1.2–)1.5–3 (–3.5) × 0.3–1(–1.3) cm, base narrowly cuneate; distals usually opposite, sometimes pseudowhorled. **Cymes** 2-flowered. **Flowers:** calyx lobes linear to linear-lanceolate, 3–6 mm; corolla white, suffused or lined with pink or yellow outside, white inside, funnelform, 7–13 mm,; lobes spreading at anthesis. **Capsules** 2–4 × 4–6 mm. **Seeds** 1–2 mm.

Flowering Apr–Jun. Open prairies or dry woods on thin, rocky or sandy soils; 10–600 m; Tex.; n, c Mexico.

Spigelia hedyotidea, of central-western Texas and northern to central Mexico, is most closely related to *S. texana* of eastern Texas (K. R. Gould and R. K. Jansen 1999). It can be distinguished from *S. texana* by its bushy growth habit with more than three stems from the base (versus one to three stems), its usually profusely scabrous (versus glabrous) stems, and its paired (versus pseudowhorled) leaves that subtend the inflorescence.

2. MITREOLA Linnaeus, Opera Var., 214. 1758 • Miterwort, hornpod [Greek *mitra*, cap, and *-la*, diminutive, alluding to fruit shape]

Cynoctonum J. F. Gmelin

Herbs or subshrubs, annual [perennial]. **Stems** erect [creeping], unbranched or branched, glabrous [glabrate, scabrous, puberulent, or pilose]. **Leaves** sessile or petiolate; blade narrowly elliptic to elliptic, ovate, or suborbiculate, venation pinnate, surfaces sparsely appressed-hairy [pilose] or glabrous, scabrous or puberulent along margins and veins. **Inflorescences** terminal or axillary, dichasial, 20–100-flowered, each flower subtended by 1 bracetole. **Flowers:** sepals distinct or shortly connate at base, green centrally with whitish edges, ovate, deltate, or oblong; corolla usually white, sometimes mauve [violet], drying purple, urceolate, throat pilose or with ring of hairs; ovary superior; stigmas knoblike, 2-lobed. **Fruits** capsules, green, often drying purple, 2-valved, 2-horned, dehiscent along medial line. **Seeds** gold to dark brown, ellipsoid, obliquely ellipsoid, or depressed-subglobose [fusiform], reticulate or smooth [warty]. $x = 10$.

Species 8–10 (3 in the flora): c, se United States, Mexico, West Indies, Central America, n South America, s Asia (n India, Malaysia [Borneo]), w Africa, Indian Ocean Islands (Madagascar), n Australia.

A. J. M. Leeuwenberg (1974) and J. B. Nelson (1980) gave thorough accounts of the nomenclatural history of *Mitreola* and its synonymy.

1. Leaves petiolate or subsessile, blades 2–8 cm, base cuneate; inflorescences lax, flowers mostly shorter than internodes . . . . 1. *Mitreola petiolata*
1. Leaves subsessile or sessile, blades (0.8–)1.2–3.3 cm, base usually rounded, sometimes cuneate; inflorescences congested, flowers longer than internodes.
    2. Capsule horns papillose-warty over inner and outer faces; seeds smooth; larger leaves 1.5–2 times as long as wide . . . . 2. *Mitreola sessilifolia*
    2. Capsule horns lightly tuberculate mostly over inner faces; seeds reticulate; larger leaves 4–5 times as long as wide . . . . 3. *Mitreola angustifolia*

1. Mitreola petiolata (J. F. Gmelin) Torrey & A. Gray, Fl. N. Amer. 2: 45. 1841 • Lax hornpod [F]

Cynoctonum petiolatum J. F. Gmelin, Syst. Nat. 2: 443. 1791; *C. mitreola* (Linnaeus) Britton; *C. succulentum* R. W. Long

Herbs or subshrubs, 5–80(–100) cm. **Leaves** petiolate or subsessile; blade ovate, narrowly ovate, or elliptic, 2–8 × 0.5–4.5 cm, larger leaves 2–3.5 times as long as wide, base cuneate, surfaces sparsely appressed-hairy or glabrous. **Inflorescences** lax, flowers mostly shorter than internodes, ultimate branches 2–15 cm. **Flowers** subsessile; calyx glabrous; corolla usually white, sometimes mauve, 1.5–2.5 mm, throat pilose. **Capsules** 2–3.5(–5) mm, horns smooth or lightly tuberculate on inner faces. **Seeds** ellipsoid or depressed-subglobose, 0.5–0.7 mm, reticulate. 2*n* = 20.

Flowering Jul–Aug. Marshes, swamp forests, ditches, along roads, grassy plains, fields, often moist, sandy, or rocky places, light shade; 0–1000 m; Ala., Ark., Fla., Ga., La., Miss., Mo., N.C., Okla., S.C., Tenn., Tex., Va.; Mexico; West Indies; Central America.

Recent work (K. M. Neubig et al., unpubl.) would indicate, based on DNA and morphological data, there are at least two additional species that should be segregated from *Mitreola petiolata*, including some Mexican and Central American material and all South American and Old World material included by Leeuwenberg (1974) in his broadly defined *M. petiolata*. Thus, the distribution of *M. petiolata* in the narrow sense is primarily North American, extending south to Guatemala and in the Caribbean.

2. Mitreola sessilifolia (J. F. Gmelin) G. Don, Gen. Hist. 4: 171. 1837 • Small-leaved miterwort, swamp hornpod

Cynoctonum sessilifolium J. F. Gmelin, Syst. Nat. 2: 443. 1791; *C. sessilifolium* var. *microphyllum* R. W. Long

Herbs, 10–50 cm. **Leaves** sessile; blade usually ovate to suborbiculate, sometimes elliptic, (0.8–)1.2–2.5 × (0.5)0.7–1.6 cm, larger leaves 1.5–2 times as long as wide, base usually rounded, sometimes cuneate, surfaces glabrous. **Inflorescences** congested, flowers longer than internodes, ultimate branches 1–3 cm. **Flowers** sessile; calyx papillose outside, glabrous within; corolla white, 2–3 mm, throat with ring of hairs. **Capsules** 2–4 mm, horns papillose-warty on inner and outer faces. **Seeds** obliquely ellipsoid, 0.4 mm, smooth.

Flowering late Jun–Sep(–early Nov). Wet savannas, streamhead ecotones, bogs, swamps, pocosins, other wet depressions; 0–200 m; Ala., Ark., Fla., Ga., La., Miss., N.C., Okla., S.C., Tex., Va.; West Indies (Bahamas).

3. Mitreola angustifolia (Torrey & A. Gray) J. B. Nelson, Phytologia 46: 339. 1980 • Narrow-leaved miterwort or hornpod [E]

Mitreola sessilifolia (J. F. Gmelin) G. Don var. *angustifolia* Torrey & A. Gray, Fl. N. Amer. 2: 45. 1841

Herbs, 10–50 cm. **Leaves** usually sessile, sometimes subsessile; blade usually narrowly elliptic, sometimes elliptic, 0.9–3.3 × 0.2–0.6(–0.7) cm, larger leaves 4–5 times as long as wide, base usually rounded, sometimes cuneate, surfaces glabrous. **Inflorescences** congested, flowers longer than internodes, ultimate branches 1–1.5 cm. **Flowers** sessile; calyx slightly ciliate along margins and midrib outside, glabrous within; corolla white, 2–3 mm, throat with ring of hairs. **Capsules** 2–4 mm, horns lightly tuberculate mostly on inner faces. **Seeds** obliquely ellipsoid, 0.4 mm, reticulate.

Flowering Jun–Sep(–Oct). Coastal plain depressional wetlands; 0–30 m; Ala., Fla., Ga., Miss., S.C.

Cynoctonum angustifolium (Torrey & A. Gray) Small (1896) is a later homonym of *C. angustifolium* Decaisne (1844) and pertains here.

3. STRYCHNOS Linnaeus, Sp. Pl. 1: 189. 1753; Gen. Pl. ed. 5, 86. 1754 • [Greek *strychnon*, name for many different poisonous plants] [I]

Deciduous shrubs or trees [lianas]. Stems erect [climbing], with lateral branches and spines [unarmed or with prickles], glabrous or with simple hairs. **Leaves** petiolate [subsessile]; blade [orbiculate], suborbiculate, ovate, or [narrowly] elliptic, venation 1 or 2 pairs from near base and curved along margin, not reaching apex. **Inflorescences** terminal [and/or axillary], thyrsoid, 1–many-flowered. **Flowers:** sepals ± connate, green, linear [to orbiculate]; corolla white, pale green, green, or yellowish [orange], urceolate to campanulate [rotate to salverform], glabrous or hairy inside and out; ovary superior, 1[or 2]-celled; stigmas oblong [capitate, conic], shallowly lobed. **Fruits** berries, deep yellow, yellow-green, yellow-brown, or orange [red, brown, blue-black]. **Seeds** usually disc-shaped to spheric, smooth. $x = 22$.

Species ca. 200 (1 in the flora): introduced, Florida; Mexico, South America, Asia (India, Sri Lanka), Africa, Indian Ocean Islands (Madagascar), n Australia.

Strychnos is the largest genus in Loganiaceae. Its species are found in tropical rainforests and savannahs in both the Old World and New World Tropics, with the greatest number in tropical Africa. Alkaloids are prevalent in *Strychnos* species (N. G. Bisset and J. D. Phillipson 1971), the best known being strychnine, which is commercially extracted from the southeastern Asian *S. nux-vomica* Linnaeus. Some American species have been used as the primary or secondary ingredients in the dart poison curare (B. A. Krukoff and J. Monachino 1942; Krukoff and R. C. Barneby 1969); medicinal uses have been reported for Old World species (H. M. Burkill 1985–2004).

Brehmia Harvey (1842), a later homonym of *Brehmia* Schrank (1824), pertains here.

SELECTED REFERENCES Krukoff, B. A. and R. C. Barneby. 1969. Supplementary notes on the American species of *Strychnos*. VIII. Mem. New York Bot. Gard. 20: 1–93. Krukoff, B. A. and J. Monachino. 1942. The American species of *Strychnos*. Brittonia 2: 248–322.

1. Strychnos spinosa Lamarck in J. Lamarck and J. Poiret, Tabl. Encycl. 2: 38. 1794 • Monkey- or Natal- or wood-orange [F] [I]

Trunk branching near base; outer bark gray or brown, tending to flake in rectangular segments; inner bark yellowish with green margins. **Branchlets** glabrous or short-hairy; spines axillary and terminal, 1 cm. **Leaves:** stipules narrowly triangular, 1–2 mm; petiole 2–10 mm; blade light to dark green, elliptic, ovate, or suborbiculate, 1.5–9(–13.5) × 1.2–7.5 cm, surfaces glabrous or sparsely hairy. **Inflorescences** congested thyrses, 10–20 mm diam.; bracts linear, 3–4 mm, sparsely hairy. **Pedicels** sparsely to densely pubescent. **Flowers:** sepals 1.5–6 mm; corolla white or pale green, 3–6 mm, glabrous or sparsely hairy outside, throat with ring of hairs inside, lobes triangular, 1.2–2 mm. **Berries** spheric, pulp yellow, exocarp thick, woody, granular, 5–12.5 cm diam. **Seeds** 10–100, brown, obliquely ovate or elliptic, flattened, 2 cm, short-hairy. $2n = 44$.

Flowering Mar–May(–Sep). Disturbed, weedy areas, fencerows, railroad rights-of-way, dumps, vacant lots; 0–20 m; introduced; Fla.; Africa; Indian Ocean Islands (Madagascar, Mauritius).

Strychnos spinosa is cultivated in southern Florida and elsewhere outside its native range for its edible, sour-sweet fruit pulp, although the seeds and unripe fruit are toxic (C. Orwa et al., http://www.worldagroforestry.org/af/treedb/). Herbarium vouchers of *S. spinosa* growing and producing fruit outside of cultivation have been collected sporadically from as early as 1961, from Highlands, Hillsborough, Martin, and Miami-Dade counties. The longevity of these trees is unknown, and the production of seedlings has not been documented. Orwa et al. cited many uses of *S. spinosa* in Africa, including as food, livestock fodder, fuel, timber, musical instruments, poison, and medicine, particularly as a treatment for snakebite. *Strychnos spinosa* occurs in savanna forests throughout its native tropical Africa, as well as in open woodland and riparian fringes, from 0–2200 m elevation (A. J. M. Leeuwenberg 1969; Orwa et al.).

GELSEMIACEAE Struwe & V. A. Albert • Jessamine Family

Alexander Krings

Vines [**shrubs**], perennial, twining (creeping), non-tendrillate. **Leaves** persistent, opposite (whorled), simple, petiolate, stipulate; interpetiolar stipules reduced to ridge [well-developed]; blade membranaceous, margins entire. **Inflorescences** axillary or terminal, flowers solitary or clustered, few-flowered [compound cymes, many-flowered]. **Flowers** bisexual, heterostylous [homostylous], actinomorphic or slightly zygomorphic, 5-merous [rarely 4-merous]; sepals caducous or persistent, 5, distinct; corolla funnelform, lobes imbricate in bud; stamens 5, epipetalous; anthers latrorse; ovary 2-locular; ovules several per locule; placentation axile; stigmas twice divided, appearing 4-cleft. **Fruits** capsules, not inflated [inflated], apex beaked [not beaked]. **Seeds** several, flattened, winged or not.

Genera 2, species 12 (1 genus, 2 species in the flora): e United States, Mexico, Central America, n South America, e Asia, Africa.

1. GELSEMIUM Jussieu, Gen. Pl., 150. 1789 • [Italian *gelsemino*, from Persian-Arabic *yāsamīn*, jasmine, probably alluding to similarity of flowers]

Stems: internodes glabrous, puberulent, or retrorsely scabridulous, nodes glabrous to antrorsely short-pubescent. **Leaves:** stipular colleters linear, 1–4 on each side of petiole base. **Inflorescences:** flowers solitary or in reduced, simple cymes, pedicellate; bracteoles lanceolate, ovate, or deltate, proximal usually with single linear colleter at base on each side. **Flowers** fragrant or not; sepals lanceolate, ovate, or oblong, apex acuminate, obtuse, or rounded; corolla yellow to orange, lobe apex obtuse, rounded, or emarginate [acuminate]; stamens inserted at base to middle of tube; filaments glabrous; anthers connivent or separate, sagittate; styles glabrous. **Capsules** elliptic to oblong, dehiscence septicidal into 4 segments. **Seeds** ovate [elliptic to reniform], wings absent or unilateral [circumferential]. $x = 8$.

Species 3 (2 in the flora): e United States, Mexico, Central America, e Asia.

SELECTED REFERENCES Ornduff, R. 1970e. The systematics and breeding system of *Gelsemium* (Loganiaceae). J. Arnold Arbor. 51: 1–17. Rogers, G. K. 1986. Genera of Loganiaceae in the southeastern United States. J. Arnold Arbor. 67: 143–185.

1. Abaxial leaf surface with small patch of spreading trichomes at extreme base; sepals acuminate; distal 1/2 of pedicels ebracteolate; seeds not winged; capsule beak 2.4–4.2 mm; flowers usually not fragrant . . . . 1. *Gelsemium rankinii*
1. Abaxial leaf surface glabrous or slightly scabridulous at base; sepals obtuse to narrowly rounded; distal 1/2 of pedicels bracteolate; seeds with unilateral wing; capsule beak 1–2 mm; flowers usually fragrant . . . . 2. *Gelsemium sempervirens*

1. Gelsemium rankinii Small, Addisonia 13: 37, plate 435. 1928 • Swamp jessamine E F

Gelsemium nitidum Michaux var. *inodorum* Nuttall

Leaves: petiole 2.5–4.3 mm, glabrous or moderately pubescent, trichomes spreading; blade narrowly lanceolate, elliptic, or ovate, 2–7.1 × 0.8–2.9 cm, base rounded to cuneate, apex obtuse, acute, or acuminate, abaxial surface glabrous except for small patch of spreading trichomes basally, adaxial surface glabrous or scattered-pubescent along midvein. **Pedicels** ebracteolate on distal 1/2. **Inflorescences** cymes, (2- or) 3–5-flowered. **Flowers** usually not fragrant; sepals lanceolate to ovate, 3.3–3.5 × 0.9–1.5 mm, apex acuminate, surfaces glabrous; pin flowers: tube 11–15 mm, lobes ovate to suborbiculate, 9–12 × 8–11 mm, surfaces glabrous; filaments 7–8 mm; anthers 4–4.1 mm; styles 19–24 mm; thrum flowers: tube 16–18 mm, lobes ovate to suborbiculate, 11–12 × 8–11 mm, surfaces glabrous; filaments 19–21 mm; anthers 4.4–4.9 mm; styles 6–8.4 mm. **Capsules** 10–15 × 6–8 mm, beak 2.4–4.2 mm. **Seeds** not winged, 3.3–4.5 × 1.4–2.5 mm. $2n = 16$.

Flowering Mar–May (also sporadically in fall); fruiting Sep–Oct. Swamp forests, primarily *Taxodium-Nyssa*; 0–300 m; Ala., Fla., Ga., La., Miss., N.C., S.C.

2. Gelsemium sempervirens (Linnaeus) J. St.-Hilaire, Expos. Fam. Nat. 1: 338. 1805 • Yellow jessamine F

Bignonia sempervirens Linnaeus, Sp. Pl. 2: 623. 1753; *Gelsemium nitidum* Michaux

Leaves: petiole 2–5.1 mm, glabrous or scabridulous; blade lanceolate, elliptic, or ovate, 2–7 × 0.8–2.1 cm, base rounded to cuneate, apex obtuse, acute, or acuminate, abaxial surface glabrous or slightly scabridulous at base, adaxial surface glabrous. **Pedicels** bracteolate on distal 1/2. **Inflorescences** solitary flowers or cymes, 2- or 3-flowered. **Flowers** usually fragrant; sepals lanceolate, oblong, or ovate, 4.3–4.6 × 1.5–1.6 mm, apex obtuse to narrowly rounded, surfaces glabrous; pin flowers: tube 14–27 mm, lobes ovate to suborbiculate, 6.8–10 × 6.5–9.5 mm, surfaces glabrous; filaments 9.1–10.6 mm; anthers 3.9–4.1 mm; styles 12–22.3 mm; thrum flowers: tube 17–24 mm, lobes ovate to suborbiculate, 8.5–13 × 6.8–-12.1 mm, surfaces glabrous; filaments 19–22 mm; anthers 4–4.1 mm; styles 5–12.1 mm. **Capsules** 13–25 × 7–12 mm, beak 1–2 mm. **Seeds** unilaterally winged, 4.7–5.4 × 3.1–3.3 mm, wing 5–5.7 mm. $2n = 16$.

Flowering Mar–May (also sporadically in fall); fruiting Sep–Nov. Sandy maritime forests to dry upland forests to moist pine flatwoods to swamp forests, roadsides, thickets; 0–1900 m; Ala., Ark., Fla., Ga., La., Miss., N.C., S.C., Tenn., Tex., Va.; Mexico (Chiapas, Oaxaca, Puebla, Veracruz); Central America (Guatemala).

APOCYNACEAE Jussieu
• Milkweed and Dogbane Family

Mark Fishbein
David E. Lemke
Alexander Krings

Trees, shrubs, lianas, vines, or herbs, perennial, taprooted, rhizomatous, or tuberous; latex milky (clear). **Stems** prostrate, trailing, erect, climbing, or twining, indument absent or variously of unicellular or multicellular glandular or eglandular trichomes. **Leaves** deciduous, semipersistent, or persistent, cauline, alternate, opposite, or whorled; stipules absent, stipular colleters variously basal to petiole, interpetiolar, intrapetiolar, or absent; petiole present or absent; blade margins entire, venation pinnate, pinnipalmate, or a single vein, laminar colleters borne at adaxial base or absent. **Inflorescences** extra-axillary, axillary, terminal, or pseudoterminal, cymose, often racemiform, umbelliform, corymbiform, or paniculiform, pedunculate (sessile). **Flowers:** perianth and androecium hypogynous [perigynous, epigynous]; sepals 5, connate, calycine colleters absent or present; petals 5, connate, corolla rotate, rotate-reflexed, campanulate, funnelform, urceolate, tubular, or salverform, aestivation dextrorse, sinistrorse, or valvate; corolline corona absent, annular, or of 5 scales or pads; stamens 5, antisepalous, epipetalous, sometimes connate throughout; anthers 2-thecous, free, connivent, or adnate to gynoecium and forming a gynostegium, margins with corneous wing in species with a gynostegium, 2- or 4-locular; gynostegial corona absent or of 5 separate or united scales, filaments, cups, or other diverse forms; pollen in monads, tetrads, or massed in each theca into a rigid pollinium, translators absent or present, when present, receiving tetrads of adjacent anthers (*Cryptostegia*) or joining pollinia of adjacent anthers to a common corpusculum, together forming a pollinarium; pistils 1, 2-carpellate, connate only by style apices; ovaries superior [partly inferior], 1-locular; styles 1 or 2, apical; stigmas subapical or lateral; ovules few–numerous; nectaries variously around the base of ovaries, on gynostegium, on receptacle, or absent. **Fruits** follicles, capsules, berries, or drupes, solitary or paired, erect or pendulous, striate, spiny, winged, or smooth. **Seeds** few–numerous, brown or black, flattened, navicular, or cylindric, marginally winged or not, comose or not, arillate or not.

Genera ca. 460, species ca. 4800 (36 genera, 175 species in the flora): nearly worldwide; especially in tropics and subtropics.

From the early nineteenth into the latter half of the twentieth centuries, the pollinia- and gynostegium-bearing species were segregated as the family Asclepiadaceae Borkhausen. The structure of the combined androecium and gynoecium and their function in pollination are the most complex among eudicots (P. K. Endress 1994). Ample morphological and molecular phylogenetic evidence provided a sound basis for reuniting these families (M. E. Fallen 1986; H.-E. Wanntorp 1988; B. Sennblad and B. Bremer 1996). As yet, the sister group of Apocynaceae within Gentianales is uncertain, with some analyses indicating Gelsemiaceae and others Gentianaceae (P. F. Stevens 2001 onwards, http://www.mobot.org/MOBOT/research/APweb/). The current classification of Apocynaceae allocates genera to five subfamilies and further into tribes and, in some cases, subtribes (M. E. Endress et al. 2014). Three subfamilies that composed the former Asclepiadaceae (Asclepiadoideae Burnett, Periplocoideae R. Brown ex Endlicher, Secamonoideae Endlicher) have been repeatedly demonstrated to be monophyletic, whereas both of the remaining two subfamilies that composed Apocynaceae in the narrow sense (Apocynoideae Burnett, Rauvolfioideae Kosteletzky) are clearly paraphyletic (T. Livshultz et al. 2007). We follow the current classification here, recognizing that future revisions will be required to obtain monophyletic higher taxa. Genera of Rauvolfioideae (nos. 1–13), Apocynoideae (nos. 14–23), Periplocoideae (no. 24), and Asclepiadoideae (nos. 25–36) are present in the flora area.

Apocynaceae species possess diverse secondary compounds, some of which are potent toxins or are pharmacologically active. The indole alkaloids vincristine and vinblastine present in *Catharanthus roseus* are used in the treatment of Hodgkin's lymphoma, leukemias, and other cancers. Toxic cardiac glycosides in *Asclepias* are sequestered by several species of milkweed-feeding insects for use in their own defense, most famously by the monarch butterfly (S. B. Malcolm 1991). Species of *Vinca* have long been cultivated as evergreen ground covers with showy flowers; species of *Amsonia* and *Asclepias* are also widely cultivated, especially for the nectar-rich flowers that are attractive to butterflies and other insects. Cultivated *Vinca* and *Vincetoxicum* species readily naturalize, are considered invasive, and present serious management concerns.

The following species have been reported as naturalized or escaped from cultivation, but are excluded from this treatment because convincing proof of persisting populations is lacking: *Calotropis gigantea* (Linnaeus) W. T. Aiton, *C. procera* (Aiton) W. T. Aiton, *Gomphocarpus fruticosus* (Linnaeus) W. T. Aiton, *G. physocarpus* E. Meyer, *Metaplexis japonica* (Thunberg) Makino, *Periploca graeca* Linnaeus, and *Vincetoxicum hirundinaria* Medikus.

Calotropis gigantea and C. procera are infrequently cultivated in California and Florida. Two collections of *C. procera* are known outside of cultivation. In California, the species was collected in an agricultural area in Imperial Valley (*Laemmlen s.n.*, UCR-51373). The single individual was destroyed with no subsequent sign of naturalization in the area (A. C. Sanders, pers. comm.). In Florida, collections have been made from a soil dump (suspected to be from the Bahamas) in Hillsborough County in which numerous exotic species appeared (A. R. Franck, pers. comm.; *Dickman s.n.*, USF-273524).

Gomphocarpus fruticosus [*Asclepias fruticosa* Linnaeus] and, especially, *G. physocarpus* [*A. physocarpa* (E. Meyer) Schlechter] have been cultivated in the flora region; the popularity of *G. physocarpus* appears to have increased dramatically in recent years because of its rapid growth and the novelty of the inflated, soft-spined follicles. Neither species is winter hardy in most of our region, although they may persist for a few years as far north as USDA Plant Hardiness Zone 7. Nonetheless, these species have not been documented often by herbarium specimens from the region, and in nearly all cases it is certain that they represent cultivated plants. A recent collection of *G. physocarpus* made in Ventura County, California, in 2016 (*Provance 1116-02*, UCR-278890) may represent the beginning of an established population.

Botanists in California, Florida, and Texas should be cognizant of the possibility that these species may establish here, as they have in subtropical regions around the world.

Metaplexis japonica [*Cynanchum rostellatum* (Turcz.) Liede & Khanum] was reported from an Iowa State University agricultural research farm in the late 1950s and early 1960s (S. L. Welsh and D. E. Anderson 1962). It was suggested that the species may have been introduced during World War II through investigations of fiber sources. Eradication measures were undertaken, and there is no evidence from collections—or observations (D. A. Lewis, pers. comm.)—that the species persists in the state. An additional collection of the species was made in Canyon County, Idaho, in 1966 (*Linford s.n.*, ID-106861). However, no additional collections have been made since, and western botanists report being unaware of any populations of the species in the state at this time (B. Ertter, pers. comm.; J. Smith, pers. comm.).

Until the end of the last century, *Periploca graeca* was cultivated as an ornamental subject across the warmer parts of the flora region. It seems to have fallen out of favor in recent times. There are several collections housed at herbaria, all of which pertain to cultivated plants or plants persisting for years to perhaps decades around homesteads. However, *P. graeca* does not appear to be naturalized anywhere in our region.

Vincetoxicum hirundinaria [synonyms: *Cynanchum medium* (Decaisne) K. Schumann, not R. Brown, *C. vincetoxicum* (Linnaeus) Persoon, *V. medium* Decaisne, *V. officinale* Moench] has been rarely cultivated in the region but is widely reported as a member of regional floras. Reports of naturalized populations ascribed to *V. hirundinaria* (or one of its synonyms) pertain to *V. rossicum*.

SELECTED REFERENCES Endress, M. E., S. Liede, and U. Meve. 2014. An updated classification for Apocynaceae. Phytotaxa 159: 175–194. Fallen, M. E. 1986. Floral structure in the Apocynaceae: Morphological, functional and evolutionary aspects. Bot. Jahrb. Syst. 106: 245–286. Liede, S. et al. 2005. Phylogenetics of the New World subtribes of Asclepiadeae (Apocynaceae–Asclepiadoideae): Metastelmatinae, Oxypetalinae, and Gonolobinae. Syst. Bot. 30: 184–195. Livshultz, T. et al. 2007. Phylogeny of Apocynoideae and the APSA clade (Apocynaceae s. l.). Ann. Missouri Bot. Gard. 94: 324–359. McDonnell, A., M. Parks, and M. Fishbein. 2018. Multilocus phylogenetics of New World milkweed vines (Apocynaceae, Asclepiadoideae, Gonolobinae). Syst. Bot. 43: 77–96. Rosatti, T. J. 1989. The genera of suborder Apocynineae (Apocynaceae and Asclepiadaceae) in the southeastern United States. J. Arnold Arbor. 70: 307–401. Sennblad, B. and B. Bremer. 1996. The familial and subfamilial relationships of Apocynaceae and Asclepiadaceae evaluated with *rbc*L data. Pl. Syst. Evol. 202: 153–175. Woodson, R. E. Jr. 1941. The North American Asclepiadaceae. I. Perspective of the genera. Ann. Missouri Bot. Gard. 28: 193–244.

1. Coronas gynostegial, at least in part; androecium fused with gynoecium into a gynostegium; pollen massed into rigid pollinia [Asclepiadoideae].
   2. Stems not twining.
      3. Leaves alternate or whorled . . . . 26. *Asclepias*, p. 156 (in part)
      3. Leaves opposite or subopposite.
         4. Follicles broadly ovoid, muricate throughout with sparse, thick protuberances, too heavy for stems to support and lying on ground; stems prostrate to ascending; herbage clothed in a mixed indument of long, eglandular and minute, glandular trichomes. . . . . 30. *Matelea*, p. 239 (in part)
         4. Follicles narrowly to broadly fusiform or lance-ovoid, not muricate throughout (if muricate throughout, follicles erect, aerial); stems prostrate to erect or vining; herbage glabrous or with various indumentum, but not glandular (if glandular in part [*Pherotrichis*], follicles aerial and sharply deflexed).
            5. Follicles sharply deflexed; stems and leaves densely hirsute with long eglandular and minute glandular trichomes; corollas urceolate, cream with longitudinal green lines . . . . 34. *Pherotrichis*, p. 262
            5. Follicles erect, spreading, or pendulous; stems and leaves glabrous or with various indumentum, trichomes never glandular; corollas rotate or campanulate, variously colored, but never cream with longitudinal lines.

6. Inflorescences umbellate, extra-axillary; corollas variously colored; follicles usually erect (if pendulous, coronas white, cream, ochroleucous, or rarely red-violet in part, sometimes tinged pink, green, or yellow, or with red-violet line) . . . . 26. *Asclepias*, p. 156 (in part)
6. Inflorescences cymes, axillary; coronas pinkish tan, reddish, or dark purple; follicles pendulous. . . . . 36. *Vincetoxicum*, p. 266 (in part)

[2. Shifted to left margin.—Ed.]

2. Stems twining, at least at tips.
   7. Leaf bases cuneate to rounded (subcordate), blades linear.
      8. Coronas double, consisting of a pentagonal ring at base of staminal column and 5 discrete, vesicular segments at base of anthers; corolla lobes 3+ mm . . .28. *Funastrum*, p. 230 (in part)
      8. Coronas absent or single, consisting of 5 discrete or basally united, laminar segments arising from staminal column; corolla lobes to 4 mm.
         9. Coronas absent; corollas yellow, becoming orange with age, tubular with incurved lobes . . . . 28. *Funastrum*, p. 230 (in part)
         9. Coronas single, consisting of 5 discrete or basally united, laminar segments arising from staminal column; corollas cream, yellow, or green, not becoming orange with age, campanulate, lobes not incurved.
            10. Inflorescences axillary, sessile; corona segments united at base; corolla lobes glabrous adaxially; leaves caducous and stems often leafless. . . . . .32. *Orthosia*, p. 260
            10. Inflorescences extra-axillary, pedunculate (if sessile, corolla lobes densely puberulent or villous); corona segments discrete or united; corolla lobes glabrous, puberulent, or villous adaxially; leaves persistent.
                11. Corolla lobes puberulent to villous adaxially; leaf blades chartaceous; corona segments discrete; aestivation valvate . . . . .31. *Metastelma*, p. 256 (in part)
                11. Corolla lobes glabrous adaxially; leaf blades fleshy; corona segments united at base; aestivation contort-dextrorse . . . . .33. *Pattalias*, p. 261
   7. Leaf bases truncate to cordate, blades lanceolate to ovate or orbiculate.
      12. Stems and/or leaves with a mixed indument of minute glandular hairs and longer eglandular hairs (if glandular hairs obscure, follicles with winged angles and dorsal anther appendages present); follicles 5-angled, muricate or tuberculate (if smooth, then gray striate and apex acuminate or long-attenuate).
          13. Anthers with laminar dorsal appendages; follicles angled. . . . . .29. *Gonolobus*, p. 236
          13. Dorsal anther appendages lacking; follicles not angled, surfaces muricate, tuberculate, or smooth.
              14. Follicles smooth, muricate, or tuberculate, sometimes gray striate; glandular trichomes not accumulating white crystals in dried specimens; style apex concave, planar, or convex, not bifid . . . . 30. *Matelea*, p. 239 (in part)
              14. Follicles smooth, mottled green and gray; glandular trichomes accumulating white crystals in dried specimens; style apex beaked, bifid. . . . . 35. *Polystemma*, p. 264
      12. Stems and leaves puberulent to glabrous, eglandular; follicles lacking angles and protuberances, not mottled.
          15. Latex clear; inflorescences axillary or extra-axillary (if extra-axillary, corona segments deeply bifid).
              16 Corollas and coronas cream; inflorescences extra-axillary; coronas of 5 separate, laminar, deeply bifid segments; follicles 2.5+ cm wide. . . . . 27. *Cynanchum*, p. 228 (in part)
              16. Corollas and coronas pale reddish brown to dark purple; inflorescences axillary; coronas thick, 5-lobed rings or 5 thick, entire segments; follicles to 1 cm wide . . . . .36. *Vincetoxicum*, p. 266 (in part)
          15. Latex white; inflorescences extra-axillary (corona segments entire or 3-lobed).
              17. Corolla lobes 1.5–3.5 mm; coronas 1 whorl of 5 discrete, laminar segments 1–2.5 mm; follicles 0.3–0.7 cm wide . . . . .31. *Metastelma*, p. 256 (in part)
              17. Corolla lobes 3–13 mm; coronas 1 or 2 whorls of discrete or basally united segments or a ring, if segments discrete and laminar, then 2.5–6 mm; follicles 0.3–7 cm wide.

18. Coronas 2 whorls, a pentagonal ring at base of staminal column and 5 discrete, vesicular segments at base of anthers; follicles 0.3–1.6 cm wide . . . . 28. *Funastrum*, p. 230 (in part)
18. Coronas 1 whorl of 5 discrete or basally united, laminar segments or a tube longer than and obscuring gynostegium; follicles 1.5–7 cm wide.
19. Corollas pubescent abaxially; coronas a tube or 5 discrete, oblong or quadrate segments; seeds 5–7 × 1.5–3 mm . . . . 25. *Araujia*, p. 154
19. Corollas glabrous abaxially; coronas of 5 basally united, laminar segments; seeds 6–10 × 4–6 mm . . . . 27. *Cynanchum*, p. 228 (in part)
1. Coronas absent or corolline only; stamens distinct, anthers connivent or not, but androecium not fused with gynoecium into a gynostegium; pollen free, sometimes in tetrads [Apocynoideae, Periplocoideae, Rauvolfioideae].
20. Pollen shed onto translators; follicles fusiform, strongly 3-sided . . . . .24. *Cryptostegia*, p. 152
20. Translators absent; fruits various, if follicles, then terete or somewhat compressed in cross section, not 3-sided.
21. Stems climbing, twining, trailing, or sprawling (sometimes suberect in *Pentalinon* and *Rhabdadenia*).
22. Corollas yellow.
23. Leaves usually whorled, sometimes opposite; aestivation sinistrorse; capsules solitary, spiny . . . . 1. *Allamanda*, p. 109
23. Leaves opposite; aestivation dextrorse; follicles paired, glabrous or pubescent, but not spiny.
24. Corollas to 10 mm, lobes 3–4 mm, spreading to strongly reflexed; follicles 10–23 mm . . . . 22. *Thyrsanthella*, p. 151 (in part)
24. Corollas 15+ mm, lobes 7–25 mm, spreading to ascending; follicles 60+ mm.
25. Petioles 1–2 mm; abaxial leaf surface glabrous; calyx lobes broadly ovate, 1–2.5 mm, glabrous; corolla lobes 7–13 mm; anther connectives not appendiculate; follicles 60–90 mm . . . . 14. *Angadenia*, p. 137 (in part)
25. Petioles 5–12 mm; abaxial leaf surface pubescent; calyx lobes linear-lanceolate, 7–12 mm, pubescent or glabrate; corolla lobes 20–25 mm; anther connectives appendiculate, elongate appendages intertwined; follicles 120–180 mm. . . . . 20. *Pentalinon*, p. 148 (in part)
22. Corollas white, pinkish white, pale yellow, cream, or shades of blue or purple.
26. Aestivation sinistrorse; corollas blue-purple, blue-violet, violet, or reddish purple (rarely white or pale blue). . . . . 13. *Vinca*, p. 135 (in part)
26. Aestivation dextrorse; corollas white, pinkish white, pale yellow or cream.
27. Corollas to 10 mm, lobes 2–5 mm wide.
28. Leaves deciduous; corolla lobes 3–4 × 2–3 mm, spreading to reflexed. . . . . 22. *Thyrsanthella*, p. 151 (in part)
28. Leaves persistent; corolla lobes 7–10 × 4–5 mm, spreading. . . . . .23. *Trachelospermum*, p. 151
27. Corollas 15+ mm, lobes 6+ mm wide.
29. Corollas salverform, adjacent corolla lobes not overlapping or for at most ¼ length from the base. . . . . 17. *Echites*, p. 143
29. Corollas funnelform, adjacent corolla lobes overlapping for ½+ of their length from the base.
30. Petioles 1–2 mm; calyx lobes 1–2.5 mm; seeds 4–8 mm . . . . 14. *Angadenia*, p. 137 (in part)
30. Petioles 5–15 mm; calyx lobes 5–12 mm; seeds 10–30 mm.
31. Peduncles 20–40 mm, sparsely pubescent; calycine colleters present; seeds very narrowly oblong, 10–12 mm . . . . 20. *Pentalinon*, p. 148 (in part)
31. Peduncles 75–100 mm, glabrous; calycine colleters absent; seeds linear, 25–30 mm . . . . 21. *Rhabdadenia*, p. 149

[21. Shifted to left margin.—Ed.]

21. Stems erect or ascending (slightly spreading in *Catharanthus* and *Cycladenia*).
    32. Stems armed with paired spines . . . . 4. *Carissa*, p. 125
    32. Stems unarmed.
        33. Leaves whorled (sometimes opposite at lower nodes; subverticillate keys in alternate lead).
            34. Aestivation sinistrorse; shrubs; corollas white, lobes ± as long as wide; fruits fleshy, initially red, maturing black . . . . 9. *Rauvolfia*, p. 130
            34. Aestivation dextrorse; combination of characters otherwise.
                35. Latex milky; calycine colleters absent; corollas salverform, lobes 3–7 mm; coralline coronas absent; anthers not connivent, not adherent to stigma; nectaries annular . . . . 2. *Alstonia*, p. 110
                35. Latex clear; calycine colleters present; corollas funnelform, lobes 15–25 mm; coralline coronas lacerate; anthers connivent, adherent to stigma; nectaries absent . . . . 19. *Nerium*, p. 148 (in part)
        33. Leaves alternate or opposite (subverticillate in *Amsonia* and *Mandevilla*).
            36. Aestivation dextrorse.
                37. Herbaceous perennials; calycine colleters absent.
                    38. Stems 20+ cm; latex milky; stipular colleters interpetiolar; corolla lobes to 3 mm; coralline coronas of 5 small, sagittate squamellae; pollen in persistent tetrads . . . . 15. *Apocynum*, p. 138
                    38. Stems 6–12 cm; latex not milky; stipular colleters absent; corolla lobes 4+ mm; coralline coronas absent; pollen free . . . . 16. *Cycladenia*, p. 141
                37. Subshrubs, shrubs, or trees; calycine colleters present (absent in *Ochrosia*).
                    39. Latex clear; coralline coronas lacerate; nectaries absent . . . . 19. *Nerium*, p. 148 (in part)
                    39. Latex milky; coralline coronas absent; nectaries 2–5 or annular.
                        40. Leaves deciduous, laminar colleters present; stipular colleters interpetiolar; Arizona, New Mexico, Texas . . . . 18. *Mandevilla*, p. 144
                        40. Leaves persistent, laminar colleters absent; stipular colleters absent or intrapetiolar; Florida.
                            41. Stipular colleters intrapetiolar; corollas white or cream, nectaries annular; fruits drupaceous, red, ellipsoid to ovoid . . . 7. *Ochrosia*, p. 129
                            41. Stipular colleters absent; corollas yellow to very pale yellow, nectaries 5, alternating with stamens; fruits follicles, green, terete to slightly moniliform . . . . 14. *Angadenia*, p. 137 (in part)
            36. Aestivation sinistrorse.
                42. Leaves opposite.
                    43. Stipular colleters absent; corollas blue-purple, blue-violet, violet, or reddish purple (rarely white or pale blue) . . . . 13. *Vinca*, p. 135 (in part)
                    43. Stipular colleters intrapetiolar or both intra- and interpetiolar; corollas white, cream, pink, red, magenta, red-violet, or yellow.
                        44. Leaf surfaces, peduncles, and calyces pubescent; corollas yellow . . . . 6. *Haplophyton*, p. 128 (in part)
                        44. Leaf surfaces, peduncles, and calyces glabrous (sometimes sparsely pubescent in *Catharanthus roseus* and ciliate in *Tabernaemontana divaricata*); corollas white, cream, pink, red, magenta, or red-violet (sometimes yellow in *Tabernaemontana alba*).
                            45. Subshrubs or herbs, 3–10(–20) dm; corollas white, pink, red, magenta, or red-violet, tube (15–)20–30 mm, throat 4–5 mm, lobes broadly obovate, often mucronulate; follicles erect; seeds not arillate . . . . 5. *Catharanthus*, p. 126
                            45. Shrubs or trees, 5–30(–150) dm; corollas white or cream (sometimes yellow in *Tabernaemontana alba*), tube 3–12 mm, throat 3–15 mm, lobes obliquely elliptic, obovate, or dolabriform; follicles deflexed; seeds arillate . . . . 10. *Tabernaemontana*, p. 132
                42. Leaves alternate (to subverticillate in *Amsonia*).

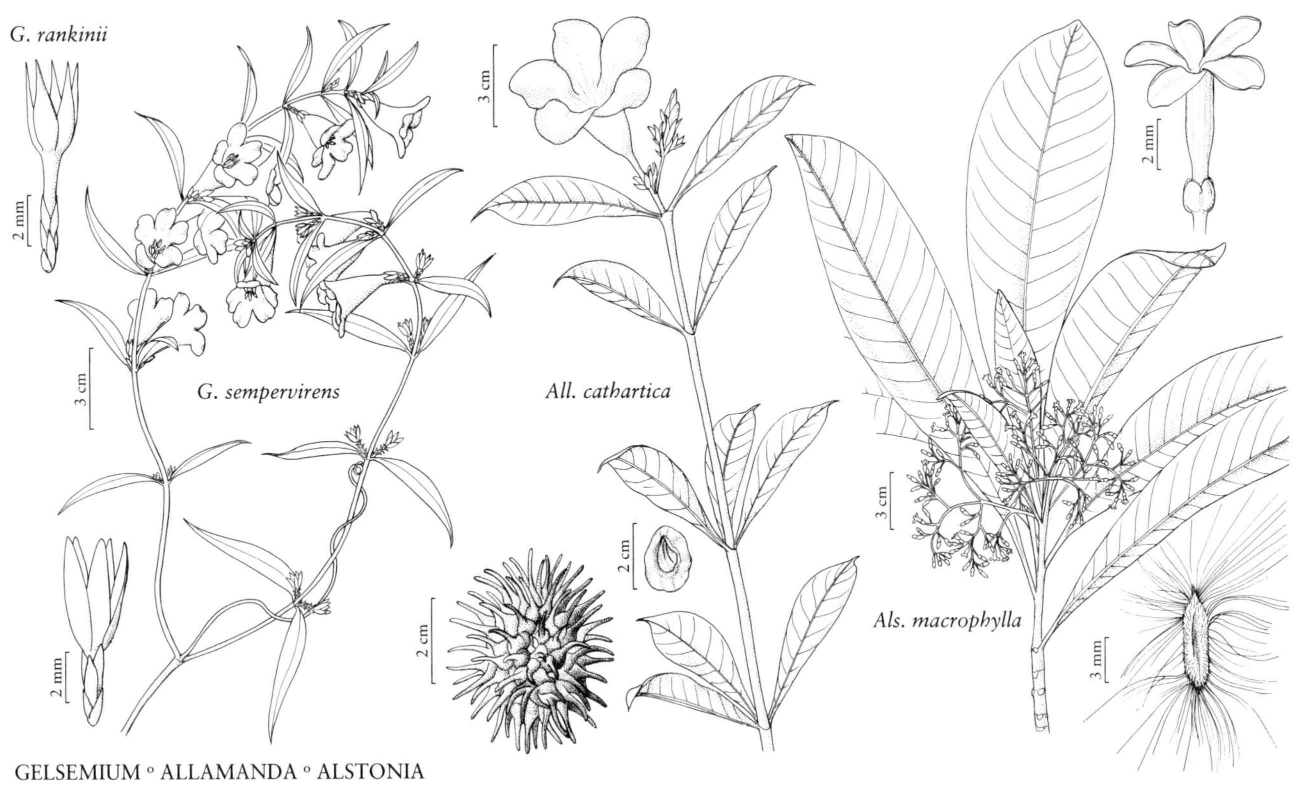

GELSEMIUM ◦ ALLAMANDA ◦ ALSTONIA

[46. Shifted to left margin.—Ed.]

46. Herbaceous perennials or shrubs/subshrubs to 60 cm.
    47. Inflorescences terminal, thyrsoid or corymbose cymes; corollas blue to purplish or lavender (infrequently white or pink); seeds not comose . . . . 3. *Amsonia*, p. 112
    47. Inflorescences axillary, solitary or occasionally 2-flowered; corollas yellow; seeds with comas on both ends. . . . . 6. *Haplophyton*, p. 128 (in part)
46. Trees or shrubs 1+ m.
    48. Corollas yellow or orange; fruits green to black, dehiscent drupes . . . . 11. *Thevetia*, p. 133
    48. Corollas white; fruits brown follicles or white, indehiscent drupes.
        49. Corolla tubes 9–20 mm, lobes (15–)25–35(–45) × 10–15 mm; stamens inserted near base of corolla tube; follicles brown; leaves deciduous . . . . 8. *Plumeria*, p. 129
        49. Corolla tubes 6–7 mm, lobes 3–5 × 1.2–1.6 mm; stamens inserted at or near orifice of corolla tube; drupes white; leaves persistent . . . . 12. *Vallesia*, p. 135

1. ALLAMANDA Linnaeus, Mant. Pl. 2: 146, 214 [as Allemanda]. 1771 • [For Frederique Louis Allamand, 1736–1803, Swiss physician and botanist, correspondent of Linnaeus] I

David E. Lemke

Shrubs [lianas, subshrubs, trees]; latex milky. **Stems** trailing [erect], unarmed, glabrous or eglandular-pubescent. **Leaves** persistent, whorled [subwhorled], rarely opposite [distally alternate], petiolate or sessile; stipular colleters present, intrapetiolar; laminar colleters absent. **Inflorescences** axillary or subterminal, cymose, pedunculate. **Flowers:** calycine colleters absent; corolla yellow [pink, red, violet], funnelform, aestivation sinistrorse; coroline corona dissected;

androecium and gynoecium not united into a gynostegium; stamens inserted near top of corolla tube; anthers not connivent, not adherent to stigma; connectives not appendiculate, locules 4; pollen free, not massed into pollinia, translators absent; nectary annular. **Fruits** capsules, solitary, erect, green or tinged with purple-red, subglobose to globose, surface spiny [smooth], glabrous. **Seeds** orbiculate to ovoid, flattened, winged, not beaked, not comose, not arillate. *x* = 9.

Species ca. 15 (1 in the flora): introduced, Florida; South America; some species introduced in tropical regions nearly worldwide.

The systematic position of *Allamanda* within Apocynaceae has long been uncertain. At anthesis the ovary is compound, leading K. M. Schumann (1895) to place *Allamanda* with other genera with compound ovaries in a supposed primitive alliance in the family. M. E. Fallen (1985) has demonstrated that the compound ovary is derived from a gynoecium with two separate ovaries and, based on both molecular and morphological evidence (A. O. Simões et al. 2007), the genus is currently placed in the Plumerieae clade of the subfamily Rauvolfioideae.

SELECTED REFERENCES Fallen, M. E. 1985. The gynoecial development and systematic position of *Allamanda* (Apocynaceae). Amer. J. Bot. 72: 572–579. Sakane, M. and G. J. Shepherd. 1986. Uma revisão do gênero *Allamanda* (Apocynaceae). Revista Brasil. Bot. 9: 125–149.

1. **Allamanda cathartica** Linnaeus, Mant. Pl. 2: 214. 1771 (as Allemanda) • Golden trumpet, brownbud allamanda [F] [I]

Shrubs sprawling. **Leaves:** petiole 0–2 mm, pubescent; blade oblanceolate, 50–150 × 15–40 mm, subcoriaceous, base attenuate, margins often slightly revolute, apex cuspidate, surfaces pubescent along midvein abaxially. **Peduncles** 10–40 mm, pubescent. **Pedicels** 4–6 mm, pubescent. **Flowers:** calyx lobes ovate to narrowly ovate, 10–12 mm, glabrous; corolla tube 20–35 × 2–3.5 mm, throat 20–40 × 15–25 mm, lobes spreading, obliquely ovate to suborbiculate, 12–35 × 10–40 mm. **Capsules** 4–5 × 3.5–4 cm. **Seeds** 13–22 mm. 2*n* = 18.

Flowering spring–fall; fruiting summer–fall. Hammocks, lakeshores, dunes, abandoned citrus groves and other disturbed sites; 0–20 m; introduced; Fla.; ne South America; introduced also in Mexico, Central America, and in tropical regions nearly worldwide.

Allamanda cathartica is native to northeastern South America (R. E. Woodson Jr. et al. 1970) but is widely cultivated and has escaped and become naturalized in both the New World and Old World Tropics. It is naturalized along much of the eastern coast of southern Mexico and Central America and has been reported to reach treelike proportions in Venezuela (T. J. Rosatti 1989). The globose capsules with spines to 15 mm are unique among the genera of Apocynaceae in the flora area.

The medicinal and toxic properties of *Allamanda cathartica* have been well documented. U. Quattrocchi (2012) reported that the latex may cause a contact dermatitis in sensitive individuals and severe catharsis on ingestion. An infusion of the leaves has been used as both an emetic and a purgative and has been demonstrated to have strong fungal toxicity against certain dermatophytes. According to G. E. Burrows and R. J. Tyrl (2013), the purgative effects result from the action of several types of iridoid compounds.

2. ALSTONIA R. Brown, Asclepiadeae, 64. 1810, name conserved • [For Charles Alston, 1683–1760, Scottish professor of botany, University of Edinburgh] [I]

David E. Lemke

Trees [**shrubs**]; latex milky. **Stems** erect, unarmed, glabrous [eglandular-pubescent]. **Leaves** persistent, whorled [opposite], petiolate; stipular colleters interpetiolar; laminar colleters absent. **Inflorescences** terminal, compound-cymose, pedunculate. **Flowers:** calycine colleters absent; corolla white, yellow, or cream, salverform, aestivation dextrorse [sinistrorse]; corolline corona absent; androecium and gynoecium not united into a gynostegium; stamens distinct, inserted at top of corolla tube; anthers not connivent, not adherent to stigma, connectives not appendiculate or enlarged, locules 4; pollen free, not massed into pollinia, translators absent; nectary annular. **Fruits** follicles, usually paired, pendulous, brown, slender, terete, surface striate, glabrous or pubescent. **Seeds** elliptic to oblong, flattened, marginally long-ciliate, not winged, not beaked, comose, not arillate. x = 11.

Species ca. 43 (2 in the flora): introduced, Florida; Asia, Africa, Indian Ocean Islands, Pacific Islands, Australia.

SELECTED REFERENCES Monachino, J. 1949. A revision of the genus *Alstonia* (Apocynaceae). Pacific Sci. 3: 133–182. Sidiyasa, K. 1998. Taxonomy, phylogeny, and wood anatomy of *Alstonia* (Apocynaceae). Blumea, Suppl. 11: 1–230.

1. Leaves in whorls of 3 or 4; corolla tube and throat together to 6.5 mm; flowering winter–spring . . . . 1. *Alstonia macrophylla*
1. Leaves in whorls of 4–8; corolla tube and throat together 7+ mm; flowering fall. . . . . 2. *Alstonia scholaris*

1. **Alstonia macrophylla** Wallich ex G. Don, Gen. Hist. 4: 87. 1837 • Deviltree [F] [I]

Trees 3–10(–50) m. **Leaves** in whorls of 3 or 4; petiole 2–25 mm, glabrous; blade narrowly elliptic to elliptic, obovate, or narrowly obovate, 4.5–25(–30) × 1.5–10 cm, coriaceous, base acute to decurrent on petiole, margins not revolute, apex rounded to narrowly acuminate, surfaces glabrous or pubescent abaxially. **Peduncles** 1–6(–8) cm, pubescent or glabrous. **Pedicels** (0–)1–4(–6) mm, pubescent or glabrous. **Flowers:** calyx lobes ovate to broadly ovate, 1–2.5 mm, pubescent or glabrous, ciliate; corolla white or cream, glabrous or very sparsely eglandular-pubescent abaxially, eglandular-pubescent adaxially, tube and throat together to 6.5 mm, tube 2–3.5 × 1–1.5 mm, throat 2–3 × 1.5–2 mm, lobes spreading, ovate or obliquely ovate, 3–7 × 1–2.5 mm. **Follicles** 25–60 × 0.2–0.4 cm. **Seeds** 5–10(–12) × 1.5–4 mm.

Flowering winter–spring; fruiting spring–summer. Disturbed pinelands and coastal hammocks; 0–10 m; introduced; Fla.; se Asia; Indian Ocean Islands; Pacific Islands.

Alstonia macrophylla has become naturalized in a few coastal hammocks and along the margins of disturbed pinelands in Miami-Dade County.

Stem, root, and leaf extracts of *Alstonia macrophylla* are commonly employed in traditional medicine in Thailand. M. S. Khyade et al. (2014) reported that leaf extracts of *A. macrophylla* exhibited antimicrobial, anti-inflammatory, and antipyretic activity, and an extract of the root bark demonstrated potent activity against a multidrug-resistant strain of the malarial parasite *Plasmodium falciparum*.

2. **Alstonia scholaris** (Linnaeus) R. Brown, Asclepiadeae, 65. 1810 • Blackboard tree, Indian deviltree [I]

Echites scholaris Linnaeus, Mant. Pl. 1: 53. 1767

Trees 2–20(–60) m. **Leaves** in whorls of 4–8; petiole 5–20(–25) mm, glabrous; blade narrowly elliptic to obovate, 5–17(–22) × 2–7(–8.5) cm, subcoriaceous, base cuneate, usually decurrent on petiole, less often acute or obtuse, margins somewhat revolute, apex obtuse, rounded, or retuse, surfaces glabrous or pubescent abaxially. **Peduncles** 0.5–5(–9) cm, pubescent or glabrate. **Pedicels** 0–2 mm, pubescent. **Flowers:** calyx lobes ovate, 1.5–2.4 mm, pubescent, ciliate; corolla white, yellow, or cream, eglandular-pubescent abaxially and adaxially, tube and throat together 7+ mm, tube 4–6 × 1–1.5 mm, throat 3–4 × 1.5–2 mm, lobes spreading, broadly ovate or suborbiculate, 3–5 × 2.5–4.5 mm. **Follicles** 20–40(–60) × 0.2–0.3 cm. **Seeds** 4–5(–7) mm. $2n$ = 44.

Flowering fall; fruiting material not seen. Coastal hardwood hammocks; 0 m; introduced; Fla.; Asia; Pacific Islands; Australia.

Alstonia scholaris is occasionally cultivated in southern Florida, where it may produce stems to 20 m in height and 1 m in diameter, and has become naturalized at a few sites in Broward, Miami-Dade, and Palm Beach counties.

M. S. Khyade et al. (2014) provided a lengthy list of the traditional medicinal uses of leaf, bark, and root extracts of *Alstonia scholaris* in India and document the antimicrobial, analgesic, anti-inflammatory, and possible anticancer potential of alkaloids isolated from the plant. In India, the wood was formerly used to make chalkboards (H. Drury 1873).

3. AMSONIA Walter, Fl. Carol., 11, 98. 1788 • Blue-star [Probably for Dr. John Amson, eighteenth-century physician in Williamsburg, Virginia, serving as its mayor 1750 and 1751]

Linh Tõ Ngô

Wendy L. Applequist

Herbs, perennial; latex milky. **Stems** usually erect and clumping, unarmed, glabrous or pubescent with eglandular hairs. **Leaves** deciduous, alternate or subverticillate, petiolate or sessile, reduced in size toward stem base, usually at least slightly heteromorphic with branch leaves proportionately narrower than main stem leaves; stipular colleters absent; laminar colleters absent. **Inflorescences** terminal, thyrsoid or corymbose cymes, distal branches occasionally terminated by small inflorescences that almost never set fruit, pedunculate. **Flowers:** calycine colleters absent; corolla blue to purplish, white, or lavender (pink), salverform, aestivation sinistrorse; corolline corona absent; androecium and gynoecium not united into a gynostegium; stamens inserted near top of corolla tube; anthers not connivent, not adherent to stigma, connectives not appendiculate, locules 4; pollen free, not massed into pollinia, translators absent; nectary absent or annular. **Fruits** follicles, paired, brown, slender, terete or moniliform, smooth, glabrous or rarely with patchy indument. **Seeds** cylindric, fusiform, or ± terete, not winged, not beaked, not comose, not arillate. $x = 11$.

Species 17 (15 in the flora): United States, n Mexico, Europe (Greece, Turkey), e Asia.

Amsonia is taxonomically problematic. Species of the southeastern and south-central United States are highly variable with, in some cases, no definitive species or varietal boundaries and persistent nomenclatural issues. Southwestern species are often variable and sometimes very similar to other species. This treatment conservatively maintains the species circumscriptions that were favored by S. P. McLaughlin (1982), except as noted. Three subgenera are recognized in this treatment, reduced from four in some recent literature. The southwestern subg. *Articularia* and *Sphinctosiphon* are most distinguishable in fruit, at which time some species of subg. *Sphinctosiphon* are hard to distinguish from one another.

SELECTED REFERENCE McLaughlin, S. P. 1982. A revision of the southwestern species of *Amsonia* (Apocynaceae). Ann. Missouri Bot. Gard. 69: 336–350.

1. Corolla tubes broadest at apex, not constricted; stigmas depressed-capitate to truncate; bracts on distal portions of inflorescences usually inconspicuous, to 2.5 mm, narrowly deltate to deltate or ovate; sc, se United States, Kansas to Virginia, s to e Texas to Florida .... 3a. *Amsonia* subg. *Amsonia*, p. 113
1. Corolla tubes slightly narrowing to conspicuously constricted at apex; stigmas apiculate with 2 small lobes; bracts on distal portions of inflorescences usually conspicuous, mostly 2–6 mm, linear to narrowly deltate (in *A. jonesii* usually short, those subtending individual flowers often less than 1 mm or absent); sw United States to c Texas.
   2. Stems usually branched for most of length (sometimes only distally); corolla tubes (7–)8–12(–13) mm, moderately to strongly constricted in narrow band just below apex (most visible just before anthesis); mature follicles moniliform, strongly constricted between seeds; seeds fusiform with acute to rounded-acute or flat-truncate (rarely diagonally truncate) ends, with mostly smooth surface .... 3b. *Amsonia* subg. *Articularia*, p. 118
   2. Stems usually branched only on distal portion (sometimes to near base, rarely unbranched); corolla tubes (6–)7.5–41(–45) mm, broadest below apex, slightly narrowing to moderately constricted at apex; mature follicles terete, at most slightly narrowed between seeds; seeds cylindric with diagonally truncate to flat-truncate ends, with mostly convoluted surface........ 3c. *Amsonia* subg. *Sphinctosiphon*, p. 120

3a. Amsonia Walter subg. Amsonia

Stems usually branched only on distal portion. **Bracts** on distal portions of inflorescences narrowly deltate to deltate or ovate, to 2.5 mm. **Flowers:** corolla tube (5–)6–9(–10) mm, widest at apex, not distally constricted; stigma depressed-capitate to truncate. **Follicles** terete, not strongly constricted between seeds. **Seeds** reddish brown, terete with usually diagonally truncate ends, with convoluted surface.

Species 7 (5 in the flora): United States, n Mexico, Europe (Greece, Turkey), e Asia.

Delimitation of species and varieties in the southeastern subg. *Amsonia* is challenging. The distinctions between the widespread and variable *Amsonia tabernaemontana* and the geographically more restricted *A. ludoviciana* and *A. rigida* are small or inconsistent. Problematic specimens identified as probable interspecific hybrids, including hybrids between *A. ciliata* and the *A. tabernaemontana* group, also exist. Outside North America, subg. *Amsonia* includes *A. orientalis* Decaisne of Europe and *A. elliptica* (Thunberg) Roemer & Schultes of Asia.

1. Abaxial leaf surfaces densely (to moderately) tomentose . . . . 3. *Amsonia ludoviciana*
1. Abaxial leaf surfaces glabrous or sparsely or moderately pubescent, not tomentose.
    2. Stem leaf blades linear.
        3. Stem leaf blades 2.5–4.6(–5.5) cm × 0.3–1(–1.6) mm (–2.4 mm if unrolled); North Carolina to Florida, Missouri, Oklahoma. . . . . .1. *Amsonia ciliata* (in part)
        3. Stem leaf blades (5–)7.5–11 cm × 2–3.2(–4.2) mm; Arkansas, Oklahoma. . . . . .2. *Amsonia hubrichtii*
    2. Stem leaf blades narrowly elliptic, elliptic, narrowly oblong, ligulate, lanceolate, ovate, or obovate.
        4. Leaf blades at least slightly heteromorphic, surfaces glabrous except margins sometimes ciliate; stem leaf blades ligulate, very narrowly elliptic to elliptic or very narrowly lanceolate, 2.5–6.5 cm × 3–15 mm; branch leaf blades linear to ligulate or narrowly elliptic, 2.5–5.6 cm × 3–6 mm. . . . . .1. *Amsonia ciliata* (in part)
        4. Leaf blades not heteromorphic, surfaces glabrous or pubescent; stem leaf blades lanceolate or narrowly elliptic to elliptic, narrowly oblong, narrowly lanceolate, ovate, or obovate, (2.5–)3.3–12.5(–14.4) cm × (4–)8–50(–65) mm; branch leaf blades similar to stem leaf blades.
            5. Stem leaf blades elliptic, narrowly oblong, lanceolate, or obovate, (3.1–)3.9–8.4(–9.3) cm × (4–)8–18(–23) mm, margins sometimes irregularly revolute giving a dentate appearance; outer surface of corollas usually glabrous (rarely sparsely short-pubescent) . . . . .4. *Amsonia rigida*
            5. Stem leaf blades lanceolate to narrowly lanceolate, ovate, narrowly elliptic, or elliptic, (2.5–)3.3–12.5(–14.4) cm × (8–)11–50(–65) mm, margins not revolute or evenly revolute; outer surface of corollas normally with 5 patches of sparse to moderate, long pubescence at base of corolla lobes and distal part of tube (very rarely glabrate) . . . . .5. *Amsonia tabernaemontana*

1. Amsonia ciliata Walter, Fl. Carol., 98. 1788 • Fringed blue-star [E]

Stems erect (basally ascending), (14–)15–49(–67) cm in flower, to 79 cm in fruit, glabrous or moderately pubescent; branches borne on distal portion of stem, often but not always exceeding infructescence. **Leaves:** petiole 0–4 mm, glabrous; blades at least slightly heteromorphic or all linear; stem leaf blades linear, ligulate, very narrowly elliptic to elliptic or very narrowly lanceolate, 2.5–6.5 cm × 3–15 mm, margins entire, slightly to strongly revolute, sometimes sparsely to moderately ciliate, apex acute to acuminate, surfaces glabrous; branch leaf blades linear to ligulate or very narrowly elliptic, 2.5–5.6 cm × 0.3–6 mm. **Flowers:** sepals deltate to very narrowly deltate, 0.8–1.8(–2) mm; corolla tube bluish, greenish above, 6–9(–10) mm, lobes blue (rarely to white, or in var. *texana* purplish blue), 5–9(–10) mm, outer surface of corolla glabrous (very rarely inconspicuously pubescent). **Follicles** erect, (5.5–)6.5–13.5cm × 2–4 mm, apex acute, glabrous. **Seeds** (5–)5.5–8(–9.3) × 1.3–2.1 mm.

Varieties 3 (3 in the flora): sc, se United States.

1. Distal stem and branch leaf blades very narrow, 0.3–1(–1.6) mm wide (rarely to 2.4 mm if unrolled), margins usually strongly revolute . . . . .1b. *Amsonia ciliata* var. *filifolia*
1. Distal stem leaf blades broader than branch leaves, at least 3 mm wide, margins slightly to strongly revolute.
   2. Stems erect, seldom somewhat ascending at base, usually pubescent, less often glabrous; stem leaf blades ligulate to very narrowly elliptic (or very narrowly lanceolate), 2.5–6.5 cm × 3–10 mm; branch leaf blades linear to ligulate (very narrowly elliptic), 3–5.6 cm × 1–3 mm; corolla lobe margins not ciliate; North Carolina to Florida and Alabama . . . . 1a. *Amsonia ciliata* var. *ciliata*
   2. Stems basally ascending or less often erect, glabrous; stem leaf blades narrowly elliptic to elliptic (ligulate), (3–)3.5–6.5 cm × (3.3–)5–15 mm; branch leaf blades linear to very narrowly elliptic, 3.2–5.6 cm × 2–6 mm; corolla lobe margins often minutely ciliate; Oklahoma, Texas. . . 1c. *Amsonia ciliata* var. *texana*

1a. Amsonia ciliata Walter var. **ciliata** E

Stems erect, seldom somewhat ascending at base, moderately to sparsely pubescent, sometimes especially around petiole bases, or glabrous; branches exceeding infructescence. **Leaves:** petiole 1–4 mm in stem leaves, 1–2 mm or absent in branch leaves; stem leaf blades ligulate to very narrowly elliptic (or very narrowly lanceolate), 2.5–6.5 cm × 3–10 mm, margins usually at least slightly revolute, ciliate or not ciliate, apex narrowly acute; branch leaf blades linear to ligulate (very narrowly elliptic), 3–5.6 cm × 1–3 mm, margins often more deeply revolute than on stem leaf blades. **Flowers:** sepals narrowly deltate, 1–1.5(–2) mm; corolla tube 6–8 mm, lobes (5–)6–8(–10) mm, margins not ciliate. **Follicles** 7.5–13.5 cm × 2–3(–3.5) mm.

Flowering spring; fruiting summer. Pine and oak woods, sand barrens, bluffs, sand hills, occasionally swampy fields, flood plains; elevation range unreported, often apparently low (0–100 m); Ala., Fla., Ga., N.C., S.C.

Rare specimens of *Amsonia ciliata* from Texas resemble morphologically the southeastern var. *ciliata* rather than the very similar var. *texana*. The occurrence of var. *ciliata* so far from its usual range is unlikely enough that the identity of these few specimens should be considered questionable.

1b. Amsonia ciliata Walter var. **filifolia** Alph. Wood, Class-book Bot. ed. s.n.(b), 589. 1861 E

Amsonia ciliata var. *tenuifolia* (Rafinesque) Woodson

Stems erect, usually glabrous or slightly pubescent, sometimes moderately pubescent; branches exceeding infructescence or not. **Leaves:** petiole 0.3–1 mm or absent; stem leaf blades linear, rarely very narrowly elliptic, 2.5–4.6(–5.5) cm × 0.3–1(–1.6) mm (–2.4 mm if unrolled), margins usually strongly revolute, ciliate, apex acute; branch leaf blades similar to stem leaf blades. **Flowers:** sepals often deltate, sometimes narrowly deltate, 0.8–1.2(–1.8) mm; corolla tube 7–8(–8.5) mm, lobes 5–7(–8) mm, not ciliate. **Follicles** (5.5–)6.5–11 (–12.5) cm × 2–3 mm. $2n = 22$.

Flowering spring; fruiting summer. Xeric pine-oak woods, rock glades, sand barrens, sand hills, coastal plain, river bluffs, gravel bars; 0–300 m; Ala., Ark., Fla., Ga., Mo., N.C., Okla., S.C.

Variety *filifolia* is distinctive for its very narrow, usually strongly revolute stem leaves.

1c. Amsonia ciliata Walter var. **texana** (A. Gray) J. M. Coulter, Contr. U.S. Natl. Herb. 2: 262. 1892 E

Amsonia angustifolia Michaux var. *texana* A. Gray in A. Gray et al., Syn. Fl. N. Amer. 2(1): 81. 1878

Stems basally ascending to erect, glabrous; branches usually at most equal to infructescence (but occasionally becoming very long). **Leaves:** petiole 1–3(–4) mm in stem leaves, 0–1 mm in branch leaves; stem leaf blades narrowly elliptic to elliptic (ligulate), (3–)3.5–6.5 cm × (3.3–)5–15 mm, margins slightly to strongly revolute, not or occasionally ciliate, apex acute to acuminate; branch leaf blades linear to very narrowly elliptic, 3.2–5.6 cm × 2–6 mm, margins often much more deeply revolute than on stem leaf blades. **Flowers:** sepals narrowly deltate, apex acuminate or acute, 1–1.5(–2) mm, sparsely ciliate with long hairs or glabrous; corolla tube (6–)7–9(–10) mm, lobes (5–)6–8(–9) mm, margins often minutely ciliate. **Follicles** (7–)11–13.5 cm × 3–4 mm.

Flowering spring; fruiting summer. Prairies, pastures, post oak woodlands, hills, rock outcrops, stream banks, along roadsides and railroad tracks; 200–900 m; Okla., Tex.

Variety *texana* is distinguished from var. *ciliata* by leaf margins that are usually not ciliate, often broader leaves (especially branch leaves), and often ciliate corolla lobes. It occurs in the hill country of Texas and

Oklahoma, geographically separated from and at higher elevations than var. *ciliata*.

2. Amsonia hubrichtii Woodson, Rhodora 45: 328. 1943 [E]

Stems erect, (35–)52–130 cm, glabrous; branches borne on distal portion of stem, much exceeding infructescence. **Leaves:** petiole 0–3(–4) mm, glabrous; blades slightly heteromorphic; stem leaf blades linear, (5–)7.5–11 cm × 2–3.2(–4.2) mm, margins entire, revolute, not ciliate, apex narrowly acute, surfaces glabrous or sparsely short-pubescent on midrib and veins adaxially; branch leaf blades linear, 5–10.5 cm × (0.8–)1–1.8(–2.4) mm. **Flowers:** sepals deltate to narrowly deltate, 0.7–1.5 mm; corolla tube blue (to purplish), 6–8 mm, lobes blue (to white, purplish), (4.5–)5.4–7.8 mm, outer surface of corolla glabrous. **Follicles** pendulous or erect, 7.5–13.8 cm × 2.5–3.4 mm, apex acute, glabrous. **Seeds** 7.5–11 × 1.6–2.2 mm.

Flowering spring; fruiting summer. Stream banks, bottoms, shaley creek beds, gravel bars and spits, moist glades; 100–400 m; Ark., Okla.

Amsonia hubrichtii resembles *A. ciliata* var. *filifolia* but is a much larger plant, with longer, adaxially lustrous leaves, and with narrower ecological preferences. It is confined to central and western Arkansas and southeastern Oklahoma.

3. Amsonia ludoviciana Vail ex Small, Fl. S.E. U.S., 935, 1336. 1903 • Louisiana blue-star [E]

Stems erect, 30–41(–78) cm, glabrous or sparsely pubescent; branches borne on distal portion of stem, much exceeding infructescence (occasionally absent). **Leaves:** petiole (3–)4–5(–6) mm, moderately to densely (sparsely) pubescent; blades not heteromorphic; stem leaf blades lanceolate, elliptic, or ovate (narrowly elliptic, broadly elliptic, or oblong-lanceolate), (4–)4.8–7.4(–9.5) cm × (10–)12–40(–45) mm, margins entire, not revolute, moderately to densely ciliate, apex acuminate (acute), surfaces densely (to moderately) tomentose abaxially, glabrous adaxially; branch leaf blades similar to stem leaf blades. **Flowers:** sepals deltate, 0.8–1.1 mm; corolla tube purplish, greenish above, 7–8 mm, lobes blue, 6–8 mm, outer surface of corolla with 5 patches of long trichomes at corolla lobe bases and distal part of tube, less often glabrous. **Follicles** erect, (6.5–)8.5–10 cm × (2–)3–4 mm, apex acuminate, pubescent at least distally or glabrous. **Seeds** 9–13 × 1.7–2 mm. **2*n*** = 22.

Flowering spring(–early summer); fruiting summer. Low, wet woods, riparian areas, roadbanks, ditches; elevation range unreported, often apparently low (0–100 m); Ga., La., Miss., Tex.

Amsonia ludoviciana is distinguishable from *A. tabernaemontana* by its tomentose leaf indument and sometimes distally pubescent follicles, but its distinctiveness has been questioned by some. Almost all of its range is in Louisiana, where it appears to be uncommon yet relatively widespread. Very few collections are known from north-central Georgia, extreme eastern Texas, and southwestern Mississippi (where it is suspected of having been extirpated; no recent collections were seen).

4. Amsonia rigida Shuttleworth ex Small, Fl. S.E. U.S., 935. 1903 [E]

Amsonia glaberrima Woodson

Stems erect (somewhat ascending at base), (34–)48–92(–107) cm, glabrous or seldom sparsely pubescent; branches borne on distal portion of stem, exceeding infructescence. **Leaves:** petiole (1–)2–5(–6) mm, glabrous or slightly to moderately pubescent; blades not heteromorphic; stem leaf blades elliptic, narrowly oblong, lanceolate, or obovate, (3.1–)3.9–8.4(–9.3) cm × (4–)8–18(–23) mm, margins entire, often slightly revolute, usually irregularly and sometimes appearing dentate, sometimes sparsely ciliate, or moderately scabrous, apex acuminate, sometimes acute, surfaces glabrous or slightly to moderately pubescent on midrib or on veins; branch leaf blades similar to stem leaf blades. **Flowers:** sepals narrowly deltate to deltate, (0.8–)1–1.5(–1.8) mm; corolla tube blue, greenish above, (5.5–)6–8(–8.5) mm, lobes blue (to white or rarely lavender), (4–)5–7.5 mm, outer surface of corolla glabrous (very sparsely short-pubescent). **Follicles** erect (sometimes partly spreading), 9–13(–15) cm × 3–4 mm, apex acute, sometimes elongated, glabrous. **Seeds** 8.5–12(–15) × 1.9–2.6(–3) mm.

Flowering spring; fruiting early summer to summer. Pond and swamp edges, stream bottoms and flats, stream banks, coastal prairie, wet meadows, bogs, ditches and flooded areas, moist woods, black prairie, roadsides; 10–200 m; Ala., Fla., Ga., La., Miss., Mo., Tex.

Amsonia rigida is very similar to narrow-leaved forms of the more common and widespread *A. tabernaemontana*, but it can be distinguished by its usually glabrous corolla and leaves with often irregularly revolute margins. Almost all populations from Alabama and Mississippi, primarily from the Black Belt Prairie region, have leaf margins that are ciliate, often with short, sturdy, curved hairs, and often pale flowers. These populations have been informally labeled as a variety or species, but no name has been validly published. Although they are not formally recognized as a variety here, for conservation purposes they should be considered possibly distinct.

5. Amsonia tabernaemontana Walter, Fl. Carol., 98. 1788 • Eastern blue-star [E]

Stems erect (ascending), 27–102 cm, glabrous (sparsely pubescent); branches borne on distal portion of stem, much exceeding infructescence. **Leaves:** petiole 1.5–7(–9) mm, sparsely to moderately pubescent or glabrous; blades not heteromorphic; stem leaf blades lanceolate to narrowly lanceolate, ovate, narrowly elliptic, or elliptic, (2.5–)3.3–12.5(–14.4) cm × (8–)11–50(–65) mm, margins entire to subentire, not to moderately revolute, ciliate or not, apex acuminate to acute (rarely with a rounded tip), surfaces sparsely pubescent abaxially (and sparsely scabrous or short-pubescent adaxially) or glabrous; branch leaf blades similar to stem leaf blades. **Flowers:** sepals deltate to narrowly deltate, (0.4–)0.5–1.8(–2.3) mm; corolla tube bluish (rarely to lilac), sometimes greenish above, (5–)6–8(–8.8) mm, lobes pale blue (rarely to dark or lilac-blue or white), (5–)6–9.8(–12) mm, outer surface of corolla with 5 patches of long trichomes at corolla lobe bases and distal part of tube (very rarely glabrate). **Follicles** erect to pendulous (spreading), (5–)7–15 cm × 2.1–4(–4.2) mm, apex acuminate (acute), glabrous. **Seeds** (5–)5.5–10.8 × (1.2–)1.5–2.4(–3.3) mm.

Varieties 4 (4 in the flora): c, e United States.

Amsonia tabernaemontana encompasses a great deal of morphological diversity, particularly in leaf size and shape and sepal indument and length, which is to some extent geographically correlated; therefore, the recognition of varieties seems desirable. However, the varieties are not well differentiated and tend to blend into one another, especially where their ranges overlap.

1. Stems ascending (erect); stem leaf blades (2.5–)3.3–7.4 cm; sepals usually sparsely pubescent; e Texas, Oklahoma . . . . . 5c. *Amsonia tabernaemontana* var. *repens*
1. Stems erect; stem leaf blades (4–)6.5–12(–14.4) cm; sepals pubescent or glabrous; widespread in se to sc United States.
    2. Sepals narrowly deltate or less often deltate, (0.5–)0.9–1.6(–1.9) mm, usually sparsely to moderately long-pubescent in flower (often becoming glabrous or glabrate in fruit); Illinois, Arkansas to Kansas, Oklahoma . . . . 5b. *Amsonia tabernaemontana* var. *illustris*
    2. Sepals deltate (seldom narrowly deltate), (0.4–)0.5–1.1(–1.6) mm, glabrous (glabrate with rare single hairs); widespread.
        3. Stem leaf blades lanceolate to ovate, narrowly elliptic, or elliptic, (5.3–)8–12(–14) cm × (17–)24–50(–65) mm . . . . 5a. *Amsonia tabernaemontana* var. *tabernaemontana*
        3. Stem leaf blades narrowly lanceolate to lanceolate (or narrowly elliptic when young), (4–)6.5–10(–11.5) cm × 10–22(–25) mm . . . . 5d. *Amsonia tabernaemontana* var. *salicifolia*

5a. Amsonia tabernaemontana Walter var. **tabernaemontana** [E]

Tabernaemontana amsonia Linnaeus

Stems erect, (38–)48–100 cm. **Leaves:** petiole (2–)4–7(–9) mm, sparsely to moderately pubescent, or glabrous; stem leaf blades lanceolate to ovate, narrowly elliptic, or elliptic, (5.3–)8–12(–14) cm × (17–)24–50(–65) mm, margins subentire, often slightly irregular, seldom revolute, almost always short-ciliate or minutely scabrous, apex acuminate (acute, rounded-acute), surfaces sparsely pubescent mostly on midrib and veins abaxially (glabrous or also sparsely scabrous to short-pubescent adaxially). **Flowers:** sepals deltate (narrowly deltate), 0.5–0.9(–1.1) mm, glabrous (rarely bearing single long hairs); corolla tube (5–)6–7.5 mm, lobes (5.5–)6.4–8.4(–9.5) mm. **Follicles** erect, (5–)7–12 cm × 2.6–3.6 mm. **Seeds** (5–)5.5–9.5 × (1.3–)1.6–2.3 mm. $2n = 22$.

Flowering spring; fruiting summer. Moist woods, stream and lake banks, hillsides, ditches, fields; 0–400 m; Ala., Ark., Ga., Ill., Ky., La., Mo., N.C., Okla., S.C., Tenn., Tex., Va.

Variety *tabernaemontana* may also occur in Indiana; the only specimens seen of the species from that state were better identified as var. *salicifolia*.

5b. Amsonia tabernaemontana Walter var. **illustris** (Woodson) J. K. Williams, Phytoneuron 2019-13: 5. 2019 [E]

Amsonia illustris Woodson, Ann. Missouri Bot. Gard. 16: 407. 1929

Stems erect, 49–88 cm. **Leaves:** petiole (3–)4–7 mm, sparsely to moderately pubescent with long, soft hairs or glabrous; stem leaf blades lanceolate (narrowly elliptic), (5.7–)7–12.5(–14.4) cm × 12–26(–31) mm, margins entire, weakly or not revolute, often ciliate with long, soft hairs (rarely weakly, irregularly crenulate), apex acuminate to acute, surfaces sparsely pubescent with long, soft hairs, especially on midrib, adaxially or on both surfaces, or glabrous. **Flowers:** sepals narrowly deltate to deltate, (0.5–)0.9–1.6(–1.9) mm, usually mostly sparsely to moderately long-pubescent, rarely all or most in an inflorescence glabrous, often glabrescent or glabrate in fruit; corolla tube 6.5–8 mm, lobes (6–)7.5–9.8(–10.5) mm. **Follicles** erect to pendulous, (7–)10–15 cm × 2.1–3.4(–4.2) mm. **Seeds** 6–8(–10.4) × (1.2–)1.5–1.7(–3.3) mm. $2n$ = 22.

Flowering spring (rarely summer); fruiting summer (rarely fall). Gravel bars, stream beds and banks, flood plains and wetlands, low woods, glades and barrens, ditches, railroad embankments, old fields; 0–400 m; Ark., Ill., Kans., Mo., Okla.

Variety *illustris* is characterized by narrow leaves and pubescent calyces; it is confined to the western portion of the range of *Amsonia tabernaemontana*.

5c. Amsonia tabernaemontana Walter var. **repens** (Shinners) J. K. Williams, Phytoneuron 2019-13: 5. 2019 [E]

Amsonia repens Shinners, Field & Lab. 19: 126. 1951

Stems ascending (erect), 27–102 cm. **Leaves:** petiole 1.5–4.5(–6) mm, sparsely to moderately long-pubescent adaxially or glabrous; stem leaf blades lanceolate, (2.5–)3.3–7.4 cm × (8–)11–21(–25) mm, margins subentire (slightly wavy), usually somewhat revolute, often sparsely ciliate especially around apex, apex acute (acuminate), surfaces sparsely pubescent on (or near) midrib adaxially at least when young, or glabrous, usually ciliate if pubescent. **Flowers:** sepals narrowly deltate (deltate), (0.6–)1.1–1.8(–2.3) mm, usually sparsely pubescent with long, soft hairs, rarely glabrous; corolla tube (5.3–)6.5–7.5(–8) mm, lobes (6.5–)7–9 (–10.5) mm. **Follicles** erect, 11–15 cm × 3.4–4 mm. **Seeds** not seen.

Flowering spring; fruiting summer. Moist coastal prairies, wet meadows and forests, water-filled ditches, sandy barrens and blacklands; 0–200 m; Okla., Tex.

Variety *repens* resembles var. *illustris* in having long, pubescent sepals. It is often a relatively small plant ascending from a horizontal rhizome, although the upper end of its size range is large, and its leaves are smaller than those of var. *illustris*. The few Oklahoma collections that appear to be best placed within this variety are confined to McCurtain County in the extreme southeast of the state, where the two varieties may intergrade.

5d. Amsonia tabernaemontana Walter var. **salicifolia** (Pursh) Woodson, Ann. Missouri Bot. Gard. 15: 406. 1928 [E]

Amsonia salicifolia Pursh, Fl. Amer. Sept. 1: 184. 1813; *A. tabernaemontana* var. *gattingeri* Woodson

Stems erect, 41–94 cm. **Leaves:** petiole 2–6(–9) mm, glabrous or sparsely pubescent; stem leaf blades narrowly lanceolate to lanceolate (or narrowly elliptic when young), (4–)6.5–10(–11.5) cm × 10–22(–25) mm, margins entire, sometimes irregular, often somewhat revolute, sometimes ciliate, apex acuminate to narrowly acute, surfaces glabrous (sparsely pubescent abaxially and rarely adaxially). **Flowers:** sepals deltate (narrowly deltate), (0.4–)0.6–1.1(–1.6) mm, glabrous; corolla tube (6.5–)7–8(–8.8) mm, lobes (5–)6–9.5(–12) mm. **Follicles** erect (rarely spreading), (5–)7–12.5 cm × 2.5–3.5 mm. **Seeds** 8–10.8 × 1.8–2.4 mm.

Flowering spring(–early summer); fruiting summer (–fall). Woods, prairies and meadows, flood plains, creek beds and gravel bars, roadsides; 30–600 m; Ala., Ark., Ga., Ill., Ind., Kans., Ky., La., Miss., Mo., N.C., Ohio, Okla., S.C., Tenn., Tex., Va.

Variety *salicifolia* is found in much of the same range as the broad-leaved var. *tabernaemontana*, which likewise has short, glabrous sepals. Although the ranges of leaf breadth in those two varieties do overlap to some extent, there is (for the most part) an apparent discontinuity between narrow-leaved and broad-leaved populations. In addition, the leaves of var. *salicifolia* are more often glabrous and more likely to have revolute margins, and the flowers on average are longer than those of var. *tabernaemontana*; therefore, continued recognition of two varieties seems appropriate. Variety *salicifolia* strongly resembles var. *illustris*, which likewise often has long, narrow leaves; var. *illustris* has typically longer, pubescent sepals, and its range is much narrower. All varieties may be dubiously distinct where they occur sympatrically.

Variety *gattingeri* was named by Woodson to include plants with pubescent leaves, especially on the midrib adaxially, although becoming glabrate in maturity. The variety putatively ranges from Indiana to Kansas and south to Tennessee, Arkansas, and Texas. It is not clearly distinguished from var. *salicifolia*.

3b. Amsonia Walter subg. Articularia Woodson, Ann. Missouri Bot. Gard. 15: 418. 1928

Stems usually branched for most of length (occasionally only distally). **Bracts** on distal portions of inflorescences linear to narrowly deltate, mostly 2–6 mm (seldom less than 2 mm). **Flowers:** corolla tube (7–)8–12(–13) mm, moderately to strongly constricted just below apex, often in distinct narrow band (most visible just before anthesis); stigma apiculate with 2 small lobes. **Follicles** moniliform, strongly constricted between seeds. **Seeds** light reddish brown to pale brown, fusiform with acute to rounded-acute or flat-truncate (rarely diagonally truncate) ends, mostly smooth except at ends and with a prominent longitudinal groove.

Species 2 (2 in the flora): sw, sc United States, n Mexico.

Within subg. *Articularia*, as many as four species, one with two varieties, have been recognized. The most recent revision (S. P. McLaughlin 1982) recognized only one species with two varieties.

1. Stem leaf blades linear (ligulate or very narrowly elliptic), 1.6–3.5(–5.5) mm wide, surfaces moderately to sparsely tomentose; flowers purple to magenta, maroon, or lavender (blue); New Mexico, Texas . . . . 6. *Amsonia arenaria*
1. Stem leaf blades ovate to narrowly lanceolate, narrowly to very narrowly elliptic, or elliptic, (3–)4–27 mm wide, surfaces glabrous or densely (rarely sparsely) tomentose; flowers bluish (violet- to lavender-tinged) to white (pinkish, bluish, or purple-tinged); Arizona, California, Nevada, Utah . . . . 7. *Amsonia tomentosa*

6. Amsonia arenaria Standley, Proc. Biol. Soc. Wash. 26: 117. 1913

Stems erect, 15–71 cm, tomentose; branches usually borne on most of stem (occasionally confined to distal portion), often exceeding main stem even in flower, always exceeding infructescence. **Leaves:** petiole 0–1.5 (–3) mm, tomentose; leaf blades slightly heteromorphic; stem leaf blades linear (ligulate or very narrowly elliptic), (2–)3–6(–7) cm × 1.6–3.5(–5.5) mm, margins entire, not revolute, not ciliate, apex acute (round-tipped), surfaces sparsely to moderately tomentose; branch leaf blades linear (ligulate, very narrowly elliptic), (2–)2.9–4.5(–6) cm × 1–2(–3.1) mm. **Flowers:** sepals subulate (to filiform, aberrantly narrowly deltate), (2–)3–4.5(–5.3) mm; corolla tube 9–11(–11.5) mm, lobes purple to magenta, maroon, or lavender (blue), (5–)6–8.3(–9.4) mm, outer surface of corolla glabrous. **Follicles** erect, 2.9–10.5 cm × 4.8–6.3 mm, apex acuminate, usually somewhat tomentose on one side and near base (glabrous). **Seeds** 14–19(–23.5) × 3.6–5 mm.

Flowering spring; fruiting summer. Dunes, sand hills, deserts, gravelly plains; 1100–1500 m; N.Mex., Tex.; Mexico (Chihuahua).

S. P. McLaughlin (1982) included *Amsonia arenaria* within *A. tomentosa* var. *stenophylla*, along with narrow-leaved populations of *A. tomentosa* from Arizona and Utah. The geographic range of the latter group is almost contiguous with that of the broader-leaved *A. tomentosa* var. *tomentosa*, which is native to southern California and Nevada and western Arizona, and both of those groups share features such as an intermingling of glabrous and densely tomentose individuals within single populations. By contrast, the populations of southern New Mexico, western Texas, and northern Mexico formerly termed *A. arenaria* are geographically disjunct. They are also morphologically distinctive, often having very long distal branches at flowering; consistent moderate leaf pubescence, rather than dense or no pubescence; more pronounced narrowing of leaves than in var. *stenophylla*; a usually different pattern of calyx indument; and, according to available label notes and photographs, usually different flower colors. Single-seeded fruits are relatively common in *A. tomentosa* but rare in *A. arenaria*, and the fruits of *A. arenaria* are usually somewhat tomentose only on one side and near the base, a state that is rare although not unknown in *A. tomentosa*. The seeds of *A. arenaria* are somewhat longer on average. These differences are at least equal in magnitude to those differentiating other pairs of commonly recognized species in this problematic genus. Thus, the continued recognition of *A. arenaria* at the species level is preferable.

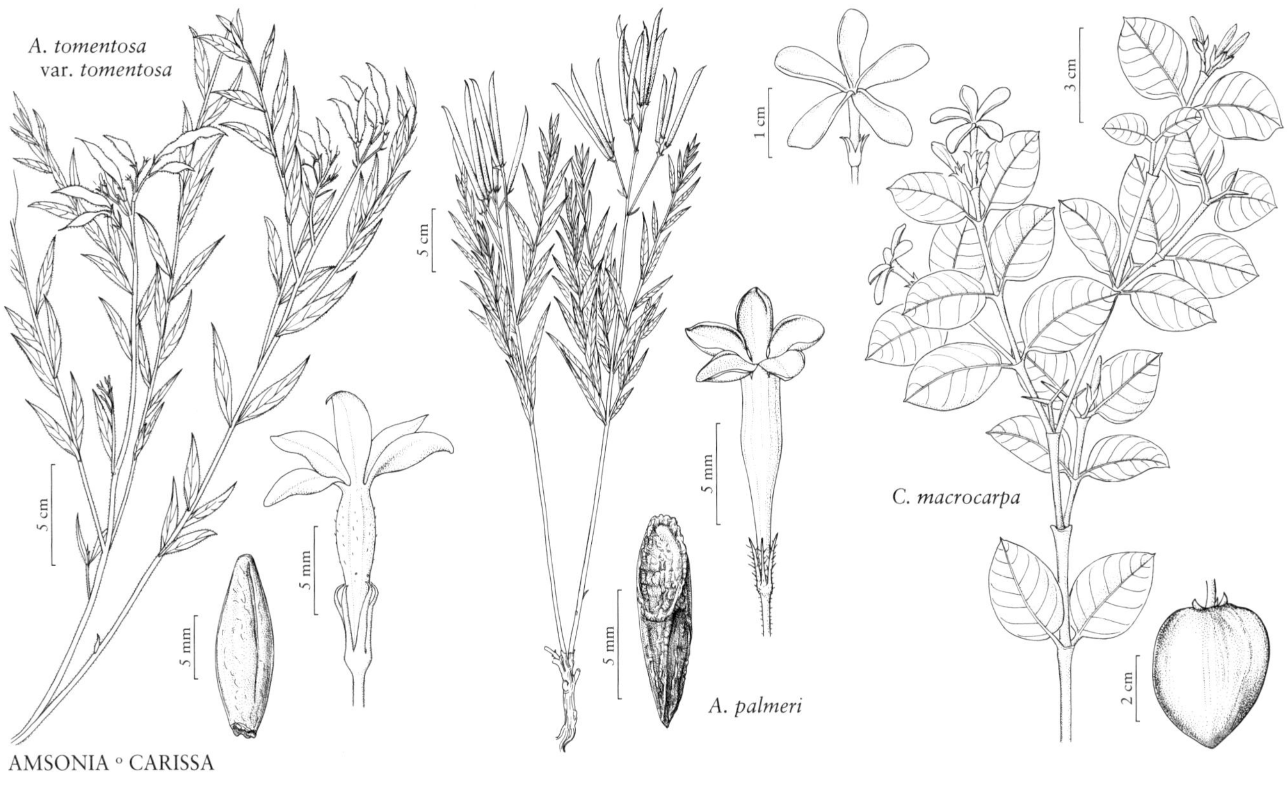

AMSONIA ◦ CARISSA

7. **Amsonia tomentosa** Torrey & Frémont in J. Frémont, Rep. Exped. Rocky Mts., 316. 1845 [E] [F]

Stems erect, 11–65 cm, glabrous or tomentose; branches usually borne on most of stem (occasionally confined to distal portion), well exceeding infructescence. **Leaves:** petiole 1–5 mm, glabrous or tomentose (leaves rarely sessile); leaf blades slightly or moderately heteromorphic; stem leaf blades ovate to narrowly lanceolate, narrowly to very narrowly elliptic, or elliptic, (2–)3–5(–6) cm × (3–)4–27 mm, margins entire, not revolute, not ciliate, apex acute (somewhat acuminate), surfaces glabrous or densely (rarely sparsely) tomentose; branch leaf blades narrower than stem leaf blades, (2.2–)3–4.7(–5.6) cm × (2–)3–6(–13) mm. **Flowers:** sepals subulate or narrowly deltate (deltate), (2–)3.5–6(–7.5) mm; corolla tube green to purplish or pinkish green, (7–)8–12(–13) mm, lobes bluish (violet- to lavender-tinged) to white especially after anthesis (pinkish, bluish, or purple-tinged), (2.8–)4.5–8 mm, outer surface of corolla glabrous. **Follicles** erect (aberrantly deformed and spreading), 2–8(–9.5) cm × (4–)4.8–6.8 mm, apex acuminate, glabrous (partly tomentose). **Seeds** (8–)11–17 × (3–)3.9–5.5(–6.3) mm.

Varieties 2 (2 in the flora): sw United States.

Both varieties of *Amsonia tomentosa* have an unusual pattern of variation in pubescence. Stems and leaves are usually either densely tomentose or glabrous, with intermediate density of pubescence rare. In some populations, the two forms are found together, suggesting that indument may be a single-gene trait.

1. Stem leaf blades ovate to lanceolate (narrowly elliptic to elliptic), (6–)8–27 mm wide, branch leaf blades markedly narrower, 4–13 mm wide; w, s Arizona, California, Nevada . . . . 7a. *Amsonia tomentosa* var. *tomentosa*
1. Stem leaf blades narrowly to very narrowly elliptic or narrowly lanceolate, (3–)4–8(–10) mm wide, branch leaf blades slightly narrower, (2–)3–6(–7) mm wide; n Arizona, Utah. . . . . 7b. *Amsonia tomentosa* var. *stenophylla*

7a. **Amsonia tomentosa** Torrey & Frémont var. **tomentosa** [E] [F]

Amsonia brevifolia A. Gray

Plants significantly heterophyllous. **Stem leaves:** petiole 1.5–5 mm; blade ovate to lanceolate (narrowly elliptic to elliptic), (2–)3–5.3 cm × (6–)8–27 mm, base convex (often attenuate basally) to rounded or cuneate. **Branch leaf blades** lanceolate to narrowly elliptic, 2.2–3.7 cm × 4–13 mm. **Corolla lobes** white to bluish, pinkish, or purplish white (blue to pink in bud) or pale blue to blue. **2*n*** = 22.

Flowering spring; fruiting late spring–summer. Sandy washes, deserts, rocky slopes and gullies, limestone or granitic alluvial fans; 600–1900 m; Ariz., Calif., Nev.

Variety *tomentosa* is known from a relatively cohesive region that includes the northwestern to north-central quarter of Arizona, southeastern Nevada, and three neighboring counties of southeastern California.

7b. Amsonia tomentosa Torrey & Frémont var. **stenophylla** Kearney & Peebles, J. Wash. Acad. Sci. 29: 487. 1939 E

Amsonia eastwoodiana Rydberg

Plants slightly heterophyllous. **Stem leaves:** petiole 1–2(–3) mm (leaves rarely sessile); blade narrowly to very narrowly elliptic or narrowly lanceolate, (2.4–)2.8–5(–6) cm × (3–)4–8 (–10) mm, base cuneate (narrowly convex). **Branch leaf blades** ligulate to very narrowly elliptic, (2.4–)3–4.7 (–5.6) cm × (2–)3–6(–7) mm. **Corolla lobes** bluish white to white, blue-violet, or blue-lavender.

Flowering spring; fruiting early summer. Sandy soils, arid shrubland and brush communities, rarely on rocky hillsides; 800–1600 m; Ariz., Utah.

Some specimens of var. *stenophylla* have characters reminiscent of *Amsonia arenaria* (patchy calyx or fruit indument or flowers said to be somewhat purplish), but on balance, these populations seem to be more naturally grouped with *A. tomentosa* (see discussion under *A. arenaria*). Two specimens identified with var. *stenophylla* from near the putative boundary between its range and that of var. *tomentosa* have very broad stem leaves (to 15 mm), indicating a likelihood of gene flow or incomplete separation between the two varieties. The synonym *A. eastwoodiana* was based on a combination of fruiting material of var. *stenophylla* and flowering material of *A. peeblesii* but was usually used for populations of var. *stenophylla* (S. P. McLaughlin 1982).

3c. AMSONIA Walter subg. SPHINCTOSIPHON (K. Schumann) Woodson, Ann. Missouri Bot. Gard. 15: 411. 1928

Amsonia sect. *Sphinctosiphon* K. Schumann in H. G. A. Engler and K. Prantl, Nat. Pflanzenfam. 120–122[IV,2]: 143. 1895

Stems branched usually only on distal portion, less often to near base (rarely unbranched). **Bracts** on distal portions of inflorescences linear to narrowly deltate, mostly 2–6 mm (occasionally shorter; in *A. jonesii* often less than 1 mm, those below individual flowers deltate or absent). **Flowers:** corolla tube (6–)7.5–41(–45) mm, slightly narrowing to moderately constricted at apex; stigma apiculate with 2 small lobes. **Follicles** terete, not strongly constricted between seeds. **Seeds** brown, cylindric and diagonally truncate to flat-truncate, with mostly convoluted surface.

Species 8 (8 in the flora): sw, sc United States, n Mexico.

The circumscription of subg. *Sphinctosiphon* in this treatment includes three long-flowered species that in previous treatments were segregated within subg. *Sphinctosiphon* as sect. *Longiflora* Woodson or removed into "subg. *Longiflora*" (not validly published) by S. P. McLaughlin (1982). These species are not otherwise distinguishable as a group from the usually shorter-flowered species of subg. *Sphinctosiphon*, so they are most conservatively treated as simply the largest-flowered species of that subgenus. The key emphasizes vegetative features where possible; fruit morphology does not distinguish most species, and fruiting specimens may be difficult to identify.

1. Stems glabrous.
   2. Stem leaf blades ovate to lanceolate, elliptic, or narrowly oblong, (6–)9–25(–32) mm wide.
      3. Stem leaf blades ovate to lanceolate or somewhat elliptic; branch leaf blades lanceolate (narrowly elliptic), 2.2–3.9 cm; sepals (0.3–)1–2.5(–4) mm; corolla tube (6–)7.5–8.5(–10) mm, lobes blue, (4–)5–7(–8.5) mm . . . . 10. *Amsonia jonesii* (in part)
      3. Stem leaf blades narrowly oblong to lanceolate; branch leaf blades very narrowly lanceolate or elliptic to linear, 3.5–6(–6.8) cm; sepals (2–)3–5(–7) mm; corolla tube 7.5–12(–17) mm, lobes blue or white, (2.5–)3–5(–7) mm . . . . 13. *Amsonia palmeri* (in part)
   2. Stem leaf blades linear to ligulate (to somewhat narrowly elliptic or lanceolate), (0.7–) 1–7(–9) mm wide.
      4. Stem leaf blades (2.3–)3–6.5(–9) cm × (0.7–)1–2.8(–5) mm; corolla tubes (20–)28–41(–45) mm . . . . 12. *Amsonia longiflora* (in part)
      4. Stem leaf blades 3.2–12(–14) cm, usually 3+ mm (in *A. palmeri* as little as 2 mm) wide; corolla tubes 7.5–21.5 mm.
         5. Corolla tubes 7.5–12(–17) mm, lobes (2.5–)3–5(–7) mm; Arizona, New Mexico, Texas . . . . 13. *Amsonia palmeri* (in part)
         5. Corolla tubes (13–)14–21.5 mm, lobes (5–)7–14(–16) mm; Arizona.
            6. Stem leaf blades (5.5–)7–12(–14) cm; corolla tubes 16–21.5 mm, lobes white to bluish, turning blue with age, 9.5–14(–16) mm; fruits 4–12(–15) cm; s Arizona. . . . . .9. *Amsonia grandiflora*
            6. Stem leaf blades 3.2–7.3(–9.3) cm; corolla tubes (13–)14–17(–19) mm, lobes white, (5–)7–9(–10) mm; fruits (2–)3–7(–10) cm; n Arizona . . . . .14. *Amsonia peeblesii*
1. Stems sparsely to moderately pubescent (glabrate, densely pubescent).
   7. Stem leaf blades linear to ligulate (very narrowly elliptic or lanceolate), (0.7–)1–2.8(–5) mm wide; corolla tubes (20–)28–41(–45) mm. . . . . .12. *Amsonia longiflora* (in part)
   7. Stem leaf blades narrowly oblong-elliptic, narrowly elliptic, narrowly lanceolate, lanceolate, ligulate, linear, or ovate, 2–25(–32) mm wide; corolla tubes (6–)7.5–23 mm.
      8. Branches few or absent, if present usually remaining shorter than infructescences; stem leaf blades ovate to lanceolate (elliptic, narrowly elliptic), (2–)3.3–5.5(–6.1) cm × (6–)9–25(–32) mm, surfaces glabrous; branch leaf blades lanceolate (narrowly elliptic), 3–9(–10) mm wide; sepals narrowly deltate to deltate (subulate, broadly deltate), (0.3–)1–2.5(–4) mm; corolla tubes (6–)7.5–8.5(–10) mm, lobes blue; Arizona, Colorado, New Mexico, Utah . . . . 10. *Amsonia jonesii* (in part)
      8. Branches present, usually elongating to slightly exceed or at least equal infructescences; stem leaf blades variable in shape and size, sometimes lanceolate but not ovate, to 18 mm wide, surfaces usually at least sparsely pubescent on midrib; branch leaf blades variable, in some linear to ligulate; sepals subulate to narrowly deltate, (2–)3–6(–7) mm; corolla tubes 7.5–23 mm, lobes white to greenish or yellowish white, cream, or blue; Arizona, New Mexico, Texas.
         9. Branch leaf blades narrowly lanceolate, (3–)6–9 mm wide; corolla tubes 12–14(–15) mm, lobes (2–)3–4 mm; seeds 3–4(–4.8) mm diam.; Baboquivari Mountains, Pima County, Arizona. . . . . 11. *Amsonia kearneyana*
         9. Branch leaf blades linear to ligulate or very narrowly lanceolate, oblong, or elliptic, 1–5(–7) mm wide; corolla tubes 7.5–23 mm, lobes (2.5–)3–11(–13) mm; seeds (1–)2–3 mm diam.; Arizona, New Mexico, Texas.
            10. Corolla tubes (16–)18–23 mm, lobes (6.5–)8–11(–13) mm; Socorro County, New Mexico . . . . 8. *Amsonia fugatei*
            10. Corolla tubes 7.5–17 mm, lobes (2.5–)3–9(–12) mm; Arizona, New Mexico, Texas.
                11. Stems 23–62(–80) cm, erect; corolla tubes 7.5–12(–17) mm; corolla lobes (2.5–)3–5(–7) mm, adaxial surface glabrous. . . . . . . 13. *Amsonia palmeri* (in part)
                11. Stems 10–20(–50) cm, usually erect, with long stems always ascending; corolla tubes (11–)12–17 mm; corolla lobes (5–)6–9(–12) mm, frequently pubescent at base of adaxial surface . . . . 15. *Amsonia tharpii*

8. **Amsonia fugatei** S. P. McLaughlin, SouthW. Naturalist 30: 563, fig. 1[center]. 1985 [C] [E]

Stems erect, 18–50 cm, sparsely to moderately pubescent (glabrate); branches borne on distal part of stem (rarely on most of length), at least slightly exceeding infructescence. **Leaves:** petiole 0–1(–2) mm, glabrate or sparsely pubescent; blades heteromorphic; stem leaf blades narrowly elliptic to narrowly lanceolate, lanceolate, or ligulate, (2.5–)3.2–5.2(–6.7) cm × (3–)5–7(–10) mm, margins entire, slightly revolute, sometimes in part sparsely ciliate, apex acute, surfaces glabrous (sparsely pubescent on midrib); branch leaf blades linear to ligulate, 2.5–5.5 cm × 1–4 mm. **Flowers:** sepals subulate to narrowly deltate, (3–)4–6 mm; corolla tube iridescent blue to purplish, (16–)18–23 mm, lobes white to cream, (6.5–)8–11(–13) mm, outer surface of corolla glabrous. **Follicles** erect, (2.4–)3.7–6(–8.3) cm × 2–4 mm, apex acuminate, glabrous. **Seeds** 8–10 × 2–3 mm.

Flowering late spring; fruiting late spring–early summer. Rocky slopes and ridges, washes, sand dunes; of conservation concern; 1100–1700 m; N.Mex.

Amsonia fugatei is endemic to Socorro County. It is most similar to *A. palmeri*, which has smaller flowers.

9. **Amsonia grandiflora** Alexander, Torreya 34: 116. 1934 [C]

Stems erect, 38–80(–90) cm, glabrous; branches borne on distal part of stem, at least slightly exceeding infructescence. **Leaves:** petiole 0–3(–5) mm, glabrous; blades somewhat heteromorphic or all linear; stem leaf blades linear to ligulate or very narrowly lanceolate to very narrowly elliptic, (5.5–)7–12(–14) cm × (2.5–)3–6(–8) mm, margins entire, sometimes slightly revolute, sparsely ciliate, apex narrowly acute, surfaces glabrous; branch leaf blades linear, (4.8–)5.5–9(–10) cm × (1–)1.5–2.6(–3.1) mm. **Flowers:** sepals subulate to narrowly deltate, often weak and curved, (3–)4.5–6.5(–7) mm, glabrous; corolla tube blue (to whitish), 16–21.5 mm, lobes bluish to white, turning blue with age (pinkish), 9.5–14(–16) mm, outer surface of corolla glabrous. **Follicles** erect, 4–12(–15) cm × 3.1–4(–4.5) mm, apex acuminate, glabrous. **Seeds** 6.4–11 × 2–3 mm.

Flowering spring (rarely summer); fruiting summer. Woodlands, hillsides, canyon bottoms and gullies, open ground, often near water; of conservation concern; 800–1400 m; Ariz.; Mexico (Sonora).

Amsonia grandiflora is primarily restricted to Pima and Santa Cruz counties and rarely extends to Maricopa County and adjacent regions of Mexico. In Mexico, it is largely or entirely restricted to the state of Sonora; it reportedly rarely occurs in Durango, but no specimen was seen. It may be distinguished from *A. palmeri* by its larger flowers. Fruiting specimens of *A. grandiflora* will typically have somewhat shorter leaves than narrow-leaved individuals of *A. palmeri*, and the sepals are on average longer.

10. **Amsonia jonesii** Woodson, Ann. Missouri Bot. Gard. 15: 414. 1928 [E]

Amsonia latifolia M. E. Jones, Contr. W. Bot. 12: 50. 1908, not Michaux 1803

Stems erect, (15–)23–60(–77) cm, glabrous or sparsely to moderately short-pubescent; branches few, confined to distal part of stem, usually remaining shorter than infructescence, or occasionally absent. **Leaves:** petiole 1–3(–4.5) mm, glabrous (base pubescent); blades heteromorphic; stem leaf blades ovate, especially proximally, to lanceolate (elliptic, rarely narrowly elliptic), (2–)3.3–5.5(–6.1) cm × (6–)9–25(–32) mm, margins entire, not revolute, not ciliate, apex slightly acuminate (acute), surfaces glabrous; branch leaf blades lanceolate (narrowly elliptic), 2.2–3.9 cm × 3–9(–10) mm. **Flowers:** sepals narrowly deltate to deltate (subulate, broadly deltate), (0.3–)1–2.5(–4) mm; corolla tube blue, (6–)7.5–8.5(–10) mm, lobes blue, (4–)5–7(–8.5) mm, outer surface of corolla glabrous or bearing villous patches on apical part of tube and basal part of lobes. **Follicles** erect, (1.4–)3.5–6.5(–9) cm × (2.5–)3–4.3(–4.8) mm, apex short-acuminate to acute, glabrous. **Seeds** 6–10 × 1.5–2.5 mm.

Flowering spring; fruiting summer. Juniper and other woodlands, sagebrush and other arid shrublands, washes, dry prairies and mesas, roadsides, frequently on sand and gravel; 900–2200 m; Ariz., Colo., N.Mex., Utah.

Amsonia jonesii is readily distinguished from other species in its subgenus by its short and relatively broad, often ovate leaves, small blue flowers with usually short sepals (subtended by short, broad bracts), and few, short, or sometimes absent stem branches. Its range extends farther north than that of any other species of subg. *Sphinctosiphon*. In Arizona, it is confined to Coconino and Mojave counties. A specimen from Utah representing a probable hybrid involving *A. jonesii* (*Thorne & Thorne s.n.*, ARIZ) has unusually narrow leaves and multiple well-developed branches on each stem.

11. Amsonia kearneyana Woodson, Ann. Missouri Bot. Gard. 15: 415. 1928 • Kearney's blue-star [C] [E]

Stems erect, 39–62(–90) cm, moderately (densely or sparsely) long-pubescent; branches borne on distal part of stem, often exceeding infructescence. **Leaves:** petiole 1–3(–5) mm, densely to moderately pubescent (glabrous); blades heteromorphic; stem leaf blades lanceolate to narrowly lanceolate (narrowly oblong-elliptic), 5–7.6 cm × 11–17 mm, margins entire, often slightly revolute, ciliate with usually long cilia (rarely not ciliate), apex acute to acuminate, surfaces long-pubescent on midrib and main veins (rarely glabrous) adaxially; branch leaf blades narrowly lanceolate, 4.3–5.7 cm × (3–)6–9 mm. **Flowers:** sepals subulate, (3–)4–5(–6) mm; corolla tube green to purplish, 12–14(–15) mm, lobes white, (2–)3–4 mm, outer surface of corolla glabrous. **Follicles** erect, 3–10 cm × 3.5–5.5 mm, apex short-acuminate, glabrous. **Seeds** 6–11 × 3–4(–4.8) mm.

Flowering spring; fruiting summer. Canyons, plains, hillsides near streams; of conservation concern; 1100–1600 m; Ariz.

Amsonia kearneyana is endemic to the Baboquivari Mountains in Pima County; it is listed as endangered. The species is notable for its often densely long-pubescent pedicels and inflorescences and unusually short corolla lobes. These features are not unknown in the widespread and variable *A. palmeri*, although specimens of that species with corolla tubes longer than 12 mm are uncommon and typically have longer corolla lobes. The seeds of *A. kearneyana* are broader than those of any other species of subg. *Sphinctosiphon*.

12. Amsonia longiflora Torrey in W. H. Emory, Rep. U.S. Mex. Bound. 2(1): 159. 1859

Stems erect, 16–60 cm, glabrous or sparsely to moderately pubescent; branches often borne on most of stem (or only distal portion), often remaining shorter than infructescence. **Leaves:** petiole 0–2.5 mm, glabrous or sparsely to moderately pubescent; blades slightly heteromorphic; stem leaf blades linear to ligulate (very narrowly elliptic or lanceolate), (2.3–)3–6.5(–9) cm × (0.7–)1–2.8(–5) mm, margins entire, sometimes somewhat revolute, ciliate or not, apex narrowly acute, surfaces glabrous or sparsely to moderately pubescent; branch leaf blades linear, (1.1–)2.5–5.2(–7) cm × 0.5–1.3(–3) mm. **Flowers:** sepals narrowly deltate (to subulate) with weak, often curved apex, (2–)3.5–7(–9.5) mm; corolla tube bluish to lead-colored, lavender or purplish below, often greenish above, (20–)28–41(–45) mm, lobes white to cream or bluish white (pale pink), (6.5–)9–13.5(–18) mm, outer surface of corolla glabrous. **Follicles** erect, (3.8–)6–11(–20) cm × (2.4–)2.9–4 mm, apex short-acuminate (acute), glabrous. **Seeds** 5.3–8.1(–10.5) × 1.5–2.7 mm.

Varieties 2 (2 in the flora): sc United States, n Mexico.

The varieties of *Amsonia longiflora* are distinguished by differing pubescence. Both are generally found on limestone substrates. Notes on a specimen collected by R. C. Sivinski (*Sivinski 4282*, UNM) indicate that var. *longiflora* was collected on dolomite and had pale blue corollas, whereas nearby var. *salpignantha* occurred on gypsum and had white to pale pink corollas (the latter is rarely reported, and corolla lobe color is not consistently different between the two varieties). It is not known whether this substrate preference may be consistent and contributes to maintaining the distinction between the two. Both are restricted to southern New Mexico and Texas, but var. *longiflora* is confined to a relatively small portion of western Texas, while var. *salpignantha* is more widely distributed in western Texas, and occurs also in several counties in a disjunct portion of central Texas.

1. Stems, leaves, petioles, and outer surface of calyces glabrous; leaf margins not ciliate, not revolute. . . . . 12a. *Amsonia longiflora* var. *longiflora*
1. Stems, leaves, petioles, and outer surface of calyces usually sparsely to moderately pubescent (seldom glabrate to glabrous); leaf margins at least sparsely ciliate, occasionally somewhat revolute . . . . 12b. *Amsonia longiflora* var. *salpignantha*

12a. Amsonia longiflora Torrey var. **longiflora**

Stems glabrous. **Leaves:** petiole glabrous; blade margins not revolute, not ciliate, surfaces glabrous. **Calyces** glabrous (sepal margins seldom sparsely ciliate or rarely papillate).

Flowering spring(–summer); fruiting (spring–)summer(–fall). Dry, rocky hills, mesas, scrub, limestone flats, roadsides, generally on limestone; 900–1500 m; N.Mex., Tex.; Mexico (Coahuila).

12b. Amsonia longiflora Torrey var. **salpignantha** (Woodson) S. P. McLaughlin, Ann. Missouri Bot. Gard. 69: 348. 1983 (as salpignatha)

Amsonia salpignantha Woodson, Ann. Missouri Bot. Gard. 15: 417, plate 52, figs. 21, 22. 1928

Stems sparsely to moderately pubescent (glabrate, glabrous). **Leaves:** petiole sparsely to moderately short-pubescent (seldom glabrate to glabrous on one or both surfaces); blade margins occasionally somewhat revolute, at least sparsely ciliate, surfaces sparsely to moderately short-pubescent (seldom glabrate or glabrous on one or both surfaces). **Calyces** sparsely pubescent (glabrous, moderately pubescent), especially on cup, sepal margins ciliate (or not).

Flowering spring; fruiting summer. Rocky hills, canyons, valley bottoms, generally on limestone; 100–1600 m; N.Mex., Tex.; Mexico (Coahuila).

13. Amsonia palmeri A. Gray, Proc. Amer. Acad. Arts 12: 64. 1876 [F]

Amsonia hirtella Standley; *A. hirtella* var. *pogonosepala* (Woodson) Wiggins

Stems erect, 23–62(–80) cm, glabrous or moderately (somewhat densely or sparsely) pubescent; branches usually borne on distal portion of stem (occasionally below midpoint), not or slightly exceeding infructescence. **Leaves:** petiole 0–2(–3) mm, glabrous or moderately (sparsely) pubescent; blades heteromorphic or all very narrow; stem leaf blades narrowly oblong-elliptic to narrowly oblong, narrowly lanceolate, or linear, 4.5–7.5(–9.2) cm × 2–18 mm, margins entire, sometimes slightly revolute, moderately short-ciliate if leaf blade is pubescent, apex acute to acuminate, surfaces glabrous or moderately (sparsely) pubescent; branch leaf blades very narrowly lanceolate or very narrowly elliptic to linear, 3.5–6(–6.8) cm × 1–5(–7) mm. **Flowers:** sepals narrowly deltate to subulate, (2–)3–5(–7) mm; corolla tube bluish or purplish green or green, 7.5–12(–17) mm, lobes white to blue or yellowish white, (2.5–)3–5(–7) mm, outer surface of corolla glabrous. **Follicles** erect, 2–10(–13) cm × 2–4 mm, apex acuminate, glabrous. **Seeds** 6–10 × 1–2.5 mm.

Flowering spring; fruiting late spring–summer. Rocky hillsides, arroyos and draws, woodlands, washes and flood plains; 600–1900 m; Ariz., N.Mex., Tex.; Mexico (Chihuahua, Sonora).

Leaf morphology and pubescence are quite variable in *Amsonia palmeri*; if stem leaves are broad, then the plant is noticeably heterophyllous at maturity with narrow branch leaves. Glabrous and pubescent individuals often occur in mixed populations (S. P. McLaughlin 1982). The species can be distinguished from *A. tharpii* in part by the adaxial base of the corolla lobes being glabrous (vs. frequently pubescent in *A. tharpii*). *Amsonia palmeri* is fairly widespread in Arizona, especially in southern and western counties, but is confined to limited portions of southwestern New Mexico and western Texas.

14. Amsonia peeblesii Woodson, Bull. Torrey Bot. Club 63: 35. 1936 [E]

Stems erect from vertical taproot, (15–)20–58(–90) cm, glabrous; branches often borne below midpoint of stem as well as distally, uppermost often exceeding infructescence, often terminated by small inflorescences. **Leaves:** petiole 0–1.2 mm, glabrous; blades moderately heteromorphic; stem leaf blades ligulate to narrowly oblong-elliptic or narrowly lanceolate, 3.2–7.3(–9.3) cm × 3–7(–9) mm, margins entire, not (slightly) revolute, not ciliate, apex acute, surfaces glabrous; branch leaf blades linear to very narrowly oblong or lanceolate, 4.1–6.4(–7.4) cm × 1–4(–7) mm. **Flowers:** sepals narrowly deltate (subulate), (2–)3–6(–9.5) mm; corolla tube purplish blue to greenish, darker and more bluish proximally, (13–)14–17(–19) mm, lobes white, (5–)7–9(–10) mm, outer surface of corolla glabrous. **Follicles** erect, (2–)3–7(–10) cm × (2.5–)3–4 mm, apex acuminate, glabrous. **Seeds** (5–)7–11 × 1.5–2.5 mm. $2n = 22$.

Flowering spring–early summer; fruiting summer. Ridge tops, valleys, washes and draws, often on sand or sandstone; 1300–1700 m; Ariz.

Amsonia peeblesii generally resembles narrower-leaved, glabrous forms of *A. palmeri*. The stems of *A. peeblesii* are much more often branched below the midpoint, and the flowers are usually larger; the distribution of *A. peeblesii* is narrower and more northerly, whereas *A. palmeri* is largely found in southern and western portions of Arizona.

15. Amsonia tharpii Woodson, Ann. Missouri Bot. Gard. 35: 237. 1948 • Tharp blue-star [C] [E]

Stems usually erect (rare long stems always ascending), 10–20(–50) cm, moderately (sparsely) pubescent; branches usually borne on distal portion, seldom to near base of stem, equaling or slightly exceeding infructescence. **Leaves:** petiole 0.5–2 mm, sparsely to moderately pubescent (glabrous or densely pubescent); blades heteromorphic; stem leaf blades narrowly oblong-elliptic, lanceolate, narrowly lanceolate, or narrowly elliptic to ligulate (markedly narrower towards distal part of stem), 2.5–5(–7.5) cm × (2–)4–10(–13) mm, margins entire, not revolute, sometimes ciliate, apex acute, surfaces sparsely pubescent abaxially primarily on midrib (or glabrous); branch leaf blades very narrowly lanceolate to very narrowly oblong or linear, 2–5.5 cm × 1–3 mm. **Flowers:** sepals subulate to narrowly deltate, (2–)3–5(–6) mm, usually ciliate; corolla tube purplish, (11–)12–17 mm, lobes white to greenish white or pale blue, (5–)6–9(–12) mm, outer surface of corolla glabrous. **Follicles** erect, 2–7(–12) cm × 2.5–5 mm, apex acuminate, glabrous. **Seeds** 7–11 × 2–3 mm.

Flowering spring; fruiting late spring–summer. Rocky limestone hills and ridges; of conservation concern; 900–1400 m; N.Mex., Tex.

The U.S. Fish & Wildlife Service concluded in 2009 that *Amsonia tharpii* should be considered for listing as federally threatened or endangered due to its limited range and threats from grazing and development, although no formal action has been taken. It is listed as endangered in New Mexico. Collections were seen only from Eddy County in southeastern New Mexico and Pecos County in western Texas. This species is notable for its usually small size and the frequent appearance of pubescence at the base of the adaxial surface of the corolla lobes, which is an extension of the indument inside the corolla tube.

4. CARISSA Linnaeus, Syst. Nat. ed. 12, 2: 135, 189. 1767; Mant. Pl. 1: 52. 1767, name conserved • [Derivation uncertain; perhaps Sanskrit *krishna*, dark blue, or Greek *charis*, grace, alluding to appearance of mature fruit] [I]

David E. Lemke

Shrubs [trees, vines]; latex milky. **Stems** erect [scandent], armed with paired spines [unarmed], glabrous or eglandular-pubescent especially on younger growth. **Leaves** persistent, opposite [whorled], petiolate; stipular colleters intrapetiolar and interpetiolar; laminar colleters absent. **Inflorescences** terminal [axillary], cymose, short-pedunculate. **Flowers:** calycine colleters present or absent; corolla white [pink], salverform, aestivation sinistrorse [dextrorse]; corolline corona absent; androecium and gynoecium not united into a gynostegium; stamens inserted at top of corolla tube; anthers not connivent, not adherent to stigma, connectives not appendiculate or enlarged, locules 4; pollen free, not massed into pollinia, translators absent; nectaries absent. **Fruits** baccate, solitary, erect [pendulous], red or purple [black], globose or ellipsoid, terete or somewhat compressed, surface smooth, glabrous. **Seeds** obliquely ovate to elliptic or orbiculate, flattened, not winged, not beaked, not comose, not arillate. $x = 11$.

Species 7 (1 in the flora): introduced; Asia, Africa, Indian Ocean Islands, Pacific Ocean Islands, Australia.

SELECTED REFERENCE Leeuwenberg, A. J. M. and F. J. H. van Dilst. 2001. Series of revisions of Apocynaceae XLIX. *Carissa* L. Wageningen Univ. Pap. 1: 3–109, 123–126.

1. Carissa macrocarpa (Ecklon) A. de Candolle in A. P. de Candolle and A. L. P. P. de Candolle, Prodr. 8: 336. 1844 • Natal plum [F] [I]

Arduina macrocarpa Ecklon, S. African Quart. J. 1: 372. 1830

Shrubs 1–2(–6) m. **Stems:** spines stout, bifurcated. **Leaves:** petiole 1–6 mm, glabrous; blade ovate, elliptic, oblong, or orbiculate, 1.3–7.2 × 0.9–5.3 cm, coriaceous, base cordate to cuneate, margins revolute, apex acute or mucronate, glabrous. **Peduncles** 3–4 mm, glabrous. **Pedicels** 3–4 mm, glabrous. **Flowers:** calyx lobes ovate to narrowly oblong, auriculate, 2–4.5(–7) mm, glabrous, colleters present or absent; corolla glabrous abaxially, eglandular-pubescent adaxially, tube 5–10 × 1.5–2 mm, throat 5–12 × 2.5–3 mm, lobes spreading, obliquely obovate, (4.5–)10–24 × 4–7 mm. **Berries** 2.7–6 × 2–3 cm. **Seeds** 4–6 × 3–4.5 mm. $2n = 22$.

Flowering spring–fall; fruiting summer–fall. Coastal hammocks, beach dunes, disturbed areas; 0–10 m; introduced; Fla., Tex.; e, se Africa; cultivated widely in warmer parts of the world.

The stout, dichotomously branched spines readily distinguish *Carissa macrocarpa* from all other North American members of Apocynaceae. Plants are widely cultivated as ornamentals and for hedges in warmer parts of the United States and have become naturalized in a few coastal counties in Florida and Texas. The flesh of the fruit is edible when fully ripe, but unripe fruits, seeds, and all vegetative parts of the plant are toxic.

5. CATHARANTHUS G. Don, Gen. Hist. 4: 71, 95. 1837 • Periwinkle [Greek *katharos*, pure, and *anthos*, flower, alluding to neatness and beauty of flowers] [I]

David E. Lemke

Subshrubs or herbs, perennial [annual]; latex milky. **Stems** erect, ascending, or decumbent, unarmed, glabrous or eglandular-pubescent. **Leaves** deciduous, opposite or occasionally subopposite, petiolate; stipular colleters interpetiolar and intrapetiolar; laminar colleters absent. **Inflorescences** axillary, solitary flowers or pairs of flowers, short-pedunculate or sessile. **Flowers:** calycine colleters absent; corolla white, pink, red, magenta, or red-violet, salverform, aestivation sinistrorse; corolline corona absent; androecium and gynoecium not united into a gynostegium; stamens inserted at top [middle] of corolla tube; anthers connivent, not adherent to stigma, connectives not appendiculate or enlarged, locules 4; pollen free, not massed into pollinia, translators absent; nectaries 2, elongate and often exceeding ovary. **Fruits** follicles, solitary or paired, erect, green, slender and weakly moniliform, terete or compressed, surface striate, pubescent. **Seeds** ovoid [oblong], flattened, not winged, not beaked, not comose, not arillate. $x = 8$.

Species 8 (1 in the flora): introduced; Asia (India), Indian Ocean Islands (Madagascar, Sri Lanka); introduced also nearly worldwide in tropical and subtropical regions.

At various times the generic names *Vinca*, *Lochnera* Reichenbach, and *Catharanthus* have been used for the species treated here, but W. T. Stearn (1966) argued convincingly that the latter name is the legitimate one for the genus. While *Catharanthus* and *Vinca* have been shown to be somewhat closely related (A. O. Simões et al. 2016), they have disjunct native geographic ranges (India, Madagascar, and Sri Lanka for *Catharanthus*, Europe and western Asia for *Vinca*) and are easily distinguished by the characteristics given in the generic key.

SELECTED REFERENCE van Bergen, M. and W. Snoeijer. 1996. Series of revisions of Apocynaceae XLII: *Catharanthus* G. Don. The Madagascar periwinkle and related species. Wageningen Agric. Univ. Pap. 96: 1–120.

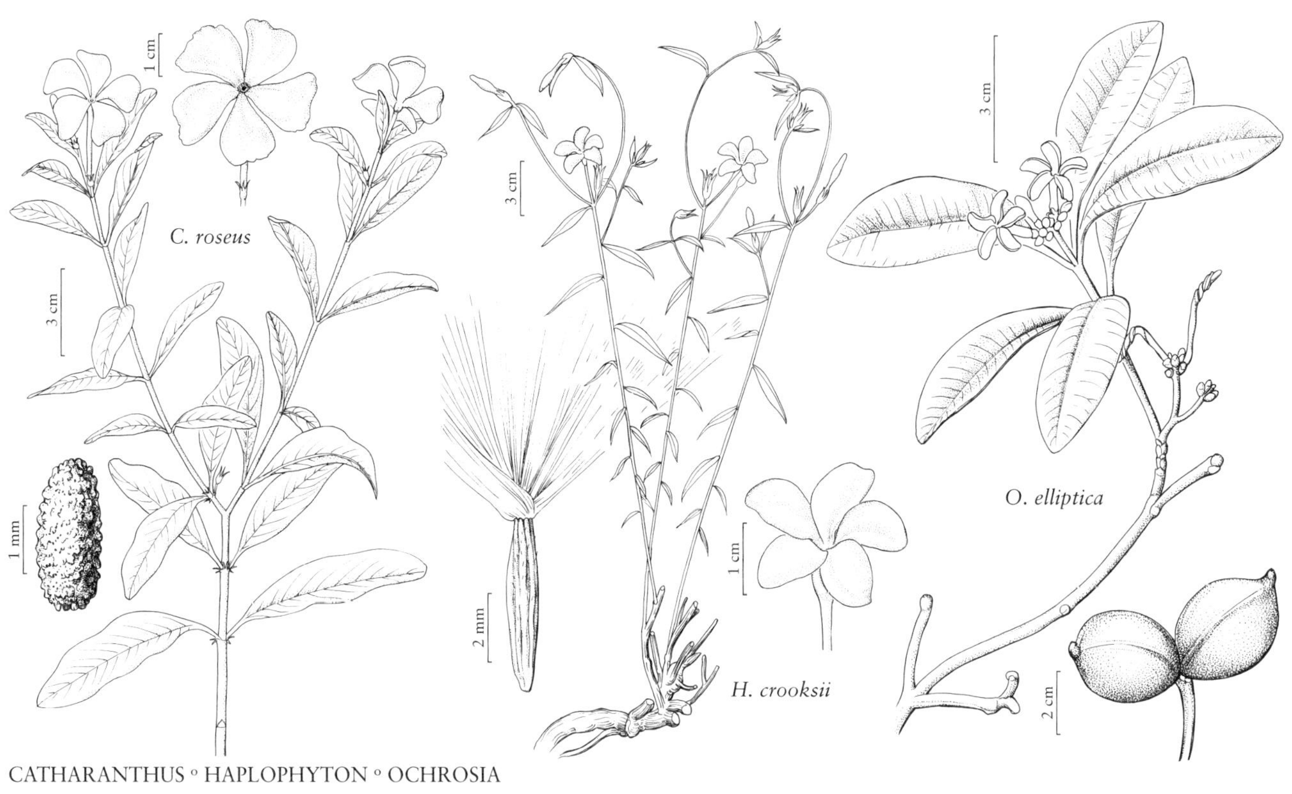

CATHARANTHUS ◦ HAPLOPHYTON ◦ OCHROSIA

1. **Catharanthus roseus** (Linnaeus) G. Don, Gen. Hist. 4: 95. 1837 • Madagascar periwinkle [F] [I] [W]

Vinca rosea Linnaeus, Syst. Nat. ed. 10, 2: 944. 1759; *Lochnera rosea* (Linnaeus) Reichenbach ex Spach

Subshrubs or herbs perennial, 3–10(–20) dm. **Leaves:** petiole (1–)3–11 mm, sparsely pubescent or glabrous; blade elliptic to obovate, oblong, or rarely lanceolate, (1–)2.5–9 × (0.6–)0.8–4 cm, membranous, base cuneate to attenuate, apex rounded to acute or obtuse, mucronulate, surfaces sparsely pubescent or glabrous. **Peduncles** 1–4 mm, sparsely pubescent or glabrous, occasionally absent. **Pedicels** 0–1 mm, sparsely pubescent or glabrous. **Flowers:** calyx lobes narrowly lanceolate, 2–6 mm, sparsely pubescent or glabrous; corolla eglandular-pubescent abaxially and adaxially, tube (15–)20–30 × 1–1.5 mm, throat 4–5 × 2–3 mm, lobes spreading, broadly obovate, often mucronulate, (5–)10–28 × 10–25 mm. **Follicles** (12–)20–50 × 1.5–2 mm. **Seeds** 1–3 × 0.5–1.5 mm. $2n$ = 16.

Flowering spring–fall; fruiting summer–fall. Disturbed areas, old homesites; 0–300 m; introduced; Ala., Calif., Fla., Ga., Kans., La., Miss., N.C., S.C., Tenn., Tex.; Indian Ocean Islands (Madagascar); widely introduced in warmer regions of the world.

Catharanthus roseus is of great pharmaceutical interest for its ability to synthesize a large number (ca. 130) of monoterpenoid indole alkaloids, the best known of which are vinblastine and vincristine (Q. Pan et al. 2016). When purified, both compounds have been shown to be useful for treating certain cancers, especially Hodgkin's disease, non-Hodgkin's lymphomas, and acute leukemia (R. Van der Heijden et al. 2004), and act by disrupting microtubules, causing dissolution of the mitotic spindle and metaphase arrest in dividing cells.

6. HAPLOPHYTON A. de Candolle in A. P. de Candolle and A. L. P. P. de Candolle, Prodr. 8: 412. 1844 • [Greek *haploos*, simple, and *phyton*, plant, alluding to absence of calycine glands and floral nectaries]

David E. Lemke

Shrubs or subshrubs; latex milky. **Stems** erect, unarmed, eglandular-pubescent. **Leaves** deciduous, alternate or occasionally opposite, petiolate; stipular colleters intrapetiolar; laminar colleters absent. **Inflorescences** axillary, solitary flowers, occasionally 2-flowered, pedunculate. **Flowers:** calycine colleters absent; corolla yellow, rotate, aestivation sinistrorse [dextrorse]; corolline corona absent; androecium and gynoecium not united into a gynostegium; stamens inserted at top of corolla tube; anthers not connivent, not adherent to stigma, connectives not appendiculate or enlarged, locules 4; pollen free, not massed into pollinia, translators absent; nectaries absent. **Fruits** follicles, solitary or paired, erect, brown, slender, terete, striate, pubescent. **Seeds** narrowly elliptic, slightly flattened, not winged, not beaked, comose, not arillate.

Species 2 (1 in the flora): sw United States, Mexico, Central America (Guatemala).

SELECTED REFERENCE Williams, J. K. 1995. Miscellaneous notes on *Haplophyton* (Apocynaceae: Plumerieae: Haplophytinae). Sida 16: 469–475.

1. Haplophyton crooksii (L. D. Benson) L. D. Benson, Amer. J. Bot. 30: 630. 1943 • Cockroach plant [F]

Haplophyton cimicidum A. de Candolle var. *crooksii* L. D. Benson, Torreya 42: 9. 1942

Shrubs or subshrubs 2–6 dm. **Leaves:** petiole 1–2 mm, pubescent; blade ovate-elliptic to narrowly oblong-elliptic, (7–)20–35 × (2–)4–8 mm, membranous, base broadly obtuse, apex acuminate to obtuse and mucronulate, surfaces pubescent. **Peduncles** 8–10 mm, pubescent. **Flowers:** calyx lobes linear-lanceolate, 5–8 mm, pubescent; corolla eglandular-pubescent abaxially and adaxially, tube 6–9 × 1.5–2 mm, lobes spreading, obliquely obovate, 10–18 × 5–10 mm. **Follicles** 50–90 × 1.5–3 mm. **Seeds** 5–7 × 0.8–1 mm.

Flowering spring–fall; fruiting summer–fall. Rocky slopes in desert scrub and desert grasslands; 700–1600 m; Ariz., N.Mex., Tex.; Mexico (Chihuahua, Coahuila, Sonora).

Many authors have treated *Haplophyton* as a monospecific genus, but J. K. Williams (1995) offered a convincing argument that plants from the southwestern United States and northern Mexico, characterized by sinistrorse corolla aestivation, smaller leaves and seeds, and shorter corolla tubes, and here referred to *H. crooksii*, represent a taxon distinct from *H. cimicidum* of central and southern Mexico and Guatemala. *Haplophyton crooksii* is widespread in southern Arizona but is known from only a few locations in Luna County, New Mexico, and El Paso and Presidio counties, Texas.

According to G. E. Burrows and R. J. Tyrl (2013), extracts of the dried leaves of *Haplophyton*, mixed with cornmeal or molasses, have been used as an insecticide and in lotions for parasite control. M. A. Mroue et al. (1996) isolated 15 indole alkaloids from *H. crooksii* and demonstrated that all showed significant in vitro inhibition of acetylcholinesterase activity.

7. OCHROSIA Jussieu, Gen. Pl., 144. 1789 • Yellowwood [Greek *okhros*, pale yellow, probably alluding to color of stem, leaf, and fruit of some species] [I]

David E. Lemke

Shrubs or small trees; latex milky. **Stems** erect, unarmed, glabrous. **Leaves** persistent, opposite [whorled], petiolate; stipular colleters intrapetiolar; laminar colleters absent. **Inflorescences** subterminal or axillary, cymose, pedunculate. **Flowers:** calycine colleters absent; corolla white or cream [pale yellow], salverform, aestivation dextrorse; corolline corona absent; androecium and gynoecium not united into a gynostegium; stamens inserted at top of corolla tube; anthers not connivent, not adherent to stigma, connectives not appendiculate or enlarged, locules 4; pollen free, not massed into pollinia, translators absent; nectary annular. **Fruits** drupaceous, paired or solitary, erect, red [yellow, orange], ellipsoid to ovoid, compressed, smooth, glabrous. **Seeds** elliptic or orbiculate, flattened, narrowly winged, not beaked, not comose, not arillate. $x = 10$.

Species ca. 40 (1 in the flora): introduced, Florida; Asia, Indian Ocean Islands (Réunion), Pacific Islands, Australia.

According to P. Boiteau (1981), the hard, fibrous endocarp of *Ochrosia* fruits facilitates flotation and dispersal by water.

1. Ochrosia elliptica Labillardière, Sert. Austro-Caledon., 25, plate 30. 1824 • Elliptic yellowwood, bloodhorn [F] [I]

Shrubs or trees 1–6(–8) m. **Leaves:** petiole 0.5–2(–3) cm, glabrous; blade elliptic to obovate, (5–)7–14 × 2–6 cm, subcoriaceous, base cuneate, often decurrent on petiole, margins revolute, apex obtuse to broadly acute or rarely emarginate, surfaces glabrous. **Peduncles** 1.3–7 cm, glabrous. **Pedicels** 1–2 mm, glabrous. **Flowers:** calyx lobes ovate, 1.5–2.5 mm, glabrous; corolla glabrous abaxially, glabrous or occasionally eglandular-pubescent adaxially, tube 5–7 × 1–1.5 mm, throat 3–4 × 1.5–2 mm, lobes spreading, obliquely oblong to oblanceolate, 6–7 × 2–2.5 mm. **Drupes** 3.5–4 × 2–2.5 cm. **Seeds** 8–11 × 6–8 mm including wing. $2n = 20$.

Flowering and fruiting year-round. Coastal strands, disturbed sites; 0–10 m; introduced; Fla.; Pacific Islands (Nauru, New Caledonia, Vanuatu); Australia.

Ochrosia elliptica is occasionally planted as an ornamental in coastal areas of southern Florida due to its salt tolerance and has become naturalized in Broward, Monroe, and Sarasota counties. The bright red fruits, which persist on the branches, are attractive but poisonous.

8. PLUMERIA Linnaeus, Sp. Pl. 1: 209. 1753; Gen. Pl, ed. 5, 99. 1854 • Frangipani [For Charles Plumier, 1646–1704, French botanist who collected extensively in the West Indies] [I]

David E. Lemke

Trees, small or medium-sized; latex clear. **Stems** erect, unarmed, glabrous or eglandular-pubescent especially on younger growth. **Leaves** deciduous, alternate, petiolate; stipular colleters intrapetiolar; laminar colleters absent. **Inflorescences** terminal or subterminal, thyrsiform, pedunculate. **Flowers:** calycine colleters absent; corolla white with yellow eye [yellow, orange, pink, red, magenta], salverform, aestivation sinistrorse; corolline corona absent; androecium and gynoecium not united into a gynostegium; stamens inserted near base of corolla tube; anthers not connivent, not adherent to stigma, connectives not appendiculate or enlarged, locules 4;

pollen free, not massed into pollinia, translators absent; nectary absent. **Fruits** follicles, paired, pendulous, brown, slender, terete or slightly compressed, truncate, surface smooth or striate, glabrous. **Seeds** narrowly ovate, flattened, winged basally, not beaked, not comose, not arillate. $x = 9$.

Species ca. 10 (1 in the flora): introduced, Florida; Mexico, West Indies, Central America, n South America; cultivated widely in warmer regions of the world.

Although it honors Plumier, the genus name has always been written as here treated.

Plumeria rubra Linnaeus, a native of Mexico and Central America, is often cultivated in frost-free regions of Arizona, California, Florida, and Texas and can be distinguished by its more open, corymbose inflorescence of pink, red, magenta, orange, or yellow (rarely white) flowers.

SELECTED REFERENCE Woodson, R. E. Jr. 1938. Studies in the Apocynaceae. VII. An evaluation of the genera *Plumeria* L. and *Himatanthus* Willd. Ann. Missouri Bot. Gard. 25: 189–224.

1. Plumeria obtusa Linnaeus, Sp. Pl. 1: 210. 1753 [F] [I]

Trees 3–10 m. **Leaves:** petiole 10–40 mm, glabrous; blade obovate or oblanceolate to obovate-oblong, 3.5–18 × 1–8.5 cm, coriaceous or subcoriaceous, base rounded, obtuse, or cuneate, apex rounded, emarginate or mucronate, surfaces glabrous. **Peduncles** 6–10 cm, glabrous. **Pedicels** 5–12 mm, glabrous. **Flowers:** calyx lobes ovate to deltate, 1–1.5 mm, glabrous; corolla glabrous abaxially, eglandular-pubescent adaxially, tube 9–20 × 1–1.5 mm, lobes ascending-spreading, obliquely obovate-oblong or obovate, (15–)25–35(–45) × 10–15 mm. **Follicles** 6.5–24 × 1–2 cm. **Seeds:** body 10–15 × 7–10 mm, wing 8–13 mm. $2n = 36$.

Flowering summer–fall; fruiting fall–winter. Coastal hammocks, pinelands; 0 m; introduced; Fla.; West Indies; Central America (Guatemala, Honduras).

Plumeria obtusa is widely cultivated as an ornamental in southern Florida but has apparently only escaped from cultivation in disturbed coastal hammocks and surrounding pinelands on Big Pine Key in Monroe County.

An interesting case of deceitful pollination in *Plumeria*, in which hawkmoths are attracted to flowers that offer no nectar reward, was described by W. A. Haber (1984).

9. RAUVOLFIA Linnaeus, Sp. Pl. 1: 208. 1753; Gen. Pl. ed. 5, 98. 1854 • [For Leonhart Rauwolff, 1535–1596, German physician and botanist] [I]

David E. Lemke

Shrubs [trees]; latex milky. **Stems** erect, unarmed, glabrous [eglandular-pubescent]. **Leaves** persistent, whorled or sometimes opposite at lower nodes, petiolate; stipular colleters intrapetiolar; laminar colleters absent. **Inflorescences** terminal and axillary, cymose, pedunculate. **Flowers:** calycine colleters absent; corolla white or cream [pale purple, pink], rotate [salverform], aestivation sinistrorse; corolline corona absent; androecium and gynoecium not united into a gynostegium; stamens inserted at top of corolla tube; anthers not connivent, not adherent to stigma; connectives not appendiculate or enlarged, locules 4; pollen free, not massed into pollinia, translators absent; nectary annular. **Fruits** drupaceous, solitary or partly to completely coherent, erect, initially red, maturing black, globose to depressed-globose, smooth, glabrous. **Seeds** ovate, not winged, not beaked, not comose, not arillate. $x = 11$.

Species ca. 80 (1 in the flora): introduced, Florida; Mexico, West Indies, Central America, South America, Asia, Africa, Indian Ocean Islands, Pacific Islands, introduced also in s Asia, Australia.

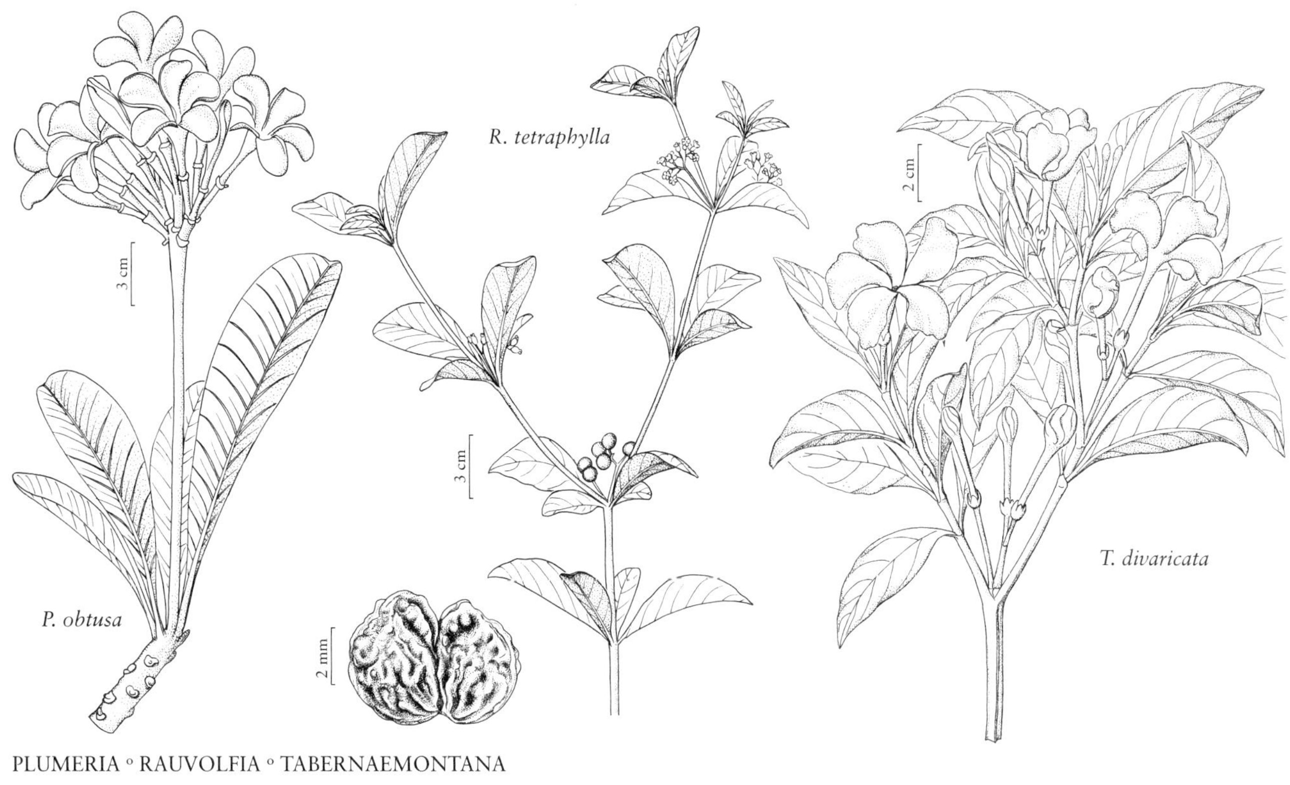

PLUMERIA ◦ RAUVOLFIA ◦ TABERNAEMONTANA

SELECTED REFERENCE Rao, A. S. 1956. A revision of *Rauvolfia* with particular reference to the American species. Ann. Missouri Bot. Gard. 43: 253–354.

1. Rauvolfia tetraphylla Linnaeus, Sp. Pl. 1: 208. 1753 [F] [I]

Shrubs 1–3(–14) m. **Leaves** anisophyllous; petiole 1–7(–20) mm, glabrous; blade narrowly oblong-elliptic to ovate-elliptic, 4–12(–17) × 1.5–5(–7) cm, membranous to subcoriaceous, base acute to rounded, apex acute to obtuse, surfaces glabrous or pubescent. **Peduncles** 1–4 cm, glabrous or sparsely pubescent. **Pedicels** 3–5 mm, glabrous or pubescent. **Flowers:** calyx lobes broadly ovate, 0.5–1.5 mm, glabrous or minutely glandular-pubescent on margins; corolla glabrous throughout or eglandular-pubescent abaxially, tube 2–3(–5) × 0.1–2 mm, throat 1.5–2 × 1.5–2 mm with a conspicuous adaxial ring of hairs, lobes spreading, obliquely obovate to suborbiculate, 2–3.5 × 1–2 mm, ± as long as wide. **Fruits** drupes, 5–7(–10) × 5–8(–13) mm. **Seeds** 3.5–6 × 2–3 mm. **2*n* = 66.**

Flowering spring–fall; fruiting summer–fall. Disturbed sites; 0–10 m; introduced; Fla.; Mexico; West Indies; Central America; n South America; introduced also in s Asia (India), Australia.

Rauvolfia tetraphylla is cultivated at several localities in Florida and has apparently become naturalized in at least one location in Palm Beach County. The fleshy fruits, with one or two small, ovoid seeds, are distinctive among North American Apocynaceae.

Rauvolfia species are well known for the production of numerous alkaloids, particularly reserpine, which are found throughout the plant and can cause reductions in blood pressure and heart rate when ingested (G. E. Burrows and R. J. Tyrl 2013). The milky latex of *R. tetraphylla* often causes irritation and reddening of the skin upon contact.

10. TABERNAEMONTANA Linnaeus, Sp. Pl. 1: 210. 1753; Gen. Pl. ed. 5, 100. 1754 • Milkwood [For Jacobus Theodorus Tabernaemontanus, 1525–1590, German physician and herbalist] [I]

David E. Lemke

Shrubs or trees; latex milky. **Stems** erect, unarmed, glabrous or eglandular-pubescent especially on younger growth. **Leaves** persistent, opposite, petiolate; stipular colleters intrapetiolar; laminar colleters absent. **Inflorescences** axillary, compound-cymose, pedunculate. **Flowers:** calycine colleters present; corolla white, cream, or yellow [pink, pale purple], salverform, aestivation sinistrorse [dextrorse]; corolline corona absent; androecium and gynoecium not united into a gynostegium; stamens inserted near top [middle] of corolla tube; anthers not connivent, not adherent to stigma, connectives not appendiculate or enlarged, locules 4; pollen free, not massed into pollinia, translators absent; nectary annular or absent. **Fruits** follicular [baccate], solitary or paired, deflexed, yellow, orange, red, or green, ellipsoid or reniform, compressed, smooth, glabrous. **Seeds** obliquely elliptic, somewhat flattened, not winged, not beaked, not comose, arillate. $x = 11$.

Species ca. 110 (2 in the flora): introduced, Florida; Mexico, West Indies, Central America, South America, Asia, Africa, Indian Ocean Islands, Pacific Islands.

Tabernaemontana was established by Linnaeus based on three species, one from tropical Asia and two from the West Indies. The genus was greatly enlarged by A. L. P. P. de Candolle (1844), who recognized more than 60 species from the New World and Old World tropics, but later largely dismembered as taxonomists such as J. Miers (1878) and O. Stapf (1902) began to establish segregate genera, these often restricted to either the Paleotropics or the Neotropics. A revision of *Tabernaemontana* by M. Pichon (1948b) proposed a broad circumscription of the genus to again include both New World and Old World species, a delimitation largely followed by A. J. M. Leeuwenberg (1991, 1994) in the most recent taxonomic treatments of the group.

A phylogenetic analysis of the tribe Tabernaemontaneae G. Don by A. O. Simões et al. (2010), based on a combination of molecular and morphological characters, found that the species assigned to *Tabernaemontana* and its segregates form a clade with a deep split into paleotropical and neotropical lineages and that this clade corresponds closely, but not exactly, to the broad circumscription of the genus by A. J. M. Leeuwenberg (1991, 1994). To render *Tabernaemontana* monophyletic, Simões et al. proposed delimiting the genus to encompass not only the New World and Old World species that had been included in *Tabernaemontana* by Leeuwenberg but also the neotropical genus *Stemmadenia* Bentham, a delimitation followed here.

SELECTED REFERENCES Leeuwenberg, A. J. M. 1990. *Tabernaemontana* (Apocynaceae): Discussion of its delimitation. In: P. Baas et al., eds. 1990. The Plant Diversity of Malesia. Dordrecht and Boston. Pp. 73–81. Leeuwenberg, A. J. M. 1991. A Revision of *Tabernaemontana*: The Old World Species. Kew. Leeuwenberg, A. J. M. 1994. A Revision of *Tabernaemontana*. Two, the New World Species and *Stemmadenia*. Kew.

1. Corolla lobes 7–14 mm; calyx lobes glabrous; seeds 10–50 . . . . 1. *Tabernaemontana alba*
1. Corolla lobes 15–27 mm; calyx lobes usually ciliate; seeds 2–10 . . . . 2. *Tabernaemontana divaricata*

1. Tabernaemontana alba Miller, Gard. Dict. ed. 8, Tabernaemontana no. 2. 1768 • White milkwood [I]

Shrubs or small trees, 1.5–3(–15) m. **Leaves:** petiole (5–)10–25 mm, glabrous; blade elliptic to obovate, (2–)7–23 × (1–)2.5–8 cm, membranous or subcoriaceous, base cuneate, margins revolute, apex acuminate or apiculate, surfaces glabrous. **Peduncles** (0.2–)1.5–7.5 cm, glabrous. **Pedicels** 4–10(–15) mm, glabrous. **Flowers:** calyx lobes ovate to broadly ovate, 1.5–3 mm, glabrous; corolla white, cream, or yellow, glabrous abaxially, eglandular-pubescent adaxially, tube 3–4 × 1.5–2.5 mm, throat 3–4 × 1.5–2 mm, lobes spreading, obliquely elliptic to dolabriform, 7–14 × 3.5–6 mm. **Follicles** orange, yellow, or green, 2–4 × 1–2.5 cm. **Seeds** 10–50, 8–10 × 4–5 mm, aril orange or red.

Flowering summer; fruiting summer–fall. Disturbed pinelands; 0–10 m; introduced; Fla.; Mexico; West Indies; Central America.

Tabernaemontana alba was reported as being rare in disturbed pinelands in Miami-Dade County of southern Florida by R. P. Wunderlin (1998); however, no recently-collected specimens have been seen.

2. Tabernaemontana divaricata (Linnaeus) R. Brown ex Roemer & Schultes in J. J. Roemer et al., Syst. Veg. 4: 427. 1819 • Pinwheel flower [F] [I]

Nerium divaricatum Linnaeus, Sp. Pl. 1: 209. 1753

Shrubs or small trees, 0.5–3 (–5) m. **Leaves:** petiole 3–10 mm, glabrous; blade narrowly elliptic to elliptic, 3–18 × 1–6 cm, membranous, base cuneate, often decurrent on petiole, margins revolute, apex acuminate, surfaces glabrous. **Peduncles** 0.5–8 cm, glabrous. **Pedicels** 3–20 mm, glabrous. **Flowers:** calyx lobes narrowly to broadly ovate, 2–3.5 mm, usually ciliate; corolla white, often with a yellow center, glabrous abaxially, eglandular-pubescent adaxially, tube 7–12 × 1–2.5 mm, throat 8–15 × 1.5–3 mm, lobes spreading, obliquely elliptic to obovate or dolabriform, 15–27 × 8–20 mm. **Follicles** orange, red, or occasionally green, 2–7 × 0.6–1.5 cm. **Seeds** 2–10, 7–8 × 3–4 mm, aril orange or red. $2n = 22$.

Flowering summer; fruiting summer–fall. Pine flatwoods; 20–30 m; introduced; Fla.; Asia.

Tabernaemontana divaricata is often grown as an ornamental in Florida but has apparently escaped and become naturalized in only a small area of Osceola County.

11. THEVETIA Linnaeus, Opera Var., 212. 1758, name conserved • [For André Thevet, sixteenth-century French priest and explorer] [I]

Patricia Sabadin

Alexander Krings

Trees or shrubs; latex milky. **Stems** erect, unarmed, glabrous [sparsely eglandular-pubescent]. **Leaves** evergreen, alternate, petiolate [sessile]; stipular colleters 2 or 3 at base of petiole on each side, continuous with intrapetiolar colleters, interpetiolar colleters absent; laminar colleters absent. **Inflorescences** terminal, corymbiform, simple or compound, pedunculate. **Flowers:** calycine colleters present [absent]; corolla yellow or orange [sometimes tinted with purple], funnelform [salverform], aestivation sinistrorse; corolline corona fingerlike [3-gonous]; androecium and gynoecium not united into a gynostegium; stamens inserted near top of corolla tube; anthers connivent, agglutinated on stigma, apical connective deltoid, locules 4; pollen free, not massed into pollinia, translators absent; nectary disciform. **Fruits** drupaceous, solitary, pendulous, green to black [red], obpyriform to subglobose [reniform], rhombic to subterete, smooth, glabrous. **Seeds** reniform, flattened, not winged [winged], not beaked, not comose, not arillate. $x = 10$.

Species ca. 8 (1 in the flora): introduced, Florida; United States, Mexico, West Indies, Central America, South America; introduced also in Bermuda, Asia, Africa, Indian Ocean Islands, Pacific Islands, Australia.

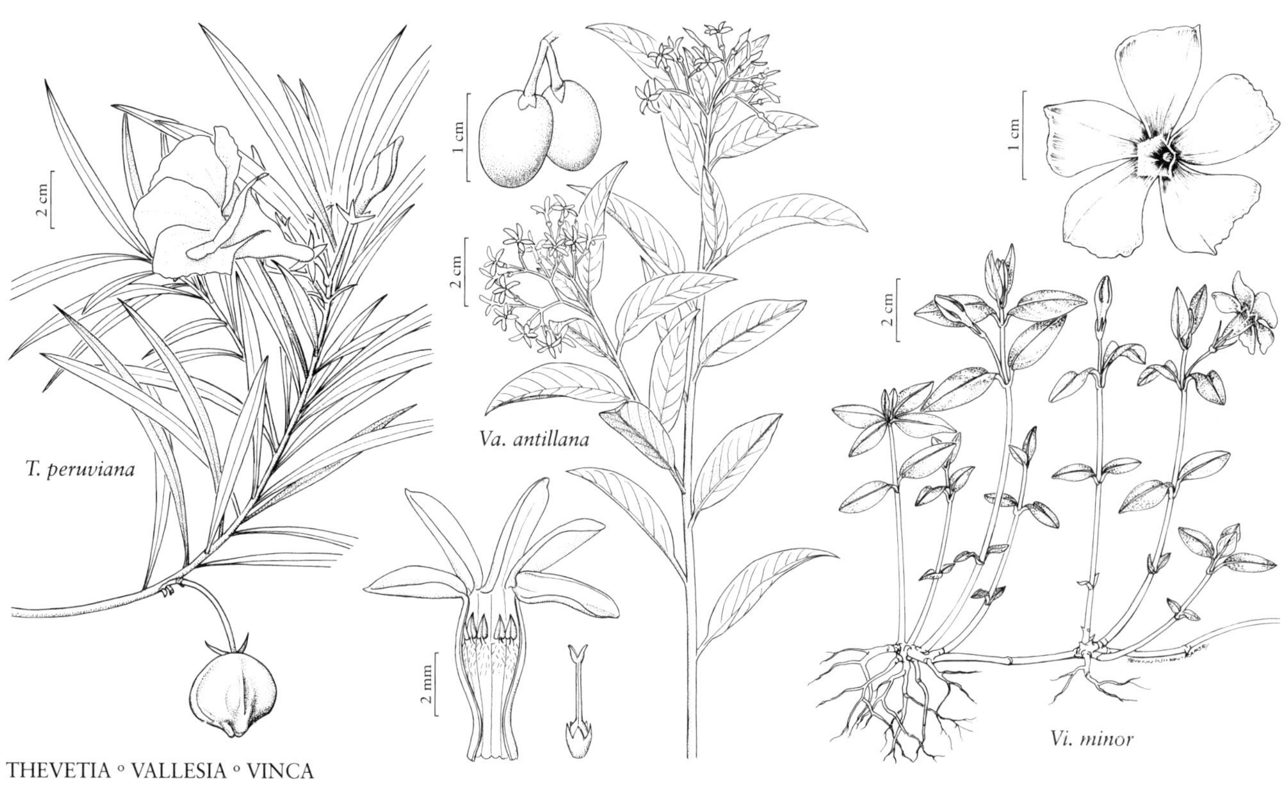

THEVETIA ◦ VALLESIA ◦ VINCA

Although some have proposed recognition of *Thevetia* in the narrow sense as distinct from *Cascabela* Rafinesque (L. O. Alvarado-Cárdenas and H. Ochoterena 2007; J. F. Morales 2009b, 2009c), the broader concept of A. O. Simões et al. (2007) of a *Thevetia* that encompasses both taxa is adopted here.

SELECTED REFERENCES Alvarado-Cárdenas, L. O. and H. Ochoterena. 2007. A phylogenetic analysis of the *Cascabela-Thevetia* species complex (Plumerieae, Apocynaceae) based on morphology. Ann. Missouri Bot. Gard. 94: 298–323. Williams, J. K. and J. K. Stutzman. 2008. Chromosome number of *Thevetia ahouai* (Apocynaceae: Rauvolfoidae: Plumerieae) with discussion on the generic boundaries of *Thevetia*. J. Bot. Res. Inst. Texas 2: 489–493.

1. **Thevetia peruviana** (Persoon) K. Schumann in H. G. A. Engler and K. Prantl, Nat. Pflanzenfam. 120–122[IV,2]): 159. 1895 • Yellow oleander, lucky nut, trumpet flower [F] [I]

Cerbera peruviana Persoon, Syn. Pl. 1: 267. 1805

Trees or shrubs 1.5–3.5(–8) m. **Leaves:** petiole 1.9–16.3 × 0.1–2.8 mm, glabrous; blade lanceolate, oblong, or oblanceolate, 4.7–15.3 × 0.4–2.2 cm, base attenuate, margins entire, apex acute, surfaces glabrous, lateral veins indistinct. **Inflorescences** to 8-flowered; peduncle 2–9 mm, glabrous; bracts deciduous, ovate, 1.8–6.3 × 0.6–2 mm, glabrous. **Pedicels** 1.3–6.2 cm, glabrous. **Flowers:** sepals spreading, ovate to lanceolate, 0.3–1.3 × 0.2–0.5 cm, apex acute to acuminate, glabrous; corolla tube 0.9–1.8 × 0.3–0.6 cm, internally pubescent, throat 0.7–1.7 × 0.5–2.2 cm, lobes ascending to spreading, obovate to oblong, (0.6–)1.4–3.5 × 1.7–2.5 cm, glabrous; filaments separate; anthers sagittate, 1.9–2.3 × 1.4–1.6 mm; ovary 1.9–4 × 1.7–3 mm, glabrous; style 10–20.1 mm; stigma 2–3 × 2.3–3 mm; nectar disc 5-lobed, 0.8–1 mm. **Drupes** dehiscent, (1.2–)2.5–4 × (1.5–)2–5 cm, sometimes lenticellate. **Seeds** 1–2 × 1–1.3 cm. **2*n*** = 20.

Flowering spring–fall; fruiting spring–fall. Shorelines; 0–90 m; introduced; Fla.; Mexico; West Indies; Central America; South America; introduced also in Bermuda, Asia, Africa, Indian Ocean Islands, Pacific Islands, Australia.

Thevetia peruviana is cultivated in Arizona, California, Florida, and Texas, as well as in the tropics around the world (L. O. Alvarado-Cárdenas and H. Ochoterena 2007). It has escaped cultivation and become naturalized in Florida in Brevard, Miami-Dade, and Sarasota counties.

12. VALLESIA Ruiz & Pavon, Fl. Peruv. Prodr., 28, plate 5[top]. 1794 • Pearl berry [For Francisco Vallés, 1524–1592, Spanish botanist, physician to Phillip II]

Eliane Meyer Norman

Shrubs, more rarely trees, to 12 m; latex sometimes present, milky [clear]. **Stems** erect, unarmed, glabrous or eglandular-pubescent. **Leaves** persistent, alternate, petiolate; stipular colleters interpetiolar; laminar colleters absent. **Inflorescences** extra-axillary, occasionally terminal, cymose, pedunculate. **Flowers:** calycine colleters absent; corolla white [yellowish orange], salverform, rarely urceolate-salverform, aestivation sinistrorse; corolline corona absent; androecium and gynoecium not united into a gynostegium; stamens inserted at or near orifice of corolla tube; anthers not connivent, adherent to stigma, connectives apiculate, locules 4; pollen free, not massed into pollinia, translators absent; nectary adherent to base of ovary. **Fruits** drupaceous, solitary (paired), erect, white, oblong-ovoid to reniform, terete to somewhat compressed, smooth, glabrous. **Seeds** oblong-ovoid, somewhat flattened, not winged, not beaked, not comose, not arillate. $x = 10$.

Species ca. 10 (1 in the flora): Florida, Mexico, West Indies, Central America, South America.

1. Vallesia antillana Woodson, Ann. Missouri Bot. Gard. 24: 13. 1937 • Tear shrub [F]

Shrubs or small trees, 1–5 m. **Stems** yellowish to light gray, furrowed; latex milky. **Leaves:** petiole 4–5 mm; blade obovate-elliptic or oblong-lanceolate, 2.5–10 × 1.2–3 cm, subcoriaceous, base broadly acute to obtuse, margins entire, apex acuminate, surfaces glabrous. **Peduncle** 8–20 mm, glabrous. **Pedicels** 4–8 mm, glabrous. **Flowers:** calyx lobes acute, 0.7–1.5 mm, glabrous; corolla tube 6–7 mm, abaxial surface puberulent below insertion of stamens, adaxial surface glabrous, lobes spreading, oblong, 3–5 × 1.2–1.6 mm, glabrous; androecium inserted 1.5 mm below orifice, 1–1.5 × 0.5–0.7 mm; gynoecium ovoid, 1.5 mm; style 2 mm; stigma 0.5 mm. **Drupes** indehiscent, oblong-ovoid, 10–12 × 5–8 mm (3–4 mm wide when dry), glistening.

Flowering and fruiting year-round. Rockland hammocks, coastal berms; 0–200 m; Fla.; Mexico; West Indies (Bahamas, Cuba, Dominican Republic, Jamaica).

Vallesia antillana, now known in the flora area only from a few populations in Monroe County including the Florida Keys, is recognized as an endangered state species (G. D. Gann et al. 2002). It is sparingly cultivated in southern Florida. In the Bahamas, the species is used to treat gastrointestinal problems.

13. VINCA Linnaeus, Sp. Pl. 1: 209. 1753; Gen. Pl. ed. 5, 98. 1754 • Periwinkle [Derivation uncertain; probably Latin *vinco*, to conquer, or *vincio*, to bind, alluding to binding and subduing other plants in its habitat] [I]

David E. Lemke

Suffrutescent herbs [subshrubs]; latex milky. **Stems** trailing or ascending [erect], unarmed, glabrous or eglandular-pubescent. **Leaves** persistent [deciduous], opposite, petiolate; stipular colleters absent; laminar colleters absent. **Inflorescences** axillary, 1(–4)-flowered, pedunculate. **Flowers:** calycine colleters absent; corolla blue-purple, blue-violet, violet, reddish purple, or white, rarely pale blue, rotate-funnelform, aestivation sinistrorse; corolline corona absent; androecium and gynoecium not united into a gynostegium; stamens inserted at top of corolla tube; anthers connivent, not adherent to stigma, connectives appendiculate, locules 4; pollen free, not massed into pollinia, translators absent; nectaries 2. **Fruits** follicles, usually paired, erect, brown, slender, terete to somewhat moniliform, surface striate, glabrous. **Seeds** narrowly ovoid to elliptic, navicular, not winged, not beaked, not comose, not arillate. $x = 23$.

Species 7 (2 in the flora): introduced; Europe, w Asia; introduced also widely.

Several species of *Vinca* are widely cultivated as ornamentals throughout Europe and North America, but only two of these are treated as naturalized elements of our flora. *Vinca herbacea* Waldstein & Kitaibel has been collected several times from northwestern Massachusetts but is here considered a waif; it can be distinguished from *V. major* and *V. minor* by the entirely herbaceous growth habit, scabrous (versus ciliate or glabrous) leaf margins, and corolla lobes less than 6 mm wide.

In Europe, species of *Vinca* have long been associated with death, being commonly planted in cemeteries and, during the Middle Ages, used in the making of wreaths that were sometimes worn as a crown by condemned persons on their way to execution.

SELECTED REFERENCE Stearn, W. T. 1973. A synopsis of the genus *Vinca* including its taxonomic and nomenclatural history. In: W. I. Taylor and N. R. Farnsworth, eds. 1973. The *Vinca* Alkaloids. Botany, Chemistry, and Pharmacology. New York. Pp. 1–94.

1. Leaf blade margins ciliate; calyx lobes ciliate, 7–15 mm; seeds 7–10 mm . . . . 1. *Vinca major*
1. Leaf blade margins not ciliate; calyx lobes not ciliate, 3–4 mm; seeds 5–7 mm . . . . 2. *Vinca minor*

1. Vinca major Linnaeus, Sp. Pl. 1: 209. 1753 • Greater periwinkle [I] [W]

Stems ascending and trailing. **Leaves:** petiole 5–15 mm, glabrous or pubescent, with 2 small glandular appendages in distal half; blade ovate to broadly ovate or lanceolate, 2.5–9 × 2–6 cm (distals), membranous, base cordate or truncate, margins ciliate, apex obtuse to acute, adaxial surface usually pubescent. **Peduncles** 1.5–4 cm, glabrous. **Flowers:** calyx lobes linear, 7–15 mm, ciliate; corolla blue-purple, rarely violet or white, glabrous abaxially, eglandular-pubescent adaxially, tube 4–5 × 3–4 mm, throat 8–12 × 4–9 mm, lobes spreading, obliquely dolabriform, 15–20 × (3–)10–20 mm. **Follicles** 25–50 × 2–3 mm. **Seeds** 7–10 × 2–2.6 mm. $2n = 92$.

Flowering spring (or year-round in the southern United States); fruiting summer. Open woodlands, stream and woodland margins, old home sites, other shaded disturbed areas; 0–1900 m; introduced; B.C.; Ala., Ariz., Ark., Calif., Conn., Del., Fla., Ga., Idaho, Ill., Ind., Ky., La., Md., Mass., Mich., Miss., Mo., N.J., N.Mex., N.Y., N.C., Ohio, Okla., Oreg., Pa., S.C., Tenn., Tex., Utah, Va., Wash., Wis.; s Europe.

Vinca major has been widely introduced as an ornamental and commonly spreads from cultivation.

2. Vinca minor Linnaeus, Sp. Pl. 1: 209. 1753 • Common periwinkle, petite pervenche [F] [I] [W]

Stems trailing. **Leaves:** petiole 1–2(–10) mm, glabrous, without glandular appendages; blade lanceolate to elliptic or sometimes ovate, 1.5–4.5 × 0.5–2.5 cm (distals), membranous, base rounded or cuneate, margins not ciliate, apex obtuse to acute, glabrous. **Peduncles** 1.5–2.5(–3.5) cm, glabrous. **Flowers:** calyx lobes narrowly lanceolate, 3–4 mm, glabrous; corolla blue-violet, rarely pale blue, reddish purple, or white, glabrous abaxially, eglandular-pubescent adaxially, tube 3–6 × 2–3 mm, throat 5–7 × 3–4 mm, lobes spreading, obliquely dolabriform, 10–15 × 6–15 mm. **Follicles** 20–30(–60) × 2–3 mm. **Seeds** 5–7 × 2–2.3 mm. $2n = 46$.

Flowering spring; fruiting summer. Roadsides, open woodlands, shaded disturbed areas, old cemeteries, homesites; 0–1300 m; introduced; B.C., N.B., N.S., Ont., Que.; Ala., Ark., Calif., Conn., Del., D.C., Fla., Ga., Idaho, Ill., Ind., Iowa, Kans., Ky., La., Maine, Md., Mass., Mich., Minn., Miss., Mo., Nebr., N.H., N.J., N.Y., N.C., Ohio, Okla., Oreg., Pa., R.I., S.C., Tenn., Tex., Utah, Vt., Va., Wash., W.Va., Wis.; Europe; w Asia.

Vinca minor has been widely introduced as an ornamental and commonly spreads from cultivation. It is more frost-hardy than *V. major* and more common in the midwestern and northeastern United States and eastern Canada, while *V. major* is abundant across the southern United States. *Vinca minor* is regarded as invasive in parts of Georgia, Kentucky, South Carolina, and Tennessee.

14. ANGADENIA Miers, Apocyn. S. Amer., 173, plate 27. 1878 • [Greek *angos*, vessel, and *adenos*, gland, alluding to urceolate nectaried disc]

David E. Lemke

Subshrubs [lianas]; latex milky. **Stems** erect, decumbent, or trailing, unarmed, eglandular-pubescent [glabrous]. **Leaves** persistent, opposite or rarely subopposite, petiolate; stipular colleters absent; laminar colleters absent. **Inflorescences** axillary, cymose, pedunculate. **Flowers:** calycine colleters present; corolla yellow to very pale yellow, funnelform, aestivation dextrorse; corolline corona absent; androecium and gynoecium not united into a gynostegium; stamens inserted at top of corolla tube; anthers connivent, adherent to stigma, connectives enlarged, 2-lobed, locules 4; pollen free, not massed into pollinia, translators absent; nectaries 5, alternating with stamens. **Fruits** follicles, usually paired, erect, green, slender, terete to slightly moniliform, surface striate, glabrous. **Seeds** narrowly lanceolate, flattened, not winged, beaked, comose, not arillate.

Species 2 (1 in the flora): Florida, West Indies.

As originally circumscribed, *Angadenia* was a genus of nearly two dozen species, most of which were transferred by R. E. Woodson Jr. (1936) to *Fernaldia* Woodson, *Mandevilla*, *Neobracea* Britton, and *Odontadenia* Bentham. Following Woodson, the genus is here treated as comprising two species, the native *A. berteroi* and *A. lindeniana* Miers of Cuba, Hispaniola, and Jamaica.

SELECTED REFERENCES Barrios, B. and S. Koptur. 2011. Floral biology and breeding system of *Angadenia berteroi* (Apocynaceae): Why do flowers of the pineland golden trumpet produce few fruits? Int. J. Pl. Sci. 172: 378–385. Barrios, B., G. Arellano, and S. Koptur. 2011. The effects of fire and fragmentation on occurrence and flowering of a rare perennial plant. Pl. Ecol. 212: 1057–1067.

1. Angadenia berteroi (A. de Candolle) Miers, Apocyn. S. Amer. 180. 1878 (as berterii) • Pineland golden trumpet [F]

Echites berteroi A. de Candolle in A. P. de Candolle and A. L. P. P. de Candolle, Prodr. 8: 447. 1844 (as berterii); *Rhabdadenia corallicola* Small; *R. sagrae* Müller Argoviensis

Subshrubs, 10–20 dm. **Stems** sparsely pubescent. **Leaves:** petiole 1–2 mm, pubescent; blade narrowly elliptic to ovate or oblong, 10–25 × 2.5–10 mm, membranous to subcoriaceous, base rounded to truncate, margins revolute, apex rounded to obtuse, surfaces glabrous. **Peduncles** 15–20 mm, glabrous. **Pedicels** 7–10 mm, glabrous. **Flowers:** calyx lobes broadly ovate, 1–2.5 mm, glabrous; corolla glabrous abaxially, usually eglandular-pubescent adaxially, tube 5–8 × 2 mm, throat 10–15 × 5–9 mm, lobes ascending, obliquely obovate, 7–13 × 6–10 mm, adjacent lobes overlapping for 1/2+ of their length from base. **Follicles** 60–90 × 1.5–4 mm. **Seeds** 4–8 mm.

Flowering spring–summer; fruiting spring–fall. Pine rocklands, rockland hammocks, marl prairies; 0–10 m; Fla.; West Indies (Bahamas, Cuba, Hispaniola).

Angadenia berteroi is known from Florida only by collections from Miami-Dade and Monroe counties. It is considered threatened in Florida.

B. Barrios and S. Koptur (2011) examined the breeding system of *Angadenia berteroi* and concluded that the low fruit set seen in natural populations may be due to a combination of self-incompatibility and low visitation by pollinators. Within the pine rocklands habitat dominated by *Pinus elliottii*, the abundance (but not flowering or fruit set) of *A. berteroi* has been shown to be correlated with habitat fragmentation and time elapsed since burning (Barrios et al. 2011).

G. E. Burrows and R. J. Tyrl (2013) reported that the milky latex of *Angadenia berteroi* causes irritation and reddening of the skin upon contact and increased mucus production by the mucosa of the digestive tract upon ingestion.

15. APOCYNUM Linnaeus, Sp. Pl. 1: 213. 1753; Gen. Pl. ed. 5, 101. 1754 • Dogbane, Indian hemp [Greek *apo*, away from, and *kynos*, dog, alluding to certain species purportedly poisonous to dogs]

David E. Lemke

Herbs, perennial, often from somewhat thickened rhizomes; latex milky. **Stems** erect or ascending, unarmed, glabrous or eglandular-pubescent. **Leaves** deciduous, opposite or rarely subopposite or subverticillate, petiolate; stipular colleters interpetiolar; laminar colleters absent. **Inflorescences** axillary or terminal, cymose, pedunculate, bracteate or ebracteate. **Flowers**: calycine colleters absent; corolla white, white with pink veins, greenish white, yellowish white, or pink, campanulate, urceolate, or cylindric, aestivation dextrorse; corolline corona of 5 small sagittate squamellae; androecium and gynoecium not united into a gynostegium; stamens inserted near base of corolla tube; translators absent; anthers connivent, adherent to stigma; connectives enlarged, narrowly 2-lobed, locules 4; pollen in tetrads, not massed into pollinia, translators absent; nectaries 5, distinct. **Fruits** follicles, usually paired, erect, brown, slender, terete, surface striate, glabrous. **Seeds** fusiform, terete, not winged, not beaked, comose, not arillate. x = 8, 11.

Species 3 (3, including 1 hybrid, in the flora): North America, n Mexico.

Apocynum is a taxonomically challenging genus in which the number of species recognized depends upon both generic and species delimitations. Nearly 250 specific and infraspecific taxa have been named in *Apocynum*, many by E. L. Greene (for example, 1904, 1912), that were later reduced by R. E. Woodson Jr. (1930) to seven species and 17 varieties, and more recent floras have recognized as few as two variable species. The present treatment conservatively views *Apocynum* as comprising three species native to North America and northern Mexico, with the Old World species placed in the segregate genera *Poacynum* Baillon and *Trachomitum* Woodson, following Woodson. A detailed discussion of the nomenclatural history, morphology, anatomy, and geography of *Apocynum* was provided by Woodson.

The species of *Apocynum* are often weedy and clonal and a single genetic individual can be spread over a large area. Although visited by a variety of potential pollinators (K. D. Waddington 1976), fruit set in *Apocynum* species is low, often less than 10%. S. R. Lipow and R. Wyatt (1999) have suggested that this is due to a late-acting self-incompatibility mechanism similar to that documented for other members of the family.

Apocynum has a long history of usage by humans, especially Native Americans, as a source of fiber for the production of thread and cord used in the manufacture of items such as netting, basketry, bowstrings, bridles, and rope. A high-quality rubber latex has been extracted from *A. androsaemifolium* and *A. cannabinum* and was used in the preparation of chewing gum by several Native American groups. The mild toxicity of *Apocynum* species is due to the presence of cardiac glycosides, and powdered roots, or root and leaf extracts, were used by Native Americans to treat a wide variety of disorders, including headache, asthma, dyspepsia, constipation, renal and cardiac conditions, and intestinal worms (S. Cheatham et al. 1995; G. E. Burrows and R. J. Tyrl 2013).

SELECTED REFERENCE Woodson, R. E. Jr. 1930. Studies in the Apocynaceae. I. A critical study of the Apocynoideae (with special reference to the genus *Apocynum*). Ann. Missouri Bot. Gard. 17: 1–172, 174–212.

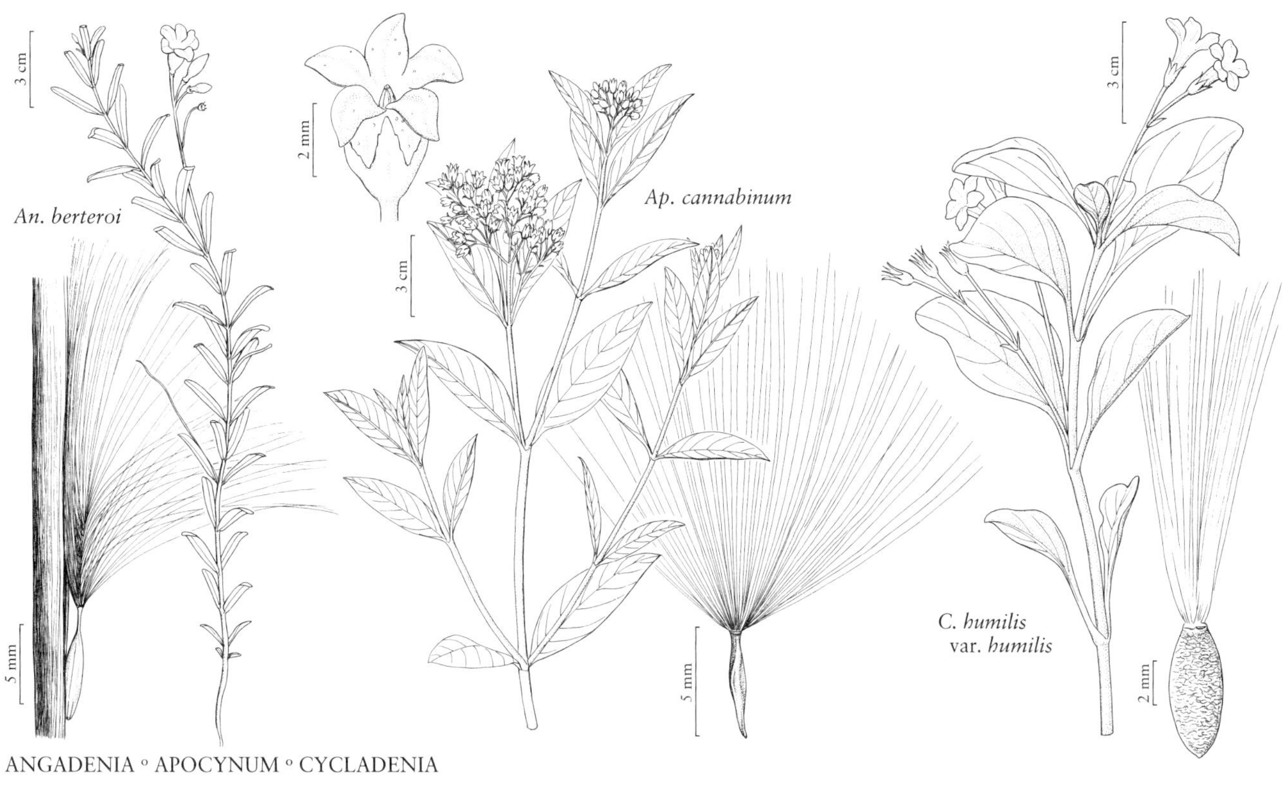

ANGADENIA ◦ APOCYNUM ◦ CYCLADENIA

1. Corollas white, greenish white, or yellowish white, tube 1.5–2.5 mm; inflorescences usually not exceeding foliage; leaf blades mostly more than 2.5 times as long as wide. . . . . 2. *Apocynum cannabinum*
1. Corollas pink, white with pink veins, or white tinged with pink, tube 4–5 mm; inflorescences usually exceeding foliage; leaf blades mostly less than 2 times as long as wide.
   2. Corolla tubes 3 times as long as calyx lobes or longer; seeds 1.5–2 mm; leaves drooping or spreading . . . . .1. *Apocynum androsaemifolium*
   2. Corolla tubes 1.5–2 times as long as calyx lobes; seeds 3–4 mm; leaves spreading to ascending . . . . .3. *Apocynum* ×*floribundum*

1. Apocynum androsaemifolium Linnaeus, Sp. Pl. 1: 213. 1753 (as fol. androsaemi) • Spreading dogbane, Apocyn à feuilles d'androsème [W]

Apocynum androsaemifolium var. *incanum* A. de Candolle; *A. androsaemifolium* var. *intermedium* Woodson

Herbs 2–10 dm. **Stems** glabrous or eglandular-pubescent. **Leaves** drooping or spreading; petiole 3–5 mm, glabrous or pubescent; blade ovate to oblong-lanceolate, (1–)3–6(–9) × (1–)1.5–3(–5) cm, mostly less than 2 times as long as wide, membranous, base rounded to cuneate, margins weakly to strongly revolute, apex acute to rounded, usually apiculate, surfaces glabrous or densely eglandular-pubescent abaxially, glabrous or sparsely eglandular-pubescent adaxially. **Inflorescences** usually exceeding foliage; peduncle 1–5 cm, glabrous or sparsely pubescent. **Pedicels** 1.5–7 mm, glabrous or sparsely pubescent. **Flowers:** calyx lobes broadly to narrowly triangular-ovate to lanceolate, 1.5–2.5 × 0.5–1 mm, glabrous or with a few scattered hairs; corolla pink or white with pink veins, glabrous abaxially and adaxially, tube 4–5 × 3–4 mm, at least 3 times as long as calyx lobes, lobes spreading or recurved, triangular-ovate, 1.5–3 × 0.7–1.2 mm. **Follicles** 60–150 × 2–3 mm. **Seeds** 1.5–2 × 0.3–0.5 mm. $2n$ = 16, 22.

Flowering spring–summer; fruiting summer–fall. Hardwood and coniferous forests and woodlands, clearings, mountain canyons, river terraces, old fields, pastures, roadsides; 10–3000 m; Alta., B.C., Man., N.B., Nfld. and Labr., N.W.T., N.S., Ont., P.E.I., Que., Sask., Yukon; Ala., Alaska, Ariz., Ark., Calif., Colo., Conn., Del., D.C., Ga., Idaho, Ill., Ind., Iowa, Ky., Maine, Md., Mass., Mich., Minn., Mo., Mont., Nebr., Nev., N.H., N.J., N.Mex., N.Y., N.C., N.Dak., Ohio, Okla., Oreg., Pa., R.I., S.C., S.Dak., Tenn., Tex., Utah, Vt., Va., Wash., W.Va., Wis., Wyo.; Mexico (Chihuahua, Coahuila).

R. E. Woodson Jr. (1930) recognized three varieties of *Apocynum androsaemifolium* based primarily on minor differences in corolla morphology and leaf pubescence; these varieties are not recognized here.

2. Apocynum cannabinum Linnaeus, Sp. Pl. 1: 213. 1753 • Prairie dogbane, Apocyn chanvrin [F] [W]

Apocynum cannabinum var. *glaberrimum* A. de Candolle; *A. cannabinum* var. *hypericifolium* (Aiton) A. Gray; *A. cannabinum* var. *pubescens* (R. Brown) A. de Candolle; *A. cordigerum* Greene; *A. hypericifolium* Aiton; *A. hypericifolium* var. *cordigerum* (Greene) Béguinot & Belosersky; *A. leuconeuron* Greene; *A. missouriense* Greene; *A. sibiricum* Jacquin var. *cordigerum* (Greene) Fernald

Herbs 3–10 dm. **Stems** glabrous or eglandular-pubescent. **Leaves** spreading to ascending; petiole 2–7 mm, often absent on lower leaves, glabrous or sparsely pubescent; blade ovate to lanceolate, 3–8(–11) × (0.4–)1.5–4 cm, mostly more than 2.5 times as long as wide, membranous, base rounded, cuneate, or auriculate, margins weakly to strongly revolute, apex acute to rounded or obtuse, usually apiculate, surfaces glabrous or densely eglandular-pubescent abaxially, glabrous or sparsely eglandular-pubescent adaxially. **Inflorescences** usually not exceeding foliage; peduncle 1–3.5 cm, glabrous or sparsely pubescent. **Pedicels** 2–3 mm, glabrous. **Flowers:** calyx lobes triangular-ovate to lanceolate, 1–2 × 0.3–0.8 mm, glabrous or infrequently pubescent; corolla white, greenish white, or yellowish white, glabrous abaxially and adaxially, tube 1.5–2.5 × 1.5–2 mm, ± as long as or slightly longer than calyx lobes, lobes erect or spreading, triangular-ovate, 1.5–2 × 0.7–1.2 mm. **Follicles** (40–)120–200 × 2–3 mm. **Seeds** 3.5–6 × 0.5–0.8 mm. $2n$ = 16, 22.

Flowering summer–fall; fruiting summer–fall. Forests, woodlands, fields, lakeshores, margins of swamps, rocky or gravelly river margins, dry creek beds, roadsides, railroad rights-of-way, other disturbed sites; 10–2500 m; Alta., B.C., Man., N.B., Nfld. and Labr., N.W.T., N.S., Ont., Que., Sask.; Ala., Ariz., Ark., Calif., Colo., Conn., Del., D.C., Fla., Ga., Idaho, Ill., Ind., Iowa, Kans., Ky., La., Maine, Md., Mass., Mich., Minn., Miss., Mo., Mont., Nebr., Nev., N.H., N.J., N.Mex., N.Y., N.C., N.Dak., Ohio, Okla., Oreg., Pa., R.I., S.C., S.Dak., Tenn., Tex., Utah, Vt., Va., Wash., W.Va., Wis., Wyo.; Mexico (Baja California, Chihuahua, Coahuila, Nuevo León, Tamaulipas).

Some American authors have treated plants with sessile or short-petiolate, lower cauline leaves and auriculate blades as *Apocynum sibiricum* Jacquin; these plants are here included in *A. cannabinum*.

Fiber derived from the stems of *Apocynum cannabinum* has long been considered to be of much higher quality than that of the other *Apocynum* species (S. Cheatham et al. 1995).

3. Apocynum ×floribundum Greene, Erythea 1: 151. 1893, as species • Hybrid dogbane, Apocyn moyen

Apocynum medium Greene

Herbs 2–5 dm. **Stems** glabrous or occasionally eglandular-pubescent. **Leaves** spreading to ascending; petiole 3–5 mm, glabrous; blade ovate to oblong-lanceolate, 3–6(–8) × 1.5–3(–5) cm, mostly less than 2 times as long as wide, membranous, base rounded to cuneate, margins weakly to strongly revolute, apex acute to rounded, usually apiculate, surfaces glabrous or eglandular-pubescent abaxially, glabrous or sparsely eglandular-pubescent adaxially. **Inflorescences** usually exceeding foliage; peduncle 1–2.5 cm, sparsely pubescent. **Pedicels** 1.5–7 mm, glabrous or sparsely pubescent. **Flowers:** calyx lobes lanceolate to oblong, 1.5–3 × 0.8–1 mm, glabrous or sparingly pubescent; corolla pink or white tinged with pink, glabrous abaxially and adaxially, tube 4–5 × 3.5–5 mm, 1.5–2 times as long as calyx lobes, lobes spreading or recurved, triangular-ovate, 2–3 × 0.8–1.2 mm. **Follicles** 70–150 × 2–3 mm. **Seeds** 3–4 × 0.4–0.6 mm.

Flowering summer–fall; fruiting summer–fall. Forests, woodlands, fields, river terraces, lakeshores, roadsides; 10–2600 m; Alta., B.C., Man., N.B., Nfld. and Labr., N.S., Ont., Que., Sask.; Ariz., Ark., Calif., Colo., Conn., Del., D.C., Ga., Idaho, Ill., Ind., Iowa, Kans., Ky., Maine, Md., Mass., Mich., Minn., Mo., Mont., Nebr., Nev., N.H., N.J., N.Mex., N.Y., N.C., N.Dak., Ohio, Oreg., Pa., R.I., S.C., S.Dak., Tenn., Tex., Utah, Vt., Va., Wash., W.Va., Wis., Wyo.; Mexico (Coahuila).

Apocynum ×floribundum (as *A. medium*) was considered by R. E. Woodson Jr. (1930) to be of hybrid origin, as were many of the taxa that he described in *Apocynum*. In a classic study, E. Anderson (1936b) investigated the possibility of a hybrid origin for *A. ×floribundum* by growing seeds from plants identified by Woodson as *A. androsaemifolium*, *A. cannabinum*, and *A. ×floribundum*. The offspring of the putative parental species bred true and had high pollen viability, while the progeny of the putative hybrids were quite variable in their morphological characteristics and had lower pollen viability, leading Anderson to conclude that *A. ×floribundum* was truly of hybrid origin. More recently, S. A. Johnson et al. (1998) found that allozyme data, as well as statistical analyses of various morphological characters, also support the hypothesis of a hybrid origin for *A. ×floribundum*.

16. CYCLADENIA Bentham, Pl. Hartw., 322. 1849 • [Greek *kyklos*, circle, and *aden*, gland, alluding to arrangement of nectaries below ovaries] [E]

David E. Lemke

Herbs, perennial, from ± woody rhizomes; latex not milky. **Stems** erect or slightly spreading, unarmed, glabrous or eglandular-pubescent. **Leaves** deciduous, opposite, petiolate; stipular colleters absent; laminar colleters absent. **Inflorescences** axillary or pseudo-terminal, few-flowered cymes, pedunculate. **Flowers:** calycine colleters absent; corolla red-violet to pink, funnelform, aestivation dextrorse; corolline corona absent; androecium and gynoecium not united into a gynostegium; stamens inserted at base of corolla throat; anthers connivent, not adherent to stigma, connectives enlarged, 2-lobed, locules 4; pollen free, not massed into pollinia, translators absent; nectary annular. **Fruits** follicles, usually paired, erect or deflexed, reddish brown, slender-fusiform, terete, striate, glabrous. **Seeds** elliptic, flattened, not winged, not beaked, comose, not arillate. $x = 7$.

Species 1: w United States.

1. Cycladenia humilis Bentham, Pl. Hartw., 323. 1849 • Waxydogbane [E] [F]

Herbs 6–12 cm. **Leaves:** petiole 5–30 mm, winged, glabrous or pubescent; blade ovate to suborbiculate, 3–7(–9.5) × 2–6.5 cm, thick-textured, base truncate to obtuse or cuneate, apex acute to obtuse, surfaces glabrous or densely pubescent, often glaucous. **Peduncles** 2.5–6 cm, glabrous or densely pubescent. **Pedicels** 7.5–12.5 mm, glabrous or densely pubescent. **Flowers:** calyx lobes ovate-lanceolate to linear, 5–7.5 mm, glabrous or eglandular-pubescent; corolla sparsely eglandular-pubescent, glabrous or eglandular-pubescent abaxially, glabrous or papillate adaxially, not clearly differentiated into tube and throat, 15–28 mm, lobes spreading, obovate-oblong, 4–9(–12) × 3–5 mm. **Follicles** 3.5–7 × 0.3–0.5 cm. **Seeds** 6–9 × 2.5–3.8 mm. $2n = 14$.

Varieties 3 (3 in the flora): w United States.

Three varieties of *Cycladenia humilis* are recognized based on morphological features and geographic distribution; however, two unpublished studies (M. P. Last 2009; H. K. Brabazon 2015) have noted inconsistencies between the morphological and genetic data and suggest a need for further study.

1. Corollas glabrous or tomentose with interwoven hairs abaxially, glabrous or papillate adaxially; n California, s Oregon .... 1a. *Cycladenia humilis* var. *humilis*
1. Corollas pubescent abaxially or marginally with hairs that are not interwoven, sparsely pubescent adaxially; n Arizona, c, s California, Utah.
   2. Corolla lobes 4–5 mm .... 1b. *Cycladenia humilis* var. *jonesii*
   2. Corolla lobes 8–12 mm .... 1c. *Cycladenia humilis* var. *venusta*

1a. Cycladenia humilis Bentham var. **humilis** • Sacramento waxydogbane [E] [F]

Cycladenia tomentosa A. Gray; *C. humilis* var. *tomentosa* (A. Gray) A. Gray

Stems glabrous or occasionally densely tomentose. **Leaves** glabrous or occasionally densely tomentose. **Corollas** glabrous or tomentose with interwoven hairs abaxially, glabrous or papillate adaxially, lobes (5–)7–9(–11) mm.

Flowering summer; fruiting summer. Talus slopes, sandy or gravelly sites, often associated with *Pinus ponderosa* or *Pseudotsuga menziesii*; 1000–3000 m; Calif., Oreg.

Variety *humilis* is known primarily from the Klamath, North Coast, and High Cascade ranges and the High Sierra Nevadas of northern California, with a single collection (*Callahan CH-MM-5 SP*, OSC-243427) from the Cascade Range of southern Oregon (Jackson County).

Plants that are densely tomentose throughout, the hairs interwoven, have been segregated as var. *tomentosa*. Such plants are found scattered within populations of the typical variety, and genetic data (H. K. Brabazon 2015) support the hypothesis that the difference in pubescence is likely to be a single-gene trait.

1b. Cycladenia humilis Bentham var. **jonesii** (Eastwood) S. L. Welsh & N. D. Atwood, Great Basin Naturalist 35: 333. 1976 • Jones's waxydogbane [C] [E]

Cycladenia jonesii Eastwood, Leafl. W. Bot. 3: 159. 1942

Stems glabrous. **Leaves** glabrous. **Corollas** pubescent abaxially or marginally with hairs that are not interwoven, sparsely pubescent adaxially, lobes 4–5 mm.

Flowering summer; fruiting summer. Steep slopes of barrens in desert scrub and pinyon-juniper woodlands; of conservation concern; 1300–2600 m; Ariz., Calif., Utah.

Variety *jonesii* is known from Utah (Emery, Garfield, Grand, and Kane counties) and Arizona (Mohave County), where it occurs on gypsiferous and nongypsiferous soils in barrens associated with desert scrub communities and pinyon-juniper woodlands, and eastern California (Inyo Mountains), where it occurs on steep shale slopes.

In Utah, plants of var. *jonesii* exhibit a floral dimorphism, with corollas 18–21 mm or 23–28 mm in length, within a single population (S. L. Welsh et al. 2003).

S. D. Sipes and V. J. Tepedino (1996) found that although var. *jonesii* is self-compatible, most reproduction is vegetative. Because a pollen vector is required, the limited fruit set observed may be due to low visitation rates by potential pollinators, or to the possibility that the original pollinator is no longer consistently found within the distribution range of the plant. These authors also observed limited fruit set in flowers that were cross-pollinated by hand, raising the possibility that fruit production could also be limited by soil nutrient availability.

Allozyme studies (S. D. Sipes and P. G. Wolf 1997) have shown var. *jonesii* to be well differentiated genetically from the typical variety despite the morphological similarity. However, the results of two unpublished studies (M. P. Last 2009; H. K. Brabazon 2015) suggest that plants from the Inyo Mountains, California, here referred to var. *jonesii*, are genetically similar to populations of var. *venusta* and may be better assigned to that variety or to an undescribed taxon.

Cycladenia humilis var. *jonesii* is listed as threatened by the United States Fish and Wildlife Service and is in the Center for Plant Conservation's National Collection of Endangered Plants.

SELECTED REFERENCES Sipes, S. D. and V. J. Tepedino. 1996. Pollinator lost? Reproduction by the enigmatic Jones cycladenia, *Cycladenia humilis* var. *jonesii*. In: J. Maschinski et al., eds. 1996. Southwestern Rare and Endangered Plants: Proceedings of the Second Conference: September 11–14, 1995, Flagstaff, Arizona. Fort Collins, Colo. Pp. 158–166. [U.S.D.A. Forest Serv., Gen. Techn. Rep. RM-283.] Sipes, S. D. and P. G. Wolf. 1997. Clonal structure and patterns of allozyme diversity in the rare endemic *Cycladenia humilis* var. *jonesii* (Apocynaceae). Amer. J. Bot. 84: 401–409.

1c. Cycladenia humilis Bentham var. **venusta** (Eastwood) Woodson ex Munz, Man. S. Calif. Bot., 379. 1935 [E]

Cycladenia venusta Eastwood, Bull. Torrey Bot. Club 29: 77. 1902

Stems glabrous. **Leaves** glabrous. **Corollas** pubescent abaxially or marginally with hairs that are not interwoven, sparsely pubescent adaxially, lobes 8–12 mm.

Flowering summer; fruiting summer. Talus and scree slopes and other gravelly sites; 1500–2500 m; Calif.

Populations of var. *venusta* are known from the San Gabriel Mountains (Los Angeles and San Bernardino counties) and the Santa Lucia Range (Monterey County), although these latter plants are genetically intermediate between varieties *humilis* and *venusta* and may be the result of hybridization between the varieties (H. K. Brabazon 2015). Collections from eastern California (Inyo County) that have been previously assigned to var. *venusta* are here treated as belonging to var. *jonesii*.

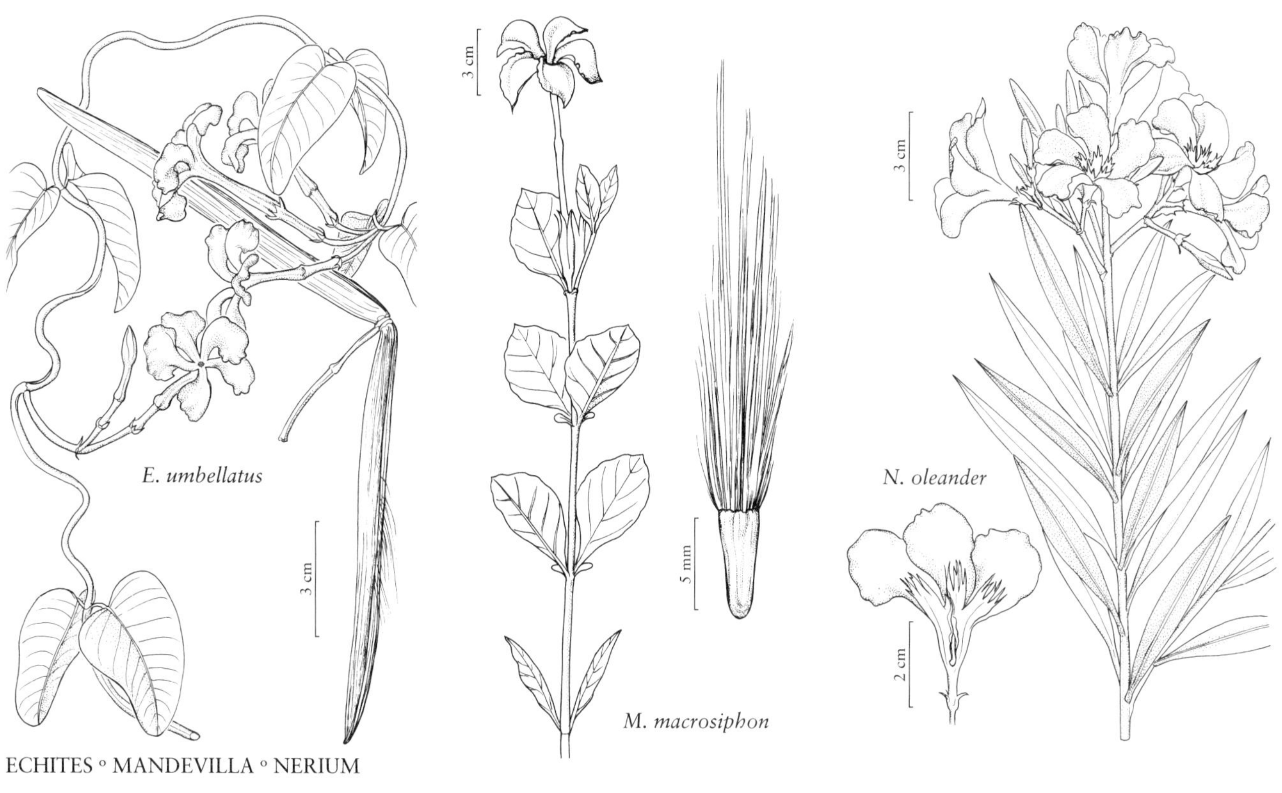

ECHITES ◦ MANDEVILLA ◦ NERIUM

17. ECHITES P. Browne, Civ. Nat. Hist. Jamaica, 182. 1756 • [Greek *echis*, viper, and *ites*, of the nature of, alluding to twining habit and deleterious quality]

David E. Lemke

Allotoonia J. F. Morales & J. K. Williams

Woody vines; latex milky. **Stems** trailing, unarmed, glabrous [eglandular-pubescent]. **Leaves** persistent, opposite, petiolate, stipular colleters interpetiolar [absent]; laminar colleters absent. **Inflorescences** axillary, terminal, or subterminal, cymose, pedunculate. **Flowers:** calycine colleters present; corolla white or pale yellow [pinkish white, yellow-orange, orange], salverform, aestivation dextrorse; corolline corona absent; androecium and gynoecium not united into a gynostegium; stamens inserted at or below middle of corolla tube; anthers connivent, adherent to stigma, connectives enlarged, 2-lobed, locules 4; pollen free, not massed into pollinia, translators absent; nectaries 5, distinct [connate and forming disc], alternating with stamens. **Fruits** follicles, usually paired, erect or deflexed, brown, slender, terete or moniliform, surface striate, glabrous. **Seeds** ovate, flattened, not winged, beaked, comose, not arillate. $x = 6$.

Species 9 or 10 (1 in the flora): Florida, Mexico, West Indies, Central America, South America (Colombia).

Echites was one of the first neotropical genera of Apocynaceae to be described. Because it was originally broadly defined, it was progressively expanded by later authors to include hundreds of species. Work by R. E. Woodson Jr. (1936) was instrumental in redefining *Echites* as a small group of species segregated into two subgenera, the remaining taxa being reduced to synonymy or transferred to other genera such as *Angadenia*, *Mandevilla*, *Mesechites* Müller Arg., *Odontadenia* Bentham, *Pentalinon*, *Prestonia* R. Brown, *Rhabdadenia*, and *Trachelospermum*, a treatment largely followed by J. F. Morales (1997). A cladistic analysis of *Echites* based on morphological characters (J. K. Williams 2004b) suggested that *Echites*, as circumscribed by both Woodson and Morales, was polyphyletic but that the species belonging to the two subgenera formed monophyletic clades. Consequently, Morales and Williams proposed the genus *Allotoonia* to include the five species of *Echites* subg. *Pseudechites* Woodson. However, a more recent phylogenetic analysis based on a combination of morphological and molecular characters (T. Livshultz et al. 2007) does not support the recognition of *Allotoonia*, and *Echites* is presently recognized as a genus of nine or ten species.

SELECTED REFERENCES Morales, J. F. 1997. A reevaluation of *Echites* and *Prestonia* sect. *Coalitae* (Apocynaceae). Brittonia 49: 328–336. Williams, J. K. 2004b. Polyphyly of the genus *Echites* (Apocynaceae: Apocynoideae: Echiteae): Evidence based on a morphological cladistic analysis. Sida 21: 117–131.

1. Echites umbellatus Jacquin, Enum. Syst. Pl., 13. 1760 (as umbellata) • Devil's potato, rubbervine [F]

Echites echites Britton

Vines twining. **Stems** glabrous. **Leaves:** petiole 5–8 mm, glabrous or sparsely [densely] pubescent; blade oblong-elliptic or ovate to suborbiculate, 3.5–7 × 1.5–4.5 cm, subcoriaceous, base obtuse to rounded or slightly cordate, margins slightly revolute, apex acuminate to apiculate or mucronate, surfaces glabrous. **Peduncles** 35–60 mm, glabrous. **Pedicels** 10–25 mm, glabrous. **Flowers:** calyx lobes ovate to narrowly triangular, 1.5–5 mm, glabrous; corolla glabrous abaxially, eglandular-pubescent adaxially, tube 20–55 × 1.5–4 mm, conspicuously widened at or below middle, lobes spreading, not overlapping or for at most 1/4 length from base, obovate, crisped distally, 11–18 × 12–16 mm. **Follicles** strongly divaricate, 60–130(–260) × 5–7 mm. **Seeds** 5–7 × 1.5–2 mm. $2n = 12$.

Flowering spring–fall; fruiting spring–early winter. Coastal pinelands, hammocks, beach dunes; 0–10 m; Fla.; Mexico; West Indies; Central America; South America (Colombia).

Echites umbellatus has been documented in Florida from Brevard County southward on the east coast and also from the keys. The species is thought to be the native host plant for the oleander caterpillar (*Syntomeida epilais*), which is common in Florida and Georgia and often causes severe defoliation of *Nerium oleander* in southern Florida (H. E. Bratley 1932).

18. MANDEVILLA Lindley, Edwards's Bot. Reg. 26: plate 7. 1840 • [For John Henry Mandeville, 1773–1861, British diplomat in Argentina and avid gardener)

David E. Lemke

Macrosiphonia Müller Arg.; *Telosiphonia* (Woodson) Henrickson

Shrubs, subshrubs, or suffrutescent perennials [woody vines]; latex milky. **Stems** erect [trailing], unarmed, eglandular-pubescent or glabrate [glabrous]. **Leaves** deciduous [persistent], opposite, subopposite, or occasionally subverticillate [whorled], petiolate; stipular colleters interpetiolar [absent]; laminar colleters present. **Inflorescences** axillary, terminal, or subterminal, cymose, pedunculate or not. **Flowers:** calycine colleters present; corolla yellow or white, often tinged with pink or red [pink, red, crimson, magenta, often with yellow eye], salverform, aestivation dextrorse; corolline corona absent; androecium and gynoecium not united into a gynostegium; stamens inserted at top of corolla tube; anthers connivent, adherent to stigma, connectives enlarged, truncate or 2-lobed, locules 4; pollen free, not massed into pollinia, translators absent; nectaries 2–5, distinct [connate and forming disc]. **Fruits** follicles, usually paired, erect or deflexed, brown to reddish brown, slender, terete or moniliform, surface smooth or striate, pubescent or glabrate [glabrous]. **Seeds** linear, flattened, not winged, not beaked, comose, not arillate. $x = 8, 10$.

Species ca. 170 (5 in the flora): sw United States, Mexico, West Indies, Central America, South America.

Mandevilla is one of the largest genera of Apocynaceae, although until recently its generic delimitation has been somewhat controversial. As noted by A. O. Simões et al. (2006), the combination of great morphological diversity and wide geographic distribution has presented challenges to workers that have been reflected in the taxonomic history of the group.

The most widely accepted circumscription of *Mandevilla* was established by R. E. Woodson Jr. (1933), who treated the genus as comprising approximately 110 neotropical species divided into two subgenera, *Exothostemon* (G. Don) Woodson and *Mandevilla*, the latter comprising five sections. M. Pichon (1948) provided a revised treatment in which he recognized the subgenera of Woodson but proposed a new classification within subg. *Mandevilla*, recognizing four sections. Pichon also included *Macrosiphonia* in *Mandevilla*, arguing that the characters used by Woodson to separate these genera were arbitrary.

Until recently, few studies have examined relationships between *Mandevilla* and putatively related genera. J. L. Zarucchi (1991) described the monotypic genus *Quiotania* from northern Colombia, indicating that he felt it to be close to *Mandevilla* but commenting that it was not clear whether it was most closely related to *Macrosiphonia*, *Mandevilla*, or other neotropical genera such as *Allomarkgrafia* Woodson, *Mesechites* Müller Arg., or *Tintinnabularia* Woodson. J. Henrickson (1996b) elevated *Macrosiphonia* subg. *Telosiphonia* to generic status, noting that all of these genera (*Allomarkgrafia*, *Macrosiphonia*, *Mandevilla*, *Mesechites*, *Quiotania*, and *Telosiphonia*) formed a distinctive and closely related group. However, in a subsequent revision of the Mexican and Central American species of *Mandevilla*, J. F. Morales (1998) chose to maintain a strict delimitation of the genus, following the circumscription by R. E. Woodson Jr. (1933).

Recent phylogenetic analyses based on a combination of molecular and morphological characters (A. O. Simões et al. 2004, 2006) have demonstrated that *Mandevilla* as circumscribed by M. Pichon (1948) is monophyletic while the genus as delimited by R. E. Woodson Jr. (1933) is paraphyletic. *Macrosiphonia* and *Telosiphonia* (as well as *Quiotania*) are nested within *Mandevilla* and are here included in the genus.

Flowering in the species of *Mandevilla* from the flora area is sporadic and typically follows rains from May through September.

Several non-native species of *Mandevilla*, especially the South American *M. sanderi* (Hemsley) Woodson, are often cultivated as ornamentals in the southern United States.

SELECTED REFERENCE Simões, A. O. et al. 2006. Is *Mandevilla* (Apocynaceae, Mesechiteae) monophyletic? Evidence from five plastid DNA loci and morphology. Ann. Missouri Bot. Gard. 93: 565–591.

1. Flowers 3+ per inflorescence; corolla yellow . . . . 5. *Mandevilla foliosa*
1. Flowers solitary or 2 or 3 per inflorescence; corolla white but often ferruginous upon drying.
    2. Corolla tube 1–3.5 cm, shorter than to 1.5 times as long as expanded corolla throat.
        3. Leaf blades ovate-lanceolate to oblong-ovate, green abaxially and adaxially; peduncles absent or to 1(–3) mm . . . . 1. *Mandevilla brachysiphon*
        3. Leaf blades linear to oblong-elliptic or oblong-ovate, white-pubescent abaxially, green adaxially; peduncles (5–)12–22(–36) mm . . . . 2. *Mandevilla hypoleuca*
    2. Corolla tube 4–10 cm, at least 2 times as long as expanded corolla throat.
        4. Shrubs; rhizomes absent; petioles mostly shorter than 3 mm . . . . 3. *Mandevilla lanuginosa*
        4. Suffrutescent perennials; rhizomes present; petioles mostly longer than 3 mm . . . . 4. *Mandevilla macrosiphon*

1. Mandevilla brachysiphon (Torrey) Pichon, Bull. Mus. Natl. Hist. Nat., sér. 2, 20: 106. 1948 • Western rocktrumpet

Echites brachysiphon Torrey in W. H. Emory, Rep. U.S. Mex. Bound. 2(1): 158. 1859; *Macrosiphonia brachysiphon* (Torrey) A. Gray; *Telosiphonia brachysiphon* (Torrey) Henrickson

Subshrubs, 2–4 dm; rhizomes present. **Stems** sparsely to densely eglandular-pubescent, especially on younger growth. **Leaves** opposite or subopposite, occasionally subverticillate; petiole 1–2 mm, pubescent; blade ovate-lanceolate to oblong-ovate, (8–)14–35 × (3.5–)6–15(–25) mm, subcoriaceous, base cuneate, rounded, or slightly cordate, margins not revolute, apex acute, acuminate, or rounded, apiculate, surfaces densely eglandular-pubescent abaxially, eglandular-pubescent adaxially. **Cymes** 1(or 2)-flowered. **Peduncles** 0–1(–3) mm, pubescent. **Pedicels** 7–15 mm, pubescent. **Flowers:** sepals reddish, oblong-ovate, 4–9 × 1–2.2 mm, pubescent; corolla white, often tinged with pink or red, often greenish below, eglandular-pubescent abaxially and adaxially, tube 13–20(–25) × 1.5 mm, throat (11–) 15–20(–25) × 4–5 mm, lobes spreading, obliquely ovate, (10–)15–25 × 10–20 mm. **Follicles** 55–120 × 4–5 mm, pubescent. **Seeds** 5–7 × 1–1.5 mm.

Flowering summer; fruiting summer–fall. Desert scrub, desert grasslands, pine-oak woodlands; 1000–1600 m; Ariz., N.Mex., Tex.; Mexico (Chihuahua, Sonora).

Mandevilla brachysiphon has the westernmost distribution of our species. Within the flora area, *M. brachysiphon* is known only from southeastern Arizona (Cochise, Graham, Pima, and Santa Cruz counties), southwestern New Mexico (Hidalgo and Luna counties), and the Franklin Mountains of El Paso County, Texas.

2. Mandevilla hypoleuca (Bentham) Pichon, Bull. Mus. Natl. Hist. Nat., sér. 2, 20: 106. 1948 • Bottomwhite rocktrumpet

Echites hypoleucus Bentham, Pl. Hartw., 23. 1839 (as hypoleuca); *Macrosiphonia hypoleuca* (Bentham) Müller Arg.; *Telosiphonia hypoleuca* (Bentham) Henrickson

Suffrutescent perennials, 1.7–7.5 dm; rhizomes present. **Stems** densely eglandular-pubescent, especially on younger growth. **Leaves** opposite, subopposite, or subverticillate; petiole 1–4 mm, pubescent; blade linear to oblong-elliptic or oblong-ovate, 25–50(–70) × (4–)8–17(–21) mm, subcoriaceous, base subcordate or rounded, margins revolute or not, apex acute or rounded, apiculate, surfaces densely white eglandular-pubescent abaxially, eglandular-pubescent adaxially. **Cymes** 1 or 2(or 3)-flowered. **Peduncles** (5–) 10–22(–36) mm, pubescent. **Pedicels** 7–15 mm, pubescent. **Flowers:** sepals reddish, oblong-linear, 7–10(–12) × 1.2–2 mm, pubescent; corolla white, often tinged with pink or red, eglandular-pubescent abaxially and adaxially, tube (8–)15–35 × 2.5–3 mm, throat (13–)17–25(–30) × 7–9 mm, lobes spreading, obliquely ovate, 15–30 × 14–28 mm. **Follicles** (80–)100–150 × 3–5 mm, pubescent. **Seeds** 6–8 × 1.5–2 mm.

Flowering summer–fall; fruiting summer–fall. Pine-oak woodlands, desert grassland ecotones, mostly on soils derived from igneous rock; 1000–1800 m; Tex.; Mexico.

Mandevilla hypoleuca has been reported from the Chinati, Chisos, and Davis mountains of western Texas but has not been well collected; most specimens date from the 1940s or earlier. The current status of the species in the flora area is uncertain, although it is widespread in northern and central Mexico.

3. Mandevilla lanuginosa (M. Martens & Galeotti) Pichon, Bull. Mus. Natl. Hist. Nat., sér. 2, 20: 106. 1948 • Woolly rocktrumpet

Echites lanuginosus M. Martens & Galeotti, Bull. Acad. Roy. Sci. Bruxelles 11(1): 357. 1844 (as lanuginosa); *Macrosiphonia lanuginosa* (M. Martens & Galeotti) Hemsley; *Telosiphonia lanuginosa* (M. Martens & Galeotti) Henrickson

Shrubs, 2–5(–10) dm; rhizomes absent. **Stems** densely eglandular-pubescent, especially on younger growth. **Leaves** opposite, rarely subopposite; petiole 1–3 mm, pubescent; blade oblong to ovate or elliptic, 15–30(–40) × 7–16(–25) mm, subcoriaceous, base rounded or subcordate, margins revolute, apex acute, rounded, or truncate, apiculate, surfaces densely grayish white eglandular-pubescent abaxially, eglandular-pubescent adaxially. **Cymes** 1(or 2)-flowered. **Peduncles** 0.5–1.5 mm, pubescent. **Pedicels** 3.5–9 mm, pubescent. **Flowers:** sepals reddish, oblong-lanceolate, 8–12.5 × 1.5–3 mm, pubescent; corolla white, often tinged with pink or red, often becoming ferruginous upon drying, eglandular-pubescent abaxially and adaxially, tube 40–80(–100) × 2.5–3 mm, throat 14–25 × 5–7 mm, lobes spreading, obliquely ovate, 15–35 × 13–23 mm. **Follicles** 50–100 × 3–4 mm, pubescent. **Seeds** 7–8 × 1.4–1.6 mm.

Flowering spring–fall; fruiting summer–fall. Grasslands and openings in thorn scrub vegetation; 0–300 m;

Tex.; Mexico (Durango, Hidalgo, Nuevo León, San Luis Potosí, Tamaulipas).

Mandevilla lanuginosa occurs at lower elevations than other *Mandevilla* species in the flora area and is characteristically found in openings in Tamaulipan thorn scrub and mesquite woodlands in southern Texas. *Mandevilla lanuginosa* is sometimes confused with *M. macrosiphon*; characters separating the two species are discussed under the latter taxon.

4. **Mandevilla macrosiphon** (Torrey) Pichon, Bull. Mus. Natl. Hist. Nat., sér. 2, 20: 106. 1948 • Plateau rocktrumpet, flor de San Juan [F]

Echites macrosiphon Torrey in W. H. Emory, Rep. U.S. Mex. Bound. 2(1): 158, plate 43. 1859; *Macrosiphonia macrosiphon* (Torrey) A. Heller; *Telosiphonia macrosiphon* (Torrey) Henrickson

Suffrutescent perennials, 1–4(–5) dm; rhizomes present. **Stems** densely eglandular-pubescent, especially on younger growth. **Leaves** opposite or rarely subopposite; petiole 3–6(–9) mm, densely pubescent; blade ovate, oblong-elliptic, or rarely orbiculate-reniform, 15–40(–60) × 10–35(–45) mm, subcoriaceous, base cuneate, rounded, or cordate, margins revolute or not, apex acute to rounded or retuse, apiculate, surfaces densely white eglandular-pubescent abaxially, eglandular-pubescent adaxially. **Cymes** 1(very rarely 2)-flowered. **Peduncles** 0.5–1 mm, pubescent. **Pedicels** 4–9(–13) mm, pubescent. **Flowers:** sepals green, often streaked with red, oblong-ovate to oblong-lanceolate, 13–18(–25) × 2.5–5 mm, pubescent; corolla white, often tinged with pink or red, often becoming ferruginous upon drying, eglandular-pubescent abaxially and adaxially, tube 50–90 × 2–3 mm, throat 15–30 × 4–6 mm, lobes spreading, obliquely ovate, 25–40 × 18–35 mm. **Follicles** 100–150 × 3.5–4 mm, pubescent. **Seeds** 5.5–7 × 1.5–2 mm.

Flowering spring–fall; fruiting summer–fall. Desert grasslands on limestone or igneous soils; 300–1700 m; Tex.; Mexico (Chihuahua, Coahuila, Durango, Zacatecas).

J. Henrickson (1996b) noted that along the eastern margin of its range, *Mandevilla macrosiphon* appears to grade into *M. lanuginosa*, both species forming small plants with densely pubescent foliage. The two species can usually be easily distinguished on the basis of growth habit, *M. macrosiphon* exhibiting a suffrutescent habit with new shoots of the season arising from a root crown or from lower branches of the previous season, while *M. lanuginosa* is a shrub with new shoots developing from its upper stems.

5. **Mandevilla foliosa** (Müller Arg.) Hemsley, Biol. Cent.-Amer., Bot. 2: 316. 1881

Amblyanthera foliosa Müller Arg., Linnaea 30: 427. 1860; *Mandevilla stans* (A. Gray) J. K. Williams; *Trachelospermum stans* A. Gray

Suffrutescent perennials [shrubs], 5–20 dm; rhizomes absent. **Stems** eglandular-pubescent to glabrate [glabrous]. **Leaves** opposite [subopposite, whorled]; petiole 3–13(–18) mm, pubescent [glabrous]; blade ovate-lanceolate to elliptic or obovate, 50–130(–150) × 15–50(–70) mm, membranous, base cuneate, obtuse, or subcordate, margins not revolute, apex acute or acuminate, surfaces eglandular-pubescent abaxially and at margins [glabrous], eglandular-pubescent to glabrate adaxially. **Cymes** 3–9(–14)-flowered. **Peduncles** 2–15 mm, pubescent [glabrous]. **Pedicels** 5–20 mm, pubescent [glabrous]. **Flowers:** sepals green, lanceolate to linear-lanceolate, (3–)5–8 × 1–1.3 mm, pubescent [glabrous]; corolla yellow, glabrous abaxially, eglandular-pubescent adaxially, tube 8–12 × 2–3 mm, throat 3–5 × 3–4 mm, lobes spreading, obliquely obovate to oblanceolate to dolabriform, often falcate, 6–10 × 3–4 mm. **Follicles** (55–)80–120 × 2–3 mm, pubescent or glabrate. **Seeds** 7–10 × 1.5–2 mm.

Flowering summer–fall; fruiting fall. Pine-juniper woodlands; 1700 m; Ariz.; Mexico.

Mandevilla foliosa is known in the United States from a single collection (*Milson 1*, ARIZ) in the Santa Rita Mountains in Santa Cruz County of southern Arizona but is widespread in Mexico.

J. K. Williams (2004c) referred populations of *Mandevilla foliosa* from northern Mexico to *M. stans*, arguing that the pubescence of stems, petioles, inflorescences, and sepals (versus glabrous in *M. foliosa*) and the disjunct geographic distribution (northern versus southern Mexico) were sufficient characters to justify recognition at the species level. This disposition was not followed in the most recent treatment of Mexican *Mandevilla* species (L. O. Alvarado-Cárdenas and J. F. Morales 2014).

19. NERIUM Linnaeus, Sp. Pl. 1: 209. 1753; Gen. Pl. ed. 5, 99. 1754 • [Ancient Greek name for oleander, perhaps from *neros*, moist or fresh, alluding to habitat and/or evergreen habit] [I]

David E. Lemke

Shrubs or small trees; latex clear. **Stems** erect, unarmed, glabrous or eglandular-pubescent especially on younger growth. **Leaves** persistent, whorled or occasionally opposite, petiolate; stipular colleters intrapetiolar; laminar colleters absent. **Inflorescences** terminal, thyrsiform, pedunculate. **Flowers:** calycine colleters present; corolla white, pink, red, purple, or rarely orange-pink, funnelform, aestivation dextrorse; corolline corona lacerate; androecium and gynoecium not united into a gynostegium; stamens inserted at top of corolla tube; anthers connivent, adherent to stigma, connectives appendiculate, elongate pubescent appendages intertwined, locules 4; pollen free, not massed into pollinia, translators absent; nectaries absent. **Fruits** follicles, solitary or paired, erect, reddish brown, slender, terete or slightly compressed, truncate, surface striate, glabrous. **Seeds** oblong, slightly flattened, not winged, not beaked, comose, not arillate. $x = 11$.

Species 1: introduced; Eurasia, Africa; introduced also nearly worldwide.

1. Nerium oleander Linnaeus, Sp. Pl. 1: 209. 1753 • Oleander, rose bay [F] [I] [W]

Leaves: petiole 2–7 mm, sparsely pubescent or glabrous; blade oblong-lanceolate, 2–15(–30) × 0.5–2.5(–3.5) cm, coriaceous, base cuneate, margins revolute, apex acuminate, surfaces pubescent abaxially, very sparsely pubescent or glabrous adaxially. **Peduncles** 3–6 cm, sparsely pubescent. **Pedicels** 5–8 mm, pubescent. **Flowers:** calyx lobes lanceolate to ovate-lanceolate, 5–7 mm, pubescent; corolla glabrous abaxially, eglandular-pubescent adaxially, tube 8–12 × 2–3 mm, throat 5–10 × 4–7 mm, lobes spreading, obliquely obovate, 15–25 × 10–20 mm. **Follicles** 8–15 × 1–1.5 cm. **Seeds** 7–10 × 1.5–2 mm, densely pubescent. $2n = 22$.

Flowering spring–summer; fruiting summer–fall. Disturbed areas, roadsides, old homesites; 0–600 m; introduced; Ala., Ariz., Calif., Ga., La., Miss., N.C., S.C., Tex.; Eurasia; Africa; widely cultivated in tropical and subtropical regions.

Nerium oleander is cultivated in warmer parts of the United States and has become sporadically naturalized from North Carolina to Florida and Texas, and in Arizona and California. The plant is widely recognized as one of the most poisonous cultivated species due to the presence of cardiac glycosides that can cause nausea, vomiting, abdominal pain, diarrhea, cardiac arrhythmia, and potassium imbalance (V. Bandara et al. 2010), although the number of human fatalities resulting from accidental ingestion of leaves and/or flowers is surprisingly small (S. D. Langford and P. J. Boor 1996).

20. PENTALINON Voigt, Hort. Suburb. Calcutt., 523. 1845 • [Greek *pente*, five, and *linon*, net, alluding to connective appendages intertwined around stigma]

David E. Lemke

Woody vines; latex milky. **Stems** twining, decumbent, or trailing, occasionally suberect, unarmed, eglandular-pubescent. **Leaves** persistent, opposite [subopposite], petiolate; stipular colleters intrapetiolar and interpetiolar, or absent; laminar colleters absent. **Inflorescences** axillary, cymose, pedunculate. **Flowers:** calycine colleters present; corolla yellow, pale yellow, or cream, funnelform, aestivation dextrorse; corolline corona absent; androecium and gynoecium not united into a gynostegium; stamens inserted near top of corolla tube; anthers connivent, adherent to stigma, connectives appendiculate, elongate appendages intertwined, locules 4; pollen free, not massed into pollinia, translators absent; nectaries 5, alternating with stamens. **Fruits** follicles, usually paired, erect to deflexed or pendulous, brown, slender, terete, surface striate, glabrous or pubescent. **Seeds** very narrowly oblong, flattened, not winged, beaked, comose, not arillate. $x = 6$.

Species 2 (1 in the flora): Florida, West Indies, Central America, n South America.

The species of *Pentalinon* have previously been treated as members of *Urechites* Müll. Arg., but B. F. Hansen and R. P. Wunderlin (1986) have convincingly argued that *Pentalinon* is legitimate and should be used instead of the more familiar, but later, *Urechites*.

1. **Pentalinon luteum** (Linnaeus) B. F. Hansen & Wunderlin, Taxon 35: 167. 1986 • Hammock viper's-tail, wild allamanda [F]

Vinca lutea Linnaeus, Cent. Pl. II, 12. 1756; *Urechites andrewsii* Small; *U. luteus* (Linnaeus) Britton; *U. luteus* var. *sericeus* R. W. Long; *U. neriandrus* Grisebach ex Miers; *U. pinetorum* Small

Woody vines, occasionally suffrutescent at base. **Stems** glabrous or densely pubescent. **Leaves** opposite; petiole 5–12 mm, pubescent; blade broadly obovate to oblong, 30–60 × 15–40 mm, membranous to subcoriaceous, base cuneate to acute or slightly cordate, margins often revolute, apex obtuse to acute or apiculate, surfaces pubescent abaxially. **Peduncles** 20–40 mm, sparsely pubescent. **Pedicels** 5–20 mm, sparsely pubescent. **Flowers:** calyx lobes linear-lanceolate, 7–12 mm, pubescent or glabrate; corolla glabrous abaxially, usually eglandular-pubescent adaxially, tube 6–15 × 2–3 mm, throat 15–35 × 8–12 mm, lobes spreading, obliquely rounded-obovate, 20–25 × 18–22 mm, adjacent lobes overlapping for 1/2+ of their length from base. **Follicles** 120–180 × 4–5 mm. **Seeds** 10–12 × 1 mm not including beak, beak ± as long as body of seed. $2n = 12$.

Flowering and fruiting year-round. Roadside thickets, pine flatwoods, coastal hammocks, mangrove swamps, beach dunes; 0–10 m; Fla.; West Indies; South America (Colombia).

In Florida, *Pentalinon luteum* occurs sporadically from St. Lucie and Lee counties southward.

G. E. Burrows and R. J. Tyrl (2013) reported that in Cuba, where *Pentalinon luteum* is locally abundant, cattle may suffer hemorrhagic diarrhea and sudden death from ventricular fibrillation upon ingestion of the plants.

21. RHABDADENIA Müller Arg. in C. F. P. von Martius et al., Fl. Bras. 6(1): 173, plate 52. 1860 • [Derivation uncertain; Greek *rhabdos*, rod, and *aden*, gland]

David E. Lemke

Woody vines [subshrubs]; latex milky. **Stems** twining to suberect, unarmed, glabrous. **Leaves** persistent, opposite, petiolate; stipular colleters present but early-deciduous, interpetiolar; laminar colleters absent. **Inflorescences** axillary or subterminal, cymose, pedunculate. **Flowers:** calycine colleters absent; corolla white to pinkish white with yellow throat, occasionally dark pink toward base [pink, magenta], funnelform, aestivation dextrorse; corolline corona absent; androecium and gynoecium not united into a gynostegium; stamens inserted near top of corolla tube; anthers connivent, adherent to stigma; connectives enlarged, 2-lobed, locules 4; pollen free, not massed into pollinia, translators absent; nectaries 5, distinct or basally connate, alternating with stamens. **Fruits** follicles, usually paired, erect, brown, slender, terete or compressed, surface striate or smooth, glabrous. **Seeds** linear, flattened, not winged, beaked, comose, not arillate.

Species 3 (1 in the flora): Florida, Mexico, West Indies, Central America, South America.

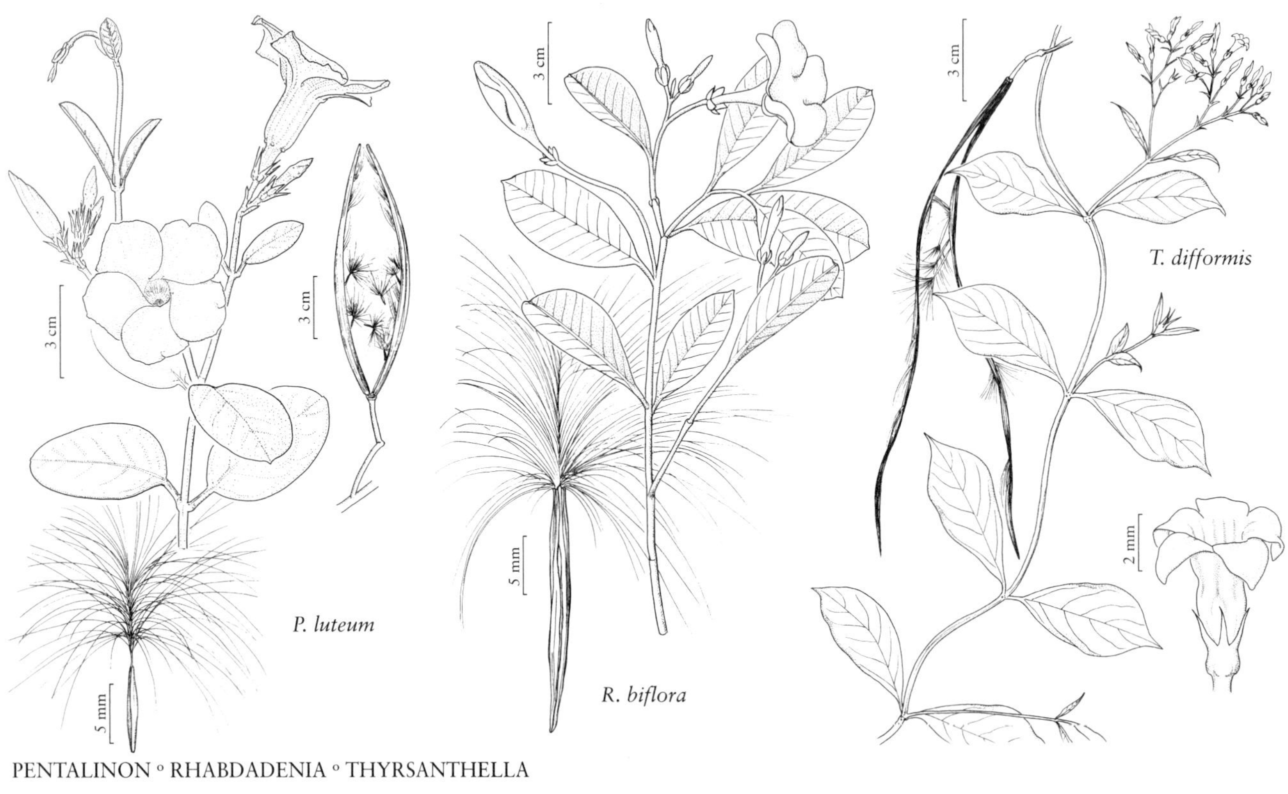

PENTALINON ◦ RHABDADENIA ◦ THYRSANTHELLA

Rhabdadenia has one of the broadest ranges of neotropical Apocynaceae, extending from the southeastern United States (Florida), West Indies, and Mexico to northern Argentina, the species typically inhabiting mangrove swamps or other flooded habitats (J. F. Morales 2009).

SELECTED REFERENCE Morales, J. F. 2009. Estudios en las Apocynaceae neotropicales XXXVII: Monografía del género *Rhabdadenia* (Apocynoideae: Echiteae). J. Bot. Res. Inst. Texas 3: 541–564.

1. **Rhabdadenia biflora** (Jacquin) Müller Arg. in C. F. P. von Martius et al., Fl. Bras. 6: 175. 1860
   • Rubbervine, mangrovevine [F]

Echites biflorus Jacquin, Enum. Syst. Pl., 13. 1760 (as biflora)

Leaves: petiole 7–15 mm, glabrous; blade obovate-oblong to elliptic or lanceolate, 30–80 × 15–30 mm, coriaceous, base acute to rounded, margins usually revolute, apex rounded to acute or apiculate, surfaces glabrous. **Peduncles** 75–100 mm, glabrous. **Pedicels** 10–15 mm, glabrous. **Flowers:** calyx lobes ovate, 5–6 mm, glabrous; corolla glabrous abaxially, eglandular-pubescent adaxially, tube 15–20 × 2–3 mm, throat 20–30 × 5–15 mm, lobes spreading, obliquely rounded-obovate, 15–20 × 15–20 mm, adjacent lobes overlapping for 1/2+ of their length from base. **Follicles** 80–100 × 3–4 mm. **Seeds** 25–30 × 1–2 mm.

Flowering spring–fall; fruiting summer–fall. Coastal hammocks, mangrove swamps; 0–10 m; Fla.; Mexico; West Indies; Central America; South America.

In Florida, *Rhabdadenia biflora* occurs from Brevard and Charlotte counties southward.

F. Lens et al. (2009) have shown that *Rhabdadenia biflora* exhibits an interesting developmental plasticity in its ability to switch abruptly between wood anatomical features characteristic of erect or suberect species (vessels in radial multiples) and those characteristic of lianas (vessels in clusters).

22. THYRSANTHELLA Pichon, Bull. Mus. Natl. Hist. Nat., sér. 2, 20: 192. 1948 • [Greek *thyrsos*, thyrse or panicle, and *anthos*, flower, and Latin *-ella*, diminutive, alluding to inflorescence] E

David E. Lemke

Vines, suffruticose; latex milky. **Stems** twining, unarmed, glabrous or sparsely eglandular-pubescent. **Leaves** deciduous, opposite, petiolate; stipular colleters interpetiolar and intrapetiolar; laminar colleters absent. **Inflorescences** axillary, thyrsiform, pedunculate. **Flowers:** calycine colleters present; corolla pale yellow, salverform, aestivation dextrorse; coralline corona absent; androecium and gynoecium not united into a gynostegium; stamens inserted near top of corolla tube; anthers connivent, adherent to stigma, connectives enlarged, 2-lobed, locules 4; pollen free, not massed into pollinia, translators absent; nectaries 5, distinct or connate. **Fruits** follicles, usually paired, erect or pendulous, brown, slender, terete, surface smooth, glabrous. **Seeds** linear, flattened, not winged, not beaked, comose, not arillate.

Species 1: c, e United States.

Pichon erected *Thyrsanthella* to accommodate a single species that has been treated traditionally as the sole native North American representative of *Trachelospermum*, a genus otherwise restricted to southeastern Asia. Recent molecular work (T. Livshultz et al. 2007) has supported the recognition of *Thyrsanthella* as distinct from *Trachelospermum*.

1. Thyrsanthella difformis (Walter) Pichon, Bull. Mus. Natl. Hist. Nat., sér. 2, 20: 192. 1948 • Climbing dogbane E F

Echites difformis Walter, Fl. Carol., 98. 1788; *Forsteronia difformis* (Walter) A. de Candolle; *Trachelospermum difforme* (Walter) A. Gray

Stems glabrous or very sparsely pubescent when young. **Leaves:** petiole 1–10(–15) mm, glabrous; blade elliptic to obovate-elliptic or occasionally linear-lanceolate or suborbiculate, 2.7–6(–12) × 0.4–3.5(–7) cm, membranous, base acute to rounded or obtuse, apex acuminate or apiculate, surfaces sparsely pubescent or glabrous. **Peduncles** 2.5–5 cm, sparsely pubescent or glabrous. **Pedicels** 4–8 mm, minutely pubescent or glabrous. **Flowers:** calyx lobes ovate-lanceolate, 3–4 mm, pubescent or glabrous; corolla glabrous or very sparsely eglandular-pubescent abaxially, eglandular-pubescent adaxially, tube 2.5–3.5 × 1–1.5 mm, throat 2.5–3.5 × 2–3 mm, lobes spreading or reflexed, obliquely obovate, 3–4 × 2–3 mm. **Follicles** 10–23 × 0.1–0.3 cm. **Seeds** 8–11 × 1.5–2 mm.

Flowering spring; fruiting spring–fall. Bottomlands, swamps, moist upland forests and woodlands; 0–400 m; Ala., Ark., Del., Fla., Ga., Ill., Ind., Ky., La., Md., Miss., Mo., N.C., Okla., S.C., Tenn., Tex., Va.

The leaves of *Thyrsanthella difformis* can exhibit a remarkable variability in width, even on branches of the same plant, most notably in young sprouts of the forest floor.

23. TRACHELOSPERMUM Lemaire, Jard. Fleur. 1: sub plate 61. 1851, name conserved • [Greek *trachelos*, collar or neck, and *sperma*, seed, alluding to elongated narrow-tipped seeds] I

David E. Lemke

Vines, suffruticose [woody]; latex milky. **Stems** twining or trailing, unarmed, glabrous or densely eglandular-pubescent. **Leaves** persistent, opposite, petiolate; stipular colleters interpetiolar and intrapetiolar; laminar colleters absent. **Inflorescences** axillary or terminal, cymose, often thyrsiform or subumbellate, pedunculate [sessile]. **Flowers:** calycine colleters present; corolla white or pale yellow, salverform, aestivation dextrorse; corolline corona absent; androecium and gynoecium not united into a gynostegium; stamens inserted near top of corolla tube; anthers connivent, adherent to stigma, connectives enlarged, 2-lobed, locules 4; pollen free, not massed into pollinia, translators absent; nectaries 5, distinct or basally connate, alternating with stamens. **Fruits** follicles, usually paired, erect or deflexed, reddish brown [brown], slender-cylindric [moniliform], surface striate, glabrous [pubescent]. **Seeds** narrowly oblong, flattened [terete], not winged, not beaked [short-beaked], comose, not arillate. x = 10.

Species ca. 15 (1 in the flora): introduced; Asia; introduced also in Mexico, West Indies, Central America, Europe, Pacific Islands, Australia.

Trachelospermum asiaticum (Siebold & Zuccarini) Nakai (Asiatic jasmine) is widely planted as a ground cover in the southern United States and has been collected from Alabama and Texas but does not appear to be naturalized. It can be distinguished from *T. jasminoides* by the growth form and by stamens that are well exserted from the corolla, in contrast to the included stamens of *T. jasminoides*.

1. **Trachelospermum jasminoides** (Lindley) Lemaire, Jard. Fleur. 1: sub plate 61. 1851 • Confederate or star jasmine [F] [I]

Rhyncospermum jasminoides Lindley, J. Hort. Soc. London 1: 74, figs. [p. 74], A. 1846

Stems glabrous or rarely ferruginous-pubescent. **Leaves:** petiole 2.5–5 mm, glabrous or sparsely pubescent; blade narrowly elliptic to broadly obovate, 20–60(–120) × 15–30(–40) mm, coriaceous, base rounded to acute, apex obtuse to broadly or abruptly acuminate, surfaces glabrous throughout or pubescent abaxially. **Peduncles** 20–50 mm, glabrous or sparsely pubescent. **Pedicels** 5–7 mm, glabrous or sparsely pubescent. **Flowers:** calyx lobes ovate to narrowly lanceolate, 2–5 mm, pubescent; corolla glabrous abaxially, eglandular-pubescent adaxially, tube 2.5–3.5 × 1–3 mm, throat 3–4 × 2–2.5 mm, lobes spreading, obliquely oblong-obovate, 7–10 × 4–5 mm; stamens included. **Follicles** 100–150 × 2.5–5 mm. **Seeds** 6(–15) × 1–1.5 mm. **2*n*** = 20.

Flowering spring; fruiting spring–summer. Disturbed habitats; 0–200 m; introduced; Ala., Fla.; Asia (China, Japan, Korea); cultivated widely and introduced also in Mexico, West Indies, Central America, Europe, Pacific Islands, Australia.

R. E. Woodson Jr. (1936b) surmised *Trachelospermum jasminoides* to be native to southeastern China and to have spread elsewhere in Asia through cultivation and escape. Although *T. jasminoides* is widely cultivated in warmer parts of the United States, only two collections have been seen that appear to represent naturalized specimens, both from disturbed woodland habitats.

24. CRYPTOSTEGIA R. Brown, Bot. Reg. 5: plate 435. 1820 • Rubbervine [Greek *kryptos*, hidden, and *stegein*, to cover, alluding to enclosure of five-scaled crown within corolla tube] [I]

Casie L. Reed

Alexander Krings

Lianas or subshrubs; latex white. **Stems** climbing or with self-supporting branches, unarmed, glabrous or eglandular-pubescent. **Leaves** persistent, opposite, petiolate; stipular colleters interpetiolar and intrapetiolar; laminar colleters absent. **Inflorescences** terminal cymes, pedunculate. **Flowers:** calycine colleters present; corolla white, pale pink, or purple-pink, infundibuliform, aestivation dextrorse; corolline corona 2-fid or entire; androecium and gynoecium not united into a gynostegium; stamens inserted at base of corolla tube; anthers connivent, adherent to stigma, connectives apiculate, locules 4; pollen in tetrads, not massed into pollinia, but shed onto translators; nectary absent. **Fruits** follicles, paired, deflexed, green to brown, fusiform, strongly 3-angled, striate or smooth, glabrous or minutely pubescent. **Seeds** oblong, flattened, not winged, not beaked, comose, not arillate. $x = 11$.

Species 2 (2 in the flora): introduced; Indian Ocean Islands (Madagascar); introduced also in Mexico, West Indies, Bermuda, Central America, South America, Asia, Africa, elsewhere in Indian Ocean Islands, Pacific Islands, Australia.

Cryptostegia is endemic to Madagascar, but its two species have been introduced pantropically. Hybrids between *C. grandiflora* and *C. madagascariensis* have been reported in cultivated landscapes in Florida. There is no evidence that these hybrids have become naturalized.

SELECTED REFERENCES Klackenberg, J. 2001. Revision of the genus *Cryptostegia* R. Br. (Apocynaceae, Periplocoideae). Adansonia 23: 205–218. Marohasy, J. and P. I. Forster. 1991. A taxonomic revision of *Cryptostegia* R. Br. (Asclepiadaceae: Periplocoideae). Austral. Syst. Bot. 4: 571–577.

1. Calyx lobes 13–20 mm, margins reflexed; corolline corona 2-fid; translator spathes orbiculate; follicles (8–)10–15.5 cm . . . . 1. *Cryptostegia grandiflora*
1. Calyx lobes 7–13(–14) mm, margins ± flat; corolline corona entire; translator spathes lanceolate to ovate; follicles 5–10 cm . . . . 2. *Cryptostegia madagascariensis*

1. Cryptostegia grandiflora R. Brown, Bot. Reg. 5: plate 435. 1820 • Palay rubbervine [F] [I]

Leaf blades 2.5–10 × 2.5–6.3 cm, glabrous. **Petioles** 3.9–15 mm, glabrous or slightly pubescent. **Flowers:** calyx lobes ovate, 13–20 mm, margins reflexed; corolla tube 18–30 mm, lobes 31.5–56 × 15–30 mm, glabrous or pubescent; corolline corona 8–11 mm, 2-fid; translator spathes orbiculate. **Follicles** (8–)10–15.5 × 2–4 cm. **Seeds** 5.2–9.7 mm; coma 18.9–38 mm. $2n = 22$.

Flowering and fruiting year-round, mostly in summer. Disturbed areas; 0–200 m; introduced; Fla., Tex.; Indian Ocean Islands (Madagascar); introduced also in Mexico, West Indies, Bermuda, Central America, South America, Asia, Africa, elsewhere in Indian Ocean Islands, Pacific Islands, Australia.

In the United States, *Cryptostegia grandiflora* is currently known outside of cultivation only in the keys of Monroe County, Florida, and along the Rio Grande (Cameron and Starr counties, Texas; T. F. Patterson and G. L. Nesom 2009).

2. Cryptostegia madagascariensis Bojer ex Decaisne in A. P. de Candolle and A. L. P. P. de Candolle, Prodr. 8: 492. 1844 • Purple or Madagascar rubbervine [I]

Leaf blades 1.5–11.2 × 1–6.5 cm, glabrous or pubescent. **Petioles** 3–10 mm, glabrous or pubescent. **Flowers:** calyx lobes ovate to elliptic, 7–13(–14) mm, margins ± flat; corolla tube (9–)15–25 mm, lobes 20–44 × 12–26 mm, glabrous or pubescent; corolline corona 6–9 mm, entire; translator spathes lanceolate to ovate. **Follicles** 5–10 × 1–4 cm. **Seeds** 5–8.9 mm; coma 18.9–38 mm.

Flowering and fruiting year-round, mostly in summer. Disturbed areas; 0–40 m; introduced; Fla.; Indian Ocean Islands (Madagascar); introduced also in Mexico, West Indies, Central America, South America, Asia, Africa, Pacific Islands, Australia.

In Florida, *Cryptostegia madagascariensis* has been found sporadically outside of cultivation in the southern third of the peninsula.

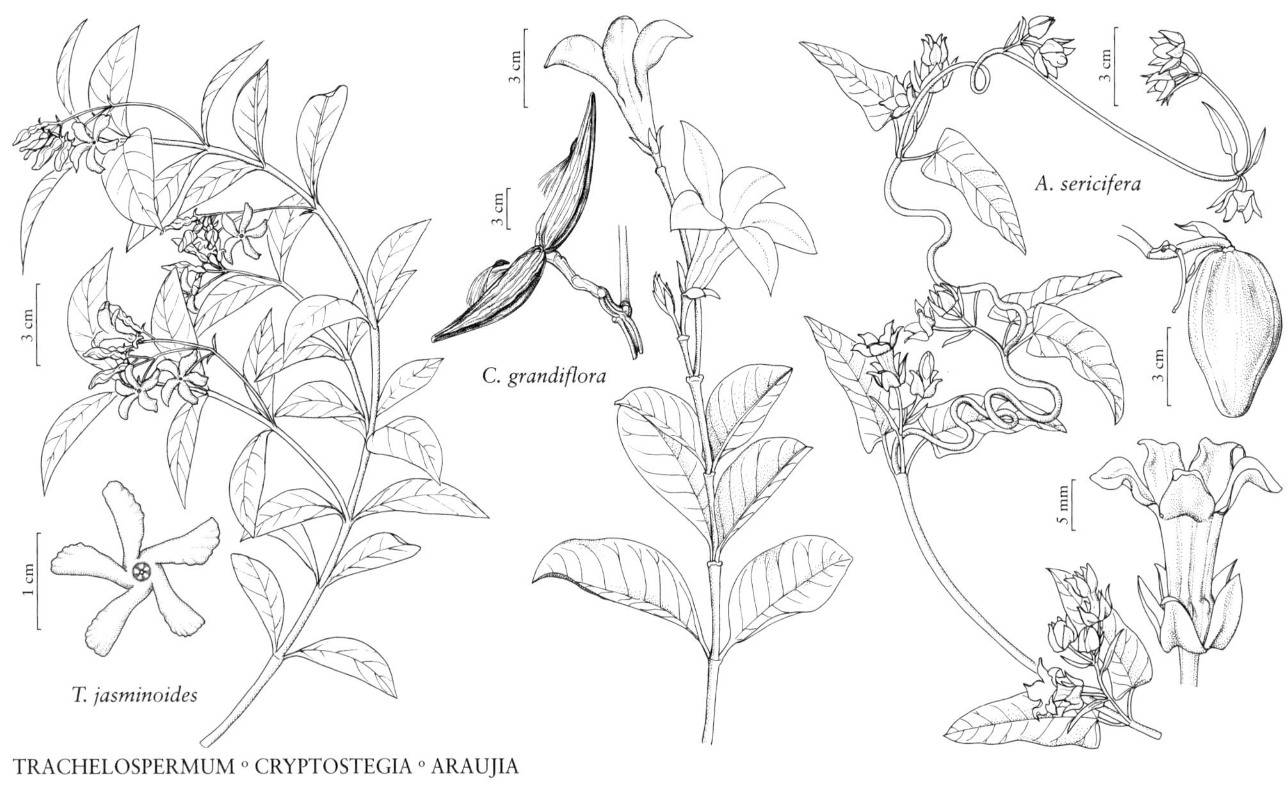

TRACHELOSPERMUM ◦ CRYPTOSTEGIA ◦ ARAUJIA

25. ARAUJIA Brotero, Trans. Linn. Soc. London 12: 62, plates 4, 5. 1818 • Bladderflower [For Antonio de Matos Araujo, nineteenth-century Portuguese plant collector] [I]

C. Lee Kimmel

Alexander Krings

Lianas; latex white. **Stems** prostrate or twining, unarmed, glabrous or eglandular-pubescent. **Leaves** persistent to semipersistent, opposite, petiolate; stipular colleters absent or 2, 1 borne at base of petiole on each side, interpetiolar and infrapetiolar colleters absent; laminar colleters present. **Inflorescences** extra-axillary, solitary, racemose or cymose, pedunculate. **Flowers:** calycine colleters absent or present; corolla white, light pink, or pale to waxy green, rotate or funnelform, aestivation valvate; corolline corona absent; androecium and gynoecium united into a gynostegium adnate to corolla tube; gynostegial corona a tube or interrupted and of cartilaginous to fleshy, irregularly toothed to entire segments; anthers adnate to style, locules 2; pollen in each theca massed into rigid, vertically oriented pollinium, pollinia lacrimiform, joined from adjacent anthers by translators to common corpusculum and together forming pollinarium. **Fruits** follicles, solitary, pendulous or somewhat deflexed, green, ovoid, terete or somewhat compressed, smooth or longitudinally furrowed, glabrous or pubescent. **Seeds** obdeltate, obpyriform, or oblong, flattened to navicular, winged or not, not beaked, comose, not arillate. x = 10, 11.

Species ca. 12 (2 in the flora): introduced; South America; introduced also in Central America, Europe, Africa, Pacific Islands, Australia.

1. Sepals lanceolate to oblong, not leaflike, 1–3 mm wide; corollas rotate, pale to waxy green, gynostegial corona of staminal and interstaminal elements fused into tube, tube 5–8 mm, obscuring gynostegium; style-head extension absent . . . . 1. *Araujia odorata*
1. Sepals ovate, leaflike, 3–7 mm wide; corollas funnelform, white or light pink, gynostegial corona of 5 distinct staminal elements, each to 3.9 mm, not fused into tube, tube 12.1–15.9 mm, not obscuring gynostegium; style-head extension 2-fid, 3.5 mm. . . . . 2. *Araujia sericifera*

1. Araujia odorata (Hooker & Arnott) Fontella & Goyder, Phytotaxa 26: 11. 2011 • Strangler or milkweed vine, latexplant [I]

Cynanchum odoratum Hooker & Arnott, J. Bot. (Hooker) 1: 294. 1835; *Morrenia odorata* (Hooker & Arnott) Lindley

Stems to 15 m. **Leaves:** petiole 0.5–4 cm, eglandular-pubescent; blade ovate, deltate, or hastate, 1.8–8 × 0.8–6 cm, base truncate to cordate, surfaces glabrous to eglandular-pubescent. **Inflorescences:** peduncle 0.1–0.7 cm, eglandular-pubescent. **Pedicels** 0.6–1.5 cm, eglandular-pubescent. **Flowers:** sepals green, lanceolate to oblong, not leaflike, 5–13 × 1–3 mm, surfaces eglandular-pubescent; calycine colleters present; corolla pale to waxy green, rotate, lobes 7–13 × 2–4 mm, abaxial surface eglandular-pubescent, adaxial surface glabrous; gynostegial corona a fused tube, 5–8 mm, obscuring gynostegium, glabrous; style-head extension absent. **Follicles** 8–12 × 2–7 cm, glabrous. **Seeds** brownish black to black, 5–6 × 1.5–2 mm; coma 2–5 cm. $2n = 22$.

Flowering summer–fall; fruiting fall–winter. Citrus groves, roadsides, waste places, dunes, beaches; 0–100 m; introduced; Fla.; South America; introduced also in Central America.

Native to central South America, *Araujia odorata* was introduced to the United States as an ornamental in the 1930s and first observed in a citrus grove in Florida in 1957 (D. P. H. Tucker and R. L. Phillips 1974; D. L. Spellman and C. R. Gunn 1976).

2. Araujia sericifera Brotero, Trans. Linn. Soc. London 12: 62, plates 4, 5. 1818 (as sericofera) • Cruel vine, moth-vine [F] [I] [W]

Stems to 12 m. **Leaves:** petiole 0.4–5 cm, eglandular-pubescent; blade hastate, deltate, oblong, or ovate, 0.8–14 × 0.2–6.4 cm, base truncate to cordate, surfaces glabrous to eglandular-pubescent. **Inflorescences:** peduncle 0.4–4.3 cm, eglandular-pubescent. **Pedicels** 0.4–1.4 cm, eglandular-pubescent. **Flowers:** sepals green, ovate, leaflike, 8–15 × 3–7 mm, surfaces eglandular-pubescent; calycine colleters absent; corolla white or light pink, funnelform, tube 12.1–15.9 × 5.5–9.8 mm, abaxial surface eglandular-pubescent, adaxial surface glabrous except eglandular-pubescent at base, lobes 6.4–9.6 × 4.5–5.4 mm, abaxial surface eglandular-pubescent, adaxial surface glabrous; gynostegial corona of 5 distinct staminal elements, revolute, oblong or quadrate, 2.5–3.9 × 1–2 mm, not fused into tube, not obscuring gynostegium, glabrous; style-head extension 2-fid, 3.5 mm. **Follicles** 4.2–12.5 × 1.6–5.7 cm, minutely eglandular-pubescent. **Seeds** brown, 5–7 × 2–3 mm; coma 2.2–5.2 cm. $2n = 20$.

Flowering summer–fall; fruiting fall–winter. Chaparral, woodlands, citrus groves, urban/suburban disturbed sites; 100–400 m; introduced; Ariz., Calif.; South America; introduced also in Central America, Europe, Africa, Pacific Islands, Australia.

The naturalization of *Araujia sericifera* in the flora area has been substantiated only for Arizona and California, although it has also been reported for Georgia (J. T. Kartesz, http://www.bonap.org/MapSwitchboard.html). A recent risk assessment estimates that about one-third of the United States presents suitable habitat for the species (Animal and Plant Health Inspection Service [U.S.D.A.] 2012).

26. ASCLEPIAS Linnaeus, Sp. Pl. 1: 214. 1753; Gen. Pl. ed. 5, 102. 1754 • Milkweed [Greek *Asklepios*, mythological chief physician, perhaps alluding to therapeutic properties of *Vincetoxicum* or a related taxon of the Old World, applied to present taxon by Linnaeus]

Mark Fishbein

Acerates Elliott; *Anantherix* Nuttall; *Asclepiodella* Small; *Asclepiodora* A. Gray; *Biventraria* Small; *Oxypteryx* Greene; *Podostemma* Greene; *Podostigma* Elliott; *Solanoa* Greene

Herbs, subshrubs, or shrubs, rhizomatous or not; latex milky (clear). **Stems** not twining, unarmed, pubescent or glabrous. **Leaves** persistent (caducous), opposite, subopposite, alternate, or whorled, petiolate or sessile; stipular colleters interpetiolar (absent); laminar colleters present or absent. **Inflorescences** extra-axillary or terminal, umbelliform, pedunculate or sessile. **Flowers:** calycine colleters present or absent; corolla green, white, cream, pink, red, maroon, purple, orange, or yellow, rotate with lobes usually reflexed to spreading (campanulate, tubular), aestivation valvate; coralline corona absent; androecium and gynoecium united into a gynostegium adnate to corolla tube; gynostegial corona of 1 whorl of 5 thick, cupulate, laminar, or clavate segments, often bearing an adaxial falcate, subulate, acicular, lingulate, or crestlike appendage; anthers adnate to style, locules 2; pollen in each theca massed into a rigid, vertically oriented pollinium, pollinia lachrimiform, joined from adjacent anthers by translators to a common corpusculum and together forming a pollinarium. **Fruits** follicles, usually solitary, erect (spreading) on upcurved or straight pedicels or pendulous, green to brown, fusiform, lance-ovoid, ovoid, or ellipsoid, smooth (ornamented), pubescent or glabrous. **Seeds** ovate or lanceolate to nearly orbiculate, flattened (naviculate in *A. albicans*, *A. linaria*; flat to somewhat naviculate in *A. subulata*), usually winged, not beaked, comose (coma absent in *A. perennis*), not arillate; comas white. $x = 11$.

Species ca. 400 (77 in the flora): North America, Mexico, Central America, South America, West Indies, Africa; introduced in Europe and pan-tropical areas.

In its broad sense, *Asclepias* is the largest genus of Apocynaceae. Attempts have been made to segregate the African species into some 20 genera, but these attempts have left a residuum of African species in *Asclepias* that are more closely related to segregate African genera than to the type-bearing American species (D. J. Goyder et al. 2007; Goyder 2009; D. Chuba et al. 2017). Hence, the broader circumscription is adopted here, although the American species as a whole (ca. 130 species) are monophyletic, and circumscription does not affect the nomenclature of the species in the flora.

Species of *Asclepias* are ecologically diverse, with habitats ranging from marshes and swamps to some of the driest regions in the North American and African deserts. They occur from sea level to over 3000 m elevation. However, the genus has diversified most dramatically in grasslands, savannas, woodlands, and forests, in seasonally dry, temperate and subtropical climates, and it is in these ecoregions that milkweeds are most conspicuous. All species are perennial, and the great majority are seasonally dormant herbs. Most species occur in environments subject to one or more of the following disturbances: extended freezing, drought, fire, or mammalian grazing. Nonetheless, the species are diverse in growth form and phenology—habit ranges from diminutive herbs less than 5 cm tall to herbs and shrubs exceeding 2 m in height, and flowering ranges from constrained two-week periods to year-round.

Two aspects of the ecology of *Asclepias* have received extensive study—reproductive and defensive strategies. As with Asclepiadoideae as a whole, *Asclepias* species have elaborate and precise insect pollination involving pollinia and coronas that mediate interactions with pollinators. *Asclepias* has been the most intensively studied genus in the subfamily with regard to the diversity of the pollinating fauna and forces that have shaped the evolution of floral traits. Whereas pollinator interactions in Asclepiadoideae genera vary widely from highly specialized to highly generalized (J. Ollerton et al. 2019), those species of *Asclepias* that have been studied quantitatively are moderately to highly generalized, including such species as *A. tuberosa*, commonly known as butterflyweed (M. Fishbein and D. L. Venable 1996). However, such studies are limited to fewer than a dozen species, and the existence of more specialized pollination seems likely for more rare and poorly known species. Species of *Asclepias* have also served as a model system for testing the hypothesis that the characteristically low fruit set in the genus is a consequence of sexual selection to increase flower number through competition among plants as pollen donors (M. F. Willson and B. J. Rathcke 1974; S. B. Broyles and R. Wyatt 1990; Fishbein and Venable 1996b).

The chemical ecology of *Asclepias* also has been widely studied because a class of secondary metabolites, cardiac glycosides (cardenolides), which is found in many Apocynaceae species, is sequestered by several specialist *Asclepias*-feeding insect herbivores and used as their own defense compounds (A. A. Agrawal et al. 2012). The interaction between *Asclepias* species and the monarch butterfly, *Danaus plexippus*, has been most intensively studied (L. P. Brower et al. 1967). Subsequently, the diversity of defensive traits of milkweeds (cardenolides, latex, trichomes, waxes, etc.) has been analyzed in an evolutionary context, considering also the functions of some of these traits as adaptations for drought resistance (Agrawal et al. 2009; M. Fishbein et al. 2018).

Species of *Asclepias* have been investigated for development of commercial products. Production of natural rubber from latex has not proved economical, even during the time of World War II shortages. There have been repeated efforts to develop an industry around harvesting seed comas as floss for stuffing and insulation. Mechanization has proved challenging, but a small industry is developing, based largely on *A. syriaca* because of its rapid growth and high seed production. *Asclepias* species have been used as a source of medicine by most Native American tribes for diverse ailments. One of the common names of *A. tuberosa* is pleurisy root, and the tuberous roots of this species have been included in herbal remedies for coughing and congestion. This seems to be the only commercialization of American milkweeds for medicinal purposes, and no clinical trials have taken place in North America involving milkweed-derived compounds. Despite toxicity, there is a tradition in native and immigrant American communities, particularly in the Appalachian region, of using young shoots of *A. syriaca* as a vegetable, following boiling with two changes of water.

Species such as *Asclepias curassavica*, *A. incarnata*, and *A. tuberosa* are widely cultivated, and horticultural varietals are available. Recently, the popularity of diverse species of *Asclepias* as components of home and commercial gardens, as well as synthetic prairies, is skyrocketing, fueled in part by interest in growing food plants for monarch butterflies, or more generally as a nectar source for the conservation of pollinator communities. Surprisingly, *A. subulata*, found naturally on arid, sandy substrates, grows rapidly in standard potting media and is becoming ubiquitous in horticultural settings in Arizona. At least 25 species are available from nurseries across the flora area.

In the key and descriptions the following conventions are used. Leaf measurements pertain to midstem leaves, usually at the node below the first inflorescence. The lowest leaves and those in the flowering portion of the plant are commonly much smaller than midstem leaves and are less useful for identification. Vestiture of stems and leaves (unless otherwise specified) is described for mature structures below the inflorescences; many species possess trichomes on stems and leaves early in development that are lost with maturity but continue to be produced on new structures in the flowering portion of the plant. The gynostegial column refers to the cylindrical structure formed by the united filaments; its length is measured from its attachment to the corolla to the base of the anthers. The corona of *Asclepias* consists largely of five discrete segments that are usually cavitate and are often referred to as hoods. In some species, these segments are strongly compressed dorsally and the cavity is greatly reduced. In others, the margins of the segments are folded in, obscuring the cavity and creating a club-shaped form. The majority of species also have an adaxial appendage of each segment, often referred to as a horn, which arises from within the cavity and is often exserted. In some species, the appendage is merely a crest within the cavity, and it may be altogether absent. The diversity of corona forms is difficult to describe, but these features are highly useful for identification. They are ideally observed in fresh material, and photographs are invaluable.

A revised infrageneric classification of *Asclepias* is needed. The species in this treatment are arranged according to phylogenetic relationships obtained from analyses of plastid genomes (M. Fishbein et al. 2018) and a set of 768 nuclear loci (J. Boutte et al. 2019). These correspond largely to the major clades presented by Fishbein et al. (2011). Species 1–3 are early-diverging lineages, species 4–6 belong to the Sonoran Desert clade, species 7–16 belong to the Incarnatae clade, species 17–22 are not placed in a major clade, species 23–27 belong to the Mexican highland clade, and species 28–77 belong to the temperate North American clade. Within the temperate North American clade, several subclades are strongly supported: species 37 in the waxy clade, species 39–42 in the Podostemma clade, species 51–56 in the Colorado Plateau clade, species 57–60 in the dwarf clade, and species 61–77 in the eastern temperate clade.

SELECTED REFERENCES Fishbein, M. et al. 2011. Phylogenetic relationships of *Asclepias* (Apocynaceae) inferred from non-coding chloroplast DNA sequences. Syst. Bot. 36: 1008–1023. Goyder, D. J., A. Nicholas, and S. Liede-Schumann. 2007. Phylogenetic relationships in subtribe Asclepiadinae (Apocynaceae: Asclepiadoideae). Ann. Missouri Bot. Gard. 94: 423–434. Kephart, S. R., R. Wyatt, and D. Parella. 1988. Hybridization in North American *Asclepias*. I. Morphological evidence. Syst. Bot. 13: 456–473. Woodson, R. E. Jr. 1954. The North American species of *Asclepias* L. Ann. Missouri Bot. Gard. 41: 1–211. Woodson, R. E. Jr. 1962. Butterflyweed revisited. Evolution 16: 168–185.

1. Leaves alternate (sometimes a few subopposite).
    2. Leaf blades filiform or needlelike to narrowly linear, 0.05–0.3 cm wide.
        3. Stems 30–160 cm.
            4. Stems woody, few–numerous; leaf blades needlelike, less than 4 cm; corona segments with rod-shaped, slightly exserted appendages . . . . 3. *Asclepias linaria*
            4. Stems herbaceous, 1–3 (rarely more); leaf blades not needlelike, greater than 4 cm; corona segment appendages absent or at most a low crest.
                5. Stems glabrous; anthers 2–3 mm.
                    6. Gynostegial column 0.5–1.5 mm; fused anthers barrel-shaped, anther wings evenly crescent-shaped and separate throughout, apical anther appendages narrowly pandurate, conduplicate, not obscuring corpuscula . . . . 49. *Asclepias engelmanniana*
                    6. Gynostegial column 0–0.5 mm; fused anthers obconic, anther wings connivent, except at the wider base, apical anther appendages deltoid, obscuring corpuscula . . . . 50. *Asclepias rusbyi*
                5. Stems puberulent with curved trichomes; anthers 1–2 mm.
                    7. Leaf blades conduplicate; corolla lobe apices minutely puberulent with curved trichomes abaxially; fused anthers obconic, 1.5–2 mm, anther wings narrowly crescent-shaped; corona segments 3–3.5 mm, appendage an internal crest, apex appearing 3-toothed; follicles erect on straight pedicels . . . . 48. *Asclepias stenophylla* (in part)
                    7. Leaf blades flat; corolla lobes glabrous; fused anthers cylindric, 1–1.5 mm, anther wings trapezoidal to triangular; corona segments 1.5–2.5 mm, appendage absent or obscure, apex not appearing toothed; follicles erect on upcurved pedicels.

8. Corona segments 1.5–2 mm, not exceeding point of anther wings; pedicels and calyces hirtellous; follicles 1–2 cm wide; flowers 34–112 per umbel. . . . . .65. *Asclepias hirtella* (in part)
8. Corona segments 2–2.5 mm, exceeding point of anther wings; pedicels and calyces puberulent with curved trichomes; follicles 0.7–1.2 cm wide; flowers 13–28(–36) per umbel. . . .66. *Asclepias longifolia* (in part)

3. Stems 5–30 cm.
9. Corona segments strongly dorsally compressed and laminar or chute-shaped, appendage absent or merely a low crest, wholly included.
10. Leaf blades conduplicate; corolla lobe apices puberulent with minute, curved trichomes abaxially; fused anthers obconic, 1.5–2 mm, anther wings narrowly crescent-shaped; corona segments 3–3.5 mm, appendage an internal crest, apex appearing 3-toothed; follicles erect on straight pedicels . . . . 48. *Asclepias stenophylla* (in part)
10. Leaf blades flat; corolla lobes glabrous; fused anthers cylindric, 1–1.5 mm, anther wings trapezoidal to triangular; corona segments 2–2.5 mm, appendage absent or obscure, apex not appearing toothed; follicles erect on upcurved pedicels . . . . .66. *Asclepias longifolia* (in part)
9. Corona segments conduplicate, cupulate, or tubular, appendage lingulate, falcate or acicular, barely to well exserted.
11. Corolla lobes 4.5–6 mm; corona segments 3.5–4.5 mm; follicles 1.5–2 cm wide, rugose. . . . . 51. *Asclepias involucrata* (in part)
11. Corolla lobes 2.5–5 mm; corona segments 1–3.5 mm; follicles 0.5–1.5 cm wide, smooth.
12. Corollas red-violet; corona segment appendages lingulate, barely exserted; follicles pendulous on spreading to declined pedicels or erect on upcurved pedicels.
13. Stems erect to ascending; leaves sessile, blades strigose to glabrate; corona segments white; follicles pendulous on spreading to declined pedicels . . . . 4. *Asclepias cutleri*
13. Stems decumbent; leaves petiolate, petioles 1–5 mm, blades puberulent with curved trichomes on midvein; corona segments red-violet at base, white to orange at apex, follicles erect on upcurved pedicels. . . . . 60. *Asclepias uncialis* (in part)
12. Corollas pink, cream, green, or tan; corona segment appendages falcate or acicular, well exserted; follicles erect on straight pedicels.
14. Leaf blades needlelike to narrowly linear, 0.05–0.1 cm wide; anthers 1–1.5 mm; corona segment appendages acicular, arching towards style apex, follicles 2.5–9.5 cm . . . . 16. *Asclepias pumila*
14. Leaf blades narrowly linear, 0.1–0.4 cm wide; anthers 1.5–3 mm; corona segment appendages falcate, sharply inflexed over style apex; follicles 7.5 12.5 cm . . . . 69. *Asclepias michauxii* (in part)

[2. Shifted to left margin.—Ed.]

2. Leaf blades linear, elliptic, oblong, oval, lanceolate, ovate, oblanceolate, obovate, or falcate, 0.3–9 cm wide.
15. Corolla lobes ascending and exceeding corona segments, 7–15 mm; corona segments clavate-tubular, deflexed at base, ascending to incurved at apex, margins connivent.
16. Leaf blade bases cuneate, apices acute to attenuate; peduncles not branched; corolla lobes 7–10 mm, puberulent abaxially at apex with curved trichomes; corona segments 4.5–7 mm, margins and appendage papillose. . . . . 55. *Asclepias asperula*
16. Leaf blade bases rounded to subcordate, apices obtuse to rounded; peduncles usually branched; corolla lobes (9–)12–15 mm, glabrous; corona segments 3–5 mm, margins and appendage hirtellous. . . . . 56. *Asclepias viridis*
15. Corolla lobes reflexed, sometimes with spreading tips, not overtopping corona segments, 3–9 mm; corona segments cupulate, tubular, conduplicate, chute-shaped, or laminar, margins separate.

[17. Shifted to left margin.—Ed.]

17. Corona segments strongly dorsally compressed and laminar or chute-shaped, appendage absent or merely a low crest.
    18. Leaf blades conduplicate; corolla lobe apices puberulent abaxially with minute, curved trichomes; fused anthers obconic, anther wings narrowly crescent-shaped; corona segment appendages crestlike, segment apices appearing 3-toothed; follicles erect on straight pedicels. . . . . 48. *Asclepias stenophylla* (in part)
    18. Leaf blades flat; corolla lobes glabrous; fused anthers cylindric, anther wings trapezoidal to triangular; corona segment appendages absent, segment apices untoothed; follicles erect on upcurved pedicels.
        19. Stems 7–20 cm, hirtellous to pilose; leaf blades hirtellous; fused anthers 1.5–2 mm; follicles hirtellous to pilose; seeds 6–7 × 4–5 mm . . . . 47. *Asclepias lanuginosa* (in part)
        19. Stems 25–125 cm, puberulent with curved trichomes; leaf blades puberulent with curved trichomes or scabridulous; fused anthers 1–1.5 mm; follicles puberulent with curved trichomes to pilosulous; seeds 10–12 × 7–8 mm.
            20. Flowers 34–112 per umbel; pedicels and calyces hirtellous; corona segments 1.5–2 mm, not exceeding point of anther wings; follicles 1–2 cm wide. . . . . .65. *Asclepias hirtella* (in part)
            20. Flowers 13–28(–36) per umbel; pedicels and calyces puberulent with curved trichomes; corona segments 2–2.5 mm, exceeding point of anther wings; follicles 0.7–1.2 cm wide . . . . .66. *Asclepias longifolia* (in part)
17. Corona segments cupulate, tubular, or conduplicate, at most slightly flattened dorsally, appendage lingulate, falcate, subulate, or conical, barely to well exserted from cavity.
    21. Stems 15–100 cm; corolla lobes 4–9 mm.
        22. Corona segments 5–7 mm.
            23. Stems, leaves, and peduncles puberulent with curved trichomes, pilosulous, tomentulose, or glabrate; corollas pink, red, or green; coronas cream, segments with reddish or purplish dorsal stripes; anthers brown; follicles pilosulous; seeds 6–7 mm. . . . . 45. *Asclepias hallii* (in part)
            23. Stems, leaves, and peduncles hirsute; corollas and coronas red, orange, or yellow; anthers yellow to yellowish green; follicles hirsute; seeds 8–9 mm . . . 70. *Asclepias tuberosa*
        22. Corona segments 2.5–3.5 mm.
            24. Pedicels 9–12 mm; corolla lobes 4–5 mm; corona segments cupulate, apex obtuse; fruiting pedicels straight, follicles 0.5–0.7 cm wide; se United States . . . . 69. *Asclepias michauxii* (in part)
            24. Pedicels 15–50 mm; corolla lobes 6–9 mm; corona segments conduplicate, apex truncate; fruiting pedicels pendulous or upcurved, follicles 1.2–3 cm wide; California or Utah.
                25. Leaf blades 2–8 cm wide, tomentose to densely puberulent with curved trichomes; anthers brown, 2.5–3 mm; follicles erect on upcurved pedicels, 5–10 cm; seeds 7–9 × 4–6 mm . . . . .36. *Asclepias eriocarpa* (in part)
                25. Leaf blades 0.5–2 cm wide, glabrous except for midvein; anthers green to yellowish green, 1.5–2 mm; follicles pendulous, 3.5–5.5 cm; seeds 12–13 × 8–9 mm . . . . 54. *Asclepias labriformis*
    21. Stems 5–15 cm; corolla lobes 3–6 mm.
        26. Corollas green, sometimes tinged red or pink; corona segment appendages well exserted, sharply inflexed towards or over style apex; follicles rugose.
            27. Leaf blades 0.2–0.8 cm wide; anthers 1.5–2 mm; corona segments cream, usually with a pink or red dorsal stripe, conduplicate, 3.5–4.5 mm; seeds 7–8 × 5–6 mm. . . . . 51. *Asclepias involucrata* (in part)
            27. Leaf blades 0.5–2 cm wide; anthers 1–1.5 mm; corona segments yellow to ochroleucous, tubular, 2–3 mm; seeds 8–12 × 6–8 mm. . . . . .52. *Asclepias macrosperma* (in part)
        26. Corollas red-violet; corona segment appendages equaling or barely exserted from segment; follicles smooth.

28. Corona segment appendages falcate, not or barely exserted; Colorado Plateau, ne Arizona, nw New Mexico, Utah.
    29. Leaf blades persistently pilosulous, more densely so on veins; pedicels and calyces densely pilosulous to tomentulose; corona segment appendages barely exserted; Utah, n of Colorado River and w of Green River, from San Rafael Reef to Waterpocket Fold. .58. *Asclepias ruthiae* (in part)
    29. Leaf blades sparsely pilosulous to glabrate, persistently hairy on veins; pedicels and calyces pilosulous; corona segment appendages included; Arizona, New Mexico, just entering Utah in Monument Valley. . . . . 59. *Asclepias sanjuanensis* (in part)
28. Corona segment appendages lingulate or conical, barely exserted; Arizona, Colorado, Nevada, New Mexico, Oklahoma, Texas, not on the Colorado Plateau.
    30. Leaf blades lanceolate to ovate, 0.6–3 cm wide, strigulose or puberulent on midvein to glabrate; flowers 2–25 per umbel; anthers 1.5–2 mm; corona appendages conical; Nevada . . . . 57. *Asclepias eastwoodiana* (in part)
    30. Leaf blades linear to lanceolate, 0.2–1 cm wide, puberulent with curved trichomes on midvein; flowers 3–7 per umbel; anthers 1–1.5 mm; corona appendages lingulate; Arizona, Colorado, New Mexico, Oklahoma, Texas . . . . 60. *Asclepias uncialis* (in part)

1. Leaves opposite, subopposite, or whorled (rarely alternate in *A. eriocarpa*).
    31. One or more nodes with leaves whorled or (in *A. quadrifolia*) apparently whorled because of a reduced internode.
        32. Shrubs, 50–400 cm; leaves ephemeral; stems often leafless or nearly so, glaucous; follicles spreading to pendulous; Sonoran Desert and se Mohave Desert, sw Arizona, se California, s Nevada.
            33. Stems 5–40 (usually 1–few), 140–400 cm; pedicels 8–16 mm; corolla lobes 4.5–6 mm; corona segments 2–3 mm, conduplicate, not exceeding style apex . . . . .5. *Asclepias albicans* (in part)
            33. Stems 3–100 (usually few–numerous), 50–175 cm; pedicels 11–21 mm; corolla lobes 7–12 mm; corona segments 7–9 mm, tubular, greatly exceeding style apex . . . . 6. *Asclepias subulata* (in part)
        32. Herbs; stems 20–150 cm; leaves persistent; stems leafy, not glaucous; follicles erect on straight or upcurved pedicels; widespread, approaching but not entering the Sonoran Desert.
            34. At least some leaves on main stem alternate or opposite, blades elliptic or lanceolate to ovate, 1–8 cm wide; corona lobes 2.5–4 mm.
                35. Plants 1–10-stemmed; whorled leaves, when present, at upper nodes, leaf blades tomentose to densely puberulent with curved trichomes; pedicels and calyces densely tomentose; corolla lobes 7–9 mm; follicles erect on upcurved pedicels, lance-ovoid, 1.5–3 cm wide; California . . . 36. *Asclepias eriocarpa* (in part)
                35. Plants single-stemmed, whorled leaves, when present, at a single, mid-stem node; leaf blades inconspicuously puberulent only on veins; pedicels puberlent on 1 side; calyces glabrous; corolla lobes 4–5 mm; follicles erect on straight pedicels, narrowly fusiform, 0.4–1 cm wide; e North America . . . . 73. *Asclepias quadrifolia* (in part)
            34. All leaves whorled (rarely opposite on vegetative branches), blades linear to linear-lanceolate or narrowly elliptic, 0.1–1.8 cm wide; corona lobes 1.5–2 mm.
                36. Leaf blades 0.2–1.8 cm wide; pedicels 9–14 mm; corollas pink (rarely green with pink tinge); California, Idaho, Nevada, Oregon, Washington . . . . 13. *Asclepias fascicularis*
                36. Leaf blades 0.1–0.4 cm wide; pedicels 5–12 mm; corollas green to cream, sometimes tinged pink or tan; widespread, s or e of Idaho, Nevada.

37. Stems 1–8, vegetative branches often present in 1+ leaf axils; leaf blades 0.1–0.4 cm wide, glabrous; anthers 1.5–2 mm; corona segment margins entire; w United States, including w Kansas, w Oklahoma, w Texas . . . . 14. *Asclepias subverticillata*
37. Stems solitary (rarely 2–3), vegetative branches absent; leaf blades 0.1–0.2 cm wide, puberulent with curved trichomes adaxially, especially on midvein to glabrate; anthers 1.2–1.5 mm; corona segment margins shallowly lobed to sharply toothed; e North America, e of Colorado, Idaho, New Mexico. . . . . 15. *Asclepias verticillata*

[31. Shifted to left margin.—Ed.]

31. Leaves opposite or subopposite, not whorled.
38. Leaf blades filiform to narrowly linear, 0.05–0.3 cm wide.
39. Shrubs, 50–400 cm; leaves ephemeral; stems often leafless or nearly so, glaucous; follicles spreading to pendulous; Sonoran Desert and se Mohave Desert, sw Arizona, se California, s Nevada.
40. Stems 5–40 (usually 1–few), 140–400 cm; pedicels 8–16 mm; corolla lobes 4.5–6 mm; corona segments 2–3 mm, conduplicate, not exceeding style apex . . . . 5. *Asclepias albicans* (in part)
40. Stems 3–100 (usually few–numerous), 50–175 cm; pedicels 11–21 mm; corolla lobes 7–12 mm; corona segments 7–9 mm, tubular, greatly exceeding style apex . . . . 6. *Asclepias subulata* (in part)
39. Subshrubs or herbs; stems 4–125 cm; leaves persistent; stems leafy, not glaucous; follicles erect on straight or upcurved pedicels; approaching but not entering the Sonoran Desert.
41. Stems prostrate to decumbent.
42. Stems 4–18 cm; petioles 1–5 mm; umbels sessile; follicles ovoid, 3–5.5 cm; Arizona, Colorado, Kansas, New Mexico, Oklahoma, Texas, Wyoming.
43. Corollas green, sometimes tinged pink or red, lobes 4.5–6 mm; corona segments 3.5–4.5 mm; follicles 1.5–2 cm wide, rugose . . . . 51. *Asclepias involucrata* (in part)
43. Corollas red-violet, lobes 3–5 mm; corona segments 1–2 mm; follicles 0.8–1.5 cm wide, smooth . . . . 60. *Asclepias uncialis* (in part)
42. Stems 15–70 cm; petioles 0–2 mm; peduncles 0–6 cm; follicles narrowly fusiform, 7.5–13.5 cm; se United States.
44. Leaf blades scabridulous to puberulent with curved trichomes; corona segments laminar, 2–2.5 mm, appendage absent or obscure; fruiting pedicels upcurved; follicles 0.7–1.2 cm wide; seeds 11–12 × 7–8 mm . . . . 66. *Asclepias longifolia* (in part)
44. Leaf blades glabrous; corona segments cupulate, 2.5–3.5 mm, appendage falcate, well exserted from cavity; fruiting pedicels straight; follicles 0.5–0.7 cm wide; seeds 7–8 × 5–6 mm . . . . 69. *Asclepias michauxii* (in part)
41. Stems ascending to erect.
45. Subshrubs, usually persistently woody at base (sometimes herbaceous to base), cespitose; stems 5–20+, sparsely to densely branched, at least at base.
46. Stems 20–60 cm; petioles 2–4 mm; flowers 3–15 per umbel; corollas cream, sometimes tinged or striped pink; corona segments tubular, glabrous at apex, appendage acicular, arching over style apex, glabrous; fruiting pedicels straight. . . . . 8. *Asclepias angustifolia* (in part)
46. Stems 5–35 cm; leaves sessile; flowers 1–7 per umbel; corollas greenish cream or greenish yellow, sometimes tinged red or purple; corona segments conduplicate or chute-shaped, hirtellous at apex, appendage falcate, arching over style apex, or an included crest, hirtellous; fruiting pedicel upcurved.

47. Flowers 2–7 per umbel; anthers 1–1.5 mm; corona segments conduplicate, 4–5 mm, apex long-caudate, appendage falcate, exserted . . . . 1. *Asclepias macrotis*
47. Flowers 1–2 per umbel; anthers 1.5–2 mm; corona segments chute-shaped, sharply inflexed, 2–3 mm, apex subcaudate, appendage an included crest. . . . . .2. *Asclepias sperryi*

[45. Shifted to left margin.—Ed.]

45. Herbs, not woody at base; stems 1(–5), rarely branched (often branched in *A. linearis*).
48. Corona segments strongly dorsally compressed or chute-shaped, margins incurved, appendages absent or a low, inlcuded crest; follicles 1–2 cm wide.
49. Flowers 9-28 per umbel; pedicels puberulent with curved trichomes to pilosulous, straight in fruit; corona segments equaling style apex . . . . 18. *Asclepias stenophylla* (in part)
49. Flowers 34-112 per umbel; pedicels hirtellous, upcurved in fruit; corona segments greatly exceeded by style apex . . . . .65. *Asclepias hirtella* (in part)
48. Corona segments conduplicate or cupulate (may be slightly flattened dorsally, but not compressed), margins not incurved, appendages evident and exserted from cavity; follicles 0.3–1 cm wide.
50. Corona segment appendages acicular, subulate, or falcate, exserted and inflexed towards or over style apex.
51. Stems 60–125 cm; leaf blades 7–25 cm; corollas red, lobes 9–10 mm; corona segments yellow to reddish orange, 5–6 mm; fruiting pedicels upcurved . . . . 67. *Asclepias lanceolata* (in part)
51. Stems 25–75 cm; leaf blades 3–9 cm; corollas green, tan, or pink, sometimes tinged brown or red, lobes 3–5 mm; coronas cream, sometimes tinged brown or green, segments 1.5–4 mm; fruiting pedicels straight.
52. Plants rhizomatous; interpetiolar ridges present; pedicels 6–8 mm; corona segments 1.5 mm; follicles 5–8.5 cm; Texas . . . . 12. *Asclepias linearis*
52. Plants not rhizomatous; interpetiolar ridges absent; pedicels 7–13 mm; corona segments 3–4 mm; follicles 9–12.5 cm; Alabama, Florida . . . . . 31. *Asclepias viridula*
50. Corona segment appendages crestlike, included, or absent.
53. Corollas and coronas yellowish green to green; corolla lobes erect, concealing gynostegium and partially obscuring corona; corona appendages absent; follicles densely puberulent with curved trichomes . . . . .33. *Asclepias pedicellata* (in part)
53. Corollas green, white, or cream, sometimes tinged gray, pink, red, lavender, or purple, lobes reflexed or spreading; coronas red, pink, violet, white, cream, gray, or a combination of these, appendage crestlike, included; follicles glabrous, glabrate, or pilosulous.
54. Corollas green, sometimes tinged red, lobes pilose abaxially; anthers tan to brown; coronas red, pink, purple, or red-violet, sometimes white at segment apices; fruiting pedicels upcurved, follicles pilosulous to glabrate; s Arizona, s New Mexico. . . . . .25. *Asclepias quinquedentata* (in part)
54. Corollas cream or white, tinged gray, pink, lavender, or purple, lobes glabrous; anthers green or lavender; corona segments cream or white, sometimes tinged gray, pink, or purple; fruiting pedicels straight, follicles glabrous; se United States.
55. Flowers erect; corollas white, sometimes tinged lavender, lobes spreading, 7–10 mm; anthers lavender; corona segments white, cupulate, apex edentate . . . . 29. *Asclepias feayi*
55. Flowers spreading to pendent; corollas and coronas cream, tinged gray, pink, or purple; corolla lobes reflexed, 4–5 mm; anthers green; corona segments conduplicate, apex toothed laterally . . . . 30. *Asclepias cinerea*

[38. Shifted to left margin.—Ed.]

38. Leaf blades linear, elliptic, oblong, oval, lanceolate, ovate, deltate, oblanceolate, obovate, oblate, or orbiculate, 0.1–15 cm wide.
    56. Both stems and leaf blades completely glabrous or glabrate, except for ciliate leaf margins in some species or abaxial midvein (*A. cryptoceras*).
        57. Stems and/or leaf blades glaucous.
            58. Corona segment appendages merely low, included crests or absent; w United States, w of Rocky Mountains plus Arizona, New Mexico, w Texas.
                59. Corollas red-violet, lobes 6–7 mm; corona segments tubular, 2–3 mm; n California, extreme w Nevada, sw Oregon . . . . 17. *Asclepias cordifolia*
                59. Corollas green to yellowish green, sometimes tinged red abaxially, lobes 8–14 mm; corona segments conduplicate, 4–8 mm; Arizona, e California, w Colorado, Idaho, Nevada, s New Mexico, e Oregon, w Texas, Utah, e Washington, s Wyoming.
                    60. Stems prostrate to decumbent, 8–25 cm; umbels sessile; corolla lobe apices minutely hirtellous adaxially; anthers brown (sometimes with green apex), 1.8–3 mm; corona segments red-violet to pinkish purple, apex with recurved, papillose teeth, appendage absent; follicles 4.5–6 cm, glabrous, glaucous . . . . 18. *Asclepias cryptoceras*
                    60. Stems erect, 30–80 cm; peduncles 5–20 cm; corolla lobes glabrous; anthers green, 3–3.5 mm; corona segments white, yellow to tan dorsally, apex edentate, appendage a low, included crest; follicles 8–13 cm, pilosulous, not glaucous . . . . 37. *Asclepias elata*
            58. Corona segment appendages exserted, falcate, subulate, or ensiform; e United States, w to Arizona, Colorado, Kansas, Nebraska, New Mexico, Texas, Utah.
                61. Peduncles, pedicels, calyx lobes, and developing leaf blades tomentose, leaves becoming glabrate with age; w United States, plains e of Rocky Mountains to c Kansas, c Nebraska, c Oklahoma, w Texas. . . 44. *Asclepias latifolia* (in part)
                61. Peduncles, pedicels, calyx lobes, and developing leaf blades glabrous, puberulent with curved trichomes, or pilosulous; e United States and Canada, w to e Kansas, e Nebraska, c Oklahoma, and e Texas.
                    62. Stems decumbent to erect; corolla lobes 5.5–6 mm; anthers brown; corona segments 3–3.5 mm, appendage ensiform, only slightly exserted. . . . . 19. *Asclepias humistrata*
                    62. Stems erect; corolla lobes 7–12 mm; anthers green; corona segments 4–7 mm, appendage falcate or subulate, well exserted.
                        63. Leaf blades 1.5–9 cm wide, apex obtuse, rounded, truncate, or emarginate (acute); pedicels 20–55 mm; anthers 2.5–3.5 mm; corona segments tubular, apex broadly obtuse and oblique or truncate and erose; follicles 1–3 cm wide.
                            64. Peduncles 5–40 cm; flowers 18–53 per umbel; pedicels puberulent with curved trichomes; corollas green, often tinged reddish purple, or bronze; coronas reddish purple to cream, segments stipitate, apex truncate, erose; follicles smooth, 9–16 cm; seeds 9–10 × 6–7 mm . . . . 61. *Asclepias amplexicaulis*
                            64. Peduncles 1–6 cm; flowers 9–29 per umbel; pedicels glabrous, glaucous; corollas dark pink, pale at base of lobes; coronas pale to dark pink, segments subsessile, apex broadly obtuse, oblique; follicles sparsely muricate, 7–11 cm; seeds 7–9 × 5–6 mm. . . . . 62. *Asclepias sullivantii*
                        63. Leaf blades 1–5 cm wide, apex acute to attenuate; pedicels 11–20 mm; anthers 2–2.5 mm; corona segments conduplicate, apex acute to obtuse and emarginate; follicles 0.9–1.6 cm wide.

65. Leaves sessile, blade base rounded to obtuse; flowers pendent; corollas green to greenish cream, sometimes tinged red; coronas green to yellowish green or greenish cream, segments sessile, apex obtuse, emarginate, appendage falcate, sharply inflexed towards style apex; seeds 6–7 × 4–5 mm; Illinois, Indiana, Iowa, Kansas, Missouri, Wisconsin . . . . .63. *Asclepias meadii*
65. Petioles 1–2 mm; leaf blade bases cordate; flowers erect; corollas pink to reddish purple; coronas pink to lavender, segments stipitate, apex acute, appendage subulate, arching above style apex; seeds 7–9 × 5–7 mm; coastal plain, e United States to e Texas. . . . . .68. *Asclepias rubra* (in part)

[57. Shifted to left margin.—Ed.]

57. Neither stems nor leaf blades glaucous.
66. Leaf blades 0.1–2 cm wide.
67. Stems 10–60 cm; corolla lobes 1.5–6 mm; corona segments 1.5–4 mm.
68. Leaf blades linear, 0.2–0.6 cm wide; flowers 4–10 per umbel; corollas green, sometimes tinged red, lobes pilose abaxially; corona segment apices truncate, appendage crestlike, barely exserted; fruiting pedicels upcurved; seeds 4–5 × 3–4 mm; s Arizona, s New Mexico . . . . .25. *Asclepias quinquedentata* (in part)
68. Leaf blades linear to elliptic, oval, oblong, or ovate, 0.3–6 cm wide; flowers 7–31 per umbel; corollas white, cream, or pink, lobes glabrous abaxially; corona segment apices obtuse to acute, appendage acicular or subulate, well exserted; fruiting pedicels straight or declined; seeds 6–15 × 4–14 mm; e United States and Canada to e Texas.
69. Subshrubs or herbs; stems cespitose; interpetiolar ridges present; petioles 10–12 mm; pedicels 7–13 mm; flowers erect to spreading; corolla lobes 3–4 mm; corona segments tubular, rounded dorsally, 1.5–2.5 mm, appendage acicular, arching above style apex; fruiting pedicels declined, follicles pendulous, 4–7 × 1–2.5 cm; seeds 12–15 × 11–14 mm, coma absent. . . . . .10. *Asclepias perennis* (in part)
69. Herbs; stems solitary; interpetiolar ridges absent; petioles 2–7 mm; pedicels 17–28 mm; flowers spreading to pendent; corolla lobes 4–5 mm; corona segments conduplicate, flattened dorsally, 2.5–4 mm, appendage subulate, inflexed over style apex; fruiting pedicels straight, follicles erect, 8–16 × 0.4–1 cm; seeds 6–7 × 4–5 mm, coma present . . . . 73. *Asclepias quadrifolia* (in part)
67. Stems 25–125 cm; corolla lobes 7–13 mm; corona segments 5–10 mm.
70. Leaf blades 2.5–8 cm; flowers spreading to pendent; corollas green, sometimes tinged brown at lobe tips, lobes 10–13 mm; anthers 3–3.5 mm; corona segments cream to pale green, 8–10 mm, clavate, incurved, appendage an included crest; follicles 11.5–15 cm . . . . 32. *Asclepias connivens* (in part)
70. Leaf blades 5–25 cm; flowers erect; corollas red, pink, or reddish purple, lobes 7–10 mm; anthers 2–2.5 mm; corona segments yellow, reddish orange, pink, or lavender, 5–7 mm, tubular or conduplicate, appendage subulate, well exserted; follicles 5.5–12 cm.
71. Petioles 0–1 mm; leaf blades linear to linear-lanceolate, 0.2–1.7 cm wide, base cuneate; flowers 4–16 per umbel; corollas red, lobes 9–10 mm; corona segments yellow to reddish orange, tubular, apex obtuse, 5–6 mm, appendage sharply inflexed over style apex; follicles 0.8–1 cm wide . . . . 67. *Asclepias lanceolata* (in part)
71. Petioles 1–2 mm; leaf blades narrowly lanceolate to ovate, 1–4.5 cm wide, base cordate; flowers 9–20 per umbel; corollas pink to reddish purple, lobes 7–9 mm; corona segments pink to lavender, conduplicate, apex acute, 6–7 mm, appendage arching above style apex; follicles 1–1.5 cm wide. . . . . .68. *Asclepias rubra* (in part)
66. Leaf blades 2–15 cm wide.

[72. Shifted to left margin.—Ed.]

72. Leaf blade bases cuneate.
    73. Petioles 0–1 mm; leaf blades linear to narrowly elliptic or oblanceolate, 0.3–2.5 cm wide; flowers 4–8 per umbel; pedicels 10–20 mm; corolla lobes 10–13 mm; corona segments 8–10 mm, clavate, appendage an included crest; coastal plain, s Alabama, Florida, s Georgia, se South Carolina . . . . 32. *Asclepias connivens* (in part)
    73. Petioles 2–15 mm; leaf blades ovate to elliptic or oblong, 1–11 cm wide; flowers 7–41 per umbel; pedicels 17–45 mm; corolla lobes 4–12 mm; corona segments 2.5–5 mm, tubular or conduplicate, appendage subulate or falcate, well exserted; inland e North America, including n Alabama, n Georgia, w South Carolina.
        74. Stems 25–60 cm; petioles 2–7 mm; leaf blades 2.5–12 × 1–6 cm; pedicels 17–28 mm; corollas pink or cream, lobes 4–5 mm; anthers tan to brown, 1–1.5 mm; corona segments conduplicate, apex obtuse, appendage subulate; follicles 0.4–1 cm wide; seeds 6–7 mm . . . . 73. *Asclepias quadrifolia* (in part)
        74. Stems 65–150 cm; petioles 5–15 mm; leaf blades 10–24 × 2–11 cm; pedicels 25–45 mm; corollas green (rarely pink-tinged), lobes 6–12 mm; anthers green, 2.5–3.5 mm; corona segments tubular, apex truncate, appendage falcate; follicles 1.5–2 cm wide; seeds 8–10 mm . . . . 74. *Asclepias exaltata* (in part)
72. Leaf blade bases truncate or rounded to cordate.
    75. Peduncles and pedicels puberulent with curved trichomes in 1 line; flowers 9–20 per umbel; corollas pink to reddish purple; coronas pink to lavender, 6–7 mm, segment apices acute; follicles fusiform, 1–1.5 cm wide; coastal plain, e United States to e Texas . . . . 68. *Asclepias rubra* (in part)
    75. Peduncles tomentose, puberulent with curved trichomes, or pilosulous; pedicels tomentose or densely pilose to glabrate; flowers 12–80 per umbel; corollas green, ochroleucous, tan, or red-violet; coronas cream to ochroleucous, 2.5–5.5 mm, segment apices truncate; follicles lance-ovoid to ovoid or ellipsoid, 1.7–3.5 cm wide; w United States.
        76. Stems 40–250 cm; leaf blades 7.5–25 cm, apex attenuate to acuminate, margins erose; umbels terminal and extra-axillary; corolla lobe apices tomentose abaxially; seeds 8–13 mm; Sonoran and Mohave deserts and surrounding areas, w Arizona, s California, s Nevada, sw Utah . . . . 35. *Asclepias erosa* (in part)
        76. Stems 25–100 cm; leaf blades 4.5–14.5 cm, apex truncate to rounded (sometimes acute), margins entire; umbels extra-axillary; corolla lobes densely tomentose throughout abaxially or glabrous; seeds 7–8 mm or 18–20 mm; e of the Sonoran and Mohave deserts, ne Arizona, s Utah to the Great Plains.
            77. Peduncles 0–2.5 cm, puberulent with curved trichomes or pilosulous; pedicels 15–35 mm; corollas green, lobes 7–9 mm, glabrous abaxially; anthers 3–3.5 mm; corona segments 3–5.5 mm, apex papillose; fruiting pedicels upcurved; seeds 7–8 × 5–6 mm; ne Arizona, s Utah to the Great Plains, substrates various including sand . . . . 44. *Asclepias latifolia* (in part)
            77. Peduncles 2.5–6 cm, tomentose; pedicels 10–15 mm; corollas ochroleucous or tan to red-violet, lobes 5.5–6.5 mm, tomentose abaxially; anthers 1.5 mm; corona segments 2.5–3.5 mm, apex glabrous; fruiting pedicels spreading; seeds 18–20 × 9–10 mm; local and rare in n Arizona, s Utah on unstabilized sand dunes . . . . 53. *Asclepias welshii* (in part)

[56. Shifted to left margin.—Ed.]

56. Either stems and/or leaf blades bearing trichomes, sometimes restricted to veins.
    78. Stems prostrate or decumbent.
        79. Stems 5–15 cm.
            80. Leaf blades 1.7–5 cm; corollas green, sometimes tinged red, pink, or brown; corona segments 2–10 mm; follicles 4.5–9.5 × 1.2–2.5 cm.
                81. Leaf blades 1.2–6.5 cm wide, apex obtuse to rounded; umbels extra-axillary; corolla lobes 7–12 mm; anthers 2–2.5 mm; corona segments green, sometimes tinged bronze, apex white or cream, sinuous-tubular, 5–10 mm . . . . 40. *Asclepias oenotheroides* (in part)
                81. Leaf blades 0.2–2 cm wide, apex acute to attenuate; umbels terminal; corolla lobes 4.5–6.5 mm; anthers 1–2 mm; corona segments cream with pink or red dorsal stripe or yellow to ochroleucous, conduplicate or tubular, but not sinuous, 2–4.5 mm.
                    82. Leaf blades 0.2–0.8 cm wide; anthers 1.5–2 mm; corona segments cream, usually with a pink or red dorsal stripe, conduplicate, 3.5–4.5 mm; seeds 7–8 × 5–6 mm . . . . 51. *Asclepias involucrata* (in part)
                    82. Leaf blades 0.5–2 cm wide; anthers 1–1.5 mm; corona segments yellow to ochroleucous, tubular, 2–3 mm; seeds 8–12 × 6–8 mm . . . . .52. *Asclepias macrosperma* (in part)
            80. Leaf blades 1.5–12 cm; corollas red-violet; corona segments 1–2 mm; follicles 3–6 × 0.5–1.5 cm.
                83. Corona segment appendages falcate, not or barely exserted; Colorado Plateau, ne Arizona, nw New Mexico, Utah.
                    84. Leaf blades persistently pilosulous, more densely so on veins; pedicels and calyces densely pilosulous to tomentulose; corona segment appendages included; Utah, n of Colorado River and w of Green River, from San Rafael Reef to Waterpocket Fold . . . . .58. *Asclepias ruthiae* (in part)
                    84. Leaf blades sparsely pilosulous to glabrate, persistently and densely hairy on veins and margins; pedicels and calyces pilosulous; corona segment appendages barely exserted; ne Arizona, nw New Mexico, just entering Utah in Monument Valley . . . . 59. *Asclepias sanjuanensis* (in part)
                83. Corona segment appendages lingulate or conical, barely exserted; Arizona, Colorado, Nevada, New Mexico, w Oklahoma, w Texas, not on the Colorado Plateau.
                    85. Leaf blades lanceolate to ovate, 0.6–3 cm wide, strigulose or pilosulous on midvein to glabrate; flowers 2–25 per umbel; anthers 1.5–2 mm; corona appendages conical; Nevada . . . . 57. *Asclepias eastwoodiana* (in part)
                    85. Leaf blades linear to lanceolate, 0.2–1 cm wide, puberulent with curved trichomes on midvein; flowers 3–7 per umbel; anthers 1–1.5 mm; corona appendages lingulate; Arizona, Colorado, New Mexico, w Oklahoma, w Texas. . . . . 60. *Asclepias uncialis* (in part)
        79. Stems 15–250 cm.
            86. Leaf blades 0.1–1 cm wide.
                87. Corolla lobes 8–11 mm, pilosulous abaxially; corona segments 5–7 mm, appendage papillose; follicles muricate ridged; s Texas . . . . . 39. *Asclepias prostrata* (in part)
                87. Corolla lobes 3–6 mm, glabrous abaxially; corona segments 1.5–4.5 mm, appendage glabrous; follicles smooth or rugose; widespread, but not s Texas.
                    88. Stems 25–70 cm; corona segments laminar, margins incurved, appendage absent or obscure; follicles 8–13.5 × 0.7–1.2 cm. . . .66. *Asclepias longifolia* (in part)
                    88. Stems 5–30 cm; corona segments cupulate or conduplicate, margins not incurved, appendage conical or falcate, barely to well exserted; follicles 3.5–6 × 0.5–2 cm or 7.5–12.5 × 0.5–0.7 cm.

89. Corollas red-violet; corona segments 1.5–2 mm, appendage conical, barely exserted, follicles spreading to pendulous, strigose . . . . 57. *Asclepias eastwoodiana* (in part)

89. Corollas green to tan, sometimes tinged pink or red; corona segments 2.5–4.5 mm, appendage falcate, well exserted, sharply inflexed over style apex; follicles erect, puberulent with curved trichomes to pilosulous.

90. Stems 5–18 cm; anthers brown; corona segments conduplicate, apex truncate, 3.5–4.5 mm; fruiting pedicels upcurved, follicles ovoid, rugose, 4.5–5.5 × 1.5–2 cm; c United States . . . . 51. *Asclepias involucrata* (in part)

90. Stems 15–30 cm; anthers green; corona segments cupulate, apex obtuse, 2.5–3.5 mm; fruiting pedicels straight, follicles narrowly fusiform, smooth, 7.5–12.5 × 0.5–0.7 cm; se United States . . . . 69. *Asclepias michauxii* (in part)

[86. Shifted to left margin.—Ed.]

86. Leaf blades 1–15 cm wide.

91. Corona segments 5–11 mm, greatly exceeding style apex, tubular, sinuous-tubular, or conduplicate-tubular, appendage papillose (rarely glabrous or glabrate).

92. Stems, leaf blades, peduncles, and pedicels pilosulous to tomentulose; petioles 2–3 mm, leaf blades 1.8–5 × 0.4–1.8 cm; flowers 3–8 per umbel; pedicels 8–14 mm; gynostegial column 3–3.5 mm; follicles pendent, muricate-ridged, tomentulose . . . . 39. *Asclepias prostrata* (in part)

92. Stems, leaf blades, peduncles, and pedicels puberulent with curved trichomes or hirtellous; petioles 2–25 mm; leaf blades 4–15 × 1.2–7.5 cm; flowers 5–32 per umbel; pedicels 10–30 mm; gynostegial column 0.3–1.5 mm; follicles erect, smooth, puberulent with curved trichomes or hirtellous.

93. Leaf blades ovate or lanceolate to oblong or elliptic, base cuneate to obtuse, apex obtuse to rounded, sometimes conduplicate; pedicels 10–20 mm; gynostegial column 1–1.5 mm; anthers 2–2.5 mm; corona segments relatively slender, green, sometimes tinged bronze, apex white or cream, deeply emarginate; se Arizona, sw Colorado, w Louisiana, New Mexico, Oklahoma, Texas . . . . 40. *Asclepias oenotheroides* (in part)

93. Leaf blades ovate to lanceolate, base cuneate or obtuse to truncate or subcordate, apex obtuse to acute, rarely conduplicate; pedicels 17–30 mm; gynostegial column 0.3–0.5 mm; anthers 1.7–2 mm; corona segments relatively stout, cream or green with cream apex, apex truncate; Arizona, se California, s Nevada, sw New Mexico. . . . . 42. *Asclepias nyctaginifolia* (in part)

91. Corona segments 1.5–6 mm, shorter than to slightly exceeding style apex, cupulate or conduplicate, appendage glabrous.

94. Corolla lobes 4–6 mm; corona segments 1.5–3 mm.

95. Leaf blades ovate to nearly orbiculate, 3–4 cm wide, base obtuse to cordate; flowers 20–55 per umbel; corolla lobes 5–6 mm; corona segments conduplicate, 2–3 mm, appendage absent; follicles 6–10 × 2–3 cm . . . . 28. *Asclepias solanoana*

95. Leaf blades narrowly lanceolate to ovate, 0.6–3 cm wide, base cuneate to obtuse; flowers 2–25 per umbel; corolla lobes 4–5 mm; corona segments cupulate, 1.5–2 mm, appendage conical, barely exserted; follicles 3.5–6 × 0.5–1 cm; Nevada. . . . . 57. *Asclepias eastwoodiana* (in part)

94. Corolla lobes 6–11 mm; corona segments 3–6 mm.

96. Flowers 5–21 per umbel; corolla lobes pilose at base adaxially; gynostegial column 2–3.5 mm; corona segments red-violet, margins connivent, papillose, appendage absent . . . . 21. *Asclepias californica* (in part)
96. Flowers 12–50 per umbel; corolla lobes papillose or glabrous at base adaxially; gynostegial column 1–1.5 mm; corona segments cream to ochroleucous or dark pink, margins separate, glabrous, appendage falcate.
97. Petioles 3–10 mm; leaf blade margins entire; peduncles 0–3.5 cm; corolla lobes densely tomentose abaxially throughout; anthers dark brown, 1.5–2 mm; corona segments sessile, apex obtuse, oblique; follicles 5–6.5 cm . . . . 22. *Asclepias vestita* (in part)
97. Petioles 0–6 mm; leaf blade margins erose; peduncles 2–10 cm; corolla lobes tomentose abaxially only at tips; anthers green, 2.5–3 mm; corona segments stipitate, apex truncate; follicles 6.5–10 cm . . . . 35. *Asclepias erosa* (in part)

[78. Shifted to left margin.—Ed.]

78. Stems spreading to erect.
98. Corolla lobes 3–5 mm.
99. Corona segments laminar or chute-shaped, margins incurved, tightly appressed to gynostegium; appendage absent or obscure.
100. Stems 7–20 cm; leaf blades oblong or lanceolate to narrowly lanceolate, 4–8 cm, hirtellous; umbels terminal; anthers 1.5–2 mm; seeds 6–7 × 4–5 mm . . . . 47. *Asclepias lanuginosa* (in part)
100. Stems 25–70 cm; leaf blades linear to linear-lanceolate, 4–18 cm, scabridulous to puberulent with curved trichomes; umbels terminal and/or extra-axillary; anthers 1–1.5 mm; seeds 10–12 × 7–8 mm.
101. Flowers 34–112 per umbel; pedicels and calyces hirtellous; corona segments 1.5–2 mm, not exceeding point of anther wings; follicles 1–2 cm wide . . . . 65. *Asclepias hirtella* (in part)
101. Flowers 13–28(–36) per umbel; pedicels and calyces puberulent with curved trichomes; corona segments 2–2.5 mm, exceeding point of anther wings; follicles 0.7–1.2 cm wide . . . . . 66. *Asclepias longifolia* (in part)
99. Corona segments conduplicate, cupulate, or tubular, margins not incurved, not appressed to gynostegium; appendage falcate, acicular, subulate, conical, lingulate, or crestlike, barely to well exserted.
102. Stems, leaf blades, and peduncles tomentose, at least initially.
103. Stems 6–15 cm; leaf blades orbiculate to obovate or oblate, 1.5–3.2 × 1.7–4 cm; peduncles 2.7–6 cm; corolla lobes glabrous abaxially; anthers 1–1.5 mm; corona segments conduplicate, appendage falcate; follicles smooth . . . . 24. *Asclepias nummularia*
103. Stems 20–40 cm; leaf blades linear-lanceolate, 5–15 × 0.3–1.3 cm; peduncles 0.2–1.5 cm; corolla lobes minutely pilosulous abaxially; anthers 2–2.5 mm; corona segments tubular, appendage lingulate; follicles ribbed . . . . 26. *Asclepias brachystephana* (in part)
102. Stems, leaf blades, peduncles, pedicels, and calyces puberulent with curved trichomes, pilose, strigulose, hirtellous, or glabrous.
104. Leaf blades linear, 0.1–0.8 cm wide.
105. Flowers pendent to spreading; corolla lobes pilose abaxially; corona segments conduplicate, sessile, apex truncate, appendage crestlike, barely exserted; fruiting pedicels upcurved . . . . 25. *Asclepias quinquedentata* (in part)
105. Flowers erect; corolla lobes glabrous abaxially; corona segments tubular or cupulate, stipitate, apex obtuse, appendage falcate or acicular, well exserted; fruiting pedicels straight.

106. Stems few–numerous, 20–60 cm; leaf blades spreading, not secund, midvein puberulent with curved trichomes; umbels all extra-axillary; corollas cream, sometimes tinged or striped pink; anthers brown; corona segments tubular, appendage acicular, arching above style apex; follicles glabrous; se Arizona . . . . 8. *Asclepias angustifolia* (in part)

106. Stems 1–4 (rarely more), 15–30 cm; leaf blades ascending, appearing secund, midvein glabrous; umbels terminal, sometimes also extra-axillary; corollas green to tan, tinged pink or red; anthers green; corona segments cupulate, appendage falcate, sharply inflexed over style apex; follicles minutely puberulent with curved trichomes; se United States . . . . 69. *Asclepias michauxii* (in part)

[104. Shifted to left margin.—Ed.]

104. Leaf blades narrowly elliptic, oval, oblong, lanceolate, ovate, oblanceolate, or obovate, 0.3–6 cm wide.

107. Corollas white, pink, or cream; corona segment appendages acicular or subulate.

108. Interpetiolar ridges absent; petioles 2–7 mm; leaf blades 1–6 cm wide; umbels terminal (sometimes also extra-axillary); flowers spreading to pendent; pedicels 17–28 mm; corona segments conduplicate, sessile, appendage subulate . . . . 73. *Asclepias quadrifolia* (in part)

108. Interpetiolar ridges present; petioles 5–20 mm; leaf blades 0.3–3 cm wide; umbels extra-axillary (may appear terminal); flowers erect to spreading; pedicels 7–14 mm; corona segments cupulate or tubular, stipitate, appendage acicular.

109. Leaf blades narrowly elliptic to oval or oblong, 5–14 cm; corona segments tubular, 1.5–2.5 mm; follicles pendulous on deflexed pedicels, lance-ovoid, 4–7 × 1–2.5 cm; seeds 12–15 × 11–14 mm, coma absent; bottomlands and ditches, e United States to e Texas . . . . . . .10. *Asclepias perennis* (in part)

109. Leaf blades ovate to lanceolate or elliptic, 2–7 cm; corona segments cupulate, 2–3 mm; follicles erect on straight pedicels, narrowly fusiform, 9–13 × 0.5–1 cm; seeds 6–7 × 4–5 mm, coma present; canyons and slopes; c, w Texas . . . . .11. *Asclepias texana*

107. Corollas green, sometimes tinged red, purple, or bronze, or red-violet; corona segment appendages falcate or conical.

110. Corollas red-violet; corona segments cupulate, 1.5–2 mm, appendage conical, barely exserted; fruiting pedicels spreading, follicles spreading to pendulous, lance-ovoid, 3.5–6 cm, strigose; Nevada . . . . 57. *Asclepias eastwoodiana* (in part)

110. Corollas green, tinged purple or bronze, rarely reddish purple; corona segments tubular or conduplicate, 2–6 mm, appendage falcate, well exserted; fruiting pedicels upcurved, follicles erect, fusiform, 5–10.5 cm, pilosulous or puberulent with curved trichomes; Florida or w Texas.

111. Stems 15–20 cm, pilosulous; leaf blades 6–8 cm, sparsely hirtellous, margins crisped; umbels terminal; peduncles 7–17 cm; pedicels 15–20 mm, pilose; anthers brown, 1.5–2 mm; corona segments tubular, 2–3 mm, apex truncate; w Texas . . . . 20. *Asclepias scaposa*

111. Stems 15–100 cm, puberulent with curved trichomes; leaf blades 1.8–5 cm, sparsely puberulent with curved trichomes on midvein only, margins entire; umbels terminal and extra-axillary; peduncles 0–4 cm; pedicels 10–14 mm, sparsely puberulent with curved trichomes; anthers green, 1–1.5 mm; corona segments conduplicate, 5–6 mm, apex attenuate; Florida. . . . . 34. *Asclepias curtissii*

[98. Shifted to left margin.—Ed.]

98. Corolla lobes 5–15 mm (rarely 4.5 mm in *A. incarnata*, with corolla and corona pink to white, corona segments 2–2.5 mm, and fruiting pedicels straight).
  112. Stems, leaf blades, peduncles, or pedicels persistently tomentose or tomentulose.
    113. Corona segments 7–15 mm.
      114. Petioles 2–6 mm; leaf blade surfaces adaxially pilosulous or tomentulose to glabrate, abaxially tomentose; corolla lobes deep maroon to green adaxially, glabrous adaxially; anthers 2–2.5 mm; coronas deep maroon to yellowish green, segments 7–9 mm, appendage absent or merely a low crest; follicles fusiform, 1.2–1.4 cm wide; extreme se Arizona and sw New Mexico. . . . . 27. *Asclepias hypoleuca*
      114. Petioles 4–12 mm; leaf blade surfaces uniformly green, tomentose to pilose; corolla lobes dark pink (rarely pale), hirtellous at base adaxially; anthers 2.5–3 mm; coronas pale pink to nearly cream, segments 9–15 mm, appendage subulate, sharply inflexed over style apex; follicles lance-ovoid, 2–3 cm wide; widespread in w United States and Canada, but not in se Arizona and sw New Mexico . . . . 76. *Asclepias speciosa*
    113. Corona segments 2–7 mm.
      115. Corolla lobes glabrous abaxially.
        116. Corollas pink, red green, or pinkish green, or reddish green; anthers brown; corona segments cream with red, pink, or purple dorsal stripe, apex acute, appendage glabrous; Rocky Mountains and associated alluvial plains, 1700–3000 m. . . . . 45. *Asclepias hallii* (in part)
        116. Corollas green to yellowish green, sometimes tinged red or purple; anthers green; corona segments green with cream apex, sometimes tinged pink or purple, or cream, greenish cream, or ochroleucous, not striped red, pink, or purple, apex truncate to rounded, appendage papillose; se United States, Great Plains, or Colorado Plateau, to 2300 m.
          117. Corona segments green with cream apex, sometimes tinged pink or purple; corona segment apices glabrous; follicles fusiform to narrowly lance-ovoid, 1–1.5 cm wide; se United States and e Texas . . . . 43. *Asclepias tomentosa* (in part)
          117. Corona segments cream to greenish cream or ochroleucous, apices papillose; follicles lance-ovoid to ovoid, 2–3 cm wide; Great Plains and Colorado Plateau to e Texas.
            118. Petioles 7–17 mm; leaf blade bases rounded, truncate, or subcordate; anthers 2–2.5 mm; corona segment apices truncate to rounded, emarginate, not oblique, appendage apex upturned; seeds 9–12 × 6–8 mm. . . 38. *Asclepias arenaria*
            118. Petioles 0–4 mm; leaf blade bases cordate; anthers 3–3.5 mm; corona segment apices truncate, not emarginate, oblique, appendage apex not upturned; seeds 7–8 × 5–6 mm . . . . 44. *Asclepias latifolia* (in part)
      115. Corolla lobes pilose, pilosulous, or tomentose abaxially, at least at tips.
        119. Leaf blades linear-lanceolate, 0.3–1.3 cm wide; corolla lobes 4–6 mm; corona segments 1.5–2 mm; follicles conspicuously striate . . . . 26. *Asclepias brachystephana* (in part)
        119. Leaf blades ovate or oval to lanceolate, elliptic, oblong, or obovate, 1.5–15 cm wide; corolla lobes 5–11 mm; corona segments 2.5–6 mm; follicles not striate.

[120. Shifted to left margin.—Ed.]

120. Corolla lobes pilose at base adaxially; gynostegial column 2–3.5 mm; corona segments red-violet, dorsally convex from base to apex, margins connivent, appendage absent . . . . 21. *Asclepias californica* (in part)

120. Corolla lobes glabrous, minutely hirtellous, or papillose at base adaxially; gynostegial column 0.2–1.5 mm; corona segments cream or ochroleucous to dark pink (sometimes reddish purple in *A. syriaca*), dorsally straight or concave from base to apex, margins separate, appendage falcate or acicular.

121. Leaf blades 3–9 × 1.5–4.5 cm, pilosulous; flowers 5–21 per umbel; s Canada (Alberta to Ontario), n United States (Montana and Wyoming e to Illinois and Michigan) . . . . 75. *Asclepias ovalifolia* (in part)

121. Leaf blades 6–30 × 2–15 cm, tomentose or puberulent with curved trichomes to glabrate (sometimes pilosulous in *A. syriaca*); flowers 12–113 per umbel; Arizona, California, Nevada, and Utah, or widespread in e United States and Canada, w to Manitoba, North Dakota, and South Dakota (*A. syriaca*).

122. Pedicels 10–15 cm; corona appendages acicular; fruiting pedicels spreading, follicles pendulous; extreme n Arizona and s Utah . . . . 53. *Asclepias welshii* (in part)

122. Pedicels 15–50 cm; corona appendages falcate; fruiting pedicels upcurved, follicles erect; w Arizona, California, s Nevada, and sw Utah, or e United States and Canada.

123. Pedicels pilose to puberulent with curved trichomes; corona segments slightly flattened dorsally, apex obtuse and usually flared, not oblique; follicles muricate (less commonly smooth); e United States and Canada . . . . 77. *Asclepias syriaca* (in part)

123. Pedicels tomentose to glabrate; corona segments rounded dorsally, apex truncate or obtuse and oblique, not flared; follicles smooth or longitudinally ridged; Arizona, California, Nevada, and Utah.

124. Peduncles 0–3.5 cm; corolla lobes tomentose abaxially throughout; anthers 1.5–2 mm; corona segments sessile, apex obtuse; follicles ovoid, 5–6.5 cm, ridged . . . . 22. *Asclepias vestita* (in part)

124. Peduncles 1–10 cm; corolla lobes tomentose abaxially only at tips; anthers 2.5–3 mm; corona segments stipitate, apex truncate; follicles lance-ovoid to ovoid, 5–10 cm, smooth.

125. Petioles 0–6 mm; leaf blade margins erose; corollas green; anthers green; coronas cream to ochroleucous, segment apices and appendages glabrous; follicles thinly tomentose; seeds 8–13 × 5–10 mm . . . . 35. *Asclepias erosa* (in part)

125. Petioles 4–15 mm; leaf blade margins entire or undulate; corollas greenish cream, ochroleucous, pink, tan, or pinkish cream; anthers dark brown; coronas cream to dark pink, segment apices and appendages papillose; follicles densely tomentose, seeds 7–9 × 4–6 mm . . . . 36. *Asclepias eriocarpa* (in part)

[112. Shifted to left margin.—Ed.]

112. Stems, leaf blades, peduncles, and pedicels variously indumented, sometimes ephemerally tomentose or tomentulose.

126. Stems, leaf blades, peduncles, and pedicels hirsute or hirtellous.

127. Stems 100–150 cm; leaf blades 3–14 cm wide; umbels terminal and extra-axillary; peduncles 6–13 cm; flowers 21–53 per umbel; corollas cream to greenish cream or ochroleucous, sometimes tinged pink; corona segments conduplicate; se Arizona . . . . 23. *Asclepias lemmonii*

127. Stems 6–70 cm (rarely to 200 cm in *A. obovata*); leaf blades 0.4–7.5 cm wide; umbels extra-axillary (sometimes appearing terminal); peduncles 0–2 cm; flowers 4–32 per umbel; corollas green, sometimes tinged red, purple, brown, or bronze; corona segments tubular; s United States, including se Arizona.

[128. Shifted to left margin.—Ed.]

128. Stems 40–70(–200) cm; flowers erect to pendent; anthers 2.5–4 mm; corona segments bronze or yellow, often tinged red, sometimes paler or cream at apex, appendage falcate, glabrous; seeds 8–9 × 6–7 mm; s Canada (Alberta to Ontario), n United States (Montana and Wyoming e to Illinois and Michigan) . . . . .64. *Asclepias obovata*

128. Stems 6–50 cm; flowers erect to ascending; anthers 1–2.5 mm; corona segments cream or green with cream or white at apex, sometimes tinged bronze, appendage lingulate, papillose or glabrous; seeds 6–8 × 4.5–6.5 mm; California to w Louisiana, n to c Oklahoma.

129. Leaf blades 0.4–2.5 cm wide, base cuneate; flowers 4–8 per umbel; corolla lobes 5–7 mm; anthers green, 1–1.5 mm; corona segments 3.5–5.5 mm, appendage flush with apex and obscuring cavity. . . . . 41. *Asclepias emoryi*

129. Leaf blades 1.2–7.5 cm wide, base cuneate to subcordate; flowers 5–32 per umbel; corolla lobes 7–13 mm; anthers brown, sometimes green at base, 1.7–2.5 mm; corona segments (5–)7–11 mm, appendage exserted below apex, exposing cavity.

130. Leaf blades ovate or lanceolate to oblong or elliptic, base cuneate to obtuse, apex obtuse to rounded, sometimes conduplicate; pedicels 10–20 mm; gynostegial column 1–1.5 mm; anthers 2–2.5 mm; corona segments green, sometimes tinged bronze, white or cream at apex, relatively slender, apex deeply emarginate; se Arizona, sw Colorado, w Louisiana, New Mexico, Oklahoma, Texas . . . . 40. *Asclepias oenotheroides* (in part)

130. Leaf blades ovate to lanceolate, base cuneate or obtuse to truncate or subcordate, apex obtuse to acute, rarely conduplicate; pedicels 17–30 mm; gynostegial column 0.3–0.5 mm; anthers 1.7–2 mm; corona segments cream or green with cream apex, relatively stout, apex truncate; Arizona, se California, s Nevada, sw New Mexico . . . . .42. *Asclepias nyctaginifolia* (in part)

[126. Shifted to left margin.—Ed.]

126. Stems, leaf blades, peduncles, and pedicels variously indumented, but trichomes not straight and spreading.

131. Corona appendages crestlike, included or barely exserted from segment, or absent.

132. Corolla lobes 10–13 mm; corona segments clavate, 8–10 mm, greatly exceeding style apex, segment apices sometimes connivent. . . . . 32. *Asclepias connivens* (in part)

132. Corolla lobes 4–10 mm; corona segments conduplicate or laminar with incurved margins, 1.5–4 mm, barely exceeding to greatly exceeded by style apex, segment apices separated by gynostegium.

133. Leaf blades linear to broadly oval or nearly orbiculate, 0.8–6 cm wide; flowers 22–60 per umbel; anthers 3–4 mm; corona segments laminar, appressed to gynostegium, margins incurved; follicles fusiform to lance-ovoid, 1.5–2 cm wide . . . . 46. *Asclepias viridiflora*

133. Leaf blades linear to narrowly elliptic or narrowly lanceolate, 0.1–0.8 cm wide; flowers 2–10 per umbel; anthers 1–2 mm; corona conduplicate, margins not incurved, not appressed to gynostegium; follicles fusiform to lance-ovoid, 0.3–1 cm wide.

134. Leaf blades linear, 6–14 cm; peduncles 1.1–2.7 cm; flowers pendent to spreading; pedicels 14–21 mm; corolla lobes reflexed, exposing corona, 4–6 mm, pilose abaxially; anthers tan to brown; corona segments red, pink, red-violet, or purple at base, white at apex, 3–4 mm; appendage crestlike, barely exserted; fruiting pedicels upcurved, follicles 0.5–1 cm wide, pilosulous to glabrate . . . . .25. *Asclepias quinquedentata* (in part)

134. Leaf blades linear to narrowly elliptic or narrowly lanceolate, 1.5–6 cm; peduncles 0.15–1 cm; flowers erect; pedicels 5–13 mm; corolla lobes erect, mostly concealing corona, 7–10 mm, glabrous abaxially; anthers green; corona segments yellowish green to green, 1.5–2.5 mm, appendage absent; fruiting pedicels straight, follicles 0.3–0.5 cm wide, puberulent with curved trichomes . . . .33. *Asclepias pedicellata* (in part)

131. Corona appendages falcate or acicular, well exserted.

[135. Shifted to left margin.—Ed.]

135. Corona segment apices truncate.
   136. Petioles 10–25 mm; corollas white with red-violet at base of lobes; anthers brown, 1.5–2 mm; seeds 5–7 × 3–5 mm . . . . 71. *Asclepias variegata*
   136. Petioles 0–15 mm; corollas green, sometimes tinged pink, red, or purple; anthers green, 2.5–4 mm; seeds 6.5–13 × 4–10 mm.
      137. Petioles 5–15 mm; leaf blades membranous, base cuneate; corona segments tubular, appendage arching above style apex; ne United States to n Alabama, n Georgia, w North Carolina, w South Carolina . . . . 74. *Asclepias exaltata* (in part)
      137. Petioles 0–9 mm; leaf blades chartaceous, coriaceous, or succulent, base obtuse to cordate; corona segments conduplicate, appendage sharply inflexed over or towards style apex; w United States or s Alabama, Florida, s Georgia, e North Carolina, e South Carolina, and e Texas.
         138. Leaf blade apices attenuate to acuminate, margins erose; umbels terminal and extra-axillary; peduncles 2–10 cm; pedicels 20–45 mm; corolla lobes tomentose abaxially at tips; anthers 2.5–3 mm; corona segment appendages glabrous; seeds 8–13 × 5–10 mm; n Arizona, s California, s Nevada, sw Utah. . . . . 35. *Asclepias erosa* (in part)
         138. Leaf blade apices acute to truncate, sometimes emarginate, margins entire or crisped; umbels only extra-axillary; peduncles 0–2.5 cm; pedicels 12–35 mm; corolla lobes glabrous abaxially; anthers 3–4 mm; corona segment appendages papillose; seeds 6.5–8 × 4.5–6 mm; se United States or c United States to n Arizona, s Utah.
            139. Petioles 2–9 mm; leaf blades 1–5 cm wide; peduncles 0–0.3 cm; pedicels 12–19 mm; corona segments green with cream apex, often tinged pink or purple; follicles fusiform to narrowly lance-ovoid, 9–18 × 1–1.5 cm; se United States and e Texas . . . . . 43. *Asclepias tomentosa* (in part)
            139. Petioles 0–4 mm; leaf blades 3–14 cm wide; peduncles 0–2.5 cm; pedicels 15–35 mm; corona segments cream, sometimes yellow dorsally; follicles ovoid, 6.5–9.5 × 2–3 cm; c United States . . . . . . 44. *Asclepias latifolia* (in part)

135. Corona segment apices acute to obtuse.
   140. Stems often branched, at least in inflorescence; interpetiolar ridges present; corona segment appendages arching above style apex; fruiting pedicels straight.
      141. Corollas red, sometimes yellow in throat (may be uniformly yellow or orange in cultivars), lobes 6–9 mm; gynostegial column 2–2.5 mm; corona segments yellow to orange, 3.5–4 mm, appendage falcate; follicles glabrous; seeds 6–7 × 4–5 mm; California, Florida, Louisiana, Texas. . . . . 7. *Asclepias curassavica*
      141. Corollas pink to white, lobes 5–6 mm; gynostegial column 1.2–1.5 mm; corona segments pink to white, 2–2.5 mm, appendage acicular; follicles sparsely puberulent with curved trichomes to pilose or pilosulous; seeds 8–9 × 5–6 mm; widespread in United States and Canada . . . . .9. *Asclepias incarnata*
   140. Stems unbranched (rarely branched in *A. syriaca*, branched only in the inflorescence in *A. hallii*, *A. purpurascens*); interpetiolar ridges absent; corona segment appendages sharply inflexed over style apex (sometimes ascending in *A. ovalifolia*); fruiting pedicels upcurved.
      142. Stems 30–70 cm; flowers 5–29 per umbel; gynostegial column 0.2–1 mm; follicles 0.7–2 cm wide; seeds 5–7 mm.
         143. Petioles 10–20 mm; corollas pale pink to red or green, tinged pink or red, lobes 7–8 mm, glabrous abaxially; anthers brown, 2–2.5 mm; corona segments cream with red, pink, or purple dorsal stripe, 5–6.5 mm; follicles 8–12 cm; Rocky Mountains and associated alluvial plains . . . . 45. *Asclepias hallii* (in part)
         143. Petioles 3–10 mm; corollas cream to yellowish, lobes 5–7 mm, pilosulous abaxially at apices; anthers green, 1.5–2 mm; corona segments cream to yellowish, 3–4 mm; follicles 5–8 cm; s Canada (Alberta to Ontario), n United States (Montana and Wyoming e to Illinois and Michigan) . . . . .75. *Asclepias ovalifolia* (in part)

[142. Shifted to left margin.—Ed.]

142. Stems 50–200 cm; flowers 17–113 per umbel; gynostegial column 1–1.5 mm; follicles 1.5–4 cm wide; seeds 7–8 mm.

144. Adaxial leaf blades sparsely pilosulous; umbels terminal and extra-axillary; flowers erect to spreading; corollas reddish purple, lobes glabrous abaxially; anthers brown; corona segments reddish purple, stipitate, conduplicate, apex acute, 5–6 mm; follicles 10–16 × 1.5–2.5 cm, smooth, puberulent with curved trichomes . . . . .72. *Asclepias purpurascens*

144. Adaxial leaf blades tomentose to glabrate; umbels only extra-axillary; flowers erect to pendent; corollas dark to pale pink or green, lobes pilose abaxially; anthers green; corona segments reddish purple to cream, sessile, tubular, apex obtuse, flared, 4–5 mm; follicles 7–12 × 2–4 cm, muricate (sometimes smooth), tomentose. . .77. *Asclepias syriaca* (in part)

1. Asclepias macrotis Torrey in W. H. Emory, Rep. U.S. Mex. Bound. 2(1): 164, plate 45, fig. B. 1859 • Longhood milkweed [F]

Subshrubs, densely cespitose. **Stems** few–numerous, erect to ascending, branched at base, 10–35 cm, minutely puberulent with curved trichomes in a line to glabrate, not glaucous, rhizomatous. **Leaves** opposite, sessile, with 1 stipular colleter on each side of leaf base; blade filiform, 2.5–7 × 0.05–0.15 cm, membranous, base cuneate, margins revolute, apex acute, mucronate, venation obscure, surfaces glabrous, margins minutely ciliate at base, laminar colleters absent. **Inflorescences** extra-axillary, sessile or pedunculate, 2–7-flowered; peduncle 0–0.8 cm, puberulent with curved trichomes, with 1 caducous bract at the base of each pedicel. **Pedicels** 6–10 mm, minutely puberulent with curved trichomes in a line. **Flowers** erect; calyx lobes lanceolate, 2–2.5 mm, apex acute, glabrous; corolla greenish cream tinged with red, lobes reflexed with spreading tips, oval, 4–5 mm, apex acute, glabrous; gynostegial column 0.5–1 mm; fused anthers brown, cylindric, 1–1.5 mm, wings right-triangular, closed, apical appendages broadly ovate; corona segments cream to greenish cream, reddish brown at base or as a dorsal stripe, subsessile, conduplicate, 4–5 mm, exceeding style apex, apex truncate, long-caudate, crisped, with a proximal tooth on each side, hirtellous, internal appendage falcate, exserted, inflexed towards style apex, hirtellous; style apex shallowly depressed, green. **Follicles** erect on upcurved pedicels, fusiform, 4.5–6 × 0.5–0.7 cm, apex attenuate, smooth, glabrous. **Seeds** ovate, 6–8 × 2.5–4 mm, margin thickly winged, faces rugulose; coma 2–2.5 cm.

Flowering May–Oct; fruiting Jun–Oct. Mesas, hills, slopes, flats, canyon rims and bottoms, arroyo margins, limestone, sandstone, shale, rhyolite, gypsum, caliche, cracks in bedrock, talus, gravel and sandy soils, oak, pine-oak, juniper, and pinyon-juniper woodlands, chaparral, shrubby grasslands, desert grasslands, prairies; 1100–2200 m; Ariz., Colo., N.Mex., Okla., Tex.; Mexico (Chihuahua, Coahuila, Sonora).

With a suffruticose, cespitose, intricately branched habit, the only other milkweed with which *Asclepias macrotis* can be confused is the rarely encountered *A. sperryi*, which is restricted to the Big Bend region of Texas in the flora area. Although *A. macrotis* ranges across western Texas, it appears to be absent from Big Bend; no mixed populations of these species are known, and hybridization is neither known nor suspected. Its distribution in Colorado is limited to the southeastern corner of the state and in Oklahoma to the extreme tip of the Panhandle, in Cimarron County. In Arizona, it is more common but limited to the three southeastern counties: Cochise, Pima, and Santa Cruz. The elongate, curled apices of the corona segments are unique, and the small, drab flowers are quite elegant.

2. Asclepias sperryi Woodson, Ann. Missouri Bot. Gard. 28: 246, fig. 2. 1941 • Sperry's milkweed

Subshrubs, densely cespitose. **Stems** few–numerous, erect to ascending, branched at base, 5–20 cm, minutely puberulent with curved trichomes to glabrate, not glaucous, rhizomatous. **Leaves** opposite, sessile, with 1 stipular colleter on each side of leaf base; blade filiform, 2–5 × 0.05–0.1 cm, membranous, base cuneate, margins revolute, apex acute, mucronate, venation obscure, surfaces glabrous, margins minutely ciliate at base, laminar colleters absent. **Inflorescences** extra-axillary, sessile, 1–2-flowered, with 1 caducous bract at the base of each pedicel. **Pedicels** 4–10 mm, minutely puberulent with curved trichomes in a line. **Flowers** erect; calyx lobes ovate, 2–3 mm, apex acute, glabrous; corolla pale greenish yellow tinged with purple, lobes spreading, ovate, 4–5 mm, apex acute, glabrous; gynostegium sessile; fused anthers brown, cylindric, 1.5–2 mm, wings deltoid, apical appendages broadly ovate; corona segments pale greenish yellow to white (sometimes tinted purple), sessile, chute-shaped, sharply inflexed, 2–3 mm, equaling to exceeded by style apex, apex truncate, subcaudate, hirtellous, internal appendage an included crest, hirtellous; style apex shallowly depressed, green. **Follicles**

erect on upcurved pedicels, fusiform, 5–7 × 0.5–0.7 cm, apex attenuate, smooth, glabrous. **Seeds** ovate, 5–6 × 3–4 mm, margin thickly winged, faces rugulose; coma 2–2.5 cm.

Flowering Apr–Sep; fruiting Aug–Sep. Slopes, ridges, canyons, limestone, rocky soils, desert scrub and desert grasslands; 1300–1800 m; Tex.; Mexico (Coahuila).

Asclepias sperryi is one of the least known milkweeds. It is exceedingly cryptic and easily mistaken for a small clump of grass, perhaps of *Bouteloua*; it is likely to be far more common than documented. However, very few observations or collections have been made in recent decades, and conservation assessment is warranted. Its range is restricted to limestone mountains of the Chihuahuan Desert in the Big Bend region of Texas, in Brewster and Pecos counties, and adjacent areas of Coahuila, entirely within the range of its sister species, *A. macrotis*. It is the only species of *Asclepias* that regularly produces a single flower per umbel.

3. **Asclepias linaria** Cavanilles, Icon. 1: 42, plate 57. 1791 • Pine-needle milkweed, hierba del cuervo

Shrubs, crown rounded. **Stems** few–numerous, erect, branched, especially distally, 30–70 cm, woody, bark brown to gray, twigs puberulent with curved trichomes, not glaucous, rhizomes absent. **Leaves** eventually caducous, alternate, spiral to irregular, sessile, with 1 stipular colleter on each side of leaf base; blade linear, needlelike, 1.5–4 × 0.1–0.15 cm, chartaceous, base cuneate, margins revolute, apex acute, mucronate, venation obscure, sparsely pilosulous to glabrate, laminar colleters absent. **Inflorescences** extra-axillary, sessile or pedunculate, 9–30-flowered; peduncle 0–2.5 cm, puberulent with curved trichomes to pilosulous, with 1 caducous bract at the base of each pedicel. **Pedicels** 10–14 mm, pilosulous. **Flowers** erect to pendent; calyx lobes lanceolate to ovate, 2–3 mm, apex acute, sparsely pilosulous to glabrate; corolla green to cream, often tinged red or purple, lobes reflexed with spreading tips, elliptic, 3.5–5 mm, apex acute, glabrous abaxially, minutely hirtellous at base adaxially, 1 margin ciliate; gynostegial column 0.2–0.5 mm; fused anthers brown, obconic, 1–1.5 mm, wings right-triangular, closed, apical appendages ovate, erose; corona segments cream, sometimes with greenish or purplish dorsal stripe, subsessile or sessile, cupulate, 2.5–3 mm, exceeding style apex, apex obtuse to rounded, glabrous, internal appendage rod-shaped, slightly exserted, glabrous; style apex shallowly depressed, green. **Follicles** erect on upcurved pedicels, ovoid, 3.5–5 × 0.6–1 cm, apex acuminate, smooth, glabrous. **Seeds** naviculate, ovate, 5–6 × 3–4 mm, margin very narrowly winged, faces rugulose, the concave one conspicuously so; coma 1.5–2 cm. $2n = 22$.

Flowering and fruiting year-round. Canyons, cliffs, arroyos, ridges, slopes, bedrock crevices, rhyolite, granite, gneiss, conglomerate, rocky, sandy, and gravel soils, pine-oak forests, oak, pinyon-juniper woodlands, chaparral, desert scrub, desert grasslands, riparian woodlands and forests; 800–1900 m; Ariz., N.Mex.; Mexico.

Asclepias linaria is arguably the most distinctive milkweed species in the Americas. It is the only species with woody stems and the only one to form hemispherical shrubs with needlelike leaves. Small plants with few stems are easily mistaken for seedling conifers. This species is widespread and occupies a great variety of habitats in Mexico. It enters the flora area in southeastern Arizona and in southwestern New Mexico only in the Peloncillo Mountains (Hidalgo County); its occurrence in the flora area is evidently relictual. In the region, it is restricted to lower reaches of protected canyons that ameliorate aridity and freezing temperatures, sites that harbor other tropical and subtropical species reaching their northern limits. The plants are often quite floriferous and attract an abundance of Hymenoptera, Lepidoptera, and Diptera.

4. **Asclepias cutleri** Woodson, Ann. Missouri Bot. Gard. 26: 263, fig. 2. 1939 • Cutler's milkweed [E]

Herbs. Stems 1–5, erect to ascending, unbranched, 7–20 cm, strigose to pilose, not glaucous, rhizomes absent. **Leaves** alternate, sessile, stipular colleters absent; blade linear to filiform, 2.5–8 × 0.1–0.2 cm, membranous, base cuneate, margins entire, apex acute, mucronate, venation obscure, surfaces strigose to glabrate, margins sparsely ciliate to glabrate, laminar colleters absent. **Inflorescences** extra-axillary at upper nodes, appearing terminal, sessile or pedunculate, 2–5-flowered (appearing greater because umbels are in close proximity); peduncle 0–0.1 cm, densely strigose, bracts few. **Pedicels** 6–15 mm, strigose to pilose. **Flowers** erect to spreading; calyx lobes lance-ovate, 1.5–2 mm, apex acute, strigose to pilose; corolla red-violet, lobes reflexed or sometimes spreading, oval, 2.5–4 mm, apex acute, pilose abaxially, glabrous adaxially; gynostegial column 0.5 mm; fused anthers brown, cylindric, 1.5–2 mm, wings right-triangular, apical appendages ovate; corona segments white, sessile, cupulate, 1.5 mm, exceeded by style apex, base saccate, apex truncate with a proximal tooth on each side, glabrous, internal appendage lingulate, barely exserted from cavity, glabrous; style apex shallowly depressed, pink to reddish. **Follicles** pendulous on spreading to declined pedicels, lance-ovoid, 3–6 × 0.5–0.8 cm,

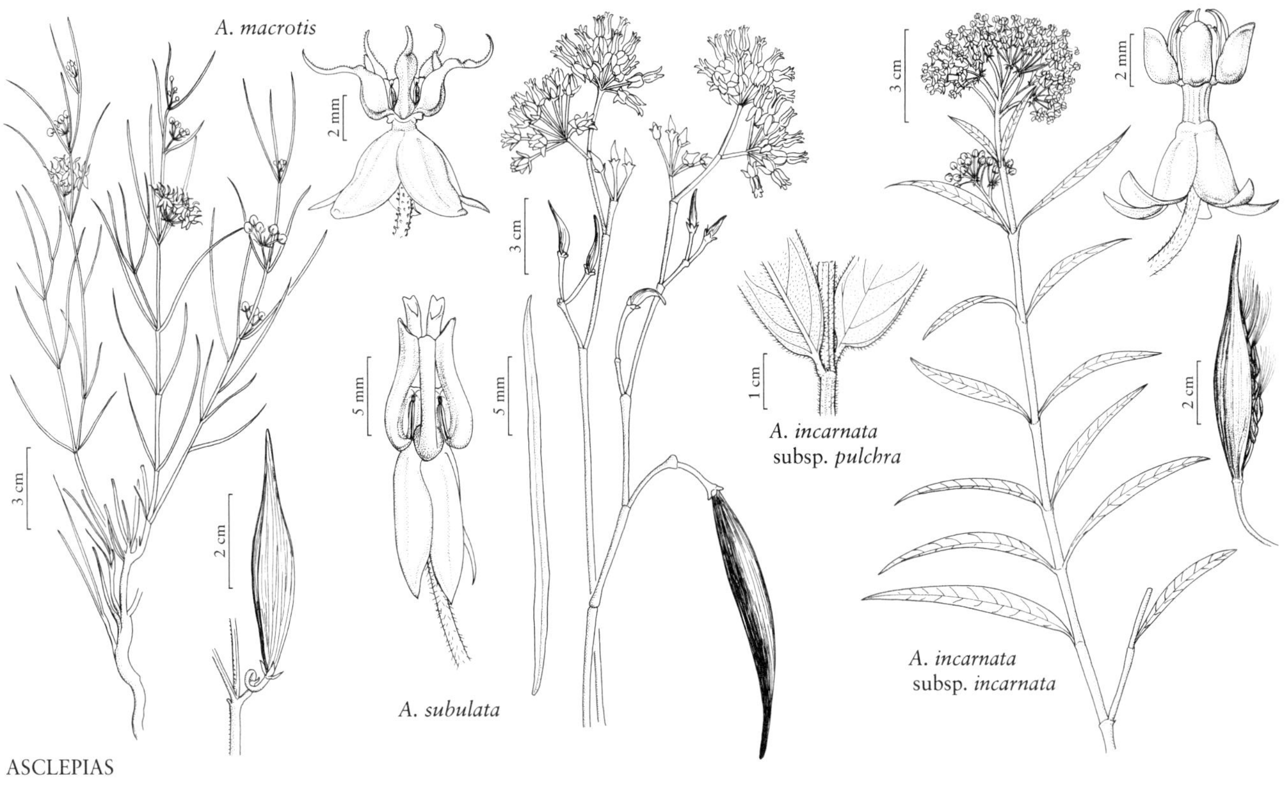

ASCLEPIAS

apex attenuate, smooth, faintly striate, strigose. **Seeds** oval, 9–11 × 4–6 mm, margin corky, winged, erose, faces ruglose-papillate, minutely hirtellous; coma 1.5 cm.

Flowering Apr–Jun; fruiting May–Jun. Sand dunes, sandy soils, grasslands, shrubby grasslands; 1400–1700 m; Ariz., Utah.

Asclepias cutleri is an edaphic endemic, limited to deep red, pink, and orange sand deposits developed on geologic units of sedimentary origin of the Colorado Plateau. Its range is restricted to Apache, Coconino, and Navajo counties in Arizona and Grand and San Juan counties in Utah. The majority of occurrences are on the Navajo reservation. A report from San Juan County, New Mexico, needs confirmation. Although rarely collected, it is highly cryptic due to its small stature and slender habit, and it is probably more common than it appears in its preferred habitat. The herbage has a bluish hue when fresh that turns green on drying, similar to *A. brachystephana* and *A. cryptoceras*. It is often erroneously described to be an annual because the very deep roots are almost never extricated, and the slender subterranean stem appears to be the root. Paired fruits from a single flower appear to be unusually common compared to other species of *Asclepias*. The flowers are remarkably similar to those of *A. brachystephana* and the so-called dwarf milkweeds, *A. eastwoodiana*, *A. ruthiae*, *A. uncialis*, and the sympatric *A. sanjuanensis*, but these species are only distantly related to *A. cutleri* (M. Fishbein et al. 2011).

5. Asclepias albicans S. Watson, Proc. Amer. Acad. Arts 24: 59. 1889 • White-stem or wax milkweed, candelilla, jumete

Shrubs. Stems 5–40 (usually 1–few), erect to ascending, branched, especially in lower half, 140–400 cm, sparsely pilose to glabrate, thickly glaucous, rhizomes absent. **Leaves** ephemeral, rarely present on flowering stems, opposite, sessile, with 0 or 1 stipular colleter on each side of leaf base; blade filiform, 1.5–2.5 × 0.1 cm, succulent, base cuneate, margins entire, apex acute, mucronate, venation obscure, surfaces pilosulous, laminar colleters absent. **Inflorescences** terminal, branched, also extra-axillary at leafless upper nodes, pedunculate, 8–50-flowered; peduncle 0.2–4.5 cm, pilose, with 1 caducous bract at the base of each pedicel. **Pedicels** 8–16 mm, pilose. **Flowers** erect to pendent; calyx lobes lanceolate to linear-lanceolate, 2.5–3 mm, apex acute, densely pilose; corolla ochroleucous to greenish cream, sometimes tinged red, faintly striate, lobes reflexed or sometimes spreading, oval, 4.5–6 mm, apex acute, glabrous; gynostegial column 1–1.8 mm; fused anthers brown, cylindric, 1.8–2 mm, wings right-triangular, closed, apical appendages ovate, erose; corona segments cream, often tinged pink, yellow, or green, shiny, sessile, conduplicate, 2–3 mm, exceeded by style apex, base saccate, apex truncate, oblique, glabrous, internal appendage falcate, exserted, sharply inflexed towards style apex, densely papillate; style apex shallowly depressed, ochroleucous to green. **Follicles** spreading to pendulous on spreading to pendulous pedicels, fusiform to lance-ovoid, 5.5–12 × 0.7–1.8 cm, apex acuminate, smooth, pilosulous. **Seeds** naviculate, lanceolate, 5–7 × 2.5–3 mm, margin narrowly winged, faces papillose and rugulose, concave face with a low keel; coma 1.5–2 cm.

Flowering and fruiting year-round. Mountain slopes, ridge tops, bajadas, flats, arroyos, granite, basalt, tuff, coarse rocky soils, cracks in boulders, sand, desert scrub; 50–800 m; Ariz., Calif.; Mexico (Baja California, Baja California Sur, Sonora).

Asclepias albicans reaches the greatest height of any *Asclepias* species in the flora area and among congeners can be confused only with *A. subulata*, the only other shrubby, leafless milkweed in the region. In the absence of flowers, it may be difficult to distinguish these species. In addition to the key characters, *A. albicans* often differs by thicker, waxier stems. The species are also ecologically divergent: *A. albicans* is most commonly found on rocky slopes and ridges (usually basalt or granite), and *A. subulata* is usually found on flats and in arroyos, often in sandy soils. However, *A. albicans* can occur on bajadas and in arroyos, often distant from the mountains and ridges harboring source populations. These species have extensively overlapping ranges but rarely hybridize at widely scattered locations in Arizona, California, and Baja California Sur. Hybrids are identified by intermediate floral morphology, especially corona size and shape. The range of *A. albicans* is within the limits of the Sonoran Desert. In Arizona, it is found in La Paz, Maricopa, Mohave, Pima, Pinal, and Yuma counties; in California only in Imperial, Riverside, San Bernardino, and San Diego counties.

6. Asclepias subulata Decaisne in A. P de Candolle and A. L. P. P. de Candolle, Prodr. 8: 571. 1844 • Rush or desert milkweed, jumete, candelilla [F]

Shrubs. **Stems** 3–100 (usually few–numerous), erect to ascending, branched, especially in lower half, 50–175 cm, sparsely pilosulous to glabrate, thickly glaucous, rhizomes absent. **Leaves** ephemeral, often present on flowering stems, opposite (rarely whorled), sessile, with 0 or 1 stipular colleter on each side of leaf base; blade filiform, 2–6 × 0.1 cm, succulent, base cuneate, margins entire, apex acute, mucronate, venation obscure, surfaces pilosulous, laminar colleters absent. **Inflorescences** extra-axillary, pedunculate, 3–21-flowered; peduncle 0.4–1.7 cm, sparsely pilose, glaucous, with 1 caducous bract at the base of each pedicel. **Pedicels** 11–21 mm, pilose. **Flowers** erect to spreading; calyx lobes lanceolate, 4–4.5 mm, apex acute, pilose; corolla green, sometimes tinged cream, yellow, or red, faintly striate, lobes reflexed, lanceolate, 7–12 mm, apex acute, glabrous; gynostegial column 1–1.5 mm; fused anthers brown, cylindric, 2–3 mm, wings right-triangular, distended at base, open at tip, apical appendages ovate; corona segments cream, often tinged pink, yellow, or green, shiny, sessile, tubular, 7–9 mm, greatly exceeding style apex, base saccate, apex truncate, spreading, with proximal flaps, glabrous, internal appendage crested, apically falcate and sharply inflexed towards style apex, barely exserted, minutely papillose; style apex shallowly depressed, cream to green. **Follicles** pendulous on spreading pedicels, fusiform, 6.5–13.5 × 1–1.8 cm, apex acuminate, smooth, sparsely pilosulous or puberulent with curved trichomes to glabrate, glaucous. **Seeds** flat to somewhat naviculate, ovate, 6–8 × 4–5 mm, margin narrowly winged, faces papillose-rugulose, concave face with a low keel; coma 1.5–3 cm. $2n$ = 22.

Flowering and fruiting year-round. Arroyos, dunes, hills, slopes, flats, depressions, bajadas, alluvial fans, basalt, granite, rhyolite, caliche, sandy, rocky, and clay soils, desert scrub; 0–1100 m; Ariz., Calif., Nev.; Mexico (Baja California, Baja California Sur, Sinaloa, Sonora).

Morphological and ecological distinctions between *Asclepias subulata* and its close relative, *A. albicans*, are discussed under the latter species. The range of *A. subulata* is more extensive than A. *albicans* and is almost exactly confluent with that of the Sonoran Desert, although it extends into the southeastern part of the Mohave Desert. In Nevada, it is restricted to Clark and Lincoln counties. The elongate, tubular, cream corona segments are remarkably similar to those of *A. nyctaginifolia*. These species were formerly considered close relatives (R. E. Woodson Jr. 1954), but they are highly dissimilar morphologically, other than the corona segments. They appear to be only distantly related (M. Fishbein et al. 2011, 2018), and the corona similarities represent a remarkable convergence. Both species are commonly visited by long-tongued tarantula hawk wasps (Pompilidae, Pepsinae), but it is not known whether they are important pollinators for these milkweeds.

7. **Asclepias curassavica** Linnaeus, Sp. Pl. 1: 215. 1753 • Blood flower, hierba de la cucaracha, scarlet or tropical milkweed I

Subshrubs or herbs. Stems 1–several, erect, sparsely to moderately branched, 30–150 cm, minutely pilosulous in a line to glabrate, not glaucous, rhizomes absent. **Leaves** persistent or gradually caducous from the base, opposite, petiolate, with 1 or 2 stipular colleters on each side of petiole on a ciliate interpetiolar ridge; petiole 4–25 mm, puberulent with curved trichomes in a line to glabrate; blade elliptic or oval to linear, 4–18 × 0.3–4.5 cm, membranous, base cuneate, margins entire, apex acute to acuminate to attenuate, venation eucamptodromous to faintly brochidodromous, surfaces sparsely puberulent with curved trichomes on veins abaxially, sparsely puberulent with curved trichomes on veins to glabrate adaxially, margins ciliate, laminar colleters absent. **Inflorescences** extra-axillary, pedunculate, 5–22-flowered; peduncle 0.5–8 cm, puberulent with curved trichomes in a line, with 1 caducous bract at the base of each pedicel. **Pedicels** 7–20 mm, puberulent with curved trichomes. **Flowers** erect; calyx lobes linear-lanceolate, 3–4 mm, apex acute, puberulent with curved trichomes; corolla red, sometimes yellow in throat (to wholly orange or yellow in cultivars), lobes reflexed with spreading tips, elliptic to oval, 6–9 mm, apex acute, glabrous abaxially, minutely papillose at base adaxially; gynostegial column 2–2.5 mm; fused anthers yellowish green to tan, cylindric, 1.5–2 mm, wings narrowly right-triangular, closed, apical appendages deltoid; corona segments yellow to orange, stipitate, tubular, dorsally somewhat flattened, 3.5–4 mm, exceeding style apex, apex obtuse to acute, glabrous, internal appendage falcate, exserted, arching over style apex, glabrous; style apex shallowly depressed, yellow. **Follicles** erect on straight pedicels, fusiform, 6–10 × 0.5–1.2 cm, apex acuminate to attenuate, smooth, glabrous. **Seeds** ovate, 6–7 × 4–5 mm, margin winged, faces minutely rugulose to smooth; coma 2.5–3 cm. $2n = 22$.

Flowering and fruiting year-round. Disturbed areas, fields, orchards, and gardens, canal banks, ditches, streamsides, wet prairies, marshes, swamps, coastal dunes, sandy soils; 0–100 m; introduced; Calif., Fla., La., Tex.; Mexico; West Indies; Central America; South America; introduced also to Old World tropics.

Asclepias curassavica is the only non-native *Asclepias* species naturalized in the flora area. It is very commonly cultivated, originally for its strikingly colored flowers and their attraction of Lepidoptera and Hymenoptera. Recently, they have been valued also as a host plant for monarch butterflies. Cultivars with pure orange or pure yellow flowers are readily available. The species develops rapidly from seed and can be grown as an annual (in the horticultural sense) anywhere in the region. Though often described as an annual, like all species of *Asclepias,* it has a perennial habit. It may persist through mild winters at least as far north as Oklahoma but has only become established in frost-free areas of the southern United States.

8. **Asclepias angustifolia** Schweigger, Enum. Pl. Hort. Regiom., 13. 1812 • Arizona or narrow-leaved milkweed

Asclepias linifolia Kunth

Subshrubs or herbs, cespitose. **Stems** few–numerous, erect, sparsely to moderately branched, 20–60 cm, puberulent with curved trichomes in a line, not glaucous, rhizomes absent. **Leaves** persistent or gradually caducous from the base, opposite, petiolate, with 1 stipular colleter on each side of petiole on a ciliate interpetiolar ridge; petiole 2–4 mm, surfaces puberulent with curved trichomes to glabrate; blade linear, 3.5–10 × 0.2–0.8 cm, chartaceous, base cuneate, margins entire, apex acute, venation faintly brochidodromous to obscure, puberulent with curved trichomes on midvein, margins ciliate, laminar colleters absent. **Inflorescences** extra-axillary at upper nodes, pedunculate, 3–15-flowered; peduncle 0.7–5 cm, puberulent with curved trichomes in a line, with 1 caducous bract at the base of each pedicel. **Pedicels** 8–18 mm, puberulent with curved trichomes in a line. **Flowers** erect; calyx lobes narrowly lanceolate, 2–3 mm, apex acute, puberulent with curved trichomes; corolla cream, sometimes pink-striped or -tinged, to pink, lobes reflexed with spreading tips, elliptic, 4–5 mm, apex acute, glabrous abaxially, minutely papillose at base adaxially; gynostegial column 0.8–1 mm; fused anthers brown, columnar, 1.5–2 mm, wings narrowly right-triangular, closed, apical appendages deltoid; corona segments cream, sometimes pink at base, stipitate, tubular, dorsally rounded, 3–3.5 mm, exceeding style apex, apex obtuse, glabrous, internal appendage acicular, exserted, arching over style apex, glabrous; style apex shallowly depressed, cream. **Follicles** erect on straight pedicels, narrowly fusiform, 4–10 × 0.4–0.7 cm, apex long-acuminate to attenuate, smooth, glabrous. **Seeds** ovate, 4.5–6 × 3–4 mm, margin narrowly winged, faces smooth; coma 2–3 cm.

Flowering Apr–Oct; fruiting Jul–Oct. Canyon bottoms, streamsides, arroyos, sandy or rocky soils, riparian forest, pine-oak-juniper forest; 1000–1800 m; Ariz.; Mexico.

The common name Arizona milkweed is misleading, as *Asclepias angustifolia* is narrowly distributed and uncommon in Arizona but widely distributed and locally common across mountainous regions of Mexico. Like *A. linaria*, this widespread species is probably relictual in its canyon habitat in southern Arizona. *Asclepias angustifolia* has been documented from the Atascosa, Dragoon, Huachuca, Pajarito, Rincon, Santa Catalina, and Santa Rita mountains (Cochise, Pima, and Santa Cruz counties). As with many milkweeds, the floral displays of well-developed plants are magnets for large numbers of Lepidoptera and Hymenoptera. Like *A. curassavica* and other members of the Incarnatae clade (species 7–16), *A. angustifolia* develops rapidly from seed and transplants well, making it a desirable species for the garden. Although its frost hardiness is limited, it can be used as a horticultural annual.

9. Asclepias incarnata Linnaeus, Sp. Pl. 1: 215. 1753
• Swamp milkweed, asclépiade incarnate [F]

Herbs. Stems 1–few, erect, unbranched to inflorescence, 30–150 cm, puberulent in a line with curved trichomes or densely pilose to glabrate, not glaucous, rhizomes absent. **Leaves** opposite, petiolate, with 1 or 2 stipular colleters on each side of petiole on a ciliate interpetiolar ridge; petiole 1–15 mm, pilosulous to pilose; blade lanceolate to linear-lanceolate or ovate, 5–15 × 0.5–4.5 cm, membranous, base obtuse to rounded or subcordate, margins entire, apex acute to acuminate or attenuate, venation eucamptodromous, surfaces sparsely puberulent with curved trichomes or pilose to glabrate, margins ciliate, 2–6 laminar colleters. **Inflorescences** extra-axillary at upper nodes, branched, pedunculate, 10–31-flowered; peduncle 1.5–7 cm, puberulent with curved trichomes to pilosulous, sometimes only on 1 side, to pilose, with 1 caducous bract at the base of each pedicel. **Pedicels** 10–15 mm, pilosulous to puberulent with curved trichomes, sometimes only on 1 side, to pilose. **Flowers** erect; calyx lobes lanceolate, 2–2.5 mm, apex acute, pilosulous to puberulent with curved trichomes; corolla pink to white, lobes reflexed with spreading tips, elliptic, (4.5–)5–6 mm, apex acute, glabrous abaxially, minutely papillose at base adaxially; gynostegial column 1.2–1.5 mm; fused anthers green to brown, columnar, 1.5–2 mm, wings narrowly right-triangular, slightly open at base, apical appendages deltoid; corona segments pink to white, often paler than corolla, stipitate, tubular, dorsally rounded to slightly flattened, 2–2.5 mm, ± equaling style apex, apex obtuse, glabrous, internal appendage acicular, exserted, arching over style apex, glabrous; style apex shallowly depressed, green, white, or pink. **Follicles** erect on straight pedicels, fusiform, 6–9 × 0.8–1.2 cm, apex long-acuminate, smooth to indistinctly ribbed, sparsely puberulent with curved trichomes to pilose or pilosulous. **Seeds** ovate, 8–9 × 5–6 mm, margin broadly winged, faces smooth; coma 1.5–2 cm.

Subspecies 2 (2 in the flora): North America, n Mexico.

Asclepias incarnata can be grown in a great variety of soil types and is surprisingly drought tolerant considering its natural predilection for hydric and mesic soils. As an easily grown, attractive, versatile species, it is one of the best options for gardening with milkweeds. It consists of two morphologically and geographically distinct, but intergrading subspecies.

1. Stems and leaf blades glabrate to sparsely puberulent with curved trichomes; petioles 7–15 mm; leaf blades lanceolate to linear-lanceolate; usually interior sites . . . . 9a. *Asclepias incarnata* subsp. *incarnata*
1. Stems and leaf blades densely pilose; petioles 1–8 mm, leaf blades ovate to lanceolate; usually coastal or on piedmont . . . . 9b. *Asclepias incarnata* subsp. *pulchra*

9a. Asclepias incarnata Linnaeus subsp. **incarnata**
• Swamp milkweed [F]

Stems 70–150 cm, puberulent in a line with curved trichomes to glabrate. **Leaves:** petiole 7–15 mm, pilosulous; blade lanceolate to linear-lanceolate, 5–15 × 0.5–4 cm, surfaces sparsely puberulent with curved trichomes, more densely on veins, to glabrate. **Peduncles** puberulent with curved trichomes to pilosulous, sometimes only on 1 side. **Pedicels** pilosulous to puberulent with curved trichomes, sometimes only on 1 side. **Flowers:** corolla bright pink to white; corona pink to white, paler than corolla; anthers green to brown. $2n = 22$.

Flowering (Apr–)Jun–Sep; fruiting Jul–Nov. Marshes, swamps, creeks, ditches, wet depressions, streamsides, pond edges, sandhills, dolomite, limestone, saturated and sandy soils, riparian woods, mixed hardwood forests, hammocks, wet meadows and prairies, pastures; 0–1700 m; Man., N.B., Ont., Que.; Ala., Ariz., Ark., Colo., Conn., D.C., Fla., Ga., Idaho, Ill., Ind., Iowa, Kans., Ky., La., Maine, Md., Mass., Mich., Minn., Mo., Mont., Nebr., Nev., N.H., N.J., N.Mex., N.Y., N.C., N.Dak., Ohio, Okla., Pa., S.C., S.Dak., Tenn., Tex., Utah, Vt., Va., W.Va., Wis., Wyo.; Mexico (Coahuila).

Subspecies *incarnata* is most common west of the Appalachian Mountains and is the only one commonly found in Florida. It has several, disjunct occurrences in wetlands in the Rocky Mountain and Intermountain states and notably along the Snake River in Idaho. It would be interesting to determine whether these occurrences represent relictual populations, recent dispersals (introductions?), or a combination of these. Subspecies *incarnata* is considered rare in Arizona (Santa Cruz County), Louisiana (Jefferson, LaFourche, St. Charles, Terrebonne, and Vernon parishes), and Montana (Carbon County). Genotypes of this subspecies are the source of several widely available cultivars, including one with pure white flowers.

9b. Asclepias incarnata Linnaeus subsp. **pulchra** (Ehrhart ex Willdenow) Woodson, Ann. Missouri Bot. Gard. 41: 53. 1954 • Swamp milkweed [E] [F]

Asclepias pulchra Ehrhart ex Willdenow, Sp. Pl. 1: 1267. 1798; *A. incarnata* var. *neoscotica* Fernald; *A. incarnata* [unranked] *pulchra* (Ehrhart ex Willdenow) Persoon

Stems 30–150 cm, densely pilose. **Leaves:** petiole 1–8 mm, pilose; blade lanceolate to ovate, 5–15 × 1.5–4.5 cm, surfaces pilose. **Peduncles** densely pilose. **Pedicels** pilose. **Flowers:** corolla dark pink to pink; corona dark to pale pink; anthers brown.

Flowering Jun–Sep; fruiting Jul–Oct. Pond and lake shores, marshes, bogs, saturated sandy and rocky soils, wet meadows, flatwoods, riparian woods, thickets; 0–900 m; N.B., N.S., P.E.I.; Conn., Del., D.C., Fla., Ga., Maine, Md., Mass., N.H., N.J., N.Y., N.C., Pa., R.I., S.C., Tex., Vt., Va.

Subspecies *pulchra* is most common on the coastal plain east of the Appalachian Mountains. There are sporadic occurrences outside this region, and the subspecies has been collected at several sites in eastern Texas. A record from western Illinois (Henry County) may represent a mistaken locality or a waif—there is no evidence that subsp. *pulchra* persists there. Only a single record is known from Prince Edward Island (*Blaney 3274* [ACAD]). The name *A. incarnata* var. *pulchra* is often seen in the literature, but this combination attributed to Persoon was unranked by him (K. Gandhi, pers. comm.).

10. Asclepias perennis Walter, Fl. Carol., 107. 1788 • Aquatic or white swamp or swamp or thin-leaf milkweed [E]

Subshrubs or herbs, cespitose. **Stems** 1–5, erect, sparsely to moderately branched, especially towards base, 30–60 cm, puberulent with curved trichomes in a line to glabrate, not glaucous, rhizomatous. **Leaves** persistent or gradually caducous from the base, opposite, petiolate, with 1 or 2 stipular colleters on each side of petiole on a ciliate interpetiolar ridge; petiole 10–12 mm, ciliate; blade narrowly elliptic to oval or oblong, 5–14 × 0.3–3 cm, membranous or chartaceous, base cuneate, margins entire, apex acute to attenuate or acuminate, minutely mucronate, venation faintly brochidodromous to eucamptodromous, surfaces sparsely puberulent with curved trichomes, more densely on veins, to glabrate, margins inconspicuously ciliate, laminar colleters absent. **Inflorescences** extra-axillary at upper nodes, sometimes appearing terminal, pedunculate, 12–30-flowered; peduncle 1.5–5 cm, puberulent with curved trichomes, sometimes only on 1 side, with 1 caducous bract at the base of each pedicel. **Pedicels** 7–13 mm, puberulent with curved trichomes, sometimes only on 1 side. **Flowers** erect to spreading; calyx lobes narrowly lanceolate to linear, 1.2–1.5 mm, apex acute, puberulent with curved trichomes; corolla white to pink-tinged, lobes reflexed with spreading tips, elliptic, 3–4 mm, apex acute to obtuse, glabrous abaxially, minutely papillose at base adaxially; gynostegial column 0.8–1.2 mm; fused anthers brown, cylindric, 1.5–2 mm, wings narrowly right-triangular, closed, apical appendages deltoid; corona segments white, sometimes faintly pink-tinged, stipitate, tubular, dorsally rounded, 1.5–2.5 mm, slightly exceeding style apex, apex obtuse to acute, glabrous, internal appendage acicular, exserted, arching over style apex, glabrous; style apex shallowly depressed, white, sometimes pink-tinged. **Follicles** pendulous on declined pedicels, lance-ovoid, 4–7 × 1–2.5 cm, apex long-acuminate, smooth, glabrous. **Seeds** broadly oval, 12–15 × 11–14 mm, margin broadly winged, faces smooth; coma absent.

Flowering (Mar–)Apr–Nov; fruiting Jun–Dec. Swamps, streamsides, ditches, bottomlands, flood plains, marshes, saturated or inundated clay, silty, and sandy soils, pine-oak, oak, and mixed hardwood forests, riparian woods, pine flatwoods; 0–300(–500?) m; Ala., Ark., Fla., Ga., Ill., Ind., Ky., La., Miss., Mo., S.C., Tenn., Tex.

Asclepias perennis is the most hydrophytic North American milkweed and is often found emerging from standing water in swamps and ditches. The pendulous fruits and hairless seeds are distinctive; it is the only milkweed in the United States with seeds lacking a coma, and only one of three such species in North America. It is widely distributed along the coastal plain; inland (Arkansas, Illinois, Indiana, Kentucky, Missouri, Tennessee), it is restricted to the valleys of the Mississippi and Ohio rivers and their tributaries.

11. Asclepias texana A. Heller, Contr. Herb. Franklin Marshall Coll. 1: 77, plate 4. 1895 • Texas milkweed

Asclepias perennis Walter var. *parvula* A. Gray

Subshrubs or herbs. Stems 1–numerous, erect, sparsely to moderately branched in upper half, 25–90 cm, puberulent with curved trichomes in a line, not glaucous, rhizomes absent. **Leaves** opposite, petiolate, with 1 or 2 stipular colleters on each side of petiole on a ciliate interpetiolar ridge; petiole 5–20 mm, puberulent with curved trichomes; blade ovate to lanceolate or elliptic, 2–7 × 0.5–3 cm, chartaceous, base cuneate to obtuse, margins entire, apex acute to acuminate, venation eucamptodromous, surfaces puberulent with curved trichomes on veins abaxially, sparsely puberulent with curved trichomes on veins to glabrate adaxially, margins ciliate, laminar colleters absent. **Inflorescences** extra-axillary at upper nodes, some appearing terminal, pedunculate, 14–31-flowered; peduncle 0.9–3 cm, puberulent with curved trichomes in a line, with 1 caducous bract at the base of each pedicel. **Pedicels** 7–14 mm, puberulent with curved trichomes in a line. **Flowers** erect; calyx lobes narrowly lanceolate, 1.5–2 mm, apex acute, puberulent with curved trichomes; corolla white, sometimes tinged green, lobes reflexed with spreading tips, elliptic, 3.5–5 mm, apex acute to obtuse, glabrous abaxially, minutely papillose at base adaxially; gynostegial column 1–1.5 mm; fused anthers brown, columnar, 1.5–2 mm, wings narrowly right-triangular, closed, apical appendages deltoid; corona segments white, sometimes tinged pink, stipitate, cupulate, dorsally rounded, 2–3 mm, equaling to slightly exceeding style apex, apex obtuse, glabrous, internal appendage acicular, exserted, arching over style apex, glabrous; style apex shallowly depressed, white. **Follicles** erect on straight pedicels, narrowly fusiform, 9–13 × 0.5–1 cm, apex long-acuminate to attenuate, smooth, glabrous. **Seeds** ovate, 6–7 × 4–5 mm, margin winged, faces smooth or sparsely papillose; coma 2–3 cm.

Flowering May–Sep; fruiting (Jul–)Aug–Oct. Canyons, arroyos, slopes, cliff bases, bluffs, streamsides, limestone, igneous rocks, rocky and clay soils, riparian, oak-juniper, and oak woods, pine-oak forest; 300–2000 m; Tex.; Mexico (Coahuila, Nuevo León).

In the flora area, *Asclepias texana* has a disjunct distribution on the Edwards Plateau and in the mountains of the Big Bend region. Although commonly occurring in canyons and riparian areas, *A. texana* is quite drought tolerant in cultivation. The tidy, bushy habit, long flowering stems topped by bright white spherical umbels, and rapid growth from seed make this a suitable candidate for horticultural use. It is known to be hardy to at least USDA Zone 7.

12. Asclepias linearis Scheele, Linnaea 21: 758. 1849 • Slim milkweed

Herbs. Stems 1–4, erect, often branched, arrested vegetative branches absent, 30–75 cm, puberulent with curved trichomes in lines, not glaucous, rhizomatous. **Leaves** opposite, sessile, with 1 or 2 stipular colleters on each side of leaf base on a ciliate interpetiolar ridge; blade linear, 3–8 × 0.1–0.3 cm, chartaceous, base cuneate, margins entire, apex acute, mucronate, venation obscure, surfaces glabrous abaxially, puberulent with curved trichomes adaxially, especially on midvein, to glabrate, margins ciliate, laminar colleters absent. **Inflorescences** extra-axillary at upper nodes, pedunculate, 7–20-flowered; peduncle 1–1.5 cm, puberulent with curved trichomes on 1 side, with 1 caducous bract at the base of each pedicel. **Pedicels** 6–8 mm, puberulent with curved trichomes on 1 side. **Flowers** erect to pendent; calyx lobes linear-lanceolate, 1.5 mm, apex acute, pilosulous; corolla green, tan, or pink, tinged red or brown, lobes reflexed with spreading tips, elliptic, 3–4 mm, apex acute, glabrous abaxially, minutely papillose at base adaxially; gynostegial column 1–1.2 mm; fused anthers green, columnar, 1.2–1.5 mm, wings narrowly right-triangular, closed, apical appendages deltoid; corona segments cream, stipitate, cupulate, dorsally flattened, 1.5 mm, exceeded by style apex, apex rounded, margin with a proximal tooth, glabrous, internal appendage acicular, exserted, arching over style apex, glabrous; style apex shallowly depressed, cream. **Follicles** erect on straight pedicels, narrowly fusiform, 5–8.5 × 0.4–0.7 cm, apex acuminate to attenuate, smooth, sparsely pilosulous to glabrate. **Seeds** ovate, 6–7 × 4–5 mm, margin winged, faces smooth; coma 2–3 cm.

Flowering (Mar–)Apr–Oct; fruiting Jul–Oct. Plains, ditches, valleys, marshes, alluvium, sandy and clay soils, prairies, savannas, mesquite grasslands, pastures; 0–200 m; Tex.; Mexico (Tamaulipas).

Although the range of *Asclepias linearis* is quite restricted, it is very common in grassland habitats on the coastal plain of southeastern Texas. Disjunct populations occur in the Dallas-Fort Worth area. It is similar in most respects to *A. verticillata* with the normally whorled leaves of that species the most reliable character distinguishing it from *A. linearis*. Although the ranges of these species overlap in the vicinity of Houston and Dallas-Fort Worth, and mixed populations have been reported, putative hybrids have not been documented, although they would be difficult to discern.

13. Asclepias fascicularis Decaisne in A. P. de Candolle and A. L. P. P. de Candolle, Prodr. 8: 569. 1844 • Narrow-leaved or Mexican or Mexican whorled milkweed

Herbs. Stems few–numerous, erect, sparsely to moderately branched, 50–150 cm, glabrous, not glaucous, rhizomes absent. **Leaves** 3–5-whorled, sessile or petiolate, with 1–3 stipular colleters on each side of petiole on a ciliate interpetiolar ridge; petiole 0–4 mm, margins puberulent with curved trichomes; blade linear to linear-lanceolate or narrowly elliptic, often somewhat conduplicate, 4.5–13 × 0.2–1.8 cm, membranous, base cuneate, margins entire, apex acute or attenuate to obtuse, mucronate, venation obscure to faintly eucamptodromous, surfaces glabrous, margins eciliate, laminar colleters absent. **Inflorescences** terminal and extra-axillary at upper nodes, sometimes branched, pedunculate, 10–37-flowered; peduncle 0.4–5.5 cm, puberulent with curved trichomes in a line to glabrate, with 1 caducous bract at the base of each pedicel. **Pedicels** 9–14 mm, puberulent with curved trichomes to pilosulous. **Flowers** erect; calyx lobes lanceolate, 1.5–2 mm, apex acute, puberulent with curved trichomes to pilosulous; corolla pale to dark pink, rarely pale green with a pink tinge, lobes reflexed with spreading tips, oval, 3–5 mm, apex acute, glabrous abaxially, minutely papillose at base adaxially; gynostegial column 1–1.5 mm; fused anthers green, cylindric, 1.5–2 mm, wings narrowly right-triangular, closed, apical appendages deltoid; corona segments cream, often tinged or striped pink, stipitate, cupulate, dorsally somewhat flattened, 1.5–2 mm, exceeded by style apex, apex obtuse, glabrous, internal appendage acicular, exserted, arching towards style apex, glabrous; style apex shallowly depressed, cream. **Follicles** erect on straight pedicels, fusiform, 6–9 × 0.5–1 cm, apex attenuate, smooth, glabrous. **Seeds** ovate, 6–7 × 4–5 mm, margin winged, faces minutely rugulose; coma 2.5–3 cm. 2*n* = 22.

Flowering Apr–Oct; fruiting (Jun–)Jul–Nov. Valleys, slopes, hills, streamsides, ditches, seeps, hot springs, wet depressions, arroyos, vernal pools, basalt, granite, limestone, clay, sandy, and silty soils, native, non-native, and shrubby grasslands, oak, pine-oak, juniper, pinyon-juniper, and riparian woodlands, chaparral, coastal sage scrub, pine and mixed-conifer forests, sometimes following fires; 0–2300 m; Calif., Idaho, Nev., Oreg., Wash.; Mexico (Baja California).

Asclepias mexicana Decaisne was misapplied to *A. fascicularis* in the past; this is the legitimate name of a related species endemic to southern and eastern Mexico. The common name Mexican (whorled) milkweed stems from this past confusion. Compared to its close relatives *A. angustifolia*, *A. linearis*, *A. pumila*, *A. subverticillata*, and *A. verticillata*, the leaves of *A. fascicularis* are not particularly narrow (despite the implication of another common name). However, very narrow leaves are found in *A. fascicularis* when it is growing at relatively dry sites, especially at the eastern limit of its range in southeastern Idaho. Such specimens (for example, *Mumford 272* [MO], *Atwood 28495* [NY]) have been attributed in the past to *A. subverticillata* in error. *Asclepias fascicularis* is completely allopatric with its Incarnatae clade relatives (species 7–16). Like these species, it is easily cultivated, and its seeds are widely available. In Washington, the range of *A. fascicularis* is restricted largely to the valleys of the Columbia and Spokane rivers and in Idaho to the Snake and Weiser rivers.

14. Asclepias subverticillata (A. Gray) Vail, Bull. Torrey Bot. Club 25: 178. 1898 • Western whorled or horsetail or poison or whorled milkweed

Asclepias verticillata Linnaeus var. *subverticillata* A. Gray, Proc. Amer. Acad. Arts 12: 71. 1876

Herbs. Stems 1–8, erect, sometimes branched, few to many arrested vegetative branches usually present, 20–90 cm, puberulent with curved trichomes in a line to glabrate, not glaucous, rhizomatous. **Leaves** 3–4-whorled, sometimes opposite on vegetative branches, sessile, with 1 stipular colleter on each side of leaf base on a ciliate interpetiolar ridge; blade linear, 3–13 × 0.1–0.4 cm, chartaceous, base cuneate, margins entire, apex acute, mucronate, venation obscure, surfaces glabrous, margins ciliate or eciliate, laminar colleters absent. **Inflorescences** extra-axillary, pedunculate, 9–25-flowered; peduncle 0.7–3.5 cm, puberulent with curved trichomes on 1 side, with 1 caducous bract at the base of each pedicel. **Pedicels** 5–12 mm, puberulent with curved trichomes. **Flowers** erect; calyx lobes lanceolate, 2–2.5 mm, apex acute, puberulent with curved trichomes to glabrate; corolla pale green to cream, sometimes pink- or tan-tinged, lobes reflexed with spreading tips, elliptic, 3.5–4.5 mm, apex acute, glabrous abaxially, minutely papillose at base adaxially; gynostegial column 0.8–1.2 mm; fused anthers green, columnar, 1.5–2 mm, wings narrowly right-triangular, closed, apical appendages deltoid; corona segments cream, sometimes green- or pink-tinged or striped, stipitate, cupulate, dorsally flattened, 1.5–2 mm, exceeded by style apex, margins entire, apex obtuse, glabrous, internal appendage acicular, exserted, arching over style apex, glabrous; style apex shallowly depressed, green to greenish cream. **Follicles** erect on straight pedicels, narrowly fusiform, 6–8.5 × 0.5–0.9 cm, apex acuminate to attenuate, smooth, minutely puberulent with curved trichomes to glabrate. **Seeds** ovate, 5–8 × 3.5–5 mm, margin winged, faces smooth; coma 2–2.5 cm.

Flowering (Apr–)May–Oct; fruiting (Jun–)Jul–Dec. Hills, ridges, mesas, slopes, flats, depressions, ciénegas, wet meadows, pastures, canyons, streamsides, arroyos, pond and lake margins, playas, bajadas, limestone, igneous rocks, sandstone, gypsum, clay, sandy, silty, and gravel soils, prairies, desert scrub, mesquite, juniper, and desert grasslands, pine savannas, chaparral, oak, pine-oak, pinyon-juniper, and riparian woodlands, pine and mixed-conifer forests; 800–2700 m; Ariz., Colo., Kans., Mo., N.Mex., Okla., Tex., Utah, Wyo.; Mexico (Chihuahua, Coahuila, Durango, Guanajuato, Nuevo León, San Luis Potosí, Sonora, Zacatecas).

Asclepias subverticillata and *A. verticillata* are amply distinct away from the zone of contact from Texas to Montana. The most reliable characters for distinguishing these species are the absence of a marginal corona segment tooth and presence of arrested vegetative branches in *A. subverticillata*. The greater frequency of multistemmed plants and completely glabrous leaves are also characteristic of *A. subverticillata*. However, absence of vegetative branches in *A. subverticillata* is common, especially in young or poorly developed plants. It can be difficult to confidently identify incomplete specimens or immature plants in the narrow zone of parapatry. Gene flow between the species has not been investigated, and the muddy species boundaries could be attributable to past or ongoing introgressive hybridization. The identity of some populations in New Mexico, the Texas Panhandle, and western Kansas have been debated, and further study is needed to determine whether they belong to one of the parental species or are advanced generation hybrids. *Asclepias subverticillata* barely enters Kansas (Grant, Hamilton, Morton, Seward, Stanton, and Stevens counties) and Oklahoma (Beaver and Cimarron counties). It also appears to be rare at the northeastern end of its range in Wyoming (Carbon County), where it is thought to be extirpated. There are few occurrences of *A. subverticillata* disjunct from the main range and within the range of *A. verticillata*. There is a single, adventive population along a railroad in St. Louis, Missouri, documented in 1962 and last observed in 1970 (V. Mühlenbach 1979); it is unknown whether this population persists. *Asclepias subverticillata* hybridizes with *A. pumila*. These hybrids are usually readily detected because the parental species are distinct in leaf arrangement and internode length (whorled and distant nodes in *A. subverticillata* versus alternate and congested in *A. pumila*). Such hybrids often have mixed phyllotaxy and have been documented in northern New Mexico. Reports of *A. subverticillata* from Idaho are based on misidentifications of *A. fascicularis* and are discussed under that species. Searches for *A. subverticillata* in southeastern Idaho have documented only *A. fascicularis* in that region (Lynn Kinter, Idaho Game and Fish, pers. comm.). Like other southwestern milkweed species with cream flowers (for example, *A. nyctaginifolia*, *A. subulata*), tarantula hawk wasps (Pompilidae, Pepsinae) are avid floral visitors to *A. subverticillata*, in spite of tiny flowers presenting minute quantities of nectar.

15. Asclepias verticillata Linnaeus, Sp. Pl. 1: 217. 1753 • Whorled or eastern whorled or horsetail milkweed, asclépiade verticillée [E]

Herbs. Stems 1 (rarely 2 or 3), erect, sparingly branched, arrested vegetative branches absent, 35–90 cm, sparsely puberulent with curved trichomes in lines, not glaucous, rhizomatous. **Leaves** 3–6-whorled, sessile, with 1 stipular colleter on each side of leaf base on a ciliate interpetiolar ridge; blade linear, 1.5–7 × 0.1–0.2 cm, chartaceous, base cuneate, margins entire, apex acute, mucronate, venation obscure, surfaces glabrous abaxially, puberulent with curved trichomes adaxially, especially on midvein, to glabrate, margins ciliate, laminar colleters absent. **Inflorescences** extra-axillary at upper nodes, pedunculate, 7–28-flowered; peduncle 0.8–4 cm, puberulent with curved trichomes on 1 side, with 1 caducous bract at the base of each pedicel. **Pedicels** 6–10 mm, puberulent with curved trichomes. **Flowers** erect to spreading; calyx lobes lanceolate, 1.5–2 mm, apex acute, puberulent with curved trichomes to glabrate; corolla pale green to cream or ochroleucous, sometimes tan-tinged, lobes reflexed with spreading tips, elliptic, 3–4 mm, apex acute, glabrous abaxially, minutely papillose at base adaxially; gynostegial column 1–1.2 mm; fused anthers green, columnar, 1.2–1.5 mm, wings narrowly right-triangular, closed, apical appendages deltoid; corona segments cream, stipitate, cupulate, dorsally flattened, 1.5–2 mm, exceeded by style apex, apex obtuse, margin shallowly lobed (sometimes obscure) to sharply toothed proximally, glabrous, internal appendage acicular, exserted, arching over style apex, glabrous; style apex shallowly depressed, green to greenish cream. **Follicles** erect on straight pedicels, narrowly fusiform, 6–11 × 0.4–0.8 cm, apex acuminate to attenuate, smooth, glabrous. **Seeds** ovate, 6–7 × 4–5 mm, margin winged, faces smooth; coma 2.5–3.5 cm. **2*n*** = 22.

Flowering Feb–Oct; fruiting Mar–Nov(–Dec). Ridges, slopes, flats, glades, bluffs, dunes, sandhills, streamsides, wet meadows and depressions, lake shores, sandstone, limestone, granite, serpentine, dolomite, shale, sandy, clay, and rocky soils, prairies, pine flatwoods and barrens, pine and oak scrubs, oak and oak-hickory woodlands, pine, pine-oak and pine-mixed-hardwood forests, forest edges; 0–1000 m; Man., Ont., Sask.; Ala., Ark., Conn., Del., D.C., Fla., Ga., Ill., Ind., Iowa, Kans., Ky., La., Md., Mass., Mich., Minn., Miss., Mo., Mont., Nebr., N.J., N.Y., N.C., N.Dak., Ohio, Okla., Pa., R.I., S.C., S.Dak., Tenn., Tex., Vt., Va., W.Va., Wis.

Asclepias verticillata is parapatric with the closely related and morphologically similar *A. linearis*, *A. pumila*, and *A. subverticillata*. It can be difficult to distinguish from these relatives where their ranges overlap. Similarities with *A. linearis* and *A. subverticillata* are discussed under those species; no definitive hybrids between *A. verticillata* and these species have been documented. The characteristic marginal corona segment tooth is often reduced to a shallow lobe in western populations of *A. verticillata*, which complicates distinguishing this species from *A. subverticillata*, and which suggests past introgression. A widely disjunct collection of *A. verticillata* was made in Arizona, well within the range of *A. subverticillata*, for Plants of the Hopis (*Millspaugh 176* [F]); persistence of the species in Arizona has not been documented by additional collections. Hybrids with *A. pumila* are usually readily detected because the parental species are distinct in leaf arrangement and internode length (whorled and distant nodes in *A. verticillata* versus alternate and congested in *A. pumila*). These hybrids often have mixed phyllotaxy and have been documented in Kansas, Montana, South Dakota, and Texas. *Asclepias verticillata* is strongly rhizomatous and forms dense colonies on roadsides and in prairies; in forests, however, genets are small and solitary stems are common. Like several other milkweed species, it is rare and declining at the northeastern terminus of its range (in Delaware, Massachusetts, New Jersey, Pennsylvania, Rhode Island, and Vermont). It is also considered rare in Saskatchewan, where it has been documented by few specimens. Reports from Wyoming (Crook County) all seem to pertain to *A. pumila* (B. Heidel, pers. comm.).

16. Asclepias pumila (A. Gray) Vail, Bull. Torrey Bot. Club 25: 175. 1898 • Plains or low or dwarf milkweed [E]

Asclepias verticillata Linnaeus var. *pumila* A. Gray, Proc. Amer. Acad. Arts 12: 71. 1876

Herbs. **Stems** 1–12, erect, unbranched to moderately branched below, 10–30 cm, puberulent with curved trichomes, not glaucous, rhizomatous. **Leaves** alternate, sessile, with 1 stipular colleter on each side of leaf base; blade needlelike to narrowly linear, 2.5–5.5 × 0.05–0.1 cm, chartaceous, base cuneate, margins entire, apex acute, mucronate, venation obscure, surfaces glabrous, margins ciliate, laminar colleters absent. **Inflorescences** extra-axillary at upper nodes, sometimes appearing terminal, pedunculate, 3–13-flowered; peduncle 0.1–2 cm, puberulent with curved trichomes, with 1 caducous bract at the base of each pedicel. **Pedicels** 6–16 mm, puberulent with curved trichomes. **Flowers** erect; calyx lobes narrowly lanceolate, 2–3 mm, apex acute, puberulent with curved trichomes to glabrate; corolla pink to cream with a pink tinge, lobes reflexed with spreading tips, elliptic, 3.5–4.5 mm, apex acute, glabrous abaxially, minutely papillose at base adaxially; gynostegial column 0.8–1 mm; fused anthers green, columnar, 1–1.5 mm, wings narrowly right-triangular, closed, apical appendages deltoid; corona segments cream, sometimes tinged or striped pink, stipitate, tubular, dorsally flattened, 2–2.5 mm, exceeded by to equaling style apex, apex obtuse, glabrous, internal appendage acicular, exserted, arching towards style apex, glabrous; style apex shallowly depressed, cream. **Follicles** erect on straight pedicels, narrowly fusiform, 2.5–9.5 × 0.5–1 cm, apex acuminate to attenuate, smooth, minutely puberulent with curved trichomes to glabrate. **Seeds** ovate, 5–6 × 3–5 mm, margin winged, faces smooth; coma 2–3 cm.

Flowering (May–)Jun–Sep; fruiting Aug–Oct. Slopes, bluffs, plains, sandhills, dunes, rock outcrops, playas, ditches, basalt, alluvium, sandy, clay, rocky, and gravel soils, prairies, meadows, shrubby grasslands, pine woodlands; 600–2300 m; Colo., Kans., Mont., Nebr., N.Mex., N.Dak., Okla., S.Dak., Tex., Wyo.

Asclepias linaria and *A. pumila* of southeastern Arizona and extreme southwestern New Mexico are the only milkweeds with densely spiraled, linear to needlelike leaves. Hybrids with *A. subverticillata* and *A. verticillata* have been documented; see discussion under these species. When not flowering, *A. pumila* is cryptic and easily overlooked in its grassland habitat, especially when growing among *Bouteloua* species. Specimens supposedly from Illinois and Arizona (*C. Mohr* [US]) undoubtedly have incorrect label data.

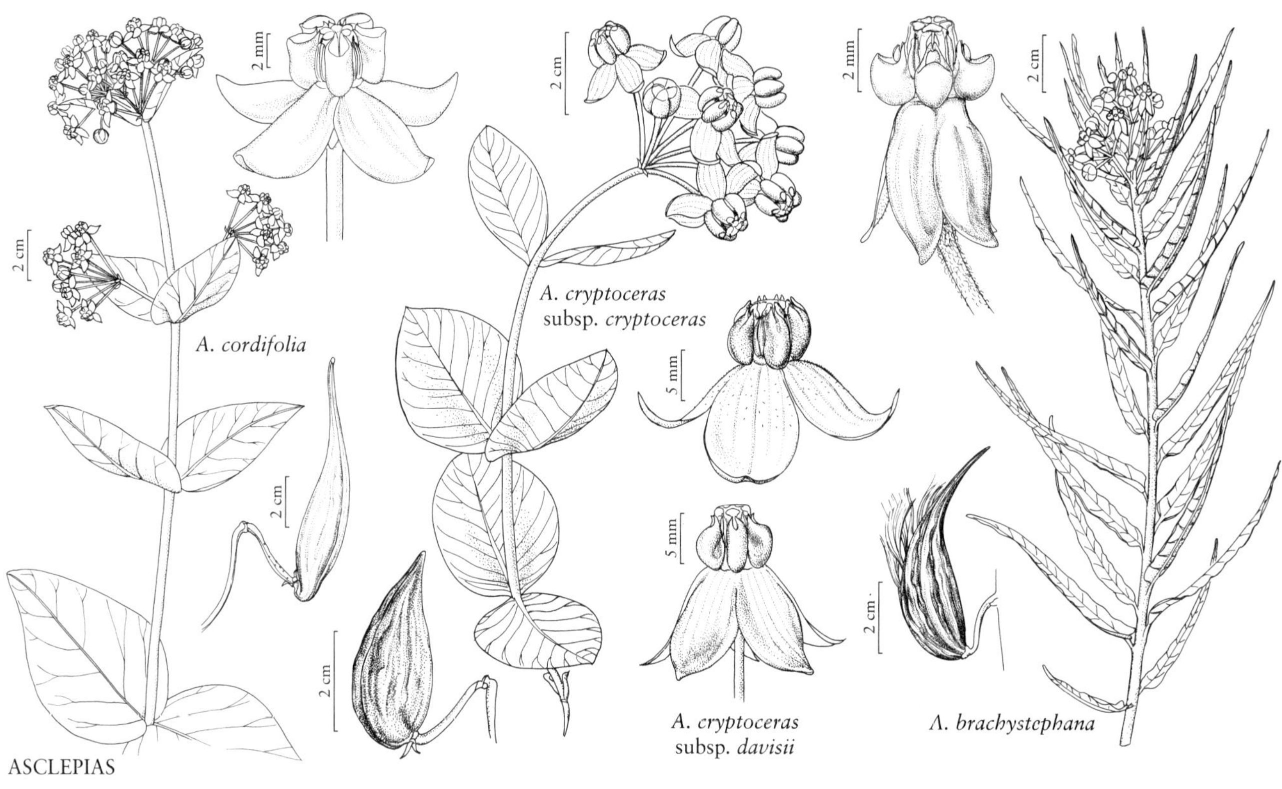

A population in the Sacramento Mountains of southern New Mexico (Otero County) appears to be a significant southern disjunction.

17. **Asclepias cordifolia** (Bentham) Jepson, Fl. W. Calif., 384. 1901 • Purple or heart-leaf milkweed [E] [F]

Acerates cordifolia Bentham, Pl. Hartw., 323. 1849; *Gomphocarpus cordifolius* (Bentham) Bentham ex A. Gray

Herbs. **Stems** 1–20, ascending to spreading, unbranched, 25–100 cm, glabrous, glaucous, rhizomes absent. **Leaves** opposite, sessile, stipular colleters absent; blade ovate to lanceolate, 3.5–11 × 1.8–8 cm, chartaceous, base cordate, clasping, margins entire, apex acute, mucronate, venation brochidodromous to eucamptodromous, surfaces glabrous, glaucous, margins minutely ciliate, laminar colleters absent. **Inflorescences** terminal, branched, sometimes also extra-axillary at upper nodes, pedunculate, 5–20-flowered; peduncle 0.3–6.5 cm, apically sparsely pilose to glabrate, with few bracts. **Pedicels** 16–37 mm, sparsely pilose. **Flowers** erect to pendent; calyx lobes elliptic to lanceolate, 3–5 mm, apex acute, pilose; corolla red-violet, lobes reflexed with spreading tips, oval, 6–7 mm, apex acute, glabrous; gynostegial column 1–1.5 mm; fused anthers brown, cylindric, 1.5–2 mm, wings right-triangular, open at tip, apical appendages deltoid; corona segments white, tinged pink to red-violet at base, sessile, tubular, 2–3 mm, exceeded by style apex, base saccate, apex truncate, oblique, glabrous, internal appendage absent; style apex shallowly depressed, pink to red-violet. **Follicles** erect on upcurved pedicels, lance-ovoid, 7.5–10.5 × 1.5–1.8 cm, apex long-acuminate, smooth, glabrous, glaucous. **Seeds** ovate, 8–9 × 5–7 mm, margin not winged, faces reticulate-rugulose; coma 3–3.5 cm. $2n = 22$.

Flowering Mar–Jul(–Aug); fruiting Jun–Sep. Hillsides, canyons, ridge tops, streamsides, seeps, basalt, serpentine, gabbro, granite, shale, limestone, talus slopes, gravel, alluvium, oak woodlands, mixed evergreen, douglas-fir, pine, pine-oak, and riparian forests, chaparral, timberline meadows, grasslands; 50–2200 (–2800) m; Calif., Nev., Oreg.

Fresh leaves and stems of *Asclepias cordifolia* are often slightly or strongly colored bluish, grayish, or purplish. This is one of the few American species of *Asclepias* with cavitate corona segments that lack adaxial appendages. Such species were segregated along with diverse African species in *Gomphocarpus* R. Brown, a polyphyletic segregate (M. Fishbein et al. 2011; D. Chuba et al. 2017). *Asclepias cordifolia* is a distinctive species unlike any other within its range. It is phylogenetically and geographically isolated, although not highly derived morphologically. In Nevada, *A. cordifolia* is restricted to the Sierra Nevada, in Carson City, Douglas, and Washoe counties. Its range extends to northern California and southwestern Oregon.

18. **Asclepias cryptoceras** S. Watson, Botany (Fortieth Parallel), 283, plate 28, figs. 1–4. 1871 • Pallid or jewel milkweed [E] [F]

Herbs. Stems 1–16, prostrate to decumbent, unbranched, 8–25 cm, glabrous, glaucous, rhizomes absent. **Leaves** opposite, sessile or petiolate, stipular colleters absent; petiole 0–8 mm, glabrous, sometimes with a few long hairs at the base; blade broadly ovate or oval to orbiculate or obovate, 3–7 × 1.8–6 cm, subsucculent, base cuneate or obtuse to cordate or truncate, margins entire, apex rounded to obtuse or acute, mucronate, venation eucamptodromous to faintly brochidodromous, surfaces glabrous, sometimes sparsely strigose on midvein abaxially, glaucous, margins ciliate, laminar colleters absent. **Inflorescences** terminal, and usually extra-axillary at uppermost node, sessile, 3–10-flowered, with 1 caducous bract at the base of each pedicel. **Pedicels** 16–35 mm, glabrous. **Flowers** ascending to pendent, calyx lobes linear-lanceolate, 5–8 mm, apex acute, sparsely strigose to pilosulous; corolla pale green to yellowish green, tinged red abaxially, lobes reflexed with spreading to ascending tips, oval, 8–14 mm, apex acute, glabrous abaxially, papillose at base and minutely hirtellous at tips adaxially; gynostegium sessile; fused anthers dark brown, sometimes green at apex, broadly cylindric, 1.8–3 mm, wings right-triangular with rounded apex, closed, apical appendages deltoid; corona segments red-violet to pinkish purple, sessile, conduplicate, dorsally rounded, 5–8 mm, slightly exceeded by to exceeding style apex, base subsaccate, margins connivent, apex truncate, oblique, with recurved, papillose teeth, papillose, internal appendage absent; style apex slightly depressed, pale green. **Follicles** sometimes paired, erect on upcurved pedicels, ovoid, 4.5–6 × 1.5–1.8 cm, apex apiculate, smooth, sometimes obscurely ribbed, glabrous, glaucous. **Seeds** ovate, 8–9 × 6–7 mm, margin very narrowly winged, faces rugulose; coma 2–2.5 cm.

Subspecies 2 (2 in flora): w United States.

Asclepias cryptoceras is one of the most striking milkweeds, with oversized flowers for its small stature. It is widely distributed in the western United States, but it is common nowhere, and occurrences are sporadic. The populations of this species fall into two morphologically discrete entities—those in Arizona, Colorado, southeastern Nevada, Utah, and Wyoming, and those in Idaho, Oregon, and Washington. Plants with intermediate flowers (corona shape and size) are found in eastern California and much of Nevada. The intermediate populations were shown by K. Weitemier (2016) to be genetically similar to subsp. *davisii* and are treated as such here. Bumblebees (*Bombus* spp.) have been observed to be avid visitors to the flowers of *A. cryptoceras*.

1. Pedicels (15–)25–35 mm; corolla lobes 11–14 mm; corona segments 6–8 mm, apices (including recurved teeth) exceeding style apices . . . . 18a. *Asclepias cryptoceras* subsp. *cryptoceras*
1. Pedicels 16–25 mm; corolla lobes 8–11 mm; corona segments 5–6 mm (–7 mm in e California and Nevada), apices (including recurved teeth) exceeded by or equaling style apices . . . . 18b. *Asclepias cryptoceras* subsp. *davisii*

18a. **Asclepias cryptoceras** S. Watson subsp. **cryptoceras** • Pallid or jewel milkweed, cow cabbage [E] [F]

Leaves: petiole 0–5 mm; blade broadly ovate or oval to orbiculate, 3–7 × 2–6 cm, apex rounded to obtuse, venation eucamptodromous to faintly brochidodromous, surfaces sparsely strigose on midvein abaxially. **Inflorescences** 4–10-flowered. **Pedicels** (15–)25–35 mm. **Flowers:** calyx lobes 7–8 mm; corolla lobes 11–14 mm; fused anthers dark brown, 2–3 mm; corona segments red-violet, 6–8 mm, apices (including recurved tooth) exceeding style apices.

Flowering Apr–Jul; fruiting May–Jul. Ridge tops, mesas, slopes, flats, arroyos, canyons, plains, alluvium, sandstone, shale, gypsum, limestone, granite, ash, sand, silt, clay, or calcareous soils, juniper, pinyon-juniper, and oak woodlands, shrubby grasslands; 1300–2700 m; Ariz., Colo., Nev., Utah, Wyo.

Subspecies *cryptoceras* co-occurs with and, in vegetative morphology, is remarkably similar to *Astragalus asclepiadoides* (Fabaceae), *Cycladenia humilis* (Apocynaceae), and *Mirabilis multiflora* (Nyctaginaceae). One cannot help but speculate that they form a mysterious mimicry complex. Subspecies *cryptoceras* is rare in Arizona (Coconino, Mohave, and Navajo counties) and Nevada (Lincoln and Nye counties).

18b. Asclepias cryptoceras S. Watson subsp. **davisii** (Woodson) Woodson, Ann. Missouri Bot. Gard. 41: 180. 1954 • Humboldt Mountains or pallid or Davis's or jewel milkweed [E] [F]

Asclepias davisii Woodson, Ann. Missouri Bot. Gard. 26: 261, fig. 1. 1939; *A. cryptoceras* var. *davisii* (Woodson) W. H. Baker

Leaves: petiole 2–8 mm; blade oval to ovate or obovate or orbiculate, 3–7 × 1.8–4.8 cm, apex rounded to acute, venation eucamptodromous, surfaces glabrous. **Inflorescences** 3–8-flowered. **Pedicels** 16–25 mm. **Flowers:** calyx lobes 5–6 mm; corolla lobes 8–11 mm; fused anthers dark brown, green at apex, 1.8–2.5 mm; corona segments pinkish purple, 5–6(–7) mm, apices (including recurved tooth) exceeded by or equaling style apices.

Flowering Apr–Jul; fruiting May–Jul. Slopes, hills, arroyos, basalt, silicic tuff, limestone, chert, serpentine, gravel, sand and clay soils, juniper woodlands, shrubby grasslands, steppe; 300–1500(–2100) m; Calif., Idaho, Nev., Oreg., Wash.

Subspecies *davisii* just enters Washington in Asotin County and is exceedingly rare in the state, where it is considered to be of conservation concern.

19. Asclepias humistrata Walter, Fl. Carol., 105. 1788 • Sandhill or pinewoods or pink-veined milkweed [E]

Herbs. Stems 1–10, decumbent or ascending to erect, unbranched, 20–50 cm, glabrous, glaucous, rhizomes absent. **Leaves** opposite, sessile, with 0 or 1 stipular colleter on each side of leaf base; blade ovate, 4–11 × 2.5–8 cm, subsucculent, base cordate, clasping, margins entire, apex obtuse, mucronulate, venation eucamptodromous to brochidodromous, surfaces glabrous, glaucous, laminar colleters absent. **Inflorescences** terminal, sometimes branched, and extra-axillary at upper nodes, pedunculate, 7–37-flowered; peduncle 2.6–5.5 cm, glabrous, glaucous, with 1 bract at the base of each pedicel. **Pedicels** 18–26 mm, glabrous. **Flowers** erect to spreading; calyx lobes lanceolate, 2–3 mm, apex obtuse, glabrous; corolla pink or red to pinkish green or reddish green, lobes reflexed, sometimes with spreading tips, oval, 5.5–6 mm, apex acute to obtuse, glabrous; gynostegial column 0.5–1 mm; fused anthers brown, cylindric, 1.2–1.5 mm, wings right-triangular, tips closed, apical appendages broadly ovate; corona segments pink to nearly white at base, white at apex, yellowing with age, sessile, conduplicate, dorsally rounded, 3–3.5 mm, slightly exceeding style apex, base subsaccate, apex truncate with a proximal tooth on each side, glabrous, internal appendage ensiform, slightly incurved, slightly exserted, glabrous; style apex shallowly depressed, pink. **Follicles** erect on upcurved pedicels, lance-ovoid, 8–12 × 0.8–1.7 cm, apex long-acuminate to attenuate, smooth, glabrous, glaucous. **Seeds** ovate to oval, 8–8.5 × 5–6 mm, margin winged, faces papillose-rugulose; coma 3–3.5 cm.

Flowering (Feb–)Mar–Oct; fruiting (Mar–)Apr–Oct (–Nov). Dunes, sandhills, ridges, slopes, coastal strand, streamsides, sandy soils, pine flatwoods, pine-oak woods, oak and pine-oak scrub; 0–200 m; Ala., Fla., Ga., La., Miss., N.C., S.C.

Asclepias humistrata is a distinctive milkweed unlike any other in its range. Its decumbent habit with vertically oriented leaves, bearing strongly contrasting white or pink venation, is unmatched by any other sandhill species. It is apparently closely related to the highly disjunct *A. cordifolia* of the Pacific Northwest, suggesting an unusual biogeographic history (M. Fishbein et al. 2011). It shares with this species bluish, grayish, or purplish glaucous herbage. *Asclepias humistrata* often exhibits remarkably high fruit set and, perhaps as a consequence, often grows in large, dense populations. It is reported as possibly extirpated from Louisiana, where it was documented from Washington Parish. *Asclepias amplexicaulis* Michaux, an illegitimate synonym, created confusion between this species and *A. amplexicaulis* Smith, a similarly glaucous, cordate-leaved species.

20. Asclepias scaposa Vail, Bull. Torrey Bot. Club 25: 171. 1898 • Bear Mountain milkweed

Herbs. Stems 1–5+, erect, unbranched (rarely at base), 15–20 cm, pilosulous, not glaucous, rhizomes absent. **Leaves** opposite, sessile or petiolate, with 1 stipular colleter on each side of petiole; petiole 0–6 mm, pilosulous to glabrate; blade oval to elliptic, 6–8 × 1.5–2.5 cm, membranous, base cuneate, margins crisped, apex obtuse to acute, venation eucamptodromous, surfaces sparsely hirtellous, margins ciliate, laminar colleters absent. **Inflorescences** terminal, solitary, pedunculate, 15–30-flowered; peduncle 7–17 cm, pilose, bracts absent or few. **Pedicels** 15–20 mm, pilose. **Flowers** erect to pendent; calyx lobes narrowly lanceolate, 2–2.5 mm, apex acute, pilose; corolla green with purplish tinge (reddish purple), lobes reflexed, elliptic, 4–5 mm, apex acute, glabrous; gynostegial column 0.5–1 mm; fused anthers brown, cylindric, 1.5–2 mm, wings curved, wider at base, apical appendages ovate; corona segments cream, sometimes yellow- or red-tinged or yellow or red at base, sessile, tubular, 2–3 mm, exceeding style apex, base saccate, apex truncate, dentate, glabrous, internal appendage falcate, exserted, arching above style apex, glabrous; style apex shallowly depressed, green or cream. **Follicles** erect on upcurved pedicels, fusiform, 5–6 × 1–1.5 cm, apex long-acuminate, smooth, pilosulous. **Seeds** not seen.

Flowering Mar–Aug; fruiting May. Ridges, slopes, limestone, rocky, silty, and clay soils, pine-oak woodlands, desert scrub, thorn scrub; 600–2000 m; Tex.; Mexico (Coahuila, Nuevo León, San Luis Potosí, Zacatecas).

The long-pedunculate, terminal inflorescence combined with short stature is distinctive in *Asclepias scaposa*. Although the locality of one of the syntypes was attributed to New Mexico by E. L. Greene, that is the only report for that state. Both syntypes were collected by Charles Wright for the United States-Mexico boundary survey, but neither of his labels indicates that they were collected in New Mexico. It is very likely that both collections were made in Texas or northeastern Mexico, and New Mexico is excluded from the distribution here. The common name Bear Mountain milkweed may refer to a ridge in the northwestern portion of the Davis Mountains, although no collections are known from this area. The few collections and observations that have been made in Texas are from scattered locations in Brewster, Crockett, Pecos, Presidio, Reeves, and Terrell counties, and conservation status in the United States merits assessment. In Mexico, *A. scaposa* also has been rarely collected, except for a local area in Nuevo León (Municipio de Galeana).

21. Asclepias californica Greene, Erythea 1: 92. 1893 • California or round-hood milkweed [E]

Acerates tomentosa Torrey in W. H. Emory, Rep. U.S. Mex. Bound. 2(1): 160, plate 44 [as Asclepias tomentosa]. 1859, not *Asclepias tomentosa* Elliott 1817

Herbs. Stems 1–20+, decumbent to erect, rarely branched, 15–90 cm, densely tomentose, not glaucous, rhizomes absent. **Leaves** opposite, sessile or petiolate, stipular colleters absent; petiole 0–17 mm, densely tomentose; blade ovate to lanceolate or oval, 5–18 × 2.5–10.5 cm, chartaceous, base cordate to truncate, margins often minutely erose, apex acuminate, venation eucamptodromous to faintly brochidodromous, surfaces densely tomentose to glabrate, margins ciliate, laminar colleters absent. **Inflorescences** terminal, sometimes branched, and extra-axillary at upper nodes, sessile or pedunculate, 5–21-flowered; peduncle 0–2.5 cm, densely tomentose, with 1 caducous bract at the base of each pedicel. **Pedicels** 15–40 mm, densely tomentose. **Flowers** spreading to pendent; calyx lobes linear to narrowly lanceolate, 4–6 mm, apex acute, densely tomentose; corolla green or tan (sometimes tinged pink) to pinkish purple, red-violet at base, lobes reflexed with spreading tips, oval, 8–11 mm, apex acute, densely tomentose abaxially, pilose at base adaxially; gynostegial column 2–3.5 mm; fused anthers dark brown, columnar, 2–2.5 mm, wings right-triangular, closed, apical appendages deltoid; corona segments red-violet, sometimes pale at apex, sessile, conduplicate, dorsally rounded, 3–6 mm, exceeded by style apex, base slightly to strongly saccate, margins connivent, apex rounded to truncate, slightly to strongly oblique, papillose, internal appendage absent; style apex planar, green. **Follicles** sometimes paired, erect on upcurved pedicels, ovoid, 5–12.5 × 2–3 cm, apex apiculate to acuminate, longitudinally ridged, densely tomentose. **Seeds** broadly oval to orbiculate, 9–12 × 8–11 mm, margin very narrowly winged, faces rugulose; coma 1.5–2.5 cm.

Subspecies 2 (2 in the flora): California.

Asclepias californica is one of the showiest milkweed species in the flora, with red-violet flowers set off by the dense, white, wooly vestiture of the rest of the plant. It is available from California nurseries but can be difficult to maintain in cultivation. An old report of the species from Baja California Sur cannot be confirmed and likely stems from a misidentification or erroneous location. Although the coronas are notoriously variable within the recognized subspecies, the key characters reliably distinguish northern and southern population systems. However, intermediates can be found in the contact zone, in Kern County.

Gomphocarpus tomentosus (Torrey) A. Gray (not Burchell 1822) is an illegitimate name found in some older regional floras that pertains here.

1. Corona segments 4–6 mm, apex usually rounded, opening extending from apex to base more than halfway, often more than three-quarters . . . . 21a. *Asclepias californica* subsp. *californica*
1. Corona segments 3–4 mm, apex truncate or rounded, opening extending from apex to base less than, rarely up to, halfway . . . . .21b. *Asclepias californica* subsp. *greenei*

21a. Asclepias californica Greene subsp. californica [E]

Corona segments 4–6 mm, apex usually rounded, opening extending from apex to base more than halfway, often more than three-quarters.

Flowering (Feb–)Mar–Jul(–Aug); fruiting (Apr–)May–Aug(–Sep). Slopes, flats, ridge tops, canyons, arroyos, bajadas, granite, volcanic substrates, rhyolite, limestone, rocky, clay, and sandy soils, pinyon, oak, juniper, and oak-juniper woodlands, pine and mixed-conifer forests, shrublands, chaparral, coastal sage scrub, riparian woods, non-native grasslands; 300–2300 m; Calif.

Subspecies *californica* ranges south from Kern and San Bernardino counties.

21b. Asclepias californica Greene subsp. **greenei** Woodson, Ann. Missouri Bot. Gard. 41: 178, fig. 98. 1954 • Greene's milkweed [E]

Corona segments 3–4 mm, apex truncate or rounded, opening extending from apex to base less than, rarely up to, halfway.

Flowering Mar–Jul; fruiting May–Aug(–Sep). Slopes, hills, flats, ridge tops, stream banks, granite, shale, serpentine, igneous substrates, talus, gravel, sandy soils, grasslands, chaparral, oak woodlands, pine-oak and mixed-conifer forests; 200–2200 m; Calif.

Subspecies *greenei* ranges north from Kern and San Luis Obispo counties. The corona may range to light pink from the more common red-violet coloration.

22. Asclepias vestita Hooker & Arnott, Bot. Beechey Voy., 363. 1839 • Wooly milkweed [E]

Asclepias vestita subsp. *parishii* (Jepson) Woodson; *A. vestita* var. *parishii* Jepson

Herbs. **Stems** 1–20, prostrate to decumbent or ascending, rarely branched, 25–90 cm, tomentose, not glaucous, rhizomes absent. **Leaves** opposite, petiolate, with 0 or 1 stipular colleter on each side of petiole; petiole 3–10 mm, tomentose; blade elliptic or oval to lanceolate or ovate, 8–20 × 2–10 cm, chartaceous, base cuneate to rounded or cordate, margins entire, apex acute to attenuate or acuminate, mucronate, venation eucamptodromous to faintly brochidodromous, surfaces densely to thinly tomentose, laminar colleters absent. **Inflorescences** extra-axillary at upper nodes, sometimes appearing terminal, sessile or pedunculate, 19–45-flowered; peduncle 0–3.5 cm, densely tomentose, with 1 caducous bract at the base of each pedicel. **Pedicels** 15–35 mm, densely tomentose. **Flowers** erect to pendent; calyx lobes elliptic, 5–6 mm, apex acute, densely tomentose; corolla green to pinkish purple, lobes reflexed, sometimes with spreading tips, oval, 6–9 mm, apex acute, densely tomentose abaxially, papillose at base adaxially; gynostegial column 1–1.5 mm; fused anthers dark brown, truncately obconic, 1.5–2 mm, wings right-triangular, closed, apical appendages ovate; corona segments cream to dark pink, sessile, conduplicate, dorsally rounded, 3–3.5 mm, equaling or slightly exceeding style apex, apex obtuse, oblique, margin with proximal tooth, glabrous, internal appendage falcate, slightly exserted, glabrous; style apex shallowly depressed, cream. **Follicles** erect on upcurved pedicels, ovoid, 5–6.5 × 2–2.5 cm, apex apiculate to acuminate, longitudinally ridged, tomentulose. **Seeds** ovate, 10–12 × 7–10 mm, margin very narrowly winged, faces smooth; coma 2–2.5 cm. 2*n* = 22.

Flowering Apr–Jul; fruiting May–Sep. Flats, slopes, ridges, canyons, arroyos, foothills, alluvial fans, fields, granite, sandstone, sandy, clay, and rocky soils, desert scrub, chaparral, grasslands, oak, pine-oak, juniper, pinyon-juniper, and Joshua tree woodlands; 50–2000 m; Calif.

W. L. Jepson (1923–1925) and R. E. Woodson Jr. (1954) segregated southern populations (Inyo, Kern, Los Angeles, San Bernardino, and Ventura counties) from northern populations (Fresno, Madera, Mariposa, Merced, Monterey, San Benito, San Joaquin, and San Luis Obispo counties) as varieties or subspecies. Of their distinguishing characters, only flower color is consistently different between these segments of the range: northern populations have pale green corollas with at most a pink tinge, whereas southern populations have pale burgundy to red-violet corollas. There is a tendency for plants in southern populations to be smaller and become more evidently glabrate late in the season, but more robust and hairier plants can also be found in the south. Further research may support recognition of distinct taxa for these populations, but they are not recognized here. *Asclepias vestita* is similar to co-occurring *A. californica* in the absence of flowers or fruits, but plants of *A. vestita* tend to be more prostrate and compact and the leaves tend to be broader towards the base and more quickly glabrate.

23. Asclepias lemmonii A. Gray, Proc. Amer. Acad. Arts 19: 85. 1883 (as lemmoni) • Lemmon's or big-leaf milkweed

Herbs. **Stems** 1–3, erect to ascending, unbranched, very stout, 100–150 cm, densely hirsute, not glaucous, rhizomes absent(?). **Leaves** opposite, petiolate, with 1 stipular colleter on each side of petiole; petiole 1–5 mm, hirsute; blade oval or oblong to ovate, 7–22 × 3–14 cm, subsucculent, base truncate to subcordate, margins entire, apex obtuse to truncate or emarginate, mucronate, venation brochidodromous, secondary veins nearly orthogonal, surfaces hirsute, margins ciliate, 8–16 laminar colleters. **Inflorescences** terminal, paired, and extra-axillary, pedunculate, 21–53-flowered; peduncle 6–13 cm, densely hirsute, with 1 caducous bract at the base of each pedicel. **Pedicels** 13–22 mm, densely hirsute. **Flowers** erect to pendent; calyx lobes lanceolate, 3.5–6 mm, apex acute, hirsute; corolla cream to greenish cream or ochroleucous, sometimes tinged pink, lobes reflexed with spreading tips, elliptic, 9–11 mm, apex acute, glabrous; gynostegial column 0.5–1 mm; fused anthers greenish brown, cylindric, 2.5–3 mm, wings right-triangular, closed, apical appendages oval; corona segments cream to ochroleucous, sometimes tinged pink, shiny, subsessile, conduplicate, 6–8 mm, equaling or exceeding style apex, apex truncate, spreading and tapering, glabrous, internal appendage laterally compressed, erect, barely exserted, glabrous; style apex shallowly depressed, green or pink. **Follicles** erect on upcurved pedicels, lance-ovoid, 9.5–13.5 × 2–3 cm, apex attenuate, smooth, densely hirsute. **Seeds** ovate, 6–7 × 4–5 mm, margin winged, faces minutely rugulose; coma 4–4.5 cm.

Flowering Jun–Sep; fruiting Aug–Oct. Canyons, slopes, streamsides, rocky and clay soils, pine-oak, pine, and riparian forests, oak woodlands, marshes; 1200–2200 m; Ariz.; Mexico (Chihuahua, Durango, Jalisco, Sonora, Zacatecas).

A highly distinctive species, *Asclepias lemmonii* just barely enters the United States in southern Arizona (Cochise, Pima, and Santa Cruz counties), where it inhabits canyons in pine-oak clad sky-island ranges. *Asclepias elata* is a common co-inhabitant of these canyons. *Asclepias lemmonii* has been documented from the Baboquiviri, Chiricahua, Huachuca, and Santa Rita mountains, and it is not common in any of these. It is considered to be of conservation concern in Arizona. The large, hirsute leaves with nearly orthogonal venation and robust, hirsute stems of *A. lemmonii* are unmatched among American milkweeds. Plants may reach heights over 2 m in the main range of the species in the northern Sierra Madre Occidental.

24. Asclepias nummularia Torrey in W. H. Emory, Rep. U.S. Mex. Bound. 2(1): 163, plate 45, fig. A. 1859 • Tufted milkweed

Herbs. Stems 1–5, erect, unbranched, 6–15 cm, taller on vegetative or post-reproductive plants, densely tomentose to glabrate, not glaucous, rhizomes absent. **Leaves** opposite, petiolate, with 1 stipular colleter on either side of petiole; petiole 3–4 mm, tomentose to glabrate; blade orbiculate to obovate or oblate, 1.5–3.2 × 1.7–4 cm, much larger on vegetative or post-reproductive stems, subsucculent, base rounded to cordate, margins entire, apex truncate or emarginate to rounded or obtuse, venation eucamptodromous, surfaces tomentose to glabrate, laminar colleters 0–10. **Inflorescences** terminal, sometimes also extra-axillary, pedunculate, 5–28-flowered; peduncle 2.7–6 cm, tomentose, with 1 bract at the base of each pedicel. **Pedicels** 12–20 mm, densely pilosulous to tomentulose. **Flowers** erect; calyx lobes ovate to lanceolate, 1.5–2.5 mm, apex acute, pilose to tomentulose; corolla pinkish violet to tan, striate, lobes reflexed, oval, 4–5 mm, apex acute, glabrous abaxially, minutely papillate at base adaxially; gynostegial column 0.2–0.5 mm; fused anthers brown, cylindric, 1–1.5 mm, wings curved, closed, apical appendages ovate, erose; corona segments white apically, red-violet basally, sessile, conduplicate, dorsally rounded, 2.5–3 mm, equaling or exceeding style apex, apex truncate, glabrous, internal appendage falcate, exserted, sharply inflexed over style apex, glabrous; style apex shallowly depressed, cream to pink. **Follicles** erect on upcurved pedicels, lance-ovoid, 4–7.5 × 1.2–1.5 cm, apex long-acuminate, smooth, tomentose. **Seeds** broadly ovate, 6–7 × 4–5 mm, margin winged, corky, erose at chalazal end, faces papillose-rugulose, hirtellous; coma 1.5–2.5 cm.

Flowering Mar–May; fruiting May–Aug. Hills, slopes, ridges, flats, arroyos, canyons, rhyolite, granite, sandstone, limestone, igneous outcrops, rocky, gravel, sandy, chalky, or clay soils, oak, oak-juniper, pinyon-juniper, and riparian woodlands, pine-oak forests, desert and oak grasslands; 1100–1900 m; Ariz., N.Mex., Tex.; n, c Mexico.

Asclepias nummularia is the only milkweed in the flora area with small stature and obovate to orbiculate leaves that are bluish under dense tomentum until late in the season. The plants often bear an uncanny resemblance to tiny cabbages. Fruit set may fail across broad regions in some years, perhaps due to drought, although the adequacy of pollination in this species has not been studied. Unlike many species of *Asclepias*, post-reproductive and non-reproductive plants of *A. nummularia* often persist until frost. Coupled with the early flowering of this species, vegetative plants are observed and collected far more often than reproductive ones. Some plants persisting into the fall months have much larger leaves and longer stems, especially in shady or moist habitats; these have often been mistaken for *A. latifolia*, the range of which only barely overlaps that of *A. nummularia* in Grant County, New Mexico. *Asclepias nummularia* is not common in New Mexico, where it has been found in Grant, Hidalgo, and Sierra counties, and is only locally common in Texas in the Davis Mountains (Jeff Davis County), and additionally in Brewster and Presidio counties. Populations from San Luis Potosí and south with narrower leaves have been segregated as *A. nummularioides* W. D. Stevens; recognition of this taxon requires further evaluation.

25. Asclepias quinquedentata A. Gray, Proc. Amer. Acad. Arts 12: 71. 1876 • Slim-pod milkweed

Asclepias rzedowskii W. D. Stevens

Herbs. **Stems** 1–5, erect to ascending, sometimes decumbent at base, unbranched or branched near base, 10–60 cm, sparsely puberulent with curved trichomes to glabrate, not glaucous, rhizomes absent. **Leaves** opposite, sessile or petiolate, with 1 or 2 stipular colleters on each side of petiole; petiole 0–1 mm, puberulent with curved trichomes to glabrate; blade linear, 6–14 × 0.2–0.6 cm, membranous, base cuneate, margins entire, apex acute, mucronate, venation obscure to faintly eucamptodromous, surfaces sparsely puberulent with curved trichomes on midvein to glabrate, margins ciliate, 0–2 laminar colleters. **Inflorescences** extra-axillary, the uppermost appearing terminal, pedunculate, 4–10-flowered (appearing greater because umbels are in close proximity); peduncle 1.1–2.7 cm, sparsely puberulent with curved trichomes, with 1 bract at the base of each pedicel. **Pedicels** 14–21 mm, puberulent with curved trichomes. **Flowers** pendent to spreading; calyx lobes lanceolate, 2.5–3.5 mm, apex acute, strigulose to pilosulous; corolla green, sometimes tinged red, lobes reflexed, exposing corona, oval, 4–6 mm, apex acute, pilose abaxially, glabrous adaxially; gynostegial column 0.5 mm; fused anthers tan to brown, cylindric, 1.5–2 mm, wings narrowly right-triangular, slightly open at base, apical appendages ovate; corona segments red or pink to red-violet or purple at base, white at apex, shiny, sessile, conduplicate, dorsally rounded, 3–4 mm, equaling to slightly exceeding style apex, apex truncate with a proximal tooth on each side, glabrous, internal appendage a crest, barely exserted from cavity; style apex shallowly depressed, white to greenish. **Follicles** erect on upcurved pedicels, fusiform, 8.5–16 × 0.5–1 cm, apex long-attenuate, smooth, faintly striate, pilosulous to glabrate. **Seeds** ovate, 4–5 × 3–4 mm, margin winged, faces smooth; coma 2–2.5 cm.

Flowering Jun–Aug; fruiting Jul–Nov. Slopes, canyons, limestone, rhyolite, rocky soils, chaparral, pinyon-juniper and oak woodlands, pine and pine-oak forests; 1300–2600 m; Ariz., N.Mex.; Mexico (Chihuahua, Distrito Federal, Durango, San Luis Potosí, México).

Although it is widely distributed, *Asclepias quinquedentata* is rarely encountered. The plant is cryptic, even in flower, because of the slender, few-leaved habit and nodding inflorescences. Nonetheless, it appears to truly be rare, at least in the United States. It is considered to be of conservation concern in Arizona, and its status in New Mexico requires evaluation. It has been reported from Texas, based on the presumed type locality. However, M. Fishbein et al. (2008) concluded that the type collection most likely was made in Arizona. The population in the Valle de México has been segregated as *A. rzedowskii* based on a subtle variant of the corona; it is here considered a synonym (Fishbein et al.).

26. Asclepias brachystephana Engelmann ex Torrey in W. H. Emory, Rep. U.S. Mex. Bound. 2(1): 163. 1859 • Short-crowned or shortcrown milkweed [F]

Herbs. **Stems** 4–25, erect, unbranched or branched near base, 20–40 cm, tomentulose, not glaucous, rhizomes absent. **Leaves** opposite to subopposite, petiolate, with 1 or 2 stipular colleters on each side of petiole; petiole 2–8 mm, tomentulose; blade linear-lanceolate, 5–15 × 0.3–1.3 cm, chartaceous, base cuneate to rounded, margins often obscurely crisped, apex acute, mucronate, venation brochidodromous, surfaces tomentulose to glabrate, midvein puberulent with curved trichomes, margins minutely ciliate, laminar colleters absent. **Inflorescences** extra-axillary, pedunculate, 4–15-flowered; peduncle 0.2–1.5 cm, tomentose, with 1 caducous bract at the base of each pedicel. **Pedicels** 9–15 mm, tomentose. **Flowers** erect; calyx lobes ovate-lanceolate, 2–3 mm, apex acute, tomentulose; corolla red-violet, sometimes green with red tinge, lobes reflexed, ovate, 4–6 mm, apex acute, minutely pilosulous; gynostegium subsessile; fused anthers brown, cylindric, 2–2.5 mm, wings right-triangular, closed, apical appendages ovate; corona segments red-violet to pink basally, white apically, sessile, tubular, 1.5–2 mm, greatly exceeded by style apex, apex truncate, oblique, with a proximal tooth on each side, glabrous, internal appendage lingulate, slightly exserted, sharply inflexed towards gynostegium, glabrous; style apex shallowly depressed, red-violet. **Follicles** erect on upcurved pedicels, lance-ovoid, 5–7 × 1.2–1.8 cm, apex acuminate, shallowly ribbed, conspicuously striate, tomentulose. **Seeds** oval to ovate, 6–7 × 4–6 mm, margin winged, faces papillate-tomentulose with dendritic scales; coma 2–2.5 cm.

Flowering Apr–Sep(–Oct); fruiting (May–)Jun–Oct. Plains, bajadas, pastures, arroyos, stream banks, riparian areas, limestone, igneous substrates, alluvium, gravel, clay, silty, and sandy soils, desert grasslands, desert scrub, oak-juniper, juniper, and mesquite woodlands; 900–1900 m; Ariz., N.Mex., Tex.; Mexico (Aguascalientes, Chihuahua, Coahuila, Durango, Nuevo León, Sonora, Zacatecas).

Asclepias brachystephana is a blue-gray, bushy herb with ascending foliage, few-flowered umbels of small, red and white flowers, and conspicuously striped follicles. It is unlike any other milkweed. Nonetheless, herbarium specimens are commonly confused with those of *A. asperula* because the herbage of *A. brachystephana* turns green on drying, and the leaves of *A. asperula* subsp. *asperula* are often of similar size and shape. However, leaf arrangement in *A. asperula* is alternate rather than opposite. The flowers of *A. brachystephana* are remarkably similar to, and convergent with, those of *A. cutleri*, *A. eastwoodiana*, *A. ruthiae*, *A. sanjuanensis*, and *A. uncialis* (M. Fishbein et al. 2011). In Arizona, *A. brachystephana* is restricted to the portion of the southeastern corner of the state with Chihuahuan floristic affinities, in Cochise, Graham, Pima, and Santa Cruz counties.

27. **Asclepias hypoleuca** (A. Gray) Woodson, Ann. Missouri Bot. Gard. 28: 206. 1941 • Mahogany milkweed, talayote

Gomphocarpus hypoleucus A. Gray, Proc. Amer. Acad. Arts 17: 222. 1882

Herbs. Stems 1 (rarely 2 or 3), erect, unbranched, 25–100 cm, puberulent with curved trichomes to pilosulous, not glaucous, rhizomes absent. **Leaves** opposite, petiolate, with 1 stipular colleter on each side of petiole; petiole 2–6 mm, puberulent with curved trichomes to pilosulous; blade ovate or lanceolate to oblong, elliptic, or oval, 5.5–11.5 × 1–5 cm, chartaceous, base cuneate or obtuse to truncate, margins entire, apex obtuse to rounded, mucronate, venation eucamptodromous, surfaces tomentose abaxially, pilosulous or tomentulose to glabrate adaxially, margins ciliate, 8–12 laminar colleters. **Inflorescences** terminal and extra-axillary, pedunculate, 12–35-flowered; peduncle 3.5–10.5 cm, puberulent with curved trichomes to tomentulose, with 1 caducous bract at the base of each pedicel. **Pedicels** 15–21 mm, densely puberulent with curved trichomes or pilosulous to tomentulose. **Flowers** erect to pendent; calyx lobes lanceolate, 3–4 mm, apex acute, pilose; corolla green, sometimes red-tinged abaxially, deep maroon to greenish red or green adaxially, lobes reflexed, tips usually spreading, oblong to elliptic, 8–10 mm, apex acute, pilosulous abaxially, glabrous adaxially; gynostegial column 1–1.5 mm; fused anthers brown, broadly cylindric, 2–2.5 mm, wings right-triangular, open at tip, apical appendages ovate, erose; corona segments deep maroon to greenish red or yellowish green, subsessile, conduplicate, 7–9 mm, greatly exceeding style apex, apex truncate, spreading and long-tapering with a proximal tooth on each side, glabrous, internal appendage absent or a low crest, glabrous; style apex shallowly depressed, green. **Follicles** erect on upcurved pedicels, fusiform, 9–11.5 × 1.2–1.4 cm, apex long-attenuate, smooth, sometimes faintly striate, pilosulous to tomentulose. **Seeds** ovate, 6–7 × 4–5 mm, margin winged, faces minutely rugulose; coma 3–3.5 cm.

Flowering Jun–Sep; fruiting Aug–Sep. Slopes, flats, lake shores, streamsides, granite, gneiss, andesite, rocky soils, pine, pine-oak, oak, and mixed-conifer forests; 1900–2800 m; Ariz., N.Mex.; Mexico (Chihuahua, Sonora).

Asclepias hypoleuca grows at higher elevations in the sky-island mountain ranges than any other milkweed. It has been documented from the Chiricahua, Huachuca, Rincon, Santa Catalina, Santa Rita, and White mountain ranges in Arizona (Cochise, Pima, and Santa Cruz counties) and the Mogollon Mountains and Black Range in New Mexico (Catron and Grant counties). Because of its limited, high-elevation distribution, and the threats of changing climate, its conservation status in the flora area merits assessment. The bicolored leaves exhibit coloration similar to co-occurring silverleaf oak (*Quercus hypoleucoides*).

28. **Asclepias solanoana** Woodson, Ann. Missouri Bot. Gard. 28: 207. 1941 • Serpentine milkweed [E]

Gomphocarpus purpurascens A. Gray, Proc. Amer. Acad. Arts 10: 76. 1874, not A. Richard 1850; *Solanoa purpurascens* Greene 1890, not *Asclepias purpurascens* Linnaeus 1753

Herbs. Stems 1–15, prostrate, unbranched (rarely branched), 15–40 cm, densely puberulent with curved trichomes to tomentose, not glaucous, rhizomes absent. **Leaves** opposite, petiolate, stipular colleters absent; petiole 5–10 mm, densely puberulent with curved trichomes; blade ovate to nearly orbiculate, 3.5–6 × 3–4 cm, subsucculent, base obtuse to cordate, margins entire, apex obtuse to acute or rounded, venation eucamptodromous, surfaces pilosulous, more densely so abaxially, especially on veins, becoming glabrate adaxially, margins inconspicuously ciliate, laminar colleters absent. **Inflorescences** terminal and extra-axillary, pedunculate, 20–55-flowered; peduncle 1.5–9 cm, tomentose, with 1 caducous bract at the base of each pedicel. **Pedicels** 10–13 mm, tomentose to pilose. **Flowers** erect to spreading; calyx lobes lanceolate, 2–3 mm, apex acute, pilose to tomentose; corolla pale pink to red, lobes reflexed with spreading tips, oval, 5–6 mm, apex acute, glabrous; gynostegium sessile; fused anthers yellow to brown or green, broadly barrel-shaped, 1.5–2 mm, wings deltoid, widest at middle, closed, apical appendages ovate; corona segments pinkish cream to cream, subsessile, conduplicate, dorsally rounded, spreading away from anthers, 2–3 mm, greatly exceeded by style apex, apex truncate, glabrous, internal appendage absent; style apex shallowly depressed, cream to green. **Follicles** erect on upcurved pedicels (at least until maturity), lance-ovoid, 6–10 × 2–3 cm, apex obtuse to acuminate, longitudinally ridged, pilosulose or tomentulose to glabrate. **Seeds** ovate, 6–8 × 5–6 mm, margin very narrowly winged, faces rugulose; coma 2–2.5 cm.

Flowering (Apr–)May–Aug; fruiting Jun–Jul. Slopes, streamsides, canyons, barrens, serpentine, rocky and deep soils, chaparral, cypress and mixed-conifer woodlands, pine and mixed-conifer forests, meadows; 200–2000 m; Calif.

Asclepias solanoana is a delightful and unique milkweed that is endemic to rugged, serpentine barrens in the northern Coast Range of California. The plants hug the ground, the stems seeming to crawl outward propelled by the highly unusual metallic, grayish or bluish green, ovate leaves. Bright pinkish rose balls of floral buds are held above, and are followed by variegated spheres of cream, pink, green, and brown flowers with a vague resemblance to heads of *Abronia* (Nyctaginaceae). It is often the only conspicuous plant species on highly exposed, south-facing slopes. S. P. Lynch (1977) documented Hymenoptera (carpenter bees, *Xylocopa*, bumblebees, *Bombus*, and honeybees, *Apis*) to be the main pollinators of *A. solanoana*. It is considered threatened by extractive industries and recreation at some sites. A naturally occurring population has been reported from southern Oregon and needs confirmation.

SELECTED REFERENCE Lynch, S. P. 1977. The floral ecology of *Asclepias solanoana* Woods. Madroño 24: 159–177.

29. Asclepias feayi A. Gray, Proc. Amer. Acad. Arts 12: 72. 1876 • Feay's or Florida milkweed [E]

Asclepiodella feayi (A. Gray) Small

Herbs. **Stems** solitary, erect, unbranched (rarely near base), 20–75 cm, minutely puberulent in a line with curved trichomes to glabrate, not glaucous, rhizomes absent. **Leaves** opposite, sessile, with 1 or 2 stipular colleters on each side of leaf base; blade filiform, 2.5–10 × 0.1–0.15 cm, membranous, base cuneate, margins entire, apex acute, venation obscure, surfaces glabrous, laminar colleters absent. **Inflorescences** terminal, sometimes branched, and often extra-axillary at upper nodes, sessile or pedunculate, 2–7-flowered; peduncle 0–5 cm, puberulent with curved trichomes on 1 side, with 1 caducous bract at the base of each pedicel. **Pedicels** 7–17 mm, minutely puberulent with curved trichomes on 1 side. **Flowers** erect; calyx lobes narrowly lanceolate, 1.5–2.5 mm, apex acute, glabrous; corolla white, sometimes pale lavender-tinged, inconspicuously striate, lobes spreading, lanceolate, 7–10 mm, apex acute to obtuse, glabrous; gynostegium sessile; fused anthers lavender, cylindric, 1.5–3 mm, wings narrowly right-triangular, open at base, apical appendages ovate; corona segments white, sessile with a basal collar, cupulate, 2.5–4 mm, equaling to slightly exceeding style apex, apex obtuse, glabrous, internal appendage a laterally flattened, included crest, glabrous; style apex shallowly depressed, lavender. **Follicles** erect on straight pedicels, fusiform, 9–12 × 0.3–0.6 cm, apex attenuate, smooth, glabrous. **Seeds** oval, 6–8 × 3–5 mm, margin winged, faces smooth; coma 3.5 cm.

Flowering (Feb–)Apr–Sep(–Nov); fruiting Jun–Aug. Flats, streamsides, sandy soils, pine scrub and flatwoods, pine-palmetto scrub, prairies, hammocks; 0–50 m; Fla.

Asclepias feayi is one of a trio of very slender milkweeds in the southeastern United States, along with *A. cinerea* and *A. viridula*. These species are divergent in floral morphology, but without flowers they are very difficult to distinguish (even in fruit), and they appear to be close relatives. However, *A. feayi* occurs primarily in peninsular central and southwestern Florida, from Lake to Collier counties, most commonly in scrub. A single disjunct population has been documented from Clay County in the northeastern part of the state (*Hall 1896* [FLAS]). *Asclepias cinerea* and *A. viridula* are found in northern Florida or further north, in flatwoods. All three species are cryptic in the absence of flowers and appear to respond positively to fire and rainfall events. They are likely to be more common than is apparent because they are inconspicuous and emerge episodically. Nonetheless, numerous historical locations for *A. feayi* have been developed and are no longer capable of supporting populations, and its conservation status merits evaluation. An unusual putative hybrid with *A. pedicellata* represented by a single collection is documented from Marion County (*Judd 2639* [FLAS]), suggested by the exactly intermediate floral morphology.

30. Asclepias cinerea Walter, Fl. Carol., 105. 1788 • Ashy or Carolina milkweed [E]

Herbs. **Stems** 1, erect, unbranched (rarely branched), 20–100 cm, minutely puberulent in a line with curved trichomes to glabrate, not glaucous, rhizomes absent. **Leaves** opposite, sessile, with 1 or 2 stipular colleters on each side of leaf base; blade filiform, 2–9 × 0.1–0.15 cm, membranous, base cuneate, margins entire, apex acute, mucronate, venation obscure, surfaces glabrous, laminar colleters absent. **Inflorescences** terminal, branched, and extra-axillary at upper nodes, pedunculate, 2–8-flowered; peduncle 0.5–1.7 cm, puberulent with curved trichomes on 1 side, with 1 caducous bract at the base of each pedicel. **Pedicels** 10–25 mm, minutely puberulent with curved trichomes on 1 side. **Flowers** spreading to pendent; calyx lobes lanceolate, 1.5–2 mm, apex acute, glabrous; corolla cream, tinged gray, pink, or purple, faintly striate, lobes reflexed with spreading tips, elliptic, 4–5 mm, apex acute to obtuse, glabrous; gynostegial column 0–0.5 mm; fused anthers green, cylindric, 1.5–2 mm, wings narrowly right-triangular, open, apical appendages deltate; corona segments cream, tinged gray, pink, or purple, sessile, conduplicate, dorsally flattened, 2–3 mm, equaling style apex, apex truncate with a proximal tooth on each side, glabrous, internal appendage a laterally flattened, included crest, glabrous; style apex shallowly depressed, white. **Follicles** erect on straight pedicels, fusiform, 8–12 × 0.3–0.7 cm, apex long-attenuate, smooth, glabrous. **Seeds** ovate, 6–7 × 4–5 mm, margin thickly winged, faces sparsely papillose; coma 2.5–3 cm.

Flowering May–Sep(–Nov); fruiting Jun–Sep. Ridges, flats, fields, sandstone, sandy soils, wet to dry pine flatwoods, barrens and savannas, pine-oak forest, often recently burned, bogs, swamps; 0–200 m; Ala., Fla., Ga., S.C.

Similarities between *Asclepias cinerea* and *A. feayi* are discussed under the latter species. *Asclepias cinerea* inhabits flatwoods mostly north of the range of *A. feayi*, co-occurring only in Clay County as far as is known. *Asclepias cinerea* is sympatric with another similar species, *A. viridula*, across northern Florida. That species tends to grow in wetter woods and meadows than *A. cinerea*. In flower, they are easily distinguished by the spreading to pendent flowers, ashy lavender corollas, and corona segments with included, crestlike appendages of *A. cinerea* (versus erect to spreading flowers, green to brownish corollas, and corona segments with exerted, falcate appendages of *A. viridula*). *Asclepias cinerea* barely enters southeastern Alabama (Covington, Geneva, and Houston counties), and the species is considered to be of conservation concern in that state. Emergence and flowering of this species appears to be stimulated by precipitation events and/or fire.

31. Asclepias viridula Chapman, Fl. South. U.S., 363. 1860 • Southern or green milkweed [C] [E]

Herbs, latex clear. **Stems** 1 (rarely 2), erect, unbranched, 25–75 cm, minutely puberulent in a line with curved trichomes to glabrate, not glaucous, rhizomes absent. **Leaves** opposite, sessile, with 1 or 2 stipular colleters on each side of leaf base; blade linear to filiform, 4.5–9 × 0.15–0.25 cm, membranous, base cuneate, margins revolute, apex acute, mucronate, venation obscure, surfaces glabrous, margins sparsely ciliate to glabrate, laminar colleters absent. **Inflorescences** extra-axillary at upper nodes, pedunculate, 4–15-flowered; peduncle 0.8–2 cm, minutely puberulent in a line with curved trichomes, with 1 caducous bract at the base of each pedicel. **Pedicels** 7–13 mm, minutely puberulent in a line with curved trichomes. **Flowers** erect to spreading; calyx lobes lanceolate, 1.5–2.5 mm, apex acute, glabrous; corolla green, tinged brown, lobes reflexed with spreading tips, elliptic, 3–5 mm, apex acute to obtuse, sometimes emarginate, glabrous; gynostegial column 0.8–1 mm; fused anthers green and brown, cylindric, 1.5–2 mm, wings right-triangular, closed, apical appendages ovate; corona segments cream, tinged brown or green, stipitate, conduplicate and dorsally rounded, 3–4 mm, slightly exceeding style apex, apex acute, spreading, with a proximal tooth on each side, glabrous, internal appendage falcate, exserted, arching towards style apex, glabrous; style apex shallowly depressed, green. **Follicles** erect on straight pedicels, fusiform, 9–12.5 × 0.6–0.9 cm, apex long-acuminate, smooth, glabrous. **Seeds** ovate, 8–9 × 5–6 mm, margin winged, faces minutely and sparsely rugulose; coma 2.5–3 cm.

Flowering Apr–Sep; fruiting Jun–Oct. Wet meadows, pine savannas, pine flatwoods, often following fires; of conservation concern; 0–50 m; Ala., Fla.

Similarities among *Asclepias cinerea*, *A. feayi*, and *A. viridula* are discussed under those species; all three are slender, cryptic when not in flower, and appear to emerge and flower in response to precipitation and fire events. *Asclepias viridula* is perhaps the most cryptic of the three, by virtue of its green corollas, and it is the most limited in range. It is typically found in wetter sites than co-occurring *A. cinerea*. *Asclepias viridula* is found disjunctly in northeastern Florida and the Florida Panhandle. Its range barely crosses into Alabama, where it is known from a single site in Houston County. Reports from Georgia are probably based on misidentifications—no specimens are known, and further searches for *A. viridula* in Georgia are warranted. It is considered to be of conservation concern throughout its range. Although not listed as a threatened or endangered species under the ESA in the United States, the number of populations is low and merits further study of population persistence and viability.

32. Asclepias connivens Baldwin in Elliott, Sketch Bot. S. Carolina 1: 320. 1817 • Large-flower milkweed [E]

Anantherix connivens (Baldwin) Feay

Herbs. Stems 1, erect, unbranched (rarely branched), 25–90 cm, minutely puberulent with curved trichomes or pilose to glabrate, not glaucous, rhizomes absent. **Leaves** opposite, sessile or petiolate, with 1 or 2 stipular colleters on each side of petiole; petiole 0–1 mm, puberulent with curved trichomes to glabrate; blade narrowly elliptic to linear or oblanceolate, 2.5–8 × 0.3–2.5 cm, chartaceous, base cuneate, margins entire, apex acute to rounded, mucronate, venation brochidodromous, surfaces sparsely pilose to glabrate, midvein puberulent with curved trichomes to glabrate, margins ciliate, 0–6 laminar colleters. **Inflorescences** terminal, sometimes branched, and extra-axillary at upper nodes, pedunculate, 4–8-flowered; peduncle 0.9–6 cm, densely puberulent with curved trichomes, with 1 caducous bract at the base of each pedicel. **Pedicels** 10–20 mm, densely puberulent with curved trichomes. **Flowers** spreading to pendent; calyx lobes lanceolate, 4–6 mm, apex acute, sparsely pilosulous; corollas green, sometimes tinged brown at apex, lobes reflexed with spreading tips, elliptic, 10–13 mm, apex acute to obtuse, sometimes emarginate, glabrous; gynostegial column 1–1.5 mm; fused anthers green, obconic, 3–3.5 mm, wings narrowly right-triangular, distended at base, closed, apical appendages oval; corona segments cream to pale green, stipitate, clavate, incurved, 8–10 mm, greatly exceeding style apex, apex rounded, glabrous, internal appendage a hidden crest, glabrous, apices of the 5 segments sometimes connivent; style apex depressed, green. **Follicles** erect on upcurved pedicels, fusiform, 11.5–15 × 0.5–1.4 cm, apex long-attenuate, smooth, minutely pilosulous. **Seeds** ovate, 7–9 × 5–6 mm, margin winged, faces minutely and sparsely rugulose; coma 3–3.5 cm.

Flowering May–Aug(–Sep); fruiting Jul–Sep(–Nov). Flats, sandy soils, pine flatwoods and barrens, often recently burned, wet meadows, marshes, bogs, swamps; 0–200 m; Ala., Fla., Ga., S.C.

Asclepias connivens is a singular species—the large, incurved, clavate corona segments are unlike any others in the genus. With wide leaves and large flowers, *A. connivens* is more conspicuous than most co-occurring milkweeds, such as *A. cinerea*, *A. feayi*, *A. pedicellata*, and *A. viridula*. It prefers wet soils and often occurs at the same sites as *A. viridula* and the red-orange-flowered *A. lanceolata*. *Asclepias connivens* barely enters South Carolina in Jasper and Beaufort counties and is considered rare and to be of conservation concern in that state. It has been reported from Mississippi, but there are no specimens from that state, and occurrence there seems unlikely, as *A. connivens* has not been documented from southwestern Alabama either. It would be interesting to discover what pollinates the large and unusual flower of *A. connivens*; however, there appear to be no reports of flower visitors to this species.

33. Asclepias pedicellata Walter, Fl. Carol., 106. 1788 • Savanna milkweed [E] [F]

Podostigma pedicellatum (Walter) Vail

Herbs. Stems 1, erect, unbranched, 10–45 cm, puberulent with curved trichomes, not glaucous, rhizomes absent. **Leaves** opposite, sessile, with 1 stipular colleter on each side of leaf base; blade linear to narrowly elliptic or narrowly lanceolate, 1.5–6 × 0.1–0.8 cm, chartaceous, base cuneate, margins entire, apex acute, mucronate, venation obscure to eucamptodromous, surfaces puberulent with curved trichomes or scabridulous to glabrate, margins ciliate, 2 laminar colleters. **Inflorescences** terminal and extra-axillary at upper nodes, pedunculate, 2–7-flowered; peduncle 0.15–1 cm, densely puberulent with curved trichomes, with 1 caducous bract at the base of each pedicel. **Pedicels** 5–13 mm, minutely puberulent with curved trichomes. **Flowers** erect; calyx lobes elliptic, 2–3 mm, apex acute, sparsely puberulent with curved trichomes to glabrate; corolla yellowish green to green, lobes erect, mostly concealing corona, narrowly elliptic, 7–10 mm, apex acute, glabrous; gynostegial column 4–6 mm; fused anthers green, broadly conic, 1–1.5 mm, wings right-triangular with decurrent base, closed, apical appendages ovate; corona segments yellowish green to green, sometimes dark green at apex, sessile, conduplicate, 1.5–2.5 mm, greatly exceeded by style apex, apex incurved, rounded, glabrous, internal appendage absent; style apex flat, green. **Follicles** erect on straight pedicels, fusiform, 8–14 × 0.3–0.5 cm, apex long-attenuate, smooth, densely puberulent with curved trichomes. **Seeds** ovate, 5–6 × 3–4 mm, margin winged, faces sparsely and minutely rugulose; coma 2–2.5 cm.

Flowering Mar–Nov(–Dec); fruiting May–Oct. Flats, streamsides, sandhills, sandy soils, pine flatwoods, savannas, pine-palmetto and oak-palmetto scrubs, often following fires; 0–100 m; Fla., Ga., N.C., S.C.

Asclepias pedicellata is found in drier habitats than some co-distributed milkweeds, such as *A. connivens* and *A. viridula*. It sometimes occurs in the same sites as *A. cinerea*, *A. curtissii*, and *A. feayi*. The erect petals and elongate gynostegial column are unique among

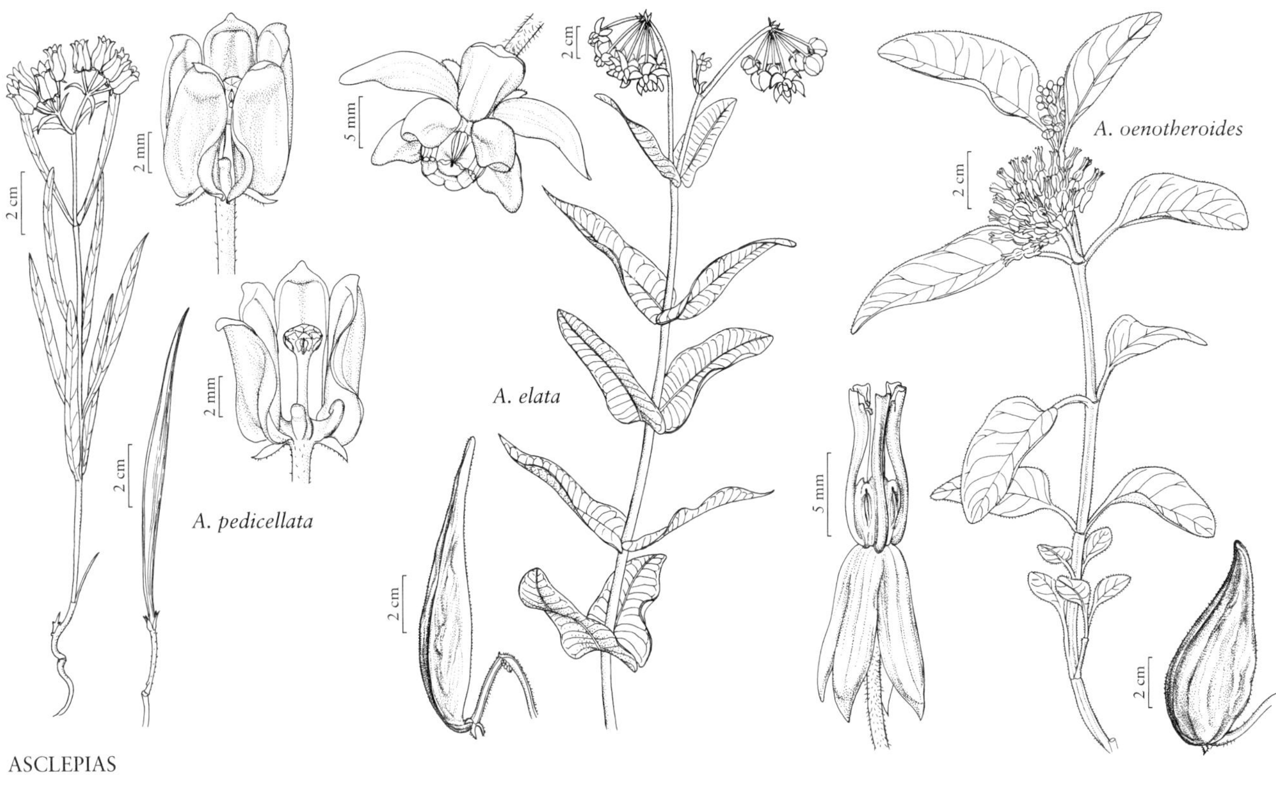

ASCLEPIAS

Asclepias species in the flora area. The green flowers with hidden coronas and low stature of the plants can make them frustratingly cryptic in their grass-dominated habitats. *Asclepias pedicellata* is common only in Florida and North Carolina and is considered rare and of conservation concern in the intervening region in Georgia and South Carolina.

34. Asclepias curtissii A. Gray, Proc. Amer. Acad. Arts 19: 85. 1883 • Curtiss's milkweed [E]

Oxypteryx curtissii (A. Gray) Small

Herbs. **Stems** 1, erect, often purplish, sometimes branched in inflorescence, 15–100 cm, puberulent with curved trichomes, not glaucous, rhizomes absent. **Leaves** opposite, petiolate, with 1 or 2 stipular colleters on each side of petiole; petiole 4–7 mm, puberulent with curved trichomes; blade oblong, elliptic, or oval to obovate, oblanceolate, ovate, or lanceolate, 1.8–5 × 0.5–2.5 cm, chartaceous, base obtuse to cuneate or rounded, margins entire, apex truncate to emarginate or acute, mucronate, venation brochidodromous to eucamptodromous, surfaces sparsely puberulent on midvein with curved trichomes, margins inconspicuously ciliate to glabrate, 2–6 laminar colleters. **Inflorescences** terminal and extra-axillary at upper nodes, sessile or pedunculate, 15–45-flowered; peduncle 0–4 cm, puberulent on 1 side with curved trichomes, with 1 caducous bract at the base of each pedicel. **Pedicels** 10–14 mm, sparsely puberulent with curved trichomes. **Flowers** erect; calyx lobes lanceolate, 2–3 mm, apex acute, glabrous; corolla green with bronze or purplish tinge, lobes reflexed with spreading tips, elliptic, 4–5 mm, apex acute, glabrous; gynostegial column 0.5–1 mm; fused anthers green, obconic, 1–1.5 mm, wings broadly triangular, widest at middle, closed, apical appendages ovate; corona segments white with a green and/or purple dorsal midline, sessile, conduplicate and dorsally flattened, 5–6 mm, greatly exceeding style apex, apex attenuate, glabrous, internal appendage falcate, exserted, sharply inflexed over style apex, glabrous; style apex shallowly depressed, green to cream. **Follicles** erect on upcurved pedicels, fusiform, 8–10.5 × 0.8–1.5 cm, apex long-acuminate, smooth, puberulent with curved trichomes. **Seeds** ovate, 8–9 × 5–6 mm, margin winged, erose, faces sparsely papillose; coma 3.5–4 cm.

Flowering Apr–Oct; fruiting Jul–Oct. Low ridges, sandy soils, oak-palmetto sand scrub, pinelands; 0–50 m; Fla.

Asclepias curtissii is endemic to white-sand substrates at interior and coastal sites on the Florida peninsula. Although the species is not considered to face imminent threat of extirpation, the scrub habitats in which it is found have been, and continue to be, heavily impacted by development. *Asclepias curtissii* cannot be mistaken for any other milkweed in its range when in flower; however, it occurs in the same habitats as *A. tomentosa*, and these species overlap considerably in vegetative features. The leaves of *A. curtissii* can be distinguished from those of *A. tomentosa* by trichomes limited to the midvein (versus sparsely to densely puberulent or tomentulose throughout).

35. **Asclepias erosa** Torrey in W. H. Emory, Rep. U.S. Mex. Bound. 2(1): 162. 1859 • Desert milkweed

Herbs. **Stems** 1–12, erect to decumbent, unbranched, 40–250 cm, thinly tomentose to glabrate, not glaucous, rhizomatous. **Leaves** opposite, sessile or petiolate, with 0 or 1 stipular colleter on each side of petiole; petiole 0–6 mm, thinly tomentose to glabrate; blade ovate to lanceolate, 7.5–25 × 2.5–15 cm, succulent, base truncate to cordate, margins minutely erose, apex attenuate to acuminate, venation eucamptodromous, surfaces tomentose to glabrate, margins ciliate, minutely erose, laminar colleters absent. **Inflorescences** terminal and extra-axillary, sometimes branched, pedunculate, 12–50-flowered; peduncle 2–10 cm, tomentose, with 1 caducous bract at the base of each pedicel. **Pedicels** 20–45 mm, densely tomentose to glabrate. **Flowers** erect to pendent; calyx lobes lanceolate, 4–5 mm, apex acute, tomentose to glabrate; corolla green, lobes reflexed, tips sometimes spreading, oval, 6–9 mm, apex acute, tomentose towards tips abaxially, glabrous adaxially; gynostegial column 1–1.5 mm; fused anthers green, cylindric, 2.5–3 mm, wings right-triangular, closed, apical appendages ovate; corona segments cream to ochroleucous, stipitate, conduplicate, dorsally rounded, 3–5.5 mm, slightly exceeding style apex, apex truncate with a proximal tooth on each side, glabrous, internal appendage falcate, exserted, sharply inflexed over style apex, glabrous; style apex shallowly depressed, cream to greenish cream. **Follicles** erect on upcurved pedicels, lance-ovoid to ovoid, 6.5–10 × 2–3.5 cm, apex acuminate to apiculate, smooth, thinly tomentose. **Seeds** ovate, 8–13 × 5–10 mm, margin narrowly winged, faces minutely rugulose, ridges papillose; coma 2–2.5 cm. $2n = 22$.

Flowering Mar–Nov; fruiting Apr–Nov. Dunes, arroyos, canyons, ridges, slopes, bajadas, flats, granite, gypsum, gravel, alluvium, volcanic substrates, sandy, saline, and silty soils, desert scrub, riparian scrub, shrubby grasslands; -50–2000 m; Ariz., Calif., Nev., Utah; Mexico (Baja California, Sonora).

Asclepias erosa is one of the most remarkable milkweeds. It inhabits the driest regions in the flora area, yet it is one of the tallest and largest-leaved species of *Asclepias*. Its leaves attain a greater size than any other sympatric milkweed, and it may possess the largest leaves of any co-occurring vascular plant species in its range. It is found most commonly in desert arroyos, and it is assumed to be deep rooted and to access reliable sources of water, which would explain its anomalously large size. Like several other milkweeds inhabiting the American deserts, *A. erosa* has white coronas and is commonly visited by tarantula hawk wasps (Pompilidae, Pepsinae). It is considered rare in Utah, where it enters the state only in Washington County. *Asclepias erosa* is often confused with *A. eriocarpa*, another robust species with an overlapping range in southern California. *Asclepias erosa* has strictly opposite, sessile to shortly petiolate leaves with erose margins and corona segments that are level at the apex and only rarely pinkish, whereas *A. eriocarpa* has leaves that may be opposite, alternate, or whorled, with longer petioles, and entire margins, and corona segments with oblique apices and that are often pinkish.

36. **Asclepias eriocarpa** Bentham, Pl. Hartw., 323. 1849 • Wooly-pod or kotolo or Indian milkweed

Asclepias eriocarpa var. *microcarpa* Munz & I. M. Johnston; *A. fremontii* Torrey; *A. kotolo* Eastwood

Herbs. **Stems** 1–10, erect to spreading, rarely branched, 30–100 cm, tomentose to puberulent with curved trichomes or glabrate, not glaucous, rhizomatous. **Leaves** opposite, or alternate or whorled at upper nodes, petiolate, with 0–2 stipular colleters on each side of petiole (sometimes additionally in the axil); petiole 4–15 mm, tomentose; blade oblong or oval to lanceolate or ovate, often conduplicate, 8–20 × 2–8 cm, chartaceous, base obtuse or truncate to cordate, margins entire or often undulate, apex acuminate to obtuse, mucronate, venation brochidodromous, surfaces tomentose to densely puberulent with curved trichomes, margins ciliate, 6–8 laminar colleters. **Inflorescences** terminal, sometimes branched, and extra-axillary at upper nodes, pedunculate, 12–57-flowered; peduncle 1–10 cm, densely tomentose, with 1 caducous bract at the base of each pedicel. **Pedicels** 15–50 mm, densely tomentose. **Flowers** erect to spreading; calyx lobes lanceolate, 3–4 mm, apex acute, densely tomentose; corolla greenish cream, ochroleucous, or pinkish cream, deep pink or tan abaxially, lobes reflexed, tips usually spreading, oval, 7–9 mm, apex acute, tomentose abaxially towards tips, glabrous adaxially; gynostegial column 1–1.5 mm; fused anthers dark brown, cylindric, 2.5–3 mm, wings right-triangular, closed, apical appendages ovate; corona segments cream to dark pink, stipitate, conduplicate, dorsally rounded, 2.5–3 mm, exceeded by style apex, apex truncate, oblique, papillose, internal appendage falcate, exserted, sharply inflexed towards style apex, papillose; style apex shallowly depressed, cream. **Follicles** erect on upcurved pedicels, lance-ovoid, 5–10 × 1.5–3 cm, apex apiculate, smooth, densely tomentose. **Seeds** ovate, 7–9 × 4–6 mm, margin narrowly winged, faces faintly rugulose; coma 2.5–3 cm.

Flowering (Apr–)May–Oct; fruiting (May–)Jun–Oct. Hills, slopes, ridge tops, flats, valleys, canyons, arroyos, stream banks, granite, rocky, alluvial, clay, and sandy soils, meadows, native and non-native grasslands, chaparral, coastal sage scrub, oak and pine-oak woodlands, pine, mixed conifer, and riparian forests, often following fires; 50–2500 m; Calif.; Mexico (Baja California).

Asclepias eriocarpa is found almost throughout the California Floristic Province in relatively dry, open sites in a variety of vegetation types. It is the only milkweed in the flora area that regularly produces alternate, opposite, and whorled leaves on a single stem. Its distribution overlaps that of several other broad-leaved, densely vestitured *Asclepias* species: *A. erosa* in desert scrub and dry grasslands, *A. californica* and *A. vestita* in chaparral and woodlands, and *A. speciosa* in woodlands. Comparison to *A. erosa* is presented under that species. *Asclepias eriocarpa* is easily distinguished from *A. californica* by the distinctive red-violet, rounded corona segments that lack appendages in the latter species. *Asclepias eriocarpa* has corona segments with truncate apices and corolla lobes that are tomentose abaxially only at the apex, in contrast to the obtuse corona segments and uniformly pubescent corolla lobes of *A. vestita*. *Asclepias speciosa* has distinctive corona segments with long, tapering apices that are much larger than and are easily distinguished from those of *A. eriocarpa*. It is possible that *A. eriocarpa* and *A. speciosa* occasionally hybridize—R. E. Woodson Jr. speculated (via annotation of the holotype, *E. Gifford s.n.* [CAS]) that *A. giffordii* Eastwood represented such a hybrid. That interpretation is accepted here. *Asclepias eriocarpa* is reported to be a resource for fiber and medicine by Native Americans.

37. **Asclepias elata** Bentham, Pl. Hartw., 290. 1849 • Nodding milkweed [F]

Asclepias glaucescens Kunth var. *elata* (Bentham) E. Fournier

Herbs. Stems 1 (rarely 2 or 3), erect, unbranched, 30–80 cm, glabrous, glaucous, rhizomatous. **Leaves** opposite, sessile, with 1 inconspicuous stipular colleter on each side of leaf base; blade oval or elliptic to oblong or lanceolate, 5.5–14 × 1.5–6 cm, subsucculent, base cordate, clasping, margins sometimes crisped, apex rounded to acute, mucronate, venation brochidodromous, surfaces glabrous, glaucous, margins minutely and remotely ciliate, laminar colleters absent. **Inflorescences** terminal and extra-axillary at upper nodes, pedunculate, 7–20-flowered; peduncle 5–20 cm, pilosulous in a line, glaucous, with 1 caducous bract at the base of each pedicel. **Pedicels** 20–32 mm, pilosulous, often in a line. **Flowers** pendent; calyx lobes elliptic to oval, 5–8 mm, apex acuminate, sparsely pilosulous to glabrate; corolla green, lobes reflexed with spreading tips, oval, 11–14 mm, apex acute, glabrous; gynostegium sessile; fused anthers green, obconic, 3–3.5 mm, wings narrowly right-triangular, thick, open just above base, apical appendages lanceolate; corona segments white, yellow to tan dorsally, subsessile, conduplicate, 4–6 mm, equaling style apex, apex truncate, glabrous, internal appendage a crest, included in segment, glabrous; style apex depressed, green to greenish cream. **Follicles** erect on upcurved pedicels, fusiform to lance-ovoid, 8–13 × 1–2 cm, apex long-attenuate, smooth, pilosulous. **Seeds** ovate to lanceolate, 6–7 × 3–6 mm, margin winged, sometimes minutely erose, faces rugulose; coma 4–4.5 cm.

Flowering Jun–Sep; fruiting Aug–Oct. Canyons, arroyos, stream banks, slopes, igneous substrates, limestone, rocky, sandy, and clay soils, pinyon-juniper and oak woodlands, pine-oak and riparian forests, meadows; 1200–2200 m; Ariz., N.Mex., Tex.; Mexico; Central America (Guatemala).

Asclepias elata is one of the few milkweeds in the flora area that possesses a mostly Mexican distribution but does not belong to the major clade that contains most of such species (species 23–27 here). Rather it is the sole representative in the flora, and the most northerly distributed member, of a small clade of glaucous, vegetatively homogeneous species ranging as far south as Costa Rica (M. Fishbein et al. 2011). *Asclepias elata* has been inconsistently recognized as distinct from *A. glaucescens* since the time of A. Gray et al. [1878–1897, vol. 2(1)]. However, at least since J. N. Rose (1892), the distinction between these species has been understood, and was presented clearly by R.E. Woodson Jr. (1954). Although indistinguishable in the absence of flowers, *A. elata* differs by fully pendent (versus erect to pendent) flowers, fewer flowers per umbel, longer corolla lobes, and corona segments that spread away from the gynostegium and are exceeded by the style apex, with the appendage merely a crest included in the segment (versus segments that are strict, overtop the style apex, and bear exserted appendages in *A. glaucescens*). Both species have extensive distributions, but they are largely allopatric, except in Mesoamerica. The northernmost populations of *A. glaucescens* are in Nayarit and San Luis Potosí. *Asclepias elata* is one of a cohort of species reaching their northwestern limits in Arizona that indicate biogeographic affinity of that region to the eastern slope of the Sierra Madre

Occidental and Altiplano, rather than the western slope of the Sierra Madre Occidental. Other examples from the milkweed flora include *A. brachystephana*, *A. involucrata*, *A. nummularia*, *A. oenotheroides*, and *A. quinquedentata*. In Arizona, *A. elata* is known from Cochise, Graham, Pima, and Santa Cruz counties, in New Mexico from Eddy, Hidalgo, and Sierra counties, and in Texas from Brewster, Culberson, Jeff Davis, and Presidio counties.

38. **Asclepias arenaria** Torrey in W. H. Emory, Rep. U.S. Mex. Bound. 2(1): 162. 1859 • Sand milkweed

Herbs. **Stems** 1–8, spreading or decumbent to erect, unbranched or rarely branched near base, 20–100 cm, tomentose, not glaucous, rhizomatous. **Leaves** opposite, petiolate, with 1 or 2 stipular colleters on each side of petiole plus 0–4 in axil; petiole 7–17 mm, tomentose; blade oblong or obovate to ovate or oval, 4.2–11.5 × 2.5–7.5 cm, subcoriaceous, base rounded or truncate to subcordate, margins often undulate or crisped, apex truncate to rounded (rarely acute), sometimes emarginate, often mucronate, venation brochidodromous, surfaces tomentose to nearly glabrate, margins ciliate, 12–24 laminar colleters. **Inflorescences** extra-axillary (sometimes appearing terminal), sessile or pedunculate, 14–51-flowered; peduncle 0–2 cm, tomentose, with 1 caducous bract at the base of each pedicel. **Pedicels** 15–25 mm, densely tomentose. **Flowers** erect to pendent; calyx lobes lanceolate, 2.5–3 mm, apex acute, densely tomentose; corolla green to yellowish green, sometimes tinged reddish or purplish, lobes reflexed with spreading tips, oval, 7–8 mm, apex acute, glabrous abaxially, minutely papillose at base adaxially; gynostegial column 1–2 mm; fused anthers green, obconic, 2–2.5 mm, wings right-triangular, closed, apical appendages ovate; corona segments cream to greenish cream or ochroleucous, subsessile, conduplicate, flaring at base, 3.5–4 mm, exceeding style apex, apex truncate to rounded, emarginate, minutely papillate, proximal margin toothed, internal appendage falcate, exserted, sharply incurved over style apex, apex upturned, minutely papillose. **Follicles** erect on upcurved pedicels, lance-ovoid, 5.5–10 × 2–2.8 cm, apex acuminate, smooth, pilosulous. **Seeds** oval, 9–12 × 6–8 mm, margin winged, faces minutely rugulose; coma 2–3 cm.

Flowering May–Aug(–Oct); fruiting Jul–Sep(–Oct). Sandhills, dunes, sandy soils, prairies, pastures, grasslands, oak scrub, riparian areas; 100–1900 m; Colo., Kans., Nebr., N.Mex., Okla., S.Dak., Tex., Wyo.; Mexico (Chihuahua).

Asclepias arenaria is the milkweed most consistently associated with pure sand soils in the western Great Plains. It is predictably found at the bases of stabilized and semi-stabilized dunes. Flowers of this species are visited by a variety of Hymenoptera, notably several species of large wasps, including tarantula hawk wasps (Pepsinae, Pompilidae) and scoliid wasps (Scoliidae), as well as by Lepidoptera. Non-flowering shoots of *A. arenaria* may produce linear leaves; they are easily overlooked and not identified as belonging to this species unless one is aware of this variation, especially when they are produced on rhizomes distant from shoots with typical foliage. This trait is found in several other broad-leaved milkweeds (for example, *A. erosa*, *A. welshii*). *Asclepias arenaria* is rare and considered to be of conservation concern in Wyoming, where it has been recorded from only two sites in Goshen County.

39. **Asclepias prostrata** W. H. Blackwell, SouthW. Naturalist 9: 178. 1964 • Prostrate milkweed [C]

Herbs. **Stems** 2–7, prostrate to decumbent, sometimes branched, 15–30 cm, pilosulous to tomentulose, not glaucous, rhizomes absent. **Leaves** opposite, petiolate, with 1 stipular colleter on either side of petiole; petiole 2–3 mm, pilosulous to tomentulose; blade linear-lanceolate to deltate, 1.8–5 × 0.4–1.8 cm, chartaceous, base truncate to rounded or subcordate, margins crisped, apex acute, mucronate, venation eucamptodromous, surfaces pilosulous to tomentulose, margins ciliate, laminar colleters absent. **Inflorescences** extra-axillary, pedunculate, 3–8-flowered; peduncle 0.4–2 cm, densely pilosulous to tomentulose, with 1 caducous bract at the base of each pedicel. **Pedicels** 8–14 mm, densely pilosulous to tomentulose. **Flowers** erect; calyx lobes lanceolate, 3–4 mm, apex acute, pilosulous; corolla green, lobes reflexed, elliptic, 8–11 mm, apex acute, pilosulous abaxially, minutely hirtellous at base adaxially; gynostegial column 3–3.5 mm; fused anthers brown, obconic, 2–2.5 mm, wings right-triangular, closed, apical appendages ovate, erose; corona segments cream, tinged dorsally yellow, green, or pinkish, subsessile, conduplicate-tubular, 5–7 mm, greatly exceeding style apex, apex truncate, glabrous, internal appendage falcate, exserted, sharply inflexed over style apex, papillose; style apex shallowly depressed, green to yellowish. **Follicles** pendent on lax pedicels, ovoid, 3.5–5.5 × 1–1.5 cm, apex acuminate, muricate-ridged, tomentulose. **Seeds** broadly ovate, 7–8 × 5–6 mm, margin corky-winged, erose, faces very sparsely papillose; coma 1–1.8 cm.

Flowering Mar–Oct; fruiting Jul–Oct(–Dec). Arroyos, flats, hills, caliche, sandy, gravel, silty, and calcareous, often compacted soils, thorn scrub; of conservation concern; 50–200 m; Tex.; Mexico (Tamaulipas).

Asclepias prostrata is one of the most unusual and poorly known milkweeds in the flora. It was first collected by A. Schott in 1853 during the United States-Mexico border survey, along the Rio Grande (Río Bravo) between Laredo and Ringgold barracks (near Rio Grande City). However, it was not described until much later, from a collection made in Tamaulipas. The species remains rarely collected in both the United States and Mexico, and it is considered extremely rare in Texas (Starr and Zapata counties) and of conservation concern. Many historically known populations in the lower Rio Grande valley have not been relocated in recent years and are presumed extirpated (A. Strong, Texas Parks and Wildlife Department, pers. comm.). Several populations are known to have been eliminated by the widening of highways; others are thought to have been impacted by the spread of the invasive grass, *Cenchrus ciliaris* Linnaeus. The prostrate habit of *A. prostrata* cannot be confused with any other species of *Asclepias*. However, *A. prostrata* exhibits a remarkable similarity in all vegetative traits, including habit, to two co-occurring asclepiads, *Matelea brevicoronata* and *M. parvifolia*, as well as species of *Acleisanthes* (Nyctaginaceae), particularly the ubiquitous *Acleisanthes longiflora*. These species form a rather curious assemblage for which there is no hypothesized explanation involving convergent evolution. Since the description of *Asclepias prostrata*, prostrate species of *Matelea* occasionally have been misidentified as this species, even far outside its range.

40. Asclepias oenotheroides Schlechtendal & Chamisso, Linnaea 5: 123. 1830 • Zizotes or longhood or sidecluster milkweed, hierba de zizotes [F]

Asclepias lindheimeri Engelmann & A. Gray; *A. longicornu* Bentham; *Podostemma helleri* Greene

Herbs. Stems 1–7, erect to spreading or decumbent, unbranched or rarely branched near base, 10–50 cm, puberulent with curved trichomes to hirsutulous, not glaucous, rhizomatous. **Leaves** persistent or gradually caducous from base, opposite, petiolate, with 1 or 2 stipular colleters on each side of petiole; petiole 2–20 mm, puberulent with curved trichomes to hirtellous; blade ovate or lanceolate to oblong or elliptic, sometimes conduplicate, 4–11 × 1.2–6.5 cm, chartaceous, base cuneate to obtuse, margins sometimes crisped, apex obtuse to rounded, venation eucamptodromous to faintly brochidodromous, surfaces hirtellous, laminar colleters absent. **Inflorescences** extra-axillary, sessile or pedunculate, 8–32-flowered; peduncle 0–1 cm, hirtellous, with 1 caducous bract at the base of each pedicel. **Pedicels** 10–20 mm, hirtellous. **Flowers** erect to ascending; calyx lobes lanceolate, 3.5–4 mm, apex acute, hirtellous; corolla green, sometime tinged red or brown, sometimes faintly striate, lobes reflexed, elliptic to linear-lanceolate, (7–)9–12 mm, apex acute, glabrous at tips abaxially or hirtellous throughout, glabrous adaxially; gynostegial column 1–1.5 mm; fused anthers brown, sometimes green proximally, obconic, 2–2.5 mm, wings triangular, widest at middle, closed, apical appendages ovate; corona segments green, sometimes tinged bronze, apex white or cream, fading yellow, sessile, sinuous-tubular, relatively slender, (5–) 7–10 mm, greatly exceeding style apex, apex slightly flared, deeply emarginate, lobed on each side, minutely papillose to glabrate, internal appendage lingulate, sharply incurved, barely exserted, exceeded by segment margin, minutely papillose to glabrate. **Follicles** erect on upcurved pedicels, lance-ovoid, 4.5–9.5 × 1.2–2.5 cm, apex attenuate to acuminate, smooth, sometimes faintly striate, puberulent with curved trichomes to hirtellous. **Seeds** ovate to oval, 6–8 × 5–6 mm, margin winged, faces smooth; coma 2–3 cm.

Flowering (Feb–)Mar–Nov; fruiting (Apr–)May–Nov. Coastal and inland dunes, salt flats, shell mounds, hills, slopes, ridges, arroyos, canyons, valleys, urban lots, ditches, limestone, sandstone, shale, basalt, volcanic ash, caliche, alluvium, sandy, clay, silty, gravel, rocky, and calcareous soils, thorn scrub, desert scrub, desert and mesquite grasslands, prairies, pastures, pinyon-juniper, juniper, oak, and riparian woodlands, pine flatwoods; 0–1900 m; Ariz., Colo., La., N.Mex., Okla., Tex.; Mexico; Central America.

Asclepias oenotheroides is one of the most widespread American milkweeds, ranging from southeastern Colorado to Nicaragua. It is very common in southern and western Texas and throughout valleys and plains across Mexico. However, it is rare at the northern limit of its range in Colorado (known only from Las Animas County), where it is considered to be of conservation concern, Louisiana (known only from Jefferson Davis Parish), and Oklahoma. In the absence of flowers, it can be difficult to distinguish from its close relatives: *A. nyctaginifolia*, in southeastern Arizona and southwestern New Mexico, and *A. emoryi* in southern Texas. Compared to *A. oenotheroides*, the leaves of *A. nyctaginifolia* tend to be broader and more consistently ovate, whereas the leaves of *A. emoryi* tend to be narrower and more consistently elliptic or linear-lanceolate. However, *A. oenotheroides* is highly variable, and the overlap with these relatives is substantial. Even in flower, pressed specimens can be challenging to distinguish. The corona segments of *A. nyctaginifolia* are thicker and wider than those of *A. oenotheroides*, which is easily observed on fresh flowers but may be obscured by drying. Similarly, in fresh flowers the flared segment apex that exceeds the exserted appendage in

A. oenotheroides is easily distinguished from the segment apex that is closed by the flush appendage in *A. emoryi*, yet this obvious distinction is frustratingly obscure in dried material. For most of the range of *A. oenotheroides* the length of the corona segments greatly exceeds that of *A. emoryi*. However, along the southern Texas coastal plain, and especially on the barrier islands, the length of corona segments of *A. oenotheroides* is shorter, overlapping slightly with *A. emoryi*. Such plants correspond to the type of *Podostemma helleri*. In addition, hybridization of *A. oenotheroides* with both *A. emoryi* and *A. nyctaginifolia* is suspected, based on a few, scattered specimens with intermediate floral morphology in the regions of overlap with each species.

41. Asclepias emoryi (Greene) Vail in J. K. Small, Fl. S.E. U.S., 948. 1903 • Emory's milkweed

Podostemma emoryi Greene, Pittonia 3: 237. 1897

Herbs. **Stems** 1–4, erect to spreading, unbranched or rarely branched near base, 6–30 cm, puberulent with curved trichomes to hirtellous, not glaucous, rhizomes absent. **Leaves** persistent or gradually caducous from base, opposite, sessile or petiolate, with 1 stipular colleter on each side of petiole; petiole 0–17 mm, puberulent with curved trichomes to hirtellous; blade elliptic to lanceolate or lance-ovate, 3–7.5 × 0.4–2.5 cm, chartaceous, base cuneate, margins often crisped, apex acute, venation eucamptodromous, surfaces hirtellous, usually conduplicate, laminar colleters absent. **Inflorescences** extra-axillary, sessile or pedunculate, 4–8-flowered; peduncle 0–2 cm, hirtellous, with 1 caducous bract at the base of each pedicel. **Pedicels** 7–10 mm, hirtellous. **Flowers** erect; calyx lobes linear-lanceolate to narrowly elliptic, 3–4 mm, apex acute, hirtellous; corolla green, sometimes tinged red or brown, faintly striate, lobes reflexed, elliptic, 5–7 mm, apex acute, hirtellous throughout or glabrate at tips abaxially, glabrous adaxially; gynostegial column 0.5–1 mm; fused anthers green, obconic, 1–1.5 mm, wings trapezoidal, closed, apical appendages ovate; corona segments proximally green, distally white or cream, sessile, tubular, 3.5–5.5 mm, greatly exceeding style apex, apex flared, deeply emarginate, minutely papillose, internal appendage lingulate, sharply incurved, at the same level as and closing the segment apex, minutely papillose. **Follicles** erect on upcurved pedicels, lance-ovoid, 5–9 × 1.2–2 cm, apex attenuate to acuminate, smooth, puberulent with curved trichomes to hirtellous, sometimes faintly striate. **Seeds** oval, 7 × 5–6 mm, margin winged, faces smooth; coma 2.5–3.5 cm.

Flowering Mar–Aug(–Oct); fruiting Jul–Nov. Plains, hills, slopes, limestone, caliche, sandy, clay, rocky, calcareous, and gravelly soils, prairies, mesquite grasslands, thorn scrub; 0–800 m; Tex.; Mexico (Coahuila, Nuevo León, San Luis Potosí, Tamaulipas).

Asclepias emoryi is distributed entirely within the range of its close relative, *A. oenotheroides*. Distinguishing them is discussed under the latter species. A few putative hybrid specimens have been collected. These can be distinguished from *A. emoryi* by slightly longer corona segments (usually shorter than in *A. oenotheroides*) with sinuate apices, slightly longer corolla lobes, and slightly broader leaves. Although not accorded conservation concern, *A. emoryi* is very rarely encountered (across its entire range) and merits study for evaluation of needed protections. Reports of *A. emoryi* from New Mexico are based upon misidentifications. It is restricted in the flora area almost entirely to southern Texas, but there are a few scattered occurrences to the northwest.

42. Asclepias nyctaginifolia A. Gray, Proc. Amer. Acad. Arts 12: 69. 1876 • Mohave milkweed

Herbs. **Stems** 1–10, spreading or decumbent to erect, unbranched or rarely branched near base, 15–40 cm, puberulent with curved trichomes or hirtellous, not glaucous, rhizomatous. **Leaves** persistent or gradually caducous from base, opposite, petiolate, with 1 or 2 stipular colleters on each side of petiole plus 0–4 in axil; petiole 6–25 mm, puberulent with curved trichomes to hirtellous; blade ovate to lanceolate, 4.5–15 × 1.5–7.5 cm, chartaceous, base cuneate or obtuse to truncate or subcordate, margins sometimes crisped, apex obtuse to acute, venation eucamptodromous to faintly brochidodromous, surfaces hirtellous, rarely conduplicate, 0–12 laminar colleters. **Inflorescences** extra-axillary, sessile or pedunculate, 5–28-flowered; peduncle 0–1 cm, hirtellous, with 1 caducous bract at the base of each pedicel. **Pedicels** 17–30 mm, hirtellous. **Flowers** erect; calyx lobes lanceolate, 3–5 mm, apex acute, hirtellous; corolla green, sometimes tinged reddish or purplish abaxially, lobes reflexed, elliptic to lanceolate, 9–13 mm, apex acute, minutely hirtellous throughout or glabrous at tips abaxially, glabrous adaxially; gynostegial column 0.3–0.5 mm; fused anthers brown, obconic, 1.7–2 mm, wings triangular, widest at middle, closed, apical appendages ovate; corona segments cream to green with cream apex, fading yellow, sessile, tubular, slightly sinuous, relatively stout, 8–11 mm, greatly exceeding style apex, apex slightly flared, truncate, minutely papillose to glabrate, internal appendage lingulate, sharply incurved, barely exserted, greatly exceeded by segment margin and exposing cavity, minutely papillose. **Follicles** erect on upcurved pedicels, lance-ovoid, 6.5–10 × 1.5–3 cm, apex acuminate, smooth, sometimes faintly striate, puberulent with curved trichomes or hirtellous. **Seeds** ovate to oval, 6–8 × 4.5–6.5 mm, margin winged, faces smooth; coma 2–4 cm.

Flowering Apr–Sep(–Nov); fruiting May–Nov. Arroyos, canyons, mesas, hills, slopes, bajadas, ridges, plains, valleys, limestone, sandstone, granite, andesite, rhyolite, volcanic ash, sandy, silty, and gravel soils, desert scrub, mesquite and oak grasslands, oak and oak-juniper, and pinyon-juniper woodlands, chaparral, pine-oak forests; 300–1800(–2000) m; Ariz., Calif., Nev., N.Mex.; Mexico (Sonora).

Asclepias nyctaginifolia is a western counterpart of the more widespread *A. oenotheroides* and differs primarily in larger leaves and more robust corona segments. Differences are discussed under the latter species. The most widely used common name, Mohave milkweed, is somewhat misleading as the species is mainly distributed along the northern and eastern margins of the Sonoran Desert and barely enters the Mohave Desert. The species is common throughout the southwestern half of Arizona and is rare in California (San Bernardino County), Nevada (Clark County), and New Mexico (Catron, Grant, and Hidalgo counties).

43. Asclepias tomentosa Elliott, Sketch Bot. S. Carolina 1: 320. 1817 • Velvetleaf or tuba milkweed [E]

Herbs. Stems solitary (rarely 2), erect, unbranched (rarely branched), 25–150 cm, densely puberulent with curved trichomes, not glaucous, rhizomes absent. **Leaves** opposite, petiolate, with 1 stipular colleter on each side of petiole; petiole 2–9 mm, densely puberulent with curved trichomes; blade lanceolate or ovate to oval or oblong or elliptic to oblanceolate or obovate, 3.5–10 × 1–5 cm, chartaceous, base obtuse to subcordate, margins crisped, apex acute or obtuse to truncate or emarginate, sometimes mucronate, venation eucamptodromous to brochidodromous, surfaces puberulent with curved trichomes to tomentulose, margins ciliate, 4–8 laminar colleters. **Inflorescences** extra-axillary, sessile or pedunculate, 5–37-flowered; peduncle 0–0.3 cm, densely puberulent with curved trichomes to tomentulose, with 1 caducous bract at the base of each pedicel. **Pedicels** 12–19 mm, densely pilose to tomentulose. **Flowers** erect; calyx lobes lanceolate, 3–4 mm, apex acute, pilose; corollas green, often tinged reddish or purplish, lobes reflexed with spreading tips, oval, 7–9 mm, apex acute, glabrous; gynostegial column 0.5–1 mm; fused anthers green, obconic, 3–4 mm, wings broadly right-triangular, closed, apical appendages broadly oval; corona segments green with cream apex, often tinged pink or purple, stipitate, conduplicate, dorsally flattened, 3–4 mm, exceeded by style apex, apex truncate, glabrous, internal appendage falcate, exserted, sharply inflexed towards style apex, papillose; style apex shallowly depressed, green, fading pink or red. **Follicles** erect on upcurved pedicels, fusiform to narrowly lance-ovoid, 9–18 × 1–1.5 cm, apex long-acuminate, smooth, puberulent with curved trichomes to pilosulous or tomentulose. **Seeds** ovate, 6.5–8 × 4.5–6 mm, margin winged, remotely erose, faces minutely and sparsely papillose and rugulose; coma 3–3.5 cm.

Flowering May–Aug(–Oct); fruiting Jun–Oct. Sandhills, dunes, sandy and marl soils, pine, pine-palmetto, pine-oak, and oak scrubs, pine flatwoods; 0–200 m; Fla., Ga., N.C., S.C., Tex.

Asclepias tomentosa is restricted largely to coastal and inland sandhills. As described by B. A. Sorrie (2016), it exhibits a disjunct distribution, with gaps of unoccupied, but suitable, habitat in eastern Georgia and from the western Florida Panhandle to Louisiana. Sorrie reports a specimen from Alabama, but this cannot be found. When not in flower, *A. tomentosa* can be confused with *A. curtissii* in peninsular Florida, where they sometimes co-occur in close proximity, and with *A. obovata* on the Gulf Coastal Plain. It can be distinguished from *A. curtissii* by the usually larger and more densely vestitured leaf blades. Both species may have purple stems. *Asclepias obovata* can be distinguished from *A. tomentosa* by the hirtellous to velutinous vestiture of the herbage. Outside of Florida, populations of *A. tomentosa* are few, but it has not been considered to be of conservation concern; evaluation of its status in Texas and Georgia (known only from Coffee County) may be warranted.

SELECTED REFERENCE Sorrie, B. A. 2016. The curious distribution of *Asclepias tomentosa* (Apocynaceae). Phytoneuron 2016-68: 1–4.

44. Asclepias latifolia (Torrey) Rafinesque, Atlantic J. 1: 146. 1832 • Corn-kernel or broad-leaf milkweed

Asclepias obtusifolia Michaux var. *latifolia* Torrey, Ann. Lyceum Nat. Hist. New York 2: 217. 1827

Herbs. Stems 1–10, erect, unbranched, 25–100 cm, puberulent with curved trichomes or thinly tomentose to glabrate, sometimes glaucous, rhizomatous. **Leaves** opposite, sessile or petiolate, with 1–4 stipular colleters on each side of petiole, sometimes also in axil; petiole 0–4 mm, thinly tomentose to glabrate; blade oval or oblong to ovate or orbiculate, 5.5–14 × 3–14 cm, subsucculent to coriaceous, base cordate, sometimes clasping, margins entire, apex truncate to rounded, sometimes emarginate, mucronate, venation brochidodromous, surfaces thinly tomentose to glabrate, sometimes glaucous, margins minutely ciliate to glabrous, 24–80 laminar colleters. **Inflorescences** extra-axillary, sessile or pedunculate, 20–59-flowered; peduncle 0–2.5 cm, puberulent with curved trichomes to pilosulous, with 1 caducous bract at the base of each pedicel. **Pedicels** 15–35 mm, densely tomentose to glabrate. **Flowers** erect to pendent; calyx lobes lanceolate, 4–5 mm, apex acute, tomentose to glabrate; corolla green, lobes reflexed, elliptic to oval, 7–9 mm, apex acute, glabrous abaxially, papillose at base adaxially; gynostegial column 1–1.5 mm; fused anthers green, cylindric, 3–3.5 mm, wings right-triangular, closed, apical appendages ovate; corona segments cream, sometimes dorsally yellow, aging yellow, stipitate, conduplicate, dorsally rounded, 3–5.5 mm, equaling to slightly exceeding style apex, apex truncate, oblique, papillose, internal appendage falcate, exserted, sharply inflexed over style apex, papillose; style apex shallowly depressed, green. **Follicles** erect on upcurved pedicels, ovoid, 6.5–9.5 × 2–3 cm, apex obtuse to apiculate, smooth, minutely pilosulous to thinly tomentulose. **Seeds** ovate, 7–8 × 5–6 mm, winged, faces minutely rugulose to smooth; coma 3–4 cm. $2n = 22$.

Flowering May–Sep; fruiting Jun–Oct. Plains, hills, slopes, dunes, canyons, arroyos, terraces, springs, ditches, limestone, shale, sandstone, caliche, silty, clay, sandy, rocky, and gravel soils, prairies, shrubby and mesquite grasslands, pastures, desert scrub, pinyon-juniper, juniper, and riparian woodlands, pine forests; 400–2300 m; Ariz., Colo., Kans., Nebr., N.Mex., Okla., Tex., Utah; Mexico (Chihuahua, Coahuila).

Asclepias latifolia is a distinctive species of the western Great Plains and Colorado Plateau, rising above short grasses and appearing as squat, leafy pagodas. It is most likely to be confused with *A. arenaria* (which is restricted to sandy substrates) due to the overlapping leaf shapes and floral colors of these species. *Asclepias latifolia* favors clayey, often rocky soils, but can be found also on sandy soils, especially on the Colorado Plateau, outside the range of *A. arenaria*. These species can be distinguished by habit (erect in *A. latifolia* versus erect to decumbent in *A. arenaria*), vestiture (more uniformly and persistently hairy in *A. arenaria*), petioles (absent or nearly so in *A. latifolia* versus present in *A. arenaria*), and the flower and seed characters included in the key. *Asclepias speciosa* in the absence of reproductive structures is also commonly confused with *A. latifolia*, but the leaves of *A. speciosa* are distinctly petiolate, persistently hairy, and typically taper to the apex. There is an apparent gap in the distribution of *A. latifolia* on the eastern Colorado Plateau, in northwestern New Mexico and southwestern Colorado, but the disjunct portions of the range are not accompanied by phenotypic divergence. *Asclepias latifolia* is limited in Nebraska to southwestern counties (Deuel, Dundy, Franklin, and Hayes), but it is apparently not uncommon there. Likewise, it is common in its limited range in Utah (Garfield, Grand, Kane, San Juan, and Wayne counties).

45. Asclepias hallii A. Gray, Proc. Amer. Acad. Arts 12: 69. 1876 • Hall's milkweed [E]

Asclepias curvipes A. Nelson

Herbs. **Stems** 1–50, erect to ascending, unbranched or occasionally branched in inflorescence, 30–70 cm, puberulent with curved trichomes or pilose to glabrate, not glaucous, rhizomatous. **Leaves** alternate to subopposite (sometimes congested into pseudo-whorls), petiolate, with 1–4 stipular colleters on each side of petiole; petiole 10–20 mm, puberulent with curved trichomes to pilosulous; blade narrowly lanceolate to ovate, 5–16 × 1.5–9 cm, chartaceous, base obtuse to truncate, often oblique or unequal, margins entire, apex acute to obtuse, mucronate, venation brochidodromous, surfaces pilosulous or tomentulose to glabrate, more densely so on veins, margins ciliate, 10–20 comparatively large laminar colleters. **Inflorescences** extra-axillary, sometimes branched, pedunculate, 9–29-flowered; peduncle 0.5–10.5 cm, pilose to tomentulose, with 1 caducous bract at the base of each pedicel. **Pedicels** 16–28 mm, densely pilose to tomentulose. **Flowers** erect to spreading; calyx lobes lanceolate, 3–4 mm, apex acute, pilose; corolla pale pink to red or green with pink or red tinge, lobes reflexed with spreading tips, lance-ovate, 7–8 mm, apex acute to obtuse, glabrous abaxially, minutely papillate at base adaxially; gynostegial column 0.5–1 mm; fused anthers brown, obconic, 2–2.5 mm, wings curved and widest at base to nearly right-triangular, closed, base distended, apical appendages ovate, erose; corona segments cream with red or pink to purple dorsal stripe, sessile, conduplicate, dorsally flattened, 5–6.5 mm, greatly exceeding style apex, apex acute, glabrous, internal appendage falcate, exserted, sharply inflexed over style apex, glabrous; style apex shallowly depressed, cream to pink. **Follicles** erect on upcurved pedicels, lance-ovoid, 8–12 × 0.7–1.5 cm, apex acuminate, smooth, pilosulous. **Seeds** narrowly ovate, 6–7 × 4 mm, margin winged, faces and wings rugulose; coma 2.5–3.5 cm.

Flowering Jun–Aug; fruiting Jul–Sep. Slopes, ridges, ditches, arroyos, field margins, shale, ash, gypsum, igneous substrates, talus, gravel, clay, silt, sandy, and rocky soils, pinyon-juniper woodlands, shrubby grasslands, meadows, pine forests; 1700–3000 m; Ariz., Colo., Nev., N.Mex., Utah, Wyo.

Asclepias hallii is the only milkweed likely to be found above 2500 m in the Rocky Mountain and Intermountain regions, where it is found mostly on rocky canyon slopes. It appears to be a shorter-statured derivative of *A. speciosa*, which it greatly resembles, and which is the only species with which it commonly co-occurs. The range of *A. hallii* extends into the plains along river valleys and there overlaps with *A. speciosa*. Hybrids between these species have been documented only rarely in Colorado, and one of the syntypes of *A. curvipes* from Wyoming may represent this hybrid. Compared to *A. speciosa*, *A. hallii* is shorter statured, the leaves are narrower and are regularly subopposite or alternate rather than strictly opposite, and the corona segments are much shorter and erect. Despite overall similarities, it appears that these two species may not be close relatives (M. Fishbein et al. 2018).

The somewhat cruciform distribution of *Asclepias hallii* is odd and may be relictual. It extends from the eastern slope of the Rocky Mountains in Colorado west to the ranges of central Nevada, and from the Wasatch Range in northern Utah south to the Sierra Ancha in central Arizona. The species is fairly common at lower elevations of the Rocky Mountains in Colorado, particularly on the eastern slope, but is quite rare across the rest of its range, and it is considered to be of conservation concern in Utah. There are few documented occurrences in Arizona (Coconino, Gila, and Navajo counties), Nevada (Elko, Eureka, Lander, and White Pine counties), and New Mexico, where it has been documented by only a single specimen from Colfax County and may be extirpated. Its conservation status in these states merits evaluation. *Asclepias hallii* appears not to have been collected in Wyoming since 1958 and has been presumed to be extirpated; however, there is recent photo documentation from Albany County.

46. Asclepias viridiflora Rafinesque, Med. Repos., hexade 2, 5: 360. 1808 • Green or green comet milkweed, asclépiade à fleurs vertes

Acerates ivesii Wooton & Standley; *Asclepias viridiflora* var. *lanceolata* Torrey; *A. viridiflora* var. *linearis* (A. Gray) Fernald

Herbs. Stems solitary, erect to ascending, unbranched (rarely), (10–)20–125 cm, puberulent with curved trichomes, not glaucous, rhizomes absent. **Leaves** opposite to subopposite, sessile or petiolate, with 1 or 2 stipular colleters on each side of petiole and also in axil; petiole 0–5 mm, puberulent with curved trichomes; blade linear to broadly oval or nearly orbiculate, 2–13 × 0.8–6 cm, chartaceous, base cuneate to rounded, margins entire or crisped, apex acute or obtuse to truncate or emarginate, mucronate, venation brochidodromous, surfaces sparsely pilosulous to glabrate, margins ciliate, laminar colleters absent. **Inflorescences** extra-axillary at upper nodes, sometimes branched at peduncle apex, sessile or pedunculate, 22–60-flowered; peduncle 0–4 cm, puberulent with curved trichomes to pilosulous, with 1 caducous bract at the base of each pedicel. **Pedicels** 7–13 mm, pilosulous. **Flowers** erect to pendent; calyx lobes narrowly lanceolate, 2–3 mm, apex acute, pilosulous; corolla green to yellowish green, sometimes tinged red, lobes reflexed, oblong, 5–7 mm, apex acute, inconspicuously pilosulous at apex abaxially, glabrous adaxially; gynostegium sessile; fused anthers green, cylindric, 3–4 mm, wings triangular, widest at middle, closed, apical appendages ovate, marginally inflexed, apically deflexed; corona segments green to cream, sometimes tinged red, sessile, laminar, margins incurved, appressed to column, 3–4 mm, greatly exceeded by style apex, apex obtuse, glabrous, internal appendage absent or obscure, glabrous; style apex shallowly depressed, green. **Follicles** erect on upcurved pedicels, fusiform to lance-ovoid, 6–10 × 1.5–2 cm, apex acuminate to attenuate, smooth, pilosulous. **Seeds** ovate, 7–8 × 4–5 mm, margin winged, faces minutely rugulose; coma 2.5–3 cm. $2n$ = 22.

Flowering Apr–Sep(–Oct); fruiting Jun–Nov. Slopes, ridges, bluffs, flats, canyons, arroyos, glades, fields, meadows, pastures, sandhills, dunes, pond edges, streamsides, playas, sandstone, limestone, gypsum, serpentine, dolomite, alluvium, silty, sandy, clay, rocky, and calcareous soils, prairies, desert grasslands, oak scrub, oak, oak-juniper, oak-hickory, pine-oak, and pine woodlands, forest openings and edges; 0–2300 m; Alta., Man., Ont., Sask.; Ala., Ariz., Ark., Colo., Conn., Del., D.C., Fla., Ga., Ill., Ind., Iowa, Kans., Ky., La., Md., Mich., Minn., Miss., Mo., Mont., Nebr., N.J., N.Mex., N.Y., N.C., N.Dak., Ohio, Okla., Pa., S.C., S.Dak., Tenn., Tex., Va., W.Va., Wis., Wyo.; Mexico (Coahuila).

Asclepias viridiflora is one of the milkweeds with spherical, greenish umbels and inconspicuous coronas (see also *A. engelmanniana*, *A. hirtella*, *A. lanuginosa*, *A. longifolia*, *A. rusbyi*, *A. stenophylla*). Prior to close examination, the tight green balls of open flowers appear to be merely in bud. The diversity in leaf morphology among individuals (linear to orbiculate) is remarkable, but has no taxonomic significance—the full range of variation may be found within single populations. This is the most widespread milkweed within the flora area, ranging across most of the United States (absent only from the westernmost states and most of New England) and southern Canada. It is nowhere abundant, but may be regularly encountered in suitable, thin-soiled prairie habitats, especially in the Great Plains. It is rare and considered to be of conservation concern on the margins of its range, in Alberta (Cypress, Forty Mile, and Warner counties), Arizona (Coconino, Gila, and Yavapai counties), Connecticut (New Haven County), Florida (Gadsden and Jackson counties), and New York (Columbia, Nassau, Richmond, and Suffolk counties).

47. Asclepias lanuginosa Nuttall, Gen. N. Amer. Pl. 1: 168. 1818 • Wooly or sidecluster or small-green milkweed, asclépiade laineuse [E]

Herbs. Stems 1 or 2 (rarely more), erect to spreading, unbranched, 7–20 cm, densely hirtellous to pilose, not glaucous, rhizomatous. **Leaves** opposite or alternate, petiolate, with 1 stipular colleter on each side of petiole; petiole 1–2 mm, hirtellous; blade oblong or lanceolate to narrowly lanceolate, 4–8 × 0.5–2.7 cm, chartaceous, base cuneate to rounded or subcordate, margins entire, apex acute to obtuse, venation eucamptodromous to reticulodromous, surfaces hirtellous, margins ciliate, laminar colleters absent. **Inflorescences** terminal, usually solitary, pedunculate, 17–50-flowered (rarely more); peduncle 1–3 cm, densely hirtellous to pilose, with 1 caducous bract at the base of each pedicel. **Pedicels** 9–13 mm, hirtellous to pilose. **Flowers** erect to spreading; calyx lobes narrowly lanceolate, 1.5–2 mm, apex acute, hirtellous; corolla greenish cream, sometimes purple-tinged, lobes reflexed, oblong, 3–5 mm, apex acute, glabrous; gynostegial column 0–0.2 mm; fused anthers green, broadly cylindric, 1.5–2 mm, wings triangular, widest below middle, closed, apical appendages ovate; corona segments cream to greenish cream, sessile, chute-shaped, margins incurved, appressed to column, 2–3.5 mm, exceeded by style apex, base saccate, apex rounded, glabrous, internal appendage absent or obscure, glabrous; style apex shallowly depressed, green to greenish cream. **Follicles** erect on upcurved pedicels, fusiform, 8–10 × 0.8–1.5 cm, apex long-acuminate, smooth, hirtellous to pilose. **Seeds** ovate, 6–7 × 4–5 mm, faces minutely rugulose; comose. **2*n*** = 22.

Flowering May–Aug; fruiting Jul–Aug. Sandhills, dunes, moraines, bluffs, slopes, sandstone, limestone, sandy, gravel, or rocky soils, prairies, pine barrens and forests, oak savannas; 200–700 m; Man.; Ill., Iowa, Kans., Minn., Nebr., N.Dak., S.Dak., Wis.

Asclepias lanuginosa is highly cryptic due to its small stature and early flowering. It resembles a short, hirtellous form of *A. viridiflora*. In addition to the differences in vestiture, *A. lanuginosa* can be distinguished from that species by the terminal inflorescence and the cream-colored corona segments. Due to severe habitat loss in the tallgrass prairie region, *A. lanuginosa* has apparently declined and is of conservation concern over much of its range, that is, in Illinois, Iowa, Kansas, North Dakota, South Dakota, and Wisconsin. It is more secure in Nebraska, but an overall re-assessment of the status of this species is warranted. The only report from Montana is from the "Yellowstone expedition," which may not have been collected in the state. R. E. Woodson Jr. (1954) considered *A. lanuginosa* Kunth (a later homonym) to be the correct name for a Mexican species, *A. otarioides* E. Fournier. He soon realized that *A. lanuginosa* Nuttall has priority, but prior usage and his extensive annotations produced lingering confusion over the correct name of the Mexican species. Torrey's replacement name for this species, *A. nuttalliana*, is illegitimate, as it was superfluous on publication.

48. Asclepias stenophylla A. Gray, Proc. Amer. Acad. Arts 12: 72. 1876 • Slimleaf or narrow-leaved milkweed [E] [F]

Polyotus angustifolius Nuttall, Trans. Amer. Philos. Soc., n. s. 5: 201. 1836, not *Asclepias angustifolia* Schweigger 1812; *Acerates angustifolia* (Nuttall) Decaisne

Herbs. Stems 1 or 2 (rarely more), erect to spreading, rarely branched, 15–85 cm, puberulent with curved trichomes, not glaucous, rhizomes absent. **Leaves** alternate, subopposite, or opposite, sessile or petiolate, with 1 stipular colleter on each side of petiole; petiole 0–1 mm, spreading to ascending, glabrate; blade linear, conduplicate, 5–16 × 0.1–0.5 cm, chartaceous, base cuneate, margins entire, apex acute, venation faintly brochidodromous to obscure, surfaces sparsely puberulent with curved trichomes, especially on midvein, to glabrate, margins ciliate, laminar colleters absent. **Inflorescences** extra-axillary, sessile or pedunculate, 9–28-flowered; peduncle 0–1.3 cm, sometimes branched at apex, puberulent with curved trichomes, with 1 caducous bract at the base of each pedicel. **Pedicels** 5–12 mm, puberulent with curved trichomes to pilosulous. **Flowers** erect to pendent; calyx lobes lanceolate, 1.5–2.5 mm, apex acute, puberulent with curved trichomes to pilosulous; corolla pale green to greenish cream, lobes reflexed with spreading tips, elliptic, 3–5 mm, apex acute, minutely puberulent with curved trichomes at apex abaxially, glabrous adaxially; gynostegial column 0–0.5 mm, fused anthers green, truncately obconic, 1.5–2 mm, wings narrowly crescent-shaped, wide open at base, apical appendages deltoid; corona segments cream, often green-tinged, sessile, chute-shaped, margins incurved, appressed to anthers, 3–3.5 mm, equaling style apex, base saccate, auriculate, apex truncate, glabrous, internal appendage a short crest, the segment appearing 3-toothed, glabrous; style apex shallowly depressed, green. **Follicles** erect on straight pedicels, fusiform, 9–13 × 1–1.2 cm, apex long-acuminate, smooth, pilosulous. **Seeds** ovate, 6–7 × 4–5 mm, margin winged, faces minutely rugulose; coma 2.5–3 cm.

Flowering May–Aug; fruiting (Jun–)Aug–Oct. Hills, ridges, bluffs, slopes, flats, glades, sandhills, streamsides, limestone, dolomite, rhyolite, sandy and clay

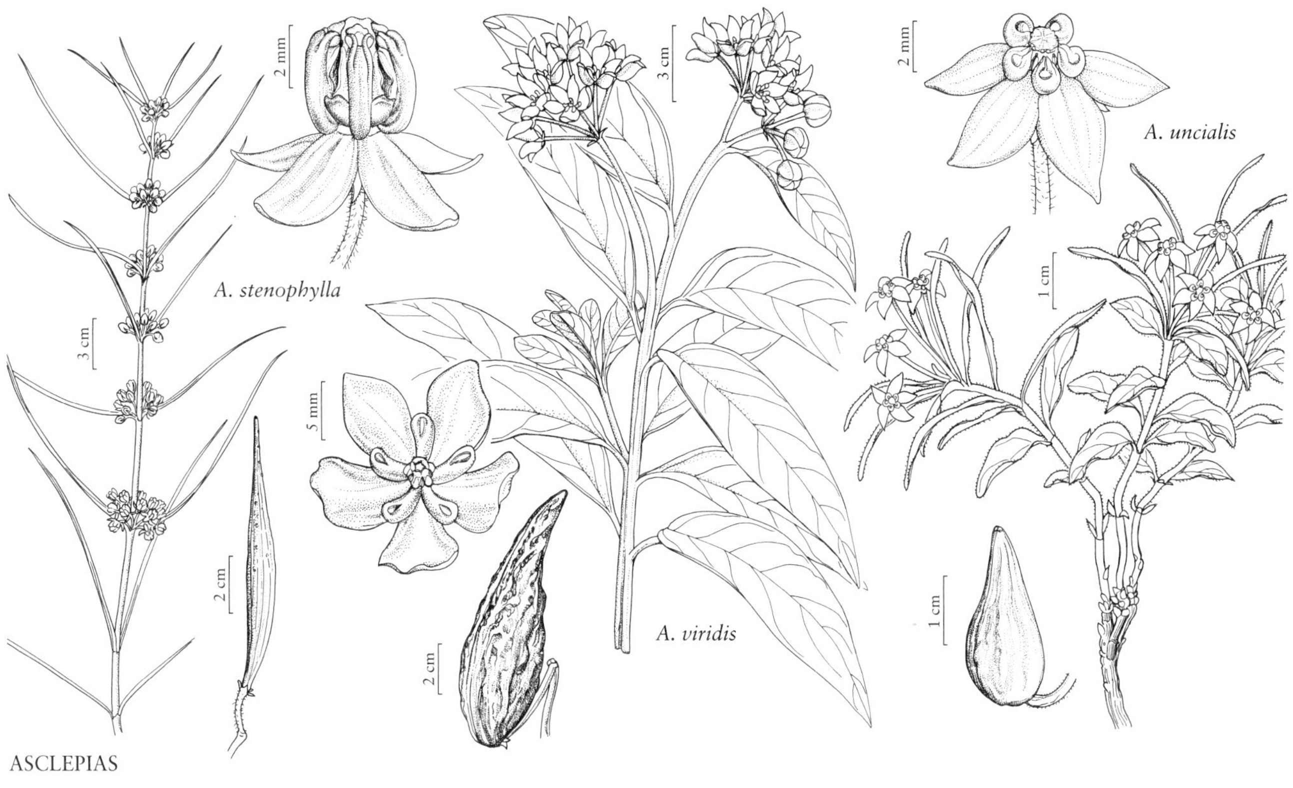

ASCLEPIAS

soils, prairies, pastures, thickets, forest openings, pine savannas; 70–1900 m; Ark., Colo., Ill., Iowa, Kans., La., Minn., Mo., Mont., Nebr., Okla., S.Dak., Tex., Wyo.

Although *Asclepias stenophylla* is a distinctive species, it is difficult to distinguish from *A. engelmanniana* in the absence of flowers or fruits, where their ranges overlap in the Great Plains. The drooping leaves of *A. engelmanniana* can reliably distinguish that species from *A. stenophylla*. *Asclepias stenophylla* is also often mistaken for *A. verticillata*, but the nearly appendageless corona segments and alternate or opposite (versus whorled) leaves readily separate *A. stenophylla* from that species. Because of its slender habit, linear leaves, and small clusters of greenish cream flowers held close to the stem, it can be overlooked in its grassland habitats. *Asclepias stenophylla* is widespread and common in its core habitat of Ozark glades and dry sites in tallgrass in Missouri, and in mixed-grass prairies from South Dakota to Texas. It is quite rare at the margins of its range in Arkansas (Baxter County), Illinois (Adams, Calhoun, and Pike counties), Iowa (Guthrie, Plymouth, and Sioux counties), Louisiana (Winn Parish), Minnesota (Houston County), Montana (Carter County), and Wyoming (Crook and Weston counties). In Colorado, it exhibits an interesting disjunction between Yuma County in the east and the Front Range of the Rocky Mountains, where it is sporadic, but impacted by development and considered to be of conservation concern. A report from North Dakota has not been confirmed.

49. Asclepias engelmanniana Woodson, Ann. Missouri Bot. Gard. 28: 207. 1941 • Engelmann's milkweed

Acerates auriculata Engelmann ex Torrey in W. H. Emory, Rep. U.S. Mex. Bound. 2(1): 160. 1859, not *Asclepias auriculata* Kunth 1819

Herbs. Stems 1 or 2 (rarely more), erect, sometimes branched, 40–160 cm, glabrous, not glaucous, rhizomes absent. **Leaves** alternate, sessile, drooping, with 1 stipular colleter on each side of leaf base; blade linear, conduplicate, 5–19 × 0.15–0.3 cm, chartaceous, base cuneate, margins entire, apex acute, venation faintly brochidodromous to obscure, surfaces sparsely puberulent with curved trichomes to glabrate, margins ciliate, laminar colleters absent. **Inflorescences** extra-axillary, sessile or pedunculate, 14–23-flowered; peduncle occasionally branched, 0–2 cm, pilosulous or puberulent with curved trichomes, with 1 caducous bract at the base of each pedicel. **Pedicels** 8–11 mm, pilose. **Flowers** erect to spreading; calyx lobes lanceolate, 3–4 mm, apex acute, pilosulous; corollas tan to russet abaxially, pale green to greenish cream or ochroleucous to tan adaxially, lobes reflexed with ascending tips, elliptic, 4–5 mm, apex acute, glabrous; gynostegial column 0.5–1.5 mm; fused anthers brown, broadly barrel-shaped, 2–2.5 mm, wings crescent-shaped and narrowly open throughout, apical appendages narrowly pandurate, conduplicate, not obscuring corpuscula; corona segments cream to tan or yellow, sessile, chute-shaped, 2–3 mm, equaling style apex, base saccate and auriculate, apex retuse to nearly truncate, glabrous, internal appendage absent or obscure, glabrous; style apex depressed, green to yellowish green. **Follicles** erect on upcurved pedicels, lance-ovoid, 6–10 × 1.2–2 cm, apex long-acuminate, smooth, pilosulous. **Seeds** ovate, 8–9 × 5–6 mm, margin winged, faces minutely papillose and rugulose; coma 2–2.5 cm.

Flowering (May–)Jun–Sep; fruiting Jul–Oct(–Nov). Hills, slopes, plains, valleys, arroyos, canyons, streamsides, ditches, sandhills, dunes, shale, sandstone, limestone, gypsum, igneous substrates, sandy, gravelly, clay, calcareous, and rocky soils, prairies, shrubby and mesquite grasslands, pastures, pinyon-juniper, juniper, oak, and oak-juniper woodlands, riparian forests; 200–2300 m; Colo., Kans., Nebr., N.Mex., Okla., Tex., Wyo.; Mexico (Chihuahua, Coahuila).

Asclepias engelmanniana is usually a tall herb with drooping leaves and spherical umbels of greenish yellow flowers rising above surrounding grassland plants. In spite of its distinctive appearance, it is quite similar to its close relatives, *A. rusbyi* and *A. stenophylla*. The yellowish green coronas, squat flowers, upcurved fruiting pedicels, and drooping leaves distinguish *A. engelmanniana* from *A. stenophylla*, which has more slender flowers with creamy coronas, straight pedicels in fruit, and spreading to ascending leaves. Despite ranges with only little overlap and few if any mixed populations, these two species are often confused, especially in the absence of flowers. Compared to its close relative, *A. engelmanniana* is distributed further west, in mixed- and short-grass prairies. A report of *A. engelmanniana* from South Dakota has not been confirmed, and reports from Iowa appear to have been based on misidentified specimens of *A. stenophylla*. Reports from Arkansas are unconfirmed and also very likely to be based on misidentifications. *Asclepias engelmanniana* is considered to be of conservation concern in Wyoming, where it has been recorded only from Goshen County. *Asclepias rusbyi* has been inconsistently distinguished from *A. engelmanniana* (for example, E. Sundell 1994), although the differences elucidated by R. E. Woodson Jr. (1954) are sound. These species are readily distinguished by the characters in the key and appear to have allopatric ranges. Reports of *A. engelmanniana* from Arizona, western New Mexico, Nevada, Utah, southwestern Colorado, and Sonora, Mexico, all pertain to *A. rusbyi*.

50. Asclepias rusbyi (Vail) Woodson, Ann. Missouri Bot. Gard. 41: 183. 1954 • Rusby's milkweed

Acerates rusbyi Vail, Bull. Torrey Bot. Club 25: 37. 1898; *Asclepias engelmanniana* Woodson var. *rusbyi* (Vail) Kearney

Herbs. Stems 1 or 2 (rarely more), erect, sometimes branched, 50–100 cm, glabrous, not glaucous, rhizomes absent. **Leaves** alternate, sessile or petiolate, drooping, with 1 stipular colleter on each side of petiole; petiole 0–0.1 cm, glabrous; blade linear, conduplicate, 9–15 × 0.2–0.3 cm, chartaceous, base cuneate, margins entire, apex acute, venation faintly brochidodromous to obscure, surfaces glabrous, margins ciliate, laminar colleters absent. **Inflorescences** extra axillary, sessile or pedunculate, 7–28-flowered; peduncle occasionally branched, 0–1.5 cm, pilosulous, with 1 caducous bract at the base of each pedicel. **Pedicels** 7–10 mm, pilose. **Flowers** erect to spreading; calyx lobes narrowly lanceolate, 3–4 mm, apex acute, pilosulous to glabrate; corollas abaxially russet or tan or bronze to pale green, adaxially pale green, lobes reflexed with ascending tips, elliptic, 4–6 mm, apex acute, glabrous; gynostegial column 0–0.5 mm; fused anthers green, obconic, 3 mm, wings crescent-shaped, connivent distally, wider and open at base, apical appendages deltoid, obscuring corpuscula; corona segments yellowish green to yellow or bronze, sessile, chute-shaped, 1.5–2.5 mm, exceeded by style apex, base saccate, auriculate, apex truncate, glabrous; internal appendage absent or a low crest, glabrous; style apex shallowly depressed, green to yellowish green. **Follicles** erect on upcurved pedicels, lance-ovoid, 9–12.5 × 1.5–2.5 cm, apex long-acuminate, smooth, faintly striate, sparsely pilosulous to glabrate. **Seeds** ovate, 6–8 × 4–6 mm, margins winged, faces minutely papillose and rugulose; coma 2–2.5 cm.

Flowering Jun–Aug; fruiting Aug–Oct. Ridges, arroyos, canyons, slopes, basalt, granite, sandstone, shale, sandy, rocky, clay, and gravel soils, oak and pinyon-juniper woodlands, pine, pine-oak, and riparian forests, forest edges; 1200–2300 m; Ariz., Colo., Nev., N.Mex., Utah; Mexico (Sonora).

As noted under *Asclepias engelmanniana*, *A. rusbyi* has been only inconsistently recognized as distinct from its more widespread relative. The species has an unusual distribution. In the north, *A. rusbyi* is encountered on the Colorado Plateau north of the Colorado River and extends west into pinyon-juniper woodlands and desert canyons at the upper margin of the Sonoran Desert. A few populations have been documented in Colorado (Archuleta, La Plata, Mesa, Montezuma, and Montrose counties), Nevada (Lincoln County), northern New

Mexico (Rio Arriba and San Juan counties), and Utah (Garfield, San Juan, and Washington counties), and should be considered to be of conservation concern in these states. South of the Colorado River, *A. rusbyi* is found throughout eastern Arizona, from the southern edge of the Colorado Plateau, through the Mogollon Rim, to the sky-island mountains of the border region, where it is found in pinyon-juniper woodlands and pine-oak forests. In this region, it is rare in New Mexico (Catron and Grant counties).

51. **Asclepias involucrata** Engelmann ex Torrey in W. H. Emory, Rep. U.S. Mex. Bound. 2(1): 163. 1859 • Dwarf milkweed

Herbs. Stems 1–15, decumbent, unbranched or branched near base, 5–18 cm, puberulent with curved trichomes to pilosulous, not glaucous, rhizomatous. **Leaves** opposite or subopposite to alternate, petiolate, with 1 or 2 stipular colleters on either side of petiole; petiole 1–2 mm, pilosulous to glabrate; blade linear to narrowly lanceolate, 1.5–12 × 0.2–0.8 cm, chartaceous, base cuneate to truncate, margins sometimes crisped, apex acute, mucronate, venation obscure to faintly eucamptodromous, surfaces sparsely pilosulous to glabrate, midvein puberulent with curved trichomes to pilosulous, margins densely ciliate, laminar colleters absent. **Inflorescences** terminal, sessile, 6–35-flowered, bracts few. **Pedicels** 12–19 mm, puberulent with curved trichomes to pilosulous. **Flowers** erect to spreading; calyx lobes elliptic, 2.5–3 mm, apex acute, pilosulous; corolla green, sometimes tinged pink or red (especially abaxially), lobes reflexed, elliptic, 4.5–6 mm, apex acute, glabrous; gynostegial column 0.5–1 mm; fused anthers brown, cylindric, 1.5–2 mm, wings right-triangular, slightly open at tip, apical appendages ovate; corona segments cream, usually with a pink or red dorsal stripe, subsessile, conduplicate, 3.5–4.5 mm, slightly exceeding style apex, apex truncate with a spreading tip, glabrous, internal appendage falcate, exserted, sharply inflexed towards the style apex, glabrous; style apex shallowly depressed, cream or greenish cream to pink. **Follicles** erect on upcurved pedicels, ovoid, 4.5–5.5 × 1.5–2 cm, apex acuminate, rugose, faintly striate, minutely puberulent with curved trichomes to pilosulous. **Seeds** ovate, 7–8 × 5–6 mm, margin thickly winged, faces papillose and rugulose, lepidote; coma 1.5–2 cm.

Flowering Mar–Jul; fruiting May–Aug. Hills, slopes, ridges, canyons, arroyos, valleys, playas, flats, dunes, limestone, sandstone, basalt, calcareous, rocky, sandy, silty, and clay soils, alluvium, prairies, mesquite, shrubby, and desert grasslands, chaparral, oak, juniper, and pinyon-juniper woodlands, pine forests, pastures; 1000–2200 m; Ariz., Colo., Kans., N.Mex., Okla., Tex.; Mexico (Chihuahua, Durango, Nuevo León, Sonora).

Despite the common name dwarf milkweed, *Asclepias involucrata* is larger than the sympatric *A. uncialis*, to which it bears a great similarity in the absence of flowers or fruit. In such conditions, *A. involucrata* is highly cryptic among the short-statured bunch grasses with which it grows. It senesces typically by summer's end, contributing to the impression that the species is not common. It has occasionally been circumscribed to include *A. macrosperma* (for example, E. Sundell 1994), although the distinctions made by R. E. Woodson Jr. (1954) were sound. Nonetheless, where the ranges of these largely parapatric species meet, in an arc from northwestern New Mexico to central Arizona, plants of intermediate morphology can be found. It is unknown whether these represent relics of the speciation process or examples of recent hybridization. *Asclepias involucrata* is rare and of conservation concern in Colorado (Baca, Bent, and Las Animas counties) and Oklahoma (Cimarron County). It is probably extirpated from Kansas (Stevens County), where it is known from a single, historical record. Reports from northern Arizona, southwestern Colorado, and Utah are based on records of *A. macrosperma*.

52. **Asclepias macrosperma** Eastwood, Bull. Torrey Bot. Club 25: 172. 1898 • Large-seed or dwarf or Eastwood's milkweed [E]

Asclepias involucrata Engelmann ex Torrey var. *tomentosa* Eastwood, Zoë 4: 120. 1893

Herbs. Stems 1–12, decumbent, unbranched or branched near base, 6–15 cm, densely puberulent with curved trichomes, not glaucous, rhizomatous. **Leaves** opposite or subopposite to alternate, petiolate, with 1 stipular colleter on either side of petiole; petiole 1–5 mm, densely puberulent with curved trichomes; blade lanceolate to linear-lanceolate or lance-ovate, 2.5–7 × 0.5–2 cm, chartaceous, base cuneate to truncate, margins crisped, apex attenuate, mucronate, venation obscure to faintly eucamptodromous, surfaces sparsely pilosulous to glabrate, midvein puberulent with curved trichomes to pilosulous, margins densely ciliate, laminar colleters absent. **Inflorescences** terminal, sessile, 12–40-flowered, bracts few. **Pedicels** 9–19 mm, densely pilose. **Flowers** erect to pendent; calyx lobes elliptic, 2.5–3 mm, apex acute, pilose; corolla green, tinged red (especially abaxially), lobes reflexed, oval, 4–5.5 mm, apex acute, glabrous; gynostegial column 0.2–0.8 mm; fused anthers brown, cylindric, 1–1.5 mm, wings right-triangular, slightly open at tip, apical appendages ovate; corona segments yellow to ochroleucous, subsessile, tubular, 2–3 mm, equaling or slightly exceeding style apex, apex truncate, glabrous, internal appendage falcate, exserted, sharply inflexed towards or over style apex, glabrous; style apex shallowly depressed, green or pink. **Follicles** erect on upcurved pedicels, ovoid, 5–6.5 × 1.2–2 cm, apex acuminate, rugose, faintly striate, densely pilosulous. **Seeds** ovate, 8–12 × 6–8 mm, margin thickly winged, faces densely rugulose; coma 1.5–2 cm.

Flowering Apr–Jun; fruiting May–Jul. Arroyos, hills, ridges, canyons, dunes, sandstone, limestone, sandy soils, juniper woodlands, shrubby grasslands, desert scrub; 900–1800 m; Ariz., Colo., N.Mex., Utah.

Asclepias macrosperma has not been consistently recognized as distinct from *A. involucrata*, as discussed under that species. It is quite homogeneous across its range on the Colorado Plateau, and intermediates with *A. involucrata* only occur where the ranges contact on the southern and eastern margins of the Plateau. It is readily distinguished from typical *A. involucrata* by broader leaves with crisped margins and smaller corona segments that are less compressed, less flared apically, and nearly uniformly yellow, as opposed to cream with a dark dorsal stripe. Also, *A. macrosperma* is largely confined to sandy, often unstabilized substrates, whereas *A. involucrata* occurs on stable, rocky, clay to sandy soils across most of its range. *Asclepias macrosperma* has been documented at few sites in Colorado (Montezuma County) and New Mexico (San Juan County) and should be considered to be of conservation concern in these states. It appears secure in Utah and on Navajo lands in Arizona.

53. Asclepias welshii N. H. Holmgren & P. K. Holmgren, Brittonia 31: 110, fig. 1. 1979 • Welsh's milkweed [C] [E]

Herbs. **Stems** 1–20, erect to ascending, unbranched, 35–100 cm, densely tomentose to glabrate, not glaucous, rhizomatous. **Leaves** subopposite to opposite (alternate), sessile or petiolate, stipular colleters absent; petiole 0–8 mm, tomentose to glabrate; blade oblong or obovate to oval or ovate, 4.5–14.5 × 2–8 cm, subsucculent, base rounded to cordate, margins entire, apex rounded or truncate to emarginate or acute, mucronate, venation brochidodromous, secondary veins orthogonal, surfaces tomentose to glabrate, margins eciliate, 0–10 laminar colleters. **Inflorescences** extra-axillary from upper nodes, pedunculate, 22–80-flowered; peduncle 2.5–6 cm, densely tomentose, with 1 caducous bract at the base of each pedicel. **Pedicels** 10–15 mm, densely pilose to tomentose. **Flowers** erect to pendent; calyx lobes linear, 5–6 mm, apex acute, densely pilose to tomentose; corolla ochroleucous or tan to red-violet, lobes reflexed with spreading tips, oval to oblong, 5.5–6.5 mm, apex acute, densely tomentose abaxially, glabrous adaxially; gynostegial column 1–1.5 mm; fused anthers green to tan, obconic, 1.5 mm, wings triangular, widest at middle, closed, apical appendages ovate; corona segments cream to ochroleucous, sometimes green at base, shiny, stipitate, tubular, dorsally flattened, 2.5–3.5 mm, exceeding style apex, apex truncate, glabrous, internal appendage falcate with acicular tip, exserted, sharply inflexed over style apex, glabrous; style apex shallowly depressed, green. **Follicles** pendulous on spreading pedicels, ellipsoid, 5–7 × 1.7–3 cm, apex acuminate, softly muricate, densely to thinly tomentose. **Seeds** oval, 18–20 × 9–10 mm, margin winged, faces minutely rugulose; coma 4–4.5 cm.

Flowering May–Sep; fruiting Jun–Oct. Active orange to red sand dunes, adjacent to pinyon-juniper and pine woodlands; of conservation concern; 1400–1900 m; Ariz., Utah.

With respect to the number of populations, *Asclepias welshii* is the most endangered *Asclepias* in the flora area and is federally listed as threatened in the United States. Most of the populations are remote and relatively secure; however, the most accessible population in a Utah park has been subject to the impacts of off-road vehicle recreation. Leaves and stems of *A. welshii* emerge densely tomentose, but the lower portions of the plants become sand blasted and smooth as the season progresses. The species is strongly rhizomatous, as befits its shifting substrate. Seedlings and sprouts from rhizomes often bear narrowly linear leaves that differ so strongly from the foliage of more robust stems that they are not easily attributed to this species. Similar heterophylly is found in *A. arenaria*, *A. erosa*, and other milkweeds of sandy substrates.

54. Asclepias labriformis M. E. Jones, Proc. Calif. Acad. Sci., ser. 2, 5: 708. 1895 • Utah or labriform or Jones's or poison milkweed [E]

Herbs. **Stems** 1–10, erect, unbranched (rarely distally), 15–70 cm, sparsely puberulent with curved trichomes to glabrate, not glaucous, rhizomatous. **Leaves** alternate (subopposite), sessile or petiolate, with 1 or 2 stipular colleters on each side of petiole plus 4–8 in axil; petiole 0–7 mm, glabrous; blade linear-lanceolate to lanceolate, often falcate, 6–14.5 × 0.5–2 cm, chartaceous, base cuneate, margins entire, apex attenuate to acute, mucronate, venation brochidodromous (often faintly), surfaces glabrous, midvein sometimes sparsely puberulent with curved trichomes abaxially, margins eciliate,

0–12 laminar colleters. **Inflorescences** extra-axillary, pedunculate, 6–28-flowered; peduncle 0.2–2.5 cm, tomentose to glabrate, with 1 caducous bract at the base of each pedicel. **Pedicels** 15–25 mm, tomentose to glabrate. **Flowers** erect to pendent; calyx lobes lanceolate, 3–4 mm, apex acute, sparsely pilose or tomentulose to glabrate; corolla ochroleucous, lobes reflexed, tips sometimes spreading, oval to elliptic, 6–7 mm, apex acute, glabrous abaxially, minutely hirtellous at base adaxially; gynostegial column 1–1.5 mm; fused anthers green to yellowish green, obconic, 1.5–2 mm, wings narrowly right-triangular, tip distended, closed, apical appendages ovate; corona segments ochroleucous to cream, substipitate, conduplicate, dorsally rounded, nearly tubular, 2.5–3.5 mm, slightly exceeding style apex, apex truncate with proximal tooth on each side, glabrous, internal appendage falcate, exserted, sharply inflexed over style apex, glabrous; style apex shallowly depressed, cream or green. **Follicles** pendulous on spreading to pendulous pedicels, ovoid, 3.5–5.5 × 1.2–2 cm, apex acuminate, smooth, tomentulose to glabrate. **Seeds** ovate to oval, 12–13 × 8–9 mm, margin winged, faces obscurely rugulose to smooth; coma 1.5–2 cm.

Flowering May–Aug; fruiting Jun–Oct. Arroyos, canyons, flats, terraces, bluffs, ditches, sandstone, shale, siltstone, gypsum, sandy, gravel, and clay soils, riparian and juniper woods, desert scrub; 1300–2000 m; Utah.

Asclepias labriformis is endemic to Utah (Emery, Garfield, Grand, Uintah, and Wayne counties) and is found in a remarkably narrow northeast–southwest band from the terraces surrounding the San Rafael Swell west of Green River, across Waterpocket Fold, to arroyo beds below the Kaiparowits Plateau southeast of Escalante. It shares pendulous follicles with several other milkweed species inhabiting sandy habitats, such as its close relative *A. welshii*, and more distant relatives *A. cutleri* and *A. subulata*. *Asclepias labriformis* is reputed to be one of the milkweeds that is most poisonous to livestock (J. M. Benson et al. 1979).

55. **Asclepias asperula** (Decaisne) Woodson, Ann. Missouri Bot. Gard. 41: 193. 1954 • Antelope horns, spider milkweed, spider antelope horns

Acerates asperula Decaisne in A. P. de Candolle and A. L. P. P. de Candolle, Prodr. 8: 522. 1844; *Asclepiodora asperula* (Decaisne) E. Fournier

Herbs. **Stems** 1–40, erect to decumbent, unbranched or branched at base, 15–60 cm, puberulent with curved trichomes, not glaucous, rhizomes absent. **Leaves** alternate to subopposite, petiolate, with 1–3 stipular colleters on each side of petiole; petiole 2–4 mm, puberulent with curved trichomes to pilosulous; blade lanceolate to linear, 5–17 × 0.4–3.7 cm, chartaceous, base cuneate, margins entire, apex attenuate to acute, mucronate, venation eucamptodromous to brochidodromous, surfaces puberulent with curved trichomes to glabrate, more densely so on veins, margins ciliate, laminar colleters absent. **Inflorescences** terminal, sessile or pedunculate, 10–60-flowered; peduncle 0–22.5 cm, puberulent with curved trichomes, with 1 caducous bract at the base of each pedicel. **Pedicels** 16–30 mm, puberulent with curved trichomes to pilose. **Flowers** erect to spreading; calyx lobes ovate to linear-lanceolate, 3–5 mm, apex acute, pilosulous to puberulent with curved trichomes; corolla pale green, sometimes tinged red abaxially, campanulate, lobes ascending and exceeding corolla segments, ovate to oval, 7–10 mm, apex acute, puberulent with curved trichomes at apex abaxially, glabrous adaxially; gynostegium sessile; fused anthers brown and green, turbinate, 2–2.5 mm, wings trapezoidal, widest above middle, closed, apical appendages ovate, erose; corona segments reddish purple and white, sessile, clavate-tubular, 4.5–7 mm, slightly exceeded by to equaling style apex, deflexed at base, margins connivent, apex incurved, rounded, upper margin and cavity papillose, internal appendage a low internal crest, papillose; style apex depressed, green. **Follicles** erect on upcurved pedicels, lance-ovoid, 6–11.5 × 1–2.5 cm, apex short- to long-acuminate, weakly to strongly arcuate, shallowly rugose-ribbed, ribs sometimes muricate, striate, pilosulous. **Seeds** ovate, 5–8 × 4–6 mm, margin winged, remotely erose, faces minutely rugulose-papillose, minutely hirtellous; coma 2.5–4 cm.

Subspecies 2 (2 in the flora): w United States, Mexico.

With terminal umbels of large, green and purple flowers, *Asclepias asperula* is highly distinctive and in flower can be confused only with its sister species, *A. viridis*. Distinguishing characteristics and the existence of interspecific hybrids are discussed under *A. viridis*. Large bees, notably *Bombus* and *Xylocopa*, are commonly observed visiting the flowers of *A. asperula*.

The subspecies of *Asclepias asperula* are strongly differentiated away from their region of contact, which extends from south of the Texas Panhandle to the extreme tip of the Oklahoma Panhandle. In the region of contact, their distinguishing traits intermix. The common occurrence of intermediates and apparent introgressants argues against elevation of the subspecies to the specific rank. There is a surprising gap in the distribution of the species as a whole on the Llano Estacado in eastern New Mexico and the Texas Panhandle that may contribute to the differentiation of the subspecies.

1. Corona segments reddish purple with white upper margin, 5–7 mm; follicle ribs inconspicuously muricate at apex or not at all . . . . 55a. *Asclepias asperula* subsp. *asperula*
1. Corona segments white, dorsally reddish purple, 4.5–6 mm; follicle ribs sparsely to densely muricate for most of the length . . . . 55b. *Asclepias asperula* subsp. *capricornu*

55a. Asclepias asperula (Decaisne) Woodson subsp. **asperula**

Asclepias capricornu Woodson subsp. *occidentalis* Woodson

Leaves blades linear-lanceolate to linear, 7.7–17 × 0.4–3.6 cm, apex attenuate. **Peduncles** 0–22.5 cm. **Corona segments** reddish purple with white upper margin, 5–7 mm. **Follicles** (6–)8–11.5 cm, ribs rarely inconspicuously muricate near apex or not at all. **Seeds** 6–8 × 5–6 mm.

Flowering Mar–Sep(–Oct); fruiting (Apr–)May–Sep(–Oct). Slopes, hillsides, mesas, ridge tops, bajadas, canyons, arroyos, streamsides, lakesides, basalt, granite, limestone, shale, sandstone, alluvium, talus, rocky, gravelly, clay, sand, and silt soils, grasslands, pinyon-juniper, oak, juniper, pine, pine-oak, mixed conifer, and mesquite woodlands, chaparral, riparian areas, desert scrub; 700–2800 m; Ariz., Calif., Colo., Idaho, Nev., N.Mex., Okla., Tex., Utah; Mexico.

Although subsp. *asperula* is uncommon in Mexico, it ranges surprisingly far south, to Oaxaca. In the flora area, subsp. *asperula* has limited distributions in California (San Bernardino County), Idaho (Franklin County), Oklahoma (Cimarron County), and Texas (trans-Pecos region).

55b. Asclepias asperula (Decaisne) Woodson subsp. **capricornu** (Woodson) Woodson, Ann. Missouri Bot. Gard. 41: 195. 1954 E

Asclepias capricornu Woodson, Ann. Missouri Bot. Gard. 32: 370. 1945, based on *Anantherix angustifolia* Rafinesque, Atlantic J. 1: 146. 1832, not *Asclepias angustifolia* Schweigger 1812; *Acerates decumbens* (Nuttall) Decaisne; *Anantherix decumbens* Nuttall; *Asclepias asperula* var. *decumbens* (Nuttall) Shinners; *Asclepiodora decumbens* (Nuttall) A. Gray

Leaves blades lanceolate, 5–14.5 × 0.7–3.7 cm, apex acute to attenuate. **Peduncles** 0–6.8 cm. **Corona segments** white, reddish purple dorsally, 4.5–6 mm. **Follicles** 6–9(–11.5) cm, ribs sparsely to densely muricate for most of length. **Seeds** 5–7 × 4–5 mm.

Flowering Mar–Sep(–Nov); fruiting Apr–Sep(–Nov). Hills, slopes, ridge tops, plains, streambanks, dunes, pastures, fields, limestone, sandstone, gypsum, igneous substrates, rocky, clay, calcareous, and sandy soils, prairies, shrubby grasslands, oak-juniper, juniper, and oak woodlands; 50–1200 m; Kans., Nebr., Okla., Tex.

Subspecies *capricornu* is considered to be of conservation concern in Nebraska, where it has been recorded only at a single location in Nuckolls County.

56. Asclepias viridis Walter, Fl. Carol., 107. 1788 • Green antelopehorn or green or spider milkweed E F

Acerates paniculata (Nuttall) Decaisne; *Anantherix paniculata* Nuttall; *A. viridis* Nuttall; *Asclepiodora viridis* (Walter) A. Gray; *Podostigma viride* (Walter) Elliott

Herbs. Stems 1–25, decumbent to erect, unbranched or rarely branched, 15–70 cm, inconspicuously puberulent with curved trichomes to glabrate, not glaucous, rhizomes absent. **Leaves** alternate to subopposite, petiolate, with 1–4 stipular colleters on each side of petiole plus 2–4 in axil; petiole 2–6 mm, puberulent with curved trichomes to glabrate; blade oblong or ovate to oval, elliptic, or lanceolate, 3–13 × 1–6 cm, chartaceous, base rounded to subcordate, margins entire, apex obtuse to rounded, often emarginate, sometimes mucronate, venation eucamptodromous to faintly brochidodromous, surfaces puberulent with curved trichomes to glabrate, more densely so on veins, margins ciliate, 6–12 laminar colleters. **Inflorescences** terminal, pedunculate, 4–23-flowered; peduncle usually branched, 0.5–6 cm, pilosulous to puberulent with curved trichomes, with 1 caducous bract at the base of each pedicel. **Pedicels** (8–)14–23 mm, puberulent with curved trichomes to pilosulous. **Flowers** erect to spreading; calyx lobes linear-lanceolate, 4–5 mm, apex acute, pilosulous to puberulent with curved trichomes; corolla pale green, campanulate, lobes ascending and exceeding corona segments, oval, (9–)12–15 mm, apex acute to obtuse, glabrous; gynostegium sessile; fused anthers brown and green, turbinate, 2.5–3 mm, wings shallowly trapezoidal, closed, apical appendages ovate, erose; corona segments pale to dark purple, upper margin usually white, sessile, clavate-tubular, 3–5 mm, greatly exceeded by style apex, deflexed at base, margins connivent, apex ascending to incurved, rounded, upper margin and cavity hirtellous, internal appendage a low internal crest, hirtellous; style apex shallowly depressed, green. **Follicles** erect on upcurved pedicels, lance-ovoid to ovoid, 6–13 × (0.5–)1.2–3 cm, apex acuminate, shallowly rugose-ribbed, inconspicuously muricate apically on ribs, striate, pilosulous. **Seeds** broadly ovate, 5.5–7.5 × 4–6 mm, margin winged, obscurely erose at chalazal end, faces rugulose, minutely hirtellous; coma 2.5–4 cm.

Flowering Jan–Nov; fruiting (Apr–)May–Nov. Slopes, flats, glades, ravines, fields, pastures, hammocks, ditches, shale, limestone, granite, sandstone, silty, sandy, rocky, clay, and calcareous soils, prairies, mesquite-juniper grasslands, oak-hickory, pine-oak, and riparian woodlands, oak forests, forest edges and openings; 0–600 m; Ala., Ark., Fla., Ga., Ill., Ind., Iowa, Kans., Ky., La., Miss., Mo., Nebr., Ohio, Okla., Tenn., Tex.

Asclepias viridis is similar only to its sister species, *A. asperula*. The broader leaves with broader apices and corona segments that are less than half the length of the corolla lobes readily distinguish *A. viridis* from *A. asperula*. These species are also largely segregated edaphically: *A. viridis* on deeper, valley soils and *A. asperula* on rocky, upland soils. Hybrids between *A. viridis* and *A. asperula* subsp. *capricornu* have been documented at several locations in northern Texas and southern Oklahoma. Both species flower early in that region (April–May) and may re-flower sporadically through the summer in response to disturbance from fire or mowing, with a second peak of flowering in the fall when weather conditions are favorable. *Asclepias viridis* displays an unusual distribution. Outside of the tall- and mixed-grass prairies of the southern Great Plains, where it is most abundant, it occupies glade habitats across the eastern United States, extending to chalk prairies in the southeastern states and pine rocklands in southern Florida. It is rare at the margins of its range and is considered to be of conservation concern in Indiana (Clark and Harrison counties) and West Virginia (Jackson and Wirt counties). Recently, it has been documented to occur in Iowa, close to the Missouri state line in Ringgold County.

57. Asclepias eastwoodiana Barneby, Leafl. W. Bot. 4: 210. 1945 • Eastwood's milkweed [E]

Herbs. Stems 1–6, ascending to decumbent, unbranched or branched near base, 6–30 cm, puberulent with curved trichomes, not glaucous, rhizomatous. **Leaves** proximally opposite, distally alternate, petiolate, with 0 or 1 stipular colleter on either side of petiole; petiole 2–10 mm, puberulent with curved trichomes; blade narrowly lanceolate to ovate, 2–4.5 × 0.6–3 cm, chartaceous, base cuneate to obtuse, margins entire, apex acute, mucronate, venation eucamptodromous, surfaces strigulose or pilosulous on midvein to glabrate, margins densely ciliate, laminar colleters absent. **Inflorescences** terminal and extra-axillary at upper nodes, sessile or pedunculate, 2–25-flowered; peduncle 0–4 cm, puberulent with curved trichomes to glabrate, bracts few. **Pedicels** 10–28 mm, pilosulous to puberulent with curved trichomes. **Flowers** erect; calyx lobes elliptic to lanceolate, 2–2.5 mm, apex acute, puberulent with curved trichomes to pilosulous; corolla red-violet, faintly striate, lobes reflexed with spreading tips, oval, 4–5 mm, apex acute, glabrous; gynostegial column 0.2–0.5 mm; fused anthers brown, cylindric, 1.5–2 mm, wings right-triangular, apical appendages ovate; corona segments red-violet dorsally, white proximally, sessile, cupulate, 1.5–2 mm, slightly exceeded by style apex, apex truncate with a proximal tooth on each side, glabrous, internal appendage conical, barely exserted from cavity, glabrous; style apex shallowly depressed, pink to red-violet. **Follicles** spreading to pendulous on spreading pedicels, lance-ovoid, 3.5–6 × 0.5–1 cm, apex acuminate, smooth, faintly striate, strigose. **Seeds** not seen.

Flowering May–Jun; fruiting Jun. Valleys, depressions, flats, slopes, arroyos, dunes, granite, gravel, sandy, calcareous, and clay soils, shrubby grasslands, desert scrub, pinyon-juniper woodlands; 1400–2200 m; Nev.

Asclepias eastwoodiana and the next three species (*A. ruthiae*, *A. sanjuanensis*, *A. uncialis*) form a complex of largely allopatric entities that have sometimes been united in a single species, for which the name *A. uncialis* holds priority (for example, E. Sundell 1994). However, each entity shows genetic, chemical, and subtle morphological distinctions, supporting recognition at the specific rank (M. B. Sady and J. N. Seibert 1991; J. P. Riser et al. 2019). *Asclepias eastwoodiana* is distinguished from the others by the differences in leaf shape, vestiture, and corona morphology indicated in the key. With *A. ruthiae* it shares a spreading to pendulous fruit that differs from the typically erect fruits of *A. sanjuanensis* and *A. uncialis*. *Asclepias eastwoodiana* is endemic to valleys in central Nevada (Esmeralda, Lander, Lincoln, Mineral, and Nye counties), where it is considered to be a species of conservation concern, potentially threatened by livestock trampling and mining development.

58. Asclepias ruthiae Maguire, Ann. Missouri Bot. Gard. 28: 245, fig. 1. 1941 • Ruth's milkweed [E]

Asclepias uncialis Greene subsp. *ruthiae* (Maguire) Kartesz & Gandhi; *A. uncialis* var. *ruthiae* (Maguire) Sundell

Herbs. Stems 1–8, decumbent, unbranched or branched near base, 9–10 cm, densely puberulent with curved trichomes, not glaucous, rhizomatous. **Leaves** proximally opposite, distally alternate, petiolate, with 0 or 1 stipular colleter on either side of petiole; petiole

2–5 mm, densely puberulent with curved trichomes; blade ovate to lanceolate, 2–3 × 0.8–2 cm, chartaceous, base truncate to cuneate, margins entire, apex acute, venation eucamptodromous, surfaces persistently pilosulous, more densely so on veins, margins densely ciliate, laminar colleters absent. **Inflorescences** terminal and extra-axillary at upper nodes, sessile, 2–6-flowered, bracts few. **Pedicels** 15–25 mm, densely pilosulous to tomentulose. **Flowers** erect; calyx lobes elliptic, 2–2.5 mm, apex acute, pilose to tomentulose; corolla red-violet, lobes reflexed with spreading tips, oval, 4–5 mm, apex acute, glabrous; gynostegial column 0.2–0.5 mm; fused anthers brown, cylindric, 1.5–2 mm, wings right-triangular, apical appendages ovate; corona segments red-violet dorsally, white proximally, sessile, cupulate, 1.5–2 mm, slightly exceeded by style apex, apex truncate with a proximal tooth on each side, glabrous, internal appendage falcate, included in cavity, glabrous; style apex shallowly depressed, pink to red-violet. **Follicles** erect to spreading on upcurved to spreading pedicels, ovoid, 3–3.5 × 0.7–1 cm, apex acuminate, smooth, pilosulous to tomentulose. **Seeds** broadly ovate, 8–9 × 6–7 mm, margin corky-winged, erose, faces rugulose, minutely hirtellous; coma 1–1.5 cm.

Flowering May–Jun; fruiting Jun–Jul. Slopes, terraces, bluffs, sandstone, basalt cobbles, limestone, sandy, clay, and rocky soils, desert scrub, shrubby grasslands; 1200–2000 m; Utah.

As discussed under *Asclepias eastwoodiana*, *A. ruthiae* is part of a complex of four species sometimes recognized as a single species. It is endemic to a small area in southeastern Utah (Emery, Sevier, and Wayne counties) on the periphery of the San Rafael Swell, the northeastern margin of Waterpocket Fold, and valleys north of the Henry Mountains. Reports from Arizona, New Mexico, and Monument Valley in Utah represent records of *A. sanjuanensis*. It commonly co-occurs with the similar *A. macrosperma* in this region and differs from the latter species by its more diminutive dimensions, red-violet corolla, and smaller, smooth fruit. Conservation status of this species requires re-assessment in light of the recent recognition of its more limited range.

59. Asclepias sanjuanensis K. D. Heil, J. M. Porter & S. L. Welsh, Great Basin Naturalist 49: 100, fig. 1. 1989 • San Juan milkweed [E]

Herbs. Stems 1–8, decumbent, unbranched or branched near base, 4–13 cm, densely puberulent with curved trichomes, not glaucous, rhizomatous. **Leaves** opposite and alternate, petiolate, with 0 or 1 stipular colleter on either side of petiole; petiole 2–5 mm, margins puberulent with curved trichomes; blade lanceolate to ovate, 2–5 × 0.5–2 cm, chartaceous, base obtuse to cuneate, margins entire, apex acute to attenuate, venation eucamptodromous, surfaces sparsely pilosulous to glabrate, persistent on veins, margins densely ciliate, laminar colleters absent. **Inflorescences** terminal and extra-axillary at upper nodes, sessile, 3–8-flowered, bracts few. **Pedicels** 11–24 mm, densely pilosulous. **Flowers** erect; calyx lobes elliptic, 2–2.5 mm, apex acute, pilose; corolla red-violet, lobes reflexed with spreading tips, oval, 4–5 mm, apex acute, glabrous; gynostegial column 0.2–0.5 mm; fused anthers brown, cylindric, 1.5–2 mm, wings narrowly right-triangular, apical appendages ovate, erose; corona segments red-violet dorsally, white to orange proximally, sessile, cupulate, 1.5–2 mm, slightly exceeded by style apex, apex truncate with a proximal tooth on each side, glabrous, internal appendage falcate, barely exserted from cavity, glabrous; style apex shallowly depressed, pink to red-violet. **Follicles** erect on upcurved pedicels, ovoid, 3–5 × 0.7–1 cm, apex acuminate, smooth, pilosulous to tomentulose. **Seeds** broadly ovate, 8–9 × 6–7 mm, margin winged, erose, faces densely rugulose, lepidote; coma 1–1.5 cm.

Flowering Apr–May; fruiting May–Jun. Slopes, ridges, arroyos, sandstone, sandy, silty, and rocky soils, juniper and pinyon-juniper woodlands, shrubby grasslands, desert scrub; 1400–1800 m; Ariz., N.Mex., Utah.

As discussed under *Asclepias eastwoodiana*, *A. sanjuanensis* is part of a complex of four species sometimes recognized as a single species. The range of *A. sanjuanensis* is larger than initially realized, extending across northwestern Arizona (Apache and Navajo counties) and barely entering southeastern Utah (San Juan County), as discussed under *A. ruthiae*. These two species are completely allopatric. In New Mexico, *A. sanjuanensis* is restricted to San Juan County.

60. Asclepias uncialis Greene, Bot. Gaz. 5: 64. 1880 • Wheel or dwarf milkweed [F]

Herbs. Stems 1–13, decumbent, unbranched or branched near base, 4–10 cm, puberulent with curved trichomes, not glaucous, rhizomatous. **Leaves** opposite and alternate, petiolate, with 1 stipular colleter on each side of petiole; petiole 1–5 mm, puberulent with curved trichomes to pilosulous; blade linear to lanceolate, 1.7–5 × 0.2–1 cm, chartaceous, base cuneate, margins entire, apex attenuate, venation obscure, surfaces puberulent on midvein with curved trichomes, margins densely ciliate, laminar colleters absent. **Inflorescences** terminal and extra-axillary at upper nodes, sessile, 3–7-flowered, bracts few. **Pedicels** 10–18 mm, densely puberulent with curved trichomes. **Flowers** erect; calyx lobes elliptic, 2–2.5 mm, apex acute, puberulent with curved trichomes; corolla red-violet, lobes reflexed with spreading tips, oval, 3–5 mm, apex acute, glabrous; gynostegial column 0.2–0.5 mm; fused anthers brown, cylindric, 1–1.5 mm, wings narrowly right-triangular, apical appendages ovate, erose; corona segments red-violet dorsally, white to orange proximally, sessile, cupulate, 1–2 mm, exceeded by style apex, apex truncate with a proximal tooth on each side, glabrous, internal appendage lingulate, barely exserted from cavity, glabrous; style apex shallowly depressed, pink to red-violet. **Follicles** erect on upcurved pedicels, ovoid, 3–5 × 0.8–1.5 cm, apex acuminate, smooth, minutely puberulent with curved trichomes. **Seeds** broadly ovate, 7–8 × 5–6 mm, margin winged, erose, faces rugulose, lepidote; coma 1.5–2 cm.

Flowering Mar–Jun; fruiting Apr–Jun. Plains, hills, ridges, canyons, bajadas, shale, alluvium, clay, sandy, and rocky soils, prairies, desert grasslands, juniper woodlands; 900–1800 m; Ariz., Colo., N.Mex., Okla., Tex.; Mexico (Sonora).

Asclepias uncialis is by far the most widespread of the four diminutive, red-violet-petaled milkweeds of western North America (species 57 60). It typically has much narrower leaves than the other three species, and is extraordinarily cryptic in the absence of flowers in its characteristic short-grass prairie habitat, where its leaves closely mimic dominant grama grasses, particularly *Bouteloua gracilis*. Although it is widespread, it is encountered commonly only in southeastern Colorado and has only been recorded at single sites in Oklahoma (Cimarron County) and Texas (Andrews County), where the species should be considered to be of conservation concern. It is considered to be of concern in Colorado and New Mexico, but it is possible that this cryptic species is more common than has been recorded. Nonetheless, it appears that it has declined in northern Colorado. An 1873 specimen (*C. C. Parry 246* [GH]) from Wyoming is the only documented record from that state and is from a highly disjunct location (attributed to Sweetwater County). It is possible that the reported location was in error; otherwise, it appears that *A. uncialis* has been extirpated from Wyoming, which is excluded from the range of the species in this treatment.

61. Asclepias amplexicaulis Smith in J. E. Smith and J. Abbott, Nat. Hist. Lepidopt. Georgia 1: 14, plate 7. 1797 • Blunt-leaved or clasping or sand milkweed [E]

Asclepias obtusifolia Michaux

Herbs. Stems 1 or 2+, erect, unbranched, 35–175 cm, glabrous, glaucous, rhizomes absent. **Leaves** opposite, sessile, with 1 stipular colleter on each side of leaf base; blade broadly ovate or oval to oblong, 6–14 × 3–7 cm, chartaceous, base cordate, clasping, margins often crisped, apex rounded to truncate, emarginate, or obtuse, sometime mucronate, venation eucamptodromous to brochidodromous, surfaces glabrous, glaucous, margins minutely ciliate, 6–16 laminar colleters. **Inflorescences** terminal (extra-axillary at upper nodes), pedunculate, 18–53-flowered; peduncle occasionally branched, 5–40 cm, glabrous, glaucous, with 1 caducous bract at the base of each pedicel. **Pedicels** 20–55 mm, puberulent with curved trichomes. **Flowers** erect to spreading; calyx lobes narrowly lanceolate, 3–5 mm, apex attenuate, sparsely pilosulous to glabrate; corolla green, often tinged red, purple, or bronze, lobes reflexed, lanceolate, 8–11 mm, apex acute, glabrous; gynostegial column 1.5–2.5 mm; fused anthers green, obconic, 2.5–3.5 mm, wings right-triangular, open at base, apical appendages rhomboid; corona segments reddish purple to cream, stipitate, tubular, 4–6 mm, exceeding style apex, apex truncate, erose, glabrous, internal appendage falcate, exserted, sharply inflexed over style apex, glabrous; style apex shallowly depressed, green. **Follicles** erect on upcurved pedicels, fusiform to narrowly lance-ovoid, 9–16 × 1–2 cm, apex long-acuminate, smooth, pilosulous. **Seeds** ovate, 9–10 × 6–7 mm, margin winged, faces minutely rugulose; coma 2.5–3 cm.

Flowering Mar–Sep; fruiting (Apr–)May–Sep. Dunes, ridges, slopes, sand hills, ravines, sandstone, rarely limestone, sandy, rocky, or silty soils, meadows, pastures, fields, railroad embankments, sand prairies, wet prairies, river banks, open oak woods, barrens, pine-oak forests, pine flatwoods and savannas, forest edges; 0–800 m; Ala., Ark., Conn., Del., D.C., Fla., Ga., Ill., Ind., Iowa, Kans., Ky., La., Md., Mass., Mich., Minn., Miss., Mo., Nebr., N.H., N.J., N.Y., N.C., Ohio, Okla., Pa., R.I., S.C., Tenn., Tex., Vt., Va., W.Va., Wis.

The common name sand milkweed refers to a strong association of *Asclepias amplexicaulis* with sandstone substrates and sandy soils. The clasping leaves and long-peduncled terminal inflorescence of *A. amplexicaulis* are distinctive among all co-occurring milkweeds. Western populations of *A. amplexicaulis*, primarily from prairies, usually have paler flowers with creamy coronas, whereas those from forest openings in the eastern and southeastern United States usually have pink to maroon coronas. The species is rare on the northwestern and northeastern margins of its range, where it is considered to be of conservation concern in Minnesota, Nebraska, New Hampshire, Rhode Island, Vermont (only in Chittenden County), and West Virginia. Hybrids with *A. exaltata*, *A. purpurascens*, and *A. syriaca* are known, but are local and not documented often. Presumed hybrids can be recognized by possessing intermediate floral and vegetative characteristics. *Asclepias* ×*intermedia* Vail probably applies to the hybrid with *A. syriaca* based on Vail's protologue (A. M. Vail 1904), but the holotype (*E. P. Bicknell s.n.* [NY]) is damaged, making the assignment tentative. The homonym *A. amplexicaulis* Michaux was applied to *A. humistrata* in the past, resulting in some taxonomic confusion between these species and the misidentification of herbarium specimens.

62. Asclepias sullivantii Engelmann ex A. Gray, Manual, 366. 1848 • Sullivant's or prairie or smooth milkweed, asclépiade de Sullivant [E]

Herbs. Stems solitary, erect, unbranched, 55–90 cm, glabrous, glaucous, rhizomatous. **Leaves** opposite, sessile, with 1 or 2 stipular colleters on each side of leaf base; blade lanceolate or ovate to oblong, 6.5–15 × 1.5–9 cm, succulent, base cordate, margins entire, apex obtuse to rounded or acute, sometimes emarginate, mucronate, venation brochidodromous, surfaces glabrous, glaucous, margins eciliate, 4–12 laminar colleters. **Inflorescences** extra-axillary at upper nodes, sometimes appearing terminal, pedunculate, 9–29-flowered; peduncle 1–6 cm, glabrous, glaucous, with 1 caducous bract at the base of each pedicel. **Pedicels** 22–36 mm, glabrous, glaucous. **Flowers** erect; calyx lobes lanceolate, 4–5 mm, apex acute, glabrous, glaucous; corolla dark pink, pale at base of lobes, lobes reflexed, sometimes with spreading tips, elliptic, 8–12 mm, apex acute, glabrous; gynostegial column 1–1.5 mm; fused anthers truncately green, obconic, 2.5–3.5 mm, wings narrowly right-triangular, open at base, apical appendages ovate; corona segments pale to dark pink, subsessile, tubular, flattened dorsally, 5–7 mm, exceeding style apex, apex broadly obtuse, oblique, glabrous, internal appendage subulate, exserted, sharply inflexed over style apex, glabrous; style apex shallowly depressed, green. **Follicles** erect on upcurved pedicels, lance-ovoid, 7–11 × 1.5–3 cm, apex acuminate, sparsely muricate, sparsely pilosulous to glabrate, glaucous. **Seeds** ovate, 7–9 × 5–6 mm, margin winged, faces minutely rugulose; coma 3.5–4.5 cm. **2*n*** = 22.

Flowering (May–)Jun–Aug(–Sep); fruiting Jun–Oct. Ditches, fields, streamsides, flood plains, alluvium, clay and sandy soils, prairies, wet prairies, shrubby grasslands, forest openings, thickets; 100–700 m; Ont.; Ill., Ind., Iowa, Kans., Mich., Minn., Mo., Nebr., N.Dak., Ohio, Okla., S.Dak., Wis.

The distribution of *Asclepias sullivantii* is coextensive with the tallgrass prairie, where it favors moist sites. The broad, smooth, glaucous, clasping leaves, often with pink venation, are similar only to *A. amplexicaulis* in the region, but the latter species favors dry sites and has a long, terminal peduncle. Because of the tremendous reduction in the extent and quality of tallgrass prairie, *A. sullivantii* is presumed to be less common than in former times and is considered to be rare and of conservation concern in Michigan, Minnesota, Wisconsin, and Ontario (Chatham-Kent, Elgin, Essex, Lambton, and Middlesex counties). It may be extirpated from North Dakota (historically in Cass and Richland counties) and South Dakota (historically in Clay, Lincoln, and Union counties). In Nebraska, it is limited to the eastern one-third of the state. Nonetheless, it is encountered commonly in suitable habitat along roadsides in the core of its range, in Illinois and Kansas. *Asclepias sullivantii* commonly co-occurs with and is often mistaken for *A. syriaca* from a distance but is easily distinguished by its smaller stature, ascending leaves, and the concentration of umbels at the apex of the stem, in addition to the glabrous and glaucous herbage that can be observed with closer examination. It is known to hybridize rarely with *A. syriaca*, and presumed hybrids can be recognized by intermediate vegetative and floral features.

63. Asclepias meadii Torrey ex A. Gray, Manual ed. 2, 704. 1856 • Mead's milkweed [C] [E]

Herbs. Stems 1 (rarely 2), erect, unbranched, 20–80 cm, glabrous, glaucous, rhizomatous. **Leaves** opposite, sessile, with 1 stipular colleter on each side of petiole; blade ovate to lanceolate, 4.5–10 × 1–5 cm, chartaceous, base rounded to obtuse, margins entire, apex acute, venation brochidodromous, surfaces glabrous, glaucous, margins inconspicuously ciliate, laminar colleters absent. **Inflorescences** terminal, rarely branched, pedunculate, 7–19-flowered; peduncle (0–)3–10 cm, sparsely pilosulous to glabrate, glaucous, with 1 caducous bract at the base of each pedicel. **Pedicels** 11–20 mm, pilosulous. **Flowers** pendent; calyx lobes narrowly lanceolate, 4–5 mm, apex acute, pilosulous; corolla green to greenish cream, sometimes tinged red, lobes reflexed with spreading tips, oval, 7–9 mm, apex acute, glabrous; gynostegial column 1.5–1.8 mm, fused anthers green, truncately obconic, 2–2.5 mm, wings right-triangular, closed, apical appendages ovate; corona segments green to yellowish green or greenish cream, sessile, conduplicate, dorsally rounded, 4–5.5 mm, greatly exceeding style apex, base saccate, apex obtuse, emarginate, slightly flared, glabrous, internal appendage falcate, exserted, sharply inflexed towards style apex, glabrous; style apex shallowly depressed, green. **Follicles** erect on upcurved pedicels, narrowly fusiform, 7–14 × 0.9–1.6 cm, apex long-acuminate, smooth to minutely rugulose, puberulent with curved trichomes. **Seeds** ovate, 6–7 × 4–5 mm, winged, faces coarsely papillose; coma 3–4.5 cm.

Flowering May–Jun(–Jul); fruiting Jun–Aug. Dry, upland prairies, chert-lime glades; of conservation concern; 100–500 m; Ill., Ind., Iowa, Kans., Mo., Wis.

With respect to historical range reduction and potential threats, *Asclepias meadii* is the most endangered North American milkweed species and is considered to be of conservation concern in each of the states in which it occurs. Nearly all of the viable populations are now restricted to Missouri and eastern Kansas; many of these are found in hay meadows, in which fruit maturation does not occur. *Asclepias meadii* is endemic to the highly-impacted tallgrass prairie ecoregion and is thought to be rare as a result of habitat loss. Consequently, concern for the continued existence of this species can be considered emblematic of concern for the tallgrass prairie as a whole. Active recovery efforts have achieved limited success, with newly established populations experiencing high mortality and slow growth of transplants. It has been re-introduced to Indiana and Wisconsin, but long-term survival of these populations is uncertain. The sessile, glaucous leaves and pendent umbels on a long peduncle suggest a diminutive version of *A. elata*; however, these species do not appear to be closely related (Fishbein et al. 2011). The terminal inflorescence of pendent umbels is unique among tallgrass prairie milkweeds.

64. Asclepias obovata Elliott, Sketch Bot. S. Carolina 1: 321. 1817 • Pineland milkweed [E]

Herbs. Stems 1 (rarely more), erect to spreading, unbranched, 40–70(–200) cm, densely hirtellous to velutinous, not glaucous, rhizomes absent. **Leaves** opposite, petiolate, with 1 or 2 stipular colleters on each side of petiole; petiole 1–4 mm, densely hirtellous to velutinous; blade oblong or elliptic to obovate or ovate, 4–9 × 1–3.5 cm, subcoriaceous, base rounded or truncate to cordate, margins sometimes crisped, apex acute to truncate, sometimes emarginate, often mucronate, venation brochidodromous, surfaces densely hirtellous to velutinous abaxially, hirtellous adaxially, margins ciliate, 8–12 laminar colleters. **Inflorescences** extra-axillary, sometimes also appearing terminal, sessile or pedunculate, 7–31-flowered; peduncle 0–0.5 cm, densely hirtellous to velutinous, with 1 caducous bract at the base of each pedicel. **Pedicels** 10–12 mm, densely hirtellous to velutinous. **Flowers** erect to pendent; calyx lobes elliptic, 5–6 mm, apex acute, densely hirtellous; corolla green, sometimes tinged reddish or bronze, lobes reflexed, sometimes with spreading tips, elliptic, 7–9 mm, apex acute, glabrous abaxially, minutely papillose at base adaxially; gynostegial column 1–1.5 mm; fused anthers green, obconic, 2.5–4 mm, wings right-triangular, open at base, apical appendages broadly ovate; corona segments bronze to yellow, often tinged red, sometimes apically cream or pale, stipitate, tubular, somewhat flattened laterally, flared at base, 5–8 mm, greatly exceeding style apex, apex rounded, flared, glabrous, internal appendage falcate, exserted, sharply incurved over style apex, glabrous; style apex shallowly depressed, green. **Follicles** erect on upcurved pedicels, narrowly to broadly fusiform, 7.5–12.5 × 1.5–2.5 cm, apex acuminate, smooth, densely hirtellous to velutinous. **Seeds** broadly ovate, 8–9 × 6–7 mm, margin winged, faces smooth; coma 2–5 cm.

Flowering May–Sep; fruiting Jul–Oct. Hills, slopes, flats, ridges, sandhills, ditches, seeps, bogs, sandstone, sandy, rocky, silty, and clay soils, pine flatwoods, pine savannas, pine, pine-oak, and bottomland hardwood forests, prairies, often following fires; 0–200 m; Ala., Ark., Fla., Ga., La., Miss., Okla., S.C., Tex.

Asclepias obovata is a common milkweed of seasonally wet, sandy soils in pine woodlands of the Gulf Coastal Plain and (rarely) the southern Atlantic Coastal Plain. It is rare and considered to be of conservation concern in Arkansas.

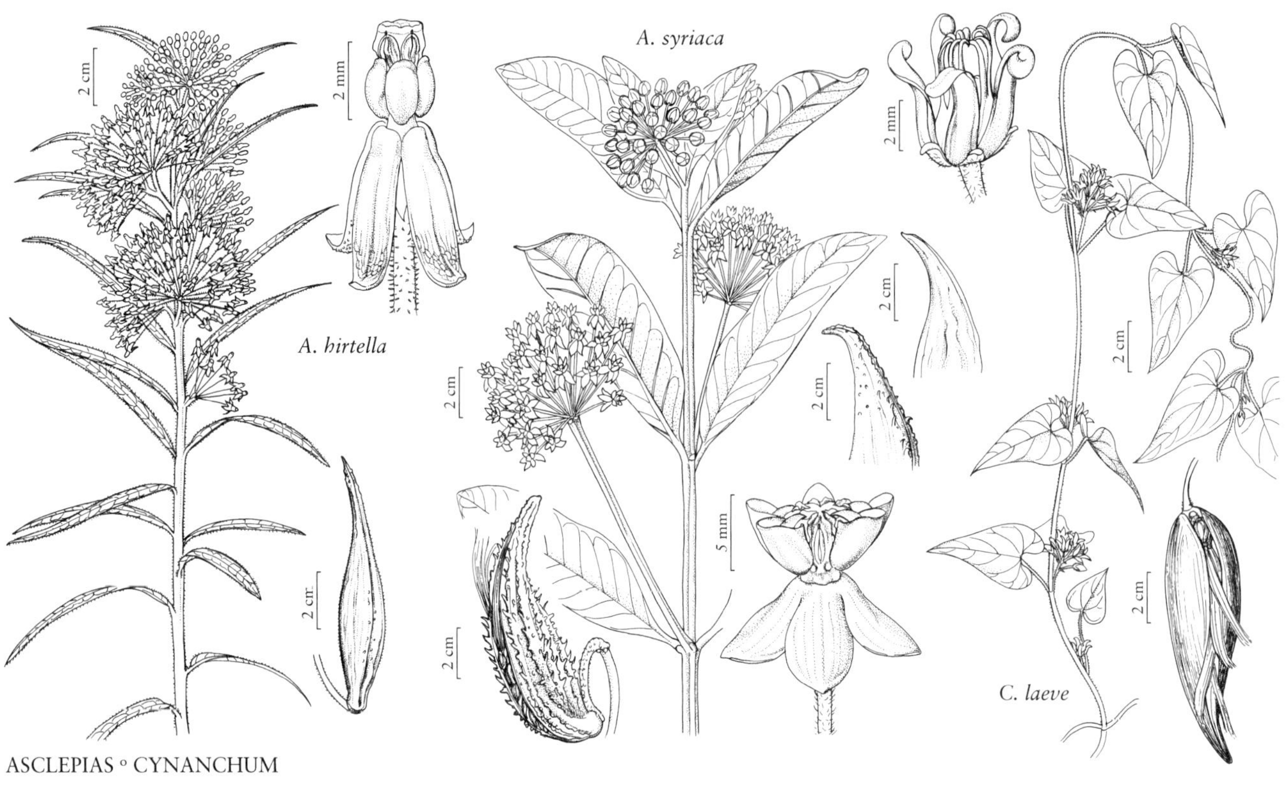

ASCLEPIAS ◦ CYNANCHUM

65. Asclepias hirtella (Pennell) Woodson, Ann. Missouri Bot. Gard. 28: 207. 1941 • Tall-green or green or prairie milkweed, asclépiade hérissée [E] [F]

Acerates hirtella Pennell, Bull. Torrey Bot. Club 46: 184. 1919; *Asclepias longifolia* Michaux subsp. *hirtella* (Pennell) J. Farmer & C. R. Bell

Herbs. **Stems** 1–3+, erect to spreading, unbranched (rarely branched), 30–125 cm, puberulent with curved trichomes, not glaucous, rhizomes absent. **Leaves** opposite or alternate, sessile or petiolate, with 1–3 stipular colleters on each side of petiole, also in axil; petiole 0–3 mm, puberulent with curved trichomes; blade linear to linear-lanceolate, 4–16 × 0.2–1.5 cm, chartaceous, base cuneate, margins entire, apex acute to attenuate, mucronate, venation faintly brochidodromous, surfaces scabridulous to puberulent with curved trichomes, especially on veins, margins ciliate, laminar colleters absent. **Inflorescences** extra-axillary, sessile or pedunculate, 34–112-flowered; peduncle 0–4 cm, hirtellous, with 1 caducous bract at the base of each pedicel. **Pedicels** 11–23 mm, hirtellous. **Flowers** erect to pendent; calyx lobes lanceolate, 1.5–2 mm, apex acute, hirtellous; corolla green to greenish cream, red-violet at tip, lobes reflexed, oblong, 3–5 mm, apex obtuse, glabrous; gynostegial column 0.5–1 mm; fused anthers green, cylindric, 1–1.5 mm, wings trapezoidal, closed, apical appendages oblate; corona segments cream to greenish cream, rarely pinkish lavender or brown, often with a dorsal red-violet stripe or at base, sessile, laminar, strongly dorsally compressed, margins incurved, appressed to column, saccate, 1.5–2 mm, not exceeding point of anther wings, greatly exceeded by style apex, apex truncate, glabrous, internal appendage absent or obscure, glabrous; style apex shallowly depressed, green. **Follicles** erect on upcurved pedicels, fusiform, 6–15 × 1–2 cm, apex attenuate, smooth, pilosulous to puberulent with curved trichomes. **Seeds** ovate, 10–11 × 7–8 mm, margin winged, faces smooth; coma 3.5–4 cm.

Flowering Apr–Oct; fruiting Jun–Oct. Plains, low hills, slopes, ditches, ravines, wet depressions, bottomlands, limestone, shale, silty, sandy, clay, and rocky soils, prairies, glades, wet meadows, oak, oak-hickory, and pine-oak forests and edges, pastures; 70–400 m; Ont.; Ala., Ark., Ga., Ill., Ind., Iowa, Kans., Ky., La., Mich., Minn., Miss., Mo., Ohio, Okla., Tenn., Tex., W.Va., Wis.

Asclepias hirtella has been included sometimes in a broadly circumscribed *A. longifolia*. The species are parapatric and usually are readily distinguished by the hirtellous pedicels of *A. hirtella* and puberulent pedicels with curved trichomes of *A. longifolia*. *Asclepias hirtella* is also typically much taller with more numerous umbels compared to *A. longifolia*. However, populations along the Gulf Coastal Plain from eastern Texas to the Mississippi River and from southern Mississippi to central Georgia have proved challenging because they include plants with the growth form of *A. longifolia* and the pedicel vestiture of *A. hirtella*. Consequently, these populations have not been consistently assigned to one species or the other. However, the previously overlooked difference in corona segment length correlates perfectly with the pedicel vestiture and with geography. Thus, Gulf Coast populations west of the Mississippi River and north of the immediate coastline are here interpreted to represent short-statured *A. hirtella*. This hypothesis warrants investigation with population genetic data—it is quite possible that populations of *A. hirtella* in southern Arkansas, Louisiana, and Texas merit recognition as a subspecies. *Asclepias hirtella* is rare at the margins of its range and is considered to be of conservation concern in Georgia, Kentucky, Louisiana, Michigan, Minnesota (Mower County), Mississippi, Tennessee, West Virginia (Jackson, Mason, and Putnam counties), and Ontario (Essex County).

66. Asclepias longifolia Michaux, Fl. Bor.-Amer. 1: 116. 1803 • Long-leaf or Florida milkweed [E]

Acerates floridana (Lamarck) Hitchcock; *A. longifolia* (Michaux) Elliott; *Asclepias floridana* Lamarck

Herbs. Stems 1–few, spreading to decumbent, unbranched, 25–70 cm, puberulent with curved trichomes, not glaucous, rhizomes absent. **Leaves** opposite or alternate, sessile or petiolate, with 1 or 2 stipular colleters on each side of petiole, also in axil; petiole 0–2 mm, puberulent with curved trichomes; blade linear to linear-lanceolate, 5–18 × 0.2–1 cm, chartaceous, base cuneate, margins entire, apex acute to attenuate, sometimes mucronate, venation brochidodromous, surfaces scabridulous to puberulent with curved trichomes, especially on veins, margins ciliate, laminar colleters absent. **Inflorescences** terminal and extra-axillary at upper nodes, sessile or pedunculate, 13–28(–36)-flowered; peduncle 0–6 cm, puberulent with curved trichomes, with 1 caducous bract at the base of each pedicel. **Pedicels** 9–16 mm, puberulent with curved trichomes. **Flowers** erect to spreading; calyx lobes narrowly lanceolate, 1.5–2 mm, apex acute, puberulent with curved trichomes; corolla white to greenish cream, purple at lobe tips, lobes reflexed, oblong, 3–5 mm, apex obtuse, glabrous; gynostegial column 0.2–0.5 mm; fused anthers green or brown, cylindric, 1–1.5 mm, wings triangular to trapezoidal, widest at middle, slightly open at tip, apical appendages ovate; corona segments pinkish lavender to red-violet, often with white apex, often a darker red-violet stripe or at base, sessile, laminar, strongly dorsally compressed, margins incurved, appressed to column, curved to subsaccate, 2–2.5 mm, exceeding point of anther wings, greatly exceeded by style apex, apex obtuse to truncate, glabrous, internal appendage absent or obscure, glabrous; style apex shallowly depressed, greenish cream to green. **Follicles** erect on upcurved pedicels, narrowly fusiform, 8–13.5 × 0.7–1.2 cm, apex attenuate, smooth, puberulent with curved trichomes. **Seeds** ovate, 11–12 × 7–8 mm, margin winged, faces smooth; coma 4–5 cm.

Flowering (Jan–)Feb–Sep(–Oct); fruiting Apr–Aug. Bogs, swamps, flats, ditches, depressions, pond edges, sandy, clay, and rocky soils, wet prairies, pine flatwoods, oak woodlands, savannas, pastures, often following fires; 0–100 m; Ala., Del., Fla., Ga., La., Miss., N.C., S.C., Va.

Asclepias longifolia has sometimes been treated to include *A. hirtella* as conspecific, as discussed under that species. As circumscribed here, the distribution of *A. longifolia* extends along the Atlantic Coastal Plain south of Delaware to Florida and westward to the Mississsppi River. Reports of *A. longifolia* from west of the Mississippi River in Louisiana and in Texas are assigned here to *A. hirtella*. Hence, the range of *A. longifolia* in Louisiana is extremely limited, and the conservation status of this species warrants assessment in that state (Livingston, Saint Helena, Saint Tammany, Tangipahoa, and Washington parishes). Reports of *A. longifolia* from Maryland and West Virginia have not been confirmed, and the species is considered extirpated from Delaware. Otherwise, it is rare at the northern extent of its range and considered to be of conservation concern in North Carolina and Virginia (Greensville and Prince George counties).

67. Asclepias lanceolata Walter, Fl. Carol., 105. 1788 • Few-flowered or smooth-orange milkweed [E]

Herbs. Stems 1, erect, unbranched, 60–125 cm, glabrous, not glaucous, rhizomes absent. **Leaves** opposite, sessile or petiolate, with 1 stipular colleter on each side of petiole on a ciliate interpetiolar ridge; petiole 0–1 mm, ciliate; blade linear to linear-lanceolate, somewhat conduplicate, 7–25 × 0.2–1.7 cm, chartaceous, base cuneate, margins entire, apex attenuate, mucronate, venation faintly brochidodromous to obscure, surfaces glabrous, margins inconspicuously ciliate, 8–12 laminar colleters, usually obscured by conduplicate petiole. **Inflorescences** terminal and often extra-axillary at an upper node, usually branched, pedunculate, 4–16-flowered; peduncle 1–7.5 cm, puberulent on 1 side with curved trichomes, with 1 caducous bract at the base of each pedicel. **Pedicels** 13–19 mm, puberulent on 1 side with curved trichomes. **Flowers** erect; calyx lobes narrowly lanceolate, 2–3 mm, apex acute, puberulent with curved trichomes to glabrate; corolla red, lobes reflexed, usually with spreading tips, elliptic, 9–10 mm, apex acute, glabrous abaxially, minutely papillose at base adaxially; gynostegial column 1.5–2 mm; fused anthers green, tinged yellow to red, truncately obconic, 2.5 mm, wings narrowly right-triangular, slightly open at base, apical appendages deltoid; corona segments yellow to reddish orange, stipitate, broadly tubular, dorsally flattened, 5–6 mm, exceeding style apex, apex obtuse, flared, glabrous, internal appendage subulate, exserted, sharply inflexed over style apex, glabrous; style apex shallowly depressed, red. **Follicles** erect on upcurved pedicels, narrowly fusiform, 5.5–10 × 0.8–1 cm, apex attenuate to long-acuminate, smooth, sparsely pilosulous or puberulent with curved trichomes to glabrate. **Seeds** broadly oval, 8–10 × 6–8 mm, margin winged, faces smooth; coma 3–3.5 cm.

Flowering year-round; fruiting Jun–Nov. Saltwater and freshwater marshes, pond edges, streamsides, bogs, swamps, ditches, glades, depressions, hill slopes, saturated sandy and silty soils, pine flatwoods, pine-oak forests, savannas, meadows, and barrens, thickets; 0–100 m; Ala., Del., Fla., Ga., La., Md., Miss., N.J., N.C., S.C., Tex., Va.

Asclepias lanceolata is a characteristic wetland milkweed of the Atlantic and Gulf coastal plains, where it occurs sometimes with its similar relative, *A. rubra*. It is distinguished from *A. rubra* by flower color (shades of orange, yellow, and red versus pink, lavender, and reddish purple), corona segment apex (obtuse versus acute), and leaf shape (linear to linear-lanceolate versus narrowly lanceolate to ovate). These species are known to hybridize locally, particularly in North Carolina, and putative hybrids can be recognized by intermediate floral and vegetative traits. The pattern of yellow-orange coronas paired with red corollas is similar to that of the introduced *A. curassavica*, and these species are sometimes confused. *Asclepias curassavica* differs by usually pure yellow (versus orangish) coronas, more flowers per umbel, much shorter peduncles, and wider, shorter leaf blades. *Asclepias lanceolata* sometimes is confused also with orange-flowered *A. tuberosa*, although that species only rarely has bicolored flowers, and the two are quite distinct in habitat, growth form, and vestiture. Through habitat loss, *A. lanceolata* has become quite rare in some areas, particularly the northeastern United States. It is considered to be of conservation concern in Delaware (Sussex County), Maryland (Dorchester County), and New Jersey. It has been reported, but not documented, from Tennessee, and its occurrence there seems unlikely. The report may be based on confusion with the name of the formerly recognized (and unrelated) *A. viridiflora* var. *lanceolata*.

68. Asclepias rubra Linnaeus, Sp. Pl. 1: 217. 1753 • Red milkweed [E]

Asclepias laurifolia Michaux

Herbs. **Stems** 1, erect, unbranched, 40–110 cm, puberulent in a single line with curved trichomes to glabrate, somewhat glaucous, rhizomes absent. **Leaves** opposite, petiolate, with 1 stipular colleter on each side of petiole on a ciliate interpetiolar ridge; petiole 1–2 mm, glabrous; blade narrowly lanceolate to ovate, 5–16 × 1–4.5 cm, chartaceous, base cordate, margins entire, apex attenuate, venation brochidodromous, surfaces glabrous, somewhat glaucous, margins ciliate, 6–8 laminar colleters, usually obscured by conduplicate petiole. **Inflorescences** terminal, usually branched, sometimes also extra-axillary at 1 upper node, pedunculate, 9–20-flowered; peduncle 0.5–13 cm, puberulent in 1 line with curved trichomes, with 1 caducous bract at the base of each pedicel. **Pedicels** 11–17 mm, puberulent on 1 side with curved trichomes. **Flowers** erect; calyx lobes lanceolate, 2–2.5 mm, apex acute, glabrous; corolla pink to reddish purple, lobes reflexed with spreading tips, elliptic, 7–9 mm, apex acute, glabrous abaxially, minutely papillose at base adaxially; gynostegial column 1.5–2 mm; fused anthers green, cylindric, 2–2.5 mm, wings right-triangular, closed, apical appendages broadly ovate; corona segments pink to lavender, stipitate, conduplicate, dorsally flattened, 6–7 mm, greatly exceeding style apex, apex acute, glabrous, internal appendage subulate, exserted, arching above style apex, glabrous; style apex shallowly depressed, green. **Follicles** erect on upcurved pedicels, narrowly fusiform, 7–12 × 1–1.5 cm, apex long-acuminate, smooth, sparsely puberulent with curved trichomes or pilosulous to glabrate. **Seeds** ovate, 7–9 × 5–7 mm, margin winged, faces minutely rugulose to smooth; coma 2–5 cm.

Flowering Apr–Sep; fruiting Jul–Oct. Streamsides, bogs, baygalls, swales, saturated soils, pine flatwoods, savannas, riparian woods, thickets; 0–300 m; Ala., Del., D.C., Fla., Ga., La., Md., Miss., N.J., N.Y., N.C., Pa., S.C., Tex., Va.

Distinguishing features and hybridization of *Asclepias rubra* and *A. lanceolata* are discussed under the latter species. The common name red milkweed is a misnomer, as the flowers are actually more commonly shades of pink or purple, whereas the similar *A. lanceolata* often has truly red corollas. *Asclepias rubra* is less common than *A. lanceolata* and considered to be of greater conservation concern in Alabama, Delaware, Georgia, Maryland, New Jersey, and Virginia. It is considered to be extirpated from the District of Columbia, New York, and Pennsylvania. It also has been reported from

Arkansas, but this remains unconfirmed and must be considered unlikely.

69. Asclepias michauxii Decaisne in A. P. de Candolle and A. L. P. P. de Candolle, Prodr. 8: 569. 1844 • Michaux's milkweed [E]

Herbs. Stems 1–4 (rarely more), spreading to decumbent or ascending, unbranched, 15–30 cm, minutely puberulent with curved trichomes, not glaucous, rhizomes absent. **Leaves** opposite or alternate, sessile, often ascending and appearing secund, with 0 or 1 stipular colleter on each side of leaf base; blade narrowly linear, 4–9 × 0.1–0.4 cm, chartaceous, base cuneate, margins entire, apex acute, venation faintly eucamptodromous to obscure, surfaces glabrous, margins remotely ciliate to glabrate, laminar colleters absent. **Inflorescences** terminal, sometimes also extra-axillary at upper nodes, sessile or pedunculate, 6–28-flowered; peduncle 0–4 cm, minutely puberulent with curved trichomes, with 1 caducous bract at the base of each pedicel. **Pedicels** 9–12 mm, minutely puberulent on 1 side with curved trichomes. **Flowers** erect, calyx lobes narrowly lanceolate, 2.5–3 mm, apex acute, sparsely puberulent with curved trichomes to glabrate; corolla green to tan, tinged pink or red, inconspicuously striate, lobes reflexed with spreading tips, elliptic, 4–5 mm, apex acute, glabrous; gynostegial column 0.5–1 mm, fused anthers green, broadly cylindric, 1.5–3 mm, wings narrowly right-triangular, open at tip, apical appendages ovate; corona segments cream, often pink-striped, to magenta, stipitate, cupulate, margins not strongly incurved, 2.5–3.5 mm, slightly exceeding style apex, apex obtuse, glabrous, internal appendage falcate, exserted, sharply inflexed over style apex, glabrous; style apex shallowly depressed, pale to deep pink. **Follicles** erect on straight pedicels, narrowly fusiform, 7.5–12.5 × 0.5–0.7 cm, apex long-attenuate, smooth, minutely puberulent with curved trichomes. **Seeds** ovate, 7–8 × 5–6 mm, margin winged, faces smooth; coma 1.5–2.5 cm.

Flowering (Mar–)Apr–Jun(–Sep); fruiting May–Aug. Flats, hills, ridges, bogs, sandhills, ravines, ditches, clay, sandy, and silty soils, pine flatwoods, oak forests, savannas, wet prairies, often following fires; 0–200 m; Ala., Fla., Ga., La., Miss., S.C.

Asclepias michauxii occurs with, and bears an overall similarity to, *A. longifolia*. They are distinguished easily in flower by the cavitate corona segments with exserted appendages of *A. michauxii*. Commonly, *A. michauxii* has decumbent stems with secund leaves, unlike *A. longifolia*. In fruit, *A. michauxii* is distinguished easily by straight pedicels. Like *A. longifolia*, *A. michauxii* barely enters Louisiana, east of the Mississippi River (Livingston, Saint Tammany, and Tangipahoa parishes), and is considered to be of conservation concern in that state.

70. Asclepias tuberosa Linnaeus, Sp. Pl. 1: 217. 1753 • Butterflyweed, butterfly milkweed, butterfly flower, pleurisy root, chigger flower, tuber root, orange milkweed, asclépiade tubéreuse

Herbs. Stems 1–numerous, erect to ascending, branched in inflorescence, 15–90 cm, densely hirsute, not glaucous, rhizomes absent. **Leaves** alternate, petiolate, with 1 or 2 stipular colleters on each side of petiole; petiole 1–4 mm, densely hirsute; blade elliptic, oblong, or oblanceolate to lanceolate or linear, 2–12 × 0.3–3 cm, chartaceous, base cuneate or obtuse to rounded, truncate, hastate, or cordate, margins entire, apex acute to attenuate or obtuse to rounded, venation brochidodromous to eucamptodromous, surfaces hirsute, more densely so on midvein abaxially, margins ciliate, 0–4 laminar colleters. **Inflorescences** corymbs of extra-axillary umbels on branches, sessile or pedunculate, 5–27-flowered; peduncle 0–4 cm, sometimes branched, hirsute, with 1 caducous bract at the base of each pedicel. **Pedicels** 9–24 mm, puberulent with curved trichomes to pilosulous. **Flowers** erect; calyx lobes narrowly lanceolate, 2–3 mm, apex acute, hirsute to puberulent with curved trichomes; corolla reddish orange (nearly red) to orange or yellow, lobes reflexed with spreading tips, narrowly elliptic, (5–)6–8 mm, apex acute, glabrous abaxially, minutely papillose at base adaxially; gynostegial column 1.2–1.5 mm, fused anthers yellow to yellowish green, cylindric, 2–3 mm, wings right-triangular, closed, apical appendages ovate; corona segments reddish orange (nearly red) to orange or yellow, substipitate, conduplicate, dorsally flattened, sulcate, 5.5–7 mm, greatly exceeding style apex, apex acute, glabrous, internal appendage subulate, exserted, arching above style apex, glabrous; style apex shallowly depressed, yellow to yellowish green. **Follicles** erect on upcurved pedicels, fusiform, 7–14 × 1.2–2 cm, apex long-acuminate or attenuate, smooth, hirsute. **Seeds** ovate, 8–9 × 4–5 mm, margin winged, faces minutely rugulose; coma 3–5 cm. **2*n*** = 22.

Subspecies 3 (3 in the flora): North America, n Mexico.

Asclepias tuberosa is one of the most familiar and beloved North American milkweeds and is a favored element of pollinator gardens because of the cheery orange flowers that attract abundant insect visitors. The clear latex is unusual in the genus and is often commented upon by collectors. The subspecies of *A. tuberosa* are highly intergrading. It is often difficult to satisfactorily place a given specimen in a particular subspecies; however, the great majority are readily assigned. It appears that the conspicuous variation in leaf morphology across the subspecies corresponds to genetically structured population variation (R. E. Woodson Jr. 1947). However, it is unknown to what extent cultivation and other human activities have blurred the distinctions among the geographic variants. Future recognition of the subspecies should be supported by genetic study with modern techniques. Although yellow-flowered plants predominate in the western plains, color variation is often pronounced in single populations, and yellow flowers may be encountered anywhere in the range. Hybridization with *A. syriaca* is documented, but is exceedingly rare. Presumed hybrids can be recognized by intermediate floral and vegetative traits.

SELECTED REFERENCE Woodson, R. E. Jr. 1947. Some dynamics of leaf variation in *Asclepias tuberosa*. Ann. Missouri Bot. Gard. 34: 353–432. Woodson, R. E. Jr. 1953. Biometric evidence of natural selection in *Asclepias tuberosa*. Proc. Natl. Acad. Sci. U.S.A. 39: 74–79.

1. Leaf bases hastate, blade margins crisped . . . . 70c. *Asclepias tuberosa* subsp. *rolfsii*
1. Leaf bases cuneate, obtuse, rounded, truncate, or subcordate, blade margins planar.
   2. Leaf bases cuneate to obtuse, rounded, or subcordate, apices rounded to acute, mostly east of the crest of the Appalachian Mountains . . . . 70a. *Asclepias tuberosa* subsp. *tuberosa*
   2. Leaf bases truncate or obtuse to cordate, apices acute to attenuate, mostly west of the crest of the Appalachian Mountains . . . . 70b. *Asclepias tuberosa* subsp. *interior*

70a. Asclepias tuberosa Linnaeus subsp. **tuberosa** E

Asclepias decumbens Linnaeus

Stems 45–80 cm. **Leaf blades** oblanceolate to narrowly elliptic or linear-lanceolate, 4–10 × 0.7–2.7 cm, base cuneate to obtuse, rounded, or subcordate, margins planar, apex acute to obtuse or rounded.

Flowering Apr–Sep(–Oct); fruiting Jul–Oct. Ridges, slopes, plains, bluffs, ditches, fields, streamsides, lake shores, salt marshes, sandhills, limestone, sandy, clay, and rocky soils, oak-hickory, pine, pine-oak, mixed-hardwood, and pine-mixed-hardwood forests, pine flatwoods, pine barrens, forest edges and openings, prairies; 0–800 m; Ont.; Ala., Ark., Conn., Del., D.C., Fla., Ga., Ill., Ind., Ky., Maine, Md., Mass., Miss., N.H., N.J., N.Y., N.C., Ohio, Pa., R.I., S.C., Tenn., Vt., Va., W.Va.

Subspecies *tuberosa* is distributed mainly on the Atlantic Coastal Plain, but is occasionally found inland in river valleys east of the Mississippi River. The plants are especially common in eastern and central Tennessee. Sporadic occurrences further inland (for example, in Arkansas, Illinois) may represent introductions. Subspecies *tuberosa* is uncommon in Rhode Island (Kent County) and is considered to be of conservation concern there.

70b. Asclepias tuberosa Linnaeus subsp. **interior** Woodson, Ann. Missouri Bot. Gard. 31: 368, plate 20 [in part]. 1944

Asclepias tuberosa var. *interior* (Woodson) Shinners; *A. tuberosa* subsp. *terminalis* Woodson

Stems 30–90 cm. **Leaf blades** narrowly elliptic or lanceolate to oblong or linear, 2–12 × 0.5–3 cm, base obtuse or truncate to cordate, margins planar, apex acute to attenuate.

Flowering (Apr–)May–Sep(–Oct); fruiting Jun–Oct. Slopes, ridges, flats, valleys, streamsides, arroyos, canyons, lake edges, cliffs, bluffs, dunes, sandhills, glades, fields, pastures, limestone, granite, sandstone, shale, sandy, silty, clay, gravel, and calcareous soils, prairies, shrubby grasslands, forest edges and openings, pine barrens, oak scrub, chaparral, oak, oak-hickory, and cedar woodlands, pine, oak, pine-oak, mixed-conifer, and riparian forests; 0–2500 m; Ont., Que.; Ala., Ariz., Ark., Colo., Conn., Fla., Ga., Ill., Ind., Iowa, Kans., Ky., La., Maine, Md., Mass., Mich., Minn., Miss., Mo., Nebr., N.H., N.J., N.Mex., N.Y., N.C., Ohio, Okla., Pa., S.C., S.Dak., Tenn., Tex., Utah, Vt., Va., W.Va., Wis.; Mexico (Chihuahua, Coahuila, Nuevo León, Sonora, Tamaulipas).

Subspecies *interior* is one of the most widespread milkweeds and is distributed primarily west of the Appalachian Mountains, although there are scattered occurrences to the Atlantic Coast. Subspecies *interior* is rare and considered to be of conservation concern in Quebec (Pontiac County). Orange and yellow corollas predominate in the western states. R. E. Woodson Jr. (1953) established subsp. *terminalis* to accommodate northern and western peripheral populations bearing leaves with cuneate bases but subsequently returned these populations to subsp. *interior* (Woodson 1962).

70c. Asclepias tuberosa Linnaeus subsp. **rolfsii** (Britton ex Vail) Woodson, Ann. Missouri Bot. Gard. 31: 368. 1944 • Rolfs's milkweed [E]

Asclepias rolfsii Britton ex Vail in J. K. Small, Fl. S.E. U.S., 943, 1336. 1903

Stems 15–70 cm. **Leaf blades** lanceolate to oblong or linear-lanceolate, 2.5–6 × 0.3–1.2 cm, base hastate, margins crisped, apex rounded to acute.

Flowering year-round; fruiting Jun–Sep. Dunes, sandhills, flood plains, limestone, sandy soils, pine flatwoods, pine, pine-oak, and oak-palmetto scrub; 0–200 m; Fla., Ga.

Subspecies *rolfsii* is nearly restricted to sandy substrates in peninsular Florida and southern Georgia and is most common on the central ridges of these states. It has been reported in error from Alabama, Mississippi, North Carolina, and Virginia. Plants of subspp. *tuberosa* and *interior* from sandy sites in southern South Carolina approach subsp. *rolfsii* and may be accommodated more properly here.

71. Asclepias variegata Linnaeus, Sp. Pl. 1: 215. 1753 • Red-ring or white milkweed, asclépiade panachée [E]

Biventraria variegata (Linnaeus) Small

Herbs. **Stems** 1–few, erect, unbranched, 30–120 cm, puberulent with curved trichomes, sometimes in a single line, sometimes glaucous, rhizomes absent. **Leaves** opposite, petiolate, with 1 or 2 stipular colleters on each side of petiole; petiole 10–25 mm, puberulent with curved trichomes; blade oval to ovate, obovate, lanceolate, or oblanceolate, 6–15 × 3–9 cm, subcoriaceous, base cuneate to obtuse or subtruncate, margins sometimes crisped, apex rounded to obtuse, apiculate or mucronate, venation eucamptodromous to faintly brochidodromous, surfaces puberulent on veins with curved trichomes, sometimes glaucous, margins ciliate, 8–12 laminar colleters. **Inflorescences** terminal, branched, also usually extra-axillary at 1 distal node, pedunculate, 11–28-flowered; peduncle 1–7 cm, densely puberulent with curved trichomes, with 1 caducous bract at the base of each pedicel. **Pedicels** 12–20 mm, densely puberulent with curved trichomes. **Flowers** erect to spreading; calyx lobes lanceolate, 2–3 mm, apex acute, pilosulous; corolla white, red-violet in throat, lobes reflexed with spreading tips, elliptic, 6–8 mm, apex acute, glabrous abaxially, minutely papillose at base adaxially; gynostegial column red-violet, 1–2 mm; fused anthers brown, truncately obconic, 1.5–2 mm, wings right-triangular, rounded, closed, apical appendages ovate; corona segments white, stipitate, conduplicate with a lateral flange on each side, 2.5–4 mm, exceeding style apex, apex truncate, glabrous, internal appendage falcate, exserted, sharply inflexed over style apex, glabrous; style apex shallowly depressed, white. **Follicles** erect on upcurved pedicels, narrowly fusiform, 10–15 × 1.5–2 cm, apex long-acuminate, smooth, pilosulous. **Seeds** ovate, 5–7 × 3–5 mm, margin winged, faces rugulose; coma 2.5–4 cm.

Flowering Mar–Aug; fruiting Jun–Nov. Ridges, slopes, bluffs, flats, glades, ravines, streamsides, lake shores, limestone, sandstone, basalt, clay, sandy, silty, and marl soils, oak-hickory, oak, mixed-hardwood, pine-mixed-hardwood, pine-oak-hickory, and pine forests, oak, pine, pine-oak, oak-hickory, and riparian woodlands, forest edges and openings; 0–900 m; Ont.; Ala., Ark., Conn., Del., D.C., Fla., Ga., Ill., Ind., Ky., La., Md., Miss., Mo., N.J., N.Y., N.C., Ohio, Okla., Pa., S.C., Tenn., Tex., Va., W.Va.

Asclepias variegata and the following species (*A. exaltata*, *A. purpurascens*, and *A. quadrifolia*) inhabit deciduous forest understories in eastern North America and often co-occur. *Asclepias variegata* has showy, snowball-like spheres of bright white flowers. On closer examination, the staminal column of each flower is colored reddish purple, forming a neat belt between the corona and corolla. Non-flowering specimens have been confused with *A. purpurascens*. The leaves of *A. variegata* have a thicker texture, and the blade apices are broader and more rounded than in *A. purpurascens*. *Asclepias variegata* has suffered serious declines at the northeastern margin of its range and is reported to have been extirpated from Ontario and Connecticut. In addition, it is considered to be of conservation concern in Delaware, New Jersey, New York, and Pennsylvania.

72. Asclepias purpurascens Linnaeus, Sp. Pl. 1: 214. 1753 • Purple milkweed, asclépiade pourpreé [E]

Herbs. **Stems** 1, erect, unbranched, 50–120 cm, puberulent in lines with curved trichomes, not glaucous, rhizomatous. **Leaves** opposite (rarely whorled at 1 midstem node), petiolate, with 1 or 2 stipular colleters on each side of petiole, also in axil; petiole 4–18 mm, puberulent with curved trichomes; blade ovate or oval to lanceolate or elliptic, 6–20 × 2–10 cm, chartaceous, base cuneate, margins entire, apex obtuse to acute, apiculate, or mucronate, venation eucamptodromous to faintly brochidodromous, surfaces pilosulous, sparsely so adaxially, margins ciliate, 12–20 laminar colleters. **Inflorescences** terminal, branched, also usually extra-axillary, sessile or pedunculate, 17–72-flowered;

peduncle 0–7 cm, puberulent with curved trichomes to pilosulous, with 1 caducous bract at the base of each pedicel. **Pedicels** 12–28 mm, puberulent with curved trichomes. **Flowers** erect to spreading; calyx lobes lanceolate, 3–5 mm, apex acute, puberulent with curved trichomes; corolla reddish purple, lobes reflexed with spreading tips, elliptic, 7–10 mm, apex acute, glabrous abaxially, papillose at base adaxially; gynostegial column 1–1.5 mm; fused anthers brown, cylindric, 2–2.5 mm, wings broadly right-triangular, closed to slightly open at tip, apical appendages ovate; corona segments reddish purple, stipitate, conduplicate with a lateral flange on each side, 5–6 mm, greatly exceeding style apex, apex acute, glabrous, internal appendage falcate, exserted, sharply inflexed over style apex, glabrous; style apex shallowly depressed, red-violet to green. **Follicles** erect on upcurved pedicels, lance-ovoid, 10–16 × 1.5–2.5 cm, apex long-acuminate, smooth, puberulent with curved trichomes. **Seeds** ovate, 7–8 × 4–5 mm, margin winged, faces rugulose; coma 3–4 cm.

Flowering May–Jul(–Sep); fruiting Jul–Oct. Slopes, ravines, fields, ditches, glades, pond and lake edges, streamsides, limestone, silty, sandy, and rocky soils, oak and riparian woods, oak-hickory and mixed-hardwood forests and edges, prairie openings; 50–400 m; Ont.; Ark., Conn., Del., D.C., Ga., Ill., Ind., Iowa, Kans., Ky., La., Md., Mass., Mich., Minn., Miss., Mo., Nebr., N.H., N.J., N.Y., N.C., Ohio, Okla., Pa., R.I., S.Dak., Tenn., Tex., Va., W.Va., Wis.

Asclepias purpurascens is most common in rocky uplands of the Ozark Mountains and the piedmont of the northern Appalachian Mountains. Its reddish purple flowers are extremely showy and the species merits cultivation. The flowers have a strong cinnamon scent. Similarities to *A. variegata* and *A. exaltata* are discussed under those species. Although widespread in eastern North America, *A. purpurascens* is now rare over most of its range and has experienced a significant loss of populations and habitat everywhere but the Ozarks. It is considered to be of conservation concern in Connecticut, Delaware, Georgia (Floyd and Murray counties), Louisiana (Caldwell and Lincoln parishes), Maryland, Massachusetts, Michigan, Mississippi (Grenada and Washington counties), Nebraska (Nemaha and Richardson counties), New York, North Carolina, Tennessee, Virginia, and Ontario (Chatham-Kent, Essex, and Lambton counties). Moreover, it is presumed extirpated from the District of Columbia, Minnesota, New Hampshire, and Rhode Island. It has been reported, but not verified, from Maine and Vermont and, if historically present, is now extirpated there as well. Hybrids with *A. amplexicaulis*, *A. exaltata*, and *A. syriaca* have been documented from the New England and mid-Atlantic regions where *A. purpurascens* is now rare or extirpated. Putative hybrids exhibit intermediate floral and vegetative characteristics.

73. **Asclepias quadrifolia** Jacquin, Observ. Bot. 2: 8, plate 33. 1767 • Four-leaf or whorled milkweed, asclépiade à quatre feuilles [E]

Herbs. **Stems** 1, erect, unbranched, 25–60 cm, puberulent on 1 side with curved trichomes to glabrate, not glaucous, rhizomes absent. **Leaves** opposite, 2 mid-stem pairs usually with shortened internode forming a pseudo-whorl, petiolate, with 1 stipular colleter on each side of petiole; petiole 2–7 mm, puberulent with curved trichomes; blade ovate to elliptic, 2.5–12 × 1–6 cm, membranous, base cuneate, margins entire, apex acute to attenuate or acuminate, venation eucamptodromous, surfaces inconspicuously puberulent on veins with curved trichomes, margins ciliate, laminar colleters absent. **Inflorescences** terminal, sometimes branched, and usually also extra-axillary at upper nodes, sessile or pedunculate, 7–31-flowered; peduncle 0–3.5 cm, puberulent on 1 side with curved trichomes, with 1 caducous bract at the base of each pedicel. **Pedicels** 17–28 mm, puberulent on 1 side with curved trichomes. **Flowers** spreading to pendent; calyx lobes lanceolate, 1.5–2 mm, apex acute, glabrous; corolla pink or cream, lobes reflexed, elliptic, 4–5 mm, apex acute, glabrous abaxially, minutely papillose at base adaxially; gynostegial column 0.5–1.5 mm; fused anthers tan to brown, cylindric, 1–1.5 mm, wings right-triangular, closed, apical appendages ovate; corona segments pink or cream, sometimes striped pink dorsally, sessile, conduplicate, flattened dorsally, 2.5–4 mm, exceeding style apex, apex obtuse, oblique, glabrous, internal appendage subulate, exserted, arched to sharply inflexed over style apex, glabrous; style apex shallowly depressed, cream to pale pink. **Follicles** erect on straight pedicels, narrowly fusiform, 8–16 × 0.4–1 cm, apex long-attenuate, smooth, sparsely puberulent with curved trichomes to glabrate. **Seeds** ovate, 6–7 × 4–5 mm, margin winged, faces rugulose; coma 3–4.5 cm.

Flowering (Mar–)Apr–Aug(–Sep); fruiting May–Nov. Ridges, slopes, valleys, flats, lake shores, canyons, limestone, chert, sandstone, oak woods, oak-hickory, pine-oak, and mixed-hardwood forests, prairie openings; 100–800 m; Ont.; Ala., Ark., Conn., Del., D.C., Ga., Ill., Ind., Iowa, Kans., Ky., Md., Mass., Mo., N.H., N.J., N.Y., N.C., Ohio, Okla., Pa., R.I., S.C., Tenn., Vt., Va., W.Va.

Asclepias quadrifolia is small statured compared to the other deciduous forest milkweeds. It is found predominantly on slopes of eroded, sedimentary rocks, especially cherty limestone. Like *A. purpurascens*, it is distributed primarily in the Appalachian and Ozark mountains. Populations with pinkish flowers are lovely in early spring; other populations have a more washed-out coloration. The characteristic pseudo-whorl of four mid-stem leaves is not always present. Like other forest milkweeds, it has experienced large population reductions at the northeastern margin of its range due to habitat loss. It is considered to be extirpated from Delaware and to be rare and of conservation concern in New Hampshire, Rhode Island (Providence County), and Ontario (Prince Edward County), as well as to the west in Kansas (Cherokee County). It has been reported, but not verified, from Minnesota. Putative hybrids with *A. exaltata* have been documented very rarely from the Appalachian Mountains and can be recognized by intermediate floral and vegetative traits.

74. Asclepias exaltata Linnaeus, Amoen. Acad. 3: 404. 1756 • Poke or tall milkweed, asclépiade très grande [E]

Asclepias bicknellii Vail; *A. phytolaccoides* G. F. Lyon ex Pursh

Herbs. Stems 1–3+, erect, unbranched, 65–150 cm, sparsely pubescent to glabrate, not glaucous, rhizomes absent. **Leaves** opposite (rarely whorled at 1 midstem node), petiolate, with 1 stipular colleter on each side of petiole; petiole 5–15 mm, minutely puberulent with curved trichomes to glabrate; blade broadly ovate to oblong or elliptic, 10–24 × 2–11 cm, membranous, base cuneate, margins entire, apex attenuate to acuminate, venation eucamptodromous to brochidodromous, surfaces pilosulous to glabrate abaxially, sparsely pilosulous to glabrate adaxially, densely so on veins, margins ciliate, 6–10 laminar colleters (often obscured in pressed specimens). **Inflorescences** extra-axillary at upper nodes (terminal), pedunculate, 11–41-flowered; peduncle 0.5–8.5 cm, puberulent on 1 side with curved trichomes, with 1 caducous bract at the base of each pedicel. **Pedicels** 25–45 mm, puberulent with curved trichomes to pilosulous on 1 side. **Flowers** spreading to drooping; calyx lobes narrowly lanceolate, 3–4.5 mm, apex attenuate, pilosulous; corolla green (rarely pink-tinged), lobes reflexed, sometimes with spreading tips, elliptic, 6–12 mm, apex acute, glabrous; gynostegial column 1.5–2 mm; fused anthers green, columnar, 2.5–3.5 mm, wings right-triangular with rounded tip, apical appendages deltoid; corona segments white to pinkish, sometimes red-purple at base, stipitate, tubular, 3–5 mm, exceeding style apex, base saccate, apex truncate with 1–2 teeth on each side, glabrous, internal appendage falcate, exserted, arching above style apex, glabrous; style apex shallowly depressed, green or cream. **Follicles** erect on upcurved pedicels, fusiform, 10–15 × 1.5–2 cm, apex long-acuminate, smooth, puberulent with curved trichomes. **Seeds** lance-ovate, 8–10 × 4–6 mm, margin winged, faces minutely rugulose; coma 2.5–3 cm. $2n = 22$.

Flowering May–Aug; fruiting May–Oct. Bluffs, summits, hills, slopes, ravines, bottomlands, stream banks, lake shores, moraines, rock outcrops, limestone, alluvium, rich, thin, rocky, and sandy soils, oak, pine-oak, mixed-hardwood, riparian, and cove forests and edges, meadows; 0–1500 m; Ont., Que.; Ala., Conn., Del., Ga., Ill., Ind., Iowa, Ky., Maine, Md., Mass., Mich., Minn., Mo., N.H., N.J., N.Y., N.C., Ohio, Pa., R.I., S.C., Tenn., Vt., Va., W.Va., Wis.

Unlike the other common deciduous forest understory milkweeds, the range of *Asclepias exaltata* does not extend to the Ozarks. Compared to these other species, *A. exaltata* seems to prefer richer soils. Non-flowering individuals are often confused with *A. purpurascens*, from which they are distinguished by leaves with thinner texture, sparser abaxial vestiture, and longer-tapered apices. Hybrids with *A. syriaca* are well established at several disjunct locations (S. R. Kephart et al. 1988), and their genetics and pollination have been studied (S. B. Broyles 2002; T. M. Stoepler et al. 2012). Hybrids with *A. purpurascens* and *A. amplexicaulis* are also known, but appear to be rare and local. Putative hybrids exhibit intermediate floral and vegetative morphology. *Asclepias exaltata* is rare at the margins of its range and is considered to be of conservation concern in Alabama (Lawrence, Madison, and Winston counties), Delaware (New Castle County), Rhode Island, and Quebec. Recently, it has been documented at a single site in Missouri (Cape Girardeau County) and should be considered to be of conservation concern in that state, too.

75. Asclepias ovalifolia Decaisne in A. P. de Candolle and A. L. P. P. de Candolle, Prodr. 8: 567. 1844 • Oval-leaf or dwarf milkweed, asclépiade à feuilles ovées [E]

Herbs. Stems 1, erect, unbranched, 30–70 cm, densely puberulent with curved trichomes or pilosulous to tomentose, not glaucous, rhizomatous. **Leaves** opposite, petiolate, with 1 or 2 stipular colleters on each side of petiole; petiole 3–10 mm, densely pilosulous to tomentose; blade broadly ovate to oval or narrowly elliptic, 3–9 × 1.5–4.5 cm, chartaceous, base obtuse or rounded to truncate, margins entire, apex acute to obtuse, mucronate, venation brochidodromous, surfaces pilosulous abaxially, sparsely so adaxially except on veins, margins ciliate, 4–16 laminar colleters. **Inflorescences** extra-axillary at upper nodes, sessile or pedunculate, 5–21-flowered; peduncle 0–7 cm, densely pilosulous to tomentose, with 1 caducous bract at the base of each pedicel. **Pedicels** 12–22 mm, densely pilosulous to tomentose. **Flowers** erect to spreading; calyx lobes lanceolate, 2–3 mm, apex acute, densely pilosulous; corolla cream to yellowish, lobes reflexed with spreading tips, elliptic, 5–7 mm, apex acute, pilosulous abaxially at apex, glabrous adaxially; gynostegial column 0.2–0.5 mm; fused anthers green, cylindric, 1.5–2 mm, wings right-triangular, closed, apical appendages ovate; corona segments cream to yellowish, subsessile, conduplicate, dorsally flattened, 3–4 mm, exceeding style apex, apex acute with proximal tooth on each side, glabrous, internal appendage falcate, exserted, sharply inflexed to ascending over style apex, glabrous; style apex shallowly depressed, cream to pinkish. **Follicles** erect on upcurved pedicels, lance-ovoid, 5–8 × 1.2–2 cm, apex acute to apiculate, smooth, densely pilosulous to tomentose. **Seeds** ovate, 5–6 × 3.5–4.5 mm, margin winged, faces rugulose; coma 2.5–3 cm.

Flowering May–Aug; fruiting Jul–Sep. Hills, slopes, ravines, bluffs, ridges, dunes, coulees, ditches, lake shores, sandstone, sandy, rocky, and clay soils, prairies, shrubby grasslands, aspen woods, oak savannas, oak woods, pine-oak and pine forests; 300–1600 m; Alta., Man., Ont., Sask.; Ill., Iowa, Mich., Minn., Mont., N.Dak., S.Dak., Wis., Wyo.

Asclepias ovalifolia is the northernmost-ranging species in the genus, and over much of its range co-occurs with at most one other species of *Asclepias*. The quality of its habitat has been degraded by woody encroachment, presumably resulting from fire suppression. It appears to be secure in the core of its range in Minnesota, North Dakota, Manitoba, and Saskatchewan. Elsewhere there are conservation concerns, as in Illinois (Cook, Kankakee, Kendall, Lake, and McHenry counties), Michigan (Lake and Menominee counties), Montana (Carter and Sheridan counties), Wyoming (Crook County), and Ontario. *Asclepias ovalifolia* was collected in 1915 in British Columbia in a valley of the Columbia Mountains (*Bain. s.n.* [UBC]), far disjunct from the species' range east of the Rocky Mountains. The occurrence has been considered to be adventive and not persistent (F. Lomer, pers. comm.). It has been reported from Nebraska based on the original determination of what became the type specimen of *A. hallii*; it is not known to have ever occurred in that state. Hybrids with *A. syriaca* are known, but appear to be rare, and can be recognized by possession of intermediate floral and vegetative traits.

76. Asclepias speciosa Torrey, Ann. Lyceum Nat. Hist. New York 2: 218. 1827 • Showy milkweed, asclépiade belle, belle asclépiade [E]

Herbs. Stems 1–few, erect, unbranched (rarely branched), 30–125 cm, tomentose to puberulent with curved trichomes, not glaucous, rhizomatous. **Leaves** opposite, petiolate, with 1 or 2 stipular colleters on each side of petiole, sometimes also in axil; petiole 4–12 mm, tomentose to pilose; blade lanceolate or ovate to oblong, 6–20 × 2–14 cm, chartaceous, base rounded to truncate or cordate, margins entire, apex acute to obtuse, sometimes mucronate, venation faintly brochidodromous, surfaces tomentose to pilose, margins ciliate, 6–32 laminar colleters. **Inflorescences** extra-axillary, pedunculate, 3–34-flowered; peduncle 1–10 cm, densely tomentose, with 1 caducous bract at the base of each pedicel. **Pedicels** 13–30 mm, densely tomentose. **Flowers** erect to pendent; calyx lobes elliptic, 4–8 mm, apex acute, tomentose; corolla dark pink (rarely pale), lobes reflexed with spreading tips, elliptic, 9–12 mm, apex acute, densely pilose abaxially, hirtellous at base adaxially; gynostegial column 0.5–1 mm; fused anthers green and brown, truncately obconic, 2.5–3 mm, wings right-triangular, open, widely so at base, apical appendages deltoid; corona segments pale pink to nearly cream, sessile, scoop-shaped, 9–15 mm, exceeding style apex, apex truncate with proximal tooth on each side and long-attenuate, flared, glabrous, internal appendage subulate, exserted, sharply inflexed over style apex, glabrous; style apex shallowly depressed, green to cream or pink. **Follicles** erect on upcurved pedicels, lance-ovoid, (5–)9–12 × 2–3 cm, apex long-attenuate, muricate or smooth, densely tomentose. **Seeds** ovate, 7–9 × 4–5 mm, margin winged, faces rugulose; coma 2.5–3 cm. 2*n* = 22.

Flowering (Apr–)May–Sep; fruiting Jul–Oct. Slopes, flats, hills, valleys, canyons, coulees, streamsides, lake and pond edges, ditches, swales, seeps, granite, basalt, schist, pumice, serpentine, alluvium, clay, sandy, silty, rocky, and saline soils, pine and mixed-conifer forests, oak and pine woodlands, chaparral, riparian woods, shrubby and non-native grasslands, prairies, meadows, agricultural fields; 0–2600 m; Alta., B.C., Man., Sask.; Ariz., Calif., Colo., Idaho, Ill., Iowa, Kans., Mich., Minn., Mo., Mont., Nebr., Nev., N.Mex., N.Dak., Okla., Oreg., S.Dak., Tex., Utah, Wash., Wis., Wyo; introduced in e Europe.

Asclepias speciosa is the western counterpart of *A. syriaca*; both are broad-leaved species with large umbels of pinkish flowers. The distinctive, large, tapering corona segments, which form the broadest corona span of any American species of *Asclepias*, immediately distinguish *A. speciosa* from *A. syriaca*. These species hybridize extensively from Minnesota and southern Manitoba to Kansas, blurring the distinctions in the zone of contact, which corresponds roughly to the transition from tallgrass to mixed-grass prairie (R. P. Adams et al. 1987b). Not every individual in this zone can be readily assigned to one species or the other. This is the most extensive hybrid zone in North American *Asclepias*. These hybrids have also been documented in Illinois, far to the east of the contact zone. Possible hybrids with *A. eriocarpa* and *A. hallii* in California are discussed under those species. Outside of its contiguous range in the West, *A. speciosa* is known from a few sporadic, mostly historical records from Illinois, Michigan, Missouri, and Wisconsin. Most, if not all, of these records represent adventive, ephemeral outposts from the native range.

77. **Asclepias syriaca** Linnaeus, Sp. Pl. 1: 214. 1753 • Common milkweed, silkweed, asclépiade commune, petits cochons, herbe à la ouate [E] [F]

Asclepias cornuti Decaisne; *A. kansana* Vail

Herbs. Stems 1 (rarely more, but forming dense colonies), erect, unbranched (rarely branched), 50–200 cm, tomentose to puberulent with curved trichomes, not glaucous, rhizomatous. **Leaves** opposite, petiolate, with 1–5 stipular colleters on each side of petiole; petiole 5–15 mm, tomentose to puberulent with curved trichomes; blade oval or ovate to oblong or elliptic, 6–30 × 2.5–11 cm, chartaceous, base cuneate to rounded or truncate, margins entire, apex obtuse to rounded or acute, often mucronate, venation brochidodromous, surfaces tomentose to pilosulous abaxially, tomentose to glabrate adaxially, margins ciliate, 4–20 laminar colleters. **Inflorescences** extra-axillary, pedunculate, 24–113-flowered; peduncle 2–12 cm, tomentulose to pilose or puberulent with curved trichomes, with 1 caducous bract at the base of each pedicel. **Pedicels** 17–40 mm, densely pilose to puberulent with curved trichomes. **Flowers** erect to pendent; calyx lobes elliptic, 3–6 mm, apex acute, tomentulose; corolla dark to pale pink or green and pink-tinged, lobes reflexed with spreading tips, oblong to oval, 6–9 mm, apex acute, pilose abaxially, minutely hirtellous at base adaxially; gynostegial column 1–1.5 mm; fused anthers green, cylindric, 2–2.5 mm, wings narrowly right-triangular, slightly open, apical appendages ovate; corona segments reddish purple to cream, sessile, tubular, slightly flattened dorsally, 4–5 mm, exceeding style apex, apex obtuse, somewhat to strongly flared, glabrous, internal appendage falcate, exserted, sharply inflexed over style apex, glabrous; style apex shallowly depressed, green or pale to dark pink. **Follicles** erect on upcurved pedicels, lance-ovoid to ovoid, 7–12 × 2–4 cm, apex acuminate, smooth or muricate, tomentose. **Seeds** narrowly ovate, 7–8 × 4–5 mm, margin winged, faces rugulose; coma 3–4 cm. $2n = 22$.

Flowering May–Sep(–Oct); fruiting Jun–Oct. Flats, slopes, ridges, valleys, fields, meadows, pastures, ditches, pond and lake edges, marshes, bogs, fens, parks, urban lots, streamsides, swales, bluffs, sandhills, limestone, clay, silty, sandy, and rocky soils, prairies, forest openings and edges, riparian woods; 0–1300 m; Man., N.B., Nfld. and Labr. (Nfld.), N.S., Ont., P.E.I., Que., Sask.; Ala., Ark., Conn., Del., D.C., Ga., Ill., Ind., Iowa, Kans., Ky., La., Maine, Md., Mass., Mich., Minn., Miss., Mo., Nebr., N.H., N.J., N.Y., N.C., N.Dak., Ohio, Okla., Pa., R.I., S.C., S.Dak., Tenn., Vt., Va., W.Va., Wis; introduced in Europe, sw Asia.

Asclepias syriaca is surely the most familiar milkweed in North America, and one that evokes ambivalence. It has been considered an undesirable species because of its prolific rhizomatous spread and ability to invade and thrive in cultivated land. However, it has been used as a food plant by indigenous and colonizing peoples, and its pleasantly fragrant and nectariferous flowers are avidly sought by diverse insects, highlighting the ecological importance of *A. syriaca*. Moreover, it has come to be appreciated because of its importance as one of the most commonly utilized host plants of the monarch butterfly, *Danaus plexippus*. There is some evidence that its population in agricultural lands in the upper midwestern United States has dramatically declined in the last several decades. However, its range and abundance prior to European settlement are not well understood and may have been much lower than in historical times, particularly in deforested areas of the eastern United States. It is considered to have been introduced to Newfoundland, Alabama, Georgia, Louisiana, Mississippi, and South Carolina and has been reported, but without documentation, from Texas. It has been documented from Salem, Oregon, but appears to not be established there (R. Halse, pers. comm.). It is considered to be of conservation concern at the northwesternmost edge of its range in Saskatchewan, where a single population is known in Estevan Municipality. *Asclepias syriaca* is the most promiscuous of milkweeds, as it is known to hybridize with at least seven other species (*A. amplexicaulis*, *A. exaltata*, *A. ovalifolia*, *A. purpurascens*, *A. speciosa*, *A. sullivantii*, *A. tuberosa*). Hybrids with *A. speciosa* are most frequent, as discussed under that species. Hybrids with *A. exaltata* are not infrequently encountered in the Appalachian Mountains and elsewhere. Other hybrids are highly localized. In all cases, putative hybrids are inferred from intermediate floral and vegetatative traits.

27. CYNANCHUM Linnaeus, Sp. Pl. 1: 212. 1753; Gen. Pl. ed. 5, 101. 1754 • Strangle-vine, swallow-wort, milkweed vine [Greek *kyon*, dog, and *ancho*, strangle, alluding to toxicity of some species]

Mark Fishbein

Ampelamus Rafinesque; *Mellichampia* A. Gray; *Rouliniella* Vail

Vines [erect herbs, subshrubs], herbaceous or somewhat woody, but not corky, at base [succulent]; latex white or clear. **Stems** twining [not twining], unarmed, puberulent in single line or glabrate [variously pubescent or glabrous]. **Leaves** persistent [reduced to inconspicuous scales], opposite, petiolate [sessile]; stipular colleters interpetiolar; laminar colleters present [absent]. **Inflorescences** extra-axillary, racemiform, corymbiform, or paniculiform [umbelliform], sessile or pedunculate. **Flowers:** calycine colleters apparently absent [present?]; corolla white, cream, or green [brown, purple, red, pink], campanulate [rotate], aestivation imbricate; corolline corona absent; androecium and gynoecium united into a gynostegium adnate to corolla tube; gynostegial corona 1-[2-]whorled; anthers adnate to style, locules 2; pollen in each theca massed into a rigid, vertically oriented pollinium, pollinia lacrimiform, joined from adjacent anthers by translators to common corpulsculum and together forming a pollinarium. **Fruits** follicles, usually solitary, variously oriented, green to brown, terete or subterete, smooth, glabrous. **Seeds** winged, not beaked, ovate, flattened, comose, not arillate. $x = 11$.

Species ca. 300 (3 in the flora): United States, Mexico, West Indies, Central America, South America, Europe, Asia, Africa, Indian Ocean Islands (Madagascar), Australia.

The influential revision of North American milkweed genera by R. E. Woodson Jr. (1941) adopted a very broad (and evidently polyphyletic) circumscription of *Cynanchum*, reducing genera recognized here to synonymy (*Metastelma*, *Orthosia*, *Pattalias*). Similarly, regional floras for Canada and the United States have included the introduced species here treated in *Vincetoxicum* under *Cynanchum*. The polyphyly of this broad concept of *Cynanchum* has been convincingly demonstrated (for example, by S. Liede and A. Täuber 2002). Another potential source of confusion is that the species of *Funastrum* have previously been included in the Old World genus *Sarcostemma* R. Brown. Although *Sarcostemma* has been synonymized with *Cynanchum* following the phylogenetic results of Liede and Täuber, the species of *Funastrum* are distantly related to those of *Cynanchum*.

SELECTED REFERENCES Liede, S. and A. Täuber. 2002. Circumscription of the genus *Cynanchum* (Apocynaceae–Asclepiadoideae). Syst. Bot. 27: 789–800. Sundell, E. 1981. The New World species of *Cynanchum* subgenus *Mellichampia* (Asclepiadaceae). Evol. Monogr. 5: 1–63.

1. Latex clear; dwarf axillary shoots common, resembling stipules (pseudostipules); corollas lacking inframarginal ridges adaxially; corona segments distinct, divided into 2 subulate lobes; c, e United States, including e Texas . . . . 1. *Cynanchum laeve*
1. Latex white; dwarf axillary shoots rare; corollas with inframarginal ridges adaxially; corona segments united at base, shallowly lobed or unlobed; Arizona, Texas.
    2. Corollas deeply campanulate, lobes 7–8 mm, cream (fading yellowish), inframarginal ridges pilose; corona segments unlobed; Arizona . . . . 2. *Cynanchum ligulatum*
    2. Corollas shallowly campanulate to campanulate-rotate, lobes 3–4 mm, green with white margins, inframarginal ridges glabrous; corona segments 3-lobed at apex; Texas . . . . 3. *Cynanchum unifarium*

1. Cynanchum laeve (Michaux) Persoon, Syn. Pl. 1: 274. 1805 • Honeyvine, sandvine, bluevine [E] [F] [W]

Gonolobus laevis Michaux, Fl. Bor.-Amer. 1: 119. 1803; *Ampelamus albidus* (Nuttall) Britton; *A. laevis* (Michaux) Krings

Latex clear. **Stems** puberulent in single line; dwarf axillary branches common (pseudostipules). **Leaves:** 1–3 early-caducous stipular colleters on each side of petiole; petiole 1–9 cm, puberulent in single line to glabrate; blade pinnipalmately veined, ovate or deltate, 2–11 × 1.5–10 cm, chartaceous, base shallowly to deeply cordate or sagittate, with 3–8 laminar colleters, margins puberulent-ciliate or glabrate, apex acute, attenuate, acuminate, or apiculate, surfaces minutely puberulent to glabrate on veins. **Inflorescences** racemiform to paniculiform, solitary (paired) at nodes, 6–30-flowered; peduncle 0.2–2(–6.5) cm, puberulent in single line; bracts caducous, 1, at base of each pedicel. **Pedicels** 3–12 mm, tomentulose. **Flowers:** calyx lobes erect, lanceolate to ovate, 1.5–2 mm, apex obtuse, acute, or acuminate, puberulent to glabrate, margins ciliate to glabrate; corolla cream, campanulate, tube 1–1.5 mm, lobes erect to spreading, plane to somewhat twisted, linear-lanceolate, 3.5–6 mm, glabrous, inframarginal adaxial ridges absent; corona united to column near base, composed of 5 distinct segments, cream, laminar, slightly exserted from corolla, ovate with apex deeply divided into 2 subulate, usually twisted lobes, 4–6 mm; style apex conic. **Follicles** ovoid to lance-ovoid, straight to falcate, 7–15 × 2–3.5 cm, apex attenuate, thick-walled. **Seeds** 50–100, tan to light brown, 8–10 × 5–7 mm, thickly winged, chalazal margin erose, faces minutely foveolate; coma white to tawny, 3–4 cm. $2n = 22$.

Flowering Apr–Nov; fruiting Jul–Dec. Streamsides, flood plains, canyons, dunes, roadsides, fields, gardens, cities and towns, especially on fences, riparian woods, thickets, deciduous forest, prairies; 0–700 m; Ala., Ark., Del., D.C., Fla., Ga., Ill., Ind., Iowa, Kans., Ky., La., Md., Mich., Minn., Miss., Mo., Nebr., N.J., N.Y., N.C., Ohio, Okla., Pa., S.C., Tenn., Tex., Va., W.Va.

Cynanchum laeve is a common and fairly weedy species from valleys west of the Appalachians to the central Great Plains. It is especially common in valleys of tributaries to the Mississippi River. It is far more evident as a weed on fences along roads, in fields, and in urban lots and gardens than in less-disturbed vegetation. Prior to massive anthropogenic environmental change in North America, the natural habitat was likely riparian vegetation. The deeply seated rhizomes are difficult to remove, and the vine is hard to eradicate from managed landscapes, where they are not necessarily unwelcome guests. The flowers are sweetly fragrant, hence the common name honeyvine. *Cynanchum laeve* has been reported from Ontario, but the author has not seen specimens to confirm its presence in the province. A report from Idaho is based on a misidentified specimen of *Metaplexis japonica* (Thunberg) Makino. Occurrences on the western and northern margins of the range and east of the Appalachians likely represent recent range expansion.

2. Cynanchum ligulatum (Bentham) Woodson, Ann. Missouri Bot. Gard. 28: 210. 1941

Enslenia ligulata Bentham, Pl. Hartw., 290. 1849; *Cynanchum sinaloense* (Brandegee) Woodson; *Mellichampia sinaloensis* (Brandegee) Kearney & Peebles

Latex white. **Stems** densely puberulent in single line; dwarf axillary branches rare. **Leaves:** 1 stipular colleter on each side of petiole; petiole 1–4 cm, densely puberulent in single line; blade pinnipalmately veined, ovate or deltate, 2.5–6.5 × 1.5–6 cm, chartaceous, base deeply cordate, with 2–8 laminar colleters, margins puberulent-ciliate or glabrate, apex acute, attenuate, acuminate, or apiculate, surfaces minutely pilosulous on veins abaxially, glabrous adaxially. **Inflorescences** racemiform to corymbiform, solitary at nodes, 3–10-flowered; peduncle 1–5.5 cm, densely puberulent in single line; bracts caducous, 1, at base of each pedicel. **Pedicels** 5–15 mm, densely puberulent in single line. **Flowers:** calyx lobes spreading to reflexed, linear-lanceolate, acute to acuminate, 4–5 mm, ciliate; corolla cream, yellowing with age, deeply campanulate, tube 2–3 mm, lobes erect to spreading with recurved tips, linear-lanceolate, 7–8 mm, with thickened, proximally pilose, inframarginal ridges adaxially, glabrous abaxially; corona united to column near base, composed of 5 segments connate at base, cream, laminar, exserted from corolla, subulate, apex incurved, unlobed, 6–11 mm; style apex convex. **Follicles** lance-ovoid, 6–9 × 1.5–3 cm, apex obtuse, thick-walled. **Seeds** 50–100, brown, 8.5–10 × 4–6 mm, narrowly winged, chalazal margin erose, faces minutely papillate; coma white, 1.5–3 cm.

Flowering Jul–Aug; fruiting Sep–Oct. Streamsides in oak grasslands and mesquite grasslands; 1100–1600 m; Ariz.; Mexico.

Cynanchum ligulatum has the showiest flowers of the species of *Cynanchum* in the flora area. Flowering can be profuse on vigorous plants that blanket supporting vegetation. Such displays exude a strong, sweet perfume. *Cynanchum ligulatum* is in the flora area only in extreme southeastern Arizona (Cochise, Pima, and Santa Cruz counties), where it is rare along perennial or ephemeral streams. Several of the few known localities have been explored repeatedly by botanists; the scarcity of collections suggests that populations may be ephemeral, that plants do not regularly emerge or flower, or that the species is in decline at the edge of its range. *Cynanchum ligulatum* ranges south in western Mexico to Guanajuato and Michoacán. Southern populations have dark reddish or purplish corollas, although pale- or white-flowered plants are regularly encountered there as well. Those in Arizona and northern Mexico (Chihuahua, Sinaloa, and Sonora), have uniformly cream-colored corollas that are somewhat smaller than those from southern populations. The conservative approach by E. Sundell (1981) is followed for now. Should additional study support recognition of two species, *C. sinaloense* is available for northern populations, including the flora area.

3. **Cynanchum unifarium** (Scheele) Woodson, Ann. Missouri Bot. Gard. 28: 210. 1941 • Talayote

Gonolobus unifarius Scheele, Linnaea 21: 760. 1849; *Cynanchum racemosum* (Jacquin) Jacquin var. *unifarium* (Scheele) Sundell; *Rouliniella unifaria* (Scheele) Vail

Latex white. **Stems** puberulent in single line or glabrate; dwarf axillary branches rare. **Leaves:** 1 stipular colleter on each side of petiole; petiole 1–7 cm, densely puberulent in single line; blade pinnipalmately veined, ovate to deltate, 3–11 × 1.5–9 cm, chartaceous, base shallowly to deeply cordate or sagittate, with 2–6 laminar colleters, margins puberulent-ciliate, apex acute, attenuate, acuminate, or apiculate, surfaces sparsely puberulent on veins abaxially and adaxially. **Inflorescences** racemiform to corymbiform, solitary at nodes, 4–20-flowered; peduncle 0.6–3.5 cm, densely puberulent in single line; bracts caducous, 1, at base of each pedicel. **Pedicels** 4–12 mm, densely puberulent in single line. **Flowers:** calyx lobes ascending to spreading, lanceolate to oblong, 2–3 mm, apex acute, ciliate; corolla green with lobe margins and apex white, shallowly campanulate to campanulate-rotate, tube 0.5–1 mm, lobes ascending to spreading with recurved tips, oblong, 3–4 mm, glabrous, with thickened inframarginal ridges adaxially; corona united to column near base, composed of 5 segments connate at base, cream, laminar, included in corolla, quadrate, apex shallowly 3-lobed, central lobe equal to lateral lobes or extended as triangular tooth up to 8 times length of lateral lobes, 1–2 mm; style apex convex. **Follicles** ovoid, 8–12 × 1.5–2.5 cm, apex obtuse, thick-walled. **Seeds** 50–100, brown, 6–8 × 4–5 mm, narrowly winged, chalazal margins erose, faces minutely papillate; coma white to tawny, 2.5–3.5 cm.

Flowering Apr–Oct; fruiting Jun–Jan. Limestone or igneous hills and valleys, alluvium, rocky slopes, canyons, streamsides, thickets, thornscrub, oak and juniper woodlands, grasslands, climbing trees, shrubs, and boulders; 0–1600 m; Tex.; e, n Mexico.

Cynanchum unifarium occurs in a great diversity of habitats from southern to western Texas and is one of the most commonly encountered milkweed vines in the state. It is most commonly found in riparian vegetation but can occur in a wide variety of plant communities on diverse substrates. *Cynanchum unifarium* was included in a broadly circumscribed *C. racemosum* (Jacquin) Jacquin by E. Sundell (1981). This approach has merit, but the narrower concept used by W. D. Stevens (2009) is adopted here until this complex receives more detailed study. The Spanish common name derives from an Aztec word that is commonly applied to diverse milkweeds across Mexico, including those in the genera *Asclepias*, *Gonolobus*, *Marsdenia* R. Brown, *Matelea*, and *Polystemma*, among others.

28. FUNASTRUM E. Fournier, Ann. Sci. Nat., Bot., sér. 6, 14: 388. 1882 • Twinevine, milkweed vine [Latin *funis*, rope, and *-astrum*, incomplete resemblance or diminutive, alluding to twining of stem tips in *F. angustissimum*, a South American species]

Mark Fishbein

Philibertella Vail; *Sarcostemma* R. Brown subg. *Ceramanthus* Kunze

Vines, herbaceous or somewhat woody and corky at base, perennial; latex white. **Stems** twining, unarmed, variously pubescent or glabrous. **Leaves** persistent or caducous, opposite, petiolate or sessile; stipular colleters interpetiolar; laminar colleters present or absent. **Inflorescences** extra-axillary, umbelliform, pedunculate or subsessile. **Flowers:** calycine colleters absent or present; corolla cream, cream tinged with pink or rose, greenish cream, reddish green, green, reddish brown, brown, bronze, pink, purple, or yellow, rotate-campanulate to rotate or tubular, aestivation contort-dextrorse; corolline corona absent; androecium and gynoecium united into a gynostegium adnate to corolla tube; gynostegial corona in 2 parts, one a raised, fleshy ring at insertion of staminal column on corolla, the other a whorl of 5 inflated segments on staminal column; anthers adnate to style, locules 2; pollen in each theca massed into a rigid, vertically oriented pollinium, pollinia lacrimiform, joined from adjacent anthers by translators to a common corpusculum and together forming a pollinarium. **Fruits** follicles, solitary or, less commonly, paired, pendulous, green to brown, lance-ovoid to fusiform, smooth, variously pubescent. **Seeds** winged, lanceolate to ovate, flattened, comose, not arillate.

Species ca. 18 (7 in the flora): s United States, Mexico, West Indies, Central America, South America.

R. E. Woodson Jr. (1941) and R. W. Holm (1950) treated the species of *Funastrum* as congeneric with those of *Sarcostemma* due to the remarkable similarity in the two-part coronas. The polyphyly of this concept was demonstrated by S. Liede and A. Täuber (2000). Subsequently, *Sarcostemma* in the narrow sense was synonymized with *Cynanchum* based on the phylogenetic results of Liede and Täuber (2002). *Funastrum* is endemic to the Americas, where the genus is recognized readily by its distinctive corona. The only discordant element in the current circumscription is *F. utahense*, which lacks a corona, but bears vegetative similarities to other species of the genus. The species of *Funastrum* are a common element of arid grasslands, coastlines, and riparian corridors in desert and scrub vegetation in subtropical North America. Several species reach their northern limits in the flora region, notably in the southwestern United States, and only *F. utahense* is endemic to the flora region.

1. Leaf blades filiform, less than 0.02 cm wide; corollas tubular, yellow, lobes incurved; coronas absent . . . . 7. *Funastrum utahense*
1. Leaf blades linear or elliptic to lanceolate, ovate, deltate, or oval, usually greater than or equal to 0.02 cm wide; corollas rotate to rotate-campanulate, cream or shades of pink, purple, red, green, or brown, lobes spreading to ascending; coronas consisting of a ring and discrete lobes.
   2. Stems, leaves, peduncles, and pedicels hirtellous or hirsute (stems rarely pilose in *F. torreyi*).
      3. Leaf blades linear, rarely linear-lanceolate, 0.07–0.5 cm wide; pedicels 5–8 mm; corolla lobes 3–4 mm; follicles 4–5.5 × 0.3–0.5 cm; Ariz., Calif., Nev. . . . . 5. *Funastrum hirtellum*
      3. Leaf blades deltate, lanceolate or ovate, 1.5–2.5 cm wide; pedicels 10–17 mm; corolla lobes 6–9 mm; follicles 8–9 × 1–1.5 cm; Tex. . . . . 6. *Funastrum torreyi*
   2. Stems, leaves, peduncles, and pedicels glabrous, glabrate, pilose, or puberulent with curved trichomes.
      4. Leaf blade margins crispate, rarely plane; corollas shades of green, red, or brown, glabrous adaxially; corona lobes with medial constriction; follicles 10–13 cm; seeds 8–9 × 3–4 mm. . . . . 2. *Funastrum crispum*
      4. Leaf blade margins plane; corollas shades of cream, pink, or purple, minutely hispidulous to glabrate adaxially; corona lobes without constriction; follicles 5–9 cm; seeds 4–6 × 2–3 mm.
         5. Leaf blades elliptic to oval; corollas cream with red or maroon ring encircling corona, lobes 6–7 mm; follicles 5–6 cm . . . . 1. *Funastrum clausum*
         5. Leaf blades ovate or lanceolate to linear; corollas shades of cream, pink, or purple, lacking red or maroon ring encircling corona, lobes 3–6 mm; follicles 6–9 cm.
            6. Leaf blades ovate to lanceolate, 0.7–4.5 cm wide, base commonly cordate, also sagittate or hastate; corollas most commonly cream or rose, often spotted or tipped with darker shades of rose; largely east of Pecos River . . . . 3. *Funastrum cynanchoides*
            6. Leaf blades linear-lanceolate to linear, 0.1–1.5 cm wide, base commonly hastate or truncate, also sagittate or cordate; corollas most commonly shades of pink or purple or cream, spotted or tipped with pink or purple; largely west of Pecos River. . . . . 4. *Funastrum heterophyllum*

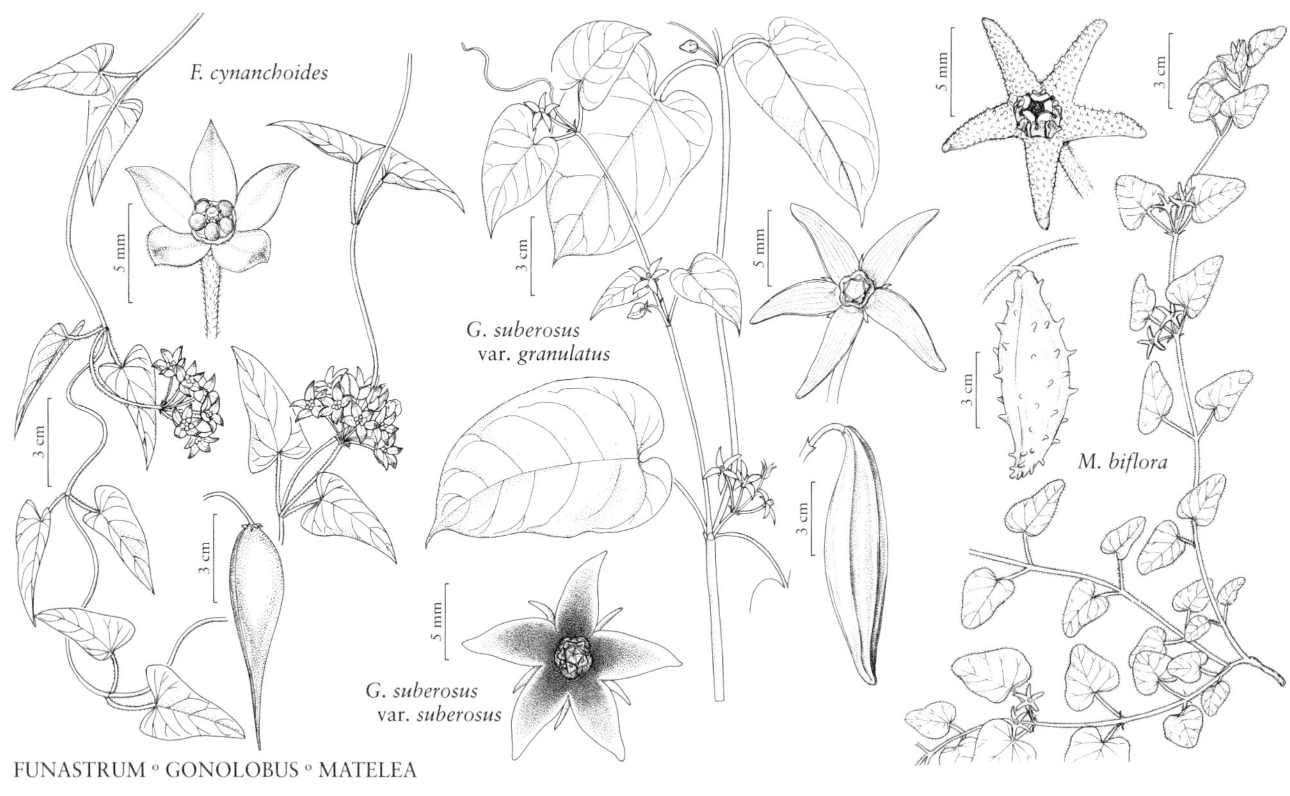

FUNASTRUM ◦ GONOLOBUS ◦ MATELEA

1. **Funastrum clausum** (Jacquin) Schlechter, Repert. Spec. Nov. Regni Veg. 13: 283. 1914 • White twinevine

Asclepias clausa Jacquin, Enum. Syst. Pl., 17. 1760; *Sarcostemma clausum* (Jacquin) Schultes

Stems sparsely pilose to glabrate, densely pilose at nodes. **Leaves** commonly caducous, stipular colleter 1 on each side of petiole; petiole 0.2–0.6 cm, pilose throughout; blade elliptic to oval, 2–5 × 0.5–2.5 cm, chartaceous to coriaceous, base obtuse to shallowly cordate, margins plane, apex acuminate, mucronate, venation pinnipalmate, surfaces sericeous to glabrate abaxially and adaxially, margins remotely ciliate with ascending hairs, laminar colleters 3–6. **Inflorescences** solitary at nodes, rarely 2 on a single peduncle, 7–15-flowered; peduncle 3.5–12 cm, glabrate; bracts caducous, 1, at base of each pedicel. **Pedicels** 15–25 mm, pilosulose. **Flowers:** calyx lobes lanceolate to ovate, 3–4 mm, apex acute, spreading, pilosulose, colleters 1–2 in sinuses; corolla cream to greenish cream, corona encircled by maroon to reddish ring, rotate to rotate-campanulate, tube 2–2.5 mm, lobes spreading to ascending, ovate, 6–7 mm, apex obtuse, pilosulose abaxially, minutely hispidulous adaxially, corona ring white, encircled by erect fringe, margins ciliate; inflated corona segments white, glossy, ovoid, 2.5–3 mm; style apex convex. **Follicles** usually solitary, lance-ovoid, 5–6 × 1–1.5 cm, apex attenuate, pilosulose. **Seeds** 30–50, tan, thickly winged, ovate, flat, 4–6 × 2–3 mm, faces coarsely papillate, chalazal margin erose to dentate; coma white, 2 cm. $2n$ = 20.

Flowering year round; fruiting Sep–Jan. Hammocks, mangroves, salt marshes, estuaries, lakeshores, streamsides, disturbed areas; 0–50 m; Fla., Tex.; Mexico; West Indies; Central America; South America.

The synonymy of *Funastrum clausum* is extensive, pertaining particularly to South America, reflecting morphological variation that greatly exceeds that present in the flora region. Flowers are deeply fragrant and are frequently visited by a diversity of insects, particularly Hymenoptera and Lepidoptera. This widespread species only enters the flora region along the Florida coastline and in the lower Rio Grande Valley of Texas.

2. Funastrum crispum (Bentham) Schlechter, Repert. Spec. Nov. Regni Veg. 13: 284. 1914 • Wavy twinevine

Sarcostemma crispum Bentham, Pl. Hartw., 291. 1849; *S. lobatum* Waterfall; *S. undulatum* Torrey

Stems retrorsely puberulent with curved trichomes to glabrate and puberulent at nodes. **Leaves** persistent, stipular colleter 1 on each side of petiole, early-caducous; petiole 0.2–1 cm, puberulent throughout with curved trichomes or in a single line; blade lanceolate to linear-lanceolate (approaching linear), 3–10 × 0.2–3 cm, chartaceous, base cordate, hastate, or truncate, margins strongly to weakly crispate (plane), apex acuminate to attenuate, single-veined to faintly pinnipalmate, surfaces sparsely to densely puberulent with curved trichomes, midvein often conspicuously densely puberulent and whitish, laminar colleters 0–4. **Inflorescences** solitary at nodes, 4–13-flowered; peduncle 0.5–4.5 cm, puberulent with curved trichomes; bracts caducous, 1, at base of each pedicel. **Pedicels** 6–25 mm, puberulent with curved trichomes to glabrate. **Flowers:** calyx lobes lanceolate, 2.5–4 mm, apex acute, erect, puberulent with curved trichomes, colleters 1–2, in sinuses; corolla green or reddish green abaxially, reddish brown, brown, bronze, or green adaxially, rotate-campanulate, tube 2–3 mm, lobes spreading to ascending, ovate, 4–9 mm, apex obtuse, sparsely to densely puberulent abaxially with curved trichomes, glabrous adaxially, margins ciliate; corona ring green to pinkish, inflated corona segments proximally green to magenta, distally glossy white to pink, oblong with medial constriction, 1.5–2.5 mm; style apex convex. **Follicles** usually solitary, lance-ovoid, 10–13 × 0.6–1.2 cm, apex attenuate, puberulent with curved trichomes. **Seeds** 20–50, light brown, thickly winged, lanceolate, shallowly naviculate, 8–9 × 3–4 mm, convex face coarsely papillate towards chalazal end, concave face nearly smooth; coma white to tawny, 3–3.5 cm. $2n$ = 44.

Flowering (Mar–)Apr–Aug(–Nov); fruiting May–Nov(–Dec). Hills, plains, cliffs, canyons, arroyos, rocky and sandy soils, limestone, granite, shale, sandstone, gypsum, grasslands, oak, pine-oak, oak-juniper, pinyon-juniper, and mesquite woodlands, chaparral, desert scrub; 200–2100 m; Ariz., Calif., Colo., N.Mex., Okla., Tex.; Mexico.

Typically, *Funastrum crispum* is easily recognized by its dark green leaves with cordate bases, white venation, and crisped margins; however, plants bearing nearly linear leaves with truncate or hastate bases and planar margins are not uncommon and are easily mistaken for *F. heterophyllum* if flowers are absent. *Funastrum crispum* is more typically found in grasslands, whereas *F. heterophyllum* is more typically found in desert scrub, especially in arroyos. R. W. Holm (1950) lectotypified *Sarcostemma heterophyllum* Engelmann ex Torrey with *Wright 1679*, resulting in the synonymy of that name with *F. crispum*; however, lectotypification of *S. heterophyllum* was predated by A. M. Vail (1897), who selected *Thomas s.n.* from Fort Yuma as the type. Consequently, that basionym applies to the species recognized as *F. heterophyllum* in this treatment (M. Fishbein and K. N. Gandhi 2018). *Funastrum crispum* is very rare at the edges of its range in California, Colorado, and Oklahoma and is considered to be of conservation concern in those states. The single population in California, in Riverside County, is quite disjunct from the rest of the range of the species.

3. Funastrum cynanchoides (Decaisne) Schlechter, Repert. Spec. Nov. Regni Veg. 13: 284. 1914 • Fringed twinevine [F] [W]

Sarcostemma cynanchoides Decaisne in A. P. de Candolle and A. L. P. P. de Candolle, Prodr. 8: 540. 1844

Stems sparsely pilosulose, puberulent with curved trichomes, or glabrate, pilosulose to glabrate at nodes. **Leaves** persistent, stipular colleter 1 on each side of petiole; petiole 0.4–2 cm, puberulent with curved trichomes in a single line; blade ovate to lanceolate, 1.2–8 × 0.7–4.5 cm, chartaceous, base cordate, sagittate, or hastate, margins plane, apex acuminate to acute, venation pinnipalmate, surfaces sparsely pilosulose, mainly on veins, or glabrous, margins inconspicuously ciliate with ascending hairs, laminar colleters 4–6. **Inflorescences** solitary (paired) at nodes, 6–31-flowered; peduncle 1–6 cm, elongating in fruit, puberulent with curved trichomes; bracts caducous, 1, at base of each pedicel. **Pedicels** 8–20 mm, puberulent with curved trichomes. **Flowers:** calyx lobes lanceolate to ovate, 2–3 mm, apex acute, erect, pilosulose, ciliate, colleters 1 or 2, in sinuses, or absent; corolla cream, often rose-tinged at lobe tips or throughout, rotate, tube 1.5–2 mm, lobes spreading with ascending tips, ovate, 3–6 mm, apex acute to obtuse, sparsely puberulent with curved trichomes to glabrate abaxially, minutely hispidulous adaxially throughout or distally or glabrate, margins ciliate; corona ring green, cream, or pinkish, inflated corona segments proximally green, distally glossy white, ovoid, 1.5–2.5 mm; style apex shallowly convex. **Follicles** usually solitary, ovoid to lance-ovoid, 6–9 × 0.8–1.6 cm, apex long-acuminate, puberulent with curved trichomes. **Seeds** 20–50, brown, winged, biconvex, lanceolate, 5–6 × 2–3 mm, faces papillate, chalazal margin erose; coma white, 2.5–3 cm. $2n$ = 20.

Flowering (Mar–)Apr–Oct; fruiting (Jun–)Aug–Nov. Canyons, cliffs, stream banks and terraces, flood plains, old fields, roadsides, sandy, silty, clayey, gravelly, or calcareous soils, sandstone, limestone, riparian, oak, and mesquite woods, grasslands, thorn scrub, desert scrub; 0–1900 m; Ariz., Ark., N.Mex., Okla., Tex.; Mexico (Chihuahua, Coahuila, Nuevo Léon, Sonora, Tamaulipas).

Funastrum cynanchoides is most similar to *F. heterophyllum*, and in the past these species were treated as conspecific (with the latter entity most commonly recognized as *Sarcostemma cynanchoides* subsp. *hartwegii*). These species are most reliably distinguished by leaf shape (ovate to lanceolate with a cordate base in *F. cynanchoides* versus linear to linear-lanceolate with a hastate to truncate base in *F. heterophyllum*), corolla color (little to no pink in *F. cynanchoides* versus faintly to strongly pink to purple in *F. heterophyllum*), flowering phenology (summer in *F. cynanchoides* versus winter to spring in *F. heterophyllum*, although summer flowering is occasional), and habitat (grasslands in *F. cynanchoides* versus desert scrub in *F. heterophyllum*); however, as emphasized in earlier treatments, the distinguishing characteristics are not constant within species, and this is especially true in the contact zone from southeastern Arizona to trans-Pecos Texas, where intermediates representing putative hybrids sporadically occur. These are most commonly encountered in Big Bend National Park. The effective isolation of these species is apparent in Tucson, Arizona, where *F. heterophyllum* occurs commonly in arroyos throughout the city and *F. cynanchoides* is adventive and common on fences; no intermediate specimens are known from this area. *Funastrum cynanchoides* is uncommon in Arkansas, and its conservation status should be evaluated in that state.

4. Funastrum heterophyllum (Engelmann ex Torrey) Standley, Contr. U.S. Natl. Herb. 23: 1170. 1924
• Fringed twinevine

Sarcostemma heterophyllum Engelmann ex Torrey in War Department [U.S.], Pacif. Railr. Rep. 5(2): 362. 1857; *Funastrum cynanchoides* (Decaisne) Schlechter var. *hartwegii* (R. W. Holm) Krings; *S. cynanchoides* Decaisne subsp. *hartwegii* R. W. Holm; *S. cynanchoides* var. *hartwegii* (R. W. Holm) Shinners

Stems glabrous to sparsely puberulent with curved trichomes, pilosulose at nodes. **Leaves** persistent to often caducous, stipular colleter 1 on each side of petiole; petiole 0.2–1.5 cm, pilosulose or puberulent with curved trichomes throughout or in a single line or glabrate; blade linear to linear-lanceolate, 1.5–8 × 0.1–1.5 cm, chartaceous, base hastate, truncate, sagittate, or cordate, margins plane, apex acute or attenuate, 1-veined or faintly pinnipalmate, surfaces pilosulose to glabrate, margins inconspicuously ciliate with ascending hairs, laminar colleters 0–4. **Inflorescences** solitary (paired) at nodes, 5–13-flowered; peduncle 0.3–4.5 cm, elongating in fruit, pilosulose to glabrate; bracts caducous, 1, at base of each pedicel. **Pedicels** 5–14 mm, hispidulous to pilosulose. **Flowers:** calyx lobes lanceolate, 1.5–2.5 mm, apex acute, erect, hispidulous to pilose, ciliate, colleter 1, in sinuses, or absent; corolla pink to purple with cream margins or cream with pink to purple blotches or streaks, rotate, tube 1–1.5 mm, lobes spreading with ascending tips, ovate, 3.5–5 mm, apex acute to obtuse, pilosulose abaxially, minutely hispidulous adaxially throughout or distally or glabrate, margins ciliate; corona ring green, cream, or pinkish, inflated corona segments proximally green, distally glossy white, ovoid, 1.5–2 mm; style apex shallowly convex. **Follicles** usually solitary, ovoid to lance-ovoid, 6–9 × 0.6–1 cm, apex long-acuminate, puberulent with curved trichomes. **Seeds** 20–50, light brown, thickly winged, biconvex, lanceolate, 5–6 × 2–3 mm, convex face coarsely papillate toward chalazal end, concave face nearly smooth; coma white, 2.5–3 cm.

Flowering and fruiting year round. Arroyos, canyons, hillsides, mountain slopes, old fields, disturbed areas, rocky, gravelly, and sandy soils, alluvium, caliche, limestone, granite, desert scrub, riparian scrub, mesquite scrub, grasslands; 100–1700 m; Ariz., Calif., Nev., N.Mex., Tex., Utah; n, c Mexico.

In the flora region, *Funastrum heterophyllum* ranges from southern California and extreme southern Nevada, where it co-occurs with *F. hirtellum*, to the Big Bend region of Texas. It is uncommon in Utah, where its conservation status merits evaluation. Differences between *F. heterophyllum* and *F. hirtellum* are discussed under the latter species. *Funastrum heterophyllum* also encounters *F. cynanchoides* at sporadic locations from southern Arizona to western Texas. The distinctions between *F. heterophyllum* and *F. cynanchoides* are discussed under the latter species. *Funastrum heterophyllum* is an exceedingly common species of Sonoran Desert arroyos and is also common in the Chihuahuan Desert. During dry periods, its leafless stems form conspicuous tangled masses on arroyo beds and banks, and riparian vegetation. Beginning with R. W. Holm (1950), the epithet *hartwegii* has been used at the specific and infraspecific levels for this entity. Holm overlooked the lectotypification of *Sarcostemma heterophyllum* Englemann ex Torrey by A. M. Vail (1897) and typified *S. heterophyllum* such that it was interpreted as a synonym of *F. crispum*. M. Fishbein and K. N. Gandhi (2018) showed that the earlier typification by Vail provides priority of *F. heterophyllum* over other epithets at this rank.

5. Funastrum hirtellum (A. Gray) Schlechter, Repert. Spec. Nov. Regni Veg. 13: 286. 1914 • Hairy milkvine

Sarcostemma heterophyllum Engelmann ex Torrey var. *hirtellum* A. Gray in W. H. Brewer et al., Bot. California 1: 478. 1876; *S. hirtellum* (A. Gray) R. W. Holm

Stems hirtellous. **Leaves** persistent or caducous, stipular colleter 1 on each side of petiole; petiole 0–1.5 cm, hirtellous throughout; blade linear (linear-lanceolate), 0.7–4 × 0.07–0.5 cm, coriaceous, base cuneate, margins plane, revolute on drying, apex acute or obtuse, 1-veined, surfaces hirtellous, laminar colleters absent. **Inflorescences** solitary at nodes, 7–13-flowered; peduncle 0.1–2.7 cm, elongating somewhat in fruit, hirtellous; bracts caducous, 1, at base of each pedicel. **Pedicels** 5–8 mm, hirtellous. **Flowers:** calyx lobes linear to narrowly elliptic, 2–2.5 mm, apex obtuse, spreading, hirtellous, colleter 1, in sinuses, or absent; corolla green, cream, or pink, usually tinged pale to deep rose toward center of lobes, rotate-campanulate, tube 0.5–1 mm, lobes ascending, lanceolate to ovate, 3–4 mm, apex acute to obtuse, hirtellous abaxially, minutely hirtellous adaxially, margins ciliate; corona ring cream or greenish cream, inflated corona segments adnate to corona ring, proximally green, distally glossy white, oblong, 1.5–2 mm; style apex shallowly convex. **Follicles** solitary or more commonly paired, lance-ovoid, 4–5.5 × 0.3–0.5 cm, apex long-attenuate, hirtellous to puberulent with curved trichomes. **Seeds** 10–20, light brown, narrowly winged, lenticular, lanceolate, 6–8 × 2.5–3 mm, convex face minutely papillate, flattened face coarsely papillate; coma white, 1.5–2 cm.

Flowering year round; fruiting May–Oct(–Nov). Arroyos, canyons, ravines, cliff bases, plains, gravel, sand, granite, gypsum, desert scrub; 30–900 m; Ariz., Calif., Nev.; Mexico (Baja California).

Funastrum hirtellum most closely resembles *F. heterophyllum*, with which it is sympatric over much of its range, occurrmg most commonly in desert arroyos. The hirtellous vestiture, smaller leaves, and corona lobes adnate to the ring distinguish *F. hirtellum* from *F. heterophyllum*, and no intermediates are known. *Funastrum hirtellum* is largely restricted to watersheds of the lower Colorado River in western Arizona, southeastern California, extreme southern Nevada, and extreme northern Baja California.

6. Funastrum torreyi (A. Gray) Schlechter, Repert. Spec. Nov. Regni Veg. 13: 287. 1914 • Soft twinevine

Philibertia torreyi A. Gray, Proc. Amer. Acad. Arts 12: 64. 1876; *Sarcostemma torreyi* (A. Gray) Woodson

Stems hirsute (pilose). **Leaves** persistent, stipular colleter 1 on each side of petiole; petiole 1–2.5 cm, hirsute throughout; blade deltate, lanceolate, or ovate, 2–5.5 × 1.5–2.5 cm, chartaceous, base sagittate or cordate, margins plane, apex acute or attenuate, mucronate, venation pinnipalmate, main veins often conspicuously pale, surfaces hirsute, margins ciliate, laminar colleters 4–10. **Inflorescences** solitary at nodes, 8–15-flowered; peduncle 2–7 cm, hirsute; bracts caducous, 1, at base of each pedicel. **Pedicels** 10–17 mm, hirsute. **Flowers:** calyx lobes narrowly lanceolate to narrowly elliptic, 4–6 mm, apex acute to acuminate, erect, hirsute, ciliate, colleter 1, in sinuses; corolla cream to pink or purplish with red or purplish blotches or stripes at bases and near tips of lobes, rotate-campanulate, tube 2–2.5 mm, lobes ascending, lanceolate to ovate, 6–9 mm, apex acute to obtuse, apiculate, hirsutulous abaxially, minutely hirtellous adaxially, margins conspicuously ciliate with flattened trichomes; corona ring cream, inflated corona segments proximally green, distally glossy white, ovoid, 2–3 mm; style apex shallowly convex. **Follicles** usually solitary, lance-ovoid, 8–9 × 1–1.5 cm, apex long-attenuate, puberulent. **Seeds** 20–50, light brown, thickly winged, biconvex, lanceolate, 6–7 × 3–4 mm, both faces papillate; coma white, 3 cm.

Flowering May–Oct; fruiting Sep–Oct. Canyons, arroyos, dry slopes, ridges, plains, igneous substrates, juniper and oak woods, thorn scrub, desert scrub; 1200–2100 m; Tex.; Mexico (Coahuila, Nuevo León, Tamaulipas).

Funastrum torreyi barely enters the flora area in the Big Bend region of Texas. It is apparently not common and re-evaluation of its conservation status is warranted. It has the largest and showiest flowers of the *Funastrum* species in the flora region.

7. Funastrum utahense (Engelmann) Liede & Meve, Nordic J. Bot. 22: 589. 2003 • Utah milkvine [E]

Astephanus utahensis Engelmann, Amer. Naturalist 9: 349. 1875; *Cynanchum utahense* (Engelmann) Woodson

Stems very sparsely pilosulose, more densely so at nodes. **Leaves** persistent, sessile, stipular colleter 1 on each side of leaf base; blade filiform, 1–4 × 0.05–0.2 cm, chartaceous, base cuneate, margins plane, apex acute or attenuate, mucronate, 1-veined, surfaces sparsely puberulent with curved trichomes, margins ciliate, laminar colleters absent. **Inflorescences** solitary at nodes, 3–6(–10)-flowered; peduncle 0.5–2 cm, sparsely puberulent with curved trichomes to glabrate; bracts caducous, 1, at base of each pedicel. **Pedicels** 3–10 mm, puberulent with curved trichomes. **Flowers:** calyx lobes linear-lanceolate, 1.3–2.5 mm, apex acute, spreading with reflexed tips, pilose, ciliate, colleters unknown; corolla yellow, becoming orange with age, green at base of tube, tubular, tube 1–3 mm, lobes incurved, closing corolla but for a pin hole, lanceolate, 0.5–1.5 mm, apex acute, puberulent to glabrate abaxially; corona absent; style apex shallowly convex. **Follicles** usually solitary, fusiform, 4–9 × 0.5–1 cm, apex acuminate, puberulent with curved trichomes to glabrate. **Seeds** not seen.

Flowering Mar–Jun(–Oct); fruiting May–Oct. Plains, arroyos, hills, bajadas, dunes, sand, silt, gravel, basalt, limestone, conglomerate, desert scrub, desert grasslands; 200–1200 m; Ariz., Calif., Nev., Utah.

Funastrum utahense is the most distinctive species in the genus and is placed here with some hesitation. Phylogenetically, it has been associated with species of *Funastrum* and *Pattalias*, but morphologically this species is not comfortably placed in either genus due to its unique, tubular corolla and lack of a corona. Whether it is best placed here, in *Pattalias*, or in a monospecific genus awaits further study. Like *F. hirtellum*, the range of *F. utahense* is restricted to the watershed of the lower Colorado River. However, *F. utahense* is far less commonly encountered. It is considered to be rare and of conservation concern in Arizona and Utah.

29. GONOLOBUS Michaux, Fl. Bor.-Amer. 1: 119. 1803 • [Greek *gonia*, angle, and *lobos*, pod, alluding to follicle shape]

Alexander Krings

Vincetoxicum Walter, Fl. Carol., 13, 104. 1788, not Wolf 1776

Vines, herbaceous [woody], perennial; latex white, rarely clear. **Stems** twining, unarmed, eglandular-pubescent or mixed eglandular- and glandular-pubescent. **Leaves** deciduous [persistent], opposite, petiolate; stipular colleters 2, 1 borne at base of petiole on each side, interpetiolar colleters absent [present], infrapetiolar colleters absent; laminar colleters present. **Inflorescences** extra-axillary, racemiform [paniculiform, umbelliform], pedunculate. **Flowers:** calycine colleters present [absent]; corolla white, light yellow, uniformly green, olive-green, yellow-green, or neon green or multicolored and generally dark maroon or brownish near base and green to yellow-green near tips, rotate, aestivation dextrorse; corolline corona ± annular, sometimes interrupted and developed only in staminal [rarely interstaminal] position, or absent; androecium and gynoecium united into a gynostegium adnate to corolla tube; gynostegial corona of fused staminal and interstaminal segments, lobed [unlobed]; anthers adnate to style, dorsal appendage laminar, locules 2; pollen in each theca massed into rigid, horizontally oriented pollinium, pollinia strongly curved, bearing sterile, navicular portion adjacent to translator, pollinia of adjacent anthers joined by translators to common corpusculum together forming pollinarium. **Fruits** follicles, solitary or paired, moderately to strongly deflexed, green to brown, ovoid, 5-angled, occasionally fewer-angled, smooth to tuberculate, glabrous or pubescent. **Seeds** obovate, flattened, winged, not beaked, comose, not arillate. $x = 11$.

Species 100–150 (3 in the flora): United States, Mexico, West Indies, Central America, South America; introduced in Africa.

The indumentum in *Gonolobus* can consist of trichomes of uniform length but more often includes a mixture of long eglandular, short eglandular, and/or short glandular capitate trichomes. Stems, petioles, peduncles, and pedicels may have indumentum in lines or completely around the stalk (ubiquitous).

SELECTED REFERENCES Krings, A. 2008. Synopsis of *Gonolobus* s.l. (Apocynaceae, Asclepiadoideae) in the United States and its territories, including lectotypification of *Lachnostoma arizonicum*. Harvard Pap. Bot. 13: 209–218. Krings, A., D. T. Thomas, and Xiang Q.-Y. 2008. On the generic circumscription of *Gonolobus* (Apocynaceae, Asclepiadoideae): Evidence from molecules and morphology. Syst. Bot. 33: 403–415.

1. Stem trichomes eglandular only; corolla white or light yellow, campanulate at base, tube 2.9–5 mm; corolline corona absent; pollinia suborbiculate to rhombic; seed distal margins entire . . . . 1. *Gonolobus arizonicus*
1. Stem trichomes eglandular and glandular-capitate; corolla uniformly green, olive-green, yellow-green, or neon green or multicolored and generally dark maroon to brownish near base and green to yellow-green near tips, subcampanulate at base, tube to 1.5 mm; corolline corona present; pollinia ovate; seed distal margins dentate.
    2. Pubescence of internodes ubiquitous, not in lines; calyx lobes 0.2–1 mm wide; corolline corona reduced to small mounds opposite anthers, glabrous. . . . . 2. *Gonolobus suberosus*
    2. Pubescence of internodes in 2 lines; calyx lobes 1–2 mm wide; corolline corona a distinctly raised ring, pubescent with unicellular hairs 0.1–0.3 mm . . . . 3. *Gonolobus taylorianus*

1. Gonolobus arizonicus (A. Gray) Woodson, Ann. Missouri Bot. Gard. 28: 243. 1941 • Arizona milkvine

Lachnostoma arizonicum A. Gray, Proc. Amer. Acad. Arts 20: 296. 1885; *Matelea arizonica* (A. Gray) Shinners

Stems densely pubescent, pubescence of internodes ubiquitous, trichomes eglandular only. **Leaves:** petiole 1.2–2.7 cm, pubescent; blade deltate, 1.9–4.5 × 0.9–2.5 cm, base cordate, apex acute to acuminate, surfaces pubescent, hairs eglandular, colleters 4–5, 0.3–0.5 mm. **Peduncles** to 32 mm, pubescent. **Pedicels** 0.8–1.3 cm, pubescent. **Flowers:** calyx lobes lanceolate, 3.7–5.3 × 0.8–1.4 mm, margins entire, apex acute to obtuse, abaxial surface pubescent, adaxial surface glabrous; corolla white or light yellow, green reticulate-veined, campanulate at base, tube 2.9–5 × 2.9–3.7 mm, lobes oblong to ovate, 3.8–5 × 2.8–3.7 mm, slightly lobed at base, glandular swelling frequently present in sinus, apex obtuse, abaxial surface pubescent, adaxial surface pubescent toward base; corolline corona absent; gynostegial corona connate approximately 1/2–3/4 distance from base and adnate to corolla; style-head ca. 3 mm diam., stipe 1.6–1.9 mm; dorsal anther appendages 0.2 × 0.9 mm, apex emarginate; pollinia suborbiculate to rhombic, 0.2 × 0.15 mm. **Follicles** narrowly ovoid, 7.5–8.4(–11) × 2.3–3.7 cm, surface minutely pubescent (appearing glabrous). **Seeds** 5.9–6.4 × 3.9–4 mm, distal margin entire, surfaces glabrous, coma white, to 1.5 cm.

Flowering Jul–Nov; fruiting Jun–Feb. Canyons, streambeds, cliff bases; 900–1400 m; Ariz.; Mexico (Baja California Sur, Sonora).

In the United States, *Gonolobus arizonicus* is restricted to south-central Arizona (Graham, Pima, and Santa Cruz counties).

2. Gonolobus suberosus (Linnaeus) R. Brown in W. Aiton and W. T. Aiton, Hortus Kew. 2: 82. 1811 • Angle-pod [E] [F]

Cynanchum suberosum Linnaeus, Sp. Pl. 1: 212. 1753; *Matelea suberosa* (Linnaeus) Shinners; *Vincetoxicum suberosum* (Linnaeus) Britton

Stems sparsely to moderately pubescent, pubescence of internodes ubiquitous, trichomes eglandular and glandular-capitate. **Leaves:** petiole 2.8–5 cm, pubescent; blade deltate to ovate-elliptic, oblong, or suborbiculate, 2.9–24.5 × 2.5–17.5 cm, base cordate, apex acuminate, abaxial surface pubescent, hairs eglandular and glandular, adaxial surface pubescent, hairs eglandular, colleters 2–4, ca. 0.3 mm. **Peduncles** to 17 mm, pubescent. **Pedicels** 0.7–3.7 cm, pubescent. **Flowers:** calyx lobes lanceolate, 1.6–5.4 × 0.2–1 mm, margins entire, apex obtuse, abaxial surface pubescent to glabrate, adaxial surface glabrous; corolla uniformly green, olive-green, yellow-green, or neon green or multicolored and generally dark maroon to brownish near base and green to yellow-green near tips, subcampanulate at base, tube 1–1.4 × 1–1.5 mm, lobes ovate to elongate-deltate, 2.6–11 × 2–4 mm, slightly lobed at base, glandular swelling frequently present in sinus, apex obtuse, abaxial surface glabrous, adaxial surface glabrous or pubescent only along right sides; corolline corona reduced to small mounds opposite anthers, glabrous; gynostegial corona prostrate-undulating; style-head 2.6–3.7 mm diam., stipe ca. 0.5 mm; dorsal anther appendages 0.9 × 0.9 mm, apex rounded to truncate, or 2-lobed to emarginate; pollinia ovate, 0.6 × 0.2 mm. **Follicles** ovoid, 8–15 × 1.5–2.8 cm, surface glabrous or minutely pubescent. **Seeds** 6–10 × 3–7 mm, distal margin dentate, surfaces somewhat muricate, glabrous, coma white, to 4.5 cm.

Varieties 2 (2 in the flora): c, e United States.

1. Corolla lobes multicolored adaxially, generally dark maroon to brownish near base and green to yellowish near tips at anthesis (or uniformly yellowish green to neon green in rare mutants), pubescent or glabrous; dorsal anther appendages dark purple or maroon tinted . . . . 2a. *Gonolobus suberosus* var. *suberosus*
1. Corolla lobes uniformly colored adaxially, olive-green at anthesis, glabrous; dorsal anther appendages yellow . . . . 2b. *Gonolobus suberosus* var. *granulatus*

2a. Gonolobus suberosus (Linnaeus) R. Brown var. **suberosus** [E] [F]

Asclepias gonocarpos (Walter) J. F. Gmelin; *Gonolobus gonocarpos* (Walter) L. M. Perry; *G. laevis* Michaux var. *macrophyllus* A. Gray; *G. tiliifolius* Decaisne; *Matelea gonocarpos* (Walter) Shinners; *Vincetoxicum gonocarpos* Walter

Flowers: corolla lobes multicolored adaxially, generally dark maroon to brownish near base and green to yellowish near tips at anthesis (or uniformly yellowish green to neon green in rare mutants), pubescent or glabrous; dorsal anther appendages dark purple or maroon tinted.

Flowering Apr–Sep; fruiting Jul–Nov. Moist woods in bottomlands, along streams, around lakes; 30–300 m; Ala., Fla., Ga., Ky., La., Miss., N.C., S.C., Tenn., Va.

2b. Gonolobus suberosus (Linnaeus) R. Brown var. **granulatus** (Scheele) Krings & Q. Y. Xiang, Harvard Pap. Bot. 10: 157. 2005 [E] [F]

Gonolobus granulatus Scheele, Linnaea 21: 759. 1849

Flowers: corolla lobes uniformly colored adaxially, olive-green at anthesis, glabrous; dorsal anther appendages yellow.

Flowering Apr–Sep; fruiting Jul–Nov. Moist woods in bottomlands, along streams, around lakes; 0–300 m; Ala., Ark., Ill., Ind., Ky., La., Miss., Mo., Okla., Tenn., Tex.

3. Gonolobus taylorianus W. D. Stevens & Montiel, Novon 12: 551. 2002 • Cuchamper [I]

Stems sparsely to moderately pubescent, pubescence of internodes in 2 lines, trichomes eglandular and glandular-capitate. **Leaves:** petiole 1–7 cm, pubescent; blade deltate, ovate, or ovate-elliptic, 2.9–11.8 × 0.9–7 cm, base cordate, or subtruncate, apex acute to acuminate, surfaces sparsely pubescent, hairs eglandular and/or glandular, colleters 2–5, 0.2–0.8 mm. **Peduncles** to 18 mm, pubescent. **Pedicels** 0.7–1.9 cm, pubescent. **Flowers:** calyx lobes lanceolate, 3–6 × 1–2 mm, margins entire, apex obtuse, abaxial surface pubescent to glabrate, adaxial surface glabrous; corolla green [purple, maroon, dark-brown], subcampanulate at base, tube 1–1.5 × 3–4 mm, lobes lanceolate to elongate-deltate, 7–11 × 2.1–3 mm, slightly lobed at base, glandular swelling frequently present in sinus, apex obtuse, abaxial surface pubescent, adaxial surface pubescent only along right side [glabrate]; corolline corona a distinctly raised ring, ca. 0.2 mm, pubescent, hairs 0.1–0.3 mm; gynostegial corona 5-lobed; style-head 3.5–4.2 mm diam., stipe ca. 0.5 mm; dorsal anther appendages 2 × 1.5 mm, apex broadly rounded to truncate, or 2-lobed to emarginate; pollinia ovate, 0.8 × 0.4 mm. **Follicles** broadly ovoid to suborbicular, 9–13.5 × 4.5–10.5 cm, surface glabrous or minutely pubescent. **Seeds** 5–6 × 3–4 mm, distal margin dentate, surfaces smooth to somewhat muricate, glabrous, coma white, to 5 cm.

Flowering Nov; fruiting Feb and May. Disturbed roadsides; 0–10 m; introduced; Fla.; Central America.

Gonolobus taylorianus is native to Central America, distributed from Costa Rica to Guatemala (W. D. Stevens and O. M. Montiel 2002). In the United States, it is currently known only from naturalized populations in Miami-Dade County (A. Krings et al. 2019). In its native range, the species is cultivated for its edible fruit (Stevens and Montiel). It may have been introduced for this purpose in Florida.

30. MATELEA Aublet, Hist. Pl. Guiane 1: 277, plate 109. 1775 • Milkvine, milkweed vine, spinypod [Derivation uncertain; possibly anagram of *amatalea*, name for riparian vine *Cissus erosa* in Wayapi language spoken in French Guiana and alluding to habitat]

Mark Fishbein

Angela McDonnell

Cyclodon Small; *Edisonia* Small; *Odontostephana* Alexander

Vines or herbs, perennial, herbaceous or woody at base; latex white. **Stems** decumbent, prostrate, or ascending, twining or not, unarmed, vestiture a mix of short glandular and long eglandular trichomes. **Leaves** persistent or tardily deciduous, opposite, petiolate; stipular colleters interpetiolar, laminar colleters present. **Inflorescences** extra-axillary or terminal, umbelliform or racemiform, sessile, subsessile, or pedunculate. **Flowers:** calycine colleters present; corolla shades of green, red, purple, brown, white, or cream, rotate or campanulate, aestivation contort-dextrorse; corolline corona absent; androecium and gynoecium united into a gynostegium adnate to corolla tube; gynostegial corona whorls 1 or 2; anthers adnate to style, locules 2; pollen in each theca massed into a rigid, horizontally oriented pollinium, pollinia lacrimiform to reniform with abaxial excavation proximal to translator, joined from adjacent anthers by translators to a common corpusculum and together forming a pollinarium. **Fruits** follicles, typically solitary, commonly pendulous, sometimes lying on ground, green, brown or purple at maturity, sometimes gray-striate, terete, smooth or variously ornamented, glabrous or with various indumentum. **Seeds** ovate, flattened, winged, not beaked, comose, not arillate. $x = 11$.

Species ca. 200 (22 in the flora): United States, Mexico, West Indies, Central America, South America.

R. E. Woodson Jr. (1941) adopted a broad (and polyphyletic) circumscription of *Matelea*, originally described as a genus of Amazonian shrubs, subsuming many small genera from North and Central America. To accommodate the diversity of an inflated *Matelea*, Woodson erected 16 subgenera. Subsequently, some distinctive elements have been removed (including *Pherotrichis* and *Polystemma* in the flora area), leaving *Matelea* in a state of temporary disarray until revision is complete. Dismemberment of *Matelea* is supported by a number of phylogenetic studies (S. Liede et al. 2005; A. Krings et al. 2008; A. McDonnell et al. 2018). In addition to difficulties in generic circumscription, *Matelea* in the flora area includes complexes with poorly understood species delimitation and a number of rare, local endemics that have been seldom collected and are poorly known. Among the Asclepiadoideae of the flora area, the species of *Matelea* are in greatest need of further taxonomic study.

The species in this treatment are ordered based on phylogenetic results (A. McDonnell et al. 2018). Within clades, species are often very difficult to distinguish, and in many cases it may be impossible to do so in the absence of flowering material. Some useful floral characters, such as color, the orientation of corolla lobes, and the form of the corona, are often distorted, obscured, or lost in herbarium material. Careful notes or photographs made at the time of collection are invaluable. Follicle morphology has not been sufficiently appreciated in differentiating species groups. Most species have muricate follicles, but in species 10–12, the follicles are typically tuberculate, whereas in species 7–9 they are typically smooth. Moreover, species 7–12 have follicles with conspicuously gray-striate coloration, similar to that of *Polystemma*, whereas the follicle color of the remaining species is more uniform. Leaf size is highly variable in most species. In this treatment, leaf measurements pertain only to leaves at nodes bearing inflorescences, at anthesis; thus, larger and smaller extremes are omitted from the ranges.

1. Stems not twining, plants decumbent, prostrate, ascending, or occasionally scandent on low vegetation (in *M. pubiflora*).
   2. Corollas densely pilose to hirsute adaxially.
      3. Corona with 2 whorls, an outer whorl of shorter, laminar segments and an inner whorl of longer, erect, subulate segments; inflorescences 2–20-flowered, pedunculate (peduncles 0.7–10 cm); s Texas . . . . 4. *Matelea parviflora* (in part)
      3. Corona with 1 whorl of united segments forming a ring, with adaxial, subulate, incurved appendages; inflorescences 1–6-flowered, sessile or subsessile; Florida, Georgia, New Mexico, Oklahoma, Texas (excluding s Texas).
         4. Inflorescences 1–2-flowered; corollas rotate-campanulate, dark maroon or brown adaxially, lobes 3–6 mm; corona exposed; New Mexico, Oklahoma, Texas . . . . 1. *Matelea biflora*
         4. Inflorescences 1–6-flowered; corollas campanulate, green, yellow, or brown adaxially, lobes 2–3.5 mm; corona included within corolla; Florida, Georgia . . . . 3. *Matelea pubiflora*
   2. Corollas glabrous adaxially or inconspicuously hirtellous at base of lobes.
      5. Corona segments overarching style apex, minutely hirtellous; corollas cream with green to pink lines that become reticulate towards lobe margins; bootheel of New Mexico . . . . 6. *Matelea chihuahuensis*
      5. Corona segments not overarching style apex, incumbent on backs of anthers or erect to spreading, glabrous; corollas green, yellow, brown, purple, or maroon, lined or not; Arkansas, Louisiana, Oklahoma, Texas.
         6. Peduncles 0–3 cm; corolla lobes 2.5–4 mm, not reticulate; corona segment appendages incumbent on anthers and exceeded by style apex, green, yellow, or brown; follicles sparsely muricate (fewer than 1 protrusion per cm of length) . . . . 2. *Matelea cynanchoides*
         6. Peduncles 0.7–10 cm; corolla lobes 1–2.5 mm, faintly to strongly reticulate; corona segment appendages not incumbent on anthers, exceeding style apex, white, greenish cream, green, or yellow; follicles moderately to densely muricate (more than 1 protrusion per cm length).
            7. Inner corona segments erect to spreading, outer segments with minute to elongate marginal lobes at apex; corolla lobes spreading to reflexed; widespread in s Texas . . . . 4. *Matelea parviflora* (in part)
            7. Inner corona segments inflexed over style apex, outer segments unlobed at apex; corolla lobes ascending to spreading; local and rare in the vicinity of Laredo, Texas . . . . 5. *Matelea brevicoronata*
1. Stems twining (at least at stem tips), usually climbing woody vegetation or fences, sometimes trailing and twining on herbs.
   8. Leaf blades 0.5–2.5 × 0.2–1.3 cm; inflorescences sessile or subsessile; pedicels 1–5 mm; follicles tuberculate (protrusions blunt) to nearly smooth.
      9. Corolla lobes 2–3 mm; leaf blades hirtellous adaxially; Arizona, California, trans-Pecos Texas. . . . . 10. *Matelea parvifolia*
      9. Corolla lobes 5–10 mm; leaf blades sparsely puberulent to glabrate adaxially; s Texas, e of the Pecos River.
         10. Corona segments 0.5–0.7 mm, exceeded by style apex, green, cream, or yellowish; corolla lobes linear; calyx lobes linear to linear-lanceolate . . . . . 11. *Matelea sagittifolia*
         10. Corona segments 2.5–3 mm, greatly exceeding style apex, white with a purple adaxial patch; corolla lobes lanceolate; calyx lobes narrowly lanceolate . . . . 12. *Matelea radiata*
   8. Leaf blades 1.5–20 × 1–18 cm; inflorescences pedunculate, peduncles (0–)0.2–13 cm; pedicels 3–40 mm; follicles muricate (protrusions sharp) or smooth.
      11. Corollas tubular-campanulate, green to reddish abaxially (rarely brown); coronas included in tube; vestiture of leaf blades wholly eglandular; Arizona, New Mexico, w Texas.
         12. Corollas glabrous adaxially, tube 3.5–8 mm; widespread in Arizona, New Mexico, w Texas . . . . 7. *Matelea producta*
         12. Corollas hirtellous adaxially, tube 1.5–2.5 mm; local in Big Bend region, Texas . . . . 8. *Matelea texensis*

[11. Shifted to left margin.—Ed.]

11. Corollas rotate, rotate-campanulate, or campanulate; coronas not concealed in tube (corona sometimes concealed by erect corolla lobes in *M. baldwyniana*, *M. carolinensis*, *M. decipiens*, and *M. obliqua*, in which case the corolla is white or maroon to brown abaxially); vestiture of leaf blades a mixture of long eglandular and inconspicuous glandular trichomes (glandular trichomes lacking in *M. edwardsensis*); e United States to w Texas.
   13. Leaf blades 2.5–5 × 1–3.5 cm, apex attenuate to long-acuminate; petioles 0.7–1.7 cm; peduncles 0–1 cm; inflorescences 1–5-flowered; coronas papillose; follicles smooth or with very sparse, low tubercules, prominently gray-striate; rare and local in Big Bend region of w Texas . . . . 9. *Matelea atrostellata*
   13. Leaf blades 1.5–20 × 1–18 cm, apex acute to acuminate; petioles 1–9.5 cm; peduncles 0.2–13 cm; inflorescences 1–40-flowered; coronas glabrous; follicles muricate, inconspicuously striate or uniform in color; e United States to w Texas.
      14. Corollas green to yellow-green, strongly reticulate with dark green lines; coronas yellow or pale green (flowers rarely maroon in *M. flavidula*).
         15. Corolla lobes with parallel, longitudinal lines, reticulate only towards margins, 3–4 mm, densely hirsute adaxially; corona segments free nearly to base; seeds 10–11 mm, chalazal end erose; stems sparsely and inconspicuously hirsute to glabrate; Edwards Plateau and surrounding area, Texas. . . . . 14. *Matelea edwardsensis*
         15. Corolla lobes reticulate throughout, 4–11 mm, glabrous or hirtellous only at base of lobes adaxially; corona a low ring or with segments free only in upper one-half; seeds 7–10 mm, chalazal end entire (scarcely erose in *M. alabamensis*); stems densely hirsute; coastal plain of se United States to w Texas.
            16. Coronas prominent, equaling or slightly exceeding style apex, segments free in upper one-half, with marginal lobes at apex; corolla lobes elliptic or oblong to oblanceolate; apical anther appendages truncate, covering less than 25% of style apex; rare and local from coastal plain of South Carolina s and e to Apalachicola watershed, Florida . . . . 16. *Matelea flavidula*
            16. Coronas a low, undulate or lobed ring, not reaching one-half the height of style, segments if present entire at apex; corolla lobes broadly ovate to nearly orbiculate; apical anther appendages deltoid or ovate, covering more than 50% of style apex; Apalachicola watershed in Florida and Georgia, w to Texas.
               17. Peduncles 1–7 cm; pedicels 12–25 mm; corolla lobes 4–5 mm; inflorescences 3–10-flowered; follicles 8–15 cm, sparsely to moderately muricate; Texas . . . . 13. *Matelea reticulata*
               17. Peduncles 0.2–1.2 cm; pedicels 5–15 mm; corolla lobes 7–10 mm; inflorescences 1–5-flowered; follicles 4.9–8 cm, densely muricate; Gulf Coastal Plain in Alabama, Florida, Georgia. . . . . 18. *Matelea alabamensis*
      14. Corollas shades of maroon, purple, brown, white, or cream (rarely green, yellow-green, pink, or orange), not or only faintly reticulate, coronas maroon, purple, white, or cream (rarely yellow, orange, or rose).
         18. Corollas and coronas white to cream or yellow (rarely purple) . . . . 22. *Matelea baldwyniana*
         18. Corollas and coronas shades of maroon, brown, or purple (rarely green, yellow-green, pink, rose, or orange).
            19. Corolla lobes 3–7.5 mm.
               20. Coronas equaling style apex, medial lobe of segments elongate, longer than lateral lobes; Florida, Georgia . . . . 17. *Matelea floridana*
               20. Coronas equaling or exceeding style apex, medial lobe of segments shorter than lateral lobes; Delaware to Georgia, w to Texas.
                  21. Corollas glabrous adaxially; coronas 2–3 mm diam.; Delaware to Georgia, w to Louisiana . . . . 15. *Matelea carolinensis* (in part)
                  21. Corollas hirtellous adaxially (rarely glabrous); coronas 3–4 mm diam.; e Texas, sw Arkansas . . . . 21. *Matelea hirtelliflora*

[19. Shifted to left margin.—Ed.]

19. Corolla lobes 7–18 mm.
  22. Corollas rotate to campanulate; coronas 2–3 mm diam. . . . . 15. *Matelea carolinensis* (in part)
  22. Corollas rotate-campanulate to campanulate; coronas 3–5 mm diam.
    23. Corollas pink, reddish, or maroon; coronas 5 mm diam., usually nearly round, segments very fleshy; Cumberland Plateau, w slope of Appalachian Mountains, and surrounding areas, Pennsylvania to North Carolina, w to Mississippi River. . . . . 19. *Matelea obliqua*
    23. Corollas purple to maroon; coronas 3–4 mm diam., usually pentagonal, segments laminar; Ozark Mountains and areas s and w, Missouri to Louisiana w to Kansas, Oklahoma, Texas . . . . 20. *Matelea decipiens*

1. **Matelea biflora** (Rafinesque) Woodson, Ann. Missouri Bot. Gard. 28: 228. 1941 • Two-flowered or star milkvine [E] [F]

Gonolobus biflorus Rafinesque, New Fl. 4: 58. 1838; *Chthamalia biflora* (Rafinesque) Decaisne; *G. biflorus* var. *wrightii* A. Gray

Herbs. **Stems** 4–10, decumbent, often branched near base, 7–40 cm, hirsute with long eglandular and minute glandular trichomes. **Leaves** 1 or 2 colleters on each side of petiole; petiole 0.5–2.5 cm, hirsute with long eglandular and minute glandular trichomes; blade ovate to deltate, 0.8–5 × 0.6–3.2 cm, base shallowly to deeply cordate, with 0–2 laminar colleters, apex acute (rounded), surfaces hirsute with long eglandular and minute glandular trichomes, especially so on veins abaxially. **Inflorescences** solitary, umbelliform, extra-axillary, sessile or subsessile, 1–2-flowered. **Pedicels** 3–11 mm, hirsute with long eglandular and minute glandular trichomes. **Flowers:** calyx lobes spreading, oval to ovate, 1.8–2.5 mm, apex rounded or acute, hirsute with long eglandular and minute glandular trichomes; corolla maroon to dark brown, not reticulate, rotate-campanulate, tube 1–1.5 mm, lobes spreading, ovate to narrowly deltate to spatulate, 3–6 mm, margins usually reflexed, pilose to hirsute adaxially; corona united to corolla and column near base, composed of 5 united segments forming a ring at base, each with an adaxial incurved appendage arching above or incumbent on anthers, equaling or exceeding style apex, maroon to dark brown, 1–1.5 mm, glabrous; apical anther appendages white, maroon to brown at base, broadly deltoid; style apex rounded, flat. **Follicles** not striate, ellipsoid to ovoid, 4.5–8.5 × 1.8–3.5 cm, apex acute, densely muricate (more than 1 protrusion per cm of length), villous to hirsute with long eglandular and minute glandular trichomes. **Seeds** tan to light brown, oval to nearly orbicular or ovate, 8–11 × 7–10 mm, margins broadly winged, chalazal end erose, faces minutely rugose; coma 2.5–4 cm.

Flowering Mar–Oct; fruiting Apr–Dec. Calcareous prairies, hillsides, pastures, fields, savannas; 100–1300 m; N.Mex., Okla., Tex.

Matelea biflora is occasionally found in grasslands and savannas of the southern Great Plains. It is most common on and around the Edwards Plateau of central Texas, where it occurs in grass-dominated habitats including disturbed areas. The range extends mostly northward and westward of that region to central Oklahoma and extreme eastern New Mexico (Lea County), where the species is much less common. Its conservation status in New Mexico merits evaluation. The plants are covered in short, glandular hairs and are malodorous when touched. The flowers occur most often in pairs, hence the common name two-flowered milkvine.

2. **Matelea cynanchoides** (Engelmann & A. Gray) Woodson, Ann. Missouri Bot. Gard. 28: 228. 1941 • Prairie milkvine [E]

Gonolobus cynanchoides Engelmann & A. Gray, Boston J. Nat. Hist. 5: 251. 1845

Herbs. **Stems** 3–10, ascending to decumbent, often branched near base, 10–70 cm, hirsute with long eglandular and minute glandular trichomes. **Leaves** with 1 or 2 colleters on each side of petiole; petiole 0.5–1.3 cm, hirsute with long eglandular and minute glandular trichomes; blade ovate to deltate (elliptic or orbiculate), 2.2–5.4 × 1–4.5 cm, base shallowly to deeply cordate, with 0–4 laminar colleters, apex acute to acuminate, surfaces hirsute with long eglandular and minute glandular trichomes, especially so on veins abaxially. **Inflorescences** solitary, umbelliform, extra-axillary, some appearing terminal, pedunculate, 2–6-flowered; peduncle 0–3 cm, hirsute with long eglandular and minute glandular trichomes. **Pedicels** 3–5 mm, hirsute with long eglandular and minute glandular trichomes. **Flowers:** calyx lobes spreading, elliptic (ovate), 2–3 mm, apex acute, hirsute with long eglandular and minute glandular trichomes;

corolla green to brown or maroon, not reticulate, rotate-campanulate, tube 0.9–1.2 mm, lobes spreading, planar, deltate to ovate, 2.5–4 mm, glabrous adaxially; corona united to corolla and column near base, composed of 5 united segments forming a ring at base, each with an adaxial incurved appendage incumbent on anthers, exceeded by style apex, green, yellow, or brown, 0.7–1 mm, fleshy, glabrous; apical anther appendages white, maroon to brown at base, rounded-truncate; style apex rounded-conic. **Follicles** not striate, ovoid to ellipsoid, 5–9.7 × 1.8–3.5 cm, apex acute, sparsely muricate (fewer than 1 protrusion per cm of length), hirsute. **Seeds** tan to light brown, broadly ovate to nearly orbicular, 10–12 × 9–10 mm, margins broadly winged, chalazal end erose, faces minutely rugose; coma 2.5–3.5 cm.

Flowering Apr–Sep; fruiting May–Oct. Sandy prairies, flood plains, stabilized dunes, hillsides, pastures, fields, savannas, pine-oak forests; 10–600 m; Ark., La., Okla., Tex.

Matelea cynanchoides occasionally occurs in sandy grasslands, savannas, and forests of the southern Great Plains. It is more abundant on the Gulf Coastal Plain of southeastern Texas in alluvial deposits overlying sandstone. The range extends eastward and northward to western Louisiana (Caddo Parish), extreme southwestern Arkansas (Miller County), and central Oklahoma, where the species is limited to deposits of deep sands. At the edge of its range, *M. cynanchoides* is considered to be of conservation concern in Arkansas and Louisiana. The plants are covered in short, glandular hairs and are malodorous when touched. *Matelea cynanchoides* is tolerant of disturbance, and resprouted, prostrate plants can be found on roadsides that are subject to regular mowing.

3. **Matelea pubiflora** (Decaisne) Woodson, Ann. Missouri Bot. Gard. 28: 230. 1941 • Trailing milkvine, trailing or sandhill spinypod [E]

Chthamalia pubiflora Decaisne in A. P. de Candolle and A. L. P. P. de Candolle, Prodr. 8: 605. 1844; *Edisonia pubiflora* (Decaisne) Small; *Gonolobus pubiflorus* (Decaisne) Engelmann & A. Gray; *Vincetoxicum pubiflorum* (Decaisne) A. Heller

Herbs. Stems 2–15, decumbent, trailing, or occasionally climbing, not twining, often branched near base, 20–110 cm, often purplish, hirsute with long eglandular and minute glandular trichomes. **Leaves** with 1 or 2 colleters on each side of petiole; petiole 0.5–3 cm, hirsute with long eglandular and minute glandular trichomes; blade ovate to deltate, 0.5–4 × 0.4–2.8 cm, base shallowly to deeply cordate, with 0–2 laminar colleters, apex acute to acuminate, surfaces hirsute with long eglandular and minute glandular trichomes, especially so on veins abaxially. **Inflorescences** solitary, umbelliform, extra-axillary, sessile or subsessile, 1–6-flowered. **Pedicels** 3–5 mm, hirsute with long eglandular and minute glandular trichomes. **Flowers:** calyx lobes spreading, ovate to lanceolate, 2–3 mm, apex acute, hirsute with long eglandular and minute glandular trichomes; corolla green, yellow, brown, and/or pink tinged, not reticulate, campanulate, tube 2.5–4 mm, lobes spreading, deltate to ovate, 2–3.5 mm, densely pilose to hirsute adaxially; corona united to corolla and column near base, composed of 5 erect, united segments forming a ring at base, each with an adaxial incurved appendage, maroon to dark brown, 1–1.5 mm, glabrous; apical anther appendages white, green to brown at base, rounded; style apex rounded, flat. **Follicles** not striate, ellipsoid to ovoid, 4.5–8.5 × 1.8–3.5 cm, apex acute, sparsely muricate, hirsute to glabrate. **Seeds** tan to light brown, orbicular to ovate, 7–11 × 6–9 mm, margins broadly winged, chalazal end erose, faces minutely rugose; coma 2–4 cm.

Flowering Mar–Aug; fruiting Apr–Sep. Dry sandhills and sand ridges, savannas, oak and pine-oak woods; 10–30 m; Fla., Ga.

Matelea pubiflora is uncommon in Georgia, where it is considered to be of conservation concern, and increasingly so in Florida due to habitat loss. It is most common in open sand on and around the central Florida ridges and uplands, where it can be found in turkey oak woods and pine and oak savannas, lending the common name of sandhill spinypod. The species tolerates moderate levels of disturbance.

4. **Matelea parviflora** (Torrey) Woodson, Ann. Missouri Bot. Gard. 28: 229. 1941 • Smallflower milkvine

Lachnostoma parviflorum Torrey in W. H. Emory, Rep. U.S. Mex. Bound. 2(1): 165. 1859; *Vincetoxicum parviflorum* (Torrey) A. Heller

Herbs. Stems 5–20, prostrate, often branched near base, 10–40 cm, hirsute with long eglandular and minute glandular trichomes. **Leaves** with 1 or 2 colleters on each side of petiole; petiole 0.2–0.9 cm, hirsute with long eglandular and minute glandular trichomes; blade ovate to deltate (lanceolate or orbiculate), 0.5–4 × 0.3–3 cm, base rounded or truncate to cordate, with 0–4 laminar colleters, apex acute to acuminate, surfaces hirsute with long eglandular and minute glandular trichomes, especially so on veins abaxially. **Inflorescences** solitary, compound racemiform, extra-axillary, pedunculate, 2–20-flowered; peduncle 0.7–10 cm, hirsute with long eglandular and minute glandular trichomes. **Pedicels** 1–10 mm, hirsute with long eglandular and minute glandular trichomes. **Flowers:** calyx lobes erect, elliptic, 1.5–2.5 mm, apex acute, hirsute with long eglandular and minute glandular trichomes; corolla green to purple, faintly to strongly reticulate, rotate, lobes spreading to reflexed, planar, ovate, lanceolate, or oblong, 1–4 mm, hirtellous adaxially at base of lobes or rarely throughout; corona united to corolla and staminal column near base, composed of 2 series, the outer of 5 laminar segments shorter than the style apex, apical margins with minute to elongate lateral lobes, the inner of 5 erect, subulate segments exceeding the style apex, white or greenish cream, 0.8–3 mm, glabrous; style apex rounded-pentagonal, flat. **Follicles** not striate, ovoid to ellipsoid, 6–10.5 × 1.5–3 cm, apex acuminate, moderately to densely muricate, hirsute. **Seeds** tan to light brown, broadly ovate to nearly orbicular, 10–16 × 9–15 mm, margins broadly winged, chalazal end minutely erose, faces minutely rugose; coma 2–4 cm.

Flowering Mar–Oct; fruiting May–Dec. Dunes, plains, valleys, hillsides, arroyos, sandstone, sandy and gravel soils, caliche, prairies, mesquite savanna, oak woodland; 10–300 m; Tex.; Mexico (Coahuila, Nuevo León, Tamaulipas).

Matelea parviflora is uncommon to occasional in the South Texas plains. The species consists of several geographic variants differing in aspects of corolla and corona morphology. The most common form has green reticulate corollas with strongly reflexed lobes that are hirtellous adaxially only at the base and coronas composed of outer segments with elongate lateral lobes at the apex and very long, erect inner segments—together forming a trident. Less common variants have olive or purple corollas, corolla lobes that are spreading and/or densely hirtellous across the adaxial surface, minutely lobed outer corona segments, or short inner corona segments. This circumscription includes specimens that have been commonly identified as *M. brevicoronata*, although they differ conspicuously in one or more attributes of the type. For distinctions with the very similar *M. brevicoronata*, see discussion under that species. An illegitimate combination in *Gonolobus* was made by A. Gray at a time when the species was treated under a broad concept of that genus.

5. **Matelea brevicoronata** (B. L. Robinson) Woodson, Ann. Missouri Bot. Gard. 28: 228. 1941 • Shortcrown milkvine [E]

Gonolobus parviflorus (Torrey) A. Gray [not Decaisne] var. *brevicoronatus* B. L. Robinson, Proc. Amer. Acad. Arts 26: 169. 1891; *Vincetoxicum brevicoronatum* (B. L. Robinson) Vail

Herbs. Stems 5–20, prostrate, often branched near base, 10–40 cm, hirsute with long eglandular and minute glandular trichomes. **Leaves** with 1 or 2 colleters on each side of petiole; petiole 0.2–0.9 cm, hirsute with long eglandular and minute glandular trichomes; blade ovate to deltate (lanceolate or orbiculate), 1.1–2.9 × 0.5–2.3 cm, base rounded or truncate (cordate), laminar colleters apparently absent, apex acute to acuminate, surfaces hirsute with long eglandular and minute glandular trichomes, especially so on veins abaxially. **Inflorescences** solitary, simple or compound umbelliform, extra-axillary, pedunculate, 2–8-flowered; peduncle 1.2–7.3 cm, hirsute with long eglandular and minute glandular trichomes. **Pedicels** 2–6 mm, hirsute with long eglandular and minute glandular trichomes. **Flowers:** calyx lobes erect, elliptic, 1.5–2.5 mm, apex acute, hirsute with long eglandular and minute glandular trichomes; corolla green, faintly to strongly reticulate, campanulate, tube 1–2 mm, lobes spreading to erect, planar, deltate, 1–3 mm, glabrous adaxially; corona united to corolla and column near base, composed of 2 series, the outer of 5 laminar segments shorter than the style apex, apical margins unlobed, the inner of 5 subulate segments inflexed over the style apex, green or yellow, 0.7–1 mm, glabrous; apical anther appendages white, green at base, deltoid; style apex rounded-conic. **Follicles** not striate, lance-ovoid, 3–6.5 × 1.5–3 cm, apex acute, densely muricate, hirsute. **Seeds** tan to light brown, ovate, 10–11 × 7–8 mm, margins broadly winged, chalazal end erose, faces minutely rugose; coma 2–2.5 cm.

Flowering Feb–Jul; fruiting Mar–Sep. Sandy substrates, flood plains, thornscrub; 100–200 m; Tex.

Matelea brevicoronata is uncommon and localized in southern Texas in the vicinity of Laredo (Webb and Zapata counties), on alluvial sands overlying sandstone. Its range is far more restricted than previously thought because variants of *M. parviflora* (as treated here) with short coronas were included in a broader concept of *M. brevicoronata*. These species are nearly identical in vegetative characters, and are distinguished only by a combination of floral characters. Compared to *M. parviflora*, the corollas of *M. brevicoronata* are campanulate with spreading to ascending lobes that are glabrous adaxially or obscurely hirtellous at the base (versus rotate with spreading to reflexed lobes that are sometimes densely hirtellous adaxially, especially when the lobes are spreading and approaching the orientation of *M. brevicoronata*). Additionally, the inner corona segments of *M. brevicoronata* are inflexed over the style apex (versus erect to spreading), and the outer segments lack marginal lobes at the apex (versus minute to more commonly elongate lobes, giving the combined corona segments the appearance of a trident). So far as known, the ranges of the two species approach each other but do not overlap. Because of the highly restricted distribution documented by fewer than ten collections, *M. brevicoronata* should be considered to be of extreme conservation concern.

6. **Matelea chihuahuensis** (A. Gray) Woodson, Ann. Missouri Bot. Gard. 28: 232. 1941 • Chihuahuan hairy milkvine

Gonolobus chihuahuensis A. Gray, Proc. Amer. Acad. Arts 21: 398. 1886

Herbs. Stems 4–20+, prostrate, 10–50 cm, hispid to hirsute with long eglandular and minute glandular trichomes. **Leaves** with 1 or 2 colleters on each side of petiole; petiole 0.2–1.6 cm, hirsute with long eglandular and minute glandular trichomes; blade deltate (ovate), 0.7–3.3 × 0.7–2.6 cm, base shallowly to deeply cordate, with 2–4 laminar colleters, apex acute, surfaces densely hirsute with long eglandular and minute glandular trichomes. **Inflorescences** solitary, umbelliform, extra-axillary, pedunculate, 2–5-flowered; peduncle 0.1–0.3 cm, hirsute with long eglandular and minute glandular trichomes. **Pedicels** 3.5–5 mm, hirsute with long eglandular and minute glandular trichomes. **Flowers:** calyx lobes spreading, elliptic, 1.9–3.5 mm, apex acute, hirsute with long eglandular and minute glandular trichomes; corolla cream with green to pink striations that become reticulate at lobe margins, rotate-campanulate, tube 0.5–1.5 mm, lobes spreading to ascending, plane, oblong to spatulate, 2.4–5 mm, hirsute adaxially at base of lobes; corona united to column near base, composed of 5 united segments forming a short ring, segments arching over style apex, translucent white to pink, ligulate, 1.5–2 mm, densely, minutely hirtellous; apical anther appendages pink, deltoid; style apex pale to dark pink, rounded, convex. **Follicles** faintly striate, lance-ovoid (ellipsoid), 4.9–5.7 × 1.7–2.5 cm, apex acuminate, densely muricate, densely villous to hirsute with long eglandular and minute glandular trichomes. **Seeds** tan to brown, oval to ovate (orbicular), 7–9 × 5–7 mm, margins broadly winged, chalazal end erose, faces rugose; coma 1.5–3 cm.

Flowering Jun–Oct; fruiting Jul–Oct. Desert grassland, hillsides, valleys; 1200–1500 m; N.Mex.; Mexico (Chihuahua).

Matelea chihuahuensis was reported in the United States apparently for the first time in 2013 by surveyors in the western piedmont of the Animas Mountains, in the bootheel of New Mexico (*Quinn s.n.*, OKLA; A. McDonnell et al. 2015). However, an earlier collection was made in 1977 in the nearby San Simon Valley (*Moir s.n.*, COLO). Because of its recent discovery in the United States, *M. chihuahuensis* has not been formally assessed for conservation status; with few known populations limited to Hidalgo County, New Mexico, this species should be considered to be of conservation concern. It is distributed across the state of Chihuahua, Mexico, but is apparently rare throughout its range. In New Mexico and in the bulk of its range in the foothills of the Sierra Madre Occidental in Chihuahua, it occurs in desert grassland in valleys and on mountain slopes. The white and green corollas, hirtellous corona segments, and small follicles are unusual in comparison to related species.

SELECTED REFERENCE McDonnell, A. et al. 2015. *Matelea chihuahuensis* (Apocynaceae): An addition to the flora of the United States and a synopsis of the species. J. Bot. Res. Inst. Texas 9: 187–194.

7. **Matelea producta** (Torrey) Woodson, Ann. Missouri Bot. Gard. 28: 230. 1941 • Texas milkvine [F]

Gonolobus productus Torrey in W. H. Emory, Rep. U.S. Mex. Bound. 2(1): 165. 1859

Vines, herbaceous (rarely suffrutescent without corky bark). **Stems** 1–5, twining (at least at stem tips), 30–200 cm, short-hirsute with spreading to retrorse eglandular and minute glandular trichomes. **Leaves** with 1 or 2 colleters on each side of petiole; petiole 0.5–3 cm, short-hirsute with eglandular and minute glandular trichomes; blade ovate to lanceolate, 2.5–8 × 1–5.5 cm, base deeply cordate, with 2–4 laminar colleters, apex long-acuminate, surfaces densely hirsute, eglandular. **Inflorescences** solitary (rarely paired), simple umbelliform or somewhat racemiform, extra-axillary, pedunculate, 2–8-flowered; peduncle 0.6–1.7 cm, short-hirsute with eglandular and minute glandular trichomes. **Pedicels** 4–9 mm, short-hirsute with eglandular and minute glandular trichomes. **Flowers:** calyx lobes ascending, elliptic to lanceolate, 2–4 mm, apex acute, short-hirsute with eglandular and minute glandular trichomes; corolla green to reddish abaxially, green to yellowish green with dark green reticulations (rarely brown) adaxially and a thin red to purple ring at top of throat, tubular-campanulate, tube

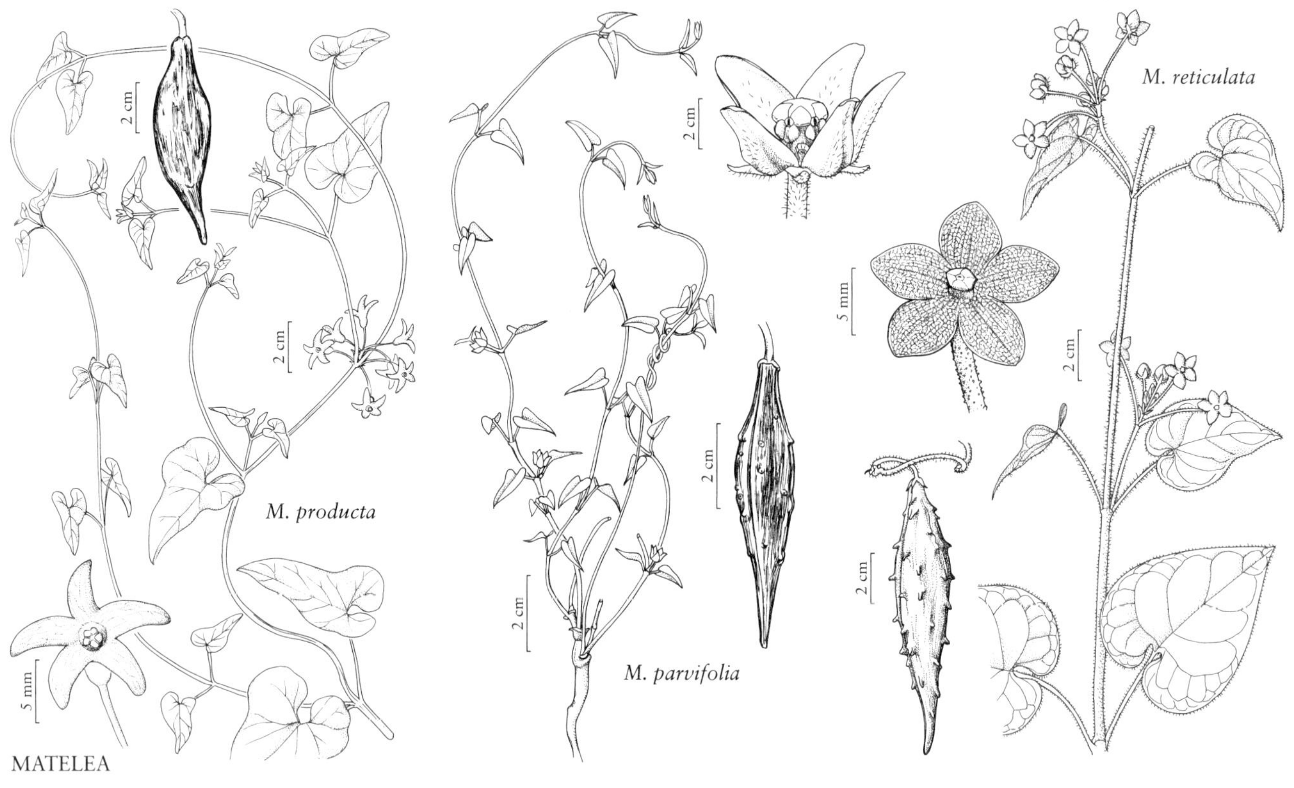

MATELEA

3.5–8 mm, lobes spreading, 5–7 mm, hirtellous abaxially, glabrous adaxially; corona concealed in throat, a ring with 5 lobes incumbent on backs of anthers, green to pinkish or lavender, 1–1.5 mm, glabrous; apical anther appendages greenish or yellowish cream, deltoid; style apex green to yellowish green, pentagonal, flat. **Follicles** gray-striate, fusiform to ellipsoid, 7–10 × 1.5–2.5 cm, apex acuminate, smooth, glabrous. **Seeds** tan, ovate, 6–8 × 4–5 mm, margins thickly winged, chalazal end minutely erose, faces smooth; coma 2–3 cm.

Flowering Apr–Sep; fruiting Jul–Oct. Hill slopes, bajadas, canyons, arroyos, gneiss, granite, limestone, sandstone, rocky, sandy, and clay soils, desertscrub, thornscrub, mesquite and juniper grasslands, chaparral, riparian and pinyon-juniper woodlands; 700–2000 m; Ariz., N.Mex., Tex.; n Mexico.

Matelea producta occurs in the flora area along a narrow arc from the Grand Canyon near Lake Mead in Arizona to the Trans-Pecos region of Texas and at scattered locations in northern New Mexico. Its tubular-campanulate green corolla is unlike that of any other milkweed vine in its range, except the very similar *M. texensis*, which differs by densely hirtellous corollas. However, *M. texensis* is narrowly endemic to a small region south of Alpine, Texas, just beyond the range of *M. producta*. The species are as of yet not known to co-occur. Many *Matelea* species have corollas that range in color from green to shades of maroon or brown. The corollas of *M. producta* are almost invariably green adaxially; however, a specimen gathered from the Davis Mountains of western Texas (*Rintz 2006-1*, SRSC) represents the only known occurrence of brown flowers in this species. Although milkweed vines in several genera are well known for ill-scented herbage, *M. producta* is particularly malodorous.

8. Matelea texensis Correll, Brittonia 18: 310. 1967 • Trans-Pecos or Texas milkvine [C] [E]

Vines, herbaceous (rarely suffrutescent without corky bark). **Stems** 1–5(–10), twining (at least at stem tips), 30–100 cm, short-hirsute with spreading to retrorse eglandular and minute glandular trichomes. **Leaves** with 1 or 2 colleters on each side of petiole; petiole 0.8–3.5 cm, short-hirsute with eglandular and minute glandular trichomes; blade deltate to ovate, 2–7.5 × 1–4.5 cm, base deeply cordate, with 2–4 laminar colleters, apex long-acuminate, surfaces densely hirsute, eglandular. **Inflorescences** simple (rarely paired), umbelliform to somewhat racemiform, extra-axillary, pedunculate, 1–6-flowered; peduncle 0.3–1.2 cm, short-hirsute with eglandular and minute glandular trichomes. **Pedicels** 3–10 mm, short-hirsute with eglandular and minute glandular trichomes. **Flowers:** calyx lobes ascending, lanceolate, 2–3 mm, apex acute, short-hirsute with eglandular and minute glandular trichomes; corolla green, not or obscurely reticulate, tubular-campanulate, densely hirtellous, tube 1.5–2.5 mm, lobes spreading, 4–6 mm; corona concealed in throat, a ring with 5 truncate lobes, each with an adaxial incurved appendage incumbent on backs of anthers, green to cream, 1–1.5 mm, glabrous; apical anther appendages cream, deltoid; style apex green, pentagonal, flat. **Follicles** gray-striate, fusiform, 9–14 × 2–2.5 cm, apex acuminate, smooth, glabrous. **Seeds** brown, ovate, 8–11 × 4–5 mm, margins thickly winged, chalazal end minutely erose, faces smooth; coma 3–4 cm.

Flowering May–Oct; fruiting Jul–Oct. Hills, slopes, arroyos, rocky igneous soils, juniper grasslands; of conservation concern; 1200–1800 m; Tex.

Endemic to the Big Bend region, *Matelea texensis* is known only from a few sites in Brewster County, south of Alpine. It is likely that some populations have been lost to road widening. *Matelea texensis* is most similar to *M. producta*, from which it differs by densely hirtellous corollas.

9. **Matelea atrostellata** Rintz, Novon. 17: 522, fig. 1. 2007 • Darkstar milkvine [C]

Vines, herbaceous. **Stems** 1–5, twining, often branched near base, 30–200 cm, short-hirsute with spreading to retrorse eglandular and minute glandular trichomes. **Leaves** with 1 colleter on each side of petiole; petiole 0.7–1.7 cm, short-hirsute with eglandular and minute glandular trichomes; blade ovate to lanceolate, 2.5–5 × 1–3.5 cm, base deeply cordate, with 2–4 laminar colleters, apex attenuate to long-acuminate, surfaces short-hirsute with eglandular and minute glandular trichomes. **Inflorescences** solitary, umbelliform to somewhat racemiform, extra-axillary, pedunculate, 1–5-flowered; peduncle 0–1 cm, short-hirsute with eglandular and minute glandular trichomes. **Pedicels** 3–10 mm, short-hirsute with eglandular and minute glandular trichomes. **Flowers:** calyx lobes spreading, lanceolate, 2–3 mm, apex acute, short-hirsute with eglandular and minute glandular trichomes; corolla pale to dark maroon or green with maroon tinge, sometimes faintly striate and reticulate towards margins, rotate, tube 1–1.5 mm, lobes spreading, narrowly lanceolate, 5–7 mm, hirtellous abaxially, hirtellous adaxially in ring at base of lobes, vestiture continuing along center of lobes with trichomes diminishing in height (reported to be rarely glabrous); corona united to corolla and column near base, composed of 5 united segments with outward flared margins, each with an adaxial incurved appendage, crimson to dark maroon, 0.5–1 mm, margins and appendages papillose; apical anther appendages white, brown at base, deltoid; style apex concave. **Follicles** gray-striate, ellipsoid to lance-ovoid, 8–11 × 1.5–2 cm, apex acuminate, smooth or with very sparse, low tubercules, glabrous. **Seeds** tan, ovate, 8 × 5–6 mm, margins thickly winged, chalazal end minutely erose, faces smooth; coma 4–5 cm.

Flowering May–Aug; fruiting Jun–Oct. Narrow canyons, slopes, arroyos, limestone, rhyolite, rocky soils, riparian woodlands; of conservation concern; 1200–1600 m; Tex.; Mexico (Coahuila).

Matelea atrostellata is endemic to Chihuahuan Desert mountain ranges from the Chinati Mountains to the Sierra del Carmen. Within the flora area, it is uncommon and restricted to Brewster County, Texas, and likely threatened. It is evidently a close relative of *M. producta* and *M. texensis* based on similarity in vegetative and fruit characters. However, the flowers diverge dramatically by their rotate corollas with outward-flared, papillose corona segments that are free from the corolla. Like *M. texensis*, *M. atrostellata* is found at the southeastern margin of the range of *M. producta*, and *M. atrostellata* does not co-occur with either species.

10. **Matelea parvifolia** (Torrey) Woodson, Ann. Missouri Bot. Gard. 28: 230. 1941 • Spearleaf [F]

Gonolobus parvifolius Torrey in W. H. Emory, Rep. U.S. Mex. Bound. 2(1): 166. 1859

Vines, suffrutesent, not corky. **Stems** 1–10, twining, 10–150 cm, retrorse-puberulent with straight or curved, eglandular trichomes. **Leaves** with 0–2 colleters on each side of petiole; petiole 0.2–1 cm, sparsely puberulent with curved, eglandular trichomes and inconspicuously glandular-hirtellous; blade ovate to lanceolate or deltate, 0.5–2 × 0.2–1 cm, base truncate to deeply cordate, with 2–4 laminar colleters, apex acute, surfaces hirtellous with eglandular and inconspicuous glandular trichomes. **Inflorescences** solitary, umbelliform, extra-axillary, sessile, 1–3-flowered. **Pedicels** 1–5 mm, hirtellous with eglandular and inconspicuous glandular trichomes. **Flowers:** calyx lobes spreading, deltate to elliptic, 1.9–3 mm, apex acute, hirtellous with eglandular and inconspicuous glandular trichomes; corolla brown or maroon to green, not reticulate, campanulate, tube 0.5–1 mm, lobes erect (spreading), deltate, 2–3 mm, abaxially hirtellous, adaxially hirtellous at base of each lobe to glabrate; corona of 5 united lobes, cup-shaped and undulate-spreading with 5 ridges opposite anthers, exceeded by style apex, maroon or brown to green (yellow), 0.5–0.7 mm, glabrous; apical anther appendages white, deltoid; style apex yellowish green to maroon, pentagonal, head flat to broadly convex. **Follicles** gray to maroon or purplish striate, narrowly lance-ovoid, 5–9 × 0.5–1.5 cm, apex acuminate, nearly smooth to sparsely tuberculate, mostly on lower half, minutely short-hirsute, glabrate. **Seeds** tan, ovate, 5–7 × 3–5 mm, margins winged, chalazal end minutely erose, faces rugulose; coma 1.5–2 cm.

Flowering and fruiting year round. Hills, slopes, bajadas, canyons, arroyos, often granitic, basaltic, or limestone substrates, rocky or sandy soils, alluvium, desertscrub; 300–1400 m; Ariz., Calif., Tex.; Mexico (Coahuila, Sonora).

Matelea parvifolia is a vigorously twining, drought-deciduous vine found disjunctly in the Sonoran and Chihuahuan deserts. To the east, it is restricted to the Big Bend region of Texas and adjacent Coahuila, Mexico. The few collections and observations made in Texas suggest that conservation status in that state merits consideration. The range is much larger in the west, extending from the lower Grand Canyon across the southwestern half of Arizona to adjacent Sonora, Mexico, and southeastern California; these two regions are disjunct by more than 500 km. The absence of *M. parvifolia* from New Mexico (and Chihuahua, Mexico) is curious, and the species may be sought in the Peloncillo Mountains, as it has been documented just across the border in Sonora. Similarly, discovery in Nevada would not be surprising. As far as known, all reports from Baja California pertain to *M. hastulata* (A. Gray) Sundell.

The closely related *Matelea sagittifolia* in southern Texas is nearly identical to *M. parvifolia* in vegetative characteristics but differs by consistently green corollas with oblong-spatulate lobes (versus shades of brown or maroon and deltate lobes), and a white, lobed style apex (versus green to maroon and pentagonal).

11. Matelea sagittifolia (A. Gray) Woodson ex Shinners, Sida 1: 363. 1964 • Arrowleaf

Gonolobus sagittifolius A. Gray, Proc. Amer. Acad. Arts 12: 77. 1876

Vines, suffrutescent, not corky. **Stems** 1–10, twining, 10–150 cm, retrorse-puberulent with curved, eglandular trichomes and inconspicuously glandular-hirtellous to glabrate. **Leaves** with 0–2 colleters on each side of petiole; petiole 0.2–1 cm, sparsely puberulent with curved, eglandular trichomes and inconspicuously glandular-hirsutulous; blade deltate to lanceolate or ovate, 0.5–2.5 × 0.2–1.3 cm, base shallowly to deeply cordate, with 2–4 laminar colleters, apex acute, surfaces sparsely puberulent with curved, eglandular trichomes and inconspicuously glandular-hirtellous (mostly on veins) to glabrate. **Inflorescences** solitary, umbelliform, extra-axillary, sessile or subsessile, 1–4-flowered. **Pedicels** 1–5 mm, puberulent with curved, eglandular trichomes and inconspicuously glandular-hirtellous. **Flowers:** calyx lobes ascending, linear to linear-lanceolate, 2–3 mm, apex acute, sparsely puberulent with curved, eglandular trichomes and inconspicuously glandular-hirtellous to glabrate; corolla green to yellow-green, not or very faintly reticulate, campanulate, tube 0.5–1 mm, lobes ascending, linear, 5–10 mm, apex often twisted, glabrous; corona cup-shaped, apex undulate, with 5 paler appendages opposite anthers, exceeded by style apex, green, cream, or yellowish, 0.5–0.7 mm, glabrous; apical anther appendages white, deltoid; style apex yellow, pentagonal-lobed, flat to broadly convex. **Follicles** gray to maroon or purplish striate, lance-ovoid, 6–7 × 1–2 cm, apex acuminate, smooth to very sparsely tuberculate, mostly on lower half, minutely puberulent to glabrate. **Seeds** tan, ovate, 5–7 × 4–5 mm, margins winged, chalazal end minutely erose, faces rugulose; coma 2–3 cm.

Flowering Mar–May(–Oct); fruiting Apr–Dec. Hills, slopes, ridges, limestone, sandstone, rocky and sandy soils, thornscrub, desertscrub; 20–700 m; Tex.; Mexico (Nuevo León, Tamaulipas).

Matelea sagittifolia is a vigorously twining, drought-deciduous vine. Its range in Texas extends from the Rio Grande Valley, (Terrell County to Starr County) northward and eastward to McMullen and San Patricio counties, where the species is uncommon. Distinctions between *M. sagittifolia* and the highly similar 10. *M. parvifolia* and 12. *M. radiata* are discussed under those species.

12. Matelea radiata Correll, Wrightia 3: 136. 1965 • Falfurrias milkvine [C] [E]

Vines, suffrutescent, not corky. **Stems** 1–10, twining, 30–200 cm, retrorse-puberulent with curved, eglandular trichomes. **Leaves** with 1 colleter on each side of petiole; petiole 0.3–0.6 cm, puberulent with curved, eglandular trichomes and inconspicuously glandular-hirtellous; blade deltate to narrowly lanceolate, 1–2 × 0.2–0.7 cm, base truncate to shallowly cordate, with 2–4 laminar colleters, apex acute, surfaces sparsely puberulent with curved, eglandular trichomes and very inconspicuously glandular-hirtellous on veins abaxially, glabrate adaxially. **Inflorescences** solitary, extra-axillary, sessile or subsessile, 1(or 2)-flowered. **Pedicels** 1–3 mm, puberulent with curved, eglandular trichomes and inconspicuously glandular-hirtellous. **Flowers:** calyx lobes spreading, narrowly lanceolate, 2–3 mm, apex acute, sparsely puberulent with curved, eglandular trichomes and inconspicuously glandular-hirtellous; corolla yellowish green abaxially, brown with green tinge adaxially, not or very faintly reticulate, campanulate, tube 1–2 mm, lobes ascending, lanceolate, 5–6 mm, glabrous; corona of 5 laminar segments opposite anthers, apex retuse, white with purple adaxial patch, 2.5–3 mm, greatly exceeding style apex, glabrous; apical anther appendages white, deltoid; style apex green, pentagonal-lobed, flat with central, bifid protrusion. **Follicles** gray-striate, lance-ovoid to fusiform, 5–8 × 0.8–1 cm, apex acuminate, smooth to moderately tuberculate, mostly on lower two-thirds, sparsely puberulent to glabrate. **Seeds** tan, ovate, 6–7 × 4–5 mm, margins winged, chalazal end scarcely erose, faces inconspicuously rugulose; coma 2–3 cm.

Flowering Apr–Aug; fruiting Jul–Dec. Low hills or plains, rocky and clay soils, thornscrub; of conservation concern; 30–100 m; Tex.

Matelea radiata is very uncommon and possibly endemic to Brooks, Hidalgo, and Starr counties in southern Texas. Only three flowering specimens are known, from near Falcon Lake and La Joya, in the Lower Rio Grande Valley, and the type collection from the imprecise location of Falfurrias. Based on the restriction of all other collections to the vicinity of the Rio Grande Valley, including fruiting specimens assigned here tentatively, it is possible that the type collection was made a good deal south of Falfurrias. Similar in most respects to close relatives *M. parvifolia* and *M. sagittifolia*, *M. radiata* is readily differentiated by elongate white corona segments that greatly exceed the style apex. Based on scant evidence, it is possible that the follicles are more densely tuberculate than in the related species. It is wholly disjunct from the range of *M. parvifolia*, which is found only as far east as the Big Bend region, and it is known to co-occur with *M. sagittifolia* only in the vicinity of Falcon Lake. Considering the rarity of *M. radiata* and development pressures in the Lower Rio Grande Valley, this species should be considered to be of extreme conservation concern.

13. Matelea reticulata (Engelmann ex A. Gray) Woodson, Ann. Missouri Bot. Gard. 28: 234. 1941 • Pearl or netted milkvine, green milkweed vine [F]

Gonolobus reticulatus Engelmann ex A. Gray, Proc. Amer. Acad. Arts 12: 75. 1876

Vines, herbaceous. **Stems** 1–5, twining, 30–200 cm, densely hirsute with long eglandular and inconspicuous glandular trichomes. **Leaves** with 0–1 colleter on each side of petiole; petiole 1–6 cm, densely hirsute with long eglandular and inconspicuous glandular trichomes; blade ovate, 2–10 × 1.5–7 cm, base deeply cordate, basal lobes rarely overlapping, with 2–4 laminar colleters, apex acute, surfaces hirsute with long eglandular and inconspicuous glandular trichomes abaxially, sparsely hirsute with eglandular trichomes adaxially. **Inflorescences** solitary, umbelliform to racemiform, extra-axillary, pedunculate, 3–10-flowered; peduncle 1–7 cm, hirsute with long eglandular and inconspicuous glandular trichomes. **Pedicels** 12–25 mm, hirsute with long eglandular and inconspicuous glandular trichomes. **Flowers:** calyx lobes spreading, lanceolate, 3–3.5 mm, apex acute, hirsute with long eglandular and inconspicuous glandular trichomes; corolla pale green with darker green reticulations on lobes, rotate, tube 1.5–2 mm, lobes spreading, broadly ovate to nearly orbiculate, 4–5 mm, hirsute with long eglandular and inconspicuous glandular trichomes abaxially, glabrous adaxially; corona united to column near base, a low, undulate-erose ring, yellow, 0.5–0.7 mm, not reaching one-half of the height of the style, glabrous; apical anther appendages silvery, reflective, deltoid, covering more than 50% of style apex; style apex nearly round, flat to broadly convex. **Follicles** sometimes striate, lance-ovoid to fusiform, somewhat thickened near base, 8–15 × 1.5–2 cm, apex acuminate, sparsely to moderately muricate, glabrate. **Seeds** brown, ovate, 9–10 × 5–9 mm, margins winged, chalazal end entire, faces rugulose; coma 3–4 cm.

Flowering Mar–Sep(–Nov); fruiting May–Feb. Hills, slopes, ridges, canyons, streamsides, limestone, granite, caliche, rocky, sandy, and clay soils, juniper-oak and riparian woodlands, thornscrub; 0–1500 m; Tex.; Mexico (Chihuahua, Coahuila, Guanajuato, Nuevo León, Querétaro, San Luis Potosí, Tamaulipas).

Matelea reticulata co-occurs with the similar *M. edwardsensis* but is more common, more widespread, and more ecologically labile than that species. Characters distinguishing these species are discussed under 14. *M. edwardsensis*. The reflective apical anther appendages of *M. reticulata* are unusual among the milkweeds of the flora region, and their role in pollinator attraction seems important but is unstudied. The species is quite uncommon and sporadic west and south of the Edwards Plateau.

14. Matelea edwardsensis Correll, Wrightia 3: 135. 1965 • Plateau milkvine [E]

Vines, herbaceous. **Stems** 1–5, twining, 30–200 cm, sparsely hirsute with short eglandular trichomes to glabrate. **Leaves** with 1 or 2 colleters on each side of petiole; petiole 1–4 cm, hirsute with eglandular and inconspicuous glandular trichomes; blade ovate, 1.5–8 × 1–6 cm, base deeply cordate, basal lobes often overlapping, with 2 laminar colleters, apex acute, surfaces sparsely hirsute with eglandular trichomes, primarily on veins, to glabrate. **Inflorescences** solitary, umbelliform to racemiform, rarely compound, extra-axillary, pedunculate, 3–12-flowered; peduncle 0.3–1.5 cm, hirsute with eglandular and inconspicuous glandular trichomes. **Pedicels** 4–10 mm, hirsute with eglandular and inconspicuous glandular trichomes. **Flowers:** calyx lobes erect, narrowly lanceolate, 2–3 mm, apex acute, hirsute with eglandular and inconspicuous glandular trichomes; corolla pale green to yellow-green with darker green lines along lobes, becoming reticulate at lobe tips, campanulate, tube 2–3 mm, lobes spreading, ovate, 3–4 mm, sparsely hirtellous abaxially, densely hirtellous adaxially; corona united to column near base, of 5 united, fleshy segments free nearly to base, each with an adaxial appendage opposite anthers, apex rounded, yellowish green to pale green, 0.7–1 mm, glabrous; apical anther appendages white, truncate; style apex green, nearly round, flat. **Follicles** sometimes striate, lance-ovoid to fusiform, 8–10 × 1.5–2 cm, apex acuminate, sparsely to moderately muricate, glabrate. **Seeds** brown, ovate, brown, 10–11 × 5–7 mm, margins broadly winged, chalazal end erose; coma 2.5–3 cm.

Flowering Mar–May(–Jun); fruiting Jun–Oct. Hills, slopes, ridges, canyons, limestone, rocky soils, juniper-oak woodlands; 100–700 m; Tex.

Matelea edwardsensis is an elusive species with numerous documented localities on and around the eastern and northern margins of the Edwards Plateau, but it appears to be common nowhere. Superficially, it is very similar to *M. reticulata*, which is almost invariably present where *M. edwardsensis* occurs. *Matelea reticulata* differs by stems, leaves, and inflorescences that are more densely vestitured with longer trichomes, rotate corollas that are glabrous adaxially and possess lobes nearly orbiculate and with a more reticulate pattern, and a highly reduced corona lacking noticeable lobes. The seeds of *M. edwardsensis* are also larger and conspicuously erose at one end. *Matelea edwardsensis* is frequent across much of its limited range and perhaps not of immediate conservation concern, but its woodland habitat is increasingly vulnerable to clearing and development.

15. Matelea carolinensis (Jacquin) Woodson, Ann. Missouri Bot. Gard. 28: 228. 1941 • Carolina or maroon Carolina milkvine, Carolina or maroon Carolina spinypod [E] [F]

Cynanchum carolinense Jacquin, Collectanea 2: 288. 1788; *Gonolobus carolinensis* (Jacquin) R. Brown ex Schultes; *Odontostephana carolinensis* (Jacquin) Alexander

Vines, herbaceous. **Stems** 1–2(–5), twining, 100–300 cm, hirsute with eglandular and inconspicuous glandular trichomes. **Leaves** with 2 colleters on each side of petiole; petiole 1–9 cm, hirsute with eglandular and inconspicuous glandular trichomes; blade ovate to orbiculate, 3–20 × 1.5–18 cm, base shallowly to deeply cordate, with 2–4 laminar colleters, apex acute to acuminate, surfaces hirsute with eglandular and inconspicuous glandular trichomes. **Inflorescences** solitary, simple or compound, umbelliform to racemiform, extra-axillary or terminal, pedunculate, 3–30(–40)-flowered; peduncle 1–13 cm, hirsute with eglandular and inconspicuous glandular trichomes. **Pedicels** 3–30 mm, hirsute with eglandular and inconspicuous glandular trichomes. **Flowers:** calyx lobes spreading, ovate to elliptic to lanceolate, 1.5–4.5 mm, apex acute to acuminate, hirsute with eglandular and inconspicuous glandular trichomes; corolla pale maroon to yellowish green tinged with maroon abaxially, dark maroon, dark brown, or dark purple (yellow and maroon or yellow with maroon tinge at base), with a cream to light rose (maroon) ring at base of corona adaxially, reticulate markings only apparent on yellow forms, rotate to campanulate, tube 1–2 mm, lobes spreading to erect or somewhat reflexed, sometimes twisted, oblong, 4–15 mm, margins plane (reflexed), minutely hirtellous abaxially, glabrous adaxially; corona united to column near base, of 5 united, fleshy segments, each with 2 lateral lobes at apex exceeding the medial lobe, forming a sheath that equals or slightly exceeds style apex, adaxial appendages incurved, incumbent on anthers, sometimes concealed when corolla lobes erect, maroon (orange-yellow or green-yellow), 0.7–2 mm, 2–3 mm diam., glabrous; apical anther appendages bright white with yellow to maroon patch at base, rhomboid; style apex pink to maroon, pentagonal, flat. **Follicles** not striate, lance-ovoid to nearly fusiform, 7–13 × 1–2.5 cm, apex acuminate, sparsely to moderately muricate, minutely hirsute. **Seeds** brown, ovate (orbicular), 8–9 × 6–8 mm, margins broadly winged, chalazal end entire, faces rugose; coma 2.5–3.5 cm.

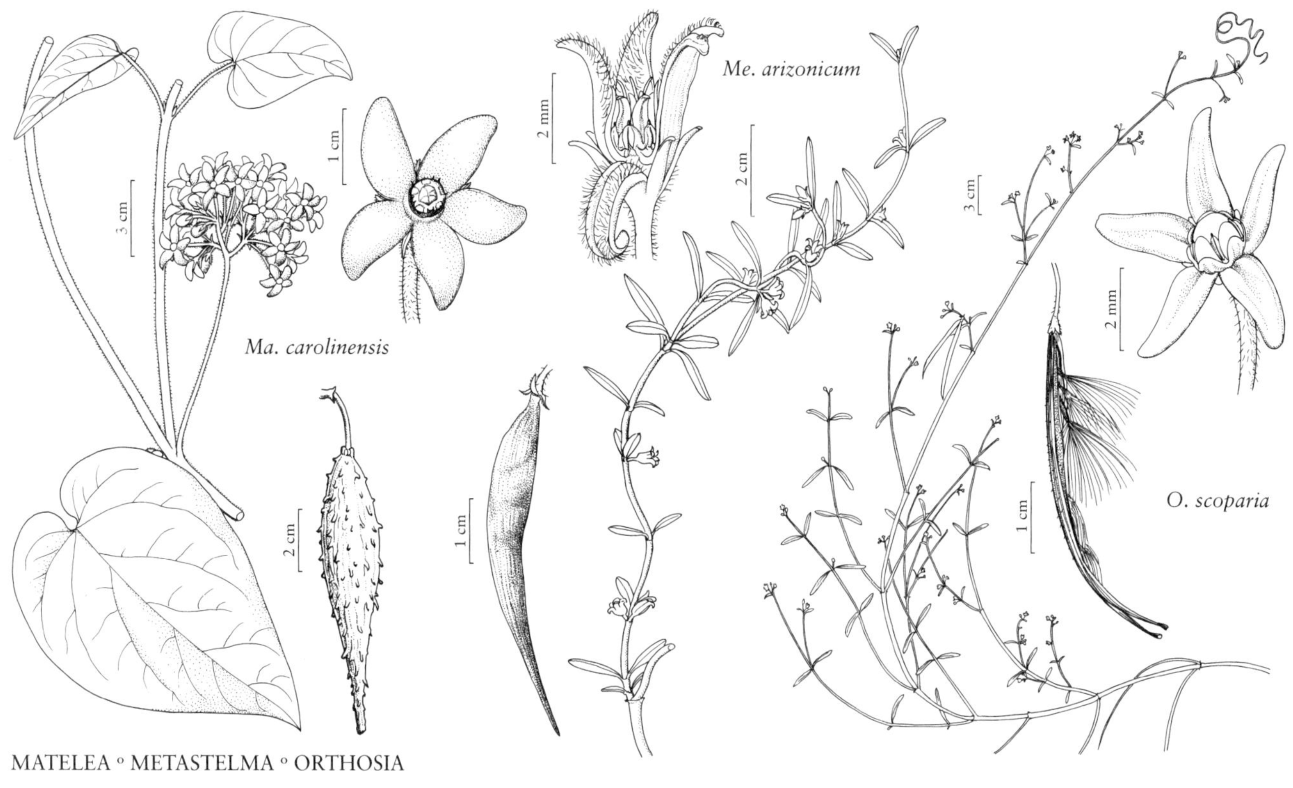

MATELEA ◦ METASTELMA ◦ ORTHOSIA

Flowering Apr–Jul(–Aug); fruiting Jun–Oct(–Nov). Hill slopes, bluffs, ridges, flats, valleys, streamsides, ditches, rock outcrops, limestone, sandstone, gabbro, serpentine, sandy and loamy soils, talus, cedar glades, barrens, oak woods, oak-hickory, oak-pine, mixed-hardwood, pine-mixed hardwood, pine, and riparian forests, old fields, secondary woods, pine plantations; 0–700 m; Ala., Del., D.C., Ga., Ky., La., Md., Miss., N.C., S.C., Tenn., Va.

Matelea carolinensis is a widespread and variable species that has long presented challenges in circumscription. The most widely distributed and common form has small, nearly rotate corollas with deltate to oval corolla lobes that are sometimes reflexed. However, Appalachian populations in the Carolinas may vary by having larger flowers with lobes that are much narrower and ascending. In the Appalachian region, there is a bewildering array of floral diversity including large and small corollas with ascending lobes and individual plants with lobes that are reflexed, spreading, and ascending, perhaps in part due to developmental changes. Plants from the Carolinas with large corollas and ascending lobes have been considered to be conspecific with those in the Ozark region treated here as *M. decipiens* since that species was first described by Alexander. These two population systems are strongly disjunct. Careful study of corona structure suggests that these populations are convergent in their narrow, campanulate corollas and that the Appalachian plants belong to the same species as the widespread populations with smaller, rotate corollas with which they intergrade. Further study using population genetics and other data types is needed to address conflicting treatments of these plants. As currently understood, *M. carolinensis* occurs almost entirely east of the Mississippi River with only a few isolated occurrences in Louisiana west of the river (for example, *Thomas 13,124* [TENN] from Catahoula Parish). Conversely, *M. decipiens* occurs entirely west of the river. *Matelea carolinensis* has suffered habitat loss at the northern edge of its range. It is presumed to be extirpated in the District of Columbia and is uncommon and of conservation concern in Delaware, Kentucky, and Maryland. Reports from Pennsylvania and West Virginia have not been confirmed and probably represent misidentifications of *M. obliqua*.

16. Matelea flavidula (Chapman) Woodson, Ann. Missouri Bot. Gard. 28: 228. 1941 • Yellow Carolina milkvine, yellow-flowered or Carolina yellow spinypod [E]

Gonolobus flavidulus Chapman, Bot. Gaz. 3: 12. 1878; *Odontostephana flavidula* (Chapman) Alexander

Vines, herbaceous. **Stems** 1, twining, 100–200 cm, hirsute with eglandular and inconspicuous glandular trichomes. **Leaves** with 2 colleters on each side of petiole; petiole 1–6 cm, hirsute with eglandular and inconspicuous glandular trichomes; blade ovate (orbiculate), 5–15 × 2–10 cm, base shallowly to deeply cordate, with (2–)4 laminar colleters, apex acute to acuminate, surfaces hirsute with eglandular and inconspicuous glandular trichomes. **Inflorescences** solitary, umbelliform, extra-axillary, pedunculate, 3–20-flowered; peduncle 0.5–4.5 cm, hirsute with eglandular and inconspicuous glandular trichomes. **Pedicels** 3–20 mm, hirsute with eglandular and inconspicuous glandular trichomes. **Flowers:** calyx lobes spreading, deltate to elliptic, 1.4–2.4 mm, apex acute, hirsute with eglandular and inconspicuous glandular trichomes; corolla pale yellow to green abaxially (maroon or maroon tinged), pale green to yellow-green (maroon or maroon tinged), with a cream to yellow ring at base of corona adaxially, frequently with green reticulations visible on both sides, rotate, tube 1–1.5 mm, lobes spreading, elliptic or oblong to oblanceolate, 4.5–11 mm, margins plane to slightly reflexed, minutely hirtellous abaxially, glabrous adaxially; corona united to column near base, of 5 segments, united in basal one-half, fleshy, each with small lateral lobes at apex, exceeded by medial lobe, forming a sheath that equals (exceeds) style apex, adaxial appendages incurved, incumbent on anthers, yellow to orange-yellow (maroon), 1.5 mm, glabrous; apical anther appendages bright white with green patch at base, truncate, covering less than 25% of style apex; style apex green, pentagonal, flat. **Follicles** not striate, lance-ovoid to ellipsoid, 8–15 × 1.5–2.5 cm, apex acuminate, muricate, sparsely and minutely hirsute to glabrous. **Seeds** dark brown, ovate (orbicular), 7–9 × 5–7 mm, margins broadly winged, chalazal end entire, faces rugose; coma 1.5–2.5 cm.

Flowering Apr–Jul; fruiting Jul–Oct. Sandy and calcareous substrates, hillsides, bluffs, and stream banks, hardwood and pine-hardwood forests; 0–300 m; Fla., Ga., S.C.

Matelea flavidula is uncommon on the Atlantic and Gulf coastal plains in Florida (Gadsden County), Georgia (Baker, Colquitt, Cook, and Decatur counties), and South Carolina (Charleston, Clarendon, and Georgetown counties). It is highly localized in Gadsden County, Florida, where its range contacts that of *M. alabamensis*, and it is considered to be endangered in that state. Reports from Alabama appear to be based on yellow-flowered plants of *M. carolinensis*. Although uncommon, its global status has not received much attention, and a current assessment is needed, patricularly since misidentification of specimens may have contributed to an overestimation of the number of extant populations. The narrower lobes of the yellow-green reticulate corollas and taller corona, along with more sparsely muricate follicles, distinguish *M. flavidula* from *M. alabamensis*.

17. Matelea floridana (Vail) Woodson, Ann. Missouri Bot. Gard. 28: 229. 1941 • Florida milkvine or spinypod [C] [E]

Vincetoxicum floridanum Vail, Bull. Torrey Bot. Club 26: 428. 1899; *Odontostephana floridana* (Vail) Alexander

Vines, herbaceous. **Stems** 1–5, twining, 100–500 cm, hirsute with eglandular and inconspicuous glandular trichomes. **Leaves** with 2 colleters on each side of petiole; petiole 1–6 cm, hirsute with eglandular and inconspicuous glandular trichomes; blade narrowly ovate, oblong, or elliptic (ovate or cordate), 3.5–11 × 2–12 cm, base shallowly to deeply cordate, with 2–6 laminar colleters, apex acute to acuminate, surfaces hirsute with eglandular and inconspicuous glandular trichomes. **Inflorescences** solitary, racemiform (umbelliform), extra-axillary, pedunculate, 3–10-flowered; peduncle 0.3–1(–5) cm, hirsute with eglandular and inconspicuous glandular trichomes. **Pedicels** 5–15 mm, hirsute with eglandular and inconspicuous glandular trichomes. **Flowers:** calyx lobes spreading, elliptic to ovate, 1–2.5 mm, apex acute to acuminate, hirsute with eglandular and inconspicuous glandular trichomes; corolla pale yellow to green abaxially (maroon or maroon tinged), maroon to brown to purple (green to yellow, sometimes maroon tinged), with a cream to yellow ring at base of corona adaxially, faintly reticulate, rotate-campanulate, tube 1–1.5 mm, lobes spreading, margins plane, lanceolate to oblong, 3–7 mm, minutely hirtellous abaxially, glabrous adaxially; corona united to column near base, of 5 united, fleshy segments, each with 2 small lateral lobes at apex, shorter than medial lobe, forming a lobed sheath that equals style apex, adaxial appendages incurved, incumbent on anthers, dark maroon, 1.5–2 mm, glabrous; apical anther appendages bright white with maroon patch at base, truncate; style apex yellow to yellow-green, maroon tinged in center, pentagonal, flat. **Follicles** not striate, lance-ovoid to ellipsoid, 7.5–11 × 1–2.5 cm, apex acuminate, sparsely to moderately muricate, sparsely and minutely hirsute. **Seeds** brown, ovate (orbicular), 7–8 × 5–6 mm, margins broadly winged, chalazal end entire, faces rugose; coma 1.5–2.2 cm.

Flowering Apr–Aug(–Oct); fruiting Jul–Oct(–Dec). Sandy and calcareous soils, sandhills, hillslopes, bluffs, sinks, ravines, hammocks, mixed-hardwood, oak-hickory, and pine forests, cypress domes, streamsides, swamps; of conservation concern; 0–70 m; Fla., Ga.

Matelea floridana is known primarily from northern peninsular Florida. Widely disjunct populations in Hendry and Miami-Dade counties are documented only by vegetative specimens that cannot be identified with any confidence; however, other than the highly dissimilar, decumbent *M. pubiflora*, no other *Matelea* occurs in peninsular Florida. A smaller disjunction is represented by an apparent gap in the distribution between Suwannee County and the Panhandle. Although widely distributed in the state, its occurrences in Florida are scattered, and *M. floridana* is considered state-endangered. The occurrence of *M. floridana* in Georgia is confirmed at a single location in Thomas County, but it has been reported from two other counties in the southwestern corner of the state. Specimens of *M. floridana* have been identified as *M. carolinensis*; however, that species is not known from Florida and differs by corona segments with the medial lobe shorter than the lateral lobes and more rounded corolla lobes (compared to the medial corona lobe much longer than the lateral lobes and acute to obtuse corolla lobes in *M. floridana*).

18. Matelea alabamensis (Vail) Woodson, Ann. Missouri Bot. Gard. 28: 234. 1941 • Alabama milkvine or spinypod [C] [E]

Vincetoxicum alabamense Vail, Bull. Torrey Bot. Club 30: 178, plate 9. 1903; *Cyclodon alabamensis* (Vail) Small

Vines, herbaceous. **Stems** 1–5, twining, 20–150 cm, hirsute with eglandular and inconspicuous glandular trichomes. **Leaves** with 0–2 colleters on each side of petiole; petiole 1–5 cm, hirsute with eglandular and inconspicuous glandular trichomes; blade ovate to nearly orbiculate, 4–12 × 2–10 cm, base shallowly to deeply cordate, with 0–2(–4) laminar colleters, apex acute to acuminate, surfaces hirsute with eglandular and inconspicuous glandular trichomes. **Inflorescences** solitary, umbelliform, extra-axillary, pedunculate, 1–5-flowered; peduncle 0.2–1.2 cm, hirsute with eglandular and inconspicuous glandular trichomes. **Pedicels** 5–15 mm, hirsute with eglandular and inconspicuous glandular trichomes. **Flowers:** calyx lobes spreading, elliptic, 1.9–3.4 mm, apex acute, hirsute with eglandular and inconspicuous glandular trichomes; corolla dark green abaxially, creamy yellow to pale green with dark green reticulations, with a deep yellow ring at base of corona adaxially, very widely rotate-campanulate, tube 0.5–1.5 mm, lobes spreading, plane, ovate, 7–10 mm, glabrous abaxially, hirtellous adaxially at base of lobes; corona united to column near base, a saucer-shaped ring with small apical teeth and 5 pointed lobes, greatly exceeded by style apex, appressed to stamens, deep yellow, 0.5–1 mm, glabrous; apical anther appendages bright white with green patch at base, ovate, covering more than 50% of style apex; style apex yellow, pentagonal, flat. **Follicles** not striate, lance-ovoid to ovoid, 4.9–8 × 1.5–2.5 cm, apex acuminate, densely muricate, minutely hirsute. **Seeds** tan to brown, broadly ovate, 7–9 × 6–8 mm, margins broadly winged, chalazal end scarcely erose, faces rugulose; coma 3.5–4 cm.

Flowering (Mar–)Apr–Jun; fruiting Jun–Dec. Sandy oak-hickory and mixed hardwoods on ravine slopes; of conservation concern; 10–100 m; Ala., Fla., Ga.

Matelea alabamensis is distributed on the Gulf Coastal Plain from the Choctawhatchee to the Chattahoochee-Apalachicola watersheds. It is uncommon and restricted to a handful of populations in Dale and Henry counties in Alabama; Gadsden, Liberty, Walton, and Washington counties in Florida; and Clay and Early counties in Georgia. A report from southeastern Georgia has not been confirmed and is doubtful. The species has experienced significant habitat loss and degradation resulting in serious concerns for conservation across its range. The greenish flowers with darker reticulations and densely muricate follicles are an unusual trait combination that is unique among the spinypods of the Gulf Coast.

Matelea alabamensis is in the Center for Plant Conservation's National Collection of Endangered Plants.

19. Matelea obliqua (Jacquin) Woodson, Ann. Missouri Bot. Gard. 28: 229. 1941 • Climbing or limerock milkvine, northern spinypod [E]

Cynanchum obliquum Jacquin, Collectanea 1: 148. 1788; *Odontostephana obliqua* (Jacquin) Alexander; *Gonolobus obliquus* (Jacquin) R. Brown ex Schultes var. *shortii* A. Gray; *Matelea shortii* (A. Gray) Woodson

Vines, herbaceous. **Stems** 1(–5), twining, 100–300 cm, hirsute with eglandular and inconspicuous glandular trichomes. **Leaves** with 2 colleters on each side of petiole; petiole 1–7 cm, hirsute with eglandular and inconspicuous glandular trichomes; blade ovate to oblong or elliptic (lanceolate, orbiculate), 3.5–15 × 2–13 cm, base shallowly to deeply cordate, with 2–4 laminar colleters, apex acute to acuminate, surfaces hirsute with eglandular and inconspicuous glandular trichomes. **Inflorescences** solitary or paired, umbelliform (rarely compound), extra-axillary, pedunculate, 5–15(–20)-flowered; peduncle 1–9 cm, hirsute with eglandular and inconspicuous glandular trichomes. **Pedicels** 5–40 mm, hirsute with eglandular and inconspicuous glandular trichomes. **Flowers:** calyx lobes spreading, elliptic to ovate (narrowly deltate), 2–4.5 mm, apex acute to acuminate, hirsute with eglandular and inconspicuous glandular trichomes; corolla pale maroon to yellowish green tinged with maroon abaxially, pink, reddish, or maroon (green, cream, or orange), with a cream to rose ring at base of corona adaxially (ring sometimes absent), not reticulate, rotate-campanulate to campanulate, tube 0.5–1.5 mm, lobes erect to spreading, twisted (coiled), oblong to linear, 8–16 mm, margins reflexed to plane, minutely hirtellous with glandular and eglandular trichomes abaxially, glabrous adaxially; corona united to column near base, nearly circular, of 5 united, very fleshy segments, each with 2 lateral lobes at apex equaling or exceeding medial lobe, forming a sheath that equals style apex, adaxial appendages incurved, incumbent on anthers, sometimes concealed when corolla lobes erect, cream to rose to maroon, 1–2 mm, 5 mm diam., glabrous; apical anther appendages bright white with maroon patch at base, truncate; style apex cream to rose to maroon, pentagonal, flat. **Follicles** not striate, lance-ovoid, 10–15 × 1.5–3 cm, apex acuminate, moderately muricate, sparsely and minutely hirsute. **Seeds** brown, ovate (orbicular), 7–9 × 4–6 mm, margins broadly winged, chalazal end entire, faces rugose; coma 2.5–3.5 cm.

Flowering (Apr–)May–Sep(–Oct); fruiting Jun–Oct. Rocky or fine soils, limestone, dolomite, sandstone, shale, hill slopes, bluffs, ridge tops, valleys, stream banks, oak and cedar woods, oak-hickory and mixed-hardwood forests, old fields, glades, barrens; 50–900 m; Ala., D.C., Ga., Ill., Ind., Ky., Md., Miss., N.C., Ohio, Pa., Tenn., Va., W.Va.

Matelea obliqua has a wide range across the Appalachian Mountains and Cumberland and Allegheny plateaus, extending into the Ohio River Valley and upper Gulf Coastal Plain, but it is common only in a few local areas. It is most common in Kentucky and Tennessee and locally in Madison County, North Carolina. The range barely enters several states, where *M. obliqua* is uncommon and is (or should be) considered to be of conservation concern, especially in Alabama (Calhoun and Madison counties) and Georgia (Catoosa and Floyd counties). Records outside the documented range are based on misidentifications pertaining to *M. carolinensis* or *M. decipiens*. Typically, *M. obliqua* can be distinguished by reddish purple to reddish brown corollas (versus maroon to purple in those species), with long, narrow, twisted corolla lobes. The most reliable character is the thick corona ring that has a diameter greater than in any other species of the eastern United States spinypods.

20. Matelea decipiens (Alexander) Woodson, Ann. Missouri Bot. Gard. 28: 228. 1941 • Oldfield milkvine, climbing milkweed [E]

Odontostephana decipiens Alexander in J. K. Small, Man. S.E. Fl., 1077. 1933; *Gonolobus decipiens* (Alexander) L. M. Perry

Vines, herbaceous. **Stems** 1(–5), twining, 50–200 cm, hirsute with eglandular and inconspicuous glandular trichomes. **Leaves** with 2 colleters on each side of petiole; petiole 1.5–9.5 cm, hirsute with eglandular and inconspicuous glandular trichomes; blade ovate to oval to orbiculate, 4–14 × 2.5–16 cm, base shallowly to deeply cordate, with 2 laminar colleters, apex acute to acuminate, surfaces hirsute with eglandular and inconspicuous glandular trichomes. **Inflorescences** solitary or paired, simple or compound umbelliform, extra-axillary, pedunculate, 10–40-flowered; peduncle 1–9 cm, hirsute with eglandular and inconspicuous glandular trichomes. **Pedicels** 10–25 mm, hirsute with eglandular and inconspicuous glandular trichomes. **Flowers:** calyx lobes spreading, elliptic to lanceolate, 2–3.8 mm, apex acute to acuminate, hirsute with eglandular and inconspicuous glandular trichomes; corolla pale maroon to green tinged with maroon abaxially, purple to maroon (greenish yellow or green tinged) adaxially, not reticulate, shallowly campanulate, tube 1.5–2.5 mm, lobes erect to spreading, slightly twisted, oblong, 7–18 mm, margins plane (recurved), glabrous abaxially, minutely hirtellous at base to glabrate adaxially; corona united to column near base, usually pentagonal, of 5 united, laminar segments, each with 2 lateral lobes at apex equaling or exceeding medial lobe, forming a ring exceeded by style apex, adaxial appendages incurved, incumbent on anthers, sometimes concealed by erect corolla lobes, maroon, 0.8–2 mm, 3–4 mm diam., glabrous; apical anther appendages bright white with pink to maroon patch at base, rhomboid; style apex yellow-green to maroon, pentagonal, flat. **Follicles** not striate, lance-ovoid, 8–12 × 1–2 cm, apex acuminate, moderately muricate, minutely hirsute. **Seeds** brown, ovate, 7–8 × 5–6 mm, margins broadly winged, chalazal end entire, faces rugose; coma 3–4.2 cm.

Flowering Apr–Jul(–Sep); fruiting Jun–Oct. Rocky and sandy soils, limestone, dolomite, granite, hill slopes, ridge tops, bluffs, valleys, stream banks, sandhills, pine and pine-oak forests, cedar glades; 20–400 m; Ark., Kans., La., Mo., Okla., Tex.

Matelea decipiens is distributed entirely west of the Mississippi River as far as presently known. Its range extends across the Ozark Mountains to the Gulf Coastal Plain in Louisiana and eastern Texas. However, it is largely absent from the western Ozarks in southwestern Missouri and northwestern Arkansas and from the Ouachita Mountains, where it is replaced by *M. baldwyniana*. *Matelea decipiens* reappears at scattered locations west of the Ozark uplift in Oklahoma and southeastern Kansas in Neosho County. It is considered to be of conservation concern in Kansas, and its status in Oklahoma and Texas merits evaluation. There are no known sites where *M. decipiens* and *M. baldwyniana* co-occur. The flowers of *M. decipiens* become smaller as the range of *M. hirtelliflora* is approached in Texas, and they barely overlap with that species (corolla lobes 7–18 mm in *M. decipiens* versus 3.2–7.5 mm in *M. hirtelliflora*). The corollas of *M. decipiens* are never hirtellous adaxially, as they very often are in *M. hirtelliflora*. E. J. Alexander (1933) and all subsequent authors considered some populations of spinypods in the Appalachian region to belong to this species, based on large, maroon corollas with ascending lobes. Appalachian populations are considered to belong to 15. *M. carolinensis*, as discussed under that species.

21. Matelea hirtelliflora McDonnell & Fishbein, Syst. Bot. 41: 781, figs. 1–3. 2016 • Hairy faced spinypod [C] [E]

Vines, herbaceous. **Stems** 1(–5), twining, 50–200 cm, hirsute with eglandular and inconspicuous glandular trichomes. **Leaves** with 2 colleters on each side of petiole; petiole 2–7 cm, hirsute with eglandular and inconspicuous glandular trichomes; blade ovate to oval, 7.5–17 × 3.5–12 cm, base shallowly to deeply cordate, with 2–4 laminar colleters, apex acute to acuminate, surfaces hirsute with eglandular and inconspicuous glandular trichomes. **Inflorescences** solitary, simple or compound umbelliform, extra-axillary, pedunculate, 1–20(–30)-flowered; peduncle 0.5–6 cm, hirsute with eglandular and inconspicuous glandular trichomes. **Pedicels** 6–13 mm, hirsute with eglandular and inconspicuous glandular trichomes. **Flowers:** calyx lobes spreading, elliptic to lanceolate, 2–2.8 mm, apex acute to acuminate, hirsute with eglandular and inconspicuous glandular trichomes; corolla pale maroon to green with a maroon tinge abaxially, purple to maroon (with a green tinge) adaxially, not reticulate, shallowly campanulate, tube 1.5–2 mm, lobes erect to spreading, slightly twisted, oblong to narrowly deltate, 3.2–7.5 mm, margins plane (recurved), minutely hirtellous to glabrate; corona united to column near base, of 5 united, fleshy segments, each with 2 lateral lobes at apex equaling or exceeding medial lobe, forming a ring exceeding style apex, adaxial appendages incurved, incumbent on anthers, maroon, 0.7–1 mm, 3–4 mm diam., glabrous; apical anther appendages bright white with yellow to green patch at base; style apex green, pentagonal, flat. **Follicles** not striate, lance-ovoid, 6–10 × 1.3–3 cm, apex acuminate, moderately muricate, minutely hirsute. **Seeds** brown, ovate, 6–7 × 5–6 mm, margins broadly winged, chalazal end entire, faces rugose; coma 3–4.2 cm.

Flowering Apr–Jun; fruiting Jun–Aug. Deep sandy soils, valleys, lake shores, hill slopes, oak-hickory and oak woodlands, pine-oak forests, often appearing following fires; of conservation concern; 80–200 m; Ark., Tex.

Matelea hirtelliflora is nearly endemic to the Piney Woods of eastern Texas. There is photographic documentation of the species in extreme southwestern Arkansas (Miller County), but we have not seen specimens. The few historical collections of this species were typically identified as *M. decipiens* prior to the description of the new species. Distinctions from *M. decipiens* are described under that species. Development with consequent habitat loss in the restricted range of *M. hirtelliflora* is cause for conservation concern.

22. Matelea baldwyniana (Sweet) Woodson, Ann. Missouri Bot. Gard. 28: 227. 1941 • Baldwyn's or climbing milkweed, Baldwyn's spinypod [E]

Gonolobus baldwinianus Sweet, Hort. Brit. ed. 2, 360. 1830; *Odontostephana baldwyniana* (Sweet) Alexander

Vines, herbaceous. **Stems** 1(–3), twining, 100–200 cm, hirsute with eglandular and inconspicuous glandular trichomes. **Leaves** with 0–2 colleters on each side of petiole; petiole 2–5 cm, hirsute with eglandular and inconspicuous glandular trichomes; blade ovate to oval (orbiculate), 3–17 × 2–15 cm, base shallowly to deeply cordate, with 2–4 laminar colleters, apex acute to acuminate, surfaces hirsute with eglandular and inconspicuous glandular trichomes. **Inflorescences** solitary or paired, simple to compound umbelliform, extra-axillary, pedunculate, 4–20-flowered; peduncle 1–9.5 cm, hirsute with eglandular and inconspicuous glandular trichomes. **Pedicels** 3–15 mm, hirsute with eglandular and inconspicuous glandular trichomes. **Flowers:** calyx lobes spreading, elliptic, 1.9–3.2 mm, apex acute, hirsute with eglandular and inconspicuous glandular trichomes; corolla white to cream, not or faintly reticulate, campanulate, tube 0.5–2 mm, lobes erect (spreading), twisted, oblong to narrowly deltate to narrowly obovate, 5–9 mm, hirtellous abaxially, glabrous adaxially; corona of 5 united lobes with lateral, subulate lobes at apex greatly exceeding medial lobe, adaxial appendages incurved, incumbent on anthers, often concealed by erect corolla lobes, white or cream to yellow (purple), 1.5–2.5 mm, glabrous; apical anther appendages cream, green at base, truncate; style apex green, rounded-pentagonal, convex. **Follicles** not striate, lance-ovoid, 9–12 × 1.7–2.5 cm, moderately muricate, minutely short-hirsute. **Seeds** brown, oval to ovate, 7–9 × 5–7 mm, margins broadly winged, chalazal end entire, faces rugose; coma 2–4 cm.

Flowering Apr–Jul; fruiting (May–)Jun–Oct(–Dec). Rocky, clay (rarely sandy) soils, limestone, shale, dolomite, hill slopes, valleys, bluffs, riparian, pine, and mixed-hardwood forests, oak-hickory woods, glades, shady meadows; 30–500 m; Ala., Ark., Fla., Mo., Okla., Tex.

Matelea baldwyniana is most common in the southwestern portion of the Ozark uplift, from southwestern Missouri to southeastern Oklahoma. A single location in northern Texas on the Oklahoma border, on the shores of Lake Texoma (an impoundment of the Red River), is disjunct by 100 km from the nearest population in the Ouachita Mountains, and is documented only by a photograph. Furthermore, the flowers on this plant are unusual, with pinkish tinged corollas and maroon coronas unknown from other populations of *M. baldwyniana*. It is possible that this population represents a pale-flowered variant of *M. decipiens*. *Matelea baldwyniana* has a well-documented disjunction of 600 km across the Mississippi Valley to southwestern Alabama, where populations extend along the Gulf Coastal Plain to the Florida Panhandle. The species is uncommon in the southeastern portion of the range, where it is considered to be of conservation concern in both Alabama (Clarke, Monroe, and Wilcox counties) and Florida (Gadsden and Jackson counties). The species is not known to occur in Georgia; however, it has been reported from the state based on uncertainty regarding typification (M. Fishbein and A. McDonnell, unpubl.). The cream corollas with ascending, twisted lobes are unique among the *Matelea* species in the flora area but recall those of *Polystemma cordifolium*, a species just entering the flora area in extreme southern Arizona.

31. METASTELMA R. Brown, Asclepiadeae, 41. 1810 • Swallow-wort, milkweed vine [Greek *meta*, change or instead, and *stemma*, girdle or crown, alluding to corona of separate scales in place of a crown]

Mark Fishbein

Basistelma Bartlett; *Cynanchum* Linnaeus subg. *Metastelma* (R. Brown) Woodson; *Epicion* Small

Vines, herbaceous or somewhat woody or corky at base, perennial; latex white. **Stems** twining, unarmed, glabrous or puberulent with eglandular trichomes. **Leaves** persistent or tardily deciduous, opposite, petiolate; stipular colleters usually present, interpetiolar; laminar colleters usually present. **Inflorescences** extra-axillary cymes, umbelliform or racemiform, sessile or pedunculate. **Flowers:** calycine colleters present or absent; corolla cream, yellowish cream, or green, campanulate, aestivation valvate; corolline corona absent; androecium and gynoecium united into gynostegium adnate to corolla tube; gynostegial corona of 1 whorl of 5 free, laminar segments; anthers adnate to style, locules 2; pollen in each theca massed into rigid, vertically oriented pollinium, pollinia lacrimiform, joined from adjacent anthers by translators to common corpusculum and together forming pollinarium. **Fruits** follicles, typically solitary, variously oriented, brown to dark brown, fusiform terete or subterete, glabrous or puberulent. **Seeds** ovate, flattened, winged, not beaked, comose, not arillate.

Species ca. 80 (7 in the flora): s United States, Mexico, West Indies, Central America, South America.

R. E. Woodson Jr. (1941) adopted a very broad (and evidently polyphyletic) circumscription of *Cynanchum*, including *Metastelma*. The microphyllous habit and minute flowers of *Metastelma* and *Orthosia* readily separate these genera from other Apocynaceae in the flora area, including

Cynanchum. The phylogenetic separation of *Cynanchum*, *Metastelma*, and *Orthosia* has been demonstrated by S. Liede et al. (2005) among others. *Orthosia scoparia* has been included in *Metastelma* by some; see the discussion of *Orthosia* for features distinguishing these genera.

All but one species of *Metastelma* in the flora area has a mixed indumentum of flattened, opaque hairs toward the adaxial apex of the corolla lobes and stiff, acicular, translucent hairs toward the center and base of the lobes pointing downward into the tube. This distinctive indumentum provides an immediately diagnostic recognition feature for *Metastelma*; however, *M. bahamense* lacks the stiff, acicular hairs. In some of the species, the apical, flattened hairs are long and form a brush; in others, these hairs are quite short and may appear granular.

SELECTED REFERENCES Liede, S. and U. Meve. 2004. Revision of *Metastelma* (Apocynaceae–Asclepiadoideae) in southwestern North America and Central America. Ann. Missouri Bot. Gard. 91: 31–86. Liede, S. et al. 2014. Phylogenetics and biogeography of the genus *Metastelma* (Apocynaceae–Asclepiadoideae–Asclepiadeae: Metastelmatinae). Syst. Bot. 39: 594–612.

1. Adaxial surface of corolla lobe apices puberulent with short, matted, flattened hairs.
   2. Leaves 1–2.5 mm wide; style apex columnar; internodes densely puberulent in single line; Arizona, New Mexico . . . . . 5. *Metastelma mexicanum*
   2. Leaves 2–15 mm wide; style apex convex or planar; internodes glabrous or glabrate; Florida, Texas.
      3. Corollas green abaxially, lobes 2–3 mm, without hispid, acicular, translucent hairs adaxially; Florida . . . . . 2. *Metastelma bahamense*
      3. Corollas cream to yellowish cream abaxially, lobes 1.5–2 mm, bearing both opaque and hispid, acicular, translucent hairs adaxially; Texas . . . . . 6. *Metastelma palmeri*
1. Adaxial surface of corolla lobe apices villous with long, erect, flattened hairs.
   4. Corolla lobes erect or slightly flared at apex, opening obscured by hairs, corona segments greater than 2 times as long as gynostegium . . . . . 7. *Metastelma pringlei*
   4. Tips of corolla lobes spreading and reflexed, opening not obscured, corona segments as long as or slightly longer than gynostegium.
      5. Internodes densely puberulent in single line; calyx lobes linear-lanceolate, puberulent; Arizona . . . . . 1. *Metastelma arizonicum*
      5. Internodes glabrous or glabrate; calyx lobes ovate, oblong, or lanceolate, glabrous; Florida, Texas.
         6. Leaf blades linear, oblong, or lanceolate, (1–)6–18 mm wide, laminar colleters present; peduncles 0–0.5 mm; corollas cream abaxially; Texas . . . . . 3. *Metastelma barbigerum*
         6. Leaf blades linear-lanceolate, 3–6 mm wide, laminar colleters absent; peduncles 0.5–1.5 mm; corollas green abaxially; Florida . . . . . 4. *Metastelma blodgettii*

1. Metastelma arizonicum A. Gray, Proc. Amer. Acad. Arts 19: 85. 1883 [F]

Cynanchum arizonicum (A. Gray) Shinners; *Metastelma watsonianum* Standley

Stems woody and corky at base, densely puberulent in single line. **Leaves:** 1 stipular colleter on each side of petiole; petiole 1–2 mm, puberulent; blade linear to narrowly lanceolate, 7–20 × 1.5–4 mm, chartaceous, single-veined, base rounded, with 1–2 laminar colleters, margins revolute, exaggerated on drying, apex acute, mucronulate, surfaces glabrous abaxially, minutely puberulent adaxially, primarily on midvein and near margins. **Inflorescences** umbelliform, sessile, 3–4-flowered. **Pedicels** 1–1.5 mm, puberulent. **Flowers:** calyx lobes linear-lanceolate, apices attenuate, 2–2.5 mm, puberulent, with 1 colleter per sinus; corolla cream, campanulate, tube 0.5–1 mm, lobes erect with spreading and recurved tips, opening not obscured, linear-lanceolate, 2.5–3.5 mm, abaxially glabrous, adaxially densely long-villous marginally and apically with erect, opaque, flattened hairs, densely hispid centrally with downward-pointing acicular translucent hairs; corona segments united to base of column, laminar, subulate, 1.5–2 mm, slightly longer than gynostegium; gynostegial column 0.5–0.7 mm; style apex convex. **Follicles** dark brown, 4–6 × 0.3–0.5 cm, apically acuminate, glabrous. **Seeds** 10–16, brown, flat to somewhat navicular, 5–6 × 2–3 mm, broadly winged, minutely tuberculate; coma white, 2–2.5 cm.

Flowering year-round, especially following winter and summer rains; fruiting year-round. Rocky substrates, valleys, canyons, alluvial plains, arroyos, hills, desert scrub (desert grasslands); 400–1400 m; Ariz.; Mexico (Sonora).

Metastelma arizonicum is a characteristic species of the upland margins of the Sonoran Desert. In Arizona, it is known from Cochise, Maricopa, Pima, and Pinal counties. Populations are widely spaced, and the plants are easily overlooked.

2. **Metastelma bahamense** Grisebach, Cat. Pl. Cub. 174. 1866 • Fragrant swallow-wort

Cynanchum northropiae (Schlechter) Alain; *Epicion northropiae* (Schlechter) Small; *Metastelma northropiae* Schlechter

Stems herbaceous, glabrate, sparsely puberulent in 1 line on new growth. **Leaves:** 1 stipular colleter on each side of petiole and 1–3 additional interpetiolar colleters; petiole 3–5 mm, glabrous or sparsely puberulent; blade lanceolate to ovate, 15–40 × 7–15 mm, chartaceous, venation faintly pinnate, base rounded, with 2 laminar colleters, margins planar, apex acute, mucronulate, surfaces glabrous, midvein adaxially or margins sometimes remotely puberulent. **Inflorescences** racemiform, 4–10-flowered; peduncle 2–6 mm, glabrous or sparsely puberulent. **Pedicels** 2–5 mm, sparsely puberulent. **Flowers:** calyx lobes lanceolate to ovate, apices obtuse to acute, 0.5–1.5 mm, margins scarious, sparsely puberulent at base to glabrate, margins sometimes remotely ciliate, colleters apparently absent; corolla green abaxially, white adaxially, campanulate, tube 1.5 mm, lobes erect with spreading tips, opening not obscured, oblong, 2–3 mm, abaxially glabrous, adaxially densely short-puberulent marginally and apically with opaque, matted, flattened hairs, acicular hairs absent; corona segments united to base of anthers, laminar, subulate, 1–1.5 mm, slightly longer than gynostegium; gynostegial column 1.5–2 mm; style apex planar. **Follicles** dark brown, 5–6 × 0.4–0.7 cm, apically acuminate, glabrous. **Seeds** 10–16, brown, flat to somewhat navicular, 5–6 × 2–3 mm, broadly winged, smooth; coma white, 2–2.5 cm.

Flowering nearly year-round; fruiting Nov–Feb, May. Beaches, sand ridges, edges of mangroves, limestone, scrub, pine woods, cactus hammocks; 0 m; Fla.; West Indies (Bahamas, Cuba).

Metastelma bahamense has been treated in the flora area most commonly as *Cynanchum northropieae.* However, the type of *M. northropiae* appears to be conspecific with *M. bahamense* (as evident by the annotations of R. Mangelsdorff). In Florida, *Metastelma bahamense* is restricted to just four counties: Brevard, Indian River, Miami-Dade, and Monroe.

3. **Metastelma barbigerum** Scheele, Linnaea 21: 760. 1849

Cynanchum barbigerum (Scheele) Shinners

Stems woody, not corky at base, glabrous. **Leaves:** 1 stipular colleter on each side of petiole; petiole 1.5–4 mm, glabrous; blade linear, oblong, or lanceolate, 10–45 × 1–18 mm, chartaceous, faintly pinnately veined, base rounded to subcordate, with 1–2 laminar colleters, margins revolute on drying, apex acute to attenuate, mucronulate, surfaces glabrous abaxially, pilosulous on midvein adaxially. **Inflorescences** umbelliform to shortly racemiform, 3–6-flowered; peduncle 0–0.5 mm. **Pedicels** 1–3 mm, glabrous. **Flowers:** calyx lobes ovate, lanceolate, or oblong, apices obtuse to acute, 0.5–1 mm, glabrous, margins scarious, with 1 colleter per sinus; corolla cream to yellowish cream, campanulate, tube 1 mm, lobes erect with spreading and recurved tips, opening not obscured, linear-lanceolate, 2.5–3.5 mm, abaxially glabrous, adaxially densely long-villous marginally and apically with erect, opaque, flattened hairs, densely hispid centrally with downward-pointing acicular translucent hairs; corona segments united to base of column, laminar, acicular, 1–1.5 mm, ± as long as gynostegium; gynostegial column 0.5–0.7 mm; style apex convex. **Follicles** dark brown, 4–6 × 0.4–0.7 cm, apically acuminate, glabrous. **Seeds** 10–16, brown, flat to somewhat navicular, 5–6 × 2–3 mm, broadly winged, smooth; coma white, 2–2.5 cm. 2*n* = 22.

Flowering Mar–Nov; fruiting Apr–Dec. Ridge tops, cliffs, stream banks, edges of marshes, coastal dunes, disturbed areas, limestone, granite, clay, silt or sand soils, thorn scrub, mesquite grasslands, oak woods; 0–800 m; Tex.; Mexico (Coahuila, Nuevo León, Tamaulipas).

Metastelma barbigerum is a morphologically and ecologically variable species. It is the most commonly encountered species of *Metastelma* in Texas. Plants on the Edwards Plateau are commonly found in more mesic microsites and have much larger leaves than plants found near the coast or in the Rio Grande valley. This name is commonly misapplied to specimens of other species in the southwestern United States and northern Mexico.

4. **Metastelma blodgettii** A. Gray, Proc. Amer. Acad. Arts 12: 73. 1876 (as blodgetti) [C]

Cynanchum blodgettii (A. Gray) Shinners

Stems herbaceous, glabrate, sparsely puberulent in 1 line on new growth. **Leaves:** 1 stipular colleter on each side of petiole; petiole 1.5–3 mm, glabrous or sparsely puberulent; blade linear-lanceolate, 10–35 × 3–6 mm, chartaceous, with single vein or obscurely pinnate, base rounded, laminar colleters absent, margins sometimes revolute on drying, apex acute to attenuate, mucronulate, surfaces glabrous, margins sometimes remotely puberulent. **Inflorescences** umbelliform, 3–5-flowered; peduncle 0.5–1.5 mm, puberulent. **Pedicels** 1.5–2 mm, sparsely puberulent to glabrate. **Flowers:** calyx lobes ovate, apices obtuse to acute, 0.5–1 mm, glabrous, margins scarious, colleters apparently absent; corolla green abaxially, white adaxially, urceolate, tube 1 mm, lobes erect with spreading tips, opening obscured by hairs, lanceolate, 2–2.5 mm, abaxially glabrous, adaxially densely villous marginally and apically with erect, opaque, flattened hairs, densely hispid centrally with downward-pointing acicular translucent hairs; corona segments united to base of column, laminar, subulate, 1–1.5 mm, slightly longer than gynostegium; gynostegium sessile; style apex convex. **Follicles** dark brown, 4–5 × 0.4–0.7 cm, apically acuminate, glabrous. **Seeds** 10–16, brown, flat to somewhat navicular, 5–6 × 2–3 mm, broadly winged, smooth; coma white, 2–2.5 cm.

Flowering nearly year-round; fruiting Mar–May, Sep–Oct. Limestone, pine rocklands, pine-palmetto scrub; of conservation concern; 0 m; Fla.; West Indies (Bahamas, Cuba).

In Florida *Metastelma blodgettii* is restricted to Miami-Dade County and Monroe County keys.

5. **Metastelma mexicanum** (Brandegee) Fishbein & R. A. Levin, Madroño 44: 270. 1997

Melinia mexicana Brandegee, Zoë 5: 216. 1905; *Basistelma angustifolium* Bartlett; *B. mexicanum* (Brandegee) Bartlett; *Cynanchum wigginsii* Shinners; *Pattalias angustifolius* S. Watson

Stems woody and corky at base, densely puberulent in single line. **Leaves:** 1 stipular colleter on each side of petiole; petiole 1–2 mm, glabrous or minutely puberulent; blade linear, 15–40 × 1–2.5 mm, chartaceous, single-veined, base cuneate to rounded, with 1–2 laminar colleters, margins revolute, exaggerated on drying, apex acute, mucronulate, surfaces glabrous abaxially, minutely puberulent adaxially, primarily on midvein. **Inflorescences** umbelliform, 2–6-flowered; peduncle 0–2 mm, densely puberulent. **Pedicels** 2–3 mm, densely puberulent. **Flowers:** calyx lobes linear, apices attenuate, 2–3 mm, puberulent, with 1 colleter per sinus; corolla cream, campanulate, tube 0.5–1 mm, lobes erect with spreading and recurved tips, opening not obscured, linear-lanceolate, 2–3 mm, abaxially glabrous, adaxially densely short-puberulent apically with opaque, matted, flattened hairs, densely hispid centrally with downward-pointing acicular translucent hairs; corona segments united to base of column, laminar, linear-elliptic, 1–1.5 mm, slightly shorter than gynostegium; gynostegium subsessile; style apex columnar. **Follicles** brown, 4–6 × 0.3–0.5 cm, apically acuminate, glabrous. **Seeds** 10–16, brown, flat, narrowly winged, minutely tuberculate; coma white, 1–2 cm.

Flowering Aug–Oct; fruiting Sep–Dec. Hillsides, steep slopes, canyons, rock outcrops, rhyolite, granite, oak woodlands, grasslands, chaparral; 1200–1800 m; Ariz., N.Mex.; Mexico (Sinaloa, Sonora).

Metastelma mexicanum is restricted in the flora area to lower mountain slopes, valleys, and hills in southeastern Arizona (Cochise and Santa Cruz counties) and extreme southwestern New Mexico (Hidalgo County). It is uncommon throughout its range, and reevaluation of its conservation status is warranted. The columnar style apex is unusual in *Metastelma*, but otherwise the morphology of the species is unremarkable in the genus.

6. **Metastelma palmeri** S. Watson, Proc. Amer. Acad. Arts 18: 115. 1883

Cynanchum maccartii Shinners

Stems woody, not corky at base, glabrous. **Leaves:** 1 stipular colleter on each side of petiole, sometimes with 1 additional interpetiolar colleter; petiole 2–5 mm, sparsely puberulent; blade linear-lanceolate, elliptic, or lanceolate, 10–30 × 2–8 mm, chartaceous, faintly pinnately veined, base rounded to cuneate, with 2 laminar colleters, margins planar, apex acute to rounded, mucronulate, surfaces glabrous abaxially, pilosulous on midvein adaxially, margins sometimes sparsely pilosulous. **Inflorescences** shortly racemiform, 3–8-flowered; peduncle 0–2 mm, puberulent. **Pedicels** 1–2 mm, glabrous. **Flowers:** calyx lobes lanceolate to ovate, apices obtuse to acute, 1–1.5 mm, margins scarious, with 1 colleter per sinus; corolla cream to yellowish cream, campanulate, tube 0.5 mm, lobes erect with spreading tips, opening not obscured, linear, 1.5–2 mm, abaxially glabrous, adaxially densely, short-puberulent marginally and apically with opaque, matted, flattened hairs, densely hispid centrally with downward-pointing acicular, translucent hairs; corona segments united to base of anthers, laminar, lanceolate, 1 mm, slightly longer than gynostegium; gynostegium sessile; style apex convex. **Follicles** dark brown, 4–7 × 0.4–0.7 cm, apically acuminate, glabrous. **Seeds** 8–12, brown, flat, 4–6 × 2–3 mm, winged, with minute papillae forming ridges, chalazal margin erose; coma white, 2–3 cm.

Flowering Apr–Oct; fruiting Jun–Oct. Valleys, plains, canyons, limestone, rocky, caliche, clay, or sandy soils, thorn scrub, oak-juniper woodland; 100–600 m; Tex.; Mexico (Coahuila, Nuevo León, Tamaulipas).

Metastelma palmeri overlaps in range with *M. barbigerum* in southern Texas. It is commonly encountered in the blackbrush thorn scrub of the Rio Grande valley and southwestern margin of the Edwards Plateau.

7. **Metastelma pringlei** A. Gray, Proc. Amer. Acad. Arts 21: 397. 1886

Cynanchum barbigerum (Scheele) Shinners var. *breviflorum* Shinners; *C. pringlei* (A. Gray) Henrickson

Stems woody, not corky at base, minutely puberulent in single line to glabrate. **Leaves:** 1 stipular colleter on each side of petiole; petiole 1.5–5 mm, sparsely puberulent; blade linear-lanceolate to lanceolate, 7–20 × 3–7 mm, chartaceous, single-veined to faintly pinnate, base cuneate to rounded, with 2 laminar colleters, margins revolute, exaggerated on drying, apex acute to attenuate, mucronulate, surfaces glabrous abaxially, minutely puberulent adaxially on midvein and near margins. **Inflorescences** umbelliform, 1–4-flowered; peduncle 0–0.5 mm. **Pedicels** 1.5–2.5 mm, puberulent. **Flowers:** calyx lobes linear-lanceolate, apices acute to attenuate, 1–1.5 mm, puberulent to glabrate, with 1 colleter per sinus; corolla cream to yellowish cream, campanulate, tube 0.5–1 mm, lobes erect or slightly flared at apex, opening obscured by hairs, lanceolate, 2–3 mm, abaxially glabrous, adaxially densely long-villous marginally and apically with erect, opaque, flattened hairs, densely hispid centrally with downward-pointing acicular translucent hairs; corona segments united to base of column, laminar, acicular, 2–2.5 mm, greater than 2 times as long as gynostegium; gynostegial column subsessile; style apex convex. **Follicles** dark brown, 4–6 × 0.4–0.5 cm, apically acuminate, glabrous. **Seeds** 10–16, brown, flat, 5–6 × 2–3 mm, broadly winged, smooth; coma white, 2–2.5 cm.

Flowering May–Oct; fruiting Jun–Oct. Mountain slopes, cliffs, ridgetops, canyons, limestone, granite, rocky soils, desert scrub, thorn scrub, oak woods, grasslands; 1000–1700 m; Tex.; Mexico.

Metastelma pringlei has the longest corona segments of any *Metastelma* in the flora area. It is the characteristic representative of the genus in the Chihuahuan Desert portion of the Big Bend region.

32. ORTHOSIA Decaisne in A. P. de Candolle and A. L. L. P. de Candolle, Prodr. 8: 526. 1844 • Swallow-wort, milkweed vine [Greek *Orthosie*, goddess of prosperity, and one of the Horae, daughters of Zeus and Themis]

Mark Fishbein

Amphistelma Grisebach

Vines herbaceous, woody only at base [scandent]; latex white. **Stems** twining, unarmed, glabrous. **Leaves** caducous [persistent], opposite, petiolate; stipular colleters interpetiolar; laminar colleters present or absent. **Inflorescences** axillary (extra-axillary) cymes, umbelliform to shortly racemiform [paniculiform], sessile [pedunculate]. **Flowers:** calycine colleters apparently absent; corolla yellow to light green (cream), campanulate [rotate or tubular], aestivation valvate; corolline corona absent; androecium and gynoecium united into a gynostegium adnate to corolla tube; gynostegial corona of 1 whorl of 5 segments connate at base; anthers adnate to style, locules 2; pollen in each theca massed into a rigid, vertically oriented pollinium, pollinia lacrimiform, joined from adjacent anthers by translators to common corpusculum and together forming a pollinarium. **Fruits** follicles, typically paired, variously oriented, brown, linear, smooth, glabrous. **Seeds** naviculate, not winged, linear, comose, not arillate. *x* = 10.

Species ca. 40 (1 in the flora): se United States, Mexico, West Indies, Central America, South America.

R. E. Woodson Jr. (1941) adopted a very broad (and evidently polyphyletic) circumscription of *Cynanchum*, including *Orthosia*. The isolated placement of *Orthosia* relative to *Cynanchum* and presumed relatives such as *Metastelma* was demonstrated by S. Liede et al. (2005). Superficially, the species are very similar to those of *Metastelma* in the small, linear leaves and tiny cream to yellow flowers. However, the often caducous leaves; deltate, glabrous corolla lobes; and commonly paired inflorescences and follicles provide a ready means in our region to distinguish *Orthosia* from *Metastelma*.

1. **Orthosia scoparia** (Nuttall) Liede & Meve, Novon 18: 207. 2008 • Leafless swallow-wort [F]

Cynanchum scoparium Nuttall, Amer. J. Sci. Arts 5: 291. 1822; *Amphistelma scoparium* (Nuttall) Small; *Metastelma scoparium* (Nuttall) Vail

Stems woody at base, not corky. **Leaves** caducous, with 1 stipular colleter on each side of petiole; petiole 1–5 mm, adaxially grooved, glabrous or sparsely puberulent; blade single-veined to faintly pinnate, linear, 10–50 × 1.5–4 mm, chartaceous, base cuneate, tapering to petiole and conduplicate, laminar colleters absent or hidden in groove, margins planar, apex acute, mucronulate, surfaces glabrous abaxially, minutely puberulent on midvein adaxially, margins remotely puberulent. **Inflorescences** 4–8-flowered. **Pedicels** 2.5–4 mm, pilosulous. **Flowers:** calyx lobes ovate, apices obtuse, 0.5–1 mm, pilosulous to usually glabrate, margins scarious; corolla tube 0.5 mm, lobes erect with spreading tips, linear, 2–2.5 mm, glabrous; corona segments basally connate and united to base of column, yellow to green, laminar, ovate, 0.5 mm, much shorter than gynostegium; gynostegium sessile; style apex planar. **Follicles** 4–7 × 0.2–0.4 cm, apically acuminate. **Seeds** 6–10, 5 mm; coma 2–2.5 cm.

Flowering nearly year-round; fruiting (Oct–)Jan–Feb(–Apr). Moist or dry hammocks, salt marshes, baygalls, swamp margins, canal banks, limestone, shell mounds, sandy soils; 0–50 m; Fla., Ga., S.C.; West Indies; South America.

Plants of *Orthosia scoparia* may appear densely leafy or completely leafless and may bear flowers in either condition. The flowers of *O. scoparia* are smaller than those of the regionally co-occurring species of *Metastelma* and *Pattalias* and, unlike those in *Metastelma*, the corollas of *O. scoparia* are completely glabrous. *Orthosia scoparia* just enters South Carolina (Beaufort County), where it is rare and considered to be of conservation concern.

33. PATTALIAS S. Watson, Proc. Amer. Acad. Arts 24: 60. 1889 • Milkvine, swallow-wort [Greek *pattalias*, pricket or stake, alluding to stout, conical beak surmounting stigma in *P. palmeri*, the type species]

Mark Fishbein

Vines, herbaceous [woody at base]; latex white. **Stems** twining vigorously, unarmed, glabrous. **Leaves** persistent, opposite, sessile; stipular colleters interpetiolar; laminar colleters apparently absent. **Inflorescences** extra-axillary, solitary at nodes, umbelliform, pedunculate. **Flowers:** calycine colleters present; corolla green or cream, often brown- or purple-tinged, campanulate, aestivation contort-dextrorse; corolline corona absent; androecium and gynoecium united into a gynostegium adnate to corolla tube; gynostegial corona of 1 whorl of 5 laminar, somewhat fleshy segments opposite stamens, connate only at very base, united to column at base of anthers; anthers adnate to style, locules 2; pollen in each theca massed into a rigid, vertically oriented pollinium, pollinia lacrimiform, joined from adjacent anthers by translators to a common corpusculum and together forming a pollinarium. **Fruits** follicles, typically solitary, variously oriented, brown, narrowly fusiform, terete, smooth, glabrous. **Seeds** winged, not beaked, ovate, flattened, comose, not arillate.

Species 2 or 3 (1 in the flora): se United States, n, se Mexico, West Indies, Central America (Belize).

Formerly included in a broadly circumscribed, polyphyletic *Cynanchum*, especially following the influential revision of North American milkweed genera by R. E. Woodson Jr. (1941), *Pattalias* is placed close to or within *Funastrum* in phylogenetic analyses (S. Liede and A. Täuber 2002). Further study may support submersion in *Funastrum*, but the morphological distinctiveness of *Pattalias* and absence of definitive phylogenetic evidence warrant maintenance of this genus (M. Fishbein and W. D. Stevens 2005). Fishbein and Stevens misinterpreted the priority of names pertaining to this genus and placed the species of *Pattalias* in *Seutera* Reichenbach, but this latter name is illegitimate (Fishbein 2017).

1. **Pattalias paluster** (Pursh) Fishbein, Phytologia 99: 87. 2017 (as palustre) • Gulf Coast or marsh swallow-wort [F]

Ceropegia palustris Pursh, Fl. Amer. Sept. 1: 184. 1813; *Cynanchum angustifolium* Persoon 1805, not *P. angustifolius* S. Watson 1889; *C. palustre* (Pursh) A. Heller; *Funastrum angustifolium* (Persoon) Liede & Meve; *Lyonia palustris* (Pursh) Small; *Seutera angustifolia* (Persoon) Fishbein & W. D. Stevens; *S. palustris* (Pursh) Vail

Leaves drooping, linear, 2–8 × 0.1–0.6 cm, fleshy, apex acute, base cuneate. **Inflorescences** 7–20-flowered; peduncles 1–5 cm. **Pedicels** 3–5 mm. **Flowers:** calyx lobes lanceolate, acute, 1–2.5 mm, ciliate, 1 colleter present adaxially at each sinus; corolla glabrous, tube ca. 0.5 mm, lobes spreading to reflexed distally, lanceolate to narrowly lanceolate, 3–4 mm, apex acute to attenuate; corona segments white, rectangular to oblong, 1.5–2 mm, apex 2-dentate to truncate; style apex 2-lobed; ovules ca. 20. **Follicles** 4–8.5 × 0.5–1 cm. **Seeds** 10–20, light brown, 5–6 × 4–5 mm, coma white, 2–3 cm.

Flowering (Feb–)Apr–Oct(–Nov); fruiting (Feb–) May–Nov. Salt marshes, coastal dunes and flats, coastal hammocks, edges of mangroves, roadsides; 0 m; Ala., Fla., Ga., La., Miss., N.C., S.C., Tex.; Mexico (Coahuila, Yucatán); West Indies; Central America (Belize).

Although *Cynanchum angustifolium* Persoon is the oldest basionym applicable to this species, the combination in *Pattalias* is precluded by *P. angustifolius* S. Watson, a synonym of *Metastelma mexicanum*. In the flora area, *P. paluster* is restricted to coastal strand, marshes, and estuaries, where it is quite likely to be the only milkweed vine present. Rarely, *Orthosia scoparia* may occur in close proximity, but that species has much smaller flowers and is commonly leafless.

34. PHEROTRICHIS Decaisne, Ann. Sci. Nat., Bot., sér. 2, 9: 322. 1838 • [Greek *phero*, bear, and *thrix*, hair, alluding to vestiture throughout, especially on corolla lobes]

Angela McDonnell

Mark Fishbein

Herbs; latex white, dilute. **Stems** usually solitary, erect, unbranched, unarmed, densely hirsute with long, eglandular and minutely glandular trichomes. **Leaves** persistent, opposite, petiolate, surfaces densely hirsute with long, eglandular and minutely glandular trichomes; stipular colleters absent; laminar colleters present. **Inflorescences** extra-axillary, cymose, umbelliform, solitary at nodes, sessile or pedunculate. **Flowers:** calycine colleters apparently absent; corolla cream with

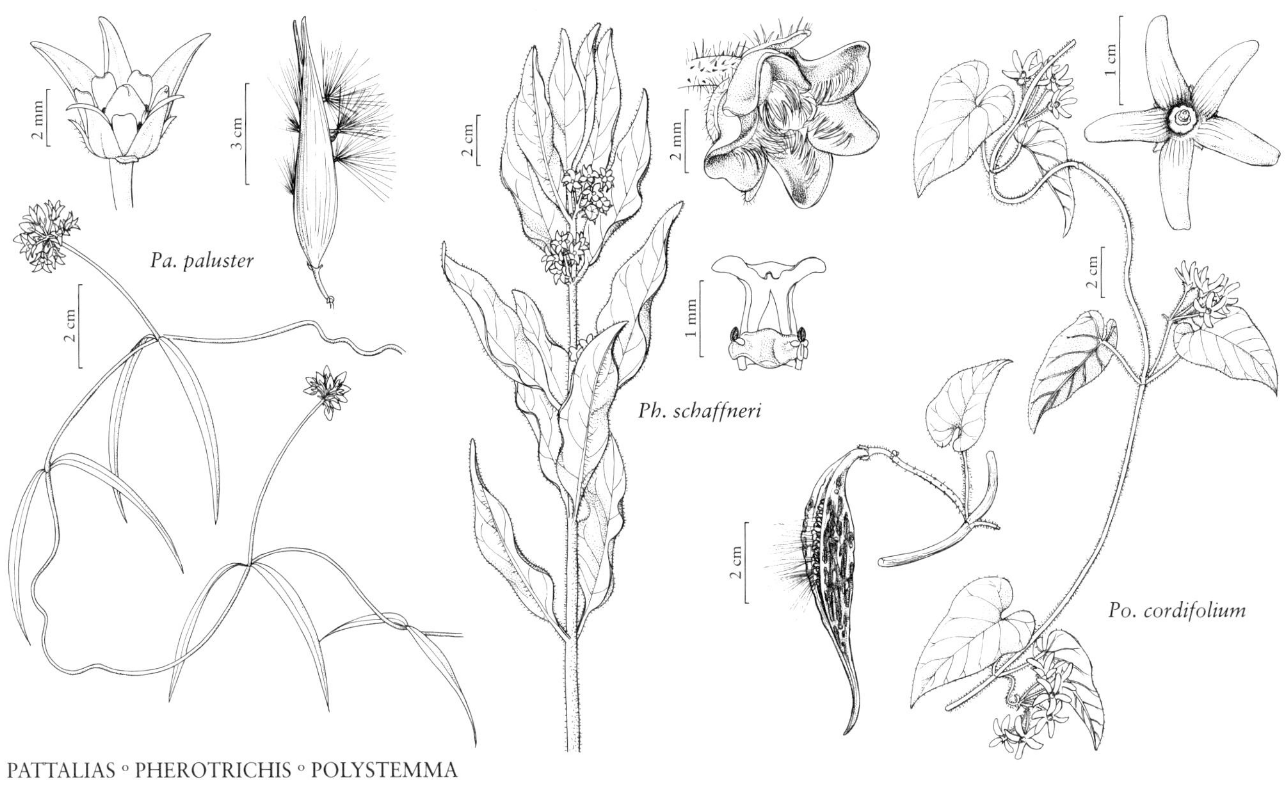

PATTALIAS ◦ PHEROTRICHIS ◦ POLYSTEMMA

longitudinal green lines, urceolate [campanulate], lobes erect to spreading, aestivation contort-dextrorse; coralline corona absent; androecium and gynoecium united into a gynostegium adnate to corolla tube; gynostegial corona of 5 laminar segments opposite stamens, distinct, fused to gynostegium at base of anthers; anthers adnate to style, apex with laminar appendages covering margins of style apex, locules 2; pollen in each theca massed into a rigid, horizontally oriented pollinium, pollinia strongly curved, bearing sterile, navicular portion adjacent to translator, pollinia of adjacent anthers joined by translators to common corpulsculum, together forming a pollinarium. **Fruits** follicles, typically solitary, sharply deflexed, fusiform, surface smooth, densely villous or hirsute, tardily glabrate. **Seeds** winged, not beaked, ovate, flattened, comose, not arillate.

Species 5–9 (1 in the flora): Arizona, Mexico, Central America (Guatemala).

Pherotrichis was formerly included in a broadly circumscribed, paraphyletic *Matelea* and is distinctive in subtribe Gonolobinae (which also includes *Gonolobus* and *Matelea*) by the erect, non-twining habit and the densely barbate corolla lobes. The removal of *Pherotrichis* from *Matelea* is but one of several generic realignments necessary to circumscribe a monophyletic *Matelea*. Species of *Pherotrichis* are very commonly mistaken for *Asclepias* but are readily distinguished by the urceloate (or campanulate in extra-regional species) corollas with villous lobes concealing the minute corona and by the sharply deflexed follicles.

SELECTED REFERENCE Lozado Pérez, L. 2003. Sistematica de *Pherotrichis* Decne. (Apocynaceae, Asclepiadoideae). M.S. thesis. Universidad Nacional Autónoma de México.

1. **Pherotrichis schaffneri** A. Gray, Syn. Fl. N. Amer. ed. 2, 2(1): 462. 1886 [F]

Herbs 2–30 cm. **Leaves:** petiole 1–2 cm; blade narrowly lanceolate to ovate, 3.5–9 × 1–5 cm, base rounded to truncate or shallowly cordate, margins ciliate, apex acute to obtuse, surfaces more densely eglandular-hairy on veins and margins, glandular hairs restricted to veins and margins, laminar colleters 6–12. **Inflorescences** 3–10-flowered; peduncle 0–7 mm. **Pedicels** 0.5–1 cm. **Flowers:** calyx lobes slightly recurved, lanceolate, 3–4 mm; corolla scarcely reticulate distally at margins, tube 1.5–2 mm, glabrous, lobes narrowly lanceolate, 3–4 mm, abaxially sparsely to moderately villous, adaxially densely villous; corona segments much shorter than style apex, apex truncate with 2 marginal teeth; apical anther appendages lanceolate; style apex elongate, yellow, domelike, 1–1.5 mm, shallowly 2-lobed. **Follicles** unknown.

Flowering Jul–Sep. Terrace of wide canyon bottom at base of mountains, pine-oak-juniper woodlands; 1700 m; Ariz.; c, n Mexico.

Pherotrichis schaffneri is known in the flora area only from the southern Huachuca Mountains in Cochise County. It had been documented by a single 1882 collection (*Lemmon 2816*), and the existence of *P. schaffneri* in the United States was questioned until it was rediscovered by J. Ksepka in 2015 (*Ksepka s.n.* [OKLA]), 133 years later. The site is an unremarkable patch of open woodland on a lower canyon slope, surrounded by thousands of acres of seemingly suitable habitat. Despite the apparent rarity of *P. schaffneri* in Arizona, it has not been afforded conservation status, in part because Lemmon's locality in the United States was unconfirmed and also because of unresolved taxonomic questions (see below). Although *P. schaffneri* is not uncommon in the northern Sierra Madre Occidental in Mexico, it is exceedingly rare in the United States, where conservation status seems warranted.

The Lemmon collection from the Huachuca Mountains served as a syntype for *Pherotrichis schaffneri*; however, this name was lectotypified by another syntype (*Schaffner 63*, lectotype GH) from San Luis Potosí, Mexico (W. D. Stevens 2005). The names *Matelea balbisii* (A. Gray) Woodson and *P. balbisii* A. Gray are most commonly applied to *P. schaffneri* throughout its range but are illegitimate as they are based on *Asclepias villosa* Balbis, a later homonym of *A. villosa* Miller, as shown by Stevens. It is questionable whether the population in Arizona and morphologically similar populations in Chihuahua, Durango, and Sonora, Mexico, are conspecific with the populations in San Luis Potosí represented by the type of *P. schaffneri*. Additional work building on the revisionary study by L. Lozado Pérez (2003) will be required to resolve this question.

35. POLYSTEMMA Decaisne in A. P. de Candolle and A. L. P. P. de Candolle, Prodr. 8: 602. 1844 • Milkweed vine [Greek *polys*, many, and *stemma*, garland or wreath, alluding to numerous filiform divisions of corona segments in *P. viridiflorum*, the type species]

Mark Fishbein

Angela McDonnell

Labidostelma Schlechter; *Rothrockia* A. Gray

Vines or scandent herbs, woody at base; latex white. **Stems** weakly to strongly twining, unarmed, mixed indumentum of long, eglandular trichomes and short, glandular trichomes that become filled with white-crystalline inclusions with age. **Leaves** drought-deciduous [persistent], opposite, petiolate; stipular colleters interpetiolar; laminar colleters present. **Inflorescences** extra-axillary, umbelliform or racemiform, pedunculate. **Flowers:** calycine colleters apparently absent; corolla cream [green, brown, purple, orange, yellow], tubular-campanulate [rotate], aestivation contort-dextrorse; corolline corona absent; androecium and gynoecium united into a gynostegium adnate to corolla tube; gynostegial corona of 1 whorl of 5 segments united at base; anthers adnate to style, locules 2; pollen in each theca massed into a rigid, horizontally oriented pollinium, pollinia lacrimiform, each with proximal, sterile, excavated region, joined from adjacent anthers by translators to common corpusculum and together forming a pollinarium. **Fruits** follicles, solitary, variously oriented, striate, lance-ovoid to fusiform, smooth, glabrous. **Seeds** winged, not beaked, ovate, flattened, comose, not arillate.

Species ca. 20 (1 in the flora): sw Arizona, Mexico, Central America.

The combination of glandular trichomes that develop crystalline inclusions with age and long, smooth, striate follicles is diagnostic for *Polystemma* and distinguishes this genus from *Matelea* (W. D. Stevens 2009). However, most species with these characteristics remain in *Matelea*, where they were placed by R. E. Woodson Jr. (1941). The genus as circumscribed here is supported by molecular phylogenetic evidence (A. McDonnell et al. 2018). A single species of *Polystemma* barely enters the flora region in southernmost Arizona.

A second species of *Polystemma* was discovered in the United States by S. Carnahan in 2019 in the southern foothills of the Santa Rita Mountains, Pima County, Arizona (*Carnahan 3807* [OKLA]). The species, which is distributed south to central Sonora, Mexico, is in the process of being described. It is easily distinguished from *P. cordifolium* by its corollas, which are deep purple to reddish brown with spreading lobes, and by its flat, pentagonal style apex. The species will key to *Polystemma* in the key to genera, except for the style apex character that points instead to *Matelea*. However, like *P. cordifolium*, this species has the characteristic glandular trichomes with crystalline inclusions that are lacking in *Matelea*.

1. Polystemma cordifolium (A. Gray) McDonnell & Fishbein, Phytologia 99: 86. 2017 • Sonoran milkweed vine [F]

Rothrockia cordifolia A. Gray, Proc. Amer. Acad. Arts 20: 295. 1885; *Matelea cordifolia* (A. Gray) Woodson

Stems: bark tan, corky, shallowly furrowed. **Leaves:** 1 stipular colleter on each side of petiole; petiole 1–4.5 cm; blade ovate, 2.5–8 × 2–6 cm, chartaceous, base deeply cordate, laminar colleters 2–4, margins planar, apex acute to acuminate, venation pinnipalmate, surfaces densely puberulent to hirsute, paler abaxially. **Inflorescences** umbelliform or racemiform, 4–6-flowered; peduncle 1–4 cm. **Pedicels** 5–10 mm. **Flowers:** calyx lobes lanceolate-ovate, apices acute, 3–5 mm, hirsutulous; corolla tube 3–4 mm, lobes ascending, sometimes twisted, linear-lanceolate to oblong, 6–12 mm, glabrous; corona segments basally connate and united to base of column, green, laminar, fleshy, ovate with minute teeth, much shorter than gynostegium; gynostegium sessile; style apex prolonged into bifid beak, papillose. **Follicles** 4.5–13.2 × 1–1.5 cm, apically attenuate. **Seeds** 100–150, reddish brown, 6 × 4 mm, granulate-tuberculate; coma white, 3–4 cm.

Flowering Jul–Sep; fruiting Oct–Mar. Arroyos, canyons, hills, slopes, plains, rock outcrops, granite, sandy and gravelly soils, desert scrub; 200–300 m; Ariz.; Mexico (Baja California, Baja California Sur, Sonora).

Polystemma cordifolium is widely distributed in the drier parts of the Sonoran Desert and reaches its northern limit in Organ Pipe Cactus National Monument (Pima County), where it may no longer be extant (R. S. Felger et al. 2014). It should be considered to be of conservation concern in the flora area.

36. VINCETOXICUM Wolf, Gen. Pl., 130. 1776 • Swallow-wort, dompte-venin [Latin *vincere*, conquer, and *toxicum*, poison, alluding to presumed medicinal benefit as a counter-poison] [I]

Mark Fishbein

Antitoxicum Pobedimova; *Cynanchum* Linnaeus sect. *Vincetoxicum* (Wolf) Tsiang & P. T. Li

Vines (erect herbs), herbaceous; latex clear. **Stems** twining, sometimes only at tips [not twining], unarmed, minutely pilosulous with eglandular trichomes in decurrent lines from nodes to glabrate [densely pubescent or glabrous]. **Leaves** persistent, opposite, petiolate (sessile); stipular colleters apparently absent; laminar colleters present or absent. **Inflorescences** axillary, cymose, pedunculate. **Flowers:** calycine colleters present; corolla pinkish tan, pale reddish brown, or dark purple [cream, yellowish], campanulate to rotate, aestivation contort-dextrorse (nearly valvate); coralline corona absent; androecium and gynoecium united into a gynostegium adnate to corolla tube; gynostegial corona annular or of 1 whorl of 5 thick, laminar to prismatic segments; anthers adnate to style, locules 2; pollen in each theca massed into a rigid, vertically oriented pollinium, pollinia lacrimiform, joined from adjacent anthers by translators to common corpusculum and together forming a pollinarium. **Fruits** follicles, solitary or paired, pendulous, narrowly lance-ovoid to fusiform, smooth, glabrous. **Seeds** winged, not beaked, ovate or lanceolate, lenticular, comose, not arillate. $x = 11$.

Species ca. 28–35 (2 in the flora): introduced; Europe, Asia.

Vincetoxicum has been treated in many regional floras in North America as a synonym of a large and polyphyletic *Cynanchum*; however, these genera have been shown to be rather distantly related (S. Liede and A. Täuber 2002; Liede et al. 2012). The circumscription adopted here corresponds in part to the *Vincetoxicum* clade of Liede et al. (2016). The later homonym *Vincetoxicum* Walter refers to distantly related species of *Gonolobus* and *Matelea* (subtribe Gonolobinae) and is a source of some taxonomic confusion because numerous American species of those genera were described under *Vincetoxicum*.

Vincetoxicum is well supported as belonging to the Tylophorinae K. Schumann (Asclepiadeae Duby). S. Liede et al. (2012, 2016) recommended uniting *Vincetoxicum* with *Tylophora* R. Brown and all other genera of Tylophorinae, except *Pentatropis* R. Brown ex Wight & Arnott, in order to resolve the non-monophyly of *Tylophora* and *Vincetoxicum*. This extreme proposal does not affect species in the flora area, but it seems prudent to exclude *Tylophora* from the synonymy of *Vincetoxicum* until a more considered approach to a revision of Tylophorinae is undertaken, weighing the advantages and disadvantages of broadly versus narrowly circumscribed genera.

Reports of *Vincetoxicum hirundinaria* Medikus [synonyms: *Cynanchum medium* (Decaisne) K. Schumann, not R. Brown, *C. vincetoxicum*, *V. medium*, *V. officinale*] are based upon plants not persisting outside cultivation or misapplication to plants of *V. rossicum*. *Vincetoxicum hirundinaria* was occasionally cultivated in the late nineteenth and early twentieth centuries at scattered sites in the northern United States and southern Canada, but none of the occurrences outside of cultivation appears to have persisted and no extant populations are known. Inconsistent recognition of the specific distinction between *V. hirundinaria* and *V. rossicum*, coupled with misapplication of the name *C. medium* to plants of *V. rossicum* in the Americas, has led to many erroneous reports of naturalized populations of *V. hirundinaria* in the flora area. For example, the concept of *V. hirundinaria* of H. A. Gleason and A. Cronquist (1991) combines attributes of both species. *Vincetoxicum hirundinaria* can be distinguished most readily from the two naturalized species in the flora area by the cream to yellowish cream corollas.

The two naturalized species in the region, *Vincetoxicum nigrum* and *V. rossicum*, are considered invasive and to be threats to native forest understory vegetation. Means to control these species are an active area of research (A. DiTomasso et al. 2005) and the continuing use of *V. nigrum* as a horticultural species should be curtailed. Monarch butterfly larvae (*Danaus plexippus*) do not complete development on *V. nigrum* or *V. rossicum*; however, there is equivocal evidence for monarch oviposition on *Vincetoxicum* (DiTomasso et al.). An unusual reproductive characteristic of many species of *Vincetoxicum*, shared by *V. nigrum* and *V. rossicum*, is polyembryonic seeds (DiTomasso et al.).

SELECTED REFERENCES DiTomasso, A., F. M. Lawlor, and S. J. Darbyshire. 2005. The biology of invasive alien plants in Canada. 2. *Cynanchum rossicum* (Kleopow) Borhidi [= *Vincetoxicum rossicum* (Kleopow) Barbar.] and *Cynanchum louiseae* (L.) Kartesz & Gandhi [= *Vincetoxicum nigrum* (L.) Moench]. Canad. J. Pl. Sci. 85: 243–263. Sheeley, S. E. and D. J. Raynal. 1996. The distribution and status of species of *Vincetoxicum* in eastern North America. Bull. Torrey Bot. Club 123: 148–156.

1. Corollas dark purple, adaxially pilosulous to hispidulous, lobes deltate, ± as long as wide, apically planar, coronas annular, shallowly lobed; peduncles 0.5–1.5 cm; seeds 6–8 mm . . . . 1. *Vincetoxicum nigrum*
1. Corollas pinkish tan to reddish brown, adaxially glabrous, lobes lanceolate, 1.5–2 times longer than wide, apically twisted, coronas of 5 prismatic segments united basally; peduncles 1.5–2.5 cm; seeds 4–6.5 mm . . . . 2. *Vincetoxicum rossicum*

1. Vincetoxicum nigrum (Linnaeus) Moench, Suppl. Meth., 313. 1802 • Black swallow-wort, dompte-venin noir [F] [I] [W]

Asclepias nigra Linnaeus, Sp. Pl. 1: 216. 1753; *Cynanchum louiseae* Kartesz & Gandhi

Stems erect to prostrate proximally, twining distally. **Leaves:** petiole 0.5–1.5 cm; blade pinnipalmately veined, lanceolate to broadly ovate, 3–12 × 1–6.5 cm, membranous, base truncate, rounded, or subcordate, with 2–8 laminar colleters, margins ciliate, apex attenuate to acuminate, surfaces pilosulous on veins abaxially, glabrous adaxially. **Inflorescences** solitary at nodes, simple or compound cymes, 4–10-flowered; peduncle 0.5–1.5 cm, pilosulous. **Pedicels** 5–6 mm, pilosulous. **Flowers:** calyx lobes lanceolate to deltate, 1–1.5 mm, apex acute, pilosulous to glabrate, margins ciliate; corolla very dark purple, rotate to rotate-campanulate, fleshy, lobes apically planar, deltate, 1.5–3 mm, ± as wide as long, apex acute, glabrous abaxially, pilosulous to hispidulous adaxially, gynostegial corona a thick, shallowly 5-lobed or crenulate ring exceeding style apex, reddish purple to dark purple; style apex depressed, umbonate, green. **Follicles** 4–8 × 0.7–1 cm, apex attenuate to acuminate. **Seeds** 7–15, brown, ovate, 6–8 × 3–4.5 mm; coma white, 2–3 cm. **2*n*** = 44.

Flowering May–Aug(–Nov); fruiting (Jun–)Jul–Oct. Disturbed areas, gardens, fences, old fields, pastures, roadsides, streamsides, ravines, slopes, beaches, railroads, limestone, igneous substrates, rocky soils, thickets, woods, grasslands; 0–400 m; introduced; Ont., Que.; Conn., Ill., Ind., Kans., Ky., Maine, Md., Mass., Mich., Mo., N.H., N.J., N.Y., Ohio, Pa., R.I., Vt., Wis.; sw Europe (France, Italy, Portugal, Spain).

Vincetoxicum nigrum is more widely known in North America by the illegitimate name *Cynanchum nigrum* (Linnaeus) Persoon, a later homonym of *C. nigrum* Cavanilles [basionym of the Mesoamerican species *Gonolobus niger* (Cavanilles) R. Brown ex Schultes]. It is the more frequently encountered and better established of the two *Vincetoxicum* species in the United States. One of the earliest and best documented escapes from cultivation occurred in Cambridge, Massachusetts, where garden plantings at Harvard University (when Asa Gray was herbarium director in the late 1800s) and elsewhere were documented as serving as invasion foci for surrounding areas. The species is still commonly encountered on fences in many places in Cambridge. In addition to the states listed above, *V. nigrum* is known from cultivation near Minneapolis, Minnesota, but is not yet known to have escaped in that state. Chromosome numbers of 2*n* = 22 and 2*n* = 44 have been reported from the native range in Europe, and it is unclear whether the introduced populations are exclusively tetraploid or also include diploids (A. DiTomasso et al. 2005). Unlike most species of Apocynaceae that have been studied, *V. nigrum* has been shown to be capable of autogamous pollination through the in situ germination of pollinia within anther thecae (DiTomasso et al.).

2. **Vincetoxicum rossicum** (Kleopow ex A. W. Hill) Barbaricz, Vyzn. Rosl. Ukrain, 346. 1950 • Pale or European swallow-wort, dompte-venin de Russie [I] [W]

Cynanchum rossicum Kleopow ex A. W. Hill in B. D. Jackson et al., Index Kew., suppl. 8, 67. 1933

Stems erect proximally, twining distally. **Leaves:** petiole 0.5–2 cm; blade pinnipalmately veined, elliptic or lanceolate to ovate, 6.5–12 × 2.5–7 cm, membranous, base truncate to rounded, with 2–8 laminar colleters, margins ciliate, apex acuminate, surfaces pilosulous on veins abaxially, glabrous adaxially. **Inflorescences** solitary at nodes, simple or compound cymes, 5–20-flowered; peduncle 1.5–2.5 cm, pilosulous in a single line. **Pedicels** 3–7 mm, pilosulous. **Flowers:** calyx lobes narrowly lanceolate, 2 mm, apex attenuate, glabrous; corollas pinkish tan to reddish brown, campanulate, not fleshy, lobes spreading, apically twisted, lanceolate, 3–5 mm, 1.5–2 times longer than wide, apex obtuse to acute, glabrous, gynostegial corona of 5 prismatic, basally united segments exceeding style apex, pinkish tan to brick red; style apex depressed, umbonate, green. **Follicles** 3.5–7 × 0.5–0.6 cm, apex attenuate to acuminate. **Seeds** 10–20, brown, lenticular, lanceolate to ovate, 4–6.5 × 2.5–3 mm; coma white, 2–3 cm. 2*n* = 22.

Flowering May–Jul(–Aug); fruiting Jun–Sep. Disturbed areas, gardens, cemeteries, old fields, pastures, powerline rights of way, coastal bluffs, basalt outcrops, limestone, shallow to deep soils, forest edges, woods, thickets, alvars, grasslands; 0–400 m; introduced; N.B., Ont., Que.; Conn., Ind., Mass., Mich., N.H., N.J., N.Y., Pa.; se Europe (sw Russia, Ukraine); introduced also in nw Europe (Norway).

The taxonomy of *Vincetoxicum rossicum* in the flora area has long been confused, as described above. This species has been reported from North America as *V. hirundinaria* (*Cynanchum vincetoxicum*) and *V. medium* (*C. medium*). *Vincetoxicum hirundinaria* is cultivated in the flora area and may persist ephemerally but appears not to have naturalized. *Vincetoxicum rossicum* has not always been distinguished from *V. hirundinaria*. Thus, early reports of *C. vincetoxicum* or *V. hirundinaria* mostly apply to *V. rossicum*. Even when the two species were distinguished, plants of *V. rossicum* were commonly called *C. medium* or *V. medium*; however, this name is now considered misapplied to *V. rossicum* and to be a synonym of *V. hirundinaria*. There have been numerous reports of *V. hirundinaria* in the northern United States and southern Canada, but the great majority of these have been shown to represent *V. rossicum*, either through examination of herbarium specimens or interpretation of descriptions of corolla color.

Vincetoxicum rossicum is the more frequently encountered and better established species of the genus in Canada. In addition to the provinces listed above, *V. rossicum* is known from cultivation near Halifax, Nova Scotia, but is not yet known to have escaped in that province. Like *V. nigrum*, *V. rossicum* has been shown to be facultatively autogamous, via in situ pollinia germination, a rare condition in Apocynaceae (A. DiTomasso et al. 2005).

CONVOLVULACEAE Jussieu

• Morning Glory Family

Daniel F. Austin†

Annuals, perennials, subshrubs, shrubs, vines, or lianas [**trees**], some with milky sap; *Cuscuta* parasitic, achlorophyllous. **Stems** decumbent, erect, procumbent, repent, trailing, or twining-climbing. **Leaves** alternate, usually simple, sometimes compound; stipules absent; petiole present or absent; blade margins entire, toothed, or lobed. **Inflorescences** axillary, ± cymose or flowers solitary; bracteate or not; bracteolate or not. **Flowers** bisexual [unisexual], actinomorphic or weakly zygomorphic; calyx persistent, sepals 5, distinct or proximally connate, equal or unequal; corolla blue, cream, green, lavender, maroon, pink, purple, rose, violet, white, or yellow, campanulate, funnelform, rotate, salverform, tubular, or urceolate, limb (3–)5-lobed or -toothed or entire, induplicate, sometimes also convolute, in bud; nectary annular or cup-shaped, sometimes 5-lobed or absent; stamens (3–)5, distinct, filaments inserted on corolla tube, anthers 2-celled, linear or oblong, dehiscent by slits; ovary superior, 1–4(–6)-locular, placentation basal or basal-axile, ovules 1–6 per locule, anatropous, bitegmic, crassi- or tenuinucellate; styles 1 or 2; stigmas 1 or 2, capitate, globose, peltate, or 2(–4)-lobed. **Fruits** usually capsular, sometimes berrylike (nutlike), dehiscent or indehiscent. **Seeds** 1–4(–6), black, brown, green, or yellow, ellipsoid, globose, obcompressed, or pyramidal, glabrous or hairy; endosperm absent or scant, cartilaginous; embryo straight, cotyledons usually folded, rarely absent.

Genera 56, species 1600–1700 (18 genera, 167 species in the flora): nearly worldwide.

A traditional circumscription of Convolvulaceae, including *Cuscuta* and *Dichondra*, is followed here. *Cuscuta* is sometimes treated as a distinct family, Cuscutaceae Dumortier, and, less commonly, *Dichondra* has been treated in a distinct family, Dichondraceae Dumortier (A. Cronquist 1981; A. B. Rendle 1959).

In addition to hundreds of horticultural cultivars, economically important members of Convolvulaceae include *Ipomoea batatas* (sweet potato) and species of *Calystegia* that are agricultural weeds. Some species of Convolvulaceae are sources of hallucinogens and medicines, especially purgatives.

SELECTED REFERENCE Stefanović, S. et al. 2003. Classification of Convolvulaceae: A phylogenetic approach. Syst. Bot. 28: 791–806.

1. Plants parasitic (lacking chlorophyll) . . . . 1. *Cuscuta*, p. 271
1. Plants not parasitic (autotrophic).
   2. Stems usually ascending, creeping, decumbent, erect, procumbent, prostrate, or trailing, seldom twining-climbing; leaves: larger blades 10–30(–100) mm; styles usually 2 (1 in *Convolvulus*).
      3. Stigmas or stigma lobes cylindric, linear, filiform, subclavate, subulate, or spatulate.
         4. Leaves sessile or subsessile; styles 2, stigma lobes 4, filiform to subclavate . . . 2. *Evolvulus*, p. 303
         4. Leaves usually petiolate, rarely sessile; styles 1, stigma lobes 2, cylindric, linear, or spatulate . . . . 14. *Convolvulus* (in part), p. 336
      3. Stigmas capitate, globose, or peltate.
         5. Styles: insertion ± basal; fruits subglobose to ± compressed, ± incised and 2-lobed, indehiscent or shattering irregularly . . . . 3. *Dichondra*, p. 306
         5. Styles: insertion not ± basal; fruits conic, fusiform, globose, oblong-ovoid, or ovoid, dehiscence valvate.
            6. Leaf blades 1–14 mm.
               7. Stems decumbent to erect, seldom, if ever, mat-forming; corollas salverform, white, 5–7 mm . . . . 4. *Cressa*, p. 308
               7. Stems decumbent, erect, prostrate, or trailing, sometimes mat-forming; corollas ± campanulate, greenish yellow to yellow, 3–4 mm . . . . . . 5. *Petrogenia*, p. 310
            6. Leaf blades (1–)10–60(–100).
               8. Sepals 9–28 mm; corollas blue, blue-purple, or white with blue limb, 35–85 mm. . . . . .6. *Bonamia*, p. 310
               8. Sepals 4–11 mm; corollas usually white, sometimes lavender, maroon, pink, purple, or red, 8–25 mm. . . . . 7. *Stylisma*, p. 311
   2. Stems usually twining-climbing, sometimes repent or trailing, rarely almost absent, ascending, decumbent, erect, or procumbent; leaves: larger blades (10–)40–270 mm; styles 1.
      9. Fruits indehiscent.
         10. Sepals lanceolate-linear, 1–2 mm; corollas white, to 8 mm. . . . . 8. *Poranopsis*, p. 314
         10. Sepals elliptic, oblong, or ovate to orbiculate, 7–20 mm; corollas lavender, purple, purplish red, red, white, or yellow, 25–95 mm.
            11. Leaf blade abaxial surfaces black-glandular-punctate; fruits dry, 25–30 mm . . . . 9. *Stictocardia*, p. 315
            11. Leaf blade abaxial surfaces glabrous, glabrate, or white-hairy; fruits fleshy or dry, 10–15 mm.
               12. Leaf blades 180–270 mm; sepals ovate, 15–20 mm; corollas lavender, 60–65 mm; fruits berrylike, fleshy . . . . .10. *Argyreia*, p. 316
               12. Leaf blades 40–100 mm; sepals oblong, 7–12 mm; corollas proximally purplish, distally ± white with greenish bands, 25–40 mm; fruits capsular or nutlike, dry . . . . 11. *Turbina*, p. 317
      9. Fruits dehiscent.
         13. Stigmas or stigma lobes cylindric, elliptic, linear, oblong, reniform, spatulate, or subulate.
            14. Stems hairy, hairs usually branched, glandular, and/or stellate, sometimes simple . . . . 12. *Jacquemontia*, p. 317
            14. Stems glabrous or hairy, hairs not branched, glandular, or stellate.
               15. Sepals (5–)8–15(–25) mm; ovary 1-locular; stigma lobes linear to oblong, apices blunt . . . . .13. *Calystegia*, p. 321
               15. Sepals 3–12 mm; ovary 2-locular; stigma lobes cylindric to spatulate, apices acute. . . . . 14. *Convolvulus* (in part), p. 336
         13. Stigmas or stigma lobes capitate or globose.
            16. Fruit dehiscence circumscissile . . . . .15. *Operculina*, p. 338
            16. Fruit dehiscence irregular or valvate.
               17. Anthers twisted after dehiscence; pollen usually 3–9-colpate, rarely aggrecolpate, not echinate . . . . 16. *Merremia*, p. 339
               17. Anthers straight after dehiscence; pollen rugate and not echinate (*Aniseia*) or pantoporate and echinate (*Ipomoea*).

[18. Shifted to left margin.—Ed.]

18. Sepals notably accrescent in fruit, outer 3 notably longer than inner 2; corolla white, campanulate, 25–30 mm . . . . 17. *Aniseia*, p. 342
18. Sepals seldom notably accrescent in fruit, outer 3 not notably longer than inner 2; corolla usually blue, lavender, pink, purple, red, violet, or white, sometimes orange, red and yellow, or red-orange, usually funnelform, sometimes campanulate or salverform, (6–)20–80 (–150+) mm. . . . . 18. *Ipomoea*, p. 342

1. CUSCUTA Linnaeus, Sp. Pl. 1: 124. 1753; Gen. Pl. ed. 5, 60. 1754 • Dodder [Aramaic and Hebrew triradical root of verb K-S-Y (Kaph, Shin, Yodh), to cover]

Mihai Costea

Guy L. Nesom

Annuals or perennials, parasitic, usually lacking chlorophyll; roots rudimentary and ephemeral. **Stems** trailing or twining, attached to host by haustoria, greenish, white, yellow, orange, or pink-purple, filiform, glabrous. **Leaves** sessile; blade reduced, scalelike, 1–2 mm, surfaces glabrous. **Inflorescences** cymes or thyrses; pedicellate or not; bracts 0–1(–11). **Flowers** (3 or)4 or 5-merous, thickened-fleshy to thin-membranous; sepals ± connate, calyx usually campanulate, cupulate, or cylindric, rarely angled or slightly zygomorphic, 2–6 mm; corolla usually white, white-cream, or yellowish, sometimes pink or purple, cupulate to cylindric, 2–6(–9) mm , limb (3 or)4- or 5-lobed; infrastaminal scales usually present, dentate or fimbriate; styles 1 or 2; stigmas 2, ± globose or ± elongate. **Fruits** capsular, depressed-globose to ovoid, indehiscent or dehiscence circumscissile. **Seeds** 1–4, glabrous, seed coat honeycombed when dry and papillate when wet or seed coat epidermis cells rectangular and not honeycombed/papillate; hilum terminal, subterminal, or lateral; embryo uniformly slender, 1–3-coiled, rarely globose-enlarged at base; cotyledons absent. $x = 4$.

Species ca. 200 (51 in the flora): nearly worldwide, most diverse in the Western Hemisphere, especially warmer regions.

Cuscuta is classified within Convolvulaceae; its outgroup relationships are not clear (S. Stefanović et al. 2002; Stefanović and R. G. Olmstead 2004). Pollen and gynoecium features suggest that *Cuscuta* is allied with Dicranostyloideae clade or Convolvuloideae Burnett (M. Welsh et al. 2010; M. A. R. Wright et al. 2011).

Some *Cuscuta* species parasitize numerous hosts; others have relatively narrow host ranges. Some of the former species can be weeds especially destructive to crops, including alfalfa, carrots, chickpeas, clover, cranberries, fava beans, flax, lespedeza, potatoes, tomatoes, and sugar beets. When growing on perennial hosts, haustoria of some *Cuscuta* species overwinter inside their hosts and generate new growth in spring (J. H. Dawson et al. 1994; M. Costea and F. J. Tardif 2006; K. Meulebrouck et al. 2009).

Seed dispersal has been considered unspecialized (J. Kuijt 1969); endozoochory by migratory birds was recently documented in some species (M. Costea et al. 2016). Worldwide dispersal of weedy *Cuscuta* species has been through contaminated seeds of forage legumes, especially alfalfa, clover, and lespedeza.

Identification of most *Cuscuta* species is usually a lengthy process: rehydration, dissection, and examination of flowers with a microscope may be necessary. For stem diameters, the following categories are used (T. G. Yuncker 1921): slender = 0.35–0.4 mm, medium = 0.4–0.6 mm, and coarse = greater than 0.6 mm. Measurements of floral parts were done on rehydrated mature flowers. Lengths of flowers were measured from base of calyx to tip of a straightened corolla lobe. Color of calyx and texture of calyx and corolla were noted only on dried flowers. Observation of papillae and multicellular appendages require magnifications of at least 50×. Orientation of corolla lobes (erect, spreading, reflexed) was noted in mature flowers. Dehiscence or indehiscence of capsules can be predicted early in development of ovaries; lines of dehiscence are readily detectable at the bases of young ovaries; the ovary wall will tear along the dehiscence line when light pressure is applied. Capsule walls sometimes rupture irregularly.

Three of the four subgenera recognized in *Cuscuta* are known from the flora area; members of subg. *Pachystigma* (Engelmann) Baker & C. H. Wright are endemic to South Africa (M. Costea et al. 2015).

SELECTED REFERENCES Costea, M. et al. 2015. A phylogenetically based infrageneric classification of the parasitic plant genus *Cuscuta* (dodders, Convolvulaceae). Syst. Bot. 40: 269–285. García, M. A. et al. 2014. Phylogeny, character evolution, and biogeography of *Cuscuta* (dodders; Convolvulaceae) inferred from coding plastid and nuclear sequences. Amer. J. Bot. 101: 670–690. Stefanović, S., M. Kuzmina, and M. Costea. 2007. Delimitation of major lineages within *Cuscuta* subgenus *Grammica* (Convolvulaceae) using plastid and nuclear DNA sequences. Amer. J. Bot. 94: 568–589. Yuncker, T. G. 1932. The genus *Cuscuta*. Mem. Torrey Bot. Club 18: 113–331. Yuncker, T. G. 1965. *Cuscuta*. In: N. L. Britton et al., eds. 1905+. North American Flora.... 47+ vols. New York. Ser. 2, part 4, pp. 1–51.

1. Inflorescences thyrsoid: paniculiform, racemiform, or spiciform; styles usually 1, sometimes distally 2-fid or separating into 2; seed coat epidermis cells rectangular, not honeycombed when dry and papillate when wet . . . . 1c. *Cuscuta* subg. *Monogynella*, p. 301
1. Inflorescences monochasial cymes: corymbiform, fasciculate, glomerulate, paniculiform, racemiform, ropelike, spiciform, or umbelliform; styles 2; seed coat epidermis cells honeycombed when dry and papillate when wet.
   2. Styles equal; stigmas clavate, cylindric, or terete . . . . 1a. *Cuscuta* subg. *Cuscuta*, p. 272
   2. Styles unequal; stigmas globose . . . . 1b. *Cuscuta* subg. *Grammica*, p. 275

1a. Cuscuta Linnaeus subg. Cuscuta [I]

Inflorescences monochasial cymes: dense, glomerulate. **Pedicels** 0–1.5 mm. **Flowers:** styles 2, equal; stigmas clavate, cylindric, or terete. **Capsules:** dehiscence circumscissile, interstylar aperture relatively small or inconspicuous. **Seeds** angled (2 flat faces and 1 convex), ellipsoid, ovoid, or subglobose; seed coat epidermis cells honeycombed when dry and papillate when wet.

Species 21 (4 in the flora): introduced; Europe, Asia, n Africa, Atlantic Islands; introduced also in Mexico, South America, s Africa, Australia.

Species of subgen. *Cuscuta* have holocentric chromosomes (B. Pazy and U. Plitmann 1994; M. A. García and S. Castroviejo 2003).

1. Flowers 4(or 5)-merous; calyx lobe apices obtuse to truncate . . . . 4. *Cuscuta europaea*
1. Flowers 5-merous; calyx lobe apices acute to acuminate.
   2. Styles plus stigmas 0.5–1.1 mm, shorter than ovary; seeds usually connate in pairs; hosts: primarily *Linum usitatissimum* . . . . 3. *Cuscuta epilinum*
   2. Styles plus stigmas 0.8–2.2 mm, equaling or longer than ovary; seeds not connate in pairs; hosts: primarily *Medicago* and *Trifolium*, sometimes other legumes and herbs from other families.
      3. Calyces golden yellow, cellular-reticulate, lobes each with a distal fleshy appendage . . . . 1. *Cuscuta approximata*
      3. Calyces creamy white or purplish, not notably reticulate, lobes without distal fleshy appendages . . . . 2. *Cuscuta epithymum*

1. Cuscuta approximata Babington, Ann. Mag. Nat. Hist. 13: 253, plate 4, fig. 3. 1844 • Alfalfa dodder, cuscute proche [I] [W]

Varieties 3 (1 in the flora): introduced; Europe, Asia, n Africa, Atlantic Islands.

Variety *episonchum* (Webb & Berthelot) Yuncker is known from western Europe, northern Africa, and Atlantic Islands (Canary Islands); var. *schiraziana* (Boissier) Yuncker is known from Europe and northern Africa.

1a. Cuscuta approximata Babington var. **approximata** [I] [W]

Stems usually yellow, rarely reddish purple, medium. **Inflorescences:** bracts elliptic to ovate, membranous, margins entire, apex acute. **Pedicels** absent. **Flowers** 5-merous, 3–4.2 mm, fleshy; papillae absent; calyx golden yellow, campanulate, ± equaling corolla tube length, divided 1/3–1/2 its length, cellular-reticulate, shiny, lobes broadly triangular to obovate, each with a distal fleshy appendage, bases overlapping, margins entire, apex acute to acuminate; corolla white, drying creamy yellow, campanulate, globose or urceolate in fruit, 2.5–4 mm, not saccate, tube 1.5–2 mm, lobes spreading, ovate-orbiculate, shorter than to equaling corolla tube length, margins entire, apex obtuse, straight; infrastaminal scales oblong to ovate, 1.4–1.8 mm, shorter than to equaling corolla tube length, bridged at 0.3–0.5 mm, base truncate to rounded, fimbriate in distal 1/2, fimbriae 0.1–0.2 mm; stamens exserted, shorter than corolla lobes; filaments 0.3–0.6 mm; anthers 0.3–0.6 × 0.3–0.4 mm; styles slightly subulate; style plus stigmas 0.8–1.6 mm, equaling or longer than ovary; stigmas cylindric to terete, 0.4–0.7 mm, shorter than to as long as styles. **Capsules** depressed-globose, 1.5–2.3 × 1.6–2 mm, not thickened or raised around interstylar aperture, translucent, capped by withered corolla. **Seeds** 3 or 4, angled, ovoid to broadly ellipsoid, 0.8–1.1 × 0.6–0.7 mm, hilum region terminal. **2*n*** = 14, 28.

Flowering May–Sep. Hosts: primarily *Medicago sativa* and *Trifolium*, sometimes other herbs; 200–2500 m; introduced; B.C.; Calif., Colo., Idaho, Mont., Nebr., Nev., N.Mex., N.Dak., Oreg., Utah, Wash., Wyo.; Europe; w Asia; n Africa.

Cuscuta approximata is distinguished from *C. epithymum* by glomerules usually with more numerous flowers, calyx golden yellow, reticulate, and shiny, and calyx lobes each with a distal fleshy appendage.

2. Cuscuta epithymum (Linnaeus) Linnaeus, Amoen. Acad. 4: 478. 1759 (as epithym.) • Clover dodder, cuscute du thym [F] [I] [W]

Cuscuta europaea Linnaeus var. *epithymum* Linnaeus, Sp. Pl. 1: 124. 1753

Varieties (subspecies) 2 (1 in the flora): introduced; Europe; introduced also in Mexico, South America, Asia, s Africa, Australia.

Cuscuta epithymum subsp. *kotschyi* (Des Moulins) Arcangeli grows in southern Europe.

2a. Cuscuta epithymum (Linnaeus) Linnaeus var. **epithymum** [F] [I] [W]

Stems usually reddish purple, rarely yellow, slender. **Inflorescences:** bracts ovate, membranous, margins entire, apex acute. **Pedicels** 0(–0.5) mm. **Flowers** 5-merous, 3–4(–4.5) mm, fleshy; papillae absent; calyx purplish or creamy white, campanulate, 1/2–2/3 corolla tube length, divided 1/2–2/3 its length, not evidently reticulate, ± shiny, lobes ovate-triangular, bases overlapping, margins entire, apex acute to acuminate; corolla usually white, sometimes pink or purple-tinged both fresh and dry, campanulate-cylindric, 2.8–4 mm, not saccate, tube 1.5–3 mm, lobes spreading, usually triangular, sometimes lanceolate, 1/2–3/4 corolla tube length, margins entire, apex acute to acuminate, straight; infrastaminal scales oblong-spatulate, 1.1–2 mm, 4/5 corolla tube length, bridged at 0.4–0.6 mm, base rounded, uniformly short-fimbriate, fimbriae 0.08–0.2 mm; stamens exserted, shorter than corolla lobes; filaments 0.4–0.7 mm; anthers 0.3–0.7 × 0.3–0.5 mm; styles terete; style plus stigmas 1.2–2.2 mm, ± equaling to 2 times ovary length; stigmas cylindric to terete, 0.6–1 mm, equaling style. **Capsules** globose, 1.6–2.2 × 1.6–2.3 mm, not thickened or raised around interstylar aperture, translucent, capped by withered corolla. **Seeds** 2–4, angled, subglobose to ovoid, 0.8–1.1 × 0.7–1 mm, hilum region terminal. **2*n*** = 14.

Flowering Jul–Oct. Hosts: herbs, especially *Medicago*, *Trifolium*, and other legumes; 0–2000 m; introduced; B.C., N.B., N.S., Ont.; Calif., Conn., Idaho, Iowa, Ky., Maine, Md., Mass., Mich., Mo., Mont., Nebr., Nev., N.J., N.Y., N.Dak., Ohio, Oreg., Pa., R.I., S.Dak., Tenn., Vt., Va., Wash., W.Va., Wyo.; Europe; introduced also in Mexico, South America, Asia, s Africa, Australia.

Ephemeral occurrences of var. *epithymum* are associated with contaminated seeds of forage legume crops.

3. Cuscuta epilinum Weihe ex Boenninghausen, Prodr. Fl. Monast. Westphal., 75. 1824 • Flax dodder, cuscute du lin [I] [W]

Stems yellow-orange, slender to medium. **Inflorescences:** bracts ovate, membranous, margins entire, apex obtuse. **Pedicels** absent. **Flowers** 5-merous, 3–4 mm, membranous; papillae absent; calyx creamy yellow, cupulate, equaling or longer than corolla tube, divided 1/2–2/3 its length, not notably reticulate or shiny, lobes broadly ovate, bases overlapping, margins entire, apex acute to

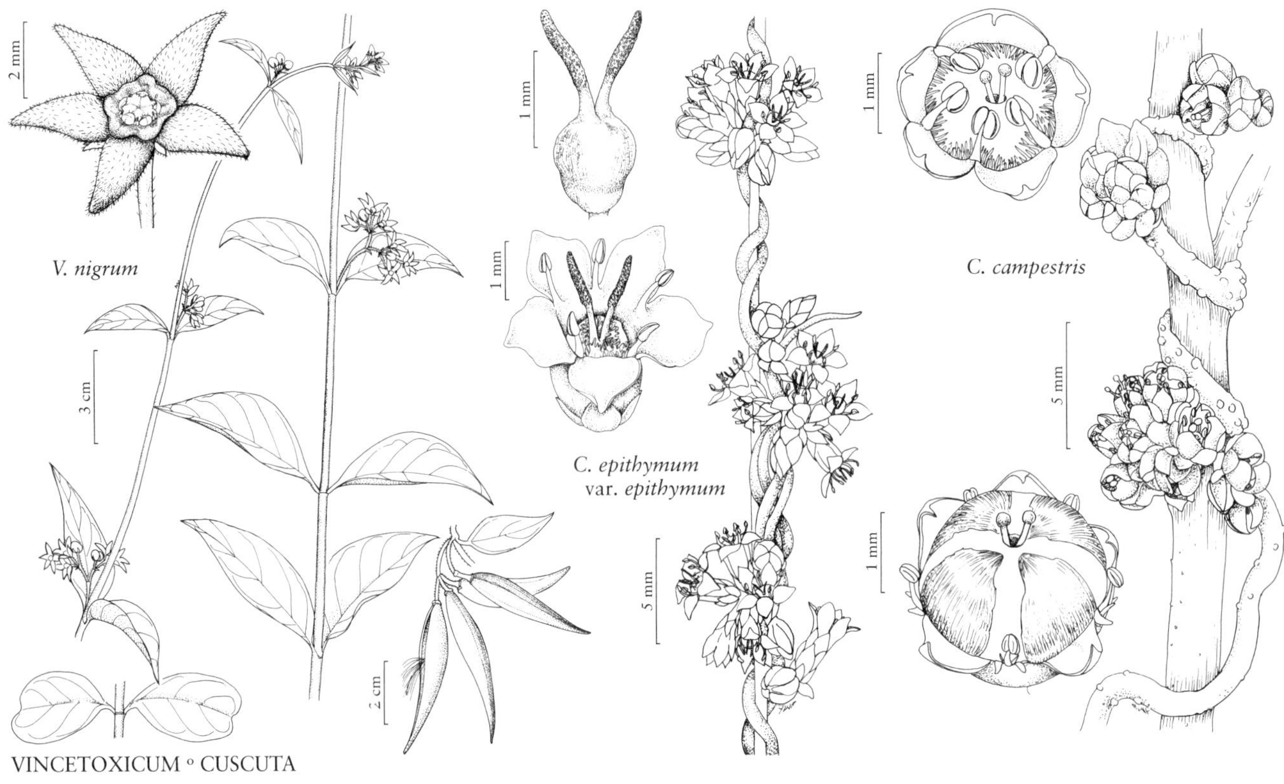

VINCETOXICUM ◦ CUSCUTA

acuminate; corolla white, drying creamy yellow, initially globose, later urceolate, 2.8–3.7 mm, tube 2.5–3 mm, not saccate, lobes erect to spreading, ovate-triangular, $^{1}/_{4}$–$^{1}/_{3}$(–$^{1}/_{2}$) corolla tube length, margins entire, apex subobtuse to acute, straight to incurved; infrastaminal scales spatulate, 0.8–1.5 mm, $^{1}/_{2}$–$^{4}/_{5}$ corolla tube length, bridged at 0.3–0.5 mm, base truncate, entire or 2-fid, short-fimbriate apically, fimbriae 0.03–0.15 mm; stamens included or barely exserted, shorter than corolla lobes; filaments 0.2–0.4 mm; anthers 0.3–0.5 × 0.3–0.4 mm; styles slightly subulate; style plus stigmas 0.5–1.1 mm, shorter than ovary; stigmas cylindric to terete, 0.3–0.6 mm, ± equaling styles. **Capsules** depressed-globose, ± angular, 2.8–3.5 × 3–4.2 mm, not thickened or raised around interstylar aperture, not translucent, capped by withered corolla. **Seeds** 4, usually connate in pairs, angled, subglobose or broadly ellipsoid to ovoid, 1.1–1.2 × 0.9–1 mm, hilum region terminal. **2*n*** = 42.

Flowering Jun–Jul. Hosts: primarily *Linum usitatissimum* but once established on flax, it may also attach to secondary hosts such as *Camelina sativa* and *Guizotia abyssinica*; 0–200 m; introduced; Que.; Del., Md., Mass., Mich., N.J., N.Y., Ohio, Pa., Vt., Wash., Wyo.; Europe; Asia.

Cuscuta epilinum is an ephemeral associated exclusively with flax. It apparently coevolved with flax since its domestication in the Nile Valley or the Middle East (M. A. García and M. P. Martín 2007).

4. Cuscuta europaea Linnaeus, Sp. Pl. 1: 124. 1753 • Greater dodder [I] [W]

Stems yellowish to reddish, medium to coarse. **Inflorescences:** bracts ovate to lanceolate, membranous, margins entire, apex obtuse to acute. **Pedicels** 0–1.5 mm. **Flowers** 4(or 5)-merous, 3–5 mm, fleshy; papillae absent; calyx creamy yellow to brownish, cupulate to obconic, shorter than to equaling corolla tube length, divided $^{1}/_{2}$–$^{3}/_{4}$ its length, not reticulate or shiny, lobes ovate to triangular-ovate, bases not overlapping, margins entire, apex obtuse to rounded; corolla white or pink, drying yellow-brown, campanulate to urceolate, 2.2–4.7 mm, tube 1.7–3.3 mm, not saccate, lobes usually erect, sometimes spreading, triangular-ovate, $^{1}/_{3}$–$^{1}/_{2}$ corolla tube length, margins entire, apex obtuse to truncate, straight; infrastaminal scales oblong, often 2-fid, 1–1.2 mm, $^{1}/_{2}$ corolla tube length, bridged at 0.4–0.6 mm, short-fimbriate apically, fimbriae 0.1–0.3 mm; stamens included, shorter than corolla lobes; filaments 0.2–0.5 mm; anthers 0.2–0.5 × 0.2–0.4 mm; styles terete; style plus stigmas 0.6–1.2 mm, shorter than ovary; stigmas cylindric to clavate, 0.3–0.6 mm, shorter than to equaling styles. **Capsules** ovoid, globose-conic, or pyriform, 1.6–4 × 2–4 mm, not thickened or raised around interstylar aperture, ± translucent, capped by withered corolla. **Seeds** 2–4, angled, subglobose, broadly ellipsoid, or ovoid, 1.2–1.5 × 0.9–1.3 mm, hilum region terminal. $2n$ = 14.

Flowering Jul–Sep. Hosts: usually *Urtica dioica*, *Humulus lupulus*, *Rubus* sp.; it can parasitize numerous other hosts both herbaceous and woody; 20–200 m; introduced; Maine, N.Y.; Europe; introduced also in South America, Asia (China, Japan), n Africa.

1b. CUSCUTA Linnaeus subg. GRAMMICA (Loureiro) Peter in H. G. A. Engler and K. Prantl, Nat. Pflanzenfam. 68[IV,3a]: 38. 1891

Grammica Loureiro, Fl. Cochinch. 1: 98, 170. 1790

Inflorescences monochasial cymes: corymbiform, fasciculate, glomerulate, paniculiform, racemiform, ropelike, spiciform, or umbelliform. **Pedicels** 0–10 mm. **Flowers** sometimes originating endogenously; styles 2, unequal; stigmas globose. **Capsules:** indehiscent or dehiscence circumscissile. **Seeds** angled (2 flat faces and 1 convex), globose, obcompressed, oblong, ellipsoid, obovoid, or ovoid; seed coat epidermis cells honeycombed when dry and papillate when wet.

Species 153 (45 in the flora): North America, Mexico, West Indies, Central America, South America, Asia, Australia; introduced nearly worldwide.

Cuscuta australis R. Brown, known from Asia and Australia, was collected in Jefferson County, New York, in 1926 (NY); it is not known to be established or recurrent in the flora area.

Cuscuta veatchii Brandegee is known only from Baja California, Mexico; it does not grow in the flora area.

Evolution of gynoecia with two unequal styles may have been the key evolutionary innovation that spurred the considerable species diversity encountered in subg. *Grammica*. The two stigmas mature sequentially and are receptive at different times (M. A. R. Wright et al. 2011). Cases of reticulate evolution have been discovered in subg. *Grammica* (S. Stefanović and M. Costea 2008; Costea and Stefanović 2010; Costea et al. 2015).

1. Capsules: dehiscence circumscissile.
    2. Styles ± subulate (tapering from base to stigma).
        3. Flowers: calyx and receptacle fleshy, corollas membranous; corolla lobes without hornlike appendages; capsules 4–7 × 4–6 mm . . . . 30. *Cuscuta mitriformis*
        3. Flowers membranous; corolla lobes each with a hornlike appendage; capsules 1.8–2.5 × 1–2.5 mm.
            4. Pedicels 0.2–0.6 mm; calyx lobes oblong, obovate, ovate, or spatulate, apex acute to obtuse, each exceeded by a hornlike appendage; seeds 0.7–1.1 × 0.7–0.9 mm . . . . 7. *Cuscuta boldinghii*
            4. Pedicels 1.5–6 mm; calyx lobes oblong-obovate or orbiculate, apex ± truncate, each not exceeded by a hornlike appendage; seeds 0.9–1.4 × 0.8–1.3 mm . . . . . 19. *Cuscuta erosa*
    2. Styles filiform (not tapered; sometimes slightly subulate in *C. dentatasquamata*).
        5. Calyces divided 1/4 length; seeds 1 . . . . 5. *Cuscuta americana*
        5. Calyces divided 1/3–3/4 length or ± to base; seeds (1 or)2–4.
            6. Infrastaminal scales usually 1/4–1/2 corolla tube length (1/2 to equaling in *C. odontolepis*).
                7. Inflorescences dense, paniculiform-glomerulate; calyx lobe bases overlapping; capsules 2.9–4 × 3–3.2 mm; seeds 1–1.2 × 0.6–0.8 mm . . . . 34. *Cuscuta odontolepis*
                7. Inflorescences loose, racemiform or umbelliform; calyx lobe bases not overlapping; capsules 1.3–2.2 × (0.7–)1.5–2.3 mm; seeds 0.6–0.9(–1.1) × 0.3–0.8 mm.

8. Flowers 5-merous; calyx lobe midveins carinate; corolla lobes erect . . . . . 46. *Cuscuta tuberculata*
8. Flowers (3 or)4-merous; calyx lobe midveins not carinate, sometimes with multicellular protuberances; corolla lobes spreading to reflexed.
9. Calyces 1/3–1/2 corolla tube length; infrastaminal scales 1/2 corolla tube length . . . . . 28. *Cuscuta leptantha*
9. Calyces equaling corolla tube length; infrastaminal scales 1/4–1/3 corolla tube length . . . . . 29. *Cuscuta liliputana*
6. Infrastaminal scales equaling or longer than corolla tube.
10. Pedicels 2–10 mm; calyx lobe midveins not carinate, without multicellular protuberances.
11. Flowers 4–5.5(–6) mm; calyx lobe apices acuminate . . . . . 27. *Cuscuta legitima*
11. Flowers 2–3 mm; calyx lobe apices obtuse to acute . . . . . 47. *Cuscuta umbellata*
10. Pedicels 0.3–2 mm; calyx lobe midveins carinate and/or with multicellular protuberances.
12. Calyx lobe bases overlapping, apices obtuse or rounded; corolla lobe apices rounded to obtuse . . . . . 12. *Cuscuta chinensis*
12. Calyx lobe bases not overlapping, apices acute; corolla lobe apices acute.
13. Flowers 2–2.6 mm, membranous; calyces straw yellow; infrastaminal scales oblong or rounded, distal 1/2 fimbriate; capsules 1.5–2.5 × 1–1.8 mm . . . . . 6. *Cuscuta azteca*
13. Flowers 2.6–3.8 mm, fleshy; calyces reddish brown; infrastaminal scales truncate or distally irregularly denticulate or 3–7-fimbriate; capsules 3–4 × 1.8–2.9 mm . . . . . 16. *Cuscuta dentatasquamata*
1. Capsules indehiscent.
14. Bracts at bases of flower clusters, pedicels, and/or flowers (or on pedicels) 2–11; calyces divided to base or nearly so.
15. Pedicels 2–5 mm; inflorescences loose, paniculiform . . . . . 15. *Cuscuta cuspidata*
15. Pedicels 0–1 mm; inflorescences glomerulate, ropelike, or spiciform.
16. Inflorescences dense, ropelike, spiraling around and closely appressed to host stem, individual flower clusters not discernible; bracts narrowly triangular to lanceolate, apex recurved . . . . . 21. *Cuscuta glomerata*
16. Inflorescences dense to loose, glomerulate or spiciform, not ropelike, individual flower clusters discernible; bracts ovate-orbiculate to ovate-triangular, apex erect.
17. Calyx lobes orbiculate, apices rounded; corolla lobe apices obtuse; seeds 1.9–2.5 × 1.7–2.3 mm . . . . . 13. *Cuscuta compacta*
17. Calyx lobes ovate, apices acute to cuspidate; corolla lobe apices usually acute, sometimes cuspidate; seeds 1.5–1.7 × 1.1–1.3 mm . . . . . 42. *Cuscuta squamata*
14. Bracts at bases of flower clusters 1(–3), at bases of pedicels and/or flowers (0 or)1(–3); calyces divided 2/5–2/3 length.
18. Infrastaminal scales absent.
19. Corolla lobes ovate-triangular, 1/3–1/2 tube length, apices inflexed . . . . . 26. *Cuscuta jepsonii*
19. Corolla lobes narrowly lanceolate to lanceolate, shorter than, equaling, or longer than tube, apices straight.
20. Styles 0.5–1(–1.5) mm; anthers 0.2–0.5 mm; capsules surrounded by, not completely enclosed by, withered corolla (capsule top visible) . . . . . 33. *Cuscuta occidentalis*
20. Styles 1.2–3 mm; anthers 0.6–1.1 mm; capsules surrounded by, sometimes also capped by, withered corolla (capsule top not or barely visible).
21. Calyces 1/4–1/2 corolla tube length; corolla lobes shorter than or equaling tube; stamen filaments 0.3–0.6 mm . . . . . 8. *Cuscuta brachycalyx*
21. Calyces 3/4 to equaling corolla tube length; corolla lobes equaling or longer than tube; stamen filaments 0.6–1.1 mm . . . . . 9. *Cuscuta californica*

[18. Shifted to left margin.—Ed.]

18. Infrastaminal scales present.

22. Infrastaminal scales relatively poorly developed, apically irregularly denticulate, or 2-fid with 1–3 fimbriae on each side of filament attachment, or 2-fid with denticulate wings.

23. Calyx lobes each with a hornlike appendage 0.5–0.7 mm . . . . 49. *Cuscuta warneri*

23. Calyx lobes without hornlike appendages.

24. Calyx lobe apices obtuse to rounded . . . . 38. *Cuscuta polygonorum* (in part)

24. Calyx lobe apices acute to long-attenuate.

25. Flowers fleshy; corolla lobe apices acute, inflexed; styles 0.7–1.8 mm, usually equaling, sometimes longer than, ovary; capsules not translucent, interstylar aperture relatively large; 50–1200 m, not California, Oregon, Washington . . . . 14. *Cuscuta coryli*

25. Flowers membranous; corolla lobe apices lance-attenuate, straight; styles 0.3–0.7 mm, $^{1}/_{4}$ ovary length; capsules translucent, interstylar aperture relatively small; 1500–2600 m, California, Oregon, Washington . . . . . 45. *Cuscuta suksdorfii*

22. Infrastaminal scales well developed, notably fimbriate.

26. Calyx and corolla lobe apices dissimilar: calyx lobe apices obtuse, rounded, or subacute; corolla lobe apices usually acuminate, acute, or subacute, sometimes cuspidate, mucronate, or truncate, usually inflexed (± cucullate in *C. draconella*).

27. Calyces $^{1}/_{2}$–$^{3}/_{4}$ corolla tube length; corolla lobes $^{1}/_{2}$ tube length . . . . 43. *Cuscuta suaveolens*

27. Calyces usually equaling corolla tube length, sometimes shorter (*C. runyonii*); corolla lobes equaling or longer than tube.

28. Flowers (3 or)4(or 5)-merous.

29. Calyces angled, lobe bases overlapping, midvein ± carinate; capsules globose to ovoid; seeds 1 or 2 . . . . 23. *Cuscuta harperi*

29. Calyces not angled, lobe bases not overlapping, midvein not carinate; capsules depressed-globose; seeds 3 or 4 . . . . 38. *Cuscuta polygonorum* (in part)

28. Flowers (4 or)5-merous.

30. Calyx lobe midveins with multicellular protuberances.

31. Calyces ± angled, lobe bases overlapping . . . . 18. *Cuscuta draconella*

31. Calyces not angled, lobe bases not overlapping . . . . 40. *Cuscuta runyonii*

30. Calyx lobe midveins without multicellular protuberances.

32. Calyces angled, lobes broadly ovate to rhombic, bases auriculate, overlapping; flowers 1.4–2.5 mm; corolla tubes 0.7–1.2 mm; anthers 0.2–0.3 mm . . . . 36. *Cuscuta pentagona*

32. Calyces not angled, lobes ovate-triangular to triangular, not auriculate, bases overlapping or not; flowers 1.9–4.6(–5) mm; corolla tubes 1.1–2.5 mm; anthers (0.3–)0.4–0.7 mm.

33. Calyx lobe bases overlapping; capsules to $^{1}/_{3}$ enveloped by withered corolla . . . . 10. *Cuscuta campestris*

33. Calyx lobe bases not or slightly overlapping; capsules $^{1}/_{2}$+ enveloped by withered corolla.

34. Perianths papillate (sometimes not obvious in calyx); corollas drying reddish brown to yellow, tube campanulate, later globose, saccate between lines of stamen attachments; infrastaminal scales equaling or longer than corolla tube . . . . 20. *Cuscuta glabrior*

34. Perianths not papillate; corollas drying creamy white, tube narrowly campanulate to cylindro-campanulate, not saccate between lines of stamen attachments; infrastaminal scales $^{3}/_{4}$–$^{4}/_{5}$ corolla tube length . . . . 37. *Cuscuta plattensis*

[26. Shifted to left margin.—Ed.]

26. Calyx and corolla lobe apices similar: calyx lobe apices either: 1) usually acuminate, acute, long-attenuate, and rarely cuspidate (*C. subinclusa*) or subacute, or 2) usually obtuse to rounded; corolla lobe apices acuminate, acute, long-attenuate, cuspidate (*C. pacifica* and *C. salina*), obtuse, or rounded, usually straight, rarely inflexed (*C. indecora*) or recurved (*C. howelliana*).
    35. Calyx lobe apices obtuse to rounded; corolla lobe apices obtuse to rounded.
        36. Calyces straw yellow, reticulate, shiny, lobe margins denticulate; seeds 1, globose to globose-ovoid . . . . 17. *Cuscuta denticulata*
        36. Calyces yellow-brown or brownish, not or finely reticulate, or not or slightly shiny, lobe margins usually entire, sometimes serrulate; seeds (1 or)2–4, ± angled, ellipsoid, obcompressed, oblong, obovoid, ovoid, or subglobose.
            37. Flowers (3 or)4(or 5)-merous . . . . 11. *Cuscuta cephalanthi*
            37. Flowers (4 or)5-merous.
                38. Calyces equaling corolla tube length; capsules depressed-globose, not raised or thickened around interstylar aperture . . . . 32. *Cuscuta obtusiflora*
                38. Calyces usually 1/3–1/2 corolla tube length, sometimes equaling corolla tube (*C. gronovii* var. *latiflora*); capsules conic-globose, ovoid, or subobpyriform, raised and thickened around interstylar aperture.
                    39. Flowers 4–6(–7) mm; capsules broadly ovoid (with stout beak 1–1.5 mm) . . . . 39. *Cuscuta rostrata*
                    39. Flowers 2–4(–4.4) mm; capsules conic-globose, ovoid, or subobpyriform (rarely apically narrowed into a neck to 1 mm).
                        40. Infrastaminal scales equaling corolla tube; styles (0.6–)1.2–2.2 mm; capsules 2.5–4.5(–5.2) mm . . . . 22. *Cuscuta gronovii*
                        40. Infrastaminal scales (1/3)–1/2 corolla tube length; styles 0.3–0.9 mm; capsules 3.5–6.5(–7) mm . . . . 48. *Cuscuta umbrosa*
    35. Calyx lobe apices usually acuminate, acute, or long-attenuate, rarely cuspidate (*C. subinclusa*); corolla lobe apices usually acuminate, acute, or long-attenuate, sometimes cuspidate (*C. pacifica* and *C. salina*).
        41. Flowers fleshy; corolla lobe apices inflexed . . . . 25. *Cuscuta indecora*
        41. Flowers membranous or base of perianth and receptacle fleshy (*C. nevadensis*); corolla lobe apices usually straight, rarely recurved (*C. howelliana*).
            42. Flowers 5–7(–9) mm; calyces 1/2 corolla tube length; corolla lobes 1/4–1/3 tube length; anthers 0.8–2 mm . . . . 44. *Cuscuta subinclusa*
            42. Flowers (2.5–)2.8–5(–6) mm; calyces equaling or longer than corolla tube; corolla lobes equaling or longer than tube; anthers 0.3–0.8 mm.
                43. Infrastaminal scales equaling corolla tube; embryo bases globose-enlarged . . . . 31. *Cuscuta nevadensis*
                43. Infrastaminal scales 1/2–2/3(–4/5) corolla tube length; embryo bases not globose-enlarged.
                    44. Flowers embedded in inflorescence of host, 4(or 5)-merous; calyx lobe apices acuminate to long-attenuate; capsules depressed-globose to globose . . . . 24. *Cuscuta howelliana*
                    44. Flowers not embedded in inflorescence of host, 5-merous; calyx lobe apices acuminate to acute; capsules ellipsoid-ovoid or ovoid.
                        45. Inflorescences subglomerulate to umbelliform; pedicels 0.5–2 mm; flowers 3.5–6 mm; corolla tubes campanulate, lobes broadly ovate to rhombic-ovate; infrastaminal scales 1/2–3/4 corolla tube length; coastal salt marshes and tidal flats, coastal interdune depressions and grasslands . . . . 35. *Cuscuta pacifica*
                        45. Inflorescences corymbiform; pedicels (0.5–)1–5 mm; flowers 2.5–4.5 mm; corolla tubes ± cylindric-campanulate to obconic, lobes lance-oblong to lance-ovate; infrastaminal scales 4/5 corolla tube length; inland salt flats . . . . 41. *Cuscuta salina*

5. Cuscuta americana Linnaeus, Sp. Pl. 1: 124. 1753 • American dodder 

Stems yellow-orange, medium. **Inflorescences** glomerulate or densely paniculiform; bracts at base of clusters 1, at base of pedicels 0 or 1, ovate to lanceolate, ± fleshy, margins entire, apex acute. **Pedicels** 0.2–0.6 mm. **Flowers** 5-merous, 2.5–4.2 mm, fleshy, not papillate; calyx brownish, cylindric, equaling or slightly shorter than corolla tube length, divided 1/4 its length, not reticulate or shiny, lobes broadly ovate, bases overlapping, margins entire, midvein not carinate, apex rounded to obtuse; corolla white, drying brown, 2–3.3 mm, tube cylindric, 1.7–2.5 mm, not saccate, lobes usually erect, sometimes spreading, ovate, 1/5–1/4 corolla tube length, margins entire, apex obtuse, ± cucullate, or straight; infrastaminal scales ovate to oblong, 1.4–2 mm, 3/4–4/5 corolla tube length, bridged at 0.6–1 mm, truncate to rounded, uniformly short-fimbriate, fimbriae 0.1–0.2 mm; stamens included, shorter than corolla lobes; filaments 0.1–0.3 mm; anthers 0.2–0.4 × 0.2–0.4 mm; styles filiform, 1.5–2.2 mm, longer than ovary. **Capsules** globose-ovoid to ovoid, 1.8–3 × 0.8–2 mm, not thickened or raised around relatively small interstylar aperture, not translucent, capped by withered corolla, dehiscence circumscissile. **Seeds** 1, subglobose to ellipsoid, 1.4–1.5 × 1–1.1 mm, hilum region terminal.

Flowering Sep–Mar. Hosts: *Bursera*, *Celtis*, *Citharexylum*, *Colubrina*, *Coursetia*, *Haematoxylum*, *Haplophyton*, *Havardia*, *Janusia*, *Jatropha*, *Karwinskia*, *Mimosa*, *Prosopis*, *Sebastiania*, *Senna*, *Vallesia*, and other herbs and woody plants; 0–40 m; Fla.; Mexico; West Indies; Central America; South America.

In Florida, *Cuscuta americana* may attack *Citrus* trees.

Cuscuta americana was used by the Aztecs to produce a yellow dye called zacatlaxcalli (B. de Sahagún 1950–1982).

6. Cuscuta azteca Costea & Stefanović, Organisms Diversity Evol. 11: 381, figs. 1h, 2g, 4. 2011 • Globe dodder

Cuscuta potosina W. Schaffner ex Engelmann var. *globifera* Yuncker

Stems orange-yellow, slender. **Inflorescences** dense, corymbiform to glomerulate; bracts at base of clusters 1, at base of pedicels 0 or 1, ovate, membranous, margins entire, apex acute. **Pedicels** 0.4–1.3 mm. **Flowers** 5-merous, 2–2.6 mm, membranous, not papillate; calyx straw yellow, cupulate, equaling corolla tube length, divided 2/3 its length, finely reticulate, shiny, lobes ovate, bases not overlapping, margins entire, midvein with multicellular protuberances, apex acute; corolla white, drying creamy, 1.5–2.1 mm, tube cylindro-campanulate, becoming globose, 0.8–1.2 mm, not saccate, lobes erect to spreading, ovate-triangular, shorter than to equaling corolla tube length, margins usually entire, sometimes irregularly toothed, apex acute, straight to slightly incurved; infrastaminal scales oblong or rounded, 0.9–1.3 mm, equaling or slightly longer than corolla tube length, bridged at 0.1–0.2 mm, uniformly fimbriate in distal 1/2, fimbriae 0.1–0.3 mm; stamens exserted, shorter than corolla lobes; filaments 0.3–0.6 mm; anthers 0.2–0.3 × 0.2–0.3 mm; styles filiform, 0.4–0.7 mm, shorter than to equaling ovary. **Capsules** depressed-globose, 1.5–2.5 × 1–1.8 mm, not thickened or raised around relatively small interstylar aperture, translucent, loosely surrounded or capped by withered corolla, dehiscence circumscissile. **Seeds** 3 or 4, angled, broadly ellipsoid to subglobose, 0.8–1 × 0.7–0.9 mm, hilum region almost terminal.

Flowering Jul–Nov. Hosts: Asteraceae, Euphorbiaceae, Fabaceae (especially *Dalea*), Malvaceae; 600–1800 m; Ariz., N.Mex.; Mexico.

7. Cuscuta boldinghii Urban, Repert. Spec. Nov. Regni Veg. 16: 38. 1919 • Caribbean dodder [W]

Stems yellow-orange, slender. **Inflorescences** glomerulate or compact-paniculiform, 15–70-flowered, flowers subsessile; bracts at base of clusters 1, at base of pedicels 0–2, ovate to lanceolate, membranous, margins entire, apex acuminate to attenuate. **Pedicels** 0.2–0.6 mm. **Flowers** 5-merous, 2.5–4 mm, membranous, not papillate; calyx brownish, campanulate, equaling corolla tube length, divided 1/2–2/3 its length, not reticulate, shiny, lobes oblong, obovate, ovate, or spatulate, bases overlapping, margins entire or finely serrulate-denticulate, midvein not carinate, apex acute to obtuse, each exceeded by hornlike appendage, 0.3–0.6 mm; corolla creamy white, drying brown, 2.2–3.2 mm, tube campanulate, 1.2–1.6 mm, not saccate, lobes spreading to reflexed, ovate to lanceolate, equaling corolla tube length, each with subapical, hornlike projection, 0.3–0.7 mm, margins entire or irregularly denticulate, apex obtuse, straight; infrastaminal scales oblong to broadly ovate or rounded, 1.2–1.6 mm, equaling corolla tube length, bridged at 0.5–0.7 mm, sparsely, uniformly short-fimbriate, fimbriae 0.04–0.1 mm; stamens exserted, shorter than corolla lobes; filaments 0.3–0.9 mm; anthers 0.4–0.5 × 0.2–0.3 mm; styles stout, ± subulate, 0.9–2 mm, longer than ovary. **Capsules**

globose to slightly depressed, 1.8–2.3 × 1–2 mm, not thickened or raised around relatively small interstylar aperture, not translucent or becoming translucent very late, capped by withered corolla, dehiscence circumscissile. **Seeds** 1–4, angled, subglobose or ovoid to broadly ellipsoid, 0.7–1.1 × 0.7–0.9 mm, hilum region subterminal.

Flowering Jul–Nov. Hosts: herbs and woody plants; 0–10 m; Fla.; Mexico; West Indies (Puerto Rico); Central America; n South America.

Cuscuta boldinghii was reported from Florida as a parasite of *Citrus* (L. C. Knorr 1949); it was collected again in Florida (Monroe County) in 2002.

Cuscuta boldinghii and *C. erosa* are the only species in the flora area bearing hornlike appendages with stomata on corolla lobes (M. Costea et al. 2011).

8. **Cuscuta brachycalyx** (Yuncker) Yuncker, Mem. Torrey Bot. Club 18: 159. 1932 [C] [E]

Cuscuta californica Hooker & Arnott var. *brachycalyx* Yuncker, Illinois Biol. Monogr. 6: 152, plate 8, fig. 45e,f, plate 11, fig. 75. 1921; *C. brachycalyx* var. *apodanthera* (Yuncker) Yuncker

Stems yellow to orange, medium. **Inflorescences** dense to loose, paniculiform; bracts at base of clusters 1, at base of pedicels 0 or 1, lanceolate, membranous, margins entire, apex acute to acuminate. **Pedicels** 1–6 mm. **Flowers** 5-merous, 4.5–6 mm, fleshy at base (receptacle and calyx base), not papillate; calyx brownish, campanulate-cupulate, 1/4–1/2 corolla tube length, divided 2/3 its length, not reticulate and shiny, lobes triangular-ovate, bases slightly overlapping, margins entire, midvein not carinate, apex acute; corolla white, drying creamy brown, 4.4–5.8 mm, tube campanulate-cylindric, becoming urceolate, 2–3.5 mm, saccate or with horizontal ridges between stamen attachments, lobes reflexed, lanceolate, somewhat shorter than to equaling corolla tube length, margins entire, apex acute, straight; infrastaminal scales absent; stamens exserted, shorter than corolla lobes; filaments 0.3–0.6 mm; anthers 0.7–1.1 × 0.3–5.5 mm; styles filiform, 1.2–3 mm, equaling or longer than ovary. **Capsules** globose to obovoid, 1.6–2.3 × 1.8–2.5 mm, apically enlarged or thickened, surrounded and capped by withered corolla, top not or barely visible, interstylar aperture not visible, indehiscent. **Seeds** 1–4, obcompressed, broadly ellipsoid to obovoid, 0.9–1.4 × 0.8–1.2 mm, hilum region lateral.

Flowering Jun–Sep. Hosts: *Allium*, *Eriogonum*, *Gilia*, *Hemizonia*, *Lagophylla*, *Linanthus*, *Trichostema*, *Trifolium*; chaparral, dry grasslands, and open places in *Pinus ponderosa* and *Abies magnifica* forests; of conservation concern; 30–2500 m; Calif., Oreg.

9. **Cuscuta californica** Hooker & Arnott, Bot. Beechey Voy., 364. 1839 • Chaparral dodder [W]

Stems yellow to orange, medium. **Inflorescences** dense, paniculiform-corymbiform or glomerulate; bracts at base of clusters 1, at base of pedicels 0 or 1, ovate to lanceolate, membranous, margins entire, apex acute to acuminate. **Pedicels** (0.5–)1–2.5 mm, sometimes papillate. **Flowers** 5-merous, 3–5(–5.5) mm, membranous or fleshy at base (receptacle and calyx base), not papillate, or receptacle, calyx, and corolla papillate; calyx yellow, turbinate-campanulate, 3/4 to equaling corolla tube length, divided 1/2–2/3 its length, finely reticulate, shiny, lobes triangular-ovate to lanceolate, bases overlapping, margins entire, midvein not carinate, apex acute to acuminate; corolla white, drying creamy white, 3–5 mm, tube cylindric-campanulate to obconic, 1.6–2.4 mm, not saccate or with horizontal ridges between stamen attachments, lobes reflexed, narrowly lanceolate, equaling or longer than corolla tube length, margins entire, apex acute, straight; infrastaminal scales absent; stamens exserted, shorter than corolla lobes; filaments 0.6–1.1 mm; anthers 0.6–1 × 0.3–0.5 mm; styles filiform, 1.2–2.2 mm, equaling or longer than ovary. **Capsules** globose, depressed-globose, or ovoid-conic, sometimes apically pointed, 1.5–2.2 × 1.8–2.5 mm, not thickened around relatively small interstylar aperture, interstylar aperture sometimes ± visible, not translucent, completely surrounded by, not capped by, withered corolla, indehiscent. **Seeds** 1–4, obcompressed, broadly ellipsoid to obovoid, 0.9–1.4 × 0.8–1.2 mm, papillate, hilum region lateral.

Varieties 3 (3 in the flora): w United States, nw Mexico.

1. Capsules ovoid-conic; seeds 1 . . . 9b. *Cuscuta californica* var. *apiculata*
1. Capsules globose to depressed-globose; seeds (1 or)2–4.
   2. Pedicels not papillate; flowers not papillate . . . . 9a. *Cuscuta californica* var. *californica*
   2. Pedicels papillate; flowers papillate . . . . 9c. *Cuscuta californica* var. *papillosa*

9a. Cuscuta californica Hooker & Arnott var. **californica** [W]

Pedicels not papillate. **Flowers** not papillate. **Capsules** globose to depressed-globose, not apically pointed. **Seeds** (1 or)2–4.

Flowering Mar–Aug(–Sep). Hosts: *Abronia*, *Acmispon*, *Adenostoma*, *Agastache*, *Ambrosia*, *Artemisia*, *Asclepias*, *Baccharis*, *Convolvulus*, *Corethrogyne*, *Croton*, *Eriodictyon*, *Eriogonum*, *Hemizonia*, *Holocarpha*, *Iva*, *Lupinus*, *Portulaca*, *Salvia*, *Spartium*, and others; 0–2500 m; Ariz., Calif., Nev., Oreg., Utah, Wash.; Mexico (Baja California, Baja California Sur).

9b. Cuscuta californica Hooker & Arnott var. **apiculata** Engelmann, Trans. Acad. Sci. St. Louis 1: 499. 1859 [C] [E]

Pedicels not papillate. **Flowers** not papillate. **Capsules** ovoid-conic, apically pointed. **Seeds** 1.

Flowering Mar–Aug. Host: *Psorothamnus spinosus*; sandy desert areas; of conservation concern; 100–600 m; Calif.

Variety *apiculata* is known with certainty only from the type specimen. Its taxonomic status is uncertain because it could not be included in molecular evolutionary studies (M. Costea and S. Stefanović 2009, 2010).

9c. Cuscuta californica Hooker & Arnott var. **papillosa** Yuncker, Illinois Biol. Monogr. 6: 152, plate 11, fig. 76. 1921 [E] [W]

Pedicels papillate. **Flowers** papillate. **Capsules** globose to depressed-globose. **Seeds** (1 or) 2–4.

Flowering Dec–Sep. Hosts: *Acmispon*, *Amelanchier*, *Artemisia*, *Ceanothus*, *Erigeron*, *Eriodictyon*, *Lepidospartum*, *Parosela*, *Penstemon*; 100–1900 m; Calif.

10. Cuscuta campestris Yuncker, Mem. Torrey Bot. Club 18: 138. 1932 • Field dodder, cuscute des champs [F] [W]

Cuscuta pentagona Engelmann var. *calycina* Engelmann, Amer. J. Sci. Arts 45: 76. 1843

Stems yellow to orange, medium. **Inflorescences** dense, corymbiform or glomerulate; bracts at base of clusters 1, at base of pedicels 0 or 1, ovate or ovate-triangular to lanceolate, membranous, margins entire, apex acute. **Pedicels** 0.3–2.5(–3.5) mm. **Flowers** (4 or)5-merous, 1.9–3.6 mm, membranous, not papillate; calyx yellow, cupulate, equaling corolla tube length, divided $^{2}/_{5}$–$^{3}/_{5}$ its length, reticulate, shiny, lobes ovate-triangular, bases overlapping, margins entire, midvein not carinate, without multicellular protuberances, apex obtuse to rounded; corolla creamy white, drying creamy or golden yellow, 2–3.5 mm, tube campanulate, (1.1–)1.5–1.9 mm, not saccate, lobes spreading, triangular to triangular-lanceolate, equaling corolla tube length, margins entire, apex acute to acuminate, inflexed; infrastaminal scales oblong-ovate to spatulate, rounded, 1.5–2 mm, equaling or exceeding corolla tube length, bridged at 0.3–0.5 mm, uniformly densely fimbriate, fimbriae 0.3–0.4(–0.5) mm; stamens exserted, shorter than corolla lobes; filaments 0.4–0.7 mm; anthers (0.3–)0.4–0.5 × 0.2–0.3 mm; styles filiform, 0.5–1.6 mm, shorter than to equaling ovary. **Capsules** depressed-globose to depressed, 1.3–2.8 × 1.9–3.8 mm, not thickened or raised around relatively large interstylar aperture, sometimes translucent, to $^{1}/_{3}$ enveloped by withered corolla, indehiscent. **Seeds** 4, angled, subglobose to broadly ellipsoid, 1.1–1.5 × 0.9–1.1 mm, hilum region subterminal. $2n = 56$.

Flowering May–Nov. Hosts: Acanthaceae, Asteraceae, Brassicaceae, Chenopodiaceae, Convolvulaceae, Euphorbiaceae, Fabaceae, Hydrophyllaceae, Polygonaceae, Solanaceae, Urticaceae, Verbenaceae, and others (M. Costea and F. J. Tardif 2006); 0–2000 m; Alta., B.C., Man., N.S., Ont., Que., Sask.; Ariz., Ark., Calif., Colo., Fla., Ga., Idaho, Ill., Ind., Iowa, Kans., Ky., La., Md., Mass., Mich., Minn., Miss., Mo., Mont., Nebr., Nev., N.J., N.Mex., N.Y., N.C., N.Dak., Ohio, Okla., Oreg., Pa., S.Dak., Tenn., Tex., Utah, Va., Wash., W.Va., Wis., Wyo.; Mexico; introduced widely.

Reports of *Cuscuta campestris* from New Brunswick, Newfoundland, and Prince Edward Island have not been verified.

Cuscuta campestris is the most widespread species of the genus in North America and perhaps the most successful and prevalent *Cuscuta* weed species worldwide; it has been recorded from South America, Europe, Asia, Africa, and Australia. It has been often referred to in North America as *C. pentagona*, which has smaller flowers and angled calyces. The two species are closely related; *C. campestris* is a hybrid species and *C. pentagona* is one of its progenitors (M. Costea et al. 2015b).

11. Cuscuta cephalanthi Engelmann, Amer. J. Sci. Arts 43: 336, plate 6, figs. 1–6. 1842 • Buttonbush dodder [E] [F] [W]

Stems yellow-orange, medium. **Inflorescences** dense to loose, spiciform or paniculiform, commonly originating endogenously; bracts at base of clusters 1 or 2, at base of pedicels and/or flowers 0 or 1, ovate, membranous, margins entire or serrulate, apex obtuse to acute. **Pedicels** 0–1 mm. **Flowers** (3 or)4(or 5)-merous, 2–3 mm, membranous, not papillate; calyx yellow-brown, shallowly cupulate, 1/2 corolla tube length, divided 2/3 its length, not reticulate, not shiny, lobes oblong-ovate, bases slightly overlapping, margins entire or serrulate, midvein not carinate, apex obtuse; corolla white, drying creamy white, 1.8–2.8 mm, tube cylindric-campanulate to cylindric, 1.1–2.2 mm, not saccate, lobes spreading, ovate, 1/3–1/2 corolla tube length, margins entire, apex obtuse, straight; infrastaminal scales oblong, rounded, 0.9–1.7 mm, shorter than to equaling corolla tube length, bridged at 0.2–0.4 mm, sparsely fimbriate, more densely distally, fimbriae 0.1–0.3 mm; stamens included to slightly exserted, equaling corolla lobes; filaments 0.2–0.4 mm; anthers 0.2–0.4 × 0.2–0.4 mm; styles filiform or narrowly terete, (0.6–)1–2 mm, equaling or longer than ovary. **Capsules** depressed-globose to globose, 2.5–3.2(–4) × 2–4 mm, not thickened or raised around relatively small interstylar aperture, not translucent, capped by withered corolla, indehiscent. **Seeds** 1 or 2, obcompressed, broadly ovoid, 1.4–2 × 1.3–1.4 mm, hilum region terminal. $2n = 60$.

Flowering Jun–Oct. Hosts: *Achillea*, *Boehmeria*, *Campsis*, *Cephalanthus*, *Chelone*, *Decodon*, *Hypericum*, *Iva*, *Justicia*, *Lycopus*, *Lysimachia*, *Lythrum*, *Persicaria*, *Physostegia*, *Potentilla*, *Pycnanthemum*, *Salix*, *Saururus*, *Scutellaria*, *Solanum*, *Solidago*, *Spiraea*, *Symphyotrichum*, *Teucrium*, *Tradescantia*, *Vernonia*, *Vicia*, and other herbs and woody plants; stream and lake shores, marshes, and floodplain forests; 0–1500 m; Alta., B.C., Man., N.B., N.S., Ont.; Ariz., Ark., Calif., Conn., D.C., Ga., Idaho, Ill., Ind., Iowa, Kans., Ky., Maine, Mass., Mich., Minn., Nebr., Nev., N.J., N.Mex., N.Y., N.C., N.Dak., Ohio, Okla., Oreg., Pa., R.I., S.C., S.Dak., Tenn., Tex., Utah, Va., Wash., Wis.

12. Cuscuta chinensis Lamarck in J. Lamarck et al., Encycl. 2: 229. 1786

Varieties 2 (1 in the flora): sw United States, Mexico, Asia, Australia.

Variety *chinensis* is known from Asia and Australia.

12a. Cuscuta chinensis Lamarck var. **applanata** (Engelmann) Costea & Stefanović, Organisms Diversity Evol. 11: 383. 2011 • Gila River dodder

Cuscuta applanata Engelmann, Trans. Acad. Sci. St. Louis 1: 479. 1859

Stems yellow to creamy, slender to medium. **Inflorescences** glomerulate to loosely paniculiform; bracts at base of clusters 1, at base of pedicels 0 or 1, broadly ovate to subround, membranous, margins entire, apex obtuse to rounded. **Pedicels** 0.4–2 mm. **Flowers** 5-merous, 2.5–3.5 mm, membranous, not papillate; calyx straw yellow, shallowly cupulate, equaling corolla tube length, divided 1/2 its length, reticulate, shiny, lobes broadly triangular-ovate, bases overlapping, margins entire or irregular, midvein carinate with multicellular protuberances, apex rounded to obtuse; corolla white, drying creamy yellow, 2–2.4 mm, tube campanulate, becoming globose to depressed-globose or urceolate, 1–1.3 mm, not saccate, lobes spreading, ovate-lanceolate, equaling corolla tube length, margins entire, apex rounded, ± incurved, not inflexed; infrastaminal scales obovate, 1.2–1.8 mm, equaling or longer than corolla tube length, bridged at 0.2–0.4 mm, long-fimbriate in distal 1/2, fimbriae 0.2–0.4 mm; stamens exserted, shorter than corolla lobes; filaments 0.3–0.6 mm; anthers 0.4–0.6 × 0.4–0.5 mm; styles filiform, 0.9–1.7 mm, equaling or longer than ovary. **Capsules** depressed-globose, slightly angular, 1.8–2.5 × 0.8–1.6 mm, not thickened or raised around relatively small interstylar aperture, translucent, surrounded by withered corolla, dehiscence circumscissile. **Seeds** (1–)3 or 4, angled, broadly ellipsoid, 0.8–1.2 × 0.8–1.1 mm, hilum region almost terminal.

Flowering Jul–Oct. Hosts: *Amaranthus*, *Ambrosia*, *Anisacanthus*, *Bahia*, *Baileya*, *Boerhavia*, *Chamaecrista*, *Chamaesaracha*, *Croton*, *Dalea*, *Flaveria*, *Ipomoea*, *Parthenium*, *Sanvitalia*, *Solanum*, *Tiquilia*, *Tragia*, *Viguiera*, and others; 100–1500 m; Ariz., N.Mex., Tex., Utah; Mexico.

Varieties *applanata* and *chinensis* are similar morphologically but geographically disjunct as a result of a long-distance dispersal event from North America to Australia/Asia (M. Costea et al. 2011b).

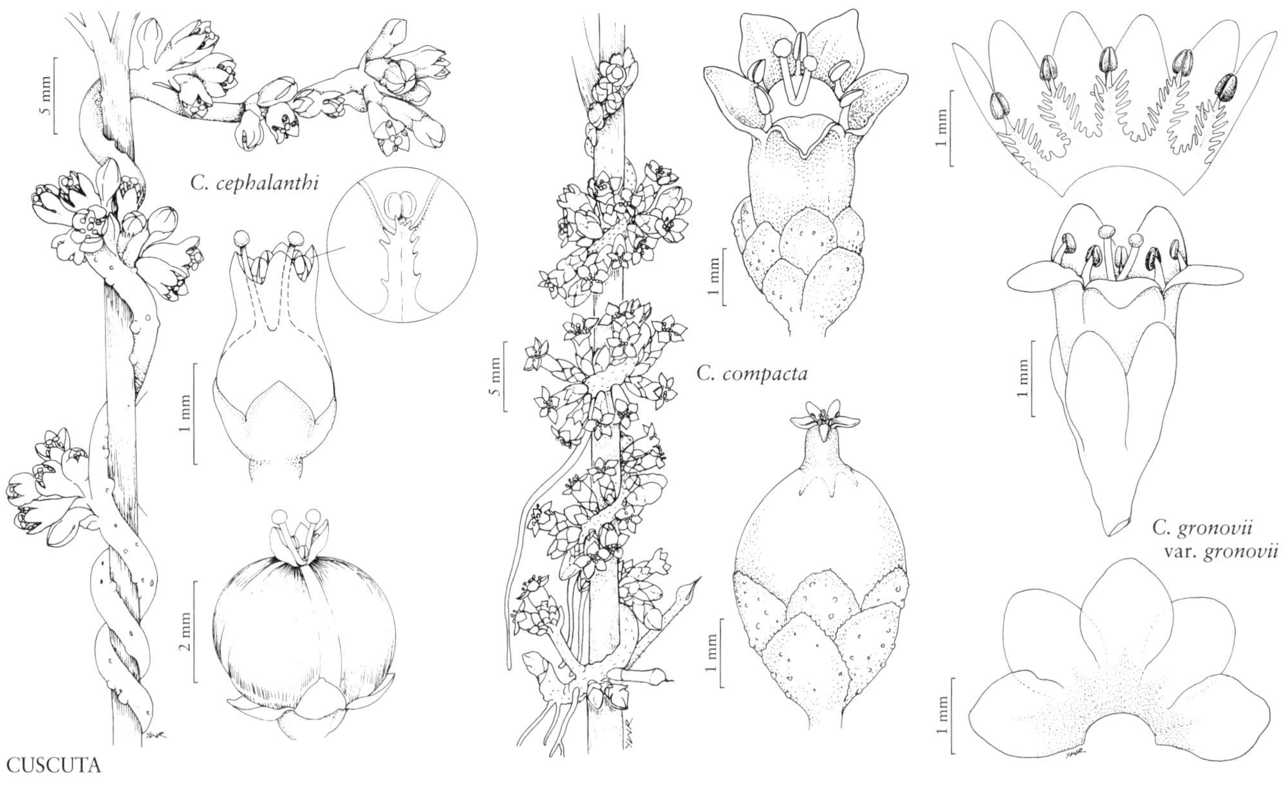

CUSCUTA

13. **Cuscuta compacta** Jussieu ex Choisy, Mém. Soc. Phys. Genève 9: 281, plate 4, fig. 2. 1842 • Compact dodder [E] [F] [W]

Cuscuta compacta var. *efimbriata* Yuncker

Stems yellow-orange, coarse. **Inflorescences** dense to loose, glomerulate, often originating endogenously; bracts at base of clusters, pedicels, and/or flowers 2–4(–6), ovate-orbiculate, fleshy, margins scarious, finely denticulate-fimbriate, apex erect, obtuse. **Pedicels** 0–1 mm. **Flowers** 5-merous, 4–6 mm, fleshy, not papillate; calyx brown-yellow, cylindro-campanulate, 3/4 corolla tube length, divided to base or nearly so, finely reticulate, shiny, lobes orbiculate, bases broadly overlapping, margins scarious, minutely denticulate-fimbriate, midvein not carinate, apex rounded; corolla white to creamy greenish, drying creamy brown, 3.8–4.5 mm, tube cylindric, 3–4 mm, not saccate, lobes spreading to reflexed, ovate, 1/3–1/4 corolla tube length, margins entire, apex obtuse, straight; infrastaminal scales oblong to obovate, 2.6–3.3 mm, 3/4 to equaling corolla tube length, bridged at 0.9–1.3 mm, usually densely fimbriate in distal 1/2–3/4, sometimes reduced, 2-fid with 1–3 fimbriae on each side of filament insertion or dentate-winged, fimbriae 0.4–0.8 mm; stamens exserted, shorter than corolla lobes; filaments 0.2–0.3 mm; anthers 0.3–0.4 × 0.3–0.4 mm; styles filiform, 1–2 mm, longer than ovary. **Capsules** conic-globose to broadly ellipsoid, 3–5(–6) × 2–2.5 mm, thickened and slightly umbonate around relatively small interstylar aperture, not translucent, capped by withered corolla, indehiscent. **Seeds** 1–4, obcompressed, broadly ovoid to broadly ellipsoid, 1.9–2.5 × 1.7–2.3 mm, hilum region subterminal. $2n = 30$.

Flowering late Jul–Nov. Hosts: *Acer*, *Alnus*, *Ampelopsis*, *Baccharis*, *Campsis*, *Carpinus*, *Carya*, *Cephalanthus*, *Cissus*, *Clethra*, *Cornus*, *Cyrilla*, *Decodon*, *Eupatorium*, *Hydrangea*, *Ilex*, *Iva*, *Lespedeza*, *Leucothoë*, *Lindera*, *Ludwigia*, *Magnolia*, *Myrica*, *Parthenocissus*, *Rhus*, *Rosa*, *Rubus*, *Sambucus*, *Sassafras*, *Smilax*, *Tecoma*, *Vaccinium*, *Viburnum*, *Vitis*, and other herbs and woody plants; wetland margins, wet woodlands; 30–300 m; Ala., Ark., Conn., Del., D.C., Fla., Ga., Ill., Ind., Iowa, Ky., La., Md., Mass., Miss., Mo., N.H., N.J., N.Y., N.C., Okla., Pa., R.I., S.C., Tenn., Tex., Va.

Cuscuta compacta var. *efimbriata* was treated as a synonym of *C. compacta* by D. F. Austin (1980). Yuncker recognized its presence in Florida (the type collection) and Arkansas; these plants apparently are populational variants, distinguished by infrastaminal scales shorter than the corolla tube, much reduced in size, bifid or winged, and reduced in marginal fimbriae.

14. Cuscuta coryli Engelmann, Amer. J. Sci. Arts 43: 337, plate 6, figs. 7–11. 1842 • Hazel dodder cuscute du noisetier [E] [W]

Stems yellow to orange, medium to slender. **Inflorescences** dense, paniculiform or glomerulate, sometimes originating endogenously; bracts at bases of clusters 1, at base of pedicels 0 or 1, ovate to lanceolate, membranous, margins entire, apex acute. **Pedicels** 0.5–3 mm. **Flowers** (3 or)4 or 5-merous, 1.7–2.6(–3) mm, fleshy; usually not papillate, perianth cells convex, domelike; calyx brownish, cupulate, usually equaling or longer than, rarely shorter than, corolla tube length, divided 1/2–2/3 its length, not reticulate or shiny, lobes triangular-ovate, bases not or slightly overlapping, margins entire, midvein ± carinate, apex acute; corolla white, usually drying dark brown, 1.5–2.5 mm, tube campanulate to suburceolate, 0.5–1.5 mm, not saccate, lobes suberect to erect, triangular-ovate, 1/3 to equaling corolla tube length, margins entire, apex acute, inflexed; infrastaminal scales relatively poorly developed, oblong, 0.7–1.5 mm, equaling or exceeding corolla tube length, bridged at 0.3–0.5 mm, 2-fid with 1–3 fimbriae on each side of filament attachment or with denticulate wings, fimbriae 0.3–0.5 mm; stamens included, shorter than corolla lobes; filaments 0.3–0.6 mm; anthers 0.2–0.4 × 0.1–0.2 mm; styles filiform, 0.7–1.8 mm, usually equaling, sometimes longer than, ovary. **Capsules** globose, ovoid, or depressed, 1.8–2.5 × 3.5–5 mm, ± thickened, not raised, around relatively large interstylar aperture, not translucent, capped or not by withered corolla, indehiscent. **Seeds** 3 or 4, heterogeneous on same plant: obcompressed to weakly angled, broadly ellipsoid, or transversely oblique, 1.3–1.6 × 1.2–1.4 mm, hilum region usually subterminal, rarely almost terminal. $2n$ = 30.

Flowering Aug–Nov. Hosts: *Desmodium*, *Glycyrrhiza*, *Hedera*, *Helianthus*, *Monarda*, *Prenanthes*, *Rubus*, *Solanum*, *Solidago*, *Symphyotrichum*, and other herbs and woody plants; woods, meadows, margins of wetlands; 50–1200 m; Man., Ont., Que., Sask.; Ala., Ariz., Ark., Conn., Del., D.C., Ill., Ind., Iowa, Kans., Ky., Md., Mass., Mich., Minn., Miss., Mo., Mont., Nebr., N.J., N.Mex., N.Y., N.C., N.Dak., Ohio, Okla., Pa., R.I., S.C., S.Dak., Tenn., Tex., Va., W.Va., Wis.

Cuscuta coryli is closely related to *C. indecora*, from which it differs by often relatively small, four-merous flowers, bifid infrastaminal scales with one to three fimbriae on each side of filament attachment or with denticulate wings, and depressed capsules (M. Costea et al. 2006).

15. Cuscuta cuspidata Engelmann, Boston J. Nat. Hist. 5: 224. 1845 • Cusp dodder [E] [W]

Stems pale yellow, medium. **Inflorescences** loose, paniculiform; bracts at bases of clusters and pedicels, and on pedicels, 2–4, ovate-orbiculate to ovate-triangular, membranous, margins ± irregularly serrulate, apex obtuse or acute to bluntly cuspidate. **Pedicels** 2–5 mm. **Flowers** 5-merous, 3.5–4.2 mm, membranous, not papillate; calyx creamy yellow, campanulate, 1/2 corolla tube length, divided to base or nearly so, finely reticulate, shiny, lobes ovate, bases overlapping, margins finely serrulate, midvein not carinate, apex acute to cuspidate; corolla white, drying creamy white, 3.3–4 mm, tube cylindric-campanulate, 2.2–3 mm, not saccate, lobes reflexed, ovate-oblong to ovate-triangular, 1/3–1/2 corolla tube length, margins entire, apex cuspidate-acute to obtuse, straight; infrastaminal scales oblong, rounded, 2–2.7 mm, 3/4–4/5 corolla tube length, bridged at 1–1.4 mm, ± uniformly densely fimbriate, fimbriae 0.3–0.6 mm; stamens exserted, slightly shorter than corolla lobes; filaments 0.6–0.9 mm; anthers 0.6–0.8 × 0.4–0.5 mm; styles filiform, 2–2.6 mm, longer than ovary. **Capsules** globose to depressed-globose, 2.5–3 × 2.8–3.2 mm, with thickened and raised ridge or collar around relatively small interstylar aperture, translucent, capped by withered corolla, indehiscent. **Seeds** 2–4, angled, subglobose to broadly ellipsoid, 1.1–1.4 × 1–1.1 mm, hilum region ± terminal. $2n$ = 30.

Flowering Jun–Oct. Hosts: Asteraceae: including *Ambrosia*, *Amphiachyris*, *Baccharis*, *Croptilon*, *Eclipta*, *Helianthus*, *Heterotheca*, *Iva*, *Liatris*, *Solidago*; less frequently genera of other families; prairies, sandy places, sometimes ruderal; 50–1500 m; Ark., Colo., Ill., Ind., Iowa, Kans., Ky., La., Miss., Mo., Nebr., N.Mex., N.Dak., Okla., S.Dak., Tex., Utah.

Reports of *Cuscuta cuspidata* from Connecticut and Wisconsin have not been verified.

16. Cuscuta dentatasquamata Yuncker, Bull. Torrey Bot. Club 49: 107, fig. 1. 1922 • Los Pinitos dodder, [C]

Stems orange, slender to medium. **Inflorescences** dense, paniculiform; bracts at base of clusters 1, at base of pedicels 0(or 1), triangular, membranous, margins entire, apex acute. **Pedicels** 0.3–2 mm. **Flowers** 5-merous, 2.6–3.8 mm, fleshy, not papillate; calyx reddish brown, campanulate, longer than corolla tube length, divided 1/3–3/4 its length, not reticulate or shiny, lobes triangular, bases not overlapping, margins entire, midvein carinate, apex acute; corolla creamy yellow, drying reddish brown, 2.5–3.4 mm, tube campanulate, 1.2–1.6 mm, later saccate, globose, lobes erect to spreading, triangular, 1/3 corolla tube length, margins entire, apex acute, straight or inflexed; infrastaminal scales oblong, 1.1–1.6 mm, equaling corolla tube length, bridged at 0.3–0.6 mm, truncate or distally irregularly denticulate or 3–7-fimbriate; stamens exserted, shorter than corolla lobes; filaments 0.3–0.4 mm; anthers 0.3–0.4 × 0.2–0.3 mm; styles filiform to slightly subulate, 0.6–1.4 mm, equaling or longer than ovary. **Capsules** depressed-globose, 3–4 × 1.8–2.9 mm, ± thickened, not raised around relatively large interstylar aperture, usually translucent, surrounded by withered corolla, dehiscence circumscissile. **Seeds** 2–4, slightly obcompressed, subglobose, 1.3–1.6 × 1–1.4 mm, hilum area subterminal.

Flowering Jul–Oct. Host: *Bouvardia*; of conservation concern; 500–1200 m; Ariz.; Mexico (Sonora).

17. Cuscuta denticulata Engelmann, Amer. Naturalist 9: 348. 1875 • Small-tooth dodder [W]

Stems light yellow, filiform. **Inflorescences** dense, glomerulate; bracts at base of clusters, pedicels, and/or flowers 1–3, subround, ovate, or rhombic to ovate-lanceolate, membranous, margins entire or denticulate, apex acute to obtuse. **Pedicels** (0–)0.5–2.2 mm. **Flowers** (4 or) 5-merous, 2–3 mm, membranous, not papillate; calyx straw yellow, campanulate to urceolate, equaling corolla tube length or nearly so, divided 2/3 its length, reticulate and shiny, lobes obovate-orbiculate, bases overlapping, margins denticulate, midvein not carinate, apex rounded; corolla white, drying straw yellow, 1.6–2.6 mm, tube campanulate, 0.6–1.5 mm, not saccate, lobes reflexed, ovate to broadly elliptic, equaling corolla tube length, margins irregularly denticulate, apex rounded, straight; infrastaminal scales ovate to oblong, 0.6–1.4 mm, equaling corolla tube length, bridged at 0.4–0.6 mm, rounded to truncate, uniformly denticulate or short-fimbriate, fimbriae 0.05–0.1 mm; stamens included to slightly exserted, shorter than corolla lobes; filaments 0.2–0.4 mm; anthers 0.2–0.4 × 0.3–0.3 mm; styles filiform, 0.3–0.5 mm, shorter than ovary. **Capsules** globose-ovoid, 1.3–2.1 × 1–2 mm, not thickened or raised around inconspicuous interstylar aperture, translucent, capped by withered corolla, indehiscent. **Seeds** 1, not angled or obcompressed or globose to globose-ovoid, 0.8–1.1 × 0.8–1.1 mm, hilum region terminal; embryo enlarged-globose at base. $2n = 30$.

Flowering Mar–Oct(–Dec). Hosts: especially *Chrysothamnus* and *Larrea*; also *Ambrosia*, *Artemisia*, *Atriplex*, *Bebbia*, *Coleogyne*, *Covillea*, *Ericameria*, *Eriogonum*, *Euphorbia*, *Gutierrezia*, *Lepidospartum*, *Lycium*, *Psorothamnus*, and other desert plants; 200–2000 m; Ariz., Calif., Colo., Idaho, Nev., Utah, Wash.; Mexico (Baja California).

18. Cuscuta draconella Costea & Stefanović, Syst. Bot. 34: 577, figs. 1B–I, 4A–C. 2009 • Dragon dodder [E]

Stems pale orange, slender. **Inflorescences** dense, glomerulate or fasciculate; bracts at base of clusters 1, at base of pedicels 1 or 2, ovate-triangular to linear, membranous, margins entire, midvein with 1–3(–5) hornlike, multicellular protuberances, apex cuspidate to long-attenuate, ± recurved. **Pedicels** 0.1–6 mm, sometimes papillate. **Flowers** 5-merous, 2.5–3.6(–4) mm, fleshy, papillate or not; calyx golden yellow to brown, campanulate, ± angled, equaling corolla tube length, divided 1/2–2/3 its length, reticulate and shiny or ± fleshy, lobes triangular-ovate or broadly ovate to subround, membranous, bases overlapping, margins irregularly serrulate, denticulate, or entire, midvein with 1 or 2(–5) hornlike, multicellular protuberances 0.1–0.3 mm, apex subacute, rounded, or obtuse; corolla white, drying creamy yellow, 2–3.5 mm, tube campanulate, 1.2–1.6 mm, not saccate, lobes spreading to reflexed, triangular-ovate, equaling corolla tube length, margins entire or irregularly denticulate, apex ± cucullate, acute, cuspidate (sometimes 2 or 3-cuspidate), mucronate, or truncate, inflexed; infrastaminal scales spatulate to obovate, 0.7–1 mm, 3/4 corolla tube length, bridged at 0.2–0.4 mm, densely fimbriate, fimbriae 0.2–0.3 mm; stamens exserted, shorter than corolla lobes; filaments 0.4–0.8 mm; anthers 0.5–0.8 × 0.4–0.5 mm; styles stout, filiform, 0.5–1.2 mm, shorter than to equaling ovary. **Capsules** not seen, lack of dehiscence line in the ovary indicates they are indehiscent. **Seeds** not seen.

Flowering Aug–Sep. Hosts: *Atriplex*, *Gutierrezia*, *Thelesperma*; 1600 m; N.Mex., Tex.

Some plants of *Cuscuta draconella* from Texas were formerly treated as *C. decipiens* Yuncker, which is known only from Zacatecas, Mexico.

19. Cuscuta erosa Yuncker, Illinois Biol. Monogr. 6: 116, plate 2, fig. 8, plate 10, fig. 61. 1921 • Sonoran dodder [W]

Stems white to creamy yellow, medium. **Inflorescences** loose to moderately dense, paniculiform or corymbiform; bracts at base of clusters 1, at base of pedicels 0 or 1, ovate-triangular to lanceolate, membranous, margins entire or serrulate-denticulate, apex obtuse to acute. **Pedicels** 1.5–6 mm. **Flowers** 5-merous, 3.5–4.5 mm, membranous, not papillate; calyx yellow, cupulate, 3/4 to ± equaling corolla tube length, divided 1/2–2/3 its length, finely reticulate, shiny, lobes orbiculate to oblong-obovate, bases overlapping, margins minutely erose or denticulate, membranous, midvein carinate, apex ± truncate, each not exceeded by a hornlike appendage, 0.1–0.2 mm; corolla creamy white, drying creamy white to reddish brown, 3.2–4 mm, tube campanulate, 1.5–2.2 mm, not saccate, sometimes with horizontal ridges between stamen attachments, lobes erect or spreading to reflexed, oblong-obovate to orbiculate, shorter than to equaling corolla tube length, margins minutely erose or denticulate, membranous, midvein unevenly carinate, apex ± truncate, each with a hornlike appendage 0.1–0.2 mm, not exceeding apex, straight; infrastaminal scales oblong to ± truncate, 1.5–2 mm, 3/4 to equaling corolla tube length, bridged at 0.4–0.6 mm, uniformly densely fimbriate, fimbriae 0.2–0.4 mm; stamens exserted, shorter than corolla lobes; filaments 0.4–1 mm; anthers 0.7–1 × 0.4–0.5 mm; styles stout, subulate, 1.8–3.2 mm, longer than ovary. **Capsules** globose, 2–2.5 × 2.2–2.5 mm, thickened, not raised around inconspicuous interstylar aperture, not translucent, withered corolla near middle or apical, dehiscence circumscissile. **Seeds** 1–4, angled, subglobose to ovoid, 0.9–1.4 × 0.8–1.3 mm, papillate, hilum region terminal.

Flowering Aug–Oct. Hosts: *Abutilon*, *Amaranthus*, *Ambrosia*, *Anisacanthus*, *Bidens*, *Carlowrightia*, *Euphorbia*, *Gomphrena*, *Ipomoea*, *Jatropha*, *Justicia*, *Kallstroemia*, *Merremia*, *Mimosa*, *Rhynchosia*, *Ruellia*, *Russelia*, *Sphinctospermum*, *Talinum*, *Tephrosia*; desert scrub; 400–1300 m; Ariz.; Mexico (Baja California, Sinaloa, Sonora).

20. Cuscuta glabrior (Engelmann) Yuncker, Mem. Torrey Bot. Club 18: 140. 1932 • Bushclover dodder [W]

Cuscuta verrucosa Engelmann var. *glabrior* Engelmann, Amer. J. Sci. Arts 43: 341. 1842; *C. glabrior* var. *pubescens* (Engelmann) Yuncker; *C. pentagona* Engelmann var. *glabrior* (Engelmann) Gandhi, R. D. Thomas & S. L. Hatch; *C. pentagona* var. *pubescens* (Engelmann) Yuncker

Stems orange, medium. **Inflorescences** loose to compact, glomerulate or corymbiform; bracts at base of clusters 1, at base of pedicels 0 or 1, ovate or ovate-triangular to lanceolate, membranous, margins entire, apex acute. **Pedicels** 0.8–4(–5) mm, sometimes papillate. **Flowers** 5-merous, 2.5–3.8 mm, membranous, perianth and ovary papillate; calyx yellow to reddish brown, cupulate, ± equaling corolla tube length, divided 1/2–2/3 its length, ± reticulate, shiny, lobes ovate-triangular, bases not overlapping, margins entire, midvein not carinate, without multicellular protuberances, apex obtuse to subacute; corolla white, drying yellow to reddish brown, 1.4–3.4 mm, tube campanulate, later globose, 1.1–1.8 mm, saccate between lines of stamen attachments, lobes spreading to reflexed, triangular, equaling corolla tube length, margins entire, apex acute to acuminate, inflexed; infrastaminal scales ovate to spatulate, 1.2–2 mm, equaling or longer than corolla tube length, bridged at 0.3–0.5 mm, rounded, uniformly densely fimbriate, fimbriae 0.3–0.7 mm; stamens exserted, shorter than corolla lobes; filaments 0.4–0.7 mm; anthers 0.4–0.7 × 0.4–0.5 mm; styles filiform, 0.9–1.6 mm, equaling or longer than ovary. **Capsules** depressed-globose to depressed, 1.5–2.8 × 2.1–3.5 mm, not thickened or raised around relatively mid-sized to large interstylar aperture, not translucent, 1/2–2/3 of base enveloped by withered corolla, indehiscent. **Seeds** 2–4, angled, subglobose to broadly ellipsoid, 0.9–1.1 × 0.8–1 mm, hilum region subterminal.

Flowering Apr–Sep. Hosts: *Amaranthus*, *Ambrosia*, *Amphiachyris*, *Asclepias*, *Convolvulus*, *Coreopsis*, *Croton*, *Dalea*, *Dyschoriste*, *Evolvulus*, *Gilia*, *Hedeoma*, *Helenium*, *Justicia*, *Lespedeza*, *Liatris*, *Machaeranthera*, *Medicago*, *Mimosa*, *Oenothera*, *Plantago*, *Polygonum*, *Prosopis*, *Ruellia*, *Solanum*, *Symphyotrichum*, *Thelesperma*, *Tragia*, *Verbena*, and other herbs; 10–1200 m; La., N.Mex., Okla., Tex., Utah; Mexico.

Cuscuta glabrior differs from *C. campestris* by papillate perianths, non-overlapping calyx lobe bases, corollas saccate between lines of stamen attachments, and capsules more than half enveloped by withered corolla.

Cuscuta glabrior is currently included in North American noxious weed lists although it is not known to attack any crops.

21. Cuscuta glomerata Choisy, Mém. Soc. Phys. Genève 9: 280, plate 4, fig. 1. 1842 • Rope dodder [E] [W]

Stems orange, medium. **Inflorescences** extremely dense, continuous (cymes indiscernible), ropelike, spiraling around and closely appressed to host stem, flowers sessile, commonly originating endogenously; bracts at base of clusters, pedicels, and/or flowers 4–11, narrowly triangular to lanceolate, folded-concave, membranous, margins serrate or lacerate, apex acute, recurved. **Pedicels** absent. **Flowers** 5-merous, 5–6 mm, membranous, not papillate; calyx stramineous yellow, campanulate-turbinate, equaling corolla tube length, divided to base or nearly so, reticulate, shiny, lobes narrowly triangular to lanceolate, bases overlapping, margins serrate to irregular, midvein not carinate, apex obtuse to acute; corolla white, drying stramineous yellow or brownish, 4.5–5.4 mm, tube cylindric, 3.6–4.4 mm, not saccate, lobes spreading to reflexed, oblong-lanceolate, 1/5–1/4 corolla tube length, margins entire, apex acute to obtuse, straight; infrastaminal scales oblong, 2.5–3 mm, 3/4 corolla tube length, bridged at 1.3–1.7 mm, rounded, ± uniformly fimbriate, fimbriae 0.3–0.5 mm; stamens exserted, shorter than corolla lobes; filaments 0.5–0.7 mm; anthers 0.6–0.8 × 0.3–0.4 mm; styles filiform, 2–3 mm, longer than ovary. **Capsules** globose to flask-shaped, 3.5–6 × 2–3.5 mm, thickened and raised in collar around style bases, not translucent, capped by withered corolla, indehiscent. **Seeds** 1 or 2, slightly obcompressed, subglobose, or ovoid, 1.4–1.7 × 1.2–1.6 mm, hilum region terminal. $2n = 30$.

Flowering Jul–Oct. Hosts: *Ambrosia*, *Asclepias*, *Helenium*, *Helianthus*, *Liatris*, *Silphium*, *Solidago*, *Symphyotrichum*, *Vernonia*, and others; 100–1000 m; Ark., Ill., Ind., Iowa, Kans., Ky., La., Mich., Minn., Miss., Mo., Nebr., N.Dak., Ohio, Okla., S.Dak., Tenn., Tex., Wis.

Inflorescences in ropelike spirals and capsules with necklike collars provide instant recognition for *Cuscuta glomerata*.

22. Cuscuta gronovii Willdenow in J. J. Roemer et al., Syst. Veg. 6: 205. 1820 • Swamp dodder, scaldweed, cuscute de gronovius [F] [W]

Stems yellow to orange, medium to coarse. **Inflorescences** loose or dense, paniculiform, sometimes originating endogenously; bracts at base of clusters 1, at base of pedicels 0 or 1, rarely 1 or 2(or 3) on pedicels, ovate to broadly triangular, membranous, margins entire or serrulate, apex acute to obtuse. **Pedicels** 1–4.5 mm. **Flowers** (4 or)5-merous, 2–4 mm, membranous, not papillate; calyx yellow-brown, cupulate or campanulate to narrowly campanulate, 1/2 to equaling corolla tube length, divided 1/2–2/3 its length, not reticulate or shiny, lobes ovate to suborbiculate or oblong, bases barely to notably overlapping, margins entire to serrulate, midvein not carinate, apex rounded or obtuse; corolla white fresh or dry, 1.8–4 mm, tube broadly to narrowly campanulate, 1–2.5(–3) mm, not saccate, lobes spreading to reflexed, mostly ovate, (1/4–)1/3 or 1/2 to equaling corolla tube length, margins entire, apex rounded to obtuse, straight; infrastaminal scales oblong, 1–2.5 mm, equaling corolla tube length, bridged at 0.4–0.8 mm, sparsely fimbriate, fimbriae 0.4–0.8 mm; stamens exserted, shorter than to equaling corolla lobes; filaments 0.4–0.7(–1) mm; anthers 0.3–0.6 × 0.3–0.4 mm; styles usually filiform, sometimes slightly subulate, (0.6–)1.2–2.2 mm, shorter than to equaling ovary. **Capsules** ovoid to conic-globose or subobpyriform, 2.5–4.5(–5.2) × 2–4(–5) mm, raised and ± thickened around relatively small to moderately large interstylar aperture, rarely apically narrowed into neck to 1 mm, not translucent, loosely surrounded or capped by withered corolla, indehiscent. **Seeds** 2–4, obcompressed to obscurely angled, subglobose to broadly ovoid, 1.3–1.7(–2.2) × 1.2–1.6 mm, hilum region subterminal.

Varieties 3 (3 in the flora): North America, West Indies; introduced in Europe.

1. Calyces equaling corolla tube length, lobes oblong to ovate, bases barely overlapping; corolla tube broadly campanulate, 1–1.5 mm, lobes 1/2 to equaling tube length . . . . 22b. *Cuscuta gronovii* var. *latiflora*
1. Calyces 1/2 corolla tube length, lobes ovate to suborbiculate, bases overlapping; corolla tube campanulate to narrowly campanulate, 1.5–2.5 (–3) mm, lobes (1/4–)1/3 tube length.
    2. Corolla tube campanulate, 1.5–2.5 mm; capsules loosely surrounded by withered corolla . . . . 22a. *Cuscuta gronovii* var. *gronovii*
    2. Corolla tube narrowly campanulate, 2–2.5 (–3) mm; capsules capped by withered corolla . . . . 22c. *Cuscuta gronovii* var. *calyptrata*

22a. Cuscuta gronovii Willdenow var. **gronovii** [F] [W]

Calyces 1/2 corolla tube length, lobes ovate to suborbiculate, bases overlapping. **Corollas:** tube campanulate, 1.5–2.5 mm, lobes 1/3 tube length. **Capsules** loosely surrounded by withered corolla. $2n = 30$.

Flowering Jun–Nov. Hosts: Acanthaceae, Anacardiaceae, Apiaceae, Asteraceae, Balsaminaceae, Betulaceae, Bignoniaceae, Brassicaceae, Caprifoliaceae, Commelinaceae, Convolvulaceae, Cornaceae, Euphorbiaceae, Fabaceae, Lamiaceae, Polygonaceae, Primulaceae, Rosaceae, Rubiaceae, Solanaceae, Urticaceae, Verbenaceae, Vitaceae, and others (M. Costea and F. J. Tardif 2006); wetland margins, wet forests; 20–300 m; Alta., Man., N.B., N.S., Ont., P.E.I., Que., Sask.; Ala., Ariz., Ark., Colo., Conn., D.C., Fla., Ga., Idaho, Ill., Ind., Iowa, Kans., Ky., La., Maine, Md., Mass., Mich., Minn., Miss., Mo., Mont., Nebr., N.H., N.J., N.Y., N.C., N.Dak., Ohio, Okla., Oreg., Pa., R.I., S.C., S.Dak., Tenn., Tex., Utah, Vt., Va., W.Va., Wis.; West Indies; introduced in Europe.

In the flora area, var. *gronovii* is the third most widespread dodder after *Cuscuta campestris* and *C. indecora*. Rarely, some plants may have capsules apically narrowed into a neck to 1 mm, reminiscent of *C. rostrata*.

Variety *gronovii* is a weed in cranberry crops in Massachusetts, New Jersey, and Wisconsin.

22b. Cuscuta gronovii Willdenow var. **latiflora** Engelmann, Trans. Acad. Sci. St. Louis 1: 508. 1859 [E] [W]

Cuscuta saururi Engelmann, Amer. J. Sci. Arts 43: 339, plate 6, figs. 17–21. 1842

Calyces equaling corolla tube length, lobes oblong to ovate, bases barely overlapping. **Corollas:** tube broadly campanulate, 1–1.5 mm, lobes 1/2 to equaling tube length. **Capsules** loosely surrounded by withered corolla.

Flowering Jul–Oct. Hosts: *Acalypha*, *Bidens*, *Boehmeria*, *Cephalanthus*, *Decodon*, *Impatiens*, *Penthorum*, *Persicaria*, *Salix*, *Saururus*, *Sium*, *Solidago*, and others; stream banks, mudflats, margins of wetlands, wet meadows, alluvial forests; 20–200 m; Ont., Que.; Ark., Conn., D.C., Ill., Ind., Iowa, Ky., La., Md., Mass., Mich., Miss., Mo., N.H., N.J., N.Y., N.C., Ohio, Okla., Pa., R.I., Tenn., Tex., Vt., Va., W.Va.

T. G. Yuncker (1932, 1965) indicated that var. *latiflora* has the same distribution as var. *gronovii*. Current herbarium data suggest that although the two varieties are sympatric over a significant portion of the flora area, var. *latiflora* has a narrower geographical distribution.

22c. Cuscuta gronovii Willdenow var. **calyptrata** Engelmann, Trans. Acad. Sci. St. Louis 1: 508. 1859 [C] [E]

Cuscuta calyptrata (Engelmann) Small

Calyces 1/2 corolla tube length, lobes ovate to suborbiculate, bases overlapping. **Corollas:** tube narrowly campanulate, 2–2.5(–3) mm, lobes (1/4–)1/3 tube length. **Capsules** capped by withered corolla.

Flowering late Aug–Oct. Hosts: herbs and woody plants; of conservation concern; 200–300 m; La., Tex.

Variety *calyptrata* is sympatric with the more widespread var. *gronovii*; it resembles *Cuscuta cephalanthi* in the way the withered corolla caps the capsule and differs in larger, usually five-merous flowers, wider infrastaminal scales, and ovoid-globose capsules.

23. Cuscuta harperi Small, Fl. S.E. U.S. ed. 2, 1361, 1375. 1913 • Harper's dodder [C] [E]

Stems orange-yellow, filiform. **Inflorescences** loose, corymbiform; bracts at base of clusters 1, at base of pedicels 0(or 1), ovate-triangular, membranous, margins entire, apex acute. **Pedicels** 0.5–2.5(–3) mm. **Flowers** 4(or 5)-merous, 0.9–1.1(–1.5) mm, fleshy, corolla lobes papillate; calyx brownish yellow, angled, cupulate, equaling corolla tube length, divided 1/2–2/3 its length, ± reticulate, rarely shiny, lobes broadly ovate-rhombic, bases ± auriculate, overlapping, forming prominent angles at sinuses, margins entire, midvein ± carinate, apex rounded; corolla white, drying cream to brownish, 0.9–1.2 mm, tube campanulate, 0.5–0.7 mm, not saccate, lobes reflexed, triangular-ovate, equaling corolla tube length, margins entire, apex subacute to acute, inflexed; infrastaminal scales narrowly oblong to obovate, 0.6–0.9 mm, equaling or longer than corolla tube length, bridged at 0.2–0.3 mm, rounded, sparsely fimbriate, fimbriae 0.1–0.3 mm; stamens exserted, slightly shorter than corolla lobes; filaments 0.1–0.2 mm; anthers 0.2–0.3 × 0.2–0.3 mm; styles filiform, 0.5–0.9 mm, shorter than ovary. **Capsules** globose to ovoid, 1.2–2.3 × 1.2–1.6 mm, not thickened around relatively small to moderately large interstylar aperture, ± translucent, withered corolla enveloping 1/4–1/3 of base, indehiscent. **Seeds** 1 or 2, subglobose, with longitudinal groove on adaxial face, 0.9–1.1 × 0.8–1.1 mm, hilum region terminal.

Flowering Sep–Nov. Hosts: *Bigelowia nuttallii*, *Croton michauxii* var. *elliptica*, *Helianthus longifolius*, *Hypericum gentianoides*, *Liatris microcephala*, and others; sandstone, less frequently, granite outcrops; of conservation concern; 200–600 m; Ala., Ga.

Cuscuta harperi is closely allied with *C. pentagona* (M. Costea et al. 2015), from which it differs in smaller, four-merous flowers.

24. Cuscuta howelliana P. Rubtzov, Leafl. W. Bot. 10: 335. 1966 • Boggs Lake dodder [E] [W]

Stems yellow to orange, slender. **Inflorescences** dense, glomerulate, 3–30-flowered, flowers sessile or subsessile, embedded in inflorescence of host; bracts at base of clusters 1, at base of pedicels and/or flowers 0 or 1, lanceolate, membranous, margins entire, apex acute. **Pedicels** 0–0.6 mm. **Flowers** 4(or 5)-merous, 3–4 mm, membranous, calyx and corolla papillate; calyx straw yellow, campanulate, equaling or longer than corolla tube length, divided 1/2–2/3 its length, finely reticulate, shiny, lobes triangular-ovate, bases not overlapping, margins entire, midvein not carinate, apex acuminate to long-attenuate, recurved; corolla drying white or creamy white to brownish, 2.8–3.5 mm, tube cylindric-campanulate to urceolate, 1.5–2 mm, not saccate, lobes suberect to spreading, triangular-ovate, equaling corolla tube length, margins entire, apex acute to long-attenuate, recurved; infrastaminal scales oblong-ovate, 1–1.3 mm, 1/2–2/3 corolla tube length, bridged at 0.2–0.3 mm, rounded, uniformly densely fimbriate, fimbriae 0.1–0.3 mm; stamens included, shorter than corolla lobes; filaments 0.1–0.3 mm; anthers 0.4–0.5 × 0.3–0.4 mm; styles filiform, 0.4–1.1 mm, 1/4 to ± equaling ovary. **Capsules** globose to depressed-globose, 1.2–1.5 × 0.8–1.2 mm, not thickened or raised around relatively small interstylar aperture, not translucent, completely enclosed or capped by withered corolla, indehiscent. **Seeds** 1–4, obcompressed to slightly angled, subglobose to broadly ellipsoid, 0.9–1.2 × 0.8–1.1 mm, hilum region subterminal.

Flowering Aug–Sep. Hosts: *Diplacus*, *Downingia*, *Epilobium*, *Eryngium*, *Navarretia*, *Polygonum*; vernal pools; 30–1000 m; Calif.

Inflorescences of *Cuscuta howelliana* develop inside inflorescences of *Epilobium densiflorum*, *Eryngium*, and *Navarretia*; the flowers apparently synchronize their anthesis with that of the host's flowers and achieve both protection from the host and access to the pollinators of the host. When it parasitizes *Diplacus* and *Downingia*, the parasite flowers but does not produce seeds.

25. Cuscuta indecora Choisy, Mém. Soc. Phys. Genève 9: 278, plate 3, fig. 3. 1842 • Large-seed dodder [F] [W]

Stems yellow to orange, slender to medium. **Inflorescences** loose to dense, paniculiform or corymbiform, sometimes originating endogenously; bracts at base of clusters 1, at base of pedicels 0 or 1, ovate to lanceolate, membranous, margins entire, apex acute. **Pedicels** 0.5–6 mm, usually papillate. **Flowers** 5-merous, 3–4.5(–5.3) mm, fleshy, perianth cells convex, domelike, perianth and ovary usually papillate; calyx creamy yellow to brownish, cupulate, 1/2–3/4 or longer than corolla tube length, divided 1/3–2/3 its length, not reticulate or shiny, lobes triangular-ovate to lanceolate, bases overlapping or not, margins entire, midvein not carinate, apex acute to attenuate; corolla white, drying creamy yellow to dark brown, 2.5–4(–5) mm, tube campanulate to campanulate-cylindric, becoming subglobose or urceolate, 1.7–3 mm, not saccate, lobes suberect to erect, triangular-ovate, 1/3 to equaling corolla tube length, margins entire, apex acute, inflexed; infrastaminal scales subspatulate to spatulate, 1.7–3 mm, equaling corolla tube length, bridged at 0.3–0.9 mm, usually rounded, rarely truncate or 2 or 3(or 4)-lobed, uniformly densely fimbriate, fimbriae 0.4–0.7 mm; stamens barely exserted or included, shorter than corolla lobes; filaments 0.3–0.7 mm; anthers 0.3–0.8 × 0.2–0.5 mm; styles filiform, 1–2.5 mm, equaling ovary. **Capsules** globose to subglobose, 2–3.5 × 1.9–4(–5) mm, thickened and raised around relatively mid-sized interstylar aperture, translucent, surrounded or capped by withered corolla, indehiscent. **Seeds** 2–4, shape heterogeneous on same plant: obcompressed to weakly angled, broadly ellipsoid to transversely oblique, 1.4–1.8 × 1.2–1.6 mm, hilum region usually subterminal, rarely almost terminal.

Varieties 3 (3 in the flora): North America, Mexico, West Indies, South America.

Cuscuta indecora is closely related to *C. coryli*; it differs by its usually five-merous, larger flowers, uniformly densely fimbriate infrastaminal scales, and more or less translucent, globose to subglobose capsules.

1. Calyces shorter than corolla tubes, divided 1/3–1/2 lengths, lobes triangular-ovate, bases overlapping . . . . 25a. *Cuscuta indecora* var. *indecora*
1. Calyces longer than corolla tubes, divided 2/3 lengths, lobes lanceolate, bases not overlapping.
   2. Flower clusters loose; calyx lobe apices acute; hosts: herbs and woody plants, including *Iva annua*. . . . . . . 25b. *Cuscuta indecora* var. *longisepala*
   2. Flower clusters dense; calyx lobe apices acute-attenuate; hosts: usually *Iva annua*, rarely *Symphyotrichum* . . . . .25c. *Cuscuta indecora* var. *attenuata*

25a. Cuscuta indecora Choisy var. **indecora** F W

Cuscuta indecora var. *bifida* Yuncker; *C. indecora* var. *neuropetala* (Engelmann) Hitchcock

Flowers: clusters loose to dense. **Calyces** 1–2 mm, shorter than corolla tube, divided 1/3–1/2 length, lobes triangular-ovate, bases overlapping, apex acute. **2*n*** = 30.

Flowering Apr–Nov. Hosts: *Acacia*, *Agalinis*, *Anulocaulis*, *Artemisia*, *Asclepias*, *Baccharis*, *Borrichia*, *Chenopodium*, *Clematis*, *Convolvulus*, *Eupatorium*, *Grindelia*, *Helenium*, *Helianthus*, *Heterotheca*, *Hypericum*, *Ipomoea*, *Kosteletzkya*, *Lactuca*, *Lepidium*, *Ligustrum*, *Malvastrum*, *Medicago*, *Mimosa*, *Myrica*, *Parthenium*, *Pithecellobium*, *Pluchea*, *Polygonum*, *Rhynchosia*, *Salsola*, *Solidago*, *Suaeda*, *Symphyotrichum*, *Tecoma*, *Tephrosia*, *Vernonia*, and others; 0–1900 m; Sask.; Ala., Ariz., Ark., Calif., Colo., Conn., Fla., Ga., Idaho, Ill., Iowa, Kans., Ky., La., Md., Mich., Minn., Miss., Mo., Mont., Nebr., Nev., N.J., N.Mex., N.C., N.Dak., Okla., S.C., S.Dak., Tenn., Tex., Utah, Va., Wash., W.Va., Wyo.; Mexico; West Indies; South America.

Variety *bifida* was noted by T. G. Yuncker (1965) to occur infrequently throughout the range of var. *indecora*. It is found that the infrastaminal scales of var. *bifida* are not truly bifid; they are spatulate with two or three, sometimes four, deeper apical incisions that create two or three, sometimes four lobes that are further fimbriate. So-called normal scales may occur in the same flower together with lobed ones. Such plants are regarded as populational variants of var. *indecora*. Bifid scales are characteristic of *C. coryli*, not *C. indecora* (M. Costea et al. 2006).

Variety *indecora* is the second most common and widespread taxon of the genus in the flora area, after *Cuscuta campestris*. It is a particularly troublesome weed in alfalfa and currently continues to spread through contaminated seeds worldwide.

25b. Cuscuta indecora Choisy var. **longisepala** Yuncker, Illinois Biol. Monogr. 6: 149, plate 8, fig. 44f, plate 11, fig. 97. 1921 F W

Flowers: clusters loose. **Calyces** 1.5–3 mm, longer than corolla tube, divided 2/3 length, lobes lanceolate, bases not overlapping, apex acute.

Flowering Jun–Sep. Hosts: *Ambrosia*, *Gutierrezia*, *Helianthus*, *Iva annua*, *Solidago*; Tex.; Mexico; introduced in South America.

Variety *longisepala* is currently included in North American lists of noxious weeds although is not known to attack any crops. It was introduced to South America, where it is more common and sometimes weedy.

25c. Cuscuta indecora Choisy var. **attenuata** (Waterfall) Costea, Sida 22: 216. 2006 C E F

Cuscuta attenuata Waterfall, Rhodora 73: 575. 1972

Flowers: clusters dense. **Calyces** 1.5–3 mm, longer than corolla tube, divided 2/3 length, lobes ovate-lanceolate to narrowly lanceolate, bases not overlapping, apex acute-attenuate. **2*n*** = 30.

Flowering late Aug–Oct. Hosts: usually *Iva annua*, rarely *Symphyotrichum*; of conservation concern; 0–400 m; Kans., Okla., Tex.

Variety *attenuata* resembles var. *longisepala* but is maintained as a separate variety of *Cuscuta indecora* because of its apparent host specialization and reproductive isolation (L. A. Prather and R. J. Tyrl 1993).

26. Cuscuta jepsonii Yuncker, Illinois Biol. Monogr. 6: 149, plate 9, fig. 52. 1921 • Jepson's dodder C E

Stems pale yellow, slender. **Inflorescences** moderately dense, corymbiform or glomerulate; bracts at base of clusters 1, at base of pedicels 0 or 1, ovate to lanceolate, membranous, margins entire, apex acute. **Pedicels** 0.5–1.5 mm, papillate. **Flowers** 5-merous, 2–3 mm, fleshy, perianth cells convex, domelike, perianth papillate; calyx yellow-brownish, shallowly cupulate, 1/2 corolla tube length, divided 1/2 its length, not reticulate or shiny,

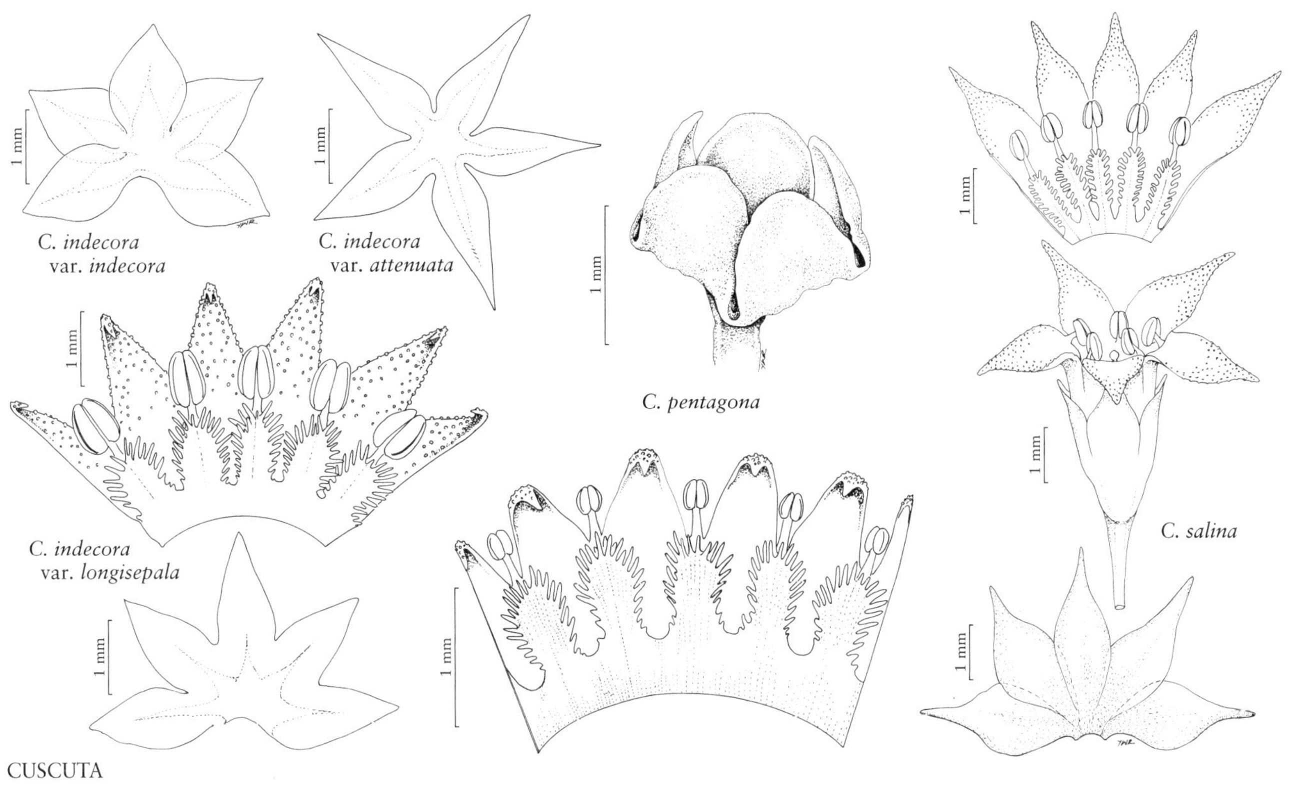

CUSCUTA

lobes triangular, bases not overlapping, margins entire, midvein not carinate, apex acute; corolla white, drying brownish, 1.8–2.8 mm, tube campanulate-globulose, 1.3–2 mm, later urceolate, not saccate, lobes erect, ovate-triangular, $^{1}/_{3}$–$^{1}/_{2}$ corolla tube length, margins entire, apex acute, inflexed; infrastaminal scales absent; stamens included, shorter than corolla lobes; filaments 0.2–0.3 mm; anthers 0.2–0.3 × 0.2–0.3 mm; styles subulate, 0.4–0.8 mm, much shorter than ovary. **Capsules** subglobose or globose to slightly depressed-globose, 2–3 × 2–3.5 mm, thickened and raised around relatively large interstylar aperture, translucent, surrounded by withered corolla, indehiscent. **Seeds** 2–4, broadly ellipsoid to subglobose, obcompressed, 0.9–1.1 × 0.8–1 mm, hilum subterminal.

Flowering Jul–Sep. Hosts: *Ceanothus diversifolius*, *C. prostratus*; mixed forests of *Pinus jeffreyi*, *P. ponderosa*, and *Pseudotsuga menziesii*; of conservation concern; 1100–1600 m; Calif.

Cuscuta jepsonii belongs to sect. *Californicae* Yunker (M. Costea and S. Stefanović 2009) and was presumed extinct until collected again from Yosemite National Park (Mariposa County) in 2009.

27. **Cuscuta legitima** Costea & Stefanović, Taxon 59: 1795, fig. 5A–D. 2010 • Large-flowered flat-globe dodder

Stems yellow-orange, slender. **Inflorescences** dense to loose, umbelliform; bracts at base of clusters 1, at base of pedicels 0 or 1, broadly triangular-ovate, membranous, finely reticulate, slightly shiny, margins entire, apex acuminate. **Pedicels** 2–10 mm. **Flowers** 5-merous, 4–5.5 (–6) mm, membranous, not papillate; calyx straw yellow, campanulate, 2.5–3.2 mm, longer than corolla tube length, divided $^{2}/_{3}$ its length, finely reticulate, slightly shiny, lobes ovate-lanceolate, bases not overlapping, margins entire, midvein not carinate, without multicellular protuberances, apex acuminate; corolla white, drying creamy white, 3.8–5.2(–5.6) mm, tube campanulate, 1.6–2.1 mm, not saccate, lobes reflexed, linear-lanceolate, longer than corolla tube length, margins entire, apex acuminate, straight; infrastaminal scales spatulate to obovate, 1.8–2.2 mm, equaling or slightly longer than corolla tube length, bridged at 0.2–0.4 mm, rounded, uniformly densely fimbriate, fimbriae 0.2–0.5 mm; stamens exserted, shorter than corolla lobes; filaments 0.6–1 mm; anthers 0.5–0.7 × 0.2–0.3 mm; styles filiform, 0.9–2.5 mm, longer than ovary.

Capsules depressed-globose, 2–3 × 1–2 mm, irregularly thickened and slightly raised around inconspicuous interstylar aperture, translucent, surrounded or capped by withered corolla, dehiscence circumscissile. **Seeds** 2–4, broadly ellipsoid to subglobose, 0.9–1.2 × 0.8–0.9 mm, hilum subterminal.

Flowering Aug–Nov. Hosts: *Allionia*, *Amaranthus*, *Boerhavia*, *Chamaesaracha*, *Evolvulus*, *Kallstroemia*, *Salsola*, *Solanum*, *Tidestromia*, *Trianthema*, *Tribulus*; 40–1200 m; Ariz., N.Mex., Tex.; Mexico (Baja California, Chihuahua, Coahuila, Sonora, Tamaulipas).

Cuscuta legitima is a hybrid species closely allied with *C. umbellata*, from which it differs in larger flowers and acuminate calyx lobes.

28. Cuscuta leptantha Engelmann, Trans. Acad. Sci. St. Louis 1: 489. 1859 • Slender dodder [W]

Cuscuta palmeri S. Watson

Stems yellow, slender. **Inflorescences** loose, umbelliform; bracts at base of clusters 1, at base of pedicels 0 or 1, ovate, membranous, margins entire, apex acute. **Pedicels** 0.7–7 mm, papillate. **Flowers** 4-merous, 3.5–4.5(–5) mm, membranous, perianth papillate; calyx straw yellow, campanulate, 1/3–1/2 corolla tube length, divided 1/2 its length, not reticulate or shiny, lobes triangular-ovate, bases not overlapping, margins entire, midvein not carinate, apex acute; corolla white, drying creamy white, 3–4 mm, tube cylindric, 1.5–2.5 mm, not saccate, lobes spreading to reflexed, lanceolate, equaling corolla tube, margins entire, often involute and appearing very narrow dry, apex acute, ± cucullate; infrastaminal scales oblong, 1.3–2.1 mm, 1/2 corolla tube length, bridged at 0.4–0.8 mm, rounded, uniformly short-fimbriate, fimbriae 0.05–0.1 mm; stamens exserted, shorter than corolla lobes; filaments 0.3–0.6 mm; anthers 0.4–0.6 × 0.3–0.4 mm; styles filiform, 1.2–2.1 mm, longer than ovary. **Capsules** globose, 1.5–2 × 1.6–1.9 mm, slightly thickened and raised around inconspicuous interstylar aperture, translucent, capped by withered corolla, dehiscence circumscissile. **Seeds** 2–4, angled, subglobose to broadly ellipsoid, 0.7–0.9 × 0.7–0.8 mm, hilum subterminal.

Flowering Feb–Nov. Hosts: *Euphorbia* subg. *Chamaesyce*; 10–100 m; N.Mex., Tex.; Mexico (Baja California, Coahuila, Sinaloa, Sonora).

29. Cuscuta liliputana Costea & Stefanović, Botany (Ottawa) 86: 802, fig. 4. 2008 [E]

Stems yellow to pale orange, slender. **Inflorescences** loose, umbelliform; bracts at base of clusters 1, at base of pedicels 0 or 1, ovate-lanceolate, membranous, ± reticulate, shiny, margins entire, apex acute. **Pedicels** (1–)2–3(–5) mm, papillate. **Flowers** (3 or)4-merous, 2.8–4 mm, calyx and corolla papillate; calyx straw yellow, cylindric, equaling corolla tube length, divided 3/4 its length, ± reticulate, shiny, lobes ovate-triangular, bases not overlapping, margins entire, midvein not carinate, sometimes with multicellular protuberances, apex acute to acuminate; corolla white, drying creamy white, 3–3.6 mm, tube cylindric, 1.5–2 mm, not saccate, lobes spreading or reflexed, lanceolate, equaling corolla tube length, margins entire, apex acute, straight; infrastaminal scales truncate to slightly obovate, 0.8–1.2 mm, 1/4–1/3 corolla tube length, bridged at 0.1–0.2 mm, distally fimbriate, fimbriae 0.1–0.2 mm; stamens exserted, shorter than corolla lobes; filaments 0.5–0.8 mm; anthers 0.3–0.5 × 0.2–0.3 mm; styles filiform, 0.8–2.5 mm, longer than ovary. **Capsules** globose to depressed-globose, 1.5–2.2 × 0.7–1.5 mm, thickened and slightly raised or with 2–4 protuberances around relatively small interstylar aperture, translucent, capped by withered corolla, dehiscence circumscissile. **Seeds** 2–4, angled, subglobose to broadly elliptic, 0.8–1.1 × 0.7–0.8 mm, hilum subterminal.

Flowering Jul–Feb. Host: *Euphorbia* subg. *Chamaesyce*; 30–1700 m; Ariz., N.Mex., Tex.

30. Cuscuta mitriformis Engelmann ex Hemsley, Diagn. Pl. Nov. Mexic. 3: 54. 1880 (as mitraeformis) • Cochise dodder [C]

Stems orange, medium to coarse. **Inflorescences** dense, corymbiform-glomerulate; bracts at base of clusters 1, at base of pedicels 0(or 1), ovate-triangular, fleshy to membranous, margins entire, apex acute. **Pedicels** 0.5–2 mm. **Flowers** 5-merous, 4–5.5 mm, receptacle and calyx fleshy, corolla membranous, not papillate; calyx brownish, shallowly cupulate, equaling corolla tube length, divided 1/3 its length, ± reticulate and shiny, lobes broadly ovate, bases overlapping, margins entire, midvein ± carinate, apex obtuse, sometimes with an appendage; corolla white, drying creamy brown, 3.6–5 mm, tube campanulate, 1.5–3.5 mm, not saccate, lobes spreading to reflexed, broadly ovate, equaling corolla tube length, without hornlike appendages, margins entire, apex obtuse, straight; infrastaminal scales ovate to oblong, 1.5–3.5 mm, equaling corolla tube length, bridged at 0.4–0.7 mm, truncate to rounded, sparsely fimbriate (sometimes only distally), fimbriae 0.1–0.4 mm; stamens slightly exserted, shorter than corolla lobes; filaments subulate, 0.5–0.9 mm; anthers 0.5–0.9 × 0.3–0.4 mm; styles stout, subulate, 0.9–2 mm, shorter than ovary. **Capsules** globose, 4–7 × 4–6 mm, thickened and raised around relatively large interstylar aperture, not translucent, surrounded by withered corolla at base, dehiscence circumscissile. **Seeds** 4, angled, subglobose to broadly ellipsoid or ovoid, 1.9–2.3 × 1.9–2.3 mm, hilum region lateral.

Flowering Jul–Sep. Hosts: *Ageratina*, *Ambrosia*, *Bouvardia*, *Commelina*, *Desmodium*, *Drymaria*, *Encelia*, *Eupatorium*, *Lupinus*, *Phaseolus*, *Salvia*, *Solanum*, *Stevia*, *Thalictrum*, *Verbena*, and others; of conservation concern; 1500–2500 m; Ariz.; Mexico.

Cuscuta mitriformis is relatively widespread in Mexico (Chihuahua, Coahuila, Durango, Estado de México, Guanajuato, Hidalgo, Michoacán, Nayarit, Nuevo León, Querétaro, Puebla, San Luis Potosí, Veracruz, and Zacatecas) but of conservation concern in the flora area, where it is known only from the Chiricahua Mountains in Cochise County, Arizona (M. Costea et al. 2013).

31. Cuscuta nevadensis I. M. Johnston, Proc. Calif. Acad. Sci., ser. 4, 12: 1133. 1924 [E] [W]

Cuscuta veatchii Brandegee var. *apoda* Yuncker, Illinois Biol. Monogr. 6: 159, plate 8, fig. 48f. 1921

Stems light yellow to orange, slender to medium. **Inflorescences** loose, umbelliform; bracts at base of clusters 1, at base of pedicels 0 or 1, lanceolate to ovate-lanceolate, membranous, margins ± entire, apex acute. **Pedicels** (0.5–)2.5–4 mm. **Flowers** (4 or)5-merous, (2.8–)3–4(–5) mm, membranous, receptacle and base of perianth fleshy, not papillate; calyx golden yellow to brownish, narrowly campanulate, equaling corolla tube length, divided 2/3 its length, reticulate and shiny, lobes lanceolate, bases overlapping, margins ± entire, midvein not carinate, apex acute to acuminate; corolla white, drying brownish yellow, 2.8–4 mm, tube campanulate, 1.3–2 mm, not saccate, lobes reflexed, ovate to oblong, equaling or longer than corolla tube length, margins entire or irregularly denticulate, apex acute, straight; infrastaminal scales ovate to oblong, 1.3–1.8 mm, equaling corolla tube length, bridged at 0.4–0.6 mm, truncate to rounded, margins uniformly fimbriate, fimbriae 0.1–0.2 mm; stamens included or slightly exserted, shorter than corolla lobes; filaments (0–)0.1–0.3 mm; anthers 0.5–0.8 × 0.3–0.4 mm; styles uniformly filiform, 0.5–1 mm, 1/2 to equaling ovary. **Capsules** globose-ovoid to ovoid, 1.4–2.1 × 1.2–2 mm, not thickened to raised around inconspicuous interstylar aperture, translucent, capped by withered corolla, indehiscent. **Seeds** 1, globose-ovoid, 0.9–1.2 × 0.8–1.08 mm, hilum region terminal, embryo base globose-enlarged.

Flowering May–Jul. Hosts: usually *Ambrosia dumosa*, *Atriplex confertifolia*, *A. hymenelytra*, *A. polycarpa*, *Psorothamnus fremontii*, *Suaeda nigra*; rarely *Acamptopappus*, *Brassica*, *Dedeckera*, *Lycium*, and *Stephanomeria*; salt-desert shrubland; 400–1500 m; Calif., Nev.

Cuscuta nevadensis is viviparous; its seeds germinate inside fruits enclosed by perianths and attached to the parent plant (M. Costea et al. 2005).

32. Cuscuta obtusiflora Kunth in A. von Humboldt et al., Nov. Gen. Sp. 3(fol.): 96; 3(qto.): 122. 1819 • Peruvian dodder [W]

Varieties 2 (1 in the flora): United States, Mexico, West Indies, South America.

Variety *obtusiflora* is known from South America.

32a. Cuscuta obtusiflora Kunth var. **glandulosa** Engelmann, Trans. Acad. Sci. St. Louis 1: 492. 1859 [W]

Cuscuta glandulosa (Engelmann) Small

Stems orange, slender to medium. **Inflorescences** dense, glomerulate; bracts at base of clusters 1, at base of pedicels and/or flowers 0 or 1, ovate, membranous, margins entire, apex acute to obtuse. **Pedicels** 0–1 mm. **Flowers** 5-merous, 1.8–2.5 mm, membranous, not papillate; calyx yellow-brown, shallowly cupulate, equaling corolla tube length, divided 1/2 its length, not reticulate or shiny, lobes ovate, bases barely overlapping, margins entire, midvein not carinate, apex obtuse; corolla creamy white, drying yellow-brown, 1.6–2.5 mm, tube campanulate, 1–1.5 mm, not saccate, lobes spreading, ovate to ovate-oblong, shorter than to equaling corolla tube length, margins entire, apex obtuse to rounded, straight; infrastaminal scales oblong to obovate, 1–1.5 mm, equaling corolla tube length, bridged at 0.2–0.5 mm, rounded, with relatively few basal fimbriae, densely fimbriate in distal 1/2 or uniformly sparsely fimbriate, fimbriae 0.2–0.6 mm; stamens exserted, shorter than corolla lobes; filaments 0.4–0.6 mm; anthers 0.3–0.4 × 0.2–0.4 mm; styles uniformly stout, 0.4–1.1 mm, equaling or shorter than ovary. **Capsules** depressed-globose, 1.5–3 × 2.5–4 mm, not raised or thickened around relatively large interstylar aperture, not translucent, base enveloped by withered corolla, indehiscent. **Seeds** (3 or)4, obcompressed, broadly ovoid to broadly ellipsoid, 1.4–1.5 × 1.2–1.3 mm, hilum area subterminal.

Flowering Jun–Sep(–Oct). Hosts: *Alternanthera*, *Dalea*, *Hygrophila*, *Justicia*, *Ludwigia*, *Lythrum*, *Persicaria*, *Xanthium*; stream banks, margins of lakes, swamps, flood plains; 20–600 m; Ala., Ark., Calif., Fla., Ga., Ky., La., Miss., N.Y., Okla., Tex.; Mexico; West Indies (Cuba, Puerto Rico).

Variety *glandulosa* is similar morphologically to *Cuscuta polygonorum*; it differs in its five-merous flowers, obtuse corolla lobes, and densely fimbriate infrastaminal scales. Both taxa are usually parasitic on *Persicaria*. Variety *glandulosa* differs from *C. gronovii* var. *latiflora* in its depressed-globose capsules that are not thickened or raised around the relatively large interstylar aperture.

33. Cuscuta occidentalis Millspaugh, Publ. Field Mus. Nat. Hist., Bot. Ser. 5: 204. 1923 [E] [W]

Cuscuta californica Hooker & Arnott var. *breviflora* Engelmann, Trans. Acad. Sci. St. Louis 1: 499. 1859

Stems yellowish to orange, medium. **Inflorescences** dense, glomerulate; bracts at base of clusters 1, at base of pedicels and/or flowers 0 or 1, lanceolate to ovate, membranous, margins entire, apex acute to acuminate. **Pedicels** 0–0.5(–1.5) mm. **Flowers** 5-merous, 2.7–3.4 mm, membranous, usually not papillate; calyx usually yellow, campanulate, somewhat shorter than to equaling corolla tube length, divided 2/5–1/2 its length, not reticulate, shiny, lobes narrowly ovate to lanceolate, bases not overlapping, margins entire, midvein not carinate, apex acuminate; corolla white, drying creamy white or yellow, 2.5–3.2 mm, tube cylindric-campanulate, 1.4–2.1 mm, saccate between lines of stamen attachments, lobes usually spreading, sometimes reflexed, lanceolate, shorter than corolla tube length, margins entire, apex acuminate, straight; infrastaminal scales absent; stamens ± exserted, shorter than corolla lobes; filaments 0.2–0.5 mm; anthers 0.2–0.5 × 0.2–0.4 mm; styles filiform, 0.5–1(–1.5) mm, shorter than ovary. **Capsules** globose to depressed-globose, 1.8–2.2 × 2–2.6 mm, slightly thickened, not raised, around relatively small interstylar aperture, translucent, surrounded by, not completely enclosed by, withered corolla (top of capsule visible), indehiscent. **Seeds** 2–4, obcompressed, subglobose to broadly ellipsoid, 0.8–1.3 × 0.8–1.1 mm, hilum region lateral.

Flowering Mar–Sep. Hosts: *Artemisia*, *Cistanthe*, *Corethrogyne*, *Diplacus*, *Epilobium*, *Ericameria*, *Eriodictyon*, *Eriogonum*, *Hemizonia*, *Iva*, *Lotus*, *Lupinus*, *Monardella*, *Oenothera*, *Polygonum*, *Salvia*, *Sisymbrium*, *Trifolium*, and others; 200–2500 m; Calif., Colo., Idaho, Nev., Oreg., Utah, Wash., Wyo.

Cuscuta occidentalis has been treated as *C. californica* var. *breviflora*, in which T. Beliz (1993) also included *C. brachycalyx* and *C. suksdorfii*. *Cuscuta occidentalis* differs from *C. californica* by its sessile or subsessile flowers, saccate corolla tube, relatively short anthers, relatively short styles, and translucent capsules that are completely enveloped by withered corollas (M. Costea and S. Stefanović 2009). Both *C. californica* and *C. occidentalis* are distinguished from *C. suksdorfii* by five-merous flowers, acute or acuminate calyx and corolla lobe apices, absence of infrastaminal scales, and capsules enclosed by withered corollas (M. Costea et al. 2006b).

34. Cuscuta odontolepis Engelmann, Trans. Acad. Sci. St. Louis 1: 486. 1859 • Santa Rita Mountain dodder [C]

Stems yellowish, slender. **Inflorescences** dense, paniculiform-glomerulate; bracts at base of clusters 1, at base of pedicels and/or flowers 0 or 1, subround to broadly ovate, membranous, margins entire, apex acute to short-acuminate, papillate. **Pedicels** 0–1 mm. **Flowers** 5-merous, 4.5–5 mm, membranous, calyx and corolla lobes papillate; calyx straw yellow, campanulate, 1/2–3/4 corolla tube length, divided 2/3 its length, finely reticulate, not shiny, lobes ovate-triangular, bases overlapping, margins entire, midvein not carinate, apex acute to short-acuminate; corolla white, drying creamy white, 3.5–4.5 mm, tube cylindric, 2.2–2.8 mm, not saccate, lobes reflexed, ovate-triangular, shorter than to equaling corolla tube length, margins entire, apex acute to short-acuminate, straight; infrastaminal scales oblong-spatulate to obovate, 2–2.5 mm, 1/2 to equaling corolla tube length, bridged at 0.2–0.5 mm, rounded, densely fimbriate in distal 1/2, fimbriae 0.2–0.3 mm; stamens barely exserted, shorter than corolla lobes; filaments 0.3–0.7 mm; anthers 0.7–1.1 × 0.2–3 mm; styles filiform, 2.8–4 mm, longer than ovary. **Capsules** globose to depressed-globose, 2.9–4 × 3–3.2 mm, thickened and raised around inconspicuous interstylar aperture, translucent, loosely surrounded and capped by withered corolla, dehiscence circumscissile. **Seeds** 3 or 4, angled, broadly ellipsoid, 1–1.2 × 0.6–0.8 mm, hilum region terminal.

Flowering Aug–Oct. Hosts: *Amaranthus*; of conservation concern; 900–1500 m; Ariz.; Mexico (Chihuahua, Sonora).

Cuscuta odontolepis was used by the Aztecs to produce a yellow dye called zacatlaxcalli (B. de Sahagún 1950–1982).

35. Cuscuta pacifica Costea & M. A. R. Wright, Syst. Bot. 34: 792, fig. 6. 2009 • Pacific salt-marsh dodder

Stems orange, slender. **Inflorescences** dense, umbelliform to subglomerulate; bracts at base of clusters 1, at base of pedicels 0 or 1, ovate to lanceolate, membranous, margins entire, apex acute to acuminate. **Pedicels** 0.5–2 mm, sometimes with domelike cells. **Flowers** 5-merous, 3.5–6 mm, membranous, papillate or corolla lobes, sometimes calyx, with domelike cells; calyx creamy yellow, usually drying dull brown, rarely drying yellow, campanulate to cupulate, equaling corolla tube length, divided 2/3 its length, not reticulate or shiny, lobes ovate-triangular, bases slightly overlapping, margins entire, midvein not carinate, apex acute to acuminate; corolla white, generally drying dark brown (rarely creamy yellow), 2.8–5.4 mm, tube campanulate, 1.5–2.6 mm, not saccate, lobes erect to spreading, broadly ovate to rhombic-ovate, equaling corolla tube length, bases overlapping, margins entire or irregular, apex usually acute to cuspidate, sometimes appearing tridentate, straight; infrastaminal scales oblong to slightly obovate, 1–1.5 mm, 1/2–3/4 corolla tube length, bridged at 0.3–0.5 mm, with few short fimbriae 0.05–0.1 mm; stamens included in completely open flower, shorter than corolla lobes; filaments 0.3–0.6 mm; anthers 0.3–0.6 × 0.3–0.5 mm; styles filiform, 0.4–0.9 mm, shorter than ovary. **Capsules** ovoid, 2–3.6 × 1.4–2.1 mm, thickened around relatively small interstylar aperture, not translucent, surrounded by withered corolla, indehiscent. **Seeds** 1 or 2, ± obcompressed, broadly ellipsoid to subglobose, 1.4–1.9 × 1.2–1.4 mm, hilum subterminal.

Varieties 2 (2 in the flora): w North America, nw Mexico.

Cuscuta pacifica was segregated from *C. salina* based on their different morphology, reproductive biology, host range, geographical distribution, and ecology (M. Costea et al. 2009).

1. Pedicels not papillate; calyces not papillate . . . . 35a. *Cuscuta pacifica* var. *pacifica*
1. Pedicels papillate; calyces papillate . . . . 35b. *Cuscuta pacifica* var. *papillata*

35a. Cuscuta pacifica Costea & M. A. R. Wright var. **pacifica**

Cuscuta salina Engelmann var. *major* Yuncker

Pedicels not papillate. **Calyces** not papillate. **2*n*** = 28, ca. 30.

Flowering Jun–Oct. Hosts: *Jaumea carnosa*, *Sarcocornia pacifica*; coastal salt marshes, tidal flats; 0–20 m; introduced; B.C.; Calif., Oreg., Wash.; Mexico (Baja California).

35b. Cuscuta pacifica Costea & M. A. R. Wright var. **papillata** (Yuncker) Costea & M. A. R. Wright, Syst. Bot. 34: 792. 2009 [C] [E]

Cuscuta salina Engelmann var. *papillata* Yuncker, Bull. Torrey Bot. Club 69: 543. 1942

Pedicels papillate. **Calyces** papillate.

Flowering Jul–Oct. Hosts: *Lupinus littoralis* var. *variicolor* and other herbs; coastal interdune depressions and grasslands; of conservation concern; 0–10 m; Calif.

36. Cuscuta pentagona Engelmann, Amer. J. Sci. Arts 43: 340, plate 6, figs. 22–24. 1842 • Five-angled dodder [E] [F] [W]

Cuscuta arvensis Beyrich ex Engelmann; *C. pentagona* var. *microcalyx* Engelmann; *Grammica pentagona* (Engelmann) W. A. Weber

Stems yellow to orange, slender to medium. **Inflorescences** dense, corymbiform to glomerulate; bracts at base of clusters 1, at base of pedicels 0 or 1, ovate or ovate-triangular to lanceolate, membranous, margins entire, apex acute. **Pedicels** 0.5–3(–4.5) mm. **Flowers** (4 or)5-merous, 1.4–2.5 mm, membranous, corolla lobes sometimes papillate; calyx yellow to brown, angled, cupulate, equaling corolla tube length, divided 1/2–2/3 its length, ± reticulate, shiny or not, lobes broadly ovate to rhombic, base auriculate, overlapping, forming prominent angles at sinuses, margins entire, midvein not carinate, without multicellular protuberances, apex rounded; corolla whitish, drying yellow to brown, 1.2–2.2 mm, tube campanulate, 0.7–1.2 mm, not saccate, lobes spreading, triangular-lanceolate, equaling corolla tube length, margins entire, apex acute to acuminate, inflexed; infrastaminal scales ovate to oblong, 0.7–1.4 mm, equaling or longer than corolla tube length, bridged at 0.3–0.5 mm, rounded, ± uniformly densely fimbriate, 0.15–0.25 mm; stamens exserted, shorter than corolla lobes; filaments 0.3–0.4 mm; anthers 0.2–0.3 × 0.2–0.3 mm; styles filiform, 0.7–1.1 mm, equaling ovary. **Capsules** depressed-globose to ovoid, 1.9–2.4 × 1.6–2.5 mm, not thickened or raised around relatively medium-sized to large interstylar aperture, translucent or not, base ± enveloped by withered corolla, indehiscent. **Seeds** 4, angled, subglobose to broadly ellipsoid, 0.9–1.1 × 0.8–1 mm, hilum region subterminal. $2n = 56$.

Flowering Apr–Nov. Hosts: herbs; 0–900 m; Man.; Ala., Ark., Del., D.C., Fla., Ga., Ill., Ind., Kans., Md., Mass., Mich., Minn., Miss., Mo., Mont., N.J., N.Y., N.C., N.Dak., Okla., Pa., S.C., S.Dak., Tenn., Tex., Va.

Cuscuta pentagona apparently has not spread outside of North America, where it is less common than *C. campestris*. It is currently included in North American noxious weeds lists although there is no evidence it attacks crops.

G. Engelmann (1859) distinguished four varieties of *Cuscuta pentagona*: var. *calycina*, var. *microcalyx*, var. *pentagona*, and var. *verrucosa* (Engelmann) Yuncker. T. G. Yuncker (1932, 1965) treated var. *calycina* and var. *verrucosa* at specific rank and provided a new name for each: *C. campestris* and *C. glabrior*, respectively. *Cuscuta glabrior* is currently accepted by all the North American overviews; *C. campestris* has been persistently considered a synonym of *C. pentagona* despite morphological and evolutionary evidence that the two are distinct (M. Costea et al. 2006c, 2015).

Cuscuta pentagona differs from *C. campestris* in its rhombic to ovate, auriculate calyx lobes with overlapping bases that form angles at sinuses and in its smaller flowers, capsules, and seeds.

37. Cuscuta plattensis A. Nelson, Bull. Torrey Bot. Club 26: 131. 1899 • Prairie dodder [C] [E]

Stems yellow to pale orange, slender to medium. **Inflorescences** loose, paniculiform; bracts at base of clusters 1, at base of pedicels 0 or 1, ovate-triangular to lanceolate, membranous, margins entire, apex acute. **Pedicels** 0.5–2.5(–3) mm. **Flowers** 5-merous, 3–4.6(–5) mm, membranous, not papillate; calyx yellow, cylindric-cupulate, equaling corolla tube length, divided 1/2–2/3 its length, reticulate, not shiny, lobes ovate-triangular, bases slightly overlapping, margins entire, midvein not carinate, without multicellular protuberances, apex obtuse to subacute; corolla white, drying creamy white, 3–4.4 mm, tube narrowly campanulate to cylindro-campanulate, 1.8–2.5 mm, not saccate, lobes spreading to reflexed, triangular, equaling corolla tube length, margins entire, apex acute, inflexed; infrastaminal scales obovate to oblong-spatulate, 1.5–2.1 mm, 3/4–4/5 corolla tube length, bridged at 0.4–0.6 mm, rounded, uniformly densely fimbriate, fimbriae 0.1–0.3 mm; stamens exserted, shorter than corolla lobes; filaments 0.3–0.4 mm; anthers 0.4–0.7 × 0.3–0.5 mm; styles filiform, 1.3–1.6 mm, equaling ovary. **Capsules** globose to depressed-globose, 1.8–3.2 × 2.2–3.6 mm, not thickened or raised around relatively large interstylar aperture, not translucent, 1/2+ enveloped by withered corolla, indehiscent. **Seeds** 1–4, obcompressed, broadly ellipsoid to obovoid, 1–1.4 × 0.9–1.2 mm, hilum region subterminal.

Flowering Aug–Sep. Hosts: *Grindelia*, *Helianthus*, *Humulus*, *Psoralea*, *Rubus*, and *Solidago*; of conservation concern; 1500–2000 m; Nebr., Wyo.

38. Cuscuta polygonorum Engelmann, Amer. J. Sci. Arts 43: 342, plate 6, figs. 26–29. 1842 • Smartweed dodder, cuscute des renouées [E] [W]

Stems yellow-orange, slender to medium. **Inflorescences** dense, glomerulate; bracts at base of clusters 1, at base of pedicels 0–1, ovate-triangular to lanceolate, membranous, margins entire, apex acute. **Pedicels** 0–1 mm. **Flowers** (3 or)4-merous, 2–2.7 mm, membranous, not papillate; calyx brownish yellow, cupulate, equaling corolla tube length, divided 1/2–2/3 its length, not reticulate or shiny, lobes triangular-ovate, bases not overlapping, margins entire, midvein not carinate, without hornlike appendages, apex obtuse to rounded; corolla white, drying yellow-brown, 1.8–2.5 mm, tube cupulate to shallowly campanulate, 1–1.4 mm, not saccate, lobes erect, triangular, equaling or longer than corolla tube length, margins entire, apex acute, inflexed; infrastaminal scales usually well developed, sometimes poorly developed, oblong, 0.8–1.2 mm, shorter than to equaling corolla tube length, bridged at 0.1–0.2 mm, mostly shallowly 2-fid with 1–3 fimbriae on each side of filament attachment or irregularly fimbriate at apex, fimbriae (0.05–)0.1–0.3 mm; stamens exserted, shorter than to ± equaling corolla lobes; filaments 0.4–0.8 mm; anthers 0.3–0.4 × 0.2–0.3 mm; styles subulate, 0.4–0.9 mm, shorter than ovary. **Capsules** depressed-globose, often appearing angled, 1.6–3 × 2.5–5 mm, not thickened or raised around relatively large interstylar aperture, not translucent, base ± enveloped by withered corolla, indehiscent. **Seeds** 3 or 4, obcompressed, subglobose or broadly ovoid to broadly ellipsoid, 1.4–1.6 × 1.2–1.3 mm, hilum region subterminal.

Flowering Jul–Oct. Hosts: usually *Persicaria*, sometimes *Impatiens*, *Ipomoea*, *Justicia*, *Laportea*, *Lycopus*, *Penthorum*, *Xanthium*, and others; margins of streams, lakes, swamps, flood plains; 30–500 m; Ont., Que.; Ark., Conn., Del., D.C., Ill., Ind., Iowa, Kans., Ky., La., Maine, Md., Mass., Mich., Minn., Mo., Nebr., N.J., N.Y., N.Dak., Ohio, Okla., Pa., R.I., Tenn., Tex., Va., Wis.

39. **Cuscuta rostrata** Shuttleworth ex Engelmann & A. Gray, Boston J. Nat. Hist. 5: 225. 1845 • Beaked dodder [E] [W]

Stems orange, coarse. **Inflorescences** dense, paniculiform to glomerulate; bracts at base of clusters 1 or 2, at base of pedicels 0 or 1, ovate, membranous, margins entire, apex acute. **Pedicels** 0.9–2 mm. **Flowers** 5-merous, 4–6(–7) mm, membranous, not papillate; calyx yellow-brown, cupulate, 1/3–1/2 corolla tube length, divided 3/5–2/3 its length, not reticulate or shiny, lobes triangular-ovate to broadly ovate, bases overlapping, margins entire, midvein not carinate, apex obtuse; corolla white, drying creamy yellow to brownish, 3.8–5.8 mm, tube campanulate, 3–5 mm, not saccate, lobes suberect to spreading or reflexed, ovate, 1/4–1/3 corolla tube length, margins entire, apex rounded, straight; infrastaminal scales oblong to obovate, 1.5–2.8 mm, 1/2–2/3 corolla tube length, bridged at 0.5–0.7 mm, rounded, sparsely fimbriate at base, more dense in distal 1/2, fimbriae 0.3–0.9 mm; stamens exserted, shorter than to equaling corolla lobes; filaments 0.5–0.7 mm; anthers 0.6–0.8 × 0.4–0.5 mm; styles filiform, 1–1.5 mm, equaling ovary. **Capsules** broadly ovoid, 4–7 × 4–5.5 mm, with stout beak 1–1.5 mm, raised and thickened around relatively small interstylar aperture, not translucent, surrounded by withered corolla, indehiscent. **Seeds** 2–4, obcompressed to obscurely angled, ovoid to oblong, 1.7–2.5 × 1.3–1.6 mm, hilum region subterminal.

Flowering Jul–Sep. Hosts: *Clematis*, *Collinsonia*, *Diervilla*, *Epilobium*, *Eupatorium*, *Euthamia*, *Hypericum*, *Impatiens*, *Laportea*, *Parthenocissus*, *Rubus*, *Rudbeckia*, *Salix*, *Solidago*, *Symphoricarpos*, *Symphyotrichum*, *Urtica*, and others; forests, especially along streams; 800–2000 m; Ga., Ky., Md., N.C., S.C., Tenn., Va., W.Va.

Cuscuta rostrata is distinguished by its relatively large flowers, beaked capsules, and Appalachian range.

40. **Cuscuta runyonii** Yuncker, Bull. Torrey Bot. Club 69: 541, fig. 1. 1942 • Runyon's dodder [E] [W]

Stems yellow-orange, slender to medium. **Inflorescences** loose, umbelliform or corymbiform; bracts at base of clusters 1, at base of pedicels 0 or 1, ovate to lanceolate, membranous, margins entire, apex acute. **Pedicels** 0.8–4(–5) mm, sometimes papillate. **Flowers** 5-merous, 2.5–3.5(–4) mm, membranous, papillate on perianth, ovary, and capsule; calyx reddish brown, cupulate, shorter than or equaling corolla tube length, divided 1/2–2/3 its length, ± reticulate, shiny, lobes triangular, bases not overlapping, margins entire, midvein sometimes ± carinate, base of calyx corresponding to each lobe with reflexed spurlike projection 0.1–0.6 mm, apex obtuse to subacute; corolla white, drying reddish brown, 2.2–3.5 mm, tube campanulate-globose, 1.1–1.8 mm, saccate between lines of stamen attachments, lobes reflexed, triangular-ovate to lanceolate, equaling corolla tube length, margins entire, apex acute to acuminate, inflexed; infrastaminal scales obovate to spatulate, 1.1–2 mm, equaling corolla tube length, bridged at 0.2–0.4 mm, rounded, uniformly densely fimbriate, fimbriae 0.3–0.7 mm; stamens exserted, shorter than corolla lobes; filaments 0.4–0.6 mm; anthers 0.4–0.6 × 0.4–0.5 mm; styles filiform, 0.9–1.8 mm, equaling or longer than ovary. **Capsules** depressed-globose, 1.6–3 × 2–3.4 mm, not thickened or raised around relatively large interstylar aperture, not translucent, 1/2+ enveloped by withered corolla, indehiscent. **Seeds** 4, angled, ovoid to broadly ellipsoid, 0.9–1.4 × 0.8–1.1 mm, hilum region subterminal.

Flowering Mar–Dec. Hosts: *Dalea*, *Dyschoriste*, *Erigeron*, *Gutierrezia*, *Hymenoxys*, *Justicia*, *Linum*, *Melampodium*, *Nama*, *Oenothera*, *Spermolepis*, *Tetraneuris*, *Thamnosma*, *Thelesperma*, *Tiquilia*; 0–200 m; Tex.

Cuscuta runyonii is closely allied with *C. glabrior* (M. Costea et al. 2015); it differs in having a spurlike appendage at the base of each calyx lobe.

41. Cuscuta salina Engelmann in W. H. Brewer et al., Bot. California 1: 536. 1876 • Inland salt-marsh dodder [F] [W]

Stems orange-yellow, slender. **Inflorescences** loose to dense, corymbiform; bracts at base of clusters 1, at base of pedicels 0 or 1, ovate-lanceolate to lanceolate, membranous, margins entire, apex acute. **Pedicels** (0.5–)1–5 mm. **Flowers** 5-merous, 2.5–4.5 mm, membranous, not papillate or corolla lobes papillate; calyx yellow, cylindric to narrowly campanulate, equaling corolla tube length, divided 1/2 its length, not reticulate, ± glossy, lobes lance-ovate to lanceolate, bases not overlapping, margins entire, midvein not carinate, apex acute to acuminate; corolla white, drying creamy white, 2.2–4 mm, tube cylindric-campanulate to obconic, 1.2–2 mm, not saccate, lobes spreading to reflexed, lance-oblong to lance-ovate, equaling corolla tube length, margins entire or irregular, apex acute to acuminate or cuspidate, sometimes appearing tridentate, straight; infrastaminal scales oblong to obovate, 1–1.7 mm, 4/5 corolla tube length, bridged at 0.2–0.4 mm, rounded, uniformly densely fimbriate, fimbriae 0.1–0.3 mm; stamens exserted at full anthesis, shorter than corolla lobes; filaments 0.3–0.7 mm; anthers 0.3–0.7 × 0.2–0.5 mm; styles uniformly filiform, 0.4–0.9 mm, shorter than ovary. **Capsules** ellipsoid-ovoid, 1.6–2.5 × 1.7–2.4 mm, thickened and raised around relatively small interstylar aperture, not translucent, surrounded or capped by withered corolla, indehiscent. **Seeds** 1, ± obcompressed, broadly ellipsoid to subglobose, 1.3–1.5 × 1.2–1.4 mm, hilum region subterminal. $2n = 30$.

Flowering Mar–Nov. Hosts: *Atriplex*, *Centromadia*, *Cressa*, *Frankenia*, *Jaumea*, *Plantago*, *Salicornia*, *Salsola*, *Suaeda*, *Trichostema*, *Wislizenia*; inland salt flats; 0–1000 m; Ariz., Calif., Nev., Utah; Mexico (Baja California, Baja California Sur, Sonora).

42. Cuscuta squamata Engelmann, Trans. Acad. Sci. St. Louis 1: 510. 1859 • Scale-flower dodder [W]

Stems yellow to orange, slender. **Inflorescences** dense, glomerulate or short-spiciform; bracts at base of clusters and flowers (2–)4 or 5(–10), ovate-orbiculate to ovate-triangular, membranous, margins denticulate, apex erect, acute to cuspidate. **Pedicels** absent. **Flowers** 5-merous, 5–6 mm, membranous, not papillate; calyx straw yellow, campanulate, 1/2–2/3 corolla tube length, divided to base or nearly so, finely reticulate, shiny, lobes ovate, broadly overlapping, margins denticulate, midvein not carinate, apex acute to cuspidate; corolla white, drying creamy, straw yellow, or light brown, 4–5.5 mm, tube cylindric, 2.4–3.5 mm, not saccate, lobes spreading to reflexed, ovate-lanceolate to oblong-ovate, 1/3–4/5 corolla tube length, margins entire, apex usually acute, sometimes cuspidate, straight; infrastaminal scales oblong, 2.4–3.4 mm, equaling corolla tube length, bridged at 0.7–1.5 mm, rounded, uniformly densely fimbriate, fimbriae 0.2–0.4 mm; stamens barely exserted, shorter than corolla lobes; filaments 0.5–0.7 mm; anthers 0.6–0.9 × 0.5–0.6 mm; styles filiform, 2.5–3.3 mm, longer than ovary. **Capsules** subglobose or ovoid to subconic, 3.4–4.5 × 2.2–3 mm, ± raised and thickened around relatively small interstylar aperture, not translucent, capped by withered corolla, indehiscent. **Seeds** 2–4, slightly obcompressed, subglobose, broadly ellipsoid to obovoid, 1.5–1.7 × 1.1–1.3 mm, hilum area subterminal.

Flowering Mar–Oct. Hosts: Asteraceae and others; 900–1500 m; N.Mex., Tex.; Mexico.

43. Cuscuta suaveolens Seringe, Ann. Sci. Phys. Nat. Lyon 3: 519. 1840 • Fimbriate dodder [I] [W]

Stems yellow-orange, slender. **Inflorescences** loose, racemiform or corymbiform; bracts at base of clusters 1, at base of pedicels 0 or 1, ovate, membranous, margins entire, apex acute. **Pedicels** 2–5(–7 mm). **Flowers** 5-merous, 3–5 mm, membranous, not papillate; calyx light brown to reddish, cupulate-turbinate, 1/2–3/4 corolla tube length, divided 1/2 its length, not reticulate or shiny, lobes triangular-ovate, bases scarcely overlapping, margins entire, revolute, midvein not carinate, apex obtuse to subacute; corolla white, drying creamy or reddish, 2.8–4.7 mm, tube campanulate, 2–3 mm, not saccate, lobes erect to spreading, triangular-ovate, 1/2 corolla tube length, margins entire, apex acute, inflexed; infrastaminal scales oblong-ovate, 1.8–2.8 mm, equaling corolla tube length, bridged at 0.8–1.4 mm, rounded, uniformly densely fimbriate, fimbriae 0.3–0.5 mm; stamens slightly exserted, shorter than corolla lobes; filaments 0.2–0.4 mm; anthers 0.6–0.8 × 0.5–0.6 mm; styles filiform, 1.6–2.5 mm, equaling or longer than ovary. **Capsules** globose-ovoid, 2.5–3.5 × 2.5–3.5 mm, not thickened around relatively small interstylar aperture, not translucent, surrounded by withered corolla, indehiscent. **Seeds** 2–4, obcompressed, subglobose to broadly ellipsoid, 1.3–2 × 1–1.7 mm, hilum region subterminal.

Flowering Jul–Oct. Hosts: *Medicago sativa*, *Trifolium*; 20–1000 m; introduced; Ala., Calif., Ohio, S.Dak., Tex.; South America; introduced also in Europe.

Cuscuta suaveolens is an ephemeral weed associated with the cultivation of forage legumes; it has not been collected in recent decades.

44. Cuscuta subinclusa Durand & Hilgard, Pl. Heermann., 42. 1854

Stems yellow to orange, medium. **Inflorescences** dense, ± glomerulate; bracts at base of clusters 1, at base of pedicels and/or flowers 0 or 1, lanceolate to ovate, membranous, margins entire, apex acute to acuminate. **Pedicels** 0–1 mm. **Flowers** 5-merous, 5–7(–9 mm), membranous, papillate on corolla lobes; calyx usually straw yellow, sometimes brown, campanulate, 1/2 corolla tube length, divided 3/5–2/3 its length, finely reticulate, shiny, lobes broadly ovate to lanceolate, bases overlapping, margins entire, midvein not carinate, apex acute, sometimes cuspidate; corolla white, drying creamy yellow or brownish, 4.5–6.6 mm, tube cylindric, 2.5–3.5(–4.5) mm, not saccate, usually with horizontal ridges between stamen attachments, lobes spreading to reflexed, ovate-triangular, 1/4–1/3 corolla tube length, margins entire, apex acute, often slightly acuminate, straight; infrastaminal scales oblong to spatulate, 1.5–2.2 mm, 1/2–2/3 corolla tube length, bridged at 0.3–0.7 mm, rounded, uniformly densely to sparsely fimbriate, fimbriae 0.2–0.4 mm; stamens slightly exserted, shorter than corolla lobes; filaments 0–0.1 mm; anthers 0.8–2 × 0.4–0.5 mm; styles 1–1.5 mm, equaling ovary. **Capsules** ovoid to ellipsoid, 1.5–3 × 1.2–2.5 mm, narrowed and thickened, forming collar around relatively small interstylar aperture, not translucent, capped by withered corolla, indehiscent. **Seeds** 1, usually subglobose to broadly ovoid, rarely slightly obcompressed, 1.3–1.7 × 1.2–1.5 mm, hilum region terminal.

Flowering Apr–Oct. Hosts: *Adenostoma*, *Amelanchier*, *Arctostaphylos*, *Artemisia*, *Asclepias*, *Ceanothus*, *Cercis*, *Citrus*, *Clematis*, *Erigeron*, *Eriogonum*, *Grindelia*, *Heteromeles*, *Monardella*, *Populus*, *Rhododendron*, *Rhus*, *Rosa*, *Salix*, *Schinus*, *Solidago*, *Vitis*, and other herbs and woody plants; forests near streams, canyon bottoms, wetlands, salt marshes; 0–2000 m; Calif.; Mexico (Baja California).

The name *Cuscuta ceanothi* Behr may pertain here. The type specimen of *C. ceanothi* was evidently destroyed, and the protologue is not sufficient to fix application of *C. ceanothi* to any one species of *Cuscuta*.

45. Cuscuta suksdorfii Yuncker, Mem. Torrey Bot. Club 18: 167. 1932 • Mountain dodder [E] [W]

Cuscuta salina Engelmann var. *acuminata* Yuncker, Illinois Biol. Monogr. 6: 162, plate 6, fig. 32f,g, plate 11, fig. 89. 1921; *C. suksdorfii* var. *subpedicellata* Yuncker

Stems yellow, slender. **Inflorescences** loose, umbelliform; bracts at base of clusters 1, at base of pedicels and/or flowers 0 or 1, ovate-lanceolate, membranous, margins entire, apex acute. **Pedicels** 0–2 mm. **Flowers** 4- or 5-merous, 2.8–3.3 mm, membranous, not papillate; calyx slightly zygomorphic, creamy yellow, broadly campanulate, 1-1/3–1-1/2 corolla tube length, divided 1/2–3/5 its length, not reticulate or shiny, lobes ovate, bases not overlapping, margins entire, midvein not carinate, apex long-attenuate; corolla white, drying creamy yellow, 2.6–3 mm, tube campanulate, 1.2–1.5 mm, not saccate, lobes suberect, triangular-ovate, longer than corolla tube length, margins entire, apex lance-attenuate, straight; infrastaminal scales relatively poorly developed, oblong, 0.6–1 mm, 1/2–3/4 corolla tube length, bridged at 0.2–0.3 mm, usually reduced to denticulate wings, rarely 2-fid with 1–3 fimbriae on each side of filament attachments, 0.1–0.2 mm; stamens included or barely visible through corolla sinuses, shorter than corolla lobes; filaments 0.2–0.5 mm; anthers 0.2–0.4 × 0.2–0.3 mm; styles terete to slightly subulate, 0.3–0.7 mm, 1/4 ovary length. **Capsules** ellipsoid-ovoid, ovoid-conic, or globose to depressed-globose, 2–3.2 × 2–3.6 mm, not thickened or raised around relatively small interstylar aperture, translucent, proximal 1/2 surrounded by withered corolla, indehiscent. **Seeds** 2–4, obcompressed, subglobose, 0.8–1.1 × 0.8–1 mm, hilum subterminal.

Flowering Jul–Sep. Hosts: mostly herbs: Asteraceae, *Cistanthe*, *Trifolium*, and others; mountain meadows; 1500–2600 m; Calif., Oreg., Wash.

46. Cuscuta tuberculata Brandegee, Univ. Calif. Publ. Bot. 3: 389. 1909 • Tubercle dodder [W]

Stems yellow-orange, filiform. **Inflorescences** loose, umbelliform-racemiform; bracts at base of clusters 1, at base of peduncles 0(or 1), ovate-lanceolate, membranous, margins entire, apex acute. **Pedicels** 2–3(–5) mm. **Flowers** 5-merous, 2.5–4 mm, membranous, papillate, especially at base of corolla tube; calyx yellow, cupulate-angular, 1/3–1/2 corolla tube length, divided ± to base, not or finely reticulate, ± glossy, lobes triangular to lanceolate, bases not overlapping, margins entire, midvein carinate and/or with multicellular protuberances, apex acute to acuminate; corolla white, drying creamy yellow, 2–3.5 mm, tube cylindric, 1.5–2.2 mm, not saccate, lobes erect, triangular-lanceolate, equaling corolla tube length, margins entire, apex acute, straight; infrastaminal scales ovate to oblong, 0.8–1.1 mm, 1/3–1/2 corolla tube length, bridged at 0.5–0.7 mm, rounded, uniformly short-fimbriate, fimbriae 0.2–0.3 mm; stamens barely exserted, shorter than to equaling corolla lobes; filaments 0.4–0.7 mm; anthers 0.5–0.8 × 0.3–0.5 mm; styles filiform, 1.5–3 mm, longer than ovary. **Capsules** globose, 1.3–2.2 × 1–2.3 mm, slightly thickened and raised around relatively small interstylar aperture, translucent, capped by withered corolla, dehiscence circumscissile. **Seeds** 2–4, angled or slightly obcompressed, ellipsoid-oblong, 0.6–0.9 × 0.3–0.5 mm, hilum area lateral.

Flowering Aug–Nov. Hosts: usually *Boerhavia*, rarely *Amaranthus*, Euphorbiaceae, or Nyctaginaceae; desert and thornscrub; 500–1000 m; Ariz., N.Mex.; Mexico (Baja California, Baja California Sur, Sonora).

47. **Cuscuta umbellata** Kunth in A. von Humboldt et al., Nov. Gen. Sp. 3(fol.): 95; 3(qto.): 121. 1819 • Flat-globe dodder [W]

Varieties 2 (1 in the flora): United States, Mexico, West Indies, South America.

Variety *desertorum* Engelmann is known from South America (Brazil).

47a. **Cuscuta umbellata** Kunth var. **umbellata** [W]

Cuscuta fasciculata Yuncker

Stems yellow-orange, slender. **Inflorescences** dense to loose, umbelliform; bracts at base of clusters 1, at base of pedicels 0 or 1, triangular-ovate, membranous, margins entire, apex acute. **Pedicels** 2–10 mm. **Flowers** 5-merous, 2–3 mm, membranous, not papillate or sometimes papillate on adaxial face of corolla lobes; calyx straw yellow, campanulate, equaling corolla tube length, divided 2/3 its length, finely reticulate, slightly shiny, lobes triangular-ovate, bases not overlapping, margins entire, midvein not carinate, without multicellular protuberances, apex obtuse to acute; corolla white, drying creamy white or brown, 2–2.5 mm, tube campanulate, 0.6–1.2 mm, not saccate, lobes reflexed, oblong to lanceolate, longer than corolla tube length, margins entire, apex obtuse to acute, straight; infrastaminal scales subspatulate to obovate, 0.8–1.2 mm, equaling or longer than corolla tube length, bridged at 0.1 mm, rounded, uniformly densely fimbriate, fimbriae 0.1–0.3 mm; stamens exserted, shorter than corolla lobes; filaments 0.4–0.7 mm; anthers 0.4–0.6 × 0.2–0.3 mm; styles filiform, 0.8–1.7 mm, longer than ovary. **Capsules** depressed-globose, 1–2.5 × 0.7–1.2 mm, thickened and slightly raised around inconspicuous interstylar aperture, translucent, surrounded or capped by withered corolla, dehiscence circumscissile. **Seeds** 2–4, angled, ovate to broadly ellipsoid or subglobose, 0.8–1.2 × 0.6–0.8 mm, hilum region terminal.

Flowering Apr–Dec. Hosts: *Acleisanthes*, *Allionia*, *Alternanthera*, *Amaranthus*, *Atriplex*, *Boerhavia*, *Gilia*, *Iresine*, *Kallstroemia*, *Salsola*, *Sesuvium*, *Suaeda*, *Tidestromia*, *Trianthema*, *Tribulus*; 10–2500 m; Ariz., Colo., Fla., Kans., La., N.Mex., Okla., Tex.; Mexico; West Indies; South America.

Cuscuta umbellata is currently included in North American noxious weed lists although there is no evidence that it attacks crops.

48. **Cuscuta umbrosa** Beyrich ex Hooker, Fl. Bor.-Amer. 2: 78. 1837 • Big-fruit dodder [E] [W]

Cuscuta curta (Engelmann) Rydberg; *C. megalocarpa* Rydberg

Stems yellow to orange, coarse. **Inflorescences** dense, paniculiform; bracts at base of clusters 1, at base of pedicels 0(or 1), ovate to broadly triangular, membranous, margins entire, apex acute to obtuse. **Pedicels** 0.9–7 mm. **Flowers** 5-merous, 2–3.5(–4.4) mm, membranous, not papillate; calyx brownish, campanulate, to 1/2 corolla tube length, divided 1/2–2/3 its length, not reticulate or shiny, lobes ovate, bases overlapping, margins entire or serrulate, midvein not carinate, apex obtuse to rounded; corolla creamy white, drying creamy brownish, 2–4 mm, tube campanulate, 1.7–2.3(–2.7) mm, not saccate, lobes spreading to reflexed, ovate to broadly triangular-ovate, 1/4–1/3 corolla tube length, margins entire, apex rounded to obtuse, straight; infrastaminal scales broadly oblong, 1.2–2 mm, (1/3–)1/2 corolla tube length, bridged at 0.5–1 mm, ± truncate to 2-lobed apically, fimbriate mostly in distal 1/2, fimbriae 0.2–0.6 mm; stamens exserted, shorter than corolla lobes; filaments 0.4–0.7 mm; anthers 0.3–0.6 × 0.2–0.4 mm; styles slender, 0.3–0.9 mm, 1/4 ovary length, slightly thickened at base. **Capsules** ovoid to globose-conic or subobpyriform, 3.5–6.5(–7) × 3–5(–6) mm, raised and thickened around relatively small interstylar aperture, apex rarely narrowed into neck to 1 mm, not translucent, surrounding capsule or capped by withered corolla, indehiscent. **Seeds** 3 or 4, obcompressed to obscurely angled, broadly ellipsoid to obovoid, 1.8–2.5(–2.8) × 1.5–1.6 mm, hilum region subterminal.

Flowering Jul–Oct. Hosts: *Ampelopsis*, *Clematis*, *Convolvulus*, *Epilobium*, *Humulus*, *Impatiens*, *Lactuca*, *Linum*, *Salix*, *Scutellaria*, *Solidago*, *Symphoricarpos*, *Urtica*, and others; 50–2100 m; Alta., Man., Ont., Sask.; Colo., Idaho, Ill., Iowa, Kans., Minn., Mont., Nebr., N.Mex., N.Dak., S.Dak., Utah, Wis., Wyo.

Navajo Indians of the southwestern United States added seeds of *Cuscuta umbrosa* to soups or stews (E. F. Castetter 1935).

49. Cuscuta warneri Yuncker, Brittonia 12: 38, fig. 1. 1960 • Warner's dodder [C] [E]

Stems yellow, filiform. **Inflorescences** dense, glomerulate or corymbiform; bracts at base of clusters 1, at base of pedicels 0(or 1), ovate, membranous, margins entire, apex acute. **Pedicels** 0.5–1 mm, papillate. **Flowers** 5-merous, 2.1–4 mm, slightly fleshy, perianth cells convex, domelike, papillate on corolla, ovary, and capsule; calyx brownish yellow, campanulate-cupulate, 1/2 corolla tube length, divided 1/2 its length, not reticulate, not shiny, lobes triangular-ovate, bases not overlapping, margins entire, midvein carinate, sometimes with multicellular projections, apex enlarged, each with a hornlike appendage 0.5–0.7 mm; corolla creamy white, drying brownish, 1.8–3.5 mm, tube campanulate-urceolate, 1.5–2.5 mm, not saccate, lobes erect, connivent, triangular-ovate, 1/4–1/3 corolla tube length, margins entire, apex acute, inflexed; infrastaminal scales relatively poorly developed, oblong, 1.3–1.2 mm, 4/5 corolla tube length, bridged at 0.2–0.3 mm, truncate, irregularly dentate or short-fimbriate distally, fimbriae 0.05–0.1 mm; stamens included, shorter than corolla lobes; filaments 0.4–0.5 mm; anthers 0.4–0.7 × 0.3–0.5 mm; styles filiform, 0.2–0.4 mm, shorter than ovary. **Capsules** globose, 1.8–2.5 × 1.9–3 mm, thickened and raised in collar around style base, translucent, surrounded or capped by withered corolla, indehiscent. **Seeds** 3 or 4, shape heterogeneous on same plant: obcompressed to weakly angled, broadly ellipsoid to transversely oblique, 1.3–1.6 × 1.2–1.4 mm, hilum region subterminal.

Flowering Jun–Sep. Hosts: known only from *Phyla cuneata* and *P. incisa*; of conservation concern; 1500–1700 m; N.Mex., Utah.

Cuscuta warneri is currently considered possibly extinct in Utah by NatureServe (http://explorer.natureserve.org); it was rediscovered in Roosevelt County, New Mexico, in 2008.

1c. Cuscuta Linnaeus subg. Monogynella (Des Moulins) Peter in H. G. A. Engler and K. Prantl, Nat. Pflanzenfam. 68[IV,3a]: 38. 1891

Monogynella Des Moulins, Étud. Cuscut., 39, 65. 1853

Inflorescences thyrsoid: paniculiform, racemiform, or spiciform. **Pedicels** 0–2 mm. **Flowers:** style usually 1, sometimes distally 2-fid or separating into 2; stigma conic, ± cuboid, depressed-globose, ellipsoid, globose, obovoid, or ovoid. **Capsules:** dehiscence circumscissile. **Seeds** ± obcompressed; seed coat epidermis cells rectangular, not honeycombed when dry and papillate when wet.

Species 15 (2 in the flora): United States, Europe, Asia, Africa.

Cuscuta cassytoides Nees ex Engelmann, known from Africa, was collected in North Carolina in 1918, probably as a waif on *Quercus phellos*; it is not known to be established or recurrent in the flora area.

Cuscuta reflexa Roxburgh, known from Asia, was collected in California in 1969; it is not known to be established or recurrent in the flora area.

1. Calyces equaling corolla tube length . . . . 50. *Cuscuta exaltata*
1. Calyces 1/3–1/2 corolla tube length . . . . 51. *Cuscuta japonica*

50. Cuscuta exaltata Engelmann, Trans. Acad. Sci. St. Louis 1: 513. 1859 • Tall dodder [E] [W]

Stems greenish yellow to purple, spotted, coarse. **Inflorescences** spikes, racemes, or loose panicles with 1–3(–5)-flowered cymes, flowers sessile or short-pedicellate; bracts at base of clusters and pedicels 1, ovate, margins entire, apex obtuse. **Pedicels** 0–2 mm. **Flowers** 5-merous, 4–5 mm, thick and fleshy, not papillate; calyx yellow-brown, globose-campanulate, equaling corolla tube length, divided 2/3 its length, not reticulate or shiny, lobes ovate to orbiculate, bases overlapping, margins entire, partly hyaline, midvein sometimes ± carinate or thickened, apex rounded; corolla white, drying yellow-brown, 3.8–4.8 mm, tube cylindric-campanulate, 2.8–3.5 mm, not saccate, lobes erect to spreading, ovate to suborbiculate, 1/4–1/3 tube length, margins entire, apex rounded; infrastaminal scales oblong to ovate, 1.5–2 mm, 1/2–2/3 corolla tube length, bridged at 0.5–1 mm, 2-fid or with coarse fimbriae distally, 0.1–0.4 mm; stamens included, shorter than corolla lobes; filaments 0.1–0.3 mm; anthers 0.3–0.5 × 0.3–0.5 mm; styles 1, sometimes distally 2-fid or separating into 2 when pulled apart, 1–1.2 mm, ± equaling or longer than ovary, longer than stigmas; stigmas depressed-globose, 0.1–0.5 mm. **Capsules** globose-ovoid to conic, 5–7(–10) × 3–5(–6) mm, not translucent, capped by withered corolla. **Seeds** 4, angled, broadly ellipsoid to ellipsoid or obovoid, 3–3.9 × 2.2–2.5 mm, hilum area lateral.

Flowering May–Oct. Hosts: *Diospyros*, *Juglans*, *Quercus*, *Rhus*, *Ulmus*, *Vitis*, and other woody plants; 0–500 m; Fla., Tex.

51. Cuscuta japonica Choisy in H. Zollinger, Syst. Verz. 2: 134. 1854 • Japanese dodder [F] [I]

Stems yellow-orange, often with purple spots, medium to coarse. **Inflorescences** spikes, racemes, or loose panicles, with 1–3(–7)-flowered cymes, flowers sessile or subsessile; bracts at base of clusters and flowers 1(or 2), broadly ovate, membranous, margins entire, apex obtuse. **Pedicels** 0–1 mm. **Flowers** 5-merous, 3–7 mm, fleshy, not papillate; calyx creamy white, drying yellow-brown, cupulate, 1/3–1/2 corolla tube length, divided 2/3 its length, not reticulate or shiny, lobes orbiculate to ovate, bases overlapping, margins entire, not partly hyaline, midvein carinate or not, apex acute to obtuse; corolla greenish white to pink, drying creamy yellow, 2.6–6.8 mm, tube cylindric, 2–4.3 mm, not saccate, lobes erect to spreading, ovate to ovate-triangular, 1/3–1/2 tube length, margins irregularly crenulate or entire, apex rounded to obtuse; infrastaminal scales oblong to ovate, 1.8–2.8 mm, 1/2 to equaling corolla tube length, bridged at 0.4–0.1 mm, rounded, with dense, thin fimbriae, 0.1–0.3 mm on margins; stamens included, shorter than corolla lobes; filaments 0–0.1 mm; anthers 0.6–1.1 × 0.5–0.7 mm; styles 1, not distally 2-fid or separable, 1–1.6 mm, equaling or longer than ovary, longer than stigma; stigmas globose, ellipsoid, ovoid, obovoid, conic, or ± cuboid, 0.1–0.6 mm. **Capsules** ovoid to helmet-shaped, 4.5–5.5(–7) × 3–3.6(–5) mm, not translucent, capped by withered corolla. **Seeds** 1–3, broadly ellipsoid to obovoid, slightly obcompressed, 2.5–2.8 × 1.1–1.7 mm, hilum region terminal.

Varieties 2 (2 in the flora): introduced; e Asia.

1. Corollas 3–5 mm; stigmas conic, ellipsoid, globose, obovoid, or ovoid . . . . 51a. *Cuscuta japonica* var. *japonica*
1. Corollas 4–7 mm; stigmas ± cuboid . . . . 51b. *Cuscuta japonica* var. *formosana*

51a. Cuscuta japonica Choisy var. **japonica** [F] [I]

Flowers: corolla 3–5 mm; stigmas conic, ellipsoid, globose, obovoid, or ovoid.

Flowering Aug–Sep. Hosts: herbs and woody plants; 0–100 m; introduced; Fla., N.J., Tex.; e Asia.

A report of var. *japonica* from South Carolina has not been verified.

51b. Cuscuta japonica Choisy var. **formosana** (Hayata) Yuncker, Mem. Torrey Bot. Club 18: 253. 1932 [I]

Cuscuta formosana Hayata, Icon. Pl. Formosan. 2: 124, plate 30. 1912

Flowers: corolla 4–7 mm; stigmas ± cuboid.

Flowering Aug–Sep. Hosts: *Citrus*, *Hedera*, *Lagerstroemia*, *Ligustrum*, *Malus*, *Nerium*, *Prunus*, *Quercus*, *Vitis*, and other woody plants; 0–200 m; introduced; Calif., Tex.; e Asia.

2. EVOLVULUS Linnaeus, Sp. Pl. ed. 2, 1: 391. 1762 • Dwarf morning glory [Latin *evolvo*, not twisting, alluding to nontwining habit, as contrasted with *Convolvulus*]

Daniel F. Austin†

Annuals, perennials, or subshrubs. Stems ascending, decumbent, erect, procumbent, or prostrate, glabrous, glabrate, or hairy. **Leaves:** sessile or subsessile; blade elliptic, lanceolate, linear, oblanceolate, oblong, orbiculate, or ovate, 2–35 mm, surfaces glabrate, glabrous, or hairy. **Inflorescences** 2 or 3+-flowered cymes or flowers solitary. **Flowers:** sepals lance-linear, lanceolate, lanceolate-ovate, oblong, or ovate, 2–6 mm; corolla usually blue, lavender, purple, or white, rarely violet, campanulate, funnelform, or rotate, 3–15+ mm, limb 5-angled or -lobed to subentire, 5–12(–22) mm diam.; styles 2, distinct or basally connate, each 2-fid 1/2+ length; stigma lobes 4, filiform to subclavate. **Fruits** capsular, globose to ovoid, dehiscence circumscissle or valvate. **Seeds** 1–4, complanate to ± globose, glabrous, surfaces smooth or verrucose. $x = 13$.

Species ca. 100 (7 in the flora): North America, Mexico, West Indies, Central America, South America; introduced in Europe, Africa, Asia, Australia.

A record of *Evolvulus nummularius* (Linnaeus) Linnaeus for Florida is based on *Rugel 108* (US), which is probably from Cuba (D. B. Ward 1968b).

SELECTED REFERENCES Austin, D. F. 1990. Comments on southwestern United States *Evolvulus* and *Ipomoea* (Convolvulaceae). Madroño 37: 124–132. Ooststroom, S. J. van. 1934. A monograph of the genus *Evolvulus*. Meded. Bot. Mus. Herb. Rijks. Univ. Utrecht 14: 1–267. Ward, D. B. 1968b. Contributions to the Flora of Florida—3. *Evolvulus*. Castanea 33: 76–79.

1. Peduncles plus pedicels filiform, (8–)12–50+ mm.
    2. Herbage glabrous or sparsely hairy and glabrescent, hairs appressed; stems usually procumbent or prostrate, sometimes decumbent . . . . 1. *Evolvulus convolvuloides*
    2. Herbage usually hairy, hairs ± appressed to spreading, sometimes tomentose; stems usually ascending, decumbent, or erect, rarely procumbent.
        3. Leaf blades lanceolate, lance-linear, or linear; sepals lanceolate to lance-linear, 3–3.5 mm, abaxially pilose to tomentose; corolla limbs (10–)12–22 mm diam. 2. *Evolvulus arizonicus*
        3. Leaf blades elliptic, lanceolate, oblong, or ovate; sepals lanceolate, 2–2.5 mm, abaxially glabrous or pilose; corolla limbs (5–)7–10 mm diam. . . . . 3. *Evolvulus alsinoides*
1. Peduncles plus pedicels stout, 0–2(–5+) mm.
    4. Leaves distichous.
        5. Stems decumbent to prostrate; leaf blades usually ovate to broadly ovate, sometimes suborbiculate, 5–8(–10) × 4–7 mm . . . . 4. *Evolvulus grisebachii*
        5. Stems ascending or procumbent; leaf blades elliptic, lanceolate, oblong, or ovate, 4–25 × 2–10 mm. . . . . 5. *Evolvulus sericeus*
    4. Leaves pentastichous.
        6. Internodes usually 4+ mm; leaf blades elliptic to linear; outer sepals ovate, 3–4 mm; rhizomatous perennial . . . . 6. *Evolvulus arenarius*
        6. Internodes rarely 4+ mm; leaf blades usually elliptic, sometimes linear-oblong or narrowly lanceolate to oblanceolate, rarely oblong; outer sepals lanceolate to narrowly lanceolate, 4–5 mm; subshrub . . . . 7. *Evolvulus nuttallianus*

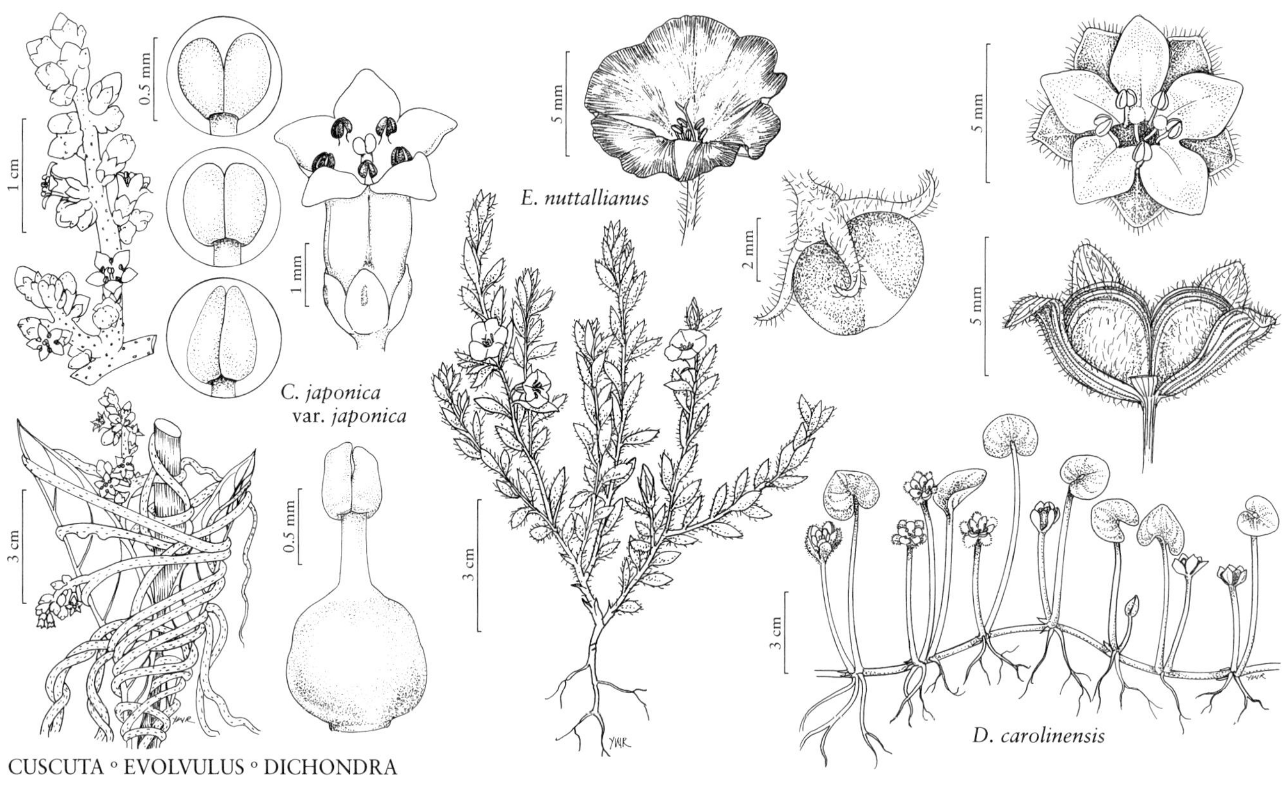

CUSCUTA ◦ EVOLVULUS ◦ DICHONDRA

1. **Evolvulus convolvuloides** (Willdenow) Stearn, Taxon 21: 649. 1972 • Bindweed dwarf morning glory

Nama convolvuloides Willdenow in J. J. Roemer et al., Syst. Veg. 6: 189. 1820; *Evolvulus glaber* Sprengel

Annuals or perennials; herbage glabrous or sparsely hairy and glabrescent, hairs appressed. **Stems** usually procumbent or prostrate, sometimes decumbent, 10–30+ cm. **Leaves:** blade usually ovate, rarely lanceolate or ± orbiculate, 8–25 × 5–15 mm. **Inflorescences** 2- or 3-flowered or flowers solitary; peduncles plus pedicels filiform, 10–25+ mm. **Flowers:** sepals lanceolate, lanceolate-ovate, or oblong, 2.5–3.5 mm; corolla pale blue or white, campanulate or rotate, limb 8–10 mm diam.

Flowering May–Feb. Saline beaches, limestone, hammocks; 0–10 m; Fla., La.; West Indies; South America.

2. **Evolvulus arizonicus** A. Gray in A. Gray et al., Syn. Fl. N. Amer. 2(1): 218. 1878 • Blue eyes, false flax

Evolvulus arizonicus var. *laetus* (A. Gray) Ooststroom

Perennials or subshrubs; herbage densely appressed-hairy to ± tomentose. **Stems** ascending, decumbent, or erect, 10–30 (–45) cm. **Leaves:** blade lanceolate, lance-linear, or linear, 10–25(–35) × 2.5–6(–14) mm. **Inflorescences** 2- or 3-flowered or flowers solitary; peduncles plus pedicels filiform, (8–)12–40+ mm. **Flowers:** sepals lanceolate to lance-linear, 3–3.5 mm, abaxially pilose to tomentose; corolla blue or blue with white stripes, rotate to broadly campanulate, limb (10–) 12–22 mm diam.

Flowering Apr–Oct. Chaparral, oak woodlands, mesquite grasslands, disturbed sites, rocky sites; 800–1900 m; Ariz., N.Mex.; Mexico (Baja California Sur, Chihuahua, Sonora); South America (Argentina).

3. Evolvulus alsinoides (Linnaeus) Linnaeus, Sp. Pl. ed. 2, 1: 392. 1762 • Slender dwarf morning glory, ojo de víbora

Convolvulus alsinoides Linnaeus, Sp. Pl. 1: 157. 1753; *Evolvulus alsinoides* var. *angustifolius* Torrey; *E. alsinoides* var. *grisebachianus* Meisner; *E. alsinoides* var. *hirticaulis* Torrey

Perennials; herbage loosely hairy to glabrate, hairs erect to spreading. **Stems** usually ascending, rarely procumbent, 6–50 cm. **Leaves:** blade elliptic, lanceolate, oblong, or ovate, 8–22 × 3.5–11 mm. **Inflorescences** 2-flowered or flowers solitary; peduncles plus pedicels filiform, (12–)25–50+ mm. **Flowers:** sepals lanceolate, 2–2.5 mm, abaxially glabrous or pilose; corolla pale blue or white, rotate, limb (5–)7–10 mm diam. $2n$ = 26 (Asia).

Flowering Feb–Dec. Pine-oak woodlands, saguaro desert scrub, desert grasslands, disturbed sites, xeric and rocky sites; 0–1900 m; Ala., Ariz., Fla., N.Mex., Tex.; Mexico; Central America; South America.

Presence of *Evolvulus alsinoides* was reported for Alabama by C. T. Mohr (1901). Its presence in Missouri has not been confirmed.

4. Evolvulus grisebachii Peter in H. G. A. Engler and K. Prantl, Nat. Pflanzenfam. 68[IV,3a]: 19. 1897 [C]

Evolvulus wrightii House

Perennials; herbage usually loosely hairy, sometimes glabrescent, hairs appressed to spreading. **Stems** decumbent to prostrate, 5–10(–25) cm. **Leaves** distichous; blade usually ovate to broadly ovate, sometimes suborbiculate, 5–8(–10) × 4–7 mm. **Inflorescences:** flowers solitary; peduncles plus pedicels stout, 0–5+ mm. **Flowers:** sepals narrowly lanceolate, 5–6 mm; corolla blue with white throat, rotate to campanulate, limb 7–10 mm diam.

Flowering Dec–Mar(–Jun). Sandy or limestone pinelands; of conservation concern; 0–10 m; Fla.; West Indies (Cuba).

5. Evolvulus sericeus Swartz, Prodr., 55. 1788 • Silver dwarf morning glory

Evolvulus macilentus Small

Perennials or subshrubs; herbage hairy, hairs appressed to ± spreading. **Stems** ascending or procumbent, 10–30 cm. **Leaves** distichous, distals ± spreading; blade elliptic, lanceolate, oblong, or ovate, 4–25 × 2–10 mm, surfaces: abaxial densely hairy, adaxial sparsely hairy or glabrate. **Inflorescences:** flowers solitary; peduncles plus pedicels stout, 0–2(–4) mm. **Flowers:** sepals oblong to lanceolate, 3–5 mm; corolla pale blue, violet, or white, rotate to broadly funnelform, limb 7–12 mm diam. $2n$ = 26 (Argentina).

Flowering Apr–Oct. Oak woodlands, desert grasslands, plains, savannas, pinelands, chaparral; 0–1900 m; Ariz., Ark., Fla., Ga., La., N.Mex., Tenn., Tex.; Mexico; West Indies; Central America; South America.

6. Evolvulus arenarius R. T. Harms, Phytoneuron 2014-20: 1, fig. 1. 2014 [E]

Perennials, rhizomatous; herbage hairy, hairs mostly appressed, 2-armed. **Stems** ascending or procumbent, 10–30+ cm, internodes usually 4+ mm. **Leaves** pentastichous, distals ± antrorse; blade elliptic to linear, 8–16 × 0.8–1.4 mm, surfaces ± hairy, hairs ± appressed. **Inflorescences:** flowers solitary; peduncles plus pedicels stout, 0–2(–4) mm. **Flowers:** outer sepals ovate, 3–4 mm; corolla lavender, fading yellow, rotate, limb 9–14 mm diam.

Flowering May–Sep. Sandy places, grasslands, prairies, scrublands; 800–1400 m; N.Mex., Tex.

7. Evolvulus nuttallianus Schultes in J. J. Roemer et al., Syst. Veg. 6: 198. 1820 • Shaggy dwarf morning glory [F]

Evolvulus argenteus Pursh, Fl. Amer. Sept. 1: 187. 1813, not R. Brown 1810

Subshrubs; herbage hairy, hairs spreading. **Stems** ascending to erect, 10–15+ cm, internodes rarely 4+ mm. **Leaves** pentastichous, distals ± antrorse; blade usually elliptic, sometimes linear-oblong or narrowly lanceolate to oblanceolate, rarely oblong, 8–20 × 1.5–5 mm, surfaces densely hairy. **Inflorescences:** flowers solitary; peduncles plus pedicels stout, 1–2(–4) mm. **Flowers:** sepals lanceolate to narrowly lanceolate, 4–5 mm; corolla blue, purple, or lavender, rotate to broadly campanulate, limb 8–12 mm diam., margins subentire.

Flowering Apr–Oct. Oak woodlands, ponderosa pine zones, sandy, rocky prairies, plains, juniper-pinyon woodlands, chaparral; 200–2500 m; Ariz., Ark., Colo., Ill., Kans., Mo., Mont., Nebr., N.Mex., Okla., S.Dak., Tenn., Tex., Utah, Wyo.; Mexico.

R. H. Mohlenbrock (1986) reported *Evolvulus nuttallianus* as established in Kane County, Illinois. Reports of it being established in North Dakota have not been confirmed.

The name *Evolvulus pilosus* of Nuttall (not validly published) pertains here.

3. DICHONDRA J. R. Forster & G. Forster, Char. Gen. Pl. ed. 2, 39, plate 20. 1776 • Ponyfoot [Greek *dis*, double, and *chondros*, grain, alluding to each flower producing two 1-seeded capsules in *D. repens*, the type species]

Daniel F. Austin†

Perennials. Stems procumbent to prostrate or trailing, usually rooting at nodes, sometimes mat-forming, glabrous or hairy. **Leaves** petiolate; blade ± cordate-orbiculate to ± reniform, 3–51 mm, to 62 mm wide, surfaces glabrate, glabrous, or hairy. **Inflorescences:** flowers usually solitary, rarely paired. **Flowers:** sepals lanceolate to obovate or spatulate, 1–4(–5.2) mm, basally connate; corolla usually cream, greenish, greenish yellow, or white, rarely purplish or reddish, campanulate to funnelform, 1.5–5 mm, limb 5-lobed; styles 2, insertion on ovary ± basal; stigmas capitate. **Fruits** capsular or utricular, subglobose to ± compressed and/or ± incised, 2-lobed, indehiscent, pericarp fragile, shattering irregularly, or dehiscence irregularly valvate. **Seeds** 1 or 2(–4), obovoid, pyriform, or subspheric, glabrous, smooth. $x = 15$.

Species 15 (8 in the flora): North America, Mexico, West Indies, Central America, South America, Pacific Islands (New Zealand), Australia; introduced also in Europe, Asia, Africa, Pacific Islands (Hawaii).

This treatment is adapted from the revision by B. C. Tharp and M. C. Johnston (1961).

SELECTED REFERENCE Tharp, B. C. and M. C. Johnston. 1961. Recharacterization of *Dichondra* (Convolvulaceae) and a revision of the North American species. Brittonia 13: 346–360.

1. Fruits slightly notched to weakly 2-lobed.
    2. Corollas purplish to reddish; abaxial leaf surfaces glabrate . . . . 1. *Dichondra occidentalis*
    2. Corollas cream; abaxial leaf surfaces sericeous.
        3. Leaf blade surfaces: abaxial and adaxial densely sericeous; pedicels 4–6 mm . . . 2. *Dichondra argentea*
        3. Leaf blade surfaces: abaxial moderately sericeous, adaxial sparsely sericeous to glabrate; pedicels 5–13(–26) mm . . . . 3. *Dichondra brachypoda*
1. Fruits notably 2-lobed.
    4. Leaf blade surfaces: abaxial densely sericeous, hairs silvery gray, adaxial sparsely sericeous to glabrate, hairs green; fruit lobes separating, each valvately dehiscent. . . 4. *Dichondra sericea*
    4. Leaf blade surfaces: abaxial densely to sparsely sericeous, adaxial sparsely sericeous, hairs sometimes patent, or glabrous; fruit lobes separating, each indehiscent (pericarp fragile, shattering irregularly).
        5. Pedicels ± straight, seldom notably recurved near tips; sepals longer than fruits . . . . .5. *Dichondra carolinensis*
        5. Pedicels recurved near tips; sepals shorter than to slightly longer than fruits.
            6. Sepals 2–2.5 mm in fruit; fruits 2–2.6 × 1.8–2.3 mm . . . . .6. *Dichondra micrantha*
            6. Sepals 2.5–3.8 mm in fruit; fruits 2.5–4 × 2.5–3.3 mm.

[7. Shifted to left margin.—Ed.]

7. Leaf blades (5–)10–15(–22) × (8–)15–25(–38) mm; corollas 2–3 mm . . . . .7. *Dichondra donelliana*
7. Leaf blades (10–)20–30(–51) × (15–)30–40(–62) mm; corollas 3.1–4 mm . . . . 8. *Dichondra recurvata*

1. Dichondra occidentalis House, Muhlenbergia 1: 130. 1906

Stems sparsely pilose. **Leaves:** petiole 12–100 mm; blade orbiculate to reniform, 8–27 × 15–42 mm, surfaces glabrate. **Pedicels** 5–31 mm, recurved near tip. **Flowers:** sepals 1.5–2 mm at anthesis, to 2–2.3 mm in fruit; corolla purplish to reddish, 3–3.5 mm. **Fruits** slightly notched to weakly 2-lobed, 4.5–5 × 2.1–2.6 mm.

Flowering Jan. Slopes, headlands, under shrubs; 0–700 m; Calif.; Mexico (Baja California).

2. Dichondra argentea Humboldt & Bonpland ex Willdenow, Hort. Berol. 2: 81, plate 81. 1806

Stems silvery-sericeous. **Leaves:** petiole 10–50 mm; blade reniform, 5–13 × 12–20 mm, surfaces: abaxial and adaxial densely sericeous, hairs silvery gray. **Pedicels** 4–6 mm, recurved near bases. **Flowers:** sepals 2–2.6 mm at anthesis, to 2.4–3 mm in fruit; corolla cream, 3.4–4 mm. **Fruits** slightly notched to weakly 2-lobed, 2.2–2.8 × 2–2.1 mm.

Flowering Apr–Nov. Desert scrub, oak woodlands; 1000–2500 m; Ariz., N.Mex., Tex.; Mexico; South America.

3. Dichondra brachypoda Wooton & Standley, Contr. U.S. Natl. Herb. 16: 160. 1913

Stems sericeous. **Leaves:** petiole 15–150 mm; blade suborbiculate to reniform, 8–40 × 10–55 mm, surfaces: abaxial moderately sericeous, adaxial sparsely sericeous to glabrate. **Pedicels** 5–13 (–26) mm, recurved near tips. **Flowers:** sepals 2.5–4 mm at anthesis, 3.8–5.2 mm in fruit; corolla cream, 3.5–5 mm. **Fruits** slightly notched to weakly 2-lobed, 6–7 × 3–5 mm.

Flowering Jun–Nov. Oak woodland, lower ponderosa pine zones; 700–1900 m; Ariz., N.Mex., Tex.; Mexico.

4. Dichondra sericea Swartz, Prodr., 54. 1788

Dichondra repens J. R. Forster & G. Forster var. *sericea* (Swartz) Choisy

Stems sparsely to densely sericeous. **Leaves:** petiole 5–35 mm; blade suborbiculate, 8–20 × 7–19 mm, surfaces: abaxial densely sericeous, hairs silvery gray, adaxial sparsely sericeous to glabrate, hairs green. **Pedicels** 5–35 mm, straight to curved or nodding, not notably recurved near tips. **Flowers:** sepals 1.5(–2.5) mm at anthesis, to 3.3 mm in fruit; corolla yellow-green, 1.5(–2) mm. **Fruits** notably 2-lobed, 2–3.5 × 1.8–2.5 mm, lobes separating, each valvately dehiscent.

Flowering: Mar–Dec [year-round]. Disturbed sites, oak woodlands, sandy soils; 500–1300 m; Ariz.; Mexico; West Indies; Central America; South America.

5. Dichondra carolinensis Michaux, Fl. Bor.-Amer. 1: 136. 1803 [F] [W]

Dichondra repens J. R. Forster & G. Forster var. *carolinensis* (Michaux) Choisy

Stems sparsely hairy, hairs appressed to erect. **Leaves:** petiole 10–50(–105) mm; blade suborbiculate to reniform, (3–)10–20(–25) × (6–)12–22 (–30) mm, surfaces: abaxial moderately to sparsely sericeous, adaxial sparsely sericeous. **Pedicels** (4–)8–20(–42) mm, ± straight or barely nodding, seldom recurved near tips. **Flowers:** sepals 1.5–3 mm at anthesis, 3.5–5 mm in fruit, longer than fruits; corolla creamy white, 1.5–3 mm at anthesis. **Fruits** notably 2-lobed, 2–3 × 1.6–2 mm, lobes separating, pericarp fragile, shattering irregularly. $2n = 30$.

Flowering year-round. Abandoned plantings, disturbed sites, pinelands, open woodlands; 0–400 m; Ala., Ark., Fla., Ga., La., Md., Miss., Mo., N.C., Okla., S.C., Tenn., Tex., Va.; Mexico.

Reports of *Dichondra carolinensis* for Illinois, Ohio, and Pennsylvania have not been confirmed.

6. Dichondra micrantha Urban, Symb. Antill. 9: 243. 1924 [I] [W]

Stems glabrous or sericeous. **Leaves:** petiole 3–25(–42) mm; blade reniform to broadly ovate-reniform, (3–)8–30 × (3–)8–30 mm, surfaces: abaxial sparsely sericeous, adaxial glabrous or sparsely sericeous, or hairs ± patent. **Pedicels** (3–)5–15 mm, recurved near tips. **Flowers:** sepals 1.5–2 mm at anthesis, 2–2.5 mm in fruit; corolla greenish to white, 1.5–2(–3) mm. **Fruits** notably 2-lobed, 2–2.6 × 1.8–2.3 mm, lobes separating, pericarp fragile, shattering irregularly. $2n$ = 28 (Asia).

Flowering Jan–Nov. Abandoned plantings, clay banks, disturbed sites, thickets; 0–500(–1000) m; introduced; Calif., Fla., Tex.; West Indies; introduced also in Mexico (Nuevo León), Europe, Asia, Africa, Pacific Islands (Hawaii, New Zealand), Australia.

Reports of *Dichondra micrantha* for Arizona, Georgia, and Maryland have not been confirmed.

7. Dichondra donelliana Tharp & M. C. Johnston, Brittonia 13: 352. 1961 [E]

Stems sericeous, hairs yellowish. **Leaves:** petiole 10–25(–75) mm; blade orbiculate-reniform, (5–)10–15(–22) × (8–)15–25(–38) mm, surfaces: abaxial densely sericeous, adaxial sparsely sericeous. **Pedicels** 4–20 mm, recurved near tips. **Flowers:** sepals 1–1.5 mm at anthesis, to 2.5–3.7 mm in fruit; corolla white to greenish white, 2–3 mm. **Fruits** notably 2-lobed, 2.5–3.9 × 2.5–3 mm, lobes separating, pericarp fragile, shattering irregularly.

Flowering Feb–Jul(–Aug). Abandoned plantings, stony, grassy slopes, moist fields; 0–300(–800) m; Calif., Oreg.

8. Dichondra recurvata Tharp & M. C. Johnston, Brittonia 13: 351. 1961 [E]

Stems densely sericeous, hairs tawny. **Leaves:** petiole 20–170 mm; blade orbiculate-reniform, (10–)20–30(–51) × (15–)30–40(–62) mm, surfaces: abaxial densely sericeous, adaxial sparsely sericeous. **Pedicels** 5–13 mm, recurved near tips. **Flowers:** sepals 2.5–3.2 mm at anthesis, 2.9–3.8 mm in fruit; corolla pale green, 3.1–4 mm. **Fruits** notably 2-lobed, 3–4 × 2.8–3.3 mm, lobes separating, pericarp fragile, shattering irregularly.

Flowering Feb–Jul. Disturbed sites, gravelly oak woodlands, lakeshores; 50–500(–700) m; Tex.

4. CRESSA Linnaeus, Sp. Pl. 1: 223. 1753; Gen. Pl. ed. 5, 104. 1754 • Alkali weed [Greek *kressa*, Cretan woman, alluding to inhabitant of island of Crete]

Daniel F. Austin†

Perennials or subshrubs, rhizomatous. **Stems** decumbent to erect, seldom, if ever, mat-forming, glabrate or hirsute to ± sericeous. **Leaves** petiolate or sessile; blade usually elliptic, lanceolate, ovate, or ovate-lanceolate, sometimes scalelike, 1–10 mm, surfaces glabrous or hirsute to ± sericeous. **Inflorescences:** flowers usually solitary, sometimes distally clustered, bracteolate. **Flowers:** sepals elliptic, oblong, obovate, or ovate, 3–6 mm; corolla white [rose], salverform, 5–7 mm, limb 5-lobed, lobes spreading to reflexed, ovate; styles 2, distinct; stigmas capitate. **Fruits** capsular, ovoid, dehiscence valvate. **Seeds** 1(–2)[–4], ovoid, glabrous, surfaces smooth or reticulate. x = 14.

Species 4 (2 in the flora): c, w United States, Mexico, South America, Asia, Indian Ocean Islands (Indonesia), Pacific Islands (Hawaii), Australia.

SELECTED REFERENCE Austin, D. F. 2000b. A revision of *Cressa* L. (Convolvulaceae). Bot. J. Linn. Soc. 133: 27–39.

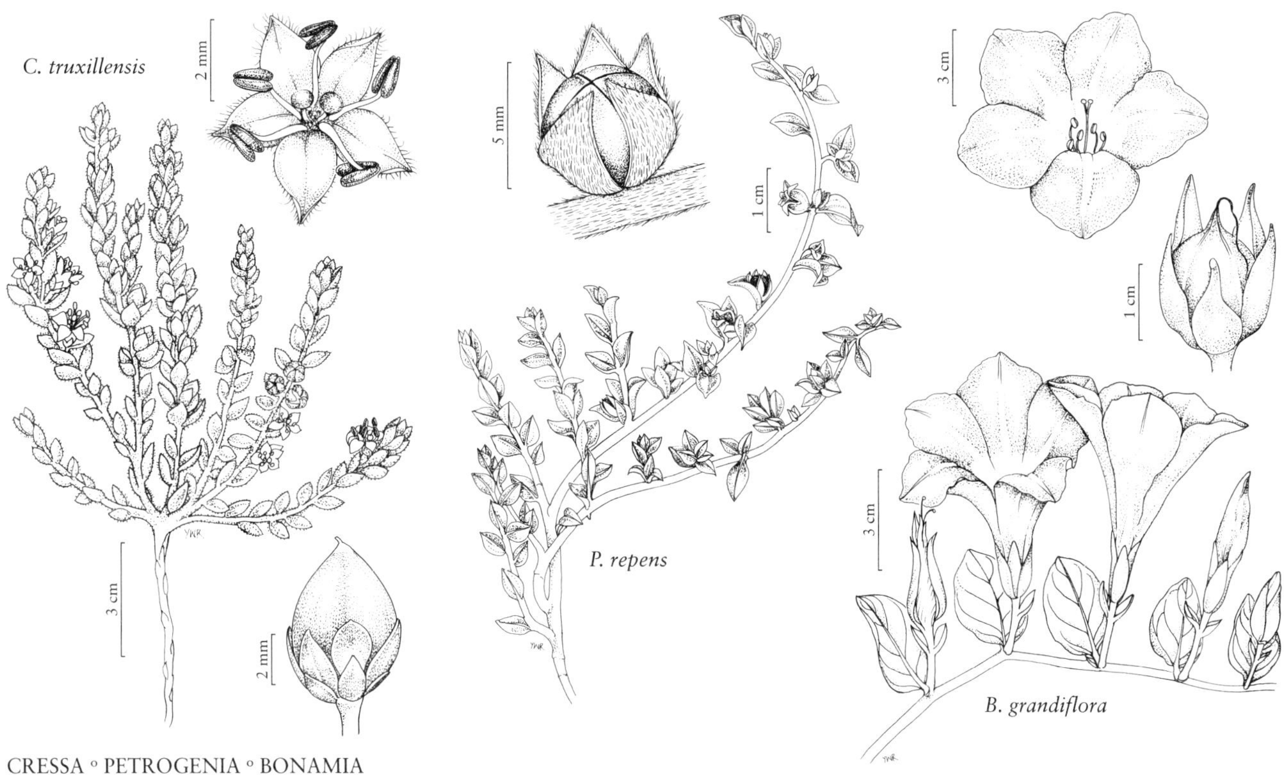

CRESSA ◦ PETROGENIA ◦ BONAMIA

1. Stems usually leafless at anthesis; leaves: blade ovate, scalelike, 1–4 × 1 mm, surfaces glabrous . . . . 1. *Cressa nudicaulis*
1. Stems leafy at anthesis; leaves: blade elliptic, lanceolate, or ovate-lanceolate, 3–10 × 1–4 mm, surfaces hirsute to ± sericeous . . . . 2. *Cressa truxillensis*

1. Cressa nudicaulis Grisebach, Abh. Königl. Ges. Wiss. Göttingen 24: 266. 1879 • Leafless cressa

Stems to 30 cm, ± sericeous, usually leafless at anthesis. **Leaves** sessile; blade ovate, scalelike, 1–4 × 1 mm, surfaces glabrous. **Inflorescences:** bracteoles elliptic to ovate, 2–3 × 1–1.5 mm. **Flowers:** sepals: outer elliptic, 4–6 × 2–2.5 mm, inner ovate, 3.5–4 × 2.5–3 mm; corolla 5–7 mm, tube 3–4 mm. **Capsules** 4–5 mm. **Seeds** 3 mm.

Flowering Apr–Jun(–Sep). Alkaline and/or saline sites; 0–20(–70) m; Tex.; Mexico (Tamaulipas); South America (Argentina).

2. Cressa truxillensis Kunth in A. von Humboldt et al., Nov. Gen. Sp. 3(fol.): 93; 3(qto): 119. 1819 [F] [W]

Cressa depressa Goodding; *C. truxillensis* var. *minima* (A. Heller) Munz; *C. truxillensis* var. *vallicola* (A. Heller) Munz

Stems to 40 cm, hirsute to sericeous; usually leafy at anthesis. **Leaves:** petioles 0.5–2 mm; blade elliptic, lanceolate, or ovate-lanceolate, 3–10 × 1–4 mm, surfaces hirsute to ± sericeous. **Inflorescences:** bracteoles ovate to ovate-lanceolate, 2–3 × 1 mm. **Flowers:** sepals: outer elliptic or obovate, 3–4 × 2.5–3 mm, inner obovate, 3–4.5 × 2–3 mm; corolla 5–6.5 mm, tube 3–3.5 mm. **Capsules** 5–6 mm. **Seeds** 3–4 mm. **2*n*** = 28.

Flowering Apr–Nov. Moist saline and/or alkaline sites; -50–1600 m; Ariz., Calif., Nev., N.Mex., Okla., Oreg., Tex., Utah; Mexico (Baja California, Chihuahua, San Luis Potosí, Sinaloa, Sonora); South America (Argentina, Chile, Ecuador, Peru); introduced in Asia (Timor), Pacific Islands (Hawaii).

5. PETROGENIA I. M. Johnston, J. Arnold Arbor. 22: 116. 1941 • [Greek *petra*, rock, and *genes*, born, alluding to habitat]

Daniel F. Austin†

Perennials or subshrubs. Stems decumbent, erect, prostrate, or trailing, sometimes rooting at nodes, mat-forming, sericeous. **Leaves** petiolate; blade elliptic to lance-elliptic, 6–14 mm, surfaces sericeous. **Inflorescences:** flowers solitary, subsessile. **Flowers:** sepals: outer ovate-lanceolate, 2.5–4(–6) mm, inner lanceolate-acuminate, 2.5–3(–5) mm; corolla greenish yellow to yellow, ± campanulate, 3–4 mm, limb 5-lobed; styles 2, distinct or connate from base to 1/2 length; stigmas globose. **Fruits** capsular, globose, dehiscence valvate. **Seeds** (1–)4, ellipsoid to trigonous, glabrous.

Species 1: Texas, n Mexico.

This treatment follows S. Stefanović et al. (2003), who recognized *Petrogenia* as a monospecific genus, whereas D. F. Austin and G. W. Staples (1985) synonymized it under *Bonamia*.

SELECTED REFERENCE Austin, D. F. and G. W. Staples. 1985. *Petrogenia* as a synonym of *Bonamia* (Convolvulaceae), with comments on allied species. Brittonia 37: 310–316.

1. Petrogenia repens I. M. Johnston, J. Arnold Arbor. 22: 116. 1941 F

Bonamia repens (I. M. Johnston) D. F. Austin & Staples

Leaves: petiole 1–2 mm; blade: base rounded to obtuse, apex usually acute, rarely retuse. **Pedicels** 0.5–1 mm. **Flowers:** sepals sericeous; corolla tube glabrous, lobes ovate, apices rounded, distally hairy. **Capsules** 4–5 mm. **Seeds** reddish brown.

Flowering Mar–Dec. Rocky, caliche and limestone ledges, slopes, or talus; 700–1300 m; Tex.; Mexico (Chihuahua, Coahuila, San Luis Potosí).

6. BONAMIA Thouars, Hist. Vég. Îles France, 33, plate 8. 1804, name conserved • Lady's nightcap [For François Bonami, 1710–1786, French physician and botanist]

Daniel F. Austin†

Perennials or subshrubs, [annuals]. Stems procumbent, prostrate, suberect, or trailing, hairy or glabrous. **Leaves** sessile or petiolate; blade oblong-ovate, orbiculate, or ovate, 10–30 mm, surfaces glabrous or hairy. **Inflorescences** usually solitary flowers, sometimes cymose. **Flowers:** sepals lanceolate, lance-ovate, oblong-ovate, or ovate, [elliptic to orbiculate], 9–28 mm, equal or unequal; corolla blue or blue-purple with white center and tube [pink, red, yellow, or white], funnelform [campanulate], 35–85 mm, limb 5-lobed [subentire or entire]; styles 2, connate from base to 1/2–3/4 length [distinct]; stigmas globose. **Fruits** capsular, conic to globose, dehiscence valvate [indehiscent]. **Seeds** 1–4(–6), globose to ovoid, glabrous or glabrate [villous or ciliate].

Species 30–45 (2 in the flora): sc, se United States, Mexico, West Indies, Central America, South America, Africa, Pacific Islands (Hawaii).

SELECTED REFERENCE Myint, T. and D. B. Ward. 1968. A taxonomic revision of the genus *Bonamia* (Convolvulaceae). Phytologia 17: 121–239.

1. Leaf blades ovate, 20–25 mm wide, surfaces glabrous or puberulent, glabrescent; sepals 15–28 mm; corollas 70–85 mm . . . . 1. *Bonamia grandiflora*
1. Leaf blades oblong-ovate, orbiculate, or ovate, 10–20 mm wide, surfaces velutinous; sepals 9–14 mm; corollas 35–50 mm . . . . 2. *Bonamia ovalifolia*

1. Bonamia grandiflora (A. Gray) Hallier f., Bull. Herb. Boissier 5: 810. 1897 • Florida lady's nightcap [C] [E] [F]

Breweria grandiflora A. Gray, Proc. Amer. Acad. Arts 15: 49. 1879

Perennials. Stems prostrate to trailing, glabrous or puberulent. **Leaves:** blade ovate, 20–30 × 20–25 mm, surfaces glabrous or puberulent, glabrescent. **Flowers:** sepals unequal, lanceolate to oblong-ovate, 15–28 × 4–10 mm; corolla 70–85 mm; styles connate 1/2 length. **Capsules** conic.

Flowering summer–early fall. Sand-pine scrub; of conservation concern; 0–30 m; Fla.

Bonamia grandiflora is in the Center for Plant Conservation's National Collection of Endangered Plants.

SELECTED REFERENCE Hartnett, D. C. and D. R. Richardson. 1989. Population biology of *Bonamia grandiflora* (Convolvulaceae): Effects of fire on plant and seed bank dynamics. Amer. J. Bot. 76: 361–369.

2. Bonamia ovalifolia (Torrey) Hallier f., Bot. Jahrb. Syst. 16: 528. 1893 [C]

Evolvulus ovalifolius Torrey in W. H. Emory, Rep. U.S. Mex. Bound. 2(1): 150. 1859; *Breweria ovalifolia* (Torrey) A. Gray

Subshrubs. Stems procumbent to suberect, sericeous or velutinous. **Leaves:** blade oblong-ovate, orbiculate, or ovate, 10–30 × 10–20 mm, surfaces velutinous. **Flowers:** sepals ± equal, ovate to ovate-lanceolate, 9–14 × 7–12 mm; corolla 35–50 mm; styles connate 3/4 length. **Capsules** globose.

Flowering summer–fall. Sandy soils; of conservation concern; 500–600 m; Tex.; Mexico (Coahuila, Tamaulipas).

Bonamia ovalifolia is known only from the Chihuahuan Desert physiographic region.

Bonamia ovalifolia is in the Center for Plant Conservation's National Collection of Endangered Plants.

SELECTED REFERENCE Austin, D. F. 1988. The rarest morning glory. Fairchild Trop. Gard. Bull. 3: 22–28.

7. STYLISMA Rafinesque, Amer. Monthly Mag. & Crit. Rev. 3: 101. 1818 • [Greek *stylos*, pillar, and *–skhísma*, division, alluding to cleft style] [E]

Charles M. Allen

Perennials. Stems usually procumbent or trailing, rarely twining, glabrous or hairy, hairs 2-armed. **Leaves** petiolate or sessile; blade linear to oblong-elliptic, 10–80 mm, surfaces glabrous or hairy. **Inflorescences** 2–7-flowered or flowers solitary. **Flowers:** sepals ovate-lanceolate to ovate, 4–11 mm; corolla usually white, sometimes lavender, maroon, pink, purple, or red, campanulate or funnelform, 8–25 mm, limb 5-angled or -lobed; styles 2, connate at base or nearly to tips; stigmas peltate. **Fruits** capsular or nutlike, dry, oblong-ovoid [ellipsoid, fusiform, turbinate], dehiscence valvate. **Seeds** 1(–4), ovoid [ellipsoid, fusiform, or globose], hairy. x = 14.

Species 6 (6 in the flora): c, se United States.

SELECTED REFERENCES Allen, C. M. 2013. Synopsis of the genus *Stylisma* (Convolvulaceae) in Louisiana. J. Bot. Res. Inst. Texas 7: 515–516. Myint, T. 1966. Revision of the genus *Stylisma* (Convolvulaceae). Brittonia 18: 97–117.

1. Leaf blades 10–18 mm; flowers: sepals 4–6 mm, corollas 8–13 mm. 1. *Stylisma abdita*
1. Leaf blades 20–80 mm; flowers: sepals 4–11 mm, corollas 10–25 mm.
   2. Corollas lavender, maroon, pink, purple, or red; stamens: filaments glabrous or glabrate . . . . 2. *Stylisma aquatica*
   2. Corollas white; stamens: filaments usually hairy, at least proximally, rarely glabrous.
      3. Leaf blades 10–30 mm wide; inflorescences usually 3–7-flowered, flowers rarely solitary.
         4. Sepals glabrous. . . . . 3. *Stylisma humistrata*
         4. Sepals densely villous . . . . .4. *Stylisma villosa*
      3. Leaf blades 0.1–8 mm wide; inflorescences: flowers usually solitary, sometimes 2–5-flowered.
         5. Leaf blades 0.1–3 mm wide; bracteoles 15–25 mm; style branches usually connate (1/2–)5/6+ length, sometimes less . . . . 5. *Stylisma pickeringii*
         5. Leaf blades 2–8 mm wide; bracteoles 1–3(–5) mm; style branches connate only at base . . . . .6. *Stylisma patens*

1. Stylisma abdita Myint, Brittonia 18: 107, fig. 4. 1966 • Showy dawnflower [E]

Bonamia abdita (Myint) R.W. Long

Stems procumbent or trailing, hairy. **Leaf blades** narrowly linear-elliptic, linear-lanceolate, or linear-oblong, 10–18 × 1.5–2 mm, surfaces densely hairy. **Inflorescences:** flowers solitary. **Flowers:** sepals ovate to ovate-lanceolate, 4–6 × 3–5 mm, abaxial surface densely villous; corolla white, 8–13 mm; stamens: filaments villous.

Flowering May–Jul. Coastal dunes; 0–10 m; Fla.

2. Stylisma aquatica (Walter) Rafinesque, Fl. Tellur. 4: 83. 1838 • Water dawnflower [E]

Convolvulus aquaticus Walter, Fl. Carol., 94. 1788; *Bonamia aquatica* (Walter) A. Gray; *Breweria aquatica* (Walter) A. Gray

Stems procumbent or trailing, hairy. **Leaf blades** narrowly oblong, oblong-elliptic, or oblong-lanceolate, 20–35 × 3–10 mm, surfaces hairy. **Inflorescences** usually 2- or 3-flowered, flowers rarely solitary. **Flowers:** sepals ovate-lanceolate, 5–8 × 3–5 mm, abaxial surface glabrous, margins ciliate; corolla lavender, maroon, pink, purple, or red, 10–15 mm; stamens: filaments glabrous or glabrate.

Flowering Jun–Aug. Coastal dunes, sandy barrens; 0–50 m; Ala., Ark., Fla., Ga., La., Miss., N.C., S.C., Tex.

3. Stylisma humistrata (Walter) Chapman, Fl. South. U.S., 346. 1860 • Southern dawnflower [E] [F]

Convolvulus humistratus Walter, Fl. Carol., 94. 1788; *Bonamia humistrata* (Walter) A. Gray; *Breweria humistrata* (Walter) A. Gray

Stems procumbent or trailing, hairy. **Leaf blades** narrowly oblong, oblong-elliptic, or oblong-lanceolate, 25–80 × 12–30 mm. **Inflorescences** usually 3–7-flowered, flowers rarely solitary. **Flowers:** sepals ovate-lanceolate, 6–9 × 3–5 mm, abaxial surface glabrous, margins ciliate; corolla white, 20–23 mm; stamens: filaments villous near base.

Flowering May–Sep. Sandy barrens, coastal dunes; 0–50 m; Ala., Ark., Fla., Ga., La., Miss., N.C., S.C., Tenn., Tex., Va.

4. Stylisma villosa (Nash) House, Bull. Torrey Bot. Club 34: 149. 1907 • Hairy dawnflower [E]

Breweria villosa Nash, Bull. Torrey Bot. Club 22: 154. 1895; *Bonamia villosa* (Nash) K. A. Wilson

Stems procumbent or trailing, glabrous or hairy. **Leaf blades** narrowly oblong-lanceolate to oblong-elliptic, 30–70 × 10–20 mm, surfaces hairy. **Inflorescences** 3–7-flowered. **Flowers:** sepals ovate-lanceolate, 7–11 × 3-5 mm, abaxial surface densely villous; corolla white, 15–25 mm; stamens: filaments glandular-villous near base.

Flowering Apr–Aug. Sandy barrens; 0–50 m; Fla., Ga., La., Miss., Tex.

5. Stylisma pickeringii (Torrey ex M. A. Curtis) A. Gray, Manual ed. 2, 335. 1856 • Pickering's dawnflower [E]

Convolvulus pickeringii Torrey ex M. A. Curtis, Boston J. Nat. Hist. 1: 129. 1835; *Bonamia pickeringii* (Torrey ex M. A. Curtis) A. Gray; *Breweria pickeringii* (Torrey ex M. A. Curtis) A. Gray

Stems procumbent or trailing, sparsely hairy. **Leaf blades** linear, 25–70 × 0.1–3 mm, surfaces glabrous or sparsely hairy. **Inflorescences** 2–5-flowered or flowers solitary; bracteoles 15–25 mm. **Flowers:** sepals ovate to ovate-lanceolate, 4–6 × 3–5 mm, abaxial surface fulvous or canescent; corolla white, 10–18 mm; stamens: filaments glabrous or proximally puberulent; style branches usually connate ($^1/_2$–)$^5/_6$+ length, sometimes less.

Varieties 2 (2 in the flora): se United States.

1. Sepal apices usually obtuse, rarely acuminate; style branches 2–3 mm, equal to subequal . . . . 5a. *Stylisma pickeringii* var. *pickeringii*
1. Sepal apices acute to subacute; style branches 1–1.5 mm, unequal. . . . . 5b. *Stylisma pickeringii* var. *pattersonii*

5a. Stylisma pickeringii (Torrey ex M. A. Curtis) A. Gray var. **pickeringii** [C] [E]

Breweria pickeringii (Torrey ex M. A. Curtis) A. Gray var. *caesariensis* Fernald & B. G. Schubert

Flowers: sepal apices usually obtuse, rarely acuminate; style branches 2–3 mm, equal to subequal. $2n = 28$.

Flowering May–Aug. Coastal plain woods, sandy soils; of conservation concern; 0–20 m; Ala., Ga., Miss., N.J., N.C., S.C.

Variety *pickeringii* is in the Center for Plant Conservation's National Collection of Endangered Plants.

5b. Stylisma pickeringii (Torrey ex M. A. Curtis) A. Gray var. **pattersonii** (Fernald & B. G. Schubert) Myint, Brittonia 18: 114. 1966 [E]

Breweria pickeringii (Torrey ex M. A. Curtis) var. *pattersonii* Fernald & B. G. Schubert, Rhodora 51: 42, plate 1129. 1949 (as pattersoni)

Flowers: sepal apices acute to subacute; style branches 1–1.5 mm, unequal. $2n = 28$.

Flowering May–Sep. Open, sandy prairies and woods, sand dunes; 0–500 m; Ala., Ark., Ill., Iowa, Kans., La., Miss., Mo., N.C., Okla., S.C., Tex.

6. Stylisma patens (Desrousseaux) Myint, Brittonia 18: 110. 1966 • Coastal plain dawnflower [E]

Convolvulus patens Desrousseaux in J. Lamarck et al., Encycl. 3: 547. 1792; *Bonamia patens* (Desrousseaux) Shinners

Stems procumbent or trailing, glabrous or hairy. **Leaf blades** elliptic or linear to lanceolate, 20–45 × 2–8 mm, surfaces hairy. **Inflorescences** 2–5-flowered or flowers solitary; bracteoles scalelike, 1–3(–5) mm. **Flowers:** sepals ovate-lanceolate, 6–9 × 3–5 mm, abaxial surface pilose, villous, glabrate, or glabrous; corolla white, 14–20 mm; stamens: filaments villous near base; style branches connate only at base.

Varieties 2 (2 in the flora): se United States.

1. Leaf blades 6–8 mm wide, length 4–6 times width; sepals: abaxial surfaces pilose or villous . . . . 6a. *Stylisma patens* var. *patens*
1. Leaf blades 2–3(–5) mm wide, length 7–15 times width; sepals: abaxial surfaces glabrous or glabrate. . . . . 6b. *Stylisma patens* var. *angustifolia*

6a. Stylisma patens (Desrousseaux) Myint var. **patens** [E]

Stylisma trichosanthes (Small) House

Leaf blades elliptic, 6–8 mm wide, length 4–6 times width. **Sepals:** abaxial surfaces pilose or villous.

Flowering May–Sep. Coastal plain woods, sandy soils; 0–100 m; Ala., Fla., Ga., Miss., N.C., S.C.

6b. Stylisma patens (Desrousseaux) Myint var. **angustifolia** (Nash) Shinners, Sida 3: 347. 1969 E

Breweria angustifolia Nash, Bull. Torrey Bot. Club 22: 155. 1895; *Bonamia angustifolia* (Nash) K. A. Wilson; *B. patens* (Desrousseaux) Shinners var. *angustifolia* (Nash) Shinners; *Stylisma angustifolia* (Nash) House; *S. patens* subsp. *angustifolia* (Nash) Myint

Leaf blades linear to lanceolate, 2–3(–5) mm wide, length 7–15 times width. **Sepals:** abaxial surfaces glabrous or glabrate, margins ciliate.

Flowering May–Sep. Coastal plain woods, sandy soils; 0–50 m; Fla., Ga., N.C., S.C.

8. PORANOPSIS Roberty, Candollea 14: 26. 1953 • [Genus *Porana* and Greek *-opsis*, resembling] I

Hayden Brislin

Alexander Krings

Perennials. Stems twining-climbing, puberulent to tomentose, glabrescent. **Leaves** petiolate; blade cordate-ovate, ovate, or suborbiculate, 40–160 mm, surfaces tomentose to villous, glabrescent. **Inflorescences** cymes, clustered in panicles, bracteoles 2, at base of calyx. **Flowers:** sepals lance-linear, 1–2 mm, outer to 24 mm in fruit; corolla white, funnelform, to 8 mm, limb 5-lobed; styles 1, nearly lacking; stigmas 2, globose. **Fruits** utricular, ellipsoid-globose, papery, puberulent [glabrous], indehiscent. **Seeds** 1, ellipsoid-globose, glabrous. $x = 13$.

Species 3 (1 in the flora): introduced, Florida; Asia; introduced also in Mexico, West Indies, Pacific Islands (Hawaii).

1. Poranopsis paniculata (Roxburgh) Roberty, Candollea 14: 26. 1953 • Bridal bouquet F I

Porana paniculata Roxburgh, Pl. Coromandel 3: 31, plate 235. 1815

Leaves: petiole 10–50(–100) mm, tomentose; blade: base cordate, margins entire, apex acuminate, acute, or obtuse. **Inflorescences** 10–30 cm. **Pedicels** 2–4 mm, tomentose. **Flowers:** corolla lobes abaxially puberulent, adaxially glabrous; anthers ellipsoid, 0.5–0.6 mm; ovary globose. **Utricles** 5–6(–7) × 4–5 mm. **Seeds** dark brown, 4–6 × 3–5 mm. $2n = 26$.

Flowering late fall–winter; fruiting early spring. Disturbed sites; 0–10 m; introduced; Fla.; Asia (Bhutan, China, India, Myanmar, Nepal, Pakistan); introduced also in Mexico, West Indies, Pacific Islands (Hawaii).

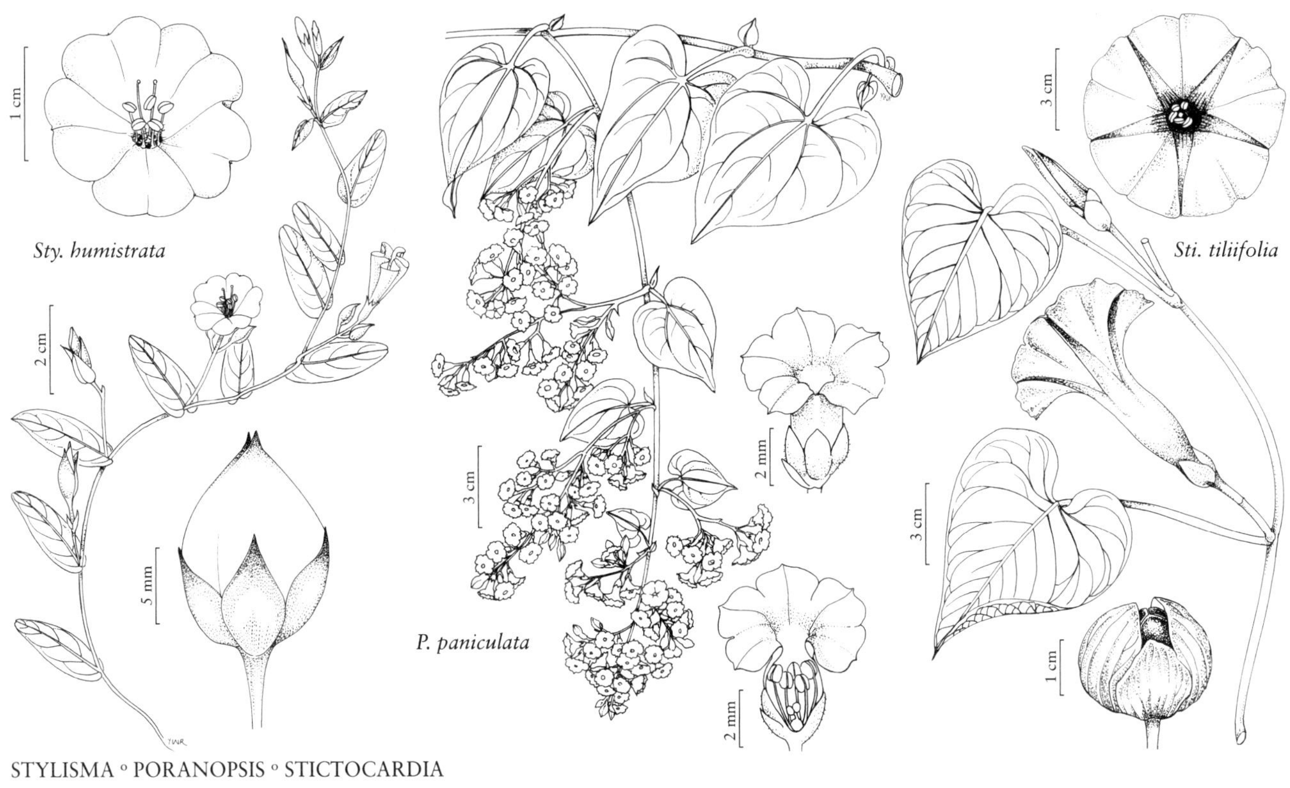

STYLISMA ◦ PORANOPSIS ◦ STICTOCARDIA

9. STICTOCARDIA Hallier f., Bot. Jahrb. Syst. 18: 159. 1893 • [Greek *stiktos*, punctured or spotted, and *kardia*, heart, alluding to glandular-punctate abaxial surfaces of heart-shaped leaf blades] I

Daniel F. Austin†

Perennials. Stems twining-climbing, to 4 m, puberulent, glabrescent. **Leaves** petiolate; blade cordate to cordate-ovate, 80–250 mm, surfaces: abaxial black-glandular-punctate, adaxial glabrous. **Inflorescences** usually solitary flowers, sometimes 2- or 3-flowered cymes. **Flowers:** sepals orbiculate, 10–20 mm; corolla purplish red, [magenta, mauve, pink-purple, red, scarlet, violet, white, yellow], funnelform, (40–)80–95 mm, limb entire; styles 1; stigmas 2, capitate. **Fruits** capsular, dry, globose, glabrous, indehiscent. **Seeds** (1–)4, ovoid, puberulent. $x = 15$.

Species 7–9+ (1 in the flora): introduced, Florida; Africa, Asia; introduced also in West Indies, Central America, South America, Pacific Islands (Guam, Hawaii, Samoa, Tahiti), Australia.

SELECTED REFERENCES Austin, D. F., D. A. Powell, and D. H. Nicolson. 1978. *Stictocardia tiliifolia* (Convolvulaceae) re-evaluated. Brittonia 30: 195–198. Austin, D. F. and Sebsebe Demissew. 1997. Unique fruits and generic status of *Stictocardia* (Convolvulaceae). Kew Bull. 52: 161–168.

1. Stictocardia tiliifolia (Desrousseaux) Hallier f., Bot. Jahrb. Syst. 18: 159. 1893 (as tiliaefolia) • Campanola, spotted-heart [F] [I]

Convolvulus tiliifolius Desrousseaux in J. Lamarck et al., Encycl. 3: 544. 1792 (as tiliaefolia)

Leaf blades: base cordate, apex acute to acuminate. **Corollas** purplish red with darker center. **Capsules** 25–30 mm, surrounded by accrescent calyx that eventually disintegrates into a vascular network. $2n = 30$.

Flowering Nov–Jan. Thicket forests; 0–10 m; introduced; Fla.; Asia; Africa; introduced also in West Indies, Central America, South America, Pacific Islands (Guam, Hawaii, Samoa, Tahiti), Australia.

Stictocardia tiliifolia is widely introduced and naturalized. The names *Rivea campanulata* (Linnaeus) House and *S. campanulata* (Linnaeus) Merrill have been misapplied to it.

10. ARGYREIA Loureiro, Fl. Cochinch. 1: 95, 134. 1790 • Elephant creeper, Hawaiian wood-rose [Greek *argyros*, silver, alluding to abaxial leaf surface appearance] [I]

Daniel F. Austin†

Perennials. Stems twining-climbing, white-hairy when young, glabrescent. **Leaves** petiolate; blade ± cordate-ovate, [40–]180–270 mm; surfaces: abaxial white-hairy, adaxial glabrous or glabrate. **Inflorescences** cymose. **Flowers:** sepals ovate [various], [3–]15–20 mm; corolla lavender [purple, red, rose, or white], funnelform [campanulate or tubular], [30–]60–65[–85] mm, limb 5-lobed to nearly entire; styles 1; stigmas 2, globose. **Fruits** berrylike, ellipsoid to globose, fleshy, indehiscent. **Seeds** 1–4, ovoid, usually glabrous, rarely hairy on hilum. $x = 15$.

Species ca. 90 (1 in the flora): introduced, Florida; Asia, Australia; introduced also in Mexico, West Indies, Central America, South America (Venezuela), Africa.

Argyreias are widely cultivated.

1. Argyreia nervosa (Burman f.) Bojer, Hortus Maurit., 224. 1837 • Woolly morning glory [F] [I]

Convolvulus nervosus Burman f., Fl. Indica, 48, plate 20, fig. 1. 1768; *Argyreia speciosa* (Linnaeus f.) Sweet

Stems to 10+ m, proximally woody. **Leaves:** blade 180–270 × 80–250 mm. **Inflorescences** clusters of 3–5+-flowered cymes; bracts conspicuous, deciduous. **Flowers:** sepals white-hairy, ± enlarged and ± red adaxially in fruit. **Fruits** red-purple or orangish to yellow, 10–15 mm. **Seeds** 8–12 mm, glabrous. $2n = 30$.

Flowering Apr–Aug. Disturbed sites, abandoned plantings; 0–10 m; introduced; Fla.; Asia; Australia; introduced also in Mexico, West Indies, Central America, South America (Venezuela), Africa.

11. TURBINA Rafinesque, Fl. Tellur. 4: 81. 1838 • [Latin *turbinata*, obconical, alluding to fruit shape] [I]

Daniel F. Austin†

Perennials or lianas, [shrubs]. Stems twining-climbing, [trailing], glabrous or hairy [glabrescent]. **Leaves** petiolate [subsessile]; blade cordate to ovate-cordate [oblong to linear], 20–100 mm, surfaces glabrous or hairy. **Inflorescences** 2–16-flowered or flowers solitary. **Flowers:** sepals oblong [ovate to lanceolate], 7–12 mm, sometimes unequal, accrescent in fruit; corolla proximally purplish, distally ± white with greenish bands, campanulate [funnelform or salverform], 25–40 [–80] mm, limb entire or 5-toothed; styles 1; stigmas 2, globose. **Fruits** capsular or nutlike, dry, oblong-ovoid [ellipsoid, fusiform, turbinate], indehiscent. **Seeds** 1(or 2), ovoid [ellipsoid, fusiform, or globose], hairy.

Species 15 (1 in the flora): introduced, Florida; Mexico, West Indies, Central America, South America, Africa, Pacific Islands (New Caledonia); introduced also in Atlantic Islands (Tenerife).

SELECTED REFERENCE Austin, D. F. and G. W. Staples. 1991. A revision of the neotropical species of *Turbina* Raf. (Convolvulaceae). Bull. Torrey Bot. Club 118: 265–280.

1. Turbina corymbosa (Linnaeus) Rafinesque, Fl. Tellur. 4: 81. 1838 • Aguinaldo blanco or de pascua, Christmas vine [F] [I]

Convolvulus corymbosus Linnaeus, Syst. Nat. ed. 10, 2: 923. 1759 (as corymbos); *Rivea corymbosa* (Linnaeus) Hallier f.

Stems proximally woody, to 5+ m. **Leaves:** petiole 40–60 mm; blade: base cordate, apex acute to acuminate, surfaces usually glabrous, rarely hairy. **Bracts** scalelike, deciduous. **Sepals** 7–12 mm, outer 2 shorter than inner 3, margins scarious. **Fruits** 10–15 mm. **Seeds** puberulent.

Flowering Dec–May. Thickets, along waterways; 0–10 m; introduced; Fla.; Mexico; West Indies; Central America; South America.

Reports of *Turbina corymbosa* from Louisiana and Texas have not been verified.

The name *Ipomoea corymbosa* (Linnaeus) Roth has been misapplied to plants of *Turbina corymbosa*. *Ipomoea sidifolia* (Kunth) Sweet, an illegitimate name, pertains here.

12. JACQUEMONTIA Choisy, Mém. Soc. Phys. Genève 6: 476. 1834 • Clustervine [For Victor Jacquemont, 1801–1832, French botanist, explorer]

Kenneth R. Robertson

Subshrubs, vines, or rarely herbs [shrubs], annual or perennial. **Stems** usually ± twining-climbing, sometimes ascending, erect, prostrate, reclining, scandent, scrambling, or trailing, hairy, hairs usually branched and/or stellate, rarely simple and/or glandular. **Leaves** petiolate; blade elliptic, obovate, ovate, or suborbiculate, (10–)60–160 mm, base cordate, cuneate, rounded, or truncate, margins usually entire, sometimes slightly repand, apex acuminate, acute, attenuate, obtuse, or retuse. **Inflorescences** usually cymes, lax to dense, 2–20+-flowered, sometimes flowers solitary; bracts usually relatively small, rarely foliaceous. **Flowers:** sepals persistent in fruit, often accrescent, shape various, equal or unequal; corollas usually blue or white, sometimes pink, lavender, or violet, campanulate to funnelform or rotate, limb entire, 5-angled, or deeply incised and 5-lobed; ovary 2-locular; styles 1, filiform; stigmas 2, each ellipsoid or oblong, flattened, tongue-shaped. **Fruits** capsular, ± globose, dehiscent by (4–)8 valves. **Seeds** (1–)4, trigonous, glabrous, the 2 abaxial edges sometimes narrowly ridged or winged. $x = 9$.

Species ca. 100 (7 in the flora): sw, c, e United States, Mexico, West Indies, Central America, South America, Asia, Africa, Pacific Islands (Hawaii), Australia.

Of the seven species of *Jacquemontia* in the flora area, four are known from southern Florida and two from southern Arizona; *J. tamnifolia* is fairly widespread in the southeastern United States. In 2012, *J. verticillata* (Linnaeus) Urban, known from Bahamas, Belize, Greater Antilles, and Mexico, was collected in Florida, Hillsborough County, on "foreign soil, probably deposited in 2011,..." (from label on University of South Florida herbarium specimen accession number 273728).

SELECTED REFERENCE Robertson, K. R. 1971. A Revision of *Jacquemontia* (Convolvulaceae) in North and Central America and the West Indies. Ph.D. dissertation. Washington University.

1. Leaf blades elliptic, obovate, ovate, or suborbiculate, bases rounded to cuneate, apices usually obtuse or retuse, sometimes acute, mucronate; corollas rotate, limbs deeply incised, 5-lobed.
    2. Leaf blades ± fleshy; outer sepals broadly obovate, rhombic, ovate, or suborbiculate, margins ciliolate . . . . 1. *Jacquemontia reclinata*
    2. Leaf blades ± herbaceous to subcoriaceous; outer sepals broadly elliptic, obovate, ovate, or spatulate, margins not ciliolate.
        3. Outer sepals obovate or spatulate; herbage hairs stellate, 4- or 5-armed. . . . 2. *Jacquemontia curtissii*
        3. Outer sepals ovate or broadly elliptic; herbage hairs stellate, 6- or 7-armed . . . . 3. *Jacquemontia havanensis*
1. Leaf blades narrowly to broadly ovate, bases cordate to truncate, apices usually acuminate, acute, or attenuate, rarely obtuse; corollas campanulate, funnelform, or subrotate, limbs entire or ± 5-angled.
    4. Inflorescences dense, capitate, bracts foliaceous, densely hairy, hairs simple, white, becoming ferruginous dry . . . . 4. *Jacquemontia tamnifolia*
    4. Inflorescences usually lax to compact, sometimes dense, bracts not foliaceous, glabrous or hairy, hairs stellate or simple, not becoming ferruginous.
        5. Annuals; herbage hairs of 2 types: 1) stellate and 3-armed, 2) simple, stalked-glandular, the latter sometimes absent; outer sepals lanceolate to lanceolate-ovate . . . . 5. *Jacquemontia agrestis*
        5. Perennials or subshrubs; herbage hairs stellate, 3- or 4–6-armed, glandular hairs absent; outer sepals ovate, broadly ovate, rhombic, or suborbiculate.
            6. Corollas white to pale blue or lavender; herbage hairs 4–6-armed; outer sepals broadly ovate to suborbiculate, base subcordate, apex acute . . . . 6. *Jacquemontia pringlei*
            6. Corollas blue; herbage hairs 3-armed; outer sepals usually rhombic, sometimes ovate, base narrowed to short stalk, apex long-attenuate . . . . 7. *Jacquemontia pentanthos*

1. Jacquemontia reclinata House ex Small, Bull. New York Bot. Gard. 3: 435. 1905 • Beach clustervine [C] [E]

Vines, perennial. **Herbage** hairy, hairs stellate, 4–7-armed, arms usually unequal, porrect. **Stems** multiple, radiating from rootstock, proximally woody, prostrate, reclining, partly twining, or ascending, to 1 m. **Leaf blades** broadly elliptic, obovate, ovate, or suborbiculate, 10–40 × 5–25 mm, ± fleshy, base cuneate to rounded, apex obtuse to retuse, mucronate. **Inflorescences** ascending, lax, 1–6-flowered. **Flowers:** sepals ± equal or unequal, outers broadly obovate, rhombic, or suborbiculate, 2.5–4 mm, margins ciliolate, inners reniform or suborbiculate, 1.5–2.5 mm, margins often scarious; corolla white or light pink, rotate, 9–15 mm, limb deeply incised, 5-lobed. **Capsules** subglobose to ovoid, 4–6 mm. **Seeds** 2.5–3 mm, outer 2 margins winged, wings 0.1–0.2 mm wide, striate, undulating.

Flowering Nov–May. Coastal sand dunes, maritime hammocks; of conservation concern; 0–10 m; Fla.

Jacquemontia reclinata is endemic to coastal sand dunes and hammocks along the eastern shore of south Florida and is federally listed as an endangered species.

Jacquemontia reclinata is in the Center for Plant Conservation's National Collection of Endangered Plants.

SELECTED REFERENCE Pinto-Torres, E. and S. Koptur. 2009. Hanging by a coastal strand: Breeding system of a federally endangered morning-glory of the south-eastern Florida coast, *Jacquemontia reclinata*. Ann. Bot. (Oxford) 104: 1301–1311.

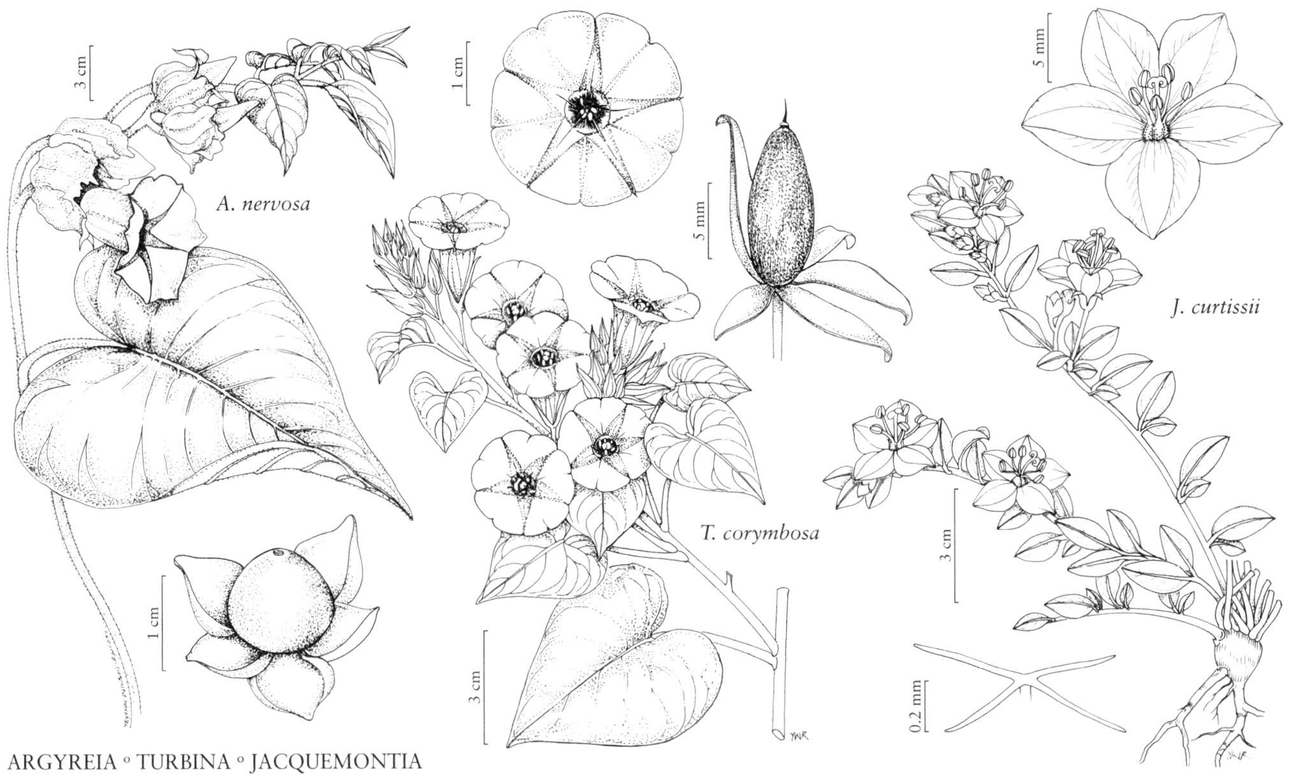

ARGYREIA ∘ TURBINA ∘ JACQUEMONTIA

2. **Jacquemontia curtissii** Peter ex Hallier f., Bausteine Monogr. Convolv., 20, plate 13, fig. 1. 1897 (as curtisii) • Pineland clustervine [C] [E] [F]

Subshrubs or vines, perennial. **Herbage** hairy, hairs stellate, 4- or 5-armed, arms often unequal, porrect. **Stems** multiple, radiating from rootstock, proximally woody, prostrate, scrambling, trailing, or ascending, to 1 m. **Leaf blades** elliptic, ovate, or suborbiculate, 10–29(–37) × 5–15(–21) mm, ± herbaceous to subcoriaceous, base cuneate, apex acute, obtuse, or retuse, mucronate. **Inflorescences** lax, 1–3(–6)-flowered. **Flowers:** sepals unequal, outers obovate or spatulate, 3–6 mm, glabrous, inners ovate to suborbiculate, 2–3 mm; corolla white or light pink, rotate, 10–17 mm, limb deeply incised, 5-lobed. **Capsules** subglobose, 5–7 mm. **Seeds** 3–3.5 mm, outer 2 margins winged, wings 0.1–0.2 mm wide, striate, undulating.

Flowering Oct–Jun(–Aug). Pinelands, openings over limestone; of conservation concern; 0–10 m; Fla.

Jacquemontia curtissii is endemic to southern Florida, especially to rocky pinelands over limestone in the Everglades, also marl prairies, mesic flatwoods, and disturbed uplands. Stout rootstocks enable *J. curtissii* plants to survive fires.

SELECTED REFERENCE Spier, L. P. and J. R. Snyder. 1998. Effects of wet- and dry-season fires on *Jacquemontia curtissii*, a South Florida pine forest endemic. Nat. Areas J. 18: 350–357.

3. **Jacquemontia havanensis** (Jacquin) Urban, Symb. Antill. 3: 342. 1902 • Havana clustervine

Convolvulus havanensis Jacquin, Observ. Bot. 2: 25, plate 45, fig. 3. 1767; *Jacquemontia jamaicensis* (Jacquin) Hallier f. ex Solereder

Vines, perennial, sometimes forming dense thickets. **Herbage** hairy, hairs stellate, 6- or 7-armed, arms unequal, porrect. **Stems** proximally woody, radiating from stout rootstock, prostrate, scrambling, or climbing, forming a dense tangle, 2–3 m. **Leaf blades** (in flora area,) elliptic or slightly ovate, 10–35 × 8–15 mm, ± herbaceous to subcoriaceous, base cuneate, apex acute to obtuse, mucronate. **Inflorescences** lax, 1–6+-flowered. **Flowers:** sepals ± equal or slightly unequal, outers ovate or broadly elliptic, 2–4 mm, glabrous, inners broadly ovate to suborbiculate, 2–3 mm; corolla white, light pink, or pale violet, rotate, 8–15 mm, limb deeply incised, 5-lobed. **Capsules** subglobose, 4–6 mm. **Seeds** 2–3 mm, outer 2 margins narrowly winged, wings 0.1–0.2 mm wide, striate, undulating. 2*n* = 20.

Flowering Aug–Jun. Coastal beach dunes, berms, rockland hammocks; 0–10 m; Fla.; Mexico; West Indies; Central America.

Jacquemontia havanensis is widespread throughout the Bahama Islands and Greater Antilles, barely reaching the Florida Keys, the Leeward Islands, Mexico, and Central America. It, *J. curtissii*, *J. reclinata*, and the West Indian *J. cayensis* Britton are closely related and form a distinctive group. All have a stout rootstock from which arise a number of radiating, horizontal stems that decline, scramble, or twine on other vegetation. Some lateral branches may be erect, and the inflorescences are also erect. Other characteristics are the mostly elliptic to suborbiculate leaf blades, 3–7-armed, stellate hairs, inflorescences usually not exceeding the leaves, deeply incised corolla limbs, and seeds with narrow, prominent wings or ridges on the two outer margins.

4. **Jacquemontia tamnifolia** (Linnaeus) Grisebach, Fl. Brit. W. I., 474. 1862 • Hairy clustervine [W]

Ipomoea tamnifolia Linnaeus, Sp. Pl. 1: 162. 1753; *Thyella tamnifolia* (Linnaeus) Rafinesque

Vines, annual. **Herbage** hairy, hairs usually simple, sometimes 2-armed, white, becoming ferruginous dry. **Stems** erect, strongly climbing, or trailing, 1–4 m, often flowering when a few dm tall. **Leaf blades** ovate, 20–160 × 10–120 mm, base cordate, margins slightly repand, apex acuminate to attenuate. **Inflorescences** dense, capitate, 6–20+-flowered, densely hirsute, bracts foliaceous, densely hairy, hairs simple, white, ferruginous dry. **Flowers:** sepals subequal or outers longer than inners, lanceolate to lance-linear, 8–10 mm, abaxial surface hirsute or sericeous; corolla blue, funnelform, 9–15 mm, limb entire or 5-angled. **Capsules** subglobose, 4–6 mm. **Seeds** 2–2.8 mm, outer 2 margins wingless or each with a very narrow ridge. $2n = 18$.

Flowering Aug–Feb. Disturbed sites, dry stream beds, fields, roadsides, swamps; 0–200 m; Ala., Ark., Fla., Ga., Ill., La., Miss., Mo., N.Y., N.C., Ohio, Okla., Pa., S.C., Tenn., Tex.; Mexico; West Indies; Central America; South America; Africa.

Within the flora area, *Jacquemontia tamnifolia* is most abundant in the southeastern United States; waif outliers are known from other states. It is the most widespread species in *Jacquemontia*. Because it is widespread and invasive in both the Old World and New World subtropics and tropics, it is uncertain where it is native and introduced.

5. **Jacquemontia agrestis** (Martius ex Choisy) Meisner in C. F. P. von Martius et al., Fl. Bras. 7: 306. 1869 • Midnight blue clustervine

Convolvulus agrestis Martius ex Choisy in A. P. de Candolle and A. L. P. P. de Candolle, Prodr. 9: 405. 1845; *Jacquemontia palmeri* S. Watson

Herbs or weak vines, rarely perennial, often delicate, 0.4–1.2 m. **Herbage** hairy, hairs of 2 types: 1) stellate and 3-armed, 2) simple, stalked-glandular, the latter sometimes absent. **Stems** slender, erect, prostrate, or scandent. **Leaf blades** broadly to narrowly ovate, 10–55 × 5–35 mm, base cordate to truncate, apex acuminate. **Inflorescences** lax, 1–6-flowered; bracts linear, inconspicuous. **Flowers:** sepals subequal or outers longer than inners, lanceolate to lance-ovate, 3–5 mm; corolla blue, narrowly funnelform to subrotate, 6–12 mm, limb 5-angled. **Capsules** subglobose, 4–5 mm. **Seeds** 2–3 mm, outer 2 margins narrowly winged.

Flowering Sep–Mar. Canyons; 1000–1300 m; Ariz.; Mexico (Baja California, Baja California Sur, Sonora); Central America; South America.

Jacquemontia agrestis reaches its northern limit of distribution in Pima County, Arizona.

6. **Jacquemontia pringlei** A. Gray, Proc. Amer. Acad. Arts 17: 227. 1882 • Pringle's clustervine

Vines, perennial, **or subshrubs. Herbage** hairy, hairs stellate, 4–6-armed, arms spreading or porrect. **Stems** trailing, twining, or ± erect, to 4 m. **Leaf blades** broadly ovate or ovate, 15–30(–65) × 10–25(–48) mm, base shallowly cordate or truncate, apex usually acute, rarely obtuse. **Inflorescences** ± lax, 1–7-flowered, bracts linear. **Flowers:** sepals unequal, outers broadly ovate to suborbiculate, 5–9 mm, longer and wider than inners, base subcordate, apex acute, surfaces hairy; corolla white to pale blue or lavender, broadly funnelform, 20–25 mm, limb entire or 5-angled. **Capsules** broadly ovoid, 5–6 mm. **Seeds** 2–2.5 mm.

Flowering May–Jan. Oak woodlands, basalt hills, rocky cliffs; 1000–1500 m; Ariz.; Mexico (Sinaloa, Sonora); Central America.

Jacquemontia pringlei is endemic to the Sonoran Desert and reaches its northern limit of distribution in Pima County, Arizona; a report for Yuma County was evidently a mistake.

7. **Jacquemontia pentanthos** (Jacquin) G. Don, Gen. Hist. 4: 283. 1837 (as pentantha) • Skyblue clustervine

Convolvulus pentanthos Jacquin, Collectanea 4: 210. 1791

Vines, perennial. **Herbage** usually hairy, sometimes glabrescent, hairs stellate, 3-armed, arms ± equal, multiangular and recurved. **Stems** from stout rootstocks, twining-climbing, to 5 m. **Leaf blades** ovate, 20–60 (–110) × 15–70(–170) mm, base cordate, margins faintly repand, apex attenuate, surfaces mostly glabrous. **Inflorescences** compact to dense, (1–)3–8+-flowered, peduncles usually longer than leaves, bracts linear, 1 × 0.5 mm. **Flowers:** sepals unequal, outers usually rhombic, sometimes ovate, 4–7 mm, base narrowed to short stalk, margins undulate, apex long-attenuate, surfaces sparsely hairy or glabrous, inners ovate, 3–7 mm, apex acute to acuminate; corollas (in flora area) blue, usually campanulate, sometimes subrotate, 10–20 mm, limb entire. **Capsules** subglobose to broadly ovoid, 4–6 mm. **Seeds** 1.5–2.5 mm. $2n = 18$.

Flowering Oct–Apr. Rockland hammocks, pine rocklands, coastal rock barrens, marl prairies, bayheads; 0–10 m; Fla.; Mexico; West Indies (mostly absent from Bahamas); Central America; South America.

Jacquemontia pentanthos is widespread in the American tropics and subtropics with its northern limit of distribution in southern Florida. It is widely cultivated. It is at the core of approximately ten species occurring in Mexico and Central America.

13. CALYSTEGIA R. Brown, Prodr., 483. 1810, name conserved • Morning glory [Greek *calyx* and *stegos,* covering, alluding to two large bracts enclosing calyx]

Richard K. Brummitt†

Perennials or subshrubs. Stems usually twining-climbing, sometimes ascending, decumbent, erect, procumbent, prostrate, or trailing, rarely almost absent, usually hairy, hairs not branched, glandular, or stellate, sometimes glabrate, glabrescent, or glabrous. **Leaves** petiolate; blade usually cordate, elliptic, linear, oblong, oblong-hastate, orbiculate, oval, ovate, reniform, triangular, or triangular-hastate, rarely palmately 7–9-lobed, (15–)20–130 mm, base usually lobed or truncate, sometimes cuneate, surfaces glabrate, glabrescent, glabrous, ± pilose, tomentose, tomentulose, or villous. **Inflorescences** usually axillary, rarely terminal, compound cymes, bracteate; peduncles 1(–4)-flowered. **Flowers:** sepals ± elliptic, lanceolate, lance-ovate, oblong, oblong-ovate, oval, or ovate, (5–)8–15(–25) mm; corolla usually white, sometimes cream, pink, purple, red, or yellow, campanulate to funnelform, (20–)25–60(–73)[–88] mm, limb entire or 5-lobed or -angled, rarely multilobed; ovary 1-locular, sometimes with partial septum; styles 1; stigma lobes 2, linear to oblong, apices blunt. **Fruits** capsular, ± globose, dehiscence irregular. **Seeds** (1–)2–4, pyramidal to subglobose or trigonous, glabrous, papillate, smooth, or reticulate. $x = 12$.

Species ca. 30 (20 in the flora): North America, Mexico, South America, Eurasia, Africa, Atlantic Islands, Pacific Islands, Australia.

H. Hallier (1893) and W. H. Lewis and R. L. Oliver (1965) summarized arguments for treating *Calystegia* and *Convolvulus* as distinct genera. In a molecular analysis, M. A. Carine et al. (2004) found *Calystegia* nested within *Convolvulus.*

Species delimitation is problematic throughout *Calystegia*, with geographic and morphological intergradation between taxa, and often arbitrary limits have to be adopted to avoid impractically broad species. Hybridization is common where species overlap geographically. It is difficult to pinpoint any species which is not taxonomically subdivided and which does not intergrade or hybridize with others.

1. Bracts (1–)2–10(–50) mm distant from sepals, margins entire, lobed, or toothed.
   2. Stems erect, sometimes intertwined; leaf blades usually linear to narrowly triangular, sometimes ovate, base not lobed or hastate-lobed and lobes ± linear, oblong, or triangular, 1-pointed . . . . 1. *Calystegia longipes*
   2. Stems ascending, decumbent, procumbent, trailing, or twining-climbing; leaf blades oblong, oblong-ovate, orbiculate, ovate, broadly to narrowly triangular, triangular-hastate, or palmately 7–9-lobed, base usually lobed and lobes rounded or 1–3-pointed, base rarely cuneate or ± truncate.
      3. Leaf blades oblong, oblong-ovate, orbiculate, or ovate, base lobed and lobes rounded or base cuneate to ± truncate . . . . 2. *Calystegia felix*
      3. Leaf blades usually broadly to narrowly triangular, ovate-triangular, or triangular-hastate (daggerlike), or palmately 7–9-lobed, sometimes ± reniform, base usually lobed and lobes rounded or 1–3-pointed, base rarely cuneate or ± truncate.
         4. Herbage glabrous.
            5. Leaf blades ± triangular to ovate-triangular; bracts linear, margins entire or proximally lobed or toothed . . . . 3. *Calystegia purpurata*
            5. Leaf blades narrowly triangular-hastate; bracts elliptic to broadly elliptic-oblong, margins entire . . . . 4. *Calystegia peirsonii* (in part)
         4. Herbage usually hairy, at least near leaf blade sinus and/or tip of peduncle, sometimes glabrate or glabrescent.
            6. Leaf blades palmately 7–9-lobed . . . . 5. *Calystegia stebbinsii*
            6. Leaf blades not palmately lobed.
               7. Bract margins entire.
                  8. Leaf blades ± triangular-hastate, middle lobe ± lance-linear; bracts 1–2 mm distant from sepals, linear, 5–16(–20) × 0.5–1.5 mm . . . . 6. *Calystegia vanzuukiae* (in part)
                  8. Leaf blades ± triangular; bracts (1–)3–7 mm distant from sepals, linear to linear-oblong, 4–13(–18) × 1–4(–5) mm . . . . 7. *Calystegia occidentalis* (in part)
               7. Bract margins proximally lobed or toothed.
                  9. Herbage tomentose to villous; leaf blades ± broadly to narrowly triangular, basal lobes 2(–3)-pointed . . . . 8. *Calystegia malacophylla* (in part)
                  9. Herbage glabrate or hairy; leaf blades narrowly triangular or triangular-hastate, basal lobes rounded or 1–2-pointed.
                     10. Bracts 1–2 mm distant from sepals, linear, 5–16(–20) × 0.5–1.5 mm. . . . . 6. *Calystegia vanzuukiae* (in part)
                     10. Bracts 2–12(–15) mm distant from sepals, lanceolate, linear, linear-oblong, oblanceolate, or narrowly to broadly triangular, 5–22(–30) × 2–4(–7) mm . . . . 7. *Calystegia occidentalis* (in part)
1. Bracts immediately subtending, less than 1 mm from, sepals, margins entire.
   11. Corollas horticultural doubles, limbs multilobed; stamens and ovaries absent . . . 9. *Calystegia pubescens*
   11. Corollas not doubles, limbs weakly 5-lobed, 5-angled, or entire; stamens and ovaries present.
      12. Bracts 1.5–3.5(–4) mm wide.
         13. Herbage tomentellous, tomentose, or villous . . . . 10. *Calystegia collina* (in part)
         13. Herbage glabrous or ± hairy, not tomentellous, tomentose, or villous.
            14. Leaf blades narrowly triangular-hastate, base lobed, lobes ± oblong to rhombic, 1-pointed . . . . 4. *Calystegia peirsonii* (in part)
            14. Leaf blades rounded-deltate to triangular-hastate, base cuneate or lobed, lobes not oblong to rhombic and 1-pointed.
               15. Herbage glabrous; bract apices acute to obtuse . . . . . 11. *Calystegia atriplicifolia* (in part)
               15. Herbage sparsely hairy; bract apices acute. . . . . 12. *Calystegia subacaulis* (in part)
      12. Bracts 4–30 mm wide.
         16. Perennials or subshrubs, rootstock woody . . . . 13. *Calystegia macrostegia*
         16. Perennials, rhizomatous.
            17. Leaf blades ± reniform, ± fleshy; corollas pink. . . . . 14. *Calystegia soldanella*

[17. Shifted to left margin.—Ed.]

17. Leaf blades not reniform, not fleshy; corollas usually pink, cream, or white, rarely purple.
    18. Stems usually erect, procumbent, or twining-climbing, sometimes trailing or proximally erect and distally twining-climbing; mostly c, e North America.
        19. Leaf blades triangular to triangular-hastate, base lobed, lobes 2(–3)-pointed; corollas usually pink, sometimes purple or white, 21–32(–35) mm. . . . . 15. *Calystegia hederacea*
        19. Leaf blades ± cordate, elliptic, elliptic-ovate, linear, oblong, ovate, broadly to narrowly triangular, or triangular-hastate, base cuneate, rounded, or lobed, lobes obtuse, rounded, or 1–2-pointed; corollas pink or white, (20–)35–65(–70) mm.
            20. Stems erect . . . . 16. *Calystegia spithamaea*
            20. Stems usually twining-climbing, sometimes trailing or proximally erect and distally twining-climbing.
                21. Leaf blades elliptic-ovate, basal lobes obtuse or rounded, surfaces sometimes whitish tomentose. . . . . 17. *Calystegia catesbeiana*
                21. Leaf blades ± cordate, linear, oblong, oblong-ovate, oval, ovate, triangular, or triangular-hastate, basal lobes usually 1–2-pointed, sometimes rounded or 1-pointed, surfaces not whitish.
                    22. Leaf blade basal sinuses usually acute to rounded, sometimes quadrate, rarely closed; bracts proximally flat or keeled, not or scarcely saccate, margins not or scarcely enfolding sepals, apices acute to subobtuse or truncate. . . . . .18. *Calystegia sepium*
                    22. Leaf blade basal sinuses ± quadrate to rounded; bracts proximally saccate, margins enfolding sepals, apices obtuse to truncate . . . . . 19. *Calystegia silvatica*
    18. Stems usually ascending-decumbent, sometimes procumbent, suberect, trailing, or proximally erect and distally weakly twining-climbing, or almost absent; mostly w North America (*C. macounii* plains and west).
        23. Herbage glabrous . . . . .11. *Calystegia atriplicifolia* (in part)
        23. Herbage moderately or sparsely hairy, puberulent, tomentellous, tomentose, or villous.
            24. Herbage sparsely hairy, hairs appressed . . . . 12. *Calystegia subacaulis* (in part)
            24. Herbage moderately or sparsely hairy, puberulent, tomentellous, tomentose, or villous, hairs not appressed.
                25. Leaf blade basal lobes ± rhombic, rounded; east of California . . . . . . 20. *Calystegia macounii*
                25. Leaf blade basal lobes 1–3-pointed; California.
                    26. Stems mostly to 60(–100) cm; leaves not in basal rosettes . . . . .8. *Calystegia malacophylla* (in part)
                    26. Stems to (2–)50 cm or almost absent; leaves usually in basal rosettes.
                        27. Herbage tomentellous, tomentose, or villous; leaf blade margins ± undulate. . . . . 10. *Calystegia collina* (in part)
                        27. Herbage moderately to sparsely hairy; leaf blade margins not notably undulate. . . . . 12. *Calystegia subacaulis* (in part)

1. Calystegia longipes (S. Watson) Brummitt, Ann. Missouri Bot. Gard. 52: 214. 1965

Convolvulus longipes S. Watson, Amer. Naturalist 7: 302. 1873

Subshrubs, rootstock woody. **Herbage** glabrous. **Stems** erect, sometimes intertwined. **Leaves:** blade usually linear to narrowly triangular, sometimes ovate, to 60 mm, base not lobed, or hastate-lobed and lobes ± linear, oblong, or triangular, 1-pointed. **Bracts** 5–20(–50) mm distant from sepals, lanceolate to linear, 3–17 × 0.2–3 mm, margins entire or lobed. **Flowers:** sepals oblong-ovate, 8–11 mm; corolla white or cream, sometimes pink-tinged, 28–36(–47) mm.

Flowering Apr–Jul. Dry, rocky sites, desert scrub; 200–2500 m; Ariz., Calif., Nev., Utah; Mexico (Baja California).

Intermediates between *Calystegia longipes* and *C. macrostegia* subsp. *tenuifolia,* and between *C. longipes* and *C. peirsonii,* occur in southern California.

2. Calystegia felix Provance & A. C. Sanders, PhytoKeys 32: 5, figs. 1–3. 2013 [E]

Perennials, rhizomatous. **Herbage** glabrous or sparsely hairy. **Stems** trailing or twining-climbing. **Leaves:** blade oblong, oblong-ovate, orbiculate, or ovate, 45–122 × 30–96 mm, base cordate and lobes rounded or base cuneate to ± truncate. **Bracts** (1–)2–3(–4) mm distant from sepals, narrowly elliptic to oblanceolate, 5–14 × 1–2.5(–3.5) mm, shorter than or equal to sepals, proximally flat, margins entire. **Flowers:** sepals lance-ovate to narrowly oblong, outers 8–11 × 2.5–5 mm, inners 11–15 × 3.5–4 mm; corolla white with yellow or purplish stripes, 27–45 mm.

Flowering Mar–Sep. Poorly drained alkali silt loams, disturbed sites; 10–300 m; Calif.

3. Calystegia purpurata (Greene) Brummitt, Ann. Missouri Bot. Gard. 52: 214. 1965 [F]

Convolvulus luteolus A. Gray var. *purpuratus* Greene, Man. Bot. San Francisco, 265. 1894; *C. occidentalis* A. Gray var. *purpuratus* (Greene) J. T. Howell

Subshrubs, rootstock woody. **Herbage** glabrous. **Stems** ascending-decumbent, trailing, or twining-climbing. **Leaves:** blade ± triangular to ovate-triangular, 15–50 mm, base lobed, lobes spreading, rounded or 2(–3)-pointed. **Bracts** 3–10(–16) mm distant from sepals, linear, (2–)3–12(–16) × 0.4–1.5 mm, margins entire or proximally lobed or toothed. **Flowers:** sepals 7–14 mm; corolla white or cream to pale purple, usually purple-striped, (23–)26–46(–52) mm.

Subspecies 2 (2 in the flora): California.

1. Stems strongly twining-climbing; leaf blades triangular, basal sinus V-shaped, apex acute; bract margins entire . . . . 3a. *Calystegia purpurata* subsp. *purpurata*
1. Stems ascending-decumbent, trailing, or weakly twining-climbing; leaf blades ovate-triangular, basal sinus ± closed, apex emarginate to rounded; bract margins lobed or toothed. . . . . 3b. *Calystegia purpurata* subsp. *saxicola*

3a. Calystegia purpurata (Greene) Brummitt subsp. **purpurata** [E]

Calystegia purpurata subsp. *solanensis* (Jepson) Brummitt; *Convolvulus occidentalis* A. Gray var. *solanensis* (Jepson) J. T. Howell

Stems strongly twining-climbing. **Leaf blades** narrowly to broadly triangular, basal lobes gradually to abruptly spreading, acutely or obscurely 2-pointed, basal sinus acute, V-shaped, apex acute. **Bracts:** margins entire.

Flowering Apr–Sep. Chaparral, coastal scrub, dry, rocky sites, shrubby slopes; 0–600(–1600) m; Calif.

3b. Calystegia purpurata (Greene) Brummitt subsp. **saxicola** (Eastwood) Brummitt, Ann. Missouri Bot. Gard. 52: 214. 1965 [E]

Convolvulus saxicola Eastwood, Bull. Torrey Bot. Club 30: 495. 1903; *C. occidentalis* A. Gray var. *saxicola* (Eastwood) J. T. Howell

Stems ascending-decumbent, trailing, or weakly twining-climbing. **Leaf blades** ovate-triangular, basal lobes rounded or obscurely 2-pointed, basal sinus ± closed, apex emarginate to rounded. **Bracts:** margins proximally lobed or toothed.

Flowering Apr–Aug. Coastal scrub, rocky headlands, sandy sites; 0–60 m; Calif.

4. Calystegia peirsonii (Abrams) Brummitt, Ann. Missouri Bot. Gard. 52: 214. 1965 [E]

Convolvulus peirsonii Abrams in L. Abrams and R. S. Ferris, Ill. Fl. Pacific States 3: 387, fig. 3866. 1951 (as piersonii)

Perennials, rhizomatous. **Herbage** glabrous, sometimes glaucous. **Stems** decumbent, procumbent, or ± twining-climbing, 15–40 cm. **Leaves:** blade narrowly triangular-hastate, to 20 mm, base lobed, lobes ± oblong to rhombic, 1-pointed, basal sinus quadrate to rounded. **Bracts** immediately subtending, or to 3 mm distant from, sepals, elliptic to broadly elliptic-oblong, 3–7 × 2.5–4 mm, margins entire. **Flowers:** sepals 9–13 mm; corolla white, 25–40 mm.

Flowering May–Jun. Chaparral, desert scrub, dry, rocky, gravelly slopes; 300–1600 m; Calif.

Calystegia peirsonii intergrades with *C. longipes* and *C. macrostegia* subspp. *cyclostegia* and *intermedia*.

5. Calystegia stebbinsii Brummitt, Kew Bull. 29: 499, fig. 1. 1974 C E

Perennials, underground stems ± woody. **Herbage** hairy, hairs appressed or spreading, whitish. **Stems** trailing or twining-climbing, to 100 cm. **Leaves:** blade palmately 7–9-lobed, lobes linear to linear-oblong, to 55 × 6 mm, base ± truncate. **Bracts** to 18 mm distant from sepals, margins palmately 3–7(–9)-lobed, lobes to 18 × 3 mm. **Flowers:** sepals 7–11 mm, basally glabrous or hairy; corolla cream, yellow, or pink-striped, 30–35 mm.

Flowering Apr–Jul. Chaparral, foothills; of conservation concern; 300–700 m; Calif.

Calystegia stebbinsii is notable for having no intermediates or hybrids with any other species or any significant infraspecific variants. In some characters, it is similar to *C. vanzuukiae*; they grow in similar habitats.

6. Calystegia vanzuukiae Brummitt & Namoff, Aliso 31: 16, figs. 2–5. 2013 C E

Perennials, rhizomatous. **Herbage** glabrate or hairy, hairs minute, appressed or ascending. **Stems** trailing or weakly twining-climbing, to 100 cm. **Leaves:** blade ± triangular-hastate (daggerlike), 20–40 mm, middle lobe ± lance-linear, (1–)2–7(–9) mm wide, base lobed, lobes 1–2-pointed. **Bracts** 1–2 mm distant from sepals, linear, 5–16(–20) × 0.5–1.5 mm, proximally flat, margins entire or proximally lobed or toothed. **Flowers:** sepals ovate to elliptic, 9–11 × 4–6 mm, apices obtuse-apiculate; corolla white, 27–36 mm.

Flowering Mar–Sep. Gabbro or serpentine soils, chaparral, mixed or coniferous woodlands, foothills; of conservation concern; 800–1200 m; Calif.

7. Calystegia occidentalis (A. Gray) Brummitt, Ann. Missouri Bot. Gard. 52: 214. 1965 E W

Convolvulus occidentalis A. Gray, Proc. Amer. Acad. Arts 11: 89. 1876

Perennials or subshrubs, rootstock woody. **Herbage** usually puberulent or pubescent, sometimes glabrescent, rarely tomentellous or ± villous. **Stems** decumbent, procumbent, or twining-climbing, to 400 cm. **Leaves:** blade ± triangular, 15–40 mm, base usually lobed, lobes rounded or 1–2-pointed, basal sinus quadrate, rounded and ± parallel-sided, or V-shaped, base sometimes ± cuneate. **Bracts** (1–)3–12(–15) mm distant from sepals, lanceolate, linear, linear-oblong, oblanceolate, or narrowly to broadly triangular, 4–22(–30) × 1–4(–7) mm, margins entire or proximally lobed or toothed. **Flowers:** sepals 9–15 mm; corolla white or cream, (20–)25–48 mm.

Subspecies 2 (2 in the flora): w United States.

Subspecies *occidentalis* and subsp. *fulcrata* are distinguished essentially by entire versus proximally lobed or toothed bract margins; the distinction is not absolute.

1. Peduncles (1–)2–4-flowered; bract margins entire . . . . .7a. *Calystegia occidentalis* subsp. *occidentalis*
1. Peduncles 1-flowered; bract margins proximally lobed or toothed . . . . .7b. *Calystegia occidentalis* subsp. *fulcrata*

7a. Calystegia occidentalis (A. Gray) Brummitt subsp. **occidentalis** E W

Calystegia malacophylla (Greene) Munz subsp. *tomentella* (Greene) Munz; *C. occidentalis* var. *tomentella* (Greene) Brummitt; *C. polymorpha* (Greene) Munz; *Convolvulus occidentalis* A. Gray subsp. *fruticetorum* (Greene) Abrams; *C. polymorphus* Greene; *C. tomentellus* Greene

Peduncles (1–)2–4-flowered. **Bracts** (1–)3–7 mm distant from sepals, usually linear to linear-elliptic or linear-oblong, sometimes oblanceolate, 4–13(–18) × 1–4(–5) mm, margins entire.

Flowering (Apr–)May–Aug(–Sep). Dry slopes, grassy, shrubby, rocky sites, conifer forests; (10–)30–1200(–1900) m; Calif., Oreg.

Plants with tomentellous to ± villous herbage from northern and southern Sierra Nevada, Greenhorn Mountains, and Mt. Pinos included here in subsp. *occidentalis* have been called *Calystegia occidentalis* var. *tomentella*, *Convolvulus tomentellus*, or *Calystegia malacophylla* subsp. *tomentella*.

7b. Calystegia occidentalis (A. Gray) Brummitt subsp. **fulcrata** (A. Gray) Brummitt, Kew Bull. 29: 502. 1974 E W

Convolvulus luteolus A. Gray var. *fulcratus* A. Gray, Proc. Amer. Acad. Arts 11: 90. 1876; *Calystegia fulcrata* (A. Gray) Brummitt; *C. malacophylla* (Greene) Munz var. *deltoidea* (Greene) Munz; *Convolvulus fulcratus* (A. Gray) Greene; *C. fulcratus* var. *deltoideus* (Greene) Jepson

Peduncles 1-flowered. **Bracts** 2–12(–15) mm distant from sepals, narrowly lanceolate to broadly triangular,

5–22(–30) × 2–4(–7) mm, margins proximally lobed or toothed.

Flowering May–Jul(–Aug). Dry slopes, rocky, shrubby sites, usually on serpentine; 30–2500 m; Calif.

8. Calystegia malacophylla (Greene) Munz, Suppl. Calif. Fl., 85. 1968 E

Convolvulus malacophyllus Greene, Pittonia 3: 326. 1898, based on *Calystegia villosa* Kellogg, Proc. Calif. Acad. Sci. 5: 17. 1873, not Rafinesque 1817

Perennials, rhizomatous. **Herbage** tomentose to villous. **Stems** ascending, decumbent, or procumbent, to 60(–100) cm. **Leaves** not in rosettes; blade usually broadly to narrowly triangular, sometimes ± reniform, 30–60 × 30–90 mm, base lobed, lobes 1–3-pointed, margins not notably undulate, apex acuminate, acute, emarginate, or obtuse. **Bracts** usually immediately subtending sepals and lanceolate to broadly ovate, rarely to 7 mm distant from sepals and narrowly triangular to subreniform, (7–)10–18(–20) × (5–)7–14(–15) mm, lengths usually equal to or longer than sepals, margins entire, lobed, or toothed. **Flowers:** sepals 9–13(–15) mm, densely hairy; corolla white, (20–)26–38(–45) mm.

Subspecies 2 (2 in the flora): California.

1. Herbage: hairs (dry) usually dull to golden brown, rarely grayish; leaf blades usually broadly triangular, sometimes ± reniform, 30–60 × 40–90 mm, basal lobes 2(–3)-pointed, apex usually acute or emarginate, sometimes acuminate; peduncles 3–5 cm . . . . 8a. *Calystegia malacophylla* subsp. *malacophylla*
1. Herbage: hairs (dry) usually gray, sometimes dull brown; leaf blades narrowly triangular, 30–45 × 30–40 mm, basal lobes 1(–2)-pointed, apex usually acute, sometimes obtuse; peduncles 6–9 cm . . . . 8b. *Calystegia malacophylla* subsp. *pedicellata*

8a. Calystegia malacophylla (Greene) Munz subsp. **malacophylla** E

Colvolvulus fulcratus A. Gray var. *berryi* (Eastwood) Jepson

Herbage: hairs (dry) usually dull to golden brown, rarely grayish. **Leaf blades** usually broadly triangular, sometimes ± reniform, 30–60 × 40–90 mm, basal lobes 2(–3)-pointed, apex usually acute or emarginate, sometimes acuminate. **Peduncles** 3–5 cm.

Flowering May–Aug. Grassy, shrubby sites; (600–)1000–2400(–3500) m; Calif.

8b. Calystegia malacophylla (Greene) Munz subsp. **pedicellata** (Jepson) Munz, Suppl. Calif. Fl., 85. 1968 E

Convolvulus villosus A. Gray var. *pedicellatus* Jepson, Man. Fl. Pl. Calif., 777. 1925 (as pedicellata); *C. malacophyllus* Greene subsp. *pedicellatus* (Jepson) Abrams

Herbage: hairs (dry) usually gray, sometimes dull brown. **Leaf blades** narrowly triangular, 30–45 × 30–40 mm, basal lobes 1(–2)-pointed, apex usually acute, sometimes obtuse. **Peduncles** 6–9 cm.

Flowering Apr–Jul. Chaparral, dry slopes, grassy, rocky sites; (80–)600–2100 m; Calif.

9. Calystegia pubescens Lindley, J. Hort. Soc. London 1: 70, fig. [p. 71]. 1846 I W

Perennials. Herbage sparsely hairy or glabrous. **Stems** trailing to twining-climbing, to 100 cm. **Leaves:** blade oblong-hastate to narrowly triangular, to 60+ mm, margins ± parallel at mid blade, base lobed, lobes abruptly spreading, ± triangular, apex acute to obtuse. **Bracts** immediately subtending sepals, lance-ovate, 15–21(–24) × 8–14 mm. **Flowers:** sepals 8–12+ mm; corolla usually pink, sometimes red or white, 40–67 mm, horticultural doubles, limb multilobed; margins entire; stamens and ovaries absent. **2*n*** = 22.

Flowering Jun–Sep. Abandoned plantings, disturbed sites; introduced; Ont., Que.; Conn., Del., D.C., Ill., Kans., Maine, Mass., Mich., Mo., N.H., N.J., N.Y., Pa., Tenn., Vt.; Asia; introduced also in Europe.

The nomenclatural type of *Calystegia pubescens* may prove to be conspecific with the type of *Convolvulus japonicus* Thunberg. Plants are sterile; reproduction is by rhizomes.

10. Calystegia collina (Greene) Brummitt, Ann. Missouri Bot. Gard. 52: 215. 1965 E

Convolvulus collinus Greene, Pittonia 3: 326. 1898; *C. malacophyllus* Greene subsp. *collinus* (Greene) Abrams

Perennials, rhizomatous. **Herbage** tomentellous, tomentose, or villous. **Stems** ascending-decumbent, to 50 cm, sometimes almost absent. **Leaves** in basal rosettes; blade usually triangular, sometimes notably

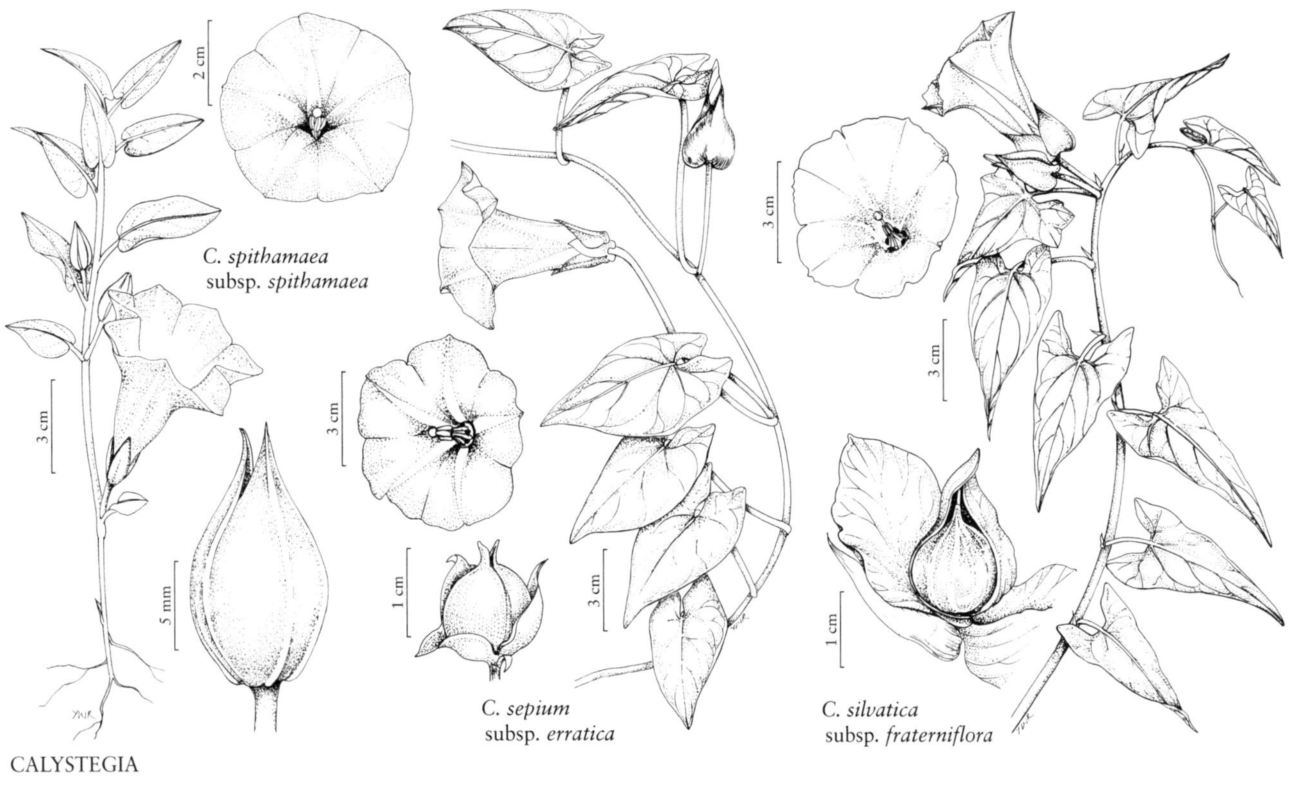

CALYSTEGIA

3-lobed, to 40 mm, base ± lobed, lobes oblong to triangular, 1–3-pointed, margins ± undulate or not, apex rounded to acute. **Bracts** immediately subtending sepals, lanceolate, linear-elliptic, or ovate, 5–17 × 1.5–14 mm. **Flowers:** sepals 8–13 mm; corolla white, (25–) 27–50(–55) mm.

Subspecies 5 (5 in the flora): California.

1. Leaf blades notably 3-lobed, lobes oblong to triangular, margins not notably undulate; bracts 1.5–3.5 mm wide . . . . 10b. *Calystegia collina* subsp. *tridactylosa*
1. Leaf blades triangular, usually basally lobed and lobes 1–3-pointed, weakly lobed in subsp. *collina*, margins undulate or not; bracts 4–14 mm wide.
   2. Sepals glabrous or medially strigose.
      3. Leaf blades ± triangular, base weakly lobed, lobes seldom well developed, apex rounded to subacute . . . . 10a. *Calystegia collina* subsp. *collina*
      3. Leaf blades broadly triangular, basal lobes well developed, apex acute . . . . 10e. *Calystegia collina* subsp. *oxyphylla*
   2. Sepals, at least outers, hairy, hairs appressed.
      4. Leaf blades mostly 40 × 40 mm, margins not undulate . . . . 10c. *Calystegia collina* subsp. *apicum*
      4. Leaf blades mostly 10–20 × 10–20 mm, margins undulate . . . . 10d. *Calystegia collina* subsp. *venusta*

10a. Calystegia collina (Greene) Brummitt subsp. **collina** E

Stems to 8(–15) cm or almost absent, densely tomentose or ± villous, hairs brownish to grayish. **Leaf blades** ± triangular, base weakly lobed, lobes seldom well developed, margins undulate, apex rounded to subacute. **Bracts** broadly ovate, 8–15 × (6–)9–14 mm. **Flowers:** sepals glabrous or medially strigose; corolla (30–)34–48(–55) mm.

Flowering Apr–Jun. Chaparral, grassy, rocky sites, woodlands; 100–1000 m; Calif.

10b. Calystegia collina (Greene) Brummitt subsp. **tridactylosa** (Eastwood) Brummitt, Ann. Missouri Bot. Gard. 52: 215. 1965 C E

Convolvulus tridactylosus Eastwood, Proc. Calif. Acad. Sci., ser. 4, 20: 151. 1931

Stems to 50 cm, densely tomentose or ± villous, hairs brownish to grayish. **Leaf blades** notably 3-lobed, middle lobe oblong to broadly to narrowly triangular, lateral lobes usually antrorse, sometimes slightly retrorse or at right angles to middle lobe, margins not notably undulate, apex acute. **Bracts** lanceolate to linear-elliptic, 5–10 × 1.5–3.5 mm, apex acute. **Flowers:** sepals, at least outers, densely hairy, hairs appressed; corolla 27–33 mm.

Flowering May–Jun. Chaparral, woodlands; of conservation concern; 600–800 m; Calif.

Subspecies *tridactylosa* may be a result of introgression between *Calystegia collina* and *C. occidentalis* subsp. *occidentalis*.

10c. Calystegia collina (Greene) Brummitt subsp. **apicum** Brummitt & Namoff, Aliso 31: 87. 2014 [E]

Stems to 30 cm or almost absent, hairy, hairs gray. **Leaf blades** ± triangular, mostly 40 × 40 mm, base lobed, lobes ± abruptly spreading, hastate, 1–3-pointed, margins not undulate, apex ± acute. **Bracts** broadly ovate to lanceolate, 8–16 × 4–10 mm. **Flowers:** sepals, at least outers, hairy, hairs appressed; corolla (25–)30–44 mm.

Flowering Apr–Jun. Chaparral, mesic sites, woodlands; 300–2000 m; Calif.

10d. Calystegia collina (Greene) Brummitt subsp. **venusta** Brummitt, Kew Bull. 35: 328. 1980 [E]

Stems 8–30 cm or almost absent, usually tomentellous, sometimes subsericeous, hairs brownish, grayish, or silvery. **Leaf blades** ± triangular, mostly 10–20 × 10–20 mm, base lobed, lobes 1–3-pointed, margins undulate, apex rounded to acute. **Bracts** lanceolate to broadly ovate, 8–16 × 4–10 mm. **Flowers:** sepals, at least outers, hairy, hairs appressed; corolla (25–)30–44 mm.

Flowering May–Jun. Chaparral, woodlands; 300–1600 m; Calif.

10e. Calystegia collina (Greene) Brummitt subsp. **oxyphylla** Brummitt, Kew Bull. 35: 328. 1980 [E]

Stems to 8(–20) cm or almost absent, densely tomentose or ± villous, hairs brown. **Leaf blades** broadly triangular, base lobed, lobes well developed, 1–3-pointed, margins undulate or not, apex acute. **Bracts** lanceolate to broadly ovate, 8–17 × 5–11 mm. **Flowers:** sepals glabrous or medially strigose; corolla 27–53 mm.

Flowering Apr–Jun. Chaparral, grassy, rocky sites, woodlands; 100–900 m; Calif.

Subspecies *oxyphylla* intergrades with subsp. *collina*.

11. Calystegia atriplicifolia Hallier f., Bull. Herb. Boissier 5: 385, plate 13, fig. 2. 1897 [E]

Perennials, rhizomatous. **Herbage** glabrous. **Stems** ascending-decumbent, erect, suberect, or trailing, to 10–50 cm or almost absent. **Leaves:** blade rounded-deltate or triangular-hastate, 20–40(–60) mm wide, base cuneate or lobed, lobes at right angles to midrib or slightly retrorse, apex acute to obtuse. **Bracts** immediately subtending sepals, elliptic, ovate, or suborbiculate, 8–19 × 3–18 mm, proximally flat or saccate, scarcely to closely enfolding sepals, apex acute, emarginate, obtuse, or rounded. **Flowers:** sepals 10–15 mm; corolla white, sometimes pink or purple-striped, 31–73 mm.

Subspecies 2 (2 in flora): w United States.

1. Stems usually suberect or trailing, to 35 cm, sometimes almost absent; leaf blades mostly 40–60 mm wide; bracts 14–19 × 9–18 mm, proximally saccate, apex emarginate to rounded; corollas 40–73 mm .... 11a. *Calystegia atriplicifolia* subsp. *atriplicifolia*
1. Stems ascending-decumbent to erect, 10–50 cm; leaf blades mostly 20–40 mm wide; bracts 8–12 × 3–9 mm, proximally flat, apex acute to obtuse; corollas 31–44 mm .... 11b. *Calystegia atriplicifolia* subsp. *buttensis*

11a. Calystegia atriplicifolia Hallier f. subsp. **atriplicifolia** [E]

Convolvulus nyctagineus Greene

Stems usually suberect or trailing, to 35 cm, sometimes almost absent. **Leaf blades** rounded-deltate or triangular-hastate, mostly 40–60 mm wide, lobes retrorse or at right angles to midrib. **Bracts** 14–19 × 9–18 mm, proximally saccate, closely enfolding sepals, apex emarginate to rounded. **Flowers:** sepals 12–14 mm; corollas 40–73 mm.

Flowering May–Aug. Grassy, wooded sites; 70–1700 m; Oreg., Wash.

11b. Calystegia atriplicifolia Hallier f. subsp. **buttensis** Brummitt, Kew Bull. 35: 327. 1980 [E]

Stems ascending-decumbent to erect, 10–50 cm. **Leaf blades** triangular-hastate, mostly 20–40 mm wide, lobes at right angles to midrib. **Bracts** 8–12 × 3–9 mm, proximally flat, not closely enfolding sepals, apex acute to obtuse. **Flowers:** sepals 10–15 mm; corollas 31–44 mm.

Flowering May–Jun. Broad-leaf, coniferous forests; 400–1200 m; Calif.

12. Calystegia subacaulis Hooker & Arnott, Bot. Beechey Voy. 363. 1839 [E]

Convolvulus subacaulis (Hooker & Arnott) Greene

Perennials, rhizomatous. **Herbage** moderately to sparsely hairy, hairs appressed, retrorse, or spreading. **Stems** ascending-decumbent, to 20–30 cm or almost absent. **Leaves** in basal rosettes or not; blade rounded-deltate to triangular-hastate, 30–40 mm, base usually cuneate, sometimes lobed, lobes retrorse, 1-pointed or rounded, margins not or weakly undulate, apex acuminate, acute, rounded, or subacute. **Bracts** immediately subtending sepals, lanceolate, oblong, or broadly ovate, 7–17 × 1.5–9 mm, shorter, or slightly longer, than sepals, proximally usually flat, sometimes keeled, apex usually rounded, sometimes acute. **Flowers:** sepals 10–13 mm; corolla white or cream, sometimes pink- to purplish-striped or -tinged, 33–62 mm.

Subspecies 2 (2 in the flora): California.

1. Herbage moderately to sparsely hairy, hairs retrorse to spreading; stems almost absent or to 2 cm; leaves in basal rosettes, blade apices usually rounded to acuminate, sometimes subacute; bracts (4–)6–9 mm wide . . . . . 12a. *Calystegia subacaulis* subsp. *subacaulis*
1. Herbage sparsely hairy, hairs appressed; stems ascending-decumbent, to 20–30 cm; leaves not in basal rosettes, blade apices acute; bracts 1.5–4 mm wide . . . . 12b. *Calystegia subacaulis* subsp. *episcopalis*

12a. Calystegia subacaulis Hooker & Arnott subsp. **subacaulis** [E]

Herbage moderately to sparsely hairy, hairs spreading to retrorse. **Stems** almost absent or to 2 cm. **Leaves** in basal rosettes; blade apices usually rounded to acuminate, sometimes subacute. **Bracts** (4–)6–9 mm wide, apex rounded.

Flowering Apr–Jun. Coastal scrub, dry hills, grassy sites, oak woodland; 0–600 m; Calif.

Subspecies *subacaulis* intergrades with *Calystegia occidentalis* subsp. *occidentalis*, *C. malacophylla* subsp. *pedicellata*, and *C. subacaulis* subsp. *episcopalis*.

12b. Calystegia subacaulis Hooker & Arnott subsp. **episcopalis** Brummitt, Kew Bull. 35: 327. 1980 [E]

Herbage sparsely hairy, hairs appressed. **Stems** ascending-decumbent, to 20–30 cm. **Leaves** not in basal rosettes; blade apices acute. **Bracts** 1.5–4 mm wide, apex acute.

Flowering Apr–Jun. Dry, grassy sites, open scrub, woodlands; 10–400 m; Calif.

Subspecies *episcopalis* intergrades with *Calystegia atriplicifolia* subsp. *buttensis* and *C. collina* subsp. *venusta*.

13. Calystegia macrostegia (Greene) Brummitt, Ann. Missouri Bot. Gard. 52: 214. 1965

Convolvulus macrostegius Greene, Bull. Calif. Acad. Sci. 1: 208, 226. 1885

Perennials or subshrubs, rootstock woody. **Herbage** glabrescent, glabrous, or ± hairy, including puberulent and/or pubescent. **Stems** prostrate, weakly trailing, or twining-climbing, to 100 cm, or strongly twining-climbing, to 900+ cm. **Leaves:** blade linear or broadly to narrowly triangular, to 130 × 1–120 mm, basally lobed, lobes 2–3-pointed or rounded, basal sinus acute, rounded, or ± quadrate. **Bracts** immediately subtending sepals, lanceolate to ovate or suborbiculate, (6–)8–30(–37) × 4–30 mm, proximally flat, keeled, or saccate. **Flowers:** sepals 7–25 mm; corolla white or cream, sometimes fading pink to purplish, 22–68 mm.

Subspecies 6 (6 in the flora): California, n Mexico.

1. Bracts 16–30 mm wide, proximally keeled or saccate.
   2. Bracts 13–23(–26); sepals 10–17(–22) mm; corollas 36–55(–60) mm; stamens 17–26 mm . . . . 13a. *Calystegia macrostegia* subsp. *macrostegia*
   2. Bracts (19–)22–30(–37); sepals 16–25 mm; corollas 47–68 mm; stamens 23–32 mm . . . . .13b. *Calystegia macrostegia* subsp. *amplissima*
1. Bracts 4–16 mm wide, proximally flat, keeled, or ± saccate.
   3. Bract apices acuminate, emarginate, or obtuse . . . . . 13c. *Calystegia macrostegia* subsp. *cyclostegia*
   3. Bract apices acute.
      4. Herbage ± grayish pubescent . . . . 13d. *Calystegia macrostegia* subsp. *arida*
      4. Herbage glabrous or puberulent, usually glabrescent.
         5. Leaf blades broadly to narrowly triangular, (7–)12–20(–30) mm wide excluding lobes . . . 13e. *Calystegia macrostegia* subsp. *intermedia*
         5. Leaf blades linear to narrowly triangular, 1–7 mm wide excluding lobes . . . . .13f. *Calystegia macrostegia* subsp. *tenuifolia*

13a. Calystegia macrostegia (Greene) Brummitt subsp. **macrostegia**

Herbage glabrous or puberulent, glabrescent. **Stems** trailing or twining-climbing, to 400+ cm. **Leaf blades** broadly triangular, lobes 2(or 3)-pointed, basal sinus broadly rounded. **Bracts** broadly ovate to suborbiculate, 13–23(–26) × 16–27 mm, proximally keeled or saccate, apex emarginate to rounded. **Flowers:** sepals 10–17(–22) mm; corolla 36–55(–60) mm; stamens 17–26 mm.

Flowering Mar–Sep. Coastal scrub, rocky canyons, shrubby hillsides; 0–1900 m; Calif.; Mexico (Baja California).

13b. Calystegia macrostegia (Greene) Brummitt subsp. **amplissima** Brummitt, Kew Bull. 35: 327. 1980 E

Herbage puberulent or pubescent, glabrescent. **Stems** trailing or twining-climbing, to 900+ cm. **Leaf blades** broadly triangular, lobes 2(or 3)-pointed, basal sinus broadly rounded. **Bracts** broadly ovate, (19–)22–30(–37) × 17–30 mm, proximally keeled or saccate, apex acuminate to rounded. **Flowers:** sepals 16–25 mm; corolla 47–68 mm; stamens 23–32 mm.

Flowering Feb–Aug. Bushy sites, grassy hillsides, rocky canyons, sand dunes; 0–500 m; Calif.

Subspecies *amplissima* intergrades with subsp. *macrostegia*.

13c. Calystegia macrostegia (Greene) Brummitt subsp. **cyclostegia** (House) Brummitt, Ann. Missouri Bot. Gard. 52: 214. 1965

Convolvulus cyclostegius House, Muhlenbergia 4: 53. 1908

Herbage glabrous or puberulent. **Stems** prostrate, trailing, or strongly twining-climbing, to 250 cm. **Leaf blades** broadly to narrowly triangular, basal lobes usually ± 2-pointed, rarely rounded, basal sinus acute, rounded, or ± quadrate. **Bracts** broadly ovate to suborbiculate, (6–)9–15(–20) × 7–13(–16) mm, proximally flat, keeled, or saccate, apex acuminate, emarginate, or obtuse. **Flowers:** sepals 9–12(–15) mm; corolla (28–)32–44(–52) mm; stamens 18–26 mm.

Flowering Apr–Oct. Pine woods, rocky slopes, sand dunes; 0–900(–1500) m; Calif.; Mexico (Baja California).

Subspecies *cyclostegia* intergrades with *Calystegia macrostegia* subsp. *intermedia*, *C. macrostegia* subsp. *macrostegia*, and *C. purpurata* subsp. *purpurata*.

13d. Calystegia macrostegia (Greene) Brummitt subsp. **arida** (Greene) Brummitt, Ann. Missouri Bot. Gard. 52: 215. 1965

Convolvulus aridus Greene, Pittonia 3: 330. 1898

Herbage ± pubescent, grayish. **Stems** trailing or twining-climbing. **Leaf blades** narrowly triangular, basal lobes 2-pointed or rounded, basal sinus rounded to quadrate. **Bracts** lanceolate, 10–16(–21) × 6–10 mm, proximally flat or ± keeled, apex acute. **Flowers:** sepals 9–12 mm; corolla white or cream, 24–34 mm; stamens 15–18 mm.

Flowering Apr–Jul. Chaparral, coastal scrub, dry sites; 0–1200 m; Calif.; Mexico (Baja California).

Subspecies *arida* intergrades with *Calystegia macrostegia* subspp. *intermedia* and *tenuifolia* and with *C. occidentalis* subsp. *fulcrata*.

13e. Calystegia macrostegia (Greene) Brummitt subsp. **intermedia** (Abrams) Brummitt, Ann. Missouri Bot. Gard. 52: 214. 1965

Convolvulus aridus Greene subsp. *intermedius* Abrams, Contr. Dudley Herb. 3: 357. 1946; *Calystegia macrostegia* (Greene) Brummitt subsp. *longiloba* (Abrams) Brummitt; *Convolvulus aridus* Greene subsp. *longilobus* Abrams

Herbage glabrous or puberulent, usually glabrescent. **Stems** trailing or twining-climbing, to 300 cm. **Leaf blades** broadly to narrowly triangular, (7–)12–20(–30) mm wide excluding lobes, lobes 2-pointed or rounded, basal sinus usually acute to rounded, sometimes ± quadrate. **Bracts** lanceolate, (10–)13–20 × 6–10(–12) mm, proximally flat or ±keeled, apex acute. **Flowers:** sepals (9–)11–14(–16) mm; corolla (24–)27–40 mm; stamens (15–)17–21 mm.

Flowering Mar–Aug. Dry, stony hillsides; 0–600 (–1100) m; Calif.; Mexico (Baja California).

Subspecies *intermedia* intergrades with subsp. *macrostegia*.

13f. Calystegia macrostegia (Greene) Brummitt subsp. **tenuifolia** (Abrams) Brummitt, Ann. Missouri Bot. Gard. 52: 215. 1965

Convolvulus aridus Greene subsp. *tenuifolius* Abrams, Contr. Dudley Herb. 3: 359. 1946

Herbage glabrous or puberulent. **Stems** trailing or weakly twining-climbing. **Leaf blades** linear to narrowly triangular, 1–7 mm wide, excluding lobes, basal lobes linear, 2-pointed or rounded, basal sinus rounded to quadrate, apex acute. **Bracts** narrowly lanceolate, 8–14 × 4–8 mm, proximally flat or ± keeled, apex acute. **Flowers:** sepals 7–10 mm; corolla white or cream, 22–30(–40) mm; stamens 13–18 mm.

Flowering Apr–Sep. Canyons, rocky hillsides, shrubby sites; 0–1400 m; Calif.; Mexico (Baja California).

Subspecies *tenuifolia* intergrades with *Calystegia longipes* and with *C. macrostegia* subspp. *arida* and *intermedia*.

14. Calystegia soldanella (Linnaeus) R. Brown, Prodr., 484. 1810

Convolvulus soldanella Linnaeus, Sp. Pl. 1: 159. 1753

Perennials, rhizomatous. **Herbage** glabrous. **Stems** usually prostrate or trailing, sometimes twining-climbing, to 100(–250) cm. **Leaves:** blade ± reniform, 15–35(–45) × 20–60(–70) mm, ± fleshy), base ± cuneate, apex emarginate or obtuse. **Bracts** immediately subtending sepals, ovate to suborbiculate, 7–16 × 5–10 mm, proximally flat, ± enfolding sepals. **Flowers:** sepals 10–16 mm; corolla pink, 32–52 mm. **2*n*** = 22 [Asia, Europe].

Flowering May–Aug. Open, sandy shores; 0(–10) m; B.C.; Calif., Oreg., Va., Wash.; South America; w Europe; Asia; Africa; Atlantic Islands (Canary Islands, Tristan da Cunha); Pacific Islands (Galapagos, New Zealand); Australia.

15. Calystegia hederacea Wallich in W. Roxburgh, Fl. Ind. 2: 94. 1824 [I] [W]

Convolvulus wallichianus Sprengel

Perennials, rhizomatous. **Herbage** glabrous. **Stems** initially erect, usually becoming procumbent or weakly twining-climbing, to 80 cm. **Leaves:** blade triangular to triangular-hastate, 20–80 × 20–70 mm, base lobed, lobes 2(–3)-pointed, basal sinus quadrate, rounded, or broadly V-shaped, apex ± acute. **Bracts** immediately subtending sepals, lance-ovate to ovate, 7–14(–18) × (4–)6–10(–12) mm, proximally flat, apex acute to subobtuse. **Flowers:** sepals 5–8(–12) mm; corolla usually pale pink, sometimes purple or white, 21–32(–35) mm. **2*n*** = 22.

Flowering May–Jul. Abandoned plantings, disturbed sites; introduced; Conn., Idaho, Ill., Ind., Kans., Ky., La., Maine, Md., Mass., Mich., Mo., N.H., N.J., N.Y., N.C., Ohio, Pa., Tenn., Vt., Va., Wis.; Asia.

Calystegia hederacea is easily recognized by its usually procumbent habit, ivylike leaves, and relatively small bracts and flowers.

16. Calystegia spithamaea (Linnaeus) Pursh, Fl. Amer. Sept. 1: 143. 1813 [E] [F]

Convolvulus spithamaeus Linnaeus, Sp. Pl. 1: 158. 1753

Perennials, rhizomatous. **Herbage** glabrate or tomentose, hairs sometimes whitish. **Stems** erect, 10–35(–50) cm. **Leaves:** blade narrowly to broadly elliptic, to 80 × 35 mm, base cuneate, rounded, or lobed, lobes rounded or ± 1-pointed, to 11 mm. **Bracts** immediately subtending sepals, lance-ovate, oblong, or oval, (13–)16–25(–29) × 8–14(–20) mm, proximally flat or ± keeled, apex acute to obtuse, closely enfolding sepals. **Flowers:** sepals 12–17 mm; corolla white, 35–63(–67) mm.

Subspecies 3 (3 in the flora): c, e North America.

1. Habit ± lax; herbage glabrate to pubescent; leaf blades ultimately ± flat . . . . . 16a. *Calystegia spithamaea* subsp. *spithamaea*
1. Habit ± compact; herbage tomentose, hairs sometimes whitish; leaf blades ± conduplicate.
    2. Leaf blade bases usually lobed, lobes to 5(–6) mm, sometimes cuneate; corollas (38–)43–59(–64) mm; plants (10–)15–35 (–40 cm) . . . . 16b. *Calystegia spithamaea* subsp. *stans*
    2. Leaf blade bases lobed, lobes 4–11 mm; corollas 35–47(–53) mm; plants (15–)25–35(–40) cm . . . . 16c. *Calystegia spithamaea* subsp. *purshiana*

16a. Calystegia spithamaea (Linnaeus) Pursh subsp. **spithamaea** [E] [F]

Plants 10–30(–50) cm; habit erect, ± lax. **Herbage** glabrate to pubescent. **Leaf blades** elliptic to broadly elliptic, ultimately ± flat, base cuneate, rounded, or lobed, lobes to 4(–6) mm. **Flowers:** corolla 41–63(–67) mm; stamens (21–)24–32(–34) mm. $2n = 22$.

Flowering May–Aug. Open grasslands, sandy and stony sites, woods; Conn., Del., Ill., Ind., Iowa, Ky., Maine, Md., Mass., Mich., Minn., Mo., N.H., N.J., N.Y., N.C., Ohio, Pa., Tenn., Vt., Va., Wis.

16b. Calystegia spithamaea (Linnaeus) Pursh subsp. **stans** (Michaux) Brummitt, Ann. Missouri Bot. Gard. 52: 215. 1965 [E]

Convolvulus stans Michaux, Fl. Bor.-Amer. 1: 136. 1803

Plants (10–)15–35(–40) cm; habit ± compact. **Herbage** tomentose, hairs sometimes whitish. **Leaf blades** elliptic, ± conduplicate, base usually lobed, lobes to 5(–6) mm, sometimes cuneate. **Flowers:** corolla (38–)43–59(–64) mm; stamens (21–)23–31 mm.

Flowering May–Aug. Alluvial gravel, glacial till, dry birch or coniferous woodlands, prairie grasslands, sandy soils; N.S., Ont., Que.; Mich., Minn., N.Y., Wis.

16c. Calystegia spithamaea (Linnaeus) Pursh subsp. **purshiana** (Wherry) Brummitt, Ann. Missouri Bot. Gard. 52: 215. 1965 [E]

Convolvulus purshianus Wherry, Proc. Pennsylvania Acad. Sci. 7: 163. 1933, based on *Calystegia tomentosa* Pursh, Fl. Amer. Sept. 1: 143. 1813, not *Convolvulus tomentosus* Linnaeus 1753

Plants (15–)25–35(–40) cm; habit ± compact. **Herbage** tomentose, hairs sometimes whitish. **Leaf blades** narrowly elliptic, ± conduplicate, base lobed, lobes 4–11 mm. **Flowers:** corolla 35–47(–53) mm; stamens 21–26 mm.

Flowering May–Aug. Clay-loam, shale barrens; Md., N.Y., Pa., Va., W.Va.

17. Calystegia catesbeiana Pursh, Fl. Amer. Sept. 2: 729. 1813 [E]

Perennials, rhizomatous. **Herbage** pubescent to tomentose, hairs usually whitish. **Stems** usually twining-climbing, sometimes proximally erect, distally twining-climbing, to 40–200(–300) cm. **Leaves:** blade elliptic-ovate, to 120 × 50 mm, base lobed, lobes obtuse or rounded, to 20 mm. **Bracts** immediately subtending sepals, lanceolate, 12–34 × 10–22 mm, proximally ± keeled, margins ± enfolding sepals, apex acute. **Flowers:** sepals 11–17 mm; corolla white, 44–64(–70) mm.

Subspecies 2 (2 in the flora): se United States.

Plants of *Calystegia catesbeiana*, especially subsp. *catesbeiana*, often have been misidentified as *C. sepium* because of their climbing habit.

1. Stems usually weakly twining-climbing, sometimes proximally erect, to 40(–100) cm; leaf blades to 60 mm, basal lobes obtuse or rounded, to 11 mm, surfaces usually pubescent, abaxial rarely whitish . . . . 17a. *Calystegia catesbeiana* subsp. *catesbeiana*
1. Stems twining-climbing to 200(–300) cm; leaf blades to 120 mm, basal lobes rounded, 9–20 mm, surfaces densely pubescent to tomentose, abaxial usually whitish . . . . 17b. *Calystegia catesbeiana* subsp. *sericata*

17a. Calystegia catesbeiana Pursh subsp. **catesbeiana** E

Convolvulus spithamaeus Linnaeus var. *pubescens* (A. Gray) Fernald

Stems usually weakly twining-climbing, sometimes proximally erect, to 40(–100) cm. **Leaf blades** to 60 mm, basal lobes obtuse or rounded, to 11 mm, surfaces usually pubescent, abaxial rarely whitish.

Flowering May–Jul. Grassy, disturbed sites; Ala., Fla., Ga., Ky., Miss., N.C., S.C., Tenn., Va., W.Va.

17b. Calystegia catesbeiana Pursh subsp. **sericata** (House) Brummitt, Ann. Missouri Bot. Gard. 52: 216. 1965 E

Convolvulus sericatus House, Torreya 6: 150. 1906; *Calystegia sericata* (House) C. R. Bell

Stems twining-climbing to 200(–300) cm. **Leaf blades** to 120 mm, basal lobes rounded, 9–20 mm, surfaces densely pubescent to tomentose, abaxial usually whitish.

Flowering Jun–Jul. Forest margins, open sites; to 1200 m; Ga., N.C., S.C.

A specimen from Indiana may be referable to subsp. *sericata.*

18. Calystegia sepium (Linnaeus) R. Brown, Prodr., 483. 1810 • Hedge bindweed F W

Convolvulus sepium Linnaeus, Sp. Pl. 1: 153. 1753

Perennials, rhizomatous. **Herbage** glabrate, glabrous, or hairy. **Stems** trailing or twining-climbing. **Leaves:** blade linear, ovate, broadly to narrowly triangular, or triangular-hastate, (24–)36–50(–150) mm, base usually lobed, sometimes nearly truncate, lobes usually 1- or 2-pointed, sometimes rounded, basal sinus usually acute to rounded, rarely almost closed. **Bracts** immediately subtending sepals, in subsp. *erratica,* intergrading with sepals, oblong, oval, or ovate, 12–34 × 5–26(–28) mm, proximally flat or keeled, not or scarcely saccate, margins not or scarcely enfolding sepals, apex acute to subobtuse or truncate. **Flowers:** sepals lanceolate, 10–19(–25) mm; corolla pink or white, (28–)30–70(–80) mm.

Subspecies 8 (6 in the flora): North America, South America, Europe, Asia, Africa.

Subspecies of *Calystegia sepium* mostly show strong geographic separation; morphologic intermediates between subspecies of *C. sepium* and between *C. sepium* and other species of *Calystegia* make taxonomy difficult. Subspecies *roseata* Brummitt is known from Atlantic coasts of Europe, temperate coasts of South America, Easter Island, New Zealand, and Australia; subsp. *spectabilis* Brummitt is known from Europe and Asia.

1. Corollas white.
   2. Leaf blade basal sinuses acute; corollas 30–50 mm; stamens 17–24 mm . . . . 18a. *Calystegia sepium* subsp. *sepium*
   2. Leaf blade basal sinuses rounded; corollas (28–)44–65(–80) mm; stamens 19–32 mm.
      3. Leaf blades broadly triangular, basal lobes spreading, ± 2-pointed . . . . 18b. *Calystegia sepium* subsp. *angulata*
      3. Leaf blades linear to narrowly triangular, basal lobes slightly to abruptly spreading, 1-pointed or rounded . . . . 18c. *Calystegia sepium* subsp. *binghamiae*
1. Corollas pink.
   4. Leaf blade basal sinuses almost closed; bracts intergrading with sepals . . . . 18d. *Calystegia sepium* subsp. *erratica*
   4. Leaf blade sinuses acute or rounded; bracts not intergrading with sepals.
      5. Herbage usually pubescent to tomentose, sometimes glabrate or glabrous; leaf blade sinuses acute . . . . 18e. *Calystegia sepium* subsp. *americana*
      5. Herbage glabrous; leaf blade sinuses rounded . . . . 18f. *Calystegia sepium* subsp. *appalachiana*

18a. Calystegia sepium (Linnaeus) R. Brown subsp. **sepium** • European bindweed [I] [W]

Herbage glabrous. **Leaf blades** ± triangular to triangular-hastate, basal lobes usually 1–2-pointed, sometimes rounded, basal sinus acute. **Bracts** 12–30 × 8–16 mm, proximally flat or slightly keeled. **Flowers:** corolla white, 30–50 mm; stamens 17–24 mm. $2n = 22$ [Fl. Eur.].

Flowering Jun–Sep. Disturbed sites; introduced; Nfld. and Labr. (Nfld.), N.S.; N.Y., Pa.; Europe; nw Asia; n Africa.

Subspecies *sepium* is distinguished by its smaller flowers. It has been sporadically recorded for the flora area since 1871 and has been over-reported owing to misidentifications.

18b. Calystegia sepium (Linnaeus) R. Brown subsp. **angulata** Brummitt, Kew Bull. 35: 328. 1980 [E] [W]

Calystegia sepium var. *angulata* (Brummitt) N. H. Holmgren; *C. sepium* var. *repens* (Linnaeus) A. Gray; *Convolvulus repens* Linnaeus; *C. sepium* Linnaeus var. *repens* (Linnaeus) A. Gray

Herbage glabrous. **Leaf blades** broadly triangular, basal lobes spreading, ± 2-pointed, basal sinus rounded. **Bracts** (12–)14–26(–32) × (6–)10–18 mm, proximally keeled. **Flowers:** corolla white, limb margin rarely pink-tinged, (28–)48–50(–80) mm; stamens (19–)23–30 mm. $2n = 22$.

Flowering Jun–Sep. Disturbed sites, edges of marshes, thickets, and woods, hedges, roadsides, stream banks, tidal swamps; 70–2400 m; Alta., B.C., Man., N.B., Ont., Que., Sask.; Ariz., Colo., Conn., D.C., Idaho, Ill., Ind., Iowa, Kans., Maine, Md., Mass., Mich., Minn., Mo., Mont., Nebr., N.H., N.Mex., N.Y., N.Dak., Ohio, Oreg., Pa., R.I., S.Dak., Utah, Vt., Wash., W.Va., Wis., Wyo.

18c. Calystegia sepium (Linnaeus) R. Brown subsp. **binghamiae** (Greene) Brummitt, Ann. Missouri Bot. Gard. 52: 216. 1965 [W]

Convolvulus binghamiae Greene, Bull. Calif. Acad. Sci. 2: 417. 1887; *Calystegia sepium* subsp. *limnophila* (Greene) Brummitt; *Convolvulus nashii* House

Herbage usually pubescent to tomentose, sometimes glabrate or glabrous. **Leaf blades** linear to narrowly triangular, basal lobes slightly to abruptly spreading, 1-pointed or rounded, basal sinus narrowly rounded. **Bracts** 14–24 × 10–16 mm, proximally flat or keeled. **Flowers:** corolla white, limb margin rarely pink-tinged, (30–)44–65 mm; stamens 26–32 mm.

Flowering mostly Mar–Jul. Marshes, stream banks; Calif., Fla., Ga., La., Nev., N.Mex., N.C., S.C., Tex.; Mexico (Baja California); South America (Peru).

When M. C. Provance and A. C. Sanders named *Calystegia felix*, they noted that Brummitt had long included specimens of *C. felix* within his circumscription of *C. sepium* subsp. *binghamiae* and that the names *C. sepium* subsp. *binghamiae* Brummitt and *C. sepium* subsp. *limnophylla* (Greene) Brummitt belong to a single subspecies circumscription. They chose to call that subspecies *C. sepium* subsp. *binghamiae* (Greene) Brummitt.

18d. Calystegia sepium (Linnaeus) R. Brown subsp. **erratica** Brummitt, Kew Bull. 35: 330. 1980 [E] [F] [W]

Herbage glabrous or hairy. **Leaf blades** ± triangular to triangular-hastate, basal lobes truncate or 2-pointed, basal sinus almost closed. **Bracts** intergrading with sepals, 16–26 × 5–12 mm, proximally ± keeled, apex acute to obtuse. **Flowers:** corolla pink, 45–60 mm; stamens 25–30 mm.

Flowering Jun–Sep. Disturbed sites, swamps; Ont., Que.; Ill., Ind., Mich., Mo., N.J., N.Y., Oreg., Pa.

18e. Calystegia sepium (Linnaeus) R. Brown subsp. **americana** (Sims) Brummitt, Ann. Missouri Bot. Gard. 52: 216. 1965 [W]

Convolvulus sepium Linnaeus var. *americanus* Sims, Bot. Mag. 19: plate 732. 1804; *C. americanus* (Sims) J. W. Loudon

Herbage usually pubescent to tomentose, sometimes glabrate or glabrous. **Leaf blades** ovate, basal lobes usually rounded or 1-pointed and not spreading, sometimes slightly spreading and 2-pointed, basal sinus acute. **Bracts** distinct from sepals, 16–25 × 10–20 mm, proximally flat or keeled, apex obtuse to rounded. **Flowers:** corolla pink, 45–70 mm; stamens 24–32 mm.

Flowering Jul–Oct. Disturbed sites, grassy banks, rocky shores, salt marshes; St. Pierre and Miquelon; N.B., Nfld. and Labr. (Nfld.), N.S., Ont., P.E.I., Que.; Conn., Del., Ill., Ind., Maine, Md., Mass., Mich., N.H., N.J., N.Y., N.C., Ohio, Pa., R.I., Vt., Va., Wis.; South America (Argentina, Uruguay); s Africa (Cape Peninsula, extinct?); Atlantic Islands (Azores, Tristan da Cunha).

18f. Calystegia sepium (Linnaeus) R. Brown subsp. **appalachiana** Brummitt, Kew Bull. 35: 329. 1980 [E] [W]

Herbage glabrous. **Leaf blades** ± triangular, basal lobes spreading, usually 2-pointed, basal sinus rounded. **Bracts** distinct from sepals, 18–34 × (12–)14–26(–28) mm, proximally strongly keeled or saccate, margins usually slightly overlapping, not closely enfolding sepals. **Flowers:** corolla pink, 47–64 mm; stamens (24–)25–30(–32) mm.

Flowering Jun–Sep. Disturbed sites, hedges, roadsides, woodland edges; N.B.; Conn., Ky., Maine, Mass., Mich., Minn., N.H., N.J., N.Y., N.C., Pa., Vt., Va., W.Va.

19. Calystegia silvatica (Kitaibel) Grisebach, Spic. Fl. Rumel. 2: 74. 1844 (as sylvatica) [F]

Convolvulus silvaticus Kitaibel, Neues J. Bot. 1: 163. 1805

Perennials, rhizomatous. **Herbage** glabrous. **Stems** twining-climbing. **Leaves:** blade ± cordate, to 50–120 mm, base rounded or lobed, lobes rounded or 1-pointed, basal sinus ± quadrate to rounded. **Bracts** immediately subtending sepals, lanceolate, to 30 × 15–35 mm, proximally saccate, margins strongly enfolding sepals, apex obtuse to truncate. **Flowers:** sepals oval to ovate, 15–23 mm; corolla white, sometimes pinkish-striped, rarely otherwise pink-tinged, 43–70[–88] mm; stamens 23–40 mm; anthers 4–6.5 mm.

Subspecies 3 (2 in the flora): North America, Europe, Asia, Pacific Islands (New Zealand); temperate regions.

Subspecies *silvatica* is native to eastern Mediterranean Europe; it has corollas (50–)55–75(–88) mm, stamens (25–)28–36(–39) mm, and bract apices emarginate to truncate.

Intermediates between *Calystegia silvatica* and *C. sepium* may be due to ancient or recent hybridization.

1. Peduncles 1–2 per axil; anthers 4–5 mm . . . . 19a. *Calystegia silvatica* subsp. *fraterniflora*
1. Peduncles 1 per axil; anthers 5–6.5 mm . . . . 19b. *Calystegia silvatica* subsp. *disjuncta*

19a. Calystegia silvatica (Kitaibel) Grisebach subsp. **fraterniflora** (Mackenzie & Bush) Brummitt, Kew Bull. 35: 332. 1980 [F]

Convolvulus sepium Linnaeus var. *fraterniflorus* Mackenzie & Bush, Man. Fl. Jackson County, 153. 1902; *Calystegia silvatica* subsp. *orientalis* Brummitt

Leaf blades: basal sinus ± quadrate. **Peduncles** 1–2 per axil. **Flowers:** corolla rarely pink-tinged, 43–70 mm; anthers 4–5 mm. $2n = 20$.

Flowering Jun–Oct. Grassy banks, prairies, roadsides; Ark., Conn., D.C., Ga., Ill., Ind., Iowa, Kans., Ky., Md., Mass., Mich., Mo., N.Y., Ohio, Pa., S.C., Tenn., Tex., Vt., Va., W.Va.; Asia (China); Pacific Islands (New Zealand).

19b. Calystegia silvatica (Kitaibel) Grisebach subsp. **disjuncta** Brummitt, Lagascalia 18: 339. 1996 • Large bindweed [I]

Leaf blades: basal sinus rounded. **Peduncles** 1 per axil. **Flowers:** corolla not pink-tinged, except mid-petaline bands, 48–65 mm; anthers 5–6.5 mm. **2*n*** = 22.

Flowering Jun–Oct. Disturbed sites, abandoned plantings; 0–100 m; introduced; B.C.; Calif., Oreg., Wash.; w Europe; nw Africa; introduced also in Pacific Islands (New Zealand), Australia.

Collections from Litchfield, Connecticut, 1941 (NY), and Long Island, New York, 1906 (PH), may be *Calystegia silvatica* subsp. *disjuncta*.

20. Calystegia macounii (Greene) Brummitt, Ann. Missouri Bot. Gard. 52: 215. 1963 [E]

Convolvulus macounii Greene, Pittonia 3: 331. 1898

Perennials, rhizomatous. **Herbage** puberulent. **Stems** usually ascending-decumbent, sometimes proximally erect, distally weakly twining-climbing, to 50–70 cm. **Leaves:** blade deltate, ovate, ovate-hastate, or ovate-lanceolate, 20–60 × 15–50 mm, base lobed, lobes ± rhombic, rounded, basal sinus rounded, apex rounded to subacute. **Bracts** immediately subtending sepals, oval, ovate, or ovate-oblong, (12–)15–21(–27) × (8–)10–16(–20) mm, proximally flat or slightly saccate, apex usually obtuse, sometimes acute. **Flowers:** sepals elliptic to ovate, 12–16 mm; corolla white, (35–)40–52(–69) mm.

Flowering May–Aug. Grassy sites, including tall grass and mixed grass prairies, prairie slopes, disturbed sites, meadows, openings in woodlands, stream banks; 400–2200 m; Alta., Man., Sask.; Ariz., Colo., Idaho, Iowa, Kans., Ky., Minn., Mo., Mont., Nebr., N.Mex., N.Dak., Okla., S.Dak., Tex., Utah, Wyo.

Calystegia macounii is morphologically and geographically intermediate between *C. catesbeiana* and *C. malacophylla* and between *C. malacophylla* and *C. spithamaea*. It differs from *C. sepium* in habit and leaf shape; in the Great Plains area, it is most readily identifiable by its hairy herbage, *C. sepium* there being glabrous.

14. CONVOLVULUS Linnaeus, Sp. Pl. 1: 153. 1753; Gen. Pl. ed. 5, 76. 1754 • Bindweed [Latin *convolvo*, to entwine, alluding to twining habit of most species]

Daniel F. Austin†

Annuals or perennials [shrubs], sometimes rhizomatous. **Stems** usually decumbent to procumbent, sometimes ascending, erect, or trailing, seldom twining-climbing, glabrous or hairy, hairs not branched, glandular, or stellate. **Leaves** usually petiolate, rarely sessile; blade deltate-ovate, oblong, oblanceolate, oblong-elliptic, elliptic, linear, ovate, ovate-lanceolate, ovate-deltate, triangular-lanceolate, or deltate, 10–100 mm, surfaces glabrous or hairy. **Inflorescences:** flowers 2–5+ per peduncle [heads] or solitary; pedicels 10–30 mm; bracts scalelike, lanceolate, lance-linear, elliptic, linear, obovate, ovate, spatulate, or subulate. **Flowers:** sepals elliptic, oblong, oblong-ovate, obovate, ovate, or suborbiculate, 3–12 mm; corolla usually pink or white, sometimes tinged or striped with blue or pink, center sometimes purplish to reddish, campanulate to ± rotate, (4–)12–30 mm, limb 5-angled to 5-lobed; ovary 2-locular; style 1; stigmas or stigma lobes 2, cylindric, linear, or spatulate, apices acute. **Fruits** capsular, ± globose, ovoid, or conic-ovoid, dehiscence valvate. **Seeds** 1–4, trigonous or rounded, glabrous, surfaces granulate, papillulate, smooth, or tuberculate. $x = 12$.

Species 190 (4 in the flora): North America, Mexico, South America, Eurasia, Africa, Atlantic Islands (Canary Islands), Australia; introduced in Pacific Islands (Hawaii).

Convolvulus althaeoides Linnaeus (collected in California in 1941, 1942, and 1950), *C. cneorum* Linnaeus, *C. sabatius* Viviani var. *mauritanicus* (Bossier) Sa'ad (cultivated as *C. mauritanicus* Boissier), and *C. tricolor* Linnaeus are widely cultivated; none of them is known to be established or recurrent in the flora area.

SELECTED REFERENCE Wood, J. R. I. et al. 2015. A foundation monograph of *Convolvulus* L. (Convolvulaceae). PhytoKeys 51: 1–282.

1. Annuals; corollas 4–6 mm, limb 5-lobed . . . . 1. *Convolvulus simulans*
1. Perennials; corollas 11–30 mm, limb 5-angled.
   2. Sepals 3–4.5 mm . . . . 2. *Convolvulus arvensis*
   2. Sepals 6–12 mm.
      3. Flowers 1(–3) per peduncle; sepals oblong to ovate, 6–12 mm; corollas (15–)25–30 mm . . . . 3. *Convolvulus equitans*
      3. Flowers (1–)3–5+ per peduncle; sepals ± elliptic, 6–7 mm; corollas 11–15(–18) mm . . . . 4. *Convolvulus crenatifolius*

1. Convolvulus simulans L. M. Perry, Rhodora 33: 76. 1931 • Small-flowered morning glory

Breweria minima A. Gray, Proc. Amer. Acad. Arts 17: 228. 1882, not *Convolvulus minimus* Aublet 1775

Annuals. Stems ascending to erect, 1–3 dm, puberulent to pubescent. **Leaf blades** oblong to oblanceolate, 15–40 × 2–8(–15) mm, surfaces sparsely villous to glabrate. **Inflorescences:** flowers solitary; bracts spatulate to subulate, 4–5 mm proximal to calyx, 3–8 mm. **Flowers:** sepals oblong-ovate, 3–4 mm, margins scarious, surfaces hairy; corolla white, tinged or striped with blue or pink, campanulate, 4–6 mm, limb 5-lobed 1/3–1/2 length. **Seeds** papillulate.

Flowering Mar–Jun. Grassy and rocky places, friable wet clay soils, serpentine ridges; 0–800 m; Ariz., Calif.; Mexico (Baja California).

Convolvulus pentapetaloides Linnaeus has been misapplied to *C. simulans*.

2. Convolvulus arvensis Linnaeus, Sp. Pl. 1: 153. 1753 • Field bindweed, liseron des champs [I] [W]

Convolvulus ambigens House; *Strophocaulos arvensis* (Linnaeus) Small

Perennials, rhizomatous. **Stems** decumbent or trailing, to 10+ dm, glabrous or glabrate. **Leaf blades** elliptic, oblong-elliptic, ovate, ovate-deltate, or ovate-lanceolate, 10–100 × 3–60 mm, length 1.6–3.3 times width, surfaces glabrous or abaxial puberulent. **Inflorescences:** flowers solitary or in 2–3-flowered cymes; bracts elliptic, linear, or obovate, 2–3(–9) mm. **Flowers:** sepals: outer elliptic, 3–4.5 mm, inner suborbiculate to obovate, 3.5–5 mm; corolla white, sometimes pink-tinged, campanulate, 12–25(–30) mm, limb 5-angled. **Seeds** tuberculate. $2n = 48, 50$.

Flowering Apr–Oct. Fields, disturbed sites, roadsides; -30–3000 m; introduced; Alta., B.C., Man., N.B., N.S., Ont., P.E.I., Que., Sask.; Ala., Ariz., Ark., Calif., Colo., Conn., Del., D.C., Fla., Ga., Idaho, Ill., Ind., Iowa, Kans., Ky., La., Maine, Md., Mass., Mich., Minn., Miss., Mo., Mont., Nebr., Nev., N.H., N.J., N.Mex., N.Y., N.C., N.Dak., Ohio, Okla., Oreg., Pa., R.I., S.C., S.Dak., Tenn., Tex., Utah, Vt., Va., Wash., W.Va., Wis., Wyo.; Europe; Asia; introduced widely in temperate and mild tropical places.

Convolvulus arvensis is a major agricultural pest and is difficult to control. Numerous medicinal uses have been attributed to *C. arvensis* (D. F. Austin 2000).

SELECTED REFERENCE Austin, D. F. 2000. Bindweed (*Convolvulus arvensis*, Convolvulaceae) in North America—From medicine to menace. J. Torrey Bot. Soc. 127: 172–177.

3. Convolvulus equitans Bentham, Pl. Hartw., 16. 1839 • Texas bindweed [F] [W]

Convolvulus carrii B. L. Turner; *C. equitans* var. *lindheimeri* J. R. I. Wood & Scotland; *C. hermannioides* A. Gray

Perennials. Stems usually decumbent to procumbent or trailing, rarely twining-climbing, to 10+ dm, densely to sparsely hairy. **Leaf blades** ± elliptic to triangular-lanceolate or linear and proximally lobed, 10–70 × 2–40 mm, length 1.7–5+ times width, base simple or palmately 2–6-lobed, margins entire or toothed, surfaces densely to sparsely hairy. **Inflorescences:** flowers 1(–3) per peduncle; bracts lance-linear to lanceolate, 1–3 mm. **Flowers:** sepals elliptic, oblong, or ovate, 6–12 mm; corolla pink or white, center sometimes purplish to reddish, campanulate to ± rotate, (15–)25–30 mm, limb 5-angled. **Seeds** granulate or smooth. $2n = 24$.

Flowering year-round. Grasslands, hills, plains; 0–2000 m; Ala., Ariz., Ark., Colo., Kans., Nebr., N.Mex., Okla., Tex., Utah; Mexico.

In the flora area, *Convolvulus hermanniae* L'Héritier and *C. incanus* Vahl were long misapplied to *C. equitans*, and plants of *C. crenatifolius* were formerly identified as *C. equitans*.

B. L. Turner (2009) indicated that *Convolvulus carrii* differs from *C. equitans* by having herbage more densely silvery-hairy, leaf blades thicker and with veins more pronounced, peduncles longer, seeds smoother, and habitat restricted to fine sands in southern Texas, but these differences are not regarded here as significant.

4. Convolvulus crenatifolius Ruiz & Pavon, Fl. Peruv. 2: 10, plate 118, fig. a. 1799

Subspecies 2 (1 in the flora): Texas, South America.

Subspecies *montevidensis* (Sprengel) J. R. I. Wood & Scotland is known from Argentina, Bolivia, Brazil, Paraguay, and Uruguay.

4a. Convolvulus crenatifolius Ruiz & Pavon subsp. **crenatifolius**

Perennials. Stems usually twining, sometimes decumbent, procumbent, or trailing, to 30+ dm, densely to sparsely hairy. **Leaf blades** ± deltate to ovate, 30–80 × 10–40 mm, length 2–3+ times width, base cordate to hastate, margins ± dentate to sinuate, surfaces densely to sparsely hairy. **Inflorescences:** flowers (1–)3–5+ per peduncle; bracts lance-linear to lanceolate, 1–3(–5) mm. **Flowers:** sepals ± elliptic, 6–7 mm; corolla pink or white, ± rotate, 11–15(–18) mm, limb 5-angled. **Seeds** smooth.

Flowering Apr–Jul. Disturbed sites, scrublands, woods; 10–30 m; Tex.; South America.

Whether subsp. *crenatifolius* is disjunct or introduced in the flora area is unclear (J. R. I. Wood et al. 2015).

15. OPERCULINA Silva Manso, Enum. Subst. Braz., 16, 49. 1836 • [Latin *operculum*, cover, alluding to distal portion of fruit separating as a lid]

Daniel F. Austin†

Perennials. Stems usually twining-climbing, sometimes procumbent, glabrous [hairy]. **Leaves** petiolate; blade ± ovate or palmately to pinnately lobed, 20–120 mm, surfaces sparsely hairy [glabrous]. **Inflorescences** 2–3-flowered cymes or flowers solitary, usually bracteate, bracts foliaceous. **Flowers:** sepals obovate, orbiculate, or ovate, 11–16 mm, equal or unequal, larger and leathery in fruit, margins sometimes dentate, glabrous; corolla white [yellow or reddish to salmon], campanulate [funnelform or salverform], 34–53 mm, limb ± entire [weakly 5-lobed]; anthers twisted after dehiscence; pollen 3-colpate, not echinate; styles 1; stigmas 2, globose. **Fruits** capsular, cuboid to globose, dehiscence circumscissile. **Seeds** 1–4, ellipsoid [ovoid], glabrous [hairy]. $x = 15$.

Species ca. 18 (1 in the flora): Texas, Mexico, worldwide in tropics and subtropics, most in Old World.

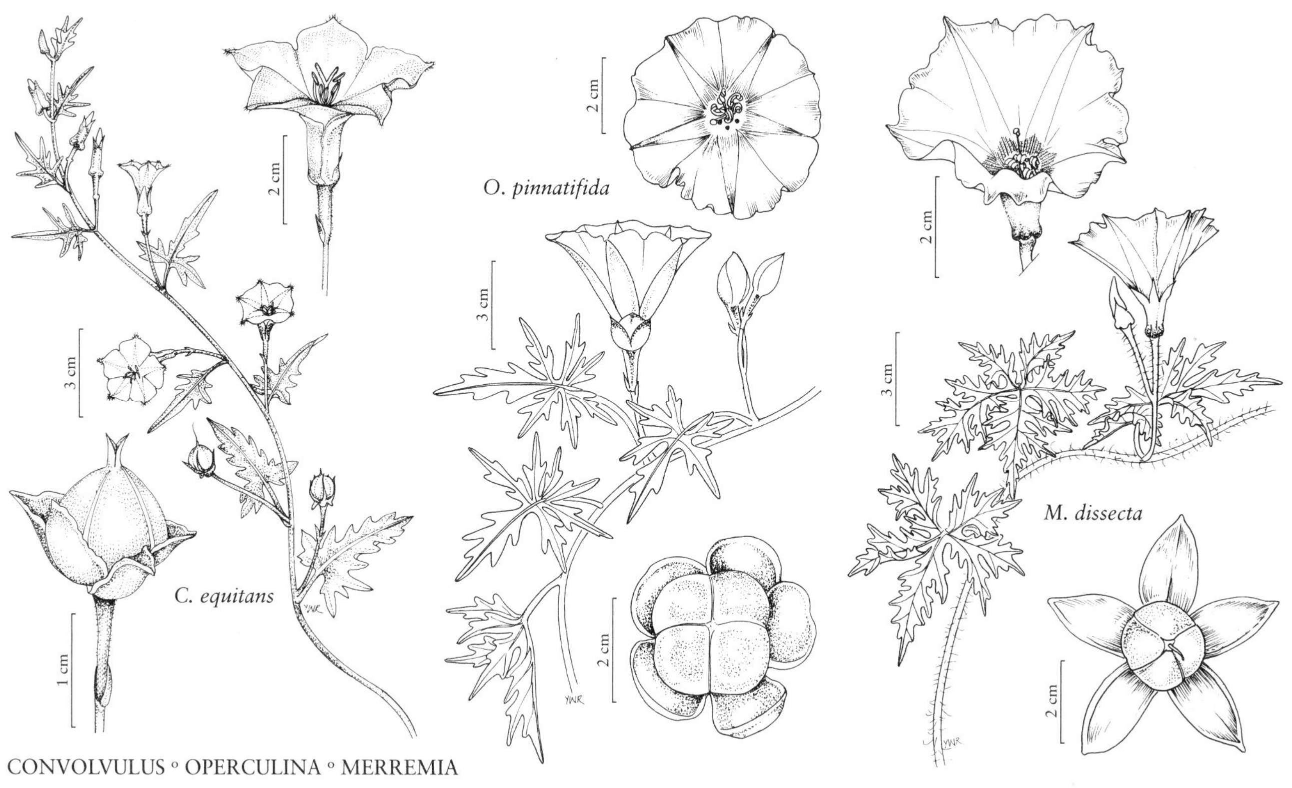

CONVOLVULUS ◦ OPERCULINA ◦ MERREMIA

1. **Operculina pinnatifida** (Kunth) O'Donell, Lilloa 23: 432. 1950 [F]

Convolvulus pinnatifidus Kunth in A. von Humboldt et al., Nov. Gen. Sp. 3(fol.): 85; 3(qto.): 108. 1819; *Ipomoea pinnatifida* (Kunth) G.Don

Plants to 6+ m. **Stems** usually terete, sometimes angular. **Leaf blades** ovate or pinnately to palmately lobed, lobes 5–9, linear, lanceolate, elliptic-obovate, or ± rhombic, base cuneate to truncate or auriculate, margins entire or ± lobed to toothed, apex obtuse-acuminate. **Peduncles** ± winged. **Flowers:** sepals stramineous or rose, accrescent in fruit; corolla limb ± entire. **Capsules** 12–20 mm, glabrous. **Seeds** 5–7 mm.

Flowering May–Oct. Pastures, scrublands; 0–20 m; Tex.; Mexico; Central America.

16. MERREMIA Dennstedt ex Endlicher, Gen. Pl. 18: 1403. 1841, name conserved • [Presumably for Blasius Merrem, 1763–1824, German professor]

Daniel F. Austin†

Perennials [shrubs]. **Stems** twining-climbing [erect or prostrate], glabrous or hairy. **Leaves** petiolate; blade usually palmately lobed or palmately compound, sometimes not lobed or compound, 15–70(–150) mm, surfaces glabrous or hairy. **Inflorescences** 2–12(–20+)-flowered cymes or umbelliform clusters or flowers solitary. **Flowers:** sepals oblong, ovate, ovate-lanceolate, or rhombic, 3–30 mm, glabrous or hairy; corolla usually cream, white, white with purplish throat, or yellow, rarely rose, campanulate to funnelform, 15–60 mm, limb subentire or 5-toothed, -angled or -lobed; anthers twisted after dehiscence; pollen usually 3–9-colpate, rarely aggrecolpate, not echinate; styles 1; stigmas globose or 2-lobed, lobes globose. **Fruits** capsular, usually ± globose, sometimes quadrangular, dehiscence irregular or valvate. **Seeds** 1–4(–6), ± trigonous, glabrous or hairy. ***x*** = 15.

Species 80+ (6 in the flora): s United States; Old World subtropics and tropics; introduced in Mexico, West Indies, Central America, South America.

Merremia hastata Hallier f. [≡ *M. tridentata* (Linnaeus) Hallier f. subsp. *hastata* Ooststroom, *Xenostegia tridentata* (Linnaeus) D. F. Austin & Staples subsp. *hastata* (Ooststroom) Parmar] was recorded from ballast in Georgia in 1902; it evidently did not persist in the flora area.

A. R. Samões and G. W. Staples (2017) treated *Merremia aegyptia* as *Distimake aegyptius* (Linnaeus) A. R. Samões & Staples, *M. cissoides* as *D. cissoides* (Lamarck) A. R. Samões & Staples, *M. dissecta* as *D. dissectus* (Jacquin) A. R. Samões & Staples, *M. quinquefolia* as *D. quinquefolius* (Linnaeus) A. R. Samões & Staples, *M. tuberosa* as *D. tuberosus* (Linnaeus) A. R. Samões & Staples, and *M. umbellata* as *Camonea umbellata* (Linnaeus) A. R. Samões & Staples.

SELECTED REFERENCE Austin, D. F. 1979. Studies of the Florida Convolvulaceae–II. *Merremia*. Florida Sci. 42: 216–222.

1. Leaf blades not lobed or compound; inflorescences usually 3–20+-flowered umbellform clusters, flowers rarely solitary . . . . 1. *Merremia umbellata*
1. Leaf blades palmately lobed or compound; inflorescences 2–9-flowered cymes or flowers solitary.
    2. Leaf blades palmately lobed.
        3. Leaf lobe margins crenate, dentate, or irregularly pinnati-sinuate; corollas white with purplish throats . . . . 2. *Merremia dissecta*
        3. Leaf lobe margins entire; corollas yellow . . . . 3. *Merremia tuberosa*
    2. Leaf blades palmately compound.
        4. Leaflet margins dentate or entire; sepal abaxial surfaces hirsute, glabrescent . . . 4. *Merremia aegyptia*
        4. Leaflet margins usually dentate, sometimes remotely serrate to subentire, rarely entire; sepal abaxial surfaces glabrous or hairy, hairs glandular and setaceous.
            5. Sepals: outer and inner 10–18 mm, apex acuminate, abaxial surface hairy, hairs glandular and setaceous . . . . 5. *Merremia cissoides*
            5. Sepals: outer 3–5 mm, inner 4–7 mm, apex obtuse, abaxial surface glabrous . . . . 6. *Merremia quinquefolia*

1. Merremia umbellata (Linnaeus) Hallier f., Bot. Jahrb. Syst. 16: 552. 1893 • Hogvine [I]

Convolvulus umbellatus Linnaeus, Sp. Pl. 1: 155. 1753; *Ipomoea polyanthes* Roemer & Schultes

Stems usually glabrous, rarely hairy. **Leaf blades** narrowly triangular to widely ovate, not lobed or compound, 15–150 × 20–100 mm, margins entire, surfaces glabrous or ± hairy. **Inflorescences** usually 3–20+-flowered umbelliform clusters, flowers rarely solitary. **Flowers:** sepals oblong, 6–8 mm, apex rounded, abaxial surface glabrous or hairy; corolla yellow, 30–35 mm.

Flowering year-round. Disturbed sites, abandoned plantings; 0–10 m; introduced; Fla.; Old World subtropics and tropics; introduced also in Mexico, West Indies, Central America, South America.

2. Merremia dissecta (Jacquin) Hallier f., Bot. Jahrb. Syst. 16: 552. 1893 (as disecta) • Alamo vine, noyau vine [F]

Convolvulus dissectus Jacquin, Observ. Bot. 2: 4, plate 28. 1767; *Ipomoea sinuata* Ortega; *Operculina dissecta* (Jacquin) House

Stems sparsely hirsute or glabrous. **Leaf blades** ± polygonal, palmately (5–)7(–9)-lobed, 40–70 × 40–100 mm; lobes ± lance-elliptic, margins usually irregularly pinnati-sinuate, sometimes crenate or dentate, surfaces usually glabrous, sometimes sparsely hairy. **Inflorescences:** flowers usually solitary, rarely 2–3+-flowered cymes. **Flowers:** sepals oblong, 18–25 mm, apex mucronate, abaxial surface glabrous; corolla white with purplish throat, 30–45 mm. **2*n*** = 32 (Asia).

Flowering Apr–Nov. Disturbed or open sites, roadsides; 0–700 m; Fla., Ga., Tex.; Mexico; West Indies; Central America; South America; introduced in Asia, Africa, Australia.

Merremia dissecta has been reported from Alabama, Arizona, Louisiana, Mississippi, and Pennsylvania based on single records and/or cultivated plants; it is apparently not established in those states.

SELECTED REFERENCE Austin, D. F. 2007. *Merremia dissecta* (Convolvulaceae)—A condiment, medicine, ornamental, and weed—A review. Econ. Bot. 61: 109–120.

3. **Merremia tuberosa** (Linnaeus) Rendle in D. Oliver et al., Fl. Trop. Afr. 4(2): 104. 1905 [I]

Ipomoea tuberosa Linnaeus, Sp. Pl. 1: 160. 1753; *Operculina tuberosa* (Linnaeus) Meisner

Stems glabrous. **Leaf blades** broadly ovate to rounded, palmately (5–)7-lobed, 80–150 × 80–150 mm; lobes lanceolate to elliptic, margins entire, surfaces glabrous. **Inflorescences** usually 4–8-flowered cymes, flowers rarely solitary. **Flowers:** sepals oblong, outers 25–30 mm, apex obtuse, mucronulate, inners 12–20 mm, apex acute, abaxial surface glabrous; corolla yellow, 50–60 mm.

Flowering Nov–Mar. Fields, hammocks, pinelands; 0–10 m; introduced; Fla.; Mexico; Central America; introduced also in West Indies, South America, Old World subtropics and tropics.

SELECTED REFERENCE Austin, D. F. 1998. Xixicamátic or wood rose (*Merremia tuberosa*, Convolvulaceae): Origins and dispersal. Econ. Bot. 52: 412–422.

4. **Merremia aegyptia** (Linnaeus) Urban, Symb. Antill. 4: 505. 1910 • Hairy wood-rose [I]

Ipomoea aegyptia Linnaeus, Sp. Pl. 1: 162. 1753

Stems usually hirsute. **Leaf blades** ± orbiculate, palmately compound, 25–150 × 25–170 mm, (3–)5(–7)-foliolate; leaflets elliptic to oblanceolate, margins dentate or entire, surfaces sparsely to densely hirsute, glabrescent. **Inflorescences** 2–3+-flowered cymes or flowers solitary. **Flowers:** sepals oblong, 20 mm, apex acute, abaxial surface hirsute, glabrescent; corolla white, 20–40 × 20–30 mm.

Flowering Oct–Feb. Disturbed sites, fields, roadsides; 0–10 m; introduced; Fla.; Mexico; West Indies; Central America; South America; introduced also in Old World subtropics and tropics.

5. **Merremia cissoides** (Lamarck) Hallier f., Bot. Jahrb. Syst. 16: 552. 1893 • Roadside wood-rose

Convolvulus cissoides Lamarck in J. Lamarck and J. Poiret, Tabl. Encycl. 1: 462. 1793

Stems hairy, hairs glandular and setaceous. **Leaf blades** ± orbiculate to pentagonal, palmately compound, 15–50 × 5–50 mm, (3–)5(–7)-foliolate; leaflets lanceolate, linear-lanceolate, or ovate-oblong, margins usually dentate, rarely entire, surfaces hairy, hairs glandular and setaceous. **Inflorescences** 3–9-flowered cymes or flowers solitary. **Flowers:** sepals rhombic, ovate, or ovate-lanceolate, 10–18 mm, apex acuminate, abaxial surface hairy, hairs glandular and setaceous; corolla usually cream or white, rarely rose, 20–25 mm.

Flowering May–Dec. Abandoned plantings, fields, lots; 0–10 m; Fla.; Mexico; Central America; South America; introduced in Old World subtropics and tropics.

6. **Merremia quinquefolia** (Linnaeus) Hallier f., Bot. Jahrb. Syst. 16: 552. 1893 • Rock rosemary

Ipomoea quinquefolia Linnaeus, Sp. Pl. 1: 162. 1753

Stems glabrous. **Leaf blades** ± orbiculate to polygonal, palmately compound, 15–50 × 25–70 mm, (3–)5(–7)-foliolate; leaflets elliptic or lanceolate to oblanceolate, margins remotely serrate to subentire, surfaces glabrous. **Inflorescences** 3–9-flowered cymes or compound cymes or flowers solitary. **Flowers:** sepals oblong, outer 3–5 mm, inner 4–7 mm, apex obtuse, abaxial surface glabrous; corolla white, 15–25 mm, glabrous.

Flowering Dec–Jun. Pinelands, disturbed sites; 0–10 m; Fla.; West Indies; Central America; South America; introduced in Old World subtropics and tropics.

17. ANISEIA Choisy, Mém. Soc. Phys. Genève 6: 481, plate 2, fig. 9. 1834 • [Greek *a*, not, and *isos* equal, alluding to unequal sepals] [I]

Daniel F. Austin†

Perennials. Stems usually twining-climbing or trailing, sometimes decumbent, sparsely hairy, glabrescent, or glabrous. **Leaves** petiolate; blade elliptic, lanceolate, or ovate, 40–80 mm, surfaces glabrous, glabrescent, or tomentose. **Inflorescences** usually solitary flowers, sometimes 2–3-flowered cymes. **Flowers:** sepals lanceolate, 12–20 mm, unequal, notably accrescent in fruit, outer 3 notably longer than inner 2; corolla white, campanulate, 25–30 mm, limb 5-toothed or entire; anthers straight after dehiscence; pollen rugate, not echinate; styles 1; stigmas 2, globose. **Fruits** capsular, ovoid [globose], dehiscence valvate. **Seeds** (3–)4, ovoid to obovoid, glabrous or hairy around margins [hairy on surfaces]. $x = 15$.

Species 3 (1 in the flora): introduced, Florida; Mexico, West Indies, Central America, South America; introduced also in Eurasia, Africa.

SELECTED REFERENCES Austin, D. F. 1973. Another adventive morning glory in Florida: *Aniseia martinicensis*. Florida Sci. 36: 197–198. Austin, D. F. 1999. The genus *Aniseia*. Syst. Bot. 23: 411–420.

1. Aniseia martinicensis (Jacquin) Choisy, Mém. Soc. Phys. Genève 8: 66. 1837 • Whitejacket [F] [I]

Convolvulus martinicensis Jacquin, Select. Stirp. Amer. Hist., 26, plate 17. 1763

Leaves: petiole to 10 mm; blade 40–80 × (5–)10–20 mm, base obtuse to attenuate, apex obtuse to acuminate, surfaces glabrous or sparsely hairy, especially along veins. **Peduncles** 3–8 cm. **Flowers:** outer sepals 12–20 × 9–12 mm, base cordate to obtuse, decurrent on pedicel or not, inner shorter, narrower. **Capsules** 15–20 mm, apex apiculate, glabrous, surrounded by ± accrescent sepals. **Seeds** dark brown to ± black, 5–6 mm. $2n = 60$.

Flowering Aug–Nov. Disturbed sites, wetlands; 0–10 m; introduced; Fla.; s Mexico; West Indies; Central America; South America; introduced also in Eurasia, Africa.

18. IPOMOEA Linnaeus, Sp. Pl. 1: 159. 1753; Gen. Pl. ed. 5, 76. 1754, name conserved • Morning glory [Greek *ipos* or *ips*, to entwine, and *homoios*, similar, alluding to twining habit]

Daniel F. Austin†

Calonyction Choisy; *Exogonium* Choisy; *Pharbitis* Choisy

Annuals, perennials, shrubs, or lianas [trees]. Stems usually decumbent, erect, trailing, or twining, sometimes ascending or repent, rarely floating; glabrous or hairy. **Leaves** petiolate; blade usually cordate, lanceolate, linear, ovate, reniform, sagittate, or ± palmately lobed, rarely cuneate-obovate, deltate, elliptic, orbiculate, pandurate, palmatisect, pentagonal, pinnatisect, sagittate, or

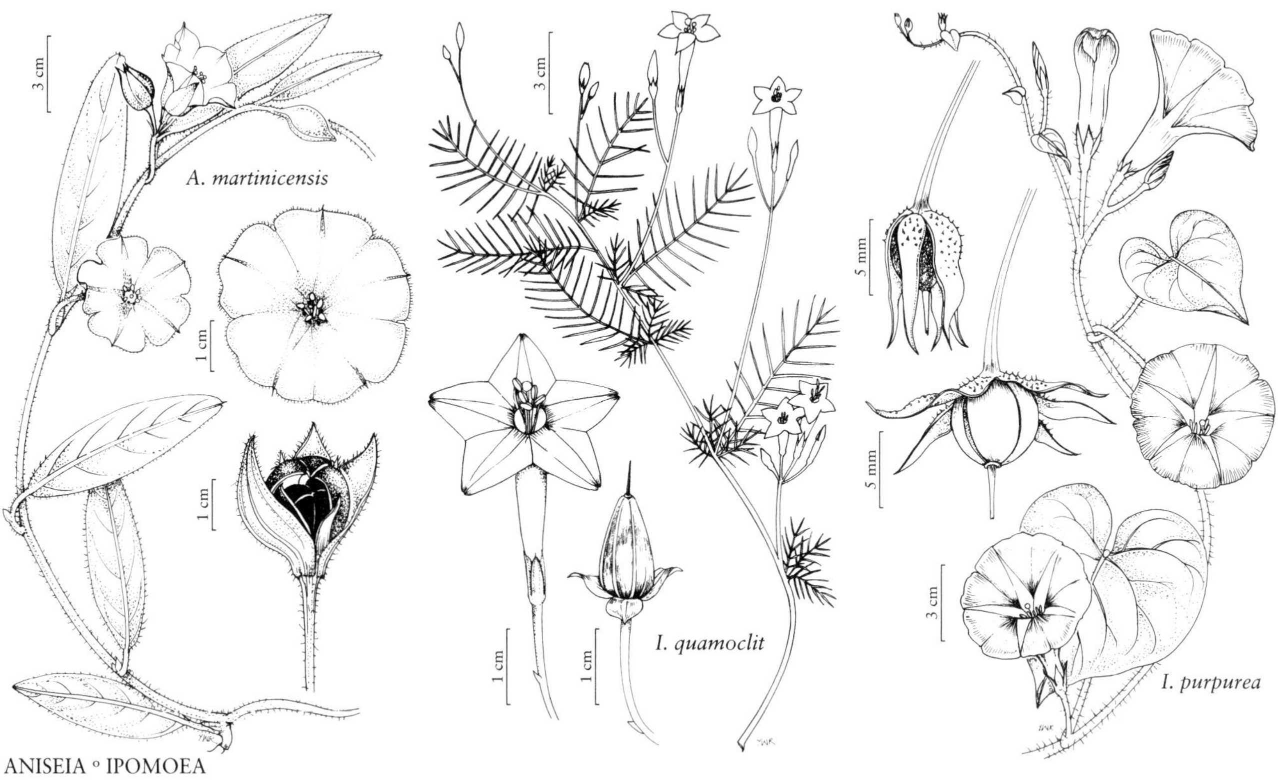

ANISEIA ◦ IPOMOEA

triangular, (10–)30–120(–250+) mm, surfaces glabrous or hairy. **Inflorescences:** flowers usually in 2–3(–25+)-flowered cymes or flowers solitary, rarely in panicles; bracts leaflike to scalelike. **Flowers:** sepals elliptic, lanceolate, oblong, ovate, or suborbiculate, (3–)8–15(–30) mm; corolla usually blue, lavender, pink, purple, red, violet, or white, sometimes orange, red and yellow, or red-orange, usually funnelform, sometimes campanulate or salverform, (6–)20–80(–150+) mm, limb entire, 5-angled, or 5-lobed; anthers straight after dehiscence, pollen pantoporate and echinate; styles 1; stigmas entire or 2(–3)-lobed, capitate or globose. **Fruits** capsular, globose to ovoid, dehiscence irregular or valvate. **Seeds** 1–4(–6), usually ellipsoid, globose, or ovoid, sometimes trigonous, glabrous or hairy. x = 15.

Species 600+ (47, including 1 hybrid, in the flora): North America, Mexico, West Indies, Central America, South America, Eurasia, Africa, Indian Ocean Islands, Pacific Islands, Australia.

Reports by J. T. Kartesz and C. A. Meacham (1999) of *Ipomoea cordifolia* Carey ex Voight from Alabama and *I. meyeri* (Sprengel) G. Don from Georgia have not been verified. *Ipomoea horsfalliae* Hooker is cultivated in Florida; it is not known to be established or recurrent in the flora area.

In protologue, *Ipomoea gilana* K. Keith & J. A. McDonald was reported to be similar to *I. lindheimeri* and to nest phylogenetically near *I. orizabensis* (G. Pelletan) Ledebour ex Steudel, which, in the broad sense, is known from Chihuahua, Mexico, and southward. *Ipomoea gilana* is known from the Black Range in southwestern New Mexico and will key here to *I. indica*.

SELECTED REFERENCE McDonald, J. A. 1995. Revision of *Ipomoea* section *Leptocallis* (Convolvulaceae). Harvard Pap. Bot. 6: 97–122.

1. Corollas ± salverform, (funnelform-salverform in *I. thurberi*; limb sometimes ± campanulate in *I. muricata* and *I. setosa*).
   2. Flowers diurnal (open all day); corollas usually orange, red, red and yellow, or red-orange, sometimes white (cultivars of *I. coccinea* and *I. quamoclit*), 14–50 mm.
      3. Perennials; corollas red . . . . 1. *Ipomoea microdactyla*
      3. Annuals; corollas usually red, red and yellow, or red-orange, sometimes white (cultivars of *I. coccinea* and *I. quamoclit*).
         4. Leaf blades palmati-pinnate or pinnatisect.
            5. Leaf blades palmati-pinnate, lobes 7–11+ . . . . 2. *Ipomoea sloteri*
            5. Leaf blades pinnatisect, lobes 19–41+ . . . . 3. *Ipomoea quamoclit*
         4. Leaf blades not palmati-pinnate or pinnatisect.
            6. Sepals ± equal, 4–4.5 mm . . . . 4. *Ipomoea hederifolia*
            6. Sepals unequal, outers 3–3.5 mm, inners 4–5.7 mm.
               7. Leaf blades usually cordate, sometimes ovate, sagittate, or triangular . . . . 5. *Ipomoea coccinea*
               7. Leaf blades usually (3–)5–7-lobed, middle lobe narrowly ± rhombic, sometimes proximal blades cordate to ovate, not lobed. . . . . 6. *Ipomoea cristulata*
   2. Flowers nocturnal (open dusk to early morning); corollas usually lavender, pink, or white, tubes sometimes purple to red, limbs or throats sometimes purple, red, or with green or yellowish bands or lines, 30–150 mm.
      8. Leaf blades usually palmatisect, sometimes sagittate.
         9. Sepals 5–12 mm; corolla limbs 30–36 mm diam. . . . . 7. *Ipomoea tenuiloba* (in part)
         9. Sepals 12–15 mm; corolla limbs 50–65 mm diam. . . . . 8. *Ipomoea thurberi*
      8. Leaf blades cordate, orbiculate, ovate, triangular, triangular-ovate, or 3–5(–7)-lobed.
         10. Leaf blade surfaces: abaxial tomentulose, adaxial glabrous; corollas white, throat lavender to purple inside . . . . 9. *Ipomoea macrorhiza*
         10. Leaf blade surfaces glabrous or hairy, not tomentulose; corollas usually lavender, pink, or white, sometimes with green or yellowish bands or lines.
             11. Sepal apices acute, outers each with ± corniform appendage.
                 12. Corollas white, throat green-banded inside; fruits 20–30 mm . . . . 10. *Ipomoea alba*
                 12. Corollas white, turning lavender; fruits 18–20 mm . . . . 11. *Ipomoea muricata*
             11. Sepal apices emarginate, obtuse, or truncate, none with corniform appendages.
                 13. Stems ± setose; corollas lavender or pink. . . . . 12. *Ipomoea setosa*
                 13. Stems not setose; corollas white, limb with yellowish lines . . . . 13. *Ipomoea violacea*
1. Corollas funnelform.
   14. Peduncles usually hairy and hairs usually ± antrorse, retrorse, or spreading, rarely appressed, rarely glabrous or glabrate (*I. barbatisepala* and *I. indica*); sepals herbaceous.
       15. Perennials.
           16. Leaf blade surfaces glabrous or ± pilose; sepals glabrous or abaxial surface sparsely hairy, hairs appressed . . . . 14. *Ipomoea indica*
           16. Leaf blade surfaces ± hirsute or sericeous; sepal surfaces: abaxial ± hispid or sericeous.
               17. Leaf blades broadly ovate to reniform, usually 3–5(–7)-lobed; sepals 18–30 mm . . . . 15. *Ipomoea lindheimeri*
               17. Leaf blades cordate, ovate, or 3–5-lobed; sepals 9–21 mm. . . . . 16. *Ipomoea pubescens*
       15. Annuals.
           18. Leaf blades (3–)5–7-lobed, palmatisect (incised nearly to petiole tip). . . 17. *Ipomoea barbatisepala*
           18. Leaf blades cordate, orbiculate, ovate, or 3(–5)-lobed, not palmatisect.
               19. Sepals elliptic, lance-oblong, or oblong, 8–15 mm, narrowed distal portion shorter to slightly longer than dilated base. . . . . 18. *Ipomoea purpurea*
               19. Sepals ± lanceolate or lance-linear, 12–30 mm, narrowed distal portion notably longer than dilated base.
                   20. Sepals ± lanceolate, 12–24 mm, proximally ovate, abruptly narrowed to ± curved or spreading distal portion; corollas 20–37(–45) mm . . . . 19. *Ipomoea hederacea*
                   20. Sepals lance-linear, 15–25(–30) mm, proximally narrowly ovate, gradually narrowed to ± straight distal portion; corollas (20–)30–60+ mm. . . 20. *Ipomoea nil*

[14. Shifted to left margin.—Ed.]

14. Peduncles usually glabrous or hairy and hairs appressed, sometimes puberulent (*I. carnea*), sparsely hispidulous (*I. costellata*), pilosulous on proximal 1–2 mm and otherwise glabrous (*I. dumetorum*), or setose; sepals usually chartaceous or coriaceous, sometimes herbaceous or membranous.
 21. Stems usually repent, (rooting at nodes), rarely twining.
  22. Leaf blades hastate, lanceolate, linear, oblong, ovate, or ± 3–5-lobed; corollas usually purple or white, throat sometimes purplish or yellow inside.
   23. Sepals elliptic-ovate to oblong-ovate, 6–8 mm, ± equal; corollas usually purple, rarely white . . . . 21. *Ipomoea aquatica*
   23. Sepals lance-oblong, 10–15 mm, unequal; corollas white, throat usually yellow, sometimes purplish, inside . . . . 22. *Ipomoea imperati*
  22. Leaf blades cordate, ± orbiculate, ovate, reniform, or rounded-cordate; corollas usually lavender, pink, or red, rarely white, throat sometimes darker inside.
   24. Leaf blade apices acute to rounded; corollas usually red, rarely white, 50–80 mm . . . . 23. *Ipomoea asarifolia*
   24. Leaf blade apices ± emarginate; corollas lavender or pink, 35–40(–70) mm . . . . 24. *Ipomoea pes-caprae*
 21. Stems not repent.
  25. Stems usually erect, sometimes ± trailing (*I. leptophylla*).
   26. Shrubs; leaf blades 40–170 mm wide, bases cordate or ± truncate . . . . 25. *Ipomoea carnea*
   26. Perennials; leaf blades 2–8(–10) mm wide, bases ± cuneate . . . . 26. *Ipomoea leptophylla*
  25. Stems usually trailing, twining, or twining only near tips, rarely ascending, decumbent, or erect.
   27. Leaf blades cuneate-obovate and distally ± incised, 3–5(–7+)-toothed . . . . 27. *Ipomoea plummerae* (in part)
   27. Leaf blades not cuneate-obovate and distally incised.
    28. Leaf blades palmatisect (incised ± to petiole tip), lobes (3–)5–9+, usually filiform, lance-linear, lanceolate, linear, or narrowly oblanceolate to spatulate, sometimes elliptic, lance-elliptic, or lance-ovate.
     29. Annuals.
      30. Corollas 10–12 mm . . . . 28. *Ipomoea costellata*
      30. Corollas 18–30(–40) mm.
       31. Leaf blade lobes 5–7(–9), usually filiform to linear, sometimes lance-linear, 0.2–2(–5) mm wide. . . . . 29. *Ipomoea ternifolia*
       31. Leaf blade lobes 5, lance-linear to lanceolate, (5–)10–15(–20) mm wide. . . . . 30. *Ipomoea wrightii*
     29. Perennials.
      32. Leaf blade lobes 5 (proximal 2 lobes sometimes 2-lobed), lance-elliptic, lanceolate, or lance-ovate, (3–)8–15(–30) mm wide . . . . 31. *Ipomoea cairica*
      32. Leaf blade lobes (3–)5–9, filiform, lanceolate, linear, or spatulate, 0.2–2.5(–6.5) mm wide.
       33. Corollas white (limb sometimes purple or pale rose), 35–65 mm . . . . 7. *Ipomoea tenuiloba* (in part)
       33. Corollas lavender, purple, or red-purple, 25–40 mm.
        34. Stems usually ± trailing, sometimes ascending, erect, or twining near tips; leaf blade lobes 3–30(–50) × (0.5–)1–2.5 mm . . . . 27. *Ipomoea plummerae* (in part)
        34. Stems usually ascending to erect, sometimes trailing; leaf blade lobes (3–)5–15(–25) × 0.2–1 mm . . . . 32. *Ipomoea capillacea*
    28. Leaf blades not palmatisect, sometimes palmately lobed.
     35. Corollas usually blue (drying pink or purple), sometimes white, sometimes throat white outside and yellow inside.
      36. Corollas 26–27 mm, limb 30–35 mm diam. . . . . 33. *Ipomoea cardiophylla*
      36. Corollas 35–60 mm, limb 50–90 mm diam. . . . . 34. *Ipomoea tricolor*

[35. Shifted to left margin.—Ed.]

35. Corollas lavender, lavender-pink, lilac, pink, pink-purple, purple, purplish, red-purple, or ± white, throat sometimes darker inside.
    37. Sepals dotted with dark spots on abaxial surface . . . . 35. *Ipomoea dumetorum*
    37. Sepals not dotted with dark spots.
        38. Leaf blades usually hastate or sagittate, ± triangular, sometimes ovate.
            39. Leaf blade surfaces glabrous; sepals elliptic, oblong, or ovate, 8–9 mm; corollas 60–90 mm . . . . .36. *Ipomoea sagittata*
            39. Leaf blade surfaces ± hairy, adaxial sometimes glabrate; sepals lance-oblong, lanceolate, or lance-ovate, 5–8 mm; corollas 30–45 mm . . . . 37. *Ipomoea tenuissima*
        38. Leaf blades usually cordate, cordate-ovate, deltate-ovate, lance-oblong, lanceolate, lance-ovate, linear, oblong-ovate, orbiculate, ovate, pandurate, or reniform, sometimes 3–5(–7)-lobed.
            40. Annuals; corollas 6–20(–25) mm.
                41. Sepals 6–7 mm, apices acute or obtuse. . . . . 38. *Ipomoea triloba*
                41. Sepals (8–)11–14 mm, apices acuminate.
                    42. Corollas white, 15–20(–25) mm; fruits 10–13 mm diam. . . . . 39. *Ipomoea lacunosa*
                    42. Corollas lavender or white, 6–15(–20) mm; fruits 7–8 mm diam. . . . . .40. *Ipomoea* ×*leucantha*
            40. Perennials; corollas 18–100 mm.
                43. Sepals 4–5.5 mm . . . . 41. *Ipomoea amnicola*
                43. Sepals 8–22 mm.
                    44. Corollas 20–38 mm. . . . . 42. *Ipomoea cordatotriloba*
                    44. Corollas (30–)40–100 mm.
                        45. Corollas usually lavender, lavender-pink, pink, or purple, sometimes white, throat usually darker inside.
                            46. Leaf blade bases cordate; corollas (30–)40–70 mm; seeds glabrous . . . . .43. *Ipomoea batatas*
                            46. Leaf blade bases cordate-hastate; corollas 70–90 mm; seeds hairy. . . . . .44. *Ipomoea rupicola*
                        45. Corollas pink or ± white, throat lavender, purple, purple-red, or red inside.
                            47. Leaf blades cordate, cordate-ovate, or pandurate . . . . . . 45. *Ipomoea pandurata*
                            47. Leaf blades deltate-ovate, lance-oblong, lanceolate, lance-ovate, or linear.
                                48. Leaf blades lance-oblong, lanceolate, or linear, 100–120 (–210) mm; Arizona. . . . . .46. *Ipomoea longifolia*
                                48. Leaf blades deltate-ovate or narrowly lance-ovate, 30–80 mm; Kansas, Oklahoma, Texas. . . . . 47. *Ipomoea shumardiana*

1. Ipomoea microdactyla Grisebach, Cat. Pl. Cub., 204. 1866 • Calcareous morning glory [C]

Exogonium microdactylum (Grisebach) House

Perennials, root relatively large, tuberlike. **Stems** trailing, twining, sometimes ± fleshy. **Leaf blades** elliptic or lanceolate, 30–100 × 10–40 mm overall, base cordate to truncate, surfaces glabrous, or ± orbiculate, (3–)5–7-lobed, incised nearly to petiole tip, lobes narrowly elliptic, linear, or oblong, 20–40 × 3–10 mm. **Peduncles** glabrous. **Flowers** diurnal; sepals oblong, orbiculate, or ovate, 6–7 mm, coriaceous, margins sometimes scarious, apex obtuse, sometimes mucronulate; corolla red, salverform, (25–)40–50 mm, limb 25–30 mm diam., weakly 5-lobed or notably 5-lobed in age.

Flowering year-round. Coppices, oölitic sites, open fields, pinelands; of conservation concern; 0–20 m; Fla.; West Indies (Bahamas, Cuba).

2. Ipomoea sloteri Macfarlane ex E. T. Reichert, Publ. Carnegie Inst. Wash. 270(2): 785. 1919 • Cardinal climber [E]

Annuals. Stems twining. **Leaf blades** deltate-ovate to oblong-ovate, 10–90 × 5–45 mm overall, palmati-pinnate, base cordate to ± truncate, lobes 7–11+, lanceolate, lance-linear, or linear, proximal ones sometimes again lobed, terminal lobe broadest, surfaces glabrous. **Peduncles** glabrous. **Flowers** diurnal; sepals elliptic to oblong, 4–8 mm, ± chartaceous, apex obtuse, mucronate, surfaces glabrous; corolla red, salverform, 40–50 mm. **2*n*** = 58.

Flowering Jun–Oct. Abandoned plantings, disturbed sites; 0–300 m; Fla., Mich., N.Y., Tex.

According to J. E. Eckenwalder (1986), the name *Ipomoea* ×*multifida* Rafinesque (as species) refers to diploid hybrid plants resulting from crosses between *I. coccinea* and *I. quamoclit*; *I. sloteri* refers to allotetraploid plants derived from diploid hybrids between *I. coccinea* and *I. quamoclit*.

3. Ipomoea quamoclit Linnaeus, Sp. Pl. 1: 159. 1753 • Cypress vine [F] [W]

Annuals. Stems twining. **Leaf blades** ± elliptic to oblong, 10–90 × 5–45 mm overall, pinnatisect, base ± truncate, lobes 19–41+, filiform to linear, surfaces glabrous. **Peduncles** glabrous. **Flowers:** diurnal; sepals elliptic to oblong, 4–8 mm, chartaceous to coriaceous, apex mucronate, surfaces glabrous; corolla usually red, sometimes white (in cultivars), salverform, 20–30 mm. **2*n*** = 30.

Flowering Jun–Oct. Abandoned plantings, forest edges, thickets; 0–1500 m; Ont.; Ala., Ark., D.C., Fla., Ga., Ill., Ind., Kans., La., Md., Miss., Mo., N.Y., N.C., Okla., Pa., S.C., Tenn., Tex., Va.; Mexico; West Indies; Central America; South America; Eurasia; Africa.

Reports for *Ipomoea quamoclit* from Kansas, Oklahoma, Texas, and Virginia may be from horticultural plantings. A report from California is from a casual garden weed; it is not naturalized there.

4. Ipomoea hederifolia Linnaeus, Syst. Nat. ed. 10, 2: 925. 1759 (as hederfol.) • Scarlet creeper [W]

Annuals. Stems twining. **Leaf blades** ± orbiculate, reniform, or 3-lobed, 20–150 × 20–150 mm, base ± cordate, surfaces glabrous or puberulent. **Peduncles** usually glabrous, sometimes sparsely hairy, hairs antrorse. **Flowers** diurnal; sepals elliptic to oblong, 4–4.5 mm, herbaceous, apex obtuse or truncate, outers with ± terminal corniform appendage, abaxial surface glabrous; corolla red to red-orange, salverform, 14–30 mm. **2*n*** = 28, 30.

Flowering Oct–Mar. Disturbed sites, fence rows, thickets; 0–1800 m; Ariz., Fla., Ga., La., Tex.; Mexico; West Indies; Central America; South America.

A report of *Ipomoea hederifolia* from Kansas was presumably based on a cultivated plant, and a report for Vermont (J. T. Atwood et al. 1973) was presumably based on waifs that did not persist. The report of *I. hederifolia* for New Mexico by W. C. Martin and C. R. Hutchins (1980) was based on misidentified specimens. *Ipomoea hederifolia* may be established in Virginia.

The names *Ipomoea coccinea* and *Quamoclit coccinea* (Linnaeus) Moench have been misapplied to plants of *I. hederifolia*.

5. Ipomoea coccinea Linnaeus, Sp. Pl. 1: 160. 1753 • Red morning glory [E] [W]

Annuals. Stems twining. **Leaf blades** usually cordate, sometimes ovate, sagittate, or triangular, 20–140 mm, base ± cordate, lobes rounded or 1–2-pointed, surfaces glabrous or proximally pilose. **Peduncles** glabrous. **Flowers** diurnal; sepals chartaceous, outers oblong to elliptic, 3–3.5 mm, apex obtuse to truncate, each with ± terminal corniform appendage 2.5–6 mm, inners oblong, 4.5–5.7 mm, chartaceous, apex obtuse to truncate, each with ± terminal, corniform appendage 2–5.5 mm; corolla usually red or red and yellow, sometimes white (in cultivars), salverform, 20–25 mm. **2*n*** = 28.

Flowering Jul–Dec. Abandoned plantings, disturbed sites; 0–300 m; Ala., Ark., Del., D.C., Fla., Ga., Ill., Ind., Kans., Ky., La., Md., Miss., Mo., N.J., N.C., Ohio, Okla., Pa., S.C., Tenn., Tex., Va., W.Va.

Ipomoea coccinea differs from *I. hederifolia* by reflexed pedicels (erect in *I. hederifolia*) and larger inner sepals (4.5–5.7 mm) than *I. hederifolia* (to 3–4 mm).

6. Ipomoea cristulata Hallier f., Meded. Rijks-Herb. 46: 20. 1922 • Star-glory

Quamoclit gracilis Hallier f., Bull. Herb. Boissier 7: 416. 1899, not *Ipomoea gracilis* R. Brown 1810

Annuals. Stems twining. **Leaf blades** usually (3–)5–7-lobed, middle lobe narrowly ± rhombic, sometimes proximal blades cordate to ovate, not lobed, 15–100 × 10–70 mm, base cordate to ± truncate, basal lobes rounded to pointed, margins ± dentate, surfaces glabrous or abaxial pilose. **Peduncles** glabrous. **Flowers** diurnal, sepals chartaceous to membranous, outers oblong, 3–3.5 mm, apex obtuse, rounded, or ± truncate, each with ± terminal corniform appendage 3–5 mm, glabrous, inners oblong, 4–5.7 mm, apex obtuse to truncate, each with ± terminal corniform appendage 2.5–3.5 mm; corolla red or red-orange, salverform, 18–26 mm, limb 10–15 mm diam. 2*n* = 30.

Flowering May–Nov. Chaparral, grasslands, oak woodlands, ponderosa pine zones; 700–2800 m; Ariz., Iowa, Kans., N.Mex., Tex.; Mexico.

Reports of *Ipomoea cristulata* from Iowa and Kansas may be based on cultivated plants; the report for Minnesota probably resulted from typographic error: MN for NM; the report for South Carolina was based on a specimen of *I. coccinea*.

7. Ipomoea tenuiloba Torrey in W. H. Emory, Rep. U.S. Mex. Bound. 2(1): 148. 1859 • Spiderleaf

Perennials, root tuberlike. **Stems** usually trailing, sometimes twining near tips. **Leaf blades** orbiculate, palmatisect, lobes 5–9, lanceolate to linear, 10–70 × 0.5–6.5 mm, surfaces glabrous. **Peduncles** glabrous. **Flowers** nocturnal; sepals chartaceous or coriaceous, outers oblong-lanceolate, 5–12 × 2–3 mm, muricate along midrib or ± smooth, margins scarious, apex mucronate, inners obovate-acuminate, 8–9 × 3–4 mm, smooth, margins scarious; corolla white, limb sometimes purple or pale rose-red, funnelform or salverform, 35–100 mm, limb 30–36 mm diam.

Varieties 2 (2 in the flora): sc, sw United States, n Mexico.

The two varieties of *Ipomoea tenuiloba* are comparatively easy to distinguish; there are intergrades (G. Yatskievych and C. T. Mason 1984), some approaching *I. plummerae*.

1. Leaf blade lobes 5–7, each to 1.2 mm wide; corollas usually white, limb sometimes pink to lavender, salverform, 65–100 mm; filaments: free portions 8–11 mm. . . . . 7a. *Ipomoea tenuiloba* var. *tenuiloba*
1. Leaf blade lobes 7–9, each to 6.5 mm wide; corollas: tube white, limb purple to red, funnelform, 35–65 mm; filaments: free portions 14–19 mm . . . . 7b. *Ipomoea tenuiloba* var. *lemmonii*

7a. Ipomoea tenuiloba Torrey var. **tenuiloba**

Leaf blade lobes 5–7, each to 1.2 mm wide. **Flowers:** corolla usually white, limb sometimes pink to lavender, salverform, 65–100 mm; filaments: free portions 8–11 mm.

Flowering Aug–Sep. Chaparral, hills, oak woodlands, ponderosa pine zone, rocky sites; 1200–2700 m; Ariz., N.Mex., Tex.; Mexico (Chihuahua, Sonora).

7b. Ipomoea tenuiloba Torrey var. **lemmonii** (A. Gray) Yatskievych & C. T. Mason, Madroño 31: 106. 1984 [E]

Ipomoea lemmonii A. Gray, Proc. Amer. Acad. Arts 19: 91. 1883 (as lemmoni)

Leaf blade lobes 7–9, each to 6.5 mm wide. **Flowers:** corolla tube white, limb purple to red, funnelform, 35–65 mm; filaments: free portions 14–19 mm.

Flowering Aug–Sep. Chaparral, hills, oak woodlands, ponderosa pine zone, rocky sites; 1500–2000 m; Ariz., Tex.

8. Ipomoea thurberi A. Gray in A. Gray et al., Syn. Fl. N. Amer. 2(1): 212. 1878 • Thurber's morning glory

Ipomoea gentryi Standley

Perennials, root elongate, tuberous. **Stems** trailing or twining. **Leaf blades** ± sagittate, 10–50 × 20–65 mm overall, base sagittate, or blades palmatisect, lobes 5–7, lanceolate, linear, or oblong, surfaces sparsely strigose. **Peduncles** glabrous. **Flowers** nocturnal; sepals lanceolate to lance-linear, 12–15 × 3–4 mm, ± herbaceous, base obscurely warty or not, apex acuminate, setaceous-caudate; corolla white, tube green, limb red, rose, drying purple, funnelform-salverform, 50–80 mm, limb 50–65 mm diam.

Flowering Aug–Sep. Oak woodlands, rocky sites; 1100–1600 m; Ariz.; Mexico (Chihuahua, Sonora).

9. Ipomoea macrorhiza Michaux, Fl. Bor.-Amer. 1: 141. 1803 • Large-root morning glory E

Perennials, root relatively large, tuberlike. **Stems** trailing or twining. **Leaf blades** ovate, triangular-ovate, or 3-lobed, 50–150 × 50–150 mm, base cordate to sagittate or truncate, margins ± crenulate, surfaces: abaxial tomentulose, adaxial glabrous, minutely beaded along veinlets. **Peduncles** tomentulose. **Flowers** nocturnal; sepals oblong-elliptic, 16–18 mm, coriaceous, sericeous; corolla white, throat lavender to purple inside, salverform, 50–80 mm.

Flowering Jun–Jul. Beaches, clearings, dunes; 0–40 m; Ala., Fla., Ga., Miss., N.C., S.C.

Ipomoea macrorhiza has been confused with the Mexican and Central American *I. jalapa* (Linnaeus) Pursh; *I. macrorhiza* differs by having nocturnal, moth-pollinated flowers with white corollas versus matinal, bee-pollinated flowers with lavender corollas.

10. Ipomoea alba Linnaeus, Sp. Pl. 1: 161. 1753 • Moonflower 

Calonyction aculeatum (Linnaeus) House; *Convolvulus aculeatus* Linnaeus

Perennials. Stems twining, usually prickly, sometimes rooting at nodes. **Leaf blades** broadly ovate to triangular or 3–5-lobed, 50–150 × 50–150 mm, base cordate, surfaces usually glabrous, rarely hairy. **Peduncles** glabrous. **Flowers** nocturnal; sepals ovate, 7–15 mm, ± coriaceous, apex acute, outers each with midrib extending as ± corniform appendage; corolla white, throat green-banded inside, salverform, 70–150 mm. **Fruits** 20–30 mm. **2*n*** = 30.

Flowering Sep–May. Forest margins, swamps, moist sites; 0–100 m; introduced; Fla., La., S.C., Tex.; Mexico; West Indies; Central America; South America; introduced also in Asia.

11. Ipomoea muricata (Linnaeus) Jacquin, Pl. Hort. Schoenbr. 3: 40. 1798 • Lilac-bell I W

Convolvulus muricatus Linnaeus, Mant. Pl. 1: 44. 1767

Annuals. Stems trailing or twining, ± warty or smooth. **Leaf blades** usually orbiculate to ovate, sometimes cordate or 3-5-lobed, 70–180 × 70–160 mm, base cordate, surfaces glabrous. **Peduncles** glabrous. **Flowers** nocturnal; sepals oblong to ovate, 6–8 mm, chartaceous to coriaceous, apex acute, outers each with ± corniform appendage 4–6 mm; corolla white, turning lavender in morning, salverform, limb sometimes ± campanulate, 30–75 mm. **Fruits** 18–20 mm. **2*n*** = 30.

Flowering Nov. Disturbed sites; 0–80 m; introduced; Ala., Ark., Fla., Ga., La., Miss., Tex.; Mexico; introduced also in South America.

Ipomoea muricata has been spread as a contaminant in soybean seeds (C. R. Gunn 1970).

The name *Ipomoea turbinata* Lagasca is illegitimate and has been misapplied to plants of *I. muricata* (G. W. Staples et al. 2006).

12. Ipomoea setosa Ker Gawler, Bot. Reg. 4: plate 335. 1818 • Brazilian morning glory I

Ipomoea melanotricha Brandegee

Perennials. Stems twining, ± setose. **Leaf blades** orbiculate or broadly ovate, usually 3–7-lobed, 100–200 × 100–200 mm overall, base cordate, lobes lanceolate to ovate, surfaces glabrous. **Peduncles** setose. **Flowers** nocturnal; sepals oblong, 10–14 mm, coriaceous, apex obtuse, abaxial surface setose; corolla lavender or pink, salverform, 60–90 mm, limb sometimes ± campanulate or rotate, 80–100 mm diam.

Flowering Nov–Jan. Abandoned plantings, disturbed sites; 0–400 m; introduced; Fla., Miss., Tex.; Mexico; Central America; South America; introduced also in Asia.

The report of *Ipomoea setosa* from Mississippi has not been verified.

13. Ipomoea violacea Linnaeus, Sp. Pl. 1: 161. 1753 • Beach moonflower

Calonyction tuba (Schlechtendal) Colla; *Ipomoea tuba* (Schlechtendal) G. Don

Perennials or lianas. Stems twining, not setose. **Leaf blades** usually cordate, orbiculate, or ovate, sometimes 3-lobed, 50–160 × 40–150 mm overall, base cordate, surfaces glabrous. **Peduncles** hairy, hairs retrorse to spreading. **Flowers** nocturnal; sepals orbiculate or ovate, 15–25 mm, herbaceous, apex emarginate, obtuse, or truncate, abaxial surfaces glabrous or proximally sparsely hirsute; corolla white, limb with yellowish lines, salverform, 50–120 mm. 2*n* = 30.

Flowering year-round. Dunes, littoral, mangrove sites; 0–50 m; Fla., Tex.; Mexico; West Indies; Central America; South America; introduced in Asia, Australia.

The name *Ipomoea violacea* has been misapplied to plants of *I. tricolor*.

14. Ipomoea indica (Burman) Merrill, Interpr. Herb. Amboin., 445. 1917 • Oceanblue morning glory [W]

Convolvulus indicus Burman, Auctuarium, index [6]. 1755; *Ipomoea mutabilis* Ker Gawler; *Pharbitis cathartica* (Poiret) Choisy

Perennials. Stems usually twining, sometimes trailing. **Leaf blades** cordate, rounded-ovate, or 3–5(–7)-lobed, 30–140 × 30–140 mm, base cordate to sagittate, surfaces glabrous or ± pilose. **Peduncles** glabrate or sparsely hairy, hairs antrorse to ± appressed. **Flowers:** sepals lance-ovate, 14–21 mm, herbaceous, apex ± acuminate, surfaces glabrous or abaxial sparsely hairy, hairs appressed; corolla usually blue to purple, rarely white, throat and tube white, funnelform, 50–70 mm. 2*n* = 30.

Flowering year-round. Roadsides, thickets; 0–1600 m; Ala., Calif., Fla., Ga., La., Miss., N.C., Pa., S.C., Tex.; Mexico; West Indies; Central America; South America; introduced in Asia.

In the flora area, *Ipomoea indica* rarely produces seeds and rarely survives winters. It is probably native in southern Florida.

SELECTED REFERENCE Fosberg, F. R. 1976. *Ipomoea indica* taxonomy: A tangle of morning glories. Bot. Not. 129: 35–38.

15. Ipomoea lindheimeri A. Gray in A. Gray et al., Syn. Fl. N. Amer. 2(1): 210. 1878 • Lindheimer's morning glory

Perennials. Stems twining. **Leaf blades** broadly ovate to reniform, 50–60 × 50–80 mm overall, usually 3–5(–7)-lobed, lobes ± lanceolate, base cordate, surfaces ± hirsute or sericeous. **Peduncles** hairy, hairs retrorse, shaggy. **Flowers:** sepals lanceolate to lance-ovate, 18–30 × 3–5(–9) mm, herbaceous, base ± dilated, abaxial surface ± hispid; corolla blue to violet, funnelform, 55–90 mm, limb 60–70 mm diam.

Flowering Apr–Nov. Oak woodlands, rocky sites, stream bottoms; 200–2300 m; N.Mex., Tex.; Mexico.

16. Ipomoea pubescens Lamarck in J. Lamarck and J. Poiret, Tabl. Encycl. 1: 465. 1793 (as Ipomaea) • Silky morning glory

Perennials, root oblong, relatively large. **Stems** twining. **Leaf blades** cordate, ovate, or 3–5-lobed, 20–80 × 20–90 mm, base cordate, lobes elliptic to ovate, surfaces ± hirsute or coarsely sericeous. **Peduncles** hairy, hairs retrorse or spreading. **Flowers:** sepals lance-ovate to ovate, 9–21 × 2–11 mm, herbaceous, abaxial surface ± hispid or coarsely sericeous; corolla blue to violet, funnelform, 55–80 mm, limb 60–70 mm diam.

Flowering Aug–Sep. Rocky sites, stream beds, oak woodlands; 100–1600 m; Ariz., N.Mex., Tex.; Mexico; South America.

17. Ipomoea barbatisepala A. Gray in A. Gray et al., Syn. Fl. N. Amer. 2(1): 212. 1878 • Canyon morning glory

Annuals. Stems twining. **Leaf blades** orbiculate-ovate, 30–80 × 15–85 mm overall, base cordate, palmatisect, incised nearly to petiole tip, lobes (3–)5–7, ± lanceolate or rhombic, surfaces glabrous, sometimes gland-dotted. **Peduncles** glabrous. **Flowers:** sepals ± lance-linear, 10–12 × 1–2 mm, herbaceous, proximally slightly dilated relative to narrowed distal portion, abaxial surface hispid-pilose; corolla blue, red-purple, or white, funnelform, 16–20(–25) mm, limb 18–20 mm diam.

Flowering Jul–Dec. Chaparral, desert scrub; 800–2500 m; Ariz., N.Mex., Tex.; Mexico.

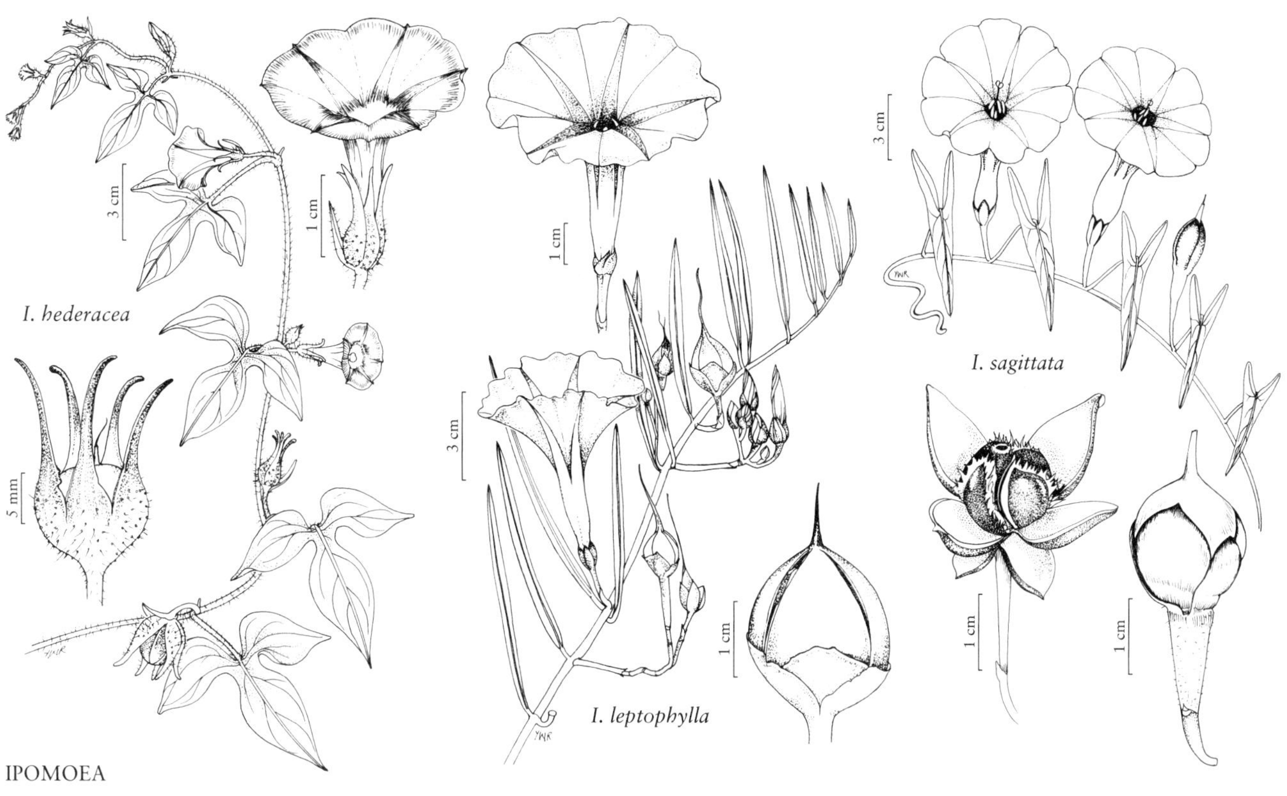

IPOMOEA

18. Ipomoea purpurea (Linnaeus) Roth, Bot. Abh. Beobacht., 27. 1787 • Tall morning glory, volubilis [F] [I] [W]

Convolvulus purpureus Linnaeus, Sp. Pl. ed. 2, 1: 219. 1762; *I. purpurea* var. *diversifolia* (Lindley) O'Donell; *Pharbitis purpurea* (Linnaeus) Voigt

Annuals. Stems twining. **Leaf blades** cordate, ovate, or 3(–5)-lobed, not palmatisect, 10–110(–180) × 10–120(–160) mm, base cordate, surfaces ± hairy, hairs ± antrorse. **Peduncles** hairy, hairs retrorse. **Flowers:** sepals elliptic, lance-oblong, or oblong, 8–15 × (1.5–)2.5–4.5 mm, herbaceous, base ± hairy, hairs dark at base, narrowed distal portion shorter to slightly longer than dilated base, apex acute to abruptly acuminate; corolla blue (purple, red, or white in cultivars), tube white inside, funnelform, (25–)40–60 mm, limb 24–48(–70) mm diam. **2*n*** = 30.

Flowering Jul–Nov. Abandoned plantings, canyons, disturbed sites, fields, stream banks; 100–2300 m; introduced; Ont., Que.; Ala., Ariz., Ark., Calif., Colo., Conn., Del., D.C., Fla., Ga., Ill., Ind., Iowa, Kans., Ky., La., Maine, Md., Mass., Mich., Minn., Miss., Mo., Mont., Nebr., Nev., N.H., N.J., N.Mex., N.Y., N.C., N.Dak., Ohio, Okla., Oreg., Pa., R.I., S.C., S.Dak., Tenn., Tex., Utah, Vt., Va., Wash., W.Va.; Mexico; introduced also in West Indies, Central America, South America, Eurasia, Africa, Australia.

In the flora area, *Ipomoea purpurea* may be native in southeastern United States and introduced elsewhere. Populations in California, Oregon, and Washington may not be truly naturalized.

19. Ipomoea hederacea Jacquin, Collecteana 1: 124, plate 36. 1787; Icon. Pl. Rar., plate 36. 1787 • Ivy-leaf morning glory [F] [W]

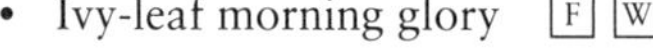

Ipomoea hederacea var. *integriuscula* A. Gray; *Pharbitis barbigera* (Sweet) G. Don

Annuals. Stems twining. **Leaf blades** ± orbiculate, ovate, or 3–5-lobed, not palmatisect, 50–120 × 50–120 mm, base cordate, basal lobes ± pointed, surfaces hairy. **Peduncles** hairy, hairs retrorse. **Flowers:** sepals ± lanceolate, 12–24 mm, herbaceous, proximally ovate, abruptly narrowed to ± curved or spreading distal portion notably longer than dilated base, abaxial surface, at least proximal 1/3, hirsute to hispid; corolla light blue, tube white or pale yellow inside, funnelform, 20–37(–45) mm, limb 17–35 mm diam. **2*n*** = 30.

Flowering May–Nov. Abandoned plantings, disturbed sites, gravel bars, roadsides; 10–1900 m; Ont.; Ala., Ariz., Ark., Calif., Conn., Del., D.C., Fla., Ga., Ill., Ind., Iowa, Kans., Ky., La., Maine, Md., Mass., Mich., Minn., Miss., Mo., Nebr., N.H., N.J., N.Mex., N.Y., N.C., N.Dak., Ohio, Okla., Pa., S.C., S.Dak., Tenn., Tex., Va., W.Va., Wis.; Mexico; South America; introduced in Europe, Africa, Pacific Islands (New Zealand), Australia.

Leaf shape in *Ipomoea hederacea*, and, perhaps, in other ipomoeas, is under simple genetic control and is useless for recognition of varieties (C. D. Elmore 1986).

20. Ipomoea nil (Linnaeus) Roth, Catal. Bot. 1: 36. 1797 • Japanese morning glory [I] [W]

Convolvulus nil Linnaeus, Sp. Pl. ed. 2, 1: 219. 1762; *Pharbitis hederacea* (Linnaeus) Choisy; *P. nil* (Linnaeus) Choisy

Annuals. Stems twining. **Leaf blades** cordate, ± orbiculate, ovate, or 3-lobed, 50–150 × 20–140 mm overall, base cordate, surfaces sparsely hirsute to sericeous. **Peduncles** hairy, hairs retrorse. **Flowers:** sepals lance-linear, 15–25(–30) mm, herbaceous, proximally narrowly ovate, densely hispid, gradually narrowed to ± straight, hispid to strigose or glabrate distal portion longer than ovate base; corolla usually blue to purplish, sometimes red or white, tube white or yellow inside, funnelform, (20–)30–60+ mm. **2*n*** = 30.

Flowering Jul–Dec. Abandoned plantings, fields, thickets; 10–2200 m; introduced; Ala., Ky., La., Md., Miss., N.C., Okla., S.C., Tenn., Tex.; Mexico; West Indies; Central America; South America; introduced also in Asia.

Reports of *Ipomoea nil* from California are based on misidentified material of *I. hederacea*.

21. Ipomoea aquatica Forsskål, Fl. Aegypt.-Arab., 44. 1775 • Water-spinach [I] [W]

Perennials. Stems usually repent, rooting at nodes, sometimes floating, rarely twining. **Leaf blades** ± hastate to lanceolate, 40–120 × 20–60 mm, base ± cordate or hastate to truncate, terminal lobe broadly to narrowly triangular or lanceolate, surfaces glabrescent. **Peduncles** glabrous. **Flowers:** sepals elliptic-ovate to oblong-ovate, 6–8 mm, chartaceous or coriaceous, apex acute or obtuse, mucronulate; corolla usually purple, rarely white, funnelform, 40–50 mm. **2*n*** = 30.

Flowering year-round. Abandoned plantings, wet sites; 0–100 m; origin uncertain; introduced; Calif., Fla.; Mexico; Central America; South America; Asia; Africa; Australia.

22. Ipomoea imperati (Vahl) Grisebach, Cat. Pl. Cub., 203. 1866 • Beach morning glory

Convolvulus imperati Vahl, Symb. Bot. 1: 17. 1790

Perennials. Stems repent, rooting at nodes and underground. **Leaf blades** lanceolate, linear, oblong, ovate, or 3–5-lobed, 15–80 × 12–60 mm, base cordate to truncate, surfaces glabrous. **Peduncles** glabrous. **Flowers:** sepals lance-oblong, 10–15 mm, outers shorter than inners, ± coriaceous, apex acute to obtuse, abaxial surface glabrous; corolla white, throat usually yellow, sometimes purplish inside, funnelform, 25–50 mm. **2*n*** = 30.

Flowering year-round. Beaches, dunes; 0–10 m; Ala., Fla., Ga., La., Miss., N.C., S.C., Tex.; Mexico; West Indies; Central America; South America; Pacific Islands (Hawaii); introduced in Europe, Asia, Africa, Australia.

Ipomoea imperati was collected once in Pennsylvania (on ballast in 1865). The names *I. littoralis* (Linnaeus) Boissier 1875, not Blume 1826, and *I. stolonifera* (Cirillo) J. F. Gmelin are illegitimate; both have been misapplied to plants of *I. imperati*.

23. Ipomoea asarifolia (Desrousseaux) Roemer & Schultes in J. J. Roemer et al., Syst. Veg. 4: 251. 1819 • Ginger-leaf morning glory [I]

Convolvulus asarifolius Desrousseaux in J. Lamarck et al., Encycl. 3: 562. 1792

Perennials. Stems repent, rooting at nodes. **Leaf blades** cordate to ± orbiculate, 30–120 × 30–120 mm, base cordate, apex acute to rounded, surfaces glabrous. **Peduncles** glabrous. **Flowers:** sepals chartaceous to coriaceous, outers elliptic, 5–9 mm, apex obtuse, inners elliptic or oblong, 10–15 mm, apex obtuse; corolla usually red, rarely white, funnelform, 50–80 mm.

Flowering year-round. Beaches, moist or swampy sites; 0–100 m; introduced; Fla.; Mexico; West Indies (Jamaica); Central America; South America.

24. Ipomoea pes-caprae (Linnaeus) R. Brown, Observ. Congo, 58. 1818 • Bayhops [W]

Convolvulus pes-caprae Linnaeus, Sp. Pl. 1: 159. 1753

Subspecies 2 (1 in the flora): tropical regions, original distribution unknown, now world-wide in subtropical and tropical climates.

Subspecies *pes-caprae* in known from coastal and island shores around and in the Indian Ocean.

24a. Ipomoea pes-caprae (Linnaeus) R. Brown subsp. **brasiliensis** (Linnaeus) Oostroom, Blumea 3: 533. 1940 [W]

Convolvulus brasiliensis Linnaeus, Sp. Pl. 1: 159. 1753; *Ipomoea pes-caprae* var. *emarginata* Hallier f.

Perennials. Stems usually repent, rooting at nodes, rarely twining, ± fleshy. **Leaf blades** ± orbiculate, ovate, or reniform, 30–100 × 50–100 mm, base cordate, rounded, or truncate, apex ± emarginate, surfaces glabrous. **Peduncles** glabrous. **Flowers:** sepals elliptic, orbiculate, or ovate, 5–11 mm, ± coriaceous; corolla lavender or pink, throat darker inside, funnelform, 35–40(–70) mm. **2*n*** = 30, 60.

Flowering year-round. Beaches, dunes; 0–10 m; place of origin uncertain; Ala., Fla., Ga., La., Miss., S.C., Tex.; Mexico; West Indies; Central America; South America; Asia; Africa; Australia.

Reports of subsp. *brasiliensis* from North Carolina and Pennsylvania have not been verified.

25. Ipomoea carnea Jacquin, Enum. Syst. Pl., 13. 1760 (as Ipomaea) [I] [W]

Subspecies 2 (1 in the flora): introduced; Mexico, West Indies, Central America, South America; introduced also in Asia, Africa.

Variety *carnea* occurs in South America.

25a. Ipomoea carnea Jacquin subsp. **fistulosa** (Martius ex Choisy) D. F. Austin, Taxon 26: 237. 1977 [I] [W]

Ipomoea fistulosa Martius ex Choisy in A. P. de Candolle and A. L. P. P. de Candolle, Prodr. 9: 349. 1845; *I. crassicaulis* (Bentham) B. L. Robinson

Shrubs [lianas]. Stems erect, [scrambling or twining]. **Leaf blades** lanceolate, ± orbiculate, or ovate, 70–250+ × 40–170 mm, base cordate or ± truncate, surfaces puberulent to glabrate. **Peduncles** puberulent. **Flowers:** sepals ± orbiculate, 3–7 mm, chartaceous or coriaceous, apex rounded; corolla usually lavender or red-purple, sometimes white, throat darker, funnelform, 40–90 mm. **2*n*** = 30.

Flowering Jan–Nov. Marshy sites, swamps; 0–1000 m; introduced; Fla., Tex.; Mexico; West Indies; Central America; South America; introduced also in Asia.

26. Ipomoea leptophylla Torrey in J. C. Frémont, Rep. Exped. Rocky Mts., 94. 1843 • Bush morning glory [E] [F]

Perennials, taproot relatively large. **Stems** usually erect, sometimes ± trailing. **Leaf blades** lance-linear to linear, 30–80(–150) × 2–8(–10) mm, base ± cuneate, surfaces glabrous. **Peduncles** glabrous. **Flowers:** sepals elliptic, orbiculate, or ovate, 5–10 mm, chartaceous or coriaceous, apex obtuse; corolla lavender-pink to purple-red, throat darker, funnelform, 50–90 mm. **2*n*** = 30.

Flowering May–Sep. Plains, prairies, sandy sites; 0–2200 m; Colo., Kans., Mont., Nebr., N.Mex., Okla., S.Dak., Tex., Wyo.

27. Ipomoea plummerae A. Gray in A. Gray et al., Syn. Fl. N. Amer. ed. 2, 2(1): 434. 1886 • Huachuca Mountain morning glory

Perennials, root globose, tuber-like. **Stems** usually ± trailing, sometimes ascending, erect, or twining near tips. **Leaf blades** of 2 forms; one form ± orbiculate, palmatisect, lobes (3–)5 (–7+), linear to spatulate, 3–30(–50) × (0.5–)1–2.5 mm, base ± truncate; the other form ± cuneate-obovate, proximally cuneate, distally ± incised, 3–5(–7+)-toothed; surfaces glabrous. **Peduncles** glabrous. **Flowers:** sepals oblong to ovate, outers 5–8 × 2–3 mm, abaxial surface ± muricate or smooth, apex acute to obtuse, mucronate, inners broadly ovate, 7–9 (–10) × 3–4 mm, apex acuminate to acute, surfaces glabrous, chartaceous to coriaceous; corolla purple to lavender, funnelform, 25–31 mm, limb 18–22 mm diam.

Varieties 3 (2 in the flora): sw United States, Mexico, South America.

Variety *cupulata* J. A. McDonald is known in Mexico from Chihuahua, Guerrero, Jalisco, and Sinaloa.

1. Leaf blades ± orbiculate, palmatisect, lobes (3–)5(–7+). . . . 27a. *Ipomoea plummerae* var. *plummerae*
1. Leaf blades ± cuneate-obovate, proximally cuneate, distally ± incised, 3–5(–7+)-toothed . . . . 27b. *Ipomoea plummerae* var. *cuneifolia*

27a. Ipomoea plummerae A. Gray var. **plummerae**

Leaf blades ± orbiculate, palmatisect, lobes (3–)5(–7+), linear to spatulate, 3–30(–50) × (0.5–) 1–2.5 mm.

Flowering Jul–Sep. Open rocky slopes; 1300–2600 m; Ariz., N.Mex., Tex.; Mexico; South America.

27b. Ipomoea plummerae A. Gray var. **cuneifolia** J. F. Macbride, Publ. Field Mus. Nat. Hist., Bot Ser. 11: 4. 1931

Ipomoea egregia House

Leaf blades ± cuneate-obovate, 10–20 × 3–12 mm, proximally cuneate, distally ± incised, 3–5(–7+)-toothed.

Flowering Sep. Open rocky slopes; 1800–2500 m; Ariz., N.Mex.; South America.

Ipomoea cuneifolia Meisner 1869, the later homonym *I. cuneifolia* A. Gray 1883, and *I. plummerae* var. *cuneifolia* apparently all refer to the same taxon.

28. Ipomoea costellata Torrey in W. H. Emory, Rep. U.S. Mex. Bound. 2(1): 149. 1859 • Crest-rib morning glory

Ipomoea costellata var. *edwardsensis* O'Kennon & G. L. Nesom

Annuals. Stems usually trailing, or twining only near tips, rarely erect. **Leaf blades** palmatisect, lobes 5–9, lance-linear, linear, oblanceolate, or spatulate, 7–28 × 0.5–3(–8) mm, surfaces glabrous or sparsely hispidulous. **Peduncles** usually glabrous, rarely sparsely hispidulous. **Flowers:** sepals lance-oblong to lanceolate, outers 3–5 × 1–2 mm, inners 4–6 × 2–3 mm, herbaceous, apex acute, abaxial surface usually ± carinate and glabrous, sometimes hispidulous on midrib; corolla pale lavender to pink, funnelform, 10–12 mm.

Flowering Jul–Nov. Chaparral, oak woodlands, ponderosa pine zone, rocky sites; 100–2200 m; Ariz., N.Mex., Tex.; Mexico; introduced in South America.

29. Ipomoea ternifolia Cavanilles, Icon. 5: 52, plate 478, fig. 1. 1799 • Tripleleaf morning glory

Varieties 3 (1 in the flora): Arizona, Mexico, Central America.

Varieties *ternifolia* and *valida* (House) J. A. McDonald are known from Mexico and Central America.

29a. Ipomoea ternifolia Cavanilles var. **villosa** (Choisy) Staples & Govaerts, Phytologia 97: 221. 2015

Ipomoea muricata Cavanilles var. *villosa* Choisy in A. P. de Candolle and A. L. P. P. de Candolle, Prodr. 9: 353. 1845; *I. ternifolia* var. *leptotoma* (Torrey) -

Annuals. Stems trailing or twining near tips. **Leaf blades** ± orbiculate to oval, 15–45(–80) × 15–40(–60) mm overall, palmatisect, lobes 5–7(–9), usually filiform to linear, sometimes lance-linear, 15–80 × 0.2–2(–5) mm, apex acute, surfaces glabrous or sparsely hairy, hairs ± erect. **Peduncles** glabrous. **Pedicels** ± hairy, hairs ± erect. **Flowers:** sepals: outers lanceolate, 8–9 × 2–3 mm, inners lance-attenuate, 10–11 × 3–4 mm, chartaceous, margins whitish, scarious, abaxial surface glabrous or hairy, hairs antrorse; corolla purple, funnelform, 25–30(–40) mm, limb 32–46 mm diam.

Flowering Jun–Oct. Desert scrub, rocky sites; 600–1400 m; Ariz.; Mexico (Baja California Sur, Chihuahua, Sinaloa, Sonora).

30. Ipomoea wrightii A. Gray in A. Gray et al., Syn. Fl. N. Amer. 2(1): 213. 1878 • Wright's morning glory [I] [W]

Annuals. Stems usually twining, sometimes trailing. **Leaf blades** ± pentagonal, 30–80 × 20–80 mm overall, palmatisect, lobes 5, lance-linear to lanceolate, (21–)35–50(–80) × (5–)10–15 (–20) mm, apex acute, surfaces glabrous. **Peduncles** usually spiraled, 50–100 mm, glabrous. **Flowers:** sepals ovate, 5–7 mm, chartaceous or coriaceous, apex obtuse to rounded, abaxial surface glabrous; corolla red, rosy, throat red-violet, funnelform, 18–30 mm.

Flowering Jul–Oct. Disturbed sites, fields, marshy sites; 0–600 m; introduced; Ala., Ark., Fla., Ga., La., Miss., Okla., Tenn., Tex.; Asia; introduced also in Mexico, West Indies, Central America, South America (Argentina, Brazil, Ecuador, Paraguay, Peru), Africa, Australia.

Ipomoea wrightii may be spread as a contaminant in seeds.

The names *Ipomoea heptaphylla* (Roxburg) Voigt and *I. pulchella* Roth have been misapplied to plants of *I. wrightii.*

31. Ipomoea cairica (Linnaeus) Sweet, Hort. Brit., 287. 1826 • Cairo morning glory [I] [W]

Convolvulus cairicus Linnaeus, Syst. Nat. ed. 10, 2: 922. 1759

Perennials. Stems usually twining, sometimes trailing. **Leaf blades** orbiculate to ovate, 30–100 × 30–100 mm overall, palmatisect, lobes 5 (proximal 2 sometimes 2-lobed), lance-elliptic, lanceolate, or lance-ovate, (5–)10–25(–70) × (3–)8–15(–30) mm, apex acute to obtuse, surfaces glabrous. **Peduncles** glabrous; pedicels straight, 10–25 mm. **Flowers:** sepals oblong to ovate, 4–6.5(–9) mm, outers slightly shorter than inners, chartaceous, margins scarious, apex obtuse to acute; corolla lavender-blue or white, throat purplish-red, funnelform, 45–60 mm. $2n = 30$.

Flowering Mar–Oct. Abandoned plantings, disturbed sites; -20–200 m; introduced; Ala., Calif., Fla., La.; Africa; introduced also in Mexico (Oaxaca), West Indies, South America.

32. Ipomoea capillacea (Kunth) G. Don, Gen. Hist. 4: 267. 1837 • Purple morning glory

Convolvulus capillaceus Kunth in A. von Humboldt et al., Nov. Gen. Sp. 3(fol.): 76; 3(qto.): 97. 1819

Perennials. Stems usually ascending to erect, sometimes trailing. **Leaf blades** palmatisect, lobes 5–9, filiform to linear, (3–)5–15(–25) × 0.2–1 mm. **Peduncles** glabrous. **Flowers:** sepals elliptic, oblong, or ovate, 5–6 × 2–3 mm, chartaceous or coriaceous, abaxial surface muricate or smooth; corolla lavender to red-purple, funnelform, 30–40 mm, limb 20–25 mm diam.

Flowering Jul–Sep. Oak woodlands, plains, ponderosa pine zones; 1500–2500 m; Ariz., N.Mex., Tex.; Mexico; Central America; South America.

The report of *Ipomoea capillacea* from Alabama (J. T. Kartesz and C. A. Meacham 1999) was probably based on a specimen of *I. muricata.*

33. Ipomoea cardiophylla A. Gray in A. Gray et al., Syn. Fl. N. Amer. 2(1): 213. 1878 • Heart-leaf morning glory

Annuals. Stems twining. **Leaf blades** cordate, 20–60 × 14–38 mm, base cordate, surfaces glabrous. **Peduncles** glabrous. **Flowers:** sepals triangular, 6 × 3–4 mm, chartaceous to coriaceous, apex acute; corolla blue (drying pink or purple), funnelform, 26–27 mm, limb 30–35 mm diam.

Flowering Aug–Oct. Desert scrub; 700–1700 m; Ariz., N.Mex., Tex.; Mexico.

The name *Ipomoea aristolochiifolia* G. Don has been misapplied to plants of *I. cardiophylla.*

34. Ipomoea tricolor Cavanilles, Icon. 3: 5, plate 208. 1795 • Heavenly blue morning glory [I] [W]

Annuals. Stems twining. **Leaf blades** ± cordate, 60–100 × 25–130 mm, base cordate, surfaces glabrous. **Peduncles** glabrous. **Flowers:** sepals lance-ovate, triangular, or oblong-triangular, (4–)6–7 mm, coriaceous, margins scarious, apex acute, abaxial surface muriculate, glabrous; corolla usually blue to deep blue, sometimes white, tube white outside, pale yellow inside, funnelform, 35–60 mm, limb 50–90 mm diam.

Flowering Oct–Dec. Abandoned plantings, thickets; 20–1900 m; introduced; Ala., Ariz., Ark., Fla., Ga., Ky., La., Miss., Mo., N.C., Pa., S.C., Tex.; Mexico; West Indies; Central America; South America.

Ipomoea tricolor is native in Mexico and has long been cultivated in North America.

The name *Ipomoea violacea* has been misapplied to plants of *I. tricolor.*

35. Ipomoea dumetorum Willdenow in J. J. Roemer et al., Syst. Veg. 4: 789. 1819 • Railway creeper

Annuals. Stems usually twining, sometimes trailing. **Leaf blades** deltate, ovate, or ovate-elongate, 24–80 × 8–87 mm, base cordate or ± sagittate to truncate, margins sometimes 3-toothed, surfaces glabrous. **Peduncles** pilosulous on proximal 1–2 mm, distally glabrous. **Flowers:** sepals elongate-ovate to ovate, 3.5–8 mm, chartaceous or coriaceous, apex acute to obtuse, abaxial surface dotted with dark spots; corolla usually dark lavender to pink, rarely white, funnelform, 15–28 mm.

Flowering Aug–Oct. Open, dry to wet sites, washes; 2000–2800 m; N.Mex., Tex.; Mexico; South America.

In the flora area, *Ipomoea dumetorum* is known from the Davis, Organ, and White mountains. The names *I. cardiophylla* and *I. pulchella* (Kunth) G. Don (not Roth) have been misapplied to plants of *I. dumetorum*.

36. Ipomoea sagittata Poiret, Voy. Barbarie 2: 122. 1789 (as Ipomea) • Saltmarsh morning glory [F]

Perennials. Stems twining. **Leaf blades** ± triangular, 40–100 × 20–60 mm overall, base hastate to sagittate, basal lobes lanceolate, linear, or narrowly triangular, 15–60(–100) × 3–8(–15) mm, surfaces glabrous. **Peduncles** glabrous. **Flowers:** sepals elliptic, oblong, or ovate, 8–9 mm, coriaceous, apex obtuse to rounded, mucronate, surfaces glabrous; corolla lavender, purple, or red-purple, funnelform, 60–90 mm, limb 60–80 mm diam.

Flowering Apr–Oct. Beaches, brackish or freshwater marshes, swamps; 0–400 m; Ala., Fla., Ga., La., Miss., N.C., S.C., Tex.; Mexico; West Indies; introduced in Eurasia, nw Africa.

37. Ipomoea tenuissima Choisy in A. P. de Candolle and A. L. P. P. de Candolle, Prodr. 9: 376. 1845 • Rockland morning glory [E]

Annuals. Stems twining. **Leaf blades** usually narrowly hastate or sagittate, sometimes ovate, 15–30 × 7–20 mm, base cordate, hastate, or sagittate, lobes usually pointed, sometimes rounded, surfaces usually ± hairy, adaxial sometimes glabrate. **Peduncles** hairy, hairs appressed. **Flowers:** sepals lance-oblong, lanceolate, or lance-ovate, 5–8 mm, chartaceous or coriaceous, margins ciliate, apex acuminate, mucronate; corolla lavender, pink, or pink-purple, throat darker inside, funnelform, 30–45 mm.

Flowering Aug–Sep. Pine flatwoods; 0–30 m; Fla.

After fires, *Ipomoea tenuissima* seeds germinate and seedlings thrive for about a year. The plants then disappear except in sites that remain open.

38. Ipomoea triloba Linnaeus, Sp. Pl. 1: 161. 1753 • Littlebell [I] [W]

Annuals. Stems usually twining, sometimes trailing. **Leaf blades** orbiculate, broadly ovate, or 3–7-lobed, 20–80 × 20–70 mm overall, base cordate, basal lobes angular, lobed, or rounded, surfaces glabrous or sparsely pilose. **Peduncles** glabrous, distally verruculose. **Flowers:** sepals narrowly elliptic-oblong, lanceolate, or oblong, 6–7 mm, chartaceous or coriaceous, margins ciliate, apex acute or obtuse, mucronulate-caudate, surface glabrous or abaxial sparsely hairy; corolla lavender, funnelform, 10–20 mm. $2n$ = 30, 60.

Flowering year-round. Disturbed sites; -40–100 m; introduced; Calif., Fla.; Mexico; West Indies; Central America; South America; introduced also in Asia.

Ipomoea triloba seeds are sometimes a contaminant in rice and other seeds.

Ipomoea trifida (Kunth) G. Don (*Convolvulus trifidus* Kunth) was incorrectly ascribed to Florida by J. K. Small (1933) on the basis of an unusual and incorrectly identified specimen of *I. triloba*.

39. Ipomoea lacunosa Linnaeus, Sp. Pl. 1: 161. 1753 • Whitestar [E] [W]

Annuals. Stems twining. **Leaf blades** cordate-ovate, deltate-ovate, ovate, or 3(–5)-lobed, 30–80 × 20–70 mm, base ± cordate, basal lobes rounded or pointed, surfaces glabrous or sparsely hairy. **Peduncles** glabrous, sometimes muricate. **Flowers:** sepals elliptic-oblong, lanceolate, or lance-ovate, (8–)11–14 mm, chartaceous or coriaceous, margins ciliate, apex acuminate, surfaces glabrous; corolla usually white, limb sometimes pink tinged, funnelform, 15–20(–25) mm. **Fruits** 10–13 mm diam. 2*n* = 30.

Flowering Apr–Oct. Disturbed sites, ditches, fields; 0–300 m; Ala., Del., D.C., Fla., Ga., Ill., Ind., Iowa, Kans., Ky., La., Md., Mass., Miss., Mo., N.J., N.C., Ohio, Okla., Pa., S.C., Tenn., Tex., Va., W.Va.

Plants of *Ipomoea lacunosa* are sometimes confused with plants of *I.* ×*leucantha*, which are derived from hybridization between *I. lacunosa* and *I. cordatotriloba* (D. F. Austin and W. E. Abel 1981). Seeds of *Ipomoea lacunosa* are 5–6 mm and seeds of *I.* ×*leucantha* are 3.2–4 mm.

Reports of *Ipomoea lacunosa* for Ontario, California, and New York are apparently based on waifs.

40. Ipomoea ×leucantha Jacquin, Icon. Pl. Rar. 2: 10, plate 318. 1788 (as species) • Whitestar morning glory

Annuals. Stems usually twining, sometimes trailing. **Leaf blades** orbiculate, ovate, or 3–5-lobed, 20–80 × 20–70 mm, base cordate, surfaces glabrous or sparsely hairy. **Peduncles** glabrous. **Flowers:** sepals lanceolate, (8–)10–14 mm, chartaceous or coriaceous, apex acuminate, surfaces glabrous; corolla lavender or white, throat usually darker, funnelform, 6–15(–20) mm, limb to 10+ mm diam. **Fruits** 7–8 mm diam.

Flowering Aug–Oct. Disturbed sites; 0–700 m; Ariz., Fla., La., Miss., S.C., Tex., Va.; Mexico; West Indies; Central America; South America.

In the southeastern United States, *Ipomoea* ×*leucantha* results from crosses between *I. cordatotriloba* and *I. lacunosa* that may be effected by honeybees (*Apis mellifera*); elsewhere, *I.* ×*leucantha* is presumably introduced.

41. Ipomoea amnicola Morong, Ann. New York Acad. Sci. 7: 170. 1893 • Red-center morning glory [I]

Perennials. Stems usually twining, sometimes decumbent. **Leaf blades** cordate-ovate, reniform, or 3-lobed, 20–125 × 20–125 mm, base cordate, surfaces glabrous. **Peduncles** glabrous, sometimes warty. **Flowers:** sepals: outers usually ± orbiculate, sometimes oblong or elliptic, rarely ovate, 4–5 mm, inners obovate to ± orbiculate, 4–5.5 mm, chartaceous or membranous, margins scarious, apex usually emarginate, mucronulate, or obtuse, sometimes truncate; corolla white with red (rose) lines or wholly lilac, throat darker, funnelform, 18–30 mm.

Flowering Apr–Jul(–Dec). Disturbed sites; 0–200 m; introduced; Tex.; Mexico; South America (Argentina, Paraguay).

In the flora area, *Ipomoea amnicola* has been reported also from a Missouri collection made in 1921.

42. Ipomoea cordatotriloba Dennstedt, Nomencl. Bot. 1: 246. 1810 • Tie-vine [F] [W]

Perennials. Stems twining. **Leaf blades** cordate-ovate, lance-ovate, ovate, or 3–5(–7)-lobed, 10–90 × 10–90 mm, base cordate, lobes usually rounded, sometimes pointed, surfaces usually hirsute, pilose, or tomentose, rarely glabrous. **Peduncles** glabrous or hairy, hairs appressed. **Flowers:** sepals lanceolate to ovate, 8–14 mm, chartaceous or coriaceous, outers lance-ovate to lanceolate, narrowed distal portion curved, glabrous or hairy, inners ovate, margins ciliate or not, abaxial surface glabrous or hairy; corolla lavender, tube darker, funnelform, 20–38 mm.

Varieties 3 (2 in the flora): United States, Mexico, South America.

Variety *australis* (O'Donnell) D. F. Austin is known from Argentina.

Varieties *cordatotriloba* and *torreyana* appear to differ by minor, trivial traits; nevertheless, they have distinctive aspects and, historically, distinct ranges and habitats. Both have been dispersed by humans and may appear sporadically in places outside their historical ranges.

1. Leaf blade surfaces usually hirsute, pilose, or tomentose, rarely glabrous; sepals hispid-pilose and/or ciliate . . . . 42a. *Ipomoea cordatotriloba* var. *cordatotriloba*
1. Leaf blade surfaces glabrous; sepals glabrous . . . . 42b. *Ipomoea cordatotriloba* var. *torreyana*

42a. Ipomoea cordatotriloba Dennstedt var. **cordatotriloba** [W]

Ipomoea trichocarpa Elliott

Leaf blade surfaces usually hirsute, pilose, or tomentose, rarely glabrous. **Sepals** hispid-pilose and/or ciliate. **2*n*** = 30.

Flowering Apr–Nov. Disturbed sites, fields, roadsides, thickets; 0–600 m; Ala., Ark., Fla., Ga., La., Miss., N.C., S.C., Tex.; Mexico (Tamaulipas); South America.

42b. Ipomoea cordatotriloba Dennstedt var. **torreyana** (A. Gray) D. F. Austin, Taxon 37: 185. 1988 [F] [W]

Ipomoea trifida (Kunth) G. Don var. *torreyana* A. Gray in A. Gray et al., Syn. Fl. N. Amer. 2(1): 212. 1878; *I. trichocarpa* Elliott var. *torreyana* (A. Gray) Shinners

Leaf blade surfaces glabrous. **Sepals** glabrous. **2*n*** = 30.

Flowering Jan–Nov. Disturbed sites, fields, roadsides, thickets; 0–1300 m; Tex.; Mexico (Chihuahua, Nuevo León, Tamaulipas).

43. Ipomoea batatas (Linnaeus) Lamarck in J. Lamarck and J. Poiret, Tabl. Encycl. 1: 465. 1793 • Sweet potato [I]

Convolvulus batatas Linnaeus, Sp. Pl. 1: 154. 1753

Perennials, root relatively large, tuberlike. **Stems** ± trailing, rarely twining. **Leaf blades** cordate, broadly ovate, or 5–7-lobed, 50–100+ × 40–100 mm overall, base cordate, surfaces glabrous or hairy. **Peduncles** glabrous or hairy, hairs appressed. **Flowers:** sepals lanceolate to oblong, 8–15 mm, chartaceous; corolla usually lavender, pink, or purplish, sometimes white, throat usually darker inside, funnelform, (30–)40–70 mm. **Seeds** glabrous. **2*n*** = 60, 84, 90.

Flowering year-round. Abandoned plantings, thickets; 0–200+ m; introduced; Fla., Kans., La., Miss., N.Y., N.C., Pa., S.C., Tex., Utah, Va.; Mexico; West Indies; Central America; South America; introduced also in Asia, Africa.

Reports of *Ipomoea batatas* from northern parts of the flora area appear to be based on ephemerals.

SELECTED REFERENCE Austin, D. F. 1978. The *Ipomoea batatas* complex–I. Taxonomy. Bull. Torrey Bot. Club 105: 114–129.

44. Ipomoea rupicola House, Ann. New York Acad. Sci. 18: 230. 1908 • Cliff morning glory

Perennials, rhizomatous. **Stems** trailing or twining. **Leaf blades** usually cordate-ovate to oblong-ovate, sometimes pandurate, 30–90 × 20–70 mm, base cordate-hastate, margins sometimes indented, rarely lobed or toothed, surfaces tomentulose, glabrescent. **Peduncles** hairy, tomentulose, hairs ± appressed. **Flowers:** sepals elliptic-oblong to oblong-ovate, 12–14 mm, coriaceous, apex obtuse, rounded, or subacute; corolla lavender-pink or purple, throat darker inside, funnelform, 70–90 mm. **Seeds** hairy.

Flowering Jun–Oct. Rocky, open sites; 20–1800 m; Tex.; Mexico (Tamaulipas).

45. Ipomoea pandurata (Linnaeus) G. Meyer, Prim. Fl. Esseq., 100. 1818 • Man-of-the-earth [E] [F] [W]

Convolvulus panduratus Linnaeus, Sp. Pl. 1: 153. 1753

Perennials, root relatively large. **Stems** usually twining, sometimes trailing. **Leaf blades** cordate, cordate-ovate, or pandurate, 30–100 × 20–90 mm, base cordate, surfaces glabrous or abaxial hairy. **Peduncles** glabrous. **Flowers:** sepals elliptic-oblong, 12–22 mm, outers sometimes shorter than inners, coriaceous, surfaces glabrous; corolla white, throat lavender or purple-red inside, funnelform, 50–80 mm. **2*n*** = 30.

Flowering Jun–Sep. Abandoned plantings, fields, prairies; 0–600 m; Ont.; Ala., Ark., Conn., Del., D.C., Fla., Ga., Ill., Ind., Iowa, Kans., Ky., La., Md., Mass., Mich., Miss., Mo., Nebr., N.J., N.Y., N.C., Ohio, Okla., Pa., S.C., Tenn., Tex., Va., W.Va.

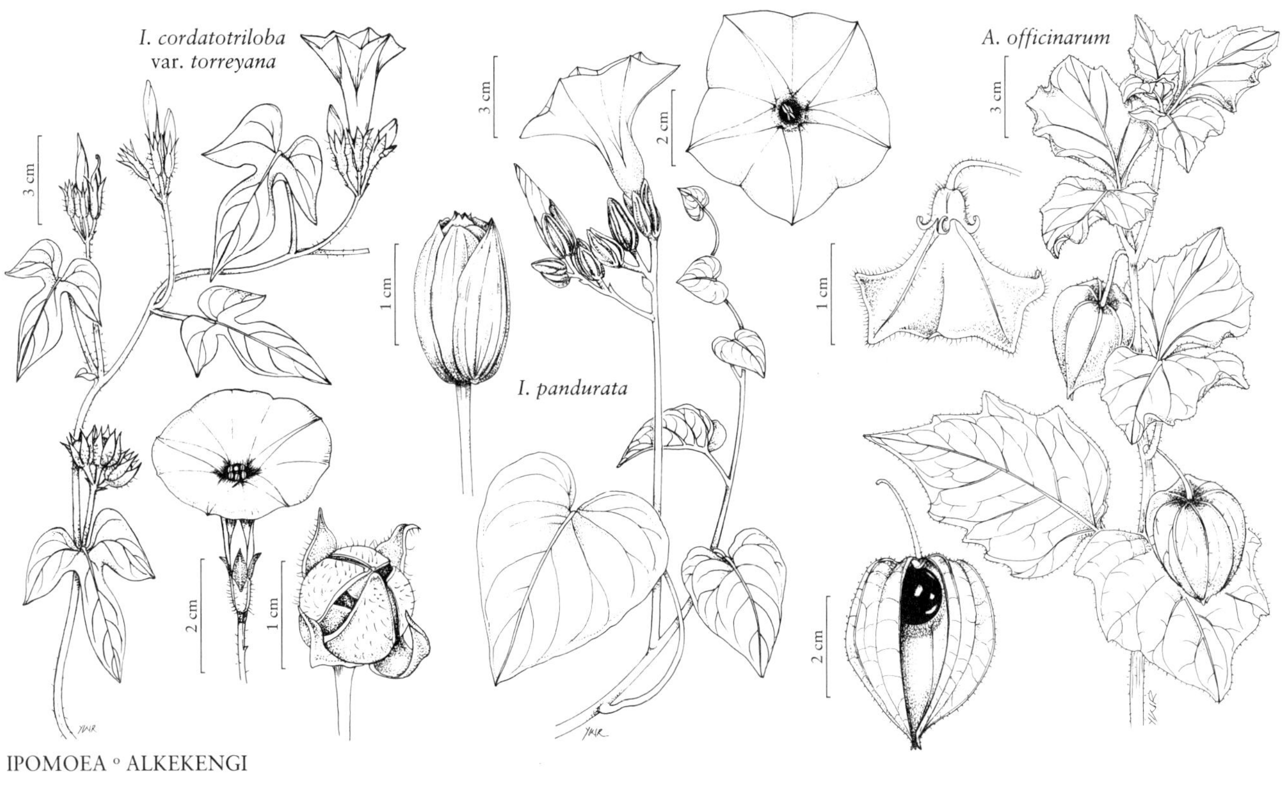

IPOMOEA ∘ ALKEKENGI

46. Ipomoea longifolia Bentham, Pl. Hartw., 16 (as Ipomaea), 345. 1839 • Pink-throat morning glory

Perennials, ± fleshy, rhizomatous. **Stems** usually trailing, rarely decumbent. **Leaf blades** lance-oblong, lanceolate, or linear, 100–120(–210) × 20–40 mm, base rounded, surfaces glabrous. **Peduncles** glabrous. **Flowers:** sepals oblong or ovate, 12–20 mm, outers 12–14(–17) × 6–7 mm, inners 15–20 × 7–8 mm, coriaceous, surfaces glabrous; corolla ± white, throat purple inside, funnelform, 70–100 mm, limb 70–80 mm diam.

Flowering Apr–Sep. Desert grasslands, oak woodlands; 900–1900 m; Ariz.; Mexico.

47. Ipomoea shumardiana (Torrey) Shinners, SouthW. Naturalist 6: 101. 1961 • Narrow-leaf morning glory [C] [E]

Convolvulus shumardianus Torrey in R. B. Marcy, Explor. Red River Louisiana, 291. 1852

Perennials. Stems trailing or twining. **Leaf blades** deltate-ovate or narrowly lance-ovate, 30–80 × 10–40 mm, base cordate to truncate, surfaces glabrous. **Peduncles** glabrous. **Flowers:** sepals not dotted with dark spots, elliptic, oblong, or orbiculate, 10–15 mm, outers shorter than inners, coriaceous, surfaces glabrous; corolla pink or white, throat purple or red inside, funnelform, 50–80 mm, limb 50–80 mm diam.

Flowering Jun–Aug. Plains, prairies; of conservation concern; 200–600 m; Kans., Okla., Tex.

Ipomoea shumardiana is known only from areas where *I. leptophylla* and *I. pandurata* are sympatric; the names *I. longifolia* and *I. pandurata* have been misapplied to plants of *I. shumardiana*.

SOLANACEAE Jussieu • Nightshade Family

Janet R. Sullivan

Herbs, annual or perennial, **vines, shrubs, or trees**, sometimes rhizomatous or tuberous, usually hermaphroditic, sometimes gynodioecious or dioecious in *Lycium* [andromonoecious or dioecious in some *Solanum*]. **Stems** erect or decumbent to prostrate, sympodially branched, glabrous or pubescent, hairs simple, dendritic, or stellate, stalked-vescicular in *Quincula*, sometimes peltate scales in *Solanum*, glandular or eglandular, glands sessile or stalked and sometimes multicellular, sometimes prickly or spiny. **Leaves** alternate and spiral or unequally paired, simple or pinnately compound [trifoliolate], estipulate, petiolate or sessile; blade margins entire, dentate, lobed, or divided. **Inflorescences** terminal, axillary, extra-axillary, in forks of dichotomous branches in some *Datura* and *Solanum*, racemose, paniculate, umbellate, or glomerulate cymes, in fascicles of 2–6, or solitary flowers. **Flowers** bisexual or rarely unisexual, radially symmetric or, less frequently, bilateral; perianth and androecium hypogynous; sepals (4 or)5, connate for almost full length to only at base, tubular, campanulate, obconic, or cyathiform [urceolate], with 3–10 equal or unequal lobes, lobules, or teeth, 5–10 veins extending into appendages that protrude below a truncate rim in *Lycianthes*, bases of lobes sagittate to cordate in *Nicandra*; calyx usually persistent in fruit and often accrescent, circumscissile and leaving a basal remnant that is slightly enlarged in some species of *Datura*, sometimes spreading, reflexed, or expanded to partially or completely enclose fruit, sometimes becoming membranous, sometimes inflated and bladderlike; petals (4 or)5, connate for almost full length to only at base, stellate, campanulate, campanulate-rotate, rotate, funnelform, salverform, or urceolate [crateriform, cylindric], margins nearly entire or shallowly to deeply 4–10-lobed, lobes sometimes widely flaring or reflexed, corona present between tube and limb in *Nectouxia*; stamens (2 or 4 plus staminodes or)5, variously adnate to corolla from near base nearly to rim, with appendage at point of fusion in *Cestrum*, equal or unequal in length, anthers sometimes connivent, basi-, dorsi-, or ventrifixed, with unequal thecae in *Bouchetia*, *Browallia*, and *Hunzikeria*, connective sometimes thickened, with 2 dorsal wings in *Nectouxia*, dehiscence longitudinal or poricidal; pistil 1, 2–5-carpellate; ovary superior, nectary disc sometimes present at base, [1 or]2–5[or 10]-locular, [1 or]2–many ovules per locule, placentation axile; style 1, stigma rounded-truncate, capitate, or 2–5-lobed. **Fruits** usually berries or 2–4-valved capsules, dehiscence septicidal, circumscissile, or irregular [septifragal, septicidal-loculicidal], sometimes drupaceous in *Lycium* [schizocarp of nutlets in

Nolana]. **Seeds** 1–many, whitish to straw-colored to dark brown or black, often flattened and discoidal or reniform, sometimes globose, ovoid, prismatic, or angulate, with hyaline margin in *Oryctes*, with white caruncle in some species of *Datura*, intermixed with sclerotic granules in *Calliphysalis* and some *Solanum*; embryo straight or curved, endosperm usually abundant.

Genera ca. 97, species ca. 2500 (27 genera, 151 species, including 2 hybrids, in the flora): nearly worldwide.

Solanaceae occur on all continents except Antarctica, but their greatest diversity is in the Neotropics. Their closest relative is the Convolvulaceae, from which they can be distinguished by growth form (when herbaceous vines, not with twining stems), the usually 2-locular ovary with numerous ovules, and the absence of laticifers with milky sap. Members of the Solanaceae exhibit a sympodial growth pattern, resulting in unusual placement of leaves, branches, and flowers (A. Child 1979). The foliage of some Solanaceae is fetid or pungent (reminiscent of rotting potatoes or with a sharp, tomato-like smell). The vegetative portions of most members of the family are poisonous due to the presence of tropane and steroidal alkaloids.

Despite the toxicity of its foliage, many Solanaceae species are of worldwide economic importance, such as tomato (*Solanum lycopersicum*), eggplant (*S. melongena*), potato (*S. tuberosum*), chili and bell peppers (*Capsicum* spp.), tobacco (*Nicotiana tabacum*), and *Petunia*. The family also includes many species of small-scale crop, medicinal, or horticultural importance. The most well known of these in the flora area are tomatillo (*Physalis philadelphica*), belladonna (*Atropa belladonna*), goji berry (*Lycium barbarum* and *L. chinense*), Chinese lantern plant (*Alkekengi officinarum*), jimsonweed (*Datura* spp.), jessamine (*Cestrum* spp.), and *Calibrachoa*.

The fruit in the Solanaceae is typically a berry or capsule (with circumscissile dehiscence in *Hyoscyamus*), but may be hardened and drupaceous in some *Lycium*. The berry-fruited taxa may have a thick, juicy pericarp, or one that is dry and thin, ultimately shattering into irregular pieces. Many genera have an accrescent calyx. The most enlarged of these may become rigid and spinescent (*Hyoscyamus*), inflated and almost closed around the berry (*Alkekengi*, *Calliphysalis*, *Physalis*, *Quincula*), loose and enveloping all (*Nicandra*) or at least the basal half of the berry (*Chamaesaracha*), or reflexed or flaring (*Jaltomata*). The brightly colored, juicy berries of many species are animal dispersed. The inflated, bladderlike calyces in several genera also aid in dispersal (short distance by water or ground level by wind).

The showy flowers of Solanaceae are pollinated by bees, flies, butterflies, moths, or hummingbirds foraging for nectar and/or pollen. *Solanum* does not produce nectar, and pollen is gathered by vibration (buzz pollination) or manipulation of the anthers. Flowers of *Cestrum*, *Datura*, and some *Nicotiana* emit a strong, sweet fragrance in the evening and are moth-pollinated.

Some Solanaceae grown as ornamentals have escaped and persisted for short periods but have not become established in the flora. *Nierembergia hippomanica* Miers var. *coerulea* R. Millan has been collected from disturbed sites in Texas (1958, 1962, 1998), and *N. scoparia* Sendtner was found along a roadside in Georgia in 1947. *Nierembergia* Ruiz & Pavon is a South American genus. *Brugmansia suaveolens* (Humboldt & Bonpland ex Willdenow) Sweet, also native to South America, was found in a ruderal yard (2001) and wooded ravine (1983) in Florida. *Brugmansia* has been treated as part of *Datura*, but it can be distinguished from that genus by its woody habit, pendulous flowers, and elongated, spineless fruits.

Since the early 1990s, our understanding of the circumscription of the Solanaceae, as well as phylogenetic relationships within the family, have been clarified using both morphological and molecular characters, particularly chloroplast DNA sequence data (W. G. D'Arcy 1991; A. T. Hunziker 2001; T. R. Martins and T. J. Barkman 2005; R. G. Olmstead and L. Bohs 2007; Olmstead and J. D. Palmer 1991, 1992; Olmstead et al. 1999, 2008; T. Särkinen et al.

2013). Based on molecular data, the family is considered monophyletic with the inclusion of *Nolana* Linnaeus f., *Schizanthus*, and *Sclerophylax* Miers. The subfamilies Solanoideae Burnett and Cestroideae Burnett, traditionally recognized on the basis of morphology alone, have been shown to be nonmonophyletic. Some generic concepts have been revised; changes relevant to the flora area are discussed under the various genera.

SELECTED REFERENCES Child, A. 1979. A review of branching patterns in the Solanaceae. In: J. G. Hawkes et al., eds. 1979. The Biology and Taxonomy of the Solanaceae. London. Pp. 345–356. D'Arcy, W. G. 1991. The Solanaceae since 1976, with a review of its biogeography. In: J. G. Hawkes et al., eds. 1991. Solanaceae III: Taxonomy, Chemistry, Evolution. Kew. Pp. 75–137. D'Arcy, W. G., ed. 1986. Solanaceae: Biology and Systematics. New York. Hawkes, J. G. et al., eds. 1979. The Biology and Taxonomy of the Solanaceae. London. Hawkes, J. G. et al, eds. 1991. Solanaceae III: Taxonomy, Chemistry, Evolution. Kew. Hunziker, A. T. 2001. Genera Solanacearum: The Genera of Solanaceae illustrated.... Ruggell. Keating, R. C., V. C. Hollowell, and T. B. Croat. 2005. A Festschrift for William G. D'Arcy: The Legacy of a Taxonomist. St. Louis. Knapp, S. 2002. Floral diversity and evolution in the Solanaceae. In: Q. C. B. Cronk et al., eds. 2002. Developmental Genetics and Plant Evolution. London. Pp. 267-297. Martins, T. R. and T. J. Barkman. 2005. Reconstruction of Solanaceae phylogeny using the nuclear gene SAMT. Syst. Bot. 30: 435–447. Nee, M. et al. 1999. Solanaceae IV: Advances in Biology and Utilization. Kew. Olmstead, R. G. et al. 1999. Phylogeny and provisional classification of the Solanaceae based on chloroplast DNA. In: M. Nee et al., eds. Solanaceae IV: Advances in Biology and Utilization. Kew. Pp. 111–137. Olmstead, R. G. et al. 2008. A molecular phylogeny of the Solanaceae. Taxon 57: 1159–1181. Olmstead, R. G. and L. Bohs. 2007. A summary of molecular systematic research in Solanaceae: 1982–2006. In: D. M. Spooner et al. eds. 2007. Solanaceae VI: Genomics Meets Biodiversity.... Leuven. [Acta Hort. 745.] Pp. 255–268. Olmstead, R. G. and J. D. Palmer. 1991. Chloroplast DNA and systematics of the Solanaceae. In: J. G. Hawkes et al., eds. 1991. Solanaceae III: Taxonomy, Chemistry, Evolution. Kew. Pp. 161–168. Olmstead, R. G. and J. D. Palmer. 1992. A chloroplast DNA phylogeny of the Solanaceae: Subfamilial relationships and character evolution. Ann. Missouri Bot. Gard. 79: 346–360. Rydberg, P. A. 1896b. The North American species of *Physalis* and related genera. Mem. Torrey Bot. Club 4: 297–372. Spooner, D. M. et al., eds. 2007. Solanaceae VI: Genomics Meets Biodiversity.... Leuven. [Acta Hort. 745.] Van den Berg, R. G. et al. 2001. Solanaceae V: Advances in Taxonomy and Utilization. Nijmegen.

1. Corollas campanulate, rotate, stellate, or urceolate; fruits berries (hardened or drupaceous in some *Lycium*).
   2. Flowers and fruits borne in subumbellate, umbellate, or cymose clusters.
      3. Flowers and fruits borne in umbellate clusters; anthers dehiscing longitudinally; calyces enlarged, with flaring lobes in fruit . . . . 14. *Jaltomata*, p. 386
      3. Flowers and fruits borne in subumbellate or cymose clusters; anthers dehiscing by pores (sometimes expanding to longitudinal slits with age); calyces only slightly enlarged in fruit.
         4. Shrubs; calyces 10-veined; leaves simple, margins entire. . . . . 16. *Lycianthes* (in part), p. 388
         4. Herbs, vines, subshrubs, shrubs, or trees; calyces 5-veined; leaves simple or pinnately compound, margins entire, dentate, or lobed . . . . 27. *Solanum*, p. 428
   2. Flowers and fruits solitary or in fascicles of 2–8.
      5. Stems creeping (rooting at nodes); anthers dehiscing by pores . . . . . . 16. *Lycianthes* (in part), p. 388
      5. Stems erect, decumbent, prostrate, or scandent; anthers dehiscing by longitudinal slits.
         6. Shrubs or scandent, perennial herbs; calyces not, or only sometimes, enlarged in fruit.
            7. Scandent, perennial herbs; flowers and fruits solitary; corollas urceolate . . . . 25. *Salpichroa*, p. 427
            7. Shrubs; flowers and fruits solitary or in fascicles of 2–8; corollas rotate, subrotate, or campanulate.
               8. Plants without spines. . . . . 7. *Capsicum*, p. 371
               8. Plants spinescent . . . . 17. *Lycium* (in part), p. 390
         6. Herbs erect or decumbent, annual or perennial; calyces enlarged in fruit.
            9. Corollas campanulate; calyx lobes enlarged and flaring in fruit, not enclosing berries . . . . 2. *Atropa*, p. 366
            9. Corollas broadly campanulate, campanulate-rotate, rotate, or urceolate; calyx or calyx lobes enlarged in fruit and completely or partly enclosing berries.

10. Calyx lobes with sagittate to cordate bases, lobes expanded and surrounding berries . . . . 19. *Nicandra*, p. 399
10. Calyx lobes without lobed bases, basal portions expanded and completely or partially enclosing berries.
  11. Plants sparsely to densely covered with stalked, white vesicles . . . . 24. *Quincula*, p. 425
  11. Plants glabrous or pubescent with simple or branched hairs.
    12. Stems erect, usually unbranched; corollas white, lobed; fruiting calyces drying orange-red or bright red, inflated and completely enclosing berries . . . . 1. *Alkekengi*, p. 365
    12. Stems erect to ascending or decumbent to prostrate, branched; corollas white, cream, yellow, greenish, or deep purple, unlobed or with obscure or shallow lobes or teeth; fruiting calyces drying brown, enlarged or inflated and completely or partially enclosing berries.
      13. Flowers and fruits solitary; fruiting calyces inflated and completely enclosing berries. . . . . 23. *Physalis*, p. 411
      13. Flowers and fruits borne singly or in fascicles of (1–)2–6; fruiting calyces expanded but not inflated, mostly or completely enclosing berries.
        14. Stems decumbent to ± prostrate; fruiting calyces shorter than berries; berries dry. . . . . 9. *Chamaesaracha*, p. 374
        14. Stems erect to ascending; fruiting calyces just shorter than or exceeding berries; berries fleshy.
          15. Corollas broadly campanulate; stamens slightly unequal; fruiting calyces enclosing berries; seeds intermixed with sclerotic granules. . . . . .6. *Calliphysalis*, p. 370
          15. Corollas rotate; stamens equal; fruiting calyces shorter than or just exceeding berries; seeds not intermixed with sclerotic granules. . . . . . .15. *Leucophysalis*, p. 386

1. Corollas funnelform, salverform, or tubular; fruits capsules or berries (hardened or drupaceous in some *Lycium*).
  16. Plants woody (subshrubs, shrubs, small trees, or lianas).
    17. Shrubs, spinescent. . . . . .17. *Lycium* (in part), p. 390
    17. Subshrubs, shrubs, small trees, or lianas, without spines.
      18. Inflorescences axillary panicles; fruits berries . . . . 8. *Cestrum*, p. 372
      18. Inflorescences axillary or terminal cymes, racemes, glomerules, or solitary flowers; fruits capsules.
        19. Shrubs or small trees, 0.2–6(–10) m; stamens 5, inserted near base of corolla . . . . 20. *Nicotiana* (in part), p. 400
        19. Shrubs or subshrubs to 1.5 m; stamens 4 (5th much reduced or absent), inserted in distal half of corolla.
          20. Corollas 5-lobed or sometimes appearing 4-lobed due to fusion of 2 abaxial lobes; capsules 2-valved; seeds prismatic or rounded on back and excavated ventrally. . . . . 4. *Browallia* (in part), p. 368
          20. Corollas 5-lobed; capsules 4-valved; seeds reniform and finely wrinkled . . . . .11. *Hunzikeria*, p. 382
  16. Plants herbaceous.

[21. Shifted to left margin.—Ed.]

21. Corollas bilaterally symmetric (or slightly bilateral); fruits capsules.
    22. Leaf margins usually deeply pinnately dissected or coarsely toothed or lobed.
        23. Leaf margins coarsely toothed to shallowly lobed; corolla lobes shorter than to as long as tube; capsules circumscissile . . . . 12. *Hyoscyamus*, p. 384
        23. Leaf margins pinnate-pinnatifid; corolla lobes longer than tube and often deeply dissected; capsules septicidal . . . . 26. *Schizanthus*, p. 427
    22. Leaf margins entire or irregularly crenate or undulate.
        24. Stems erect; stamens 5 . . . . 20. *Nicotiana* (in part), p. 400
        24. Stems erect to ascending, decumbent, or procumbent; stamens 4, sometimes with reduced or sterile 5th.
            25. Stems sprawling or procumbent; corollas funnelform; calyces not accrescent; capsules hemispheric, 2-valved . . . . 5. *Calibrachoa*, p. 369
            25. Stems erect or ascending to decumbent; corollas funnelform or salverform; calyces accrescent; capsules ellipsoid or ovoid, 2- or 4-valved.
                26. Corollas funnelform; capsules ellipsoid, 4-valved . . . . 3. *Bouchetia*, p. 366
                26. Corollas salverform; capsules ovoid, 2-valved . . . . 4. *Browallia* (in part), p. 368
21. Corollas radially symmetric; fruits capsules or berries.
    27. Corollas funnelform with 5 long-acuminate lobes; fruits 4-valved or irregularly dehiscing capsules, sometimes with prickles or tubercles . . . . 10. *Datura*, p. 378
    27. Corollas funnelform, salverform, or tubular, lobes not long-acuminate; fruits berries or 2–4-valved capsules, not prickly.
        28. Plants glabrous.
            29. Plants prostrate to ascending; inflorescences solitary or fascicled flowers; corolla lobes acute to acuminate; fruits berries . . . . 13. *Jaborosa*, p. 385
            29. Plants usually erect; inflorescences cymose, forming false racemes or glomerules; corolla lobes rounded to deltate; fruits capsules . . . . 20. *Nicotiana* (in part), p. 400
        28. Plants glandular-pubescent.
            30. Stamens equal or slightly unequal; fruits berries.
                31. Corollas salverform, with corona between tube and limb; berries narrowly ovoid, juicy . . . . 18. *Nectouxia*, p. 398
                31. Corollas tubular; berries globose, dry . . . . 21. *Oryctes*, p. 409
            30. Stamens unequal; fruits capsules.
                32. Flowers and fruits in cymose clusters (appearing as false racemes or glomerules); calyx lobes deltate or triangular . . . . 20. *Nicotiana* (in part), p. 400
                32. Flowers and fruits solitary; calyx lobes linear . . . . 22. *Petunia*, p. 409

1. ALKEKENGI Miller, Gard. Dict. Abr. ed. 4, vol. 1. 1754 • Chinese lantern plant, Japanese lantern, Jerusalem-cherry, strawberry ground-cherry, winter-cherry [Probably from Greek *halikakabon*, bladder, known in Persian as *al-kākunadj* and in Arabic as *hub-ul-kakinj*, name for a nightshade] [I]

Janet R. Sullivan

Herbs, perennial, rhizomatous, glabrous or sparsely pubescent, hairs simple, eglandular. **Stems** usually simple. **Leaves** alternate or geminate. **Inflorescences** axillary, solitary flowers. **Flowers** 5-merous; calyx campanulate, lobes 5, lanceolate, accrescent, inflated and completely enclosing berry; corolla white with pale green in throat, campanulate-rotate, limb widely flaring, lobes broad, blunt; stamens inserted at base of corolla tube, equal; anthers basifixed, oblong, dehiscing by longitudinal slits; ovary 2-carpellate; style slender, straight; stigma broadly capitate. **Fruits** berries, globose. **Seeds** reniform, flattened. $x = 12$.

Species 1: introduced; Eurasia, introduced also in Australia.

Until recently, *Alkekengi* was treated within *Physalis*; phylogenetic analysis showed that it does not fall within the strongly supported clade of New World *Physalis* (M. Whitson and P. S. Manos 2005). *Alkekengi* can be distinguished by its lobed, white corolla and orange-red or bright red fruiting calyx.

1. Alkekengi officinarum Moench, Suppl. Meth., 177. 1802 • Coqueret alkékenge [F] [I] [W]

Physalis alkekengi Linnaeus, Sp. Pl. 1: 183. 1753

Stems 2.5–9 dm. **Leaves:** petiole 0.6–4.4 cm; blade broadly ovate, (4–)6–11 × (2.5–)4–8.5 cm, margins entire or irregularly dentate. **Flowering pedicels** 9–13 mm. **Flowers:** calyx 4–7 mm, lobes 2–3.5 mm, tomentose; corolla 10–15 mm; anthers yellow, 2.5–3 mm. **Fruiting pedicels** 20–40 mm. **Berries** red, 1 cm diam., enclosed in papery, orange-red to bright red calyx, 3–5.5 × 2.5–4.5 cm. $2n = 24$.

Flowering Jun–Aug. Fence rows, thickets, vacant lots, cemeteries, roadsides, railroad tracks, stream banks; 0–600 m; introduced; B.C., Man., N.B., Ont., Que.; Conn., Del., Idaho, Ill., Iowa, Maine, Md., Mass., N.Y., Pa., Tenn., Vt., W.Va.; Eurasia; introduced also in Australia.

Early herbals mention the winter-cherry as a diuretic and a treatment for gout; it is used as a febrifuge in China. The ripe berries are edible. In the flora area, *Alkekengi officinarum* is most often cultivated as an ornamental. Stems with inflated, orange-red or bright red fruiting calyces are used in floral arrangements. The dried calyces maintain their color for long periods in bouquets. Escaped plants can spread aggressively via rhizomes, and specimen data indicate reproduction and dispersal by seed.

In its native range, two varieties of *Alkekengi officinarum* are recognized: var. *officinarum* and var. *franchetii* (Masters) R. J. Wang. Plants in the flora area, derived from cultivation and likely from multiple sources, may exhibit characteristics of both on a single specimen.

2. ATROPA Linnaeus, Sp. Pl. 1: 181. 1753; Gen. Pl. ed. 5, 85. 1754 • Belladonna [Greek *atropos*, inexorable or unchangeable, alluding to one of the three Moirai, goddesses of fate and destiny in Greek mythology] [I]

Zheng Li

Michael A. Vincent

Herbs, perennial, rhizomatous, glabrous or pubescent. **Stems** branching. **Leaves** alternate (sometimes geminate). **Inflorescences** axillary, solitary flowers. **Flowers** 5-merous; calyx accrescent, campanulate, 5-lobed, lobes acute to acuminate, widely flaring in fruit; corolla yellow or purple, radial, campanulate, lobes flaring to reflexed; stamens inserted at base of corolla tube, equal; anthers basifixed, ovate, (basally lobed), dehiscing by longitudinal slits; ovary 2-carpellate; style arched, curved; stigma capitate. **Fruits** berries, globose, fleshy. **Seeds** globose-reniform. $x = 12$.

Species 3 (1 in the flora): introduced; Europe, w, c Asia, n Africa; introduced also in South America.

1. Atropa belladonna Linnaeus, Sp. Pl. 1: 181. 1753 • Belle-dame 

Herbs 0.5–1.3 m, pubescent. **Leaves:** petiole 0.3–3 cm; blade broadly lanceolate to ovate, 2–23.7 × 0.8–7.8 cm, base cuneate, margins entire, apex acute. **Flowering pedicels** 0.8–2.1 cm. **Flowers:** calyx 0.6–1.5 × 0.5–2 cm; corolla purple, 2–3 × 1.2–2.2 cm; stamens 16–18 mm; anthers introrse; ovary 3 × 2 mm; style 14 mm; stigma 0.5 × 1 mm. **Fruiting pedicels** 2.7–3.9 cm. **Fruiting calyces** 1.2–1.7 × 2–2.6 cm. **Berries** usually black, rarely yellow, 1.1–1.2 × 1.1–1.4 cm. **Seeds** golden brown, 2 × 1.5 mm, testa alveolate. $2n = 72$.

Flowering Jun–Aug. Waste places; 0–600 m; introduced; Mich., N.J., Wash.; Europe; w Asia; n Africa; introduced also in South America.

The whole plant of *Atropa belladonna* is poisonous, containing hyoscyamine, scopolamine, and atropine (J. M. Rowson 1950). In herbal medicine, belladonna has been used for the treatment of headache, menstrual symptoms, peptic ulcer disease, inflammation, and motion sickness and may have some efficacy in treating such complaints as irritable bowel syndrome (C. Ulbricht et al. 2004). Atropine is used in ophthalmic medicine and was used as a nerve-gas antidote in the first Gulf War (T. R. Forbes 1977; D. J. Mabberley 2008).

Reports of *Atropa belladonna* from Newfoundland (P. J. Scott 1991), New York (R. S. Mitchell 1986), and Oregon (C. L. Hitchcock et al. 1955–1969) were not verified. Belladonna is known from California as an agricultural weed but is not established in that state.

3. BOUCHETIA de Candolle ex Dunal in A. P. de Candolle and A. L. P. P. de Candolle, Prodr. 13(1): 589. 1852 • Painted-tongue [For Dominique Bouchet, 1770–1845, French botanist]

Alexander Krings

Herbs, perennial, taprooted, pubescent, hairs glandular-capitate and eglandular. **Stems** decumbent or ascending, branched near base and at distal nodes. **Leaves** alternate, sessile; blade margins entire. **Inflorescences** terminal or pseudoterminal, appearing axillary, solitary flowers. **Flowers** 5-merous, slightly bilaterally symmetric; calyx accrescent, oblong-campanulate, lobes 5, narrowly deltate to oblong, tightly enclosing and slightly exceeding fruit; corolla white or white tinged with lavender, slightly bilateral, funnelform, lobes broadly ovate, suborbiculate, or rhombic, apically rounded; stamens 4, sometimes plus a 5th, inserted in proximal 1/2 of corolla

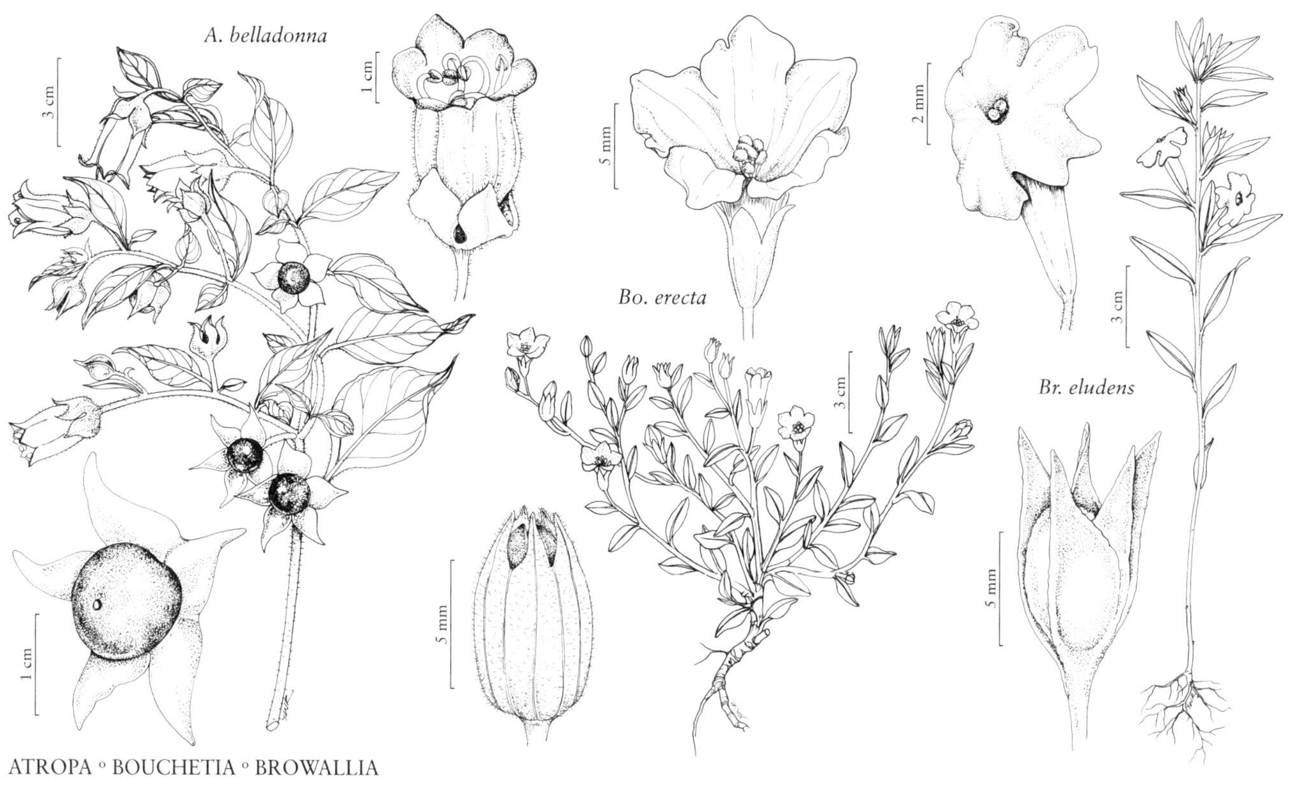

ATROPA ◦ BOUCHETIA ◦ BROWALLIA

tube, ± didynamous; anthers basifixed, ovate, ± unequal, that of 5th stamen sometimes smaller, dehiscing by longitudinal slits; ovary 2-locular; style filiform; stigma subreniform. **Fruits** capsules, ellipsoid, (4-valved). **Seeds** prismatic, faces irregular. $x = 8$.

Species 3 (1 in the flora): Texas, Mexico, Central America, South America (Argentina).

1. **Bouchetia erecta** de Candolle ex Dunal in A. P. de Candolle and A. L. P. P. de Candolle, Prodr. 13(1): 589. 1852 [F]

Salpiglossis erecta (de Candolle ex Dunal) D'Arcy

Herbs antrorsely strigillose, hairs of 2 types: glandular-capitate, spreading, translucent to white, frequently with darker tip, 0.1 mm, and eglandular, white, 0.1–0.5 mm. **Stems** to 25 cm. **Leaves** various, decreasing in size from stem base to apex, sparsely to moderately pubescent, basal leaf blades spatulate to oblanceolate, 1.6–4 × 0.4–1.4 cm, base narrowly attenuate, apex obtuse to acute, lateral veins in 2–3(–4) alternate to opposite pairs, arcuate to straight, distal cauline leaf blades linear to narrowly elliptic, 0.6–2.5 × 0.01–0.6 cm, base attenuate to cuneate, apex obtuse to acute, lateral veins not distinct. **Flowering pedicels** 6–22(–30) mm. **Flowers:** calyx tube 3.1–6.1 × 2.5–3.2 mm, to 4.9 mm diam. in fruit, inner surface glabrous near base, becoming as pubescent as outer surface near apex and on lobes, outer surface moderately to densely pubescent, lobes 2.1–5 × 1–2.3 mm, apically obtuse; corolla tube 8–13.9 × 5.4–11.2 mm, inner surface glabrous, outer surface sparsely to moderately pubescent, lobes 3.6–4.2 × 3.6–6.2 mm, inner surface glabrous, outer surface sparsely to moderately pubescent; filaments adnate to corolla tube 3.7–4.8 mm beyond base, 4.2–7.5 mm, glabrous; anthers connivent, 0.9–1.1 × 0.8–1 mm; ovary ellipsoid, 1.9–2.3 × 0.9–1.1 mm, glabrous; style 6.7–7.8 mm, glabrous; stigma greenish or yellowish, 0.6–1 × 1.3–2 mm. **Capsules** 6.8–8 × 4.3–4.9 mm, glabrous or sparsely to moderately pubescent, hairs glandular-capitate. **Seeds** brown, 0.7–1 × 0.6–0.8 mm, surfaces distinctly alveolate. $2n = 16$.

Flowering Mar–Nov. Woodland openings, alluvial terraces, coastal prairies, disturbed or managed pastures, roadsides, hay meadows, usually on well-drained calcareous substrates; 0–400 m; Tex.; n Mexico.

Bouchetia erecta is known from central and southern Texas. The single putative record of *B. erecta* from Mississippi (*Warren s.n.*, USMS) was found to be *Jacquemontia tamnifolia* (Convolvulaceae).

4. BROWALLIA Linnaeus, Sp. Pl. 2: 631. 1753; Gen. Pl. ed. 5, 278. 1754 • [For John Browall, 1707–1755, Bishop of Åbo, Sweden, and friend of Linnaeus]

Philip D. Jenkins†

Herbs, annual or perennial, **or shrubs,** taprooted or with fibrous roots, hairs simple or with multicellular glandular heads. **Stems** erect to decumbent and sprawling, branched. **Leaves** alternate, petiolate or sessile; blade simple, margins entire. **Inflorescences** axillary and terminal, solitary flowers or racemoid. **Flowers** 5-merous, bilaterally symmetric; calyx accrescent, tubular-campanulate, lobes 5, triangular, as long as or slightly longer than capsule, not inflated; corolla white, creamy white to yellowish, blue, or violet, bilateral (sometimes appearing 4-parted by fusion of abaxial 2 lobes), salverform, lobes spreading or reflexed, rounded; stamens 4, unequal, (abaxial 2 with 2 fertile thecae each, adaxial 2 with 1 fertile and 1 abortive theca each), sometimes also with 5th smaller stamen, abaxial 2 inserted in distal $1/2$ of corolla tube, adaxial 2 at mouth of corolla; anthers basifixed, spheric or subspheric, dehiscing by longitudinal slits; ovary 2-carpellate; style proximally slender, distally sigmoidally curved and broadened, (wrinkled); stigma capitate-bilobed. **Fruits** capsules, ovoid, (2-valved). **Seeds** (4–50), prismatic or concave abaxially and convex adaxially. $x = 10, 11$.

Species 5 (2 in the flora): Arizona, Mexico, Central America, South America, introduced in Asia, Africa, Indian Ocean Islands, Pacific Islands, Australia.

J. F. Macbride (1962) commented on the difficulty of circumscribing morphological species in *Browallia*, leading him to question the genetic status and taxonomy of species in the genus. He noted that the annuals, especially, had weedy tendencies whose considerable morphological diversity appeared to be influenced by environmental conditions. Plants of some species grown in the garden from wild-collected seed exhibit characteristics found in the widely cultivated *B. americana* after a single generation.

1. Plants glabrous or viscid-pubescent; corollas blue or violet with white or yellow center or entirely white, limb spreading; seeds to 1 mm, prismatic; escaped garden plants . . . . . 1. *Browallia americana*
1. Plants scabrous, hairs eglandular; corollas creamy white to yellowish, limb reflexed; seeds 1–1.3 mm, concave abaxially and convex adaxially; native and local in Arizona. . . . . . . .2. *Browallia eludens*

1. Browallia americana Linnaeus, Sp. Pl. 2: 631. 1753 • Jamaican forget-me-not, no-me-olvidas [I]

Stems 0.3–150 cm, glabrous or viscid-pubescent, hairs usually fine, simple. **Leaves:** petiole 0–5 cm; blade ovate to cordate. **Pedicels** 0–15 mm. **Flowers:** calyx length $1/4$–$1/2$ times corolla tube, glabrous or viscid-pubescent; corolla blue or violet with white or yellow center or entirely white, without hairs enclosing mouth, 1–12 cm, limb spreading. **Seeds** prismatic, to 1 mm, reticulate-foveolate. $2n = 22$.

Flowering Jun–Aug. Disturbed sites; 0–1000 m; introduced; Conn., Fla., Ga., La., Mass., Miss.; Central America; South America; introduced also in Mexico, Asia, Africa, Indian Ocean Islands, Pacific Islands, Australia.

Browallia americana is highly variable. It may be an ephemeral annual or a shrub. Native to the Andes of northwestern South America, and perhaps Central America, it has been introduced in tropical and subtropical areas around the world. The variable vegetative morphology in Peru and Ecuador has challenged taxonomists and resulted in over 30 synonyms (J. F. Macbride 1962; A. H. Gentry 1993; D. J. Mabberley 2008).

In the flora area, *Browallia americana* sometimes escapes from cultivation. The cultivated plants do not retain the variability found where they are native, and only blue-, violet-, and white-flowered forms are in the nursery trade.

2. **Browallia eludens** Van Devender & P. D. Jenkins, Madroño 40: 214, figs. 1–3. 1993 • Bush violet [C] [F]

Stems 1–25 cm, slightly scabrous, eglandular. **Leaves:** petiole 2–5 mm; blade rhombic. **Pedicels** 0–10 mm. **Flowers:** calyx length 9/10 corolla tube; corolla creamy white to yellowish, with hairs enclosing mouth, 0.5–0.8 cm, limb reflexed. **Seeds** concave abaxially and convex adaxially, 1–1.3 mm, foveolate.

Flowering Jul–Sep. Oak woodland-savanna; of conservation concern; 1500–1800 m; Ariz.; Mexico (Chihuahua, Sonora).

Browallia eludens is isolated geographically from the other species of *Browallia* and is known in the flora area only from Santa Cruz County.

Chloroplast phylogenies have placed *Browallia eludens* as more closely related to *Streptosolen jamesonii* (Bentham) Miers than to other (blue-flowered) *Browallia* (R. G. Olmstead and J. D. Palmer 1992; Olmstead et al. 2008). Occurrence of *B. eludens* in oak woodland-savanna is enigmatic when compared with other *Browallia* species; it is the only *Browallia* native to northwestern Mexico and the flora area. The reflexed corolla limb is unique in *Browallia*. Pollination may be by syrphid flies, especially species with nonretractable tubelike mouthparts, and, perhaps, by sphinx moths. The flowers are lightly scented. The plants are known to self-pollinate in drought conditions.

5. CALIBRACHOA Cervantes in P. de la Llave and J. M. de Lexarza, Nov. Veg. Descr. 2: 3. 1825 • [For Antonio de la Cal y Bracho, 1764/1766–1833, Spanish-born Mexican botanist and pharmacologist] [I]

Philip D. Jenkins†

Herbs, annual or perennial, viscid-pubescent [eglandular], roots fibrous or woody. **Stems** sprawling or procumbent, branched. **Leaves** alternate, (subopposite immediately proximal to flowers), petiolate [sessile]; blade ([membranous] fleshy), margins entire. **Inflorescences** axillary, solitary flowers. **Flowers** 5-merous, usually bilaterally symmetric; calyx not accrescent, tubular-obconic to campanulate, lobes 5, lanceolate; corolla white, white with blue to violet limb, blue, or violet [yellow], ± bilateral, funnelform [salverform or tubular], lobes rounded; stamens 4, inserted in abaxial 1/2 of corolla tube, didynamous plus 5th smaller, sterile filament; anthers oblong, ventrifixed, dehiscing by longitudinal slits; ovary 2-carpellate; style (not exserted), sigmoidally curved, proximally slender, distally expanded; stigma 2-lobed [capitate]. **Fruits** capsules, hemispheric, (2-valved). **Seeds** (130–1200), ovoid [spheric to subreniform] (foveolate-reticulate). $x = 9$.

Species ca. 30 (1 in the flora): introduced; South America, introduced also in Mexico, Central America, elsewhere in South America.

Diversity of *Calibrachoa* is greatest in South America, especially southern Brazil. Members of *Calibrachoa* were previously incorporated in *Petunia* and can be found there in some references.

1. **Calibrachoa parviflora** (Jussieu) D'Arcy, Phytologia 67: 465. 1989 • Seaside petunia [F] [I] [W]

Petunia parviflora Jussieu, Ann. Mus. Natl. Hist. Nat. 2: 216, plate 47, fig. 1. 1803

Stems 0.1–1 m, internodes sometimes relatively long. **Leaf blades** deflexed in fruit, elliptic-spatulate, 2–6(–10) mm, fleshy. **Flowers:** calyx 5–10 mm; corolla 5–15 mm. **Capsules** 3–5 mm, calyx associated with developing fruit. **Seeds** pale brown, 0.5–0.8 mm.

Flowering Mar–Nov. Disturbed, sandy soils, coastal dunes, sandy to muddy margins of seasonal wetlands, reservoirs, ballast; 0–1500 m; introduced; Ala., Ariz., Calif., Colo., Fla., Ga., Kans., La., Md., Miss., Nev., N.J., N.Mex., N.C., S.C., Tex., Utah, Va.; South America (Argentina, Paraguay, Uruguay); introduced also in Mexico, Central America (Belize, Costa Rica, El Salvador, Guatemala, Honduras, Panama), elsewhere in South America (Bolivia, Brazil).

Stems, leaves, and flowers of *Calibrachoa parviflora* are occasionally encrusted with sand or soil particles captured by the glandular indument.

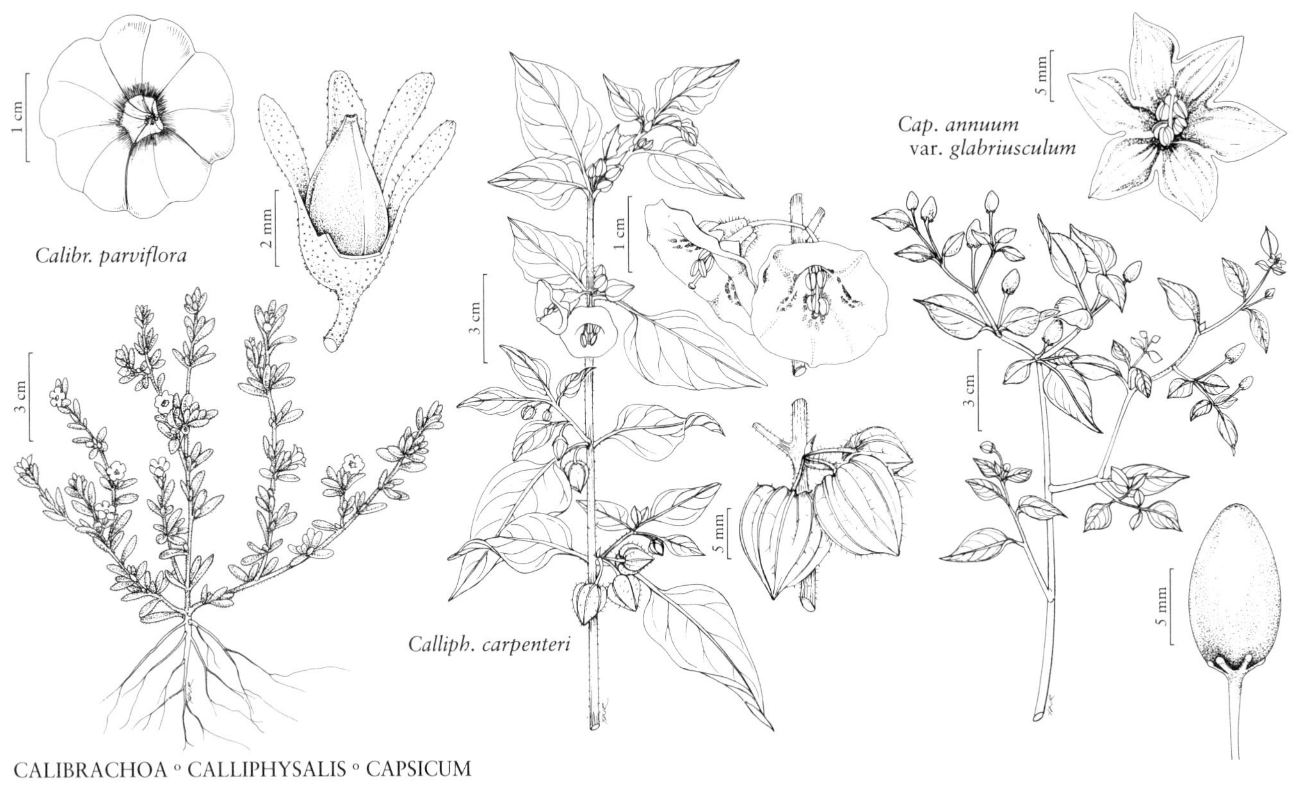

CALIBRACHOA ◦ CALLIPHYSALIS ◦ CAPSICUM

6. CALLIPHYSALIS Whitson, Rhodora 114: 137, figs. 1–3. 2012 • Carpenter's groundcherry [Greek *calli*, beautiful, and genus *Physalis*, alluding to resemblance] E

Maggie Whitson

Herbs, perennial, pubescent, hairs simple, mostly glandular; taproot fleshy. **Stems** branching. **Leaves** alternate. **Inflorescences** axillary, fasciculate, 2–6-flowered. **Flowers** 5-merous; calyx campanulate, enclosing berry, lobes 5, accrescent, broadly triangular; corolla yellow with 5 olive green spots at base of throat, radial, broadly campanulate; stamens inserted at base of corolla tube; anthers basifixed, oblong, dehiscing by longitudinal slits; ovary 2-carpellate; style filiform; stigma capitate. **Fruits** berries, globose, fleshy. **Seeds** discoid to reniform (intermixed with sclerotic granules). $x = 12$.

Species 1: se United States.

Calliphysalis was originally considered a morphologically odd species of *Physalis*. It shares an enlarged fruiting calyx and yellow corollas with *Physalis*, but its perennial taproot is unique and the clustered flowers atypical. Molecular data (M. Whitson and P. S. Manos 2005; R. G. Olmstead et al. 2008) do not place *Calliphysalis* within the group of North American *Physalis* species, supporting its recognition as a distinct physaloid genus.

1. Calliphysalis carpenteri (Riddell) Whitson, Rhodora 114: 330. 2012 [E] [F]

Physalis carpenteri Riddell, New Orleans Med. Surg. J. 9: 610. 1853 (as carpenterii)

Stems terete, 6–7.5(–10) dm, pubescent, viscid, becoming slightly woody at base. **Leaves:** petiole 1.5–5.5(–9.5) cm; blade ovate to ovate-lanceolate or elliptic, 5–10(–14) × 3–6(–9.5) cm, margins entire or sinuate-dentate or irregularly dentate with 1–3(–5) teeth per side, pubescent. **Flowering pedicels** 5–10 mm. **Flowers:** calyx shallowly incised, 4–7.5 mm; corolla 0.8–1.5 × 1–1.5(–2.5) cm; stamens included; anthers yellow; style 8–10 mm. **Berries** yellow, 2-locular, 6–10 mm diam., enclosed in papery, expanded calyx, 1.2–2.1 × 1–1.9 cm. **Seeds** dark brown, 1–1.5 mm. $2n = 24$.

Flowering (late Apr–)Jun–Aug(–Sep). Disturbed woodland areas, bluffs, clearings, trailsides, dump sites, animal diggings; 0–200 m; Ala., Fla., Ga., La., Miss.

7. CAPSICUM Linnaeus, Sp. Pl. 1: 188. 1753; Gen. Pl. ed. 5, 86. 1754 • Chili pepper, aji [Greek *kapsa*, box or capsule, alluding to fruit]

W. Hardy Eshbaugh

Shrubs [rarely trees], rhizomatous, usually glabrous or glabrescent, sometimes sparsely or densely pubescent. **Stems** erect or spreading, dichotomously branched, often hollow. **Leaves** alternate. **Inflorescences** axillary, solitary flowers or 2–3[4–6]-flowered fascicles. **Flowers** 5–7-merous; calyx not expanded or accrescent, cyathiform or campanulate, margins truncate, entire, ± undulated, or with 5–10 teeth of 2 lengths, base somewhat enlarged in fruit, not expanded or accrescent; corolla pure white, greenish white, bluish white, yellow, or purple, sometimes with tan or greenish yellow markings, radial, rotate, subrotate, or campanulate, stellate or 5-angled or -lobed; stamens inserted at base of corolla, equal [unequal], [with or without paired appendages at base of each filament]; anthers basifixed, oblong to lanceolate, dehiscing by longitudinal slits; ovary 2-carpellate; style cylindric, elongate, slender, usually longer than ovary [except in species with stylar heteromorphism]; stigma capitate. **Fruits** berries, globose or ovoid and elongated, dry to fleshy. **Seeds** (5–45), yellow to cream or brown, flattened, reniform or subreniform. $x = 12, 13$.

Species 40 (1 in the flora): s United States, Mexico, West Indies, Central America, South America (north of Amazon River); introduced nearly worldwide.

Capsicum is included in the subtribe Capsicinae (T. Yamazaki 1993), which is characterized chiefly by its prominent stapet (auriculate, bidenticulate, or winged appendages at the filament bases).

Capsicum is native in the Americas; some species are cultivated worldwide for food, spice, or medicine. In cultivation, *Capsicum* is typically a herbaceous annual with five- to eight-merous flowers. Two domesticated species have been introduced into the United States: *C. annuum* and *C. baccatum* var. *pendulum* (Willdenow) Eshbaugh. A. T. Hunziker (2001) considered the cultivated *C. annuum*, *C. chinense*, and *C. frutescens* species complex to be a single species, *C. annuum*. W. H. Eshbaugh (2012) discussed this issue at length and adopted Hunziker's position.

SELECTED REFERENCES Eshbaugh, W. H. 2012. The taxonomy of the genus *Capsicum* (Solanaceae). In: V. M. Russo, ed. 2012. Peppers: Botany, Production and Uses. Wallingford. Pp. 14–22. Walsh, B. M. and S. B. Hoot. 2001. Phylogenetic relationships of *Capsicum* (Solanaceae) using DNA sequences from two noncoding regions: The chloroplast *atpB-rbcL* spacer region and the nuclear *waxy* introns. Int. J. Pl. Sci. 162: 1409–1418.

1. Capsicum annuum Linnaeus, Sp. Pl. 1: 188. 1753 [F]

Varieties ca. 20 (1 in the flora): s United States, Mexico, West Indies, Central America, South America (north of Amazon River); introduced nearly worldwide.

1a. Capsicum annuum Linnaeus var. **glabriusculum** (Dunal) Heiser & Pickersgill, Baileya 19: 156. 1975 • Bird pepper, chiltepin, chili piquin, tepin [F]

Capsicum hispidum Dunal var. *glabriusculum* Dunal in A. P. de Candolle and A. L. P. P. de Candolle, Prodr. 13(1): 420. 1852; *C. annuum* var. *aviculare* (Dierbach) D'Arcy & Eshbaugh; *C. annuum* var. *minus* (Fingerhuth) Shinners; *C. minimum* Miller

Stems 0.3–2 m, branched, frequently purple-striate, slender, usually glabrous, rarely puberulent. **Leaves:** petiole (4–)6–7(–11) mm; blade ovate or elliptic-ovate, (20–)25–37(–45) × (10–)14–16(–20) mm. **Flowering pedicels** usually 1 per node, rarely 2–3 per node, (4–)10–14(–25) mm. **Flowers:** calyx mostly truncate with small umbos in place of teeth; corolla usually white, rarely greenish, 8–12 mm diam., lobes 5–7 mm, sinuses 1–5 mm deep; anthers violet to blue; ovary globose; style short-capitate. **Berries** deciduous, erect, red, globose to ovoid, 5–10 mm diam., rarely 14+ mm long. **Seeds** cream to yellow. $2n$ = 24, 48.

Flowering Mar–Oct, sporadically year-round. Fence rows, pastures, shell mounds, hammocks, waste places, well-drained soils, silty and sandy loams [coffee plantations]; 0–300 m; Ala., Ariz., Fla., La., Tex.; Mexico; West Indies; Central America; South America (north of the Amazon River).

Variety *glabriusculum* is considered to be the progenitor of domesticated var. *annuum*. The center of origin of the bird pepper is believed to be in southern Mexico, and its native range is from the southeastern and southwestern United States and the Caribbean to northern Peru. Domesticated var. *annuum* (tabasco pepper) is grown throughout warm temperate North America. Plants sometimes escape and persist for years (for example, California, Florida, Missouri, and New Mexico) but are unlikely to become established in the flora area.

The epithet *baccatum* has been associated erroneously with what is recognized here as var. *glabriusculum*. *Capsicum baccatum* Linnaeus is an entirely different species native in South America that has distinct, paired, yellow to tan to greenish markings on each lobe of the corolla.

In 1999, Native Seeds/SEARCH and the United States Forest Service established a 1000-hectare wild chili botanical area and reserve located in Rock Corral Canyon near Tumacacori, Arizona. The preserve protects a large population of var. *glabriusculum* as an in-situ genetic (germplasm) reserve.

8. CESTRUM Linnaeus, Sp. Pl. 1: 191. 1753; Gen. Pl. ed. 5, 88. 1754 • Jessamine [Etymology uncertain, perhaps Greek *kestra*, a kind of hammer, alluding to corolla shape] [I]

Alexandre K. Monro

Shrubs, trees, or lianas, glabrous or pubescent, hairs simple or branched. **Stems** erect or lax, sparsely branched from base. **Leaves** alternate. **Inflorescences** axillary (sometimes clustered in leaf axils, often bracteate or bracteolate) [terminal], paniculate [racemose]. **Flowers** 5-merous, radially symmetric to slightly bilateral; calyx campanulate or tubular, lobes 3–5, acute to linear (equal or unequal), expanding slightly in fruit; corolla white, pale yellow, pale green, or yellow-green, [red, pink, or orange], radial, tubular (tube frequently expanded around anthers), lobes 4 or 5, deltate to acute; stamens equal or subequal, inserted at varying levels in corolla tube, filaments frequently pubescent, frequently with an appendage at point of fusion to corolla; anthers dorsifixed, oblong to ellipsoid, dehiscing by longitudinal slits; ovary 2-carpellate (2- or 4-locular); style slender, usually surpassing stamens; stigma entire or 2-lobed, rarely exserted. **Fruits** berries, often juicy, globose, ovoid, or oblong. **Seeds** oblong to angulate. x = 8.

Species ca. 175 (3 in the flora): introduced; Mexico, Central America, South America, West Indies (Greater Antilles).

Some species of *Cestrum* are cultivated as ornamentals in warmer parts of the United States. Three are established in the flora area, and *C. aurantiacum* Lindley and *C. fasciculatum* (Schlechtendal) Miers may become established; they are shrubs or trees (to 4–5 m) and have brightly colored corollas (orange in *C. aurantiacum* and pink or red in *C. fasciculatum*). Berries of *C. aurantiacum* are white; those of *C. fasciculatum* are red. *Cestrum fasciculatum* may also be known as *C. elegans* Francey, an illegitimate homonym.

SELECTED REFERENCE Francey, P. 1935. Monographie du genre *Cestrum* L. Candollea 6: 46–398.

1. Axillary branches not subtended by minor leaves; berries ripening white. . . . . 2. *Cestrum nocturnum*
1. Axillary branches usually subtended by 1–3 minor leaves; berries ripening dark purple or black.
   2. Calyces 3–4 mm, lobes 0.5–0.8 mm; corollas 11–16 mm, lobes 1.7–2.3 mm . . . . 1. *Cestrum diurnum*
   2. Calyces 4.5–5.5 mm, lobes 1.2–1.5 mm; corollas 18–22 mm, lobes 4–4.5 mm . . . . 3. *Cestrum parqui*

1. Cestrum diurnum Linnaeus, Sp. Pl. 1: 191. 1753 • Day-blooming jessamine F I W

Shrubs, 0.5–4 m; stems densely to sparsely pubescent, glabrescent; axillary branches usually subtended by minor leaf. **Leaves:** petiole 5–24 mm; blade elliptic, oblong, ovate, oblong-ovate, or obovate, 2.5–12 × 0.6–3.6 cm. **Inflorescences** 1–3 per axil, each cluster 3–11-flowered. **Flowers:** calyx 3–4 × 1.7–2.3 mm, lobes 5, erect, 0.5–0.8 mm; corolla pale yellow or white, 11–16 mm, lobes 1.7–2.3 mm. **Berries** purple to black, 6–11 × 4.5–9 mm. **2*n*** = 16.

Flowering late spring–mid-summer. Secondary scrub, forest edges, roadsides; 0–10 m; introduced; Fla.; Mexico (Chiapas, Quintana Roo, Yucatán); West Indies (Cuba); Central America (Guatemala).

Cestrum diurnum has been introduced as an ornamental shrub into most of tropical and subtropical America, and is considered to be an environmental weed (R. P. Randall 2012). It is listed as invasive or potentially invasive in Florida.

2. Cestrum nocturnum Linnaeus, Sp. Pl. 1: 191. 1753 • Night-blooming jessamine I

Shrubs or trees, 1–12 m; young stems sparsely pubescent, hairs glandular; axillary branches not subtended by minor leaf. **Leaves:** petiole 5–18 mm; blade ovate, elliptic, or ovate-elliptic, 3.7–21 × 1.4–8.5 cm. **Inflorescences** 1–3 per axil, each cluster 1–4-flowered. **Flowers:** calyx 2.2–3.5 × 1.2–1.8 mm, lobes 5, erect or spreading, 0.5–1 mm; corolla pale yellow to pale green, 16–24 mm, lobes 2–4.5 mm. **Berries** white, 5–11 × 5–9 mm. **2*n*** = 16.

Flowering mid-summer–winter. Secondary scrub, forest edges, roadsides; 0–200 m; introduced; Calif., Fla., La.; Mexico (Chiapas, Morelos, Oaxaca, San Luis Potosí, Yucatán); Central America (Nicaragua, Panama); South America (Brazil, Colombia, Venezuela).

Cestrum nocturnum is considered to be an agricultural and environmental weed (R. P. Randall 2002).

3. Cestrum parqui L'Héritier, Stirp. Nov. 4: 73, plate 36. 1788 • Chilean jessamine I

Shrubs, to 2 m; stems glabrous or pubescent; axillary branches usually subtended by 1–3 minor leaves. **Leaves:** petiole 6–10 mm; blade narrowly ovate, narrowly elliptic, narrowly ovate-elliptic, or narrowly oblong-elliptic, 4.1–8.7 × 1.2–3 cm. **Inflorescences** 1 per axil, each cluster 1–6-flowered. **Flowers:** calyx 4.5–5.5 × 2–2.8 mm, lobes 3 or 5, erect, 1.2–1.5 mm; corolla pale yellow to pale green, 18–22 mm, lobes 4–4.5 mm. **Berries** black, 7–9 × 3–6.5 mm. **2*n*** = 16.

Flowering mid-summer–winter. Disturbed forest, secondary scrub, riversides; 60–400 m; introduced; Calif., Fla., Tex.; South America (Argentina, Bolivia, Chile, Paraguay, Peru, Uruguay).

Cestrum parqui has the potential to become an invasive species where it has escaped in the United States.

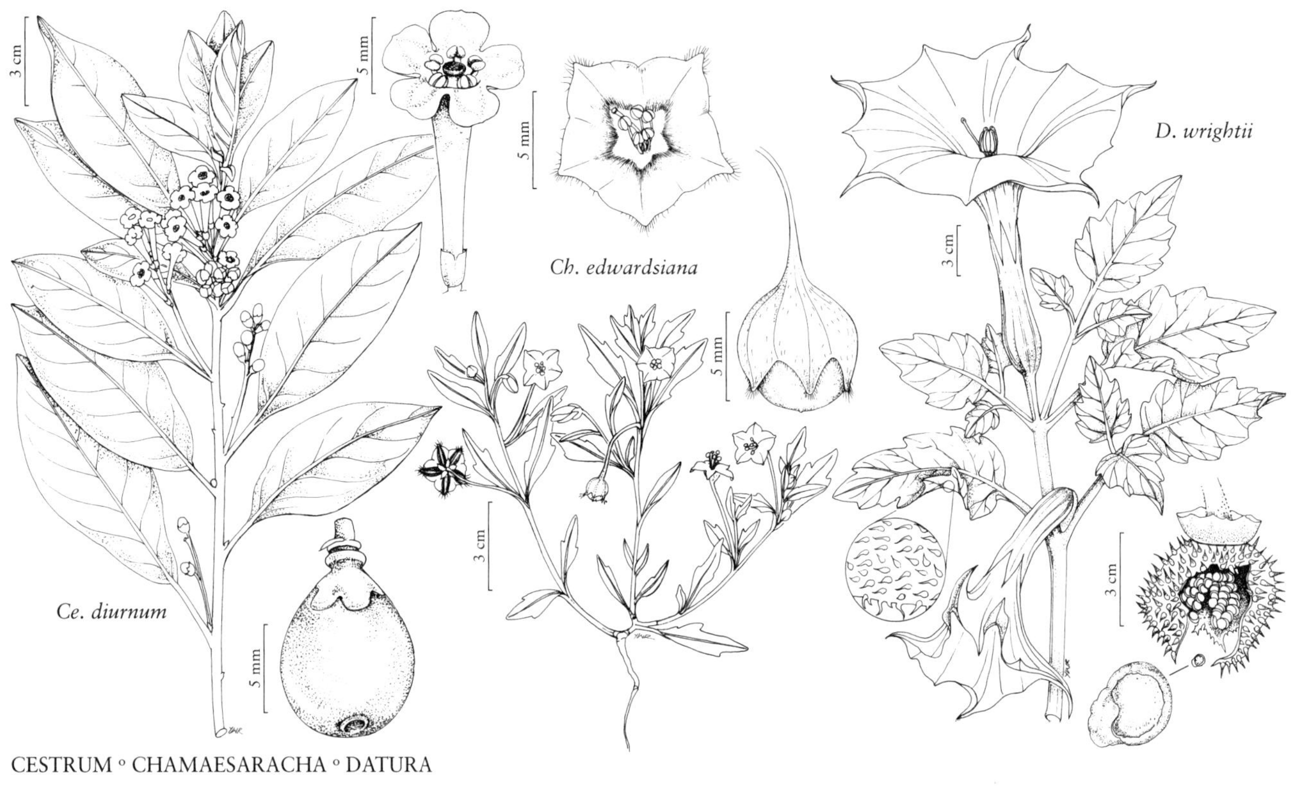

CESTRUM ◦ CHAMAESARACHA ◦ DATURA

9. CHAMAESARACHA (A. Gray) Bentham & Hooker f., Gen. Pl. 2: 891. 1876 • Five eyes [Greek *chamai*, on the ground, and genus *Saracha*, alluding to low habit and similarity]

John E. Averett†

Saracha Ruiz & Pavon [unranked] *Chamaesaracha* A. Gray, Proc. Amer. Acad. Arts 10: 62. 1874

Herbs, perennial, spreading, rhizomatous, glabrous or variously pubescent, hairs eglandular or glandular. **Stems** decumbent to ± prostrate or suberect, branching from base and nodes. **Leaves** alternate, subsessile or petiolate; blade simple, ± undulate, entire to deeply lobed. **Inflorescences** axillary, (1–)2–4(–5)-flowered clusters; pedicels slender, 1–3 cm, elongating to 2.5–3.5 and becoming curved in fruit. **Flowers** 5-merous; calyx accrescent, campanulate, 5-lobed, in fruit not inflated and shorter than berry; corolla creamy white to light yellow, rotate; stamens equal; filaments inserted near base of corolla tube; anthers basifixed, oblong, dehiscing by longitudinal slits; ovary 2-carpellate; style straight to slightly curved, slender; stigma capitate. **Fruits** dry berries, globose, tightly invested by, not enclosed by, accrescent calyx. **Seeds** flattened, reniform rugose-reticulate. $x = 12$.

Species 10 (8 in the flora): sc, sw United States, n Mexico.

Chamaesaracha is commonly encountered in the Sonoran and Chihuahuan deserts and arid grasslands of the southwestern United States and northern Mexico. It is distinguished from other Physalideae Miers by the dry berry and closely appressed, accrescent calyx that does not completely enclose the fruit, and by rugose-reticulate seeds. *Chamaesaracha nana* (A. Gray) A. Gray was transferred to *Leucophysalis* (J. E. Averett 1970). *Chamaesaracha geohintonii* Averett & B. L. Turner and *C. rzedowskiana* Hunziker occur in Mexico.

SELECTED REFERENCES Averett, J. E. 1973. Biosystematic study of *Chamaesaracha* (Solanaceae). Rhodora 75: 325–365. Turner, B. L. 2015. Taxonomy of *Chamaesaracha* (Solanaceae). Phytologia 97: 226–245.

1. Leaf blades 4–10 times as long as wide; herbage glabrous or sparsely pubescent, hairs white and short-stellate . . . . 1. *Chamaesaracha coronopus*
1. Leaf blades 2–5 times as long as wide; herbage glabrous to densely pubescent, hairs simple, 1-branched, or dendritic.
   2. Herbage glabrous or glabrate . . . . 7. *Chamaesaracha edwardsiana*
   2. Herbage pubescent (rarely glabrous in *C. pallida*).
      3. Herbage with hairs dendritic, eglandular (rarely glabrous in *C. pallida*).
         4. Leaf blades broadly lanceolate to rhombic, margins entire to sinuate or lobed . . . . 8. *Chamaesaracha pallida*
         4. Leaf blades linear-lanceolate or oblanceolate to rhombic, margins shallowly to deeply lobed (sometimes only a few lobes).
            5. Leaf blade margins deeply lobed; Oklahoma, nc Texas . . . . 3. *Chamaesaracha darcyi*
            5. Leaf blade margins shallowly to deeply lobed; Colorado, Kansas, New Mexico, Oklahoma, Texas. . . . . 4. *Chamaesaracha coniodes* (in part)
      3. Herbage pubescent with hairs simple or 1-branched, glandular or eglandular.
         6. Stems 1–3 dm; leaf blades 0.4–2 cm wide; Arizona, California, Colorado, Kansas, New Mexico, Oklahoma, Texas.
            7. Herbage densely glandular-pubescent; leaf blade margins sinuate or lobed . . . . 2. *Chamaesaracha sordida*
            7. Herbage densely pubescent, hairs glandular and eglandular; leaf blade margins ± lobed . . . . 4. *Chamaesaracha coniodes* (in part)
         6. Stems (1.5–)2–5 dm; leaf blades 1.5–3.5 cm wide; Big Bend region of Texas.
            8. Herbage not villous, hairs simple, long and eglandular intermixed with shorter glandular; petioles 1/3 total leaf length . . . . 5. *Chamaesaracha crenata*
            8. Herbage villous, hairs mostly elongate, eglandular, 1-branched at tip; petioles to 1/4 total leaf length . . . . 6. *Chamaesaracha villosa*

1. Chamaesaracha coronopus (Dunal) A. Gray in W. H. Brewer et al., Bot. California 1: 540. 1876 • Greenleaf five eyes [W]

Solanum coronopus Dunal in A. P. de Candolle and A. L. P. P. de Candolle, Prodr. 13(1): 64. 1852; *Chamaesaracha arida* Henrickson; *C. felgeri* B. L. Turner; *Saracha coronopus* (Dunal) A. Gray

Stems decumbent to ± erect, often cinerescent or purplish, 1–5 dm. **Herbage** glabrous or sparsely pubescent, hairs white and short-stellate, eglandular. **Leaves** subsessile; blade linear-lanceolate, 2–10 × 0.5–1.5 cm, length 4–10 times width, margins sinuate to deeply lobed. **Inflorescences** 1–2-flowered. **Flowers:** calyx 4–5 mm, pubescent, especially along lobe margins; corolla 10–15 mm diam. **Berries** 5–8 mm diam. $2n$ = 48, 72.

Flowering Mar–Oct (mostly late spring–early summer, depending on rain). Dry, open grasslands, coniferous woodlands, sagebrush scrub; 0–2300 m; Ariz., Calif., Colo., N.Mex., Okla., Tex., Utah; Mexico (Chihuahua, Coahuila, Nuevo León, Sonora, Tamaulipas, Zacatecas).

Chamaesaracha coronopus is widespread and frequent within its range. The species is highly variable in leaf shape, vestiture, and stature; it is characterized by the long-linear, deeply incised leaf blades four or more times as long as wide. Plants are glabrous or slightly pubescent, with short-stellate hairs that appear as tufts emerging from the leaf surface. Most populations of *C. coronopus* have $2n$ = 48; populations with $2n$ = 72 are known from southern Texas, the Big Bend region of western Texas, and southern Arizona. J. Henrickson (2009) described *C. arida* from New Mexico, citing only the type; that entity and populations comparable to it are here included in *C. coronopus*. B. L. Turner (2015) described *C. felgeri* from one population in southern Arizona and cited another collection from southern Texas; he suggested that some might consider *C. felgeri* an edaphic ecotype. With the data available, that suggestion is favored and those populations are included within *C. coronopus*.

2. Chamaesaracha sordida (Dunal) A. Gray in A. Gray et al., Syn. Fl. N. Amer. 2(1): 232. 1878 • Hairy five eyes

Withania sordida Dunal in A. P. de Candolle and A. L. P. P. de Candolle, Prodr. 13(1): 456. 1852

Stems decumbent to suberect, green, 1–3 dm (1.5 mm diam.). **Herbage** densely pubescent, viscid, hairs simple, mostly glandular. **Leaves:** subsessile; blade oblanceolate to rhombic, 1.5–4 × 0.4–0.8(–1) cm, length 4–5 times width, margins sinuate or lobed. **Inflorescences** 1–2-flowered. **Flowers:** calyx 4–5 mm, pubescent, especially along lobe margins; corolla 10–15 mm diam. **Berries** 5–8 mm diam. **2*n*** = 24.

Flowering Mar–Oct (mostly late spring–early summer, depending on rain). Deserts, dry, open grasslands; 0–2000 m; Ariz., Calif., N.Mex., Tex.; Mexico (Chihuahua, Coahuila, Nuevo León, Tamaulipas).

Chamaesaracha sordida is known from western Texas, New Mexico, and Arizona. It is most closely related to *C. coronopus* and differs in having broader, mostly entire leaves. It is also similar to *C. pallida*; it lacks the relatively dense dendritic hairs on the leaves and has a generally more eastern distribution. Herbarium specimens from 2007 and 2008 indicate that it has been introduced into southern California (Clark Mountain Range, San Bernardino County) and may be naturalizing.

3. Chamaesaracha darcyi Averett, Monogr. Syst. Bot. Missouri Bot. Gard. 104: 350. 2005 [E]

Stems ± prostrate to decumbent, pale green, 1–3 dm. **Herbage** pubescent, hairs dendritic, eglandular. **Leaves** subsessile; blade linear-lanceolate to rhombic, 2–5 × 1–2 cm, length 2–2.5 times width, margins deeply lobed. **Inflorescences** 1–5-flowered. **Flowers:** calyx 4–5 mm, densely pubescent, hairs mostly relatively long; corolla 5–15 mm diam. **Berries** 8–10 mm diam. **2*n*** = 48.

Flowering Mar–Oct (mostly late spring–early summer, depending on rain). Dry grasslands, prairies; 100–400 m; Okla., Tex.

Chamaesaracha darcyi is known from the Rolling Plains of north-central Texas and extends to the Cross Timbers region of Texas and Oklahoma. It is very close to *C. coniodes*, having a dense vestiture of branched, dendritic hairs like those found on the type of *C. coniodes*; *C. darcyi* typically has more deeply incised leaf margins and a nearly prostrate habit.

4. Chamaesaracha coniodes (Moricand ex Dunal) Bentham & Hooker f. ex B. D. Jackson in B. D. Jackson et al., Index Kew. 1(1): 505. 1893 • Gray five eyes

Solanum coniodes Moricand ex Dunal in A. P. de Candolle and A. L. P. P. de Candolle, Prodr. 13(1): 64. 1852; *Chamaesaracha texensis* Henrickson

Stems decumbent to ± prostrate, pale green, 1–3 dm. **Herbage** pubescent, viscid, hairs simple, dendritic in some populations, especially in s Texas, eglandular and glandular. **Leaves** subsessile; blade linear-lanceolate or oblanceolate to rhombic, 2–6 × 0.5–2 cm, length 3–4 times width, margins shallowly to deeply lobed (occasionally only a few shallow lobes). **Inflorescences** 1–5-flowered. **Flowers:** calyx 4–5 mm, densely pubescent, hairs mostly relatively long; corolla 5–15 mm diam. **Berries** 8–10 mm diam. **2*n*** = 48.

Flowering Mar–Oct (mostly late spring–early summer, depending on rain). Deserts, grasslands; 0–2000 m; Colo., Kans., N.Mex., Okla., Tex.; Mexico (Chihuahua, Coahuila, Nuevo León, Tamaulipas).

Chamaesaracha coniodes is highly variable in leaf shape, vestiture, and stature. It is similar to *C. darcyi* and *C. coronopus*. Some populations are similar to *C. sordida*. The variation and its significance were discussed by J. E. Averett (2010b). J. Henrickson (2009) recognized *C. texensis*, which has simple hairs and may also have an understory of glandular hairs, as distinct from the type of *C. coniodes*, which has only dendritic hairs; this is part of the variability across the range of the species.

5. Chamaesaracha crenata Rydberg, Mem. Torrey Bot. Club 4: 368. 1896 (as Camaesaracha) • Toothed five eyes

Stems decumbent, pale green, (1.5–)3–4 dm. **Herbage** pubescent, viscid, hairs simple, long and eglandular intermixed with shorter glandular. **Leaves:** petiole 1/3 total leaf length; blade ovate to broadly rhombic, 4–6 × 1.5–3.5 cm, length 2–2.5 times width, margins crenate or sinuate. **Inflorescences** 1–5-flowered. **Flowers:** calyx 5–7 mm, densely pubescent, hairs mostly relatively long; corolla 5–15 mm diam. **Berries** 8–10 mm diam. **2*n*** = 24.

Flowering Mar–Oct (mostly late spring–early summer, depending on rain). Deserts, dry grasslands, frequently on roadsides or dry desert washes; 500–900 m; Tex.; Mexico (Coahuila).

Chamaesaracha crenata occurs primarily in the Big Bend region of Texas and adjacent Coahuila, Mexico. Plants are robust, forming dense mounds. J. Henrickson (2009) included *C. crenata* within *C. villosa*. The two species are very similar; populations in and around Big Bend National Park compare more closely to the type of *C. crenata*.

6. **Chamaesaracha villosa** Rydberg, Mem. Torrey Bot. Club 4: 368. 1896 • Trans-Pecos five eyes

Stems decumbent, pale green, (1.5–)2–5 dm. **Herbage** villous, hairs mostly elongate, frequently 1-branched at tip, eglandular. **Leaves:** petiole to 1/4 total leaf length; blade ovate to broadly rhombic, 4–6 × 1.5–3.5 cm, length 2–2.5 times width, margins crenate or entire. **Inflorescences** 1–5-flowered. **Flowers:** calyx 5–7 mm, densely pubescent, hairs mostly relatively long; corolla 5–15 mm diam. **Berries** 8–10 mm diam. $2n$ = 24.

Flowering Mar–Oct (mostly late spring–early summer, depending on rain). Deserts, dry grasslands, roadsides, dry desert washes; 400–1300 m; Tex.; Mexico (Chihuahua, Coahuila, Durango).

Chamaesaracha villosa occurs within 30–40 miles of the Rio Grande River in Trans-Pecos Texas and adjacent Mexico. It is robust and villous, with branching stems forming mounds 1 m across; it is most similar to *C. crenata*, with which it is easily confused. The principal characters to distinguish the two species are leaf shape and vestiture. The similarities and differences were discussed by J. E. Averett (2010).

7. **Chamaesaracha edwardsiana** Averett, Sida 5: 48. 1972 • Edwards Plateau five eyes [E] [F]

Stems decumbent to suberect, purplish or greenish gray, 0.7–3 dm. **Herbage** glabrous or glabrate, hairs usually dendritic, sometimes simple, eglandular. **Leaves** subsessile; blade linear-lanceolate to rhombic, 2.5–7 × 0.7–1.5 cm, length 3.5–4.5 times width, margins occasionally few-lobed. **Inflorescences** 1–2-flowered. **Flowers:** calyx 4–5 mm, pubescent, especially along lobe margins; corolla 10–15 mm diam. **Berries** 5–8 mm diam. $2n$ = 48.

Flowering Mar–Oct (mostly late spring–early summer, depending on rain). Roadsides, limestone soils; 0–700 m; Tex.

Chamaesaracha edwardsiana occurs in the Edwards Plateau region of central Texas. J. E. Averett (1973) mapped a few populations in northern Mexico but most of those may belong with *C. pallida*. *Chamaesaracha edwardsiana* is most closely related to *C. coronopus* but differs in having broader, entire or nearly entire leaves. The species is similar also to *C. pallida* but lacks the relatively dense dendritic hairs on the leaves and has a generally more eastern distribution.

8. **Chamaesaracha pallida** Averett, Sida 5: 49. 1972 • Pale five eyes

Stems decumbent to suberect, pale green, 0.5–1.5 dm. **Herbage** usually pubescent, rarely glabrous, hairs dendritic, eglandular. **Leaves** subsessile; blade broadly lanceolate to rhombic, 2–3 × 0.6–1.5 cm, length 2–3 times width, margins entire to sinuate. **Inflorescences** 1–2-flowered. **Flowers:** calyx 3–4 mm, pubescent, especially along lobe margins; corolla 10–13 mm diam. **Berries** 5–7 mm diam. $2n$ = 72.

Flowering Mar–Oct (mostly late spring–early summer, depending on rain). Deserts, high grasslands; 300–2000 m; N.Mex., Tex.; Mexico (Chihuahua, Durango, Nuevo León).

Chamaesaracha pallida is found in southeastern New Mexico and western Texas. It is most closely related to *C. edwardsiana*; it differs in having a relatively dense vestiture of dendritic hairs (except for a few populations in the Guadalupe Mountains that are largely glabrous), a more western distribution, and a hexaploid chromosome complement.

10. DATURA Linnaeus, Sp. Pl. 1: 179. 1753; Gen. Pl. ed. 5, 83. 1754 • Jimsonweed, thorn-apple, toloache [Sanskrit *dhattura*, illusion, alluding to hallucinogenic properties; Latin *dator*, giver, alluding to extract supposedly given to enhance sexual potency]

Robert A. Bye

Herbs, annual or perennial, taprooted, usually tuberous in perennial species, usually pubescent, sometimes glabrous. **Stems** dichotomously branching. **Leaves** alternate, sometimes appearing subopposite on flowering branches; blade entire to sinuate-dentate or lobed. **Inflorescences** terminal, often appearing leaf-opposed, solitary flowers. **Flowers** 5-merous (erect); calyx cylindric, 5-toothed or splitting irregularly to produce a variable number of unequal teeth, circumscissile in fruit leaving a basal remnant that is slightly accrescent or not; corolla white to purple, radial, funnelform or trumpet-shaped, with 5 acuminate lobes (each subtended by 3 prominent veins) alternating with either lobules or shallow sinuses; stamens equal, inserted in proximal 1/2 of corolla tube; anthers basifixed, linear-oblong, dehiscing by longitudinal slits; ovary 2-carpellate (2- or 4-locular); style filiform, equaling, slightly longer, or shorter than stamens; stigma subcapitate, 2-lobed. **Fruits** capsules, 4-valved or irregularly dehiscing, ovoid, (2- or 4-locular, smooth or with prickles or tubercles). **Seeds** 40–120(–400), reniform or subreniform (black or tan-brown, with or without convex marginal ridge, some species with a white caruncle). $x = 12$.

Species 14 (8 in the flora): North America, Mexico, West Indies, Central America, n South America; introduced nearly worldwide.

The large flowers and evening fragrance have assured the cultivation of *Datura* spp., in particular, *D. innoxia*, *D. metel*, *D. stramonium*, and *D. wrightii*, throughout the world. Nocturnal anthesis lasts only one night in wild species. Some species of *Datura* are a sacred component of Native American ritual passage and have been employed since pre-Columbian times (W. E. Safford 1922; W. J. Litzinger 1981; C. E. Boyd 2003). Indigenous and Hispanic peoples employ some species in traditional healing practices for treating wounds and inflammations and for psychotropic effects. Tropane alkaloids, of which more than 30 have been reported in *Datura* (E. Eich 2008), cause delirium. In particular, atropine, hyoscyamine, and scopolamine are responsible for the anticholinergic properties of some pharmaceutical preparations used in treating motion sickness, broncho- and vasoconstriction, and other ailments.

Some species of *Datura* have become worldwide weeds (K. Hammer et al. 1983). The commerce of various crop seeds contaminated by *D. stramonium* is one of the principal causes of its spread during the last century.

Poisonings and deaths from consumption of seeds and/or foliage of *Datura* spp. are reported for livestock (cattle, horses, swine, and chickens) and humans.

Mexico is considered the center of origin and diversification of *Datura* (D. E. Symon and L. A. R. Haegi 1991). Prior to human settlement, species of *Datura* native in what is today southwestern United States included *D. discolor*, *D. innoxia*, *D. quercifolia*, and *D. wrightii*. Pre-Columbian dispersion of *D. wrightii* (as a sacred and medicinal plant) and of *D. stramonium* (as a medicinal plant and a weed associated with Mesoamerican agriculture) is related to settlement histories and migration of various indigenous peoples. Pre-Columbian presence of *D. metel* (R. Geeta and W. Gharaibeh 2007) and *D. stramonium* (A. Touwaide 1998) in Eurasia is supported by iconographic, literature, and linguistic sources.

The polymorphism in trichome morphology (for example, glandular versus non-glandular hairs) and density has been shown to have a genetic basis that responds to selection pressures of insect herbivores (N. M. van Dam et al. 1999). Certain seed characters of taxonomic importance such as caruncles (external food bodies rich in amino acids and sugars, also known as elaiosomes) are key to ant-*Datura* mutualism and maintenance of wild plants populations (D. J. O'Dowd and M. E. Hay 1980). Because of distinctive patterns of variation, certain species of *Datura* serve as experimental organisms contributing to understanding chromosome diversity, host plant and herbivore interactions, and relationships between various floral characters and hawkmoth pollinators (A. G. Avery et al. 1959; P. L. Valverde et al. 2001; J. L. Bronstein et al. 2009; respectively).

SELECTED REFERENCES Barclay, A. S. 1959. New considerations in an old genus: *Datura*. Bot. Mus. Leafl. 18: 245–272. Barclay, A. S. 1959b. Studies in the Genus *Datura* (Solanaceae). I. Taxonomy of the Subgenus *Datura*. Ph.D. thesis. Harvard University. Bye, R. A. and V. Sosa. 2013. Molecular phylogeny of the jimsonweed genus *Datura* (Solanaceae). Syst. Bot. 38: 818–829. Hammer, K., A. Romeike, and C. Tittel. 1983. Vorarbeiten zur monographischen Darstellung von wildpflanzensortimenten: *Datura* L., sections *Dutra* Bernh., *Ceratocaulis* Bernh. et *Datura*. Kulturpflanze 31: 13–75. Safford, W. E. 1921. Synopsis of the genus *Datura*. J. Wash. Acad. Sci. 11: 173–189. Symon, D. E. and L. A. R. Haegi. 1991. *Datura* (Solanaceae) is a New World genus. In: J. G. Hawkes et al., eds. 1991. Solanaceae III: Taxonomy, Chemistry, Evolution. Kew. Pp. 197–210.

1. Corollas usually 4–11 cm, limb with sinuses alternating with acuminate lobes; capsules erect.
   2. Leaf blades elliptic to narrowly ovate, margins usually pinnately lobed, sometimes sinuate-dentate . . . . 6. *Datura quercifolia*
   2. Leaf blades broadly ovate, margins usually sinuate-dentate, sometimes pinnately lobed.
      3. Corollas 4–6 cm; capsule prickles unequal, some 15+ mm . . . . 3. *Datura ferox*
      3. Corollas (5–)6–11 cm; capsule prickles ± equal, to 15 mm . . . . 7. *Datura stramonium*
1. Corollas usually greater than 10 cm, limb with lobules alternating with acuminate lobes; capsules pendent.
   4. Corolla limbs with lobules larger than acuminate lobes, throat with purple ring; capsules with prickles, dehiscing by 4 valves, pericarp dry . . . . 2. *Datura discolor*
   4. Corolla limbs with lobules smaller than or of similar size as acuminate lobes, throat without purple ring; capsules with or without prickles, dehiscing irregularly, pericarp fleshy.
      5. Leaf blades narrowly ovate to lanceolate; corolla surfaces usually waxy; calyces glabrous, tube split along 1 side, (appearing spathe-like), with poorly defined teeth; capsules without prickles or tubercles. . . . . 1. *Datura ceratocaula*
      5. Leaf blades ovate; corolla surfaces not waxy; calyces hairy, tube cylindric, 5-toothed; capsules with prickles or tubercles.
         6. Stems sparsely hairy, glabrescent; corollas white, purple, or yellow, with single, double, or triple whorls; capsules tuberculate. . . . . 5. *Datura metel*
         6. Stems usually hairy, sometimes glabrescent or glabrous; corollas white, sometimes lavender or purple-tinged, with a single whorl; capsules with prickles.
            7. Corolla limbs with lobules ± equal to acuminate lobes; calyces villous (especially along veins); abaxial leaf surface villous (especially along veins), hairs spreading. . . . . 4. *Datura innoxia*
            7. Corolla limbs with lobules smaller than acuminate lobes; calyces canescent (especially along veins); abaxial leaf surface canescent (especially along veins), hairs appressed or curved . . . . 8. *Datura wrightii*

1. Datura ceratocaula Ortega, Nov. Pl. Descr. Dec., 11. 1797 • Latin or Mexican thorn-apple, tornaloca [I]

Herbs annual, to 8 dm. **Stems** glabrous. **Leaf blades** narrowly ovate to lanceolate, to 15 × 8 cm, margins sinuate to pinnately lobed, abaxial surface tomentose, adaxial surface glabrous. **Flowers:** calyx glabrous, tube split along 1 side, appearing spathe-like and with poorly defined unequal teeth; corolla white with red-purple hues, trumpet-shaped surface usually waxy, 11.5–20 cm, acuminate lobes alternating with smaller lobules. **Capsules** pendent, irregularly dehiscent, pericarp fleshy, glabrous, without prickles or tubercles; calyx remnant not accrescent. **Seeds** black, 3.5–5 mm, convex marginal ridge absent, testa finely pitted; caruncle present. $2n$ = 24.

Flowering Jul–Sep. Seasonal ponds, livestock ponds, ditches, desert grassland-shrublands; 1300–1400 m; introduced; N.Mex.; Mexico.

The caruncle of *Datura ceratocaula* usually does not detach from the seed (as it does in the case of seeds of other species), and swells and becomes sticky when wetted. In this way, the seeds are adapted to dispersal by aquatic birds and livestock. In New Mexico, *D. ceratocaula* is known only from Hildago County.

2. Datura discolor Bernhardi, Neues J. Pharm. Aerzte 26: 149. 1833; Linnaea 8: Litt. Ber. 138. 1833 • Desert thorn-apple, small datura [W]

Datura thomasii Torrey

Herbs annual or short-lived perennial, to 10 dm. **Stems** usually hairy, sometimes glabrous. **Leaf blades** ovate, to 18 × 16 cm, margins entire or dentate, abaxial surface hairy, adaxial surface glabrous. **Flowers:** calyx hairy along veins, tube cylindric, 5-toothed; corolla white throat with purple ring, trumpet-shaped, 8–15 cm, acuminate lobes alternating with larger lobules. **Capsules** pendent, regularly dehiscing by 4 valves, pericarp dry, hairy, with prickles to 3.2 cm; calyx remnant accrescent (sometimes reflexed). **Seeds** black, 3–4.5 mm, convex marginal ridge absent, testa rugose; caruncle present. $2n$ = 24.

Flowering Mar–Oct. Streamsides, irrigation ditches, road and trail margins, waste places, desert shrublands, grasslands, pinyon-juniper-oak woodlands; 0–600 (–1800) m; Ariz., Calif.; Mexico; Central America; introduced nearly worldwide.

Outside of its typical flowering period, *Datura discolor* flowers sporadically after rains.

3. Datura ferox Linnaeus, Demonstr. Pl., 6. 1753 • Fierce thorn-apple [I] [W]

Herbs annual, to 10 dm. **Stems** puberulent, sometimes glabrescent. **Leaf blades** broadly ovate, to 13 × 8 cm, margins usually sinuate-dentate, sometimes pinnately lobed, surfaces glabrescent. **Flowers:** calyx with minute pubescence along veins, tube cylindric, 5-toothed; corolla white, trumpet-shaped, 4–6 cm, acuminate lobes alternating with sinuses. **Capsules** erect, dehiscent by 4 valves, pericarp dry, sparsely hairy, with prickles unequal, some 15+ mm, proximals shorter than distals; calyx remnant not accrescent. **Seeds** black, 4–4.5 mm, convex marginal ridge absent, testa rugose; caruncle absent. $2n$ = 24.

Flowering Jun–Sep. Cultivated fields, irrigation ditches, road and trail margins, waste places; 0–150 m; introduced; Ala., Ark., Calif., Ga., Nev., N.Y., N.C., Pa.; occasional nearly worldwide.

The origin and native status of *Datura ferox* is unresolved, although China has been cited as the country of origin since Linnaeus. The most extensive phytogeographic distribution of this species is in northern Argentina. Its association with ship ballast and seed stock of monocultural crops may explain its local abundance when introduced.

4. Datura innoxia Miller, Gard. Dict. ed. 8, Datura no. 5. 1768 (as inoxia) • Angel's-trumpet, downy thorn-apple, herbe aux sorciers, Indian-apple, moonflower, pomme épineuse, pricklyburr, tlapatl, stramoine innofensive [W]

Datura meteloides de Candolle ex Dunal

Herbs perennial, to 10 dm, roots tuberous. **Stems** usually villous-pubescent, sometimes glabrous. **Leaf blades** ovate, to 22 × 16 cm, margins entire or irregularly sinuate-dentate, surfaces villous to glabrescent, (trichomes spreading, often more dense along veins, sometimes glandular). **Flowers:** calyx villous along veins, hairs spreading, tube cylindric, 5-toothed; corolla white, sometimes lavender- or purple-tinged, funnelform, 10–22 cm, sparsely hairy, glabrescent, acuminate lobes alternating with lobules of similar size. **Capsules** pendent, irregularly dehiscent, pericarp fleshy, hairy, with prickles 10 mm; calyx remnant slightly accrescent. **Seeds** brown, 4–6 mm, convex marginal ridge present, testa smooth; caruncle present. $2n$ = 24.

Flowering Jul–Oct. Streamsides, road and trail margins, waste places, desert shrublands, grasslands;

0–2000 m; Ont., Que., Sask.; Ala., Ark., Conn., D.C., Fla., Ga., Ill., Ind., Kans., Ky., Md., Mich., Miss., Mo., N.J., N.Mex., N.Y., N.C., Ohio, Okla., Pa., S.C., Tenn., Tex., W.Va., Wis.; Mexico; West Indies; Central America; n South America (Colombia, Ecuador, Venezuela); introduced nearly worldwide.

Datura innoxia is native to Texas and possibly New Mexico. Elsewhere in the flora area, it is widely introduced as an ornamental and, inadvertently, as a weed.

5. **Datura metel** Linnaeus, Sp. Pl. 1: 179. 1753 • Angel's-trumpet, herbe aux sorciers, Indian-apple, moonflower [I] [W]

Herbs perennial, to 20 dm, roots tuberous. **Stems** purple in some cultivars, sparsely puberulent, glabrescent. **Leaf blades** ovate, to 24 × 20 cm, margins entire or irregularly sinuate-dentate, surfaces puberulent, glabrescent. **Flowers:** calyx hairy along veins, tube cylindric, 5-toothed; corolla white, yellow, or purple, broadly funnelform, with single, double, or triple whorls, finely puberulent along veins, 11–22 cm, acuminate lobes alternating with smaller lobules (or emarginate). **Capsules** pendent, irregularly dehiscent, pericarp fleshy, puberulent, glabrescent, tuberculate; calyx remnant slightly accrescent. **Seeds** brown, 4–6 mm, convex marginal ridge present, testa smooth; caruncle present. $2n = 24$.

Flowering Jun–Sep. Waste places; 0–1000 m; introduced; Ont., Que.; Calif., Fla., Ill., Kans., La., Mass., N.C., Okla., Tex.; Mexico; introduced and cultivated nearly worldwide.

Datura metel is an ornamental and ritual plant that was domesticated in the region of southern Mexico and Central America prior to European contact; it is derived from a common ancestor shared with *D. innoxia*. Using old Arabic and Indic references as well as iconographic representations from southern India, R. Geeta and W. Gharaibeh (2007) supported the hypothesis that *D. metel* was transferred to the Old World at least a millennium ago. Plants escaped from cultivation may persist for only a few years.

6. **Datura quercifolia** Kunth in A. von Humboldt et al., Nov. Gen. Sp., 3(fol.): 6; 3(qto.): 7. 1818 • Oak-leaf jimsonweed or thorn-apple [W]

Herbs annual, to 8 dm. **Stems** hairy, sometimes villous. **Leaf blades** elliptic to narrowly ovate, to 16 × 10 cm, margins usually pinnately lobed, sometimes sinuate-dentate, abaxial surface hairy, adaxial surface glabrescent. **Flowers:** calyx hairy along veins, tube cylindric, 5-toothed; corolla whitish to purple, trumpet-shaped, 4–8 cm, acuminate lobes alternating with sinuses. **Capsules** erect, dehiscent by 4 valves, pericarp dry, glabrous or hairy, with prickles unequal, proximals shorter than distals; calyx remnant not accrescent. **Seeds** black, 3–5 mm, convex marginal ridge absent, testa rugose; caruncle absent. $2n = 24$.

Flowering Jul–Oct. Gardens, cultivated fields, irrigation ditches, margins of roads and trails, waste places in various types of vegetation; 0–2200 m; Ariz., Ark., Calif., Ga., Kans., La., Md., Mass., N.Mex., N.C., Okla., Oreg., Pa., S.C., Tex.; Mexico.

Datura quercifolia is native to the Chihuahuan Desert, where it hybridizes occasionally with introduced *D. stramonium*. In the flora area, it is native to Texas and possibly New Mexico. Its geographic range is expanding especially in agricultural habitats. Although recently documented in central California (former orange orchard in Riverside; 1984, 1996), it has not expanded its range in that state.

7. **Datura stramonium** Linnaeus, Sp. Pl. 1: 179. 1753 • Devil's apple or weed, herbe aux sorciers, Jamestown weed, mad-apple, pomme épineuse, stink-wort, stramonium, stramoine commune [I] [W]

Datura stramonium var. *tatula* (Linnaeus) Torrey; *D. tatula* Linnaeus

Herbs annual, to 15 dm. **Stems** sometimes purple, sparsely puberulent, glabrescent. **Leaf blades** broadly ovate, to 22 × 12 cm, margins coarsely sinuate-dentate, surfaces glabrescent. **Flowers:** calyx hairy along veins, tube cylindric, 5-toothed; corolla usually white, sometimes purplish, trumpet-shaped, (5–)6–11 cm, acuminate lobes alternating with sinuses. **Capsules** erect, dehiscent by 4 valves, pericarp dry, glabrous or hairy, with prickles ± equal, to 15 mm; calyx remnant not accrescent. **Seeds** black, 3–4 mm, convex marginal ridge absent, testa rugose; caruncle absent. $2n = 24$.

Flowering summer. Gardens, cultivated fields, irrigation ditches, pastures, road and trail margins, waste places; 0–1800 m; introduced; Alta., B.C., N.B., N.S., Ont., P.E.I., Que., Sask.; Ala., Ariz., Ark., Calif., Colo., Conn., Del., D.C., Fla., Ga., Idaho, Ill., Ind., Iowa, Kans., Ky., La., Maine, Md., Mass., Mich., Minn., Miss., Mo., Mont., Nebr., Nev., N.H., N.J., N.Mex., N.Y., N.C., N.Dak., Ohio, Okla., Oreg., Pa., R.I., S.C., S.Dak., Tenn., Tex., Utah, Vt., Va., Wash., W.Va., Wis.; Mexico; introduced nearly worldwide.

Although a weed found throughout the world, *Datura stramonium* is probably native to central and southern Mexico and accompanied the expansion of Mesoamerican agriculture. Based upon a revised interpretation of ancient Latin and Greek texts, A. Touwaide (1998) argued that it was known in the Old World prior to the discovery of the New World in 1492. The delirious consequences of the British soldiers' consumption of young leaves at Jamestown, Virginia, in 1676 led to the application of the common name of jimsonweed to *D. stramonium* (R. Beverley 1705).

8. **Datura wrightii** Regel, Gartenflora 8: 193, plate 260. 1859 • Angel's trumpet, Indian-apple, sacred datura [F] [W]

Datura metel Linnaeus var. *quinquecuspida* Torrey

Herbs perennial, to 12 dm, roots tuberous. **Stems** usually canescent, sometimes glabrescent. **Leaf blades** ovate, to 22 × 16 cm, margins entire or irregularly sinuate-dentate, abaxial surface canescent (especially along veins), hairs appressed or curved, sometimes glandular, adaxial surface puberulent to glabrescent. **Flowers:** calyx canescent along veins, tube cylindric, 5-toothed; corolla white, sometimes tinged pale lavender, broadly funnelform, usually puberulent along veins, 14–26 cm, acuminate lobes alternating with smaller lobules. **Capsules** pendent, irregularly dehiscent, pericarp fleshy, puberulent, with prickles usually less than 10 mm; calyx remnant slightly accrescent. **Seeds** brown, 4–6 mm, convex marginal ridge present, testa smooth; caruncle present. $2n$ = 24.

Flowering Jun–Oct. Streamsides, irrigation ditches, road and trail margins, waste places, desert and desert-margin shrublands, grasslands; 0–2100 m; Ala., Ariz., Ark., Calif., Colo., Conn., Fla., Ga., Idaho, Ill., Ind., Iowa, Kans., Ky., La., Maine, Md., Mass., Mich., Minn., Miss., Mo., Nebr., Nev., N.H., N.J., N.Mex., N.Y., N.C., Ohio, Okla., Oreg., Pa., R.I., S.C., Tex., Utah, Va., Wash., W.Va., Wis., Wyo.; Mexico (Baja California, Chihuahua, Sonora); introduced nearly worldwide.

Datura wrightii has been introduced worldwide both as an ornamental and unintentionally. In the flora area, it is native in Texas and possibly New Mexico. The combination *D. innoxia* subsp. *quinquecuspida* (Torrey) A. S. Barclay is an invalidly published synonym of *D. wrightii*.

11. HUNZIKERIA D'Arcy, Phytologia 34: 283. 1976 • [For Armando Theodoro Hunziker, 1919–2001, Argentinean botanist]

Philip D. Jenkins†

Subshrubs, pubescent, hairs simple, glandular or eglandular. **Stems** decumbent, branched. **Leaves** alternate. **Inflorescences** axillary and terminal, solitary or paired flowers. **Flowers** 5-merous, bilaterally symmetric; calyx not accrescent, campanulate, lobes 5, triangular; corolla pink to purple, bilateral, salverform, (slightly enlarged at summit of tube), lobes broadly spreading, rounded; stamens 4, inserted near summit of corolla tube, unequal, sometimes with a 5th represented by a staminode; anthers ventrifixed, spheric, dehiscing by longitudinal slits; ovary 2-carpellate; style proximally slender, distally ± curved; stigma capitate-bilobed. **Fruits** capsules, subglobose, (4-valved). **Seeds** reniform. x = 8.

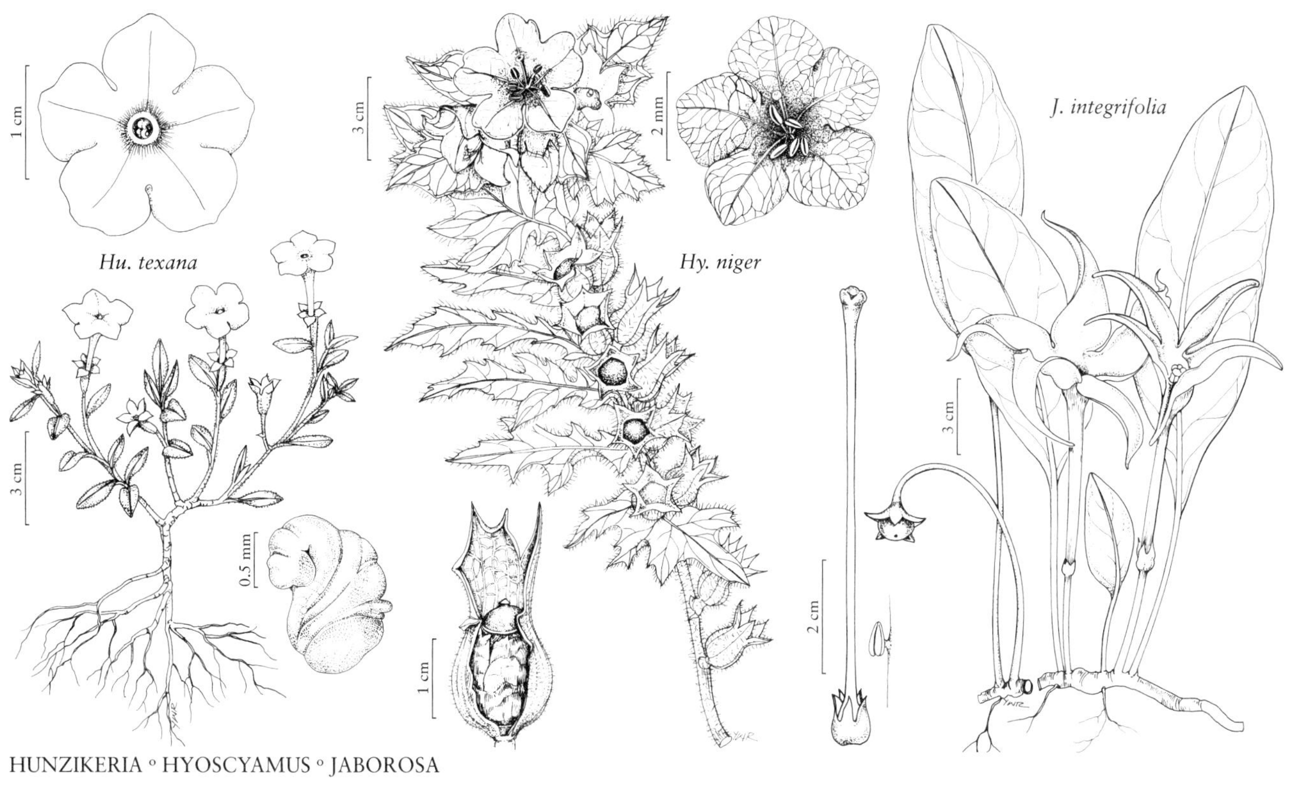

HUNZIKERIA ◦ HYOSCYAMUS ◦ JABOROSA

Species 3 (1 in the flora): Texas, Mexico, Central America, South America (to Venezuela).

Hunzikeria embraces subshrub plants of Texas, Mexico, and Venezuela that A. Gray placed in *Leptoglossis* Bentham. The flowers of *Hunzikeria* resemble those of *Nierembergia*, which is cultivated in the flora area, but the latter have five fertile stamens. A most unusual feature of *Hunzikeria* is the seed, which is large (ca. 1.5 mm) and wrinkled or furrowed. The obconic calyx with five shortened, deltate lobes, plus the long salverform corolla with a distal enlargement of the tube and a broadly spreading limb, serve to distinguish *Hunzikeria* from other Solanaceae.

1. **Hunzikeria texana** (Torrey) D'Arcy, Phytologia 34: 283. 1976 • Texas cup flower [F]

Browallia texana Torrey in W. H. Emory, Rep. U.S. Mex. Bound. 2(1). 156. 1859; *Leptoglossis texana* (Torrey) A. Gray

Stems to 20 cm, densely viscid-pubescent. **Leaves:** petiole 2–5 mm; blade ovate to obovate, 10–20 × 5–8 mm, base sometimes cuneate, margins entire, apex obtuse or acute, veins obscure, surfaces viscid-pilose, costas ciliate. **Flowering pedicels** 7 mm, pilose. **Flowers:** calyx 10 mm, lobes 3 mm; corolla 15–20 mm, scarcely expanded upwards, mouth tightly constricted around anthers, limb rotate, 20–23 mm diam., held slightly oblique to tube; style 2-winged proximal to stigma; stigma included, compressed by anthers, crescentric, 2-armed. **Capsules** included within calyx tube. **Seeds** reddish, darker on ridges, 1.5 mm, wrinkled with 4–8 ridges or furrows. $2n$ = 32.

Flowering Feb–May. On limestone; 300–600 m; Tex.; Mexico (Coahuila, Hidalgo, Nuevo León, San Luis Potosí, Tamaulipas).

Hunzikeria texana is limited to Edwards, Kinney, and Val Verde counties at the western edge of the Edwards Plateau. In this species, the anthers are of two sizes due to abortion of one of the thecae in each of the adaxial stamens. There are typically only five seeds per capsule.

12. HYOSCYAMUS Linnaeus, Sp. Pl. 1: 179. 1753; Gen. Pl. ed. 5, 84. 1754 • Henbane [Greek *hyos*, hog, and *kyamos*, bean pod, alluding to use of fruits as hog's food] [I]

Michael A. Vincent

Herbs, annual, biennial, or perennial, taprooted [rhizomatous], glandular-pubescent [glabrous]. **Stems** simple or branched. **Leaves** alternate (basal rosettes and cauline), petiolate (cauline leaves may be sessile or petiolate proximally, sessile distally); blade toothed or lobed. **Inflorescences** terminal [axillary], racemose [cymose], bracts leaflike. **Flowers** 5-merous, bilaterally symmetric or slightly irregular; calyx not to slightly accrescent, enclosing fruit, tubular-campanulate [urceolate or cup-shaped], 5-lobed, lobes triangular (lobes shorter than to as long as tube), becoming rigid with rigid or spinescent lobe-tips; corolla yellow or white, throat and veins sometimes purple, bilateral, funnelform; lobes spreading, triangular-ovate, shorter than tube, adaxial lobe usually ± larger; stamens slightly unequal, inserted near base or middle of corolla tube; anthers dorsifixed, ellipsoidal, dehiscing by longitudinal slits (introrse); ovary 2-carpellate, (2-locular); style filiform; stigma capitate, ± 2-lobed. **Fruits** capsules, ellipsoidal, dehiscence circumscissile. **Seeds** ovate-subreniform, flattened, (surface reticulate to honeycombed, dull or shiny). $x = 14, 17$.

Species 23 (2 in the flora): introduced; Eurasia; n Africa; introduced also in South America, s Africa, Australia.

Species of *Hyoscyamus* sometimes are cultivated as ornamentals. They contain toxic alkaloids, including hyoscyamine and scopolamine, which may have medical applications (M. F. Roberts and M. Wink 1998; W. H. Lewis and M. P. F. Elvin-Lewis 2003).

1. Mid and distal cauline leaves sessile, blade bases clasping; corolla veins purple. . . . . 1. *Hyoscyamus niger*
1. Cauline leaves all petiolate, blade bases not clasping; corolla veins pale . . . . 2. *Hyoscyamus albus*

1. Hyoscyamus niger Linnaeus, Sp. Pl. 1: 179. 1753 • Black henbane, jusquiame noire [F] [I] [W]

Herbs annual or biennial, 0.3–1.2 m, aromatic. **Stems** simple or branched. **Leaves:** proximal cauline leaves petiolate, mid and distal cauline leaves sessile; blade broadly lanceolate to ovate, 5–30 × 2–15 cm, base clasping, margins coarsely toothed or shallowly to deeply lobed, lobes pinnate, narrowly triangular, 0.5–5 cm, apex acute. **Flowers:** calyx 8–15 × 5–10 mm, lobes triangular, apically acute-acuminate; corolla pale yellow to greenish yellow with purple throat and veins, 25–45 × 23–43 mm, tube narrow, 8–12 × 4–5 mm, lobes broadly triangular-ovate, 5–7 × 6–10 mm, apically rounded, acute, or emarginate, abaxially often glandular-hairy in proximal half; stamens 7–8 mm; filaments 5–7 mm; anthers 2.1 × 1.1 mm; ovary 2–3 × 2–2.5 mm; style 10–12 mm.

Flowering May–Sep. Waste places, roadsides, fields; 0–2900 m; introduced; Alta., B.C., Man., N.B., N.S., Ont., P.E.I., Que., Sask.; Colo., Conn., Del., Idaho, Ill., Ind., Iowa, Maine, Md., Mass., Mich., Mont., Nebr., Nev., N.H., N.J., N.Mex., N.Y., N.Dak., Oreg., Pa., S.Dak., Utah, Vt., Wash., Wis., Wyo.; Eurasia; introduced also in South America, s Africa, Australia.

Hyoscyamus niger is highly toxic and should not be ingested. It contains the alkaloids hyoscyamine and scopolamine in all parts of the plant (W. H. Blackwell 1990). These compounds, as well as atropine and others, have medicinal applications, due in part to their activity as acetylcholine depressors, perhaps for treatment of certain heart conditions, as antispasmodics, to reduce symptoms of emphysema, and to relieve toothache (G. M. Hocking 1947; W. H. Lewis and M. P. F. Elvin-Lewis 2003).

Collections of *Hyoscyamus niger* from Nova Scotia and Prince Edward Island appear to be waifs.

2. Hyoscyamus albus Linnaeus, Sp. Pl. 1: 180. 1753 • White henbane [I]

Herbs biennial or perennial, 0.3–0.9 m, not aromatic. **Stems** branched. **Leaves:** cauline leaves all petiolate; blade broadly lanceolate to ovate, 4–7.3 × 3–7 cm, base not clasping, margins coarsely toothed to shallowly lobed, lobes pinnate, broadly triangular, 0.5–1.2 cm, apex acute. **Flowers:** calyx 8–15 × 5–10 mm, lobes triangular, apex acute-apiculate; corolla creamy white to pale yellow with greenish yellow, or purple throat greenish, veins pale, 10–15 × 12–18 mm, tube narrow, 6–9 × 3–4 mm, lobes broadly triangular-ovate, 4–6 × 4–8 mm, apex rounded, rarely emarginate, often glandular-hairy abaxially in lower half; stamens 15–20 mm; filaments 13–18 mm; anthers 2 × 1.1 mm; ovary 1.5–2 × 1.5–2 mm; style 15–16 mm. **2*n*** = 34, 68.

Flowering May–Sep. Waste places, roadsides, ballast piles; 0–200 m; introduced; Ont.; Fla., N.J., Pa.; Eurasia; n Africa; introduced also in Australia.

Hyoscyamus albus collections from Pennsylvania are from ballast and waste ground in Philadelphia; the most recent collections date from 1921 (A. F. Rhoads and W. M. Klein 1993). The report of *H. albus* from Florida is based on a record from ballast in Pensacola by C. T. Mohr (1878).

13. JABOROSA Jussieu, Gen. Pl., 125. 1789 • [Arabic name for a species of mandrake] [I]

Michael A. Vincent

Herbs, perennial [annual, biennial], rhizomatous, glabrous [pubescent]. **Stems** prostrate to ascending, branched. **Leaves** alternate; blade margins entire or denticulate [lobed, pinnatisect, bipinnatisect]. **Inflorescences** axillary, solitary or fascicled flowers. **Flowers** 5-merous; calyx ± accrescent, campanulate, lobes 5, as long as or longer than tube; corolla white [yellow or purple], radial, salverform [rotate, campanulate, urceolate, cylindric], lobes acute to acuminate, flaring to reflexed; stamens inserted centrally or distally on corolla tube, equal; anthers dorsifixed, ellipsoidal, dehiscing by longitudinal slits; ovary 2–5-carpellate; style filiform; stigma capitate or 2–5-lobed. **Fruits** berries, subglobose-compressed, juicy. **Seeds** globose-reniform. ***x*** = 12.

Species 23 (1 in the flora): introduced; South America.

Species of *Jaborosa* are sometimes cultivated as ornamentals, dating back to the mid-nineteenth century (J. D. Wilson 1851). *Jaborosa* is a source of withanolides, which may be useful as insecticides (J. C. Oberti 1998).

1. Jaborosa integrifolia Lamarck in J. Lamarck et al., Encycl. 3: 189. 1789 • Springblossom [F] [I]

Herbs 0.3 m. **Stems** subterranean. **Leaves:** (erect); petiole 0.3–1.3 cm; blade broadly lanceolate to ovate, 5–30 × 1–8 cm, base cuneate, apex rounded. **Flowering pedicels** 4–17 cm. **Flowers:** calyx 0.6–1.2 × 0.7–1 cm; corolla fragrant, 5.5–11.5 × 6–11 cm, lobes often glandular-hairy abaxially; stamens 2.5–3 mm; filaments nearly completely adnate to corolla; anthers introrse; ovary 4–5 × 3–4 mm; style 4.5–8 cm; stigma 3–5 × 3–5 mm. **Fruiting pedicels** 6–19 cm. **Fruiting calyces** 1.2–1.9 × 1.6–2.3 cm. **Berries** greenish yellow, 1.5–2.7 × 2–3 cm. **Seeds** dark brown, 3–3.5 × 2.8–3.2 mm, testa alveolate. **2*n*** = 24.

Flowering Apr–Sep. Waste places; 0–100 m; introduced; Ala., La.; South America (Argentina, Brazil, Paraguay, Uruguay).

Jaborosa integrifolia was first reported for North America as a ballast plant in Mobile County, Alabama (C. T. Mohr 1901), and later in Louisiana as a weed in wet areas of Plaquemines Parish (R. D. Thomas and C. M. Allen 1993–1998).

14. JALTOMATA Schlechtendal, Index Seminum (Halle) 1838: 8. 1838 • [Possibly Jaltomate, name of a small pueblo in Zacatecas, Mexico]

Thomas Mione

Herbs, perennial, [**shrubs**], glabrous [densely pubescent], roots tuberous [lacking expanded root]. **Stems** erect to procumbent, branched. **Leaves** alternate. **Inflorescences** axillary, umbellate. **Flowers** 5-merous; calyx rotate, stellate, deeply incised, lobes accrescent, triangular [narrowly triangular or obtuse], flaring, planar, or reflexed [bowl-shaped] in fruit; corolla pale green [green, whitish, bluish, blue-purple, purple, sometimes red at base], radial, rotate, 5-lobed or lobes alternating with lobules and then totaling 10 [crateriform, broadly campanulate, tubular, urceolate]; stamens inserted at base of corolla tube, equal; anthers ventrifixed [basifixed], ovate [lanceolate], dehiscing by longitudinal slits; ovary 2-carpellate; style slender; stigma capitate [truncate, style narrowing from thicker base]. **Fruits** berries, subspheric, juicy. **Seeds** ovate to reniform. $x = 12$.

Species ca. 60 (1 in the flora): Arizona, Mexico, West Indies (Cuba, Hispaniola, Jamaica, Puerto Rico), Central America, South America.

SELECTED REFERENCE Mione, T. 1992. Systematics and Evolution of *Jaltomata* (Solanaceae). Ph.D. dissertation. University of Connecticut.

1. Jaltomata procumbens (Cavanilles) J. L. Gentry, Phytologia 27: 287. 1973 [F]

Atropa procumbens Cavanilles, Icon. 1: 53, plate 72. 1791; *Saracha procumbens* (Cavanilles) Ruiz & Pavon

Stems 4- or 5-sided [angular], to 8 [13] dm, glabrous [pubescent]. **Leaves:** petiole winged or cuneate, to 4.5 cm, wing tapering to base; blade ovate, to 13 × 8.5 cm [21 × 11 cm], margins entire or toothed, surfaces glabrate. **Inflorescences** 6[–18]-flowered; peduncle rarely absent, green to purplish green, to 3[–7.5] cm. **Pedicels** green, with raised longitudinal ridges, to 2[–3] cm. **Flowers:** calyx green [purple in fruit], 9–13 mm diam., 1.8–2.5 cm diam. in fruit; corolla 20–31 mm diam., adaxial face pilosulous; stamens with expanded bases; filaments ventrifixed, straight. **Berries** black to dark purple [rarely green], to 12 × 14 [14 × 17] mm. $2n = 24$.

Flowering Jul–Sep. Canyons, along streams, shade of oaks, shelter of rocks, alluvium, rhyolite, or other rich soils [disturbed habitats]; 1000–1900[–2900] m; Ariz.; Mexico; Central America; South America (Colombia, Ecuador, Venezuela).

Jaltomata procumbens is protogynous and self-pollination occurs in late-stage flowering. Plants cultivated in the absence of pollinators (Connecticut) abundantly set fruit.

In Mexico, plants of *Jaltomata procumbens* are deliberately not weeded out of agricultural fields, and the fruits are gathered, consumed uncooked, and can be purchased in some markets (Tilton Davis and R. A. Bye 1982; D. E. Williams 1985; Davis 1986).

Jaltomata procumbens has been reported from Maryland, on a chrome ore pile at a seaport (C. F. Reed 1964). It may never have become naturalized in that state.

15. LEUCOPHYSALIS Rydberg, Mem. Torrey Bot. Club 4: 365. 1896 • [Greek *leucos*, white, and genus *Physalis*, alluding to large, white corolla and resemblance] [E]

John E. Averett†

Herbs, annual or perennial, from fleshy or subligneous taproot, variously pubescent. **Stems** erect or ascending, branched, (stems and branches striate, sometimes angled). **Leaves** alternate or geminate. **Inflorescences** axillary, solitary flowers or clusters of 2–4. **Flowers** 5-merous;

calyx accrescent, campanulate, lobes 5, broadly triangular, exceeding or just shorter than berry, not inflated in fruit; corolla cream-white to pale yellow, radial, rotate, weakly lobed; stamens inserted near base of corolla throat, equal, (shorter than style); anthers basifixed, oblong, dehiscing by longitudinal slits; ovary 2-carpellate; style slender, straight, slightly curved; stigma blunt. **Fruits** berries, globose to ovoid, fleshy. **Seeds** flattened, reniform. $x = 12$.

Species 2 (2 in the flora): North America.

Leucophysalis consists of two species: *L. grandiflora* occurring from Alberta to Quebec and adjacent states in the United States and *L. nana* occurring in California, Nevada, and Oregon. The northern distribution is unusual among related Solanaceae.

The relationships of the two species and of the genus were more fully discussed by J. E. Averett (2009).

SELECTED REFERENCE Averett, J. E. 2009. Taxonomy of *Leucophysalis* (Solanaceae, tribe Physaleae). Rhodora 111: 209–217.

1. Plants annual, erect, 3–9 dm; stems pubescent, hairs glandular and simple; fruiting calyces exceeding berries . . . . 1. *Leucophysalis grandiflora*
1. Plants perennial, spreading, mounds to 2.5 dm; stems strigose-hispidulous; fruiting calyces not or rarely exceeding berries . . . . 2. *Leucophysalis nana*

1. Leucophysalis grandiflora (Hooker) Rydberg, Mem. Torrey Bot. Club 4: 366. 1896 • Large false groundcherry, coqueret à grandes fleurs E F

Physalis grandiflora Hooker, Fl. Bor.-Amer. 2: 90. 1837; *Chamaesaracha grandiflora* (Hooker) Fernald

Herbs annual, erect, 3–9 dm. **Roots** fleshy. **Stems** pubescent, hairs glandular and simple. **Leaves:** petiole 2–5 cm; blade broadly ovate to ovate-lanceolate, 4–10 × 2.5–8 cm, margins entire, surfaces glabrous or slightly pubescent. **Pedicels** 2–4 per node. **Flowers:** calyx 5–10 mm, hairy, especially at base, lobes sharply acute to attenuate; corolla cream-white to pale yellow, with darker yellow or yellow-green markings in throat, 2–6 cm diam. **Berries** globose to ovoid, 10–15 mm diam. **Fruiting calyces** accrescent, exceeding berries. $2n = 48$.

Flowering Jun–Jul. Disturbed or burned areas, usually open forest; 0–600 m; Alta., Man., Ont., Que., Sask.; Mich., Minn., N.Y., Vt., Wis.

Leucophysalis grandiflora has been collected from Alberta to Quebec and the adjacent states in the United States; there are few recent collections from the eastern part of its range. It seems relatively rare, appearing for a short while and then disappearing.

2. Leucophysalis nana (A. Gray) Averett, Ann. Missouri Bot. Gard. 57: 380. 1971 • Dwarf false groundcherry E

Saracha nana A. Gray, Proc. Amer. Acad. Arts 10: 62. 1874; *Chamaesaracha nana* (A. Gray) A. Gray

Herbs perennial, spreading, mounds to 2.5 dm. **Roots** fleshy to subligneous. **Stems** strigose-hispidulous. **Leaves:** petiole 0.5–3.5 cm; blade ovate-lanceolate to rhombic, 1–7 × 2–4 cm, margins entire or slightly undulate, abaxial surface slightly pubescent. **Pedicels** 1–2 per node. **Flowers:** calyx 3–4 mm, densely pubescent, lobes acuminate or sharply acute; corolla cream-white to pale yellow, with yellow-green markings in throat, 2 cm diam. **Berries** ovoid, 8 mm diam. **Fruiting calyces** accrescent, not or rarely exceeding berries. $2n = 24$.

Flowering May–Aug. Sandy flats, thickets, rocky meadows; 900–2600 m; Calif., Nev., Oreg.

Leucophysalis nana is found in the Sierra Nevada of California and Nevada to the Cascade Mountains and Great Basin of Oregon.

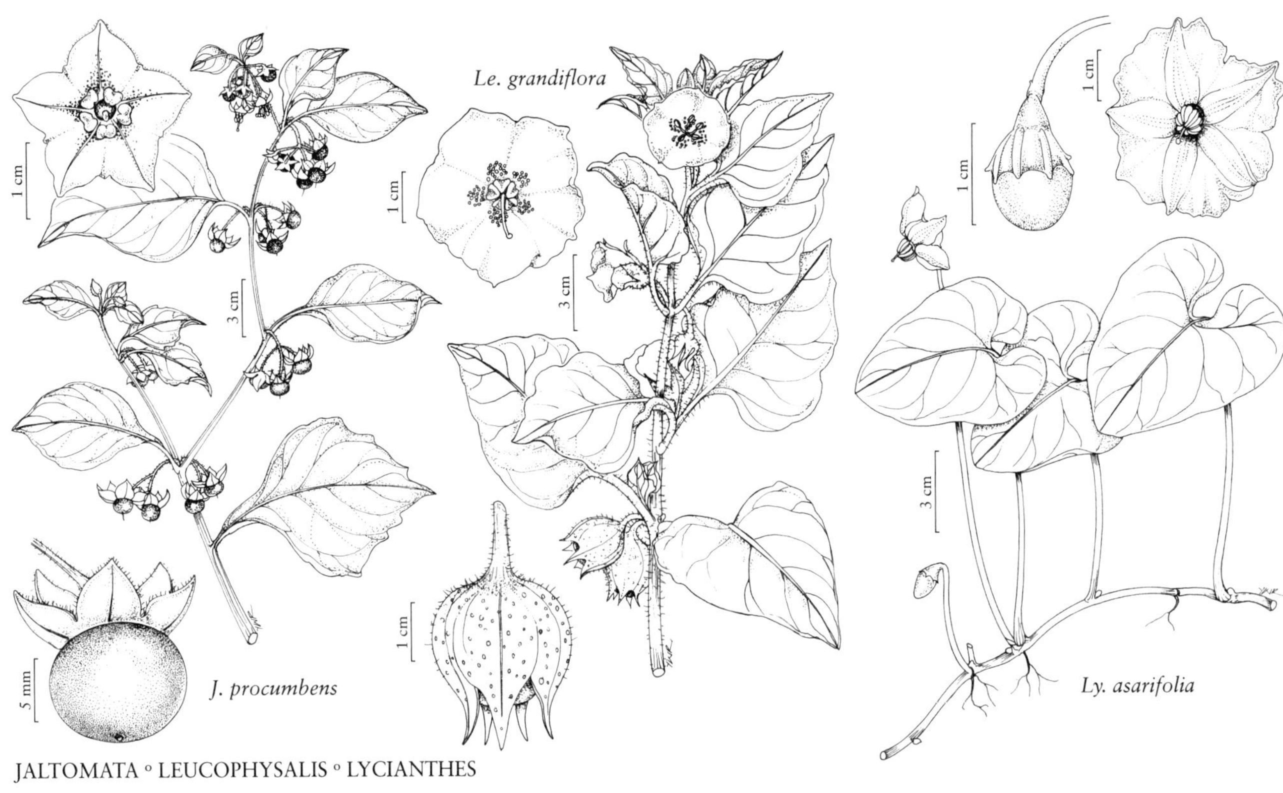

JALTOMATA ◦ LEUCOPHYSALIS ◦ LYCIANTHES

16. LYCIANTHES (Dunal) Hassler, Annuaire Conserv. Jard. Bot. Genève 20: 180. 1917, name conserved • Potato bush, ginger leaf [Genus *Lycium* and Greek *anthos*, flower, presumably alluding to resemblance of spinescent *L. lycioides*] [I]

Ellen A. Dean

Solanum Linnaeus subsect. *Lycianthes* Dunal in A. P. de Candolle and A. L. P. P. de Candolle, Prodr. 13(1): 29. 1852

Herbs, perennial, **or shrubs [vines, epiphytes],** rhizomatous or stoloniferous [tuberous roots], hairs simple or branched [stellate], glandular or eglandular. **Stems** creeping, erect, or ascending, branched. **Leaves** alternate or geminate; blade margins usually entire and undulate-wavy [remotely dentate]. **Inflorescences** axillary, solitary flowers or subumbellate. **Flowers** 5-merous, radially symmetric except for unequal stamens; calyx accrescent or somewhat accrescent, campanulate to obconic (becoming rotate to bowl-shaped in fruit), 10-veined, 5 or 10 veins often extending into 5 or 10 appendages that protrude from calyx below truncate rim; corolla white to pale yellow, or purplish to blue (with a yellow center in *L. rantonnetii*) [sometimes with contrasting green or purple markings], radial, rotate, or reflexed [campanulate], entire or lobed; stamens inserted near base of corolla limb, equal or unequal; anthers basifixed, lanceolate to elliptic, dehiscing by pores [by longitudinal slits]; ovary 2-carpellate; style straight or curved; stigma capitate to slightly lobed [deeply lobed]. **Fruits** berries, globose to ellipsoid, fleshy or juicy. **Seeds** lenticular-compressed to round or angular, (sclerotic granules sometimes present in outer part of mesocarp [these rarely enclosing seeds]). x = 12.

Species 150–200 (2 in the flora): introduced; Mexico, Central America, South America, se Asia, Pacific Islands, Australia.

The flowers of *Lycianthes* often open and close daily for two to five days in a row; outside the flora area, they are sometimes open only in early morning.

SELECTED REFERENCE Bitter, F. A. G. 1919. Die Gattung *Lycianthes*. Bremen. [Preprinted from Abh. Naturwiss. Vereins Bremen 24: 292–520. 1920.]

1. Herbs; corollas greenish white or yellow-white to white . . . . 1. *Lycianthes asarifolia*
1. Shrubs; corollas violet-purple to blue, with yellow center . . . . 2. *Lycianthes rantonnetii*

1. Lycianthes asarifolia (Kunth & Bouché) Bitter, Gatt. Lycianthes, 423. 1929 [F] [I] [W]

Solanum asarifolium Kunth & Bouché, Index Seminum (Berlin), App. 1845: 10. 1845

Herbs, stoloniferous or rhizomatous, hairs simple, eglandular. **Stems** creeping, rooting at nodes, pubescent to villous. **Leaves** mostly geminate (2d leaf often reduced, minute, caducous), glabrate; petioles of non-reduced leaves erect, 3–14(–23) cm; non-reduced leaf blades widely ovate to deltate, 2–12.5 × 1.6–12(–15) cm, base cordate to reniform, apex rounded to acute. **Inflorescences** solitary flowers. **Pedicels** erect, 4–9(–14) cm. **Flowers:** calyx campanulate, puberulent or glabrous, tube angled, 3–5 × 3–4 mm (accrescent in fruit to 12–14 mm diam.); corolla greenish white or yellow-white to white, 6–12 (–15) mm; stamens slightly unequal; filaments 1–3 mm, glabrous; anthers yellow, 2.5–3.5 mm; ovary subglobose to conic, 1–2.7 mm; style exserted beyond anthers, 5–6 mm. **Berries** orange-red, compressed-globose or ellipsoid, 0.7–2.5 cm. **Seeds** 70–85 per fruit, 2–2.5 mm, (sclerotic granules not present).

Flowering year-round. Residential areas, urban parks; 0–30 m; introduced; La., Tex.; South America.

In the flora area, *Lycianthes asarifolia* is known only from Houston, Texas, and New Orleans, Louisiana, where it has become an invasive weed. The New Orleans population does not fruit and may be a self-sterile clone. The Houston populations bear fruit, outcompete other ground covers and lawn plants, and have the potential to invade other areas. *Lycianthes asarifolia* is native in South America, where it grows in moist forest or open habitats, including being weedy along roadsides and in croplands.

2. Lycianthes rantonnetii (Carrière) Bitter, Gatt. Lycianthes, 332. 1929 • Blue potato bush, Paraguay nightshade [I]

Solanum rantonnetii Carrière, Rev. Hort. 32: 135, fig. 32. 1859 (as rantonnei)

Shrubs, 1–3(–4) m, branched from base, hairs whitish, simple or branched, eglandular and glandular. **Stems** ascending to erect. **Leaves** proximally alternate, distally geminate or solitary; petiole (0.1–)0.8–2.5(–4) cm; blade widely ovate or rhombic-lanceolate to narrowly lanceolate, 1–15.5 × 0.5–7.5 cm, base cuneate, sometimes unequal, apex acute to acuminate. **Inflorescences** subumbellate, 1–7-flowered. **Pedicels** spreading, straight, 0.5–2.5 cm. **Flowers:** calyx campanulate to obconic, puberulent, tube 1.5–4 × 2.5–4.5 mm (somewhat accrescent in fruit, becoming rotate and to 9 mm diam.); corolla violet-purple to blue, with yellow center, 6–18 mm; stamens unequal; three filaments 2–3 mm, two 0.8–1.5 mm, densely to sparsely puberulent at juncture with corolla tube; anthers orange, 2.5–4 mm; ovary ovoid, 1.8–2.3 mm; style not exserted beyond anthers, 3.5–5.5 mm. **Berries** rarely seen in flora region, immature fruits sometimes seen, yellow, subglobose to ellipsoid, 1–2[–3.2] cm. **Seeds** 3–12 per fruit, 2.5–3.5 mm, (15+ irregularly shaped sclerotic granules sometimes attached to seeds). $2n = 24$.

Flowering year-round. Chaparral; 700–800 m; introduced; Calif.; South America.

Lycianthes rantonnetii is a popular horticultural plant, flowering prolifically throughout the year depending on climate. It is known in the flora area only from San Diego County, where it persists near abandoned homesteads. Mature fruits are rarely seen in the flora area; when present, they are much smaller than the size reported from the native range. *Lycianthes rantonnetii* is native in South America, where it grows naturally in thickets and woodlands and has been documented as becoming weedy in disturbed areas along roadsides.

17. LYCIUM Linnaeus, Sp. Pl. 1: 191. 1753; Gen. Pl. ed. 5, 88. 1754 • Wolfberry, boxthorn [Greek *lykion*, name used by Dioscorides and Pliny for a spiny shrub, probably a species of *Rhamnus* supposedly from Lycia, ancient region of Asia Minor, alluding to resemblance]

Rachel A. Levin

Jill S. Miller

Shrubs, glabrous or hairy, leaves sometimes glaucous. **Stems** erect to prostrate, spinescent, with single (rarely) or multiple branches (often with divaricate branching). **Leaves** alternate, usually in fascicles (often drought-deciduous), petiolate or sessile, sometimes succulent; blade simple. **Inflorescences** axillary, fasciculate or solitary flowers. **Flowers** bisexual or unisexual, 4–5(–6)-merous, radially symmetric or calyx occasionally ± bilateral; calyx cupulate, tubular, or campanulate, sometimes accrescent in fruit; corolla white, greenish, yellowish, or lavender to deep purple lobes sometimes white with purple veins, tubular, funnelform, campanulate, or campanulate-rotate, lobes spreading or reflexed; stamens inserted at or proximal to midpoint of corolla tube, equal or unequal; anthers dorsifixed, ovate, dehiscing by longitudinal slits; ovary 2-carpellate; style filiform; stigma slightly 2-lobed. **Fruits** berries, juicy, occasionally hardened or drupaceous, globose to ovoid, rarely with constrictions (*L. cooperi*, *L. macrodon*, and *L. puberulum*). **Seeds** discoid to auriform, flattened. $x = 12$.

Species ca. 90 (18 in the flora): North America, Mexico, West Indies, South America, Eurasia, Africa, Atlantic Islands, Indian Ocean Islands, Pacific Islands, Australia.

Species of *Lycium* typically inhabit subtropical regions, often growing in desert, coastal, or saline environments. Some species can spread vegetatively via root suckering; plants have also been known to sprout from roots. Most species of *Lycium* are hermaphroditic; some are gynodioecious or dioecious. At least two species, *L. californicum* and *L. carolinianum*, are polymorphic for sexual strategy, having either hermaphroditic or dimorphic (gynodioecious or functionally dioecious) populations. Most species are diploid; some are polyploid. Polyploidy is positively correlated with sexual dimorphism.

Lycium appears to have evolved in South America, with subsequent dispersal to North America and a single long-distance dispersal event to the Old World. *Grabowskia* Schlechtendal and *Phrodus* Miers were formerly treated as separate genera; they have been transferred to *Lycium*.

SELECTED REFERENCES Chiang Cabrera, F. 1981. A Taxonomic Study of the North American Species of *Lycium* (Solanaceae). Ph.D. Dissertation, University of Texas. Chiang Cabrera, F. and L. R. Landrum. 2009 Vascular plants of Arizona: Solanaceae part three: *Lycium*. Canotia 5: 17–26. Hitchcock, C. L. 1932. A monographic study of the genus *Lycium* of the western hemisphere. Ann. Missouri. Bot. Gard. 19: 179–348, 350–375. Levin, R. A. and Jill S. Miller. 2005. Relationships within tribe Lycieae (Solanaceae): Paraphyly of *Lycium* and multiple origins of gender dimorphism. Amer. J. Bot. 92: 2044–2053. Miller, Jill S. 2002. Phylogenetic relationships and the evolution of gender dimorphism in *Lycium* (Solanaceae). Syst. Bot. 27: 416–428.

1. Berries not fleshy, green to yellow, orange, or brown, seeds 2–10.
    2. Berries not constricted; flowers 4(–5)-merous; plants 0.3–0.6 m, bark pale tan to white . . . . 1. *Lycium shockleyi*
    2. Berries constricted; flowers (4–)5-merous; plants 0.6–3 m, bark usually dark brown, sometimes reddish, purple, or black.
        3. Berries constricted at or distal to middle; calyx lobe lengths 0.5–1 times tube; leaf surfaces usually densely glandular-pubescent, not glaucous . . . . 2. *Lycium cooperi*
        3. Berries constricted proximal to middle, calyx lobe lengths 1–2 times tube; leaf surfaces glabrous or pubescent, glaucous.
            4. Calyx lobes linear; Sonoran Desert. . . . . 3. *Lycium macrodon*
            4. Calyx lobes ovate; Chihuahuan Desert. . . . . 4. *Lycium puberulum*

1. Berries ± fleshy, red to orange to orange-yellow; seeds 2–50+.
   5. Seeds 2, each enclosed by a hard layer forming a pyrene .... 5. *Lycium californicum*
   5. Seeds 4–50+, each not enclosed by a hard layer.
      6. Leaves glaucous; corollas (8–)12–25 mm, funnelform .... 6. *Lycium pallidum*
      6. Leaves rarely glaucous; corollas 4–16(–20) mm, tubular, funnelform, campanulate, or campanulate-rotate.
         7. Calyx cupulate (to tubular in *L. torreyi*).
            8. Corollas narrowly tubular or narrowly tubular-funnelform.
               9. Corolla lobe margins glabrous or sparsely ciliate .... 7. *Lycium andersonii*
               9. Corolla lobe margins densely ciliate-lanate .... 8. *Lycium torreyi*
            8. Corollas tubular to funnelform or campanulate-rotate.
               10. Corollas campanulate-rotate; coastal or wetland areas .... 9. *Lycium carolinianum*
               10. Corollas tubular to funnelform; mainly desert areas.
                   11. Leaf surfaces glabrous .... 10. *Lycium berlandieri*
                   11. Leaf surfaces pubescent .... 11. *Lycium texanum*
         7. Calyx tubular, campanulate, or tubular-campanulate.
            12. Calyx tubular or tubular-campanulate, 2–10 mm.
                13. Leaf surfaces glabrous; plants with bisexual flowers only ..... 12. *Lycium ferocissimum*
                13. Leaf surfaces glandular-pubescent; plants with either pistillate flowers or bisexual flowers.
                    14. Corollas deep lavender to purple, 8–20 mm; stamens included or slightly exserted; saline desert flats .... 13. *Lycium fremontii*
                    14. Corollas greenish white to lavender, 7–14 mm; stamens exserted 2–3+ mm from corolla in bisexual flowers; desert washes and bajadas .... 14. *Lycium exsertum*
            12. Calyx campanulate, 2–6 mm.
                15. Leaf surfaces glabrous; pedicels 10–20 mm; corollas funnelform; occurring mainly near habitation.
                    16. Corolla lobes equaling or longer than tube; leaves subsessile ...15. *Lycium chinense*
                    16. Corolla lobes shorter than or equaling tube; leaves petiolate ... .16. *Lycium barbarum*
                15. Leaf surfaces glabrous or puberulent to densely pubescent; pedicels 1–10 mm; corollas campanulate to tubular or funnelform; not restricted to human-modified areas.
                    17. Leaf surfaces glabrous or puberulent; corollas lavender or white with purple markings, campanulate to tubular; berries 10 mm; coastal desert areas, including by the Salton Sea .... 17. *Lycium brevipes*
                    17. Leaf surfaces densely pubescent; corollas pale lavender to purple, narrowly campanulate to funnelform; berries 4–7 mm; inland, along desert washes and bajadas .... 18. *Lycium parishii*

1. Lycium shockleyi A. Gray, Proc. Amer. Acad. Arts 22: 311. 1887 • Shockley's desert-thorn [E]

Lycium rickardii C. H. Muller

Shrubs prostrate, 0.3–0.6 m; bark pale tan to white; stems glabrous. **Leaves:** blade spatulate to oblanceolate-ovate, 4–23 × 1.5–6 mm, succulent, surfaces glabrous or sparsely glandular-puberulent. **Inflorescences** solitary flowers. **Pedicels** to 1 mm. **Flowers** 4(–5)-merous; calyx tubular to campanulate, 5–15 mm, lobe lengths 0.5–1 times tube; corolla white to greenish white or pale purple with bluish veins, tubular-funnelform, 8–14 mm, lobes 1.5–3 mm; stamens included to exserted. **Berries** green, ovoid, 5–6 mm, dry, hard, with strongly accrescent calyx. **Seeds** 2–6.

Flowering Apr–Jun. Sandy to rocky desert flats (Great Basin Desert); 900–1700 m; Nev.

Lycium shockleyi, endemic to western Nevada, is sometimes confused with the more widespread *L. cooperi*. *Lycium shockleyi* can be readily differentiated by its prostrate growth form, four-merous flowers, glabrous leaves, and pale bark. The ovary has a large, red nectar disc that persists until the fruits are almost mature.

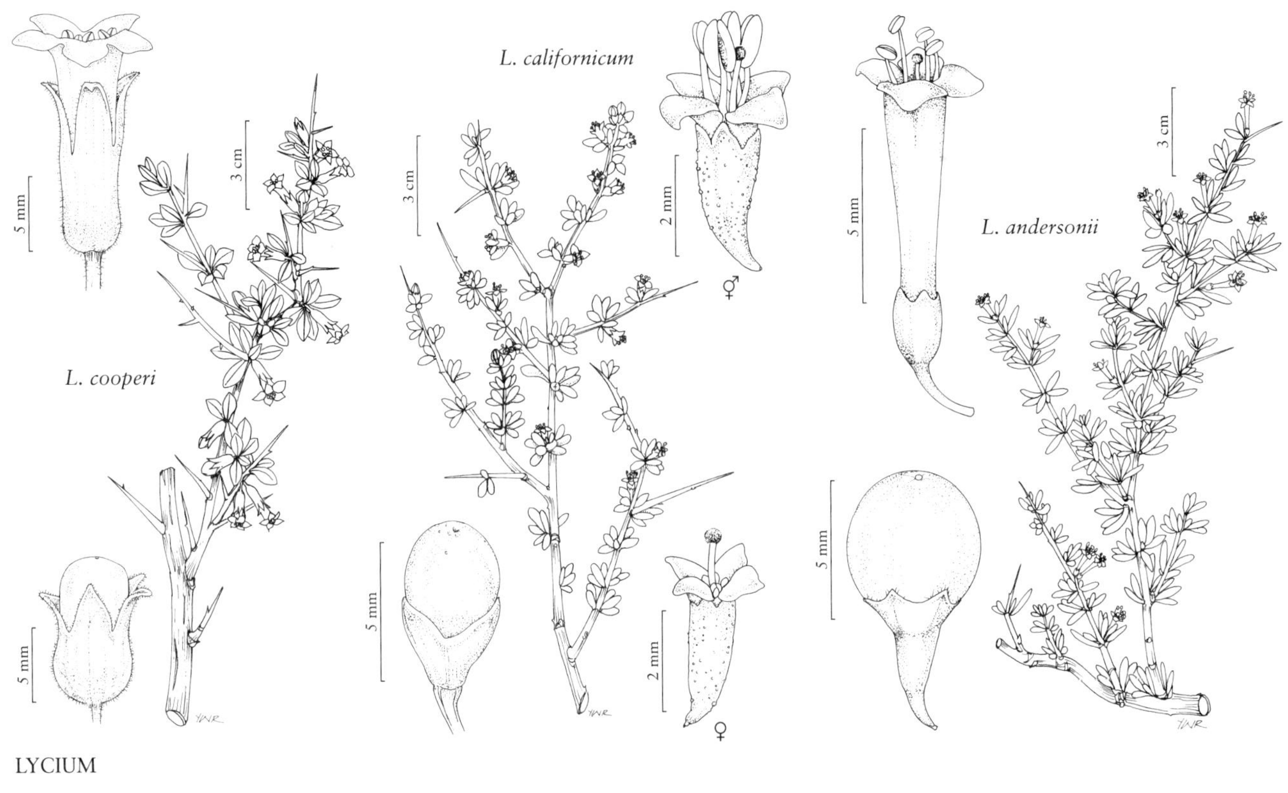

2. **Lycium cooperi** A. Gray, Proc. Amer. Acad. Arts 7: 388. 1868 • Peach-thorn, Cooper's desert-thorn E F

Shrubs erect, 0.6–2.5 m; bark usually purplish to reddish; stems densely glandular-pubescent. **Leaves:** blade spatulate to oblanceolate, 10–35 × 2–23 mm, surfaces usually densely glandular-pubescent. **Inflorescences** 2–3-flowered fascicles or solitary flowers. **Pedicels** 2–8 mm. **Flowers** (4–)5-merous; calyx narrowly campanulate, 4–14 mm, lobe lengths 0.5–1 times tube; corolla white or greenish yellow, sometimes purple-veined, tubular to funnelform, 8–15 mm, lobes 1.5–3 mm; stamens included to exserted. **Berries** greenish yellow to orange, ovoid, constricted at or distal to middle, 5–10 mm, dry, hard, strongly accrescent calyx usually rupturing with fruit growth. **Seeds** 6–10. $2n$ = 24.

Flowering Mar–May. Sandy washes to slopes (Mojave and Colorado deserts); 100–2000 m; Ariz., Calif., Nev., Utah.

Lycium cooperi occurs in western Arizona, southeastern California, southern Nevada (Clark and Esmeralda counties), and southwestern Utah (Washington County). It can be distinguished from the similar species *L. pallidum* and *L. shockleyi* by its dense, glandular pubescence and hard, constricted fruit.

3. **Lycium macrodon** A. Gray, Proc. Amer. Acad. Arts 6: 45. 1862 • Desert wolfberry, mahogany desert-thorn

Lycium dispermum Wiggins; *L. macrodon* var. *dispermum* (Wiggins) F. Chiang

Shrubs erect, 1–3 m; bark dark brown to black; stems glabrate. **Leaves:** blade spatulate, 5–35 × 2–12 mm, glaucous, surfaces glabrous or sparsely puberulent. **Inflorescences** 2-flowered fascicles or solitary flowers. **Pedicels** 1–7 mm. **Flowers** (4–)5-merous; calyx campanulate, 4–9 mm, sparsely glandular-puberulent, lobes linear, length 1.5–2 times tube; corolla white to greenish white to pale lilac on lobes, tubular, 6–12 mm, lobes 2–5 mm; stamens included or exserted. **Berries** yellowish to brown, ovoid, constricted proximal to middle, 6–10 mm, dry, hard. **Seeds** 2–4. $2n$ = 24.

Flowering Feb–Apr. Sandy washes and flats (Sonoran Desert); 200–800 m; Ariz.; Mexico (Sonora).

4. Lycium puberulum A. Gray, Proc. Amer. Acad. Arts 6: 46. 1862 • Downy desert-thorn or wolfberry

Lycium berberioides Correll; *L. puberulum* var. *berberidoides* (Correll) F. Chiang

Shrubs erect, 0.7–2 m; bark chocolate brown or reddish purple to black; stems glabrate. **Leaves:** blade spatulate, 5–40 × 2.5–11 mm, glaucous, surfaces glabrous or pubescent. **Inflorescences** 2-flowered fascicles or solitary flowers. **Pedicels** 1–4 mm. **Flowers** (4–)5-merous; calyx campanulate, 4–8 mm, lobes ovate, length 1–2 times tube; corolla pale purple to white with greenish lobes, tubular to funnelform, 7–13 mm, lobes 2–3 mm; stamens included. **Berries** pale orange-yellow, ovoid, constricted proximal to middle, 4–9 mm, glaucous, dry, hard. **Seeds** 2–4. $2n = 24$.

Flowering Mar–Apr. Desert hills, flats; 500–1200 m; Tex.; Mexico (Chihuahua, Coahuila, Durango).

Lycium puberulum is similar to *L. cooperi* and *L. macrodon*; it is restricted to the Chihuahuan Desert, occurring in western Texas and adjacent northern Mexico. In that region, *L. pallidum* is the most similar species, but it occurs at higher elevations, has much larger flowers, and its fruit is a fleshy berry versus the hardened fruit of *L. puberulum*.

5. Lycium californicum Nuttall ex A. Gray in W. H. Brewer et al., Bot. California 1: 542. 1876 • California box-thorn [F]

Lycium californicum var. *arizonicum* A. Gray; *L. californicum* var. *interior* F. Chiang; *L. carinatum* S. Watson

Shrubs prostrate to erect, 0.1–2 m; bark yellowish to gray; stems glabrous. **Leaves:** blade spheric to spatulate, 2–25 × 1–3 mm, fleshy to succulent, surfaces glabrous. **Inflorescences** 2–3-flowered fascicles or solitary flowers. **Pedicels** 1–5 mm. **Flowers** bisexual or pistillate, 4-merous; calyx campanulate, 2–4-lobed, 2–3 mm, lobe lengths 0.3 times tube; corolla white to pale purple, campanulate, 2–6 mm, lobe lengths to 1 times tube; stamens exserted (bisexual flowers) or included, with nonfunctional anthers (pistillate flowers). **Berries** orange to red, ovoid, 2–6 mm, thinly fleshy. **Seeds** 2, each enclosed by a hard layer. $2n = 24, 36, 48$.

Flowering Jan–Mar(–Sep). Saline flats, coastal flats and bluffs (Mediterranean floristic province, Sonoran Desert, and restricted locations in the Chihuahuan Desert); 0–500 m; Ariz., Calif.; Mexico (Baja California, Baja California Sur, Coahuila, San Luis Potosí, Sinaloa, Sonora, Zacatecas).

Within the flora area, *Lycium californicum* occurs in southern Arizona and southwestern California. Populations of *L. californicum* vary in both ploidy (diploid to tetraploid) and sexual system (hermaphroditic to gender dimorphic). Gender dimorphic populations are morphologically gynodioecious (functionally dioecious), and plants are sexually dimorphic for flower size, with smaller pistillate flowers.

SELECTED REFERENCE Miller, Jill S. et al. 2016. Correlated polymorphism in cytotype and sexual system within a monophyletic species, *Lycium californicum*. Ann. Bot. (Oxford) 117: 307–317.

6. Lycium pallidum Miers, Ann. Mag. Nat. Hist., ser. 2, 14: 131. 1854 • Pale wolfberry

Shrubs usually erect, sometimes prostrate, 1–2.5 m; bark yellowish, gray to reddish, or black; stems glabrous or sparsely puberulent. **Leaves:** blade spatulate to oblanceolate, 10–50 × 3–25 mm, glaucous, surfaces glabrous. **Inflorescences** 2–3-flowered fascicles or solitary flowers. **Pedicels** 4–16 mm. **Flowers** 5-merous; calyx cupulate to campanulate, 2.5–8 mm, lobe lengths 1–2 times tube; corolla greenish white to lavender, often with purple veins, funnelform, (8–)12–25 mm, lobes 3–5 mm; stamens exserted. **Berries** red, ovoid, 10 mm, glaucous, fleshy, apex sometimes hard. **Seeds** 4–50. $2n = 24$.

Lycium pallidum is known from throughout Arizona and New Mexico, southeastern California, southern Colorado, south-central Nevada (Nye County), western Oklahoma (Cimarron County), western Texas, southern Utah, and northeastern Mexico. Although the fruits of *L. pallidum* are fleshy, they occasionally have a hardened apex. The range of *L. pallidum* overlaps with those of several other *Lycium* species; however, its large, glaucous leaves and long, funnelform flowers are very distinctive.

Varieties 2 (2 in the flora): sw United States, Mexico.

1. Seeds 20–50; corollas 12–25 mm . . . . 6a. *Lycium pallidum* var. *pallidum*
1. Seeds 4–8; corollas (8–)12–20 mm . . . . 6b. *Lycium pallidum* var. *oligospermum*

6a. Lycium pallidum Miers var. **pallidum**

Corollas 12–25 mm. **Seeds** 20–50.

Flowering Mar–Jun. Desert grasslands and shrublands, submontane washes, flats, rocky slopes; 800–2300 m; Ariz., Colo., N.Mex., Okla., Tex., Utah; Mexico (Coahuila, Nuevo León, San Luis Potosí, Zacatecas).

6b. Lycium pallidum Miers var. **oligospermum** C. L. Hitchcock, Ann. Missouri Bot. Gard. 19: 304. 1932

Corollas (8–)12–20 mm. **Seeds** 4–8.

Flowering Mar–May. Desert washes, flats, rocky slopes (Mohave Desert, rarely nw Sonoran Desert); 200–1300 m; Calif., Nev.

7. Lycium andersonii A. Gray, Proc. Amer. Acad. Arts 7: 388. 1868 • Water jacket, redberry desert-thorn [F]

Lycium andersonii var. *deserticola* (C. L. Hitchcock) Jepson; *L. andersonii* var. *pubescens* S. Watson; *L. andersonii* var. *wrightii* A. Gray

Shrubs erect, 0.5–3 m; bark silvery tan to dark brown; stems glabrous or pubescent. **Leaves:** blade spatulate, 3–35 × 1–8 mm, fleshy to succulent, sometimes glaucous, surfaces glabrous or pubescent. **Inflorescences** 2-flowered fascicles or solitary flowers. **Pedicels** 1–10 mm. **Flowers** 4–5-merous; calyx cupulate, 1.5–3 mm, lobe lengths 0.25 times tube; corolla white to light purple, narrowly tubular-funnelform, 4–16 mm, lobes 1–2.5 mm, margins glabrous or sparsely ciliate; stamens included or exserted. **Berries** orange to red, ovoid, 3–8 mm, fleshy. **Seeds** 50+. **2*n*** = 24.

Flowering Feb–May. Desert washes, flats, grasslands; 100–1900 m; Ariz., Calif., Nev., N.Mex., Utah; Mexico (Baja California, Baja California Sur, Sinaloa, Sonora).

F. Chiang Cabrera (1981) recognized three varieties of *Lycium andersonii* (var. *andersonii*, var. *deserticola*, var. *wrightii*) in the flora area; these varieties are based on corolla lobe number (four or five), flower size, and leaf size and shape. Distinctiveness of these varieties is unclear; these characters generally vary within most species and indeed within individuals.

8. Lycium torreyi A. Gray, Proc. Amer. Acad. Arts 6: 47. 1862 • Squaw-thorn, squaw desert-thorn, Torrey's wolfberry

Shrubs erect, 1–3 m; bark yellowish tan to brown; stems glabrous. **Leaves:** blade spatulate to obovate, 10–50 × 1.5–15 mm, ± fleshy, surfaces glabrous. **Inflorescences** 2–8-flowered fascicles or solitary flowers. **Pedicels** 5–20 mm. **Flowers** (4–)5-merous; calyx cupulate to tubular, 2.5–6 mm, lobe lengths 0.25-0.5 times tube; corolla white to greenish lavender, narrowly tubular, 5–15 mm, lobes spreading, 1–4 mm, margins densely ciliate-lanate; stamens slightly exserted. **Berries** orange to red, ovoid, 6–12 mm, fleshy. **Seeds** 8–30. **2*n*** = 24.

Flowering Mar–May. Desert washes, alluvial flats, along streams and canals; 50–1000 m; Ariz., Calif., Nev., N.Mex., Tex., Utah; Mexico (Chihuahua).

In the flora area, *Lycium torreyi* occurs in Arizona, southeastern California, eastern Nevada (Clark and Lincoln counties), western New Mexico, western Texas, and southern Utah. It can be distinguished from *L. andersonii* by its densely ciliate-lanate corolla lobes, and the mouth of the corolla is not quite as narrow. Further, *L. torreyi* usually occurs by streams or canals, with branches more cascading than upright. C. L. Hitchcock (1932) reported the fruits to be juicy and sweet.

9. Lycium carolinianum Walter, Fl. Carol., 84. 1788 • Carolina wolfberry, Christmas berry

Lycium carolinianum var. *gaumeri* C. L. Hitchcock; *L. carolinianum* var. *quadrifidum* (Mociño & Sessé ex Dunal) C. L. Hitchcock; *L. carolinianum* var. *sandwicense* (A. Gray) C. L. Hitchcock; *L. sandwicense* A. Gray

Shrubs erect to prostrate, 0.4–3 m; bark tan to gray; stems glabrous. **Leaves:** blade spatulate to oblanceolate, 10–35 × 1–6 mm, fleshy, surfaces glabrous. **Inflorescences** 2–3-flowered fascicles or solitary flowers. **Pedicels** 5–30 mm. **Flowers** bisexual or pistillate, 4(–5)-merous; calyx cupulate, 3 mm, lobe lengths ± 1 times tube; corolla white to purple, campanulate-rotate, 7–12 mm, lobe lengths ± 1 times tube; stamens exserted. **Berries** red, globose, 10 mm, fleshy. **Seeds** 50+. **2*n*** = 24 (48).

Flowering Jan–Dec. Coastal dunes, tidal flats, wetlands, salt marshes; 0–5 m; Ala., Fla., Ga., La., Miss., Tex.; Mexico; West Indies (Cuba); e Asia (Japan: Daiton

Islands, Ogasawara Islands); Pacific Islands (Easter Island, Hawaii, French Polynesia, Pitcairn Islands, Tonga).

Lycium carolinianum is the most widespread species of the genus and shows considerable variation in habit, from prostrate to relatively short and unbranched, to taller and intricately branched. Within the flora area, *L. carolinianum* is most common in the Gulf Coast areas of Florida and Texas; it also occurs in Louisiana, southwestern Alabama (Mobile County), southeastern Georgia (Camden County), and southeastern Mississippi (Jackson County). Historically, *L. carolinianum* has been reported from coastal South Carolina. Fruits (especially in Florida and Texas) tend to be pleasantly sweet and juicy.

SELECTED REFERENCE Blank, C. M., R. A. Levin, and J. S. Miller. 2014. Intraspecific variation in gender strategies in *Lycium* (Solanaceae): Associations with ploidy and changes in floral form following the evolution of gender dimorphism. Amer. J. Bot. 101: 2160–2168.

10. Lycium berlandieri Dunal in A. P. de Candolle and A. L. P. P. de Candolle, Prodr. 13(1): 520. 1852 • Berlandier's wolfberry [W]

Lycium berlandieri var. *longistylum* C. L. Hitchcock; *L. berlandieri* var. *parviflorum* (A. Gray) A. Terracciano; *L. berlandieri* var. *peninsulare* (Brandegee) C. L. Hitchcock

Shrubs erect, 0.7–2.5 m; bark tan to gray to reddish or almost black; stems glabrous or pubescent. **Leaves:** blade linear to spatulate, 1.5–15 × 1–4.5 mm, surfaces glabrous. **Inflorescences** 2–3-flowered fascicles or solitary flowers. **Pedicels** 3–20 mm. **Flowers** 4–5-merous; calyx cupulate, 1–3 mm, lobe lengths 0.3 times tube; corolla white to pale lavender, tubular to funnelform, 4–9 mm, lobe lengths 0.17–0.3 times tube; stamens included to exserted. **Berries** orange to red, globose, 5 mm, fleshy. **Seeds** 50+. **2*n*** = 24.

Flowering Jul–Sep. Desert washes, rocky slopes, flats (Sonoran and Chihuahuan deserts); 300–900 m; Ariz., N.Mex., Okla., Tex.; Mexico (Baja California Sur, Chihuahua, Coahuila, Durango, Hidalgo, Nuevo León, San Luis Potosí, Sinaloa, Sonora, Tamaulipas, Zacatecas).

Within the flora area, *Lycium berlandieri* occurs in Arizona, New Mexico, and Texas; it has been reported from western Oklahoma (Harmon and Jackson counties).

11. Lycium texanum Correll, Wrightia 3: 139. 1965 • Texas wolfberry [C] [E]

Shrubs erect, 1–2 m; bark silvery tan to dark brown; stems hispidulous. **Leaves:** blade linear to spatulate, to 20 × 3 mm, surfaces hispidulous-puberulous. **Inflorescences** 2-flowered fascicles or solitary flowers. **Pedicels** 1.5–9 mm. **Flowers** 4–5-merous; calyx cupulate, 1.5–3 mm, minutely lobed; corolla lavender to white, tubular to funnelform, 7–8 mm, lobes 1.5–2.5 mm; stamens slightly exserted. **Berries** orange-red, ovoid, 3–8 mm, fleshy. **Seeds** 50+.

Flowering Mar–Oct. Rocky and sandy soils, desert canyons, semidesert grasslands, thorn scrub (Trans-Pecos region); of conservation concern; 1000–1400 m; Tex.

F. Chiang Cabrera (1981) noted that *Lycium texanum* is similar to *L. andersonii*, differing mainly in the type of pubescence (short, straight hairs versus longer, curved hairs). Data from at least one nuclear gene region suggest a close relationship with *L. andersonii* (R. A. Levin et al. 2009), and it is possible that *L. texanum* is simply the Texas variant of *L. andersonii*.

12. Lycium ferocissimum Miers, Ann. Mag. Nat. Hist., ser. 2, 14: 187. 1854 • African boxthorn [I] [W]

Shrubs erect, 2–3 m; bark pale gray to pinkish brown; stems glabrous. **Leaves:** blade bright green, obovate, 12–35 × 4–10 mm, fleshy, surfaces glabrous. **Inflorescences** 2-flowered fascicles or solitary flowers. **Pedicels** 5–15 mm. **Flowers** 5-merous; calyx tubular, 5–7 mm, lobe lengths ± 0.25 times tube; corolla white with purple veins, funnelform, 6–8 mm, lobes 3–4 mm; stamens exserted. **Berries** red, ovoid, 8–10 mm, fleshy. **Seeds** 50+. **2*n*** = 24.

Flowering year-round following rain. Coastal salt marshes, dunes, hedgerows, waste places; introduced; Calif.; Africa (South Africa); introduced also in Europe (Cyprus, Spain), n Africa (Morocco, Tunisia), Pacific Islands (New Zealand), Australia.

Within the flora area, *Lycium ferocissimum* is known only from Los Angeles County. It has the potential to become an invasive weed and its import and cultivation are restricted.

13. Lycium fremontii A. Gray, Proc. Amer. Acad. Arts 6: 46. 1862 (as fremonti) • Frémont's desert-thorn

Lycium fremontii var. *congestum* C. L. Hitchcock

Shrubs erect, 1–3 m; bark tan, gray, or brown; stems densely glandular-pubescent. **Leaves:** blade bright green, spatulate, 8–35 × 2–15 mm, fleshy, surfaces densely glandular-pubescent. **Inflorescences** 2–3-flowered fascicles or solitary flowers. **Pedicels** 4–25 mm. **Flowers** bisexual or pistillate, 5-merous; calyx tubular, 2–10 mm, lobes to 1–2 mm, glandular-puberulent; corolla deep lavender to purple, tubular to funnelform, 8–20 mm, lobes 2–8 mm; stamens included to slightly exserted. **Berries** red, ovoid, 5–9 mm, fleshy. **Seeds** 40–60. **2*n*** = 96, 120.

Flowering Jan–Apr. Sandy washes, saline flats (Sonoran Desert); 100–1300 m; Ariz., Calif.; Mexico (Baja California, Baja California Sur, Sonora).

Lycium fremontii occurs in the Sonoran Desert of Arizona, southern California, and northwestern Mexico. Populations of *L. fremontii* are morphologically gynodioecious (functionally dioecious), and plants are sexually dimorphic for flower size. Pistillate plants are often covered with orange-red berries in March and April. Plants are robust; in southern Arizona they often thrive at the edges of agricultural fields, where there is water run-off. This species co-occurs with *L. andersonii*, *L. berlandieri*, and *L. californicum*. However, the combination of larger bright green leaves, deep lavender flowers, floral dimorphism, and considerable glandular pubescence differentiates this species.

14. Lycium exsertum A. Gray, Proc. Amer. Acad. Arts 20: 305. 1885 • Arizona desert-thorn

Shrubs erect, 1–4 m; bark dark gray to brown; stems densely glandular-pubescent. **Leaves:** blade spatulate, 5–25 × 3–10 mm, surfaces densely glandular-pubescent. **Inflorescences** 2–3-flowered fascicles or solitary flowers. **Pedicels** 3–6 mm. **Flowers** bisexual or pistillate, 5-merous; calyx tubular-campanulate, 2.5–6 mm, lobe lengths 0.25–0.5 times tube; corolla greenish white to lavender, funnelform, 7–14 mm, lobes 1–2 mm; stamens exserted 2–3+ mm in bisexual flowers. **Berries** red, ovoid, 6–8 mm, fleshy. **Seeds** 20–35. **2*n*** = 48.

Flowering Jan–Apr. Desert washes, bajadas (Sonoran Desert); 300–1400 m; Ariz.; Mexico (Baja California, Sinaloa, Sonora).

Populations of *Lycium exsertum* are morphologically gynodioecious (functionally dioecious), and plants are sexually dimorphic for flower size. *Lycium exsertum* can be differentiated from *L. fremontii* by its light purple pendent flowers, often considerably exserted stamens or stigma, and a more upland habitat.

15. Lycium chinense Miller, Gard. Dict. ed. 8, Lycium no. 5. 1768 • Chinese wolfberry, matrimony vine, goji [I]

Lycium chinense var. *potaninii* (Pojarkova) A. M. Lu

Shrubs erect, 0.5–3 m; bark pale gray; stems glabrous. **Leaves** subsessile; blade ovate to linear-lanceolate, 15–100 × 5–40 mm, surfaces glabrous. **Inflorescences** 2–4-flowered fascicles or solitary flowers. **Pedicels** 10–20 mm. **Flowers** 5-merous; calyx campanulate, 2–4 mm, lobe lengths 0.3–1 times tube; corolla pale purple, funnelform, 9–14 mm, lobe lengths ± equal to tube; stamens slightly included to exserted. **Berries** red, ovoid, 7–22 mm, fleshy. **Seeds** 50+. **2*n*** = 24.

Flowering May–Aug. Waste places, roadsides; to 1000 m; introduced; Ont.; Calif., Conn., Ga., Ill., Ky., Maine, Md., Mass., Mich., Mo., N.H., N.Y., N.C., Ohio, Oreg., Pa., Tenn., Vt., Va.; Asia (China); introduced also in Europe and elsewhere in Asia (Japan, Korea, Mongolia, Nepal, Pakistan, Taiwan, Thailand).

Lycium chinense is difficult to differentiate from *L. barbarum*. It is likely that records for *L. chinense* in the flora area include both *L. barbarum* and *L. chinense*; they have been treated as conspecific in some floras. Jill S. Miller et al. (2011), using material collected in China, demonstrated that they are distinct. They are distinguished mainly by the length of the corolla lobes relative to the corolla tube. In China, *L. chinense* has been cultivated for centuries for its medicinal properties, as well as for use in cooking. Young shoots, leaves, and berries (both fresh and dried) are commonly consumed. The root bark is used for relief of fever and cough, and the seed oil is used in cooking and as a lubricant.

16. Lycium barbarum Linnaeus, Sp. Pl. 1: 192. 1753 • Matrimony vine, Ningxia goji, goji berry, lyciet de Barbarie [I] [W]

Lycium barbarum var. *auranticarpum* K. F. Ching; *L. halimifolium* Miller

Shrubs erect, 0.8–3 m; bark silvery tan; stems glabrous. **Leaves:** blade lanceolate to oblong, 20–60 × 3–35 mm, surfaces glabrous. **Inflorescences** 2–4-flowered fascicles or solitary flowers. **Pedicels** 10–20 mm. **Flowers** 4–6-merous; calyx campanulate, often 2-lobed, 3–5 mm, lobes 1–2 mm; corolla lavender to purple, funnelform, 8–13 mm, lobe lengths 0.5–1 times tube; stamens exserted. **Berries** red or orange-yellow, ovoid, 4–20 mm, fleshy. **Seeds** 4–20. **2*n*** = 24.

Flowering Mar–Oct. Waste places, roadsides, fields; 0–2300 m; introduced; Alta., B.C., N.S., Ont., Que., Sask.; Ala., Ark., Calif., Colo., Conn., Del., D.C., Ga., Idaho, Ill., Ind., Iowa, Kans., Ky., La., Maine, Md., Mass., Mich., Minn., Mo., Mont., Nebr., Nev., N.H., N.J., N.Mex., N.Y., N.C., N.Dak., Ohio, Okla., Oreg., Pa., R.I., S.C., S.Dak., Tenn., Tex., Utah, Vt., Va., Wash., W.Va., Wis., Wyo.; Asia (China); introduced also in Eurasia, Pacific Islands (New Zealand), Australia.

Lycium barbarum is naturalized across North America, Europe, and Asia, as well as Australia and New Zealand. It is commonly cultivated in northern China, especially in Ningxia province. The plants have uses from medicinal to tea and wine. See discussion of 15. *L. chinense* for confusion between these two introduced species, especially in the flora area.

17. Lycium brevipes Bentham, Bot. Voy. Sulphur, 40. 1844 • Baja desert-thorn

Lycium brevipes var. *hassei* (Greene) C. L. Hitchcock; *L. cedrosense* Greene; *L. hassei* Greene; *L. richii* A. Gray; *L. verrucosum* Eastwood

Shrubs usually erect, sometimes prostrate, 1–3 m; bark tan to brown; stems glandular-puberulent. **Leaves:** blade spatulate to obovate, 3–30 × 3–19 mm, fleshy, surfaces glabrous or puberulent. **Inflorescences** 2–3-flowered fascicles or solitary flowers. **Pedicels** 1–10 mm. **Flowers** 5-merous; calyx campanulate, 2–6 mm, lobe lengths to 0.5–1 times tube; corolla lavender to white, sometimes with deep purple markings, campanulate to tubular, 4–10 mm, lobes 3–5 mm; stamens exserted. **Berries** red, ovoid, 10 mm, fleshy. **Seeds** 50+. **2*n*** = 24.

Flowering Mar–Apr. Coastal dunes, flood plains (mainly in Sonoran Desert); 0–600 m; Calif.; Mexico (Baja California, Baja California Sur, Sinaloa, Sonora).

Corolla color varies across the range of *Lycium brevipes* and plants may be spreading or erect. Within the flora area, *L. brevipes* is restricted to southern California, mainly along the northern coastline of the Salton Sea. It is rare in, or extirpated from, the California Channel Islands (historically it was collected on the southern islands).

18. Lycium parishii A. Gray, Proc. Amer. Acad. Arts 20: 305. 1885 • Parish's desert-thorn

Lycium modestum I. M. Johnston; *L. parishii* var. *modestum* (I. M. Johnston) F. Chiang

Shrubs erect, 1–3.5 m; bark silvery to brown; stems glandular-pubescent. **Leaves:** blade spatulate, 3–12 × 1–5 mm, surfaces densely pubescent. **Inflorescences** 2-flowered fascicles or solitary flowers, erect. **Pedicels** 2–10 mm. **Flowers** 5-merous; calyx campanulate, 2–6 mm, lobe lengths 0.5–1 times tube; corolla pale lavender to purple, narrowly campanulate to funnelform, 6–10 mm, lobes 2–3 mm; stamens exserted. **Berries** red, ovoid, 4–7 mm, fleshy. **Seeds** 7–15. **2*n*** = 24.

Flowering Feb–Apr. Desert washes, bajadas (Sonoran Desert); 200–1200 m; Ariz., Calif., Nev.; Mexico (Baja California, Baja California Sur, Coahuila, San Luis Potosí, Sonora).

Within the flora area, *Lycium parishii* occurs in Arizona, southern California, and southern Nevada (Clark County).

18. NECTOUXIA Kunth in A. von Humboldt et al., Nov. Gen. Sp. 3(fol.): 8; 3(qto.): 10; plate 193. 1818 • Trans-Pecos stinkleaf, puckering nightshade [For Hippolyte Nectoux, 1759–1836, French botanist who accompanied Napoleon to Egypt]

Ellen A. Dean

Herbs, perennial, rhizomatous, pubescent, hairs simple, glandular. **Stems** erect or ascending, branched. **Leaves** alternate or geminate. **Inflorescences** axillary, solitary flowers. **Flowers** 5-merous; calyx campanulate, incised nearly to base, lobes 5, linear-subulate, lengthening in fruit; corolla yellow-green, black when dried, radial, salverform, corona present between tube and limb; stamens inserted beyond middle of corolla tube, equal; anthers basifixed, oblong to lanceolate, (connective with 2 wings), dehiscing by longitudinal slits; ovary 2-carpellate; style filiform; stigma ± expanded, capitate or 2-lobed. **Fruits** berries, narrowly ovoid, juicy. **Seeds** reniform-lenticular.

Species 1: Texas, Mexico.

Nectouxia is closely related to the South American genus *Salpichroa*; the corollas of both genera turn black upon drying (A. T. Hunziker 2001). C. Carrizo García et al. (2018) concluded that *Nectouxia* and *Salpichroa* should be combined into a single genus. In that case *Nectouxia* would have priority, requiring name changes for over 20 *Salpichroa* species. A proposal is under consideration to conserve the name *Salpichroa* over *Nectouxia* (G. E. Barboza et al. 2016).

Stems of *Nectouxia* exhibit sympodial growth. The branching from the nodes of the first sympodium is poorly to very well developed.

SELECTED REFERENCE Carrizo García, C. et al. 2018. Unraveling the phylogenetic relationships of *Nectouxia* (Solanaceae): Its position relative to *Salpichroa*. Pl. Syst. Evol. 304: 177–183.

1. **Nectouxia formosa** Kunth in A. von Humboldt et al., Nov. Gen. Sp. 3(fol.): 8; 3(qto.): 11; plate 193. 1818 [F]

Rhizomes elongate. **Stems** to 4[–7] dm; first above-ground sympodium terminating after 5–11[–14] spirally arranged leaves, subsequent sympodia with 1 or 2 leaves, largest of these distal sympodia to 8 cm × 4 mm; vegetative parts ± glandular-puberulent throughout, hairs erect [appressed], fetid-smelling. **Leaves:** blade widely to narrowly ovate, largest 1–8[–9.5] × 0.6–5.5[–6] cm (smaller toward tips and base of plant), base truncate or cordate [rounded], sometimes unequal, apex acute [acuminate]. **Flowering pedicels** ascending to curved, 2–20[–33] mm, shorter than calyx. **Flowers:** calyx densely hairy, tube 1–3 × [1.7–]2–3 mm, lobes [5–]6–13[–22] × 0.5–1[–1.5] mm; corolla glabrate, tube 8–25[–30] × [2–]2.5–4[–5] mm, limb reflexed at maturity, lobes [3–]5–14[–23] × 2–6[–14] mm, corona [1–]1.5–2.5 × 2–2.5[–4] mm; filaments [0–]1–2 mm, filiform; anthers yellow, 3[–4] mm; nectariferous disc present; ovary conic to narrowly oblong, 2–3.5[–4] mm; style often kinked near middle, sometimes exserted to 3 mm beyond corona, 15–22[–28] mm. **Berries** 15–30 × 6–9 mm. **Seeds** 10–25 per fruit, tan, round to oval, flattened, 2–2.5 × 1.5–2 mm, minutely pitted.

Flowering late spring–early fall. Rocky slopes, sometimes pinyon pine forest understory; 2100–2300 m; Tex.; Mexico.

In the flora area, *Nectouxia formosa* is known only from Emory Peak in the Chisos Mountains of Big Bend National Park. Its distribution extends to southern Mexico. In Mexico, it has often been collected in disturbed areas such as pastures, roadsides, and cultivated fields.

The fruits have been reported to be edible (J. W. Harshberger 1898). Reports that *Nectouxia formosa* is annual in Texas are erroneous. The chromosome number is unknown.

19. NICANDRA Adanson, Fam. Pl. 2: 219, 582. 1763, name conserved • Apple of Peru, shoo-fly plant [For Nicander of Colophon, second century B.C.E., Greek physician and poet known for his works on toxicology and natural history] [I]

Maggie Whitson

Herbs, annual, glabrous or sparsely pubescent, hairs simple, mostly eglandular, taproot slender, somewhat fibrous. **Stems** branched. **Leaves** alternate; blade simple or margins slightly lobed. **Inflorescences** axillary, solitary flowers. **Flowers** 5-merous; calyx accrescent, campanulate, incised 2/3 its length, lobes 5, bases sagittate to cordate; corolla light purple to nearly white, radial, broadly campanulate, shallowly incised; stamens inserted at base of corolla tube, equal; anthers basifixed, oblong, dehiscing by longitudinal slits; ovary 3–5-carpellate; style filiform; stigma capitate-lobed. **Fruits** berries, globose, dry. **Seeds** discoidal to reniform. $x = 10$.

Species 3 (1 in the flora): introduced; South America, introduced also in Mexico, Central America, Africa, Pacific Islands.

Nicandra has been considered a close relative of *Physalis*. The resemblance is superficial; current molecular data neither place this genus with *Physalis* and its relatives nor indicate a particularly close relationship to any other genus. Although *Nicandra* comes out near *Exodeconus* Rafinesque and *Solandra* Swartz (R. G. Olmstead et al. 2008), it is morphologically distinct; A. T. Hunziker (2001) left it in its own tribe (the Nicandreae Lowe).

1. **Nicandra physalodes** (Linnaeus) Gaertner, Fruct. Sem. Pl. 2: 237. 1791 (as physaloides) • Nicandre faux-coqueret [F] [I] [W]

Atropa physalodes Linnaeus, Sp. Pl. 1: 181. 1753; *Boberella nicandra* E. H. L. Krause; *Calydermos erosus* Ruiz & Pavon; *Pentagonia physalodes* (Linnaeus) Hiern

Stems hollow, ridged, angular in cross section, 5–10(–20) dm, glabrous or sparsely pubescent at nodes. **Leaves:** blade ovate to oblong or elliptic, 3–20(–31) × 2–10(–20) cm, margins sinuate-dentate, irregularly dentate, or lobed, surfaces sparsely pubescent. **Flowering pedicels** 20–30 mm, puberulent. **Flowers:** calyx 1–2 cm, lobes partially connate; corolla (1.5–)2–3 × 2–3.5 cm, throat white, often with 5 purple spots at base; stamens included; anthers yellow; style 3–5 mm, sparsely pubescent. **Berries** tan, 3–5-locular, 1–2 cm diam., very thin-walled, cracking easily when dry, enclosed in papery, expanded calyx, 2.5–3.5 × 2.5–3.5 cm. **Seeds** dark brown, 1–2 mm. $2n = 20$.

Flowering Jul–Oct (year-round in tropical areas). Cultivated fields, waste ground, dumps, old garden sites, rocky balds, roadsides; 0–1100 m; introduced; B.C., N.S., Ont., P.E.I., Que.; Ala., Ark., Calif., Colo., Conn., Del., D.C., Fla., Ga., Idaho, Ill., Ind., Iowa, Kans., Ky., Maine, Md., Mass., Mich., Minn., Miss., Mo., Nebr., N.H., N.J., N.Y., N.C., N.Dak., Ohio, Okla., Pa., R.I., S.C., Tenn., Vt., Va., Wash., W.Va., Wis.; South America; introduced also in Mexico, Central America, Africa, Pacific Islands (Hawaii).

Nicandra physalodes is cultivated for its ornamental flowers and fruits. Historically, *Nicandra* was grown around farmsteads and used as a source of fly poison, thus the common name shoo-fly plant.

Nicandra is most likely to naturalize in areas with ample precipitation and warm climates. Many of the collections seen are old (pre-1950); they are indicative of areas in which *N. physalodes* can naturalize. Although scattered wild populations of *N. physalodes* have been documented from across the United States and Canada, these plants are more common in the eastern half of the United States and are probably uncommon in most of the rest of the range. In most areas, *N. physalodes* may reseed itself in garden sites and persist in this way for a year or two. Escaped plants of this type should occasionally be found throughout the continental United States and southern Canada.

20. NICOTIANA Linnaeus, Sp. Pl. 1: 180. 1753; Gen. Pl. ed. 5, 84. 1754 • Tobacco [For Jean Nicot, 1530–1600, French ambassador at Lisbon, who sent tobacco plants and/or seeds to French court ca. 1560]

Sandra Knapp

Herbs, shrubs, or small trees, with soft wood, annual, biennial, or short-lived perennial, often rosette-forming, from taproot or rarely horizontal rootstock (forming colonies), sparsely to densely viscid-pubescent with simple hairs, rarely glabrous. **Stems** usually erect, sometimes branching from base. **Leaves** alternate, densely clustered in rosette-forming species, sessile or petiolate; petiole often winged; blade simple, margins entire or irregularly and obscurely crenate and sometimes undulate. **Inflorescences** terminal or apparently axillary, cymose, usually forming false racemes, or glomerulate. **Flowers** bisexual, 5-merous (fasciated with increased parts only in cultivars), radially symmetric or somewhat bilaterally symmetric, especially in androecium; calyx tubular or narrowly campanulate, 5-lobed, lobes persistent, usually deltate or triangular, equal or unequal, usually slightly accrescent and mostly enclosing capsule; corolla white to cream, variously marked or tinged with pink or purple, or yellow-green, radial or more commonly at least somewhat bilateral, tubular, funnelform, or salverform, limb deeply 5-lobed to ± entire, lobes, if present, rounded to deltate, sometimes emarginate; stamens 5, inserted variously from near base of corolla tube to near apex, sometimes unequal (2 + 2 + 1), on equal filaments or filaments of unequal length with one inserted at a different level and usually shorter than the other 4; anthers dorsifixed (or appearing basifixed), ellipsoid to globose, dehiscing by longitudinal slits; ovary 2-carpellate (irregularly more in some cultivars); style filiform, straight or curved; stigma capitate or slightly 2-lobed. **Fruits** capsules, usually ovoid, sometimes narrowly so, 2–4-valved (occasionally many-valved in cultivars), dehiscent apically with long septicidal cleft and shorter loculicidal cleft. **Seeds** angular to oblong (minute), occasionally somewhat reniform. x = (9, 10), 12, (16, 18, 19), 24.

Species ca. 75 (12 in the flora): North America, Mexico, West Indies, Central America, South America, s Africa, Australia; introduced widely.

Nicotiana is recognizable by its tubular flowers and usually sticky pubescence. Of the species occurring in North America, only *N. glauca* is not sticky-pubescent. The genus has long been of interest due the high number of allopolyploid species (T. H. Goodspeed 1954; M. W. Chase et al. 2003; E. W. McCarthy et al. 2015).

Most species of *Nicotiana* in the flora area have distinct basal rosettes until flowering occurs, and cauline leaves usually differ in morphology from those in the basal rosette. Many species occur as short-lived perennials in areas without frost, but occur as annuals where winters are harder. The tree tobacco (*N. glauca*) is the only truly woody species occurring in North America.

Species of *Nicotiana* that bloom in the evening usually have white or cream flowers and are pollinated by moths, while day-flowering species have yellow, pink, or cream flowers and are pollinated by bees; however, this overall pattern does not hold strictly true. *Nicotiana attenuata* has become a model system for the study of the complex interplay between pollinators and herbivores in the evolution of floral and other traits (for example, C. Diezel et al. 2011; D. Kessler et al. 2015).

Nicotiana species are often cultivated and occur occasionally as garden or greenhouse escapes. *Nicotiana alata* Link & Otto and *N.* ×*sanderae* W. Watson have been reported for the flora area but are known only as ephemerals or from cultivation. Some records of *N. alata* are possible misidentifications for *N. longiflora*, from which *N. alata* differs in its strongly decurrent cauline leaves and larger corollas with wider tubes. *Nicotiana* ×*sanderae* has striking red salverform corollas; various horticultural hybrids of this long-flowered species are possible ephemeral occurrences on waste heaps.

The two tobaccos of commerce, *Nicotiana rustica* and *N. tabacum*, are cultivated as ornamentals or as sources of leaves for human use. Although most herbarium specimens from the flora area are clearly from cultivation, both taxa are sometimes adventive.

The small seeds and weedy habits of *Nicotiana* species predispose them to become invasive; for example, the non-native species *N. glauca* is a common component of ecosystems in western North America. This propensity to grow in disturbed areas means cultivated species are likely to become established briefly in areas where soil is loose or in old garden sites.

SELECTED REFERENCES Chase, M. W. et al. 2003. Molecular systematics, GISH and the origin of hybrid taxa in *Nicotiana* (Solanaceae). Ann. Bot. (Oxford) 92: 107–127. Clarkson, J. J. et al. 2004. Phylogenetic relationships in *Nicotiana* (Solanaceae) inferred from multiple plastid DNA regions. Molec. Phylogen. Evol. 33: 75–90. Goodspeed, T. H. 1954. The genus *Nicotiana*. Chron. Bot. 16. Knapp, S., M. W. Chase, and J. J. Clarkson. 2004. Nomenclatural changes and a new sectional classification in *Nicotiana* (Solanaceae). Taxon 53: 73–82. McCarthy, E. W. et al. 2015. The effect of polyploidy and hybridization in the evolution of flower colour in *Nicotiana* (Solanaceae). Ann. Bot. (Oxford) 115: 1117–1131.

1. Tubular portion of corollas (tube + throat) greenish yellow, pale yellow, or bright yellow, often with green limb; cauline leaves petiolate.
   2. Leaf blades glaucous, glabrous; corolla tubes bright yellow to greenish yellow, limb usually a distinct color from tube, often bright green; small trees or shrubs . . . . 4. *Nicotiana glauca*
   2. Leaf blades not glaucous, viscid-pubescent; corolla tubes pale yellow to greenish yellow, limb same color as tube; short-lived perennials or annual herbs . . . . 10. *Nicotiana rustica*
1. Tubular portion of corollas (tube + throat) ivory, cream, white (often tinged green, gray, or purple), pink, or occasionally red (*N. tabacum*); cauline leaves petiolate, sessile, or variously clasping.
   3. At least some cauline leaves with a distinct petiole (very short in *N. clevelandii* and *N. quadrivalvis*).
      4. Tubular portion of corollas greater than 2 times calyx length (including lobes); inflorescences usually branched, somewhat (not densely) leafy.
         5. Corolla limbs circular, lobes deltate or emarginate; styles longer than longest stamens; calyces with dark midveins, trichomes without enlarged bases . . . . 1. *Nicotiana acuminata*
         5. Corolla limbs pentagonal (occasionally circular), lobes broadly triangular; styles shorter than longest stamens; calyces uniformly green, trichomes with enlarged bases . . . . 2. *Nicotiana attenuata*
      4. Tubular portion of corollas less than or equaling 2 times calyx length (including lobes), if greater than 2 times calyx length then limb greater than 2 cm diam.; inflorescences usually unbranched, densely leafy to bracteate.
         6. Stamens included in throat; corolla limbs to 1 cm diam.; calyx lobes unequal, one longer than tubular portion of calyx . . . . 3. *Nicotiana clevelandii*
         6. Stamens exserted from throat; corolla limbs 2+ cm diam.; calyx lobes ± equal . . . . 8. *Nicotiana quadrivalvis*
   3. Cauline leaves sessile or with variously clasping bases (proximalmost near rosette sometimes appearing petiolate).
      7. Corolla tubes pale greenish cream to pink or red, throat 0.5 cm diam., tubular portion straight or strongly curved and dilated distally. . . . . 12. *Nicotiana tabacum*
      7. Corolla tubes white or cream (occasionally appearing pale yellow in very old flowers), often tinged with green, gray, or purple, throat 0.1–0.6 cm diam., tubular portion straight, variously shaped.

[8. Shifted to left margin.—Ed.]

8. Corollas usually 5 cm or shorter.
    9. Corolla length (excluding limb) to 2 times limb diam.; basal leaves not in distinct rosette . . . . 6. *Nicotiana obtusifolia*
    9. Corolla length (excluding limb) 2+ times limb diam.; basal leaves in distinct rosette.
        10. Anthers to 0.1 mm (filaments free for at least some of their length); corollas glabrous or minutely puberulent (not cobwebby-pubescent) internally; distal cauline leaf blades lanceolate to linear-lanceolate . . . . 7. *Nicotiana plumbaginifolia*
        10. Anthers appearing sessile (filaments fused to corolla throughout their entire length); corollas cobwebby-pubescent internally; distal cauline leaf blades oblong-ovate or pandurate . . . . 9. *Nicotiana repanda* (in part)
8. Corollas usually longer than 5 cm.
    11. Tubular portion of corollas inflated and ventricose in middle to distal 1/3, slightly curved, corollas white; inflorescences panicles with congested branches, appearing moplike . . . . 11. *Nicotiana sylvestris*
    11. Tubular portion of corollas gradually widening distally, straight, corollas white or grayish white; inflorescences false racemes, occasionally few-branched.
        12. Corolla throat constricted; filaments unequal, free for at least some of their length (anthers not sessile); cauline leaves sessile, blade lanceolate or linear, base auriculate . . . . 5. *Nicotiana longiflora*
        12. Corolla throat gaping; filaments equal, fused to corolla along their entire length (anthers appearing sessile); cauline leaves sessile, blade pandurate or oblong-ovate, base clasping. . . . . 9. *Nicotiana repanda* (in part)

1. Nicotiana acuminata (Graham) Hooker, Bot. Mag. 56: plate 2919. 1829 • Manyflower tobacco [I] [W]

Petunia acuminata Graham, Edinburgh New Philos. J. 5: 378. 1828; *Nicotiana acuminata* var. *multiflora* Reiche

Herbs, annual, robust, from loose basal rosette. **Stems** branched from base (proximal branches longer), 5–20 dm, viscid-pubescent. **Rosette leaves:** petiole length equaling blade; blade ovate or orbiculate, 6–12 cm, surfaces viscid-pubescent, somewhat scabrous adaxially. **Cauline leaves:** petiole length 1/3–1/6 blade; blade elliptic to lanceolate, becoming very narrow near inflorescence, 10–25 cm, apex acuminate, surfaces viscid-pubescent, somewhat scabrous adaxially. **Inflorescences** few-branched, somewhat leafy; flowering crepuscular. **Pedicels** 0.5–2 cm. **Flowers:** calyx strongly purple-veined, 1–2 cm, membranous, viscid-pubescent (hairs without swollen bases), lobes long-triangular, unequal, longest equaling tube; corolla straight, 2.5–10 cm (excluding limb), viscid-puberulent externally, tube white or white tinged with greenish purple, sometimes striped, 0.8–4 cm × 2–3 mm, widening to throat 1–4 × 5 mm, glabrous or minutely puberulent internally, limb spreading, white, circular, 2–4 cm diam., lobes shallow, deltate or emarginate; stamens inserted near base of throat, included; filaments unequal, in 2 equal or unequal pairs, 4 cm, 1 filament shorter than either pair, pubescent proximally; style straight, just surpassing longest stamen pair. **Capsules** broadly ovoid, 1–1.2 cm. **Fruiting calyces** tearing along membranous sinuses, covering ca. 1/2 of mature capsule. **Seeds** 0.9–1 mm. $2n = 24$.

Flowering Dec–Jul. Open sandy or gravelly areas; 0–2000 m; introduced; Calif., Nev., Oreg., Wash.; South America (Chile); introduced also in Mexico (Baja California).

Nicotiana acuminata is native to Chile and is naturalized on the West Coast from Baja California to Washington. It is easy to confuse with the native *N. attenuata* but differs in its much longer corolla tube and usually emarginate limb. The flowers are often clustered near the tips of the few inflorescence branches. Plants from the western United States are often identified as var. *multiflora*.

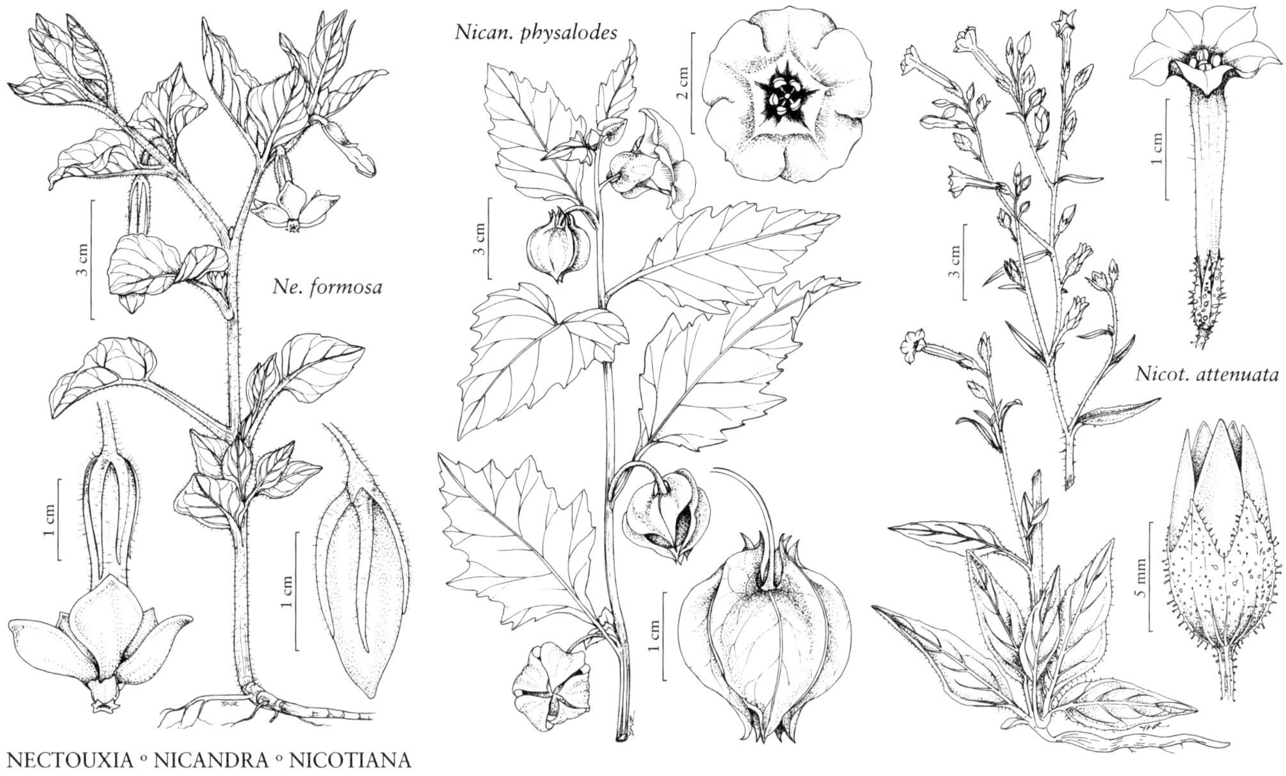

NECTOUXIA ◦ NICANDRA ◦ NICOTIANA

2. **Nicotiana attenuata** Torrey ex S. Watson, Botany (Fortieth Parallel), 276, plate 27, figs. 1, 2. 1871
• Coyote tobacco [F]

Herbs, annual, robust, from basal rosette. **Stems** single or with few weak lateral branches, 5–20 dm, viscid-pubescent or glabrate with few hairs with swollen bases. **Rosette leaves:** petiole length shorter than or almost equaling blade (2–3 cm); blade elliptic or oblong-elliptic, 5–10 cm, surfaces viscid-pubescent. **Cauline leaves:** petiole length to 1/2 blade, distal leaves sessile; blade lanceolate to linear, progressively narrower distally, 2–8(–10.5) cm, apex acute to acuminate, surfaces viscid-pubescent or glabrate. **Inflorescences** unbranched or with few short lateral branches, somewhat leafy; flowering crepuscular (early morning). **Pedicels** 0.2–0.4 cm. **Flowers:** calyx uniformly green or with weakly developed veins, 0.6–1 cm, densely viscid-pubescent (hairs with swollen bases), lobes narrowly triangular, shorter than tube, unequal; corolla straight, 2–3.5 cm (excluding limb), viscid-puberulent externally, tube creamy white or tinged with purple-green or gray-green, 0.5–0.7 cm × 1.5–2 mm, widening to throat 2–3 × 5 mm (asymmetrically dilated distally), glabrous or minutely puberulent within, limb spreading or slightly reflexed, white or cream, pentagonal to ± circular and often asymmetrically spreading, 0.4–0.8 cm diam., lobes shallow and rounded or obtuse, broadly triangular (proximal lobes reflexed); stamens inserted near base of throat, included; filaments unequal, 4 of 3 cm (2 of these slightly longer), 1 shorter, 1.5–2 cm, glabrous or minutely pubescent proximally; style straight, just shorter than longest stamen pair. **Capsules** ovoid, 0.8–1.2 cm. **Fruiting calyces** not tearing at sinuses, covering to 1/2 of mature capsule. **Seeds** 0.8 mm. **2*n*** = 24.

Flowering May–Nov. Sandy slopes, banks and rocky outcrops, disturbed places, often appearing after fire; (50–)1000–2600 m; B.C.; Ariz., Calif., Colo., Idaho, Mont., Nev., N.Mex., Oreg., Tex., Utah, Wash., Wyo.; Mexico (Baja California, Sinaloa, Sonora).

Nicotiana attenuata is the most common species of the genus in the Great Basin, and often forms large colonies after fires and other disturbance. It has been the subject of intensive study over many years by the Max Planck Institute for Chemical Ecology at the Lytle Ranch Preserve in southwestern Utah (for example, C. Diezel et al. 2011; D. Kessler et al. 2015). There are some records of its use by Native American peoples as a smoking or chewing tobacco; it has also been reported as a medicinal plant used by the Zuni people (M. C. Stevenson 1915). *Nicotiana torreyana* A. Nelson & J. F. Macbride is an illegitimate name that has been applied to this species.

3. Nicotiana clevelandii A. Gray in A. Gray et al., Syn. Fl. N. Amer. 2(1): 242. 1878 (as clevelandi) • Cleveland's tobacco

Nicotiana greeneana Rose

Herbs, annual, from basal rosette. **Stems** single or with few lateral branches (several robust basal branches on older plants), 2–6 dm, villous and viscid-pubescent. **Rosette leaves:** petiole length 1/8–1/2 blade; blade broadly elliptic to rhombic-ovate, 6–20 cm, surfaces softly viscid-pubescent. **Cauline leaves** sessile to short-petiolate; blade ovate, 1–6 cm, becoming smaller distally, (fleshy), apex acute, surfaces scabrous with persistent swollen bases of short, patent trichomes. **Inflorescences** unbranched, few-flowered, densely leafy; flowering crepuscular. **Pedicels** 0.2–0.5 cm. **Flowers:** calyx uniformly green with poorly developed membranous sinuses, 0.8–1 cm, viscid-pubescent, lobes subulate, unequal (4 as long as tube, 1 longer); corolla straight, 1.4–2 cm (excluding limb), minutely viscid-pubescent externally, tube white or tinged with purple externally, 0.3–0.4 cm × 1 mm, widening to throat 10–17 × 2–4 mm (oblique, often bent at junction of tube and throat), glabrous or minutely puberulent within, limb spreading, white, pentagonal to stellate, 0.6–0.8(–1) cm diam., lobes broadly triangular, unequal, acute; stamens inserted at base of throat, extending nearly to corolla mouth; filaments 1–1.5 cm, unequal, 1 much shorter (curved), glabrous; style straight, just exceeding 4 long stamens. **Capsules** ovoid, 0.4–0.6 cm. **Fruiting calyces** not tearing at sinuses, covering capsule. **Seeds** 0.5 mm. **2*n*** = 48.

Flowering Feb–Jul. Sandy areas, dunes, sea cliffs, washes, desert slopes; 0–500 m; Ariz., Calif.; Mexico (Baja California, Baja California Sur, Sonora).

4. Nicotiana glauca Graham, Bot. Mag. 55: plate 2837. 1828 • Tree tobacco [I] [W]

Small trees or shrubs. Stems usually branched near base, occasionally with distinct trunk (branches drooping), 10–60 (–100) dm, glabrous, somewhat glaucous. **Cauline leaves:** petiole length 1/2 blade (not winged); blade ovate to lanceolate, 5–25 cm, base acute or cordate, apex rounded, (rubbery), surfaces glabrous, glaucous. **Inflorescences** branched, not leafy; flowering diurnal. **Pedicels** 0.3–1 cm. **Flowers:** calyx green, (evenly cylindric), 1–1.5 cm, without membranous sinuses, glabrous or minutely pubescent, lobes sharply triangular, equal, much shorter than tube; corolla straight, 2.5–4.5 cm (excluding limb), glabrous or finely pubescent externally, tube bright yellow to greenish yellow, (cylindric to clavate, slightly constricted apically), 0.5–0.8 cm × 3 mm, widening slightly to throat 1.5–4 cm × 6–8 mm, glabrous within, limb assurgent, greenish yellow or bright green (usually distinct color from tube in young flowers), turning yellow and same color as tube with age, circular or pentagonal, 0.6–0.8 cm diam., lobes rounded, broadly triangular, equal; stamens inserted at base of throat, extending to corolla mouth; filaments subequal 2.5–4.5 cm, (geniculate at base), glabrous; style straight, exceeding stamens and exserted from corolla mouth. **Capsules** ovoid, 0.7–1.5 cm. **Fruiting calyces** not tearing along sinuses, covering mature capsule. **Seeds** 0.5 mm. **2*n*** = 24.

Flowering year-round. Open areas along roads, disturbed habitats, often in Mediterranean vegetation; 0–2600 m; introduced; Ala., Ariz., Calif., Fla., Ga., Miss., Nev., N.Mex., Tex.; South America (Argentina, Bolivia, Chile); introduced also in Mexico, Europe (France, Greece, Italy, Spain, Turkey), sw Asia (Israel, Lebanon), Africa, Pacific Islands (New Zealand), Australia.

Nicotiana glauca is registered as an invasive plant in the United States (www.invasives.org). It can form monodominant stands due to high seed set and germination success. It was originally introduced from Argentina to Mexico, thence to the United States and worldwide (T. H. Goodspeed 1954). *Siphaulax glabra* Rafinesque is an illegitimate, superfluous name for this species.

5. Nicotiana longiflora Cavanilles, Descr. Pl., 106. 1802 • Longflower tobacco [I]

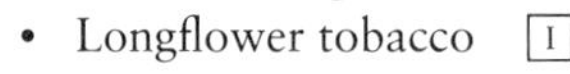

Nicotiana acuta Grisebach; *N. acutiflora* A. Saint-Hilaire

Herbs, annual or biennial, from basal rosette. **Stems** single or with few spreading lateral branches, 5–8(–10) dm, sparsely pubescent, usually not viscid, tuberculate. **Rosette leaves** sessile; blade elliptic to oblanceolate, 10–30(–50) cm, base narrowed and winged, surfaces coarsely viscid-pubescent. **Cauline leaves** sessile; blade ovate to lanceolate or linear, 1–5 cm, progressively smaller and more linear towards inflorescence, base auriculate, apex acute to attenuate, surfaces coarsely viscid-pubescent. **Inflorescences** false racemes, occasionally with few weak branches, not leafy; flowering crepuscular. **Pedicels** 0.5–1.3 cm. **Flowers:** calyx green or occasionally somewhat purplish-tinged, 1.5–2.5 cm, tube elliptic, 10-ribbed, sinus membranes long and transparent, minutely pubescent, sometimes viscid, lobes usually spreading, subulate, ± equal, equaling tube; corolla straight, 4–12 cm (excluding limb), puberulent externally, tube and throat not well differentiated, straight, white or often grayish white, 2 mm diam. at base, gradually widening and somewhat broader in distal 1/4, abruptly swollen to 6 mm diam. just below constricted mouth, glabrous or minutely puberulent internally, limb spreading, sometimes with purplish-gray veins abaxially, adaxially white or ivory, stellate, 2–5 cm diam., lobes triangular to deltate, acute; stamens inserted in upper part of tube just below mouth, included; filaments unequal, free for at least some of their length (anthers not sessile),four 0.1–0.8 cm (2 of these slightly longer), 1 shorter, ca. 0.1 cm, glabrous; style straight, just exceeding stamens, exserted from corolla mouth. **Capsules** ovoid, 1.1–1.6 cm. **Fruiting calyces** not tearing at scarious sinuses, nearly covering capsule, lobe tips spreading. **Seeds** 0.5 mm. **2*n*** = 20.

Flowering Apr–Aug. Open fields, stream banks, wet places, ballast sites near ports; 0–200 m; introduced; Ont., Que.; Ala., Fla., Ga., Ill., Ind., La., Mass., Miss., Mo., Tex., W.Va.; South America (Argentina, Bolivia, Paraguay); introduced also in Europe (Germany, Sweden), Africa (South Africa).

Nicotiana longiflora is a relatively rare weed along rivers and in waste places. It could be confused with *N. plumbaginifolia*, with which it is sympatric along the Gulf Coast, but differs from that species in its much larger flowers and its strongly 10-ribbed calyx. In a vegetative state, the two species are very difficult to distinguish.

6. **Nicotiana obtusifolia** M. Martens & Galeotti, Bull. Acad. Roy. Soc. Bruxelles 12(1): 129. 1845 • Desert tobacco [F] [W]

Nicotiana glandulosa Buckley; *N. multiflora* Nuttall ex Torrey; *N. palmeri* A. Gray; *N. trigonophylla* Dunal

Herbs, annual or short-lived perennial, without marked basal rosette. **Stems** branched from base, (slender, brittle), 5–10 dm, viscid-tomentose. **Cauline leaves:** petiole short and winged or leaves sessile; proximal blades oblanceolate to narrowly elliptic, distal blades panduriform to trigonate, 5–20 cm, smaller very near inflorescence, base somewhat clasping, apex acute or acuminate, surfaces densely viscid-pubescent. **Inflorescences** secund false racemes, occasionally few-branched, somewhat leafy; flowering diurnal. **Pedicels** 0.2–0.5 cm. **Flowers:** calyx green, cup-shaped, 10-ribbed, 0.8–2 cm, with minute membranous sinuses, pubescent and somewhat rough, lobes triangular to long-triangular, ± unequal, as long or longer than tube, tips slightly recurved; corolla straight, 1.2–3.5 cm (excluding limb), viscid-pubescent externally, tube gray-green, cream, or ± yellowish cream, 0.3–0.6 cm × 2–4 mm, widening to throat 6–12 × 5 mm, glabrous or minutely puberulent within, limb slightly reflexed to spreading, cream or whitish green, pentagonal, 0.6–8 cm diam., lobes broadly triangular and rounded apically, equal; stamens inserted at base of throat, included; filaments unequal, 4 straight, 1.2–3.4 cm, sometimes kneed, extending to corolla mouth, 1 shorter, 1–2 cm, not kneed, curved away from corolla, all slightly pubescent just above insertion point; style straight, just exceeding longest stamens. **Capsules** broadly ovoid (acute), 0.8–1.1 cm. **Fruiting calyces** not tearing at sinuses, covering lower 2/3 of capsule. **Seeds** 0.5 mm. **2*n*** = 24.

Flowering year-round. Rocky or gravelly areas in deserts; 0–2500(–2900) m; Ariz., Calif., Nev., N.Mex., Okla., Tex., Utah; Mexico.

7. **Nicotiana plumbaginifolia** Viviani, Elench. Pl., 26, plate 1. 1802, name conserved • Tex-Mex tobacco

Herbs, annual or occasionally biennial, from basal rosette. **Stems** single (slender and wiry), with long basal branches, 2–10 dm, tuberculate-hispid. **Rosette leaves:** petiole short and broad-winged or leaves sessile; blade spatulate, obovate, or oblanceolate, 5–30 cm, surfaces hispid. **Cauline leaves** sessile; proximal blades rounded to ovate, distal blades lanceolate to linear-lanceolate, 1–5 cm, base clasping, apex acuminate and often twisted, surfaces hispid. **Inflorescences** simple or few-forked to more rarely many times branched, few-flowered, not leafy; flowering crepuscular. **Pedicels** 0.3–0.7 cm. **Flowers:** calyx green or purplish green, elliptic to ovate, 10-ribbed, 0.8–1.3 cm, sinus membranes long, minutely hispid, lobes linear-subulate, equal, length ± equaling tube, tips somewhat spreading; corolla straight, 2.5–3.5 cm (excluding limb), puberulent, tube and throat not clearly differentiated, white or greenish-gray tinged, 2.5–3.5 cm, gradually widening from 1 mm to 2 mm diam., abruptly swollen to 4 mm diam., just below contracted mouth, glabrous or minutely puberulent internally, limb spreading or slightly reflexed, cream or white, stellate, 1 cm diam., lobes white adaxially, ivory or greenish purple or with purplish veins abaxially, ovate-acute; stamens inserted just below mouth (4 inserted 0.4 cm below mouth, one 0.4 cm lower), included; filaments free for at least some of their length (anthers not sessile), 4 nearly shorter than 0.1 mm and sometimes unequal, one 1 mm, glabrous; anthers to 0.1 mm; style straight, equaling or just exceeding stamens. **Capsules** narrowly ovoid, 0.8–1.1 cm. **Fruiting calyces** not tearing at sinuses, just covering capsule. **Seeds** 0.5 mm. **2*n*** = 24.

Flowering May–Jan. Moist ground, semishade, widespread in disturbed habitats; 0–2000 m; Fla., La., Tex.; Mexico; Central America; South America; introduced in s Asia (India, Taiwan).

8. Nicotiana quadrivalvis Pursh, Fl. Amer. Sept. 1: 141. 1813 • Indian or Bigelow's or Wallace's tobacco [E] [W]

Amphipleis quadrivalvis (Pursh) Rafinesque; *Dictyocalyx quadrivalvis* (Pursh) Hooker f. ex Walpers; *Nicotiana bigelovii* (Torrey) S. Watson; *N. bigelovii* var. *wallacei* A. Gray; *N. multivalvis* Lindley; *N. plumbaginifolia* Viviani var. *bigelovii* Torrey; *N. quadrivalvis* var. *bigelovii* (Torrey) DeWolf; *Polydiclis quadrivalvis* (Pursh) Miers

Herbs, annual, from basal rosette. **Stems** single or with multiple robust branches, 3–20 dm, moist, patent, viscid-pubescent. **Rosette leaves:** petiole length to 1/2 blade; blade elliptic to narrowly ovate, 1–1.5 cm, surfaces usually viscid-pubescent abaxially, glabrous adaxially. **Cauline leaves** sessile or short-petiolate, congested toward inflorescence; blade ovate to lanceolate, 1–4 cm, gradually decreasing in size and narrower distally, apex acute to acuminate, surfaces coarsely viscid-pubescent. **Inflorescences** unbranched or few-branched, leafy to bracteate; flowering crepuscular. **Pedicels** 0.4–1 cm (longer in fruit). **Flowers:** calyx green, elliptic, 10-ridged (or more in fasciated plants), 0.9–3.5 cm, densely viscid-pubescent, (trichomes occasionally with swollen bases), lobes linear, ± equal, length equaling or sometimes exceeding tube, sinus membranes long, transparent; corolla straight, 2–5 cm (excluding limb), minutely viscid-pubescent externally, tube and throat not well differentiated, white or ivory, or tinged purple externally and oily-glossy, broadly trumpet-shaped, tubular portion from 2 mm diam. at base to 7 mm diam. at mouth (much wider in fasciated varieties), glabrous or minutely puberulent internally, limb spreading, cream to white flushed with grayish purple (externally), stellate to pentagonal, 2–5 cm diam., lobes broadly triangular, acute; stamens unequal, 4 inserted just below mouth, exserted, 1 inserted ca. 1 cm deeper in throat, included; filaments ± equal, 0.3–1 cm, glabrous; style straight, just exceeding stamens. **Capsules** narrowly ovoid, 1–2.5 cm. **Fruiting calyces** not tearing at sinuses, completely enclosing capsule. **Seeds** 0.9 mm. **2*n*** = 48.

Flowering Apr–Oct(–Dec). Dry river beds, washes, gravel bars, mesas, plains, burned areas; 0–600 (–2000) m; B.C.; Ariz., Calif., Nev., Oreg., Wash.

Nicotiana quadrivalvis was widely cultivated by western Native American peoples, and prior to European expansion to the west, was grown from Haida Gwaii in British Columbia across the Great Plains to North Dakota and Missouri. The tobacco collected by Meriwether Lewis and William Clark in North Dakota was *N. quadrivalvis*; however, recent collections from this area are only from cultivated plants (for example, *Reveal 8376*, NY). Plants derived from cultivated forms often have supernumerary flower parts (fasciation) and flowers with more than five petals. These forms are most often collected from northern California northwards; the first collection of this type was made along the Columbia River in Oregon by David Douglas.

9. Nicotiana repanda Willdenow in J. G. C. Lehmann, Gen. Nicot. Hist., 40, plate 3. 1818 • Fiddleleaf tobacco

Eucapnia repanda (Willdenow) Rafinesque; *Nicotiana doniana* Dunal; *N. lyrata* Kunth

Herbs, annual, from basal rosette. **Stems** loosely branched, 5–15 dm, rough with minute tubercules. **Rosette leaves:** petiole short and winged or leaves sessile; blade spatulate to oblong-ovate, 10–15 cm, surfaces softly viscid-pubescent. **Cauline leaves** sessile; blade pandurate, becoming smaller distally, distal leaves oblong-ovate or pandurate, 1–5 cm, base clasping, apex acute, surfaces softly viscid-pubescent. **Inflorescences** long, flexuous false racemes, occasionally with shorter lateral branches, not leafy; flowering crepuscular. **Pedicels** 0.3–0.7 cm. **Flowers:** calyx green, 1.5 cm, minutely hispid, tube globose, strongly 10-ribbed, lobes erect, linear, length equaling tube, subequal, viscid-pubescent; corolla white, straight, 4–6 cm (excluding limb), sparsely pubescent or glabrous at base, tube white, straight, 3–5.5 cm × 1 mm, gradually widening to gaping throat 5–6 × 3 mm, cobwebby-pubescent internally, limb spreading or (in daytime) slightly assurgent, white, pentagonal, 2–2.4 cm diam., lobes acute apically; stamens inserted at base of throat, included; filaments equal (fused to corolla tube for their entire length, anthers appearing sessile), glabrous; style straight, equaling stamens. **Capsules** ovoid, 1 cm. **Fruiting calyces** not tearing at sinuses, covering entire capsule. **Seeds** 0.6 mm. **2*n*** = 48.

Flowering Feb–Sep. Moist ground along streams, open areas; 0–800 m; Ala., Tex.; Mexico (Nuevo Léon, Tamaulipas, Veracruz); West Indies (Cuba).

Nicotiana repanda occurs along the Gulf Coast and in Texas in southern, central, and the eastern edge of west Texas, in adjacent Mexico, and on the island of Cuba. It overlaps in distribution with *N. plumbaginifolia* but is easy to distinguish from that species by its pandurate (fiddle-shaped) cauline leaves and longer, thinner flowers.

10. Nicotiana rustica Linnaeus, Sp. Pl. 1: 180. 1753 • Aztec tobacco [I]

Nicotiana pavonii Dunal; *N. rugosa* Miller

Herbs, annual or occasionally perennial, coarse and robust, without basal rosette. **Stems** single or less often with weak lateral branches, 5–20 dm, viscid-pubescent. **Cauline leaves:** petiole much shorter than blade, not winged; blade ovate, elliptic, or sometime lanceolate near inflorescence, (7.5–)10–15 cm, base often oblique, apex acute to rounded, surfaces viscid-pubescent, not glaucous. **Inflorescences** branched with distinct central axis, usually somewhat leafy; flowering diurnal. **Pedicels** 0.3–0.5 cm (longer in fruit). **Flowers:** calyx green, cylindric, 0.8–1.5 cm, viscid, lobes erect, broadly triangular, acute, ± equaling tube, 1 much longer than others; corolla straight, 1.2–1.7 cm (excluding limb), puberulent externally, tube yellow to greenish yellow, broadly obconic with slight constriction at mouth, 0.3 cm × 2 mm, widening to throat 10 × 6–8 mm, glabrous or minutely puberulent internally, limb spreading to slightly reflexed, yellowish green, pentagonal, 0.6–0.9 cm diam., lobes greenish yellow, apiculate, equal, very short; stamens inserted at base of throat, included; filaments unequal, 4 sigmoid, 1.2–1.6 cm, extending to corolla mouth, 1 shorter, ca. 1.1 cm, all cottony-pubescent at base; style straight or slightly curved, equaling or slightly exceeding longer 4 stamens. **Capsules** ellipsoid-ovoid to subglobose, 0.7–1.6 cm. **Fruiting calyces** not markedly tearing at sinuses, almost covering capsule. **Seeds** 0.7–1.1 mm. $2n = 48$.

Flowering year-round. Disturbed areas, field edges, roadsides, escaped from cultivation; 0–1000 m; introduced; B.C., Ont.; Ill., Md., Mass., N.Y., Oreg., Tex.; South America (Bolivia, Peru); introduced nearly worldwide except for Antarctica.

Nicotiana rustica is one of the two commercially cultivated species of tobacco and was likely the dominant species on the east coast of North America before the introduction of *N. tabacum* by European settlers. It has been recorded as the so-called sacred Indian tobacco of the Iroquois nation and the authentic pre-settlement stock of the Onondagas. It is usually found in cultivation, often associated with towns and villages of Native American peoples and it can be difficult to tell from herbarium labels (unless specified) if plants are spontaneous or specifically cultivated. There are a number of apparently wild-collected specimens of *N. rustica* from the late 19th and early 20th centuries that are not obviously labeled as cultivated (for example, from New Mexico in association with pueblos); *N. rustica* could easily escape anywhere it is cultivated, but probably does not persist. Most records from more northern latitudes (for example, Massachusetts) are historical specimens taken from rubbish heaps or dumps.

11. Nicotiana sylvestris Spegazzini, Gartenflora 47: 131, fig. 38. 1898 • South American tobacco [I]

Herbs, perennial, [or soft-wooded small trees], from basal rosette that soon merges with cauline leaves. **Stems** single or multiple, stout, branches erect, 4–30 dm, viscid-pubescent. **Rosette leaves** sessile; blade elliptic, to 50 cm, base decurrent and often auriculate or clasping, surfaces viscid-pubescent. **Cauline leaves** sessile; proximal blades elliptic, distal elliptic to elliptic-ovate, decreasing in size distally, 20–50 cm, base auriculate, apex acute to acuminate, surfaces viscid-pubescent. **Inflorescences** branched panicles with congested branches, appearing moplike, leafy; flowering crepuscular. **Pedicels** 0.5–1.5 cm (nodding, flowers pendent). **Flowers:** calyx green, oblong or subglobose (somewhat twisted), 1–1.8 cm, viscid-pubescent, lobes erect, deltate to triangular, equal or nearly so, much shorter than tube; corolla white, straight or very slightly curved, 6–9 cm (excluding limb), minutely viscid-pubescent without, tube white, cylindric, straight or slightly curved, 2 cm × 1–2 mm, throat 40–70 mm, glabrous or minutely puberulent internally, ventricose and inflated in middle to distal 1/3 (often somewhat asymmetrically), narrowing toward apex with constriction at mouth, 5 mm diam., limb spreading, tips somewhat reflexed, white, stellate, 1.5–2 cm diam., lobes white, broadly triangular, acute apically; stamens inserted subequally at base of throat, included; filaments unequal, 4 longer, 4–7 cm, just reaching corolla mouth, 1 slightly shorter, 4–6 cm, pubescent at insertion point; style straight, just exceeding the 4 longer stamens, slightly exserted. **Capsules** ovoid, 1.5–1.8 cm. **Fruiting calyces** not tearing at sinuses, almost completely covering capsule. **Seeds** 0.5 mm. $2n = 24$.

Flowering year-round. Disturbed areas, abandoned gardens; 0–2000 m; introduced; Ariz., Calif.; South America (Argentina, Bolivia); introduced also in Europe (Germany, Sweden, United Kingdom).

Nicotiana sylvestris is widely cultivated and has been recorded as escaping, self-sowing, and persisting in disturbed areas. Most herbarium specimens have been collected from gardens or greenhouses, but the ease with which the species self-sows means it is likely to become at least ephemerally naturalized in areas with little frost.

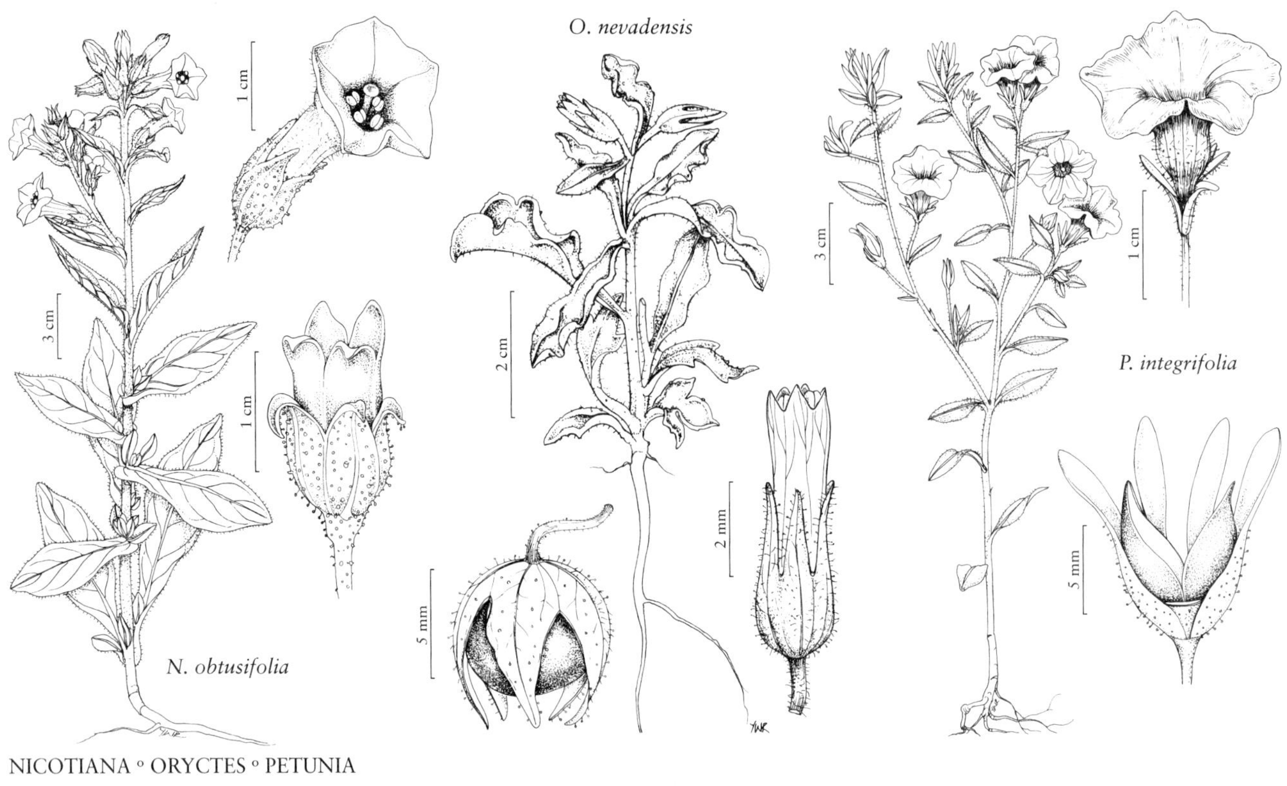

NICOTIANA ◦ ORYCTES ◦ PETUNIA

12. Nicotiana tabacum Linnaeus, Sp. Pl. 1: 180. 1753 [I]

Nicotiana angustifolia Miller; *N. fruticosa* Linnaeus

Herbs, perennial, **or shrubs to small trees,** soft-wooded, without basal rosette. **Stems** single, usually unbranched, woody at base (hollow), 10–30 dm, viscid-pubescent. **Cauline leaves** sessile; blade elliptic to lanceolate, 5–50 cm, becoming smaller distally, base tapering or decurrent, apex acute to acuminate, surfaces viscid-pubescent. **Inflorescences** branched with distinct central axis (branches themselves branched and shorter than central axis), usually somewhat leafy; flowering diurnal. **Pedicels** (spreading), 0.5–1.5 cm. **Flowers:** calyx uniformly green, 1.2–2.5 cm, viscid-pubescent, lobes long-triangular, equaling or shorter than tube, unequal; corolla straight or strongly curved in distal 1/2, 3–5 cm (excluding limb), viscid-puberulent externally, tube pale greenish cream to pink or red, slightly curved or straight, 0.7–1.5 cm × 2–2.5 mm, widening to throat 25–40 × 5 mm, somewhat dilated distally, glabrous or minutely puberulent internally, limb spreading to somewhat reflexed, pale pink to reddish pink, occasionally white, pentagonal, 2–3 cm diam., lobes pale pink to reddish pink, occasionally white, acute; stamens inserted near base of throat; filaments unequal, 4 slightly exserted, 3–5 cm, 1 included, 3 cm (shorter than the other 4), pubescent on proximal 1/2; style straight or slightly curved, ± equaling longer stamens. **Capsules** narrowly ellipsoid, ovoid, or globose, 1.2–2 cm. **Fruiting calyces** often tearing at sinuses (especially in cultivars), covering 1/2 mature capsule. **Seeds** 0.5 mm. **2*n*** = 48.

Flowering year-round. Disturbed areas, field edges; 0–1000 m; introduced; Ont.; Fla., Ga., Ky., La., Md., Mich., Mo., N.C., S.C., Tenn., Va., W.Va.; South America; cultivated nearly worldwide except Antarctica.

Nicotiana tabacum is the principal tobacco of commerce and was the mainstay of the economy of the Chesapeake Bay region during Colonial times. It replaced *N. rustica* as the main cultivated species in North America in the early 1600s. Most herbarium specimens of *N. tabacum* come from gardens or research greenhouses, but it occasionally escapes and is an ephemeral weed where the climate is mild. Commercial tobacco cultivars are grown for their large leaves, and flowers are removed to allow further growth of top leaves before harvest, but if marginal individuals are not harvested and are collected, they could be mistaken for naturalized weeds. *Nicotiana tabacum* is widely cultivated across North America.

21. ORYCTES S. Watson, Botany (Fortieth Parallel), 274, plate 28, figs. 5–10. 1871 • [Greek *orykter*, digger, alluding to a name once applied to some of the indigenous people within the range of the genus, now considered derogatory] E

John E. Averett†

Herbs, annual, viscid-pubescent, hairs sparse, scurfy, taproot fleshy. **Stems** branched. **Leaves** alternate, clustered at tips of branches, petiolate to subsessile. **Inflorescences** axillary, solitary flowers or 2–4-flowered clusters. **Flowers** 5-merous; calyx accrescent, tubular, lobes 5, narrowly triangular, slightly unequal, investing, not enclosing berry; corolla cream with purplish tinge, radial, tubular, 5-lobed; stamens inserted at base of corolla tube, ± unequal; anthers basifixed, globose, dehiscing by longitudinal slits; ovary 2-carpellate; style slender, slightly curved; stigma blunt. **Fruits** berries, globose, dry. **Seeds** flattened, reniform, margin hyaline.

Species 1: w United States.

Oryctes is monospecific and, because of the accrescent fruiting calyx, has been included among the genera surrounding *Physalis*. It is distinctive in having a tubular corolla and hyaline-margined seeds. A. T. Hunziker (2001) included *Oryctes* in Physalideae Miers subtribe Iochrominae Reveal; molecular data (R. G. Olmstead et al. 2008) support its relationship to genera near *Physalis* in Physalideae.

SELECTED REFERENCE Averett, J. E. and W. G. D'Arcy. 1983. Flavonoids in *Oryctes* (Solanaceae). Phytochemistry 22: 2325–2326.

1. **Oryctes nevadensis** S. Watson, Botany (Fortieth Parallel), 274, plate 28, figs. 5–10. 1871 C E F

Stems leafy, 15–20 cm. **Leaf blades** linear to ovate, 1–3 cm, base narrowing to petiole, margins entire or lobed, undulate, surfaces glandular-pubescent. **Flowers:** calyx 2–4 mm; corolla 5–8 mm; stamens slightly exserted. **Berries** 6–7 mm diam. **Seeds** brown, 2 mm.

Flowering May–Jun. Sand, stabilized dunes, washes, valley flats; of conservation concern; 1200–1800 m; Calif., Nev.

Oryctes nevadensis is relatively rare and known from approximately 50 populations. The species is restricted to Inyo County in California, and Churchill, Esmeralda, Humboldt, Mineral, Pershing, and Washoe counties in Nevada.

22. PETUNIA Jussieu, Ann. Mus. Natl. Hist. Nat. 2: 215, plate 47. 1803, name conserved • [Tupi-Guarani (Brazilian) *petun*, tobacco, alluding to affinity with *Nicotiana*] I

Kathryn L. Fox

Janet R. Sullivan

Stimoryne Rafinesque

Herbs, annual [perennial], taprooted, sparsely to densely viscid-glandular pilose. **Stems** erect or ascending to decumbent [prostrate], branched at base and distal nodes. **Leaves** alternate, geminate subtending flowers at distal nodes, petiolate. **Inflorescences** axillary, solitary flowers. **Flowers** 5-merous, radially symmetric; calyx campanulate, lobes 5, linear [lanceolate], not accrescent (lobes erect or reflexed); corolla white to rose-purple (drying deep violet) [red], throat reticulated rose-purple or greenish white, radial, salverform to funnelform with shallow, rounded lobes; stamens slightly unequal (2 long, 2 medium, 1 short [4 equal, 1 longer]), inserted at base or near midpoint of corolla tube; anthers ventrifixed, oblong to elliptic, dehiscing by longitudinal slits; ovary 2-carpellate; style slender to thick, curved to straight; stigma obconic to capitate-truncate, 2-lobed. **Fruits** capsules (2-valved, apex bidentate or entire, dehiscence septicidal), ovoid, length 1/2–3/4 calyx lobes. **Seeds** globose-reniform (red-brown; foveolate-reticulate with coarse, wavy middle lamellae and anticlinal walls). $x = 7$.

Species 17 (3, including 1 hybrid, in the flora): introduced; South America (Argentina, Bolivia, Brazil, Paraguay, Uruguay); introduced also widely.

The hybrid of *Petunia axillaris* and *P. integrifolia*, *P.* ×*atkinsiana*, is one of the most popular ornamental garden plants today and has been grown in cultivation since its creation around 1834 (T. Ando et al. 2005). Natural hybridization does not occur, even in the species' native ranges, due to differences in pollination syndrome (Ando et al. 1999, 2001; M. E. Hoballah et al. 2007; T. Gübitz et al. 2009). Artificial hybridization results in viable seed when the carpellate parent is *P. axillaris* (Ando et al. 1999, 2001, 2005b; Gübitz et al. 2009; T. L. Sims and T. P. Robbins 2009).

Since the late eighteenth century, the parent species have been popular garden plants in their own rights; today, they are grown much less commonly than the hybrid. Three species occur in the flora area as naturalized populations in disturbed areas; they probably do not persist for more than a few years.

SELECTED REFERENCES Ando, T. et al. 2005. Phylogenetic analysis of *Petunia* sensu Jussieu (Solanaceae) using chloroplast DNA RFLP. Ann. Bot. (Oxford) 96: 289–297. Stehmann, J. R. et al. 2009. The genus *Petunia*. In: T. Gerats and J. Strommer, eds. 2009. *Petunia*: Evolutionary, Developmental and Physiological Genetics, ed. 2. New York. Pp. 1–28. Watanabe, H. et al. 1999. Three groups of species in *Petunia* sensu Jussieu (Solanaceae) inferred from the intact seed morphology. Amer. J. Bot. 86: 302–305.

1. Corollas ivory to white, salverform . . . . 1. *Petunia axillaris*
1. Corollas white to pale pink or rose-purple (drying deep violet), ± funnelform.
    2. Corollas rose-purple (drying deep violet), with slight abaxial bulge in the tube, tube 1–3 cm, limb 1–4 cm diam. . . . . 2. *Petunia integrifolia*
    2. Corollas white to pale pink, without bulge, tube 1.1–5.5 cm, limb 1.3–7 cm diam . . . . 3. *Petunia* ×*atkinsiana*

1. Petunia axillaris (Lamarck) Britton, Sterns & Poggenburg, Prelim. Cat., 38. 1888 • White-flowered petunia [I]

Nicotiana axillaris Lamarck in J. Lamarck & J. Poiret, Tabl. Encycl. 2(3[1]): 7. 1794

Stems 1.5–7 dm. **Leaf blades:** proximalmost oblanceolate, distalmost lanceolate to ovate, 2–8.5 (including petiole) × 0.8–3.5 cm, margins entire. **Pedicels** 1.5–8 cm. **Flowers:** calyx 8–17 mm, lobes 4–14 mm; corolla ivory to white (lobe apex sometimes drying pale pink), veins often green or dark purple (drying brownish), salverform, tube 2.8–5.2 cm, limb 2.5–5.5 cm diam.; stamens inserted near midpoint of corolla tube, not surpassing style; anthers and pollen yellow; filaments green; pistil 3.2–4.5 cm. **Capsules** 5–12 mm. $2n = 14$.

Flowering May–Oct. Waste places, along railroads and roadsides, poor soil; 0–400 m; introduced; Ala., Conn., Del., D.C., Fla., Ga., Ill., Ind., Iowa, Mass., Mich., Minn., Mo., N.J., N.Y., N.C., Ohio, Pa., S.C., Tenn., Tex., W.Va.; South America (Argentina, Bolivia, Brazil, Paraguay); introduced also in Australia.

Specimens of *Petunia axillaris* are often misidentified as *P.* ×*atkinsiana*, especially if there is some slight color in the corolla. *Petunia* ×*atkinsiana* with white corollas sometimes has the blue pollen and anther color of *P. integrifolia*.

2. **Petunia integrifolia** (Hooker) Schinz & Thellung, Vierteljahrsschr. Naturf. Ges. Zürich 60: 361. 1915 • Violet-flowered petunia F I

Salpiglossis integrifolia Hooker, Bot. Mag. 58: plate 3113. 1831; *Petunia violacea* Lindley

Stems 1–7 dm. **Leaf blades:** proximalmost oblanceolate, distalmost lanceolate to ovate, 1.4–7.2 (including petiole) × 0.3–3 cm, margins entire. **Pedicels** 1–6 cm. **Flowers:** calyx 5–14 mm, lobes 3–11 mm; corolla rose-purple (drying deep violet), veins sometimes darker rose-purple or violet (not distinct from rest of corolla when dried), funnelform with slight abaxial bulge in the tube, tube 1–3 cm, limb 1–4 cm diam.; stamens inserted at base of corolla tube, longest 2 surpassing style; anthers and pollen blue to violet; filaments light purple to green; pistil 1–2.2 cm. **Capsules** 3–9 mm. $2n = 14$.

Flowering May–Oct. Waste places, along railroads and roadsides, poor soil or sand; 0–400 m; introduced; Ala., Calif., Conn., D.C., Fla., Ill., Maine, Md., Mass., N.H., N.J., N.Y., N.C., Ohio, Pa., R.I., Tex., W.Va., Wis.; South America (Argentina, Brazil, Paraguay).

Naturalized populations of *Petunia integrifolia* appear to have been more common prior to the 1960s, based on herbarium specimens.

SELECTED REFERENCE Ando, T. et al. 2005b. A morphological study of the *Petunia integrifolia* complex (Solanaceae). Ann. Bot. (Oxford) 96: 887–900.

3. **Petunia ×atkinsiana** (Sweet) D. Don ex W. H. Baxter in J. C. Loudon, Hort. Brit. ed. 3, 655. 1839 • Common garden or hybrid petunia I

Nierembergia ×atkinsiana Sweet, Brit. Fl. Gard. 6: plate 268. 1834; *Petunia ×hybrida* E. Vilmorin

Stems 0.9–10 dm. **Leaf blades:** proximalmost oblanceolate, distalmost lanceolate to ovate, 1–12 (including petiole) × 0.2–5 cm, margins entire. **Pedicels** 1–6 cm. **Flowers:** calyx 7–21 mm, lobes 4–16 mm; corolla white to pale pink with white tube (drying white to pale violet), veins green to deep pink or dark purple (drying brownish to deep violet), ± funnelform, tube 1.1–5.5 cm, limb 1.3–7 cm wide; stamens inserted at base or near midpoint of corolla tube, longest 2 just shorter than to just surpassing style; anthers and pollen blue, violet, or yellow; filaments blue, violet, or yellow; pistil 2–4.5 cm. **Capsules** 5–15 mm. $2n = 14$.

Flowering Apr–Oct(–Dec). Waste places, along railroads and roadsides, poor soil, gravel, or sand; 0–1600 m; introduced; Ont., Que.; Ala., Ark., Calif., Conn., Fla., Ga., Ill., Iowa, Kans., Ky., La., Md., Mass., Mich., Minn., Miss., Mo., N.J., N.Y., N.C., Ohio, Pa., S.C., Tex., Utah, Vt., Va., W.Va., Wis.; introduced also in Europe, Asia, Australia.

Petunia ×atkinsiana is the most commonly naturalized of the three species and is often robust. When fertile, it can spread by seed from cultivation and become established in disturbed areas. Some herbarium specimens probably represent waifs; it is often difficult to determine from label data. Cultivated plants are available in a wide range of colors and color patterns. Only white to pale pink morphs were seen in specimens representing naturalized populations; other color forms may be fertile and may occur outside of cultivation.

23. PHYSALIS Linnaeus, Sp. Pl. 1: 182. 1753; Gen. Pl. ed. 5, 85. 1754 • Ground-cherry, husk-tomato, tomaté [Greek *physa*, bladder or bellows, and *alis*, belonging to, alluding to inflated fruiting calyx]

Janet R. Sullivan

Margaranthus Schlechtendal

Herbs [shrubs], annual or perennial, taprooted or rhizomatous, glabrous or pubescent, hairs simple, forked, or dendroid-stelliform, sometimes multicellular and appearing jointed, glandular or eglandular. **Stems** erect to weakly decumbent, branching. **Leaves** alternate, sometimes geminate, petiolate or sessile; blade simple, margins entire, dentate, or sinuate. **Inflorescences**

axillary, solitary flowers [fascicles of 2–5]. **Flowers** 5-merous, (pendent or nodding); calyx campanulate with 5 broadly to narrowly triangular lobes, accrescent and inflated in fruit, becoming reticulate-membranous and bladderlike with narrow orifice and completely enclosing berry; corolla yellow or pale cream-yellow to nearly white, often with 5 large spots or smudges or star-shaped tinge of color in throat, and mat of white hairs at base of throat, radial, rotate or campanulate-rotate, limb sometimes widely flaring or reflexed at maturity, 5-angulate or obscurely 5-lobed, (*P. solanacea* dark purple or rarely yellowish or greenish with large purple spots, urceolate, with 5 shallow teeth) [campanulate with open limb]; stamens inserted at base of corolla tube, equal; anthers basifixed, oblong to narrowly elliptic, dehiscing by longitudinal slits; ovary 2-carpellate; style straight, slender, sometimes expanding distally; stigma minutely capitate or truncate. **Fruits** berries, globose, juicy [somewhat dry], completely enclosed by inflated fruiting calyx. **Seeds** reniform, flattened [oblique-triangular]. x = 12, 24.

Species ca. 90 (24, including 1 hybrid, in the flora): North America, Mexico, West Indies, Bermuda, Central America, South America; introduced in Asia, Africa, Pacific Islands, Australia.

Physalis is recognizable by the fruiting calyx that enlarges and inflates to completely enclose the berry, and pendent or nodding flowers borne singly at each node; most members also have an unlobed, yellow or cream-yellow, campanulate-rotate corolla. *Physalis solanacea* is atypical in having an urceolate corolla that is mostly or completely lurid purple; the molecular phylogenetic analysis by M. Whitson and P. S. Manos (2005) showed that it is part of a clade of more typical species. Their study also supported treatment of *Calliphysalis* and *Alkekengi* as genera distinct from *Physalis*.

The variable morphology of *Physalis* species has resulted in many misidentified herbarium specimens. Hair morphology is important in identifying most taxa.

Two species occur as waifs in the flora area: *Physalis minima* Linnaeus, native to the New World tropics, has been collected as a weed in sugarcane fields and sandy open pastures in Louisiana; and *P. ixocarpa* Brotero ex Hornemann [including var. *parviflora* (Waterfall) Kartesz & Gandhi and var. *immaculata* (Waterfall) Kartesz & Gandhi], native to Mexico, has been collected sporadically from disturbed sites in the United States and Canada.

Several species of *Physalis* are commonly cultivated for their edible berries: *P. philadelphica* (tomatillo) has fruits with a flavor reminiscent of tomato and is used in Mexican-style salsa; *P. minima* (pygmy groundcherry) has fruits similar to a cherry tomato; and *P. peruviana* Linnaeus (cape gooseberry) and *P. grisea* (strawberry tomato) have sweet fruits that are used in pies, jams, sauces, and as a garnish. The berries of many wild-growing species are eaten as well. Fruits of *P. minima* and *P. peruviana* are also used worldwide as an antispasmodic, diuretic, antiseptic, sedative, and analgesic. Unripe fruits and foliage of *Physalis* species contain solanine and other solanidine alkaloids and are toxic if ingested in large quantities.

After this manuscript was completed, *Physalis macrosperma* Pyne, E. L. Bridges & Orzell was published describing plants similar to *P. heterophylla* but with larger fruiting calyces and seeds. Further study is needed to clarify the disposition of this taxon.

SELECTED REFERENCES Martínez, M. 1998. Revision of *Physalis* section *Epeteiorhiza* (Solanaceae). Anales Inst. Biol. Univ. Nac. Autón. Méx., Bot. 69: 71–117. Martínez, M. 1999. Infrageneric taxonomy of *Physalis*. In: M. Nee et al., eds. 1999. Solanaceae IV: Advances in Biology and Utilization. Kew. Pp. 275–283. Menzel, M. Y. 1951. The cytotaxonomy and genetics of *Physalis*. Proc. Amer. Philos. Soc. 95: 132–183. Rydberg, P. A. 1896b. The North American species of *Physalis* and related genera. Mem. Torrey Bot. Club 4: 297–374. Seithe, A. and J. R. Sullivan. 1990. Hair morphology and systematics of *Physalis* (Solanaceae). Pl. Syst. Evol. 170: 193–204. Sullivan, J. R. 1985. Systematics of the *Physalis viscosa* complex (Solanaceae). Syst. Bot. 10: 426–444. Sullivan, J. R. 2004. The genus *Physalis* (Solanaceae) in the southeastern United States. Rhodora 106: 305–326. Sullivan, J. R., V. P. Patel, and W. Chissoe. 2005. Palynology and systematics of *Physalis* (Solanaceae). In: R. C. Keating et al., eds. 2005. A Festschrift for William G. D'Arcy: The Legacy of a Taxonomist. St. Louis. Pp. 287–300. Waterfall, U. T. 1958. A taxonomic study of the genus *Physalis* in North America north of Mexico. Rhodora 60: 107–114, 128–142, 152–173. Whitson, M. and P. S. Manos. 2005. Untangling *Physalis* (Solanaceae) from the physaloids: A two-gene phylogeny of the *Physalinae*. Syst. Bot. 30: 216–230.

1. Corollas urceolate, 2.5–4.5 mm; pedicels 1.5–3 mm in flower, 3–5 mm in fruit; fruiting calyx 5-ribbed . . . . 22. *Physalis solanacea*
1. Corollas rotate to campanulate-rotate, 5–20 mm; pedicels 2–46 mm in flower, 3–60 mm in fruit; fruiting calyx 10-ribbed or sharply 5-angled.
    2. Anthers strongly twisted after dehiscence; fruiting calyx filled, and often burst, by berry . . . . 19. *Physalis philadelphica*
    2. Anthers not twisted after dehiscence; fruiting calyx loosely enclosing, or nearly filled, by berry.
        3. Plants pubescent, hairs 2- or 3-branched or dendroid-stelliform, or glabrous except for dendroid-stelliform hairs on leaf margins and calyx.
            4. Hairs predominantly 2- or 3-branched.
                5. Hairs to 0.5 mm, appressed, giving plants a grayish appearance; pedicels 3–11(–17) mm in flower, 10–15(–20) mm in fruit . . . . 10. *Physalis fendleri* (in part)
                5. Hairs 0.5–2 mm, divergent; pedicels 8–46 mm in flower, 15–55 mm in fruit . . . . 21. *Physalis pumila* (in part)
            4. Hairs predominantly dendroid-stelliform.
                6. Plants glabrous except for leaf margins and calyx.
                    7. Leaves sessile; blade linear-lanceolate (sometimes folded along midrib), 0.2–0.8(–1) cm wide . . . . 3. *Physalis angustifolia*
                    7. Leaves sessile or petiole to 1/10 blade; blade narrowly spatulate to linear-lanceolate, 0.2–1.5(–2) cm wide . . . . 9. *Physalis ×elliottii* (in part)
                6. Plants pubescent.
                    8. Leaves sessile or petiole to 1/10 blade; blade elliptic or spatulate to linear-lanceolate.
                        9. Leaf blades 1–4 cm wide; corollas with dark purple-black spots; coastal Louisiana, Texas . . . . 6. *Physalis cinerascens* (in part)
                        9. Leaf blades 0.2–1.5(–2) cm wide; corollas with pale brown, ochre, or green spots or smudges; Florida . . . . 9. *Physalis ×elliottii* (in part)
                    8. Leaves petiolate, petioles 1/5 to as long as blade; blades orbiculate to broadly ovate or elliptic.
                        10. Leaf blades broadly elliptic to ovate, base rounded, margins usually entire or rarely shallowly sinuate . . . . 24. *Physalis walteri*
                        10. Leaf blades orbiculate to ovate, base truncate to slightly attenuate, margins usually dentate or sinuate, sometimes entire.
                            11. Plants sparsely to somewhat densely pubescent, hairs to 1 mm . . . . 6. *Physalis cinerascens* (in part)
                            11. Plants densely pubescent, hairs to 1 mm (obscuring plant surface on younger growth), sometimes also with 2–4 mm branched or simple hairs . . . . 17. *Physalis mollis*
        3. Plants pubescent, hairs simple, or glabrous except for simple hairs on pedicels and calyx, or glabrous.
            12. Annuals, taprooted; fruiting calyces sharply 5 angled or 10-ribbed.
                13. Corollas rotate; flowering pedicels (13–)20–34(–40) mm. . . . . 1. *Physalis acutifolia*
                13. Corollas campanulate-rotate; flowering pedicels 2–17(–22) mm.
                    14. Corollas without spots or smudges, or only tinged purple; fruiting calyces 10-ribbed.
                        15. Plants glabrous or sparsely pubescent, hairs eglandular; leaf blades narrowly elliptic-ovate to linear-lanceolate; pedicels 7–17(–22) mm in flower, 15–30 mm in fruit . . . . 2. *Physalis angulata*
                        15. Plants sparsely to densely pubescent, hairs intermixed glandular and eglandular; leaf blades broadly ovate to orbiculate; pedicels 4–7 mm in flower, 5–10 mm in fruit . . . . 16. *Physalis missouriensis*
                    14. Corollas with 5 large, dark purple-black spots or smudges; fruiting calyces sharply 5-angled.

16. Leaf margins coarsely dentate, teeth 10+ per side; pedicels (10–) 15–35 mm in fruit . . . . 7. *Physalis cordata*
16. Leaf margins entire, irregularly crenate-dentate, or coarsely dentate, teeth fewer than 10 per side; pedicels 5–15 mm in fruit.
17. Plants villous, hairs intermixed with stalked and sessile glands; leaf blades gray-green, usually drying orange or with orange patches . . . . 11. *Physalis grisea*
17. Plants ± glabrous to villous, hairs glandular and/or eglandular; leaf blades green, drying green or grayish-brownish.
18. Pedicels noticeably stout, especially in fruit; fruiting calyces nearly spheric . . . . 18. *Physalis neomexicana*
18. Pedicels slender; fruiting calyces always noticeably longer than wide . . . . 20. *Physalis pubescens*

[12. Shifted to left margin.—Ed.]

12. Perennials, rhizomatous; fruiting calyces 10-ribbed.
19. Plants becoming suffrutescent, pubescent, hairs divergent, to 0.5 mm; flowering calyces (3–)4–7(–8) mm . . . . 8. *Physalis crassifolia*
19. Plants remaining herbaceous, glabrous to densely pubescent, hairs appressed or not, 0.5+ mm; flowering calyces 5–14 mm.
20. Leaf blades broadly ovate to orbiculate; plants often glandular.
21. Pedicels 4–8(–13) mm in flower, 5–15 mm in fruit . . . . 12. *Physalis hederifolia*
21. Pedicels (8–)9–17(–25) mm in flower, 15–30(–35) mm in fruit.
22. Plants glabrous to villous, hairs antrorse, to 1 mm, sometimes also simple, jointed, divergent, 1–2 mm; often with slender, shallowly buried rhizomes . . . . 4. *Physalis arenicola*
22. Plants villous, hairs divergent, 1–2 mm, sometimes also with shorter glandular hairs; all rhizomes stout and deeply buried . . . . 13. *Physalis heterophylla*
20. Leaf blades ovate to elliptic, broadly lanceolate, or oblanceolate; plants not glandular.
23. Plants pubescent, hairs 1–3 mm; anthers usually dark purple to blue, rarely yellow . . . . 5. *Physalis caudella*
23. Plants glabrous or pubescent, hairs to 0.5 mm; anthers yellow or with blue or purple tinge.
24. Plants pubescent, hairs mostly simple with some 2- or 3-branched intermixed . . . . 10. *Physalis fendleri* (in part)
24. Plants glabrous or pubescent, hairs simple only.
25. Plants glabrous or sparsely strigose, hairs antrorse, to 0.5 mm . . . 15. *Physalis longifolia*
25. Plants sparsely to densely pubescent, hairs divergent, 1–1.5 mm, and antrorse or retrorse, to 0.5 mm.
26. Stems erect; leaf blade margins coarsely to shallowly dentate or entire; hairs divergent and retrorse . . . . 23. *Physalis virginiana*
26. Stems erect to decumbent; leaf blade margins entire or sinuate; hairs divergent and antrorse.
27. Leaf blades oblanceolate; eastern coastal plain . . . . 14. *Physalis lanceolata*
27. Leaf blades elliptic-ovate to ovate-lanceolate; Great Plains . . . . 21. *Physalis pumila* (in part)

1. Physalis acutifolia (Miers) Sandwith, Kew Bull. 14: 232. 1960 [W]

Saracha acutifolia Miers, Ann. Mag. Nat. Hist., ser. 2, 3: 449. 1849; *Physalis wrightii* A. Gray

Herbs annual, taprooted, sparsely pubescent to ± glabrous, hairs simple, appressed, antrorse, to 0.5 mm. **Stems** erect to decumbent, branching at most nodes, branches spreading and sometimes decumbent, 1–5 dm. **Leaves** petiolate; petiole mostly 1/2–2/3 blade; blade narrowly elliptic-ovate to lanceolate, (1.5–)2.5–6.8(–8.3) × (0.7–)1–2.5(–5.4) cm, base attenuate to rounded, margins coarsely, deeply, irregularly dentate, teeth acuminate. **Pedicels** (13–)20–34(–40) mm, (20–)25–35(–39) mm in fruit. **Flowers:** calyx (3–)4–5(–6) mm, lobes (1–)2–4 mm, (acute to acuminate); corolla pale yellow to nearly white with green or darker yellow tinge, rotate, 5–15 mm; anthers usually blue-tinged, rarely all blue or yellow, not twisted after dehiscence, 1–3 mm. **Fruiting calyces** nearly filled by berry, 10-ribbed, 15–25(–30) × 13–20(–22) mm. $2n = 24$.

Flowering (May–)Jul–Nov. Disturbed areas along streams and roadsides, gravel and sand, cultivated fields, parks; 100–2000 m; Ala., Ariz., Calif., Ga., Miss., N.Mex., Tex.; Mexico (Baja California, Baja California Sur, Chihuahua, Sinaloa, Sonora).

Corollas of *Physalis acutifolia* are nearly rotate with a very short floral tube and somewhat reflexed, widely flaring limb when fully open. Unless it is in flower, *P. acutifolia* is difficult to distinguish from narrow-leaved *P. angulata*, which has corollas that are more campanulate-rotate, without a reflexed limb.

2. Physalis angulata Linnaeus, Sp. Pl. 1: 183. 1753 [W]

Physalis angulata var. *lanceifolia* (Nees) Waterfall; *P. angulata* var. *pendula* (Rydberg) Waterfall; *P. lanceifolia* Nees; *P. pendula* Rydberg

Herbs annual, taprooted, glabrous or sparsely pubescent, hairs simple, jointed, to 0.5 mm. **Stems** erect (angulate, at least proximally), branching at most nodes, branches spreading, 1–20 dm. **Leaves** petiolate; petiole 1/3–2/3 blade; blade narrowly elliptic-ovate to linear-lanceolate, 3–10(–14) × 1–8 cm, base rounded to attenuate, margins coarsely, deeply, irregularly dentate, teeth acuminate. **Pedicels** 7–17(–22) mm, 15–30 mm in fruit. **Flowers:** calyx 3–5 mm, sparsely hairy or glabrous except for margins, lobes 1–3 mm; corolla yellow, without spots or smudges or rarely tinged purple, campanulate-rotate, 6–10 mm; anthers blue or blue-tinged, not twisted after dehiscence, 1–3 mm. **Fruiting calyces** loosely enclosing berry, 10-ribbed, 20–40 × 15–25 mm. $2n = 24, 48$.

Flowering year-round in areas without frost, mostly Jun–Nov. Hardwood and pine woods, woodland borders, stream margins, floodplains, marshy areas, fields, pastures, waste places; 0–1600 m; Ont.; Ala., Ariz., Ark., Calif., Conn., Fla., Ga., Ill., Ind., Kans., Ky., La., Md., Mass., Miss., Mo., N.Mex., N.C., Okla., S.C., Tenn., Tex., Va.; Mexico; West Indies; Bermuda; Central America; South America; introduced in Asia, Africa, Pacific Islands, Australia.

Populations of *Physalis angulata* with linear to lanceolate, sinuate leaf blades can be found in Arizona, California, Florida, New Mexico, Oklahoma, and Texas. When not in flower, narrow-leaved *P. angulata* is difficult to distinguish from *P. acutifolia*. The latter species has nearly rotate, widely flaring corollas that are pale yellow to nearly white with a green or yellow star-shaped tinge in the throat.

3. Physalis angustifolia Nuttall, J. Acad. Nat. Sci. Philadelphia 7: 113. 1834 [E]

Herbs perennial, rhizomatous, rhizomes deeply buried, often also with slender, shallow rhizomes, glabrous except for sparse dendroid-stelliform hairs to 1 mm on leaf margins and calyx. **Stems** erect to decumbent, branching at most nodes, proximal branches spreading and decumbent, 0.5–1.5(–2.5) dm. **Leaves** sessile; blade linear-lanceolate, sometimes folded along midrib, 2.5–9 × 0.2–0.8(–1) cm, base tapering to stem, margins entire. **Pedicels** 14–21 mm, 15–35(–42) mm in fruit. **Flowers:** calyx 6–8 mm, lobes (2–)3–4 mm; corolla yellow with 5 ochre smudges, campanulate rotate, (8–)11–15 mm; anthers yellow, not twisted after dehiscence, 2–2.5 mm. **Fruiting calyces** orange drying brown, loosely enclosing berry, 10-ribbed, (15–)20–30(–40) × 15–25 mm. $2n = 24$.

Flowering year-round in areas without frost. Sand, beach dunes, disturbed coastal areas in sand; 0 m; Ala., Fla., La., Miss.

In Florida, plants occur along the panhandle east to Franklin County. Narrow-leaved plants of *Physalis* ×*elliottii* var. *glabra* occurring in peninsular Florida are sometimes mistakenly keyed to *P. augustifolia* (J. R. Sullivan 1985, 2013).

4. Physalis arenicola Kearney, Bull. Torrey Bot. Club 21: 485. 1894 [E]

Physalis arenicola var. *ciliosa* (Rydberg) Waterfall; *P. ciliosa* Rydberg

Herbs perennial, rhizomatous, rhizomes deeply buried, slender, typically also with shallowly buried, slender rhizomes, glabrous to villous, hairs simple, antrorse, to 1 mm, sometimes also with simple, jointed, divergent hairs, 1–2 mm, sometimes glandular. **Stems** erect, few-branched, 0.5–3 dm. **Leaves** petiolate; petiole 1/4–2/3 blade; blade ovate to suborbiculate, 1.5–6(–6.5) × 1–5 cm, base truncate to cordate, margins entire or coarsely, irregularly dentate with few teeth. **Pedicels** (8–)11–17(–25) mm, 15–30 (–35) mm in fruit. **Flowers:** calyx 6–12 mm, villous, lobes 2–5 mm; corolla yellow with 5 pale reddish-brown smudges or not, campanulate-rotate, 10–17 mm; anthers yellow, not twisted after dehiscence, 2.5–4 mm. **Fruiting calyces** loosely enclosing berry, 10-ribbed, 20–35 × 15–25 mm. $2n$ = 24.

Flowering year-round in areas without frost. Sandy soil, pine-oak woods, hammocks, fields, pastures, roadsides; 0–50 m; Ala., Fla., Ga., Miss.

Physalis arenicola is found throughout Florida; only a few records exist from the other states in its range.

5. Physalis caudella Standley, Publ. Field Mus. Nat. Hist., Bot. Ser. 17: 273. 1937

Herbs perennial, rhizomatous, rhizomes deeply buried and seldom collected, sparsely pubescent to densely villous, hairs simple, jointed, 1–3 mm. **Stems** erect, branching infrequently, branches ascending to spreading, 1–3(–4) dm. **Leaves** petiolate; petiole to 1/3 blade at proximal nodes, appearing ± sessile at distal nodes; blade lanceolate or lanceolate-ovate, (2.5–)4.5–7.5 (–9.5) × 1.2–2.5(–4) cm, base rounded and tapering to petiole, margins entire, saliently few-toothed, or repand. **Pedicels** 8–13(–15) mm, 10–20(–25) mm in fruit. **Flowers:** calyx 6–10 mm, lobes 2–5(–7) mm; corolla yellow with dark purple-black spots, campanulate-rotate, 14–16 mm; anthers dark purple to blue, rarely yellow, not twisted after dehiscence, (2–)3–3.5 mm. **Fruiting calyces** loosely enclosing berry, 10-ribbed, (20–)30–50 × 20–30(–35) mm, lobes attenuate. $2n$ = 24.

Flowering (sporadically Jun–)Aug. Loose, gravelly soil near streams, slopes, rocky ridges, pinyon-oak-juniper woodlands; 1200–2800 m; Ariz., N.Mex.; Mexico (Chihuahua, Sonora).

Only a few herbarium specimens of *Physalis caudella* have been seen from the flora area (Apache, Cochise, Pima, and Santa Cruz counties in Arizona, and Catron County in New Mexico).

6. Physalis cinerascens (Dunal) Hitchcock, Key Spring Fl. Manhattan, 32. 1894

Physalis pensylvanica Linnaeus var. *cinerascens* Dunal in A. P. de Candolle and A. L. P. P. de Candolle, Prodr. 13(1): 435. 1852; *P. viscosa* Linnaeus var. *cinerascens* (Dunal) Waterfall

Herbs perennial, rhizomatous, rhizomes deeply buried, stout, sparsely to ± densely pubescent, hairs dendroid-stelliform, to 1 mm. **Stems** erect to decumbent, branching at most nodes, proximal branches spreading and decumbent, 0.5–5 dm. **Leaves** sessile or petiolate; petiole 1/5 to as long as blade; blade orbiculate to broadly ovate or elliptic to spatulate, 1.5–8(–9) × 1–6(–8) cm, base truncate to attenuate, margins coarsely dentate, sinuate, or entire. **Pedicels** 10–33 mm, 15–60 mm in fruit. **Flowers:** calyx (3.5–) 5–9 mm, lobes 1.5–4 mm; corolla yellow with 5 dark purple-black spots, campanulate-rotate, (8–)10–16 mm; anthers yellow, rarely purple-tinged, not twisted after dehiscence, 2–5 mm. **Fruiting calyces** loosely enclosing berry, 10-ribbed, 15–35(–45) × 10–35 mm.

Varieties 2 (2 in the flora): c, sc United States, Mexico.

1. Leaf blades orbiculate to broadly ovate, 1–6(–8) cm wide, base truncate to slightly attenuate, margins dentate, sinuate, or entire; corolla limbs reflexed when fully open . . . . . . .6a. *Physalis cinerascens* var. *cinerascens*
1. Leaf blades elliptic to spatulate, 1–4 cm wide, base attenuate, margins entire; corolla limbs not reflexed when fully open . . . . . . 6b. *Physalis cinerascens* var. *spathulifolia*

6a. Physalis cinerascens (Dunal) Hitchcock var. **cinerascens**

Leaf blades orbiculate to broadly ovate, 1–6(–8) cm wide, base truncate to slightly attenuate, margins dentate, sinuate, or entire. **Pedicels** 15–60 mm in fruit. **Corollas:** limbs reflexed when fully open. **Fruiting calyces** 15–30 × 10–20 mm. $2n$ = 24.

Flowering year-round in areas without frost. Prairies, fields, roadsides, disturbed habitats; 100–1500 m; Ala., Ark., Kans., La., Mo., N.Mex., Okla., Tex.; Mexico.

In Mexico, var. *cinerascens* is widespread in central and eastern regions south to Yucatán.

SELECTED REFERENCE Sullivan, J. R. 1984. Pollination biology of *Physalis viscosa* var. *cinerascens* (Solanaceae). Amer. J. Bot. 71: 815–820.

6b. Physalis cinerascens (Dunal) Hitchcock var. **spathulifolia** (Torrey) J. R. Sullivan, Syst. Bot. 10: 444. 1985 (as spathulaefolia)

Physalis lanceolata Michaux var. *spathulifolia* Torrey in W. H. Emory, Rep. U.S. Mex. Bound. 2(1): 153. 1859 (as spathulaefolia); *P. spathulifolia* (Torrey) B. L. Turner; *P. viscosa* Linnaeus var. *spathulifolia* (Torrey) A. Gray

Leaf blades elliptic to spatulate, 1–4 cm wide, base attenuate, margins entire. **Pedicels** 25–60 mm in fruit. **Corollas:** limbs not reflexed when fully open. **Fruiting calyces** 25–45 × (15–)20–35 mm. **2*n*** = 24.

Flowering year-round in areas without frost. Gulf dunes, disturbed habitats near Gulf Coast in sand; 0 m; La., Tex.; Mexico (Tamaulipas).

In the past, var. *spathulifolia* was considered to be more closely related to *Physalis walteri* (as varieties of *P. viscosa* subsp. *maritima*). Both are plants of coastal sand dunes and have similar vegetative morphology. Relationships among the taxa with dendroid-stelliform hairs were examined in detail by J. R. Sullivan (1985).

7. Physalis cordata Miller, Gard. Dict. ed. 8, Physalis no. 14. 1768

Herbs annual, taprooted, glabrous or sparsely pubescent, hairs simple, appressed, to 0.5 mm. **Stems** erect, branching at most nodes, branches spreading, 1.5–5(–20) dm. **Leaves** petiolate; petiole 2/3 to as long as blade; blade broadly ovate to orbiculate, 4.5–8.5 × 3.5–7.5 cm, base rounded to truncate or cordate, margins coarsely dentate, teeth 10+ per side. **Pedicels** (4.5–)6–11 mm, (10–)15–35 mm in fruit. **Flowers:** calyx 3.5–6.5 mm, lobes lanceolate, 2–4.5 mm; corolla yellow with 5 large purple-brown-black spots, campanulate-rotate, 6.5–9.5 mm; anthers blue or blue-tinged, not twisted after dehiscence, 1.5–2.5 mm. **Fruiting calyces** loosely enclosing berry, sharply 5-angled, (25–)30–40 × 20–30 mm. **2*n*** = 24.

Flowering Jul–Oct. Sandy or clay soils, along streams, pine woods, disturbed habitats; 0–100 m; Ala., Ark., Fla., Ga., La., Miss., Mo., N.C., Okla., S.C., Tenn., Tex.; Mexico; West Indies; Bermuda; Central America; South America (to Brazil); introduced in Asia.

Herbarium specimens of *Physalis cordata* often consist of only the distal portions of the plants, with label data stating that they are quite tall. The upper limit in this description is taken from M. Martínez (1998).

8. Physalis crassifolia Bentham, Bot. Voy. Sulphur, 40. 1844 [F]

Physalis crassifolia var. *cardiophylla* (Torrey) A. Gray; *P. crassifolia* var. *versicolor* (Rydberg) Waterfall; *P. versicolor* Rydberg

Herbs perennial, becoming suffrutescent, rhizomatous, rhizomes often just below soil surface, vertical, stout, puberulent, hairs divergent, to 0.5 mm, some glandular, appearing ± glabrous without magnification. **Stems** erect, branching from near base and at most nodes, branches widely spreading, distinctly zigzag, slender, 1–4(–10) dm. **Leaves** petiolate; petiole mostly as long as blade; blade broadly ovate to deltate, 0.8–3.3(–4.5) × 0.8–3.3 (–4.5) cm, base cordate, sometimes slightly unequal, margins entire to unevenly coarsely dentate, sometimes thick and slightly succulent. **Pedicels** 8–24(–33) mm, (11–)14–30(–35) mm in fruit. **Flowers:** calyx (3–)4–7(–8) mm, lobes 1–3 mm; corolla pale yellow with yellow or greenish-brown smudges or tinge, campanulate-rotate, 8–14 mm; anthers yellow, not twisted after dehiscence, 1.5–3 mm. **Fruiting calyces** loosely enclosing berry, 10-ribbed, 14–30(–40) × (10–)15–20(–25) mm. **2*n*** = 24.

Flowering year-round in areas without frost, mostly Mar–Apr. Gravelly or sandy slopes, washes, roadsides, mesas, canyons; 100–1700 m; Ariz., Calif., Nev., Utah; Mexico (Baja California, Baja California Sur, Chihuahua, Sinaloa, Sonora); introduced in Australia.

In *Physalis crassifolia*, the corolla limb is widely flaring and reflexed when the flower is fully open, and the flowers are more nodding than fully pendent. Some herbarium specimen labels indicate that plants flower the first year. *Physalis greenei* Vasey & Rose, not validly published, has been misapplied to some representatives of *P. crassifolia*. *Physalis crassifolia* is widespread in Arizona, but it is restricted in California to south-southeastern counties as far north as Inyo, in Nevada to Clark and Lincoln counties, and in Utah to Washington County.

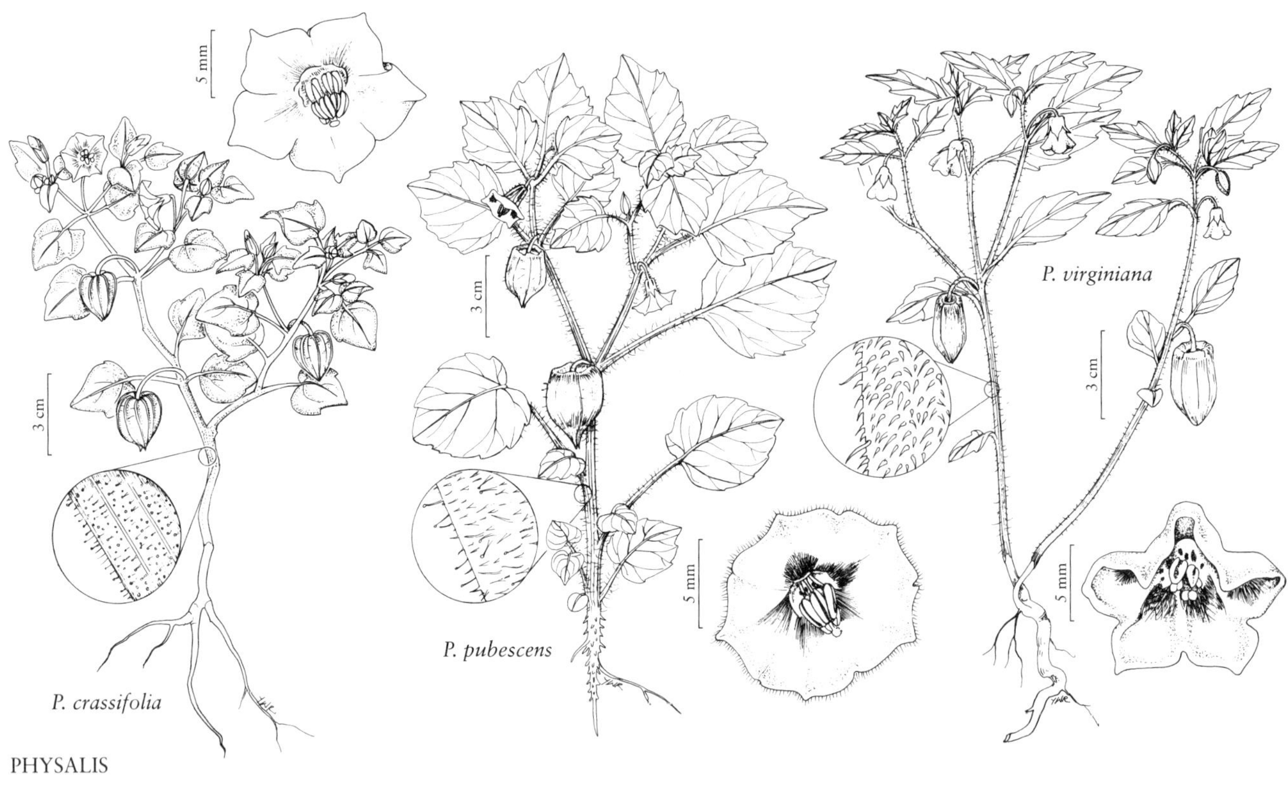

PHYSALIS

9. **Physalis ×elliottii** Kunze, Linnaea 20: 33. 1847, as species [E]

Physalis viscosa Linnaeus var. *elliottii* (Kunze) Waterfall

Herbs perennial, rhizomatous, rhizomes deeply buried, often also with slender, shallow rhizomes, glabrous or sparsely to densely pubescent, hairs dendroid-stelliform, to 1 mm. **Stems** erect to decumbent, branching at most nodes, proximal branches spreading and decumbent, 0.5–3(–4) dm. **Leaves** sessile or petiolate; petiole to 1/10 blade; blade narrowly spatulate to linear-lanceolate, 3–10(–12) × 0.2–1.5(–2) cm, base tapering to slightly winged petiole, margins entire. **Pedicels** 10–27 mm, 15–35 mm in fruit. **Flowers:** calyx 5–10 mm, lobes 2–4 mm; corolla yellow with 5 pale brown, ochre, or green spots or smudges, campanulate-rotate, 12–20 mm; anthers yellow, not twisted after dehiscence, 2.5–3.5 mm. **Fruiting calyces** loosely enclosing berry, 10-ribbed, 25–35 × 15–25 mm. $2n = 24$.

Nothovarieties 2 (2 in the flora): Florida.

Physalis ×elliottii, its parentage, and its nothovarieties were discussed in detail by J. R. Sullivan (1985, 2013).

1. Plants sparsely to densely dendritic-pubescent. . . . . 9a. *Physalis ×elliottii* nothovar. *elliottii*
1. Plants glabrous except along leaf margins and calyx . . . . 9b. *Physalis ×elliottii* nothovar. *glabra*

9a. **Physalis ×elliottii** Kunze nothovar. **elliottii** [E]

Herbs sparsely to densely dendritic-pubescent. **Leaf blades** narrowly spatulate to lanceolate. $2n = 24$.

Flowering year-round in areas without frost. Beach dunes, edges of pine woods, disturbed coastal areas in sand; 0 m; Fla.

Nothovariety *elliottii* occurs from Wakulla County east and south, typically along the Gulf Coast, also in Broward, Hendry, and Miami-Dade counties. Plants sometimes appear to be glabrous without magnification.

9b. **Physalis ×elliottii** Kunze nothovar. **glabra** (Waterfall) J. R. Sullivan, Rhodora 115: 291. 2013 [E]

Physalis viscosa Linnaeus forma *glabra* Waterfall, Rhodora 60: 135. 1958; *P. walteri* Nuttall var. *glabra* (Waterfall) D. B. Ward

Herbs glabrous except along leaf margins and calyx. **Leaf blades** narrowly spatulate to narrowly lanceolate. $2n = 24$.

Flowering year-round in areas without frost. Beach dunes, edges of pine woods, disturbed coastal areas in sand; 0 m; Fla.

Nothovariety *glabra* occurs along the Gulf Coast from Pinellas County southward. Leaf width can vary considerably on a single plant; populations in Monroe County (Big Pine Key, Little Pine Key, No-Name Key) tend to have at least a few stems that resemble *Physalis angustifolia*.

10. Physalis fendleri A. Gray, Proc. Amer. Acad. Arts 10: 66. 1874

Physalis fendleri var. *cordifolia* A. Gray; *P. hederifolia* A. Gray var. *cordifolia* (A. Gray) Waterfall; *P. hederifolia* var. *fendleri* (A. Gray) Cronquist

Herbs perennial, rhizomatous, rhizome stout, pubescent, hairs simple, forked, or 3-branched, to 0.5 mm, most branching at base and branches appressed to surface, giving plants a grayish appearance. **Stems** erect, usually branching from base and most nodes, branches spreading, 0.5–5 dm. **Leaves** petiolate; petiole $^1/_3$–$^2/_3$ blade; blade ovate-lanceolate to ovate, 1–5.5 × 1–3.5 cm, base deltate-truncate to slightly cordate, sometimes unequal, margins sinuate to coarsely and irregularly dentate, teeth few, sometimes only 1 tooth per side near base. **Pedicels** 3–11(–17) mm, 10–15(–20) mm in fruit. **Flowers:** calyx 5–8 mm, lobes 2–4 mm; corolla yellow with 5 greenish to brown smudges, campanulate-rotate, (7–)10–12 mm; anthers yellow or purple-tinged, not twisted after dehiscence, 1.5–3 mm. **Fruiting calyces** loosely enclosing berry, 10-ribbed, 17–30(–35) × 15–25 mm. $2n = 24$.

Flowering May–Sep. Rocky to sandy soil, loose slopes, pinyon-juniper-ponderosa zones; 1300–2300 m; Ariz., Calif., Colo., Nev., N.Mex., Okla., Tex., Utah; Mexico (Baja California, San Luis Potosí, Tamaulipas).

In *Physalis fendleri*, the corolla limb is reflexed at maturity. Some populations in New Mexico have mostly retrorse, and very few branched, hairs. In the flora area, *P. fendleri* can be found as far west as southern California, northward into southern Nevada, Utah, and Colorado, and eastward into western Oklahoma and Texas.

11. Physalis grisea (Waterfall) M. Martínez, Taxon 42: 104. 1993 [E] [W]

Physalis pubescens Linnaeus var. *grisea* Waterfall, Rhodora 60: 167. 1958

Herbs annual, taprooted, villous, hairs simple, jointed, 0.5–1 mm, intermixed with stalked glands to 0.5 mm and sessile glands. **Stems** erect, branching at most nodes, branches spreading, 1–6 dm. **Leaves** petiolate; petiole $^2/_5$–$^4/_5$ blade; blade gray-green, drying orange or with orange patches, broadly ovate, 3.5–11 × 2.5–10 cm, base broadly rounded to slightly cordate, margins coarsely dentate with fewer than 10 teeth per side. **Pedicels** 4–6 mm, 5–12 mm in fruit. **Flowers:** calyx 3–5 mm, short-pubescent, lobes 1.5–2.5 mm; corolla yellow with 5 large, dark purple-black spots, campanulate-rotate, 5–8 mm; anthers blue or blue-tinged, not twisted after dehiscence, 1–2 mm. **Fruiting calyces** loosely enclosing berry, sharply 5-angled, 20–35 × 15–25 mm. $2n = 24$.

Flowering Jun–Oct. Open areas, meadows, pastures, disturbed woodlands, stream bottoms, cultivated sites; 50–200 m; B.C., Ont.; Ala., Calif., Conn., Del., D.C., Ga., Ill., Ind., Iowa, Maine, Md., Mass., Minn., Mo., N.J., N.Y., N.C., Ohio, Pa., S.C., S.Dak., Tenn., Vt., Va., Wash., W.Va.

Physalis grisea can be distinguished from *P. pubescens* by the often larger leaves that have a distinctive gray-green color and usually exhibit orange patches on drying. The orange-yellow fruit of *P. grisea* is sweet, and the species is offered in seed catalogs as "strawberry tomato." Most herbarium specimens of *P. grisea* from outside of cultivation probably represent short-lived populations derived from garden escapes; the geographic distribution given here is likely to change over time (the only Oregon population is historical). The name *P. pruinosa* Linnaeus has been misapplied to *P. grisea* (M. Martínez 1993).

12. Physalis hederifolia A. Gray, Proc. Amer. Acad. Arts 10: 65. 1874 (as hederaefolia)

Physalis comata Rydberg; *P. hederifolia* var. *comata* (Rydberg) Waterfall; *P. hederifolia* var. *palmeri* (A. Gray) C. L. Hitchcock; *P. hederifolia* var. *puberula* A. Gray; *P. palmeri* A. Gray; *P. puberula* Fernald; *P. rotundata* Rydberg

Herbs perennial, rhizomatous, rhizome stout, densely pubescent, hairs simple, sometimes jointed, glandular, 0.5–1 mm, sometimes also with sessile glands. **Stems** erect to decumbent, usually branching from base and at most nodes, branches spreading, 0.5–3 dm. **Leaves** petiolate; petiole 1/2 to ± as long as blade; blade broadly ovate to orbiculate, 1.5–3.5 × 1–3 cm, base cordate to rounded, margins ± entire or coarsely dentate, teeth sharp to blunt. **Pedicels** 4–8(–13) mm, 5–15 mm in fruit. **Flowers:** calyx 5–7(–10) mm, lobes 1.5–3.5(–5) mm; corolla yellow with 5 dark brown spots, campanulate-rotate, 7–12 mm; anthers yellow, not twisted after dehiscence, 2–4 mm. **Fruiting calyces** loosely enclosing berry, 10-ribbed, 20–30 × 15–25(–30) mm. **2*n*** = 24.

Flowering Apr–Aug. Dry open gravelly sites, rocky ledges, open plains; 200–2600 m; Ariz., Calif., Colo., Kans., Mont., Nebr., Nev., N.Mex., Okla., S.Dak., Tex., Utah, Wyo.; Mexico (Baja California, Chihuahua, Coahuila, Durango, Nuevo León, San Luis Potosí, Sonora, Tamaulipas, Zacatecas).

In *Physalis hederifolia*, the corolla limb is reflexed at maturity. This is a widespread species of the southwestern United States and the Great Plains (as far west as southern California and southeastern Nevada, and east into the western half of Iowa, Nebraska, Oklahoma, South Dakota, and Texas). Several varieties have been recognized based primarily on indument characters, but these features vary considerably over the range of the species. Plants from the more southern part of the range tend to have shorter hairs and to be more densely glandular; plants from northern Oklahoma and New Mexico northward tend to have longer hairs. All are clearly distinguishable from *P. fendleri*, which is eglandular, typically has forked or few-branched hairs, a distinctive leaf shape, and corollas with greenish-brownish smudges rather than distinct brown spots.

13. Physalis heterophylla Nees, Linnaea 6: 463. 1831 E W

Physalis ambigua (A. Gray) Britton; *P. heterophylla* var. *ambigua* (A. Gray) Rydberg; *P. heterophylla* var. *clavipes* Fernald; *P. heterophylla* var. *nyctaginea* (Dunal) Rydberg; *P. heterophylla* var. *villosa* Waterfall; *P. nyctaginea* Dunal; *P. sinuata* Rydberg

Herbs perennial, rhizomatous, rhizomes deeply buried, stout, densely villous, hairs simple, jointed, divergent, 1–2 mm, sometimes also with shorter glandular hairs. **Stems** erect to decumbent, branching at most nodes, branches spreading and decumbent, 1.5–10 dm. **Leaves** petiolate; petiole 1/3–2/3 blade; blade broadly ovate to suborbiculate, (2–)4–11(–13) × 3–9(–10) cm, base truncate to slightly cordate, margins deeply and irregularly dentate to ± entire. **Pedicels** 9–15(–20) mm, 20–30 mm in fruit. **Flowers:** calyx 6–12 mm, lobes 3–6 mm; corolla yellow with 5 large purple-brown smudges, campanulate-rotate, 10–17 mm; anthers yellow, rarely blue-tinged, not twisted after dehiscence, 2.5–4.5 mm. **Fruiting calyces** loosely enclosing berry, 10-ribbed, 25–40 × 15–30 mm. **2*n*** = 24.

Flowering May–Sep. Openings in hardwood forests, edges of pine woods, grasslands, fields, roadsides, disturbed sites; 10–400 m; Ont., Que., Sask.; Ala., Ark., Colo., Conn., Del., D.C., Fla., Ga., Ill., Ind., Iowa, Kans., Ky., La., Maine, Md., Mass., Mich., Minn., Miss., Mo., Mont., Nebr., N.H., N.J., N.Y., N.C., N.Dak., Ohio, Okla., Pa., R.I., S.C., S.Dak., Tenn., Tex., Vt., Va., W.Va., Wis., Wyo.

Physalis heterophylla is widespread east of the Rocky Mountains. Herbarium specimens from Manitoba, Oregon, and Utah represent historical collections. Morphological variation has been recognized taxonomically in some manuals, although intergradation occurs among varieties and they often cannot be identified reliably. The cultivated *P. peruviana* (cape gooseberry) is similar to *P. heterophylla* except that it is not glandular and has shorter pedicels (6–8 mm in flower, 13–15 mm in fruit). W. F. Hinton (1975b) reported the uncommon occurrence of a population in North Carolina representing natural hybridization between *P. heterophylla* and *P. virginiana*.

14. Physalis lanceolata Michaux, Fl. Bor.-Amer. 1: 149. 1803 E

Herbs perennial, rhizomatous, rhizomes stout, sparsely pubescent, hairs simple, antrorse, to 0.5 mm, or simple, jointed, divergent, 1–1.5 mm. **Stems** decumbent or weakly ascending, infrequently branching, branches spreading and decumbent or parallel to ground, 2–4 dm. **Leaves** petiolate; petiole 1/25–1/3 blade; blade oblanceolate, 4–10 × 2–6 cm, base attenuate, margins entire to slightly sinuate. **Pedicels** 10–20 mm, 10–30 mm in fruit. **Flowers:** calyx 6–10 mm, hispid, lobes 2–5 mm; corolla yellow with 5 pale brown smudges, campanulate-rotate, 10–15 mm; anthers yellow, not twisted after dehiscence, 2.5–3.5 mm. **Fruiting calyces** loosely enclosing to nearly filled by berry, 10-ribbed, 20–35 × 15–30 mm. **2*n*** = 24.

Flowering Apr–Sep. Dry to xeric pine-oak-grass communities of the Sandhills Region; 100–200 m; Ga., N.C., S.C.

Physalis lanceolata occurs as populations of 1 to 20 plants scattered within suitable habit, notably where fire management is practiced. W. F. Hinton (1970, 1976) showed that *P. lanceolata* is not a hybrid and that the name had been misapplied to plants of the Great Plains.

15. Physalis longifolia Nuttall, Trans. Amer. Philos. Soc., n. s. 5: 193. 1836

Herbs perennial, rhizomatous, rhizomes deeply buried, stout, glabrous or sparsely strigose, hairs simple, antrorse, to 0.5 mm. **Stems** erect or erect to decumbent, branching frequently at distal nodes or several-branched from base, branches spreading or ascending, 1–6 dm. **Leaves** petiolate; petiole 1/5–2/5 blade; blade ovate to ovate-lanceolate or broadly lanceolate, 2.5–10(–13) × 0.5–6(–7) cm, base truncate to rounded, margins entire to coarsely dentate or irregularly crenate-dentate with only a few teeth. **Pedicels** 5–18 mm, 12–35 mm in fruit. **Flowers:** calyx (5–)7–12 mm, sparsely strigose with short, antrorse hairs, lobes 3–6 mm; corolla yellow with 5 purple-brown smudges, campanulate-rotate, 10–20 mm; anthers yellow or blue-tinged, not twisted after dehiscence, 2–4 mm. **Fruiting calyces** loosely enclosing berry, 10-ribbed, 20–40 × 15–30 mm. **2*n*** = 24.

Varieties 3 (3 in the flora): North America, n Mexico; introduced in Australia.

1. Stems erect to decumbent, branching at base; leaf blades glabrous . . . . . 15c. *Physalis longifolia* var. *texana*
1. Stems erect, branching at distal nodes; leaf blades glabrous or sparsely strigose.
    2. Leaf blades narrowly ovate-lanceolate to lanceolate; anthers yellow. . . . . .15a. *Physalis longifolia* var. *longifolia*
    2. Leaf blades ovate to ovate-lanceolate or broadly lanceolate; anthers blue-tinged . . . . 15b. *Physalis longifolia* var. *subglabrata*

15a. Physalis longifolia Nuttall var. **longifolia** [W]

Physalis polyphylla Greene; *P. rigida* Pollard & C. R. Ball; *P. virginiana* Miller var. *polyphylla* (Greene) Waterfall; *P. virginiana* var. *sonorae* (Torrey) Waterfall

Herbs glabrous or sparsely strigose, hairs simple, antrorse, to 0.5 mm. **Stems** erect, branching frequently at distal nodes, 1–6 dm. **Leaf blades** narrowly ovate-lanceolate to lanceolate, 2.5–7.5 × 0.5–2.5 cm. **Flowers:** corolla 10–20 mm; anthers yellow. **2*n*** = 24.

Flowering May–Oct. Fields, open woods, sandy areas, disturbed sites, sagebrush and pinyon-juniper communities; 200–2700 m; B.C.; Ariz., Ark., Calif., Colo., Idaho, Ill., Iowa, Kans., Mich., Minn., Miss., Mo., Mont., Nebr., Nev., N.Mex., Okla., Oreg., S.Dak., Tex., Utah, Wash., Wis., Wyo.; Mexico (Chihuahua, Coahuila, Hidalgo, San Luis Potosí, Sonora); introduced in Australia.

Variety *longifolia* is widespread in the United States from the prairie states westward. In the westernmost states, its distribution is restricted (Lincoln County, Nevada; Siskiyou County, California; Baker, Malheur, and Wallowa counties in Oregon; Asotin County, Washington), and these plants are slightly smaller than in the rest of the range. It reportedly does not persist in British Columbia.

15b. Physalis longifolia Nuttall var. **subglabrata** (Mackenzie & Bush) Cronquist in C. L. Hitchcock et al., Vasc. Pl. Pacif. N.W. 4: 286. 1959 [E] [W]

Physalis subglabrata Mackenzie & Bush, Trans. Acad. Sci. St. Louis 12: 86. 1902; *P. macrophysa* Rydberg; *P. virginiana* Miller var. *subglabrata* (Mackenzie & Bush) Waterfall

Herbs glabrous or sparsely strigose, hairs simple, antrorse, to 0.5 mm. **Stems** erect, branching frequently at distal nodes, 1–6 dm. **Leaf blades** ovate to ovate-lanceolate or broadly lanceolate, 3.5–10(–13) × 2–6(–7) cm. **Flowers:** corolla 10–18 mm; anthers blue-tinged. **2*n*** = 24.

Flowering May–Oct. Open woods, fields, stream bottoms, roadsides, disturbed or cultivated sites; 10–400 m; Ont.; Ala., Ark., Conn., Del., D.C., Fla., Ga., Ill., Ind., Iowa, Kans., Ky., La., Md., Mass., Mich., Miss., Mo., N.J., N.Y., N.C., Ohio, Okla., Pa., R.I., S.C., Tenn., Tex., Vt., Va., W.Va., Wis.

Plants of var. *subglabrata* with large fruiting calyces have been distinguished as forma *macrophysa* (Rydberg) Steyermark, but they appear to represent an extreme of a continuum and do not warrant taxonomic recognition.

15c. Physalis longifolia Nuttall var. **texana** (Rydberg) J. R. Sullivan, Rhodora 115: 291. 2013 [E]

Physalis texana Rydberg, Mem. Torrey Bot. Club 4: 339. 1896; *P. virginiana* Miller var. *texana* (Rydberg) Waterfall

Herbs glabrous except for short, sparse, antrorse hairs on flowering calyx and pedicel. **Stems** erect to decumbent, branching at base, 1–1.5 dm. **Leaf blades** ovate, 3.5–5.5 × 1.7–4.5 cm. **Flowers:** corolla 11 mm; anthers yellow. **2*n*** = 24.

Flowering Mar–Nov. Alluvial sand or clay, partially shaded stream banks, woodlands, thickets, disturbed or cultivated sites; 0–20 m; Tex.

Variety *texana* is known from central to southern Texas.

16. Physalis missouriensis Mackenzie & Bush, Trans. Acad. Sci. St. Louis 12: 84. 1902 [E]

Physalis pubescens Linnaeus var. *missouriensis* (Mackenzie & Bush) Waterfall

Herbs annual, taprooted, sparsely to densely pubescent, hairs simple, jointed, glandular and eglandular, to 0.5 mm. **Stems** erect, branching at most nodes, branches spreading, 1.5–5.5 dm. **Leaves** petiolate; petiole 2/5 to as long as blade; blade broadly ovate to orbiculate, 2.5–5.5 × 1.5–5 cm, base rounded, sometimes truncate, margins irregularly, shallowly crenate-dentate. **Pedicels** 4–7 mm, 5–10 mm in fruit. **Flowers:** calyx 2.5–4 mm, densely glandular-pubescent, lobes 1–2 mm; corolla yellow, without spots or smudges, campanulate-rotate, 5–7 mm; anthers yellow, not twisted after dehiscence, 1–1.5 mm. **Fruiting calyces** loosely enclosing berry, 10-ribbed, 10–20 × 10–20 mm. **2*n*** = 24.

Flowering Jun–Sep(–Oct). Rocky bluffs, dolomite ledges, cliffs, wooded slopes and stream banks primarily on the Ozark Plateau; 50–500 m; Ark., Kans., Mo., Nebr., Okla.

Physalis missouriensis is uncommon and most easily confused with *P. pubescens*, from which it can be distinguished by its unspotted corolla and ten-ribbed fruiting calyx.

17. Physalis mollis Nuttall, Trans. Amer. Philos. Soc., n. s. 5: 194. 1836 [E]

Physalis viscosa Linnaeus subsp. *mollis* (Nuttall) Waterfall

Herbs perennial, rhizomatous, rhizomes deeply buried, stout, often also with shallowly buried, slender rhizomes, densely pubescent, hairs dendroid-stelliform, to 1 mm, obscuring plant surface on younger growth, occasionally also jointed, branched or simple, 2–4 mm, glandular or eglandular. **Stems** erect, branching occasionally, branches ascending, 1.5–5 dm. **Leaves** petiolate; petiole 1/3–4/5 blade; blade ovate, 2.5–7 × 1.5–6(–7) cm, base truncate, margins coarsely dentate or irregular to ± entire. **Pedicels** 10–25(–35) mm, 20–40 (–52) mm in fruit. **Flowers:** calyx 6–10(–12) mm, lobes 2.5–5.5 mm; corolla yellow with 5 pale to dark brown smudges or dark purple-black spots, campanulate-rotate, 9.5–15(–17) mm; anthers yellow, rarely blue- or purple-tinged, not twisted after dehiscence, 3–4 mm. **Fruiting calyces** loosely enclosing berry, 10-ribbed, 25–40(–50) × 15–35 mm.

Varieties 2 (2 in the flora): sc United States.

1. Plants eglandular; corolla with pale to dark brown smudges. . . . . 17a. *Physalis mollis* var. *mollis*
1. Plants glandular; corolla with deep purple-black spots . . . . 17b. *Physalis mollis* var. *variovestita*

17a. Physalis mollis Nuttall var. **mollis** [E]

Hairs dendroid-stelliform, to 1 mm, eglandular, occasionally also jointed, branched or simple, 2–4 mm, eglandular. **Corollas** with pale to dark brown smudges. **Fruiting calyces** 15–30 mm wide. **2*n*** = 24.

Flowering Mar–Oct. Sandy soil, prairies, roadsides, disturbed habitats; 50–200 m; Ark., La., Okla., Tex.

Variety *mollis* can be distinguished by its dense, eglandular tomentum, which often obscures the leaf surface.

17b. Physalis mollis Nuttall var. **variovestita** (Waterfall) J. R. Sullivan, Syst. Bot. 10: 442. 1985 [E]

Physalis variovestita Waterfall, Rhodora 60: 137. 1958; *P. cinerascens* (Dunal) Hitchcock var. *variovestita* (Waterfall) B. L. Turner

Hairs dendroid-stelliform, to 1 mm, glandular, occasionally also jointed, branched or simple, 2–4 mm, glandular. **Corollas** with deep purple-black spots. **Fruiting calyces** 25–35 mm wide. **2*n*** = 24.

Flowering Mar–Oct. Disturbed areas in sand; 0–20 m; Tex.

Variety *variovestita* is found in extreme southern Texas, north along the coast to Aransas County.

18. Physalis neomexicana Rydberg, Mem. Torrey Bot. Club 4: 325. 1896 (as neo-mexicana) [E]

Physalis foetens Poiret var. *neomexicana* (Rydberg) Waterfall; *P. subulata* Rydberg var. *neomexicana* (Rydberg) Waterfall ex Kartesz & Gandhi

Herbs annual, taprooted, densely glandular-pubescent, hairs simple, mostly 0.5(–1 mm), grayish brown in appearance when dry. **Stems** erect, angulate and blue-tinged, at least distally, branching at most nodes, internodes noticeably long, branches spreading, 1–5 dm. **Leaves** petiolate; petiole 1/3–1/2 blade; blade broadly ovate to orbiculate, 2–6 × 1.5–5 cm, base deltate to rounded-attenuate, margins coarsely, irregularly crenate-dentate. **Pedicels**

stout, 2–5 mm, 5–10(–12) mm in fruit. **Flowers:** calyx 3–5 mm, lobes 1–2.5(–3) mm, long-attenuate; corolla yellow with 5 large, dark purple-blue-black spots, campanulate-rotate, 6–10 mm; anthers blue or blue-tinged, not twisted after dehiscence, 1 mm. **Fruiting calyces** loosely enclosing berry, sharply 5-angled (ribs often deep purple), nearly spheric, 20–25(–30) × 15–20 (–30) mm. **2*n*** = 24.

Flowering May–Sep. Sandy soil, pinyon-juniper associations, disturbed grasslands, roadsides, cultivated fields, gardens; 1500–2500 m; Ariz., Colo., N.Mex., Tex.

Physalis neomexicana can be distinguished from *P. pubescens* by its stout pedicels, nearly spheric fruiting calyces, and grayish brown appearance when dry. Some herbarium specimen labels mention that the plants are ill-smelling. M. Martínez (1998) determined the name *P. subulata* Rydberg to be a synonym of *P. patula* Miller, which is a Mexican species.

19. Physalis philadelphica Lamarck in J. Lamarck et al., Encycl. 2: 101. 1786 • Tomatillo, Mexican ground-cherry or husk-tomato [I] [W]

Herbs annual, taprooted, glabrous or sparsely hairy, hairs simple, appressed, mostly 0.5 mm. **Stems** erect, branching mostly at distal nodes, branches spreading, sometimes streaked with purple, 1.5–10 dm. **Leaves** petiolate; petiole 1/2 to as long as blade; blade ovate to ovate-lanceolate, 2–7 × 2–4 cm, base rounded to attenuate, margins dentate to entire. **Pedicels** 3–6 mm, 3–8(–11) mm in fruit. **Flowers:** calyx 5–7(–10) mm, lobes 2–4 mm; corolla yellow with 5 blue-tinged spots or smudges, campanulate-rotate, 7–15 mm; anthers blue, strongly twisted after dehiscence, 3 mm. **Fruiting calyces** filled, or burst, by berry, 10-ribbed, 20–30 × 20–30 mm. **2*n*** = 24.

Flowering year-round in areas without frost. Disturbed sites, fence rows, edges of cultivated fields, roadsides; 0–200 m; introduced; B.C., Sask.; Ariz., Calif., Idaho, Ill., Md., Mass., Minn., Mo., N.Mex., Oreg., Pa., Tenn., Tex., Vt., Va., Wash., W.Va.; Mexico; introduced also in Australia.

Physalis philadelphica is native to Mexico and, possibly, the southwestern United States; it is cultivated for its fruits, which are used in Mexican-style salsa. It frequently escapes cultivation and can become established in disturbed habitats. Considerable morphological diversity has been documented in this species (M. Y. Menzel 1951; W. D. Hudson 1986); the measurements given here reflect only wild-growing populations in the flora area. The mature berry is pale green to purplish or purple-streaked. Seeds can remain viable in the soil for several years.

SELECTED REFERENCE Hudson, W. D. 1986. Relationships of domesticated and wild *Physalis philadelphica*. In: W. G. D'Arcy, ed. 1986. Solanaceae: Biology and Systematics. New York. Pp. 416–432.

20. Physalis pubescens Linnaeus, Sp. Pl. 1: 183. 1753 [F] [W]

Physalis barbadensis Jacquin; *P. barbadensis* var. *glabra* (Michaux) Fernald; *P. floridana* Rydberg; *P. latiphysa* Waterfall; *P. pubescens* var. *glabra* (Michaux) Waterfall; *P. pubescens* var. *integrifolia* (Dunal) Waterfall; *P. turbinata* Medikus

Herbs annual, taprooted, ± glabrous to villous, hairs simple, jointed, glandular and eglandular, of varying lengths, all shorter than 0.5 mm, plants from southwestern United States all glandular, green in appearance when dry. **Stems** erect, branching at most nodes, branches spreading, 0.5–8 dm. **Leaves** petiolate; petiole 1/5 to as long as blade; blade broadly ovate to orbiculate, (1.6–)2.5–8(–9.5) × (1–)2–7 cm, base rounded to slightly cordate, margins entire or coarsely dentate, teeth fewer than 8 per side. **Pedicels** slender, 3.5–9 mm, 5–15 mm in fruit. **Flowers:** calyx 3–6(–7) mm, lobes 1–3.5 mm; corolla yellow with 5 large, dark purple-brown-black spots, campanulate-rotate, 6–11 mm; anthers blue, rarely yellow or blue-tinged, not twisted after dehiscence, 1–2 mm. **Fruiting calyces** loosely enclosing berry, sharply 5-angled, 20–35 × 15–25(–30) mm, always noticeably longer than wide. **2*n*** = 24.

Flowering year-round in areas without frost, mostly May–Oct. Low woods, edges of swamps, stream banks, floodplains, hammocks, disturbed habitats; 0–900 m; Ala., Ariz., Ark., Calif., D.C., Fla., Ga., Ill., Ind., Iowa, Kans., Ky., La., Md., Mich., Miss., Mo., N.J., N.Mex., N.C., Ohio, Okla., Pa., S.C., Tenn., Tex., Utah, Va., W.Va., Wis.; Mexico; West Indies; Central America; South America; introduced in Australia.

Fresh plants of *Physalis pubescens* reportedly have a strong fetid odor (M. Martínez 1998). This widespread species exhibits considerable variability in the character of the leaf margins and degree of indument. The fruits of *P. pubescens* are reportedly gathered for food.

21. Physalis pumila Nuttall, Trans. Amer. Philos. Soc., n. s. 5: 193. 1836 [E]

Herbs perennial, rhizomatous, rhizomes deeply buried, stout, hispid, hairs simple or 2- or 3-branched, divergent and antrorse, jointed, 0.5–2 mm. **Stems** erect to decumbent, branching at most nodes or infrequently and only at distal nodes, branches ascending, 1.5–4 dm. **Leaves** petiolate; petiole 1/10–2/5 blade; blade elliptic-ovate to ovate-lanceolate, 3–8(–10) × 2–4(–5) cm, base rounded to attenuate and narrowing to petiole, margins entire to sinuate, rarely shallowly, irregularly sinuate-dentate. **Pedicels** hispid, 8–46 mm, 15–55 mm in fruit. **Flowers:** calyx 6–12 mm, lobes 2.5–6 mm; corolla yellow with pale brown, ochre, or green tinge or smudges, campanulate-rotate, 9–17 mm; anthers yellow, rarely blue-tinged, not twisted after dehiscence, 1–3 mm. **Fruiting calyces** loosely enclosing berry, 10-ribbed, 20–40 × 15–30 mm. **2*n*** = 24.

Varieties 2 (2 in the flora): c, sc United States.

Varieties *hispida* and *pumila* are quite distinctive in the field but are often difficult to distinguish in the herbarium.

SELECTED REFERENCE Hinton, W. F. 1976. The systematics of *Physalis pumila* subsp. *hispida*. Syst. Bot. 1: 188–193.

1. Stems erect; plants hispid throughout, hairs simple and 2- or 3-branched . . . . 21a. *Physalis pumila* var. *pumila*
1. Stems erect to decumbent; plants hispid on pedicels and calyx, hairs simple, rarely with a few branched hairs . . . . . . 21b. *Physalis pumila* var. *hispida*

21a. Physalis pumila Nuttall var. **pumila** [E]

Herbs hispid, hairs simple and 2- or 3-branched, jointed, 0.5–2 mm. **Stems** erect. **Pedicels** 13–46 mm, 25–55 mm in fruit. **Flowers:** calyx 6–12 mm; corolla 10–17 mm; anthers yellow, rarely blue-tinged, 1–3 mm. **Fruiting calyces** 25–40 mm. **2*n*** = 24.

Flowering Mar–Sep. Dry, rocky soil, prairies, fields, disturbed habitats; 100–600 m; Ark., Ill., Iowa, Kans., La., Mo., Nebr., Okla., Tex.

Variety *pumila* is morphologically distinguishable by the abundance of two- or three-branched hairs. It is found in dry rocky soil on the eastern Great Plains.

21b. Physalis pumila Nuttall var. **hispida** (Waterfall) J. R. Sullivan, Rhodora 115: 291. 2013 [E]

Physalis virginiana Miller var. *hispida* Waterfall, Rhodora 60: 154. 1958; *P. hispida* (Waterfall) Cronquist; *P. longifolia* Nuttall var. *hispida* (Waterfall) Steyermark; *P.*≈*pumila* subsp. *hispida* (Waterfall) W. F. Hinton

Herbs pubescent, hairs on stem simple, rarely with a few branched hairs, jointed, antrorse, 0.5–1 mm, on pedicels and calyx spreading-hispid, simple, jointed, 0.5–1.5 mm. **Stems** erect to decumbent. **Pedicels** 8–15(–22) mm, 15–35 mm in fruit. **Flowers:** calyx 6–10 mm; corolla 9–14 mm; anthers yellow, 2–3 mm. **Fruiting calyces** 20–30 mm. **2*n*** = 24.

Flowering May–Aug (mostly Jun). Sandhills, dunes, sandy floodplains, sandy prairies and roadsides; 500–2000 m; Colo., Kans., Mo., Nebr., Okla., Tex., Wyo.

Variety *hispida* is morphologically distinguishable by the lack or rarity of branched hairs. It is found in sandy soil in the Great Plains. The name *Physalis lanceolata* Michaux has been misapplied to var. *hispida* (W. F. Hinton 1976).

22. Physalis solanacea (Schlechtendal) Axelius, Phytologia 79: 11. 1995 (as solanaceous)

Margaranthus solanaceus Schlechtendal, Index Seminum (Halle) 1838: 8. 1838.

Herbs annual, taprooted, sparsely pubescent to ± glabrous, hairs simple, antrorse, appressed, to 0.5 mm. **Stems** erect, branching from near base and at most nodes, branches spreading, distal internodes congested, 1–18 dm. **Leaves** petiolate; petiole mostly 1/2 as long as blade; blade ovate to ovate-lanceolate, (1–)2–4.5(–7) × 1–3(–4) cm, base unequal, margins entire to irregularly sinuate or with 1+ large teeth. **Pedicels** 1.5–3 mm, 3–5 mm in fruit. **Flowers:** calyx 1.5–2.5 mm, lobes 0.5–1 mm; corolla deep purple, rarely greenish or yellowish with 5 very large purple spots visible through corolla, urceolate, bulging beyond calyx, 2.5–4.5 mm; anthers yellow to purple, not twisted after dehiscence, 1 mm; style sometimes exserted. **Fruiting calyces** nearly filled by berry, 5-ribbed, not sharply angled, 8–17 × 6–13 mm. **2*n*** = 24.

Flowering (Jun–)Aug–Oct(–Dec). Grasslands, pinyon-oak-juniper woodlands, disturbed areas, weedy fields, along streams; 10–2100 m; Ariz., N.Mex., Tex.; Mexico.

Congested internodes on the distal portions of plants of *Physalis solanacea* give the superficial appearance of multiple leaves or flowers at a single node. It is widespread in Mexico except for Baja California (north and south).

23. Physalis virginiana Miller, Gard. Dict. ed. 8, Physalis no. 4. 1768 E F W

Physalis monticola C. Mohr; *P. virginiana* var. *campaniforma* Waterfall

Herbs perennial, rhizomatous, rhizomes deeply buried, stout, hispid, hairs simple, jointed, divergent, mostly 1 mm, and retrorse, to 0.5 mm. **Stems** erect, branching infrequently and only at distal nodes, branches ascending, sometimes with multiple aerial stems arising from apex of rhizome, 1–4 dm. **Leaves** petiolate; petiole 1/5–1/2 blade; blade ovate to broadly lanceolate, 2–7(–9) × 1–5(–6) cm, base truncate to obtuse or rounded, margins entire or coarsely to shallowly dentate with few teeth. **Pedicels** (6–)9–19(–27) mm, 12–30(–33) mm in fruit. **Flowers:** calyx 6–12(–14) mm, lobes 3–6 mm; corolla yellow with 5 dark purple-brown-black smudges, campanulate-rotate, 9–17(–20) mm; anthers yellow or blue-tinged, not twisted after dehiscence, 2–3 mm. **Fruiting calyces** loosely enclosing berry, 10-ribbed, pyramidal, narrowing to lobes, 20–40 × 15–30 mm. $2n = 24$.

Flowering Apr–Oct. Sandy soils, prairies, fields, thickets, pine-oak-hickory woodlands, gravelly pinyon-juniper slopes, disturbed habitats, sandy or gravelly roadsides, cultivated ground, waste places, along railroads; 50–2500 m; Man., Ont.; Ala., Ark., Colo., Conn., Del., D.C., Fla., Ga., Ill., Ind., Iowa, Kans., Ky., La., Maine, Md., Mass., Mich., Minn., Miss., Mo., Nebr., N.H., N.J., N.Mex., N.C., N.Dak., Ohio, Okla., Pa., S.C., S.Dak., Tenn., Tex., Vt., Va., W.Va., Wis., Wyo.

Physalis virginiana occurs primarily in the Midwest and central plains states and southern and eastern Canadian prairies. It is found sporadically in disturbed habitats in the eastern United States and on wooded, gravelly slopes in the foothills of the Rocky Mountains. In Manitoba, the fruits of *P. virginiana* are gathered and preserved (canned) for winter use.

SELECTED REFERENCE Hinton, W. F. 1975b. Natural hybridization and extinction of a population of *Physalis virginiana* (Solanaceae). Amer. J. Bot. 62: 198–202.

24. Physalis walteri Nuttall, J. Acad. Nat. Sci. Philadelphia 7: 112. 1834 E F

Physalis maritima M. A. Curtis; *P. viscosa* Linnaeus subsp. *maritima* (M. A. Curtis) Waterfall; *P. viscosa* var. *maritima* (M. A. Curtis) Rydberg

Herbs perennial, rhizomatous, rhizomes deeply buried, stout, pubescent, hairs dendroid-stelliform, to 1 mm. **Stems** erect to decumbent, branching at most nodes, proximal branches spreading and decumbent, (0.5–)1.5–3(–4) dm. **Leaves** petiolate; petiole 1/5–1/3 blade; blade broadly elliptic to ovate (2–)3–7(–11) × 1.5–4(–7.5) cm, base rounded, margins entire, rarely irregularly shallowly sinuate. **Pedicels** 10–25(–30) mm, 15–35(–45) mm in fruit. **Flowers:** calyx 5–9 mm, lobes 2–4 mm; corolla yellow with 5 dark purple-brown spots, campanulate-rotate, 10–16 mm; anthers yellow, rarely purple-tinged, not twisted after dehiscence, 2.5–3.5 mm. **Fruiting calyces** loosely enclosing berry, 10-ribbed, 20–35 × 15–25 mm. $2n = 24$.

Flowering year-round in areas without frost. Beach dunes, maritime woodlands, inland sandhills, disturbed areas in sand; 0–60 m; Ala., Fla., Ga., Miss., N.C., S.C., Va.

Physalis walteri occurs in inland, sandy areas in Alabama, Florida, Georgia, and Mississippi and on beach dunes from Florida north along the Atlantic Coast to southern Virginia.

24. QUINCULA Rafinesque, Atlantic J. 1: 145. 1832 • Purple ground-cherry [Latin *quinc-*, five, and *-ula*, diminutive, alluding to opaque spots on corolla as compared to dark spots in *Physalis*]

Janet R. Sullivan

Herbs, perennial, rhizomatous, sparsely to densely covered with stalked white vesicles. **Stems** erect, weakly decumbent, or prostrate, branched from base and at most nodes. **Leaves** alternate; blade margins entire or sinuate to deeply pinnately incised. **Inflorescences** axillary,

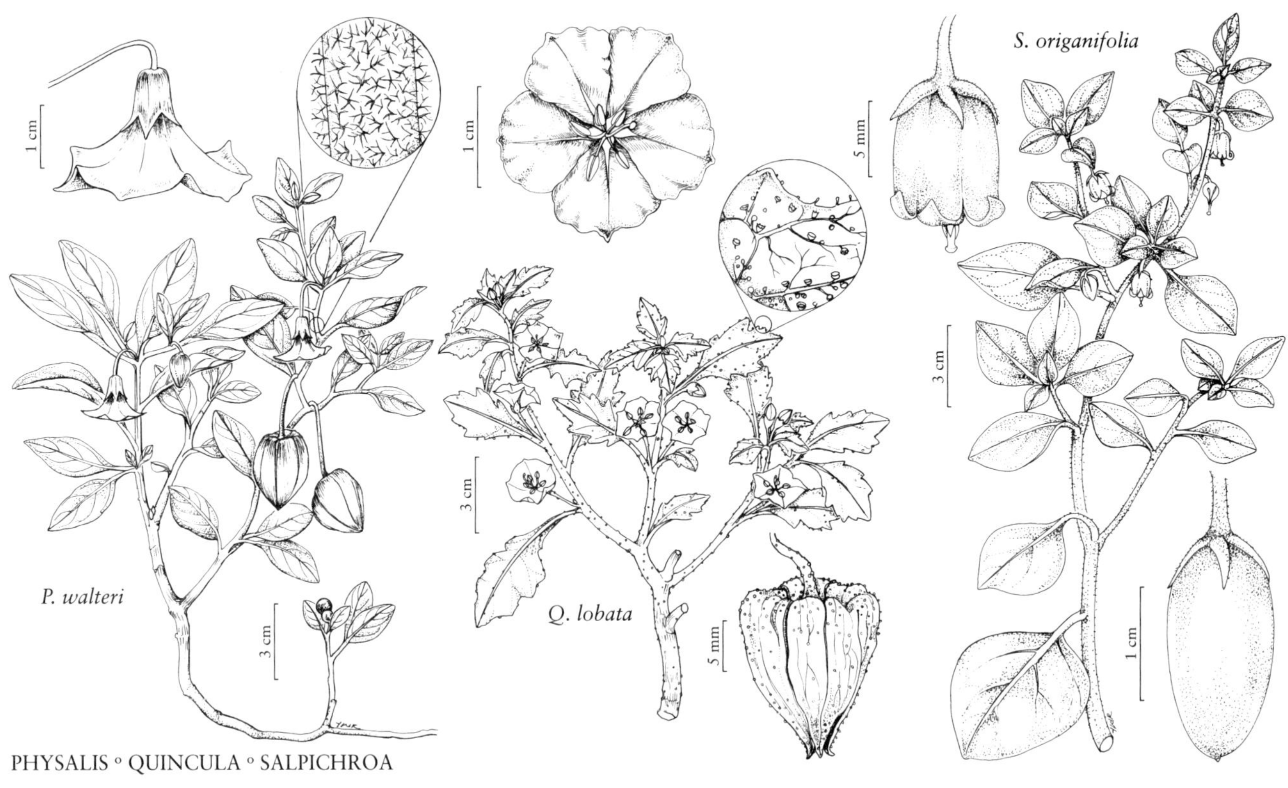

PHYSALIS ◦ QUINCULA ◦ SALPICHROA

usually 2–4-flowered clusters, sometimes solitary flowers. **Flowers** 5-merous; calyx accrescent, campanulate, lobes 5, broadly triangular, completely and loosely enclosing berry; corolla usually purple, rarely white, radial, rotate, lobes relatively short; stamens inserted near base of corolla tube, equal; anthers basifixed, narrowly elliptic, dehiscing by longitudinal slits; ovary 2-carpellate; style slender, slightly curved to 1 side; stigma capitate. **Fruits** berries, globose, dry. **Seeds** angular. $x = 11$.

Species 1: sw, sc United States, n Mexico.

Quincula can be distinguished from other North American physaloid genera (*Calliphysalis*, *Chamaesaracha*, *Leucophysalis*, *Oryctes*, *Physalis*) by its purple, rarely white, corollas with five opaque spots, and stalked white vesicles scattered on the stems, foliage, and calyces.

1. **Quincula lobata** (Torrey) Rafinesque, Atlantic J. 1: 145. 1832 [F] [W]

Physalis lobata Torrey, Ann. Lyceum Nat. Hist. New York 2: 226. 1827

Stems 2–30 cm. **Leaf blades** oblanceolate to spatulate, (2–)4–6(–7.5) (including petiole) × (0.5–)1.5–2.5 cm, slightly fleshy. **Flowering pedicels** (5–)10–25(–30) mm. **Flowers:** calyx (2.5–)3–6(–7) mm, lobes 1–2(–3) mm; corolla 11–26(–28) mm diam., with 5 darker purple patches along main veins and mat of white hairs in throat; filaments filiform; anthers yellow, rarely purple-tinged, 1–2(–2.5) mm. **Fruiting pedicels** (16–)20–30(–39) mm. **Fruiting calyces** green, occasionally with purple along main veins, triangular overall, 15–20(–25) × 15–20 mm, very sharply 5-angled and invaginated at base. **Berries** green, 0.5–1 cm diam. $2n = 22, 44$.

Flowering Mar–Oct (sporadically year-round). Roadsides, grasslands, grazed areas, prairies, disturbed habitats, stream bottoms, washes, sand, gravel, clay, silt, loam; 300–1900 m; Ariz., Calif., Colo., Kans., Nev., N.Mex., Okla., Tex.; Mexico (Chihuahua, Coahuila, Nuevo León, Sonora, Tamaulipas).

Forms of *Quincula lobata* with purple and white corollas can be found in the same population. In some publications, the forms have been recognized taxonomically as *Physalis lobata* forma *lobata* and forma *albiflora* Waterfall. Occasional aneuploidy was reported by M. Y. Menzel (1950).

25. SALPICHROA Miers, London J. Bot. 4: 321. 1845 • [Greek *salpinx*, trumpet, and *chroa*, color or complexion, alluding to flowers] [I]

Philip D. Jenkins†

Herbs, perennial, [**shrubs**], rhizomatous, pubescent, hairs simple, short, unicellular, or glabrate. **Stems** scandent [prostrate, decumbent, or pendent], branched. **Leaves** subopposite or geminate. **Inflorescences** axillary, solitary flowers [rarely paired]. **Flowers** 5-merous; calyx not accrescent, campanulate, lobes 5, linear [acute-triangular or narrowly ovate]; corolla greenish yellow to white [sulphur yellow, pinkish], radial, urceolate [cylindric or urceolate-cylindric], lobes triangular [linear, cuspidate, subulate, or narrowly ovate], revolute [spreading or reflexed]; stamens 5, inserted in adaxial 1/2 of tube, or near mouth of corolla, equal; anthers dorsifixed, oblong, dehiscing by longitudinal slits; ovary 2-carpellate; style (exserted or not), slender, straight; stigma capitate. **Fruits** berries, ovoid-oblong [ellipsoid], juicy. **Seeds** (10–20), reniform. x = 12.

Species 21 (1 in the flora): introduced; South America, introduced also in Europe, Africa, Australia.

All species of *Salpichroa* except *S. origanifolia* are Andean. A 2018 paper by C. Carrizo García et al. indicates that *Nectouxia* and *Salpichroa* should be combined into a single genus. In that case *Nectouxia* would have priority, requiring name changes for over 20 *Salpichroa* species. A proposal is under consideration to conserve the name *Salpichroa* over *Nectouxia* (G. E. Barboza et al. 2016).

1. **Salpichroa origanifolia** (Lamarck) Thellung, Fl. Advent. Montpellier, 452. 1912 • Cock's eggs, lily-of-the-valley vine [F] [I] [W]

Physalis origanifolia Lamarck, Tabl. Encycl. 2: 28. 1794; *Perizoma rhomboidea* (Gillies & Hooker) Small; *Salpichroa rhomboidea* (Gillies & Hooker) Miers

Herbs with strong odor, (0.2–) 0.4–3(–5) m, root sometimes fleshy. **Stems** ± lignified, usually 1–2(–4)-winged, turning dark when dry. **Leaves:** petiole shorter than blade; blade ovate-rhombic to suborbiculate, 1.5–4(–6) × 1.5–4(–5) cm, fleshy. **Pedicels** pendent, slender. **Flowers:** calyx 2–3.5 mm, incised nearly to base; corolla 3.5–10 mm, inside with dense, wooly, annular band of hairs; stamens not exserted, connivent. **Berries** pale yellowish white, nearly translucent. $2n$ = 24.

Flowering Mar–Jul. Cultivated fields, waste ground; 0–2000 m; introduced; Ala., Ariz., Calif., Fla., Ga., La., Miss., N.C., S.C., Tex., Va.; South America (Argentina, se Bolivia, s Brazil, Paraguay, Uruguay); introduced also in w, s Europe (England, France, Italy, Portugal, Spain), Africa (Algeria, Egypt), Australia.

Salpichroa origanifolia can escape cultivation and persist for short periods. The rhizomes are a source of alkaloids (W. C. Evans et al. 1972), and whole plants (growing in Argentina) are a source of withanolides.

26. SCHIZANTHUS Ruiz & Pavon, Fl. Peruv. Prodr., 6. 1794 • Poor man's orchid [Greek *schizos*, divide, and *anthos*, flower, alluding to irregularly divided corolla] [I]

Michael A. Vincent

Herbs, annual [biennial], taprooted [rhizomatous], glandular-pubescent. **Stems** branched. **Leaves** alternate; petiolate (or petiolate proximally, sessile distally); blade pinnate or pinnate-pinnatifid [toothed]. **Inflorescences** axillary and terminal, paniculate. **Flowers** 5-merous, bilaterally symmetric (resupinate); calyx not accrescent, campanulate, deeply 5-lobed nearly to base, lobes lanceolate-linear; corolla white, pink, red, purple, or blue [yellow, orange, or brownish], often mottled, bilateral, funnelform, lobes spreading, longer than tube, adaxial larger and often deeply dissected; stamens inserted at base of corolla tube, unequal, 2 fertile and lateral, 3 sterile; anthers dorsifixed, ellipsoidal, dehiscing (explosively) by longitudinal slits (introrse); ovary 2-carpellate (2-locular); ovules 10–50; style filiform; stigma discoid, unlobed. **Fruits** capsules, ellipsoidal (slightly exceeding calyx), dehiscence septicidal. **Seeds** 8–40, ovate-reniform, flattened, (muricate-reticulate, dull). $x = 10$.

Species 12 (1 in the flora): introduced; s South America (Argentina, Chile).

Schizanthus is sister to the rest of the Solanaceae in phylogenetic studies, possibly pointing to a Southern Hemisphere origin of the family (R. G. Olmstead et al. 2008). Indument of *Schizanthus* is a mixture of unicellular hairs and glandular, multiseriate, multicellular capitate hairs that are seemingly unique in the Solanaceae (A. T. Hunziker 2001). Species of *Schizanthus* are cultivated as ornamentals for their butterfly- or orchid-shaped flowers (L. H. Bailey 1924).

SELECTED REFERENCE Grau, J. and E. Gronbach. 1984. Untersuchungen zur Variabilität in der Gattung *Schizanthus* (Solanaceae). Mitt. Bot. Staatssamml. München 20: 111–203.

1. **Schizanthus pinnatus** Ruiz & Pavon, Fl. Peruv. 1: 13, plate 17. 1798 • Butterfly-flower [F] [I]

Herbs 0.3–1.3 m, sparsely pubescent. **Leaves:** petiole 0.3–1 cm; blade broadly lanceolate to ovate, 1.5–10 × 1–4.5 cm, lobes: base cuneate, margins entire, denticulate, or pinnate, apex acute. **Inflorescence bracts** lanceolate-ovate, 1.5–3 mm, margins entire. **Flowering pedicels** 1.5–2.1 cm. **Flowers:** calyx 0.6–0.9 cm, lobes distinct for 7/8 their length, unequal; corolla 2–3.5 cm diam., tube narrow, 2.5–3 × 1.8–2.1 mm, lobe shape highly variable, adaxial narrowly elliptic-ovate, lateral and abaxial further subdivided into narrower lobes, 1–1.6 × 0.3–1 cm, apex broadly rounded to acute, glabrate, often with various large or small spots of darker or brighter colors; fertile stamens 7.5–8.5 mm; filaments nearly completely distinct; ovary 1.5–3 × 0.8–1.3 mm; style 5–7 mm. **Capsules** tan, 4–6 × 7–9 mm. **Seeds** brown, 1.8–2 × 1.6–1.8 mm. $2n = 20$.

Flowering Apr–Sep. Waste places; 50–300 m; introduced; N.Y., Tex.; s South America (Chile).

Schizanthus pinnatus is commonly cultivated as an ornamental bedding plant (C. D. Brickell and J. D. Zuk 1997). Reports of the species from Maine (C. D. Richards et al. 1983) could not be confirmed; it was excluded from the Maine flora by A. Haines and T. F. Vining (1998).

27. SOLANUM Linnaeus, Sp. Pl. 1: 184. 1753; Gen. Pl. ed. 5, 85. 1754 • Nightshade [Derivation uncertain; possibly Latin *solis*, sun, and anus, connection, alluding to sunny habitat, or *solor*, soothe, and *anus*, connection, alluding to soothing narcotic property]

Lynn Bohs[1]

[1]The author wishes to acknowledge co-authorship with David M. Spooner† on *S. jamesii* and *S. stoloniferum* and with Sandra Knapp and Tiina Särkinen on the black nightshade species.

Lycopersicon Miller

Herbs, vines, subshrubs, shrubs, or trees, annual or perennial, sometimes rhizomatous or with tubers, unarmed or prickly, glabrous or pubescent, hairs simple, dendritically branched, or stellate, glandular or eglandular [peltate scales]. **Stems** erect to prostrate or twining, with sympodial growth. **Leaves** alternate, petiolate or sessile; blade simple or pinnately compound, margins entire, sinuate-dentate, dentate, or toothed to pinnately lobed or divided. **Inflorescences** terminal, leaf-opposed, extra-axillary, or in branch forks, branched or unbranched. **Flowers** usually bisexual, 5-merous; calyx sometimes accrescent, campanulate, truncate to 5-lobed; corolla white, cream, green, yellow, pink, blue, violet, or purple, radially or bilaterally symmetric, rotate, rotate-stellate, campanulate, stellate-pentagonal, or stellate, shallowly to deeply 5-lobed; stamens inserted near base of corolla, equal or unequal; anthers oblong, ellipsoidal, or lanceolate, broad or narrow and tapering toward the apex, dehiscing by pores, these sometimes expanding into longitudinal slits; ovary 2-carpellate (sometimes multicarpellate in cultivated species); style straight or curved, stigma truncate to capitate. **Fruits** berries, globose to ellipsoidal or ovoid, fleshy or juicy (occasionally dry), with sclerotic granules in some species. **Seeds** usually reniform, flattened, plump, or rounded, occasionally angled. $x = 12$.

Species ca. 1500 (52 in the flora): nearly worldwide on all continents except Antarctica.

Solanum is the largest genus in the Solanaceae and one of the largest genera of angiosperms. It is the most speciose genus in the family in North America. The North American taxa are composed of natives, most of which have ranges that extend into Mexico or the Caribbean, and a number of species that have been introduced deliberately or unintentionally. Some of these have spread and become naturalized, occasionally as pests, whereas others are only sporadically or ephemerally escaped from cultivation.

Solanum is one of the world's most economically important genera. It includes crop plants such as the potato, tomato, and eggplant as well as species used on a small scale as edible fruits or as ornamentals. Most *Solanum* plants contain high levels of alkaloids and can be toxic. In general, the foliage or unripe fruits of *Solanum* species should not be consumed by humans or other animals due to danger of poisoning.

Sclerotic granules (also called stone cells) occur in fruits of some *Solanum* species. These are hardened masses within the fruit pulp that are usually spherical, 0.5–5 mm in diameter, and visible with 10× magnification. In dissected fruits they are generally paler, rounded, and more homogeneous in appearance than the seeds. The number of sclerotic granules per fruit can be diagnostic in species of the black nightshade group. They are very numerous in *S. laciniatum* and *S. triflorum* and can be seen in pressed fruits without dissection.

Some *Solanum* species are cultivated as foods or ornamentals and may occasionally escape, while other non-native species are known only from historical records.

The potato (*Solanum tuberosum* Linnaeus) is the world's leading root crop and second only to the cereal grains in food production. Potatoes are widely cultivated in agricultural fields and home gardens. Volunteer plants of potato can grow from discarded tubers but do not appear to persist or spread. Eggplants (*S. melongena* Linnaeus) are commonly grown in home gardens or as specialty vegetables in small-scale agriculture. They may occasionally escape from cultivation in southern Florida but have not become naturalized. *Solanum retroflexum* Dunal (sunberry, wonderberry) and *S. scabrum* Miller (garden huckleberry) are sometimes grown for their edible fruits but apparently do not escape from cultivation. *Solanum glaucophyllum* Desfontaines is known in the United States only from two specimens collected in Pensacola, Florida, over a hundred years ago. *Solanum pilcomayense* Morong, native to Argentina and Paraguay, is only known from two collections made in port areas in New Jersey and Texas, but it has also not been collected since the early 20th century and is not established. *Solanum villosum* Miller (hairy nightshade, woolly nightshade) is an occasional introduction from the Old World but has not persisted or become naturalized in North America. It was first recorded in the United States in 1899 in Pensacola, Florida, perhaps arriving in ships' ballast. After this manuscript was completed, *S. houstonii* Martyn (syn. *S. tridynamum* Dunal) was discovered in Santa Cruz County, Arizona. This species was formerly considered endemic to Mexico (S. Knapp et al. 2017). It is most similar to *S. hindsianum* but differs in its dimorphic hermaphroditic and staminate flowers with strongly unequal anthers in the latter.

The North American *Solanum* flora includes a number of species in the black nightshade group (*Solanum* sect. *Solanum*, or the Morelloid clade). These can form hybrid and polyploid complexes and are notoriously difficult to identify. Previous floras have used erroneous names or conflated several species under a single name, often referring to all black nightshades in an area as *S. americanum* or *S. nigrum*. Many of the older determinations of specimens in herbaria should be considered suspect. The North American black nightshade species comprise natives and endemics (for example, *S. emulans*, *S. interius*, *S. pseudogracile*) as well as introduced or adventive species from Central and South America and the Old World. This is still a difficult group under active study. T. Särkinen et al. (2018) includes complete descriptions, nomenclature, and synonymy for taxa treated here that have been introduced from the Old World. S. Knapp et al. (2019) similarly treated the black nightshade species of North and Central America and the Caribbean.

SELECTED REFERENCES Knapp, S. 2002b. *Solanum* section *Geminata* (Solanaceae). In: Organization for Flora Neotropica. 1968+. Flora Neotropica. 121+ nos. New York. No. 84. Knapp, S. 2013. A revision of the Dulcamaroid clade of *Solanum* Linnaeus (Solanaceae). PhytoKeys 22: 1–432. Knapp, S. et al. 2017. A revision of the *Solanum elaeagnifolium* clade (Elaeagnifolium clade; subgenus *Leptostemonum*, Solanaceae). PhytoKeys 84: 1–104. Knapp, S. et al. 2019. A revision of the Morelloid Clade of *Solanum* L. (Solanaceae) in the Caribbean and North and Central America. PhytoKeys 123: 1–144, doi: 10.3897/phytokeys.123.31738. Nee, M. 1991. Synopsis of *Solanum* section *Acanthophora*: A group of interest for glycoalkaloids. In: J. G. Hawkes et al,. eds. 1991. Solanaceae III: Taxonomy, Chemistry, Evolution. Kew. Pp. 257–266. Peralta, I. E., D. M. Spooner, and S. Knapp. 2008. Taxonomy of wild tomatoes and their relatives (*Solanum* sect. *Lycopersicoides*, sect. *Juglandifolia*, sect. *Lycopersicon*; Solanaceae). Syst. Bot. Monogr. 84: 1–186. Roe, K. E. 1967. A revision of *Solanum* sect. *Brevantherum* (Solanaceae) in North and Central America. Brittonia 19: 353–373. Särkinen, T. et al. 2018. A revision of the Old World black nightshades (Morelloid clade of *Solanum* L., Solanaceae). PhytoKeys 106: 1–223. Spooner, D. M. et al. 2004. Wild potatoes (*Solanum* section *Petota*; Solanaceae) of North and Central America. Syst. Bot. Monogr. 68: 1–209. Stern, S. R. et al. 2013. A revision of *Solanum* section *Gonatotrichum*. Syst. Bot. 38: 471–496. Wahlert, G. A., F. E. Chiarini, and L. Bohs. 2015. A revision of *Solanum* section *Lathyrocarpum* (the Carolinense clade, Solanaceae). Syst. Bot. 40: 853–887. Whalen, M. D. 1979. Taxonomy of *Solanum* section *Androceras*. Gentes Herbarum 11: 359–426.

1. Plants usually with prickles (at least on basal parts or on young plants) and usually with stellate hairs (except in *S. capsicoides*); anthers narrow and tapered, dehiscent by terminal pores.
    2. Leaves 1–3 times deeply pinnatifid (divided more than halfway to midrib) to compound; fruits nearly to completely covered by accrescent prickly calyx.
        3. Corollas radially symmetric; stamens equal or nearly so; calyces loosely covering red, juicy fruits . . . . 43. *Solanum sisymbriifolium*
        3. Corollas bilaterally symmetric; stamens dimorphic, with one much longer than others; calyces tightly covering brown, dry fruits.
            4. Stem hairs stellate or otherwise branched; corollas yellow . . . . 39. *Solanum rostratum*
            4. Stem hairs mostly simple, often glandular (sometimes mixed with sparse stellate hairs); corollas violet, blue, white, or yellow.
                5. Corollas white or yellow.
                    6. Stems with well-spaced, needlelike prickles; corollas white, 2–2.5 cm diam., rotate-stellate, with abundant interpetalar tissue; w Texas . . . . 9. *Solanum cordicitum*
                    6. Stems with dense, bristlelike prickles; corollas yellow, 1.3–1.8 cm diam., stellate, with sparse interpetalar tissue; s Arizona . . . 27. *Solanum lumholtzianum*
                5. Corollas violet or blue.
                    7. Corollas 1–1.5 cm diam.; long anther 2–5 mm.
                        8. Stems densely pubescent with simple, glandular hairs 0.2–0.4 mm; stems sparsely to moderately prickly, with 20 or fewer prickles per cm; corollas pentagonal-stellate, with sparse interpetalar tissue; endemic to nc New Mexico . . . . 34. *Solanum novomexicanum*
                        8. Stems sparsely pubescent with simple, glandular hairs ca. 0.2 mm; stems densely prickly, with 30+ prickles per cm; corollas pentagonal, with abundant interpetalar tissue; w Texas to c New Mexico, Arizona . . . . 42. *Solanum setigeroides*
                    7. Corollas 1.4–3.5 cm diam.; long anther 5.5–20 mm.

9. Plants perennials; seeds plump, 2.8–3.6 mm . . . . 46. *Solanum tenuipes*
9. Plants annuals; seeds flattened, 2.3–3 mm.
10. Long anther 11–16 mm; corollas 2.5–3.5 cm diam.; large leaves usually only 2 times pinnatifid, with obtuse to rounded ultimate lobes. . . . . 8. *Solanum citrullifolium*
10. Long anther 5.5–8.5 mm; corollas 1.4–2 cm diam.; large leaves often 3 times pinnatifid, with acute ultimate lobes . . . . . . 10. *Solanum davisense*

[2. Shifted to left margin.—Ed.]

2. Leaves entire or lobed, rarely deeply pinnatifid or compound; fruiting calyces not accrescent or, if accrescent, not almost completely covering fruits.
11. Leaves sessile or with petiole to 1 cm, blade rhombic; calyx lobes linear; corollas to 1–2 cm diam.; c Florida . . . . 23. *Solanum jamaicense*
11. Leaves with petiole (0.1–)1–10(–13) cm, blade lanceolate to oblong, elliptic, ovate, obovate, or suborbiculate; calyx lobes linear-lanceolate, lanceolate, triangular, deltate, ovate-lanceolate, elliptic-acuminate, or broadly deltate; corollas (1–)1.5–5 cm diam.
12. Pubescence of stems and upper leaf surfaces of unbranched hairs or plants glabrate.
13. Pubescence eglandular or plants glabrate; berries orange to red; seeds winged . . . . .5. *Solanum capsicoides*
13. Pubescence glandular and eglandular; berries yellow, often mottled with green when young; seeds not winged . . . . .51. *Solanum viarum*
12. Pubescence of stem and upper leaf surfaces of stellate hairs.
14. Pubescence dense and silvery or bright white, particularly on lower leaf surfaces.
15. Stellate hairs scalelike, with rays fused at center . . . . 17. *Solanum elaeagnifolium*
15. Stellate hairs not scalelike, rays all separate.
16. Leaves (0.5–)1–3(–4.5) cm wide . . . . 21. *Solanum hindsianum*
16. Leaves 3–15 cm wide.
17. Calyces not prickly; inflorescences usually much-branched; corollas usually blue to purple, rarely whitish; berries 0.7–1.5 cm diam., yellow to orange . . . . .26. *Solanum lanceolatum*
17. Calyces prickly; inflorescences unbranched or forked; corollas white to pale purple; berries 3.5–4(–5) cm diam., yellow. . .29. *Solanum marginatum*
14. Pubescence sparse to dense, of various colors, but not noticeably silvery or bright white.
18. Scandent shrubs with branches 1–2+ m; stems glabrate to sparsely pubescent; wet areas in Florida and Texas . . . . .45. *Solanum tampicense*
18. Erect or spreading herbs, shrubs, or small trees; stems noticeably stellate-pubescent; widespread, including Florida and Texas.
19. Leaf margins entire to very shallowly lobed; corollas stellate, without interpetalar tissue; berries red.
20. Inner surface of anther tube densely stellate-pubescent; inflorescences usually unbranched; pedicels recurved to one side of axis in fruit . . . . .3. *Solanum bahamense*
20. Inner surface of anther tube glabrous; inflorescences branched; pedicels erect in fruit. . . . . 14. *Solanum donianum*
19. Leaf margins usually shallowly to deeply lobed, rarely entire, subentire, or sinuate; corollas rotate-campanulate, rotate-stellate, stellate-pentagonal, or stellate, with sparse to abundant interpetalar tissue; berries green, yellow, or purplish.
21. Trees or shrubs 1–4 m; inflorescence pubescence of unbranched glandular hairs . . . . .47. *Solanum torvum*
21. Erect or spreading annual or perennial herbs or shrubs 0.2–1.2 m; inflorescence pubescence stellate.

22. Stems, inflorescences, and calyces densely armed; annual herbs, usually spreading and to 0.5(–1) m; major leaf lobes with acute teeth or shallow lobes . . . . 4. *Solanum campechiense*
22. Stems, inflorescences, and calyces sparsely to moderately armed; erect perennial herbs or shrubs 0.2–1.2 m; major leaf lobes, if present, entire to coarsely lobed.
    23. Stems and petioles stellate-pubescent with central ray 1-celled and equal to or shorter than lateral rays . . . . 12. *Solanum dimidiatum*
    23. Stems and petioles stellate-pubescent with central ray 1–5-celled and longer than lateral rays.
        24. Plants to 0.2 m; leaf margins entire, sinuate, or shallowly lobed; inflorescences 1–4-flowered; endemic to dolomite outcrops in Alabama and possibly Georgia . . . . 38. *Solanum pumilum*
        24. Plants to 1.2 m; leaf margins subentire, sinuate, or shallowly to deeply lobed; inflorescences 2–15-flowered; plants not confined to dolomite outcrops and distributions more widespread.
            25. Stems armed with prickles to 6 mm; inflorescences unbranched or rarely forked; corollas 2–3 cm diam.; widespread in North America . . . . . .6. *Solanum carolinense*
            25. Stems armed with prickles to 15 mm; inflorescences forked to several times branched; corollas 2–4.4 cm diam.; mainly Alabama, Florida, and Georgia, rarely Mississippi . . . 35. *Solanum perplexum*

1. Plants without prickles, with or without stellate hairs; anthers of various forms, dehiscent by terminal pores that often open into longitudinal slits or by introrse longitudinal slits.
    26. Leaves compound; pedicels articulated above base.
        27. Leaflets lobed; corollas yellow; plants without tubers; berries usually red, orange, or yellow. . . . . 28. *Solanum lycopersicum*
        27. Leaflets entire; corollas white to pink, blue, or purple; plants with underground tubers; berries green.
            28. Corollas white, stellate; pseudostipules (when present) pinnatifid . . . . 24. *Solanum jamesii*
            28. Corollas purple, blue, pale pink, or rarely white, pentagonal to rotate; pseudostipules entire . . . . 44. *Solanum stoloniferum*
    26. Leaves simple, entire, undulate, sinuate-dentate to toothed, deeply lobed, pinnatifid, or hastate, simple to pinnately compound in *S. seaforthianum*; pedicels articulated at base.
        29. Plants climbing or scrambling vines.
            30. Stamens (filaments) unequal; corollas purple; most leaves lobed or compound with up to 4 pairs of leaflets . . . . 41. *Solanum seaforthianum*
            30. Stamens equal; corollas white or purple; leaves unlobed or with 1–3 basal lobes.
                31. Corollas purple (rarely white) with green and white shiny spots at base of each lobe; n 2/3 of United States, s Canada . . . . 16. *Solanum dulcamara*
                31. Corollas white or tinged with purple, often with shiny green or greenish white eye; Texas. . . . . 49. *Solanum triquetrum*
        29. Plants herbs, subshrubs, shrubs, or small trees, not climbing.
            32. Shrubs or small trees, 2–12 m; ovaries and berries pubescent.
                33. Corollas white; without small axillary leaves. 19. *Solanum erianthum*
                33. Corollas purple; larger leaves often with smaller axillary leaves. . . 30. *Solanum mauritianum*
            32. Herbs, subshrubs, shrubs, or small trees, 0.1–2(–4) m; ovaries and berries glabrous.

[34. Shifted to left margin.—Ed.]

34. Corollas 3–5 cm diam., usually pink to blue, violet, or purple, rarely whitish; shrubs or small trees, 1–4 m.
    35. Plants densely pubescent with usually glandular hairs; Santa Catalina Island, California . . . . 52. *Solanum wallacei*
    35. Plants glabrous; California, Oregon.
        36. Corolla lobes acute at apex; berries orange to red, sclerotic granules inconspicuous to absent. . . . . 2. *Solanum aviculare*
        36. Corolla lobes notched at apex; berries yellow to orange, sclerotic granules abundant . . . . 25. *Solanum laciniatum*
34. Corollas 2.5 cm or less diam., usually white to pale purple (pale to deep purple in *S. umbelliferum*); small herbs, subshrubs, or shrubs, to 1.5 m, rarely to 3 m.
    37. Corollas pale to deep purple or occasionally white, with green spots edged with white at base of lobes; anthers 3.5–4.5 mm . . . . 50. *Solanum umbelliferum*
    37. Corollas white or tinged with purple, often with central star of different color but not with pronounced green spots at base of lobes; anthers 0.7–3.5(–5) mm.
        38. Leaf margins deeply and regularly pinnatifid . . . . 48. *Solanum triflorum* (in part)
        38. Leaf margins entire, or sinuate, sinuate-dentate, to shallowly toothed or lobed.
            39. Plants noticeably glandular-pubescent, sticky to the touch.
                40. Fruiting calyces not or only slightly accrescent, covering only base of fruit; plants glabrate to moderately pubescent; hairs mostly eglandular but occasionally glandular, to 1 mm. . . . . 32. *Solanum nigrum* (in part)
                40. Fruiting calyces strongly accrescent, covering at least half of fruit; plants moderately to densely pubescent; hairs glandular, 1.5–2 mm.
                    41. Fruiting calyces covering ca. half of fruit; berries shiny greenish to purplish brown, with (0–)2–3 sclerotic granules; inflorescences usually extra-axillary, 4–8(–10)-flowered . . . . 33. *Solanum nitidibaccatum*
                    41. Fruiting calyces nearly covering fruit; berries dull pale green, with 4–6 sclerotic granules; inflorescences usually leaf-opposed, 2–5(–7)-flowered . . . . 40. *Solanum sarrachoides*
            39. Plants glabrous or eglandular-pubescent (sometimes with a few glandular hairs), not sticky to the touch.
                42. Berries yellow, orange, or red when ripe; fruiting pedicels erect; inflorescences leaf-opposed.
                    43. Shrubs 1–2 m, glabrous or minutely puberulent; corollas 0.7–1 cm diam.; anthers 1.5–2 mm; berries orange when ripe . . . . 13. *Solanum diphyllum*
                    43. Shrubs to 1 m, glabrous or densely pubescent with branched hairs; corollas 1–1.5(–2.5) cm diam.; anthers 3–4 mm; berries yellow, orange, or red when ripe . . . . 36. *Solanum pseudocapsicum*
                42. Berries green, white, yellowish green, or purple to purplish black when ripe; fruiting pedicels erect, spreading, recurved, reflexed, or nodding; inflorescences leaf-opposed or extra-axillary.
                    44. Inflorescences nearly sessile; berries white to greenish and semi-transparent . . . . 11. *Solanum deflexum*
                    44. Inflorescences 0.5–4 cm; berries green, dark green, yellowish green, or purple to purplish black.
                        45. Anthers 0.7–1.5 mm.
                            46. Berries shiny purplish black, with (0–)2–4(–6) sclerotic granules; fruiting pedicels erect or spreading; calyx lobes strongly reflexed in fruit. . . . . 1. *Solanum americanum*
                            46. Berries dull or slightly shiny purple-black, with 6–9 sclerotic granules; fruiting pedicels reflexed or recurved; calyx lobes appressed to spreading in fruit . . . . 18. *Solanum emulans*
                        45. Anthers (1.8–)2–4.5 mm.
                            47. Fruiting peduncles sharply reflexed from base . . . . . 7. *Solanum chenopodioides*
                            47. Fruiting peduncles spreading or curved downwards.

[48. Shifted to left margin.—Ed.]

48. Berries without sclerotic granules (*S. pseudogracile* rarely with 2).
    49. Fruiting pedicels usually spreading, occasionally recurved; inflorescences racemelike; seeds 1.8–2 × 1.5–1.6 mm . . . . 32. *Solanum nigrum* (in part)
    49. Fruiting pedicels recurved to reflexed; inflorescences umbel-like; seeds 1–1.3 × 0.8–0.9 mm . . . . 37. *Solanum pseudogracile*
48. Berries with sclerotic granules.
    50. Inflorescences forked, 6–14-flowered; w coast from California to Washington . . . . 20. *Solanum furcatum*
    50. Inflorescences unbranched, 1–14-flowered; w coast to c and se United States, w, c Canada.
        51. Plants decumbent to prostrate, fleshy; sclerotic granules 13–30 per fruit . . . . 48. *Solanum triflorum* (in part)
        51. Plants erect to sprawling, not fleshy; sclerotic granules 2–13 per fruit.
            52. Anthers (2.5–)3–4.5 mm, slightly tapered towards the tips; corollas 1–2 cm diam. . . . . 15. *Solanum douglasii*
            52. Anthers 1.8–3 mm, not tapered; corollas 0.5–1.5 cm diam.
                53. Sclerotic granules 2–4 per fruit; seeds 1.8–2 × 1.5–1.6 mm . . . . 22. *Solanum interius*
                53. Sclerotic granules (4–)5–6(–13) per fruit; seeds 1.2–1.5 × 1–1.1 mm . . . 31. *Solanum nigrescens*

1. Solanum americanum Miller, Gard. Dict. ed. 8 Solanum no. 5. 1768 • American black or common or West Indian nightshade [W]

Solanum nigrum Linnaeus var. *americanum* (Miller) O. E. Schulz; *S. nigrum* var. *nodiflorum* (Jacquin) A. Gray; *S. nodiflorum* Jacquin; *S. ptychanthum* Dunal

Herbs to subshrubs, annual to perennial, erect, unarmed, to 1.5 m, glabrate to moderately pubescent, hairs whitish, unbranched, to 1 mm, eglandular. **Leaves** petiolate; petiole 1–4 cm; blade simple, ovate to ovate-elliptic, 2–10.5 × 1–4.5 cm, margins entire or shallowly sinuate-dentate, base decurrent. **Inflorescences** extra-axillary or leaf-opposed, unbranched, usually umbel-like, 3–10-flowered, 0.5–3 cm. **Pedicels** erect or spreading, 0.3–1 cm in flower, to 1.5 cm in fruit. **Flowers** radially symmetric; calyx somewhat accrescent, unarmed, 1–3 mm, lobes deltate, strongly reflexed in fruit; corolla white, sometimes with yellowish central star, stellate, 0.4–0.8 cm diam., without interpetalar tissue; stamens equal; anthers ellipsoidal, 0.7–1.5 mm, dehiscent by terminal pores that open into longitudinal slits; ovary glabrous. **Berries** shiny purplish black, globose, 0.5–1 cm diam., glabrous, with (0–)2–4(–6) sclerotic granules, usually 2–4 larger and 2 smaller. **Seeds** pale yellow to brown, 1–1.5 × 0.5–1.5 mm, minutely pitted. **2*n*** = 24.

Flowering May–Nov (year-round in Fla.). Weedy habitats, secondary forest, disturbed areas; 0–1000 (–2000) m; B.C.; Ala., Ariz., Calif., Fla., Ga., La., Miss., Mo., Oreg., S.C., Tex., Utah, Wash.; Mexico; West Indies; Central America; South America; introduced in Europe, Asia, Africa, Pacific Islands, Australia.

Solanum americanum is a morphologically variable and globally distributed weedy species. It has often been confused with other species in the black nightshade group such as *S. emulans*, *S. nigrescens*, *S. nigrum*, and *S. pseudogracile* and has often been referred to as *S. nodiflorum* in floristic treatments. It is distinguished by its very short anthers and shiny black mature fruits with strongly reflexed calyx lobes and usually two to four (rarely none or as many as six) sclerotic granules per fruit. Leaf shape and pubescence can vary considerably throughout its range, and there are some suspected cases of introgression with other species such as *S. nigrescens* or *S. pseudogracile*. The name *S. nigrum* has been misapplied to *S. americanum* (for example, A. E. Radford et al. 1968).

2. Solanum aviculare G. Forster, Pl. Esc., 42. 1786 • New Zealand nightshade, poroporo [I]

Shrubs, erect, unarmed, 1–4 m, glabrous. **Leaves** petiolate; petiole 1–1.5 cm; blade simple, elliptic, 10–30 × 2–15 cm, margins entire or coarsely pinnatifid with 1–3 lobes per side, lobe margins entire, base cuneate. **Inflorescences** leaf-opposed or in branch fork, unbranched or forked, to 10-flowered, to 15 cm. **Pedicels** 1.5–2 cm in flower and fruit. **Flowers** radially symmetric; calyx somewhat accrescent, unarmed, 3–4 mm, glabrous, lobes deltate; corolla blue to deep purple, rotate-stellate, lobes acute at apex, 3–4 cm diam., with abundant interpetalar tissue; stamens equal; anthers oblong, slightly tapered, 3–4 mm, dehiscent by terminal pores that sometimes open into longitudinal slits; ovary glabrous. **Berries** bright orange to red, obovoid to ellipsoidal,

1.5–2 × 1–1.5 cm, glabrous, with sclerotic granules inconspicuous to absent. **Seeds** reddish brown, flattened, 1.5–2 × 1.5–2 mm, finely reticulate. **2*n*** = 46.

Flowering Jan–Jul. Open, disturbed sites; 0–2000 m; introduced; Calif., Oreg.; Pacific Islands (New Guinea, New Zealand); Australia.

The name *Solanum laciniatum* Aiton has been misapplied to *S. aviculare* (for example, M. Nee 1993). Although only *S. aviculare* is listed in that treatment as occurring in California, both *S. aviculare* and *S. laciniatum* apparently are found there, and most photos labeled *S. aviculare* on the CalFlora website are of *S. laciniatum*. *Solanum laciniatum* has notched and ruffled corolla lobes with abundant interpetalar tissue (versus acute and entire corolla lobes with little interpetalar tissue in *S. aviculare*) and yellow fruits (versus red) with numerous stone cells (versus stone cells inconspicuous to absent).

Solanum aviculare and *S. laciniatum* have been introduced from Australia and New Zealand as ornamentals and now are found in scattered localities in California, especially near the coast. Some plants have escaped and become naturalized, and these species have the potential to be invasive in the future.

3. Solanum bahamense Linnaeus, Sp. Pl. 1: 188. 1753 • Bahama or Rugel's nightshade, cankerberry

Solanum bahamense var. *luxurians* D'Arcy; *S. bahamense* var. *rugelii* D'Arcy; *S. racemosum* Jacquin

Shrubs or small trees, erect, sparsely to densely armed (at least when young), to 4 m, prickles yellow or orange, straight, to 10 mm, sparsely to densely pubescent, hairs sessile, stellate, 6–8-rayed, central ray usually shorter than lateral rays, sometimes absent, occasionally as long as or longer than lateral rays. **Leaves** petiolate; petiole 0.1–2.7 cm; blade simple, narrowly elliptic, 2–21 × 0.5–5.5 cm, margins entire or shallowly lobed, base attenuate to rounded. **Inflorescences** extra-axillary or leaf-opposed, usually unbranched, 30+-flowered, to 15 cm. **Pedicels** 0.8–1.5 cm in flower, 1–2 cm and recurved to one side of the inflorescence in fruit. **Flowers** radially symmetric; calyx not accrescent, unarmed, 1.5–2.5 mm, moderately stellate-pubescent, lobes triangular; corolla white or violet, stellate, 1.5–2 cm diam., without interpetalar tissue; stamens equal; anthers narrow and tapered, 5.5–8 mm, dehiscent by terminal pores, anther tube densely stellate-pubescent within; ovary glabrous or sparsely pubescent with short-glandular hairs. **Berries** bright shiny red, globose, 0.5–0.7 cm diam., glabrous, without sclerotic granules. **Seeds** pale tan, flattened, 2–2.5 × 1–1.5 mm, minutely pitted. **2*n*** = 24.

Flowering year-round. Coastal forests, forest and beach margins, dunes, often on coral or calcareous soils; 0–100 m; Fla.; West Indies.

Solanum bahamense is morphologically variable, especially in leaf shape, hair morphology, prickle density, and corolla color. Using morphological and molecular data, R. Strickland-Constable et al. (2010) established that several formerly recognized taxa are encompassed within its range of variability and should be considered as synonyms of *S. bahamense*. It is unique among the North American spiny solanums in having stellate hairs on the inner (adaxial) surface of the anthers. *Solanum bahamense* occurs in southern Florida and the Keys and throughout the Caribbean.

SELECTED REFERENCE Strickland-Constable, R. et al. 2010. Species identity in the *Solanum bahamense* species group (Solanaceae, *Solanum* subgenus *Leptostemonum*). Taxon 59: 209–226.

4. Solanum campechiense Linnaeus, Sp. Pl. 1: 187. 1753 • Redberry nightshade

Herbs, annual, usually spreading, densely armed, to 0.5(–1) m, prickles pale yellow, straight, to 10 mm, sparsely to densely tomentose, hairs nearly sessile to thick-stalked, stellate, 4–9-rayed, central ray equal to or slightly longer than lateral rays. **Leaves** petiolate; petiole 1–5 cm; blade simple, ovate, 4–13 × 3–12 cm, margins deeply lobed with 2–4 lobes per side, these with additional coarse, acute teeth or shallow lobes, base cordate. **Inflorescences** extra-axillary, unbranched, 1–3(–6)-flowered, 0.5–2(–5) cm. **Pedicels** 0.5–1 cm in flower, 1–2 cm in fruit. **Flowers** radially symmetric; calyx somewhat accrescent and partially covering fruit, densely prickly, 4–6 mm, densely stellate-pubescent, lobes narrowly triangular; corolla white, bluish white, blue, or lilac, rotate-campanulate, 1–1.5 cm diam., with abundant interpetalar tissue; stamens equal; anthers narrow and tapered, 2–5 mm, dehiscent by terminal pores; ovary glabrous. **Berries** purplish, greenish, or yellowish, globose, 1.5–2 cm diam., glabrous, without sclerotic granules. **Seeds** yellowish brown, flattened, 1.8–3 × 1.4–2 mm, pustulate.

Flowering Feb–Dec. Muddy edges of ephemeral lakes and streams, tropical and subtropical dry forest; 0–100 m; Tex.; Mexico (Chiapas, Jalisco, Oaxaca, Tamaulipas, Veracruz, Yucatán); West Indies; Central America (Belize, Costa Rica, El Salvador, Guatemala, Honduras); South America (Ecuador, Peru).

In the flora region, *Solanum campechiense* is found in southernmost Texas. Although D. S. Correll and M. C. Johnston (1970) stated that the fruits are cherry-red at maturity, the specimens seen indicate that they turn purplish when ripe.

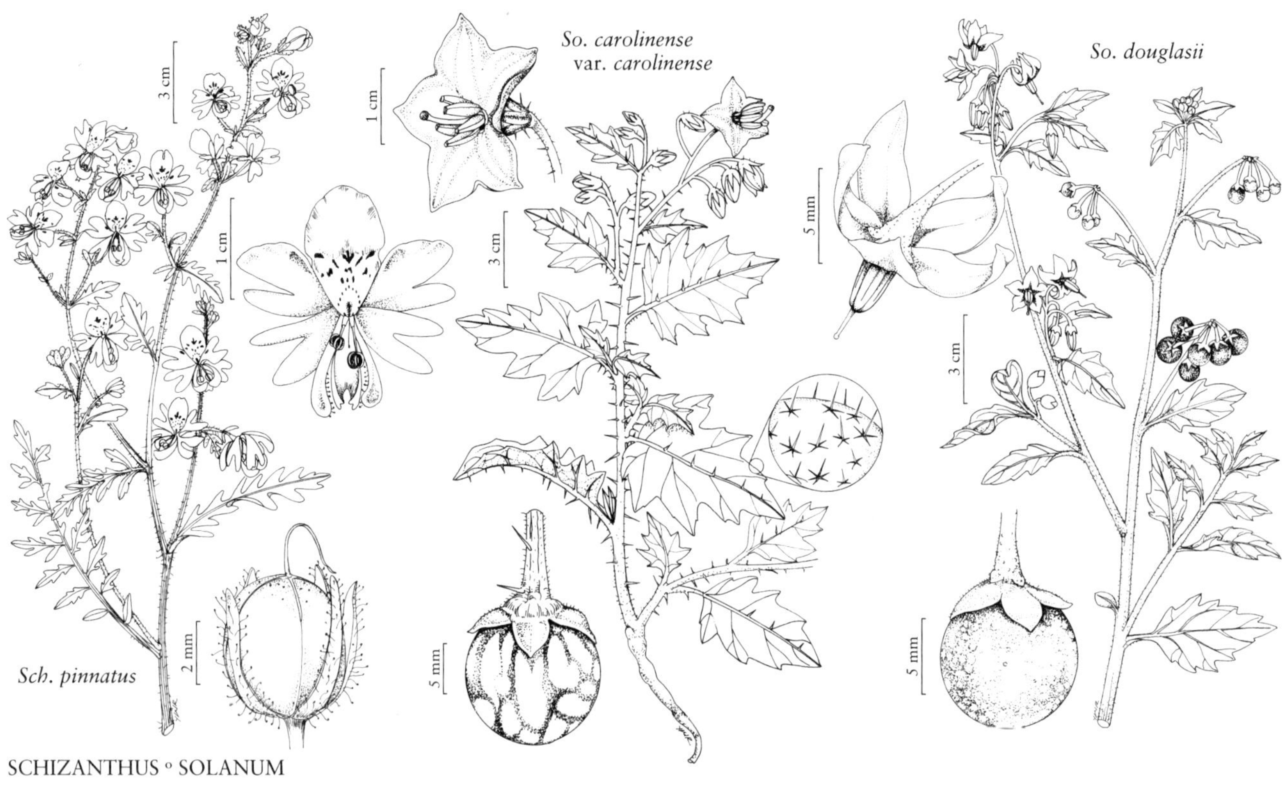

SCHIZANTHUS ◦ SOLANUM

5. **Solanum capsicoides** Allioni, Auct. Syn. Meth. Stirp. Hort. Regii Taurin., 12. 1773 • Cockroach berry, soda- or Sodom-apple [I] [W]

Solanum ciliatum Lamarck

Shrubs, erect, sparsely to densely armed, 0.3–1(–2) m, prickles yellowish, straight or slightly reflexed, to 15 mm, glabrate to moderately pubescent, hairs spreading, unbranched, eglandular. **Leaves** petiolate; petiole 2–10(–13) cm; blade simple, broadly ovate, 4–15 × 4–15 cm, margins shallowly to deeply lobed with 2–3 lobes per side, lobe margins entire to coarsely lobed, base cordate. **Inflorescences** extra-axillary, ± sessile, unbranched, 1–7-flowered. **Pedicels** ca. 1 cm in flower, 1.5–2 cm in fruit. **Flowers** radially symmetric; calyx somewhat accrescent, sometimes prickly, 4–6 mm, pubescent with minute glands and longer, simple, eglandular hairs, lobes triangular; corolla white, stellate, 1.5–2 cm diam., without interpetalar tissue; stamens equal; anthers narrow and tapered, 5–6.5 mm, dehiscent by terminal pores; ovary glabrous. **Berries** dull orange to red, globose, 2–4 cm diam., glabrous, without sclerotic granules. **Seeds** yellow, flattened, winged at maturity, 4–6 × 4–6 mm, minutely pitted. **2*n*** = 24.

Flowering Apr–Oct (year-round in Fla.). Disturbed areas, sandy soils; 0–1000 m; introduced; Fla., La., Miss., N.C., S.C., Tex.; South America (Brazil); introduced also in West Indies, Central America, elsewhere in South America, Asia, Africa, Pacific Islands, Australia.

Solanum capsicoides is presumed native to coastal Brazil and is grown as an ornamental for its showy red fruits. It has spread from cultivation and become naturalized in tropical and subtropical climates. In the flora region, it is found particularly in Florida. The name *S. aculeatissimum* Jacquin has been misapplied to *S. capsicoides* (for example, J. K. Small 1913; A. E. Radford et al. 1968; D. S. Correll and M. C. Johnston 1970; R. W. Long and O. Lakela 1971).

6. **Solanum carolinense** Linnaeus, Sp. Pl. 1: 187. 1753 • Carolina horsenettle, horsenettle, bull nettle [E] [F] [W]

Herbs or shrubs, perennial, erect, sparsely to moderately armed, to 1.2 m, prickles cream to yellowish, straight, to 6 mm, sparsely to densely pubescent, hairs white to cream, sessile to short-stalked, stellate, 4(–8)-rayed, central ray 1–3(–5)-celled and to 3 mm, longer than lateral rays. **Leaves** petiolate; petiole 0.4–4 cm; blade simple, ovate, lanceolate, or elliptic, 2–15 × 2–10 cm, margins subentire, sinuate, or lobed with 1–4 lobes per side, sometimes very deeply lobed almost to midrib, lobe margins entire to coarsely lobed, base cuneate.

Inflorescences extra-axillary, unbranched or rarely forked, 2–12-flowered, 2–9 cm. **Pedicels** 0.5–1 cm in flower, 1.2–1.8 cm and curved downward in fruit. **Flowers** radially symmetric; calyx somewhat accrescent, unarmed or with sparse prickles, 5–8 mm, moderately stellate-pubescent, lobes lanceolate to elliptic-acuminate; corolla white to pale blue or violet, stellate to stellate-pentagonal or rotate-stellate, 2–3 cm diam., with interpetalar tissue at margins and bases of lobes; stamens equal; anthers narrow and tapered, 4.5–6.5 mm, dehiscent by terminal pores; ovary glabrous or sparsely to moderately glandular-puberulent, rarely moderately pubescent, hairs white, stellate or simple. **Berries** light green with darker green mottling or pale greenish white, turning bright yellow, subglobose to depressed-globose, 1–2 × 1–1.8 cm, glabrous, without sclerotic granules. **Seeds** yellow, flattened, 1.7–2.4 × 1.6–1.8 mm, minutely pitted. $2n = 24$.

Varieties 2 (2 in the flora): North America; introduced in Europe, Asia, Pacific Islands (New Zealand), Australia.

1. Leaf margins subentire, sinuate, or lobed, sinuses of lobes, when present, reaching less than 1/2 distance to midvein; apex of leaf lobes subacute to acute, sometimes rounded . . . . 6a. *Solanum carolinense* var. *carolinense*
1. Leaf margins deeply lobed, sinuses of lobes reaching more than 1/2 distance to midvein or almost to midvein; apex of leaf lobes typically rounded . . . . . . 6b. *Solanum carolinense* var. *floridanum*

6a. Solanum carolinense Linnaeus var. **carolinense** E F W

Solanum carolinense var. *albiflorum* Kuntze

Leaves with margins subentire, sinuate, or lobed less than 1/2 distance to midvein; lobes typically subacute to acute at apex, sometimes rounded. $2n = 24$.

Flowering Apr–Oct. Disturbed or urban areas, cultivated fields, light or sandy soils; 0–900(–1200) m; N.S., Ont., Que.; Ala., Ariz., Ark., Calif., Colo., Conn., Del., D.C., Fla., Ga., Idaho, Ill., Ind., Iowa, Kans., Ky., La., Maine, Md., Mass., Mich., Minn., Miss., Mo., Nebr., N.H., N.J., N.Mex., N.Y., N.C., Ohio, Okla., Oreg., Pa., R.I., S.C., S.Dak., Tenn., Tex., Utah, Vt., Va., Wash., W.Va., Wis.; introduced in Europe, Asia, Pacific Islands (New Zealand), Australia.

In the United States, var. *carolinense* is apparently native east of the Rocky Mountains and south of the Great Lakes. It is introduced and weedy north and west of its native range in North America. In its native range and elsewhere, var. *carolinense* is extremely difficult to control in cultivated areas and pastures once established. It has been classified as a noxious weed in several states. All parts of the plant are considered to be toxic and cases of poisoning have been reported in cattle, sheep, deer, and humans (G. A. Wahlert et al. 2015). Some authors recognize two forms: forma *carolinense* and forma *albiflorum* (Kuntze) Benke based on violet versus white corollas, but mixed populations occur with both flower colors, so this distinction is not maintained in the latest taxonomic revision (Wahlert et al.).

6b. Solanum carolinense Linnaeus var. **floridanum** Chapman, Fl. South. U.S., 349. 1860 (as floridana) • Florida horsenettle E

Solanum floridanum Shuttleworth ex Dunal in A. P. de Candolle and A. L. P. P. de Candolle, Prodr. 13(1): 306. 1852, not Rafinesque 1840; *S. godfreyi* Shinners

Leaves with margins deeply lobed more than 1/2 distance to midvein, often lobed almost to midvein; lobes typically rounded at apex. $2n = 24$.

Flowering May–Sep. Moist areas, sandy riverbanks, slash pine–palmetto woodlands, roadsides; 0–10 m; Fla., Ga.

Variety *floridanum* is found in the vicinity of the Apalachicola, Aucilla, and Suwannee rivers in the Gulf Coast region of Florida and Georgia and sparingly in the coastal plain of Georgia.

7. Solanum chenopodioides Lamarck in J. Lamarck and J. Poiret, Tabl. Encycl. 2: 18. 1794 • Whitetip or velvety nightshade I

Solanum americanum Miller var. *baylisii* D'Arcy; *S. ottonis* Hylander

Herbs or shrubs, annual to short-lived perennial, erect or somewhat sprawling, unarmed, to 1 m, glabrescent to densely pubescent, hairs unbranched, to 1 mm, eglandular. **Leaves** petiolate; petiole 1–3 cm; blade simple, narrowly ovate to elliptic, 1.5–5(–7) × 0.5–3.5 cm, margins entire or sinuate, base cuneate to decurrent. **Inflorescences** extra-axillary or leaf-opposed, unbranched or rarely forked, umbel-like, 3–7(–10)-flowered, 1–3(–4) cm, fruiting peduncles sharply reflexed from base. **Pedicels** 0.5–1 cm in flower and fruit, reflexed downward in fruit. **Flowers** radially symmetric; calyx not accrescent, unarmed, 2–3.5 mm, sparsely pubescent, lobes deltate, appressed in fruit; corolla white or purplish, with greenish, yellowish, or brown central star, stellate, 0.8–1.5 cm diam., with sparse interpetalar tissue; stamens equal; anthers ellipsoidal, 2–3 mm, dehiscent by terminal pores that open into longitudinal slits; ovary glabrous. **Berries** dull purplish black, globose, 0.5–1 cm diam., glabrous, without sclerotic granules. **Seeds** pale yellow, flattened, 1–1.5 × 1–1.5 mm, minutely pitted. **2*n*** = 24.

Flowering May–Oct (year-round in Fla.). Sandy soil, disturbed areas; 0–2000 m; introduced; Calif., Fla., Ga., Md., Mo., N.C., Wis.; South America (Argentina, Brazil, Paraguay, Peru, Uruguay); introduced also in Europe, Africa, Pacific Islands (New Zealand), Australia.

Solanum chenopodioides has been introduced sporadically and is occasionally adventive in North America. It is distinctive in having the fruiting peduncles strongly reflexed downward, but is otherwise difficult to distinguish from *S. pseudogracile*, with which it may be conspecific.

The illegitimate superfluous name *Solanum gracile* Dunal has often been used for *S. chenopodioides* (for example, J. K. Small 1913; A. E. Radford et al. 1968). W. G. D'Arcy (1974) included *S. gracile* (and its replacement name *S. ottonis*) in the synonymy of *S. nigrescens* but the taxa are distinct.

8. Solanum citrullifolium A. Braun, Index Seminum (Friburg) 1849: [3]. 1849 • Watermelon or melon-leaf nightshade [W]

Herbs, annual, spreading, sparsely to moderately armed, 0.3–0.8 m, prickles yellowish, straight, needlelike, 3–7 mm, sparsely to densely pubescent, hairs short, unbranched, glandular, occasionally with a few longer, unbranched, eglandular hairs, abaxial leaf surfaces usually also with sessile to short-stalked, few-rayed, stellate hairs, central ray equal to or longer than lateral rays. **Leaves** petiolate; petiole 2–7 cm; blade simple to compound, broadly ovate, 4–10(–15) × 3–8 cm, margins bipinnately lobed or divided with 3–4 main leaflets per side, these with obtuse or rounded lobes, base truncate. **Inflorescences** extra-axillary, unbranched, 4–10-flowered, 3–11 cm. **Pedicels** 1–2 cm in flower, 1–2 cm and erect in fruit. **Flowers** bilaterally symmetric; calyx accrescent and tightly covering fruit, densely prickly, 2.5–3.8 mm, densely glandular-pubescent, lobes linear-lanceolate; corolla violet or blue, pentagonal-stellate, 2.5–3.5 cm diam., with interpetalar tissue at margins and bases of lobes; stamens unequal, lowermost much longer and curved; anthers narrow and tapered, dehiscent by terminal pores, short anthers yellow, 6–10 mm, longer anther purplish, 11–16 mm; ovary glabrous. **Berries** brown, globose, 0.8–1.2 cm diam., glabrous, dry, without sclerotic granules. **Seeds** dark brown, flattened, 2.3–3 × 2–2.5 mm, reticulately wrinkled, ridged, or undulate. **2*n*** = 24.

North American plants identified as *Solanum heterodoxum* Dunal are largely misidentifications of *S. citrullifolium*. *Solanum heterodoxum* in the current sense is now restricted to Mexico.

Varieties 2 (2 in the flora): sc, se United States, Mexico.

1. Stems scattered-prickly with fewer than 20 prickles per cm of stem; prickles often to 1 mm diam. at base; stems densely glandular-pubescent . . . . .8a. *Solanum citrullifolium* var. *citrullifolium*
1. Stems densely bristly with 25+ bristles per cm of stem; bristles mostly less than 0.5 mm diam. at base; stems sparsely glandular-pubescent . . . . 8b. *Solanum citrullifolium* var. *setigerum*

8a. Solanum citrullifolium A. Braun var. **citrullifolium** [W]

Stems with spreading prickles 3–7 mm and often to 1 mm diam. at base, fewer than 20 per cm of stem, densely pubescent with short-stipitate glands and often with unbranched, spreading hairs to 1 mm. **Calyx lobes** 2.5–3.8 mm. **Seeds** reticulately wrinkled or merely undulate.

Flowering Apr–Oct. Well-drained, often igneous, rocky or sandy soils, sparsely vegetated mountainsides, dry grasslands, disturbed places; (100–)1300–1900 m; Fla., Tex.; Mexico (Coahuila).

Variety *citrullifolium* is found in central and western Texas south to Mexico. The plants from central Texas differ from those of western Texas and Mexico in having more rounded ultimate leaf lobes, leaf undersides often lacking stellate hairs, and longer fruiting inflorescences. In Florida, var. *citrullifolium* has escaped from cultivation and become naturalized. Massachusetts records (F. C. Seymour 1982) are historical (1885, 1913, 1949) as introductions that did not persist.

8b. Solanum citrullifolium A. Braun var. **setigerum** Bartlett, Proc. Amer. Acad. Arts 44: 628. 1909 [W]

Stems densely bristly with spreading or slightly descending prickles 4–8 mm and mostly less than 0.5 mm diam. at base, 25+ per cm of stem, sparsely pubescent with short-stipitate glands. **Calyx lobes** 3.3–3.5 mm. **Seeds** with low, radially oriented ridges.

Flowering Jul–Oct. Limestone, lava, sand, alluvial deposits, calcareous, gypseous, and saline flats, around wet depressions; 1100–1400 m; Tex.; Mexico (Chihuahua, Coahuila).

Variety *setigerum* is found occasionally in Presidio County, where it is becoming a serious nuisance in overgrazed and irrigated lands.

9. Solanum cordicitum S. R. Stern, J. Bot. Res. Inst. Texas 8: 2, figs. 1, 2. 2014 • Valentine nightshade [E]

Herbs, annual, erect, moderately armed, to 0.35 m, prickles whitish or yellowish, needlelike, to 5 mm, usually less than 20 per cm of stem, moderately pubescent with stipitate glands 0.5–1 mm mixed with sparse, unbranched, eglandular hairs 1–2 mm, abaxial leaf surfaces with sparse, sessile to short-stalked, stellate hairs, 2–4-rayed, central ray equal to lateral rays. **Leaves** petiolate; petiole 0.5–3 cm; blade simple, ovate to elliptic, 3–8 × 1.5–4 cm, margins deeply lobed to pinnatifid with 3–4 lobes per side, these shallowly lobed, base obtuse. **Inflorescences** extra-axillary, unbranched, 5–8-flowered, 8–12 cm. **Pedicels** 0.4–1 cm in flower, 1–1.8 cm and erect in fruit. **Flowers** bilaterally symmetric; calyx accrescent and tightly covering fruit, moderately prickly, 4–6 mm, moderately pubescent, lobes narrowly triangular; corolla white, rotate-stellate, 2–2.5 cm diam., with abundant interpetalar tissue; stamens unequal, lowermost much longer and curved; anthers narrow and tapered, dehiscent by terminal pores, short anthers 5–6 mm, longer anther 9–11 mm; ovary glabrous. **Berries** brown, globose, 1–1.2 cm diam., glabrous, dry, without sclerotic granules. **Seeds** dark brown, flattened, reniform, ca. 1.5 × 1 mm, reticulately ridged.

Flowering Sep–Nov. Open and disturbed areas; 1300–1900 m; Tex.

Solanum cordicitum is currently known only from three collections from Jeff Davis County.

10. Solanum davisense Whalen, Wrightia 5: 234, fig. 35. 1976 • Davis horsenettle

Herbs, annual, erect, moderately armed, 0.4–0.8 m, prickles whitish or yellowish, straight, needlelike, 3–15 mm, moderately to densely pubescent, hairs unbranched, glandular and eglandular, abaxial leaf surfaces also with sessile, few-rayed, stellate hairs, central ray equal to or longer than lateral rays. **Leaves** petiolate; petiole 2–6 cm; blade simple to compound, broadly ovate, 5–10 × 2.5–8 cm, margins 2–3 times lobed or divided with 3–4 main leaflets per side, leaflets with acute lobes, base truncate. **Inflorescences** extra-axillary, unbranched, 5–9-flowered, 4–7 cm. **Pedicels** 1–1.5 cm in flower, 1–1.5 cm and erect in fruit. **Flowers** bilaterally symmetric; calyx accrescent and tightly covering fruit, densely prickly, 3–5 mm, densely pubescent, lobes linear; corolla violet or blue, pentagonal-stellate, 1.4–2 cm diam., with interpetalar tissue at the margins and bases of lobes; stamens unequal, lowermost much longer and curved; anthers narrow and tapered, dehiscent by terminal pores, short anthers yellow, 4–5.5 mm, longer anther purplish, 5.5–8.5 mm; ovary glabrous. **Berries** brown, globose, 0.8–1 cm diam., glabrous, dry, without sclerotic granules. **Seeds** dark brown, flattened, 2.6–3 × 2–2.5 mm, minutely pitted. $2n = 24$.

Flowering Jun–Sep. Igneous soils, sand or gravel streambeds; 900–2100 m; Tex.; Mexico (Coahuila).

In Texas, *Solanum davisense* is known only from the Chinati, Chisos, and Davis mountains.

11. Solanum deflexum Greenman, Proc. Amer. Acad. Arts 32: 301. 1897 • Sonoita nightshade

Salpichroa wrightii A. Gray 1886, not *Solanum wrightii* Bentham 1861

Herbs, annual, erect, unarmed, 0.1–0.3(–0.4) m, sparsely to densely pubescent, hairs 1–2-celled, unbranched, 1–2 mm, eglandular. **Leaves** petiolate; petiole 0.5–2 cm; blade simple, elliptic to elliptic-ovoid, 1–4.5 × 0.5–2.5 cm, margins entire, base rounded to obtuse and often decurrent. **Inflorescences** nearly sessile, extra-axillary or sub-opposite leaves, unbranched, racemelike, 1–5-flowered. **Pedicels** 0.5–1.2 cm in flower, 1–2 mm and spreading or nodding in fruit. **Flowers** radially symmetric; calyx not accrescent, unarmed, 3–9 mm, moderately to densely pubescent, lobes linear-lanceolate; corolla white, rotate, 0.5–1 cm diam., with abundant interpetalar tissue; stamens equal; anthers oblong, slightly tapered, 1.5–3 mm, dehiscent by terminal pores that open into longitudinal slits; ovary glabrous. **Berries** white to greenish and semitransparent, globose, 5–12 mm diam., glabrous, without sclerotic granules. **Seeds** light brown, somewhat flattened, ca. 2.5 × 1.5 mm, notched where connected to placenta, ridged.

Flowering Aug–Sep. Sandy soils in grazed areas, roadsides, disturbed areas in dry forests; 1000–1700 m; Ariz.; Mexico; Central America (Costa Rica, El Salvador, Guatemala, Honduras, Nicaragua).

Solanum deflexum occurs sporadically in southeastern Arizona (Cochise, Pima, Pinal, and Santa Cruz counties). It is unique in its fruits with explosive dehiscence. As the berries mature, they build up turgor pressure until they burst, propelling seeds up to several feet from the parent plant. Plants of *S. deflexum* were often identified as *S. adscendens* Sendtner (M. Nee 1989), a distinct species found only in South America (S. Stern et al. 2013).

12. Solanum dimidiatum Rafinesque, Autik. Bot., 107. 1840 • Torrey's nightshade, western or robust horsenette [W]

Solanum torreyi A. Gray

Herbs, perennial, erect, sparsely to moderately armed, to 1 m, prickles cream to yellowish, straight or slightly curved, to 6.5 mm, sparsely to densely pubescent, hairs whitish, sessile to short-stalked, stellate, (4–)6–10-rayed, central ray 1-celled and equal to or shorter than lateral rays. **Leaves** petiolate; petiole 1–4 cm; blade simple, ovate, 6–15 × 3–10 cm, margins sinuate or shallowly to deeply lobed with 2–4 lobes per side, lobe margins entire to coarsely lobed, base truncate to cuneate and often oblique. **Inflorescences** extra-axillary, 1–several times branched, to ca. 20-flowered, 6–14 cm. **Pedicels** 1–2.5 cm in flower, 1.5–3 cm and curved downward in fruit. **Flowers** radially symmetric; calyx not accrescent, unarmed or with sparse prickles, 6–14 mm, densely stellate-pubescent, lobes ovate-lanceolate; corolla lavender, pale blue, or sometimes white, stellate to stellate-pentagonal or rotate-stellate, 2–4.6 cm diam., with abundant interpetalar tissue at margins and bases of lobes; stamens equal; anthers narrow and tapered, 5–9 mm, dehiscent by terminal pores; ovary minutely pubescent, hairs simple and stellate, rarely densely stellate-pubescent, glandular and eglandular. **Berries** yellow, subovoid to depressed-globose, 1–2 × 1–2.5 cm, glabrous, without sclerotic granules. **Seeds** yellow, flattened, 1.9–3 × 1.7–2.5 mm, minutely pitted. $2n$ = 72.

Flowering Apr–Oct. Prairies, woodlands, disturbed areas; 20–700(–2000) m; Ark., Calif., Ill., Kans., La., Mo., N.Mex., Okla., S.C., Tex.; Mexico (Nuevo León); introduced in Australia.

Solanum dimidiatum is found mainly in the south-central United States, with outlier populations in Illinois, Missouri, New Mexico, and South Carolina. The species is introduced in California, where it is considered a noxious weed by the California Department of Agriculture.

13. Solanum diphyllum Linnaeus, Sp. Pl. 1: 184. 1753 • Twoleaf nightshade [I] [W]

Shrubs, erect, unarmed, 1–2 m, glabrous or occasionally minutely puberulent, hairs white, unbranched, eglandular. **Leaves** petiolate; petiole 0.2–0.5 cm; blade simple, elliptic, 0.9–6.8 × 0.6–2.2 cm, margins entire, base acute to attenuate or decurrent. **Inflorescences** leaf-opposed, unbranched, 5–20-flowered, 0.3–1.2 cm. **Pedicels** ca. 0.5 cm in flower, ca. 1.2 cm and erect in fruit. **Flowers** radially symmetric; calyx somewhat accrescent, unarmed, 1.5–2 mm, glabrous, lobes deltoid; corolla white, often tinged with lavender, stellate, 0.7–1 cm diam., without interpetalar tissue; stamens equal; anthers oblong, 1.5–2 mm, dehiscent by terminal pores that open into longitudinal slits; ovary glabrous. **Berries** yellow to orange, globose, 0.7–1.2 cm diam., glabrous, without sclerotic granules. **Seeds** pale yellow or tan, flattened, ca. 3 × 2.5 mm, minutely pitted. $2n$ = 24.

Flowering year-round. Dry lowland areas, hammocks, disturbed sites; 0–300 m; introduced; Fla.; Mexico; West Indies; Central America (Belize, Costa Rica, El Salvador, Guatemala, Honduras, Nicaragua); South America (Brazil); introduced also in Europe (s France, Italy), Asia, Pacific Islands (Java, Philippines).

Solanum diphyllum is often cultivated for its brightly colored fruits and can escape from cultivation in tropical and subtropical areas. It occurs sporadically and does not appear to be common, but in other areas where it has escaped it has become naturalized (Asia; S. Knapp 2002b).

14. Solanum donianum Walpers, Repert. Bot. Syst. 3: 54. 1844 • Mullein nightshade

Solanum blodgettii Chapman

Shrubs, erect, sparsely armed when young, older growth unarmed, 0.5–2.5 m, prickles brownish, 1–3 mm, straight, moderately to densely pubescent, hairs sessile to short-stalked, stellate, 6–9-rayed, central ray absent or shorter than lateral rays. **Leaves** petiolate; petiole 1–3 cm; blade simple, ovate to elliptic, 4.5–13 × 2–5.5 cm, margins entire, base rounded to acute. **Inflorescences** terminal to extra-axillary, much-branched, with numerous flowers, 2–8 cm. **Pedicels** erect and 0.7–1 cm in flower and fruit. **Flowers** radially symmetric; calyx not accrescent, unarmed, 2–4 mm, moderately stellate-pubescent, lobes triangular; corolla white, stellate, 1.5–2 cm diam., without interpetalar tissue; stamens equal; anthers narrow and tapered, 3–4.5 mm, dehiscent by terminal pores; ovary glabrous or sparsely glandular-pubescent. **Berries** red, globose, 0.5–1 cm diam., glabrous or sparsely glandular-pubescent, without sclerotic granules. **Seeds** yellow, flattened, 2.5–3.5 × 1.5–2.5 mm, minutely pitted. $2n = 24$.

Flowering year-round. Seasides, hammocks, pine forests, limestone soils; 0 m; Fla.; s Mexico; West Indies (Bahamas); Central America (Belize, Guatemala).

Solanum donianum is occasional in southern Florida. The oldest name for this species is *S. verbascifolium* Linnaeus, but it has been widely misapplied to *S. erianthum* and is now rejected.

15. Solanum douglasii Dunal in A. P. de Candolle and A. L. P. P. de Candolle, Prodr. 13(1): 48. 1852 • Greenspot nightshade [F] [W]

Solanum arizonicum Parish

Herbs or shrubs, perennial, erect, unarmed, to 1.5(–3) m, sparsely to moderately pubescent, hairs white, curved, unbranched, 0.5–1 mm, eglandular. **Leaves** petiolate; petiole 0.5–3(–7) cm; blade simple, ovate, 1–5(–9) × 0.5–3(–6) cm, margins entire to coarsely and irregularly toothed, base truncate to acute and decurrent. **Inflorescences** extra-axillary or leaf-opposed, unbranched, racemelike, 2–7(–14)-flowered, 2–4 cm. **Pedicels** 0.5–1 cm in flower and fruit, nodding or deflexed downward in fruit. **Flowers** radially symmetric; calyx not accrescent, unarmed, 2–3 mm, sparsely pubescent, lobes deltate; corolla white with yellow-green to brownish central star, stellate, 1–2 cm diam., without interpetalar tissue; stamens equal; anthers ellipsoidal and slightly tapered towards the tips, (2.5–) 3–4.5 mm, dehiscent by terminal pores that open into longitudinal slits; ovary glabrous. **Berries** dull purplish black, globose, 0.5–1 cm diam., glabrous, with (2–)6–8 sclerotic granules per fruit. **Seeds** pale yellow to tan, flattened, 1.5–2 × 1–1.5 mm, finely reticulate. $2n = 24$.

Flowering Mar–Nov (nearly year-round in Calif.). Dry shrubland, woodland, rocky slopes, stream banks, canyons; 0–2500 m; Ariz., Calif., N.Mex., Tex.; Mexico; West Indies (Guadeloupe); Central America (El Salvador, Guatemala, Honduras, Nicaragua).

Solanum douglasii is most commonly found west of the Rocky Mountains. It can be difficult to distinguish from *S. nigrescens*, but its longer, slightly tapered anthers on very short (relative to anther length) filaments is a good distinguishing character for plants in flower. The buds of *S. douglasii* are more pointed than those of *S. nigrescens*.

16. Solanum dulcamara Linnaeus, Sp. Pl. 1: 185. 1753 • Woody or climbing or bitter nightshade, bittersweet, morelle douce-amère [F] [W]

Solanum dulcamara var. *villosissimum* Desvaux

Vines, climbing or scrambling, herbaceous or woody, unarmed, to 8–10 m, sparsely to densely pubescent, hairs unbranched and/or dendritic, rarely glabrous. **Leaves** petiolate; petiole 0.5–5 cm; blade simple, elliptic or ovate to cordate, 2.5–12 × 1.2–9 cm, margins entire to deeply pinnatifid and usually 3-lobed near base, lobe margins entire, base truncate to cordate. **Inflorescences** terminal or lateral, extra-axillary, much-branched, 7–40-flowered, (1–)4–15 cm. **Pedicels** inserted into a small sleeve on the inflorescence axis, 0.6–1.2 cm in flower and fruit. **Flowers** radially symmetric; calyx not accrescent, unarmed, 1–2 mm, glabrous to densely pubescent, lobes triangular, shallow; corolla purple (rarely white), with green and white shiny spots at base of each lobe, deeply stellate, 1.5–2 cm diam., without interpetalar tissue; stamens equal; anthers oblong, slightly tapered, 4.5–6 mm, dehiscent by terminal pores that often open into longitudinal slits; ovary glabrous. **Berries** bright shiny red, globose to ellipsoidal, 0.5–1.5 × 0.5–1 cm, glabrous, without sclerotic granules. **Seeds** pale yellow or tan, flattened, 2–3 mm diam., minutely pitted. $2n = 24$.

Flowering May–Nov. Weedy, in a wide variety of habitats, often associated with water; 0–2000 m; B.C., Man., N.B., Nfld. and Labr. (Nfld.), N.S., Ont., P.E.I., Que., Sask.; Ala., Ariz., Calif., Colo., Conn., D.C., Fla., Ga., Idaho, Ill., Ind., Iowa, Kans., Ky., Maine, Md., Mass., Mich., Minn., Mo., Mont., Nebr., Nev., N.H., N.J., N.Mex., N.Y., N.C., N.Dak., Ohio, Oreg., Pa., R.I., S.Dak., Tenn., Utah, Vt., Va., Wash., W.Va., Wis., Wyo.; Eurasia.

Solanum dulcamara is widely distributed across Eurasia and boreal North America. The North American populations are thought to be introductions, but it is possible that the species has a truly circumboreal distribution. A white-flowered form has been recognized by some authors as *S. dulcamara* forma *albiflorum* House and an especially pubescent form as *S. dulcamara* var. *villosissimum*, but variation in a number of morphological features is continuous across the range of the species and these and other variants are not recognized in the latest monograph of the group (S. Knapp 2013).

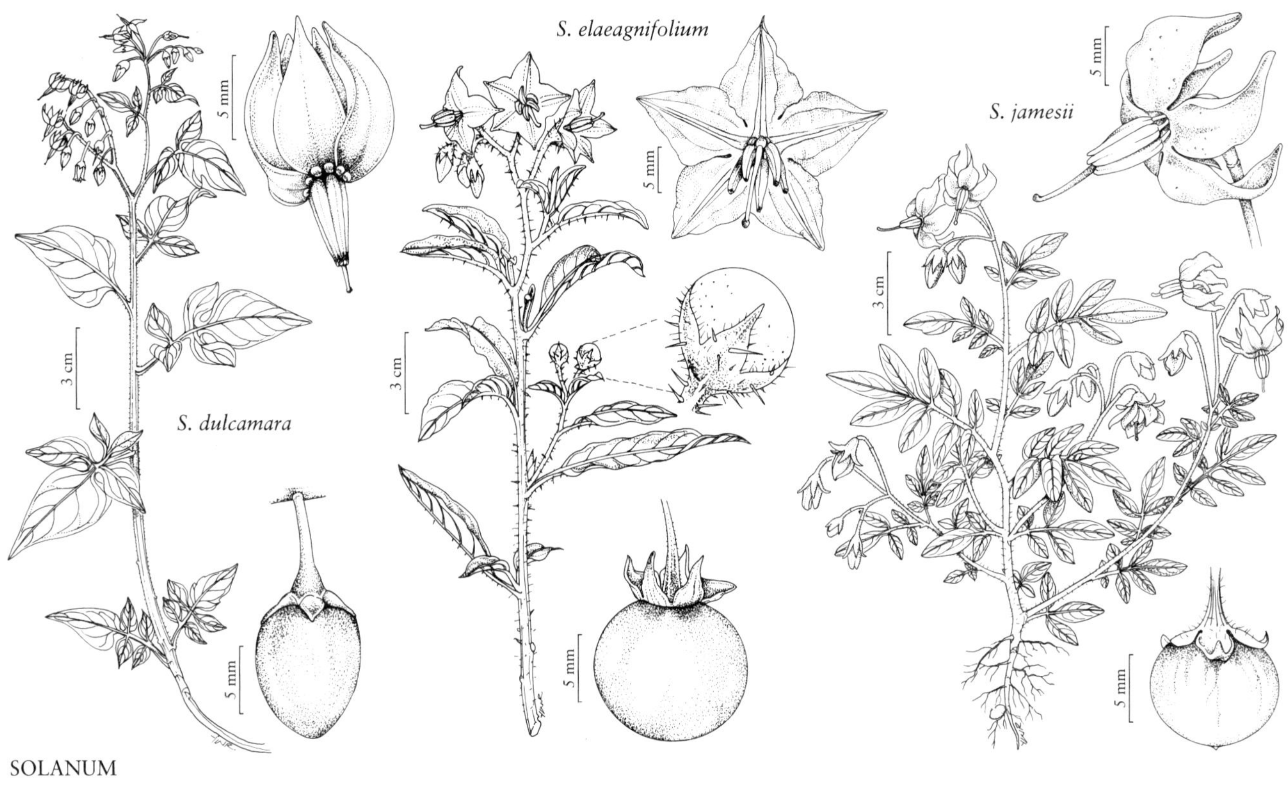

SOLANUM

17. **Solanum elaeagnifolium** Cavanilles, Icon. 3: 22, plate 243. 1795 • Silverleaf nightshade, white horsenettle, trompillo [F] [W]

Herbs or shrubs, perennial, erect, rhizomatous, sparsely to densely armed, to 1 m, prickles orange to brown, straight, to 5 mm, densely silvery-pubescent, hairs sessile or subsessile, stellate, scalelike, 10–15-rayed, central ray shorter than lateral rays, lateral rays fused at center. **Leaves** petiolate; petiole 1–5 cm; blade simple, lanceolate to oblong, 5–15 × 0.5–3 cm, margins undulate, subentire to sinuate or shallowly lobed, lobe margins entire, base truncate to cuneate. **Inflorescences** extra-axillary, unbranched, 3–5(–7)-flowered, 3–5 cm. **Pedicels** 1–3 cm in flower, reflexed and 1–3 cm in fruit. **Flowers** radially symmetric; calyx not accrescent, unarmed or with sparse prickles, 5–10 mm, densely silvery-stellate-pubescent, lobes linear-lanceolate; corolla pale to deep blue or violet, rarely white, pentagonal-stellate, 2–3(–3.5) cm diam., with abundant interpetalar tissue; stamens equal; anthers narrow and tapered, 6–9 mm, dehiscent by terminal pores; ovary glabrous to densely stellate-pubescent. **Berries** yellow to orange, drying brown or black, globose, 0.5–1.5 cm diam., glabrous, without sclerotic granules. **Seeds** yellowish, flattened, 3–5 × 2–4 mm, minutely pitted. **2*n*** = 24, 48, 72.

Flowering Mar–Nov. Dry sites, open woods, disturbed areas, roadsides, railroads, fields; 0–2100 m; Ala., Ariz., Ark., Calif., Colo., Fla., Ga., Idaho, Ill., Ind., Kans., Ky., La., Md., Miss., Mo., Nebr., Nev., N.Mex., N.C., Okla., Oreg., S.C., Tenn., Tex., Utah, Wash.; Mexico; West Indies; South America (Argentina, Chile, Paraguay, Uruguay); introduced elsewhere in South America (Colombia, Peru), Eurasia (Mediterranean, Middle East, India, Pakistan), Africa, Australia.

Solanum elaeagnifolium has a disjunct native distribution. It occurs in arid regions of the southwestern United States and Mexico and also in Argentina, Chile, Paraguay, and Uruguay. North American plants are diploid, whereas those in Argentina are diploid, tetraploid, or hexaploid. It is invasive and considered a noxious weed in 21 states in the flora area as well as in many tropical and subtropical regions worldwide. It is toxic to livestock and can form large, rhizomatous patches that are difficult to eradicate.

A white-flowered form has been recognized as *Solanum elaeagnifolium* forma *albiflorum* Cockerell.

18. Solanum emulans Rafinesque, Autik. Bot., 107. 1840 • Eastern nightshade, morelle noire de l'Est [E]

Solanum nigrum Linnaeus var. *virginicum* Linnaeus

Herbs or shrubs, annual or perennial, erect, unarmed, to 1 m, glabrous to sparsely or rarely densely pubescent, hairs unbranched, to 1 mm, eglandular. **Leaves** petiolate; petiole 1–5 cm; blade simple, ovate to elliptic, 4.5–10.5 × 2–6 cm, margins entire to sinuate-dentate, base attenuate to rounded. **Inflorescences** extra-axillary, unbranched, umbel-like, (2–)3–6-flowered, 1–2.5 cm. **Pedicels** straight and spreading in flower and recurved to reflexed in fruit, 0.5–1 cm. **Flowers** radially symmetric; calyx not accrescent, unarmed, 2–3 mm, glabrous to sparsely pubescent, lobes appressed in fruit, deltate; corolla white, sometimes with yellow central star, rarely purplish, stellate, 0.5–1 cm diam., without interpetalar tissue; stamens equal; anthers ellipsoidal, 1–1.5 mm, dehiscent by terminal pores that open into longitudinal slits; ovary glabrous. **Berries** dull or slightly shiny purplish black, globose, 0.5–1 cm diam., glabrous, with 6–9 sclerotic granules per fruit. **Seeds** yellowish, flattened, 1.5–2 × 1–1.5 mm, finely reticulate. **2*n*** = 24.

Flowering May–Oct. Moist, open woodlands, stream banks, fields, roadsides, disturbed areas; 0–700 (–1700) m; B.C., Man., N.B., Ont., Que., Sask.; Ala., Ariz., Ark., Colo., Conn., Del., D.C., Fla., Ga., Ill., Ind., Iowa, Kans., Ky., La., Maine, Md., Mass., Mich., Minn., Miss., Mo., Nebr., N.H., N.J., N.Mex., N.Y., N.C., N.Dak., Ohio, Okla., Pa., R.I., S.C., S.Dak., Tenn., Tex., Vt., Va., W.Va., Wis., Wyo.

Solanum emulans has often been called *S. ptychanthum* Dunal (with the variant spelling ptycanthum), but that name is a synonym of *S. americanum*.

Solanum emulans is the most common species in the black nightshade group in northeastern North America. It can be distinguished from other North American species in the black nightshade group by its unbranched inflorescences, short anthers, appressed fruiting calyx lobes, and numerous sclerotic granules in the fruits.

19. Solanum erianthum D. Don, Prodr. Fl. Nepal., 96. 1825 • Potato tree, salvadora

Shrubs or small trees, erect, unarmed, 2–8 m, densely pubescent, hairs sessile to short-stalked, stellate to echinoid. **Leaves** petiolate; petiole 1–10 cm; blade simple, elliptic to ovate, 10–25 × 3–15 cm, margins entire, base rounded or acute. **Inflorescences** terminal, becoming leaf-opposed, much-branched, 10–50-flowered, 5–20 cm. **Pedicels** 0.2–0.6 cm in flower, erect and 0.4–10 cm in fruit. **Flowers** radially symmetric; calyx accrescent and subtending fruit, unarmed, 5–7 mm, densely pubescent, hairs stellate to echinoid, lobes broadly triangular; corolla white, stellate, 1–2 cm diam., without interpetalar tissue; stamens equal; anthers oblong, 2.5–3.5 mm, dehiscent by terminal pores that open into longitudinal slits; ovary tomentose, hairs stellate or echinoid. **Berries** yellow to orange, globose, 1–2 cm diam., densely pubescent, without sclerotic granules. **Seeds** yellowish brown, flattened, 1.5–2 × 1–1.5 mm, minutely pitted. **2*n*** = 24.

Flowering Apr–Oct (year-round in Fla.). Hammocks, pinelands, disturbed sites; 0–100 m; Fla., Tex.; Mexico; West Indies; Central America; South America (Colombia); introduced in Asia, Africa, Pacific Islands (including the Galapagos Islands), Australia.

In the United States, *Solanum erianthum* is common only in central to southern Florida and in extreme southern Texas near the Gulf of Mexico.

The name *Solanum verbascifolium* Linnaeus has been widely misapplied to *S. erianthum* (K. E. Roe 1968), but is a synonym of *S. donianum* that has now been rejected.

20. Solanum furcatum Dunal in J. Lamarck et al., Encycl., suppl. 3: 750. 1814 • Forked nightshade [I] [W]

Herbs, annual or perennial, erect to sprawling, unarmed, to 1 m, sparsely pubescent, hairs unbranched, to 0.5 mm, eglandular. **Leaves** petiolate; petiole 1–3.5 cm; blade simple, ovate-lanceolate, 3–10 × 2–5 cm, margins entire to sinuate-dentate, base cuneate to truncate. **Inflorescences** extra-axillary, forked, umbel-like or racemelike, 6–14-flowered, 1.5–3 cm. **Pedicels** straight and spreading and 0.5–1 cm in flower, strongly reflexed and 0.5–1 cm in fruit. **Flowers** radially symmetric; calyx not accrescent, unarmed, 3–4 mm, sparsely pubescent, lobes obtuse; corolla white to pale purple with yellowish or greenish central star, stellate,

1–2 cm diam., with sparse interpetalar tissue; stamens equal; anthers ellipsoidal, 2.5–3.5 mm, dehiscent by terminal pores that open into longitudinal slits; ovary glabrous. **Berries** dull green to purple, globose, 0.5–0.9 cm diam., glabrous, with 6–14 sclerotic granules per fruit. **Seeds** pale yellow to light brown, flattened, 1.5–2 × 1–1.5 mm, finely reticulate. $2n = 72$.

Flowering May–Oct. Open and disturbed areas near sea cliffs, bluffs, and on sand dunes; 0–500 m; introduced; Calif., Oreg., Wash.; South America (Argentina, Chile); introduced also in Pacific Islands (New Zealand), Australia.

Solanum furcatum is found in coastal environments in the western United States. M. Nee (1993) stated that the name *S. gayanum* (J. Remy) F. Philippi has been misapplied to plants of *S. furcatum*, but no basis can be found for this assertion and the two species are morphologically very different. *Solanum gayanum*, a synonym of *S. crispum* Ruiz & Pavon, and native to Chile, is cultivated and perhaps naturalized in San Francisco, California (P. A. Munz 1968).

Solanum furcatum can be distinguished from the similar and sympatric *S. douglasii* by its usually forked inflorescences and fruits with usually more than ten sclerotic granules. A distinctive character of *S. furcatum* is the long style that is about twice the length of the anthers.

21. **Solanum hindsianum** Bentham, Bot. Voy. Sulphur, 39. 1844 • Hinds's or Baja or Sonoran nightshade

Shrubs, erect, unarmed to sparsely armed, 0.5–3 m, prickles reddish brown, straight, 2–15 mm, densely silvery-pubescent, hairs sessile to short-stalked, stellate, 8–12-rayed, central ray shorter than or equal to lateral rays. **Leaves** petiolate; petioles 0.5–1.5 cm; blade simple, ovate to elliptic, (1–)2–6.5 × (0.5–)1–3(–4.5) cm, margins entire or undulate, base rounded to truncate and usually oblique. **Inflorescences** extra-axillary, unbranched, 3–4-flowered, 3–4 cm. **Pedicels** 0.4–1.5 cm in flower, erect and 1–2 cm in fruit. **Flowers** radially symmetric; calyx not accrescent, unarmed, 8–10 (–20) mm, densely stellate-pubescent, lobes long-triangular to linear-lanceolate; corolla violet, pentagonal, 2.5–4(–5) cm diam., with abundant interpetalar tissue; stamens equal or slightly unequal; anthers narrow and tapered, 6–10 mm, dehiscent by terminal pores; ovary glabrous. **Berries** light green, sometimes with darker mottling, drying dark brown or reddish brown, globose, 1–1.5(–2) cm diam., glabrous, cracking open to expose seeds, without sclerotic granules. **Seeds** dark brown, flattened, 2–3 mm diam., minutely pitted.

Flowering Jan–Mar. Rocky soils, hillsides; 500–600 m; Ariz.; Mexico (Baja California, Baja California Sur, Sinaloa, Sonora).

Solanum hindsianum is endemic to the Sonoran Desert of extreme southern Arizona and northern Mexico. In Arizona, it is known only from Organ Pipe Cactus National Monument.

22. **Solanum interius** Rydberg, Bull. Torrey Bot. Club 31: 641. 1905 • Plains black or deadly nightshade [E]

Solanum nigrum Linnaeus [unranked] *interius* (Rydberg) F. C. Gates

Herbs or shrubs, annual to short-lived perennial, erect, unarmed, to 1 m, sparsely to densely pubescent, hairs unbranched, usually to 1 mm, eglandular. **Leaves** petiolate; petiole 0.5–3.5 cm; blade simple, ovate to ovate-lanceolate, 4.5–11 × 2.5–7 cm, margins entire to sinuate-dentate, base cuneate to rounded or slightly decurrent. **Inflorescences** extra-axillary, unbranched, (2–)3–8-flowered, 2.5–3.5 cm. **Pedicels** spreading in flower, recurved to reflexed in fruit, 0.5–1 cm in flower and fruit. **Flowers** radially symmetric; calyx not accrescent, unarmed, 2–5 mm, sparsely pubescent, lobes lanceolate, sometimes reflexed in fruit; corolla white, sometimes tinged with purple, with yellowish central star, stellate, 0.5–1 cm diam., without interpetalar tissue; stamens equal; anthers ellipsoidal, 1.8–2.5 mm, dehiscent by terminal pores that open into longitudinal slits; ovary glabrous. **Berries** shiny purplish black, globose, 1–1.5 cm diam., glabrous, with 2–4 sclerotic granules. **Seeds** yellowish to brown, flattened, 1.8–2 × 1.5–1.6 mm, finely reticulate. $2n = 24$.

Flowering Jun–Oct. Pastures, open woodlands, stream valleys, thickets, disturbed areas, sandy soils; (100–)500–2500 m; Colo., Idaho, Iowa, Kans., Mont., Nebr., Nev., N.Mex., N.Dak., Okla., S.Dak., Tex., Utah, Wyo.

Solanum interius is endemic to North America and is most common in the Great Plains and eastern Rocky Mountains. Distinctive characters are the basal flower with its pedicel articulated above the base and the very large seeds. In Texas, *S. interius* can be very difficult to distinguish from *S. nigrescens*, but *S. interius* usually has longer calyx lobes. Records of *S. interius* from Saskatchewan are actually *S. emulans*.

23. Solanum jamaicense Miller, Gard. Dict. ed. 8, Solanum no. 17. 1768 • Jamaican nightshade [I] [W]

Shrubs, erect to scandent, 1–2 m, moderately armed, prickles yellow to green, recurved, to 8 mm, moderately to densely white-pubescent, hairs short-stalked, stellate, 6–8-rayed, central ray shorter than or equal to lateral rays. **Leaves** petiolate or sessile; petiole to 1 cm; blade simple, rhombic, 4–13 × 3–8 cm, margins entire or with 2–5 shallow lobes per side, lobe margins entire, base cuneate and decurrent. **Inflorescences** extra-axillary, unbranched, 5–15-flowered, 1–3 cm. **Pedicels** 0.5–1 cm in flower, 1–1.5 cm in fruit. **Flowers** radially symmetric; calyx not accrescent, unarmed, 2–7 mm, moderately to densely stellate-pubescent, hairs long-stalked, lobes linear; corolla white, stellate, 1–2 cm diam., without interpetalar tissue; stamens equal; anthers narrow and tapered, 3.5–5 mm, dehiscent by terminal pores; ovary glabrous to very sparsely glandular-puberulent. **Berries** bright shiny red to orange, globose, 0.4–1.2 cm diam., glabrous, without sclerotic granules. **Seeds** yellow, flattened, 1–1.5 × 0.5–1 mm, minutely pitted and ridged. $2n = 24$.

Flowering Jul–Sep. Lakesides, shaded hammocks; 0–10 m; introduced; Fla.; Mexico; West Indies; Central America; South America (Bolivia, Brazil, Colombia, Ecuador, French Guiana, Guyana, Peru, Venezuela).

Solanum jamaicense is thought to have been spread to Florida by birds that eat the bright red berries. It was first seen in the state in 1930 and, although locally invasive in hammocks of central Florida, has not become a widespread pest.

SELECTED REFERENCE Diaz, R., W. A. Overholt, and K. Langeland. 2008. Jamaican nightshade (*Solanum jamaicense*): A threat to Florida's hammocks. Invasive Pl. Sci. Managem. 1: 422–425.

24. Solanum jamesii Torrey, Ann. Lyceum Nat. Hist. New York 2: 227. 1827 • Wild potato [F]

Herbs, perennial, erect, unarmed, bearing tubers to 2 cm long, to 0.5 m, glabrous or sparsely pubescent, hairs unbranched, gland-tipped. **Leaves** petiolate; petiole 1.5–3.5 cm, sometimes with pair of pinnatifid pseudostipules at base; blade compound, elliptic to ovate, 7–15 × 4–9 cm, margins divided into 1–4(–5) pairs of leaflets, leaflet margins entire, base attenuate. **Inflorescences** terminal, extra-axillary, generally forked or 3-fid, 4–10(–20)-flowered, to 3 cm. **Pedicels** articulated near middle, 1.6–3 cm in flower and fruit. **Flowers** radially symmetric; calyx not accrescent, unarmed, 4–6 mm, glabrous to sparsely pubescent, lobes deltate-acuminate; corolla white, stellate, 2.8–3.5 cm diam., without interpetalar tissue; stamens equal; anthers oblong, slightly tapered, 5–6 mm, dehiscent by terminal pores that open into longitudinal slits; ovary glabrous. **Berries** green, globose, ca. 1 cm diam., glabrous, without sclerotic granules. **Seeds** dark reddish brown, rounded, 1–2 mm diam., rugose. $2n = 24$.

Flowering Jun–Oct. Hillsides, stream bottoms, sandy soils, disturbed grasslands, pinyon-juniper forests, oak thickets, coniferous and deciduous forests; 1300–2900 m; Ariz., Colo., N.Mex., Tex., Utah; Mexico (Chihuahua, Querétaro, San Luis Potosí, Sonora).

The tubers of *Solanum jamesii* have been gathered as food by Native Americans, and starch grains identified as *S. jamesii* from stone tools in Utah form the earliest evidence for the use of potatoes in North America (L. A. Louderback and B. M. Pavlik 2017). All other parts of the plant are toxic.

25. Solanum laciniatum Aiton, Hort. Kew. 1: 247. 1789 • New Zealand nightshade, kangaroo-apple [I]

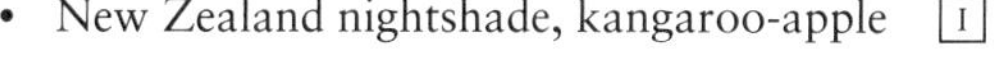

Shrubs, erect, unarmed, 1–3 m, glabrous. **Leaves** petiolate; petiole 1–1.5 cm; blade simple, elliptic, 10–30 × 1.5–15 cm, margins entire to coarsely pinnatifid with 1–3 lobes per side, lobe margins entire, base cuneate. **Inflorescences** leaf-opposed or in branch forks, unbranched or forked, to 10-flowered, 5–15 cm. **Pedicels** 1.5–3 cm in flower and fruit. **Flowers** radially symmetric; calyx somewhat accrescent, unarmed, 3–4 mm, glabrous, lobes deltate; corolla blue to deep purple, rotate-stellate, 3–5 cm diam., lobes notched at apex, with abundant interpetalar tissue; stamens equal; anthers oblong, slightly tapered, 3–4 mm, dehiscent by terminal pores that sometimes open into longitudinal slits; ovary glabrous. **Berries** yellow to orange-yellow, ovoid to obovoid, 1.5–2 × 1–1.5 cm, glabrous, with abundant sclerotic granules. **Seeds** reddish brown, flattened, 2.5–3 × 2–2.5 mm, concentrically reticulate. $2n = 92$.

Flowering Jan–Jul. Open, disturbed sites; 0–100 m; introduced; Calif.; Pacific Islands (New Zealand); Australia.

Solanum laciniatum has often been confused with *S. aviculare* in floras. See discussion under 2. *S. aviculare*.

26. Solanum lanceolatum Cavanilles, Icon. 3: 23, plate 245. 1795 • Orangeberry or lanceleaf nightshade [I] [W]

Shrubs or small trees, erect, sparsely armed, 1–2(–3.5) m, prickles brownish, straight or recurved, to 10 mm, moderately to densely bright white-pubescent, hairs sessile to short-stalked, stellate, 5–8-rayed, central ray shorter than lateral rays. **Leaves** petiolate; petiole 1.5–4 cm; blade simple, ovate to elliptic, 5–25 × 3–15 cm, margins entire to shallowly lobed with 2–4 lobes per side, lobe margins entire to undulate, base obtuse to cordate. **Inflorescences** extra-axillary, usually much-branched, 10–20-flowered, to 10 cm. **Pedicels** 0.5–1 cm in flower, erect and 1–2 cm in fruit. **Flowers** radially symmetric; calyx not accrescent, unarmed, 5–8 mm, densely stellate-pubescent, lobes triangular; corolla usually blue to purple, rarely whitish, stellate to rotate-stellate, 2.5–4 cm diam., with abundant interpetalar tissue; stamens equal; anthers narrow and tapered, 6–8 mm, dehiscent by terminal pores; ovary glabrous to stellate-pubescent. **Berries** yellow to orange, globose, 0.7–1.5 cm diam., stellate-pubescent at apex, without sclerotic granules. **Seeds** yellow to light brown, flattened, 2.5–3 × 2–2.5 mm, minutely pitted.

Flowering Apr–Aug. Disturbed sites; 0–400 m; introduced; Calif.; Mexico; Central America.

Solanum lanceolatum is naturalized along the California coast and inland in central and southern California, where it is listed as a noxious weed by the California Department of Agriculture.

27. Solanum lumholtzianum Bartlett, Proc. Amer. Acad. Arts 44: 629. 1909 • Sonoran nightshade

Herbs, annual, erect, moderately to densely armed, 0.2–0.7 m, prickles whitish or yellowish, straight, bristlelike, 4–8 mm, 20+ per cm of stem, sparsely pubescent, hairs sessile to short-stalked, unbranched, glandular, sometimes with a few unbranched, eglandular hairs, abaxial leaf surfaces with sparse, sessile, 4–6-rayed, stellate hairs, central ray equal to lateral rays. **Leaves** petiolate; petiole 2.5–6.5 cm; blade simple to compound, broadly ovate, 5–13 × 2.5–9 cm, margins 2–3-times lobed or divided with 4–5 main leaflets per side, leaflets with deep, acute lobes, base truncate. **Inflorescences** extra-axillary, unbranched, 6–10-flowered, 3–8 cm. **Pedicels** 0.5–1.5 cm in flower, erect and 0.5–1.5 cm in fruit. **Flowers** bilaterally symmetric; calyx accrescent and tightly covering fruit, densely prickly, 3.5–4.5 mm, sparsely to moderately glandular-pubescent, lobes linear-lanceolate; corolla yellow, stellate, 1.3–1.8 cm diam., with ovate or narrowly deltate lobes, with sparse interpetalar tissue; stamens unequal, lowermost longer and curved, adjacent pair somewhat shorter, uppermost pair shortest; anthers narrow and tapered, dehiscent by terminal pores, shortest anthers 4.5–6 mm, medium-sized anthers 5.6–7.5 mm, longest anther 6.5–8.6 mm; ovary glabrous. **Berries** brown, globose, 1–1.4 cm diam., glabrous, dry, without sclerotic granules. **Seeds** dark brown, plump, 3–3.5 × 2.5–3 mm, radially ridged with hilum sunken in a deep notch. 2*n* = 24.

Flowering Aug–Oct. Sandy or gravelly soils, washes, low ground near wet depressions, along stream banks, roadsides; 900–1400 m; Ariz.; Mexico (Sinaloa, Sonora).

In the flora area, *Solanum lumholtzianum* is found only in Pima and Santa Cruz counties.

28. Solanum lycopersicum Linnaeus, Sp. Pl. 1: 185. 1753 • Tomato, tomate [I]

Lycopersicon esculentum Miller, name conserved; *L. lycopersicum* (Linnaeus) H. Karsten; *L. esculentum* var. *cerasiforme* Alefeld; *Solanum lycopersicum* var. *cerasiforme* (Alefeld) Voss

Herbs, annual, erect or scandent, unarmed, ca. 0.5–1.5 m, moderately to densely pubescent, hairs simple, glandular and eglandular. **Leaves** petiolate; petiole 2–10 cm; blade compound, elliptic, 10–30+ × 5–25 cm, margins divided with 3(–5) pairs of leaflets, interspersed with smaller interjected leaflets, leaflet margins toothed, base truncate to cordate. **Inflorescences** lateral, extra-axillary, simple or rarely forked, 4–15-flowered, to 10 cm. **Pedicels** articulated near middle, 0.5–2 cm in flower, to 3 cm in fruit. **Flowers** radially symmetric; calyx accrescent, unarmed, 5–25 mm, lobes 5–7, lanceolate-acuminate; corolla yellow, pentagonal to stellate, 1–3 cm diam., with interpetalar tissue at margins and bases of lobes; stamens equal; anthers narrow and tapered, 6–11 mm, each with a sterile apical appendage, dehiscent by oblong pores that open into introrse, longitudinal slits; ovary glabrous or glandular-puberulent. **Berries** usually red, orange, or yellow, globose to ellipsoid or obovoid, 1.5–12 cm diam., glabrous, without sclerotic granules. **Seeds** pale brown, flattened, 2–4 × 1.5–2 mm, pubescent. 2*n* = 24.

Flowering year-round in Fla., May–frost elsewhere. Disturbed habitats, rich soils; 0–2000 m; introduced; widely cultivated; South America; introduced also in temperate and tropical countries worldwide.

The tomato is a major agricultural crop and is commonly grown in home gardens. The fruits can have a variety of shapes, sizes, colors, and flavors and are used raw or cooked in a vast array of dishes. Other parts of the plants are considered to be poisonous. The flowers can commonly be six- to eight-merous, especially in varieties selected for very large fruits.

The ancestors of *Solanum lycopersicum* were originally from western South America, but the species is now known only from cultivation. Tomatoes frequently escape from cultivation or germinate in compost piles or garbage dumps but are very susceptible to frost, rarely persist, and are not invasive, even in warm climates.

29. **Solanum marginatum** Linnaeus f., Suppl. Pl., 147. 1782 • White-margined or purple African nightshade I W

Herbs or shrubs, erect, moderately armed, 1–2 m, prickles pale orange, straight or slightly curved, 5–12 mm, densely bright white-pubescent, hairs short-stalked, stellate, 10–20-rayed, central ray equal to lateral rays. **Leaves** petiolate; petiole 1.5–6 cm; blade simple, ovate, 8–23 × 7–13 cm, margins coarsely lobed with 3–4 lobes per side, lobe margins entire to coarsely lobed, base cordate. **Inflorescences** extra-axillary or leaf-opposed, occasionally sessile, unbranched or forked, 6–15(–30)-flowered, 3.5–8 cm. **Pedicels** 0.5–2 cm in flower, pendent and 2–3 cm in fruit. **Flowers** radially symmetric; calyx accrescent and subtending fruit, sparsely to moderately armed, 7–15 mm, densely stellate-pubescent, lobes broadly deltate; corolla white to pale purple, rotate-stellate, 2.5–4 cm diam., with abundant interpetalar tissue; stamens equal; anthers narrow and tapered, 5.5–7 mm, dehiscent by terminal pores; ovary moderately stellate-pubescent. **Berries** yellow, dark green mottled with white when young, globose, 3.5–4(–5) cm diam., glabrous, without sclerotic granules. **Seeds** light brown, flattened, 2.5–3.5 × 2–2.5 mm, minutely pitted. $2n = 24$.

Flowering May–Aug. Disturbed sites; 0–1000 m; introduced; Calif.; Asia; Africa (Eritrea, Ethiopia); introduced also in South America, Europe, Atlantic Islands (Canary Islands), Australia.

Solanum marginatum is naturalized along the central and southern coast of California, where it has been listed as a noxious weed by the California Department of Food and Agriculture. There are few recent collections.

30. **Solanum mauritianum** Scopoli, Delic. Fl. Faun. Insubr. 3: 16, plate 8. 1788 • Earleaf nightshade I W

Solanum auriculatum Aiton

Shrubs or small trees, erect, unarmed, 2–12 m, densely pubescent, hairs white, sessile to long-stalked, stellate to echinoid. **Leaves** petiolate; petiole 1–8 cm; blade simple, elliptic to ovate, 11–31 × 4–14 cm, margins entire, base acute, often with smaller axillary leaves. **Inflorescences** terminal, becoming leaf-opposed, much-branched, 50–100-flowered, 5–24 cm. **Pedicels** erect and 0.2–0.5 cm in flower and fruit. **Flowers** radially symmetric; calyx slightly accrescent, unarmed, 4–7.5 mm, densely pubescent, lobes deltate; corolla purple, stellate-pentagonal, 1–1.5 cm diam., with abundant interpetalar tissue; stamens equal; anthers oblong, 2–3.5 mm, dehiscent by terminal pores that open into longitudinal slits; ovary tomentose. **Berries** yellow, globose, 1–1.5 cm diam., tomentose, without sclerotic granules. **Seeds** yellowish brown, flattened, 1.5–2.5 × 1.5–2 mm, minutely pitted. $2n = 24$.

Flowering Mar–Jul. Disturbed sites; 0–500 m; introduced; Calif., Fla.; South America (Brazil, Uruguay); introduced also in Asia (India), Africa, Atlantic Islands, Indian Ocean Islands, Pacific Islands, Australia.

In Florida, *Solanum mauritianum* has become naturalized and common only at one site in Pasco County. It also occurs frequently in southern California from Santa Barbara south to San Diego with urban waifs in the Bay Area. It appears to be spreading into relatively undisturbed riparian areas in the San Gabriel Mountains and may become a widespread pest.

31. **Solanum nigrescens** M. Martens & Galeotti, Bull. Acad. Roy. Sci. Bruxelles 12(1): 140. 1845 • Divine nightshade

Herbs, perennial, erect to somewhat sprawling, unarmed, to 3 m, nearly glabrous to moderately pubescent, hairs unbranched, to 1 mm, eglandular. **Leaves** petiolate; petiole 0.5–2 cm; blade simple, ovate to ovate-elliptic, 4–10.5 × 2–5 cm, margins entire or shallowly sinuate-dentate, base decurrent. **Inflorescences** extra-axillary or leaf-opposed, unbranched, umbel-like to racemelike, (2–)5–10-flowered, 1–3.5 cm. **Pedicels** spreading and 0.5–1 cm in flower, spreading and 1–1.5 cm in fruit. **Flowers** radially symmetric; calyx not accrescent, unarmed, 1–2 mm, sparsely pubescent, lobes deltate; corolla white, rarely purplish, often with green or purplish central star, stellate, 1–1.5 cm diam., with sparse interpetalar tissue; stamens equal; anthers ellipsoidal, 2–3 mm, dehiscent by terminal pores that open into longitudinal slits; ovary glabrous. **Berries** dull green or purplish, globose, 0.5–0.8 cm diam., glabrous, with (4–)5–6(–13) sclerotic granules. **Seeds** tan, flattened, 1.2–1.5 × 1–1.1 mm, finely pitted. 2*n* = 24.

Flowering year-round. Deciduous and coniferous forests, fields, swampy areas; 0–1500 m; Ala., Fla., La., Miss., N.Mex., N.C., Tex.; Mexico; West Indies; Central America; South America.

Solanum nigrescens is widespread in Central and northern South America and the Caribbean and extends northward into the southeastern United States along the Gulf Coast and slightly inland. Where sympatric with *S. americanum*, it can be distinguished by its longer anthers and dull green or purplish berries with appressed to spreading calyx lobes. Plants collected as weeds in rice and sugarcane fields of Louisiana and provisionally identified as the Chinese species *S. merrillianum* T. N. Liou are somewhat intermediate between *S. americanum* and *S. nigrescens* and could represent recent hybrid populations (S. Knapp et al. 2019). *Solanum nigrescens* differs from *S. douglasii* in its shorter anthers and longer filaments relative to anther length; moreover, *S. douglasii* is usually found west of the Rocky Mountains, whereas *S. nigrescens* occurs in the southeastern United States. The ranges of *S. nigrescens* and *S. interius* overlap (for example, in Texas). *Solanum nigrescens* may be distinguished from *S. interius* by its usually acute calyx lobes, smaller seeds, and more numerous sclerotic granules in the fruits. *Solanum nigrescens* differs from *S. nigrum* in its more slender peduncles and pedicels, smaller seeds, and fruits with sclerotic granules.

32. Solanum nigrum Linnaeus, Sp. Pl. 1: 186. 1753 • Black nightshade, morelle noire [I] [W]

Solanum nigrum subsp. *schultesii* (Opiz) Wessely

Herbs, annual or perennial, erect or sprawling, unarmed, to 1 m, glabrescent to moderately pubescent, hairs unbranched, to 1 mm, usually eglandular or occasionally glandular. **Leaves** petiolate; petiole 0.5–3 cm; blade simple, ovate, 3.5–7 × 2.2–5 cm, margins entire to coarsely toothed with 3–5 lobes per side, base truncate to cuneate. **Inflorescences** extra-axillary, unbranched or occasionally forked, racemelike, (3–)4–10-flowered, 1–2 cm. **Pedicels** 0.5–1 cm, spreading to occasionally recurved in flower and fruit. **Flowers** radially symmetric; calyx not accrescent, unarmed, 1.5–2 mm, sparsely pubescent, lobes deltate, spreading to reflexed in fruit; corolla white with yellowish central star, stellate, 1–1.5 cm diam., without interpetalar tissue; stamens equal; anthers ellipsoidal, (1.8–)2–2.5 mm, dehiscent by terminal pores that open into longitudinal slits; ovary glabrous. **Berries** dull or slightly shiny purple-black or green to yellowish green, globose, 0.5–1 cm diam., glabrous, without sclerotic granules. **Seeds** yellow, flattened, 1.8–2 × 1.5–1.6 mm, minutely pitted. 2*n* = 72.

Flowering May–Oct. Disturbed areas, irrigated fields; 0–2200 m; introduced; B.C., N.S., Ont.; Alaska, Calif., D.C., Fla., Ga., Idaho, Iowa, Maine, Md., Mass., Mo., Mont., Nev., N.J., N.Y., N.C., Okla., Oreg., Pa., Tex., Utah, Va., Wash.; Eurasia, n Africa; introduced also in Pacific Islands (New Zealand), Australia.

Many regional floras have used *Solanum nigrum* as the name for various species in the black nightshade group, and it can be difficult to distinguish this species from *S. emulans* and *S. nigrescens*. *Solanum nigrum* can be distinguished from the native North American species of the black nightshade group (*S. americanum*, *S. douglasii*, *S. emulans*, *S. interius*, *S. nigrescens*) by its thicker peduncles and pedicels, larger seeds, and fruits lacking sclerotic granules. *Solanum nigrum* was probably introduced from northern Europe and has been locally naturalized in North America.

33. Solanum nitidibaccatum Bitter, Repert. Spec. Nov. Regni Veg. 11: 208. 1912 • Hoe or hairy nightshade

Solanum physalifolium Rusby var. *nitidibaccatum* (Bitter) Edmonds

Herbs, annual, erect or prostrate, unarmed, to 0.2(–0.4) m, moderately to densely pubescent, hairs unbranched, 1.5–2 mm, glandular. **Leaves** petiolate; petiole 0.5–3 cm; blade simple, ovate to lanceolate, 2–10 × 1–5 cm, margins entire to sinuate-dentate, base cuneate to decurrent. **Inflorescences** usually extra-axillary, occasionally leaf-opposed, unbranched, 4–8(–10)-flowered, 1–2 cm. **Pedicels** spreading to reflexed and 0.4–1 cm in flower and fruit. **Flowers** radially symmetric; calyx accrescent and covering ca. one-half berry, unarmed, 3–4 mm, sparsely to moderately pubescent, lobes broadly triangular; corolla white with yellowish central star edged with reddish purple to dark brown, rotate-stellate, 0.5–1 cm diam., with sparse interpetalar tissue; stamens equal; anthers ellipsoidal, 1–1.4 mm, dehiscent by terminal pores that open into longitudinal slits; ovary glabrous. **Berries** shiny greenish to purplish brown, globose, 0.5–1 cm diam., glabrous, with (0–)2–3 sclerotic granules. **Seeds** yellow to brown, flattened, 1.5–2.5 × 1.5–2 mm, minutely pitted. 2*n* = 24.

Flowering May–Oct. Disturbed areas, fields; (0–) 1200–2500 m; B.C., Man., N.B., Ont., Que.; Alaska, Ariz., Ark., Calif., Colo., Idaho, Mass., Minn., Mo., Mont., Nev., N.Mex., N.Y., N.C., N.Dak., Oreg., Pa., Tex., Utah, Wash., Wis., Wyo.; South America (Argentina, Chile); introduced in Europe, Africa, Pacific Islands (New Zealand), Australia.

Solanum nitidibaccatum has often been confused with and misidentified as *S. sarrachoides*, which has a much longer fruiting calyx that nearly covers the mature berry. *Solanum nitidibaccatum* also differs from *S. sarrachoides* in its smaller leaves, larger number of flowers per inflorescence (four to eight versus three or four in *S. sarrachoides*), and fruits with usually two or three sclerotic granules (versus four to six in *S. sarrachoides*).

Most references to *Solanum sarrachoides* in North American floras are actually *S. nitidibaccatum*. *Solanum nitidibaccatum* has also been confused with *S. villosum* Miller (R. L. McGregor 1986). J. M. Edmonds (1986) regarded *S. nitidibaccatum* as a variety of *S. physalifolium*, but the two taxa are now recognized as distinct species, with *S. physalifolium* restricted to South America. The name *S. physalifolium*, however, has been used for *S. nitidibaccatum* in a number of North American floras.

Solanum nitidibaccatum is currently considered to be native to both North and South America. It is a common weed in cultivated fields in the Great Plains, Pacific Northwest, and adjacent parts of Canada.

SELECTED REFERENCE Edmonds, J. M. 1986. Biosystematics of *Solanum sarrachoides* Sendtner and *S. physalifolium* Rusby (*S. nitidibaccatum* Bitter). Bot. J. Linn. Soc. 92: 1–38.

34. Solanum novomexicanum (Bartlett) S. R. Stern, J. Bot. Res. Inst. Texas 8: 6. 2014 • New Mexico nightshade [E]

Solanum heterodoxum Dunal var. *novomexicanum* Bartlett, Proc. Amer. Acad. Arts 44: 628. 1909; *Androcera novomexicana* (Bartlett) Wooton & Standley

Herbs, annual, spreading, sparsely to moderately armed, 0.3–0.7 m, prickles whitish or yellowish, straight, tapered, 3–8 mm, usually 20 or fewer per cm of stem, densely pubescent with stipitate-glandular hairs 0.2–0.4 mm, abaxial leaf surfaces also with scattered, sessile, 4–6-rayed, stellate hairs, central ray equal to lateral rays. **Leaves** petiolate; petiole 2–7 cm; blade simple to compound, broadly ovate to deltate, 4–11 × 4–8 cm, margins bipinnately lobed to divided with 2–3 main leaflets per side, leaflets with obtuse or rounded lobes, base truncate. **Inflorescences** extra-axillary, unbranched, 5–9-flowered, 4–10 cm. **Pedicels** 1–1.5 cm in flower, erect and 1–1.5 cm in fruit. **Flowers** bilaterally symmetric; calyx accrescent and tightly covering fruit, densely prickly, 4.5–6.5 mm, densely glandular-pubescent, lobes lanceolate; corolla violet or blue, pentagonal-stellate, with narrowly deltate lobes, 1–1.5 cm diam., with sparse interpetalar tissue; stamens unequal, lowermost much longer and curved; anthers narrow and tapered, dehiscent by terminal pores, short anthers 2–4 mm, longer anther 3.5–5 mm; ovary glabrous. **Berries** brown, globose, 1–1.2 cm diam., glabrous, dry, without sclerotic granules. **Seeds** dark brown, flattened, 2.5–3 × 2–2.5 mm, reticulately wrinkled or merely undulate.

Flowering Jun–Sep. Gravelly or sandy soils, open hillsides, arroyo banks, roadsides; 1900–2300 m; N.Mex.

Solanum novomexicanum is uncommon and endemic to the mountains of northcentral New Mexico.

35. Solanum perplexum Small, Man. S. E. Fl., 1115, 1508. 1933 [E]

Herbs, perennial, erect, sparsely to moderately armed, to 1 m, prickles cream to yellowish, straight or slightly curved, to 15 mm, nearly glabrous or sparsely to densely pubescent, hairs yellowish, sessile to short-stalked, stellate, (4–)6–8-rayed, central ray 1–2-celled and longer than lateral rays. **Leaves** petiolate; petiole 1–6 cm; blade simple, broadly ovate, 7–22 × 8–18 cm, margins shallowly to deeply lobed with 2–5 lobes per side, lobe margins entire to coarsely lobed, base truncate to cuneate and often oblique. **Inflorescences** extra-axillary, forked to several times branched, to 15-flowered, 7–15 cm. **Pedicels** 1–2 cm in flower, curved downward and to ca. 2.4 cm in fruit. **Flowers** radially symmetric; calyx not accrescent, unarmed or sparsely prickled, 7–13 mm, densely stellate-pubescent, lobes ovate-lanceolate; corolla lavender, stellate to stellate-pentagonal or rotate-stellate, 2–4.4 cm diam., with sparse to moderate interpetalar tissue at margins and base of lobes; stamens equal; anthers narrow and tapered, 4–10 mm, dehiscent by terminal pores; ovary glabrous. **Berries** yellow, subglobose, 1.8–3.5 × 2–4 cm, glabrous, without sclerotic granules. **Seeds** yellow, flattened, ca. 2 × 2.5 mm, minutely pitted. $2n$ = ca. 72.

Flowering May–Aug. Disturbed areas, peanut and cotton fields, roadsides, grazed pastures, urban waste areas; 90–200 m; Ala., Fla., Ga., Miss.

Solanum perplexum is similar to *S. dimidiatum* and was placed in synonymy with *S. dimidiatum* by W. G. D'Arcy (1974). The two species can be distinguished by their indumentum [golden stellate hairs with six to eight (rarely as few as four) lateral rays with the central ray one- or two-celled and longer than lateral rays in

S. perplexum versus whitish stellate hairs with six to ten (rarely as few as four) lateral rays with the central ray one-celled and equal to or shorter than lateral rays in *S. dimidiatum*], the larger prickles on the stems and leaves (up to 15 mm in *S. perplexum* versus up to 6.5 mm in *S. dimidiatum*), and the larger leaves (up to 22 × 18 cm in *S. perplexum* versus up to 16 × 10 cm in *S. dimidiatum*).

Solanum perplexum occurs mainly in the region where the borders of Alabama, Florida, and Georgia meet, with a single outlying population known from western Mississippi.

36. Solanum pseudocapsicum Linnaeus, Sp. Pl. 1: 184. 1753 • Jerusalem- or winter-cherry [I] [W]

Solanum capsicastrum Link ex Schauer

Shrubs, erect, unarmed, to 1 m, glabrous to densely pubescent, hairs dendritically branched. **Leaves** petiolate; petiole 0.2–1 cm; blade simple, elliptic, 1–9 × 0.5–4.5 cm, margins entire, base acute to attenuate. **Inflorescences** leaf-opposed, unbranched, 1–8-flowered, 0.2–1 cm. **Pedicels** 0.3–0.7 cm in flower, 0.8–1 cm and erect in fruit. **Flowers** radially symmetric; calyx somewhat accrescent, unarmed, 2.5–6 mm, glabrous to densely pubescent with dendritic hairs, lobes long-triangular; corolla white, stellate, 1–1.5(–2.5) cm diam., without interpetalar tissue; stamens equal; anthers oblong, 3–4 mm, dehiscent by terminal pores that open into longitudinal slits; ovary glabrous. **Berries** yellow to orange or red, globose, 1–2 cm diam., glabrous, without sclerotic granules. **Seeds** yellowish, flattened with thickened margins, 3–4 × 2.5–3 mm, minutely pitted. $2n = 24$.

Flowering May–Sep. Disturbed sites; 0–1000 m; introduced; Fla., Tex.; Mexico; Central America; South America; often escaped in tropical and subtropical countries worldwide.

Solanum pseudocapsicum is native from Mexico to Argentina, southern Brazil, and Uruguay. It is grown as an ornamental for its showy fruits, especially around Christmas. It occasionally escapes from cultivation in southern Florida and Texas. In Texas, it has become established and fairly common in Austin, in the Lower Rio Grande Valley, and in Goliad and Caldwell counties. Cultivated forms are usually glabrous, but some can have branched pubescence. The fruits are mildly poisonous when ingested by humans but can be highly toxic to dogs and some birds.

Solanum pseudocapsicum, along with *S. diphyllum*, has a distinctive leaf arrangement in which a longer, narrower leaf is paired with a shorter, often more rounded one.

37. Solanum pseudogracile Heiser, Bot. J. Linn. Soc. 76: 294. 1978 • Glowing nightshade [E]

Herbs or shrubs, annual or perennial, erect, unarmed, to 1 m, sparsely to moderately pubescent, hairs unbranched, to 1 mm, eglandular. **Leaves** petiolate; petiole 0.5–3 cm; blade simple, elliptic to lanceolate, 1.5–8 × 1–4 cm, margins entire or nearly so, base cuneate to attenuate. **Inflorescences** extra-axillary, unbranched or rarely forked, umbel-like, 3–8-flowered, 1–2 cm. **Pedicels** 0.5–1 cm in flower and fruit, recurved to reflexed in fruit. **Flowers** radially symmetric; calyx not accrescent, unarmed, 1.5–3 mm, sparsely pubescent, lobes deltate, reflexed in fruit; corolla white with yellowish central star, stellate, 1–1.5 cm diam., with sparse interpetalar tissue; stamens equal; anthers ellipsoidal, 2–3 mm, dehiscent by terminal pores that open into longitudinal slits; ovary glabrous. **Berries** dull purplish black, globose, 0.5–1.5 cm diam., glabrous, without (or rarely with 2) sclerotic granules. **Seeds** pale yellow, flattened, 1–1.3 × 0.8–0.9 mm, minutely pitted. $2n = 24$.

Flowering May–Oct (year-round in Fla.). Coastal dunes, margins of maritime forests, brackish marshes; 0–400 m; Ala., Fla., Ga., La., Miss., N.C., S.C., Tex.

Solanum pseudogracile is very similar to and perhaps not distinct from *S. chenopodioides*. It is ecologically distinctive, occurring in sand dunes and salt marshes of the Atlantic and eastern Gulf Coastal Plain and inland in some parts of Florida and Georgia.

38. Solanum pumilum Dunal in A. P. de Candolle and A. L. P. P. de Candolle, Prodr. 13(1): 287. 1852 • Dwarf horsenettle [C] [E]

Solanum hirsutum Nuttall, J. Acad. Nat. Sci. Philadelphia 7: 109. 1834, not Dunal 1813; *S. carolinense* Linnaeus var. *hirsutum* A. Gray

Herbs, perennial, erect, sparsely armed, to 0.2 m, prickles cream to yellowish, straight, to 3.5 mm, moderately to densely pubescent, hairs whitish, sessile, stellate, 4–8-rayed, central ray (1–)2–5-celled and longer than lateral rays. **Leaves** petiolate; petiole 0.2–1 cm; blade simple, elliptic to obovate, 2.2–8.6 × 1.1–5.1 cm, margins entire, sinuate, or shallowly lobed with 2–6 lobes per side, lobe margins entire, base cuneate to attenuate. **Inflorescences** extra-axillary, unbranched, 1–4-flowered, 3–7 cm. **Pedicels** 1–3.5 cm in flower and fruit. **Flowers** radially symmetric; calyx not accrescent, unarmed or sparsely prickly, 6–7 mm, densely stellate-pubescent, lobes triangular; corolla white, stellate to

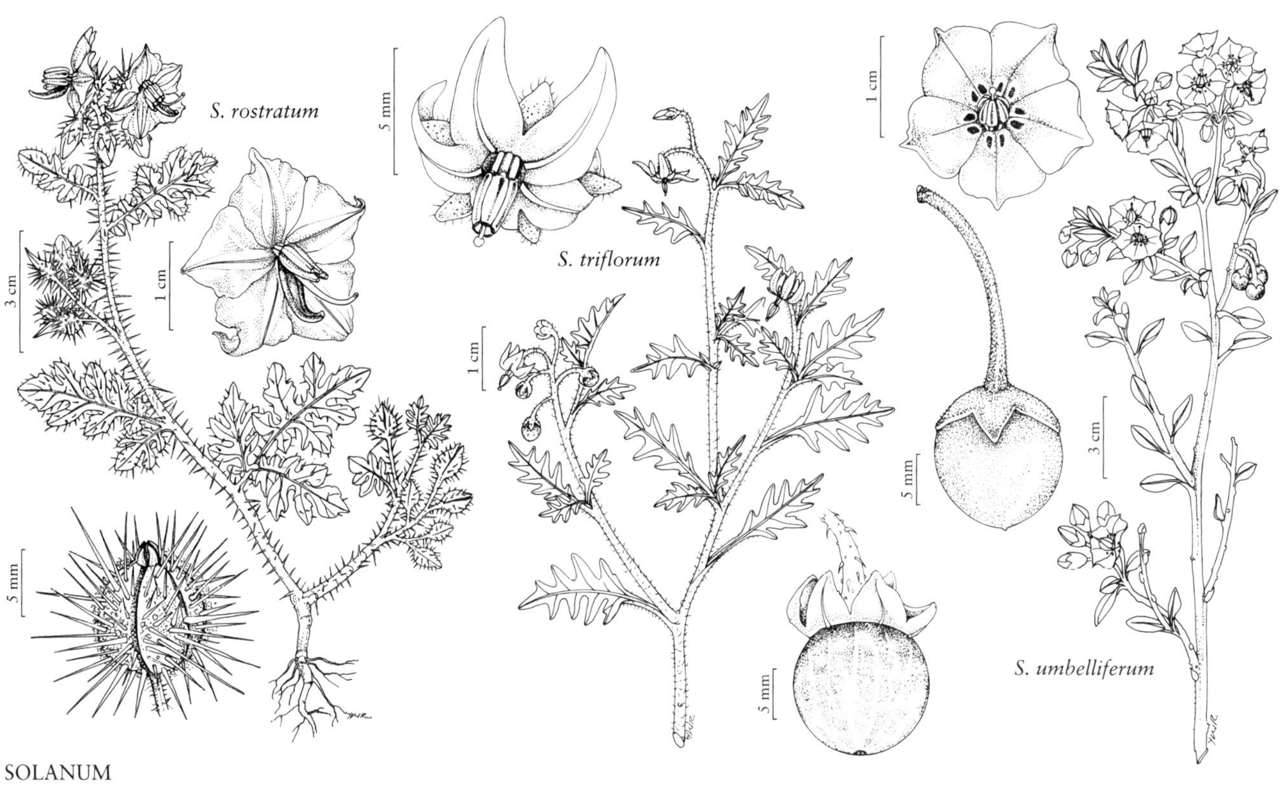

stellate-pentagonal, 1.8–3 cm diam., with abundant interpetalar tissue at margins and base of lobes; stamens equal; anthers narrow and tapered, 6–7 mm, dehiscent by terminal pores; ovary glabrous. **Berries** unknown. **Seeds** unknown.

Flowering Apr–May; of conservation concern; 80–200 m; Ala., Ga.

Solanum pumilum is endemic to Ketona dolomite outcroppings near the Little Cahaba River in Bibb County and on amphibolite outcroppings near the Coosa River in Chilton and Coosa counties in Alabama. It was collected originally from Georgia in the 1830s from Baldwin and Muskogee counties.

39. Solanum rostratum Dunal, Hist. Nat. Solanum, 234, plate 24. 1813 • Buffalobur nightshade, buffalobur, Kansas thistle, mala mujer [F] [W]

Androcera rostrata (Dunal) Rydberg; *Solanum heterandrum* Pursh

Herbs, annual, erect, moderately to densely armed, to 1 m, prickles yellow, straight, to 12 mm, moderately to densely pubescent, hairs sessile to long-stalked, stellate, 6–10-rayed, central ray equal to or longer than lateral rays. **Leaves** petiolate; petiole 2–10 cm; blade simple to twice-compound, ovate to elliptic, (2–)4–16 × 3–12 cm, margins lobed to 1–2 times divided with 2–4 main leaflets per side, leaflets with deep, rounded lobes, base truncate to subcordate. **Inflorescences** extra-axillary, unbranched, 5–12-flowered, 4–11 cm. **Pedicels** 0.5–1.5 cm in flower, erect and 0.5–1.5 cm in fruit. **Flowers** bilaterally symmetric; calyx accrescent and tightly covering fruit, densely prickly or bristly, 7.5–12.5 mm, densely stellate-pubescent, lobes linear to lanceolate; corolla yellow, rotate-pentagonal, 1.5–3.5 cm diam., with abundant interpetalar tissue; stamens unequal, lowermost much longer and curved; anthers narrow and tapered, dehiscent by terminal pores, short anthers 4, yellow, 6–8 mm, longer anther reddish or purplish, 10–14 mm; ovary glabrous. **Berries** brown, globose, 1–1.2 cm diam., glabrous, dry, without sclerotic granules. **Seeds** dark brown, flattened, 2–3 × 1.8–2 mm, minutely pitted and irregularly ridged. **2*n*** = 24.

Flowering year-round. Disturbed sites, versatile in soil tolerance, roadsides, pasturelands; 0–2500 m; B.C., N.B., N.S., Ont., Que.; Ala., Alaska, Ariz., Ark., Calif., Colo., Conn., Del., D.C., Ga., Idaho, Ill., Ind., Iowa, Kans., Ky., La., Maine, Md., Mass., Mich., Minn., Miss., Mo., Mont., Nebr., Nev., N.H., N.J., N.Mex., N.Y., N.C., N.Dak., Ohio, Okla., Oreg., Pa., R.I., S.C., S.Dak., Tenn., Tex., Utah, Vt., Va., Wash., W.Va., Wis., Wyo.; Mexico.

Solanum rostratum is widespread in the central Mexican highlands from Chihuahua and Coahuila to

Puebla and Oaxaca, and its native range likely extends from Mexico City north to the United States Great Plains. It is widely introduced outside this presumed area of origin.

Solanum rostratum is considered a noxious weed in several states. It is often invasive in gardens, pastures, and disturbed areas. The plants are extremely spiny, and there are reports of pigs being poisoned by eating the berries and roots. This species is thought to be the original host of the Colorado potato beetle (*Leptinotarsa decemlineata*) before potatoes were widely cultivated in the western and central United States. The beetle then adopted potatoes as its primary host and rapidly spread eastward.

40. Solanum sarrachoides Sendtner in C. F. P. von Martius et al., Fl. Bras. 10: 18, plate 1, figs. 1–8. 1846 • Viscid nightshade [I] [W]

Herbs, annual, erect to decumbent, unarmed, to 1 m, moderately to densely pubescent, hairs unbranched, to 2 mm, glandular. **Leaves** petiolate; petiole 0.5–3 cm; blade simple, ovate to elliptic, 3–7.5 × 3–6 cm, margins entire to sinuate-dentate, base truncate to cordate. **Inflorescences** leaf-opposed or occasionally extra-axillary, unbranched, umbel-like, 2–5(–7)-flowered, 0.5–1.5 cm. **Pedicels** 0.5–1 cm in flower and fruit, spreading in flower, reflexed in fruit. **Flowers** radially symmetric; calyx accrescent and nearly covering berry, unarmed, 2–3 mm, sparsely to moderately pubescent, lobes narrowly triangular; corolla white with yellowish or greenish central star, rotate-stellate, 0.5–1 cm diam., with abundant interpetalar tissue; stamens equal; anthers ellipsoidal, 1.2–2 mm, dehiscent by terminal pores that open into longitudinal slits; ovary glabrous. **Berries** dull pale green, globose, 0.5–1 cm diam., glabrous, with 4–6 sclerotic granules. **Seeds** pale yellow, flattened, 1.5–2 × 1–1.5 mm, nearly smooth. $2n = 24$.

Flowering May–Oct. Farmyards, fields, open woodlands, roadsides, disturbed areas; 0–500 m; introduced; Ark., Conn., Fla., Ill., Kans., Md., Mo., N.C., Okla., R.I., S.C., Va., Wash.; s South America (Argentina, Brazil, Paraguay); introduced also in Europe, Africa (South Africa).

Many accounts of *Solanum sarrachoides* in North America actually refer either to *S. nitidibaccatum* or to a mixture of the two species. In North America, *S. sarrachoides* is much less widespread and common than *S. nitidibaccatum*.

SELECTED REFERENCE Edmonds, J. M. 1986. Biosystematics of *Solanum sarrachoides* Sendtner and *S. physalifolium* Rusby (*S. nitidibaccatum* Bitter). Bot. J. Linn. Soc. 92: 1–38.

41. Solanum seaforthianum Andrews, Bot. Repos. 8: plate 504. 1808 • Brazilian nightshade [I]

Climbing or scrambling vines, woody, unarmed, to ca. 3 m, glabrous or sparsely pubescent, hairs white, unbranched, ca. 0.2 mm. **Leaves** petiolate; petioles twining around supports, 1–4 cm; blade simple to compound, elliptic to broadly ovate, (2–)3.5–10(–13) × (1–)2–9(–11) cm, margins entire to divided with up to 4 pairs of leaflets, leaflet margins entire, base truncate or slightly cordate. **Inflorescences** terminal, becoming lateral, extra-axillary, much-branched, to 100+-flowered, to 25+ cm. **Pedicels** inserted into small sleeve on inflorescence axis, 0.8–1.4 cm in flower and fruit. **Flowers** radially symmetric; calyx not accrescent, unarmed, ca. 0.5 mm, nearly truncate, glabrous or sparsely pubescent on lobe tips; corolla purple, stellate, 1–2.5 cm diam., with sparse interpetalar tissue; stamens unequal due to unequal filaments; anthers ellipsoidal, 2–3 mm, dehiscent by terminal pores; ovary glabrous. **Berries** bright shiny red, globose, 0.8–1.5 cm diam., glabrous, without sclerotic granules. **Seeds** pale yellowish tan, flattened, 4–4.5 × 2.5-3 mm, minutely pitted. $2n = 24$.

Flowering year-round. Disturbed sites; 0–200 m; introduced; Fla.; Mexico; West Indies; Central America (Belize, Costa Rica, Guatemala, Honduras, Nicaragua, Panama); South America (Brazil, Colombia, Ecuador, Venezuela); introduced also elsewhere in South America (Argentina, Paraguay, Peru), Asia, Africa, Pacific Islands, Australia.

Solanum seaforthianum is widely cultivated as an ornamental and sporadically escapes in Florida. A similar species with twining petioles, *S. laxum* Sprengel, is occasionally cultivated in California. It is distinguished from *S. seaforthianum* by having tufts of hairs in the vein axils of the abaxial leaf surfaces, white rather than violet corollas, and equal stamens.

42. Solanum setigeroides (Whalen) S. R. Stern, J. Bot. Res. Inst. Texas 8: 5. 2014 • Bristly nightshade

Solanum heterodoxum Dunal var. *setigeroides* Whalen, Wrightia 5: 237. 1976

Herbs, annual, spreading, branching from near base, densely armed, 0.3–0.7 m, prickles straight, 4–8 mm, 30+ per cm of stem, sparsely pubescent, hairs ca. 0.2 mm, stipitate-glandular, abaxial leaf surfaces also with scattered, sessile, 4–6-rayed, stellate hairs, central ray equal to lateral rays. **Leaves** petiolate; petiole 2–7 cm;

blade simple to compound, broadly ovate to deltate, 4–11 × 4–8 cm, margins twice-lobed to twice-divided with 2–3 main leaflets per side, leaflets with obtuse or rounded lobes, base truncate. **Inflorescences** extra-axillary, unbranched, 5–9-flowered, 4–10 cm. **Pedicels** 1–2 cm in flower, erect and 1–2 cm in fruit. **Flowers** bilaterally symmetric; calyx accrescent and tightly covering fruit, densely prickly, 4.5–6.5 mm, sparsely glandular-pubescent, lobes lanceolate; corolla violet or blue, pentagonal, 1–1.5 cm diam., with abundant interpetalar tissue; stamens unequal, lowermost much longer and curved; anthers narrow and tapered, dehiscent by terminal pores, short anthers 2–4 mm, longer anther 3.5–5 mm; ovary glabrous. **Berries** brown, globose, 1–1.2 cm diam., glabrous, dry, without sclerotic granules. **Seeds** dark brown, flattened, 2.5–3 × 2–2.5 mm, minutely pitted and weakly ridged or faceted.

Flowering Jun–Oct. Silty, sandy, or gravelly soils, playas, dunes, streambeds, arroyos, open hillsides; 600–2000 m; Ariz., N.Mex., Tex.; Mexico (Chihuahua).

Solanum setigeroides is a weed of disturbed and overgrazed places ranging from central Arizona and New Mexico to extreme western Texas.

43. Solanum sisymbriifolium Lamarck, Tabl. Encycl. 2: 25. 1794 (as sisymbrifolium) • Sticky nightshade [I] [W]

Herbs, annual, erect, sparsely to moderately armed, 1–1.5 m, prickles yellow, straight or curved, 1–15 mm, densely pubescent, hairs unbranched, glandular and stellate, sessile, 4–7-rayed, central ray glandular or eglandular, longer than lateral rays. **Leaves** petiolate; petiole 2–5 cm; blade simple to compound, broadly ovate, 8–15 × 3–8.5 cm, margins deeply lobed to divided with 4–7 lobes or leaflets per side, lobes or leaflets with rounded to acute lobes, base acute or cordate. **Inflorescences** extra-axillary, unbranched, 4–11-flowered, 4–15 cm. **Pedicels** 0.5–1 cm in flower, 1.5–2.5 cm in fruit. **Flowers** radially symmetric; calyx accrescent, moderately prickly, 6–9 mm, densely pubescent, hairs simple or sessile and stellate, glandular and eglandular, lobes subtending to almost completely and loosely covering fruit at maturity, deltate; corolla white or pale blue, rotate-pentagonal, 2–3 cm diam., with abundant interpetalar tissue; stamens equal or nearly so; anthers narrow and tapered, 8–10 mm, dehiscent by terminal pores; ovary glabrous to sparsely glandular-pubescent. **Berries** bright red, globose, 1–2 cm diam., glabrous to sparsely glandular-pubescent, juicy, without sclerotic granules. **Seeds** pale yellow, plump, 3–3.5 × 2–3 mm, minutely pitted. $2n = 24$.

Flowering Feb–Oct. Disturbed sites; 0–100 m; introduced; Ont.; Ala., Ariz., Calif., Del., Fla., Ga., La., Mass., Miss., N.J., N.Y., N.C., Pa., S.C., Tex., Va.; South America (Argentina, Bolivia, Brazil, Paraguay); introduced also in Mexico, Central America (Costa Rica), nw South America, Europe, Asia (China, India), Africa.

The bright red fruits of *Solanum sisymbriifolium* are edible, and the plants are used in pest control and as a nematode and beetle trap in Europe and the United Kingdom; however, cultivation of this species should be discouraged due to its invasive potential. Reports of this species from Oregon are old; it is not naturalized there.

44. Solanum stoloniferum Schlechtendal, Linnaea 8: 255. 1833 • Wild potato

Solanum fendleri A. Gray; *S. fendleri* subsp. *arizonicum* Hawkes; *S. fendleri* var. *texense* Correll; *S. leptosepalum* Correll

Herbs, perennial, erect, bearing tubers to 3 cm, unarmed, to 0.7 m, sparsely to densely pubescent to strigose, hairs unbranched, eglandular. **Leaves** petiolate; petiole 1.5–4 cm, with pair of entire lunate pseudostipules at base; blade compound, elliptic to ovate, 7.5–20 × 3.5–8 cm, margins divided with 1–4 pairs of leaflets, these sometimes interspersed with smaller, interjected leaflets, lowermost leaflets sometimes greatly reduced in size, leaflet margins entire, base cuneate to cordate. **Inflorescences** terminal, extra-axillary or leaf-opposed, generally forked or 3-fid, 3–26-flowered, to 10 cm. **Pedicels** articulated near middle, 1.1–3.7 cm in flower and fruit. **Flowers** radially symmetric; calyx not accrescent, unarmed, 4–8 mm, lobes deltate-acuminate; corolla purple, blue, pale pink, or rarely white, pentagonal to rotate, 1.8–3.3 cm diam., with abundant interpetalar tissue; stamens equal; anthers oblong, slightly tapered, 3.5–6.5 mm, dehiscent by terminal pores that open into longitudinal slits; ovary glabrous. **Berries** green, sometimes with dark green stripes or white spots, globose or slightly ovoid, 0.9–1.7 cm diam., glabrous, without sclerotic granules. **Seeds** greenish white, rounded, 1–2 mm diam., rugose. $2n = 48$.

Flowering Jul–Oct(–Nov). Hillsides, stream bottoms, sandy soils, disturbed areas in grasslands, pinyon-juniper forests, alpine meadows, coniferous and deciduous forests; 1400–3100 m; Ariz., N.Mex., Tex.; Mexico.

Solanum stoloniferum is widespread in highland Mexico. Its northern range extends into New Mexico, southeastern Arizona, and southwestern Texas. It is one of the most common and polymorphic species of wild potatoes in North America and Mexico.

45. Solanum tampicense Dunal in A. P. de Candolle and A. L. P. P. de Candolle, Prodr. 13(1): 284. 1852 • Scrambling or wetland nightshade, aquatic soda apple [I] [W]

Shrubs, scandent, branches 1–2+ m, moderately to densely armed, prickles yellow, recurved, 2–8 mm, glabrate to sparsely pubescent, hairs tan to reddish, stellate, 3–6-rayed, central ray equal to lateral rays. **Leaves** petiolate; petiole 1–4 cm; blade simple, ovate, 4–16 × 2–6 cm, margins shallowly to moderately lobed with 2–5 lobes per side, lobe margins entire, base cuneate and slightly decurrent. **Inflorescences** extra-axillary, unbranched, 3–10-flowered, 1–4 cm. **Pedicels** 0.5–2 cm in flower and fruit. **Flowers** radially symmetric; calyx not accrescent, sometimes prickly, 2–5 mm, glabrous, lobes narrowly triangular; corolla white to cream, stellate, 1.5–2 cm diam., without interpetalar tissue; stamens equal; anthers narrow and tapered, 4–6 mm, dehiscent by terminal pores; ovary glabrous. **Berries** red, globose, 0.5–0.8 cm diam., glabrous, without sclerotic granules. **Seeds** yellow, flattened, 1.5–2 × 1–1.5 mm, minutely pitted and ridged.

Flowering Feb–Dec. Swamps, riverbanks, wet areas; 0–200 m; introduced; Fla., Tex.; Mexico; West Indies (Cuba, Lesser Antilles); Central America; South America (Colombia, Venezuela).

In the flora area, *Solanum tampicense* is found in central and southern Florida. A single population was found in 2016 in Cameron County, Texas. It was first collected in Florida in 1983, and although not common, it has the potential to become invasive. It is listed by the United States federal government and several states as a noxious weed. It is a species of riverbanks and swamps, where it can form impenetrable, spiny thickets.

SELECTED REFERENCE Wunderlin, R. et al. 1993. *Solanum viarum* and *S. tampicense* (Solanaceae): Two weedy species new to Florida and the United States. Sida 15: 605–611.

46. Solanum tenuipes Bartlett, Proc. Amer. Acad. Arts 44: 629. 1909 • Fancy nightshade

Herbs, perennial, spreading, moderately to densely armed, 0.2–0.5 m, prickles pale, straight, to 8 mm, glabrate to sparsely or moderately pubescent, hairs unbranched, glandular or eglandular, abaxial leaf surfaces with some sessile stellate hairs, 4–6-rayed, central ray equal to lateral rays. **Leaves** petiolate; petiole 2–6 cm; blade simple to 2–3 times lobed to compound, broadly ovate to elliptic, 4–9 × 2–7 cm, margins 2–3 times lobed to 2–3 times divided with 2–3 main leaflets per side, leaflets lobed to pinnately dissected, base truncate. **Inflorescences** extra-axillary, unbranched, 6–9-flowered, 4–9 cm. **Pedicels** 1–2 cm in flower, erect and 2–3 cm in fruit. **Flowers** bilaterally symmetric; calyx accrescent and almost completely and tightly covering fruit, moderately to densely prickly, 4–6 mm, sparsely to moderately pubescent, lobes linear to lanceolate; corolla violet or blue, stellate, 2.5–3.5 cm diam., with sparse interpetalar tissue; stamens unequal, lowermost much longer and curved; anthers narrow and tapered, dehiscent by terminal pores, short anthers 7–10 mm, longer anther 12–20 mm; ovary glabrous. **Berries** brown, globose to depressed-globose, 0.7–1 cm diam., glabrous, dry, without sclerotic granules. **Seeds** dark brown, plump, 2.8–3.6 × 2–3 mm, minutely pitted and irregularly ridged. $2n = 24$.

Varieties 2 (2 in the flora): Texas, nc Mexico.

1. Larger leaves 3 times pinnatifid; seeds 3.1–3.6 mm . . . . 46a. *Solanum tenuipes* var. *tenuipes*
1. Leaves usually only 2 times pinnatifid; seeds to 3 mm . . . . .46b. *Solanum tenuipes* var. *latisectum*

46a. Solanum tenuipes Bartlett var. **tenuipes**

Herbs usually much-branched from base, glandular-pubescent but often glabrate. **Leaves:** larger ones 3 times pinnatifid, ultimate lobes narrow, rounded, and often elongate. **Seeds** 3.1–3.6 mm.

Flowering Jun–Nov. Open desert, semidesert, calcareous or gypsum hills, bajadas, flats; (70–)600–1300 m; Tex.; Mexico (Coahuila, Durango, Nuevo León).

Variety *tenuipes* is found in Brewster, Crockett, Maverick, and Terrell counties.

46b. Solanum tenuipes Bartlett var. **latisectum** Whalen, Wrightia 5: 238. 1976

Herbs branched from corky, perennial base or with erect stems and branched above, viscid-glandular, hairs scattered along young stems, straight, spreading, uniseriate, ca. 0.5 mm. **Leaves:** larger ones 2 times pinnatifid, ultimate lobes obtuse or rounded and usually ± as broad as long. **Seeds** to 3 mm.

Flowering Apr. Open desert, semidesert, calcareous or gypsum hills, bajadas, flats; 800 m; Tex.; Mexico (Chihuahua, Coahuila, Durango).

Variety *latisectum* has rarely been collected in the United States, with only one collection known from 30 miles east of Presidio.

47. **Solanum torvum** Swartz, Prodr., 47. 1788, name proposed for conservation • Turkey berry [W]

Shrubs or trees, erect, sparsely to moderately armed, 1–4 m, prickles brownish, straight to recurved, 3–7 mm, moderately to densely pubescent, hairs sessile to short-stalked, stellate, 4–8-rayed, central ray shorter than or equal to lateral rays, moderately pubescent with unbranched, glandular hairs on inflorescences and calyces. **Leaves** petiolate; petiole 1–5 cm; blade simple, ovate to elliptic, 7–23 × 4–14 cm, margins subentire to coarsely lobed with 3–4 lobes per side, lobe margins entire to coarsely lobed, base truncate to subcordate and asymmetrical. **Inflorescences** extra-axillary, unbranched or branched, 10–20-flowered, to 6 cm. **Pedicels** 1–1.5 cm in flower, erect and 1.5–2.5 cm in fruit. **Flowers** radially symmetric; calyx slightly accrescent, unarmed, 4–8 mm, sparsely to moderately pubescent with unbranched, gland-tipped hairs, lobes lanceolate; corolla white, stellate, 2–3 cm diam., with sparse interpetalar tissue; stamens equal; anthers narrow and tapered, 6–9 mm, dehiscent by terminal pores; ovary glabrous or glandular-puberulent at apex. **Berries** green to yellow, globose, 1–1.5 cm diam., glabrous, without sclerotic granules. **Seeds** light brown, flattened, 2.5–3 × 2–2.5 mm, minutely pitted. $2n = 24$.

Flowering year-round in frost-free areas. Disturbed sites; 0–10 m; Ala., Fla.; Mexico; West Indies; Central America; South America (Brazil, Colombia, Ecuador, French Guiana, Guyana, Venezuela); introduced in Asia, Africa, Indian Ocean Islands, Pacific Islands, Australia.

Solanum torvum is listed by the United States federal government and several states as a noxious weed. It is occasional to rare in peninsular Florida and could possibly become invasive elsewhere in subtropical climates. The green fruits are used in Asian and West Indian cuisine, often as an addition to soups and curries.

48. **Solanum triflorum** Nuttall, Gen. N. Amer. Pl. 1: 128. 1818 • Cutleaf nightshade, morelle à trois fleurs [F] [W]

Herbs, annual, decumbent to prostrate, unarmed, to 0.4 m, fleshy, nearly glabrous to moderately pubescent, hairs unbranched, to 2 mm, eglandular, rarely glandular. **Leaves** petiolate; petiole 0.5–2.5 cm; blade simple, elliptic to oblong, 2–5 × 1–3 cm, margins shallowly lobed to deeply and regularly pinnatifid with 3–6 lobes per side, lobe margins entire or occasionally coarsely lobed, base cuneate and decurrent. **Inflorescences** extra-axillary, unbranched, umbel-like, 1–6-flowered, 1–3 cm. **Pedicels** spreading and 0.5–1.5 cm in flower, reflexed and 0.5–1.5 cm in fruit. **Flowers** radially symmetric; calyx accrescent and covering base of berry, unarmed, 2–4(–7) mm, moderately pubescent, lobes deltate, reflexed; corolla white or light purple with green or purplish central star, stellate, 0.5–1 cm diam., with sparse interpetalar tissue; stamens equal; anthers narrowly ellipsoidal, 2.5–4 mm, dehiscent by terminal pores that open into longitudinal slits; ovary glabrous. **Berries** shiny dark green to purplish black, globose, 0.8–2 cm diam., glabrous, with 13–30 sclerotic granules. **Seeds** yellow, plump, 2–3 × 2–2.5 mm, minutely pitted. $2n = 24$.

Flowering Apr–Sep. Disturbed areas, roadsides, stream banks, along railroad tracks, prairie dog towns; (0–)700–2900 m; Alta., B.C., Man., Sask.; Ariz., Calif., Colo., Idaho, Iowa, Kans., Mass., Mich., Minn., Mo., Mont., Nebr., Nev., N.Mex., N.Dak., Okla., Oreg., S.Dak., Tex., Utah, Wash., Wyo.; South America (Argentina); introduced in Europe, Africa, Australia.

Solanum triflorum is found in South America (Argentina) and is also considered to be native to central and western North America. It is occasionally adventive in the eastern United States. It is poisonous to livestock and can become a serious weed in cultivated fields, especially in the Great Plains.

49. **Solanum triquetrum** Cavanilles, Icon. 3: 30, plate 259. 1795 • Texas nightshade

Solanum lindheimerianum Scheele

Vines, semiwoody or scramblers with enlarged woody base, to 2 m, occasionally erect subshrubs to 0.5 m, unarmed, glabrous to densely pubescent, hairs usually ascending and pointing distally on stems, weak, unbranched, to 0.5 mm. **Leaves** petiolate; petiole 0.3–1.2 cm; blade simple, deltate to hastate or triangular, sometimes linear, (1–)1.8–5 × (0.3–)1–3.5 cm, margins entire to basally 2-lobed, lobe margins entire, base truncate to subcordate or hastate. **Inflorescences** terminal or lateral, leaf-opposed or occasionally extra-axillary, unbranched or occasionally forked, 3–6-flowered, 1–3 cm. **Pedicels** inserted into small sleeve on inflorescence axis, 0.6–1.2 cm in flower, 1–1.5 cm in fruit. **Flowers** radially symmetric; calyx not accrescent, unarmed, 2.5–3.5 mm, glabrous to sparsely pubescent, lobes triangular-acuminate; corolla white or tinged with purple, often with shiny green or greenish white eye, stellate, 1.5–2 cm diam., without interpetalar tissue; stamens equal; anthers oblong, slightly tapered, 3.5–4 mm, dehiscent by terminal pores that open into longitudinal slits; ovary glabrous. **Berries** bright shiny red, globose, 1–1.5 cm diam., glabrous, without sclerotic granules. **Seeds** reddish brown, plump-reniform to flattened, ca. 4 × 2.5 mm, minutely pitted. $2n = 24$.

Flowering year-round. Slopes, thickets, moist places; 0–1400 m; Tex.; Mexico (Coahuila, Durango, Hidalgo, Nuevo León, San Luis Potosí, Tamaulipas, Zacatecas).

Solanum triquetrum is widespread in central, southern, and western Texas. It could be confused with *S. dulcamara*, which also has shiny green dots at the corolla lobe bases, but the flowers of *S. triquetrum* are white and the leaves more sharply triangular. Leaf shape and size in *S. triquetrum* are extremely variable (S. Knapp 2013).

50. **Solanum umbelliferum** Eschscholtz, Mém. Acad. Imp. Sci. St. Pétersbourg Hist. Acad. 10: 283. 1826 • Blue witch, blue witch or chaparral nightshade F

Solanum clokeyi Munz; *S. obispoense* Eastwood; *S. parishii* A. Heller; *S. tenuilobatum* Parish; *S. umbelliferum* var. *clokeyi* (Munz) D. J. Keil; *S. umbelliferum* var. *glabrescens* Torrey; *S. umbelliferum* var. *hoffmannii* (Munz) D. J. Keil; *S. umbelliferum* var. *incanum* Torrey; *S. umbelliferum* var. *intermedium* (Parish) D. J. Keil; *S. umbelliferum* var. *montanum* (Munz) D. J. Keil; *S. umbelliferum* var. *obispoense* (Eastwood) D. J. Keil; *S. umbelliferum* var. *xanti* (A. Gray) D. J. Keil; *S. wallacei* (A. Gray) Parish var. *clokeyi* (Munz) McMinn; *S. xanti* A. Gray; *S. xanti* var. *glabrescens* Parish; *S. xanti* var. *hoffmannii* Munz; *S. xanti* var. *intermedium* Parish; *S. xanti* var. *montanum* Munz

Shrubs or subshrubs, erect or somewhat spreading, unarmed, to 1.5 m, glabrous to densely pubescent, hairs unbranched, to 2 mm, glandular or eglandular and dendritic. **Leaves** petiolate; petiole 0.2–1.5(–3) cm; blade simple, lanceolate to ovate or obovate, (0.5–)1–4(–9) × 0.5–2(–6.5) cm, margins entire to pinnatifid with 1(–3) pairs of lobes at base, lobe margins entire to undulate, base attenuate to truncate, occasionally subcordate. **Inflorescences** terminal or lateral, leaf-opposed or extra-axillary, simple or once-branched, 5–20-flowered, 1–8 cm. **Pedicels** inserted into small sleeve on inflorescence axis, 0.5–1.5 cm in flower, 1.2–2 cm in fruit. **Flowers** radially symmetric; calyx slightly accrescent, unarmed, 2.5–5.5 mm, glabrous to densely pubescent, hairs unbranched or dendritic, lobes broadly deltate; corolla pale to deep purple or occasionally white, with green spots edged with white at base of lobes, spots separate or confluent, rotate, (1–)1.3–2.5 cm diam., with abundant interpetalar tissue; stamens equal; anthers ellipsoidal, slightly tapered, 3.5–4.5 mm, dehiscent by terminal pores that open into longitudinal slits; ovary glabrous. **Berries** green, greenish black, or black, globose, 1–2 cm diam., glabrous, without sclerotic granules. **Seeds** reddish brown, flattened, ca. 2 × 1.5 mm, minutely pitted.

Flowering Feb–Nov (most of the year in California). Sand dunes, chaparral, coastal sage scrub, rocky slopes, pine forests; 0–2000 m; Ariz., Calif., Nev., Oreg., Wash.; Mexico (Baja California).

Solanum umbelliferum is common in the western part of North America from Washington to Baja California. It is found throughout California except for Modoc Plateau, Desert Province, and Central Valley.

Past treatments have divided *Solanum umbelliferum* into a number of taxa based on leaf size and shape and pubescence type and density, but the most recent monograph (S. Knapp 2013) regarded it as one highly variable and widespread species in which no character discontinuities can be seen. Glabrous populations from northern California have been called *S. parishii*, sticky-glandular populations from central and southern California *S. xanti*, glabrous populations from southern California have been called var. *glabrescens*, and densely pubescent eglandular populations from central California have been called *S. californicum* Dunal. Island populations with larger leaves have been called *S. clokeyi* (but see 52. *S. wallacei*, a distinct endemic on Santa Catalina Island). A number of new varietal combinations were published by D. J. Keil (2018) to accommodate much of this regional and local variation, but the group needs thorough study using both molecular and morphological analysis across its range to assess the taxonomic validity of these segregants. Some of the variation may be environmental.

51. Solanum viarum Dunal in A. P. de Candolle and A. L. P. P. de Candolle, Prodr. 13(1): 240. 1852 • Tropical soda apple [F] [I] [W]

Shrubs, erect, sparsely armed, 0.5–2 m, prickles white or yellowish, straight or recurved, 1–25 mm, densely pubescent, hairs unbranched, short-glandular and longer-eglandular, with sessile, stellate hairs on abaxial leaf surface, these 4(–5)-rayed, central ray shorter than lateral rays. **Leaves** petiolate; petiole 3–6 cm; blade simple, ovate to suborbiculate, 7–10(–20) × 6–8(–15) cm, margins coarsely lobed with 3–5 lobes per side, lobe margins entire to coarsely toothed, base truncate to cordate. **Inflorescences** extra-axillary, sessile or nearly so, unbranched, 3–5-flowered. **Pedicels** 0.7–1.1 cm in flower, 1–2 cm in fruit. **Flowers** radially symmetric; calyx somewhat accrescent, unarmed or sparsely prickly, 3–4 mm, densely pubescent, lobes triangular; corolla greenish or whitish, stellate, 1.5–2.5 cm diam., without interpetalar tissue; stamens equal; anthers narrow and tapered, 5.5–7(–10) mm, dehiscent by terminal pores; ovary densely pubescent, hairs glandular and eglandular. **Berries** light green mottled with dark green when young, yellow when ripe, globose, (1.5–)2–3 cm diam., glabrous, without sclerotic granules. **Seeds** reddish brown, flattened, 2–3 × 2–2.5 mm, minutely pitted. $2n = 24$.

Flowering May–frost (year-round in Fla.). Pastures, roadsides, disturbed areas; 0–1000 m; introduced; Ala., Fla., Ga., La., Miss., N.C., Pa., S.C., Tenn., Tex.; South America (Argentina, Brazil, Paraguay, Uruguay); introduced also in Asia (India), Africa.

In the United States, *Solanum viarum* was first collected in Florida in 1988 and has subsequently become an aggressive and invasive species in the Southeast. It is on the Federal Noxious Weeds List and is classified as a noxious weed or plant pest in many states. Cattle and other animals eat the fruits and spread the seeds through their feces, and the seeds are coated with a sticky substance that makes them adhere to farm equipment when the plants are mowed. It can form large patches that are difficult to eradicate due to their extensive root systems and sharp prickles. It is a major agricultural pest and a threat to native ecosystems.

SELECTED REFERENCE Wunderlin, R. P. et al. 1993. *Solanum viarum* and *S. tampicense* (Solanaceae): Two weedy species new to Florida and the United States. Sida 15: 605–611.

52. Solanum wallacei (A. Gray) Parish, Proc. Calif. Acad. Sci., ser. 3, 2: 166. 1901 • Greasy or Santa Catalina or Wallace's nightshade [C] [E]

Solanum xanti A. Gray var. *wallacei* A. Gray, Proc. Amer. Acad. Arts 11: 91. 1876; *S. umbelliferum* Eschscholtz var. *wallacei* (A. Gray) D. J. Keil

Shrubs or small trees, erect to spreading, unarmed, 1–1.5 (–3) m, densely pubescent, hairs transparent, unbranched, to ca. 3 mm, usually glandular. **Leaves** petiolate; petiole 1–2.5(–4) cm; blade simple, elliptic to obovate, 3–11 (–14) × 1.6–5.5(–9) cm, margins entire or slightly undulate, occasionally with 2 small lobes at base, lobe margins entire, base truncate or acute. **Inflorescences** terminal or lateral, leaf-opposed or extra-axillary, usually once-branched, occasionally more, 20–30-flowered, (2–)4–10 cm. **Pedicels** inserted into small sleeve on inflorescence axis, 1.5–2 cm in flower and fruit. **Flowers** radially symmetric; calyx not accrescent, unarmed, 6–7 mm, densely pubescent, lobes deltate; corolla violet to purple with or without green spots at base of lobes, spots usually small and not confluent, rotate, 3–4.5 cm diam., with abundant interpetalar tissue; stamens equal; anthers ellipsoidal, 4.5–5 mm, dehiscent by terminal pores that open into longitudinal slits; ovary glabrous. **Berries** shiny green, turning yellow then black, globose, 3–4 cm diam., glabrous, without sclerotic granules. **Seeds** reddish brown, plump to flattened, 1.5–2 × 1–1.5 mm, minutely pitted.

Flowering Mar–Aug. Chaparral, open areas, canyon bottoms; of conservation concern; 0–300 m; Calif.

Solanum wallacei is endemic to Santa Catalina Island.

SPHENOCLEACEAE Baskerville

• Gooseweed Family

J. Richard Carter
Jordan C. Jones

Herbs, annual. **Stems** mostly branched. **Leaves** alternate, estipulate, petiolate; blade: margins entire, often undulate, venation pinnate. **Inflorescences** terminal spikes. **Flowers** sessile, epigynous, actinomorphic; calyx lobes 5, imbricate, connivent, proximally connate; corolla short-tubular, lobes 5, alternate with calyx lobes, proximally connate; stamens 5, alternate with corolla lobes, adnate to corolla tube; filaments 0.1–0.2 mm; anthers ± quadrate, 2-locular; ovary inferior, 2-locular; ovules 100+, anatropous; placentation axile; styles 0.2–0.3 mm; stigmas discoid-capitate, depressed. **Fruits** capsules (pyxides), ± obconic, proximally laterally compressed and wedge-shaped, chartaceous, lustrous, dehiscence circumscissile; lids ± flat, discoid, subcoriaceous, covered by persistent, ± connivent, appressed calyx lobes. **Seeds** 100+, tan, oblong, ± lustrous, alveolate; endosperm cellular; embryos straight, filling the seed.

Genus 1: introduced; Asia, Africa; introduced also in Mexico, West Indies, Central America, South America.

Sphenoclea has commonly been included within Campanulaceae. H. K. Airy Shaw (1948), noting similarities with both Phytolaccaceae (habit, anatomy) and Primulaceae, treated it in monogeneric Sphenocleaceae. T. J. Rosatti (1986) also placed it in Sphenocleaceae, citing differences between *Sphenoclea* and Campanulaceae (for example, spicate inflorescences, circumscissile capsules, tetracyclic stomata, pericyclic stem sclerenchyma, cluster crystals, and absence of laticiferous phloem canals). Molecular evidence (B. Bremer et al. 2002) supports treatment in Sphenocleaceae, and that family in Solanales of the Lamiid clade (Angiosperm Phylogeny Group 2009).

A root exudate of *Sphenoclea zeylanica* has been shown to be an effective nematocide (C. Mohandas et al. 1981). In Java, nascent shoots are consumed as a condiment of rice, imparting a bitter flavor (H. K. Airy Shaw 1948), and it is intensively cultivated in Bali as a highly nutritious vegetable (I. W. A. Permadi et al. 2016). *Sphenoclea zeylanica* has been cited as an agricultural weed of rice in the southeastern United States (C. T. Bryson and M. S. DeFelice 2009) and in rice-growing areas throughout the world (T. J. Rosatti 1986). Herbicide-resistant biotypes of *S. zeylanica* have been reported in Asia (K. Itoh and K. Ito 1994).

SELECTED REFERENCE Rosatti, T. J. 1986. The genera of Sphenocleaceae and Campanulaceae in the southeastern United States. J. Arnold Arbor. 67: 1–64.

1. SPHENOCLEA Gaertner, Fruct. Sem. Pl. 1: 113, plate 24, fig. 5. 1788, name conserved • [Greek *sphen*, wedge, and *kleio*, to inclose, probably alluding to persistent calyx covering cuneiform fruits] [I]

Herbs glabrous. **Roots** coarse, usually whitish. **Stems** erect, hollow, soft. **Leaf blades** elliptic to oblanceolate, base cuneate, decurrent nearly to base of petiole, apex acute to rounded, sometimes mucronate. **Inflorescences** dense (rachis hidden by flowers and fruits), acropetal, pedunculate, with 1 small bract subtending each pedicel; bracts greenish white, spatulate, base carinate, apex incurved, acuminate, ± erose; bracteoles 2 per pedicel, greenish white, spatulate, apex incurved, acute to rounded, ± erose. **Flowers:** calyx persistent, dark green with whitish margins; corolla white. **Seeds** longitudinally striate. $x = 12$.

Species 2 (1 in the flora): introduced; Asia, Africa; introduced also in Mexico, West Indies, Central America, South America.

1. Sphenoclea zeylanica Gaertner, Fruct. Sem. Pl. 1: 113, plate 24, fig. 5. 1788 • Gooseweed, chickenspike [F] [I]

Herbs 18.5–97 cm. **Stems** ± terete, green, 2–10 mm diam. at mid stem, often proximally spongy and rooting from proximal nodes where submerged. **Leaves:** petiole 0.3–2.7 cm; blade 1–13.8 × 0.3–5.7 cm. **Spikes** narrowly ovoid to cylindric, 0.5–10.5 × 0.3–1.3 cm. **Peduncles** 0.4–10.3 cm. **Flowers:** calyx lobes broadly ovate to deltate, 1.2–1.5 × 1.5–1.8 mm, enlarging in fruit, obtuse to rounded, ± erose; corolla caducous, 1.8–2.3 mm, lobes oblong, length 1–2 times tube, apex obtuse to acute, ± erose; filaments 0.1–0.2 mm; anthers 0.5 × 0.6 mm; styles 0.3–0.4 mm; stigmas 0.4 mm diam. **Capsules** sessile, 2–3 × 3–4 mm, lids 3–4 mm diam. **Seeds** 0.4–0.5 mm. $2n = 24$.

Flowering Jun–Nov. Rice fields, ditches, shallow margins of ponds and lakes, stream banks, wet disturbed soils, sometimes emergent; 0–300 m; introduced; Ala., Ark., Fla., Ga., La., Miss., Mo., N.C., Okla., S.C., Tex.; Asia; Africa; introduced also in Mexico, West Indies, Central America, South America.

Sphenoclea zeylanica is a common weed of rice fields and other wet disturbed sites in the coastal prairies of southern Louisiana and southeastern Texas. It occurs sporadically in other southeastern states, and its dispersal is correlated with rice agriculture. The earliest records in the United States were collected in Louisiana about 1850, where it was probably introduced as a contaminant of rice seed (J. R. Carter et al. 2014).

West African plants with shorter oblong spikes, pink corollas, and stamens with longer filaments are sometimes segregated as the endemic *Sphenoclea dalzielii* N. E. Brown (S. M. H. Jafri, http://www.tropicos.org/Name/5504127).

HYDROLEACEAE R. Brown

• Hydrolea Family

Lawrence J. Davenport

Herbs or small shrubs, perennial, occasionally annual, semi-aquatic, erect or decumbent, unbranched or branched, to 6 dm, often gregarious, with new culms arising from prostrate stems. **Stems** succulent to woody, base often swollen; thorns 1 or 2 per node or absent. **Leaves** alternate, sessile or tapering to short (to 1 cm) petiole, estipulate. **Inflorescences** terminal corymbs, short-pedicellate, in clusters at branch tips, terminal leafy panicles, axillary fascicles (from compacted branchlets) on peduncles to 4 cm, or flowers solitary in leaf axils. **Pedicels** to 1 cm. **Flowers:** sepals persistent in fruit, 5, distinct, equal; petals 5, imbricate in bud, connate basally, equal; stamens 5, exserted or included, distinct, alternate with petals, inserted on short corolla tube, equal; anthers dorsifixed, 4-lobed, dehiscing longitudinally; disc absent; ovary superior; placentation axile, placentas adnate to thin septum, spongy, entire or 2-fid in cross section, bearing 100–200 anatropous ovules; styles persistent in fruit, 2–4, separate, equal. **Fruits** capsular, erect or drooping at maturity, dehiscence septicidal, loculicidal, or irregular. **Seeds** 100–200. x = 9, 10, or 12.

Genus 1, species 11 (5 in the flora): sc, e United States, Mexico, Central America, South America, s, se Asia, Africa, Indian Ocean Islands (Madagascar), Pacific Islands (Philippines), n Australia.

The genus *Hydrolea* has been segregated in its own family (R. Brown 1818) or included within the Hydrophyllaceae (A. Gray 1875; A. Brand 1913; L. J. Davenport 1988). Modern molecular data (M. W. Chase et al. 1993; M. E. Cosner et al. 1994; D. M. Ferguson 1998) and floral development studies (C. Erbar et al. 2005) support maintaining the Hydroleaceae as a distinct family.

SELECTED REFERENCES Davenport, L. J. 1988. A monograph of *Hydrolea* (Hydrophyllaceae). Rhodora 90: 169–208. Erbar, C., S. Porembski, and P. Leins. 2005. Contributions to the systematic position of *Hydrolea* (Hydroleaceae) based on floral development. Pl. Syst. Evol. 252: 71–83.

1. HYDROLEA Linnaeus, Sp. Pl. ed. 2, 1: 328. 1762; Gen. Pl. ed. 6, 124. 1764, name conserved • [Greek *hydor*, water, and *elaia*, olive, probably alluding to habitat and oiliness and/or form of leaves]

Herbs or small shrubs, usually perennial. **Stems** green, brown, or purple, glabrous or hispid-hirsute, with or without glandular trichomes or long, jointed hairs; thorns occasionally bearing small leaves toward tips. **Leaf blades** ovate, lanceolate, or linear, occasionally orbiculate, pinnately veined, base attenuate, acute, round, or obtuse, margins entire or serrulate, often undulate, apex acuminate to acute, surfaces glabrous or densely pubescent, sometimes gland-dotted, with or without glandular trichomes or long, jointed hairs, occasionally pubescent only along main veins. **Inflorescences** terminal or axillary, pedunculate, highly branched or unbranched. **Flowers:** sepals lanceolate or ovate, margins entire, surfaces glabrous or hispid-hirsute, with or without glandular trichomes or long, jointed hairs; corolla blue or white, campanulate; anthers pale pink, white, or blue; filaments white or blue, abruptly dilated at base, glabrous; pollen white or yellow; ovary green or blue, globose or subglobose, proximal 1/2 glabrous, distal 1/2 glabrous, puberulent, or glandular-pubescent; styles white, brown, or blue, often curved inward to summits, glabrous, puberulent, or glandular-pubescent (especially at base); stigmas funnelform. **Capsules** brown or purple, globose or ovoid, occasionally truncated at style bases, 3–7 × 2.5–7 mm, proximal 1/2 glabrous, distal 1/2 glabrous, puberulent, or glandular-pubescent. **Seeds** tan to dark brown, ovoid to cylindric, symmetric or occasionally asymmetric, 0.4–0.7 × 0.2–0.4 mm.

Species 11 (5 in the flora): sc, e United States, Mexico, Central America, South America, s, se Asia, Africa, Indian Ocean Islands (Madagascar), Pacific Islands (Philippines), n Australia.

Species in the flora area are placed in *Hydrolea* sect. *Hydrolea.*

1. Inflorescences axillary fascicles.
    2. Stems and sepals with long, jointed hairs . . . . 1. *Hydrolea quadrivalvis*
    2. Stems and sepals without long, jointed hairs . . . . 2. *Hydrolea uniflora*
1. Inflorescences terminal corymbs or leafy panicles.
    3. Stems pubescent or hispid-hirsute, usually densely covered with glandular trichomes, occasionally glabrous . . . . 5. *Hydrolea spinosa*
    3. Stems pubescent, sometimes with few glandular trichomes.
        4. Leaf blades ovate, 1–2.5 cm wide, margins entire, surfaces pubescent; styles 10–15 mm . . . . 3. *Hydrolea ovata*
        4. Leaf blades lanceolate, 0.3–1 cm wide, margins serrulate, surfaces glabrous or pubescent; styles 5–10 mm . . . . 4. *Hydrolea corymbosa*

1. Hydrolea quadrivalvis Walter, Fl. Carol., 110. 1788 • Water-pod E F

Nama quadrivalvis (Walter) Kuntze

Herbs, erect or decumbent, to 6 dm, unbranched, with short reproductive branches or with branches arising from prostrate stems. **Stems** green to brown, with long, jointed hairs; thorns 1 or 2 per node, rarely absent, 5–12 × 0.6–1.2 mm. **Leaf blades** lanceolate, 4–10 × 1–2.5 cm, base attenuate to acute, margins entire or serrulate, surfaces hairy, with occasional long trichomes restricted to veins. **Inflorescences** axillary, fasciculate, 1–10-flowered, or on short, leafy branches. **Flowers:** sepals lanceolate, 6–10 × 1.5–2.5 mm, with long, jointed hairs; corolla blue, rarely white, petals 8–11 × 3.5–6 mm; ovary glabrous; styles 2, 3.5–5 mm, glabrous. **Capsules** globose, 5.5–7 × 5–7 mm, glabrous. **Seeds** cylindric, 0.6–0.7 × 0.2–0.3 mm. $2n = 20$.

Flowering Jul–Sep. Pond margins, stream banks and mudflats; 30–400 m; Ala., Fla., Ga., La., Miss., N.C., S.C., Tenn., Va.

In morphology, *Hydrolea quadrivalvis* is closest to *H. uniflora* but differs in having long, jointed hairs on stems and sepals.

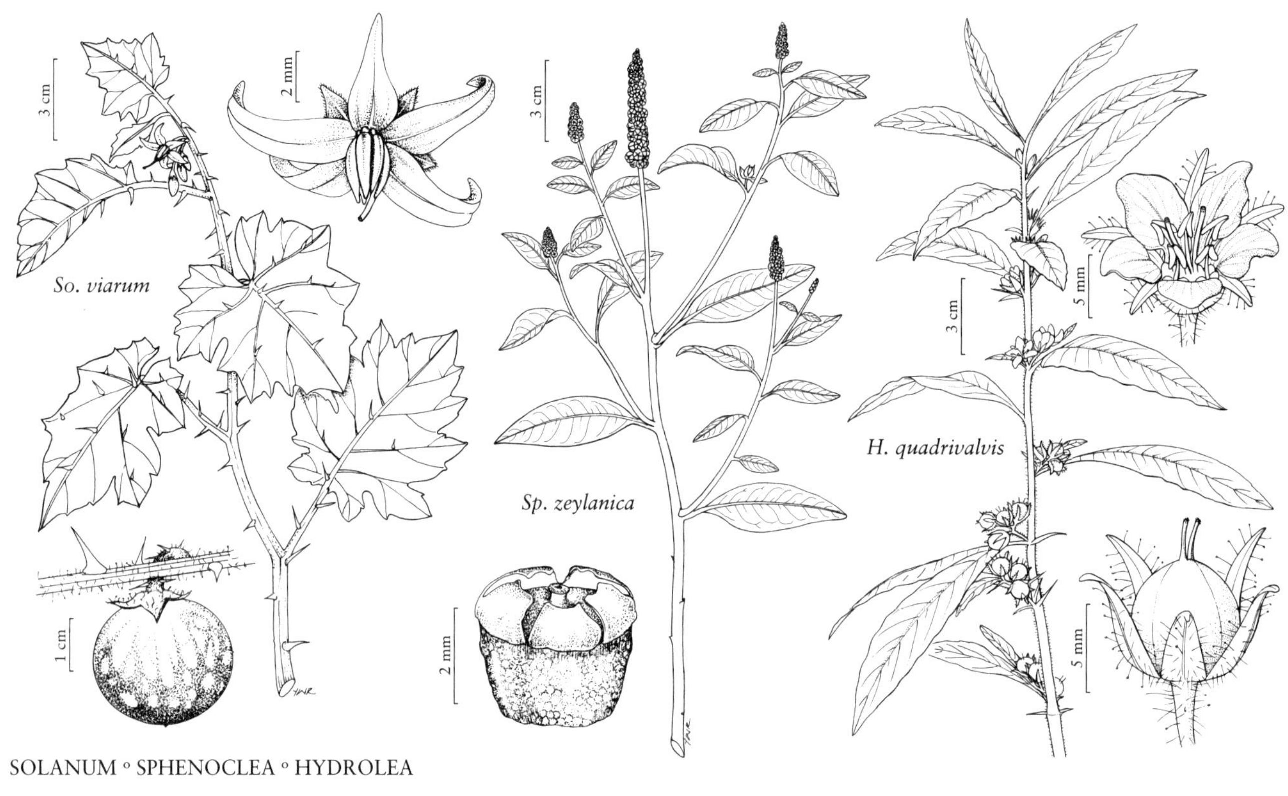

SOLANUM ◦ SPHENOCLEA ◦ HYDROLEA

2. **Hydrolea uniflora** Rafinesque, Autik. Bot., 34. 1840 • One-flower false fiddleleaf [E] [W]

Hydrolea affinis A. Gray; *Nama affinis* (A. Gray) Kuntze

Herbs, erect or decumbent, to 6 dm, unbranched, with short reproductive branches or with branches arising from prostrate stems. **Stems** green to brown, glabrous or puberulent, without long, jointed hairs; thorns 1 or 2 per node, 6–15 × 0.5–1.2 mm. **Leaf blades** lanceolate, 3–10 × 1–2 cm, base attenuate to acute, margins entire or serrulate, surfaces glabrous. **Inflorescences** axillary, fasciculate, 1–10-flowered, or on short, leafy branches. **Flowers:** sepals broadly lanceolate to ovate, 5–8 × 1.5–4 mm, glabrous or puberulent, without glandular trichomes; corolla blue, occasionally white, petals 7–11 × 4–6 mm; ovary glabrous; styles 2, 3.5–5 mm, glabrous. **Capsules** globose, 4–6 mm wide, glabrous. **Seeds** ovoid, 0.5–0.6 × 0.2–0.3 mm. **2*n*** = 20.

Flowering Jun–Sep. Stream banks and pond margins; 0–300 m; Ala., Ark., Ill., Ky., La., Miss., Mo., Okla., Tenn., Tex.

Hydrolea uniflora is very similar morphologically to *H. quadrivalvis* but differs from it in lacking long, jointed hairs and by having flower clusters borne on long (to 4 cm) peduncles.

3. **Hydrolea ovata** Nuttall ex Choisy, Mém. Soc. Phys. Genève 6: 109, plate 1. 1833 • Ovate false fiddleleaf [E]

Hydrolea ovata var. *georgiana* Brand; *Nama ovata* (Nuttall ex Choisy) Britton

Herbs or small shrubs, erect, to 10 dm, usually broadly branched. **Stems** green to brown, densely pubescent, occasionally with few longer or glandular trichomes; thorns 1 or 2 per node, 5–12 × 0.3–0.8 mm. **Leaf blades** ovate, occasionally orbiculate, 1.5–7 × 1–2.5 cm, base attenuate, acute, or obtuse, margins entire, surfaces pubescent. **Inflorescences** terminal, paniculate, leafy, broadly branching, 25–40-flowered. **Flowers:** sepals narrowly lanceolate, 6–9 × 1–2.5 mm, hispid-hirsute, with glandular trichomes; corolla blue, occasionally white, petals 11–17 × 5–9 mm; ovary glabrous or puberulent, upper 1/2 often with glandular trichomes; styles 2, 10–15 mm, glandular-pubescent toward bases. **Capsules** globose, 4.5–5.5 × 4–5.5 mm, upper 1/2 puberulent or glandular-pubescent. **Seeds** cylindric, symmetric, 0.5–0.6 × 0.2–0.3 mm. **2*n*** = 20.

Flowering Jun–Sep. Edges of sloughs, marshes, and ponds; 2–200 m; Ala., Ark., Fla., Ga., Ky., La., Miss., Mo., Okla., Tenn., Tex.

Hydrolea ovata is much more robust than *H. corymbosa*, with densely pubescent stems and leaves. Its leaves are ovate, while those of *H. corymbosa* are lanceolate.

Some forms of *Hydrolea spinosa* have small, rounded leaves similar to those of *H. ovata*; however, the stems of *H. ovata* are only rarely glandular-pubescent, and the styles are much longer than those of *H. spinosa*.

A number of specimens collected in Louisiana and eastern Texas appear to result from hybridization between *Hydrolea ovata* and *H. uniflora*. These are marked by several intermediate characteristics: leaves are ovate-lanceolate and sparsely pubescent; stems and sepals are also pubescent, with the latter sometimes glandular-pubescent; petals are either the same size as in one of the species or intermediate. The putative hybrids generally are extensively branched, with branches from the lower nodes producing a much bushier and more sprawling plant than is typical of either species. Clusters of flowers are borne either at the tips or at the nodes of these lateral branches.

4. **Hydrolea corymbosa** J. Macbride ex Elliott, Sketch Bot. S. Carolina 1: 336. 1817 • Skyflower [E]

Nama corymbosa (J. Macbride ex Elliott) Kuntze

Herbs, erect, to 6 dm, unbranched or with short reproductive branches. **Stems** green, brown, or purple, densely pubescent, without glandular trichomes; thorns rare, 1 per node or absent, 4–11 × 0.2–0.6 mm. **Leaf blades** lanceolate, 2–5.5 × 0.3–1 cm, base acute to rounded, margins serrulate, surfaces glabrous or pubescent. **Inflorescences** terminal, leafy panicles or corymbs, 15–30-flowered. **Flowers:** sepals lanceolate, 4.5–7 × 1–2 mm, hispid-hirsute, with glandular trichomes; corolla blue, petals 10–15 × 5–8 mm; ovary glabrous or puberulent, upper 1/2 often with glandular trichomes; styles 2, 5–10 mm, glandular-pubescent toward bases. **Capsules** globose to slightly ovoid, 3–4.5 × 2.5–4 mm, upper 1/2 puberulent or glandular-pubescent. **Seeds** broadly ovoid, symmetric, 0.6–0.7 × 0.3–0.4 mm.

Flowering Jul–Sep. Wet roadsides and ditches; 0–20 m; Fla., Ga., S.C.

Hydrolea corymbosa is morphologically very similar to the more western *H. ovata*, sharing the paniculate or corymbose type of inflorescence and similar stem and sepal pubescence. However, *H. corymbosa* is a much smaller and more slender plant with fewer (if any) thorns, stems that are dark brown or purple, and lanceolate leaves.

5. **Hydrolea spinosa** Linnaeus, Sp. Pl. ed. 2, 1: 328. 1762

Nama spinosa (Linnaeus) Kuntze

Varieties 3 (1 in the flora): Texas, Mexico, Central America, South America, Asia.

5a. **Hydrolea spinosa** Linnaeus var. **spinosa** • Spiny false fiddleleaf

Hydrolea extra-axillaris C. Morren; *H. tetragynia* Sessé & Mociño; *H. trigyna* Swartz; *Nama extra-axillaris* (C. Morren) Kuntze

Herbs or small shrubs, erect or decumbent, to 20 dm, unbranched to broadly branched. **Stems** green, brown, or purple, pubescent or hispid-hirsute, occasionally glabrous, usually densely covered with short, glandular trichomes; thorns 1 or 2 per node or absent, 4–30 × 0.4–2 mm. **Leaf blades** ovate to lanceolate, occasionally linear, 1–12 × 0.2–3 cm, base attenuate to acute, margins entire or serrulate, surfaces puberulent to hispid-hirsute, with or without glandular trichomes. **Inflorescences** terminal, narrow or broadly branching, leafy panicles or clustered at branch tips, 20–100-flowered. **Flowers:** sepals lanceolate, 6–14 × 1.5–3.5 mm, puberulent to hispid-hirsute, with glandular trichomes; corolla blue, rarely white, petals 5–17 × 2–12 mm; ovary puberulent, upper 1/2 usually glandular-pubescent; styles 2–4, 1.5–13 mm, glandular-pubescent toward bases. **Capsules** globose to ovoid, 3.5–8 × 3–7 mm, upper 1/2 puberulent or glandular-pubescent. **Seeds** ovoid to cylindric, symmetric, 0.4–0.7 × 0.2–0.3 mm. $2n$ = 20, 40.

Flowering year-round. Wet pond margins, open flood plains; 0–10 m; Tex.; Mexico; Central America; South America (south to Argentina).

In the flora area, var. *spinosa* is known from Cameron County. Specimens vary in pubescence, thorniness, and leaf shape and size.

Literature Cited

Robert W. Kiger, Editor

This is a consolidated list of all works cited in volume 14, whether as selected references, in text, or in nomenclatural contexts. In citations of articles, both here and in the taxonomic treatments, and also in nomenclatural citations, the titles of serials are rendered in the forms recommended in G. D. R. Bridson and E. R. Smith (1991). When those forms are abbreviated, as most are, cross references to the corresponding full serial titles are interpolated here alphabetically by abbreviated form. In nomenclatural citations (only), book titles are rendered in the abbreviated forms recommended in F. A. Stafleu and R. S. Cowan (1976–1988) and Stafleu et al. (1992–2009). Here, those abbreviated forms are indicated parenthetically following the full citations of the corresponding works, and cross references to the full citations are interpolated in the list alphabetically by abbreviated form. Two or more works published in the same year by the same author or group of coauthors will be distinguished uniquely and consistently throughout all volumes of *Flora of North America* by lower-case letters (b, c, d, ...) suffixed to the date for the second and subsequent works in the set. The suffixes are assigned in order of editorial encounter and do not reflect chronological sequence of publication. The first work by any particular author or group from any given year carries the implicit date suffix "a"; thus, the sequence of explicit suffixes begins with "b". There may be citations in this list that have dates suffixed "b," "c," "d," etc. but that are not preceded by citations of "[a]," "b," and/or "c," etc. works for that year. In such cases, the missing "[a]," "b," and/or "c," etc. works are ones cited (and encountered first from) elsewhere in the *Flora* that are not pertinent in this volume.

Abh. Königl. Ges. Wiss. Göttingen = Abhandlungen der Königlichen Gesellschaft der Wissenschaften zu Göttingen.

Abrams, L. and R. S. Ferris. 1923–1960. Illustrated Flora of the Pacific States: Washington, Oregon, and California. 4 vols. Stanford. (Ill. Fl. Pacific States)

Acta Bot. Neerl. = Acta Botanica Neerlandica.

Acta Horti Gothob. = Acta Horti Gothoburgensis; Meddelanden från Göteborgs Botaniska Trädgård.

Acta Phytotax. Sin. = Acta Phytotaxonomica Sinica. [Chih Wu Fen Lei Hsüeh Pao.]

Adams, R. P. et al. 1987b. Investigation of hybridization between *Asclepias speciosa* and *A. syriaca* using alkanes, fatty acids and triterpenoids. Biochem. Syst. & Ecol. 15: 395–399.

Adanson, M. 1763[–1764]. Familles des Plantes. 2 vols. Paris. [Vol. 1, 1764; vol. 2, 1763.] (Fam. Pl.)

Adansonia = Adansonia; Recueil Périodique d'Observations Botaniques.

Addisonia = Addisonia; Colored Illustrations and Popular Descriptions of Plants.

Agrawal, A. A. et al. 2009. Phylogenetic ecology of leaf surface traits in the milkweeds (*Asclepias* spp.): Chemistry, ecophysiology, and insect behavior. New Phytol. 183: 848–867.

Agrawal, A. A. et al. 2012. Toxic cardenolides: Chemical ecology and coevolution of specialized plant-herbivore interactions. New Phytol. 194: 28–45.

Agrotrop = Agrotrop; Journal on Agricultural Science.

Airy Shaw, H. K. 1948. Sphenocleaceae. In: C. G. G. J. van Steenis, ed. 1948+. Flora Malesiana.... Series I. Spermatophyta. 18+ vols., some in parts. Djakarta and Leiden. Vol. 4, pp. 27–28.

Aiton, W. 1789. Hortus Kewensis; or, a Catalogue of the Plants Cultivated in the Royal Botanic Garden at Kew. 3 vols. London. (Hort. Kew.)

Aiton, W. and W. T. Aiton. 1810–1813. Hortus Kewensis; or a Catalogue of the Plants Cultivated in the Royal Botanic Garden at Kew. 5 vols. London. (Hortus Kew.)

Albach, D. C., H. M. Meudt, and B. Oxelman. 2005. Piecing together the "new" Plantaginaceae. Amer. J. Bot. 92: 297–315.

Albert, V. A. and L. Struwe. 1997. Phylogeny and classification of *Voyria* (saprophytic Gentianaceae). Brittonia 49: 466–479.

Alexander, E. J. 1933. *Odontostephana*. In: J. K. Small. 1933. Manual of the Southeastern Flora, Being Descriptions of the Seed Plants Growing Naturally in Florida, Alabama, Mississippi, Eastern Louisiana, Tennessee, North Carolina, South Carolina and Georgia. New York. Pp. 1076–1078.

Allen, C. K. 1933. A monograph of the American species of the genus *Halenia*. Ann. Missouri Bot. Gard. 20: 119–222.

Allen, C. M. 2013. Synopsis of the genus *Stylisma* (Convolvulaceae) in Louisiana. J. Bot. Res. Inst. Texas 7: 515–516.

Allioni, C. 1773. Auctuarium ad Synopsim Methodicam Stirpium Horti Regii Taurinensis. Turin. [Preprinted from Mélanges Philos. Math. Soc. Roy. Turin 5: 53–96. 1774.] (Auct. Syn. Meth. Stirp. Taurin.)

Allred, K. W. 1999. New plant distribution records. New Mexico Bot. Newslett. 13: 7.

Alvarado-Cárdenas, L. O. and J. F. Morales. 2014. El género *Mandevilla* (Apocynaceae: Apocynoideae, Mesechiteae) en México. Bot. Sci. 92: 59–79.

Alvarado-Cárdenas, L. O. and H. Ochoterena. 2007. A phylogenetic analysis of the *Cascabela-Thevetia* species complex (Plumerieae, Apocynaceae) based on morphology. Ann. Missouri Bot. Gard. 94: 298–323.

Amer. J. Bot. = American Journal of Botany.

Amer. J. Sci. Arts = American Journal of Science, and Arts.

Amer. Midl. Naturalist = American Midland Naturalist; Devoted to Natural History, Primarily That of the Prairie States.

Amer. Monthly Mag. & Crit. Rev. = American Monthly Magazine and Critical Review.

Amer. Naturalist = American Naturalist....

Amoen. Acad.—See: C. Linnaeus 1749[–1769]

Anales Inst. Biol. Univ. Nac. Autón. México, Bot. = Anales del Instituto de Biológia de la Universidad Nacional Autónoma de México. Série Botánica.

Anales Jard. Bot. Madrid = Anales del Jardín Botánico de Madrid.

Anderson, E. 1936b. An experimental study of hybridization in the genus *Apocynum*. Ann. Missouri Bot. Gard. 23: 159–168.

Ando, T. et al. 1999. Floral anthocyanins in wild taxa of *Petunia* (Solanaceae). Biochem. Syst. & Ecol. 27: 623–650.

Ando, T. et al. 2001. Reproductive isolation in a native population of *Petunia* sensu Jussieu (Solanaceae). Ann. Bot. (Oxford) 88: 403–413.

Ando, T. et al. 2005. Phylogenetic analysis of *Petunia* sensu Jussieu (Solanaceae) using chloroplast DNA RFLP. Ann. Bot. (Oxford) 96: 289–297.

Ando, T. et al. 2005b. A morphological study of the *Petunia integrifolia* complex (Solanaceae). Ann. Bot. (Oxford) 96: 887–900.

Angiosperm Phylogeny Group. 2009. An update of the Angiosperm Phylogeny Group classification for the orders and families of flowering plants: APG III. Bot. J. Linn. Soc. 161: 105–121.

Animal and Plant Health Inspection Service [U.S.D.A.]. 2012. Weed Risk Assessment for *Araujia sericifera* Brot. (Apocynaceae)—Cruel Plant. Raleigh.

Ann. Bot. (Oxford) = Annals of Botany. (Oxford.)

Ann. Lyceum Nat. Hist. New York = Annals of the Lyceum of Natural History of New York.

Ann. Mag. Nat. Hist. = Annals and Magazine of Natural History, Including Zoology, Botany, and Geology.

Ann. Missouri Bot. Gard. = Annals of the Missouri Botanical Garden.

Ann. Mus. Natl. Hist. Nat. = Annales du Muséum National d'Histoire Naturelle. = ["National" dropped after vol. 5.]

Ann. New York Acad. Sci. = Annals of the New York Academy of Sciences.

Ann. Sci. Nat., Bot. = Annales des Sciences Naturelles. Botanique.

Ann. Sci. Phys. Nat. Lyon = Annales des Sciences Physiques et Naturelles, d'Agriculture et de l'Industrie; Publiées par la Société Royale d'Agriculture, d'Histoire Naturelle et Arts Utiles de Lyon.

Annuaire Conserv. Jard. Bot. Genève = Annuaire du Conservatoire et Jardin Botaniques de Genève.

Apocyn. S. Amer.—See: J. Miers 1878

Arch. Bot. (Leipzig) = Archiv für die Botanik. (Leipzig.)

Asclepiadeae—See: R. Brown 1810b

Atlantic J. = Atlantic Journal, and Friend of Knowledge.

Atwood, J. T. et al. 1973. Checklist of Vermont Plants: Including All Vascular Plants Growing without Cultivation. Burlington.

Aublet, J. B. 1775. Histoire des Plantes de la Guiane Françoise.... 4 vols. Paris. [Vols. 1 and 2: text, paged consecutively; vols. 3 and 4: plates.] (Hist. Pl. Guiane)

Aubréville, A. et al., eds. 1967+. Flore de la Nouvelle-Calédonie et Dependances. 26+ vols. Paris.

Auct. Syn. Meth. Stirp. Taurin.—See: C. Allioni 1773

Auctuarium—See: J. Burman 1755

Austin, D. F. 1973. Another adventive morning glory in Florida: *Aniseia matinicensis.* Florida Sci. 36: 197–198.

Austin, D. F. 1978. The *Ipomoea batatas* complex—I. Taxonomy. Bull. Torrey Bot. Club 105: 114–129.

Austin, D. F. 1979. Studies of the Florida Convolvulaceae—II. *Merremia.* Florida Sci. 42: 216–222.

Austin, D. F. 1980. Studies of the Florida Convolvulaceae—III. *Cuscuta.* Florida Sci. 43: 294–302.

Austin, D. F. 1988. The rarest morning glory. Fairchild Trop. Gard. Bull. 3: 22–28.

Austin, D. F. 1990. Comments on southwestern United States *Evolvulus* and *Ipomoea* (Convolvulaceae). Madroño 37: 124–132.

Austin, D. F. 1998. Xixicamátic or wood rose (*Merremia tuberosa,* Convolvulaceae): Origins and dispersal. Econ. Bot. 52: 412–422.

Austin, D. F. 1999. The genus *Aniseia.* Syst. Bot. 23: 411–420.

Austin, D. F. 2000. Bindweed (*Convolvulus arvensis,* Convolvulaceae) in North America—From medicine to menace. J. Torrey Bot. Soc. 127: 172–177.

Austin, D. F. 2000b. A revision of *Cressa* L. (Convolvulaceae). Bot. J. Linn. Soc. 133: 27–39.

Austin, D. F. 2007. *Merremia dissecta* (Convolvulaceae)—A condiment, medicine, ornamental, and weed—A review. Econ. Bot. 61: 109–120.

Austin, D. F. and W. E. Abel. 1981. Introgressive hybridization between *Ipomoea trichocarpa* and *I. lacunosa.* Bull. Torrey Bot. Club 108: 231–239.

Austin, D. F. and S. Demissew. 1997. Unique fruits and generic status of *Stictocardia* (Convolvulaceae). Kew Bull. 52: 161–168.

Austin, D. F., D. A. Powell, and D. H. Nicolson. 1978. *Stictocardia tiliifolia* (Convolvulaceae) re-evaluated. Brittonia 30: 195–198.

Austin, D. F. and G. W. Staples. 1985. *Petrogenia* as a synonym of *Bonamia* (Convolvulaceae), with comments on allied species. Brittonia 37: 310–316.

Austin, D. F. and G. W. Staples. 1991. A revision of the neotropical species of *Turbina* Raf. (Convolvulaceae). Bull. Torrey Bot. Club 118: 265–280.

Austral. Syst. Bot. = Australian Systematic Botany.

Autik. Bot.—See: C. S. Rafinesque 1840

Averett, J. E. 1970. New combinations in the Solaneae (Solanaceae) and comments regarding the taxonomic status of *Leucophysalis.* Ann. Missouri Bot. Gard. 57: 380–381.

Averett, J. E. 1973. Biosystematic study of *Chamaesaracha* (Solanaceae). Rhodora 75: 325–365.

Averett, J. E. 2009. Taxonomy of *Leucophysalis* (Solanaceae, tribe Physaleae). Rhodora 111: 209–217.

Averett, J. E. 2010. A new species of *Chamaesaracha* (Solanaceae) from Mexico and the separation of *C. crenata* from *C. villosa.* Phytologia 92: 435–441.

Averett, J. E. 2010b. The status of *Chamaesaracha coniodes* and *C. coronopus* (Solanaceae). Phytoneuron 2010-57: 1–5.

Averett, J. E. and W. G. D'Arcy. 1983. Flavonoids in *Oryctes* (Solanaceae). Phytochemistry 22: 2325–2326.

Avery, A. G., S. Satina, and J. Rietsema. 1959. Blakeslee: The Genus *Datura.* New York.

B. M. C. Evol. Biol. = B M C Evolutionary Biology.

Baas, P., K. Kalkman, and R. Geesink, eds. 1990. The Plant Diversity of Malesia.... Dordrecht and Boston.

Bailey, L. H. 1924. Manual of Cultivated Plants.... New York and London.

Baileya = Baileya; a Quarterly Journal of Horticultural Taxonomy.

Bandara, V. et al. 2010. A review of the natural history, toxicology, diagnosis and clinical management of *Nerium oleander* (common oleander) and *Thevetia peruviana* (yellow oleander) poisoning. Toxicon 56: 273–281.

Barbaricz, A. I. 1950. Vyznacnyk Roslyn Ukrajiny. Kiev. (Vyzn. Rosl. Ukrain.)

Barboza, G. E., R. Deanna, and P. González. 2016. Proposal to conserve the name *Salpichroa* against *Nectouxia* (Solanaceae). Taxon 65: 1433–1434.

Barclay, A. S. 1959. New considerations in an old genus: *Datura.* Bot. Mus. Leafl. 18: 245–272.

Barclay, A. S. 1959b. Studies in the Genus *Datura* (Solanaceae). I. Taxonomy of the Subgenus *Datura.* Ph.D. thesis. Harvard University.

Barrios, B., G. Arellano, and S. Koptur. 2011. The effects of fire and fragmentation on occurrence and flowering of a rare perennial plant. Pl. Ecol. 212: 1057–1067.

Barrios, B. and S. Koptur. 2011. Floral biology and breeding system of *Angadenia berteroi* (Apocynaceae): Why do flowers of the pineland golden trumpet produce few fruits? Int. J. Pl. Sci. 172: 378–385.

Barton, B. S. 1812. Flora Virginica.... Philadelphia. (Fl. Virgin.)

Bausteine Monogr. Convolv.—See: H. Hallier 1897

Beliz, T. 1993. *Cuscuta.* In: J. C. Hickman, ed. 1993. The Jepson Manual. Higher Plants of California. Berkeley, Los Angeles, and London. Pp. 538–539.

Bell, N. B. and L. J. Lester. 1978. Genetic and morphological detection of introgression in a clinal population of *Sabatia* section *Campestria* (Gentianaceae). Syst. Bot. 3: 87–104.

Benson, J. M. et al. 1979. Effects on sheep of the milkweeds *Asclepias eriocarpa* and *A. labriformis* and of cardiac glycoside-containing derivative material. Toxicon 17: 155–165.

Bentham, G. 1839[–1857]. Plantas Hartwegianas Imprimis Mexicanas.... London. [Issued by gatherings with consecutive signatures and pagination.] (Pl. Hartw.)

Bentham, G. 1844[–1846]. The Botany of the Voyage of H.M.S. Sulphur, under the Command of Captain Sir Edward Belcher...during the Years 1836–1842. 6 parts. London. [Parts paged consecutively.] (Bot. Voy. Sulphur)

Bentham, G. and J. D. Hooker, J. D. 1862–1883. Genera Plantarum ad Exemplaria Imprimis in Herbariis Kewensibus Servata Definita. 3 vols. London. (Gen. Pl.)

Biochem. Syst. & Ecol. = Biochemical Systematics and Ecology.

Biol. Cent.-Amer., Bot.—See: W. B. Hemsley 1879–1888

Bisset, N. G. and J. D. Phillipson. 1971. The African species of *Strychnos*. Part II. The alkaloids. Lloydia 34: 1–60.

Bitter, F. A. G. 1919. Die Gattung *Lycianthes*. Bremen. [Preprinted from Abh. Naturwiss. Vereins Bremen 24: 292–520. 1920.] (Gatt. Lycianthes)

Blackwell, W. H. 1990. Poisonous and Medicinal Plants. Englewood Cliffs.

Blake, S. F. 1915. Notes on the genus *Sabatia*. Rhodora 17: 50–57.

Blank, C. M., R. A. Levin, and J. S. Miller. 2014. Intraspecific variation in gender strategies in *Lycium* (Solanaceae): Associations with ploidy and changes in floral form following the evolution of gender dimorphism. Amer. J. Bot. 101: 2160–2168.

Blumea = Blumea; Tidjschrift voor die Systematiek en die Geografie der Planten (A Journal of Plant Taxonomy and Plant Geography).

Blumea, Suppl. = Blumea. Supplement.

Boenninghausen, C. F. M. von. 1824. Prodromus Florae Monasteriensis Westphalorum.... Münster. (Prodr. Fl. Monast. Westphal.)

Boiteau, P. 1981. Apocynaceae. In: A. Aubréville et al., eds. 1967+. Flore de la Nouvelle-Calédonie et Dependances. 26+ vols. Paris. Vol. 10, pp. 1–302.

Bojer, W. 1837. Hortus Mauritianus.... Mauritius. (Hortus Maurit.)

Bonplandia (Hannover) = Bonplandia; Zeitschrift für die gesammte Botanik.

Börner, C. [1912]b. Eine Flora für das deutsche Volk.... Leipzig. (Fl. Deut. Volk)

Boston J. Nat. Hist. = Boston Journal of Natural History.

Bot. Abh. Beobacht.—See: A. W. Roth 1787

Bot. Acta = Botanica Acta; Berichte der Deutschen botanischen Gesellschaft.

Bot. Beechey Voy.—See: W. J. Hooker and G. A. W. Arnott [1830–]1841

Bot. California—See: W. H. Brewer et al. 1876–1880

Bot. Gaz. = Botanical Gazette; Paper of Botanical Notes.

Bot. J. Linn. Soc. = Botanical Journal of the Linnean Society.

Bot. Jahrb. Syst. = Botanische Jahrbücher für Systematik, Pflanzengeschichte und Pflanzengeographie.

Bot. Mag. = Botanical Magazine; or, Flower-garden Displayed.... [Edited by Wm. Curtis.] [With vol. 15, 1801, title became Curtis's Botanical Magazine; or....]

Bot. Mag. (Tokyo) = Botanical Magazine. [Shokubutsu-gaku Zasshi.] (Tokyo.)

Bot. Mus. Leafl. = Botanical Museum Leaflets. [Harvard University.]

Bot. Not. = Botaniska Notiser.

Bot. Reg. = Botanical Register....

Bot. Repos. = Botanists' Repository, for New, and Rare Plants.

Bot. Sci. = Botanical Sciences.

Bot. Voy. Sulphur—See: G. Bentham 1844[–1846]

Bot. Zhurn. (Moscow & Leningrad) = Botanicheskii Zhurnal. (Moscow and Leningrad.)

Botany (Fortieth Parallel)—See: S. Watson 1871

Botany (Ottawa) = Botany.

Boutte, J. et al. 2019. NGS-Indel Coder: A new pipeline to code indel characters in phylogenetic data with an example of its application in milkweeds *(Asclepias)*. Molec. Phylogen. Evol. 139.

Boyd, C. E. 2003. Rock Art of the Lower Pecos. College Station, Tex.

Brabazon, H. K. 2015. Delimiting Species and Varieties of *Cycladenia humilis* (Apocynaceae). M.S. thesis. Brigham Young University.

Brand, A. 1913. Hydrophyllaceae. In: H. G. A. Engler, ed. 1900–1953. Das Pflanzenreich.... 107 vols. Berlin. Vol. 59[IV,251], pp. 1–210.

Bratley, H. E. 1932. The oleander caterpillar, *Syntomeida epilais,* Walker. Florida Entomol. 15: 57–64.

Bremer, B. et al. 2002. Phylogenetics of asterids based on 3 coding and 3 non-coding chloroplast DNA markers and the utility of non-coding DNA at higher taxonomic levels. Molec. Phylogen. Evol. 24: 274–301.

Brewer, W. H. et al. 1876–1880. Geological Survey of California.... Botany.... 2 vols. Cambridge, Mass. (Bot. California)

Brickell, C. D. and J. D. Zuk, eds. 1997. The American Horticultural Society A–Z Encyclopedia of Garden Plants. New York.

Bridson, G. D. R. 2004. BPH-2: Periodicals with Botanical Content. 2 vols. Pittsburgh.

Bridson, G. D. R. and E. R. Smith. 1991. B-P-H/S. Botanico-Periodicum-Huntianum/Supplementum. Pittsburgh.

Brit. Fl. Gard.—See: R. Sweet 1823–1837

Brit. Herb.—See: J. Hill 1756[–1757]

Britton, N. L. et al., eds. 1905+. North American Flora.... 47+ vols. New York. [Vols. 1–34, 1905–1957; ser. 2, parts 1–13+, 1954+.]

Britton, N. L., E. E. Sterns, J. F. Poggenburg, et al. 1888. Preliminary Catalogue of Anthophyta and Pteridophyta Reported As Growing Spontaneously within One Hundred Miles of New York City. New York. [Authorship often attributed as B.S.P. in nomenclatural contexts.] (Prelim. Cat.)

Brittonia = Brittonia; a Journal of Systematic Botany....

Bronstein, J. L., T. H. Huxman, and G. Davidowitz. 2009. Reproductive biology of *Datura wrightii:* The benefits of associating with an herbiverous pollinator. Ann. Bot. (Oxford) 103: 1435–1443.

Broome, C. R. 1973. Systematics of *Centaurium* (Gentianaceae) of Mexico and Central America. Ph.D. dissertation. Duke University.

Broome, C. R. 1978. Chromosome numbers and meiosis in North and Central American species of *Centaurium* (Gentianaceae). Syst. Bot. 3: 299–312.

Broome, C. R. 1981. A new variety of *Centaurium namophilum* (Gentianaceae) from the Great Basin. Great Basin Naturalist 41: 192–197.

Brower, L. P., J. van Brower, and J. M. Corvino. 1967. Plant poisons in a terrestrial food chain. Proc. Natl. Acad. Sci. U.S.A. 57: 893–898.

Brown, R. 1810. Prodromus Florae Novae Hollandiae et Insulae van-Diemen.... London. (Prodr.)

Brown, R. 1810b. On the Asclepiadeae.... London. [Preprinted from Mem. Wern. Nat. Hist. Soc. 1: 12–78. 1811.] (Asclepiadeae)

Brown, R. 1818. Observations, systematical and geographical, on Professor Christian Smith's collection of plants from the vicinity of the river Congo. In: J. H. Tuckey. 1818. Narrative of an Expedition to Explore the River Zaire.... London. Pp. 420–485.

Brown, R. 1818b. Observations...on the Herbarium Collected by Prof. Christian Smith in the Vicinity of the Congo.... London. (Observ. Congo)

Browne, P. 1756. The Civil and Natural History of Jamaica.... London. (Civ. Nat. Hist. Jamaica)

Broyles, S. B. 2002. Hybrid bridges to gene flow: A case study in milkweeds *(Asclepias)*. Evolution 56: 1943–1953.

Broyles, S. B. and J. Wyatt. 1990. Paternity analysis in a natural population of *Asclepias exaltata:* Multiple paternity, functional gender, and the "pollen-donation hypothesis." Evolution 44: 1454–1468.

Brummitt, R. K. and C. E. Powell, eds. 1992. Authors of Plant Names. A List of Authors of Scientific Names of Plants, with Recommended Standard Forms of Their Names, Including Abbreviations. Kew.

Bryson, C. T. and M. S. DeFelice, eds. 2009. Weeds of the South. Athens, Ga.

Bull. Acad. Roy. Sci. Bruxelles = Bulletins de l'Académie Royale des Sciences et Belles-lettres de Bruxelles.

Bull. Calif. Acad. Sci. = Bulletin of the California Academy of Sciences.

Bull. Herb. Boissier = Bulletin de l'Herbier Boissier.

Bull. Josselyn Bot. Soc. Maine = Bulletin of the Josselyn Botanical Society of Maine.

Bull. Mus. Natl. Hist. Nat. = Bulletin du Muséum National d'Histoire Naturelle.

Bull. New York Acad. Med. = Bulletin of the New York Academy of Medicine.

Bull. New York Bot. Gard. = Bulletin of the New York Botanical Garden.

Bull. New York State Mus. Sci. Serv. = Bulletin of the New York State Museum and Science Service.

Bull. S. Calif. Acad. Sci. = Bulletin of the Southern California Academy of Sciences.

Bull. Soc. Neuchâtel. Sci. Nat. = Bulletin de la Société Neuchâteloise de Sciences Naturelles.

Bull. Torrey Bot. Club = Bulletin of the Torrey Botanical Club.

Bull. Univ. New Mexico, Biol. Ser. = Bulletin of the University of New Mexico. Biological Series.

Burkill, H. M. 1985–2004. The Useful Plants of West Tropical Africa, ed. 2. 6 vols. Kew.

Burman, J. 1755. Het Auctuarium.... Amsterdam. (Auctuarium)

Burman, N. L. 1768. Flora Indica ... Nec Non Prodromus Florae Capensis. Leiden and Amsterdam. (Fl. Indica)

Burrows, G. E. and R. J. Tyrl. 2013. Toxic Plants of North America, ed. 2. Oxford.

Bye, R. A. and V. Sosa. 2013. Molecular phylogeny of the jimsonweed genus *Datura* (Solanaceae). Syst. Bot. 38: 818–829.

Callihan, R. H., S. L. Carson, and R. T. Dobbins. 1995. NAWEEDS, Computer-aided Weed Identification for North America. Illustrated User's Guide plus Computer Floppy Disk. Moscow, Idaho.

Cameron, D. D. and J. F. Bolin. 2010. Isotopic evidence of partial mycoheterotrophy in Gentianaceae: *Bartonia viginica* and *Obolaria virginica* as case studies. Amer. J. Bot. 97: 1272–1277.

Canad. Field-Naturalist = Canadian Field-Naturalist.

Canad. J. Bot. = Canadian Journal of Botany.

Canad. J. Pl. Sci. = Canadian Journal of Plant Science.

Candolle, A. L. P. P. de. 1844. Apocynaceae. In: A. P. de Candolle and A. L. P. P. de Candolle, eds. 1823–1873. Prodromus Systematis Naturalis Regni Vegetabilis.... 17 vols. Paris etc. Vol. 8, pp. 317–489.

Candolle, A. P. de and A. L. P. P. de Candolle, eds. 1823–1873. Prodromus Systematis Naturalis Regni Vegetabilis.... 17 vols. Paris etc. [Vols. 1–7 edited by A. P. de Candolle, vols. 8–17 by A. L. P. P. de Candolle.] (Prodr.)

Candollea = Candollea; Organe du Conservatoire et du Jardin Botaniques de la Ville de Genève.

Canotia = Canotia; a New Journal of Arizona Botany.

Card, H. H. 1931. A revision of the genus *Frasera.* Ann. Missouri Bot. Gard. 18: 245–280, plate 14.

Carine, M. A. et al. 2004. Relationships of the Macaronesian and Mediterranean floras: Molecular evidence for multiple colonizations into Macaronesia and back-colonization of the continent in *Convolvulus* (Convolvulaceae). Amer. J. Bot. 91: 1070–1085.

Carrizo García, C. et al. 2018. Unraveling the phylogenetic relationships of *Nectouxia* (Solanaceae): Its position relative to *Salpichroa.* Pl. Syst. Evol. 304: 177–183.

Carter, J. R., J. C. Jones, and R. H. Goddard. 2014. *Sphenoclea zeylanica* (Sphenocleaceae) in North America—Dispersal, ecology and morphology. Castanea 79: 33–50.

Caryologia = Caryologia; Giornale di Citologia, Citosistematica e Citogenetica.

Castanea = Castanea; Journal of the Southern Appalachian Botanical Club.

Castetter, E. F. 1935. Ethnobiological studies in the American Southwest I. Uncultivated native plants used as sources of food. Bull. Univ. New Mexico, Biol. Ser. 4(1): 1–44.

Cat. Edwards's Nat. Hist.—See: C. Linnaeus 1776

Cat. Pl. Amer. Sept.—See: G. H. E. Muhlenberg 1813

Cat. Pl. Cub.—See: A. H. R. Grisebach 1866

Catal. Bot.—See: A. W. Roth 1797–1806

Cavanilles, A. J. 1791–1801. Icones et Descriptiones Plantarum, Quae aut Sponte in Hispania Crescunt, aut in Hortis Hospitantur. 6 vols. Madrid. (Icon.)

Cavanilles, A. J. [1801–]1802. Descripción de las Plantas.... Madrid. (Descr. Pl.)

Cent. Pl. II—See: C. Linnaeus [1756]

Chambers, K. L. and J. Greenleaf. 1989. *Gentiana setigera* is the correct name for *G. bisetaea.* Madroño 36: 49–50.

Chapman, A. W. 1860. Flora of the Southern United States.... New York. (Fl. South. U.S.)

Char. Gen. Pl. ed. 2—See: J. R. Forster and G. Forster 1776

Chase, M. W. et al. 1993. Phylogenetics of seed plants: An analysis of nucleotide sequences from the plastid gene *rbc*L. Ann. Missouri Bot. Gard. 80: 528–580.

Chase, M. W. et al. 2003. Molecular systematics, GISH and the origin of hybrid taxa in *Nicotiana* (Solanaceae). Ann. Bot. (Oxford) 92: 107–127.

Chassot, P. et al. 2001. High paraphyly of *Swertia* L. (Gentianaceae) in the *Gentianella*-lineage as revealed by nuclear and chloroplast DNA sequence variation. Pl. Syst. Evol. 229: 2–21.

Cheatham, S. et al. 1995. The Useful Wild Plants of Texas, the Southeastern and Southwestern United States, the Southern Plains, and Northern Mexico. 2 vols. Austin.

Chiang Cabrera, F. 1981. A Taxonomic Study of the North American Species of *Lycium* (Solanaceae). Ph.D. dissertation. University of Texas.

Chiang Cabrera, F. and L. R. Landrum. 2009. Vascular plants of Arizona: Solanaceae part three: *Lycium*. Canotia 5: 17–26.

Child, A. 1979. A review of branching patterns in the Solanaceae. In: J. G. Hawkes et al., eds. 1979. The Biology and Taxonomy of the Solanaceae. London. Pp. 345–356.

Chron. Bot. = Chronica Botanica....

Chuba, D. et al. 2017. Phylogenetics of the African *Asclepias* complex (Apocynaceae) based on three plastid DNA regions. Syst. Bot. 42: 148–159.

Ciotir, C., C. Yesson, and J. R. Freeland. 2013. The evolutionary history and conservation value of disjunct *Bartonia paniculata* subsp. *paniculata* (branched bartonia) populations in Canada. Botany (Ottawa) 91: 605–613.

Civ. Nat. Hist. Jamaica—See: P. Browne 1756

Cladistics = Cladistics; the International Journal of the Willi Hennig Society.

Clarkson, J. J. et al. 2004 Phylogenetic relationships in *Nicotiana* (Solanaceae) inferred from multiple plastid DNA regions. Molec. Phylogen. Evol. 33: 75–90.

Class-book Bot. ed. s.n.(b)—See: A. Wood 1861

Collectanea—See: N. J. Jacquin 1786[1787]–1796[1797]

Compan. Bot. Mag. = Companion to the Botanical Magazine....

Contr. Arctic Inst. = Contributions of the Arctic Institute, Catholic University of America.

Contr. Dudley Herb. = Contributions from the Dudley Herbarium of Stanford University.

Contr. Herb. Franklin Marshall Coll. = Contributions from the Herbarium of Franklin and Marshall College.

Contr. Lab. Bot. Univ. Montréal = Contributions du Laboratoire de Botanique de l'Université de Montréal.

Contr. Life Sci. Roy. Ontario Mus. = Contributions, Life Sciences, Royal Ontario Museum.

Contr. U.S. Natl. Herb. = Contributions from the United States National Herbarium.

Contr. W. Bot. = Contributions to Western Botany.

Correll, D. S. and M. C. Johnston. 1970. Manual of the Vascular Plants of Texas. Renner, Tex.

Cosner, M. E., R. K. Jansen, and T. G. Lammers. 1994. Phylogenetic relationships in the Campanulales based on *rbc*L sequences. Pl. Syst. Evol. 190: 79–95.

Costea, M. et al. 2011. Systematics of "horned" dodders: Phylogenetic relationships, taxonomy, and two new species within the *Cuscuta chapalana* complex (Convolvulaceae). Botany (Ottawa) 89: 715–730.

Costea, M. et al. 2011b. Systematics of *Cuscuta chinensis* species complex (subgenus *Grammica*, Convolvulaceae): Evidence for long-distance dispersal and one new species. Organisms Diversity Evol. 11: 373–386.

Costea, M. et al. 2013. More problems despite bigger flowers: Systematics of *Cuscuta tinctoria* clade (subgenus *Grammica*, Convolvulaceae) with description of six new species. Syst. Bot. 38: 1160–1187.

Costea, M. et al. 2015. A phylogenetically based infrageneric classification of the parasitic plant genus *Cuscuta* (dodders, Convolvulaceae). Syst. Bot. 40: 269–285.

Costea, M. et al. 2015b. Entangled evolutionary history of *Cuscuta pentagona* clade: A story involving hybridization and Darwin in the Galapagos. Taxon 64: 1225–1242.

Costea, M. et al. 2016. Waterfowl endozoochory: An overlooked long-distance dispersal mode for *Cuscuta* (dodder). Amer. J. Bot. 103: 957–962.

Costea, M., G. L. Nesom, and S. Stefanović. 2006. Taxonomy of *Cuscuta indecora* complex in North America. Sida 22: 176–195.

Costea, M., G. L. Nesom, and S. Stefanović. 2006b. Taxonomy of *Cuscuta salina-californica* complex in North America. Sida 22: 197–207.

Costea, M., G. L. Nesom, and S. Stefanović. 2006c. Taxonomy of *Cuscuta pentagona* complex in North America. Sida 22: 151–175.

Costea, M., G. L. Nesom, and F. J. Tardif. 2005. Taxonomic status of *Cuscuta nevadensis* and *C. veatchii* (Convolvulaceae). Brittonia 57: 264–272.

Costea, M. and S. Stefanović. 2009. *Cuscuta jepsonii* (Convolvulaceae), an invasive weed or an extinct endemic? Amer. J. Bot. 96: 1744–1750.

Costea, M. and S. Stefanović. 2010. Molecular phylogeny of the *Cuscuta californica* complex and a new species from New Mexico and Trans-Pecos. Syst. Bot. 34: 570–579.

Costea, M. and F. J. Tardif. 2006. The biology of Canadian weeds. 133. *Cuscuta campestris* Yuncker, *C. gronovii* Willd. ex Schult., *C. umbrosa* Beyr. ex Hook., *C. epithymum* (L.) L. and *C. epilinum* Weihe. Canad. J. Pl. Sci. 86: 293–316.

Costea, M., M. A. R. Wright, and S. Stefanović. 2009. Untangling the systematics of salt marsh dodders. Syst. Bot. 34: 787–795.

Courson, F. 1998. La Situation du Gentianopsis de Victorin *(Gentianopsis victorinii)* au Québec. Québec.

Cronk, Q. C. B. et al., eds. 2002. Developmental Genetics and Plant Evolution. London.

Cronquist, A. 1981. An Integrated System of Classification of Flowering Plants. New York.

Cronquist, A. et al. 1972–2017. Intermountain Flora. Vascular Plants of the Intermountain West, U.S.A. 7 vols. in 9. New York and London. (Intermount. Fl.)

Crum, H. A. and L. E. Anderson. 1980–1983. Mosses of North America. 4 vols. Durham, N.C. [Text to accompany exsiccata.]

Curr. Med. Chem. = Current Medicinal Chemistry.

Curtis, J. T. 1959. The Vegetation of Wisconsin: An Ordination of Plant Communities. Madison.

Cycl.—See: A. Rees [1802–]1819–1820

D'Arcy, W. G. 1974. *Solanum* and its close relatives in Florida. Ann. Missouri Bot. Gard. 61: 819–867.

D'Arcy, W. G., ed. 1986. Solanaceae: Biology and Systematics. New York.

D'Arcy, W. G. 1991. The Solanaceae since 1976, with a review of its biogeography. In: J. G. Hawkes et al., eds. 1991. Solanaceae III: Taxonomy, Chemistry, Evolution. Kew. Pp. 75–137.

Danm. Holst. Fl.—See: C. G. Rafn 1796–1800

Darwiniana = Darwiniana; Carpeta del "Darwinion."

Davenport, L. J. 1988. A monograph of *Hydrolea* (Hydrophyllaceae). Rhodora 90: 169–208.

Davidse, G., M. Sousa S., A. O. Chater, et al., eds. 1994+. Flora Mesoamericana. 6+ vols. in parts. Mexico City, St. Louis, and London.

Davis, Tilton. 1986. *Jaltomata* in the Tarahumara Indian region of northern Mexico. In: W. G. D'Arcy, ed. 1986. Solanaceae: Biology and Systematics. New York. Pp. 405–411.

Davis, Tilton and R. A. Bye. 1982. Ethnobotany and progressive domestication of *Jaltomata* (Solanaceae) in Mexico and Central America. Econ. Bot. 36: 225–241.

Dawson, J. H. et al. 1994. Biology and control of *Cuscuta*. Rev. Weed Sci. 6: 265–317.

Delic. Fl. Faun. Insubr.—See: J. A. Scopoli 1786[–1788]

Delpinoa = Delpinoa; Nuova Serie del Bulletino dell' Orto Botanico della Università di Napoli.

Demonstr. Pl.—See: C. Linnaeus [1753]

Dennstedt, A. W. 1810. Nomenclator Botanicus.... 2 vols. Eisenberg. (Nomencl. Bot.)

Des Moulins, C. R. A. 1853. Études Organiques sur les Cuscutes. Toulouse. (Étud. Cuscut.)

Descr. Pl.—See: A. J. Cavanilles [1801–]1802

Diagn. Pl. Nov. Mexic.—See: W. B. Hemsley 1878–1880

Diaz, R., W. A. Overholt, and K. Langeland. 2008. Jamaican nightshade *(Solanum jamaicense)*: A threat to Florida's hammocks. Invasive Pl. Sci. Managem. 1: 422–425.

Diezel, C., D. Kessler, and I. T. Baldwin. 2011. Pithy protection: *Nicotiana attenuata*'s jasmonic acid-mediated defenses are required to resist stem-boring weevil larvae. Pl. Physiol. (Lancaster) 155: 1936–1946.

DiTomasso, A., F. M. Lawlor, and S. J. Darbyshire. 2005. The biology of invasive alien plants in Canada. 2. *Cynanchum rossicum* (Kleopow) Borhidi [= *Vincetoxicum rossicum* (Kleopow) Barbar.] and *Cynanchum louisieae* (L.) Kartesz & Gandhi [= *Vincetoxicum nigrum* (L.) Moench]. Canad. J. Pl. Sci. 85: 243–263.

Don, D. 1825. Prodromus Florae Nepalensis.... London. (Prodr. Fl. Nepal.)

Don, G. 1831–1838. A General History of the Dichlamydeous Plants.... 4 vols. London. (Gen. Hist.)

Drury, H. 1873. The Useful Plants of India, ed. 2. London.

Dunal, M. F. 1813. Histoire Naturelle, Médicale et Économique des *Solanum*.... Paris etc. (Hist. Nat. Solanum)

Durand, E. M. and T. C. Hilgard. 1854. Plantae Heermannianae.... Philadelphia. [Preprinted from J. Acad. Nat. Sci. Philadelphia, n. s. 3: 37–46. 1855.] (Pl. Heermann.)

Dutilly, A., E. Lepage, and M. G. Duman. 1958. Contribution à la flore des îles (T.N.G.) et du versant oriental (Qué.) de la Baie James. Contr. Arctic Inst. 9F.

Eckenwalder, J. E. 1986. Nomenclature of the cardinal climber (Convolvulaceae) reconsidered. Taxon 35: 169–170.

Ecology = Ecology, a Quarterly Journal Devoted to All Phases of Ecological Biology.

Econ. Bot. = Economic Botany; Devoted to Applied Botany and Plant Utilization.

Edinburgh New Philos. J. = Edinburgh New Philosophical Journal.

Edmonds, J. M. 1986. Biosystematics of *Solanum sarrachoides* Sendtner and *S. physalifolium* Rusby (*S. nitidibaccatum* Bitter). Bot. J. Linn. Soc. 92: 1–38.

Edwards's Bot. Reg. = Edwards's Botanical Register....

Eich, E. 2008. Solanaceae and Convolvulaceae: Secondary Metabolites: Biosynthesis, Chamotaxonomy, Biological and Economic Significance (a Handbook). Berlin.

Elench. Pl.—See: D. Viviani 1802

Elliott, S. [1816–]1821–1824. A Sketch of the Botany of South-Carolina and Georgia. 2 vols. in 13 parts. Charleston. (Sketch Bot. S. Carolina)

Elmore, C. D. 1986. Mode of reproduction and inheritance of leaf shape in *Ipomoea hederacea*. Weed Sci. 34: 391–395.

Emory, W. H. 1857–1859. Report on the United States and Mexican Boundary Survey, Made under the Direction of the Secretary of the Interior. 2 vols. in parts. Washington. (Rep. U.S. Mex. Bound.)

Encycl.—See: J. Lamarck et al. 1783–1817

Endlicher, S. L. 1836–1840[–1850]. Genera Plantarum Secundum Ordines Naturales Disposita. 18 parts + 5 suppls. in 6 parts. Vienna. [Paged consecutively through suppl. 1(2); suppls. 2–5 paged independently.] (Gen. Pl.)

Endlicher, S. L. and E. Fenzl. 1839. Novarum Stirpium Decades. 10 parts. Vienna. [Parts (decades) paged consecutively.] (Nov. Stirp. Dec.)

Endress, M. E., S. Liede, and U. Meve. 2014. An updated classification for Apocynaceae. Phytotaxa 159: 175–194.

Endress, P. K. 1997. Diversity and Evolutionary Biology of Tropical Flowers. Cambridge and New York.

Engelmann, G. 1859. Systematic arrangement of the species of the genus *Cuscuta* with critical remarks on old species and descriptions of new ones. Trans. Acad. Sci. St. Louis 1: 453–523.

Engelmann, G. 1863. New species of *Gentiana*, from the alpine regions of the Rocky Mountains. Trans. Acad. Sci. St. Louis 2: 214–220.

Engelmann, G. 1879. Gentianeae. In: J. T. Rothrock. 1878[1879]. Report upon United States Geographical Surveys West of the One Hundredth Meridian, in Charge of First Lieut. Geo. M. Wheeler.... Vol. 6—Botany. Washington. Pp. 189–197.

Engler, H. G. A., ed. 1900–1953. Das Pflanzenreich.... 107 vols. Berlin. [Sequence of vol. (Heft) numbers (order of publication) is independent of the sequence of series and family (Roman and Arabic) numbers (taxonomic order).]

Engler, H. G. A. et al., eds. 1924+. Die natürlichen Pflanzenfamilien..., ed. 2. 26+ vols. Leipzig and Berlin.

Engler, H. G. A. and K. Prantl, eds. 1887–1915. Die natürlichen Pflanzenfamilien.... 254 fascs. Leipzig. [Sequence of fasc. (Lieferung) numbers (order of publication) is independent of the sequence of division (Teil) and subdivision (Abteilung) numbers (taxonomic order).] (Nat. Pflanzenfam.)

English, C. S. 1934. Notes on northwestern flora. Proc. Biol. Soc. Wash. 47: 189–192.

Enum. Pl. Hort. Regiom.—See: A. F. Schweigger 1812

Enum. Subst. Braz.—See: A. L. P. da Silva Manso 1836

Enum. Syst. Pl.—See: N. J. Jacquin 1760

Environment Canada. 2011. Recovery Strategy for the Victorin's Gentian (*Gentianopsis virgata* subsp. *virgata*) in Canada. Ottawa.

Erbar, C., S. Porembski, and P. Leins. 2005. Contributions to the systematic position of *Hydrolea* (Hydroleaceae) based on floral development. Pl. Syst. Evol. 252: 71–83.

Erythea = Erythea; a Journal of Botany, West American and General.

Eshbaugh, W. H. 2012. The taxonomy of the genus *Capsicum* (Solanaceae). In: V. M. Russo, ed. 2012. Peppers: Botany, Production and Uses Wallingford. Pp. 14–28.

Étud. Cuscut.—See: C. R. A. Des Moulins 1853

Evans, W. C. et al. 1972. Alkaloids of *Salpichroa origanifolia*. Phytochemistry 11: 469.

Evol. Monogr. = Evolutionary Monographs.

Evolution = Evolution; International Journal of Organic Evolution.

Explor. Red River Louisiana—See: R. B. Marcy 1853

Expos. Fam. Nat.—See: J. St.-Hilaire 1805

Fairchild Trop. Gard. Bull. = Fairchild Tropical Garden Bulletin.

Fallen, M. E. 1985. The gynoecial development and systematic position of *Allamanda* (Apocynaceae). Amer. J. Bot. 72: 572–579.

Fallen, M. E. 1986. Floral structure in the Apocynaceae: Morphological, functional and evolutionary aspects. Bot. Jahrb. Syst. 106: 245–286.

Fam. Pl.—See: M. Adanson 1763[–1764]

Favre, A. et al. 2016. Out-of-Tibet: The spatio-temporal evolution of *Gentiana* (Gentianaceae). J. Biogeogr. 43: 1967–1978.

Favre, A. et al. 2020. Phylogenetic relationships and sectional delineation within *Gentiana* (Gentianaceae). Taxon 69: 1221–1238.

Felger, R. S., S. Rutman, and J. Malusa. 2014. Ajo Peak to Tinajas Altas: Flora of southwestern Arizona: Part 8. Eudicots: Acanthaceae–Apocynaceae. Phytoneuron 2014-85: 1–71.

Ferguson, D. M. 1998. Phylogenetic analysis and relationships in Hydrophyllaceae based on *ndh*F sequence data. Syst. Bot. 23: 253–268.

Fernald, M. L. 1902. Some little known plants from Florida and Georgia. Bot. Gaz. 33: 154–157.

Fernald, M. L. 1917b. Some forms of American gentians. Rhodora 19: 149–152.

Field & Lab. = Field & Laboratory.

Fishbein, M. 2017. Taxonomic adjustments in North American Apocynaceae. Phytologia 99: 86–88.

Fishbein, M. et al. 2011. Phylogenetic relationships of *Asclepias* (Apocynaceae) inferred from non-coding chloroplast DNA sequences. Syst. Bot. 36: 1008–1023.

Fishbein, M. et al. 2018. Evolution at the tips: *Asclepias* phylogenomics and new perspectives on leaf surfaces. Amer. J. Bot. 105: 514–524.

Fishbein, M. and K. N. Gandhi. 2018. Typification of *Sarcostemma heterophyllum* and nomenclatural notes in North American *Funastrum* (Apocynaceae). Novon 26: 165–167.

Fishbein, M., V. Juárez-Jaimes, and L. O. Alvarado-Cárdenas. 2008. Resurrection of *Asclepias schaffneri* (Apocynaceae, Asclepiadadoideae), a rare, Mexican milkweed. Madroño 55: 69–75.

Fishbein, M. and W. D. Stevens. 2005. Resurrection of *Seutera* Reichenbach (Apocynaceae, Asclepiadoideae). Novon 15: 531–533.

Fishbein, M. and D. L. Venable. 1996. Diversity and temporal change in the effective pollinators of *Asclepias tuberosa*. Ecology 77: 1061–1073.

Fishbein, M. and D. L. Venable. 1996b. Evolution of inflorescence design: Theory and data. Evolution 50: 2165–2177.

Fl. Adv. Montpellier—See: A. Thellung 1912

Fl. Aegypt.-Arab.—See: P. Forsskål 1775

Fl. Amer. Sept.—See: F. Pursh [1813]1814

Fl. Bor.-Amer.—See: W. J. Hooker [1829–]1833–1840; A. Michaux 1803

Fl. Bras.—See: C. F. P. von Martius et al. 1840–1906

Fl. Brit. W. I.—See: A. H. R. Grisebach [1859–]1864

Fl. Carol.—See: T. Walter 1788

Fl. Cochinch.—See: J. de Loureiro 1790

Fl. Deut. Volk—See: C. Börner [1912]b

Fl. Ind.—See: W. Roxburgh 1820–1824

Fl. Indica—See: N. L. Burman 1768

Fl. N. Amer.—See: J. Torrey and A. Gray 1838–1843

Fl. N. Middle United States—See: J. Torrey [1823–]1824

Fl. Peruv.—See: H. Ruiz López and J. A. Pavon 1798–1802

Fl. Peruv. Prodr.—See: H. Ruiz López and J. A. Pavon 1794

Fl. Portug.—See: J. C. Hoffmannsegg and H. F. Link 1809 [–1840]

Fl. Ross.—See: P. S. Pallas 1784–1788[–1831]

Fl. S.E. U.S.—See: J. K. Small 1903

Fl. S.E. U.S. ed. 2—See: J. K. Small 1913

Fl. South. U.S.—See: A. W. Chapman 1860

Fl. Tellur.—See: C. S. Rafinesque 1836[1837–1838]

Fl. Trop. Afr.—See: D. Oliver et al. 1868–1937

Fl. Virgin.—See: B. S. Barton 1812

Fl. W. Calif.—See: W. L. Jepson 1901

Flexner, S. B. and L. C. Hauck, eds. 1987. The Random House Dictionary of the English Language, ed. 2 unabridged. New York.

Flora = Flora; oder (allgemeine) botanische Zeitung. = [Vols. 1–16, 1818–1833, include "Beilage" and "Ergänzungsblätter"; vols. 17–25, 1834–1842, include "Beiblatt" and "Intelligenzblatt."]

Florida Entomol. = Florida Entomologist.

Florida Sci. = Florida Scientist.

Folia Geobot. Phytotax. = Folia Geobotanica et Phytotaxonomica.

Forbes, T. R. 1977. Why is it called "beautiful lady"? A note on belladonna. Bull. New York Acad. Med. 53: 403–406.

Forsskål, P. 1775. Flora Aegyptiaco-Arabica. Copenhagen. (Fl. Aegypt.-Arab.)

Forster, G. 1786b. De Plantis Esculentis Insularum Oceani Australis Commentatio Botanica. Berlin. (Pl. Esc.)

Forster, J. R. and G. Forster. 1776. Characteres Generum Plantarum, Quas in Itinere ad Insulas Maris Australis..., ed. 2. London. (Char. Gen. Pl. ed. 2)

Fosberg, F. R. 1976. *Ipomoea indica* taxonomy: A tangle of morning glories. Bot. Not. 129: 35–38.

Francey, P. 1935. Monographie du genre *Cestrum* L. Candollea 6: 46–398.

Franklin, J. et al. 1823. Narrative of a Journey to the Shores of the Polar Sea, in the Years 1819, 20, 21 and 22. London. [Richardson: Appendix VII. Botanical appendix, pp. [729]–768, incl. bryophytes by Schwägrichen, algae and lichens by Hooker.] (Narr. Journey Polar Sea)

Frasier, C. L. 2008. Evolution and Systematics of the Angiosperm Order Gentianales with an In-depth Focus on Loganiaceae and Its Species-rich and Toxic Genus *Strychnos*. Ph.D. dissertation. Rutgers, The State University of New Jersey.

Frémont, J. C. 1843–1845. Report of the Exploring Expedition to the Rocky Mountains in the Year 1842, and to Oregon and North California in the Year 1843–44. 2 parts. Washington. [Parts paged consecutively.] (Rep. Exped. Rocky Mts.)

Fries, E. M. 1845–1849. Summa Vegetabilium Scandinaviae.... 2 sects. Uppsala, Stockholm, and Leipzig. [Sections paged consecutively.] (Summa Veg. Scand.)

Froelich, J. A. von. [1796]. De *Gentiana* Dissertatio.... Erlangen. (Gentiana)

Fruct. Sem. Pl.—See: J. Gaertner 1788–1791[–1792]

Gaertner, J. 1788–1791[–1792]. De Fructibus et Seminibus Plantarum.... 2 vols. Stuttgart and Tübingen. [Vol. 1 in 1 part only, 1788. Vol. 2 in 4 parts paged consecutively: pp. 1–184, 1790; pp. 185–352, 353–504, 1791; pp. 505–520, 1792.] (Fruct. Sem. Pl.)

Gann, G. D., K. A. Bradley, and S. W. Woodmansee. 2002. Rare Plants of South Florida: Their History, Conservation and Restoration. Miami.

García, M. A. et al. 2014. Phylogeny, character evolution, and biogeography of *Cuscuta* (dodders, Convolvulaceae) inferred from coding plastid and nuclear sequences. Amer. J. Bot. 101: 670–690.

García, M. A. and S. Castroviejo. 2003. Estudios citotaxonómicos en las especies Ibéricas del género *Cuscuta* (Convolvulaceae). Anales Jard. Bot. Madrid 60: 33–44.

García, M. A. and M. P. Martín. 2007. Phylogeny of *Cuscuta* subgenus *Cuscuta* (Convolvulaceae) based on nrDNA ITS and chloroplast *trn*L intron sequences. Syst. Bot. 32: 899–916.

Gard. Dict. ed. 8—See: P. Miller 1768

Gard. Dict. Abr. ed. 4—See: P. Miller 1754

Gardner, H. W. 2011. Tallgrass Prairie Restoration in the Midwestern and Eastern United States: A Hands-on Guide. New York.

Gartenflora = Gartenflora; Monatsschrift für deutsche und schweizerische Garten- und Blumenkunde.

Gatt. Lycianthes—See: F. A. G. Bitter 1919

Geeta, R. and W. Gharaibeh. 2007. Historical evidence for a pre-Columbian presence of *Datura* in the Old World and implications for a first millennium transfer from the New World. J. Biosci. 32: 1227–1244.

Gen. Hist.—See: G. Don 1831–1838

Gen. N. Amer. Pl.—See: T. Nuttall 1818

Gen. Nicot. Hist.—See: J. G. C. Lehmann 1818

Gen. Pl.—See: G. Bentham and J. D. Hooker 1862–1883; A. L. de Jussieu 1789; S. L. Endlicher 1836–1840[–1850]; N. M. Wolf 1776

Gen. Pl. ed. 5—See: C. Linnaeus 1754

Gen. Pl. ed. 6—See: C. Linnaeus 1764

Gen. Sp. Gent.—See: A. H. R. Grisebach 1839

Genet. Resources Crop Evol. = Genetic Resources and Crop Evolution; an International Journal.

Gentes Herbarum = Gentes Herbarum; Occasional Papers on the Kinds of Plants.

Gentiana—See: J. A. von Froelich [1796]

Gentry, A. H. 1993. A Field Guide to the Families and Genera of Woody Plants of Northwest South America.... Chicago.

Gerats, T. and J. Strommer, eds. 2009. *Petunia:* Evolutionary, Developmental and Physiological Genetics, ed. 2. New York.

Ges. Naturf. Freunde Berlin Neue Schriften = Der Gesellschaft naturforschender Freunde zu Berlin, neue Schriften.

Gillett, J. M. 1957. A revision of the North American species of *Gentianella* Moench. Ann. Missouri Bot. Gard. 44: 195–269.

Gillett, J. M. 1959. A revision of *Bartonia* and *Obolaria* (Gentianaceae). Rhodora 61: 43–62.

Gleason, H. A. 1952. The New Britton and Brown Illustrated Flora of the Northeastern United States and Adjacent Canada. 3 vols. New York.

Gleason, H. A. and A. Cronquist. 1991. Manual of Vascular Plants of Northeastern United States and Adjacent Canada, ed. 2. Bronx.

Gmelin, J. F. 1791[–1792]. Caroli à Linné...Systema Naturae per Regna Tria Naturae.... Tomus II. Editio Decima Tertia, Aucta, Reformata. 2 parts. Leipzig. (Syst. Nat.)

Goodspeed, T. H. 1954. The genus *Nicotiana*. Chron. Bot. 16.

Gould, K. R. 1997. Systematic Studies in *Spigelia*. Ph.D. dissertation. University of Texas.

Gould, K. R. and R. K. Jansen. 1999. Taxonomy and phylogeny of a Gulf Coast disjunct group of *Spigelia* (Loganiaceae sensu lato). Lundellia 2: 1–13.

Goyder, D. J. 2009. A synopsis of *Asclepias* (Apocynaceae: Asclepiadoideae) in tropical Africa. Kew Bull. 64: 369–399.

Goyder, D. J., A. Nicholas, and S. Liede-Schumann. 2007. Phylogenetic relationships in subtribe Asclepiadinae (Apocynaceae: Asclepiadoideae). Ann. Missouri Bot. Gard. 94: 423–434.

Grana Palynol. = Grana Palynologica; an International Journal of Palynology.

Grau, J. and E. Gronbach. 1984. Untersuchungen zur Variabilität in der Gattung *Schizanthus* (Solanaceae). Mitt. Bot. Staatssamml. München 20: 111–203.

Gray, A. 1848. A Manual of the Botany of the Northern United States.... Boston, Cambridge, and London. (Manual)

Gray, A. 1848b. *Obolaria virginica,* Linn. Mem. Amer. Acad. Arts, ser. 2, 3: 21–31.

Gray, A. 1856. A Manual of the Botany of the Northern United States..., ed. 2. New York. (Manual ed. 2)

Gray, A. 1875. A conspectus of the North American Hydrophyllaceae. Proc. Amer. Acad. Arts 10: 312–332.

Gray, A. et al. 1886. Synoptical Flora of North America: The Gamopetalae, Being a Second Edition of Vol. i Part ii, and Vol. ii Part i, Collected. 2 vols. New York, London, and Leipzig. [Reissued 1888 as Smithsonian Misc. Collect. 591.] (Syn. Fl. N. Amer. ed. 2)

Gray, A., S. Watson, B. L. Robinson, et al. 1878–1897. Synoptical Flora of North America. 2 vols. in parts and fascs. New York etc. [Vol. 1(1,1), 1895; vol. 1(1,2), 1897; vol. 1(2), 1884; vol. 2(1), 1878.] (Syn. Fl. N. Amer.)

Great Basin Naturalist = Great Basin Naturalist.

Great Plains Flora Association. 1986. Flora of the Great Plains. Lawrence, Kans.

Greene, E. L. 1894. Manual of the Botany of the Region of San Francisco Bay.... San Francisco. (Man. Bot. San Francisco)

Greene, E. L. 1904. Seven new apocynums. Leafl. Bot. Observ. Crit. 1: 56–59.

Greene, E. L. 1912. Accessions to *Apocynum*. Leafl. Bot. Observ. Crit. 2: 164–189.

Grisebach, A. H. R. 1839. Genera et Species Gentianearum.... Stuttgart and Tübingen. (Gen. Sp. Gent.)

Grisebach, A. H. R. 1843–1845[–1846]. Spicilegium Florae Rumelicae et Bithynicae.... 2 vols. in 6 parts. Braunschweig. [Vols. paged independently but parts numbered consecutively.] (Spic. Fl. Rumel.)

Grisebach, A. H. R. [1859–]1864. Flora of the British West Indian Islands. 7 parts. London. [Parts paged consecutively.] (Fl. Brit. W. I.)

Grisebach, A. H. R. 1866. Catalogus Plantarum Cubensium Exhibens Collectionem Wrightianam Aliasque Minores ex Insula Cuba Missas. Leipzig. (Cat. Pl. Cub.)

Groff, P. A. 1989. Studies in Whole-plant Morphology. Ph.D. dissertation. University of California, Berkeley.

Gübitz, T. et al. 2009. *Petunia* as a model system for the genetics and evolution of pollination syndromes. In: T. Gerats and J. Strommer, eds. 2009. *Petunia:* Evolutionary, Developmental and Physiological Genetics, ed. 2. New York. Pp. 29–50.

Gunn, C. R. 1970. History and taxonomy of the purple moonflower, *Ipomoea turbinata* Lagasca y Segura. Proc. Assoc. Off. Seed Analysts N. Amer. 59: 116–123.

Haber, W. A. 1984. Pollination by deceit in a mass-flowering tropical tree *Plumeria rubra* L. (Apocynaceae). Biotropica 16: 269–275.

Hagen, K. B. von and J. W. Kadereit. 2001. The phylogeny of *Gentianella* (Gentianaceae) and its colonization of the Southern Hemisphere as revealed by nuclear and chloroplast DNA sequence variation. Organisms Diversity Evol. 1: 61–79.

Hagen, K. B. von and J. W. Kadereit. 2002. Phylogeny and flower evolution of the Swertiinae (Gentianaceae, Gentianeae): Homoplasy and the principle of variable proportions. Syst. Bot. 27: 592–597.

Haines, A. 1998. Flora of Maine.... Bar Harbor.

Haines, A. 2011. New England Wildflower Society's Flora Novae Angliae.... New Haven.

Halda, J. J. 1996. The Genus *Gentiana*. Dobré.

Hallier, H. 1897. Bausteine zu einer Monographie der Convolvulaceen. Geneva. (Bausteine Monogr. Convolv.)

Hallier, H. 1893. Versuch einer natürlichen Gliederung der Convolvulceen auf morphologischer und anatomischer Grundlage. Bot. Jahrb. Syst. 16: 453–591.

Hammer, K., A. Romeike, and C. Tittel. 1983. Vorarbeiten zur monographischen Darstellung von wildpflanzensortimenten: *Datura* L., sections *Dutra* Bernh., *Ceratocaulis* Bernh. et *Datura*. Kulturpflanze 31: 13–75.

Hansen, B. F. and R. P. Wunderlin. 1986. *Pentalinon* Voigt, an earlier name for *Urechites* Müll. Arg. (Apocynaceae). Taxon 35: 166–168.

Harshberger, J. W. 1898. Botanical observations on the Mexican flora, especially on the flora of the Valley of Mexico. Proc. Acad. Nat. Sci. Philadelphia 50: 372–413.

Hartnett, D. C. and D. R. Richardson. 1989. Population biology of *Bonamia grandiflora* (Convolvulaceae): Effects of fire on plant and seed bank dynamics. Amer. J. Bot. 76: 361–369.

Harvard Pap. Bot. = Harvard Papers in Botany.

Haston, E. M. et al. 2007. A linear sequence of Angiosperm Phylogeny Group II families. Taxon 56: 7–12.

Hawkes, J. G. et al., eds. 1979. The Biology and Taxonomy of the Solanaceae. London.

Hawkes, J. G. et al., eds. 1991. Solanaceae III: Taxonomy, Chemistry, Evolution. Kew.

Hayata, B. 1911–1921. Icones Plantarum Formosanarum.... 10 vols. Taihoku. (Icon. Pl. Formos.)

Hemsley, W. B. 1878–1880. Diagnoses Plantarum Novarum... Mexicanarum et Centrali-americanarum. 3 parts. London. [Parts paged consecutively.] (Diagn. Pl. Nov. Mexic.)

Hemsley, W. B. 1879–1888. Biologia Centrali-Americana.... Botany.... 5 vols. London. (Biol. Cent.-Amer., Bot.)

Henrickson, J. 1996. Notes on *Spigelia* (Loganiaceae). Sida 17: 89–103.

Henrickson, J. 1996b. Studies in *Macrosiphonia* (Apocynaceae): Generic recognition of *Telosiphonia.* Aliso 14: 179–195.

Henrickson, J. 2009. New names in *Chamaesaracha* (Solanaceae). Phytologia 91: 186–187.

Hereditas (Lund) = Hereditas. (Lund.)

Hickman, J. C., ed. 1993. The Jepson Manual. Higher Plants of California. Berkeley, Los Angeles, and London.

Hill, J. 1756[–1757]. The British Herbal: An History of Plants and Trees.... 52 fascs. London. [Fascicles paged and plates numbered consecutively.] (Brit. Herb.)

Hinton, W. F. 1970. The taxonomic status of *Physalis lanceolata* (Solanaceae) in the Carolina sandhills. Brittonia 22: 14–19.

Hinton, W. F. 1975b. Natural hybridization and extinction of a population of *Physalis virginiana* (Solanaceae). Amer. J. Bot. 62: 198–202.

Hinton, W. F. 1976. The systematics of *Physalis pumila* subsp. *hispida* (Solanaceae). Syst. Bot. 1: 188–193.

Hist. Nat. Solanum—See: M. F. Dunal 1813

Hist. Pl. Guiane—See: J. B. Aublet 1775

Hist. Vég. Îles France—See: L. M. A. du P. Thouars 1804

Hitchcock, A. S. 1894. A Key to the Spring Flora of Manhattan.... Manhattan, Kans. (Key Spring Fl. Manhattan)

Hitchcock, C. L. 1932. A monographic study of the genus *Lycium* of the western hemisphere. Ann. Missouri Bot. Gard. 19: 179–348, 350–375.

Hitchcock, C. L. 1959. Gentianaceae. In: C. L. Hitchcock et al. 1955–1969. Vascular Plants of the Pacific Northwest. 5 vols. Seattle. Vol. 4, pp. 57–76..

Hitchcock, C. L. et al. 1955–1969. Vascular Plants of the Pacific Northwest. 5 vols. Seattle. [Univ. Wash. Publ. Biol. 17.] (Vasc. Pl. Pacif. N.W.)

Ho, T. N. and Liu S. W. 2001. A Worldwide Monograph of *Gentiana.* Beijing and New York.

Ho, T. N., Liu S. W., and Lu X. F. 1996. A phylogenetic analysis of *Gentiana* (Gentianaceae). Acta Phytotax. Sin. 34: 505–530.

Hoballah, M. E. et al. 2007. Single gene-mediated shift in pollinator attraction in *Petunia.* Pl. Cell 19: 779–790.

Hocking, G. M. 1947. Henbane: Healing herb of Hercules and of Apollo. Econ. Bot. 1: 306–316.

Hoffmannsegg, J. C. and H. F. Link. 1809[–1840]. Flore Portugaise.... 2 vols in 22 parts. Berlin. [Vols. paged independently, plates numbered consecutively.] (Fl. Portug.)

Holm, R. W. 1950. The American species of *Sarcostemma* R. Br. (Asclepiadaceae). Ann. Missouri Bot. Gard. 37: 477–560.

Holm, T. 1897. *Obolaria virginica* L.: A morphological and anatomical study. Ann. Bot. (Oxford) 11: 369–383.

Holmgren, N. H. 1984b. Gentianaceae. In: A. Cronquist et al. 1972–2017. Intermountain Flora. Vascular Plants of the Intermountain West, U.S.A. 7 vols. in 9. New York and London. Vol. 4, pp. 4–23.

Hooker, W. J. [1829–]1833–1840. Flora Boreali-Americana; or, the Botany of the Northern Parts of British America.... 2 vols. in 12 parts. London, Paris, and Strasbourg. (Fl. Bor.-Amer.)

Hooker, W. J. and G. A. W. Arnott. [1830–]1841. The Botany of Captain Beechey's Voyage; Comprising an Account of the Plants Collected by Messrs Lay and Collie, and Other Officers of the Expedition, during the Voyage to the Pacific and Bering's Strait, Performed in His Majesty's Ship Blossom, under the Command of Captain F. W. Beechey...in the Years 1825, 26, 27, and 28. 10 parts. London. [Parts paged and plates numbered consecutively.] (Bot. Beechey Voy.)

Hort. Berol.—See: C. L. Willdenow 1803–1816

Hort. Brit.—See: R. Sweet 1826

Hort. Brit. ed. 2—See: R. Sweet 1830

Hort. Brit. ed. 3—See: J. C. Loudon 1839

Hort. Kew.—See: W. Aiton 1789

Hort. Suburb. Calcutt.—See: J. O. Voigt 1845

Horticulture (Boston) = Horticulture; an Illustrated Journal [later the Magazine (later Art) of American Gardening].

Hortus Kew.—See: W. Aiton and W. T. Aiton 1810–1813

Hortus Maurit.—See: W. Bojer 1837

Hudson, W. D. 1986. Relationships of domesticated and wild *Physalis philadelphica.* In: W. G. D'Arcy, ed. 1986. Solanaceae: Biology and Systematics. New York. Pp. 416–432.

Hultén, E. 1968. Flora of Alaska and Neighboring Territories: A Manual of the Vascular Plants. Stanford.

Humboldt, A. von, A. J. Bonpland, and C. S. Kunth. 1815[1816]–1825. Nova Genera et Species Plantarum Quas in Peregrinatione Orbis Novi Collegerunt, Descripserunt.... 7 vols. in 36 parts. Paris. (Nov. Gen. Sp.)

Hunziker, A. T. 2001. Genera Solanacearum: The Genera of Solanaceae Illustrated.... Ruggell.

Hurley, H. 1968. A Taxonomic Revision of the Genus *Spigelia* (Loganiaceae). Ph.D. dissertation. George Washington University.

Icon.—See: A. J. Cavanilles 1791–1801

Icon. Pl. Formos.—See: B. Hayata 1911–1921

Icon. Pl. Rar.—See: N. J. Jacquin 1781–1793[–1795]

Ill. Fl. Pacific States—See: L. Abrams and R. S. Ferris 1923–1960

Illinois Biol. Monogr. = Illinois Biological Monographs.

Iltis, H. H. 1965. The genus *Gentianopsis* (Gentianaceae): Transfers and phytogeographic comments. Sida 2: 129–154.

Imhoff, S. 1997. Root anatomy and mycotrophy of the achlorophyllous *Voyria tenella* Hook. (Gentianaceae). Bot. Acta 110: 298–305.

Imhoff, S. 1999. Root morphology, anatomy and mycotrophy of the achlorophyllous *Voyria aphylla* (Jacq.) Pers. (Gentianaceae). Mycorrhiza 9: 33–39.

Index Kew.—See: B. D. Jackson et al. [1893–]1895+

Index Seminum (Berlin), App. = Index Seminum (Berlin), Appendix.

Inouye, D. W. and O. R. Taylor. 1980. Variation in generation time in *Frasera speciosa* (Gentianaceae), a long-lived perennial monocarp. Oecologia 47: 171–174.

Int. J. Pl. Sci. = International Journal of Plant Sciences.

Intermount. Fl.—See: A. Cronquist et al. 1972–2017

Interpr. Herb. Amboin.—See: E. D. Merrill 1917

Invasive Pl. Sci. Managem. = Invasive Plant Science and Management.

Itoh, K. and K. Ito. 1994. Weed ecology and its control in south-east tropical countries. Jap. J. Trop. Agric. 38: 369–373.

Iwatsuki, K. et al., eds. 1993+. Flora of Japan. 3+ vols. in 6+. Tokyo.

J. Acad. Nat. Sci. Philadelphia = Journal of the Academy of Natural Sciences of Philadelphia.

J. Arizona-Nevada Acad. Sci. = Journal of the Arizona-Nevada Academy of Science.

J. Arnold Arbor. = Journal of the Arnold Arboretum.

J. Biogeogr. = Journal of Biogeography.

J. Biosci. = Journal of Biosciences.

J. Bot. (Hooker) = Journal of Botany, (Being a Second Series of the Botanical Miscellany), Containing Figures and Descriptions....

J. Bot. Res. Inst. Texas = Journal of the Botanical Research Institute of Texas.

J. Ethnopharmacol. = Journal of Ethnopharmacology; Interdisciplinary Journal Devoted to Bioscientific Research on Indigenous Drugs.

J. Evol. Biol. = Journal of Evolutionary Biology.

J. Fac. Sci. Hokkaido Imp. Univ., Ser. 5, Bot. = Journal of the Faculty of Science of the Hokkaido Imperial University. Series 5, Botany. [Rigaku-bu Kiyo.]

J. Herbal Pharmacotherapy = Journal of Herbal Pharmacotherapy; Innovations in Clinical & Applied Evidence Based Herbal Medicinals.

J. Heredity = Journal of Heredity.

J. Hort. Soc. London = Journal of the Horticultural Society of London.

J. Linn. Soc., Bot. = Journal of the Linnean Society. Botany.

J. Nat. Prod. (Lloydia) = Journal of Natural Products (Lloydia).

J. Torrey Bot. Soc. = Journal of the Torrey Botanical Society.

J. Wash. Acad. Sci. = Journal of the Washington Academy of Sciences.

Jackson, B. D. et al. comps. [1893–]1895+. Index Kewensis Plantarum Phanerogamarum.... 2 vols. + 21+ suppls. Oxford. (Index Kew.)

Jacquin, N. J. 1760. Enumeratio Systematica Plantarum, Quas in Insulis Caribaeis Vicinaque Americes Continente Detexit Novas.... Leiden. (Enum. Syst. Pl.)

Jacquin, N. J. 1763. Selectarum Stirpium Americanarum Historia.... Vienna. (Select. Stirp. Amer. Hist.)

Jacquin, N. J. 1764–1771. Observationum Botanicarum.... 4 parts. Vienna. (Observ. Bot.)

Jacquin, N. J. 1781–1793[–1795]. Icones Plantarum Rariorum. 3 vols. in fascs. Vienna etc. [Vols. paged independently, plates numbered consecutively.] (Icon. Pl. Rar.)

Jacquin, N. J. 1786[1787]–1796[1797]. Collectanea ad Botanicam, Chemiam, et Historiam Naturalem Spectantia.... 5 vols. Vienna. (Collectanea)

Jacquin, N. J. 1797–1804. Plantarum Rariorum Horti Caesarei Schoenbrunnensis Descriptiones et Icones. 4 vols. Vienna, London, and Leiden. (Pl. Hort. Schoenbr.)

Jap. J. Trop. Agric. = Japanese Journal of Tropical Agriculture.

Jard. Fleur.—See: C. Lemaire 1851–1854

Jepson, W. L. 1901. A Flora of Western Middle California.... Berkeley. (Fl. W. Calif.)

Jepson, W. L. [1923–1925.] A Manual of the Flowering Plants of California.... Berkeley. (Man. Fl. Pl. Calif.)

Johnson, S. A., L. P. Bruederle, and D. F. Tomback. 1998. A mating system conundrum: Hybridization in *Apocynum* (Apocynaceae). Amer. J. Bot. 85: 1316–1323.

Jussieu, A. L. de. 1789. Genera Plantarum.... Paris. (Gen. Pl.)

Kartesz, J. T. 1994. A Synonymized Checklist of the Vascular Flora of the United States, Canada, and Greenland, ed. 2. 2 vols. Portland.

Kartesz, J. T. and C. A. Meacham. 1999. Synthesis of the North American Flora, ver. 1.0. Chapel Hill. [CD-ROM.] (Synth. N. Amer. Fl.)

Keating, R. C., V. C. Hollowell, and T. B. Croat, eds. 2005. A Festschrift for William G. D'Arcy: The Legacy of a Taxonomist. St. Louis.

Keating, W. H. 1824. Narrative of an Expedition to the Source of St. Peter's River...under the Command of Stephen H. Long.... 2 vols. Philadelphia. (Narrat. Exp. St. Peter's River)

Keil, D. J. 2018. New combinations and nomenclatural notes in the *Solanum umbelliferum* complex (Solanaceae). Phytoneuron 2018-61: 1–4.

Kephart, S. R., R. Wyatt, and D. Parella. 1988. Hybridization in North American *Asclepias*. I. Morphological evidence. Syst. Bot. 13: 456–473.

Kessler, D. et al. 2015. How scent and nectar influence floral antagonists and mutualists. eLife 4: doi:10.7554/eLife.07641

Kew Bull. = Kew Bulletin.

Key Spring Fl. Manhattan—See: A. S. Hitchcock 1894

Khyade, M. S., D. M. Kasote, and N. P. Vaikos. 2014. *Alstonia scholaris* (L.) R. Br. and *Alstonia macrophylla* Wall. ex G. Don: A comparative review on traditional uses, phytochemistry and pharmacology. J. Ethnopharmacol. 153: 1–18.

Kiger, R. W. and D. M. Porter. 2001. Categorical Glossary for the Flora of North America Project. Pittsburgh.

Klackenberg, J. 2001. Revision of the genus *Cryptostegia* R. Br. (Apocynaceae, Periplocoideae). Adansonia 23: 205–218.

Knapp, S. 2002. Floral diversity and evolution in the Solanaceae. In: Q. C. B. Cronk et al., eds. 2002. Developmental Genetics and Plant Evolution. London. Pp. 267–297.

Knapp, S. 2002b. *Solanum* section *Geminata* (Solanaceae. In: Organization for Flora Neotropica. 1968+. Flora Neotropica. 121+ nos. New York. No. 84.

Knapp, S. 2013. A revision of the Dulcamaroid clade of *Solanum* Linnaeus (Solanaceae). PhytoKeys 22: 1–432.

Knapp, S. et al. 2017. A revision of the *Solanum elaeagnifolium* clade (Elaeagnifolium clade; subgenus *Leptostemonum*, Solanaceae). PhytoKeys 84: 1–104.

Knapp, S. et al. 2019. A revision of the Morelloid clade of *Solanum* L. (Solanaceae) in the Caribbean and North and Central America. PhytoKeys 123: 1–144.

Knapp, S., M. W. Chase, and J. J. Clarkson. 2004. Nomenclatural changes and a new sectional classification in *Nicotiana* (Solanaceae). Taxon 53: 73–82.

Knorr, L. C. 1949. Parasitism of *Citrus* in Florida by various species of dodder, including *Cuscuta boldinghii* Urb., a species newly reported for the United States. Phytopathology 39: 411–412.

Köhlein, F. 1991. Gentians (English translation by D. Winstanley). Portland.

Kongl. Vetensk. Acad. Nya Handl. = Kongl[iga]. Vetenskaps Academiens Nya Handlingar.

Kral, R., ed. 1983. A Report on Some Rare, Threatened, or Endangered Forest-related Vascular Plants of the South. 2 vols. Washington. [U.S.D.A. Forest Serv., Techn. Publ. R8-TP 2.]

Krings, A. 2008. Synopsis of *Gonolobus* s.l. (Apocynaceae, Asclepiadoideae) in the United States and its territories, including lectotypification of *Lacnostoma arizonicum*. Harvard Pap. Bot. 13: 209–218.

Krings, A. et al. 2019. *Gonolobus taylorianus* (Apocynaceae, Asclepiadoideae, Gonolobinae) in Florida, U.S.A. J. Bot. Res. Inst. Texas 13: 315–317.

Krings, A., D. T. Thomas, and Xiang Q. Y. 2008. On the generic circumscription of *Gonolobus* (Apocynaceae, Asclepiadoideae): Evidence from molecules and morphology. Syst. Bot. 33: 403–415.

Krukoff, B. A. and R. C. Barneby. 1969. Supplementary notes on the American species of *Strychnos*. VIII. Mem. New York Bot. Gard. 20: 1–93.

Krukoff, B. A. and J. Monachino. 1942. The American species of *Strychnos*. Brittonia 2: 248–322.

Kubitzki, K. et al., eds. 1990+. The Families and Genera of Vascular Plants. 15+ vols. Berlin etc.

Kuijt, J. 1969. The Biology of Parasitic Flowering Plants. Berkeley.

Kulturpflanze = Kulturpflanze. Berichte und Mitteilungen aus dem Institut für Kulturpflanzenforschung der Deutschen Akademie der Wissenschaften zu Berlin in Gatersleben Krs. Aschersleben.

Kuntze, O. 1891–1898. Revisio Generum Plantarum Vascularium Omnium atque Cellularium Multarum.... 3 vols. Leipzig etc. [Vol. 3 in 3 parts paged independently; parts 1 and 3 unnumbered.] (Revis. Gen. Pl.)

La Llave, P. de and J. M. de Lexarza. 1824–1825. Novorum Vegetabilium Descriptiones. 2 fascs. Mexico City. [Fasc. 2 includes Orchidianum Opusculum, paged separately.] (Nov. Veg. Descr.)

Labillardière, J. J. H. de. 1824–1825. Sertum Austro-Caledonicum.... 2 parts. Paris. (Sert. Austro-Caledon.)

Lamarck, J. et al. 1783–1817. Encyclopédie Méthodique. Botanique.... 13 vols. Paris and Liège. [Vols. 1–8, suppls. 1–5.] (Encycl.)

Lamarck, J. and J. Poiret. 1791–1823. Tableau Encyclopédique et Méthodique des Trois Règnes de la Nature. Botanique.... 6 vols. Paris. [Vols. 1–2 = tome 1; vols. 3–5 = tome 2; vol. [6] = tome 3. Vols. paged consecutively within tomes.] (Tabl. Encycl.)

Langford, S. D. and P. J. Boor. 1996. Oleander toxicity: An examination of human and animal toxic exposures. Toxicology 109: 1–13.

Last, M. P. 2009. Intraspecific Phylogeography of *Cycladenia humilis* (Apocynaceae). M.S. thesis. Brigham Young University.

Lauderback, L. A. and B. M. Pavlik. 2017. Starch granule evidence for the earliest potato use in North America. Proc. Natl. Acad. Sci. U.S.A. 114: 7606–7610.

Leafl. Bot. Observ. Crit. = Leaflets of Botanical Observation and Criticism.

Leafl. W. Bot. = Leaflets of Western Botany.

Leeuwenberg, A. J. M. 1969. The Loganiaceae of Africa VIII. *Strychnos* III: Revision of the African species with notes on the extra-African. Meded. Landbouwhoogeschool 69: 1–316.

Leeuwenberg, A. J. M. 1974. The Loganiaceae of Africa XII. A revision of *Mitreola* L. Meded. Landbouwhoogeschool 74: 1–28.

Leeuwenberg, A. J. M. 1990. *Tabernaemontana* (Apocynaceae): Discussion of its delimitation. In: P. Baas et al., eds. 1990. The Plant Diversity of Malesia.... Dordrecht and Boston. Pp. 73–81.

Leeuwenberg, A. J. M. 1991. A Revision of *Tabernaemontana:* The Old World Species. Kew.

Leeuwenberg, A. J. M. 1994. A Revision of *Tabernaemontana*. Two, the New World Species and *Stemmadenia*. Kew.

Leeuwenberg, A. J. M. and P. W. Leenhouts. 1980. Loganiaceae. Taxonomy. In: H. G. A. Engler et al., eds. 1924+. Die natürlichen Pflanzenfamilien..., ed. 2. 26+ vols. Leipzig and Berlin. Vol. 29b(1), pp. 8–96.

Leeuwenberg, A. J. M. and F. J. H. van Dilst. 2001. Series of revisions of Apocynaceae XLIX. *Carissa* L. Wageningen Univ. Pap. 1: 3–109, 123–126.

Lehmann, J. G. C. 1818. Generis Nicotiniarum Historia.... Hamburg. (Gen. Nicot. Hist.)

Lemaire, C. 1851–1854. Le Jardin Fleuriste.... 4 vols. Gand. (Jard. Fleur.)

Lens, F. et al. 2009. Vessel grouping patterns in subfamilies Apocynoideae and Periplocoideae confirm phylogenetic value of wood structure within Apocynaceae. Amer. J. Bot. 96: 2168–2183.

Levin, R. A., J. Blanton, and Jill S. Miller. 2009. Phylogenetic utility of nuclear nitrate reductase: A milti-locus comparison of nuclear and chloroplast sequence data for inference of relationships among American Lycieae (Solanaceae). Molec. Phylogen. Evol. 50: 608–617.

Levin, R. A. and Jill S. Miller. 2005. Relationships within tribe Lycieae (Solanaceae): Paraphyly of *Lycium* and multiple origins of gender dimorphism. Amer. J. Bot. 92: 2044–2053.

Lewis, W. H. and M. P. F. Elvin-Lewis. 2003. Medical Botany: Plants Affecting Human Health, ed. 2. Hoboken.

Lewis, W. H. and R. L. Oliver. 1965. Realignment of *Calystegia* and *Convolvulus* (Convolvulaceae). Ann. Missouri Bot. Gard. 52: 217–222.

L'Héritier de Brutelle, C.-L. 1784[1785–1805]. Stirpes Novae aut Minus Cognitae.... 2 vols. in 9 fascs. Paris. [Fascicles paged and plates numbered consecutively.] (Stirp. Nov.)

Liede, S. et al. 2005. Phylogenetics of the New World subtribes of Asclepiadeae (Apocynaceae–Asclepiadoidcae): Metastelmatinae, Oxypetalinae, and Gonolobinae. Syst. Bot. 30: 184–195.

Liede, S. et al. 2012. *Vincetoxicum* and *Tylophora* (Apocynaceae: Asclepiodoideeae: Asclepiadeae)—Two sides of the same medal: Independent shifts from tropical to temperate habitats. Taxon 61: 803–825.

Liede, S. et al. 2014. Phylogenetics and biogeography of the genus *Metastelma* (Apocynaceae–Asclepiadoideae–Asclepiadeae: Metastelmatinae). Syst. Bot. 39: 594-612.

Liede, S. et al. 2016. Going west—A subtropical lineage (*Vincetoxicum*, Apocynaceae: Asclepiadoideae) expanding into Europe. Molec. Phylogen. Evol. 94: 436–446.

Liede, S. and U. Meve. 2004. Revision of *Metastelma* (Apocynaceae–Asclepiadoideae) in southwestern North America and Central America. Ann. Missouri Bot. Gard. 91: 31–86.

Liede, S. and A. Täuber. 2000. *Sarcostemma* R. Br. (Apocynaceae–Asclepiadoideae)—a controversial generic circumscription reconsidered: Evidence from *trn*L-F spacers. Pl. Syst. Evol. 225: 133–140.

Liede, S. and A. Täuber. 2002. Circumscription of the genus *Cynanchum* (Apocynaceae–Asclepiadoideae). Syst. Bot. 27: 789–800.

Lilloa = Lilloa; Revista de Botánica.

Linnaea = Linnaea; ein Journal fur die Botanik in ihrem ganzen Umfange.

Linnaeus, C. 1749[–1769]. Amoenitates Academicae seu Dissertationes Variae Physicae, Medicae Botanicae.... 7 vols. Stockholm and Leipzig. (Amoen. Acad.)

Linnaeus, C. 1753. Species Plantarum.... 2 vols. Stockholm. (Sp. Pl.)

Linnaeus, C. [1753.] Demonstrationes Plantarum in Horto Upsaliensi.... Uppsala. (Demonstr. Pl.)

Linnaeus, C. 1754. Genera Plantarum..., ed. 5. Stockholm. (Gen. Pl. ed. 5)

Linnaeus, C. [1756.] Centuria II. Plantarum.... Uppsala. (Cent. Pl. II)

Linnaeus, C. 1758. Opera Varia...Fundamenta Botanica, Sponsalia Plantarum, et Systema Naturae.... Lucca. (Opera Var.)

Linnaeus, C. 1758[–1759]. Systema Naturae per Regna Tria Naturae..., ed. 10. 2 vols. Stockholm. (Syst. Nat. ed. 10)

Linnaeus, C. 1762–1763. Species Plantarum..., ed. 2. 2 vols. Stockholm. (Sp. Pl. ed. 2)

Linnaeus, C. 1764. Genera Plantarum..., ed. 6. Stockholm. (Gen. Pl. ed. 6)

Linnaeus, C. 1766–1768. Systema Naturae per Regna Tria Naturae..., ed. 12. 3 vols. Stockholm. (Syst. Nat. ed. 12)

Linnaeus, C. 1767[–1771]. Mantissa Plantarum. 2 parts. Stockholm. [Mantissa [1] and Mantissa [2] Altera paged consecutively.] (Mant. Pl.)

Linnaeus, C. and G. C. Edwards. 1776. A Catalogue of the Birds, Beasts, Fishes, Insects, Plants, &C Contained in Edwards's Natural History.... London. (Cat. Edwards's Nat. Hist.)

Linnaeus, C. f. 1781[1782]. Supplementum Plantarum Systematis Vegetabilium Editionis Decimae Tertiae, Generum Plantarum Editionis Sextae, et Specierum Plantarum Editionis Secundae. Braunschweig. (Suppl. Pl.)

Lipow, S. R. and R. Wyatt. 1999. Floral morphology and late-acting self-incompatibility in *Apocynum cannabinum* (Apocynaceae). Pl. Syst. Evol. 219: 99–109.

Litzinger, W. J. 1981. Ceramic evidence for prehistoric *Datura* use in North America. J. Ethnopharmacol. 4: 57–74.

Liu, J. Q. and Ho T. N. 1996. The embryological studies of *Comastoma pulmonarium*. Acta Phytotax. Sin. 34: 577–585.

Liu, S. W. and Ho T. N. 1992. Systematic study on *Lomatogonium* A. Br. (Gentianaceae). Acta Phytotax. Sin. 30: 289–319.

Livshultz, T. et al. 2007. Phylogeny of Apocynoideae and the APSA clade (Apocynaceae s.l.). Ann. Missouri Bot. Gard. 94: 324–359.

Lloydia = Lloydia; a Quarterly Journal of Biological Science.

London J. Bot. = London Journal of Botany.

Long, R. W. and O. Lakela. 1971. A Flora of Tropical Florida: A Manual of the Seed Plants and Ferns of Southern Peninsular Florida. Coral Gables. [Reprinted 1976, Miami.]

Loudon, J. C. 1839. Hortus Britannicus. A Catalogue of All the Plants Indigenous, Cultivated in, or Introduced to Britain, ed. 3. London. (Hort. Brit. ed. 3)

Loureiro, J. de. 1790. Flora Cochinchinensis.... 2 vols. Lisbon. [Vols. paged consecutively.] (Fl. Cochinch.)

Löve, Á. and D. Löve. 1956. Cytotaxonomical conspectus on the Gentianaceae. Acta Horti Gothob. 20: 65–291.

Löve, Á. and D. Löve. 1982b. In: IOPB chromosome number reports LXXV. Taxon 31: 342–368.

Löve, Á. and D. Löve. 1986. In: IOPB chromosome number reports XCIII. Taxon 35: 897–903.

Löve, D. 1953. Cytotaxonomical remarks on the Gentianaceae. Hereditas (Lund) 33: 421–422.

Lozado Pérez, L. 2003. Sistematica de *Pherotrichis* Decne. (Apocynaceae, Asclepiadoideae). M.S. thesis. Universidad Nacional Autónoma de México.

Luna-Cavazos, M., R. A. Bye, and M. Jiao. 2009. The origin of *Datura metel* (Solanaceae): Genetic and phylogenetic evidence. Genet. Resources Crop Evol. 56: 263–275.

Lundellia = Lundellia; Journal of the Plant Resources Center of the University of Texas at Austin.

Lynch, S. P. 1977. The floral ecology of *Asclepias solanoana* Woods. Madroño 24: 159–177.

Maas, P. J. M. and P. Ruyters. 1986. *Voyria* and *Voyriella* (saprophytic Gentianaceae). In: Organization for Flora Neotropica. 1968+. Flora Neotropica. 121+ nos. New York. No. 41.

Mabberley, D. J. 2008. The Plant-book: A Portable Dictionary of the Vascular Plants..., ed. 3. Cambridge.

Macbride, J. F. 1962. Solanaceae. In: J. F. Macbride et al. 1936+. Flora of Peru. 6+ parts. Chicago. Part 5B(1).

Macbride, J. F. et al. 1936+. Flora of Peru. 6+ parts. Chicago. [Published in numerous fascs. constituting 6 nominal parts (together designated as vol. 13 of Field Mus. Nat. Hist., Bot. Ser.) plus later unnumbered increments (designated as individual issues of Fieldiana, Bot.).]

Mackenzie, K. K. and B. F. Bush. 1902. Manual of the Flora of Jackson County, Missouri. Kansas City. (Man. Fl. Jackson County)

Madroño = Madroño; Journal of the California Botanical Society [from vol. 3: a West American Journal of Botany].

Malcolm, S. B. 1991. Cardenolide-mediated interactions between plants and herbivores. In: G. A. Rosenthal and M. R. Berenbaum, eds. 1991. Herbivores: Their Interactions with Secondary Plant Metabolites, ed. 2. Volume 1. San Diego. Pp. 251–296.

Man. Bot. San Francisco—See: E. L. Greene 1894

Man. Fl. Jackson County—See: K. K. Mackenzie and B. F. Bush 1902

Man. Fl. Pl. Calif.—See: W. L. Jepson [1923–1925]

Man. S. Calif. Bot.—See: P. A. Munz 1935

Man. S.E. Fl.—See: J. K. Small 1933

Mansion, G. 2004. A new classification of the polyphyletic genus *Centaurium* Hill (Chironiinae, Gentianaceae): Description of the New World endemic *Zeltnera,* and reinstatement of *Gyrandra* Griseb. and *Schenkia* Griseb. Taxon 53: 719–740.

Mansion, G. and L. Struwe. 2004. Generic delimitation and phylogenetic relationships within the subtribe Chironiinae (Chironieae: Gentianaceae), with special reference to *Centaurium:* Evidence from nrDNA and cpDNA sequences. Molec. Phylogen. Evol. 42: 951–977.

Mansion, G. and L. Zeltner. 2004. Phylogenetic relationships within the New World endemic *Zeltnera* (Gentianaceae-Chironiinae) inferred from molecular and karyological data. Amer. J. Bot. 91: 2069–2086.

Mansion, G., L. Zeltner, and F. Bretagnolle. 2005. Phylogenetic patterns and polyploid evolution within the Mediterranean genus *Centaurium* (Gentianaceae-Chironieae). Taxon 54: 931–950.

Mant. Pl.—See: C. Linnaeus 1767[–1771]

Manual—See: A. Gray 1848

Manual ed. 2—See: A. Gray 1856

Marcy, R. B. 1853. Exploration of the Red River of Louisiana, in the Year 1852.... Washington. (Explor. Red River Louisiana)

Marohasy, J. and P. I. Forster. 1991. A taxonomic revision of *Cryptostegia* R. Br. (Asclepiadaceae: Periplocoideae). Austral. Syst. Bot. 4: 571–577.

Martin, W. C. and C. R. Hutchins. 1980. A Flora of New Mexico. 2 vols. Vaduz.

Martínez, M. 1993. The correct application of *Physalis pruinosa* L. (Solanaceae). Taxon 42: 103–104.

Martínez, M. 1998. Revision of *Physalis* section *Epeteiorhiza* (Solanaceae). Anales Inst. Biol. Univ. Nac. Autón. México, Bot. 69: 71–117.

Martínez, M. 1999. Infrageneric taxonomy of *Physalis.* In: M. Nee et al., eds. 1999. Solanaceae IV: Advances in Biology and Utilization. Kew. Pp. 275–283.

Martins, T. R. and T. J. Barkman. 2005. Reconstruction of Solanaceae phylogeny using the nuclear gene SAMT. Syst. Bot. 30: 435–447.

Martius, C. F. P. von, A. W. Eichler, and I. Urban, eds. 1840–1906. Flora Brasiliensis. 15 vols. in 40 parts, 130 fascs. Munich, Vienna, and Leipzig. [Vols. and parts numbered in systematic sequence, fascs. numbered independently in chronological sequence.] (Fl. Bras.)

Maschinski, J., H. D. Hammond, and L. Holter, eds. 1996. Southwestern Rare and Endangered Plants: Proceedings of the Second Conference: September 11–14, 1995, Flagstaff, Arizona. Fort Collins. [U.S.D.A. Forest Serv., Gen. Techn. Rep. RM-283.]

Mason, C. T. 1991. A new *Gentiana* (Gentianaceae) from northern California and southern Oregon. Madroño 37: 289–292.

Mason, C. T. and H. H. Iltis. 1966. Preliminary reports on the flora of Wisconsin no. 53. Gentianaceae and Menyanthaceae—gentian and buckbean families. Trans. Wisconsin Acad. Sci. 54: 295–329.

Mathews, K. G. et al. 2009. A phylogenetic analysis and taxonomic revision of *Bartonia* (Gentianaceae: Gentianeae), based on molecular and morphological evidence. Syst. Bot. 34: 162–172.

Mathews, K. G., M. S. Ruigrok, and G. Mansion. 2015. Phylogeny and biogeography of the eastern North American rose gentians (*Sabatia,* Gentianaceae). Syst. Bot. 40: 81–85.

McCarthy, E. W. et al. 2015. The effect of polyploidy and hybridization in the evolution of flower colour in *Nicotiana* (Solanaceae). Ann. Bot. (Oxford) 115: 1117–1131.

McDonald, J. A. 1995. Revision of *Ipomoea* section *Leptocallis* (Convolvulaceae). Harvard Pap. Bot. 6: 97–122.

McDonnell, A. et al. 2015. *Matelea chihuahuensis* (Apocynaceae): An addition to the flora of the United States and a synopsis of the species. J. Bot. Res. Inst. Texas 9: 187–194.

McDonnell, A., M. Parks, and M. Fishbein. 2018. Multilocus phylogenetics of New World milkweed vines (Apocynaceae, Asclepiadoideae, Gonolobinae). Syst. Bot. 43: 77–96.

McGregor, R. L., J. L. Gentry, and R. E. Brooks. 1986. Solanaceae In: Great Plains Flora Association. 1986. Flora of the Great Plains. Lawrence, Kans. Pp. 637–651.

McLaughlin, S. P. 1982. A revision of the southwestern species of *Amsonia* (Apocynaceae). Ann. Missouri Bot. Gard. 69: 336–350.

Med. Fl.—See: C. S. Rafinesque 1828[–1830]

Med. Repos. = Medical Repository.

Meded. Bot. Mus. Herb. Rijks Univ. Utrecht = Mededeelingen van het Botanisch Museum en Herbarium van de Rijks Universiteit te Utrecht.

Meded. Landbouwhoogeschool = Mededeelingen van de Landbouwhoogeschool te Wageningen.

Meded. Rijks-Herb. = Mededeelingen van 's Rijks-Herbarium.

Melderis, A. 1972. Taxonomic studies on the European species of the genus *Centaurium* Hill. J. Linn. Soc., Bot. 65: 224–250.

Mém. Acad. Imp. Sci. St. Pétersbourg Hist. Acad. = Mémoires de l'Académie Impériale des Sciences de St. Pétersbourg. Avec l'Histoire de l'Académie.

Mém. Acad. Imp. Sci. St.-Pétersbourg, Sér. 6, Sci. Math. = Mémoires de l'Académie Impériale des Sciences de St.-Pétersbourg. Sixième Série. Sciences Mathématiques, Physiques et Naturelles.

Mem. Amer. Acad. Arts = Memoirs of the American Academy of Arts and Science.

Mem. New York Bot. Gard. = Memoirs of the New York Botanical Garden.

Mém. Soc. Phys. Genève = Mémoires de la Société de Physique et d'Histoire Naturelle de Genève.

Mem. Torrey Bot. Club = Memoirs of the Torrey Botanical Club.

Mennega, A. M. W. 1980. Loganiaceae. Anatomy of the secondary xylem. In: H. G. A. Engler et al., eds. 1924+. Die natürlichen Pflanzenfamilien..., ed. 2. 26+ vols. Leipzig and Berlin. Vol. 28b(1), pp. 112–161.

Menzel, M. Y. 1950. Cytotaxonomic observations on some genera of the Solanae: *Margaranthus, Saracha,* and *Quincula.* Amer. J. Bot. 37: 25–30.

Menzel, M. Y. 1951. The cytotaxonomy and genetics of *Physalis.* Proc. Amer. Philos. Soc. 95: 132–183.

Merriam-Webster. 1988. Webster's New Geographical Dictionary. Springfield, Mass.

Merrill, E. D. 1917. An Interpretation of Rumphius's Herbarium Amboinense.... Manila. (Interpr. Herb. Amboin.)

Methodus—See: C. Moench 1794

Meulebrouck, K. et al. 2009. Hidden in the host—Unexpected vegetative hibernation of the holoparasitic *Cuscuta epithymum* (L.) L. and its implications for population persistence. Flora 204: 306–315.

Meyer, G. F. W. 1818. Primitiae Florae Essequeboensis.... Göttingen. (Prim. Fl. Esseq.)

Michaux, A. 1803. Flora Boreali-Americana.... 2 vols. Paris and Strasbourg. (Fl. Bor.-Amer.)

Michigan Bot. = Michigan Botanist.

Miers, J. 1878. On the Apocynaceae of South America.... London and Edinburgh. (Apocyn. S. Amer.)

Miller, Jill S. 2002. Phylogenetic relationships and the evolution of gender dimorphism in *Lycium* (Solanaceae). Syst. Bot. 27: 416–428.

Miller, Jill S. et al. 2011. Out of America to Africa or Asia: Inference of dispersal histories using nuclear and plastid DNA and the S-RNase self-incompatibility locus. Molec. Biol. Evol. 28: 793–801.

Miller, Jill S. et al. 2016. Correlated polymorphism in cytotype and sexual system within a monophyletic species, *Lycium californicum.* Ann. Bot. (Oxford) 117: 307–317.

Miller, P. 1754. The Gardeners Dictionary.... Abridged..., ed. 4. 3 vols. London. (Gard. Dict. Abr. ed. 4)

Miller, P. 1768. The Gardeners Dictionary..., ed. 8. London. (Gard. Dict. ed. 8)

Mink, J. N. et al. 2011b. *Centaurium tenuiflorum* (Gentianaceae) new to Oklahoma and *Centaurium texense* in Mexico. Phytoneuron 2011-49: 1–3.

Mione, T. 1992. Systematics and Evolution of *Jaltomata* (Solanaceae). Ph.D. dissertation. University of Connecticut.

Mitchell, R. S. 1986. A checklist of New York State plants. Bull. New York State Mus. Sci. Serv. 458.

Mitt. Bot. Staatssamml. München = Mitteilungen (aus) der Botanischen Staatssammlung München.

Mitt. Naturwiss. Vereins Univ. Wien = Mitteilungen des Naturwissenschaftlichen Vereins der Universität Wien.

Moench, C. 1794. Methodus Plantas Horti Botanici et Agri Marburgensis.... Marburg. (Methodus)

Moench, C. 1802. Supplementum ad Methodum Plantas.... Marburg. (Suppl. Meth.)

Mohandas, C., Y. S. Rao, and S. C. Sahu. 1981. Cultural control of rice root nematodes (*Hirschmanniella* spp.) with *Sphenoclea zeylanica.* Proc. Indian Natl. Sci. Acad., B 90: 373–376.

Mohlenbrock, R. H. 1986. Guide to the Vascular Flora of Illinois, rev. ed. Carbondale.

Mohr, C. T. 1878. Foreign plants introduced to the gulf states. Bot. Gaz. 3: 42–46.

Mohr, C. T. 1901. Plant life of Alabama. Contr. U.S. Natl. Herb. 6.

Molec. Biol. Evol. = Molecular Biology and Evolution.

Molec. Phylogen. Evol. = Molecular Phylogenetics and Evolution.

Monachino, J. 1949. A revision of the genus *Alstonia* (Apocynaceae). Pacific Sci. 3: 133–182.

Monogr. Syst. Bot. Missouri Bot. Gard. = Monographs in Systematic Botany from the Missouri Botanical Garden.

Morales, J. F. 1997. A reevaluation of *Echites* and *Prestonia* sect. *Coalitae* (Apocynaceae). Brittonia 49: 328–336.

Morales, J. F. 1998. A synopsis of the genus *Mandevilla* (Apocynaceae) in Mexico and Central America. Brittonia 50: 214–232.

Morales, J. F. 2009. Estudios en las Apocynaceae neotropicales XXXVII: Monografía del género *Rhabdadenia* (Apocynoideae: Echiteae). J. Bot. Res. Inst. Texas 3: 541–564.

Morales, J. F. 2009b. Estudios en las Apocynaceae neotropicales XXXIX: Revisión de las Apocynoideae y Rauvolfioideae de Honduras. Anales Jard. Bot. Madrid 66: 217–262.

Morales, J. F. 2009c. La familia Apocynaceae (Apocynoideae, Rauvolfioideae) en Guatemala. Darwiniana 47: 140–184.

Mroue, M. A. et al. 1996. Indole alkaloids of *Haplophyton crooksii*. J. Nat. Prod. (Lloydia) 59: 890–893.

Muhlenbach, V. 1979. Contributions to the synanthropic (adventive) flora of the railroads in St. Louis, Missouri, U.S.A. Ann. Missouri Bot. Gard. 66: 1–108.

Muhlenberg, G. H. E. 1813. Catalogus Plantarum Americae Septentrionalis.... Lancaster, Pa. (Cat. Pl. Amer. Sept.)

Muhlenbergia = Muhlenbergia; a Journal of Botany.

Munz, P. A. 1935. A Manual of Southern California Botany.... San Francisco. (Man. S. Calif. Bot.)

Munz, P. A. 1968. Supplement to A California Flora. Berkeley and Los Angeles. (Suppl. Calif. Fl.)

Myint, T. 1966. Revision of the genus *Stylisma* (Convolvulaceae). Brittonia 18: 97–117.

Myint, T. and D. B. Ward. 1968. A taxonomic revision of the genus *Bonamia* (Convolvulaceae). Phytologia 17: 121–239.

Narr. Journey Polar Sea—See: J. Franklin et al. 1823

Narrat. Exp. St. Peter's River—See: W. H. Keating 1824

Nat. Areas J. = Natural Areas Journal; Quarterly Publication of the Natural Areas Association.

Nat. Hist. Lepidopt. Georgia—See: J. E. Smith and J. Abbot 1797

Nat. Pflanzenfam.—See: H. G. A. Engler and K. Prantl 1887–1915

Nee, M. 1989. Notes on *Solanum* sect. *Gonatotrichum*. Solanaceae Newslett. 3: 80–82.

Nee, M. 1991. Synopsis of *Solanum* section *Acanthophora:* A group of interest for glycoalkaloids. In: J. G. Hawkes et al., eds. 1991. Solanaceae III: Taxonomy, Chemistry, Evolution. Kew. Pp. 257–266.

Nee, M. 1993. Solanaceae. In: J. C. Hickman, ed. 1993. The Jepson Manual. Higher Plants of California. Berkeley, Los Angeles, and London. Pp. 1068–1077.

Nee, M. et al., eds. 1999. Solanaceae IV: Advances in Biology and Utilization. Kew.

Negrón-Ortiz, V. 2011. Technical/Agency Draft Recovery Plan for *Spigelia gentianoides* (Gentian Pinkroot). Atlanta.

Nelson, J. B. 1980. *Mitreola* vs. *Cynoctonum,* and a new combination for the southeastern United States. Phytologia 46: 338–340.

Nesom, G. L. 1991g. Taxonomy of *Gentianella* (Gentianaceae) in Mexico. Phytologia 70: 1–20.

Neues J. Bot. = Neues Journal für die Botanik.

Neues J. Pharm. Aerzte = Neues Journal der Pharmacie für Aerzte, Apotheker und Chemiker.

New Fl.—See: C. S. Rafinesque 1836[–1838]

New Mexico Bot. Newslett. = The New Mexico Botanist Newsletter.

New Orleans Med. Surg. J. = New Orleans Medical and Surgical Journal.

New Phytol. = New Phytologist; a British Botanical Journal.

Nilsson, S. 1967. Pollen morphological studies in the Gentianaceae-Gentianinae. Grana Palynol., ser. 2, 7: 46–145.

Nomencl. Bot.—See: A. W. Dennstedt 1810

Nordic J. Bot. = Nordic Journal of Botany.

Notul. Syst. (Paris) = Notulae Systematicae, Herbier du Muséum de Paris. Phanérogamie.

Nov. Gen. Sp.—See: A. von Humboldt et al. 1815[1816]–1825

Nov. Pl. Descr. Dec.—See: C. G. Ortega 1797–1800

Nov. Stirp. Dec.—See: S. L. Endlicher and E. Fenzl 1839

Nov. Veg. Descr.—See: P. de La Llave and J. M. de Lexarza 1824–1825

Novon = Novon; a Journal for Botanical Nomenclature.

Nuttall, T. 1818. The Genera of North American Plants, and Catalogue of the Species, to the Year 1817.... 2 vols. Philadelphia. (Gen. N. Amer. Pl.)

Oberti, J. C. 1998. Withanolidos en Solanaceae con especial referencia a la tribu Jaborosae. Monogr. Syst. Bot. Missouri Bot. Gard. 68: 47–52.

Observ. Bot.—See: N. J. Jacquin 1764–1771

Observ. Congo—See: R. Brown 1818b

O'Dowd, D. J. and M. E. Hay. 1980. Mutualism between harvester ants and a desert ephemeral: Seed escape from rodents. Ecology 61: 531–540.

Oesterr. Bot. Z. = Oesterreichische botanische Zeitschrift. Gemeinützиges Organ für Botanik.

Oliver, D. et al., eds. 1868–1937. Flora of Tropical Africa.... 10 vols. in sects. London. [Sections in each volume paged independently.] (Fl. Trop. Afr.)

Ollerton, J. et al. 2019. The diversity of pollination systems in large plant clades: Apocynaceae as a case study. Ann. Bot. (Oxford) 123: 311–325.

Olmstead, R. G. et al. 1999. Phylogeny and provisional classification of the Solanaceae based on chloroplast DNA. In: M. Nee et al., eds. 1999. Solanaceae IV: Advances in Biology and Utilization. Kew. Pp. 111–137.

Olmstead, R. G. et al. 2008. A molecular phylogeny of the Solanaceae. Taxon 57: 1159–1181.

Olmstead, R. G. and L. Bohs. 2007. A summary of molecular systematic research in Solanaceae: 1982–2006. In: D. M. Spooner et al., eds. 2007. Solanaceae VI: Genomics Meets Biodiversity.... Leuven. Pp. 255–268.

Olmstead, R. G. and J. D. Palmer. 1991. Chloroplast DNA and systematics of the Solanaceae. In: J. G. Hawkes et al., eds. 1991. Solanaceae III: Taxonomy, Chemistry, Evolution. Kew. Pp. 161–168.

Olmstead, R. G. and J. D. Palmer. 1992. A chloroplast DNA phylogeny of the Solanaceae: Subfamilial relationships and character evolution. Ann. Missouri Bot. Gard. 79: 346–360.

Ooststroom, S. J. van. 1934. A monograph of the genus *Evolvulus*. Meded. Bot. Mus. Herb. Rijks Univ. Utrecht 14: 1–267.

Opera Bot. = Opera Botanica a Societate Botanice Lundensi.

Opera Var.—See: C. Linnaeus 1758

Organisms Diversity Evol. = Organisms, Diversity and Evolution; Journal of the Gesellschaft für Biologische Systematik.

Organization for Flora Neotropica. 1968+. Flora Neotropica. 121+ nos. New York.

Ornduff, R. 1970e. The systematics and breeding system of *Gelsemium* (Loganiaceae). J. Arnold Arbor. 51: 1–17.

Orrell Ellison, L. C. 2006. The Natural History, Genetics and Population Biology of *Sabatia kennedyana* (Plymouth Gentian), an Endangered Plant of Atlantic Coastal Pondshores. Ph.D. dissertation. University of Massachusetts, Boston.

Ortega, C. G. 1797–1800. Novarum, aut Rariorum Plantarum Horti Reg. Botan. Matrit. Descriptionum Decades.... 10 decades in 4 parts. Madrid. [Parts paged consecutively.] (Nov. Pl. Descr. Dec.)

Ottawa Naturalist = Ottawa Naturalist; Transactions of the Ottawa Field-Naturalists' Club.

Pacif. Railr. Rep.—See: War Department 1855–1860

Pacific Sci. = Pacific Science; a Quarterly Devoted to the Biological and Physical Sciences of the Pacific Region.

Pallas, P. S. 1784–1788[–1831]. Flora Rossica seu Stirpium Imperii Rossici par Europam et Asiam Indigenarum.... 2 vols. in 3 parts. St. Petersburg. [Parts 1 and 2 of vol. 1 paged independently.] (Fl. Ross.)

Pan, Q. et al. 2016. Monoterpenoid indole alkaloids biosynthesis and its regulation in *Catharanthus roseus:* A literature review from genes to metabolites. Phytochem. Rev. 15: 221–250.

Parad. Lond.—See: R. A. Salisbury 1805[–1808]

Parke, M. 1986. The new look of lisianthus. Horticulture (Boston) 64(3): 32–34.

Patterson, D. T. et al. 1989. Composite List of Weeds. Champaign.

Patterson, T. F. and G. L. Nesom. 2009. *Cryptostegia grandiflora* (Apocynaceae: Asclepiadoideae), a new non-native weed for Texas. J. Bot. Res. Inst. Texas 3: 461–463.

Pazy, B. and U. Plitmann. 1994. Holocentric chromosome behavior in *Cuscuta* (Cuscutaceae). Pl. Syst. Evol. 191: 105–109.

Peralta, I. E., D. M. Spooner, and S. Knapp. 2008. Taxonomy of wild tomatoes and their relatives (*Solanum* sect. *Lycopersicoides,* sect. *Juglandifolia,* sect. *Lycopersicon;* Solanaceae). Syst. Bot. Monogr. 84: 1–186.

Permadi, I. A. W. et al. 2016. Identifikasi karakter morfologi dan agronomi tanaman gonda (*Sphenoclea zeylanica* Gaertn) di Kabupaten Jembrana, Bali. Agrotrop 5(1): 43–54.

Perry, J. D. 1971. Biosystematic studies in the North American genus *Sabatia.* Rhodora 73: 309–369.

Persoon, C. H. 1805–1807. Synopsis Plantarum.... 2 vols. Paris and Tubingen. (Syn. Pl.)

Phytochem. Rev. = Phytochemistry Reviews; Fundamentals and Perspectives of Natural Products Research.

Phytologia = Phytologia; Designed to Expedite Botanical Publication.

Phytopathology = Phytopathology; Official Organ of the American Phytopathological Society.

Pichon, M. 1948. Classification des Apocynacées: X. Genre *Mandevilla.* Bull. Mus. Natl. Hist. Nat., sér. 2, 20: 211–216.

Pichon, M. 1948b. Classification des Apocynacées: VI. Genre *Tabernaemontana.* Notul. Syst. (Paris) 13: 230–253.

Pinto-Torres, E. and S. Koptur. 2009. Hanging by a coastal strand: Breeding system of a federally endangered morning-glory of the south-eastern Florida coast, *Jacquemontia reclinata.* Ann. Bot. (Oxford) 104: 1301–1311.

Pl. Cell = The Plant Cell.

Pl. Coromandel—See: W. Roxburgh 1795–1820

Pl. Ecol. = Plant Ecology.

Pl. Esc.—See: G. Forster 1786b

Pl. Hartw.—See: G. Bentham 1839[–1857]

Pl. Heermann.—See: E. M. Durand and T. C. Hilgard 1854

Pl. Hort. Schoenbr.—See: N. J. Jacquin 1797–1804

Pl. Physiol. (Lancaster) = Plant Physiology.

Pl. Syst. Evol. = Plant Systematics and Evolution.

Poiret, J. 1789. Voyage en Barbarie.... 2 vols. Paris. (Voy. Barbarie)

Popovkin, A. V. et al. 2011. *Spigelia genuflexa* (Loganiaceae), a new geocarpic species from the Atlantic forest of northeastern Bahia, Brazil. PhytoKeys 6: 47–65.

Prather, L. A. and R. J. Tyrl. 1993. The biology of *Cuscuta attenuata* Waterfall (Cuscutaceae). Proc. Oklahoma Acad. Sci. 73: 7–13.

Prelim. Cat.—See: N. L. Britton et al. 1888

Prim. Fl. Esseq.—See: G. F. W. Meyer 1818

Pringle, J. S. 1967. Taxonomy and distribution of *Gentiana,* section *Pneumonanthae* [sic], in eastern North America. Brittonia 19: 1–32.

Pringle, J. S. 1968. The status and distribution of *Gentiana linearis* and *G. rubricaulis* in the Upper Great Lakes region. Michigan Bot. 7: 99–112.

Pringle, J. S. 2004. Notes on the distribution and nomenclature of North American *Gentianopsis* (Gentianaceae). Sida 21: 525–530.

Pringle, J. S. 2008. Nomenclature of the spurred-gentian of the southwestern United States and northwestern Mexico, *Halenia rothrockii* (Gentianaceae). Madroño 54: 326–328.

Pringle, J. S. 2010b. The identity and nomenclature of the Pacific North American species *Zeltnera muhlenbergii* (Gentianaceae) and its distinction from *Centaurium tenuiflorum* and other species with which it has been confused. Madroño 57: 184–202.

Pringle, J. S. 2013. Nomenclatural notes on North American *Gyrandra* and *Lomatogonium* (Gentianaceae). Rhodora 115: 96–111.

Proc. Acad. Nat. Sci. Philadelphia = Proceedings of the Academy of Natural Sciences of Philadelphia.

Proc. Amer. Acad. Arts = Proceedings of the American Academy of Arts and Sciences.

Proc. Amer. Philos. Soc. = Proceedings of the American Philosophical Society.

Proc. Assoc. Off. Seed Analysts N. Amer. = Proceedings of the Association of Official Seed Analysts of North America.

Proc. Biol. Soc. Wash. = Proceedings of the Biological Society of Washington.

Proc. Calif. Acad. Sci. = Proceedings of the California Academy of Sciences.

Proc. Indian Natl. Sci. Acad., B = Proceedings of the Indian National Science Academy. Part B, Biological Sciences.

Proc. Natl. Acad. Sci. U.S.A. = Proceedings of the National Academy of Sciences of the United States of America.

Proc. Oklahoma Acad. Sci. = Proceedings of the Oklahoma Academy of Science.

Proc. Pennsylvania Acad. Sci. = Proceedings of the Pennsylvania Academy of Science.

Prodr.—See: R. Brown 1810; A. P. de Candolle and A. L. P. P. de Candolle 1823–1873; O. P. Swartz 1788

Prodr. Fl. Monast. Westphal.—See: C. M. F. von Boenninghausen 1824

Prodr. Fl. Nepal.—See: D. Don 1825

Publ. Carnegie Inst. Wash. = Publications of the Carnegie Institution of Washington.

Publ. Field Columbian Mus., Bot. Ser. = Publications of the Field Columbian Museum. Botanical Series.

Publ. Field Mus. Nat. Hist., Bot. Ser. = Publications of the Field Museum of Natural History. Botanical Series.

Pursh, F. [1813]1814. Flora Americae Septentrionalis; or, a Systematic Arrangement and Description of the Plants of North America. 2 vols. London. (Fl. Amer. Sept.)

Quattrocchi, U. 2012. CRC Dictionary of Medicinal and Poisonous Plants. Boca Raton.

Radford, A. E., H. E. Ahles, and C. R. Bell. 1968. Manual of the Vascular Flora of the Carolinas. Chapel Hill.

Rafinesque, C. S. 1828[–1830]. Medical Flora; or, Manual of the Medical Botany of the United States of North America. 2 vols. Philadelphia. (Med. Fl.)

Rafinesque, C. S. 1836[–1838]. New Flora and Botany of North America.... 4 parts. Philadelphia. [Parts paged independently.] (New Fl.)

Rafinesque, C. S. 1836[1837–1838]. Flora Telluriana.... 4 vols. Philadelphia. (Fl. Tellur.)

Rafinesque, C. S. 1840. Autikon Botanikon. 3 parts. Philadelphia. [Parts paged consecutively.] (Autik. Bot.)

Rafn, C. G. 1796–1800. Danmarks og Holsteens Flora.... 2 vols. Copenhagen. (Danm. Holst. Fl.)

Randall, R. P. 2002. A Global Compendium of Weeds. Melbourne.

Rao, A. S. 1956. A revision of *Rauvolfia* with particular reference to the American species. Ann. Missouri Bot. Gard. 43: 253–354.

Reed, C. F. 1964. A flora of the chrome and manganese ore piles at Canton, in the port of Baltimore, Maryland, and at Newport News, Virginia, with descriptions of genera and species new to the flora of eastern United States. Phytologia 10: 321–406.

Rees, A. [1802–]1819–1820. The Cyclopaedia; or, Universal Dictionary of Arts, Sciences, and Literature.... 39 vols. in 79 parts. London. [Pages unnumbered.] (Cycl.)

Rendle, A. B. 1959. The Classification of Flowering Plants, ed. 2. Cambridge and New York.

Rep. (Annual) Board Regents Smithsonian Inst. = Report (Annual) of the Board of Regents, Smithsonian Institution.

Rep. (Annual) Bur. Amer. Ethnol. = Annual Report of the Bureau of American Ethnology.

Rep. Bot. Exch. Club Soc. Brit. Isles = (Report,) Botanical Exchange Club and Society of the British Isles.

Rep. Exped. Rocky Mts.—See: J. C. Frémont 1843–1845

Rep. For. N. America—See: C. S. Sargent 1884

Rep. U.S. Geogr. Surv., Wheeler—See: J. T. Rothrock 1878[1879]

Rep. U.S. Mex. Bound.—See: W. H. Emory 1857–1859

Repert. Bot. Syst.—See: W. G. Walpers 1842–1847

Repert. Spec. Nov. Regni Veg. = Repertorium Specierum Novarum Regni Vegetabilis.

Rev. Hort. = Revue Horticole; Journal d'Horticulture Pratique.

Rev. Weed Sci. = Reviews of Weed Science.

Revis. Gen. Pl.—See: O. Kuntze 1891–1898

Revista Brasil. Bot. = Revista Brasileira de Botánica; Publicação Oficial da Sociedade Botánica do Brasil, Regional de São Paulo.

Rhoads, A. F. and W. M. Klein. 1993. The Vascular Flora of Pennsylvania: Annotated Checklist and Atlas. Philadelphia.

Rhodora = Rhodora; Journal of the New England Botanical Club.

Richards, C. D., F. Hyland, and L. M. Eastman. 1983. Checklist of the vascular plants of Maine, rev. 2. Bull. Josselyn Bot. Soc. Maine 11.

Riley, J. L. and S. M. McKay. 1980. The vegetation and phytogeography of coastal southwestern James Bay. Contr. Life Sci. Roy. Ontario Mus. 124.

Riser, J. P. 2019. Genetics and ecological niche define species boundaries in the dwarf milkweed clade (*Asclepias:* Asclepiadoideae: Apocynaceae). Int. J. Pl. Sci. 180: 160–170.

Roberts, M. F. and M. Wink. 1998. Alkaloids: Biochemistry, Ecology, and Medicinal Applications. New York.

Robertson, K. R. 1971. A Revision of *Jacquemontia* (Convolvulaceae) in North and Central America and the West Indies. Ph.D. dissertation. Walshington University.

Roe, K. E. 1967. A revision of *Solanum* sect. *Brevantherum* (Solanaceae) in North and Central America. Brittonia 19: 353–373.

Roe, K. E. 1968. *Solanum verbascifolium* L. misidentification and misapplication. Taxon 17: 176–179.

Roemer, J. J., J. A. Schultes, and J. H. Schultes. 1817[–1830]. Caroli a Linné...Systema Vegetabilium...Editione XV.... 7 vols. Stuttgart. (Syst. Veg.)

Rogers, G. K. 1986. The genera of Loganiaceae in the southeastern United States. J. Arnold Arbor. 67: 143–185.

Rosatti, T. J. 1986. The genera of Sphenocleaceae and Campanulaceae in the southeastern United States. J. Arnold Arbor. 67: 1–64.

Rosatti, T. J. 1989. The genera of suborder Apocynineae (Apocynaceae and Asclepiadaceae) in the southeastern United States. J. Arnold Arbor. 70: 307–401.

Rose, J. N. 1892. Notes on *Asclepias glaucescens* and *A. elata*. Bot. Gaz. 17: 193–194.

Rosenthal, G. A. and M. R. Berenbaum, eds. 1991. Herbivores: Their Interactions with Secondary Plant Metabolites, ed. 2. Volume 1. San Diego.

Roth, A. W. 1787. Botanische Abhandlungen und Beobachtungen.... Nuremberg. (Bot. Abh. Beobacht.)

Roth, A. W. 1797–1806. Catalecta Botanica.... 3 parts. Leipzig. [Parts paged independently.] (Catal. Bot.)

Rothrock, J. T. 1878[1879]. Report upon United States Geographical Surveys West of the One Hundredth Meridian, in Charge of First Lieut. Geo. M. Wheeler.... Vol. 6—Botany. Washington. (Rep. U.S. Geogr. Surv., Wheeler)

Rousseau, J. 1932. Contribtion a l'étude du *Gentiana victorinii*. Contr. Lab. Bot. Univ. Montréal 23.

Roxburgh, W. 1795–1820. Plants of the Coast of Coromandel.... 3 vols. in parts. London. [Volumes paged independently, plates numbered consecutively.] (Pl. Coromandel)

Roxburgh, W. 1820–1824. Flora Indica; or Descriptions of Indian Plants.... 2 vols. Serampore. (Fl. Ind.)

Ruiz López, H. and J. A. Pavon. 1794. Flora Peruvianae, et Chilensis Prodromus.... Madrid. (Fl. Peruv. Prodr.)

Ruiz López, H. and J. A. Pavon. 1798–1802. Flora Peruviana, et Chilensis, sive Descripciones, et Icones Plantarum Peruvianarum, et Chilensium.... 3 vols. [Madrid.] (Fl. Peruv.)

Russo, V. M., ed. 2012. Peppers: Botany, Production and Uses Wallingford.

Rydberg, P. A. 1896b. The North American species of *Physalis* and related genera. Mem. Torrey Bot. Club 4: 297–374.

S. African Quart. J. = South African Quarterly Journal.

Sady, M. B. and J. N. Seibert. 1991. Chemical differences between species of *Asclepias* from the intermountain region of North America. Phytochemistry 30: 3001–3003.

Safford, W. E. 1921. Synopsis of the genus *Datura*. J. Wash. Acad. Sci. 11: 173–189.

Safford, W. E. 1922. Daturas of the Old World and New: An account of their narcotic properties and their use in oracular and initiatory ceremonies. Rep. (Annual) Board Regents Smithsonian Inst. 1920: 537–567.

Sahagún, B. de. 1950–1982. Florentine Codex: General History of the Things of New Spain. Books 1–12. Transl. A. J. O. Anderson and C. E. Dibble. Salt Lake City.

Sakane, M. and G. J. Shepherd. 1986. Uma revisão do gênero *Allamanda* (Apocynaceae). Revista Brasil. Bot. 9: 125–149.

Salisbury, R. A. 1805[–1808]. The Paradisus Londinensis.... 2 vols. London. [Plates numbered consecutively.] (Parad. Lond.)

Samões, A. R. and G. W. Staples. 2017. Dissolution of Convolvulaceae tribe Merremieae and a new classification of the constituent genera. Bot. J. Linn. Soc. 183: 561–586.

Sargent, C. S. 1884. Department of the Interior, Census Office...Report on the Forests of North America (Exclusive of Mexico).... Washington. [47 Congr., 2 Sess., House Misc. Doc. 42(9).] (Rep. For. N. America)

Sargentia = Sargentia; Continuation of the Contributions from the Arnold Arboretum of Harvard University.

Särkinen, T. et al. 2013. A phylogenetic framework for evolutionary study of the nightshades (Solanaceae): A dated 1000-tip tree. B. M. C. Evol. Biol. 13: 214.

Särkinen, T. et al. 2018. A revision of the Old World black nightshades (Morelloid clade of *Solanum* L., Solanaceae). PhytoKeys 106: 1–223.

Scamman, E. 1940. A list of plants from interior Alaska. Rhodora 42: 309–343.

Schumann, K. M. 1895. Apocynaceae. In: H. G. A. Engler and K. Prantl, eds. 1887–1915. Die natürlichen Pflanzenfamilien.... 254 fascs. Leipzig. Fascs. 120–122[IV,2], pp. 109–189.

Schweigger, A. F. 1812. Enumeratio Plantarum Horti Botanici Regiomontani. Königsberg. (Enum. Pl. Hort. Regiom.)

Scopoli, J. A. 1786[–1788]. Deliciae Florae et Faunae Insubricae.... 3 vols. Pavia. (Delic. Fl. Faun. Insubr.)

Scott, P. J. 1991. Deadly nightshade, *Atropa belladonna* (Solanaceae), in Newfoundland. Canad. Field-Naturalist 105: 112.

Seithe, A. and J. R. Sullivan. 1990. Hair morphology and systematics of *Physalis* (Solanaceae). Pl. Syst. Evol. 170: 193–204.

Select. Stirp. Amer. Hist.—See: N. J. Jacquin 1763

Sennblad, B. and B. Bremer. 1996. The familial and subfamilial relationships of Apocynaceae and Asclepiadaceae evaluated with *rbc*L data. Pl. Syst. Evol. 202: 153–175.

Sert. Austro-Caledon.—See: J. J. H. de Labillardière 1824–1825

Seymour, F. C. 1982. The Flora of New England: A Manual for the Identification of All Vascular Plants...Growing without Cultivation..., ed. 2. Plainfield, N.J.

Shah, J. 1984. Taxonomic Studies in the Genus *Swertia* (Gentianaceae). Ph.D. thesis. University of Aberdeen.

Sheeley, S. E. and D. J. Raynal. 1996. The distribution and status of species of *Vincetoxicum* in eastern North America. Bull. Torrey Bot. Club 123: 148–156.

Shinners, L. H. 1957. Synopsis of the genus *Eustoma* (Gentianaceae). SouthW. Naturalist 2: 38–43.

Sida = Sida; Contributions to Botany.

Sidiyasa, K. 1998. Taxonomy, phylogeny, and wood anatomy of *Alstonia* (Apocynaceae). Blumea, Suppl. 11: 1–230.

Silva Manso, A. L. P. da. 1836. Enumeração das Substancias Brazileiras.... Rio de Janeiro. (Enum. Subst. Braz.)

Simões, A. O. et al. 2004. Tribal and intergeneric relationships of Mesechiteae (Apocynoideae, Apocynaceae): Evidence from three noncoding plastid DNA regions and morphology. Amer. J. Bot. 91: 1409–1418.

Simões, A. O. et al. 2006. Is *Mandevilla* (Apocynaceae, Mesechiteae) monophyletic? Evidence from five plastid DNA loci and morphology. Ann. Missouri Bot. Gard. 93: 565–591.

Simões, A. O. et al. 2007. Phylogeny and systematics of the Rauvolfioideae (Apocynaceae) based on molecular and morphological evidence. Ann. Missouri Bot. Gard. 94: 268–297.

Simões, A. O. et al. 2016. Systematics and character evolution of Vinceae (Apocynaceae). Taxon 65: 99–122.

Simões, A. O., M. E. Endress, and E. Conti. 2010. Systematics and character evolution of Tabernaemontaneae (Apocynaceae, Rauvolfioideae) based on molecular and morphological evidence. Taxon 59: 772–790.

Sims, T. L. and T. P. Robbins. 2009. Gametophytic self-incompatibility in *Petunia.* In: T. Gerats and J. Strommer, eds. 2009. *Petunia:* Evolutionary, Developmental and Physiological Genetics, ed. 2. New York. Pp. 85–106.

Sipes, S. D. and V. J. Tepedino. 1996. Pollinator lost? Reproduction by the enigmatic Jones cycladenia, *Cycladenia humilis* var. *jonesii.* In: J. Maschinski et al., eds. 1996. Southwestern Rare and Endangered Plants: Proceedings of the Second Conference: September 11–14, 1995, Flagstaff, Arizona. Fort Collins. Pp. 158–166.

Sipes, S. D. and P. G. Wolf, P. G. 1997. Clonal structure and pattern of allozyme diversity in the rare endemic *Cycladenia humilis* var. *jonesii* (Apocynaceae). Amer. J. Bot. 84: 401–409.

Sketch Bot. S. Carolina—See: S. Elliott [1816–]1821–1824

Skr. Kiøbenhavnske Selsk. Laerd. Elsk. = Skrifter, Som Udi det Kiøbenhavnske Selskab af Laerdoms og Videnskabers Elskere ere Fremlagte og Oplaeste.

Small, J. K. 1903. Flora of the Southeastern United States.... New York. (Fl. S.E. U.S.)

Small, J. K. 1913. Flora of the Southeastern United States..., ed. 2. New York. (Fl. S.E. U.S. ed. 2)

Small, J. K. 1933. Manual of the Southeastern Flora, Being Descriptions of the Seed Plants Growing Naturally in Florida, Alabama, Mississippi, Eastern Louisiana, Tennessee, North Carolina, South Carolina and Georgia. New York. (Man. S.E. Fl.)

Smith, J. E. and J. Abbot. 1797. The Natural History of the Rare Lepidopterous Insects of Georgia. 2 vols. London. (Nat. Hist. Lepidopt. Georgia)

Solanaceae Newslett. = Solanaceae Newsletter.

Sorrie, B. A. 2016. The curious distribution of *Asclepias tomentosa* (Apocynaceae). Phytoneuron 2016-68: 1–4.

SouthW. Naturalist = Southwestern Naturalist.

Sp. Pl.—See: C. Linnaeus 1753; C. L. Willdenow et al. 1797–1830

Sp. Pl. ed. 2—See: C. Linnaeus 1762–1763

Spellman, D. L. and C. R. Gunn. 1976. *Morrenia odorata* and *Araujia sericofera* (Asclepiadaceae): Weeds in citrus groves. Castanea 41: 139–148.

Spic. Fl. Rumel.—See: A. H. R. Grisebach 1843–1845[–1846]

Spier, L. P. and J. R. Snyder. 1998. Effects of wet- and dry-seaxon fires on *Jacquemontia curtisii,* a South Florida pine forest endemic. Nat. Areas J. 18: 350–357.

Spooner, D. M. et al. 2004. Wild potatoes (*Solanum* section *Petota:* Solanaceae) of North and Central America. Syst. Bot. Monogr. 68: 1–209.

Spooner, D. M. et al., eds. 2007. Solanaceae VI: Genomics Meets Biodiversity.... Leuven. [Acta Hort. 745.]

St. John, H. 1941. Revision of the genus *Swertia* (Gentianaceae) and the reduction of *Frasera.* Amer. Midl. Naturalist 26: 1–29.

St.-Hilaire, J. 1805. Exposition des Familles Naturelles.... 2 parts. Paris and Strasbourg. [Parts paged independently.] (Expos. Fam. Nat.)

Stafleu, F. A. et al. 1992–2009. Taxonomic Literature: A Selective Guide to Botanical Publications and Collections with Dates, Commentaries and Types. Supplement. 8 vols. Königstein.

Stafleu, F. A. and R. S. Cowan. 1976–1988. Taxonomic Literature: A Selective Guide to Botanical Publications and Collections with Dates, Commentaries and Types, ed. 2. 7 vols. Utrecht etc.

Stapf, O. 1902. Apocynaceae. In: D. Oliver et al., eds. 1868–1937. Flora of Tropical Africa.... 10 vols. in sects. London. Vol. 4, pp. 24–192.

Staples, G. W. et al. 2006. The restoration of *Ipomoea muricata* (L.) Jacq. (Convolvulaceae). Taxon 54: 1075–1079.

Stearn, W. T. 1966. *Catharanthus roseus,* the correct name for the Madagascar periwinkle. Lloydia 29: 196–200.

Stearn, W. T. 1973. A synopsis of the genus *Vinca* including its taxonomic and nomenclatural history. In: W. I. Taylor and N. R. Farnsworth, eds. 1973. The *Vinca* Alkaloids. Botany, Chemistry, and Pharmacology. New York. Pp. 1–94.

Steenis, C. G. G. J. van, ed. 1948+. Flora Malesiana.... Series I. Spermatophyta. 18+ vols., some in parts. Djakarta and Leiden.

Stefanović, S., D. F. Austin, and R. G. Olmstead. 2003. Classification of Convolvulaceae: A phylogenetic approach. Syst. Bot. 28: 791–806.

Stefanović, S. and M. Costea. 2008. Reticulate evolution in the parasitic genus *Cuscuta* (Convolvulaceae): Over and over and over again. Botany (Ottawa) 86: 791–808.

Stefanović, S., L. Krueger, and R. G. Olmstead. 2002. Monophyly of the Convolvulaceae and circumscription of their major lineages based on DNA sequences of multiple chloroplast loci. Amer. J. Bot. 89: 1510–1522.

Stefanović, S., M. Kuzmina, and M. Costea. 2007. Delimitation of major lineages within *Cuscuta* subgenus *Grammica* (Convolvulaceae) using plastid and nuclear DNA sequences. Amer. J. Bot. 94: 568–589.

Stefanović, S. and R. G. Olmstead. 2004. Testing the phylogenetic position of a parasitic plant (*Cuscuta,* Convolvulaceae, Asteridae): Bayesian inference and the parametric bootstrap on data drawn from three genomes. Syst. Biol. 53: 384–399.

Stehmann, J. R. et al. 2009. The genus *Petunia.* In: T. Gerats and J. Strommer, eds. 2009. *Petunia:* Evolutionary, Developmental and Physiological Genetics, ed. 2. New York. Pp. 1–28.

Stern, S. R. et al. 2013. A revision of *Solanum* section *Gonatotrichum.* Syst. Bot. 38: 471–496.

Stevens, W. D. 2001. Asclepiadaceae. In: W. D. Stevens et al., eds. 2001. Flora de Nicaragua. 4 vols. St. Louis. Vol. 1, pp. 234–270.

Stevens, W. D. 2005. New and interesting milkweeds (Apocynaceae, Asclepiadoideae). Novon 15: 602–619.

Stevens, W. D. 2009. Asclepiadaceae. In: G. Davidse et al., eds. 1994+. Flora Mesoamericana. 6+ vols. in parts. Mexico City, St. Louis, and London. Vol. 4(1), pp. 703–768.

Stevens, W. D. et al., eds. 2001. Flora de Nicaragua. 4 vols. St. Louis.

Stevens, W. D. and O. M. Montiel. 2002. A new species of *Gonolobus* (Apocynaceae, Asclepiadoideae) from Mesoamerica. Novon 12: 551–554.

Stevenson, M. C. 1915. Ethnobotany of the Zuñi Indians. Rep. (Annual) Bur. Amer. Ethnol. 30: 35–102. [Reprinted 1993, New York.]

Stirp. Nov.—See: C.-L. L'Héritier de Brutelle 1784[1785–1805]

Stoepler, T. M. et al. 2012. Differential pollinator effectiveness and importance in a milkweed (*Asclepias*, Apocynaceae) hybrid zone. Amer. J. Bot. 99: 448–458.

Strickland-Constable, R. et al. 2010. Species identity in the *Solanum bahamense* species group (Solanaceae, *Solanum* subgenus *Leptostemonum*). Taxon 59: 209–226.

Struwe, L. et al. 2002. Systematics, character evolution, and biogeography of Gentianaceae. In: L. Struwe and V. A. Albert, eds. 2002. Gentianaceae: Systematics and Natural History. Cambridge. Pp. 21–309.

Struwe, L. and V. A. Albert, eds. 2002. Gentianaceae: Systematics and Natural History. Cambridge.

Struwe, L., V. A. Albert, and B. Bremer. 1994. Cladistics and family level classification of the Gentianales. Cladistics 10: 175–206.

Struwe, L. and J. S. Pringle. 2019. Gentianaceae. In: K. Kubitzki et al., eds. 1990+. The Families and Genera of Vascular Plants. 15+ vols. Berlin etc. Vol. 15, pp. 453–503.

Sullivan, J. R. 1984. Pollination biology of *Physalis viscosa* var. *cinerascens* (Solanaceae). Amer. J. Bot. 71: 815–820.

Sullivan, J. R. 1985. Systematics of the *Physalis viscosa* complex (Solanaceae). Syst. Bot. 10: 426–444.

Sullivan, J. R. 2004. The genus *Physalis* (Solanaceae) in the southeastern United States. Rhodora 106: 305–326.

Sullivan, J. R. 2013. Nomenclatural innovations in North American *Physalis* (Solanaceae). Rhodora 115: 290–292.

Sullivan, J. R., V. P. Patel, and W. Chissoe. 2005. Palynology and systematics of *Physalis* (Solanaceae). In: R. C. Keating et al., eds. 2005. A Festschrift for William G. D'Arcy: The Legacy of a Taxonomist. St. Louis. Pp. 287–300.

Summa Veg. Scand.—See: E. M. Fries 1845–1849

Sundell, E. 1981. The New World species of *Cynanchum* subgenus *Mellichampia* (Asclepiadaceae). Evol. Monogr. 5: 1–63.

Sundell, E. 1994. Vascular plants of Arizona: Asclepiadaceae, milkweed family. J. Arizona-Nevada Acad. Sci. 27: 169–187.

Sunyatsenia = Sunyatsenia. Journal of the Botanical Institute; College of Agriculture, Sun Yatsen University.

Suppl. Calif. Fl.—See: P. A. Munz 1968

Suppl. Meth.—See: C. Moench 1802

Suppl. Pl.—See: C. Linnaeus f. 1781[1782]

Swartz, O. P. 1788. Nova Genera & Species Plantarum seu Prodromus.... Stockholm, Uppsala, and Åbo. (Prodr.)

Sweet, R. 1823–1837. The British Flower Garden.... 7 vols. London. [Vols. 5–7 also issued as vols. 1–3 of series 2.] (Brit. Fl. Gard.)

Sweet, R. 1826. Hortus Britannicus.... 2 parts. London. [Parts paged consecutively.] (Hort. Brit.)

Sweet, R. 1830. Hortus Britannicus..., ed. 2. London. (Hort. Brit. ed. 2)

Symb. Antill.—See: I. Urban 1898–1928

Symb. Bot.—See: M. Vahl 1790–1794

Symon, D. E. and A. R. Haegi. 1991. *Datura* (Solanaceae) is a New World genus. In: J. G. Hawkes et al., eds. 1991. Solanaceae III: Taxonomy, Chemistry, Evolution. Kew. Pp. 197–210.

Syn. Fl. N. Amer.—See: A. Gray et al. 1878–1897

Syn. Fl. N. Amer. ed. 2—See: A. Gray et al. 1886

Syn. Pl.—See: C. H. Persoon 1805–1807

Synth. N. Amer. Fl.—See: J. T. Kartesz and C. A. Meacham 1999

Syst. Biol. = Systematic Biology.

Syst. Bot. = Systematic Botany; Quarterly Journal of the American Society of Plant Taxonomists.

Syst. Bot. Monogr. = Systematic Botany Monographs; Monographic Series of the American Society of Plant Taxonomists.

Syst. Nat.—See: J. F. Gmelin 1791[–1792]

Syst. Nat. ed. 10—See: C. Linnaeus 1758[–1759]

Syst. Nat. ed. 12—See: C. Linnaeus 1766[–1768]

Syst. Veg.—See: J. J. Roemer et al. 1817[–1830]

Syst. Verz.—See: H. Zollinger 1854–1855

Tabl. Encycl.—See: J. Lamarck and J. Poiret 1791–1823

Taxon = Taxon; Journal of the International Association for Plant Taxonomy.

Taylor, W. I. and N. R. Farnsworth, eds. 1973. The *Vinca* Alkaloids. Botany, Chemistry, and Pharmacology. New York.

Tharp, B. C. and M. C. Johnston. 1961. Recharacterization of *Dichondra* (Convolvulaceae) and a revision of the North American species. Brittonia 13: 346–360.

Thellung, A. 1912. La Flore Adventice de Montpellier. Cherbourg. [Preprinted from Mém. Soc. Sci. Nat. Math. Cherbourg 38: 57–728. 1912.] (Fl. Adv. Montpellier)

Thomas, R. D. and C. M. Allen. 1993–1998. Atlas of the Vascular Flora of Louisiana. 3 vols. Baton Rouge.

Thouars, L. M. A. du P. 1804. Histoire des Végétaux Recueillis sur les Îles de France, la Réunion (Bourbon) et Madagascar. Première Partie. Paris. (Hist. Vég. Îles France)

Threadgill, P. F., J. M. Baskin, and C. C. Baskin. 1981. The ecological life cycle of *Frasera caroliniensis,* a long-lived monocarpic perennial. Amer. Midl. Naturalist 105: 277–288.

Torrey, J. [1823–]1824. A Flora of the Northern and Middle Sections of the United States.... 1 vol. in 3 fascs. New York. (Fl. N. Middle United States)

Torrey, J. and A. Gray. 1838–1843. A Flora of North America.... 2 vols. in 7 parts. New York, London, and Paris. (Fl. N. Amer.)

Torreya = Torreya; a Monthly Journal of Botanical Notes and News.

Touwaide, A. 1998. *Datura stramonium* L.: Old or New World? Delpinoa 39/40: 29–43.

Toxicology = Toxicology; International Journal Concerned with the Effects of Chemicals on Living Systems.

Toyokuni, H. 1963. Conspectus Gentianacearum Japonicarum. J. Fac. Sci. Hokkaido Imp. Univ., Ser. 5, Bot. 7: 137–259.

Trans. Acad. Sci. St. Louis = Transactions of the Academy of Science of St. Louis.

Trans. Amer. Philos. Soc. = Transactions of the American Philosophical Society Held at Philadelphia for Promoting Useful Knowledge.

Trans. Linn. Soc. London = Transactions of the Linnean Society of London.

Trans. Wisconsin Acad. Sci. = Transactions of the Wisconsin Academy of Sciences, Arts and Letters.

Tucker, D. P. and R. L. Phillips. 1974. The strangler vine: A major weed pest in Florida citrus groves. Weeds Today 5(4): 6–8.

Tuckey, J. H. 1818. Narrative of an Expedition to Explore the River Zaire.... London.

Turner, B. L. 1993d. The Texas species of *Centaurium* (Gentianaceae). Phytologia 75: 259–275.

Turner, B. L. 2009. *Convolvulus carrii,* a localized endemic from southernmost Texas. Phytologia 91: 394–400.

Turner, B. L. 2015. Taxonomy of *Chamaesaracha* (Solanaceae). Phytologia 97: 226–245.

Ulbricht, C. et al. 2004. An evidence-based systematic review of belladonna by the National Standard Research Collaboration. J. Herbal Pharmacotherapy 4: 61–90.

Univ. Calif. Publ. Bot. = University of California Publications in Botany.

University of Chicago Press. 1993. The Chicago Manual of Style, ed. 14. Chicago.

Uppsala Univ. Årsskr. = Uppsala Universitets Årsskrift.

Urban, I., ed. 1898–1928. Symbolae Antillanae seu Fundamenta Florae Indiae Occidentalis.... 9 vols. Berlin etc. (Symb. Antill.)

Vahl, M. 1790–1794. Symbolae Botanicae.... 3 vols. Copenhagen. (Symb. Bot.)

Vail, A. M. 1897. Studies in the Asclepiadaceae–I. Notes on the genus *Philibertella* in the United States. Bull. Torrey Bot. Club 24: 305–310.

Vail, A. M. 1904. Studies in the Asclepiadaceae–VIII. A new species of *Asclepias* from Kansas and two possible hybrids from New York. Bull. Torrey Bot. Club 31: 457–460.

Valverde, P. L., J. Fornoni, and J. Nuñez-Farfán. 2001. Defensive role of leaf trichomes in resistance to herbivorous insects in *Datura stramonium.* J. Evol. Biol. 14: 424–432.

van Bergen, M. and W. Snoeijer. 1996. Series of revisions of Apocynaceae XLII: *Catharanthus* G. Don. Wageningen Agric. Univ. Pap. 96: 1–120.

van Dam, N. M., J. D. Hare, and E. Elle. 1999. Inheritance and distribution of trichome phenotypes in *Datura wrightii.* J. Heredity 91: 220–227.

Van den Berg, R. G. et al. 2001. Solanaceae V: Advances in Taxonomy and Utilization. Nijmegen.

van der Heijden, R. et al. 2004. The *Catharanthus* alkaloids: Pharmacognosy and biotechnology. Curr. Med. Chem. 11: 607–628.

Vasc. Pl. Pacif. N.W.—See: C. L. Hitchcock et al. 1955–1969

Vierteljahrsschr. Naturf. Ges. Zürich = Vierteljahrsschrift der Naturforschenden Gesellschaft in Zürich.

Viviani, D. 1802. Elenchus Plantarum.... Genoa. (Elench. Pl.)

Voigt, J. O. 1845. Hortus Suburbanus Calcuttensis. Calcutta. (Hort. Suburb. Calcutt.)

Voy. Barbarie—See: J. Poiret 1789

Vyzn. Rosl. Ukrain.—See: A. I. Barbaricz 1950

W. N. Amer. Naturalist = Western North American Naturalist.

Waddington, K. D. 1976. Pollination of *Apocynum sibiricum* (Apocynaceae) by Lepidoptera. SouthW. Naturalist 21: 31–36.

Wageningen Agric. Univ. Pap. = Wageningen Agricultural University Papers.

Wageningen Univ. Pap. = Wageningen University Papers

Wagstaff, S. J. 2004. Tetrachondraceae. In: K. Kubitzki et al., eds. 1990+. The Families and Genera of Vascular Plants. 15+ vols. Berlin etc. Vol. 7, pp. 441–444.

Wahlert, G. A., F. E. Chiarini, and L. Bohs. 2015. A revision of *Solanum* section *Lathyrocarpum* (the Carolinense clade, Solanaceae). Syst. Bot. 40: 853–887.

Walpers, W. G. 1842–1847. Repertorium Botanices Systematicae.... 6 vols. Leipzig. (Repert. Bot. Syst.)

Walsh, B. M. and S. B. Hoot. 2001. Phylogenetic relationships of *Capsicum* (Solanaceae) using DNA sequences from two noncoding regions: The chloroplast *atp*B-*rbc*L spacer region and the nuclear *waxy* introns. Int. J. Pl. Sci. 162: 1409–1418.

Walter, T. 1788. Flora Caroliniana, Secundum Systema Vegetabilium Perillustris Linnaei Digesta.... London. (Fl. Carol.)

Wanntorp, H.-E. 1988. The genus *Microloma* (Asclepiadaceae). Opera Bot. 98: 1–69.

War Department [U.S.]. 1855–1860. Reports of Explorations and Surveys, to Ascertain the Most Practicable and Economical Route for a Railroad from the Mississippi River to the Pacific Ocean. Made under the Direction of the Secretary of War, in 1853[–1856].... 12 vols. in 13. Washington. (Pacif. Railr. Rep.)

Ward, D. B. 1968b. Contributions to the flora of Florida—3. *Evolvulus.* Castanea 33: 76–79.

Ward, D. B. 2007. Thomas Walter typification project, IV: Neotypes and epitypes for 43 Walter names, of genera A through C. J. Bot. Res. Inst. Texas 1: 1091–1105.

Watanabe, H. et al. 1999. Three groups of species in *Petunia* sensu Jussieu (Solanaceae) inferred from the intact seed morphology. Amer. J. Bot. 86: 302–305.

Waterfall, U. T. 1958. A taxonomic study of the genus *Physalis* in North America north of Mexico. Rhodora 60: 107–114, 128–142, 152–173.

Watson, S. 1871. United States Geological Expolration [sic] of the Fortieth Parallel. Clarence King, Geologist-in-charge. [Vol. 5] Botany. By Sereno Watson.... Washington. [Botanical portion of larger work by C. King.] [Botany (Fortieth Parallel)]

Weaver, J. E. and F. E. Clements. 1929. Plant Ecology. New York and London.

Weber, W. A. and R. C. Wittmann. 1985. Additions to the flora of Colorado—XI. Phytologia 58: 385–388.

Weed Sci. = Weed Science; Journal of the Weed Science Society of America.

Weitemier, K. 2016. Genomic Investigations of Diversity within the Milkweed Genus *Asclepias*, at Multiple Scales. Ph.D. dissertation. Oregon State University.

Welsh, M., S. Stefanović, and M. Costea. 2010. Pollen evolution and its taxonomic significance in *Cuscuta* (dodders, Convolvulaceae). Pl. Syst. Evol. 285: 83–101.

Welsh, S. L. and D. E. Anderson. 1962. *Metaplexis japonica:* An oriental milkweed from an Iowa cornfield. Brittonia 14: 186–188.

Wettstein, R. 1896. Die Gattungsgehörigkeit und systematische Stellung der *Gentiana tenella* Rottb. und *G. nana* Wulf. Oesterr. Bot. Z. 46: 172–176.

Whalen, M. D. 1979. Taxonomy of *Solanum* section *Androceras*. Gentes Herbarum 11: 359–426.

Whitson, M. and P. S. Manos. 2005. Untangling *Physalis* (Solanaceae) from the physaloids: A two-gene phylogeny of the Physalinae. Syst. Bot. 30: 216–230.

Wilbur, R. L. 1955. A revision of the North American genus *Sabatia* (Gentianaceae). Rhodora 57: 1–33, 43–71, 78–104.

Wilbur, R. L. 1970b. Infraspecific classification in the Carolina flora. Rhodora 72: 51–65.

Wilbur, R. L. 1984. A synopsis of the genus *Halenia* (Gentianaceae) in Mexico. Rhodora 86: 311–337.

Willdenow, C. L. 1803–1816. Hortus Berolinensis.... 2 vols. in 10 fascs. Berlin. [Fascs. and plates numbered consecutively.] (Hort. Berol.)

Willdenow, C. L., C. F. Schwägrichen, and J. H. F. Link. 1797–1830. Caroli a Linné Species Plantarum.... Editio Quarta.... 6 vols. Berlin. [Vols. 1–5(1), 1797–1810, by Willdenow; vol. 5(2), 1830, by Schwägrichen; vol. 6, 1824–1825, by Link.] (Sp. Pl.)

Williams, D. E. 1985. Tres Arvenses Solanáceas Comestibles y su Proceso de Domesticación en el Estado de Tlaxcala, México. Master's thesis. Institución de Enseñanza e Investigación en Ciencias Agrícolas.

Williams, J. K. 1995. Miscellaneous notes on *Haplophyton* (Apocynaceae: Plumerieae). Sida 16: 469–475.

Williams, J. K. 2004b. Polyphyly of the genus *Echites* (Apocynaceae: Apocynoideae: Echiteae): Evidence based on a morphological cladistic analysis. Sida 21: 117–131.

Williams, J. K. 2004c. A new combination in Mexican *Mandevilla* (Apocynaceae subfamily Apocynoideae) III. Phytologia 86: 178–183.

Williams, J. K. and J. K. Stutzman. 2008. Chromosome number of *Thevetia ahouai* (Apocynaceae: Rauvolfoidae: Plumerieae) with discussion on the generic boundaries of *Thevetia*. J. Bot. Res. Inst. Texas 2: 489–493.

Willson, M. F. and B. J. Rathcke. 1974. Adaptive design of the floral display in *Asclepias syriaca* L. Amer. Midl. Naturalist 92: 47–57.

Wilson, B. L., V. Hipkins, and T. Kaye. 2010. One taxon or two: Are *Frasera umpquaensis* and *F. fastigiata* (Gentianaceae) distinct species? Madroño 57: 106–119.

Wilson, J. D. 1851. The Rural Cyclopedia, or a General Dictionary of Agriculture. Edinburgh.

Wolf, N. M. 1776. Genera Plantarum.... [Danzig.] (Gen. Pl.)

Wood, A. 1861. A Class-book of Botany.... New York. [Class-book Bot. ed. s.n.(b)]

Wood, C. E. Jr. and R. E. Weaver. 1982. The genera of Gentianaceae in the southeastern United States. J. Arnold Arbor. 63: 441–487.

Wood, J. R. I. et al. 2015. A foundation monograph of *Convolvulus* L. (Convolvulaceae). PhytoKeys 51: 1–282.

Woodson, R. E. Jr. 1930. Studies in the Apocynaceae. I. A critical study of the Apocynoideae (with special reference to the genus *Apocynum*). Ann. Missouri Bot. Gard. 17: 1–172, 174–212.

Woodson, R. E. Jr. 1933. Studies in the Apocynaceae. IV. The American genera of Echitoideae. Ann. Missouri Bot. Gard. 20: 605–790.

Woodson, R. E. Jr. 1936. Studies in the Apocynaceae. IV. The American genera of Echitoideae. Ann. Missouri Bot. Gard. 23: 169–438.

Woodson, R. E. Jr. 1936b. Studies in the Apocynaceae. V. A revision of the Asiatic species of *Trachelospermun* Lem. Sunyatsenia 3: 65–105.

Woodson, R. E. Jr. 1938. Studies in the Apocynaceae. VII. An evaluation of the genera *Plumeria* L. and *Himatanthus* Willd. Ann. Missouri Bot. Gard. 25: 189–224.

Woodson, R. E. Jr. 1941. The North American Asclepiadaceae. I. Perspective of the genera. Ann. Missouri Bot. Gard. 28: 193–244.

Woodson, R. E. Jr. 1947. Some dynamics of leaf variation in *Asclepias tuberosa*. Ann. Missouri Bot. Gard. 34: 353–432.

Woodson, R. E. Jr. 1953. Biometric evidence of natural selection in *Asclepias tuberosa*. Proc. Natl. Acad. Sci. U.S.A. 39: 74–79.

Woodson, R. E. Jr. 1954. The North American species of *Asclepias* L. Ann. Missouri Bot. Gard. 41: 1–211.

Woodson, R. E. Jr. 1962. Butterflyweed revisited. Evolution 16: 168–185.

Woodson, R. E. Jr., R. W. Schery, et al., eds. 1943–1981. Flora of Panama. 41 fascs. St. Louis. [Fascs. published as individual issues of Ann. Missouri Bot. Gard. and aggregating 8 nominal parts + introduction and indexes.]

Woodson, R. E. Jr., R. W. Schery, and J. W. Nowicke. 1970. Apocynaceae. In: R. E. Woodson Jr. et al., eds. 1943–1981. Flora of Panama. 41 fascs. St. Louis. [Ann. Missouri Bot. Gard. 57: 59–130.]

Wright, M. A. R., M. Welsh, and M. Costea. 2011. Diversity and evolution of gynoecium in *Cuscuta* (dodders, Convolvulaceae) in relation to their reproductive biology: Two styles are better than one. Pl. Syst. Evol. 296: 51–76.

Wrightia = Wrightia; a Botanical Journal.

Wunderlin, R. P. 1998. Guide to the Vascular Plants of Florida. Gainesville.

Wunderlin, R. P. et al. 1993. *Solanum viarum* and *S. tampicense* (Solanaceae): Two weedy species new to Florida and the United States. Sida 15: 605–611.

Yamazaki, T. 1993. Solanaceae. In: K. Iwatsuki et al., eds. 1993+. Flora of Japan. 3+ vols. in 6+. Tokyo. Vol. 3a, pp. 183–194.

Yatskievych, G. and C. T. Mason. 1984. A taxonomic study of *Ipomoea tenuiloba* Torrey (Convolvulaceae), with notes on related species. Madroño 31: 102–108.

Yuan, Y. M. and P. Küpfer. 1993. Karyological studies of *Gentianopsis* and some related genera of Gentianaceae from China. Caryologia 58: 115–123.

Yuan, Y. M. and P. Küpfer. 1995. Molecular phylogenetics of the subtribe Gentianinae (Gentianaceae) inferred from the sequences of internal transcribed spacers (ITS) of nuclear ribosomal DNA. Pl. Syst. Evol. 196: 207–226.

Yuncker, T. G. 1921. Revision of the North American and West Indian species of *Cuscuta.* Illinois Biol. Monogr. 6: 91–231.

Yuncker, T. G. 1932. The genus *Cuscuta.* Mem. Torrey Bot. Club 18: 113–331.

Yuncker, T. G. 1965. *Cuscuta.* In: N. L. Britton et al., eds. 1905+. North American Flora.... 47+ vols. New York. Ser. 2, part 4, pp. 1–51.

Zarucchi, J. L. 1991. *Quiotania:* A new genus of Apocynaceae-Apocynoideae from northern Colombia. Novon 1: 33–36.

Zeltner, L. 1970. Recherches biosystématique sur les genres *Blackstonia* Huds. et *Centaurium* Hill. Bull. Soc. Neuchâtel. Sci. Nat. 93: 1–164.

Zoë = Zoë; a Biological Journal.

Zollinger, H. 1854–1855. Systematisches Verzeichniss der im indischen Archipel.... 3 vols. Zürich. (Syst. Verz.)

Zuev, V. V. 1990. K sistematike semeistva Gentianaceae v Sibiri. Bot. Zhurn. (Moscow & Leningrad) 75: 1296–1305.

Index

Names in *italics* are synonyms, casually mentioned hybrids, or plants not established in the flora. Part numbers are shown in parentheses followed by a colon. Page numbers in **boldface** indicate the primary entry for a taxon. Page numbers in italics indicate an illustration. Roman type is used for all other entries, including author names, vernacular names, and accepted scientific names for plants treated as established members of the flora.

Flora of North America — Index to Families/Volumes of Vascular Plants

Boldface denotes published volume: page number, current as of October 2022.